Chemical Safety Sheets

Chemical Safety Sheets

working safely
with hazardous chemicals

1991

Published by: Kluwer Academic Publishers
Samsom Chemical Publishers
Dutch Institute for the Working Environment
Dutch Chemical Industry Association

ISBN 0-7923-1258-9

Published by Samsom Chemical Publishers
and by Kluwer Academic Publishers
P.O. Box 17, 3300 AA Dordrecht, The Netherlands.

Kluwer Academic Publishers incorporates
the publishing programmes of
D. Reidel, Martinus Nijhoff, Dr W. Junk and MTP Press.

Sold and distributed in the U.S.A. and Canada
by Kluwer Academic Publishers
101 Philip Drive, Norwell, MA 02061, U.S.A.

Sold and distributed in the U.K.
by Graham & Trotman Limited
Sterling House, 66 Wilton Road, London SWIV DE

In all other countries sold and distributed
by Kluwer Academic Publishers Group
P.O. Box 322, 3300 AH Dordrecht, The Netherlands

Printed in The Netherlands

ACKNOWLEDGEMENTS

Chemical Safety Sheets is the international edition of 'Chemiekaarten', the authoritative work of reference on the safe handling of chemicals, prepared under the auspices of the Dutch Institute for the Working Environment and the Dutch Chemical Industry Association, with the assistance of a committee of chemical and safety experts and a committee of medical experts. Dr. J. Zawierko acted as coordinating editor.

Valuable advice and criticism was provided by Dr. R. Horner of the Environmental Protection Agency, Washington, USA and by P. Colman and G. Walsh of the Electricity Supply Board, Dublin, Ireland.

Permission was granted by Dr. W.D. Kelley of the American Conference of Governmental Industrial Hygienists to include the Threshold Limit Values from the 1990-1991 ACGIH List of TLV's, and the Recommendations Regarding Gloves and Protective Clothing to be used while handling hazardous chemicals from the *Guidelines for the Selection of Chemical Protective Clothing*, 3rd ed., 1987.

For the NFPA Standard System to the Identification of Fire Hazards the 1991 edition of the *Fire Protection Guide of Hazardous Materials* was used.

Editor and publishers wish to thank all who contributed to the preparation of the international edition.

The format of 'Chemiekaarten' has been widely acclaimed by professionals from the major industrial countries and has been selected by experts from the World Health Organization, the International Labour Office, the United Nations Environment Program and the European Community to serve as an example for the preparation of the *International Chemical Safety Cards*.

Comments and suggestions regarding Chemical Safety Sheets will be welcomed at our editorial address:

Samsom Chemical Publishers
P.O. Box 316
2400 AH Alphen aan den Rijn
Netherlands

Tel. 31 1720 66 420
Fax. 31 1720 93 270

TABLE OF CONTENTS

HOW TO USE THIS BOOK

Chemical Safety Sheets has been designed as a work of reference on the safe handling of chemicals in laboratories, industrial workplaces and other institutions.

Each sheet presents data of interest to experts as well as data of interest to the employees in charge of handling the respective chemicals.

The information presented covers such aspects as
- identification and recognition of the substance
- occupational exposure levels
- effects of exposure
- safe handling and personal protection
- hazard prevention
- dealing with accidents
- safe storage and dealing with spillages
- labeling
- transportation
- fire hazards

All these data are presented in a uniform format in order to facilitate quick reference and comparisons of the hazardous properties of various chemicals.

Each heading and all frequently occurring standard phrases are explained in the chapter Explanatory Notes.

Finally, the book contains two indexes: an alphabetic index specifying CAS-no., EEC-no., RTECS-no., Hazard Identification-no. and UN-no for each substance and an index in order of the UN-numbers.

The editors prepared the information contained in Chemical Safety Sheets with great care and to the best of their knowledge, but no warranty, express or implied, is made.
The publishers accept no responsibility for damage or liability of any kind alleged to result from the use of this book.

ethanal
acetic aldehyde
acetylaldehyde
ethyl aldehyde

CH_3CHO

ACETALDEHYDE

PHYSICAL PROPERTIES	IMPORTANT CHARACTERISTICS

PHYSICAL PROPERTIES	
Boiling point, °C	21
Melting point, °C	− 123
Flash point, °C	− 40
Auto-ignition temperature, °C	140
Relative density (water = 1)	0.8
Relative vapor density (air = 1)	1.5
Relative density at 20 °C of saturated mixture vapor/air (air = 1)	1.5
Vapor pressure, mm Hg at 20 °C	749
Solubility in water	∞
Explosive limits, vol% in air	4-60
Minimum ignition energy, mJ	0.37
Electrical conductivity, pS/m	1.2×10^8
Relative molecular mass	44.1
Log P octanol/water	0.4

Gross formula: C_2H_4O

IMPORTANT CHARACTERISTICS

COLORLESS LIQUID WITH CHARACTERISTIC ODOR
Vapor is heavier than air and spreads at ground level, with risk of ignition at a distance. Lye can cause violent polymerization. Can autoxidize, forming explosive acetaldehyde monoperacetate. Do not use compressed air when filling, emptying or processing. Strong reducing agent which reacts violently with oxidants. Reacts violently with various organic substances, halogens, sulfuric acid and amines, with risk of fire and explosion.

TLV-TWA	100 ppm	180 mg/m³
STEL	150 ppm	270 mg/m³

Absorption route: Can enter the body by inhalation or ingestion. Harmful atmospheric concentrations can build up very rapidly on evaporation at 20°C.
Immediate effects: Irritates the skin. Corrosive to the eyes and respiratory tract. Inhalation of vapor/fumes can cause severe breathing difficulties (lung edema). In serious cases risk of death.
Effects of prolonged/repeated exposure: Prolonged or repeated contact can cause skin disorders.

HAZARDS/SYMPTOMS	PREVENTIVE MEASURES	FIRE EXTINGUISHING/FIRST AID
Fire: extremely flammable.	keep away from open flame, sparks and hot surfaces (e.g. steam lines), no smoking.	powder, alcohol-resistant foam, large quantities of water, carbon dioxide, (halons).
Explosion: forms explosive air-vapor mixtures.	sealed machinery, ventilation, explosion-proof electrical equipment and lighting, non-sparking tools.	in case of fire: keep tanks/drums cool by spraying with water.
Inhalation: *corrosive*; sore throat, cough, severe breathing difficulties, unconsciousness.	ventilation, local exhaust or respiratory protection.	fresh air, rest, place in half-sitting position, take to hospital.
Skin: redness, pain.	gloves.	remove contaminated clothing, flush skin with water or shower, refer to a doctor.
Eyes: *corrosive*; redness, pain, impaired vision.	face shield, or combined eye and respiratory protection.	flush with water, take to a doctor.
Ingestion: *corrosive*; abdominal pain, diarrhea, vomiting.		rinse mouth, take immediately to hospital.

SPILLAGE	STORAGE	LABELING / NFPA
ventilate, collect leakage in sealable containers, flush away remainder with water, do not use sawdust or other combustible absorbents; (additional individual protection: breathing apparatus).	keep in cool, fireproof place, separate from oxidants, halogens, lye, organic substances, amines and concentrated sulfuric acid; add an inhibitor.	R: 12-36/37-40 S: 16-33-36/37 Highly flammable Harmful 4 2 2

NOTES
Log P octanol/water is estimated. Flush contaminated clothing with water (fire hazard). Lung edema symptoms usually develop several hours later and are aggravated by physical exertion: rest and hospitalization essential. As first aid, a doctor or authorized person should consider administering a corticosteroid spray. Unbreakable packaging preferred; if breakable, keep in unbreakable container.

Transport Emergency Card TEC(R)-620

HI: 33; UN-number: 1089

1

β-ACETALDOXIME

PHYSICAL PROPERTIES	IMPORTANT CHARACTERISTICS
Boiling point, °C 115 Melting point, °C 12 Flash point, °C <22 1) Relative density (water = 1) 0.97 Relative vapor density (air = 1) 2.0 Relative density at 20 °C of saturated mixture vapor/air (air = 1) 1.01 Vapor pressure, mm Hg at 26 °C 9.9 Solubility in water very good Explosive limits, vol% in air see notes Relative molecular mass 59.1 Log P octanol/water − 0.1 Gross formula: C_2H_5NO	**COLORLESS LIQUID WITH CHARACTERISTIC ODOR** Vapor mixes readily with air. Able to form peroxides (by autoxidation). Distilling can cause explosion. Toxic → *nitrous vapors* are a product of combustion. Reacts with acids, giving off → *acetaldehyde* and → *hydroxylamine*. Reacts violently with strong oxidants. Attacks many base metals. TLV-TWA not available **Absorption route:** Can enter the body by inhalation or ingestion. Insufficient data on the rate at which harmful concentrations can build up.

HAZARDS/SYMPTOMS	PREVENTIVE MEASURES	FIRE EXTINGUISHING/FIRST AID
Fire: extremely flammable.	keep away from open flame and sparks, no smoking.	powder, alcohol-resistant foam, large quantities of water, carbon dioxide, (halons).
Explosion: forms explosive air-vapor mixtures.	sealed machinery, ventilation, explosion-proof electrical equipment and lighting.	in case of fire: keep tanks/drums cool by spraying with water.
Inhalation:	ventilation, local exhaust or respiratory protection.	fresh air, rest, call a doctor.
Skin:	gloves.	remove contaminated clothing, flush skin with water or shower, call a doctor if necessary.
Eyes:	face shield, or combined eye and respiratory protection.	flush with water, take to a doctor if necessary.
Ingestion:		rinse mouth, take to a doctor.

SPILLAGE	STORAGE	LABELING / NFPA
collect leakage in sealable containers, soak up with sand or other inert absorbent and remove to safe place; (additional individual protection: breathing apparatus).	keep in dark, fireproof place, separate from oxidants.	

NOTES
1) Melting point of α-isomer 47° C. Flash point and data on fire and explosion hazards refer to technical grade. Pure substance has a flash point of 40° C and is thus much less of a fire hazard. Before distilling check for peroxides; if found, render harmless. Insufficient toxicological data on harmful effects to humans.

Transport Emergency Card TEC(R)-30G36

UN-number: 2332

ethanoic acid
ethylic acid
methanecarboxylic acid
vinegar

CH_3COOH

ACETIC ACID
(100%)

PHYSICAL PROPERTIES		IMPORTANT CHARACTERISTICS		
Boiling point, °C	118	**COLORLESS LIQUID WITH PUNGENT ODOR**		
Melting point, °C	17	Vapor mixes readily with air. Reacts violently with oxidants, with risk of fire and explosion. Reacts		
Flash point, °C	40	violently with bases, evolving heat. Attacks many metals, giving off flammable gas (→ *hydrogen*).		
Auto-ignition temperature, °C	450			
Relative density (water = 1)	1.0	TLV-TWA	10 ppm	25 mg/m³
Relative vapor density (air = 1)	2.1	STEL	15 ppm	37 mg/m³
Relative density at 20 °C of saturated mixture				
vapor/air (air = 1)	1.02	**Absorption route:** Can enter the body by inhalation or ingestion. Harmful atmospheric		
Vapor pressure, mm Hg at 20 °C	12	concentrations can build up fairly rapidly on evaporation at approx. 20°C - even more rapidly in		
Solubility in water	∞	aerosol form.		
Explosive limits, vol% in air	4.0-17	**Immediate effects:** Corrosive to the eyes, skin and respiratory tract. Inhalation of vapor/fumes		
Electrical conductivity, pS/m	6 x 10⁵	can cause severe breathing difficulties (lung edema). In serious cases risk of death.		
Relative molecular mass	60.0			
Log P octanol/water	−0.2			
Gross formula:	$C_2H_4O_2$			

HAZARDS/SYMPTOMS	PREVENTIVE MEASURES	FIRE EXTINGUISHING/FIRST AID
Fire: flammable.	keep away from open flame and sparks, no smoking.	powder, water spray, alcohol-resistant foam, carbon dioxide, (halons).
Explosion: above 40°C: forms explosive air-vapor mixtures.	above 40°C: sealed machinery, ventilation, explosion-proof electrical equipment.	in case of fire: keep tanks/drums cool by spraying with water.
Inhalation: *corrosive*; sore throat, cough, shortness of breath, severe breathing difficulties.	ventilation, local exhaust or respiratory protection.	fresh air, rest, place in half-sitting position, take to hospital.
Skin: *corrosive*; redness, pain, serious burns.	gloves.	remove contaminated clothing, flush skin with water or shower, refer to a doctor.
Eyes: *corrosive*; pain, impaired vision.	face shield.	flush with water, take to a doctor.
Ingestion: *corrosive*; sore throat, diarrhea, abdominal cramps, vomiting.		rinse mouth, call a doctor or take to hospital.

SPILLAGE	STORAGE	LABELING / NFPA
collect leakage in sealable containers, flush away remainder with water; (additional individual protection: breathing apparatus).	keep in fireproof place, separate from oxidants and strong bases.	R: 10-35 S: 2-23-26 Note B ⚠ Corrosive

NFPA diamond: 2 (top), 2 (left), 0 (right)

NOTES

Lung edema symptoms usually develop several hours later and are aggravated by physical exertion: rest and hospitalization essential. As first aid, a doctor or authorized person should consider administering a corticosteroid spray. Does not attack aluminum or stainless steel. UNDER NO CIRCUMSTANCES add water to acid; when diluting ALWAYS add acid to water.

Transport Emergency Card TEC(R)-614

HI: 83; UN-number: 2789

CAS-No: [64-19-7]
ethanoic acid
ethylic acid
methanecarboxylic acid
vinegar

CH_3COOH

ACETIC ACID
(85% in water)

PHYSICAL PROPERTIES	IMPORTANT CHARACTERISTICS
Boiling point, °C 103 Melting point, °C −15 Flash point, °C 50 Auto-ignition temperature, °C 485 Relative density (water = 1) 1.05 Relative vapor density (air = 1) 2.0 Relative density at 20 °C of saturated mixture vapor/air (air = 1) 1.02 Vapor pressure, mm Hg at 20°C 11 Solubility in water ∞ Explosive limits, vol% in air 4-17 Relative molecular mass 60.1 Log P octanol/water −0.2 Gross formula: $C_2H_4O_2$	**COLORLESS LIQUID WITH PUNGENT ODOR** Vapor mixes readily with air. Medium strong acid. Reacts violently with strong oxidants. Reacts with bases, evolving large quantities of heat. Attacks many metals, giving off flammable gas ($\rightarrow$ *hydrogen*). TLV-TWA 10 ppm 25 mg/m³ STEL 15 ppm 37 mg/m³ **Absorption route:** Can enter the body by inhalation or ingestion. Harmful atmospheric concentrations can build up fairly rapidly on evaporation at approx. 20°C - even more rapidly in aerosol form. **Immediate effects:** Corrosive to the eyes, skin and respiratory tract. Inhalation of vapor/fumes can cause severe breathing difficulties (lung edema). In serious cases risk of death.

HAZARDS/SYMPTOMS	PREVENTIVE MEASURES	FIRE EXTINGUISHING/FIRST AID
Fire: flammable.	keep away from open flame and sparks, no smoking.	powder, water spray, AFFF, carbon dioxide, (halons).
Explosion: above 50°C: forms explosive air-vapor mixtures.	above 50°C: sealed machinery, ventilation, explosion-proof electrical equipment.	in case of fire: keep tanks/drums cool by spraying with water.
Inhalation: *corrosive*; sore throat, cough, shortness of breath, severe breathing difficulties.	ventilation, local exhaust or respiratory protection.	fresh air, rest, place in half-sitting position, take to hospital.
Skin: *corrosive*; redness, pain, burns.	gloves.	remove contaminated clothing, flush skin with water or shower, call a doctor.
Eyes: *corrosive*; redness, pain, impaired vision.	face shield.	flush with water, take to doctor.
Ingestion: *corrosive*; diarrhea, stomach cramps, vomiting.		rinse mouth, take immediately to hospital.

SPILLAGE	STORAGE	LABELING / NFPA
collect leakage in sealable containers, flush away remainder with water; (additional individual protection: breathing apparatus).	keep in fireproof place, separate from oxidants and strong bases.	R: 34 S: 2-23-26 Note B Corrosive

NOTES
Lung edema symptoms usually develop several hours later and are aggravated by physical exertion: rest and hospitalization essential. As first aid, a doctor or authorized person should consider administering a corticosteroid spray. UN No. 2789 applies to acetic acid (10-80% aqueous solution).

Transport Emergency Card TEC(R)-614

HI: 83; UN-number: 2789 (10-80%)

4

CAS-No: [108-24-7]
acetic acid anhydride
acetyl anhydride
acetyl ether
acetyl oxide
ethanoic anhydrate

$(CH_3CO)_2O$

ACETIC ANHYDRIDE

PHYSICAL PROPERTIES		IMPORTANT CHARACTERISTICS	
Boiling point, °C	140	**COLORLESS LIQUID WITH PUNGENT ODOR**	
Melting point, °C	−73	Vapor mixes readily with air. Highly corrosive, esp. in the presence of water. Reacts violently with hot water, alcohols, amines, strong bases, strong oxidants and many other compounds. Reacts slowly with cold water. Attacks many plastics.	
Flash point, °C	49		
Auto-ignition temperature, °C	316		
Relative density (water = 1)	1.1		
Relative vapor density (air = 1)	3.5	TLV-TWA	5 ppm	21 mg/m³ C
Relative density at 20 °C of saturated mixture vapor/air (air = 1)	1.01	**Absorption route:** Can enter the body by inhalation or ingestion. Harmful atmospheric concentrations can build up fairly rapidly on evaporation at approx. 20°C - even more rapidly in aerosol form.	
Vapor pressure, mm Hg at 20 °C	3.8		
Solubility in water	reaction	**Immediate effects:** Corrosive to the eyes, skin and respiratory tract. Inhalation of vapor/fumes can cause severe breathing difficulties (lung edema). In serious cases risk of death.	
Explosive limits, vol% in air	2.0-10.3		
Electrical conductivity, pS/m	7.5x10⁷	**Effects of prolonged/repeated exposure:** Prolonged or repeated contact can cause skin disorders.	
Relative molecular mass	102.1		
Gross formula:	$C_4H_6O_3$		

HAZARDS/SYMPTOMS	PREVENTIVE MEASURES	FIRE EXTINGUISHING/FIRST AID
Fire: flammable.	keep away from open flame and sparks, no smoking.	powder, carbon dioxide, (halons), DO NOT USE WATER-BASED EXTINGUISHERS.
Explosion: above 49°C: forms explosive air-vapor mixtures.	above 49°C: sealed machinery, ventilation, explosion-proof electrical equipment.	in case of fire: keep tanks/drums cool by spraying with water, but DO NOT spray substance with water.
Inhalation: *corrosive*; sore throat, cough, shortness of breath, severe breathing difficulties.	ventilation, local exhaust or respiratory protection.	fresh air, rest, place in half-sitting position, artificial respiration if necessary, take to hospital.
Skin: *corrosive*; redness, pain, burns.	gloves, protective clothing.	remove contaminated clothing, flush skin with water or shower, call a doctor if necessary.
Eyes: *corrosive*; redness, pain, impaired vision.	face shield.	flush with water, take to a doctor.
Ingestion: *corrosive*; sore throat, abdominal pain, vomiting.		rinse mouth, immediately take to hospital.

SPILLAGE	STORAGE	LABELING / NFPA
collect leakage in sealable containers, soak up with sand or other inert absorbent and remove to safe place, flush away remainder with water; (additional individual protection: breathing apparatus).	keep in dry, fireproof place, separate from oxidants, strong bases and alcohols.	R: 10-34 S: 26 Corrosive

NOTES

Acids promote reaction with water. Fight large-scale fires with large quantities of water. TLV is maximum and must not be exceeded. Lung edema symptoms usually develop several hours later and are aggravated by physical exertion: rest and hospitalization essential. As first aid, a doctor or authorized person should consider administering a corticosteroid spray. Use airtight packaging.

Transport Emergency Card TEC(R)-63

HI: 83; UN-number: 1715

CAS-No: [67-64-1]
2-propanone
dimethyl formaldehyde
dimethyl ketone
ketone propane
methyl ketone

CH₃COCH₃

ACETONE

PHYSICAL PROPERTIES		IMPORTANT CHARACTERISTICS
Boiling point, °C	56	**COLORLESS LIQUID WITH CHARACTERISTIC ODOR**
Melting point, °C	− 95	Vapor is heavier than air and spreads at ground level, with risk of ignition at a distance. Do not use
Flash point, °C	− 19	compressed air when filling, emptying or processing. Reacts with strong oxidants, forming an
Auto-ignition temperature, °C	540	explosive peroxide. Reacts violently with chlorofom in basic environment, with risk of fire and
Relative density (water = 1)	0.8	explosion. Attacks many plastics.
Relative vapor density (air = 1)	2.0	

Relative density at 20 °C of saturated mixture		TLV-TWA	750 ppm	1780 mg/m³
vapor/air (air = 1)	1.2	STEL	1000 ppm	2380 mg/m³

Vapor pressure, mm Hg at 20 °C	177	
Solubility in water	∞	**Absorption route:** Can enter the body by inhalation or ingestion. Harmful atmospheric
Explosive limits, vol% in air	2.3-13	concentrations can build up fairly rapidly on evaporation at approx. 20°C - even more rapidly in
Minimum ignition energy, mJ	1.15	aerosol form.
Electrical conductivity, pS/m	4.9x10⁵	**Immediate effects:** Irritates the eyes, skin and respiratory tract. Liquid destroys the skin's natural
Relative molecular mass	58.1	oils. Affects the nervous system.
Log P octanol/water	− 0.2	

Gross formula: C₃H₆O

HAZARDS/SYMPTOMS	PREVENTIVE MEASURES	FIRE EXTINGUISHING/FIRST AID
Fire: extremely flammable.	keep away from open flame and sparks, no smoking.	powder, alcohol-resistant foam, large quantities of water, carbon dioxide, (halons).
Explosion: forms explosive air-vapor mixtures, reaction with (1) strong oxidants or (2) chloroform in basic environment can cause explosion.	sealed machinery, ventilation, explosion-proof electrical equipment and lighting.	in case of fire: keep tanks/drums cool by spraying with water.
Inhalation: sore throat, cough, headache, dizziness, drowsiness.	ventilation, local exhaust or respiratory protection.	fresh air, rest, call a doctor.
Skin: redness, pain.	gloves.	remove contaminated clothing, flush skin with water or shower.
Eyes: redness, pain.	safety glasses.	flush with water, take to a doctor.
Ingestion: sore throat, headache, dizziness, drowsiness.		rinse mouth, call a doctor or take to hospital.

SPILLAGE	STORAGE	LABELING / NFPA
collect leakage in sealable containers, soak up with sand or other inert absorbent and remove to safe place, flush away remainder with water; (additional individual protection: breathing apparatus).	keep in fireproof place, separate from oxidants.	R: 11 S: 9-16-23-33 🔥 Flammable 3 / 1 / 0

NOTES
Alcohol consumption increases toxic effects.

Transport Emergency Card TEC(R)-30

HI: 33; UN-number: 1090

CAS-No: [75-86-5]
2-hydroxy-2-methylpropanenitrile
2-cyano-2-propanol
2-hydroxyisobutyronitril
2-hydroxy-2-methylpropionitrile
2-methyllactonitril

$(CH_3)_2C(OH)C \equiv N$

ACETONE CYANOHYDRIN

PHYSICAL PROPERTIES	IMPORTANT CHARACTERISTICS
Boiling point, °C 95 Melting point, °C −20 Flash point, °C 74 Auto-ignition temperature, °C 688 Relative density (water = 1) 0.9 Relative vapor density (air = 1) 2.9 Relative density at 20 °C of saturated mixture vapor/air (air = 1) 1.0 Vapor pressure, mm Hg at 20 °C 0.76 Solubility in water ∞ Explosive limits, vol% in air 2.3-11 Relative molecular mass 85.1	**COLORLESS LIQUID** Vapor mixes readily with air. Decomposes when heated above 120°C or due to traces of base, giving off toxic gas (→ *hydrogen cyanide*) and flammable vapor (→ *acetone*). Reacts violently with concentrated acids and strong bases, with risk of fire and explosion. Reacts with strong oxidants, with risk of fire and explosion.
	TLV-TWA not available
	Absorption route: Can enter the body by inhalation or ingestion or through the skin. Harmful atmospheric concentrations can build up fairly rapidly on evaporation at approx. 20°C - even more rapidly in aerosol form. **Immediate effects:** Irritates the eyes, skin and respiratory tract. Impedes tissue respiration.
Gross formula: C_4H_7NO	

HAZARDS/SYMPTOMS	PREVENTIVE MEASURES	FIRE EXTINGUISHING/FIRST AID
Fire: combustible.	keep away from open flame, no smoking.	powder, alcohol-resistant foam, water spray, carbon dioxide, (halons).
Explosion: above 74°C: forms explosive air-vapor mixtures.	above 74°C: sealed machinery, ventilation.	in case of fire: keep tanks/drums cool by spraying with water.
Inhalation: sore throat, shortness of breath, headache, nausea, vomiting.	ventilation, local exhaust or respiratory protection.	fresh air, rest, take to hospital.
Skin: *is absorbed*; redness, pain.	gloves, protective clothing.	remove contaminated clothing, wash skin with soap and water, call a doctor.
Eyes: redness, pain.	face shield, or combined eye and respiratory protection.	flush with water, take to doctor if necessary.
Ingestion: sore throat, headache, shortness of breath, nausea, vomiting.		rinse mouth, take immediately to hospital.

SPILLAGE	STORAGE	LABELING / NFPA
collect leakage in sealable containers, soak up with sand or other inert absorbent and remove to safe place; (additional individual protection: breathing apparatus).	keep in cool, fireproof place, separate from oxidants, strong acids and strong bases.	R: 26/27/28 S: 7/9-27-45 ☠ Toxic NFPA: 3 / 2 / 2

NOTES
Limited storage life, unless slightly acidified. Special first aid required in the event of poisoning: antidotes must be available (with instructions). Use airtight packaging.

Transport Emergency Card TEC(R)-172

HI: 66; UN-number: 1541

CAS-No: [75-05-8]
cyanomethane
ethane nitrile
ethyl nitril
methanecarbonitril
methyl cyanide

$CH_3C \equiv N$

ACETONITRILE

PHYSICAL PROPERTIES	IMPORTANT CHARACTERISTICS

PHYSICAL PROPERTIES

Boiling point, °C	81
Melting point, °C	−45
Flash point, °C	2
Auto-ignition temperature, °C	525
Relative density (water = 1)	0.8
Relative vapor density (air = 1)	1.4
Relative density at 20 °C of saturated mixture vapor/air (air = 1)	1.04
Vapor pressure, mm Hg at 20 °C	73.7
Solubility in water	∞
Explosive limits, vol% in air	3-16
Electrical conductivity, pS/m	6x10⁴
Relative molecular mass	41.1
Log P octanol/water	−0.3

| Gross formula: | C_2H_3N |

IMPORTANT CHARACTERISTICS

COLORLESS LIQUID WITH CHARACTERISTIC ODOR
Vapor mixes readily with air, forming explosive mixtures. Do not use compressed air when filling, emptying or processing. Decomposes when heated, giving off combustible and toxic vapors. Reacts violently with strong oxidants. Hydrolizes in alkaline or acid water, forming → *acetic acid* and → *anhydrous ammonia*.

TLV-TWA	40 ppm	67 mg/m³
STEL	60 ppm	101 mg/m³

Absorption route: Can enter the body by inhalation or ingestion or through the skin. Harmful atmospheric concentrations can build up fairly rapidly on evaporation at approx. 20°C - even more rapidly in aerosol form.
Immediate effects: Irritates the eyes, skin and respiratory tract. Impedes tissue respiration. Exposure to high concentrations can cause unconsciousness and death. Effects can be delayed. Keep under medical observation.

HAZARDS/SYMPTOMS	PREVENTIVE MEASURES	FIRE EXTINGUISHING/FIRST AID
Fire: extremely flammable.	keep away from open flame and sparks, no smoking.	powder, alcohol-resistant foam, large quantities of water, carbon dioxide, (halons).
Explosion: forms explosive air-vapor mixtures.	sealed machinery, ventilation, explosion-proof electrical equipment and lighting.	in case of fire: keep tanks/drums cool by spraying with water.
	STRICT HYGIENE	IN ALL CASES CALL A DOCTOR
Inhalation: severe breathing difficulties, headache, dizziness, cramps, feeling of weakness.	ventilation, local exhaust or respiratory protection.	fresh air, rest, special treatment, take to hospital.
Skin: *is absorbed*; blue lips or fingernails; see also 'Inhalation'.	gloves, protective clothing.	remove contaminated clothing, flush skin with water or shower.
Eyes: redness, pain.	face shield.	flush with water, take to a doctor if necessary.
Ingestion: headache, severe breathing difficulties, dizziness, cramps, feeling of weakness.		rinse mouth, special treatment, call a doctor or take to hospital.

SPILLAGE	STORAGE	LABELING / NFPA
collect leakage in sealable containers, soak up with sand or other inert absorbent and remove to safe place, render remainder harmless with sodium hypochlorite solution; (additional individual protection: breathing apparatus).	keep in fireproof place, separate from oxidants.	R: 11-23/24/25 S: 16-27-44 Flammable Toxic 3 / 3 / 0

NOTES
Odor limit is above TLV. Special first aid required in the event of poisoning: antidotes must be available (with instructions).

Transport Emergency Card TEC(R)-148

HI: 336; UN-number: 1648

ACETONYLACETONE

PHYSICAL PROPERTIES		IMPORTANT CHARACTERISTICS	
Boiling point, °C	192	**COLORLESS LIQUID WITH CHARACTERISTIC ODOR**	
Melting point, °C	−9	Vapor mixes readily with air. Reacts violently with strong oxidants, with risk of fire and explosion.	
Flash point, °C	79		
Auto-ignition temperature, °C	490	TLV-TWA	not available
Relative density (water = 1)	0.97		
Relative vapor density (air = 1)	3.9	**Absorption route:** Can enter the body by inhalation or ingestion. Insufficient data on the rate at which harmful concentrations can build up.	
Relative density at 20 °C of saturated mixture			
vapor/air (air = 1)	1.01	**Immediate effects:** Irritates the eyes. Liquid destroys the skin's natural oils. Affects the nervous system. In serious cases risk of unconsciousness.	
Vapor pressure, mm Hg at 20 °C	1.3		
Solubility in water	∞		
Relative molecular mass	114.2		
Log P octanol/water	0.0		
Gross formula:	$C_6H_{10}O_2$		

HAZARDS/SYMPTOMS	PREVENTIVE MEASURES	FIRE EXTINGUISHING/FIRST AID
Fire: combustible.	keep away from open flame, no smoking.	powder, alcohol-resistant foam, water spray, carbon dioxide, (halons).
Explosion: above 79°C: forms explosive air-vapor mixtures.	above 79°C: sealed machinery, ventilation.	in case of fire: keep tanks/drums cool by spraying with water.
Inhalation: cough, drowsiness.	ventilation, local exhaust or respiratory protection.	fresh air, rest, call a doctor.
Skin:	gloves.	remove contaminated clothing, flush skin with water or shower.
Eyes: redness, pain.	safety glasses.	flush with water, take to a doctor.
Ingestion: cough, drowsiness.		rinse mouth, rest, call a doctor.

SPILLAGE	STORAGE	LABELING / NFPA
collect leakage in sealable containers, soak up with sand or other inert absorbent and remove to safe place, flush away remainder with water.	keep separate from oxidants; ventilate at floor level.	1 / 1 / 0

NOTES
Log P octanol/water is estimated.

CAS-No: [98-86-2]
acetylbenzene
methylphenylketone
1-phenylethanone
phenylmethylketone

$C_6H_5COCH_3$

ACETOPHENONE

PHYSICAL PROPERTIES		IMPORTANT CHARACTERISTICS
Boiling point, °C	202	**COLORLESS LIQUID OR WHITE CRYSTALS WITH CHARACTERISTIC ODOR**
Melting point, °C	20	Vapor mixes readily with air. Reacts with oxidants.
Flash point, °C	77	
Auto-ignition temperature, °C	535	TLV-TWA not available
Relative density (water = 1)	1.02	
Relative vapor density (air = 1)	4.15	
Relative density at 20 °C of saturated mixture		**Absorption route:** Can enter the body by inhalation or ingestion or through the skin. Insufficient data on the rate at which harmful concentrations can build up.
vapor/air (air = 1)	1.0	**Immediate effects:** Irritates the eyes, skin and respiratory tract. Liquid destroys the skin's natural oils. Affects the nervous system.
Vapor pressure, mm Hg at 20 °C	0.99	
Solubility in water, g/100 ml at 20 °C	0.7	
Electrical conductivity, pS/m	3.1x10⁵	
Relative molecular mass	120.1	
Log P octanol/water	1.6	
Gross formula: C_8H_8O		

HAZARDS/SYMPTOMS	PREVENTIVE MEASURES	FIRE EXTINGUISHING/FIRST AID
Fire: combustible.	keep away from open flame, no smoking.	powder, AFFF, foam, carbon dioxide, (halons).
Explosion: above 77°C: forms explosive air-vapor mixtures.	above 77°C: sealed machinery, ventilation.	
Inhalation: cough, headache, dizziness.	ventilation, local exhaust or respiratory protection.	fresh air, rest, call a doctor.
Skin: *is absorbed*; redness, pain.	gloves.	remove contaminated clothing, flush skin with water or shower.
Eyes: redness, pain.	safety glasses.	flush with water, take to a doctor if necessary.
Ingestion: sore throat, abdominal pain, nausea; see also 'Inhalation'.		rinse mouth, immediately call a doctor or take to hospital.

SPILLAGE	STORAGE	LABELING / NFPA
collect leakage in sealable containers, soak up with sand or other inert absorbent and remove to safe place; if solid: clean up spillage and remove to safe place; (additional individual protection: P2 respirator).		

NFPA: 2 / 1 / 0

NOTES
Alcohol consumption increases toxic effects.

Transport Emergency Card TEC(R)-30G37

2,4-pentanedione
diacetyl methane

$CH_3COCH_2COCH_3$

ACETYL ACETONE

PHYSICAL PROPERTIES	IMPORTANT CHARACTERISTICS
Boiling point, °C　140 Melting point, °C　−23 Flash point, °C　34 Auto-ignition temperature, °C　340 Relative density (water = 1)　0.98 Relative vapor density (air = 1)　3.5 Relative density at 20 °C of saturated mixture vapor/air (air = 1)　1.03 Vapor pressure, mm Hg at 20 °C　7.1 Solubility in water, g/100 ml　16 Explosive limits, vol% in air　?-? Relative molecular mass　100.1 Log P octanol/water　2.1 Gross formula:　$C_5H_8O_2$	**COLORLESS LIQUID WITH CHARACTERISTIC ODOR** Vapor mixes readily with air. Reacts violently with strong oxidants. Reacts with organic acids, amines and isocyanates. TLV-TWA　　　　　　　not available **Absorption route:** Can enter the body by inhalation or ingestion. Insufficient data on the rate at which harmful concentrations can build up. **Immediate effects:** Irritates the eyes, skin and respiratory tract. Liquid destroys the skin's natural oils. Affects the nervous system.

HAZARDS/SYMPTOMS	PREVENTIVE MEASURES	FIRE EXTINGUISHING/FIRST AID
Fire: flammable.	keep away from open flame and sparks, no smoking.	powder, AFFF, foam, carbon dioxide, (halons).
Explosion: above 34°C: forms explosive air-vapor mixtures.	above 34°C: sealed machinery, ventilation, explosion-proof electrical equipment.	in case of fire: keep tanks/drums cool by spraying with water.
Inhalation: cough, drowsiness.	ventilation, local exhaust or respiratory protection.	fresh air, rest, call a doctor if necessary.
Skin: redness.	gloves.	remove contaminated clothing, flush skin with water or shower.
Eyes: redness, pain.	face shield.	flush with water, take to a doctor if necessary.
Ingestion: abdominal pain, nausea, drowsiness.		rinse mouth, rest, call a doctor.

SPILLAGE	STORAGE	LABELING / NFPA
collect leakage in sealable containers, soak up with sand or other inert absorbent and remove to safe place.	keep in fireproof place, separate from oxidants.	R: 10-22 S: 21-23-24/25 ✖ Harmful

NFPA: 2 / 2 / 0

NOTES
Explosive limits not given in the literature. Log P octanol/water is estimated. Packaging: special material. Unbreakable packaging preferred; if breakable, keep in unbreakable container.

Transport Emergency Card TEC(R)-591

HI: 30; UN-number: 2310

ACETYLBROMIDE

PHYSICAL PROPERTIES	IMPORTANT CHARACTERISTICS
Boiling point, °C 77 Melting point, °C −96 Relative density (water = 1) 1.6 Relative vapor density (air = 1) 4.2 Relative density at 20 °C of saturated mixture vapor/air (air = 1) 1.4 Vapor pressure, mm Hg at 20 °C 101 Solubility in water reaction Relative molecular mass 123.0	**COLORLESS FUMING LIQUID WITH PUNGENT ODOR** Vapor is heavier than air. Decomposes in flame or on hot surface, giving off corrosive vapors (→ *hydrogen bromide*). Reacts violently with water, alcohols and many other compounds, forming corrosive bromide water mist. Reacts with air to form corrosive vapors which are heavier than air and spread at ground level (→ *hydrogen bromide*).

TLV-TWA	not available

Absorption route: Can enter the body by inhalation or ingestion. Harmful atmospheric concentrations can build up very rapidly on evaporation at 20°C.
Immediate effects: Corrosive to the eyes, skin and respiratory tract. Inhalation of vapor/fumes can cause severe breathing difficulties (lung edema). In serious cases risk of death.

Gross formula: C$_2$H$_3$BrO

HAZARDS/SYMPTOMS	PREVENTIVE MEASURES	FIRE EXTINGUISHING/FIRST AID
Fire: non-combustible, many chemical reactions can cause fire and explosion.	avoid contact with water, alcohols and many other substances.	in case of fire in immediate vicinity: DO NOT USE WATER-BASED EXTINGUISHERS.
Explosion:		in case of fire: keep tanks/drums cool by spraying with water, but DO NOT spray substance with water.
	STRICT HYGIENE	
Inhalation: *corrosive*; sore throat, cough, shortness of breath, severe breathing difficulties.	ventilation, local exhaust or respiratory protection.	fresh air, rest, place in half-sitting position, artificial respiration if necessary, take to hospital.
Skin: *corrosive*; redness, pain, burns.	gloves, protective clothing.	remove contaminated clothing, flush skin with water or shower, call a doctor.
Eyes: *corrosive*; redness, pain, impaired vision.	face shield, or combined eye and respiratory protection.	flush with water, take to a doctor.
Ingestion: *corrosive*; sore throat, abdominal pain, vomiting.		rinse mouth, take to hospital.

SPILLAGE	STORAGE	LABELING / NFPA
evacuate area, call in an expert, collect leakage in sealable containers, soak up with sand or other inert absorbent and remove to safe place; (additional individual protection: CHEMICAL SUIT).	keep dry; ventilate at floor level, separate from alcohols.	

NOTES
Reacts violently with extinguishing agents such as foam. Lung edema symptoms usually develop several hours later and are aggravated by physical exertion: rest and hospitalization essential. As first aid, a doctor or authorized person should consider administering a corticosteroid spray. Under no circumstances use near flame or hot surface or when welding. Unbreakable packaging preferred; if breakable, keep in unbreakable container.

Transport Emergency Card TEC(R)-80G14

HI: 80; UN-number: 1716

ACETYL CAPROLACTAM

PHYSICAL PROPERTIES		IMPORTANT CHARACTERISTICS
		COLORLESS LIQUID WITH CHARACTERISTIC ODOR
Boiling point, °C	> 250	Presumed to be able to form peroxides. Decomposes when heated or burned, giving off toxic vapors (→ *nitrous vapors*). Reacts with strong oxidants.
Melting point, °C	− 13	
Flash point, °C	> 110	
Relative density (water = 1)	1.09	**TLV-TWA** not available
Vapor pressure, mm Hg at 130 °C	13	
Solubility in water	good	**Absorption route:** Can enter the body by inhalation of vapor or ingestion. Harmful atmospheric concentrations build up very slowly, if at all, on evaporation at approx. 20° C, but much more rapidly in aerosol form.
Minimum ignition energy, mJ	1.1	
Relative molecular mass	155.2	
		Immediate effects: Irritates the eyes and respiratory tract. If liquid is swallowed, droplets can enter the lungs, with risk of pneumonia. Exposure to high concentrations can cause convulsions.
Gross formula:	$C_8H_{13}NO_2$	

HAZARDS/SYMPTOMS	PREVENTIVE MEASURES	FIRE EXTINGUISHING/FIRST AID
Fire: combustible.	keep away from open flame, no smoking.	powder, alcohol-resistant foam, water spray, carbon dioxide, (halons).
Inhalation: slight irritation, cough, headache.	ventilation.	fresh air, rest, consult a doctor.
Skin: redness, slight irritation.	gloves.	remove contaminated clothing, flush skin with water and wash with soap and water.
Eyes: irritation, redness.	safety glasses.	flush thoroughly with water (remove contact lenses if easy), take to a doctor.
Ingestion: burning sensation, nausea.		rinse mouth, consult a doctor.

SPILLAGE	STORAGE	LABELING / NFPA
collect leakage in sealable containers, soak up with sand or other inert absorbent and remove to safe place.	keep in dark place, separate from oxidants; ventilate at floor level.	

NOTES

Boiling point at 26.6mm Hg approx. 135° C. Before distilling check for peroxides; if found, render harmless. Insufficient data on harmful effects to humans: exercise great caution.

13

ACETYL CHLORIDE

PHYSICAL PROPERTIES		IMPORTANT CHARACTERISTICS
Boiling point, °C	52	**COLORLESS LIQUID WITH PUNGENT ODOR**
Melting point, °C	−112	Vapor is heavier than air and spreads at ground level, with risk of ignition at a distance.
Flash point, °C	4	Decomposes when heated, giving off toxic gas (→ *phosgene*) and corrosive vapor
Auto-ignition temperature, °C	390	(→ *hydrochloric acid*). Do not use compressed air when filling, emptying or processing. Reacts
Relative density (water = 1)	1.1	violently with water, alcohols, bases and many other compounds. Reacts with air to form
Relative vapor density (air = 1)	2.7	corrosive → *hydrochloric acid* fumes and → *acetic acid.*
Relative density at 20 °C of saturated mixture		
vapor/air (air = 1)	1.5	TLV-TWA not available
Vapor pressure, mm Hg at 20 °C	213	
Solubility in water	reaction	**Absorption route:** Can enter the body by inhalation or ingestion. Harmful atmospheric
Explosive limits, vol% in air	5-?	concentrations can build up very rapidly on evaporation at 20°C.
Electrical conductivity, pS/m	4x10^7	**Immediate effects:** Corrosive to the eyes, skin and respiratory tract. Inhalation of vapor/fumes
Relative molecular mass	78.5	can cause severe breathing difficulties (lung edema). In serious cases risk of death.
Gross formula:	C$_2$H$_3$ClO	

HAZARDS/SYMPTOMS	PREVENTIVE MEASURES	FIRE EXTINGUISHING/FIRST AID
Fire: extremely flammable, many chemical reactions can cause fire and explosion.	keep away from open flame and sparks, no smoking.	powder, carbon dioxide, (halons), DO NOT USE WATER-BASED EXTINGUISHERS.
Explosion: forms explosive air-vapor mixtures.	sealed machinery, ventilation, explosion-proof electrical equipment and lighting.	in case of fire: keep tanks/drums cool by spraying with water, but DO NOT spray substance with water.
	STRICT HYGIENE	IN ALL CASES CALL IN A DOCTOR
Inhalation: *corrosive*; sore throat, cough, severe breathing difficulties.	ventilation, local exhaust or respiratory protection.	fresh air, rest, artificial respiration if necessary, take to hospital.
Skin: *corrosive*; redness, pain, burns.	gloves, protective clothing.	remove contaminated clothing, flush skin with water or shower, take to hospital.
Eyes: *corrosive*; redness, pain, impaired vision.	face shield, or combined eye and respiratory protection.	flush with water, take to a doctor.
Ingestion: *corrosive*; sore throat, abdominal pain.		rinse mouth, call a doctor or take to hospital.

SPILLAGE	STORAGE	LABELING / NFPA
evacuate area, call in an expert, collect leakage in sealable containers, soak up with dry sand or other inert absorbent and remove to safe place, combat gas cloud with water curtain; (additional individual protection: CHEMICAL SUIT).	keep in dry, fireproof place, separate from strong bases and alcohols.	R: 11-14-34 S: 9-16-26 Flammable Corrosive 3 / 3 2 / W

NOTES
Lung edema symptoms usually develop several hours later and are aggravated by physical exertion: rest and hospitalization essential. As first aid, a doctor or authorized person should consider administering a corticosteroid spray. Use airtight packaging. Unbreakable packaging preferred; if breakable, keep in unbreakable container.

Transport Emergency Card TEC(R)-55

HI: X338; UN-number: 1717

ACETYLENE

<table>
<tr><th colspan="2">PHYSICAL PROPERTIES</th><th colspan="2">IMPORTANT CHARACTERISTICS</th></tr>
<tr>
<td colspan="2">
Sublimation temperature, °C   −84
Melting point, °C   −82
Flash point, °C   flammable gas
Auto-ignition temperature, °C   305
Relative vapor density (air = 1)   0.9
Vapor pressure, mm Hg at 21 °C   33,896
Solubility in water   poor
Explosive limits, vol% in air   1.5-100
Minimum ignition energy, mJ   0.017
Relative molecular mass   26
</td>
<td colspan="2">
COLORLESS GAS WITH CHARACTERISTIC ODOR, USUALLY DISSOLVED UNDER PRESSURE IN ACETONE
Mixes readily with air, forming explosive mixtures. Reacts with copper, silver and mercury to form shock-sensitive compounds. Decomposes when heated above 160° C or if subjected to shock or rise in pressure, with risk of fire and explosion. Strong reducing agent which reacts violently with oxidants. Reacts violently with fluorine, chlorine, bromine, hypochlorite solutions and many other compounds, with risk of fire and explosion.
</td>
</tr>
<tr>
<td colspan="2"></td>
<td colspan="2">TLV-TWA   not available</td>
</tr>
<tr>
<td colspan="2">Gross formula: C_2H_2</td>
<td colspan="2">
Absorption route: Can enter the body by inhalation. Can saturate the air if released, with risk of suffocation.
Immediate effects: Liquid can cause frostbite due to rapid evaporation. Affects the nervous system.
</td>
</tr>
<tr><th>HAZARDS/SYMPTOMS</th><th colspan="2">PREVENTIVE MEASURES</th><th>FIRE EXTINGUISHING/FIRST AID</th></tr>
<tr>
<td>**Fire:** extremely flammable.</td>
<td colspan="2">keep away from open flame and sparks, no smoking.</td>
<td>shut off supply; if impossible and no danger to surrounding area, allow fire to burn itself out; otherwise extinguish with powder, carbon dioxide, (halons).</td>
</tr>
<tr>
<td>**Explosion:** forms explosive air-gas mixtures,, contact with copper can cause explosion (see notes).</td>
<td colspan="2">sealed machinery, ventilation, explosion-proof electrical equipment and lighting, non-sparking tools, do not subject to shock or friction.</td>
<td>in case of fire: keep cylinders cool by spraying with water, fight fire from sheltered location.</td>
</tr>
<tr>
<td>**Inhalation:** dizziness, nausea, drowsiness, unconsciousness.</td>
<td colspan="2">ventilation, local exhaust or respiratory protection.</td>
<td>fresh air, rest, artificial respiration if necessary, call a doctor and take to hospital.</td>
</tr>
<tr>
<td>**Skin:**</td>
<td colspan="2">insulating gloves.</td>
<td></td>
</tr>
<tr>
<td>**Eyes:**</td>
<td colspan="2">acid goggles, or combined eye and respiratory protection.</td>
<td></td>
</tr>
<tr><th>SPILLAGE</th><th colspan="2">STORAGE</th><th>LABELING / NFPA</th></tr>
<tr>
<td>evacuate area, call in an expert, ventilate; (additional individual protection: breathing apparatus).</td>
<td colspan="2">keep in fireproof place, separate from oxidants and many other substances.</td>
<td>
R: 5-6-12
S: 9-16-33

🔥
Flammable

NFPA: 1 / 4 / 3
</td>
</tr>
<tr><th colspan="4">NOTES</th></tr>
<tr>
<td colspan="4">Material used for lines must not exceed 63% copper content. In confined spaces combustion can cause explosion. High atmospheric concentrations, e.g. in poorly ventilated spaces, can cause oxygen deficiency, with risk of unconsciousness. Use special acetylene detector. Turn leaking cylinder so that leak is on top to prevent flammable acetylene/acetone mixture. Close valve when not in use; check hoses and lines regularly; test for leaking connections with soap solution.</td>
</tr>
</table>

Transport Emergency Card TEC(R)-20G25

HI: 239; UN-number: 1001

ACETYLENE
(cylinder)

PHYSICAL PROPERTIES	IMPORTANT CHARACTERISTICS
Sublimation temperature, °C −84 Melting point, °C −81 Flash point, °C flammable gas Auto-ignition temperature, °C 305 Relative density (water = 1) 0.4 Relative vapor density (air = 1) 0.9 Vapor pressure, mm Hg at 21 °C 13,072 Solubility in water none Explosive limits, vol% in air 2.4-83 Minimum ignition energy, mJ 0.019 Relative molecular mass 26	**GAS DISSOLVED UNDER PRESSURE IN ACETONE WITH CHARACTERISTIC ODOR** Mixes readily with air, forming explosive mixtures. Reacts with copper, silver and mercury to form shock-sensitive compounds. Can produce thermal explosion when heated. Decomposes explosively when moderately heated or subjected to rise in pressure. Strong reducing agent which reacts violently with oxidants. Reacts violently with fluorine, chlorine and bromine, with risk of fire and explosion. Phosphine can occur as contaminant.
	TLV-TWA not available
	Absorption route: Can enter the body by inhalation. **Immediate effects:** Affects the nervous system.
Gross formula: C_2H_2	

HAZARDS/SYMPTOMS	PREVENTIVE MEASURES	FIRE EXTINGUISHING/FIRST AID
Fire: extremely flammable.	keep away from open flame and sparks, no smoking.	shut off supply; if impossible and no danger to surrounding area, allow fire to burn itself out; otherwise extinguish with powder, carbon dioxide, (halons).
Explosion: forms explosive air-gas mixtures, heating in cylinder can cause explosion, avoid contact with copper, mercury, silver, chlorine and fluorine.	sealed machinery, ventilation, explosion-proof electrical equipment and lighting, non-sparking tools, anti-blowback device on torch.	in case of fire: keep cylinders cool by spraying with water, fight fire from sheltered location, cylinders which become warm should be allowed to cool off in safe place under water.
Inhalation: dizziness, nausea, drowsiness, unconsciousness.	local exhaust; ventilation.	fresh air, rest, call a doctor or take to hospital.

SPILLAGE	STORAGE	LABELING / NFPA
evacuate area, call in an expert; ventilate.	keep in cool, fireproof, separate storage area.	◇ 4 / 1 3

NOTES

Material used for lines must not exceed 63% copper content. High atmospheric concentrations, e.g. in poorly ventilated spaces, can cause oxygen deficiency, with risk of unconsciousness. Sublimation point and melting point are for pure acetylene (not solution in acetone). Close valve when not in use; check hoses and lines regularly; test for leaking connections with soap solution. Use cylinders with porous filler. Use special fittings.

CAS-No: [107-02-8]
acrylic aldehyde
acraldehyde
acroleine
acrylaldehyde
2-propanal

$CH_2 = CHCHO$

ACROLEIN

PHYSICAL PROPERTIES		IMPORTANT CHARACTERISTICS	
Boiling point, °C	53	**COLORLESS LIQUID WITH PUNGENT ODOR**	
Melting point, °C	−88	Vapor is heavier than air and spreads at ground level, with risk of ignition at a distance. Presumed to be able to form peroxides. Readily able to polymerize. Do not use compressed air when filling, emptying or processing. Strong reducing agent which reacts violently with oxidants, acids and bases. Reacts violently with many substances, with risk of fire and explosion. Attacks zinc and cadmium.	
Flash point, °C	−26		
Auto-ignition temperature, °C	220		
Relative density (water = 1)	0.8		
Relative vapor density (air = 1)	1.9		
Relative density at 20 °C of saturated mixture			
vapor/air (air = 1)	1.27	**TLV-TWA** 0.1 ppm 0.23 mg/m³	
Vapor pressure, mm Hg at 20 °C	223	STEL 0.3 ppm 0.69 mg/m³	
Solubility in water, g/100 ml at 20 °C	20.6		
Explosive limits, vol% in air	2.8-31	**Absorption route:** Can enter the body by inhalation or ingestion. Harmful atmospheric concentrations can build up very rapidly on evaporation at 20°C. **Immediate effects:** Corrosive to the eyes, skin and respiratory tract. Inhalation of vapor/fumes can cause severe breathing difficulties (lung edema). In serious cases risk of death.	
Minimum ignition energy, mJ	0.13		
Electrical conductivity, pS/m	1.55x10⁷		
Relative molecular mass	56.1		
Gross formula:	C_3H_4O		

HAZARDS/SYMPTOMS	PREVENTIVE MEASURES	FIRE EXTINGUISHING/FIRST AID
Fire: extremely flammable.	keep away from open flame and sparks, no smoking.	powder, AFFF, foam, carbon dioxide, (halons).
Explosion: forms explosive air-vapor mixtures.	sealed machinery, ventilation, explosion-proof electrical equipment and lighting, non-sparking tools, do not subject to shock or friction.	in case of fire: keep tanks/drums cool by spraying with water.
	STRICT HYGIENE	IN ALL CASES CALL A DOCTOR
Inhalation: *corrosive*; sore throat, cough, shortness of breath, severe breathing difficulties.	ventilation, local exhaust or respiratory protection.	fresh air, rest, place in half-sitting position, take to hospital.
Skin: *corrosive*; redness, pain, burns.	gloves, protective clothing.	remove contaminated clothing, flush skin with water or shower, refer to a doctor.
Eyes: *corrosive*; redness, pain, impaired vision.	face shield, or combined eye and respiratory protection.	flush with water, take to a doctor.
Ingestion: *corrosive*; sore throat, abdominal pain, diarrhea.		rinse mouth, take immediately to hospital.

SPILLAGE	STORAGE	LABELING / NFPA
evacuate area, call in an expert, render small quantities harmless with 10% sodium bisulfite solution, soak up with sand or other inert absorbent and remove to safe place, do not use sawdust or other combustible absorbents; (additional individual protection: CHEMICAL SUIT).	keep in fireproof place, separate from oxidants, acids and bases; add an inhibitor.	R: 11-23-36/37/38 S: 29-33-44 Note D Flammable Toxic 3 / 3 / 3

NOTES

Before distilling check for peroxides; if found, render harmless. Odor limit is above TLV. Lung edema symptoms usually develop several hours later and are aggravated by physical exertion: rest and hospitalization ssential. As first aid, a doctor or authorized person should consider administering a corticosteroid spray. Strong irritation warns of concentrations below acute danger level (watering of the eyes). Unbreakable packaging preferred; if breakable, keep in unbreakable container.

Transport Emergency Card TEC(R)-118

HI: 336; UN-number: 1092

acrylic amide
ethylenecarboxamide
propenamide
propenoic acid amide
vinyl amide

$$CH_2 = CHCONH_2$$

ACRYLAMIDE

PHYSICAL PROPERTIES	IMPORTANT CHARACTERISTICS	
Melting point, °C 84 Relative density (water = 1) 1.1 Relative density at 20 °C of saturated mixture vapor/air (air = 1) 1.0 Vapor pressure, mm Hg at 20 °C 6.8x10^{-3} Solubility in water, g/100 ml at 30 °C 215 Relative molecular mass 71.1	**WHITE CRYSTALS OR WHITE POWDER** Polymerizes readily when heated above 84°C or exposed to light. Decomposes when heated or burned, giving off toxic gases (→ *nitrous vapors*). Reacts violently with strong oxidants.	
	TLV-TWA (skin)	0.03 mg/m³ A2
	Absorption route: Can enter the body by inhalation or ingestion or through the skin. Harmful atmospheric concentrations build up fairly slowly on evaporation at 20°C, but more rapidly in the form of airborne particles or aerosol. **Immediate effects:** Irritates the eyes, skin and respiratory tract. Affects the nervous system. **Effects of prolonged/repeated exposure:** Can cause liver damage.	
Gross formula: C_3H_5NO		

HAZARDS/SYMPTOMS	PREVENTIVE MEASURES	FIRE EXTINGUISHING/FIRST AID
Fire: combustible.	keep away from open flame, no smoking.	water spray, powder.
Explosion: finely dispersed particles form explosive mixtures on contact with air.	keep dust from accumulating; sealed machinery, explosion-proof electrical equipment and lighting, grounding.	
	STRICT HYGIENE, KEEP DUST UNDER CONTROL	IN ALL CASES CALL IN A DOCTOR
Inhalation: sore throat, cough, tiredness, feeling of weakness.	local exhaust or respiratory protection.	fresh air, rest, place in half-sitting position, call a doctor.
Skin: *is absorbed*; redness, pain.	gloves, protective clothing.	remove contaminated clothing, flush skin with water or shower, refer to a doctor.
Eyes: redness, pain.	face shield, or combined eye and respiratory protection.	flush with water, take to a doctor.
Ingestion: abdominal pain, slackness of muscles, shaking.	do not eat, drink or smoke while working.	rinse mouth, immediately take to hospital.

SPILLAGE	STORAGE	LABELING / NFPA
clean up spillage, carefully collect remainder; (additional individual protection: breathing apparatus).	keep in cool, dark place, separate from oxidants; add an inhibitor.	R: 23/24/25-33 S: 27-44 ☠ Toxic

NFPA: 2 (top) 3 (left) 2 (right)

NOTES
Alcohol consumption increases toxic effects. Depending on the degree of exposure, regular medical checkups are advisable.

CAS-No: [79-10-7]
vinylformic acid
acroleic acid
ethylenecarboxylic acid
propeneacid
propenoic acid

$CH_2 = CHCOOH$

ACRYLIC ACID

PHYSICAL PROPERTIES	IMPORTANT CHARACTERISTICS
Boiling point, °C　　　　　　　　　141 Melting point, °C　　　　　　　　　13 Flash point, °C　　　　　　　　　　54 Auto-ignition temperature, °C　　　375 Relative density (water = 1)　　　　1.06 Relative vapor density (air = 1)　　　2.5 Relative density at 20 °C of saturated mixture vapor/air (air = 1)　　　　　　　　1.01 Vapor pressure, mm Hg at 20 °C　　3.3 Solubility in water　　　　　　　　∞ Explosive limits, vol% in air　　　5.3-26 Relative molecular mass　　　　　72.1 Log P octanol/water　　　　　　　0.4 Gross formula:　　　　　　　　$C_3H_4O_2$	**COLORLESS LIQUID WITH CHARACTERISTIC ODOR** Vapor mixes readily with air. Moderate heating, exposure to light or peroxides or other activators can cause violent polymerization, with risk of fire and explosion. Medium strong acid. Reacts violently with strong bases. Reacts violently with oxidants, with risk of fire and explosion. Attacks many metals (incl. nickel and copper), giving off flammable gas (→ *hydrogen*). TLV-TWA (skin)　　　　2 ppm　　　　5.9 mg/m³ **Absorption route:** Can enter the body by inhalation or ingestion. Harmful atmospheric concentrations can build up fairly rapidly on evaporation at approx. 20°C - even more rapidly in aerosol form. **Immediate effects:** Corrosive to the eyes, skin and respiratory tract. **Effects of prolonged/repeated exposure:** Can cause kidney damage.

HAZARDS/SYMPTOMS	PREVENTIVE MEASURES	FIRE EXTINGUISHING/FIRST AID
Fire: flammable.	keep away from open flame and sparks, no smoking.	powder, alcohol-resistant foam, water spray, carbon dioxide, (halons).
Explosion: above 54°C: forms explosive air-vapor mixtures.	above 54°C: sealed machinery, ventilation, explosion-proof electrical equipment.	in case of fire: keep tanks/drums cool by spraying with water.
Inhalation: *corrosive*; sore throat, cough, shortness of breath.	ventilation, local exhaust or respiratory protection.	fresh air, rest, place in half-sitting position, call a doctor.
Skin: *corrosive*; redness, pain, burns.	gloves, protective clothing.	remove contaminated clothing, flush skin with water or shower, call a doctor.
Eyes: *corrosive*; redness, pain, impaired vision.	face shield, or combined eye and respiratory protection.	flush with water, take to doctor.
Ingestion: *corrosive*; sore throat, abdominal pain, vomiting.		rinse mouth, call a doctor or take to hospital.

SPILLAGE	STORAGE	LABELING / NFPA
collect leakage in sealable containers, flush away remainder with water; (additional individual protection: breathing apparatus).	keep in cool, fireproof place, separate from oxidants and strong bases; add an inhibitor.	R: 10-34 S: 26-36 Corrosive 3　2　2

NOTES
Log P octanol/water is estimated. If stored below melting point: first melt entire mass, then stir and add inhibitor before drawing off. Vapor contains no inhibitor; condensate can therefore polymerize in exhaust or ventilation facilities, with risk of breakdown.

Transport Emergency Card TEC(R)-706

HI: 89; UN-number: 2218

CAS-No: [107-13-1]
AN
cyanoethylene
2-propenenitrile
vinyl cyanide

$CH_2 = CHC \equiv N$

ACRYLONITRILE

PHYSICAL PROPERTIES		IMPORTANT CHARACTERISTICS
Boiling point, °C	77	**COLORLESS LIQUID WITH CHARACTERISTIC ODOR**
Melting point, °C	−82	Vapor mixes readily with air, forming explosive mixtures. Readily able to polymerize when
Flash point, °C	−5	moderately heated and exposed to light. Polymerizes violently in the presence of bases or
Auto-ignition temperature, °C	480	peroxides. Do not use compressed air when filling, emptying or processing. Reacts violently with
Relative density (water = 1)	0.8	strong oxidants. Attacks copper and its alloys.
Relative vapor density (air = 1)	1.83	

Relative density at 20 °C of saturated mixture vapor/air (air = 1)   1.1

TLV-TWA (skin)   2 ppm   4.3 mg/m³ A2

Vapor pressure, mm Hg at 20 °C   83.6
Solubility in water, g/100 ml at 20 °C   7.3
Explosive limits, vol% in air   2.8-28
Minimum ignition energy, mJ   0.16
Relative molecular mass   53.1
Log P octanol/water   −0.9

Absorption route: Can enter the body by inhalation or ingestion or through the skin. Harmful atmospheric concentrations can build up very rapidly on evaporation at 20°C.
Immediate effects: Irritates the eyes and respiratory tract. Impedes tissue respiration. Corrosive to the skin, but sometimes not until several hours later. Affects the nervous system. In serious cases risk of unconsciousness and death.
Effects of prolonged/repeated exposure: Has been found to cause a type of cancer in humans under certain circumstances.

Gross formula: C_3H_3N

HAZARDS/SYMPTOMS	PREVENTIVE MEASURES	FIRE EXTINGUISHING/FIRST AID
Fire: extremely flammable.	keep away from open flame and sparks, no smoking.	powder, AFFF, foam, carbon dioxide, (halons).
Explosion: forms explosive air-vapor mixtures.	sealed machinery, ventilation, explosion-proof electrical equipment and lighting, grounding, non-sparking tools.	in case of fire: keep tanks/drums cool by spraying with water.
	STRICT HYGIENE	IN ALL CASES CALL A DOCTOR
Inhalation: sore throat, severe breathing difficulties, dizziness, nausea, feeling of weakness.	ventilation, local exhaust or respiratory protection.	fresh air, rest, special treatment, take to hospital.
Skin: *is absorbed*; redness; see also 'Inhalation'.	gloves, protective clothing.	remove contaminated clothing, flush skin with water or shower, refer to a doctor.
Eyes: redness, pain.	face shield, or combined eye and respiratory protection.	flush with water, take to a doctor.
Ingestion: sore throat, severe breathing difficulties, nausea, dizziness, feeling of weakness.		rinse mouth, special treatment, take immediately to hospital.

SPILLAGE	STORAGE	LABELING / NFPA
evacuate area, call in an expert, collect leakage in non-sealable containers, soak up with sand or other inert absorbent and remove to safe place, render remainder harmless with sodium hypochlorite solution; (additional individual protection: breathing apparatus).	keep in cool, dark, fireproof place, separate from oxidants and strong bases; add an inhibitor.	R: 45-11-23/24/25-38 S: 53-16-27-44 Note D + E Flammable Toxic

NOTES

Evaporation of acrylonitrile from leather is negligible and with time causes burning of the skin: avoid leather clothing (incl. shoes). Odor limit is above TLV. Special first aid required in the event of poisoning: antidotes must be available (with instructions).

Transport Emergency Card TEC(R)-61

HI: 336; UN-number: 1093

ADIPIC ACID

PHYSICAL PROPERTIES		IMPORTANT CHARACTERISTICS
Boiling point (decomposes), °C	337	**WHITE CRYSTALLINE POWDER**
Melting point, °C	152	In dry state can form electrostatic charge if stirred, transported pneumatically, poured etc.
Flash point, °C	196	Reacts with strong oxidants.
Auto-ignition temperature, °C	420	
Relative density (water = 1)	1.36	TLV-TWA not available
Vapor pressure, mm Hg at 20 °C	<0.00076	
Solubility in water, g/100 ml at 20 °C	1.5	**Absorption route:** Can enter the body by inhalation or ingestion. Evaporation negligible at 20°C, but unpleasant atmospheric concentrations can build up rapidly in aerosol form.
Relative molecular mass	146.1	**Immediate effects:** Irritates the eyes and respiratory tract.
Log P octanol/water	0.1	
Gross formula:	$C_6H_{10}O_4$	

HAZARDS/SYMPTOMS	PREVENTIVE MEASURES	FIRE EXTINGUISHING/FIRST AID
Fire: combustible.	keep away from open flame, no smoking.	water spray, powder.
Explosion: finely dispersed particles form explosive mixtures on contact with air.	keep dust from accumulating, sealed machinery, explosion-proof electrical equipment and lighting, grounding.	
Inhalation: sore throat, cough.	local exhaust or respiratory protection.	fresh air, rest.
Skin: redness.	gloves.	remove contaminated clothing, flush skin with water or shower.
Eyes: redness, pain.	goggles.	flush with water, take to a doctor if necessary.
Ingestion: sore throat, abdominal pain.		rinse mouth, call a doctor.

SPILLAGE	STORAGE	LABELING / NFPA
clean up spillage, flush away remainder with water; (additional individual protection: P1 respirator).		R: 36 ✖ Irritant

NOTES

Transport Emergency Card TEC(R)-61G05

adipic acid dinitrile
adipylnitrile
1,4-dicyanobutane
tetramethylene cyanide

$CN(CH_2)_4C \equiv N$

ADIPONITRILE

PHYSICAL PROPERTIES	IMPORTANT CHARACTERISTICS
Boiling point, °C .... 295 Melting point, °C .... 2 Flash point, °C .... 93 Auto-ignition temperature, °C .... 460 Relative density (water = 1) .... 0.96 Relative vapor density (air = 1) .... 3.7 Relative density at 20 °C of saturated mixture vapor/air (air = 1) .... 1.0 Vapor pressure, mm Hg at 20 °C ... 0.0023 Solubility in water, g/100 ml at 20 °C ... 4.5 Relative molecular mass 108.1	**COLORLESS OILY LIQUID** Vapor mixes readily with air. Toxic vapors are a product of combustion. Decomposes when heated above 93°C, giving off toxic vapors (→ *hydrogen cyanide*). Reacts violently with strong oxidants.
	TLV-TWA not available
	Absorption route: Can enter the body by inhalation or ingestion or through the skin. Harmful atmospheric concentrations build up very slowly, if at all, on evaporation at approx. 20°C, but much more rapidly in aerosol form. **Immediate effects:** Irritates the eyes, skin and respiratory tract. Impedes tissue respiration. In serious cases risk of seizures, unconsciousness and death.
Gross formula: $C_6H_8N_2$	

HAZARDS/SYMPTOMS	PREVENTIVE MEASURES	FIRE EXTINGUISHING/FIRST AID
Fire: combustible.	keep away from open flame, no smoking.	powder, AFFF, foam, carbon dioxide, (halons).
Explosion: above 93°C: forms explosive air-vapor mixtures.	above 93°C: sealed machinery, ventilation.	
Inhalation: sore throat, shortness of breath, headache, cramps, unconsciousness.	ventilation, local exhaust or respiratory protection.	fresh air, rest, special treatment, take to hospital.
Skin: *is absorbed*; redness, pain.	gloves.	remove contaminated clothing, flush skin with water or shower, call a doctor.
Eyes: redness, pain, impaired vision.	face shield.	flush with water, take to a doctor.
Ingestion: abdominal pain; see also 'Inhalation'.	do not eat, drink or smoke while working.	rinse mouth, special treatment, call a doctor or take to hospital.

SPILLAGE	STORAGE	LABELING / NFPA
collect leakage in sealable containers, soak up with sand or other inert absorbent and remove to safe place.	keep separate from oxidants; ventilate at floor level.	NFPA diamond: Health 2, Fire 1, Reactivity (blank), 4

NOTES

Flash point applies to the technical grade. Pure adiponitrile has a flash point of 163°C. Special first aid required in the event of poisoning: antidotes must be available (with instructions).

AIR
(liquid, refrigerated)

PHYSICAL PROPERTIES	IMPORTANT CHARACTERISTICS
Boiling point, °C −190 Melting point, °C −216 Relative density (water = 1) see notes Relative vapor density (air = 1) see notes Solubility in water, g/100 ml at 20 °C 0.02 Relative molecular mass 29.0 (mean)	**COLORLESS, EXTREMELY COLD LIQUID** Mixture of approx. 80% nitrogen and 20% oxygen. Cold vapor given off by liquid is heavier than surrounding air. When stored for long periods, nitrogen evaporates first, leaving liquid with very high oxygen content, with increased risk of fire (see liquid oxygen).
	TLV-TWA not available
	Immediate effects: Cold liquid can cause frostbite.

HAZARDS/SYMPTOMS	PREVENTIVE MEASURES	FIRE EXTINGUISHING/FIRST AID
Fire: non-combustible.	avoid contact with combustible substances.	in case of fire in immediate vicinity: preferably no water.
Skin: *is absorbed*; redness, pain, blisters.	insulated gloves, protective clothing.	*in case of frostbite*: DO NOT remove clothing, flush skin with water or shower, take to a doctor.
Eyes: *in case of frostbite*: redness, pain, impaired vision.	face shield.	flush with water, take to a doctor.

SPILLAGE	STORAGE	LABELING / NFPA
ventilate, under no circumstances spray liquid with water.		

NOTES
Density of liquid at boiling point: 0.9 kg/l. Relative density in relation to air at the same temperature is, of course, 1.00, but very cold vapor is much heavier than surrounding air. Faster evaporation of nitrogen component causes rise in oxygen content of remaining liquid, with increased risk of fire. Top up frequently. Use special insulated pressure vessel.

Transport Emergency Card TEC(R)-707   **HI: 225; UN-number: 1003**

2-propen-1-ol
1-propen-3-ol
2-propenyl alcohol
vinyl carbinol

$$CH_2 = CHCH_2OH$$

ALLYL ALCOHOL

PHYSICAL PROPERTIES	IMPORTANT CHARACTERISTICS
Boiling point, °C 97 Melting point, °C −129 Flash point, °C 21 Auto-ignition temperature, °C 375 Relative density (water = 1) 0.9 Relative vapor density (air = 1) 2.0 Relative density at 20 °C of saturated mixture vapor/air (air = 1) 1.02 Vapor pressure, mm Hg at 20 °C 18 Solubility in water ∞ Explosive limits, vol% in air 2.5-18 Electrical conductivity, pS/m 6.5×10^8 Relative molecular mass 58 Log P octanol/water 0.2 Gross formula: C_3H_6O	**COLORLESS LIQUID WITH PUNGENT ODOR** Vapor mixes readily with air, forming explosive mixtures. Presumed to be able to form peroxides. Can polymerize when heated above 100°C or exposed to light. Strong reducing agent which reacts violently with oxidants, with risk of fire and explosion. TLV-TWA (skin) 2 ppm 5.8 mg/m³ STEL 4 ppm 9.5 mg/m³ **Absorption route:** Can enter the body by inhalation or ingestion or through the skin. Harmful atmospheric concentrations can build up fairly rapidly on evaporation at approx. 20°C - even more rapidly in aerosol form. **Immediate effects:** Corrosive to the eyes, skin and respiratory tract. Affects the nervous system, liver and kidneys. Inhalation of vapor/fumes can cause severe breathing difficulties (lung edema). In serious cases risk of death.

HAZARDS/SYMPTOMS	PREVENTIVE MEASURES	FIRE EXTINGUISHING/FIRST AID
Fire: flammable.	keep away from open flame and sparks, no smoking.	powder, alcohol-resistant foam, large quantities of water, carbon dioxide, (halons).
Explosion: above 21°C: forms explosive air-vapor mixtures.	above 21°C: sealed machinery, ventilation, explosion-proof electrical equipment.	in case of fire: keep tanks/drums cool by spraying with water.
	STRICT HYGIENE	IN ALL CASES CALL A DOCTOR
Inhalation: *corrosive*; sore throat, cough, shortness of breath, severe breathing difficulties, headache.	ventilation, local exhaust or respiratory protection.	fresh air, rest, place in half-sitting position, take to hospital.
Skin: *corrosive, is absorbed*; redness, pain, burns.	gloves, protective clothing.	remove contaminated clothing, flush skin with water or shower, take to hospital.
Eyes: *corrosive*; redness, pain, impaired vision, light-sensitivity.	face shield, or combined eye and respiratory protection.	flush with water, take to doctor if necessary.
Ingestion: *corrosive*; sore throat, abdominal pain, diarrhea, nausea, vomiting.		rinse mouth, refer to a doctor.

SPILLAGE	STORAGE	LABELING / NFPA
call in an expert, collect leakage in sealable containers, soak up with sand or other inert absorbent and remove to safe place; (additional individual protection: breathing apparatus).	keep in fireproof place, separate from oxidants.	R: 11-26-36/37/38 S: 16-39-45 🔥 Flammable ☠ Highly toxic NFPA: 3/3/1

NOTES
Before distilling check for peroxides; if found, render harmless. Odor limit is above TLV. Fumes or vapors can also damage eyes. Lung edema symptoms usually develop several hours later and are aggravated by physical exertion: rest and hospitalization essential. As first aid, a doctor or authorized person should consider administering a corticosteroid spray. Use airtight packaging. Unbreakable packaging preferred; if breakable, keep in unbreakable container.

Transport Emergency Card TEC(R)-116

HI: 663; UN-number: 1098

3-amino-1-propene
3-aminopropylene
2-propenamide
2-propenylamine

$CH_2 = CHCH_2NH_2$

ALLYLAMINE

PHYSICAL PROPERTIES		IMPORTANT CHARACTERISTICS
Boiling point, °C	53	**COLORLESS TO LIGHT YELLOW LIQUID WITH PUNGENT ODOR**
Melting point, °C	−88	Vapor is heavier than air and spreads at ground level, with risk of ignition at a distance.
Flash point, °C	−29	Decomposes when heated or burned, giving off toxic vapors (→ *nitrous vapors*). Medium strong
Auto-ignition temperature, °C	370	base which reacts violently with acids and corrodes aluminum, tin, zinc etc. Also attacks copper.
Relative density (water = 1)	0.8	Reacts violently with strong oxidants, with risk of fire and explosion.
Relative vapor density (air = 1)	2.0	
Relative density at 20 °C of saturated mixture		
vapor/air (air = 1)	1.3	TLV-TWA not available
Vapor pressure, mm Hg at 20 °C	195	
Solubility in water	∞	**Absorption route:** Can enter the body by inhalation of vapor or ingestion. Harmful atmospheric
Explosive limits, vol% in air	2.2-22	concentrations can build up very rapidly on evaporation at 20°C.
Electrical conductivity, pS/m	5.7×10^9	**Immediate effects:** Corrosive to the eyes, skin and respiratory tract. Inhalation of vapor/fumes
Relative molecular mass	57.1	can cause lung edema. If liquid is swallowed, droplets can enter the lungs, with risk of pneumonia.
		Exposure to high concentrations can be fatal. Effects can be delayed. Keep under medical
		observation.
Gross formula:	C_3H_7N	

HAZARDS/SYMPTOMS	PREVENTIVE MEASURES	FIRE EXTINGUISHING/FIRST AID
Fire: extremely flammable.	keep away from open flame and sparks, no smoking.	powder, alcohol-resistant foam, large quantities of water, carbon dioxide, (halons).
Explosion: forms explosive air-vapor mixtures.	sealed machinery, ventilation, explosion-proof electrical equipment and lighting; do not use compressed air when filling, emptying or processing.	in case of fire: keep tanks/drums cool by spraying with water.
	PREVENT VAPORIZATION, STRICT HYGIENE	IN ALL CASES CALL A DOCTOR
Inhalation: *corrosive*; cough, shortness of breath, sore throat.	ventilation, local exhaust or respiratory protection.	fresh air, rest, place in half-sitting position, take immediately to hospital.
Skin: *corrosive*; redness, pain.	gloves, protective clothing.	remove contaminated clothing, flush skin with water or shower, refer to a doctor.
Eyes: *corrosive*; redness, pain, impaired vision.	face shield, or combined eye and respiratory protection.	flush thoroughly with water (remove contact lenses if easy), take to a doctor.
Ingestion: *corrosive*; stomach cramps, nausea, sore throat, vomiting.		rinse mouth, DO NOT induce vomiting, take immediately to hospital.

SPILLAGE	STORAGE	LABELING / NFPA
evacuate area, call in an expert, ventilate, collect leakage in sealable containers, soak up with sand or other inert absorbent with added sodium bisulfate and remove to safe place; (additional individual protection: breathing apparatus).	keep in fireproof place, separate from oxidants and acids.	R: 11-23/24/25 S: 9-16-24/25-44 Flammable Toxic

NOTES
Lung edema symptoms usually develop several hours later and are aggravated by physical exertion: rest and hospitalization essential. As first aid, a doctor or authorized person should consider administering a corticosteroid spray. Unbreakable packaging preferred; if breakable, keep in unbreakable container.

Transport Emergency Card TEC(R)-165

HI: 336; UN-number: 2334

CAS-No: [106-95-6]
1-bromo-2-propene
3-bromopropene
3-bromopropylene

$BrCH_2CH = CH_2$

ALLYL BROMIDE

<table>
<tr>
<th colspan="2">PHYSICAL PROPERTIES</th>
<th colspan="2">IMPORTANT CHARACTERISTICS</th>
</tr>
<tr>
<td colspan="2">
Boiling point, °C   71
Melting point, °C   −119
Flash point, °C   −1
Auto-ignition temperature, °C   295
Relative density (water = 1)   1.4
Relative vapor density (air = 1)   4.2
Relative density at 20 °C of saturated mixture
vapor/air (air = 1)   1.5
Vapor pressure, mm Hg at 20 °C   ca. 112
Solubility in water   none
Explosive limits, vol% in air   4.3-7.3
Relative molecular mass   121.0
</td>
<td colspan="2">
COLORLESS LIQUID WITH PUNGENT ODOR
Vapor is heavier than air and spreads at ground level, with risk of ignition at a distance. Do not use compressed air when filling, emptying or processing. Decomposes when heated or burned or exposed to light, giving off corrosive vapors (→ *hydrogen bromide*). Reacts violently with oxidants.

TLV-TWA   not available

Absorption route: Can enter the body by inhalation or ingestion or through the skin. Harmful atmospheric concentrations can build up very rapidly on evaporation at 20°C.
Immediate effects: Corrosive to the eyes, skin and respiratory tract. Inhalation of vapor/fumes can cause severe breathing difficulties (lung edema). In serious cases risk of death.
Effects of prolonged/repeated exposure: Prolonged or repeated contact with the vapor can cause skin disorders. Can cause liver and kidney damage.
</td>
</tr>
<tr>
<td colspan="2">Gross formula: C_3H_5Br</td>
<td colspan="2"></td>
</tr>
<tr>
<th>HAZARDS/SYMPTOMS</th>
<th>PREVENTIVE MEASURES</th>
<th colspan="2">FIRE EXTINGUISHING/FIRST AID</th>
</tr>
<tr>
<td>**Fire:** extremely flammable.</td>
<td>keep away from open flame and sparks, no smoking, avoid contact with oxidants.</td>
<td colspan="2">powder, AFFF, foam, carbon dioxide, (halons).</td>
</tr>
<tr>
<td>**Explosion:** forms explosive air-vapor mixtures.</td>
<td>sealed machinery, ventilation, explosion-proof electrical equipment and lighting.</td>
<td colspan="2">in case of fire: keep tanks/drums cool by spraying with water.</td>
</tr>
<tr>
<td></td>
<td>STRICT HYGIENE</td>
<td colspan="2">IN ALL CASES CALL IN A DOCTOR</td>
</tr>
<tr>
<td>**Inhalation:** *corrosive*; sore throat, cough, shortness of breath, severe breathing difficulties.</td>
<td>ventilation, local exhaust or respiratory protection.</td>
<td colspan="2">fresh air, rest, take to hospital.</td>
</tr>
<tr>
<td>**Skin:** *corrosive*; redness, pain, serious burns.</td>
<td>gloves, protective clothing.</td>
<td colspan="2">remove contaminated clothing, flush skin with water or shower, call a doctor.</td>
</tr>
<tr>
<td>**Eyes:** *corrosive*; redness, pain, impaired vision.</td>
<td>face shield, or combined eye and respiratory protection.</td>
<td colspan="2">flush with water, take to a doctor.</td>
</tr>
<tr>
<td>**Ingestion:** *corrosive*; abdominal pain, diarrhea, vomiting.</td>
<td></td>
<td colspan="2">rinse mouth, immediately take to hospital.</td>
</tr>
<tr>
<th>SPILLAGE</th>
<th>STORAGE</th>
<th colspan="2">LABELING / NFPA</th>
</tr>
<tr>
<td>evacuate area, call in an expert, collect leakage in sealable containers, soak up with sand or other inert absorbent and remove to safe place; (additional individual protection: CHEMICAL SUIT).</td>
<td>keep in dark, fireproof place, separate from oxidants.</td>
<td colspan="2">
3
3   1
</td>
</tr>
<tr>
<th colspan="4">NOTES</th>
</tr>
<tr>
<td colspan="4">Depending on the degree of exposure, regular medical checkups are advisable. Lung edema symptoms usually develop several hours later and are aggravated by physical exertion: rest and hospitalization essential. As first aid, a doctor or authorized person should consider administering a corticosteroid spray. Unbreakable packaging preferred; if breakable, keep in unbreakable container.</td>
</tr>
<tr>
<td colspan="2">**Transport Emergency Card TEC(R)-30G32**</td>
<td colspan="2">**HI: 336; UN-number: 1099**</td>
</tr>
</table>

3-chloro-1-propene
3-chloropropylene

$ClCH_2CH = CH_2$

ALLYL CHLORIDE

PHYSICAL PROPERTIES	IMPORTANT CHARACTERISTICS
Boiling point, °C 45 Melting point, °C −135 Flash point, °C −29 Auto-ignition temperature, °C 390 Relative density (water = 1) 0.9 Relative vapor density (air = 1) 2.6 Relative density at 20 °C of saturated mixture vapor/air (air = 1) 1.6 Vapor pressure, mm Hg at 20 °C 300 Solubility in water poor Explosive limits, vol% in air 3.2-11.2 Relative molecular mass 76.5 Gross formula: C_3H_5Cl	**COLORLESS LIQUID WITH PUNGENT ODOR** Vapor is heavier than air and spreads at ground level, with risk of ignition at a distance. Presumed to be able to form peroxides. Able to polymerize on exposure to light or due to peroxide formation. Do not use compressed air when filling, emptying or processing. Toxic and corrosive gases are a product of combustion. Decomposes slowly on contact with water, forming → *hydrochloric acid*. Reacts violently with strong oxidants. Attacks steel and many other metals, esp. in the presence of moisture.

TLV-TWA	1 ppm	3 mg/m³
STEL	2 ppm	6.0 mg/m³

Absorption route: Can enter the body by inhalation. Harmful atmospheric concentrations can build up very rapidly on evaporation at 20°C.
Immediate effects: Corrosive to the eyes, skin and respiratory tract. Inhalation of vapor/fumes can cause severe breathing difficulties (lung edema). In serious cases risk of death.
Effects of prolonged/repeated exposure: Can cause liver and kidney damage.

HAZARDS/SYMPTOMS	PREVENTIVE MEASURES	FIRE EXTINGUISHING/FIRST AID
Fire: extremely flammable.	keep away from open flame and sparks, no smoking.	powder, AFFF, foam, carbon dioxide, (halons).
Explosion: forms explosive air-vapor mixtures.	sealed machinery, ventilation, explosion-proof electrical equipment and lighting.	in case of fire: keep tanks/drums cool by spraying with water.
	STRICT HYGIENE	
Inhalation: *corrosive*; sore throat, cough, shortness of breath, severe breathing difficulties, headache, nausea, vomiting, abdominal cramps.	ventilation, local exhaust or respiratory protection.	fresh air, rest, place in half-sitting position, take to hospital.
Skin: *corrosive*; redness, pain, serious burns.	gloves, protective clothing.	remove contaminated clothing, wash skin with soap and water, take to hospital.
Eyes: *corrosive*; redness, pain, impaired vision.	face shield, or combined eye and respiratory protection.	flush with water, take to a doctor if necessary.
Ingestion: *corrosive*; see Notes.		rinse mouth, immediately take to hospital.

SPILLAGE	STORAGE	LABELING / NFPA
evacuate area, call in an expert, collect leakage in sealable containers, soak up with sand or other inert absorbent and remove to safe place; (additional individual protection: CHEMICAL SUIT).	keep in dry, dark, fireproof place, separate from oxidants.	R: 11-26 S: 16-29-33-45 Note D 🔥 Flammable ☠ Toxic NFPA: 3 / 3 / 1

NOTES

Odor limit is above TLV. Depending on the degree of exposure, regular medical checkups are advisable. Lung edema symptoms usually develop several hours later and are aggravated by physical exertion: rest and hospitalization essential. As first aid, a doctor or authorized person should consider administering a corticosteroid spray. Ingestion is virtually impossible, due to nauseous odor. Effects on the skin can take some hours to become apparent, in the form of a deep-seated severe pain at the point of contact. Packaging: special material.

Transport Emergency Card TEC(R)-106

HI: 336; UN-number: 1100

ALLYL ETHER

PHYSICAL PROPERTIES		IMPORTANT CHARACTERISTICS
Boiling point, °C	94	**COLORLESS LIQUID WITH CHARACTERISTIC ODOR**
Melting point, °C	?	Vapor is heavier than air and spreads at ground level, with risk of ignition at a distance. Readily
Flash point, °C	−7	able to form peroxides. Flow, agitation etc. can cause build-up of electrostatic charge due to
Relative density (water = 1)	0.8	liquid's low conductivity. Do not use compressed air when filling, emptying or processing. Reacts
Relative vapor density (air = 1)	3.4	violently with oxidants, with risk of fire and explosion.
Relative density at 20°C of saturated mixture		
vapor/air (air = 1)	1.14	TLV-TWA not available
Vapor pressure, mm Hg at 20°C	44	
Solubility in water	none	**Absorption route:** Can enter the body by inhalation or ingestion or through the skin. Insufficient
Relative molecular mass	98.2	data on the rate at which harmful concentrations can build up.
Log P octanol/water	0.7	**Immediate effects:** Irritates the eyes and respiratory tract. Liquid destroys the skin's natural oils.
		Affects the nervous system.
Gross formula: $C_6H_{10}O$		

HAZARDS/SYMPTOMS	PREVENTIVE MEASURES	FIRE EXTINGUISHING/FIRST AID
Fire: extremely flammable.	keep away from open flame and sparks, no smoking.	powder, water spray, foam, carbon dioxide, (halons).
Explosion: forms explosive air-vapor mixtures.	sealed machinery, ventilation, explosion-proof electrical equipment and lighting, grounding.	in case of fire: keep tanks/drums cool by spraying with water.
Inhalation: cough, drowsiness, sleepiness.	ventilation, local exhaust or respiratory protection.	fresh air, rest.
Skin: *is absorbed*; redness, pain.	gloves.	remove contaminated clothing, flush skin with water or shower.
Eyes: redness, pain.	face shield.	flush with water, take to doctor if necessary.
Ingestion: nausea, drowsiness.		rinse mouth, call a doctor.

SPILLAGE	STORAGE	LABELING / NFPA
collect leakage in sealable containers, soak up with sand or other inert absorbent and remove to safe place; (additional individual protection: breathing apparatus).	keep in dark, fireproof place, separate from oxidants; add an inhibitor.	3 3 2

NOTES
Log P octanol/water is estimated. Before distilling check for peroxides; if found, render harmless. Alcohol consumption increases toxic effects.

Transport Emergency Card TEC(R)-30G32

HI: 336; UN-number: 2360

$C_2H_5OCH_2CH=CH_2$

ALLYL ETHYL ETHER

PHYSICAL PROPERTIES	IMPORTANT CHARACTERISTICS
Boiling point, °C 66 Flash point, °C <21 Relative density (water = 1) 0.8 Relative vapor density (air = 1) 3.0 Relative density at 20 °C of saturated mixture vapor/air (air = 1) 1.35 Vapor pressure, mm Hg at 20 °C 136 Solubility in water none Explosive limits, vol% in air ?-? Relative molecular mass 86.1 Log P octanol/water 0.7	**COLORLESS LIQUID** Vapor is heavier than air and spreads at ground level, with risk of ignition at a distance. Presumed to be able to form peroxides. Flow, agitation etc. can cause build-up of electrostatic charge due to liquid's low conductivity. Do not use compressed air when filling, emptying or processing. Reacts violently with strong oxidants.
	TLV-TWA not available
	Absorption route: Can enter the body by inhalation or ingestion or through the skin. **Immediate effects:** Irritates the eyes and respiratory tract. Liquid destroys the skin's natural oils. Affects the nervous system.
Gross formula: $C_5H_{10}O$	

HAZARDS/SYMPTOMS	PREVENTIVE MEASURES	FIRE EXTINGUISHING/FIRST AID
Fire: extremely flammable.	keep away from open flame and sparks, no smoking.	powder, AFFF, foam, carbon dioxide, (halons).
Explosion: forms explosive air-vapor mixtures.	sealed machinery, ventilation, explosion-proof electrical equipment and lighting, grounding.	in case of fire: keep tanks/drums cool by spraying with water.
Inhalation: cough, headache, drowsiness, sleepiness.	ventilation, local exhaust or respiratory protection.	fresh air, rest.
Skin: *is absorbed*; redness.	gloves.	remove contaminated clothing, flush skin with water or shower.
Eyes: redness, pain.	face shield.	flush with water, take to a doctor if necessary.
Ingestion: abdominal pain, headache, nausea, drowsiness.		rinse mouth, call a doctor.

SPILLAGE	STORAGE	LABELING / NFPA
collect leakage in sealable containers, soak up with sand or other inert absorbent and remove to safe place; (additional individual protection: breathing apparatus).	keep in fireproof place, separate from oxidants.	

NOTES
Explosive limits not given in the literature. Log P octanol/water is estimated. Before distilling check for peroxides; if found, render harmless. Alcohol consumption increases toxic effects.

Transport Emergency Card TEC(R)-30G32

HI: 336; UN-number: 2335

CAS-No: [106-92-3]
AGE
allyl 2,3-epoxypropyl ether
1-allyloxy-2,3-epoxypropane
((2-propenyloxy)methyl)oxirane

$$CH_2{=}CH{-}CH_2{-}O{-}CH_2{-}\overset{\displaystyle O}{CH}{-}CH_2$$

ALLYL GLYCIDYL ETHER

PHYSICAL PROPERTIES		IMPORTANT CHARACTERISTICS	
Boiling point, °C	154	**COLORLESS LIQUID WITH CHARACTERISTIC ODOR**	
Melting point, °C	− 100	Vapor mixes readily with air. Able to form peroxides. Able to polymerize on contact with bases or	
Flash point, °C	57	acids. Reacts violently with amines and oxidants.	
Relative density (water = 1)	0.96		
Relative vapor density (air = 1)	3.9	TLV-TWA 5 ppm 23 mg/m³	
Relative density at 20 °C of saturated mixture		STEL 10 ppm 47 mg/m³	
vapor/air (air = 1)	1.02		
Vapor pressure, mm Hg at 20 °C	3.7	**Absorption route:** Can enter the body by inhalation or ingestion or through the skin. Harmful	
Solubility in water, g/100 ml	14.1	atmospheric concentrations can build up fairly rapidly on evaporation at approx. 20°C - even	
Relative molecular mass	114.2	more rapidly in aerosol form.	
		Immediate effects: Irritates the eyes, skin and respiratory tract. Affects the nervous system.	
		Inhalation of vapor/fumes can cause severe breathing difficulties (lung edema).	
		Effects of prolonged/repeated exposure: Prolonged or repeated contact can cause skin	
		disorders.	
Gross formula:	$C_6H_{10}O_2$		

HAZARDS/SYMPTOMS	PREVENTIVE MEASURES	FIRE EXTINGUISHING/FIRST AID
Fire: combustible.	keep away from open flame, no smoking.	powder, AFFF, foam, carbon dioxide, (halons).
Explosion: above 57°C: forms explosive air-vapor mixtures.	above 57°C: sealed machinery, ventilation.	
Inhalation: cough, shortness of breath, drowsiness, sleepiness.	ventilation, local exhaust or respiratory protection.	fresh air, rest, call a doctor.
Skin: *is absorbed*; redness, pain.	gloves.	remove contaminated clothing, flush skin with water or shower, refer to a doctor if necessary.
Eyes: redness, pain, impaired vision.	face shield.	flush with water, take to a doctor if necessary.
Ingestion: headache, shortness of breath, nausea, vomiting, drowsiness, sleepiness.		rinse mouth, rest, call a doctor or take to hospital.

SPILLAGE	STORAGE	LABELING / NFPA
collect leakage in non-sealable containers, soak up with sand or other inert absorbent and remove to safe place; (additional individual protection: breathing apparatus).	keep in fireproof place, separate from oxidants, amines, bases and acids; add an inhibitor.	R: 20-43 S: 24/25 ✖ Harmful

NOTES
Before distilling check for peroxides; if found, render harmless. Alcohol consumption increases toxic effects. Depending on the degree of exposure, regular medical checkups are advisable. Lung edema symptoms usually develop several hours later and are aggravated by physical exertion: rest and hospitalization essential. As first aid, a doctor or authorized person should consider administering a corticosteroid spray.

Transport Emergency Card TEC(R)-113

HI: 30; UN-number: 2219

phenyl allyl ether
(2-propenyloxy)benzene

$C_6H_5OCH_2CH=CH_2$

ALLYL PHENYL ETHER

PHYSICAL PROPERTIES		IMPORTANT CHARACTERISTICS
Boiling point, °C	192	**COLORLESS LIQUID**
Flash point, °C	62	Vapor mixes readily with air. Presumed to be able to form peroxides. Reacts violently with strong oxidants.
Relative density (water = 1)	0.98	
Relative vapor density (air = 1)	4.6	
Relative density at 20 °C of saturated mixture vapor/air (air = 1)	1.0	TLV-TWA not available
Vapor pressure, mm Hg at 20 °C	ca. 0.53	
Solubility in water	none	**Absorption route:** Can enter the body by inhalation or ingestion or through the skin. Insufficient data on the rate at which harmful concentrations can build up.
Relative molecular mass	134.2	**Immediate effects:** Irritates the eyes and respiratory tract. Affects the nervous system.
Log P octanol/water	2.9	
Gross formula:	$C_9H_{10}O$	

HAZARDS/SYMPTOMS	PREVENTIVE MEASURES	FIRE EXTINGUISHING/FIRST AID
Fire: combustible.	keep away from open flame, no smoking.	powder, AFFF, foam, carbon dioxide, (halons).
Explosion: above 62°C: forms explosive air-vapor mixtures.	above 62°C: sealed machinery, ventilation.	
Inhalation: cough, headache, drowsiness, sleepiness.	ventilation, local exhaust or respiratory protection.	fresh air, rest.
Skin: *is absorbed*; redness.	gloves.	remove contaminated clothing, flush skin with water or shower.
Eyes: redness, pain.	face shield.	flush with water, take to a doctor if necessary.
Ingestion: abdominal pain, headache, nausea, drowsiness.		rinse mouth, call a doctor.

SPILLAGE	STORAGE	LABELING / NFPA
collect leakage in non-sealable containers, soak up with sand or other inert absorbent and remove to safe place.	keep separate from oxidants.	

NOTES
Before distilling check for peroxides; if found, render harmless.

Transport Emergency Card TEC(R)-30G15

ALUMINUM
(powder)

PHYSICAL PROPERTIES	IMPORTANT CHARACTERISTICS	
Boiling point, °C 2470 Melting point, °C 660 Auto-ignition temperature, °C 590 Relative density (water = 1) 2.7 Vapor pressure, mm Hg at 20 °C 0 Solubility in water none Explosive limits, vol% in air ?-? Relative molecular mass 27.0	**GREY POWDER** In dry state can form electrostatic charge if stirred, transported pneumatically, poured etc. Do not use compressed air when filling, emptying or processing. Strong reducing agent which reacts violently with oxidants. Reacts violently with strong bases, acids and some halogenated hydrocarbons, giving off flammable gas (→ *hydrogen*) and corrosive fumes. Reacts violently with many compounds (incl. nitrates, sulfates, halogens), metal oxides (thermic reaction in welding) and mercury and its compounds.	
	TLV-TWA 10 mg/m³	
	Absorption route: Can enter the body by inhalation or ingestion. Evaporation negligible at 20°C, but unpleasant concentrations of airborne particles can build up. **Immediate effects:** Slightly irritates the respiratory tract. **Effects of prolonged/repeated exposure:** Prolonged inhalation of particles can cause lung disorders.	
Gross formula: Al		

HAZARDS/SYMPTOMS	PREVENTIVE MEASURES	FIRE EXTINGUISHING/FIRST AID
Fire: combustible, many chemical reactions can cause fire and explosion.	keep away from open flame and sparks, no smoking, avoid contact with oxidants, strong acids and bases.	dry sand, special extinguishing powder, NO OTHER EXTINGUISHING AGENTS.
Explosion: finely dispersed particles form explosive mixtures on contact with air.	keep dust from accumulating, sealed machinery, explosion-proof electrical equipment and lighting, grounding.	
Inhalation: cough, shortness of breath, (vapor), in some cases metal fume fever.	local exhaust or respiratory protection.	fresh air, rest, consult a doctor.
Skin:	gloves.	flush skin with water or shower.
Eyes: redness.	goggles.	flush with water, take to doctor if necessary.
Ingestion:		rinse mouth, consult a doctor if necessary.

SPILLAGE	STORAGE	LABELING / NFPA
clean up spillage; (additional individual protection: P1 respirator).	keep in dry, fireproof place, separate from oxidants, strong acids, strong bases, metal oxides and metal salts, halogens and halogenated hydrocarbons.	R: 10-15 S: 7/8-43

NOTES

Besides standard grade (phlegmatized) there is pyrophoric aluminum powder, which can ignite spontaneously on exposure to air; process under inert gas. Aluminum powder is susceptible to dust explosion, depending on concentration and the presence of ignition sources. Aluminum-based paint on steel can ignite if subjected to shock or welding. Burning aluminum powder reacts violently with common extinguishing agents such as water, foam, sodium bicarbonate and halons. Depending on the degree of exposure, regular medical checkups are advisable. TLV of pyrophoric aluminum: 5mg/m³.

Transport Emergency Card TEC(R)-40

HI: 43; UN-number: 1396

ALUMINUM CHLORIDE
(anhydrous)

PHYSICAL PROPERTIES	IMPORTANT CHARACTERISTICS	
Sublimation temperature, °C 178 Relative density (water = 1) 2.5 Vapor pressure, mm Hg at 50 °C 0.0030 Solubility in water reaction Relative molecular mass 133.4	**WHITE TO YELLOW HYGROSCOPIC CRYSTALS OR POWDER WITH PUNGENT ODOR** Decomposes when heated, giving off corrosive vapors (→ *hydrochloric acid*). In aqueous solution is a medium strong acid which reacts with bases. Reacts violently with water, forming corrosive → *hydrochloric acid*. At higher temperatures reacts violently with alcohols and various other organic substances. Reacts with air to form corrosive → *hydrochloric acid* fumes. Attacks many metals, giving off flammable gas (→ *hydrogen*).	
	TLV-TWA	2 mg/m³
	Absorption route: Can enter the body by inhalation or ingestion. Evaporation negligible at 20°C, but harmful atmospheric concentrations can build up rapidly in the form of airborne particles or due to formation of hydrochloric acid fumes. **Immediate effects:** Corrosive to the eyes, skin and respiratory tract. Inhalation can cause lung edema. In serious cases risk of death.	
Gross formula: AlCl₃		

HAZARDS/SYMPTOMS	PREVENTIVE MEASURES	FIRE EXTINGUISHING/FIRST AID
Fire: non-combustible.		in case of fire in immediate vicinity: do not use water.
Explosion: contact with water, finely powdered metals and various organic substances can cause explosion.		
	KEEP DUST UNDER CONTROL	
Inhalation: *corrosive*; sore throat, cough, shortness of breath, severe breathing difficulties.	local exhaust or respiratory protection.	fresh air, rest, place in half-sitting position, take to hospital.
Skin: *corrosive*; redness, pain.	gloves, protective clothing.	remove contaminated clothing, flush skin with water or shower.
Eyes: *corrosive*; redness, pain, impaired vision.	face shield, or combined eye and respiratory protection.	flush with water, take to a doctor.
Ingestion: *corrosive*; sore throat, abdominal pain.	do not eat, drink or smoke while working.	rinse mouth, immediately take to hospital.

SPILLAGE	STORAGE	LABELING / NFPA
clean up spillage, flush away remainder with water; (additional individual protection: CHEMICAL SUIT).	keep dry, separate from strong bases; ventilate at floor level.	R: 34 S: 7/8-28 Corrosive

NOTES
Reacts violently with extinguishing agents such as water and foam. Lung edema symptoms usually develop several hours later and are aggravated by physical exertion: rest and hospitalization essential. As first aid, a doctor or authorized person should consider administering a corticosteroid spray. Use airtight packaging.

Transport Emergency Card TEC(R)-56

HI: 80; UN-number: 1726

ALUMINUM NITRATE
(nonahydrate)

PHYSICAL PROPERTIES	IMPORTANT CHARACTERISTICS	
Decomposes below boiling point, °C 135 à 150 Melting point, °C 70 Solubility in water, g/100 ml at 25 °C 64 Relative molecular mass 375.1	**WHITE CRYSTALS** Decomposes when heated above 135°C, giving off → *aluminumoxide*, → *nitrous vapors* and → *oxygen*, with increased risk of fire. Strong oxidant which reacts violently with combustible substances and reducing agents. In aqueous solution is a strong acid which reacts violently with bases and is corrosive.	
	TLV-TWA	2 mg/m³
	Absorption route: Can enter the body by inhalation or ingestion. **Immediate effects:** Irritates the eyes, skin and respiratory tract. In serious cases risk of unconsciousness. **Effects of prolonged/repeated exposure:** Can affect the blood if ingested. Prolonged or repeated contact can cause skin disorders (eczema).	
Gross formula: $AlN_3O_9.9H_2O$		

HAZARDS/SYMPTOMS	PREVENTIVE MEASURES	FIRE EXTINGUISHING/FIRST AID
Fire: not combustible but enhances combustion of other substances.	avoid contact with combustible substances.	in case of fire in immediate vicinity: use any extinguishing agent.
Inhalation: sore throat, cough, shortness of breath.	local exhaust or respiratory protection.	fresh air, rest, call a doctor.
Skin: redness, pain.	gloves, protective clothing.	remove contaminated clothing, flush skin with water or shower.
Eyes: redness, pain.	goggles or face shield.	flush with water, take to a doctor.
Ingestion: abdominal cramps, nausea, blue skin, feeling of weakness.		rinse mouth, call a doctor or take to hospital.

SPILLAGE	STORAGE	LABELING / NFPA
clean up spillage, flush away remainder with water; (additional individual protection: P2 respirator).	keep separate from combustible substances, reducing agents and strong bases.	

NOTES
Apparent boiling point 100°C due to loss of water of crystallization. Contaminated aluminum nitrate can be self-heating and cause fire. Flush contaminated clothing with water (fire hazard). Effects on blood due to formation of methemoglobin. Special first aid required in the event of aluminum nitrate poisoning: antidotes must be available (with instructions).

Transport Emergency Card TEC(R)-51G01

UN-number: 1438

ALUMINUM OXIDE

PHYSICAL PROPERTIES	IMPORTANT CHARACTERISTICS	
Boiling point, °C ca. 3000 Melting point, °C 2030 Relative density (water = 1) 3.7 Solubility in water none Relative molecular mass 102.0	**WHITE POWDER OR WHITE GRANULES**	
	TLV-TWA	10 mg/m³
	Absorption route: Can enter the body by inhalation or ingestion. **Immediate effects:** Irritates the eyes and respiratory tract. **Effects of prolonged/repeated exposure:** Prolonged exposure to particles can cause lung disorders.	
Gross formula: Al_2O_3		

HAZARDS/SYMPTOMS	PREVENTIVE MEASURES	FIRE EXTINGUISHING/FIRST AID
Fire: non-combustible.		in case of fire in immediate vicinity: use any extinguishing agent.
Inhalation: cough, shortness of breath.	local exhaust or respiratory protection.	fresh air, rest, call a doctor if necessary.
Eyes: redness, pain.	goggles.	flush with water, take to a doctor.
Ingestion:		rinse mouth, consult a doctor if necessary.

SPILLAGE	STORAGE	LABELING / NFPA
clean up spillage; (additional individual protection: P1 respirator).		

NOTES

ALUMINUM PHOSPHIDE

PHYSICAL PROPERTIES	IMPORTANT CHARACTERISTICS
Boiling point, °C ? Melting point, °C > 1000 Relative density (water = 1) 2.9 Solubility in water reaction Relative molecular mass 58.0 Gross formula: AlP	**DARK GRAY OR DARK YELLOW CRYSTALS** Can ignite spontaneously on exposure to air by forming phosphine with atmospheric moisture. Reacts slowly with water or bases, but violently with steam or dilute acids, giving off → *phosphine* gas, which can ignite spontaneously on exposure to air. TLV-TWA not available **Absorption route:** Can enter the body by inhalation or ingestion. **Immediate effects:** Irritates the eyes and respiratory tract. Affects metabolism and the nervous system. In serious cases risk of death. **Effects of prolonged/repeated exposure:** Prolonged or repeated contact can cause skin disorders (eczema). Can cause liver and kidney damage.

HAZARDS/SYMPTOMS	PREVENTIVE MEASURES	FIRE EXTINGUISHING/FIRST AID
Fire: non-combustible, but risk of fire and explosion on contact with moisture.	avoid contact with water, steam, bases and acids.	if involved in fire: dry sand, special powder extinguisher, DO NOT USE WATER-BASED EXTINGUISHERS.
Explosion: can combine with moist air to form → *phosphine* gas, with risk of explosion.		
Inhalation: shortness of breath, headache, dizziness, nausea; after some hours: vomiting, diarrhea, abdominal cramps, muscle cramps.	local exhaust or respiratory protection.	fresh air, rest, take to hospital.
Skin: redness.	gloves.	remove contaminated clothing, flush skin with water or shower.
Eyes: redness, pain.	goggles, or combined eye and respiratory protection.	flush with water, take to a doctor if necessary.
Ingestion: diarrhea, stomach cramps, vomiting; see also 'Inhalation'.	do not eat, drink or smoke while working.	rinse mouth, take immediately to hospital.

SPILLAGE	STORAGE	LABELING / NFPA
evacuate area, clean up spillage, carefully collect remainder, do not flush away with water (to avoid formation of phosphine); (additional individual protection: breathing apparatus).	keep in dry, fireproof place, separate from acids and bases; ventilate at floor level.	R: 15/29-28 S: 1/2-22-43-45 Flammable Toxic 4 / 3 2 / W

NOTES
Reacts violently with extinguishing agents such as water, foam, carbon dioxide. Use airtight packaging.

CAS-No: [10043-01-3]
aluminum alum
aluminum trisulfate
alunogenite
dialuminum sulfate

$Al_2(SO_4)_3$

ALUMINUM SULFATE

PHYSICAL PROPERTIES	IMPORTANT CHARACTERISTICS	
Melting point (decomposes), °C 770 Relative density (water = 1) 2.7 Solubility in water good Relative molecular mass 342.2	**WHITE POWDER** Decomposes when heated above 770°C, giving off corrosive vapors (→ *sulfur trioxide*). In aqueous solution is a strong acid which reacts violently with bases and is corrosive.	
	TLV-TWA	2 mg/m³
	Absorption route: Can enter the body by inhalation or ingestion. **Immediate effects:** Irritates the eyes and respiratory tract. **Effects of prolonged/repeated exposure:** Prolonged or repeated contact can cause skin disorders (eczema).	
Gross formula: $Al_2O_{12}S_3$		

HAZARDS/SYMPTOMS	PREVENTIVE MEASURES	FIRE EXTINGUISHING/FIRST AID
Fire: non-combustible.		in case of fire in immediate vicinity: use any extinguishing agent.
Inhalation: sore throat, cough.	local exhaust or respiratory protection.	fresh air, rest, call a doctor if necessary.
Skin:	gloves.	
Eyes: redness, pain.	face shield.	flush with water, take to a doctor if necessary.
Ingestion: abdominal pain.		rinse mouth, call a doctor.

SPILLAGE	STORAGE	LABELING / NFPA
clean up spillage, flush away remainder with water; (additional individual protection: P2 respirator).	keep separate from strong bases.	

NOTES

H_2NCH_2COOH

AMINOACETIC ACID

PHYSICAL PROPERTIES		IMPORTANT CHARACTERISTICS	
Melting point (decomposes), °C	232	**WHITE CRYSTALS**	
Relative density (water = 1)	1.2	Decomposes when heated giving off → *nitrous vapors*.	
Solubility in water, g/100 ml at 25 °C	25		
Relative molecular mass	75.1	TLV-TWA not available	
Log P octanol/water	−2.3		
		Absorption route: Can enter the body by inhalation or ingestion. Evaporation negligible at 20°C, but unpleasant concentrations of airborne particles can build up.	
Gross formula:	$C_2H_5NO_2$		

HAZARDS/SYMPTOMS	PREVENTIVE MEASURES	FIRE EXTINGUISHING/FIRST AID
Fire: combustible.	keep away from open flame, no smoking.	water spray, powder.
Inhalation: cough.	local exhaust or respiratory protection.	fresh air, rest, place in half-sitting position, call a doctor if necessary.
Skin:		remove contaminated clothing, flush skin with water or shower.
Eyes: redness.	goggles.	flush with water, take to a doctor if necessary.
Ingestion: in large quantities: nausea.		rinse mouth, take to a doctor if necessary.

SPILLAGE	STORAGE	LABELING / NFPA
clean up spillage, flush away remainder with water; (additional individual protection: P2 respirator).		

NOTES
Log P octanol/water is estimated. Insufficient toxicological data on harmful effects to humans.

$$CH_2-CH_2$$

$$HN \qquad N-CH_2-CH_2-NH_2$$

$$CH_2-CH_2$$

N-AMINOETHYLPIPERAZINE

PHYSICAL PROPERTIES	IMPORTANT CHARACTERISTICS
Boiling point, °C 222 Melting point, °C − 18 Flash point, °C 93 Auto-ignition temperature, °C > 300 Relative density (water = 1) 1.0 Relative vapor density (air = 1) 4.4 Relative density at 20 °C of saturated mixture vapor/air (air = 1) 1.0 Vapor pressure, mm Hg at 20 °C 0.076 Solubility in water very good Explosive limits, vol% in air 1.6-6.5 Relative molecular mass 129.2 Gross formula: $C_6H_{15}N_3$	**LIGHT YELLOW LIQUID** Vapor mixes readily with air. Decomposes when heated or burned, giving off toxic vapors (→ *nitrous vapors*). In aqueous solution is a strong base which reacts violently with acids and corrodes aluminum, zinc etc. Reacts with strong oxidants. Attacks copper, nickel and cobalt. TLV-TWA not available **Absorption route:** Can enter the body by inhalation or ingestion. Insufficient data on the rate at which harmful concentrations can build up. **Immediate effects:** Irritates the eyes, skin and respiratory tract.

HAZARDS/SYMPTOMS	PREVENTIVE MEASURES	FIRE EXTINGUISHING/FIRST AID
Fire: combustible.	keep away from open flame, no smoking.	powder, alcohol-resistant foam, water spray, carbon dioxide, (halons).
Explosion: above 93°C: forms explosive air-vapor mixtures.	above 93°C: sealed machinery, ventilation.	
Inhalation: sore throat, cough, shortness of breath.	ventilation, local exhaust or respiratory protection.	fresh air, rest, call a doctor.
Skin: redness, pain.	gloves.	remove contaminated clothing, flush skin with water or shower, refer to a doctor.
Eyes: redness, pain.	face shield.	flush with water, take to a doctor if necessary.
Ingestion: abdominal pain, diarrhea, vomiting.		rinse mouth, call a doctor or take to hospital.

SPILLAGE	STORAGE	LABELING / NFPA
collect leakage in sealable containers, soak up with sand or other inert absorbent and remove to safe place.	keep separate from oxidants and strong acids; ventilate at floor level.	R: 21/22-34-43 S: 26-36/37/39 Corrosive

NOTES
Unbreakable packaging preferred; if breakable, keep in unbreakable container.

o-AMINOPHENOL

PHYSICAL PROPERTIES	IMPORTANT CHARACTERISTICS	
Melting point, °C 174 Relative density (water = 1) 1.33 Vapor pressure, mm Hg at 153 °C 11 Solubility in water, g/100 ml 2 Relative molecular mass 109.1 Log P octanol/water 0.6	**COLORLESS OR WHITE CRYSTALS, TURNING BROWN ON EXPOSURE TO AIR** Sublimates. Decomposes when heated or burned, giving off toxic vapors (→ *nitrous vapors*). Reacts violently with oxidants.	
	TLV-TWA not available	
	Absorption route: Can enter the body by inhalation or ingestion or through the skin. Evaporation negligible at 20° C, but harmful concentrations of airborne particles can build up rapidly. **Immediate effects:** Irritates the eyes, skin and respiratory tract. Affects the nervous system. Frequent prolonged contact can cause lung disorders (asthma). In serious cases risk of unconsciousness and death. **Effects of prolonged/repeated exposure:** Prolonged or repeated contact can cause skin disorders. Can affect the blood. Can cause liver, kidney and brain damage.	
Gross formula: C_6H_7NO		

HAZARDS/SYMPTOMS	PREVENTIVE MEASURES	FIRE EXTINGUISHING/FIRST AID
Fire: combustible.	keep away from open flame, no smoking.	water spray, powder.
	KEEP DUST UNDER CONTROL	
Inhalation: shortness of breath, headache, dizziness.	local exhaust or respiratory protection.	fresh air, rest, call a doctor or take to hospital.
Skin: *is absorbed*; redness, pain; see also 'Inhalation'.	gloves, protective clothing.	remove contaminated clothing, flush skin with water or shower, call a doctor.
Eyes: redness, pain.	face shield.	flush with water, take to a doctor if necessary.
Ingestion: abdominal pain, unconsciousness; see also 'Inhalation'.	do not eat, drink or smoke while working.	rinse mouth, take immediately to hospital.

SPILLAGE	STORAGE	LABELING / NFPA
clean up spillage, carefully collect remainder; (additional individual protection: P2 respirator).	keep in dry, dark place, separate from oxidants; ventilate at floor level.	R: 20/21/22 S: 28 Note C ✖ Harmful

NOTES

Decomposition temperature not given in the literature. Depending on the degree of exposure, regular medical checkups are advisable. Asthma symptoms usually develop several hours later and are aggravated by physical exertion: rest and hospitalization essential. Persons with history of asthma symptoms should under no circumstances come in contact again with substance. Effects on blood due to formation of methemoglobin. Special first aid required in the event of poisoning: antidotes must be available (with instructions). The measures on this card also apply to meta-aminophenol and para-aminophenol. Melting point: meta-isomer: 122° C; para-isomer: 186° C.

Transport Emergency Card TEC(R)-885

HI: 60; UN-number: 2512

AMMONIA
(25% solution of ammonia in water)

PHYSICAL PROPERTIES	IMPORTANT CHARACTERISTICS	
Melting point, °C −55 Auto-ignition temperature, °C see notes Relative density (water = 1) 0.9 Relative vapor density (air = 1) 1.2 Relative density at 20 °C of saturated mixture vapor/air (air = 1) 1.1 Vapor pressure, mm Hg at 21 °C 334 Solubility in water ∞ Explosive limits, vol% in air see notes Minimum ignition energy, mJ see notes Relative molecular mass 35.1 Log P octanol/water −1.3	**COLORLESS LIQUID WITH PUNGENT ODOR** Vapor is heavier than air. Medium strong base which corrodes aluminum and zinc. Reacts violently with acids. Forms shock-sensitive compounds with halogens, mercury oxide and silver oxide.	
	TLV-TWA 25 ppm 17 mg/m³ ✶ STEL 35 ppm 24 mg/m³	
	Absorption route: Can enter the body by inhalation or ingestion. Harmful atmospheric concentrations can build up very rapidly on evaporation at 20°C. **Immediate effects:** Corrosive to the eyes, skin and respiratory tract. Inhalation of vapor/fumes can cause severe breathing difficulties (lung edema).	
Gross formula: H_5NO		

HAZARDS/SYMPTOMS	PREVENTIVE MEASURES	FIRE EXTINGUISHING/FIRST AID
Fire: see Notes.	keep away from open flame, no smoking.	in case of fire in immediate vicinity: use any extinguishing agent.
Inhalation: *corrosive*; sore throat, cough, severe breathing difficulties.	ventilation, local exhaust or respiratory protection.	fresh air, rest, place in half-sitting position, take to hospital.
Skin: *corrosive*; redness, pain, serious burns.	gloves.	remove contaminated clothing, flush skin with water or shower, refer to a doctor.
Eyes: *corrosive*; redness, pain, impaired vision.	face shield, or combined eye and respiratory protection.	flush with water, take to a doctor.
Ingestion: *corrosive*; sore throat, abdominal pain, nausea.		rinse mouth, immediately take to hospital.

SPILLAGE	STORAGE	LABELING / NFPA
collect leakage in sealable containers, flush away remainder with water; (additional individual protection: CHEMICAL SUIT).	keep cool, separate from acids; ventilate.	R: 36/37/38 S: 2-26 Note B ✖ Irritant

NOTES

Log P octanol/water is estimated. Any tanks or drums which have contained ammonia and are to be rinsed with water should be well ventilated (risk of implosion). Can form combustible air-vapor mixtures under certain circumstances (15-29% by vol.), which are difficult to ignite. ✶ TLV equals that of ammonia (NH_3). Lung edema symptoms usually develop several hours later and are aggravated by physical exertion: rest and hospitalization essential. As first aid, a doctor or authorized person should consider administering a corticosteroid spray. Use airtight packaging.

HI: 80; UN-number: 2672

AMMONIA (ANHYDROUS)
(cylinder)

PHYSICAL PROPERTIES	IMPORTANT CHARACTERISTICS
Boiling point, °C −33 Melting point, °C −78 Flash point, °C flammable gas Auto-ignition temperature, °C 630 Relative density (water = 1) 0.77 Relative vapor density (air = 1) 0.6 Vapor pressure, mm Hg at 20 °C 6460 Solubility in water, g/100 ml at 20 °C 53 Explosive limits, vol% in air 15-29 Minimum ignition energy, mJ 680 Electrical conductivity, pS/m 1.9×10^7 Relative molecular mass 17.0 Log P octanol/water −1.3	**COMPRESSED LIQUEFIED GAS WITH PUNGENT ODOR** Gas is lighter than air. Dissolves in water, evolving heat. Reacts with mercuric and silver oxide to form shock-sensitive compounds. Reacts violently with halogens, strong oxidants and acids. Attacks copper, zinc, aluminium and their alloys.

	TLV-TWA	25 ppm	17 mg/m³
	STEL	35 ppm	24 mg/m³

Absorption route: Can enter the body by inhalation. Harmful atmospheric concentrations can build up very rapidly if gas is released.
Immediate effects: Corrosive to the eyes, skin and respiratory tract. Can cause frostbite due to rapid evaporation. Inhalation of vapor/fumes can cause severe breathing difficulties (lung edema). In serious cases risk of death.

Gross formula: H_3N

HAZARDS/SYMPTOMS	PREVENTIVE MEASURES	FIRE EXTINGUISHING/FIRST AID
Fire: combustible.	keep away from open flame and sparks, no smoking.	shut off supply; if impossible and no danger to surrounding area, allow fire to burn itself out; otherwise extinguish with powder, carbon dioxide, (halons). .
Explosion: forms explosive air-gas mixtures.	sealed machinery, ventilation, explosion-proof electrical equipment and lighting.	in case of fire: keep cylinder cool by spraying with water.
Inhalation: *corrosive*; sore throat, cough, shortness of breath, severe breathing difficulties.	ventilation, local exhaust or respiratory protection.	fresh air, rest, place in half-sitting position, call a doctor.
Skin: *corrosive*; *in case of frostbite*: redness, pain, serious burns.	insulated gloves, protective clothing.	*in case of frostbite*: DO NOT remove clothing, flush skin with water or shower, refer to a doctor.
Eyes: *corrosive*; redness, pain, impaired vision.	face shield, or combined eye and respiratory protection.	flush with water, take to a doctor.

SPILLAGE	STORAGE	LABELING / NFPA
evacuate area, call in an expert, under no circumstances spray liquid with water, combat gas cloud with water curtain, shut off supply if possible; (additional individual protection: CHEMICAL SUIT).	if stored indoors, keep in fireproof place, separate from oxidants, acids and halogens.	R: 10-23 S: 7/9-16-38 ☠ Toxic ◇ 1 / 3 0

NOTES
Hardly combustible and difficult to ignite. Solubility applies to gas at pressure of 760mm Hg. Log P octanol/water is estimated. Odor limit is above TLV. Lung edema symptoms usually develop several hours later and are aggravated by physical exertion: rest and hospitalization essential. As first aid, a doctor or authorized person should consider administering a corticosteroid spray. Turn leaking cylinder so that leak is on top to prevent liquid ammonia escaping. Use cylinder with special fittings.

Transport Emergency Card TEC(R)-1

HI: 268; UN-number: 1005

AMMONIUM BICARBONATE

PHYSICAL PROPERTIES	IMPORTANT CHARACTERISTICS
Melting point (decomposes), °C 35 Relative density (water = 1) 1.58 Solubility in water, g/100 ml at 20 °C 17 Relative molecular mass 79.1	**WHITE CRYSTALS WITH CHARACTERISTIC ODOR** Decomposes when heated above 35°C, forming → *ammonia* and → *carbon dioxide*. When heated slowly, decomposition is complete at 60°C. Reacts violently with acids, giving off → *carbon dioxide*. Reacts with strong bases, forming → *ammonia*.

TLV-TWA	not available

Absorption route: Can enter the body by inhalation or ingestion. Evaporation negligible at 20°C, but unpleasant concentrations of airborne particles can build up.
Immediate effects: Irritates the eyes and respiratory tract.

Gross formula: CH_5NO_3

HAZARDS/SYMPTOMS	PREVENTIVE MEASURES	FIRE EXTINGUISHING/FIRST AID
Fire: non-combustible.		in case of fire in immediate vicinity: use any extinguishing agent.
Explosion:		in case of fire: keep tanks/drums cool by spraying with water.
Inhalation: cough, shortness of breath.	local exhaust or respiratory protection.	fresh air, rest, place in half-sitting position, call a doctor if necessary.
Eyes: redness, pain.	goggles.	flush with water, take to a doctor if necessary.
Ingestion: in large quanties: stomachache.		call a doctor if necessary.

SPILLAGE	STORAGE	LABELING / NFPA
clean up spillage, flush away remainder with water; (additional individual protection: P1 respirator).	keep cool; separate from strong acids and strong bases, ventilate at floor level.	

NOTES
When heated rapidly the melting point is 108°C.

AMMONIUMBICHROMATE

PHYSICAL PROPERTIES	IMPORTANT CHARACTERISTICS	
Decomposes below melting point, °C 170 Relative density (water = 1) 2.15 Solubility in water, g/100 ml 36 Relative molecular mass 252.1	**YELLOW TO ORANGE-RED CRYSTALS** Can produce thermal explosion when heated. Decomposes in a self-maintaining matter when heated above 225°C, giving off → *nitrogen* and voluminous green chromium (III) oxide, with risk of fire and explosion. Strong oxidant which reacts violently with combustible substances and reducing agents. In aqueous solution is a strong acid which reacts violently with bases and is corrosive.	
	TLV-TWA	0.05 mg/m³ ✷
Gross formula: $CrH_8N_2O_7$	**Absorption route:** Can enter the body by inhalation or ingestion. Evaporation negligible at 20°C, but harmful atmospheric concentrations of airborne particles can build up rapidly. **Immediate effects:** Corrosive to the eyes, skin and respiratory tract. Inhalation can cause lung edema. In serious cases risk of death. **Effects of prolonged/repeated exposure:** Prolonged or repeated contact can cause skin disorders. Has been found to cause a type of lung cancer in certain animal species under certain circumstances.	

HAZARDS/SYMPTOMS	PREVENTIVE MEASURES	FIRE EXTINGUISHING/FIRST AID
Fire: many chemical reactions cause fire and explosion, decomposes with appearance of fire.	keep away from open flame and sparks, no smoking, avoid contact with combustible substances and hot surfaces (steam lines).	water spray, powder.
Explosion: heating in confined space can cause explosion.		
	STRICT HYGIENE, KEEP DUST UNDER CONTROL	
Inhalation: *corrosive*; sore throat, cough, shortness of breath, severe breathing difficulties.	local exhaust or respiratory protection.	fresh air, rest, place in half-sitting position, take to hospital.
Skin: *corrosive*; redness, pain, burns.	gloves.	remove contaminated clothing, flush skin with water or shower, call a doctor.
Eyes: *corrosive*; redness, pain, impaired vision.	face shield, or combined eye and respiratory protection.	flush with water, take to a doctor.
Ingestion: *corrosive*; sore throat, abdominal pain, vomiting.	do not eat, drink or smoke while working.	rinse mouth, immediately take to hospital.

SPILLAGE	STORAGE	LABELING / NFPA
clean up spillage, carefully collect remainder, do not use sawdust or other combustible absorbents; (additional individual protection: P3 respirator).	keep in fireproof place, separate from combustible substances, reducing agents and strong bases.	R: 1-8-36/37/38-43 S: 28-35 Explosive Irritant

NOTES
✷ TLV equals that of chromium (chromium (VI) compounds). Flush contaminated clothing with water (fire hazard). In confined spaces combustion can cause explosion. Depending on the degree of exposure, regular medical checkups are advisable. Lung edema symptoms usually develop several hours later and are aggravated by physical exertion: rest and hospitalization essential. As first aid, a doctor or authorized person should consider administering a corticosteroid spray. Packaging: special material.

CAS-No: [12124-99-1]
ammonium hydrogen sulfide
ammonium sulfide

NH$_4$HS

AMMONIUMBISULFIDE

PHYSICAL PROPERTIES	IMPORTANT CHARACTERISTICS
Decomposes below melting point, °C 118 Relative density (water = 1) 1.2 Vapor pressure, mm Hg at 22 °C 395 Solubility in water, g/100 ml at 20 °C 128 Relative molecular mass 51.1	**COLORLESS HYGROSCOPIC CRYSTALS, TURNING YELLOW ON EXPOSURE TO AIR** Sublimates readily in connection with high vapor pressure. Decomposes at room temperature forming → *hydrogen sulfide* and → *ammonia*. Decomposes when heated or burned, giving off toxic and corrosive vapors (→ *sulfur dioxide*, → *nitrous vapors*, → *anhydrous ammonia*). Reacts with acids, forming → *sulfur dioxide* and/or → *hydrogen sulfide*. Reacts violently with strong oxidants, with risk of fire and explosion.
	TLV-TWA not available
	Absorption route: Can enter the body by inhalation or ingestion or through the skin. Harmful atmospheric concentrations can build up very rapidly on evaporation at 20°C. **Immediate effects:** Corrosive to the eyes, skin and respiratory tract. Inhalation of vapor/fumes can cause severe breathing difficulties (lung edema). In serious cases risk of unconsciousness and death.
Gross formula: H$_5$NS	

HAZARDS/SYMPTOMS	PREVENTIVE MEASURES	FIRE EXTINGUISHING/FIRST AID
Fire: extremely flammable, many chemical reactions can cause fire and explosion.	keep away from open flame and sparks, no smoking.	water spray, powder.
Explosion: many chemical reactions can cause explosion.	sealed machinery, ventilation.	in case of fire: keep tanks/drums cool by spraying with water.
Inhalation: *corrosive*; sore throat, cough, severe breathing difficulties, drowsiness.	local exhaust or respiratory protection.	fresh air, rest, place in half-sitting position, artificial respiration if necessary, take to hospital.
Skin: *is absorbed*; redness, pain, burns.	gloves, protective clothing.	remove contaminated clothing, flush skin with water or shower, call a doctor if necessary.
Eyes: *corrosive*; redness, pain, impaired vision.	face shield, or combined eye and respiratory protection.	flush with water, take to a doctor.
Ingestion: *corrosive*; abdominal cramps, vomiting, drowsiness, unconsciousness.		rinse mouth, immediately take to hospital.

SPILLAGE	STORAGE	LABELING / NFPA
clean up spillage, carefully collect remainder; (additional individual protection: breathing apparatus).	keep in cool, dry, fireproof place, separate from acids, special packaging.	

NOTES
Ammonium sulfide (NH$_4$)$_2$S is only stable below 0°C. The normal trade product is ammonium bisulfide which also exists as 40% solution in water. This solution is more stable than the anhydrous substance. Lung edema symptoms usually develop several hours later and are aggravated by physical exertion: rest and hospitalization essential. As first aid, a doctor or authorized person should consider administering a corticosteroid spray.

UN-number: 2683

45

sal ammoniac

AMMONIUM CHLORIDE

PHYSICAL PROPERTIES		IMPORTANT CHARACTERISTICS	
Sublimation temperature, °C	338	**WHITE HYGROSCOPIC CRYSTALS**	
Relative density (water = 1)	1.5	Reacts with silver salts to form shock-sensitive compounds. Decomposes when heated, forming → *nitrogen oxides*, → *anhydrous ammonia* and → *hydrochloric acid*. Reacts with strong acids, giving off → *hydrochloric acid* gas, and with strong bases, giving off → *anhydrous ammonia*. Reacts violently with strong oxidants. Reacts violently with → *ammonium nitrate*, → *potassium chlorate*, boron trifluoride and iodine heptafluoride, with risk of explosion. Attacks copper and copper compounds.	
Relative vapor density (air = 1)	1.8		
Vapor pressure, mm Hg at 160 °C	0.99		
Solubility in water, g/100 ml at 0 °C	26		
Relative molecular mass	53.5		
		TLV-TWA	10 mg/m³ ✳
		Absorption route: Can enter the body by inhalation or ingestion. Evaporation negligible at 20° C, but unpleasant concentrations of airborne particles can build up. **Immediate effects:** Irritates the eyes and respiratory tract.	
Gross formula:	ClH_4N		

HAZARDS/SYMPTOMS	PREVENTIVE MEASURES	FIRE EXTINGUISHING/FIRST AID
Fire: non-combustible.		in case of fire in immediate vicinity: use any extinguishing agent.
Inhalation: sore throat, cough.	local exhaust or respiratory protection.	fresh air, rest.
Skin: redness.	gloves.	remove contaminated clothing, flush skin with water or shower.
Eyes: redness, pain.	goggles.	flush with water, take to a doctor if necessary.
Ingestion: nausea, vomiting.		rinse mouth, call a doctor.

SPILLAGE	STORAGE	LABELING / NFPA
clean up spillage, flush away remainder with water; (additional individual protection: respirator with A/P2 filter).	keep separate from oxidants, strong acids and strong bases; ventilate.	R: 22-36 S: 22 ✖ Harmful

NOTES
✳ TLV equals that of ammonium chloride fume.

AMMONIUMFLUORIDE

PHYSICAL PROPERTIES	IMPORTANT CHARACTERISTICS	
Melting point, °C　　　　　　　sublimes Relative density (water = 1)　　　　1.01 Solubility in water, g/100 ml at 0 °C　100 Relative molecular mass　　　　37.0	**COLORLESS DELIQUESCENT CRYSTALS** Decomposes when heated, corrosive gases (→ *hydrogen fluoride* and → *ammonia*). Reacts with acids, giving off corrosive → *hydrogen fluoride*. Reacts with bases, forming corrosive → *ammonia*. Attacks glass and many metals.	
	TLV-TWA	2.5 mg/m³　✳
	Absorption route: Can enter the body by inhalation or ingestion. **Immediate effects:** Corrosive to the eyes, skin and respiratory tracts. Inhalation of vapor/fumes can cause severe breathing difficulties (lung edema). In serious cases risk of death. **Effects of prolonged/repeated exposure:** Can cause bone disorders.	
Gross formula:　　　　　　　FH_4N		

HAZARDS/SYMPTOMS	PREVENTIVE MEASURES	FIRE EXTINGUISHING/FIRST AID
Fire: non-combustible.		in case of fire in immediate vicinity: use any extinguishing agent.
	STRICT HYGIENE, KEEP DUST UNDER CONTROL	
Inhalation: *corrosive*; sore throat, cough, shortness of breath, severe breathing difficulties.	local exhaust or respiratory protection.	fresh air, rest, place in half-sitting position, take to hospital.
Skin: *corrosive*; redness, pain, serious burns.	gloves, protective clothing.	remove contaminated clothing, flush skin with water or shower, refer to a doctor.
Eyes: *corrosive*; redness, pain, impaired vision.	face shield.	flush with water, take to a doctor.
Ingestion: *corrosive*; diarrhea, abdominal cramps, vomiting.	do not eat, drink or smoke while working.	rinse mouth, immediately take to hospital.

SPILLAGE	STORAGE	LABELING / NFPA
clean up spillage, flush away remainder with water; (additional individual protection: breathing apparatus).	keep dry, keep separate from acids and bases, special packaging.	R: 23/24/25 S: 1/2-26-44 ☠ Toxic

NFPA diamond: 3 (left), 0 (top), 0 (right)

NOTES
✳ TLV as fluorine. Depending on the degree of exposure, regular medical checkups are advisable. Lung edema symptoms usually develop several hours later and are aggravated by physical exertion: rest and hospitalization essential. As first aid, a doctor or authorized person should consider administering a corticosteroid spray. Under no circumstances use near flame or hot surface or when welding.

Transport Emergency Card TEC(R)-80G11　　　　　　　　　　　　　　　　　　　　　　**HI: 60; UN-number: 2505**

AMMONIUM NITRATE

PHYSICAL PROPERTIES	IMPORTANT CHARACTERISTICS
Decomposes below boiling point, °C 210 Melting point, °C 170 Relative density (water = 1) 1.7 Solubility in water, g/100 ml at 20 °C 190 Relative molecular mass 80.0	**WHITE HYGROSCOPIC CRYSTALS** Reacts with contaminants to form shock-sensitive mixtures. Decomposes when heated, giving off toxic vapors (→ *nitrous vapors*) and → *laughing gas*. Strong oxidant which reacts violently with combustible substances and reducing agents, with risk of fire and explosion. Reacts violently with metal powders, with risk of fire and explosion.
	TLV-TWA not available
	Absorption route: Can enter the body by inhalation or ingestion. **Immediate effects:** Irritates the eyes, skin and respiratory tract. In serious cases risk of unconsciousness. **Effects of prolonged/repeated exposure:** Can affect the blood.
Gross formula: H$_4$N$_2$O$_3$	

HAZARDS/SYMPTOMS	PREVENTIVE MEASURES	FIRE EXTINGUISHING/FIRST AID
Fire: non-combustible, but makes other substances more combustible.	keep away from open flame and sparks, no smoking, avoid contact with combustible substances and reducing agents, prevent contamination.	large quantities of water.
Explosion: can explode if mixed with combustible substances.		in case of fire: keep tanks/drums cool by spraying with water.
	KEEP DUST UNDER CONTROL	
Inhalation: sore throat, cough, shortness of breath.	local exhaust or respiratory protection.	fresh air, rest, call a doctor.
Skin: redness, pain.	gloves.	remove contaminated clothing, flush skin with water or shower.
Eyes: redness, pain.	goggles.	flush with water, take to doctor.
Ingestion: abdominal cramps, nausea, blue skin, feeling of weakness.		rinse mouth, call a doctor or take to hospital.

SPILLAGE	STORAGE	LABELING / NFPA
clean up spillage, flush away remainder with water, do not use sawdust or other combustible absorbents; (additional individual protection: P2 respirator).	keep in cool, dry, fireproof place, separate from combustible substances and reducing agents.	0 / 1 / 3 / oxy

NOTES
Flush contaminated clothing with water (fire hazard). Becomes shock-sensitive when contaminated with organic substances. In confined spaces combustion can cause explosion. Effects on blood due to formation of methemoglobin. Special first aid required in the event of poisoning: antidotes must be available (with instructions).

Transport Emergency Card TEC(R)-51G09   HI: 50; UN-number: 1942

$(COONH_4)_2.H_2O$

AMMONIUM OXALATE
(monohydrate)

PHYSICAL PROPERTIES	IMPORTANT CHARACTERISTICS
Melting point (decomposes), °C 70 Relative density (water = 1) 1.5 Solubility in water, g/100 ml 5 Relative molecular mass 142.1	**COLORLESS CRYSTALS** Decomposes when heated, giving off corrosive gas (→ *ammonia* and → *nitrogen oxides*). Attacks many metals.
	TLV-TWA not available
	Absorption route: Can enter the body by inhalation or ingestion. Evaporation negligible at 20°C, but harmful concentrations of airborne particles can build up rapidly. **Immediate effects:** Corrosive to the eyes, skin and respiratory tract. Inhalation can cause lung edema. In serious cases risk of death. **Effects of prolonged/repeated exposure:** Can cause kidney damage.
Gross formula: $C_2H_8N_2O_4.H_2O$	

HAZARDS/SYMPTOMS	PREVENTIVE MEASURES	FIRE EXTINGUISHING/FIRST AID
Fire: combustible.	keep away from open flame, no smoking.	water spray, powder.
	KEEP DUST UNDER CONTROL	
Inhalation: *corrosive*; sore throat, cough, severe breathing difficulties.	local exhaust or respiratory protection.	fresh air, rest, take to hospital.
Skin: *corrosive*; redness, pain, serious burns.	gloves, protective clothing.	remove contaminated clothing, flush skin with water or shower, call a doctor.
Eyes: *corrosive*; redness, pain, impaired vision.	face shield.	flush with water, take to a doctor.
Ingestion: *corrosive*; sore throat, abdominal pain, vomiting.		rinse mouth, allow to drink milk with water, rest, take immediately to hospital.

SPILLAGE	STORAGE	LABELING / NFPA
clean up spillage, carefully collect remainder; (additional individual protection: P2 respirator).		R: 21/22 S: 2-24/25 Note A ✖ Harmful

NOTES
Lung edema symptoms usually develop several hours later and are aggravated by physical exertion: rest and hospitalization essential. As first aid, a doctor or authorized person should consider administering a corticosteroid spray.

Transport Emergency Card TEC(R)-61G11C

HI: 60; UN-number: 2449

$(NH_4)_2S_2O_8$

AMMONIUM PERSULFATE

PHYSICAL PROPERTIES	IMPORTANT CHARACTERISTICS	
Decomposes below melting point, °C 120 Relative density (water = 1) 2.0 Solubility in water good Relative molecular mass 228.2	**WHITE ODORLESS CRYSTALS** Decomposes when heated to 120°C, giving off → *oxygen* and toxic vapors (incl. → *sulfur dioxide*). Strong reducing agent which reacts violently with combustible materials and reducing agents. Reacts violently with finely dispersed metals, with risk of fire and explosion.	
	TLV-TWA not available	
	Absorption route: Can enter the body by inhalation or ingestion. Evaporation negligible at 20°C, but harmful atmospheric concentrations of airborne particles can build up rapidly. **Immediate effects:** Corrosive to the eyes, skin and respiratory tract. Inhalation can cause lung edema. **Effects of prolonged/repeated exposure:** Prolonged or repeated contact can cause skin disorders.	
Gross formula: $H_8N_2O_8S_2$		

HAZARDS/SYMPTOMS,	PREVENTIVE MEASURES	FIRE EXTINGUISHING/FIRST AID
Fire: non-combustible, but makes other substances more combustible.	avoid contact with combustible substances.	in case of fire in immediate vicinity: use any extinguishing agent.
Explosion: reaction with reducing agents or metal powders can cause explosion.		
	KEEP DUST UNDER CONTROL	
Inhalation: *corrosive*; sore throat, cough, shortness of breath, severe breathing difficulties.	local exhaust or respiratory protection.	fresh air, rest, take to hospital.
Skin: *corrosive*; redness, pain, burns.	gloves.	remove contaminated clothing, flush skin with water or shower, take to hospital.
Eyes: *corrosive*; redness, pain, impaired vision.	face shield.	flush with water, take to a doctor.
Ingestion: *corrosive*; abdominal pain, nausea, vomiting.	do not eat, drink or smoke while working.	rinse mouth, take immediately to hospital.

SPILLAGE	STORAGE	LABELING / NFPA
clean up spillage, flush away remainder with water; (additional individual protection: P2 respirator).	keep separate from combustible substances and reducing agents.	

NOTES
Lung edema symptoms usually develop several hours later and are aggravated by physical exertion: rest and hospitalization essential. As first aid, a doctor or authorized person should consider administering a corticosteroid spray.

Transport Emergency Card TEC(R)-51G01

UN-number: 1444

diammonium hydrogen phosphate
diammonium orthophosphate
diammonium phosphate

$(NH_4)_2HPO_4$

sec-AMMONIUM PHOSPHATE

PHYSICAL PROPERTIES	IMPORTANT CHARACTERISTICS
Decomposes below melting point, °C ca. 100 Relative density (water = 1) 1.6 Solubility in water, g/100 ml at 10 °C 58 Relative molecular mass 132.1	**COLORLESS TO WHITE CRYSTALS OR POWDER** Gives off → *anhydrous ammonia* slowly on contact with air. Decomposes when heated above 100°C, giving off toxic and corrosive vapors (→ *anhydrous ammonia*, nitrous gases and phosphorus oxides). Reacts violently with some strong oxidants (esp. → *potassium chlorate*), with risk of fire and explosion. Reacts with strong bases, giving off toxic gas (→ *anhydrous ammonia*).
	TLV-TWA not available
	Absorption route: Can enter the body by inhalation of particles or ingestion. Evaporation negligible at 20°C, but harmful concentrations of airborne particles can build up rapidly. **Immediate effects:** Irritates the eyes, skin and respiratory tract. Can affect the upper respiratory tract.
Gross formula: $H_9N_2O_4P$	

HAZARDS/SYMPTOMS	PREVENTIVE MEASURES	FIRE EXTINGUISHING/FIRST AID
Fire: non-combustible.		in case of fire in immediate vicinity: use any extinguishing agent.
	KEEP DUST UNDER CONTROL	
Inhalation: cough, shortness of breath, sore throat.	local exhaust or respiratory protection.	fresh air, rest, consult a doctor.
Skin: redness, pain.	gloves.	remove contaminated clothing, flush skin with water and wash with soap and water, consult a doctor.
Eyes: redness, pain.	goggles.	flush thoroughly with water (remove contact lenses if easy), take to a doctor.
Ingestion: abdominal cramps, nausea, sore throat, vomiting, feeling of weakness.		rinse mouth, immediately take to hospital.

SPILLAGE	STORAGE	LABELING / NFPA
clean up spillage, flush away remainder with water; (additional individual protection: P2 respirator).	keep separate from oxidants and strong bases; ventilate.	

NOTES
Insufficient data on harmful effects to humans: exercise caution.

AMMONIUM RHODANIDE

PHYSICAL PROPERTIES	IMPORTANT CHARACTERISTICS	
Decomposes below boiling point, °C 170 Melting point, °C 149 Relative density (water = 1) 1.3 Solubility in water good Relative molecular mass 76.1	**COLORLESS HYGROSCOPIC CRYSTALS** Decomposes when heated, giving off toxic vapors (→ *cyanides*). Decomposes when exposed to light. Reacts violently with strong oxidants.	
	TLV-TWA not available	
	Absorption route: Can enter the body by ingestion. Evaporation negligible at 20°C, but harmful concentrations of airborne particles can build up rapidly. **Immediate effects:** In high doses affects the central nervous system and blood pressure.	
Gross formula: CH₄N₂S		

HAZARDS/SYMPTOMS	PREVENTIVE MEASURES	FIRE EXTINGUISHING/FIRST AID
Fire: hardly combustible.		in case of fire in immediate vicinity: use any extinguishing agent.
Explosion:		in case of fire: keep tanks/drums cool by spraying with water.
Skin: redness.	gloves.	remove contaminated clothing, flush skin with water or shower.
Eyes: redness, pain.	safety glasses.	flush with water, take to a doctor.
Ingestion: headache, nausea, dizziness, vomiting, feeling of weakness.		rinse mouth, call a doctor or take to hospital.

SPILLAGE	STORAGE	LABELING / NFPA
clean up spillage, flush away remainder with water; (additional individual protection: P2 respirator).	keep in dark place, separate from oxidants.	

NOTES

AMMONIUMSILICOFLUORIDE

PHYSICAL PROPERTIES	IMPORTANT CHARACTERISTICS	
Decomposes below boiling point, °C ? Decomposes below melting point, °C ? Relative density (water = 1) 2.0 Vapor pressure, mm Hg at 20 °C < 0.076 Solubility in water, g/100 ml at 17 °C 18.6 Relative molecular mass 178.1	**COLORLESS CRYSTALS** Decomposes when heated, giving off corrosive vapors (→ *nitrous vapors and* → *fluorine*). Corrosive in aqueous solution.	
	TLV-TWA not available	
	Absorption route: Can enter the body by inhalation or ingestion. Evaporation negligible at 20°C, but harmful atmospheric concentrations can build up rapidly in aerosol form. **Immediate effects:** Corrosive to the eyes, skin and respiratory tract. Inhalation of vapor/fumes can cause severe breathing difficulties (lung edema). In serious cases risk of death. **Effects of prolonged/repeated exposure:** Can cause bone disorders.	
Gross formula: $F_6H_8N_2Si$		

HAZARDS/SYMPTOMS	PREVENTIVE MEASURES	FIRE EXTINGUISHING/FIRST AID
Fire: non-combustible.		in case of fire in immediate vicinity: use any extinguishing agent.
	STRICT HYGIENE, KEEP DUST UNDER CONTROL	
Inhalation: *corrosive*; sore throat, cough, shortness of breath, severe breathing difficulties.	local exhaust or respiratory protection.	fresh air, rest, place in half-sitting position, take to hospital.
Skin: *corrosive*; redness, pain, serious burns.	gloves, protective clothing.	remove contaminated clothing, flush skin with water or shower, refer to a doctor.
Eyes: *corrosive*; redness, pain, impaired vision.	face shield, or combined eye and respiratory protection.	flush with water, take to a doctor.
Ingestion: *corrosive*; diarrhea, abdominal cramps, vomiting.	do not eat, drink or smoke while working.	rinse mouth, immediately take to hospital.

SPILLAGE	STORAGE	LABELING / NFPA
clean up spillage, flush away remainder with water; (additional individual protection: P3 respirator).	keep dry.	

NOTES
The solid sublimes. Decomposition temperature not given in the literature. Depending on the degree of exposure, regular medical checkups are advisable. Lung edema symptoms usually develop several hours later and are aggravated by physical exertion: rest and hospitalization essential. As first aid, a doctor or authorized person should consider administering a corticosteroid spray. The measures on this card also apply to sodium silicofluoride.

Transport Emergency Card TEC(R)-61G11

HI: 60; UN-number: 2854

AMMONIUM SULFATE
(40% solution)

PHYSICAL PROPERTIES	IMPORTANT CHARACTERISTICS	
Boiling point, °C 105 Melting point, °C <0 Relative density (water = 1) 1.2 Solubility in water ∞ Relative molecular mass 132.1	**COLORLESS AQUEOUS SOLUTION OF AMMONIUM SULFATE** Reacts with bases to form → *anhydrous ammonia* gas. Attacks many metals.	
	TLV-TWA not available	
	Absorption route: Can enter the body by inhalation or ingestion. **Immediate effects:** Irritates the eyes, skin and respiratory tract.	
Gross formula: $H_8N_2O_4S$		

HAZARDS/SYMPTOMS	PREVENTIVE MEASURES	FIRE EXTINGUISHING/FIRST AID
Fire: non-combustible.		in case of fire in immediate vicinity: use any extinguishing agent.
Inhalation: sore throat, cough, shortness of breath.	ventilation.	fresh air, rest, call a doctor.
Skin: redness, pain.	gloves.	remove contaminated clothing, flush skin with water or shower.
Eyes: redness, pain.	acid goggles.	flush with water, take to a doctor if necessary.
Ingestion: sore throat, abdominal pain, nausea.		rinse mouth, call a doctor and take to hospital if necessary.

SPILLAGE	STORAGE	LABELING / NFPA
collect leakage in sealable containers, flush away remainder with water.	keep separate from strong bases.	

NOTES

AMMONIUM SULFATE

PHYSICAL PROPERTIES	IMPORTANT CHARACTERISTICS	
Decomposes below melting point, °C 235 Relative density (water = 1) 1.8 Solubility in water, g/100 ml at 25 °C 77 Relative molecular mass 132.1	**COLORLESS CRYSTALS OR WHITE POWDER** Decomposes when heated above 235°C, giving off toxic and corrosive vapors (→ *ammonia*, → *sulfur oxides*). Reacts violently when heated with chlorates, nitrates and nitrites. Slightly acidic in aqueous solution.	
	TLV-TWA not available	
	Absorption route: Can enter the body by inhalation or ingestion. **Immediate effects:** Irritates the eyes, skin and respiratory tract. Inhalation of high concentrations of vapor/fumes can cause severe breathing difficulties (lung edema).	
Gross formula: $H_8N_2O_4S$		

HAZARDS/SYMPTOMS	PREVENTIVE MEASURES	FIRE EXTINGUISHING/FIRST AID
Fire: non-combustible.		in case of fire in immediate vicinity: use any extinguishing agent.
Inhalation: sore throat, cough, shortness of breath.	local exhaust or respiratory protection.	fresh air, rest, call a doctor.
Skin: redness.	gloves.	remove contaminated clothing, flush skin with water or shower.
Eyes: redness, pain.	goggles.	flush with water, take to a doctor if necessary.
Ingestion: sore throat, abdominal pain, nausea, vomiting.		rinse mouth, call a doctor.

SPILLAGE	STORAGE	LABELING / NFPA
clean up spillage, flush away remainder with water; (additional individual protection: P2 respirator).		

NOTES
Lung edema symptoms usually develop several hours later and are aggravated by physical exertion: rest and hospitalization essential. As first aid, a doctor or authorized person should consider administering a corticosteroid spray.

CAS-No: [628-63-7]
acetic acid, pentyl ester
acetic acid, amyl ester
amyl acetic ester
1-pentanol acetate
n-pentyl acetate

$CH_3COOC_5H_{11}$

n-AMYLACETATE

PHYSICAL PROPERTIES	IMPORTANT CHARACTERISTICS
Boiling point, °C 147 Melting point, °C −71 Flash point, °C 37 Auto-ignition temperature, °C 375 Relative density (water = 1) 0.9 Relative vapor density (air = 1) 4.5 Relative density at 20 °C of saturated mixture vapor/air (air = 1) 1.02 Vapor pressure, mm Hg at 20 °C 7.1 Solubility in water, g/100 ml 0.2 Explosive limits, vol% in air 1.0-7.5 Relative molecular mass 130.2 Gross formula: $C_7H_{14}O_2$	**COLORLESS LIQUID WITH CHARACTERISTIC ODOR** Vapor mixes readily with air. Reacts with strong oxidants. Attacks many plastics. TLV-TWA 100 ppm 532 mg/m³ **Absorption route:** Can enter the body by inhalation or ingestion. Harmful atmospheric concentrations build up fairly slowly on evaporation, but much more rapidly in aerosol form. **Immediate effects:** Irritates the eyes, skin and respiratory tract. Liquid destroys the skin's natural oils. Affects the nervous system. In serious cases risk of unconsciousness.

HAZARDS/SYMPTOMS	PREVENTIVE MEASURES	FIRE EXTINGUISHING/FIRST AID
Fire: flammable.	keep away from open flame and sparks, no smoking.	powder, AFFF, foam, carbon dioxide, (halons).
Explosion: above 37°C: forms explosive air-vapor mixtures.	above 37°C: sealed machinery, ventilation, explosion-proof electrical equipment.	in case of fire: keep tanks/drums cool by spraying with water.
Inhalation: sore throat, cough, headache, dizziness, drowsiness, increased puls rate, tiredness.	ventilation, local exhaust or respiratory protection.	fresh air, rest, call a doctor.
Skin: redness, pain.	gloves.	remove contaminated clothing, flush skin with water or shower, call a doctor if necessary.
Eyes: redness, pain.	safety glasses.	flush with water, take to a doctor.
Ingestion: sore throat, abdominal pain, nausea; see also 'Inhalation'.		rinse mouth, call a doctor or take to hospital.

SPILLAGE	STORAGE	LABELING / NFPA
collect leakage in sealable containers, soak up with sand or other inert absorbent and remove to safe place.	keep in fireproof place, separate from oxidants.	R: 10 S: 23 Note C

NFPA diamond: Health 1, Flammability 3, Reactivity 0

NOTES
Alcohol consumption increases toxic effects.

Transport Emergency Card TEC(R)-581

HI: 30; UN-number: 1104

2-acetoxypentane
1-methylbutyl acetate
2-pentanol acetate
2-pentyl acetate

$CH_3COOC_5H_{11}$

sec-AMYL ACETATE

PHYSICAL PROPERTIES		IMPORTANT CHARACTERISTICS	
Boiling point, °C	121	**COLORLESS LIQUID WITH CHARACTERISTIC ODOR**	
Melting point, °C	−100	Vapor mixes readily with air. Reacts with strong oxidants. Attacks many plastics.	
Flash point, °C	32		
Auto-ignition temperature, °C	?	TLV-TWA 100 ppm 532 mg/m³	
Relative density (water = 1)	0.85		
Relative vapor density (air = 1)	4.5	**Absorption route:** Can enter the body by inhalation or ingestion. Harmful atmospheric concentrations build up fairly slowly on evaporation at 20°C, but much more rapidly in aerosol form.	
Relative density at 20 °C of saturated mixture vapor/air (air = 1)	1.03		
Vapor pressure, mm Hg at 20 °C	7.1	**Immediate effects:** Irritates the eyes, skin and respiratory tract. Liquid destroys the skin's natural oils. Affects the nervous system. In serious cases risk of unconsciousness.	
Solubility in water	none		
Explosive limits, vol% in air	?-?		
Relative molecular mass	130.2		
Gross formula: $C_7H_{14}O_2$			

HAZARDS/SYMPTOMS	PREVENTIVE MEASURES	FIRE EXTINGUISHING/FIRST AID
Fire: flammable.	keep away from open flame and sparks, no smoking.	powder, AFFF, foam, carbon dioxide, (halons).
Explosion: above 32°C: forms explosive air-vapor mixtures.	above 32°C: sealed machinery, ventilation, explosion-proof electrical equipment.	in case of fire: keep tanks/drums cool by spraying with water.
Inhalation: sore throat, cough, headache, dizziness, drowsiness.	ventilation, local exhaust or respiratory protection.	fresh air, rest, call a doctor.
Skin: redness, pain.	gloves.	remove contaminated clothing, flush skin with water or shower, call a doctor if necessary.
Eyes: redness, pain.	safety glasses.	flush with water, take to a doctor.
Ingestion: sore throat, abdominal pain, nausea; see also 'Inhalation'.		rinse mouth, call a doctor or take to hospital.

SPILLAGE	STORAGE	LABELING / NFPA
collect leakage in sealable containers, soak up with sand or other inert absorbent and remove to safe place.	keep in fireproof place, separate from oxidants.	R: 10 S: 23 Note C

NOTES
Explosive limits not given in the literature. Alcohol consumption increases toxic effects.

Transport Emergency Card TEC(R)-581

HI: 30; UN-number: 1104

1-pentanol
n-butyl carbinol
pentyl alcohol

$CH_3(CH_2)_3CH_2OH$

prim-n-AMYL ALCOHOL

PHYSICAL PROPERTIES		IMPORTANT CHARACTERISTICS
Boiling point, °C	138	**COLORLESS LIQUID WITH CHARACTERISTIC ODOR**
Melting point, °C	−78	Vapor mixes readily with air. Reacts violently with oxidants, with risk of fire and explosion. Reacts
Flash point, °C	33	violently with alkaline-earth and alkali metals, giving off flammable gas (→ *hydrogen*).
Auto-ignition temperature, °C	300	
Relative density (water = 1)	0.8	
Relative vapor density (air = 1)	3.0	TLV-TWA not available
Relative density at 20 °C of saturated mixture		
vapor/air (air = 1)	1.01	**Absorption route:** Can enter the body by inhalation or ingestion. Harmful atmospheric
Vapor pressure, mm Hg at 20 °C	2.3	concentrations build up very slowly, if at all, on evaporation at approx. 20°C, but much more
Solubility in water, g/100 ml	2.7	rapidly in aerosol form.
Explosive limits, vol% in air	1.3-10.5	**Immediate effects:** Irritates the eyes, skin and respiratory tract. In substantial concentrations
Electrical conductivity, pS/m	2.6×10^7	can impair consciousness.
Relative molecular mass	88.1	**Effects of prolonged/repeated exposure:** Can cause liver damage.
Log P octanol/water	1.3	
Gross formula:	$C_5H_{12}O$	

HAZARDS/SYMPTOMS	PREVENTIVE MEASURES	FIRE EXTINGUISHING/FIRST AID
Fire: flammable.	keep away from open flame and sparks, no smoking.	powder, AFFF, alcohol-resistant foam, carbon dioxide, (halons).
Explosion: above 33°C: forms explosive air-vapor mixtures.	above 33°C: sealed machinery, ventilation, explosion-proof electrical equipment.	in case of fire: keep tanks/drums cool by spraying with water.
Inhalation: cough, headache, nausea, drowsiness.	ventilation.	fresh air, rest, call a doctor.
Skin: redness.	gloves.	remove contaminated clothing, flush skin with water or shower.
Eyes: redness, pain.	safety glasses.	flush with water, take to a doctor.
Ingestion: cough, headache, dizziness, vomiting.		rinse mouth, call a doctor.

SPILLAGE	STORAGE	LABELING / NFPA
collect leakage in sealable containers, soak up with sand or other inert absorbent and remove to safe place.	keep in fireproof place, separate from oxidants.	R: 10-20 S: 24/25 Note C ✖ Harmful

NFPA diamond: 3 (top), 1 (left), 0 (right)

NOTES
Log P octanol/water is estimated. Alcohol consumption increases toxic effects. Depending on the degree of exposure, regular medical checkups are advisable.

Transport Emergency Card TEC(R)-582

HI: 30; UN-number: 1105

2-pentanol
sec-amyl alcohol
2-hydroxypentane
methyl propyl carbinol
sec-pentyl alcohol

$CH_3CH_2CH_2CHOHCH_3$

sec-n-AMYL ALCOHOL

PHYSICAL PROPERTIES	IMPORTANT CHARACTERISTICS
Boiling point, °C 119	**COLORLESS LIQUID WITH CHARACTERISTIC ODOR**
Melting point, °C −50	Vapor mixes readily with air. Reacts violently with strong oxidants. Reacts violently with alkali
Flash point, °C 34	metals, and to a lesser extent with alkaline-earth metals, giving off → *hydrogen* gas.
Auto-ignition temperature, °C 340	
Relative density (water = 1) 0.8	TLV-TWA not available
Relative vapor density (air = 1) 3.0	
Relative density at 20 °C of saturated mixture	**Absorption route:** Can enter the body by inhalation or ingestion. Harmful atmospheric
vapor/air (air = 1) 1.01	concentrations build up very slowly, if at all, on evaporation at approx. 20°C, but much more
Vapor pressure, mm Hg at 20 °C 2.3	rapidly in aerosol form.
Solubility in water, g/100 ml 13.5	**Immediate effects:** Irritates the eyes, skin and respiratory tract.
Explosive limits, vol% in air 1.2-8.0	**Effects of prolonged/repeated exposure:** Can affect the blood. Can cause liver damage.
Relative molecular mass 88.1	
Log P octanol/water 1.4	
Gross formula: $C_5H_{12}O$	

HAZARDS/SYMPTOMS	PREVENTIVE MEASURES	FIRE EXTINGUISHING/FIRST AID
Fire: flammable.	keep away from open flame and sparks, no smoking.	powder, AFFF, alcohol-resistant foam, carbon dioxide, (halons).
Explosion: above 34°C: forms explosive air-vapor mixtures.	above 34°C: sealed machinery, ventilation, explosion-proof electrical equipment.	in case of fire: keep tanks/drums cool by spraying with water.
Inhalation: cough, headache, nausea, drowsiness.	ventilation, local exhaust or respiratory protection.	fresh air, rest, call a doctor.
Skin: redness.	gloves.	remove contaminated clothing, flush skin with water or shower.
Eyes: redness, pain.	safety glasses.	flush with water, take to a doctor.
Ingestion: cough, headache, dizziness, vomiting.		rinse mouth, call a doctor.

SPILLAGE	STORAGE	LABELING / NFPA
collect leakage in sealable containers, flush away remainder with water.	keep in fireproof place, separate from oxidants and alkaline/alkaline-earth metals.	R: 10-20 S: 24/25 Note C ✖ Harmful

NOTES
Alcohol consumption increases toxic effects. Depending on the degree of exposure, regular medical checkups are advisable.

Transport Emergency Card TEC(R)-582

HI: 30; UN-number: 1105

CAS-No: [75-85-4]
2-methyl-2-butanol
t-amyl alcohol
amylene hydrate
dimethylethylcarbinol
tert-pentanol
t-pentyl alcohol

$CH_3CH_2COH(CH_3)_2$

tert-AMYL ALCOHOL

PHYSICAL PROPERTIES		IMPORTANT CHARACTERISTICS
Boiling point, °C	102	**COLORLESS LIQUID WITH CHARACTERISTIC ODOR**
Melting point, °C	− 8	Vapor mixes readily with air, forming explosive mixtures. Do not use compressed air when filling, emptying or processing. Reacts with alkali metals, giving off → *hydrogen*. Reacts violently with strong oxidants.
Flash point, °C	19	
Auto-ignition temperature, °C	435	
Relative density (water = 1)	0.8	
Relative vapor density (air = 1)	3.0	
Relative density at 20 °C of saturated mixture		**TLV-TWA** not available
vapor/air (air = 1)	1.03	
Vapor pressure, mm Hg at 20 °C	12	**Absorption route:** Can enter the body by inhalation or ingestion or through the skin. Harmful atmospheric concentrations can build up fairly rapidly on evaporation at approx. 20°C - even more rapidly in aerosol form.
Solubility in water	moderate	
Explosive limits, vol% in air	1.2-9	**Immediate effects:** Irritates the eyes, skin and respiratory tract. Affects the nervous system. In serious cases risk of unconsciousness.
Relative molecular mass	88.1	
Log P octanol/water	0.9	**Effects of prolonged/repeated exposure:** Can affect the blood.
Gross formula:	$C_5H_{12}O$	

HAZARDS/SYMPTOMS	PREVENTIVE MEASURES	FIRE EXTINGUISHING/FIRST AID
Fire: extremely flammable.	keep away from open flame and sparks, no smoking.	powder, AFFF, alcohol-resistant foam, carbon dioxide, (halons).
Explosion: forms explosive air-vapor mixtures.	sealed machinery, ventilation, explosion-proof electrical equipment and lighting.	in case of fire: keep tanks/drums cool by spraying with water.
Inhalation: cough, shortness of breath, dizziness, drowsiness, blue skin, feeling of weakness.	ventilation, local exhaust or respiratory protection.	fresh air, rest, call a doctor.
Skin: *is absorbed*; redness, pain; see also 'Inhalation'.	gloves.	remove contaminated clothing, wash skin with soap and water, call a doctor.
Eyes: redness, pain.	face shield.	flush with water, take to a doctor if necessary.
Ingestion: abdominal pain, diarrhea, vomiting; see also 'Inhalation'.		rinse mouth, call a doctor.

SPILLAGE	STORAGE	LABELING / NFPA
collect leakage in sealable containers, soak up with sand or other inert absorbent and remove to safe place, flush away remainder with water; (additional individual protection: breathing apparatus).	keep in fireproof place, separate from oxidants and alkali metals.	R: 11-20 S: 9-16-24/25 Flammable Harmful

NOTES
Alcohol consumption increases toxic effects. Depending on the degree of exposure, regular medical checkups are advisable. Effects on blood due to formation of methemoglobin. Special first aid required in the event of poisoning: antidotes must be available (with instructions).

Transport Emergency Card TEC(R)-512

HI: 33; UN-number: 1105

AMYLENE-DIMER

PHYSICAL PROPERTIES		IMPORTANT CHARACTERISTICS
Boiling point, °C	150-170	**COLORLESS LIQUID WITH CHARACTERISTIC ODOR**
Melting point, °C	<30	Vapor mixes readily with air. Flow, agitation etc. can cause build-up of electrostatic charge due to
Flash point, °C	48	liquid's low conductivity.
Auto-ignition temperature, °C	250	
Relative density (water = 1)	0.8	TLV-TWA not available
Relative vapor density (air = 1)	4.8	
Relative density at 20 °C of saturated mixture		**Absorption route:** Can enter the body by inhalation or ingestion. Harmful atmospheric
vapor/air (air = 1)	1.02	concentrations build up very slowly, if at all, on evaporation at approx. 20°C, but much more
Vapor pressure, mm Hg at 20 °C	ca. 3.8	rapidly in aerosol form.
Solubility in water	none	**Immediate effects:** Irritates the eyes and respiratory tract. Liquid destroys the skin's natural oils.
Explosive limits, vol% in air	1.1-6.8	If liquid is swallowed, droplets can enter the lungs, with risk of pneumonia.
Relative molecular mass	140.3	
Gross formula:	$C_{10}H_{10}$	

HAZARDS/SYMPTOMS	PREVENTIVE MEASURES	FIRE EXTINGUISHING/FIRST AID
Fire: flammable.	keep away from open flame and sparks, no smoking.	powder, AFFF, foam, carbon dioxide, (halons).
Explosion: above 48°C: forms explosive air-vapor mixtures.	above 48°C: sealed machinery, ventilation, explosion-proof electrical equipment, grounding.	in case of fire: keep tanks/drums cool by spraying with water.
Inhalation: sore throat, cough.	ventilation, local exhaust or respiratory protection.	fresh air, rest.
Skin: redness.	gloves.	remove contaminated clothing, wash skin with soap and water.
Eyes: redness, pain.	safety glasses.	flush with water, take to a doctor if necessary.
Ingestion: abdominal pain, nausea.		rinse mouth, immediately take to hospital.

SPILLAGE	STORAGE	LABELING / NFPA
collect leakage in sealable containers, soak up with sand or other inert absorbent and remove to safe place.		

NOTES
Technical grade contains approx. 80% dimer, otherwise mainly higher polymers.

Transport Emergency Card TEC(R)-30G08   **UN-number: 1142**

AMYL ETHER

PHYSICAL PROPERTIES	IMPORTANT CHARACTERISTICS
Boiling point, °C 190 Melting point, °C −69 Flash point, °C 57 Auto-ignition temperature, °C 170 Relative density (water = 1) 0.8 Relative vapor density (air = 1) 5.5 Relative density at 20 °C of saturated mixture vapor/air (air = 1) 1.0 Vapor pressure, mm Hg at 20 °C ca. 0.53 Solubility in water none Explosive limits, vol% in air 2-8 Electrical conductivity, pS/m 10^7 Relative molecular mass 158.3 Gross formula: $C_{10}H_{22}O$	**COLORLESS TO LIGHT YELLOW LIQUID** Vapor mixes readily with air. Able to form peroxides. Reacts violently with strong oxidants. TLV-TWA not available **Absorption route:** Can enter the body by inhalation or ingestion or through the skin. Harmful atmospheric concentrations build up very slowly, if at all, on evaporation at approx. 20°C, but much more rapidly in aerosol form. **Immediate effects:** Irritates the eyes, skin and respiratory tract. Liquid destroys the skin's natural oils. In substantial concentrations can impair consciousness. In serious cases risk of seizures and unconsciousness.

HAZARDS/SYMPTOMS	PREVENTIVE MEASURES	FIRE EXTINGUISHING/FIRST AID
Fire: combustible.	keep away from open flame, no smoking.	powder, AFFF, foam, carbon dioxide, (halons).
Explosion: above 57°C: forms explosive air-vapor mixtures.	above 57°C: sealed machinery, ventilation.	
Inhalation: sore throat, cough, drowsiness, sleepiness.	ventilation.	fresh air, rest, call a doctor.
Skin: *is absorbed*; redness, pain.	gloves.	remove contaminated clothing, flush skin with water or shower.
Eyes: redness, pain.	face shield.	flush with water, take to a doctor if necessary.
Ingestion: abdominal pain, nausea.		rinse mouth, call a doctor.

SPILLAGE	STORAGE	LABELING / NFPA
collect leakage in sealable containers, soak up with sand or other inert absorbent and remove to safe place.	keep in dark place, separate from oxidants; add an inhibitor.	2 1 0

NOTES
Before distilling check for peroxides; if found, render harmless. The measures on this card apply to all isomers and mixtures.

Transport Emergency Card TEC(R)-30G15

AMYL NITRITE

PHYSICAL PROPERTIES	IMPORTANT CHARACTERISTICS	
Boiling point, °C .... 99 Flash point, °C .... 3 Auto-ignition temperature, °C .... 210 Relative density (water = 1) .... 0.9 Relative vapor density (air = 1) .... 4.0 Relative density at 20 °C of saturated mixture vapor/air (air = 1) .... 1.1 Vapor pressure, mm Hg at 20 °C ...... ca. 27 Solubility in water .... poor Explosive limits, vol% in air .... see notes Relative molecular mass 117.2	**LIGHT YELLOW CLEAR LIQUID WITH CHARACTERISTIC ODOR** Vapor is heavier than air and spreads at ground level, with risk of ignition at a distance. Decomposes when heated or burned, giving off toxic vapors (→ *nitrous vapors*). Decomposes slowly on prolonged exposure to light, air or water, also giving off → *nitrous vapors*. Strong oxidant which reacts with combustible substances and reducing agents; also reacts violently with oxidants, with risk of fire and explosion. Attacks metals in the presence of moisture.	
	TLV-TWA not available	
Gross formula: $C_5H_{11}O_2$	**Absorption route:** Can enter the body by inhalation of vapor, through the skin or by ingestion of liquid. Harmful atmospheric concentrations can build up fairly rapidly on evaporation at approx. 20°C - even more rapidly in aerosol form. **Immediate effects:** Irritates the eyes and respiratory tract. Can affect the blood vessels, causing red face, palpitations and low blood pressure. Exposure to high concentrations can cause unconsciousness. Keep under medical observation. **Effects of prolonged/repeated exposure:** Exposure to high concentrations can affect hemoglobin, causing formation of methemoglobin.	

HAZARDS/SYMPTOMS	PREVENTIVE MEASURES	FIRE EXTINGUISHING/FIRST AID
Fire: extremely flammable.	keep away from open flame and sparks, no smoking.	powder, AFFF, carbon dioxide, (halons).
Explosion: forms explosive air-vapor mixtures.	sealed machinery, ventilation, explosion-proof electrical equipment and lighting; do not use compressed air when filling, emptying or processing.	in case of fire: keep tanks/drums cool by spraying with water.
	PREVENT MIST FORMATION, STRICT HYGIENE, AVOID EXPOSURE OF PREGNANT WOMEN	
Inhalation: disorientation, dizziness, drowsiness, blue lips and fingernails, unconsciousness.	ventilation, local exhaust or respiratory protection.	fresh air, rest, special treatment, consult a doctor and take to hospital if necessary.
Skin: *is absorbed*; see also 'Inhalation'.	gloves.	remove contaminated clothing, flush skin with water and wash with soap and water, refer to doctor.
Eyes: irritation, redness, pain.	safety glasses.	flush thoroughly with water (remove contact lenses if easy), take to a doctor.
Ingestion: burning sensation; see also 'Inhalation'.		rinse mouth, give no liquids, special treatment, take immediately to hospital.

SPILLAGE	STORAGE	LABELING / NFPA
ventilate, collect leakage in sealable containers, soak up with sand or other inert absorbent and remove to safe place, do not use sawdust or other combustible absorbents, do not flush into sewer; (additional individual protection: breathing apparatus).	keep in dry, dark fireproof place, separate from combustible substances, reducing agents and oxidants.	◇ 1 / 2

NOTES
One source gives estimated lower explosive limit of 1% by vol. Special first aid required in the event of poisoning: antidotes must be available (with instructions). Flush contaminated clothing with water (fire hazard).

Transport Emergency Card TEC(R)-30G06

UN-number: 1113

CAS-No: [80-46-6]
(dimethyl-1,1- propyl)-4 phenol
4-(1,1-dimethylpropyl)phenol
2-methyl-2-p-hydroxyphenylbutane
pentaphene

$(CH_3)_2(C_2H_5)CC_6H_4OH$

p-tert-AMYLPHENOL

PHYSICAL PROPERTIES	IMPORTANT CHARACTERISTICS
Boiling point, °C 249 Melting point, °C 93 Flash point, °C 112 Relative density (water = 1) 0.9 Relative vapor density (air = 1) 5.7 Relative density at 20 °C of saturated mixture vapor/air (air = 1) 1 Vapor pressure, mm Hg at 125 °C 9.9 Relative molecular mass 164.2	**WHITE CRYSTALS** In dry state can form electrostatic charge if stirred, transported pneumatically, poured etc. Reacts with strong oxidants.
	TLV-TWA not available
	Absorption route: Can enter the body by inhalation or ingestion. Harmful atmospheric concentrations build up very slowly, if at all, on evaporation at approx. 20°C, but harmful concentrations of airborne particles can build up much more rapidly. **Immediate effects:** Irritates the eyes, skin and respiratory tract. Inhalation of vapor/fumes can cause severe breathing difficulties (lung edema). **Effects of prolonged/repeated exposure:** Prolonged or repeated contact can cause patchy discoloration of the skin, preceded by redness and itching. The discoloration is temporary. Can affect the blood. Can cause liver and kidney damage.
Gross formula: $C_{11}H_{16}O$	

HAZARDS/SYMPTOMS	PREVENTIVE MEASURES	FIRE EXTINGUISHING/FIRST AID
Fire: combustible.	keep away from open flame, no smoking.	water spray, powder.
Explosion: finely dispersed particles form explosive mixtures on contact with air.	keep dust from accumulating, sealed machinery, explosion-proof electrical equipment and lighting, grounding.	
	STRICT HYGIENE, KEEP DUST UNDER CONTROL; AVOID EXPOSURE OF PREGNANT WOMEN AS FAR AS POSSIBLE	
Inhalation: sore throat, cough, nausea, vomiting, diarrhea, abdominal cramps, drowsiness, blue skin, intense thirst, heavy perspiration.	local exhaust or respiratory protection.	fresh air, rest, place in half-sitting position, artificial respiration if necessary, call a doctor and take to hospital.
Skin: irritation, redness, itching.	gloves.	remove contaminated clothing, flush skin with water or shower, call a doctor if necessary.
Eyes: redness, pain, impaired vision.	goggles.	flush with water, take to a doctor.
Ingestion: *corrosive*; sore throat, cough, abdominal pain, diarrhea, drowsiness, blue skin.	do not eat, drink or smoke while working.	rinse mouth, take immediately to hospital.

SPILLAGE	STORAGE	LABELING / NFPA
clean up spillage, carefully collect remainder; (additional individual protection: P2 respirator).	keep separate from oxidants.	1 2 0

NOTES

Depending on the degree of exposure, regular medical checkups are advisable. Lung edema symptoms usually develop several hours later and are aggravated by physical exertion: rest and hospitalization essential. As first aid, a doctor or authorized person should consider administering a corticosteroid spray. Effects on blood due to formation of methemoglobin.

Transport Emergency Card TEC(R)-61G12C

HI: 60; UN-number: 2430

benzenamine
amino benzene
phenylamine

ANILINE

PHYSICAL PROPERTIES		IMPORTANT CHARACTERISTICS
Boiling point, °C	184	**OILY COLORLESS TO LIGHT BROWN LIQUID WITH CHARACTERISTIC ODOR, TURNING BROWN ON EXPOSURE TO LIGHT**
Melting point, °C	−6	Vapor mixes readily with air. Decomposes when heated above 190°C or burned, giving off toxic (→ *nitrous vapors*) and flammable vapors. Medium strong base which reacts with strong acids. Reacts violently with strong oxidants, with risk of fire and explosion. Reacts with alkaline-earth and alkali metals, giving off → *hydrogen*. Attacks copper and its alloys.
Flash point, °C	70	
Auto-ignition temperature, °C	630	
Relative density (water = 1)	1.02	
Relative vapor density (air = 1)	3.2	
Relative density at 20 °C of saturated mixture vapor/air (air = 1)	1.0	

TLV-TWA (skin)   2 ppm   7.6 mg/m³

Vapor pressure, mm Hg at 20°C	0.30
Solubility in water, g/100 ml at 20 °C	3.4
Explosive limits, vol% in air	1.2-11
Electrical conductivity, pS/m	2.4x10⁶
Relative molecular mass	93.1
Log P octanol/water	0.9

Absorption route: Can enter the body by inhalation or ingestion or through the skin. Harmful atmospheric concentrations build up fairly slowly on evaporation at 20°C, but much more rapidly in aerosol form.
Immediate effects: Affects red blood cells. In serious cases risk of unconsciousness.
Effects of prolonged/repeated exposure: Can affect the blood.

Gross formula: C₆H₇N

HAZARDS/SYMPTOMS	PREVENTIVE MEASURES	FIRE EXTINGUISHING/FIRST AID
Fire: combustible.	keep away from open flame, no smoking.	powder, AFFF, foam, carbon dioxide, (halons).
Explosion: above 70°C: forms explosive air-vapor mixtures.	above 70°C: sealed machinery, ventilation.	
	STRICT HYGIENE	
Inhalation: shortness of breath, dizziness, blue skin, feeling of weakness.	ventilation, local exhaust or respiratory protection.	fresh air, rest, call a doctor.
Skin: *is absorbed*; see also 'Inhalation'.	gloves, protective clothing.	remove contaminated clothing, wash skin with soap and water, call a doctor.
Eyes: redness, pain.	face shield, or combined eye and respiratory protection.	flush with water, take to a doctor if necessary.
Ingestion: shortness of breath, dizziness, blue skin, feeling of weakness.	do not eat, drink or smoke while working.	rinse mouth, take immediately to hospital.

SPILLAGE	STORAGE	LABELING / NFPA
collect leakage in sealable containers, soak up with sand or other inert absorbent and remove to safe place.	keep in dark place, separate from oxidants and strong acids.	R: 23/24/25-33 S: 28-36/37-44 ☠ Toxic NFPA: 2 3 0

NOTES

Odor gives inadequate warning of exceeded TLV. Depending on the degree of exposure, regular medical checkups are advisable. Effects on blood due to formation of methemoglobin. Special first aid required in the event of poisoning: antidotes must be available (with instructions). Unbreakable packaging preferred; if breakable, keep in unbreakable container.

Transport Emergency Card TEC(R)-62

HI: 60; UN-number: 1547

ANILINE HYDROCHLORIDE

PHYSICAL PROPERTIES	IMPORTANT CHARACTERISTICS
Boiling point, °C 245 Melting point, °C 198 Flash point, °C 193 Relative density (water = 1) 1.2 Relative vapor density (air = 1) 4.5 Relative density at 20 °C of saturated mixture vapor/air (air = 1) 1.0 Vapor pressure, mm Hg at 20 °C <0.076 Solubility in water, g/100 ml at 25 °C 107 Relative molecular mass 129.6	**WHITE HYGROSCOPIC CRYSTALS, TURNING GREY ON EXPOSURE TO AIR** Vapor mixes readily with air. Toxic gases are a product of combustion (incl. → *nitrous vapors*). Decomposes when heated or on contact with acids, giving off toxic vapors and gases (→ *aniline* and → *hydrochloric acid*). Reacts violently with oxidants, with risk of fire and explosion.
	TLV-TWA not available
	Absorption route: Can enter the body by inhalation or ingestion or through the skin. Harmful atmospheric concentrations build up fairly slowly on evaporation at 20°C, but harmful concentrations of airborne particles can build up much more rapidly. **Immediate effects:** Irritates the respiratory tract. Corrosive to the eyes. In serious cases risk of unconsciousness and death. **Effects of prolonged/repeated exposure:** Can affect the blood.
Gross formula: C_6H_8ClN	

HAZARDS/SYMPTOMS	PREVENTIVE MEASURES	FIRE EXTINGUISHING/FIRST AID
Fire: combustible.	keep away from open flame, no smoking.	water spray, powder.
	STRICT HYGIENE, KEEP DUST UNDER CONTROL	IN ALL CASES CALL A DOCTOR
Inhalation: sore throat, cough, shortness of breath, headache, dizziness, drowsiness, unconsciousness, blue skin, feeling of weakness.	local exhaust or respiratory protection.	fresh air, rest, take to hospital.
Skin: *is absorbed*; see also 'Inhalation'.	gloves, protective clothing.	remove contaminated clothing, flush skin with water or shower, call a doctor.
Eyes: *corrosive*; redness, pain, impaired vision.	face shield, or combined eye and respiratory protection.	flush with water, take to a doctor.
Ingestion: abdominal pain, nausea; see also 'Inhalation'.	do not eat, drink or smoke while working.	rinse mouth, take immediately to hospital.

SPILLAGE	STORAGE	LABELING / NFPA
clean up spillage, carefully collect remainder, remove to safe place; (additional individual protection: A/P2 combination filter).	keep separate from oxidants and strong acids; ventilate at floor level.	R: 23/24/25-33 S: 28-36/37-44 Note A ☠ Toxic

NOTES
TLV for aniline is usually a good guide; TLV-TWA is 2ppm and 7.6 mg/m³ (skin). Depending on the degree of exposure, regular medical checkups are advisable. Effects on blood due to formation of methemoglobin. Special first aid required in the event of poisoning: antidotes must be available (with instructions). Unbreakable packaging preferred; if breakable, keep in unbreakable container.

Transport Emergency Card TEC(R)-61G12

UN-number: 1548

ANISOLE

PHYSICAL PROPERTIES	IMPORTANT CHARACTERISTICS
Boiling point, °C 154 Melting point, °C −37 Flash point, °C 43 Auto-ignition temperature, °C 475 Relative density (water = 1) 0.99 Relative vapor density (air = 1) 3.7 Relative density at 20 °C of saturated mixture vapor/air (air = 1) 1.01 Vapor pressure, mm Hg at 25 °C 3.1 Solubility in water none Explosive limits, vol% in air ?-? Electrical conductivity, pS/m 10 Relative molecular mass 108.1 Log P octanol/water 2.1	**COLORLESS LIQUID WITH CHARACTERISTIC ODOR** Vapor mixes readily with air. Able to form peroxides. Flow, agitation etc. can cause build-up of electrostatic charge due to liquid's low conductivity. Reacts with oxidants.
	TLV-TWA not available
	Absorption route: Can enter the body by inhalation or ingestion or through the skin. Harmful atmospheric concentrations build up fairly slowly on evaporation at 20° C, but much more rapidly in aerosol form. **Immediate effects:** Irritates the eyes, skin and respiratory tract. Liquid destroys the skin's natural oils.
Gross formula: C_7H_8O	

HAZARDS/SYMPTOMS	PREVENTIVE MEASURES	FIRE EXTINGUISHING/FIRST AID
Fire: flammable.	keep away from open flame and sparks, no smoking.	powder, AFFF, foam, carbon dioxide, (halons).
Explosion: above 43°C: forms explosive air-vapor mixtures.	above 43°C: sealed machinery, ventilation, explosion-proof electrical equipment, grounding.	in case of fire: keep tanks/drums cool by spraying with water.
Inhalation: sore throat, cough, shortness of breath.	ventilation, local exhaust or respiratory protection.	fresh air, rest, place in half-sitting position, call a doctor.
Skin: *is absorbed*; redness.	gloves.	remove contaminated clothing, flush skin with water or shower.
Eyes: redness, pain.	face shield.	flush with water, take to a doctor if necessary.
Ingestion: abdominal cramps, vomiting.		rinse mouth, call a doctor.

SPILLAGE	STORAGE	LABELING / NFPA
collect leakage in sealable containers, soak up with sand or other inert absorbent and remove to safe place.	keep in dark, fireproof place, separate from oxidants; add an inhibitor.	 2 1 0

NOTES
Before distilling check for peroxides; if found, render harmless.

anthracin
paranaphthalene

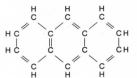

ANTHRACENE

PHYSICAL PROPERTIES		IMPORTANT CHARACTERISTICS	
Boiling point, °C	340	**YELLOW/COLORLESS CRYSTALS WITH BLUE FLUORESCENCE**	
Melting point, °C	217	Sublimates readily. Reacts with strong oxidants.	
Flash point, °C	121		
Auto-ignition temperature, °C	540	TLV-TWA not available	
Relative density (water = 1)	1.24		
Relative vapor density (air = 1)	6.2	**Absorption route:** Can enter the body by inhalation or ingestion. Harmful atmospheric concentrations build up very slowly, if at all, on evaporation at approx. 20°C, but harmful concentrations of airborne particles can build up much more rapidly.	
Vapor pressure, mm Hg at 145°C	0.99		
Solubility in water	none		
Explosive limits, vol% in air	0.6-?	**Immediate effects:** Irritates the eyes, skin and respiratory tract.	
Relative molecular mass	178.2	**Effects of prolonged/repeated exposure:** Prolonged or repeated contact and exposure to sunlight can cause skin disorders.	
Log P octanol/water	4.5		
Gross formula:	$C_{14}H_{10}$		

HAZARDS/SYMPTOMS	PREVENTIVE MEASURES	FIRE EXTINGUISHING/FIRST AID
Fire: combustible.	keep away from open flame, no smoking.	water spray, powder.
Explosion: finely dispersed particles form explosive mixtures on contact with air.		
Inhalation: sore throat, cough, shortness of breath.	local exhaust or respiratory protection.	fresh air, rest, call a doctor.
Skin: redness, pain.	gloves, protective clothing.	remove contaminated clothing, wash skin with soap and water.
Eyes: redness, pain, impaired vision.	goggles, or combined eye and respiratory protection.	flush with water, take to a doctor if necessary.
Ingestion: sore throat, abdominal pain, diarrhea, nausea.	do not eat, drink or smoke while working.	rinse mouth, call a doctor or take to hospital.

SPILLAGE	STORAGE	LABELING / NFPA
clean up spillage; (additional individual protection: P2 respirator).		

NOTES
Data apply only to pure anthracene: crude anthracene and anthracene oil can cause severe skin disorders.

ANTIMONY PENTACHLORIDE

PHYSICAL PROPERTIES	IMPORTANT CHARACTERISTICS	
Decomposes below boiling point, °C 77 Melting point, °C 4 Relative density (water = 1) 2.3 Relative vapor density (air = 1) 10.2 Relative density at 20 °C of saturated mixture vapor/air (air = 1) 1.01 Vapor pressure, mm Hg at 20 °C 0.84 Solubility in water reaction Relative molecular mass 299	**YELLOW LIQUID, FUMING ON EXPOSURE TO AIR, WITH PUNGENT ODOR** Vapor mixes readily with air. Decomposes when heated above 77°C, giving off toxic gas (*chlorine*) and → *antimony trichloride*. Reacts with water, initially forming stable hydrates; with large quantities of water forms → *hydrochloric acid* and antimony oxide. Reacts violently with bases, ammonia and many other substances, with risk of fire and explosion. Reacts with air to form corrosive vapors which are heavier than air and spread at ground level. Attacks many metals in the presence of moisture, giving off flammable gas (→ *hydrogen*).	
	TLV-TWA	0.5 mg/m³ ✳
	Absorption route: Can enter the body by inhalation or ingestion. Harmful atmospheric concentrations build up fairly rapidly due to formation of → *hydrochloric acid* fumes on exposure to air. **Immediate effects:** Corrosive to the eyes, skin and respiratory tract. Inhalation of vapor/fumes can cause severe breathing difficulties (lung edema). In serious cases risk of death. **Effects of prolonged/repeated exposure:** Can cause heart, liver and kidney damage.	
Gross formula: Cl$_5$Sb		

HAZARDS/SYMPTOMS	PREVENTIVE MEASURES	FIRE EXTINGUISHING/FIRST AID
Fire: non-combustible.		in case of fire in immediate vicinity: use any extinguishing agent.
	STRICT HYGIENE	IN ALL CASES CALL IN A DOCTOR
Inhalation: *corrosive*; sore throat, cough, shortness of breath, severe breathing difficulties.	ventilation, local exhaust or respiratory protection.	fresh air, rest, place in half-sitting position, take to hospital.
Skin: *corrosive*; redness, pain, burns.	gloves, protective clothing.	remove contaminated clothing, flush skin with water or shower, take to a doctor.
Eyes: *corrosive*; redness, pain, impaired vision.	face shield, or combined eye and respiratory protection.	flush with water, take to a doctor.
Ingestion: *corrosive*; sore throat, abdominal pain, vomiting.	do not eat, drink or smoke while working.	rinse mouth, take to hospital.

SPILLAGE	STORAGE	LABELING / NFPA
evacuate area, collect leakage in sealable containers, soak up with dry sand or other inert absorbent and remove to safe place; (additional individual protection: CHEMICAL SUIT).	keep dry, separate from bases; ventilate at floor level.	R: 34-37 S: 26 Corrosive 3 0 1

NOTES
✳ TLV equals that of antimony. Depending on the degree of exposure, regular medical checkups are advisable. Lung edema symptoms usually develop several hours later and are aggravated by physical exertion: rest and hospitalization essential. As first aid, a doctor or authorized person should consider administering a corticosteroid spray. The measures on this card also apply to antimony pentafluoride: melting point: 7°C; relative density: 2.9; relative molecular mass: 216.7. Use airtight packaging. Unbreakable packaging preferred; if breakable, keep in unbreakable container.

Transport Emergency Card TEC(R)-127

HI: 80; UN-number: 1730

ANTIMONY TRIBROMIDE

PHYSICAL PROPERTIES	IMPORTANT CHARACTERISTICS	
Boiling point, °C 280 Melting point, °C 97 Relative density (water = 1) 4.1 Relative vapor density (air = 1) 12.1 Vapor pressure, mm Hg at 20°C ≤0.076 Solubility in water reaction Relative molecular mass 361.5	**YELLOW HYGROSCOPIC CRYSTALS** Decomposes when exposed to light and water. Reacts with water, moist air or alcohol, giving off → *hydrogen bromide.*	
	TLV-TWA	0.5 mg/m³ ✱
	Absorption route: Can enter the body by inhalation or ingestion. Evaporation negligible at 20°C, but harmfull concentrations of airborne particles can build up rapidly. **Immediate effects:** Corrosive to the eyes, skin and respiratory tract. Inhalation can cause lung edema. In serious cases risk of death. **Effects of prolonged/repeated exposure:** Can cause heart, liver and kidney damage.	
Gross formula: Br$_3$Sb		

HAZARDS/SYMPTOMS	PREVENTIVE MEASURES	FIRE EXTINGUISHING/FIRST AID
Fire: non-combustible.		in case of fire in immediate vicinity: use any extinguishing agent.
	STRICT HYGIENE	IN ALL CASES CALL IN A DOCTOR
Inhalation: *corrosive*; sore throat, cough, shortness of breath, severe breathing difficulties.	ventilation, local exhaust or respiratory protection.	fresh air, rest, place in half-sitting position, take to hospital.
Skin: *corrosive*; redness, pain, burns.	gloves, protective clothing.	remove contaminated clothing, flush skin with water or shower, take to a doctor.
Eyes: *corrosive*; redness, pain, impaired vision.	face shield.	flush with water, take to a doctor.
Ingestion: *corrosive*; sore throat, abdominal pain, nausea, vomiting.	do not eat, drink or smoke while working.	rinse mouth, immediately take to hospital.

SPILLAGE	STORAGE	LABELING / NFPA
call in an expert, clean up spillage, soak up remainder with dry sand and remove to safe place; (additional individual protection: breathing apparatus).	keep in dry, dark place, separate from bases.	R: 20/22 S: 22 Note A ✖ Harmful

NOTES
✱ TLV equals that of antimony. Depending on the degree of exposure, regular medical checkups are advisable. Lung edema symptoms usually develop several hours later and are aggravated by physical exertion: rest and hospitalization essential. As first aid, a doctor or authorized person should consider administering a corticosteroid spray. Use airtight packaging.

ANTIMONY TRICHLORIDE

PHYSICAL PROPERTIES	IMPORTANT CHARACTERISTICS	
Boiling point, °C 223 Melting point, °C 73 Relative density (water = 1) 3 Relative vapor density (air = 1) 7.9 Relative density at 20 °C of saturated mixture vapor/air (air = 1) 1.00 Vapor pressure, mm Hg at 20 °C 0.12 Solubility in water, g/100 ml at 25 °C 10 Relative molecular mass 228.1	**COLORLESS HYGROSCOPIC CRYSTALS WITH PUNGENT ODOR** Vapor mixes readily with air. Decomposes slowly on contact with water, forming → *hydrochloric acid* and toxic antimony oxychloride. Reacts violently with strong bases and ammonia. Reacts with air to form corrosive fumes (→ *hydrochloric acid*). Attacks many metals in the presence of moisture, giving off flammable gas (→ *hydrogen*).	
	TLV-TWA	0.5 mg/m³ ✱
	Absorption route: Can enter the body by inhalation or ingestion. Harmful atmospheric concentrations of hydrochloric acid fumes can build up fairly rapidly. **Immediate effects:** Corrosive to the eyes, skin and respiratory tract. Inhalation of vapor/fumes can cause severe breathing difficulties (lung edema). In serious cases risk of death. **Effects of prolonged/repeated exposure:** Can cause heart, liver and kidney damage.	
Gross formula: Cl$_3$Sb		

HAZARDS/SYMPTOMS	PREVENTIVE MEASURES	FIRE EXTINGUISHING/FIRST AID
Fire: non-combustible.		in case of fire in immediate vicinity: use any extinguishing agent.
	STRICT HYGIENE	IN ALL CASES CALL IN A DOCTOR
Inhalation: *corrosive*; sore throat, cough, shortness of breath, severe breathing difficulties.	local exhaust or respiratory protection.	fresh air, rest, place in half-sitting position, take to hospital.
Skin: *corrosive*; redness, pain, burns.	gloves, protective clothing.	remove contaminated clothing, flush skin with water or shower, take to a doctor.
Eyes: *corrosive*; redness, pain, impaired vision.	face shield.	flush with water, take to a doctor.
Ingestion: *corrosive*; sore throat, abdominal pain, nausea, vomiting.	do not eat, drink or smoke while working.	rinse mouth, immediately take to hospital.

SPILLAGE	STORAGE	LABELING / NFPA
clean up spillage, carefully collect remainder; (additional individual protection: breathing apparatus).	keep dry, separate from strong bases; ventilate at floor level.	R: 34-37 S: 26 Corrosive

NOTES
✱ TLV equals that of antimony. Depending on the degree of exposure, regular medical checkups are advisable. Lung edema symptoms usually develop several hours later and are aggravated by physical exertion: rest and hospitalization essential. As first aid, a doctor or authorized person should consider administering a corticosteroid spray. Use airtight packaging.

Transport Emergency Card TEC(R)-80G11   **HI: 80; UN-number: 1733**

ANTIMONY TRIOXIDE

PHYSICAL PROPERTIES	IMPORTANT CHARACTERISTICS	
Sublimation temperature, °C ... 1550 Melting point, °C ... 655 Relative density (water = 1) ... 5.5 Vapor pressure, mm Hg at 574 °C ... 0.99 Solubility in water, g/100 ml at 30 °C ... 0.0014 Relative molecular mass ... 291.5	**WHITE CRYSTALS OR POWDER** On reduction with hydrogen gives off highly toxic gas (antimony hydride).	
	TLV-TWA	0.5 mg/m³ ✻
	Absorption route: Can enter the body by inhalation or ingestion. Evaporation negligible at 20°C, but harmful concentrations of airborne particles can build up rapidly. **Immediate effects:** Irritates the eyes, skin and respiratory tract. **Effects of prolonged/repeated exposure:** Can cause heart and lung disorders. Prolonged exposure to particles can cause lung disorders.	
Gross formula: O_3Sb_2		

HAZARDS/SYMPTOMS	PREVENTIVE MEASURES	FIRE EXTINGUISHING/FIRST AID
Fire: non-combustible.		
	STRICT HYGIENE, KEEP DUST UNDER CONTROL	
Inhalation: sore throat, cough, nausea, vomiting.	local exhaust or respiratory protection.	fresh air, rest, call a doctor if necessary.
Skin: redness, pain.	gloves.	remove contaminated clothing, flush skin with water or shower, refer to a doctor if necessary.
Eyes: redness, pain.	goggles, or combined eye and respiratory protection.	flush with water, take to a doctor.
Ingestion: sore throat, abdominal pain, nausea, vomiting, diarrhea.	do not eat, drink or smoke while working.	rinse mouth, call a doctor if necessary.

SPILLAGE	STORAGE	LABELING / NFPA
clean up spillage, carefully collect remainder; (additional individual protection: P2 respirator).		

NOTES
✻ TLV equals that of antimony. Depending on the degree of exposure, regular medical checkups are advisable.

Transport Emergency Card TEC(R)-61G11

HI: 60; UN-number: 1549

ARGON

(cylinder)

PHYSICAL PROPERTIES	IMPORTANT CHARACTERISTICS
Boiling point, °C −186 Melting point, °C −189 Relative vapor density (air = 1) 1.38 Solubility in water none Relative atomic mass 39.9 Log P octanol/water 0.9	**COLORLESS, ODORLESS COMPRESSED GAS** Gas is heavier than air and can accumulate close to ground level, causing oxygen deficiency, with risk of unconsciousness.
	TLV-TWA not available
	Absorption route: Can saturate the air if released, with risk of suffocation. **Immediate effects:** Inhalation can cause severe breathing difficulties. In serious cases risk of unconsciousness.
Gross formula: Ar	

HAZARDS/SYMPTOMS	PREVENTIVE MEASURES	FIRE EXTINGUISHING/FIRST AID
Fire: non-combustible.		in case of fire in immediate vicinity: use any extinguishing agent.
Explosion:		in case of fire: keep cylinders cool by spraying with water.
Inhalation: severe breathing difficulties, headache, dizziness, unconsciousness.	ventilation, local exhaust or respiratory protection.	fresh air, rest, artificial respiration if necessary, take to hospital.

SPILLAGE	STORAGE	LABELING / NFPA
ventilate; (additional individual protection: breathing apparatus).	if stored indoors, keep in cool, fireproof place.	

NOTES
Log P octanol/water is estimated. High atmospheric concentrations, e.g. in poorly ventilated spaces, can cause oxygen deficiency, with risk of unconsciousness.

Transport Emergency Card TEC(R)-20G01

HI: 20; UN-number: 1006

ARGON
(liquid, cooled)

PHYSICAL PROPERTIES	IMPORTANT CHARACTERISTICS	
Boiling point, °C − 186 Melting point, °C − 189 Relative density (water = 1) see notes Relative vapor density (air = 1) 1.38 Vapor pressure, mm Hg at − 186 °C 760 Solubility in water none Relative atomic mass 39.9 Log P octanol/water 0.9	**COLORLESS, ODORLESS, EXTREMELY COLD LIQUID** Gas is heavier than air and can accumulate close to ground level, causing oxygen deficiency, with risk of unconsciousness.	
	TLV-TWA not available	
	Absorption route: If liquid is released, air can become saturated due to evaporation, with risk of suffocation. **Immediate effects:** Cold liquid can cause frostbite.	
Gross formula: Ar		

HAZARDS/SYMPTOMS	PREVENTIVE MEASURES	FIRE EXTINGUISHING/FIRST AID
Fire: non-combustible.		in case of fire in immediate vicinity: use any extinguishing agent, but do not spray liquid with water.
Inhalation: see Notes.	ventilation, local exhaust or respiratory protection.	fresh air, rest, artificial respiration if necessary, call a doctor, take to hospital if necessary.
Skin: *in case of frostbite*: redness, pain, blisters, sores.	insulating gloves.	*in case of frostbite:* DO NOT remove clothing, flush skin with water, call a doctor.
Eyes: *in case of frostbite*: redness, pain, impaired vision.	face shield, or combined eye and respiratory protection.	flush with water, take to a doctor.

SPILLAGE	STORAGE	LABELING / NFPA
ventilate, under no circumstances spray liquid with water; (additional individual protection: breathing apparatus).	ventilate at floor level, keep refrigerated.	◇

NOTES
Density of liquid at boiling point 1.38kg/l. Critical temperature − 122° C. Log P octanol/water is estimated. High atmospheric concentrations, e.g. in poorly ventilated spaces, can cause oxygen deficiency, with risk of unconsciousness. Use special insulated drums.

Transport Emergency Card TEC(R)-20G22   **HI: 22; UN-number: 1951**

ARSENIC TRICHLORIDE

PHYSICAL PROPERTIES	IMPORTANT CHARACTERISTICS
Boiling point, °C 130 Melting point, °C −16 Relative density (water = 1) 2.2 Relative vapor density (air = 1) 6.3 Relative density at 20 °C of saturated mixture vapor/air (air = 1) 1.06 Vapor pressure, mm Hg at 20 °C 8.4 Solubility in water reaction Relative molecular mass 181.3	**COLORLESS, OILY, FUMING LIQUID WITH PUNGENT ODOR** Decomposes when exposed to light, giving off toxic → *chlorine* gas and → *arsenic trioxide*. Reacts violently with bases and anhydrous ammonia. Reacts violently with strong oxidants, giving off → *chlorine* gas, and with water, forming → *hydrochloric acid* and arsenic acid, evolving heat. Reacts with air to form corrosive → *hydrochloric acid* fumes. Attacks many metals, esp. in the presence of moisture, giving off flammable gas (→ *hydrogen*).

TLV-TWA	0.2 mg/m^3 ✽

Absorption route: Can enter the body by inhalation or ingestion or through the skin. Harmful atmospheric concentrations can build up very rapidly on evaporation at 20°C. **Immediate effects:** Corrosive to the eyes, skin and respiratory tract. Affects the nervous system. Inhalation of vapor/fumes can cause severe breathing difficulties (lung edema). In serious cases risk of death. **Effects of prolonged/repeated exposure:** Prolonged or repeated contact can cause skin disorders. Can cause damage to the liver, kidneys and nasal septum. Has been found to cause a type of skin, lung and liver cancer in humans under certain circumstances.

Gross formula: AsCl$_3$

HAZARDS/SYMPTOMS	PREVENTIVE MEASURES	FIRE EXTINGUISHING/FIRST AID
Fire: non-combustible.		in case of fire in immediate vicinity: use any extinguishing agent.
	STRICT HYGIENE	IN ALL CASES CALL IN A DOCTOR
Inhalation: *corrosive*; sore throat, cough, severe breathing difficulties, vomiting, diarrhea, abdominal cramps.	ventilation, local exhaust or respiratory protection.	fresh air, rest, place in half-sitting position, call a doctor or take to hospital.
Skin: *corrosive, is absorbed*; redness, pain, serious burns.	gloves, protective clothing.	remove contaminated clothing, flush skin with water or shower, call a doctor or take to hospital.
Eyes: *corrosive*; redness, pain, impaired vision.	face shield, or combined eye and respiratory protection.	flush with water, take to a doctor.
Ingestion: *corrosive*; sore throat, diarrhea, abdominal cramps, vomiting.	do not eat, drink or smoke while working.	rinse mouth, take immediately to hospital.

SPILLAGE	STORAGE	LABELING / NFPA
evacuate area, call in an expert, collect leakage in sealable containers, soak up with sand or other inert absorbent and remove to safe place; (additional individual protection: CHEMICAL SUIT).	keep in dry, dark place, separate from oxidants and bases; ventilate at floor level.	R: 23/25 S: 1/2-20/21-28-44 Note A ☠ 〽 Toxic

NOTES
✽ TLV equals that of arsenic (arsenic and soluble compounds). Depending on the degree of exposure, regular medical checkups are advisable. Lung edema symptoms usually develop several hours later and are aggravated by physical exertion: rest and hospitalization essential. As first aid, a doctor or authorized person should consider administering a corticosteroid spray. Use airtight packaging. Unbreakable packaging preferred; if breakable, keep in unbreakable container.

Transport Emergency Card TEC(R)-61G01

HI: 66; UN-number: 1560

ARSENIC TRIOXIDE

PHYSICAL PROPERTIES	IMPORTANT CHARACTERISTICS	
Boiling point, °C　　　　　　　　400 Melting point, °C　　　　　　　　312 Relative density (water = 1)　　　3.8 Solubility in water, g/100 ml at 25°C　2.1 Relative molecular mass　　　197.8	**WHITE POWDER** Reacts with acids.	
	TLV-TWA	0,2 mg/m³　A2　✱
	Absorption route: Can enter the body by inhalation or ingestion or through the skin. Evaporation negligible at 20°C, but harmful concentrations of airborne particles can build up rapidly. **Immediate effects:** Corrosive to the eyes, skin and respiratory tract. Affects the nervous system. Inhalation can cause lung edema. In serious cases risk of seizures, unconsciousness and death. **Effects of prolonged/repeated exposure:** Prolonged or repeated contact can cause skin disorders. Can cause damage to liver, kidneys and nasal septum. Has been found to cause a type of skin, lung and liver cancer in humans under certain circumstances.	
Gross formula:　　　　　As$_2$O$_3$		

HAZARDS/SYMPTOMS	PREVENTIVE MEASURES	FIRE EXTINGUISHING/FIRST AID
Fire: non-combustible.		in case of fire in immediate vicinity: use any extinguishing agent.
	STRICT HYGIENE, KEEP DUST UNDER CONTROL	
Inhalation: *corrosive*; sore throat, cough, severe breathing difficulties, abdominal cramps, vomiting, diarrhea.	local exhaust or respiratory protection.	fresh air, rest, place in half-sitting position, take to hospital.
Skin: *corrosive, is absorbed*; redness, pain.	gloves, protective clothing.	remove contaminated clothing, wash skin with soap and water, take to hospital.
Eyes: *corrosive*; redness, pain, impaired vision.	face shield, or combined eye and respiratory protection.	flush with water, take to a doctor.
Ingestion: *corrosive*; sore throat, abdominal cramps, vomiting, diarrhea.	do not eat, drink or smoke while working.	rinse mouth, take immediately to hospital.

SPILLAGE	STORAGE	LABELING / NFPA
clean up spillage, carefully collect remainder; (additional individual protection: P3 respirator).	keep separate from acids.	R: 45-28-34 S: 53-45 Note E ☠ Highly toxic

NFPA diamond: Health 2, Fire 0, Reactivity 0

NOTES
Melting point depends on crystalline form. ✱ TLV equals that of arsenic. Do not take work clothing home. Depending on the degree of exposure, regular medical checkups are advisable. Lung edema symptoms usually develop several hours later and are aggravated by physical exertion: rest and hospitalization essential. As first aid, a doctor or authorized person should consider administering a corticosteroid spray. Unbreakable packaging preferred; if breakable, keep in unbreakable container.

Transport Emergency Card TEC(R)-61G11

HI: 60; UN-number: 1561

ARSINE
(cylinder)

PHYSICAL PROPERTIES	IMPORTANT CHARACTERISTICS
Boiling point, °C −62 Melting point, °C −116 Flash point, °C flammable gas Auto-ignition temperature, °C ? Relative density (water = 1) 3.5 Relative vapor density (air = 1) 2.7 Vapor pressure, mm Hg at 20 °C 7828 Solubility in water, g/100 ml 0.07 Explosive limits, vol% in air ?-? Relative molecular mass 77.9	**COLORLESS COMPRESSED LIQUEFIED GAS WITH CHARACTERISTIC ODOR** Gas is heavier than air and spreads at ground level, with risk of ignition at a distance. Flow, agitation etc. can cause build-up of electrostatic charge due to liquid's low conductivity. In moist state decomposes when exposed to light. Decomposes when heated above 300°C, giving off toxic → *arsenic* vapor. Reacts violently with chlorine, with risk of fire and explosion.

	TLV-TWA	0.05 ppm	0.16 mg/m³

Absorption route: Can enter the body by inhalation. Harmful atmospheric concentrations can build up very rapidly if gas is released.
Immediate effects: Can cause frostbite due to rapid evaporation. Affects the nervous system. Inhalation can cause lung edema. In serious cases risk of death.
Effects of prolonged/repeated exposure: Can affect the blood. Can cause liver, kidney and heart damage.

Gross formula: AsH₃

HAZARDS/SYMPTOMS	PREVENTIVE MEASURES	FIRE EXTINGUISHING/FIRST AID
Fire: extremely flammable.	keep away from open flame and sparks, no smoking, avoid contact with chlorine.	shut off supply; if impossible and no danger to surrounding area, allow fire to burn itself out; otherwise extinguish with powder, carbon dioxide, (halons); in case of fire in immediate vicinity: use any extinguishing agent.
Explosion: forms explosive air-gas mixtures.	sealed machinery, ventilation, explosion-proof electrical equipment and lighting, grounding.	in case of fire: keep cylinders cool by spraying with water.
	STRICT HYGIENE	IN ALL CASES CALL A DOCTOR
Inhalation: headache, dizziness, vomiting, unconsciousness, dark-colored urine.	ventilation, local exhaust or respiratory protection.	fresh air, rest, take immediately to hospital.
Skin: *in case of frostbite*: redness, pain, burns.	insulated gloves, protective clothing.	*in case of frostbite*: DO NOT remove clothing, flush skin with water or shower, call a doctor.
Eyes: *in case of frostbite*: redness, pain, impaired vision.	face shield, or combined eye and respiratory protection.	flush with water, take to doctor.

SPILLAGE	STORAGE	LABELING / NFPA
evacuate area, call in an expert, ventilate, remove cylinder to safe place if possible, under no circumstances spray liquid with water; (additional individual protection: CHEMICAL SUIT).	if stored indoors, keep in cool, fireproof place; ventilate at floor level.	NFPA diamond: Blue 4, Red 4, Yellow 2

NOTES
Explosive limits not given in the literature. Odor limit is above TLV. Depending on the degree of exposure, regular medical checkups are advisable. Lung edema symptoms usually develop several hours later and are aggravated by physical exertion: rest and hospitalization essential. As first aid, a doctor or authorized person should consider administering a corticosteroid spray. Turn leaking cylinder so that leak is on top to prevent liquid arsine escaping. Use cylinder with special fittings.

Transport Emergency Card TEC(R)-20G12

HI: 236; UN-number: 2188

CAS-No: [12001-29-5]
amosite (brown)
chrysotile (white)
crocidolite (blue)

ASBESTOS

PHYSICAL PROPERTIES	IMPORTANT CHARACTERISTICS	
Melting point, °C 1000-1500 Relative density (water = 1) 2.5 Solubility in water none	**WHITE, BROWN OR BLUE SOLID WITH FIBROUS CRYSTALLINE STRUCTURE** Unable to react with virtually any other substance. Becomes powdery (amorphous) and loses its hazardous properties when heated above approx. 1200°C. Processes entailing the release of fine fibers into the air should be avoided wherever possible.	
	TLV-TWA ✱	
	Absorption route: Can enter the body by inhalation or ingestion. **Effects of prolonged/repeated exposure:** Risk of asbestosis if asbestos fibers are inhaled over a long period. Risk of malignant disorders of pleura, peritoneum and digestive organs, esp. if crocidolite is inhaled. Also risk of malignant disorders of the respiratory tract (esp. in smokers).	

HAZARDS/SYMPTOMS	PREVENTIVE MEASURES	FIRE EXTINGUISHING/FIRST AID
Fire: non-combustible.		in case of fire in immediate vicinity: use any extinguishing agent.
	STRICT HYGIENE, KEEP DUST UNDER CONTROL	
Inhalation:	local exhaust or respiratory protection.	
Skin:	special clothing.	remove contaminated clothing, flush skin with water or shower.
Eyes:	safety glasses, combined eye and respiratory protection if required.	
Ingestion:	do not eat, drink or smoke while working.	

SPILLAGE	STORAGE	LABELING / NFPA
use special vacuum cleaner or dampen spilled asbestos with water and collect carefully; (additional individual protection: P3 respirator).	limit to minimum quantity necessary.	◇◇

NOTES
✱ TLV-TLV for asbestos: amosite 0.5 fiber/cc (A1), chrysotile 2 fibers/cc (A1), crocidolite 0.2 fibers/cc (A1) and other forms 2 fibers/cc (A1). Do not take work clothing home. The measures on this card also apply to products containing asbestos. Use dustproof packaging. UN no. 2212 applies to blue asbestos, no. 2590 to white asbestos.

UN-number: 2212, 2590

CAS-No: [103-33-3]
azobenzide
azobenzol
azobisbenzene
1,2-diphenyldiazene
diphenyldiimide

$C_6H_5N = NC_6H_5$

AZOBENZENE

PHYSICAL PROPERTIES		IMPORTANT CHARACTERISTICS
Boiling point, °C	293	**ORANGE CRYSTALS**
Melting point, °C	68	Decomposes when heated or burned, giving off toxic gases ($\rightarrow$ *nitrous vapors*).
Relative density (water = 1)	1.2	
Vapor pressure, mm Hg at 20 °C	<0.00076	TLV-TWA not available
Solubility in water	none	
Relative molecular mass	182.2	**Absorption route:** Can enter the body by inhalation or ingestion or through the skin. Evaporation negligible at 20°C, but harmful concentrations of airborne particles can build up rapidly. **Immediate effects:** Irritates the eyes, skin and respiratory tract. In serious cases risk of unconsciousness and death. **Effects of prolonged/repeated exposure:** Can affect the blood. Can cause liver damage.
Log P octanol/water	3.8	
Gross formula:	$C_{12}H_{10}N_2$	

HAZARDS/SYMPTOMS	PREVENTIVE MEASURES	FIRE EXTINGUISHING/FIRST AID
Fire: combustible.	keep away from open flame, no smoking.	water spray, powder.
Inhalation: cough, shortness of breath, dizziness, drowsiness, blue skin, feeling of weakness.	local exhaust or respiratory protection.	fresh air, rest.
Skin: *is absorbed*; redness; see also 'Inhalation'.	gloves, protective clothing.	remove contaminated clothing, wash skin with soap and water.
Eyes: redness, pain.	face shield, or combined eye and respiratory protection.	flush with water, take to a doctor if necessary.
Ingestion: abdominal pain; see also 'Inhalation'.		rinse mouth, immediately take to hospital.

SPILLAGE	STORAGE	LABELING / NFPA
clean up spillage, carefully collect remainder and remove to safe place; (additional individual protection: P2 respirator).		R: 20/22 S: 28 ✖ Harmful

NOTES
Alcohol consumption increases toxic effects. Depending on the degree of exposure, regular medical checkups are advisable. Special first aid required in the event of poisoning: antidotes must be available (with instructions).

BARIUM

PHYSICAL PROPERTIES	IMPORTANT CHARACTERISTICS	
Boiling point, °C 1640 Melting point, °C 725 Relative density (water = 1) 3.5 Solubility in water reaction Relative atomic mass 137.3	**WHITE LUSTROUS LIGHT** Can ignite spontaneously on exposure to air. Strong reducing agent which reacts violently with oxidants. Reacts violently with acids and water, giving off flammable gas (→ *hydrogen*). Reacts violently with halogenated solvents, with risk of fire and explosion.	
	TLV-TWA	0.5 mg/m³
	Absorption route: Can enter the body by inhalation or ingestion. Evaporation negligible at 20°C, but in powder form harmful atmospheric concentrations of airborne particles can build up rapidly. **Immediate effects:** Irritates the eyes, skin and respiratory tract. Stimulates the smooth and striated muscles, including the heart, with subsequent paralysis. Can cause weakness, paralysis of limbs and respiratory tract. In serious cases risk of seizures, cardiac arrest and death.	
Gross formula: Ba		

HAZARDS/SYMPTOMS	PREVENTIVE MEASURES	FIRE EXTINGUISHING/FIRST AID
Fire: combustible (in powder form; extremly flammable).	keep away from open flame and sparks, no smoking.	special powder extinguisher, dry sand, NO other extinguishing agents.
Explosion: finely dispersed particles form explosive mixtures on contact with air.	keep dust from accumulating; sealed machinery, explosion-proof electrical equipment and lighting, grounding.	
	STRICT HYGIENE, KEEP DUST UNDER CONTROL	
Inhalation: sore throat, cough, shortness of breath; see also 'Ingestion'.	local exhaust or respiratory protection.	fresh air, rest, call a doctor or take to hospital.
Skin: redness, pain.	gloves, protective clothing.	remove contaminated clothing, flush skin with water or shower.
Eyes: redness, pain.	safety glasses, or combined eye and respiratory protection.	flush with water, take to a doctor if necessary.
Ingestion: abdominal pain, diarrhea, trembling, vomiting, feeling of weakness, paralysis in arms and legs.		rinse mouth, immediately take to hospital.

SPILLAGE	STORAGE	LABELING / NFPA
clean up spillage, carefully collect remainder; (additional individual protection: P2 respirator).	keep in fireproof, dry place; keep separate from oxidants, acids and halogenated solvents.	

NOTES
The measures for respiratory protection only apply to barium in powder form. Reacts violently with extinguishing agents such as water, halons, carbon dioxide and bicarbonate powder.

Transport Emergency Card TEC(R)-43G01

UN-number: 1400

$$Ba(C_2H_3O_2)_2.H_2O$$

BARIUM ACETATE

PHYSICAL PROPERTIES	IMPORTANT CHARACTERISTICS	
Melting point, °C 110 Relative density (water = 1) 2.2 Solubility in water, g/100 ml 67 Relative molecular mass 273.4	**WHITE CRYSTALS**	
	TLV-TWA	0.5 mg/m³ ✱
	Absorption route: Can enter the body by inhalation or ingestion. Evaporation negligible at 20°C, but harmful concentrations of airborne particles can build up rapidly. **Immediate effects:** Irritates the eyes, skin and respiratory tract. Stimulates the smooth and striated muscles, including the heart, with subsequent paralysis. Can cause weakness, paralysis of limbs and respiratory tract. In serious cases risk of seizures, cardiac arrest and death.	
Gross formula: $C_4H_6BaO_4.H_2O$		

HAZARDS/SYMPTOMS	PREVENTIVE MEASURES	FIRE EXTINGUISHING/FIRST AID
Fire: non-combustible.		in case of fire in immediate vicinity: use any extinguishing agent.
	STRICT HYGIENE, KEEP DUST UNDER CONTROL	
Inhalation: sore throat, cough, shortness of breath; see also 'Ingestion'.	local exhaust or respiratory protection.	fresh air, rest, take to a doctor.
Skin: redness, pain.	gloves.	remove contaminated clothing, flush skin with water or shower.
Eyes: redness, pain.	goggles, or combined eye and respiratory protection.	flush with water, take to a doctor if necessary.
Ingestion: abdominal pain, diarrhea, vomiting, trembling, feeling of weakness, paralysis in arms and legs.		rinse mouth, take immediately to hospital.

SPILLAGE	STORAGE	LABELING / NFPA
clean up spillage, carefully collect remainder; (additional individual protection: P2 respirator).		R: 20/22 S: 28 Note A ❌ Harmful

NOTES
Apparent melting point due to loss of water of crystallization. ✱ TLV equals that of barium (soluble compounds).

BARIUM AZIDE
(anhydrous)

PHYSICAL PROPERTIES	IMPORTANT CHARACTERISTICS	
Melting point (decomposes), °C 120 Relative density (water = 1) 2.9 Solubility in water, g/100 ml at 17 °C 17.3 Relative molecular mass 221.4	**WHITE CRYSTALS** Finely dispersed in air can cause dust explosion. Can decompose explosively if subjected to shock. Reacts with lead compounds and other heavy metals, to form extremely shock-sensitive compounds. Can produce explosion when heated. Decomposes when heated above 120°C, giving off → *nitrogen* and → *nitrous vapors*. Reacts with acids to form toxic and corrosive → *hydrogen azide*. Reacts violently with oxidants.	
	TLV-TWA	0.5 mg/m³ ✳
	Absorption route: Can enter the body by inhalation. Evaporation negligible at 20°C, but harmful concentrations of airborne particles can build up rapidly. **Immediate effects:** Irritates the eyes, skin and respiratory tract. Keep under medical observation.	
Gross formula: BaN$_6$		

HAZARDS/SYMPTOMS	PREVENTIVE MEASURES	FIRE EXTINGUISHING/FIRST AID
Fire: extremely flammable.	keep away from open flame and sparks, no smoking.	large quantities of water.
Explosion: finely dispersed particles form explosive mixtures on contact with air; subjecting to friction/shock or heating can cause explosion.	do not subject to shock or friction, keep dust from accumilating, sealed machinery, explosion-proof electrical equipment and lighting, grounding, preferably moisten before use (at least 50% water).	in case of fire: keep tanks/drums cool by spraying with water, fight fire from sheltered location.
	PREVENT MIST FORMATION, STRICT HYGIENE, AVOID ALL CONTACT	IN ALL CASES CALL IN A DOCTOR
Inhalation: sore throat, cough, shortness of breath; see also 'Ingestion'.	local exhaust or respiratory protection.	fresh air, rest, consult a doctor, take to hospital.
Skin: redness, pain.	gloves.	remove contaminated clothing, flush skin with water and wash with soap and water.
Eyes: redness, pain, impaired vision.	goggles.	flush thoroughly with water (remove contact lenses if easy), take to a doctor.
Ingestion: abdominal cramps, abdominal pain, diarrhea, vomiting, feeling of weakness, trembling or paralysis in arms and legs.		rinse mouth, give no liquids, take immediately to hospital.

SPILLAGE	STORAGE	LABELING / NFPA
evacuate area, call in an expert, dampen spillage with water and clean up, remove to safe place; (additional individual protection: P2 respirator).	keep in fireproof place, dampened with water (>50%); keep separate from acids.	R: 20/22 S: 28 Note A ✖ Harmful

NOTES
Anhydrous barium azide in practice seldom used. For phlegmatizing at least 50% water required. This form has UN no. 1571. ✳ TLV equals that of barium.

UN-number: 0224

BARIUM CARBONATE

PHYSICAL PROPERTIES	IMPORTANT CHARACTERISTICS	
Boiling point (decomposes), °C 1450 Melting point, °C 811 Relative density (water = 1) 4.4 Solubility in water none Relative molecular mass 197.4	**WHITE POWDER**	
	TLV-TWA	0.5 mg/m³ ✳
	Absorption route: Can enter the body by inhalation or ingestion. Evaporation negligible at 20°C, but harmful atmospheric concentrations of airborne particles can build up rapidly. **Immediate effects:** Irritates the eyes, skin and respiratory tract. Stimulates the smooth and striated muscles, including the heart, with subsequent paralysis. Can cause weakness, paralysis of limbs and respiratory tract. In serious cases risk of seizures, cardiac arrest and death.	
Gross formula: BaCO$_3$		

HAZARDS/SYMPTOMS	PREVENTIVE MEASURES	FIRE EXTINGUISHING/FIRST AID
Fire: non-combustible.		in case of fire in immediate vicinity: use any extinguishing agent.
	STRICT HYGIENE, KEEP DUST UNDER CONTROL	
Inhalation: sore throat, cough, shortness of breath; see also 'Ingestion'.	local exhaust or respiratory protection.	fresh air, rest, call a doctor or take to hospital.
Skin: redness, pain.	gloves.	remove contaminated clothing, flush skin with water or shower.
Eyes: redness, pain.	safety glasses.	flush with water, take to a doctor if necessary.
Ingestion: abdominal pain, diarrhea, vomiting, trembling, feeling of weakness or paralysis in arms and legs.		rinse mouth, take immediately to hospital.

SPILLAGE	STORAGE	LABELING / NFPA
clean up spillage, carefully collect remainder; (additional individual protection: P2 respirator).		R: 20/22 S: 28 Note A ✖ Harmful

NOTES
✳ TLV equals that of barium (soluble compounds).

BARIUM CHLORATE
(monohydrate)

PHYSICAL PROPERTIES	IMPORTANT CHARACTERISTICS
Melting point, °C 414 Relative density (water = 1) 3.2 Solubility in water, g/100 ml at 20 °C 26 Relative molecular mass 322.3	**COLORLESS OR WHITE CRYSTALS** Decomposes when heated above 250°C, giving off oxygen, with increased risk of fire. Strong oxidant which reacts violently with combustible substances and reducing agents, with risk of fire or explosion. Mixtures with reducing agents, ammonium compounds or powdered metals can explode if subjected to shock or friction. Reacts violently with concentrated sulfuric acid, with risk of fire and explosion.
	TLV-TWA 0.5 mg/m³ ✶
	Absorption route: Can enter the body by inhalation or ingestion. Evaporation negligible at 20°C, but harmful concentrations of airborne particles can build up rapidly. **Immediate effects:** Irritates the eyes, skin and respiratory tract. Stimulates the smooth and striated muscles, including the heart, with subsequent paralysis. Can cause weakness, paralysis of limbs and respiratory tract. In serious cases risk of seizures, cardiac arrest and death.
Gross formula: $BaCl_2O_6.H_2O$	

HAZARDS/SYMPTOMS	PREVENTIVE MEASURES	FIRE EXTINGUISHING/FIRST AID
Fire: non-combustible, but makes other substances more combustible: many chemical reactions can cause fire and explosion.	avoid contact with combustible substances, reducing agents and concentrated acids.	in case of fire in immediate vicinity: use any extinguishing agent.
Explosion: many chemical reactions can cause explosion.		
	STRICT HYGIENE, KEEP DUST UNDER CONTROL	
Inhalation: sore throat, cough, shortness of breath; see also 'Ingestion'.	local exhaust or respiratory protection.	fresh air, rest, call a doctor.
Skin: redness, pain.	gloves.	flush with water and only then remove clothing, flush skin with water or shower.
Eyes: redness, pain.	goggles, or combined eye and respiratory protection.	flush with water, take to a doctor if necessary.
Ingestion: abdominal pain, diarrhea, vomiting, trembling, feeling of weakness or paralysis in arms and legs.		rinse mouth, take immediately to hospital.

SPILLAGE	STORAGE	LABELING / NFPA
clean up spillage, flush away remainder with water; (additional individual protection: P2 respirator).	keep separate from combustible substances, reducing agents and strong acids.	R: 9-20/22 S: 13-27 🔥 Oxidant ✖ Harmful

NOTES

Apparent melting point at 120°C due to loss of water of crystallization. ✶ TLV equals that of barium. Flush contaminated clothing with water (fire hazard).

BARIUM CHLORIDE

PHYSICAL PROPERTIES	IMPORTANT CHARACTERISTICS	
Boiling point, °C 1560 Melting point, °C 960 Relative density (water = 1) 3.9 Solubility in water, g/100 ml 36 Relative molecular mass 208.3	**WHITE CRYSTALLINE POWDER**	
	TLV-TWA	0.5 mg/m^3 ✱
	Absorption route: Can enter the body by inhalation or ingestion. Evaporation negligible at 20° C, but harmful concentrations of airborne particles can build up rapidly. **Immediate effects:** Irritates the skin. Corrosive to the eyes and respiratory tract. Affects the gastro-intestinal tract. Stimulates the smooth and striated muscles, including the heart, with subsequent paralysis. Can cause weakness, paralysis of limbs and respiratory tract. In serious cases risk of seizures, unconsciousness and death.	
Gross formula: BaCl$_2$		

HAZARDS/SYMPTOMS	PREVENTIVE MEASURES	FIRE EXTINGUISHING/FIRST AID
Fire: non-combustible.		in case of fire in immediate vicinity: use any extinguishing agent.
	STRICT HYGIENE, KEEP DUST UNDER CONTROL	
Inhalation: *corrosive*; sore throat, cough, severe breathing difficulties; see also 'Ingestion'.	local exhaust or respiratory protection.	fresh air, rest, call a doctor and take to hospital.
Skin: redness, pain.	gloves.	remove contaminated clothing, flush skin with water or shower.
Eyes: *corrosive*; redness, pain, impaired vision.	goggles.	flush with water, take to a doctor.
Ingestion: *corrosive*; abdominal pain, diarrhea, vomiting, trembling, feeling of weakness, paralysis in arms and legs.		rinse mouth, take immediately to hospital.

SPILLAGE	STORAGE	LABELING / NFPA
clean up spillage, carefully collect remainder; (additional individual protection: P2 respirator).		R: 20/22 S: 28 Note A ✖ Harmful

NOTES
✱ TLV equals that of barium (soluble compounds).

BARIUM DIPHENYLAMINE SULFONATE

PHYSICAL PROPERTIES	IMPORTANT CHARACTERISTICS	
Melting point (decomposes), °C > 300 Solubility in water moderate Relative molecular mass 633.9	**WHITE CRYSTALS** Decomposes when heated, giving off toxic vapors.	
	TLV-TWA	0.5 mg/m³ ✱
	Absorption route: Can enter the body by inhalation or ingestion. Evaporation negligible at 20°C, but harmful concentrations of airborne particles can build up rapidly. **Immediate effects:** Irritates the eyes, skin and respiratory tract. Stimulates the smooth and striated muscles, including the heart, with subsequent paralysis. Can cause weakness, paralysis of limbs and respiratory tract. In serious cases risk of seizures, cardiac arrest and death. **Effects of prolonged/repeated exposure:** Prolonged or repeated contact can cause skin disorders.	
Gross formula: $C_{24}H_{20}BaN_2O_6S_2$		

HAZARDS/SYMPTOMS	PREVENTIVE MEASURES	FIRE EXTINGUISHING/FIRST AID
Fire: combustible.	keep away from open flame, no smoking.	powder, carbon dioxide, (halons); in case of fire in immediate vicinity: use any extinguishing agent.
	STRICT HYGIENE, KEEP DUST UNDER CONTROL	
Inhalation: sore throat, cough, shortness of breath; see also 'Ingestion'.	local exhaust or respiratory protection.	fresh air, rest, call a doctor.
Skin: redness, pain.	gloves.	remove contaminated clothing, flush skin with water or shower, take to a doctor.
Eyes: redness, pain.	goggles, or combined eye and respiratory protection.	flush with water, take to a doctor.
Ingestion: abdominal pain, diarrhea, vomiting, trembling, feeling of weakness, paralysis in arms and legs.		rinse mouth, take immediately to hospital.

SPILLAGE	STORAGE	LABELING / NFPA
clean up spillage, carefully collect remainder; (additional individual protection: P2 respirator).		R: 20/22 S: 28 Note A ✖ Harmful

NOTES
✱ TLV equals that of barium (soluble compounds).

BARIUM HYDROXIDE
(octahydrate)

PHYSICAL PROPERTIES	IMPORTANT CHARACTERISTICS	
Boiling point, °C 408 Melting point, °C 78 Relative density (water = 1) 2.2 Solubility in water, g/100 ml at 15 °C 5.6 Relative molecular mass 315.5	**COLORLESS CRYSTALS OR WHITE POWDER** Decomposes when heated, giving off toxic and corrosive vapors. Strong base which reacts violently with acids and corrodes aluminum, zinc etc. Reacts violently with chlorine rubber (when heated), with risk of fire and explosion.	
	TLV-TWA	0.5 mg/m³ ✱
	Absorption route: Can enter the body by inhalation or ingestion. Evaporation negligible at 20°C, but harmful concentrations of airborne particles can build up rapidly. **Immediate effects:** Irritates the eyes, skin and respiratory tract. Stimulates the smooth and striated muscles, including the heart, with subsequent paralysis. Can cause weakness, paralysis of limbs and respiratory tract. In serious cases risk of seizures, cardiac arrest and death.	
Gross formula: $BaH_2O_2.8H_2O$		

HAZARDS/SYMPTOMS	PREVENTIVE MEASURES	FIRE EXTINGUISHING/FIRST AID
Fire: non-combustible.		in case of fire in immediate vicinity: use any extinguishing agent.
	STRICT HYGIENE, KEEP DUST UNDER CONTROL	
Inhalation: sore throat, cough, shortness of breath, see also: 'Ingestion'.	local exhaust or respiratory protection.	fresh air, rest, call a doctor or take to hospital.
Skin: redness, pain.	gloves.	remove contaminated clothing, flush skin with water or shower.
Eyes: redness, pain.	goggles, or combined eye and respiratory protection.	flush with water, take to a doctor if necessary.
Ingestion: abdominal pain, diarrhea, stiffening of neck and mouth, vomiting, feeling of weakness, paralysis in arms and legs.		rinse mouth, immediately take to hospital.

SPILLAGE	STORAGE	LABELING / NFPA
clean up spillage, carefully collect remainder; (additional individual protection: P2 respirator).	keep separate from strong acids.	R: 20/22 S: 28 Note A ✖ Harmful

NOTES

Boiling point of anhydrous barium hydroxide. Apparent melting point due to loss of water of crystallization. ✱ TLV equals that of barium.

BARIUM NITRATE

PHYSICAL PROPERTIES	IMPORTANT CHARACTERISTICS	
Melting point, °C 590 Relative density (water = 1) 3.2 Solubility in water, g/100 ml at 20 °C 9 Relative molecular mass 261.4	**COLORLESS CLEAR CRYSTALS** Decomposes when heated above melting point, forming → *barium nitrite*, → *barium peroxide*, → *barium oxide*, → *nitrous vapors*, → *nitrogen* and → *oxygen*, depending on temperature and speed of rise in temperature. Strong oxidant which reacts violently with combustible substances and reducing agents, with risk of fire or explosion in some cases. Mixtures with sulfur powder and powders of light metals are friction and shock-sensitive.	
	TLV-TWA 0.5 mg/m³ ✶	
	Absorption route: Can enter the body by inhalation or ingestion. Evaporation negligible at 20°C, but harmful concentrations of airborne particles can build up rapidly. **Immediate effects:** Irritates the eyes, skin and respiratory tract. Stimulates the smooth and striated muscles, including the heart, with possibility of subsequent paralysis. Can cause weakness, paralysis of limbs and respiratory tract. In serious cases risk of seizures, cardiac arrest and death.	
Gross formula: BaN$_2$O$_6$		

HAZARDS/SYMPTOMS	PREVENTIVE MEASURES	FIRE EXTINGUISHING/FIRST AID
Fire: not combustible but enhances combustion of other substances.	avoid contact with combustible substances and reducing agents.	in case of fire in immediate vicinity: use any extinguishing agent.
Explosion: some reactions can cause explosion.		
	STRICT HYGIENE, KEEP DUST UNDER CONTROL	
Inhalation: sore throat, cough, shortness of breath; see also 'Ingestion'.	local exhaust or respiratory protection.	fresh air, rest, call a doctor.
Skin: redness, pain.	gloves.	remove contaminated clothing, flush skin with water or shower, call a doctor.
Eyes: redness, pain.	goggles, or combined eye and respiratory protection.	flush with water, take to a doctor if necessary.
Ingestion: abdominal pain, diarrhea, vomiting, trembling, feeling of weakness or paralysis in arms and legs.		rinse mouth, take immediately to hospital.

SPILLAGE	STORAGE	LABELING / NFPA
clean up spillage, carefully collect remainder; (additional individual protection: P2 respirator).	keep separate from combustible substances and reducing agents.	R: 20/22 S: 28 Note A ✖ Harmful

NOTES
✶ TLV equals that of barium. Flush contaminated clothing with water (fire hazard).

Transport Emergency Card TEC(R)-51G02

HI: 50; UN-number: 1446

BARIUM PERCHLORATE
(trihydrate)

PHYSICAL PROPERTIES	IMPORTANT CHARACTERISTICS	
Melting point (decomposes), °C 400 Relative density (water = 1) 2.7 Solubility in water, g/100 ml at 33 °C 198 Relative molecular mass 390.4	**COLORLESS CRYSTALS** Decomposes when heated, giving off toxic vapors and oxygen, with increased risk of fire. Strong oxidant which reacts violently with combustible substances and reducing agents. Mixtures with organic substances, ammonium compounds, sulfur and metal powders can explode if subjected to shock or friction. Reacts with strong acids, forming explosive → *perchloric acid*.	
	TLV-TWA	0.5 mg/m³ ✱
	Absorption route: Can enter the body by inhalation or ingestion. **Immediate effects:** Irritates the eyes, skin and respiratory tract. Stimulates the smooth and striated muscles, including the heart, with subsequent paralysis. Can cause weakness, paralysis of limbs and respiratory tract. In serious cases risk of seizures, cardiac arrest and death. **Effects of prolonged/repeated exposure:** Prolonged or repeated contact can cause skin disorders.	
Gross formula: BaCl$_2$O$_8$·3H$_2$O		

HAZARDS/SYMPTOMS	PREVENTIVE MEASURES	FIRE EXTINGUISHING/FIRST AID
Fire: non-combustible, but makes other substances more combustible: many chemical reactions can cause fire and explosion.	avoid contact with combustible substances.	in case of fire in immediate vicinity: use any extinguishing agent.
Explosion: mixing with combustible substances can cause explosion.		in case of fire: keep tanks/drums cool by spraying with water, fight fire from sheltered location.
	STRICT HYGIENE, KEEP DUST UNDER CONTROL	
Inhalation: sore throat, cough, shortness of breath; see also 'Ingestion'.	local exhaust or respiratory protection.	fresh air, rest, call a doctor.
Skin: redness, pain.	gloves.	flush with water and only then remove clothing.
Eyes: redness, pain.	goggles.	flush with water, take to a doctor if necessary.
Ingestion: abdominal pain, diarrhea, vomiting, trembling, muscle weakness or paralysis in arms and legs.		rinse mouth, call a doctor or take immediately to hospital.

SPILLAGE	STORAGE	LABELING / NFPA
clean up spillage, carefully collect remainder; (additional individual protection: P2 respirator).	keep separate from combustible substances, reducing agents and strong acids.	R: 9-20/22 S: 27 🔥 Oxidant ✖ Harmful

NOTES

Melting point of anhydrous barium perchlorate 505° C. Apparent melting point due to loss of water of crystallization. ✱ TLV equals that of barium. Flush contaminated clothing with water (fire hazard).

BARIUM PEROXIDE

PHYSICAL PROPERTIES	IMPORTANT CHARACTERISTICS	
Boiling point (decomposes), °C 800 Melting point, °C 450 Relative density (water = 1) 5.0 Solubility in water poor Relative molecular mass 169.4	**GRAYISH-WHITE POWDER** Decomposes when heated, giving off oxygen, with increased risk of fire. Strong oxidant which reacts violently with combustible substances and reducing agents. Mixtures with organic substances can ignite or explode if subjected to shock or friction. Reacts with water and acids, giving off oxygen. Attacks metals slowly.	
	TLV-TWA	0.5 mg/m³ ✶
Gross formula: BaO_2	**Absorption route:** Can enter the body by inhalation or ingestion. **Immediate effects:** Irritates the eyes, skin and respiratory tract. Stimulates the smooth and striated muscles, including the heart, with subsequent paralysis. Can cause slight paralysis of limbs and respiratory muscles. In serious cases risk of seizures, cardiac arrest and death. **Effects of prolonged/repeated exposure:** Prolonged or repeated contact can cause skin disorders.	

HAZARDS/SYMPTOMS	PREVENTIVE MEASURES	FIRE EXTINGUISHING/FIRST AID
Fire: non-combustible, but makes other substances more combustible: many chemical reactions can cause fire and explosion.	avoid contact with combustible substances.	in case of fire in immediate vicinity: use any extinguishing agent.
Explosion: mixing with combustible substances can cause explosion.		in case of fire: keep tanks/drums cool by spraying with water, fight fire from sheltered location.
	STRICT HYGIENE, KEEP DUST UNDER CONTROL	
Inhalation: sore throat, cough, shortness of breath; see also 'Ingestion'.	local exhaust or respiratory protection.	fresh air, rest, place in half-sitting position, call a doctor.
Skin: redness, pain.	gloves.	flush with water and only then remove clothing.
Eyes: redness, pain.	goggles, or combined eye and respiratory protection.	flush with water, take to a doctor if necessary.
Ingestion: abdominal pain, diarrhea, vomiting, trembling, muscle weakness or paralysis in arms and legs.		rinse mouth, take immediately to hospital.

SPILLAGE	STORAGE	LABELING / NFPA
clean up spillage, carefully collect remainder, do not use sawdust or other combustible absorbents; (additional individual protection: P2 respirator).	keep dry, separate from combustible substances and reducing agents.	R: 8-20/22 S: 13-27 Oxidant Harmful

NOTES
✶ TLV equals that of barium. Flush contaminated clothing with water (fire hazard). Reacts slowly with cold water; dissolves only slightly. Packaging: special material.

Transport Emergency Card TEC(R)-51G02   **HI: 50; UN-number: 1449**

BARIUM SULFIDE

PHYSICAL PROPERTIES	IMPORTANT CHARACTERISTICS
Melting point, °C 1200 Relative density (water = 1) 4.3 Solubility in water moderate Relative molecular mass 169.4	**GRAYISH-WHITE POWDER WITH CHARACTERISTIC ODOR** Decomposes on contact with acids or moist air, giving off flammable and toxic gas (→ *hydrogen sulfide*).
	TLV-TWA 0.5 mg/m³ ✱
	Absorption route: Can enter the body by inhalation or ingestion. Evaporation negligible at 20°C, but harmful atmospheric concentrations can build up rapidly in aerosol form. **Immediate effects:** Corrosive to the respiratory tract. Irritates the eyes and skin. Stimulates the smooth and striated muscles, including the heart, with possibility of subsequent paralysis. Affects the nervous system. Can cause weakness, paralysis of limbs and respiratory tract. Inhalation can cause lung edema. In serious cases risk of seizures, cardiac arrest and death.
Gross formula: BaS	

HAZARDS/SYMPTOMS	PREVENTIVE MEASURES	FIRE EXTINGUISHING/FIRST AID
Fire: non-combustible.		in case of fire in immediate vicinity: preferably not water.
	STRICT HYGIENE, KEEP DUST UNDER CONTROL	
Inhalation: *corrosive*; sore throat, cough, shortness of breath, severe breathing difficulties; see also 'Ingestion'.	local exhaust or respiratory protection.	fresh air, rest, place in half-sitting position, take to hospital.
Skin: redness, pain.	gloves.	remove contaminated clothing, flush skin with water or shower.
Eyes: redness, pain.	goggles, or combined eye and respiratory protection.	flush with water, take to a doctor if necessary.
Ingestion: abdominal pain, diarrhea, vomiting, feeling of weakness, trembling or paralysis in arms and legs.		rinse mouth, take immediately to hospital.

SPILLAGE	STORAGE	LABELING / NFPA
clean up spillage, carefully collect remainder; (additional individual protection: breathing apparatus).	keep dry, separate from acids.	R: 20/22-31 S: 28 ✖ Harmful

NOTES

✱ TLV equals that of barium (soluble compounds). Lung edema symptoms usually develop several hours later and are aggravated by physical exertion: rest and hospitalization essential. As first aid, a doctor or authorized person should consider administering a corticosteroid spray. Use airtight packaging.

Transport Emergency Card TEC(R)-61G11   **HI: 60; UN-number: 1564**

BARIUM SULFITE

PHYSICAL PROPERTIES	IMPORTANT CHARACTERISTICS	
Melting point, °C decomposes Relative density (water = 1) > 2 Solubility in water, g/100 ml at 20°C 0.02 Relative molecular mass 217.4	**COLORLESS CRYSTALS** Decomposes when heated, giving off toxic gas (→ *sulfur dioxide*). Reacts with acids, giving off toxic gas (→ *sulfur dioxide*).	
	TLV-TWA	0.5 mg/m³ ✱
	Absorption route: Can enter the body by inhalation or ingestion. Evaporation negligible at 20°C, but harmful concentrations of airborne particles can build up rapidly. **Immediate effects:** Irritates the eyes, skin and respiratory tract. Stimulates the smooth and striated muscles, including the heart, with subsequent paralysis. Can cause weakness and paralysis of limbs and respiratory muscles. In serious cases risk of seizures, cardiac arrest and death.	
Gross formula: BaO₃S		

HAZARDS/SYMPTOMS	PREVENTIVE MEASURES	FIRE EXTINGUISHING/FIRST AID
Fire: non-combustible.		in case of fire in immediate vicinity: use any extinguishing agent.
	STRICT HYGIENE, KEEP DUST UNDER CONTROL	
Inhalation: sore throat, cough, shortness of breath; see also 'Ingestion'.	local exhaust or respiratory protection.	fresh air, rest, call a doctor.
Skin: redness, pain.	gloves.	remove contaminated clothing, flush skin with water or shower.
Eyes: redness, pain.	goggles.	flush with water, take to a doctor if necessary.
Ingestion: abdominal pain, diarrhea, vomiting, feeling of weakness or paralysis in arms and legs, trembling.		rinse mouth, take immediately to hospital.

SPILLAGE	STORAGE	LABELING / NFPA
clean up spillage, carefully collect remainder; (additional individual protection: P2 respirator).	keep separate from strong acids.	R: 20/22 S: 28 Note A ✖ Harmful

NOTES
✱ TLV equals that of barium (soluble compounds).

BENZAL CHLORIDE

PHYSICAL PROPERTIES	IMPORTANT CHARACTERISTICS	
Boiling point, °C 205 Melting point, °C − 16 Flash point, °C 67 Auto-ignition temperature, °C 585 Relative density (water = 1) 1.3 Relative vapor density (air = 1) 5.6 Relative density at 20 °C of saturated mixture vapor/air (air = 1) 1.0 Vapor pressure, mm Hg at 20 °C 0.30 Solubility in water reaction Relative molecular mass 161.0	**COLORLESS LIQUID WITH PUNGENT ODOR, FUMES ON EXPOSURE TO AIR** Vapor mixes readily with air. Decomposes when heated or burned or on contact with acids and bases, giving off toxic and corrosive vapors (→ *hydrochloric acid* and → *phosgene*). Reacts violently with many metals with resin formation and formation of → *hydrochloric acid*. Reacts with strong oxidants, with risk of fire and explosion. Reacts with air to form → *hydrochloric acid* fumes. Attacks many plastics.	
	TLV-TWA	not available
	Absorption route: Can enter the body by inhalation or ingestion. Harmful atmospheric concentrations can build up fairly rapidly on evaporation at approx. 20°C - even more rapidly in aerosol form. **Immediate effects:** Corrosive to the eyes, skin and respiratory tract. Affects the nervous system. Inhalation of vapor/fumes can cause severe breathing difficulties (lung edema). In serious cases risk of unconsciousness.	
Gross formula: $C_7H_6Cl_2$		

HAZARDS/SYMPTOMS	PREVENTIVE MEASURES	FIRE EXTINGUISHING/FIRST AID
Fire: combustible.	keep away from open flame, no smoking.	powder, water spray, foam, carbon dioxide, (halons).
Explosion: above 67°C: forms explosive air-vapor mixtures.	above 67°C: sealed machinery, ventilation.	
Inhalation: *corrosive*: sore throat, cough, severe breathing difficulties, drowsiness, unconsciousness.	ventilation, local exhaust or respiratory protection.	fresh air, rest, place in half-sitting position, take to hospital.
Skin: *corrosive*; redness, pain, burns.	gloves, protective clothing.	remove contaminated clothing, flush skin with water or shower, call a doctor.
Eyes: *corrosive*; redness, pain, impaired vision.	face shield, or combined eye and respiratory protection.	flush with water, take to a doctor.
Ingestion: sore throat, diarrhea, stomach cramps, vomiting.		rinse mouth, immediately take to hospital.

SPILLAGE	STORAGE	LABELING / NFPA
collect leakage in sealable containers, render harmless with caustic soda and soak up with sand or other inert absorbent; (additional individual protection: breathing apparatus).	keep dry, separate from oxidants, separate from all metals, ventilate at floor level.	R: 36/37/38 S: 39 ✖ Irritant

NOTES
Lung edema symptoms usually develop several hours later and are aggravated by physical exertion: rest and hospitalization essential. As first aid, a doctor or authorized person should consider administering a corticosteroid spray. Under no circumstances use near flame or hot surface or when welding. Use airtight packaging made of special material.

Transport Emergency Card TEC(R)-887

HI: 68; UN-number: 1886

CAS-No: [100-52-7]
benzenecarbonal
benzene carboxaldehyde
benzenemethylal
benzoic aldehyde
phenylmethanal

C_6H_5CHO

BENZALDEHYDE

PHYSICAL PROPERTIES	IMPORTANT CHARACTERISTICS
Boiling point, °C 179 Melting point, °C −26 Flash point, °C 64 Auto-ignition temperature, °C 190 Relative density (water = 1) 1.05 Relative vapor density (air = 1) 3.7 Relative density at 20 °C of saturated mixture vapor/air (air = 1) 1.00 Vapor pressure, mm Hg at 20 °C 0.99 Solubility in water, g/100 ml at 20 °C 0.3 Explosive limits, vol% in air 1.4-? Relative molecular mass 106.1 Log P octanol/water 1.5 Gross formula: C_7H_6O	**COLORLESS LIQUID WITH CHARACTERISTIC ODOR, TURNING BROWN ON EXPOSURE TO AIR** Vapor mixes readily with air. Reacts violently with aluminum, iron, bases and phenol, with risk of fire and explosion. Reacts violently with strong oxidants. Finely dispersed over a large area has a tendency to self-ignition. Reacts with air to form peroxides. TLV-TWA not available **Absorption route:** Can enter the body by inhalation or ingestion or through the skin. Harmful atmospheric concentrations build up fairly slowly on evaporation, but much more rapidly in aerosol form. **Immediate effects:** Irritates the eyes, skin and respiratory tract. Liquid destroys the skin's natural oils. Affects the nervous system. Inhalation can cause severe breathing difficulties (asthma). In serious cases risk of seizures and unconsciousness. **Effects of prolonged/repeated exposure:** Prolonged or repeated contact can cause skin disorders. Can cause kidney damage.

HAZARDS/SYMPTOMS	PREVENTIVE MEASURES	FIRE EXTINGUISHING/FIRST AID
Fire: combustible.	keep away from open flame, no smoking.	powder, AFFF, foam, carbon dioxide, (halons).
Explosion: above 64°C: forms explosive air-vapor mixtures.	above 64°C: sealed machinery, ventilation.	
Inhalation: sore throat, cough, shortness of breath, headache, drowsiness.	ventilation, local exhaust or respiratory protection.	fresh air, rest, call a doctor.
Skin: *is absorbed*; headache, drowsiness, redness, pain.	gloves.	remove contaminated clothing, flush skin with water or shower, call a doctor if necessary.
Eyes: redness, pain.	face shield.	flush with water, take to doctor if necessary.
Ingestion: sore throat, abdominal pain, nausea, drowsiness.		rinse mouth, call a doctor or take to hospital.

SPILLAGE	STORAGE	LABELING / NFPA
collect leakage in sealable containers, soak up with sand or other inert absorbent and remove to safe place.	keep separate from oxidants; ventilate at floor level.	R: 22 S: 24 ✖ Harmful

NFPA diamond: 2 (top), 2 (left), 0 (right)

NOTES
Flush contaminated clothing with water (fire hazard). Alcohol consumption increases toxic effects. Asthma symptoms usually develop several hours later and are aggravated by physical exertion: rest and hospitalization essential. Persons with history of asthma symptoms should under no circumstances come in contact again with substance.

Transport Emergency Card TEC(R)-30G15

HI: 30; UN-number: 1990

CAS-No: [532-28-5]
laetrile
mandelonitrile

$C_6H_5CHOHC \equiv N$

BENZALDEHYDE CYANOHYDRIN

PHYSICAL PROPERTIES	IMPORTANT CHARACTERISTICS
Boiling point (decomposes), °C 170 Melting point, °C − 10 Flash point, °C ? Relative density (water = 1) 1.1 Relative vapor density (air = 1) 4.7 Relative density at 20 °C of saturated mixture vapor/air (air = 1) 1.01 Vapor pressure, mm Hg at 20 °C ca. 1.1 Solubility in water none Relative molecular mass 133.2	**YELLOW VISCOUS LIQUID WITH CHARACTERISTIC ODOR** Vapor mixes readily with air. Decomposes when heated, giving off toxic vapors (→ *nitrous vapors* and → *hydrogen cyanide*). Reacts with oxidants.
	TLV-TWA not available
	Absorption route: Can enter the body by inhalation or ingestion or through the skin. Insufficient data on the rate at which harmful concentrations can build up. **Immediate effects:** Irritates the eyes, skin and respiratory tract.
Gross formula: C_8H_7NO	

HAZARDS/SYMPTOMS	PREVENTIVE MEASURES	FIRE EXTINGUISHING/FIRST AID
Fire: combustible.	keep away from open flame, no smoking.	powder, AFFF, foam, carbon dioxide, (halons).
Inhalation: cough, shortness of breath, headache, dizziness.	ventilation, local exhaust or respiratory protection.	fresh air, rest, place in half-sitting position, call a doctor.
Skin: *is absorbed*; redness.	gloves, protective clothing.	remove contaminated clothing, wash skin thoroughly with soap and water, call a doctor.
Eyes: redness, pain.	face shield, or combined eye and respiratory protection.	flush with water, take to a doctor.
Ingestion: sore throat; see also 'Inhalation'.	do not eat, drink or smoke while working.	rinse mouth, call a doctor.

SPILLAGE	STORAGE	LABELING / NFPA
evacuate area, call in an expert, collect leakage in sealable containers, carefully collect remainder and remove to safe place; (additional individual protection: breathing apparatus).	keep separate from oxidants; ventilate at floor level.	

NOTES
Combustible, but flash point not given in the literature. Insufficient toxicological data on harmful effects to humans, but short-term laboratory experiments on animals indicate high toxicity.

Transport Emergency Card TEC(R)-61G06A

CAS-No: [71-43-2]
benzol
carbon oil
coal naphtha
cyclohexatriene

C_6H_6

BENZENE

PHYSICAL PROPERTIES	IMPORTANT CHARACTERISTICS
Boiling point, °C — 80 Melting point, °C — 6 Flash point, °C — −11 Auto-ignition temperature, °C — 555 Relative density (water = 1) — 0.9 Relative vapor density (air = 1) — 2.7 Relative density at 20°C of saturated mixture vapor/air (air = 1) — 1.2 Vapor pressure, mm Hg at 20°C — 76 Solubility in water, g/100 ml — 0.18 Explosive limits, vol% in air — 1.2-8 Minimum ignition energy, mJ — 0.2 Electrical conductivity, pS/m — 3.8 Relative molecular mass — 78.1 Log P octanol/water — 1.9 Gross formula: C_6H_6	**COLORLESS LIQUID WITH CHARACTERISTIC ODOR** Vapor is heavier than air and spreads at ground level, with risk of ignition at a distance. Flow, agitation etc. can cause build-up of electrostatic charge due to liquid's low conductivity. Do not use compressed air when filling, emptying or processing. Reacts violently with strong oxidants, with risk of fire and explosion. Attacks rubber and plastics.

TLV-TWA	10 ppm	32 mg/m³ A2

Absorption route: Can enter the body by inhalation or ingestion or through the skin. Harmful atmospheric concentrations can build up fairly rapidly on evaporation at approx. 20°C - even more rapidly in aerosol form.
Immediate effects: Irritates the eyes and respiratory tract. Liquid destroys the skin's natural oils. Affects the nervous system. Affects the organs of blood cell production. In serious cases risk of unconsciousness and death.
Effects of prolonged/repeated exposure: Can affect the blood. Can cause liver and kidney damage. Exposure to high concentrations over long periods can cause serious, sometimes fatal, blood diseases and even a particular form of leukemia.

HAZARDS/SYMPTOMS	PREVENTIVE MEASURES	FIRE EXTINGUISHING/FIRST AID
Fire: extremely flammable.	keep away from open flame and sparks, no smoking.	powder, AFFF, foam, carbon dioxide, (halons).
Explosion: forms explosive air-vapor mixtures.	sealed machinery, ventilation, explosion-proof electrical equipment and lighting, grounding, non-sparking tools.	in case of fire: keep tanks/drums cool by spraying with water.
	STRICT HYGIENE	
Inhalation: headache, dizziness, agitation, nausea, drowsiness, unconsciousness.	ventilation, local exhaust or respiratory protection.	fresh air, rest, artificial respiration if necessary, call a doctor or take to hospital.
Skin: *is absorbed.*	gloves, protective clothing.	remove contaminated clothing, flush skin with water or shower, refer to a doctor.
Eyes: redness, pain.	face shield, or combined eye and respiratory protection.	flush with water, take to a doctor.
Ingestion: sore throat, abdominal pain, headache, nausea, dizziness, drowsiness, unconsciousness.		rinse mouth, take immediately to hospital.

SPILLAGE	STORAGE	LABELING / NFPA
collect leakage in sealable containers, soak up with sand or other inert absorbent and remove to safe place; (additional individual protection: breathing apparatus).	keep in fireproof place, separate from oxidants.	R: 45-11-23/24/25-48 S: 53-16-29-44 Note E Flammable Toxic

NFPA: 3 / 2 / 0

NOTES
Odor limit is above TLV. Alcohol consumption increases toxic effects. Depending on the degree of exposure, regular medical checkups are advisable. Unbreakable packaging preferred; if breakable, keep in unbreakable container. Should not be used as a solvent, cleaning agent or thinner, except in a closed system or under conditions offering at least equivalent protection against poisoning.

Transport Emergency Card TEC(R)-7

HI: 33; UN-number: 1114

CAS-No: [98-09-9]
benzenesulfonic chloride ·
benzenesulfonyl chloride

$C_6H_5SO_2Cl$

BENZENESULFONYL CHLORIDE

PHYSICAL PROPERTIES	IMPORTANT CHARACTERISTICS	
Boiling point (decomposes), °C 251 Melting point, °C 15 Flash point, °C 130 Auto-ignition temperature, °C 460 Relative density (water = 1) 1.4 Relative vapor density (air = 1) 6.13 Vapor pressure, mm Hg at 66 °C 0.99 Solubility in water moderate Relative molecular mass 176.6	**COLORLESS VISCOUS LIQUID WITH PUNGENT ODOR** Corrosive vapors (→ *hydrogen* and → *sulfur dioxide*) are a product of combustion. In aqueous solution is a strong acid which reacts violently with bases and is corrosive. Reacts with strong oxidants. Reacts slowly with water, forming benzenesulfonic acid and → *hydrochloric acid*).	
	TLV-TWA not available	
	Absorption route: Can enter the body by inhalation or ingestion. Insufficient data on the rate at which harmful concentrations can build up. **Immediate effects:** Corrosive to the eyes, skin and respiratory tract. Inhalation of vapor/fumes can cause severe breathing difficulties (lung edema).	
Gross formula: C_6HClO_2S		

HAZARDS/SYMPTOMS	PREVENTIVE MEASURES	FIRE EXTINGUISHING/FIRST AID
Fire: combustible.	keep away from open flame, no smoking.	powder, AFFF, foam, carbon dioxide, (halons), in case of fire in, immediate vicinity: DO NOT USE WATER-BASED EXTINGUISHERS.
Inhalation: *corrosive*; sore throat, cough, severe breathing difficulties.	ventilation, local exhaust or respiratory protection.	fresh air, rest, place in half-sitting position, take to hospital.
Skin: *corrosive*; redness, pain, burns.	gloves.	remove contaminated clothing, wash skin with soap and water, call a doctor.
Eyes: *corrosive*; redness, pain, impaired vision.	face shield.	flush with water, take to a doctor.
Ingestion: *corrosive*, abdominal cramps, vomiting.	do not eat, drink or smoke while working.	rinse mouth, immediately take to hospital.

SPILLAGE	STORAGE	LABELING / NFPA
collect leakage in sealable containers, flush away remainder with water.	keep separate from oxidants and strong bases.	

NOTES
Lung edema symptoms usually develop several hours later and are aggravated by physical exertion: rest and hospitalization essential. As first aid, a doctor or authorized person should consider administering a corticosteroid spray.

Transport Emergency Card TEC(R)-858

HI: 80; UN-number: 2225

(1,1'-bifenyl)-4,4'-diamine
p-diaminodiphenyl
4,4'-dianiline
4,4'-diphenylenediamine

$H_2NC_6H_4-C_6H_4NH_2$

BENZIDINE

PHYSICAL PROPERTIES	IMPORTANT CHARACTERISTICS
Boiling point, °C ca. 400 Melting point, °C 127 Relative density (water = 1) 1.3 Solubility in water, g/100 ml at 12 °C 0.04 Relative molecular mass 184.2	**WHITE CRYSTALS OR POWDER, TURNING BROWNISH-RED ON EXPOSURE TO AIR** Decomposes when heated or burned, giving off toxic vapors (→ *nitrous vapors*). Reacts violently with strong oxidants.

	IMPORTANT CHARACTERISTICS (cont.)
	TLV-TWA (skin) A1
Gross formula: $C_{12}H_{12}N_2$	**Absorption route:** Can enter the body by inhalation or ingestion or through the skin. Evaporation negligible at 20°C, but harmful concentrations of airborne particles can build up rapidly. **Immediate effects:** Irritates the eyes. Corrosive tot the skin and respiratory tract. Inhalation can cause lung edema. In serious cases risk of seizures and unconsciousness. **Effects of prolonged/repeated exposure:** Prolonged or repeated contact can cause skin disorders. Can affect the blood. Can cause liver and kidney damage.

HAZARDS/SYMPTOMS	PREVENTIVE MEASURES	FIRE EXTINGUISHING/FIRST AID
Fire: combustible.	keep away from open flame, no smoking.	water spray, powder.
	STRICT HYGIENE, KEEP DUST UNDER CONTROL	IN ALL CASES CALL IN A DOCTOR
Inhalation: *corrosive*; sore throat, cough, severe breathing difficulties.	sealed machinery.	fresh air, rest, place in half-sitting position, take to hospital.
Skin: *is absorbed*; redness, pain.	gloves, protective clothing.	remove contaminated clothing, wash skin with soap and water, call a doctor.
Eyes: *corrosive*; redness, pain, impaired vision.	face shield, or combined eye and respiratory protection.	flush with water, take to a doctor.
Ingestion: *corrosieve*, abdominal pain, pain with swallowing, nausea, vomiting.	do not eat, drink or smoke while working.	rinse mouth, immediately take to hospital.

SPILLAGE	STORAGE	LABELING / NFPA
evacuate area, call in an expert, clean up spillage, carefully collect remainder and remove to safe place; (additional individual protection: P3 respirator).	keep in dark place, separate from oxidants.	R: 45-22 S: 53-44 Note E ☠ Toxic

NOTES

Can cause cancer in humans. Depending on the degree of exposure, regular medical checkups are advisable. Lung edema symptoms usually develop several hours later and are aggravated by physical exertion: rest and hospitalization essential. As first aid, a doctor or authorized person should consider administering a corticosteroid spray. Unbreakable packaging preferred; if breakable, keep in unbreakable container.

Transport Emergency Card TEC(R)-530

HI: 60; UN-number: 1885

CAS-No: [65-85-0]
benzenecarboxylic acid
benzeneformic acid
carboxybenzene
dracylic acid
phenylcarboxylic acid

C_6H_5COOH

BENZOIC ACID

PHYSICAL PROPERTIES	IMPORTANT CHARACTERISTICS
Boiling point, °C 249 Melting point, °C 122 Flash point, °C 121 Auto-ignition temperature, °C 570 Relative density (water = 1) 1.3 Relative vapor density (air = 1) 4.2 Relative density at 20 °C of saturated mixture vapor/air (air = 1) 1.0 Vapor pressure, mm Hg at 20 °C < 0.076 Solubility in water, g/100 ml at 20 °C 0.3 Relative molecular mass 122.1 Log P octanol/water 1.9 Gross formula: $C_7H_6O_2$	**WHITE CRYSTALS WITH CHARACTERISTIC ODOR** Sublimation begins at 100°C. Reacts with strong oxidants. TLV-TWA not available **Absorption route:** Can enter the body by inhalation or ingestion. Harmful atmospheric concentrations build up very slowly, if at all, on evaporation at approx. 20°C, but harmful concentrations of airborne particles can build up much more rapidly. **Immediate effects:** Irritates the eyes, skin and respiratory tract.

HAZARDS/SYMPTOMS	PREVENTIVE MEASURES	FIRE EXTINGUISHING/FIRST AID
Fire: combustible.	keep away from open flame, no smoking.	water spray, powder.
Explosion: finely dispersed particles form explosive mixtures on contact with air.	keep dust from accumulating; sealed machinery, explosion-proof electrical equipment and lighting, grounding.	
Inhalation: sore throat, cough.	local exhaust or respiratory protection.	fresh air, rest, place in half-sitting position.
Skin: redness.	gloves.	remove contaminated clothing, flush skin with water or shower.
Eyes: redness, pain.	goggles.	flush with water, take to a doctor if necessary.
Ingestion: sore throat, abdominal pain.		rinse mouth, call a doctor or take to hospital.

SPILLAGE	STORAGE	LABELING / NFPA
clean up spillage, carefully collect remainder; (additional individual protection: P2 respirator).	keep separate from oxidants; ventilate at floor level.	

NOTES

BENZONITRILE

PHYSICAL PROPERTIES	IMPORTANT CHARACTERISTICS
Boiling point, °C 191 Melting point (decomposes), °C − 13 Flash point, °C 70 Auto-ignition temperature, °C 550 Relative density (water = 1) 1.01 Relative vapor density (air = 1) 3.6 Relative density at 20 °C of saturated mixture vapor/air (air = 1) 1.00 Solubility in water, g/100 ml at 100 °C 1 Explosive limits, vol% in air 1.4-7.2 Electrical conductivity, pS/m 2×10^6 Relative molecular mass 103.1 Log P octanol/water 1.6 Gross formula: C_7H_5N	**COLORLESS LIQUID WITH CHARACTERISTIC ODOR** Vapor mixes readily with air. Decomposes when heated, giving off → *hydrogen cyanide* and → *nitrous vapors*. Reacts violently with strong acids, giving off → *hydrogen cyanide*. TLV-TWA not available **Absorption route:** Can enter the body by inhalation or ingestion or through the skin. Harmful atmospheric concentrations build up very slowly, if at all, on evaporation at approx. 20°C, but much more rapidly in aerosol form. **Immediate effects:** Impedes tissue respiration. In serious cases risk of unconsciousness.

HAZARDS/SYMPTOMS	PREVENTIVE MEASURES	FIRE EXTINGUISHING/FIRST AID
Fire: combustible.	keep away from open flame, no smoking.	powder, AFFF, foam, carbon dioxide, (halons).
Explosion: above 70°C: forms explosive air-vapor mixtures.	above 70°C: sealed machinery, ventilation.	
	STRICT HYGIENE	
Inhalation: headache, dizziness, nausea, sleepiness.	ventilation, local exhaust or respiratory protection.	fresh air, rest, call a doctor or take to hospital.
Skin: *is absorbed*; redness.	gloves, protective clothing.	remove contaminated clothing, wash skin with soap and water, call a doctor if necessary.
Eyes: redness, pain.	face shield.	flush with water, take to a doctor if necessary.
Ingestion: abdominal pain; see also 'Inhalation'.	do not eat, drink or smoke while working.	rinse mouth, call a doctor or take to hospital.

SPILLAGE	STORAGE	LABELING / NFPA
collect leakage in sealable containers, render remainder harmless with sodium hypochlorite solution; (additional individual protection: breathing apparatus).	keep separate from strong acids.	R: 21/22 S: 23 ❌ Harmful

NOTES
Special first aid required in the event of poisoning: antidotes must be available (with instructions).

Transport Emergency Card TEC(R)-61G17

HI: 60; UN-number: 2224

CAS-No: [98-07-7]
benzenyl chloride
benzenyl trichloride
phenylchloroform
phenyltrichloromethane
toluene trichloride

$C_6H_5CCl_3$

BENZOTRICHLORIDE

PHYSICAL PROPERTIES	IMPORTANT CHARACTERISTICS
Boiling point, °C 214 Melting point, °C −5 Flash point, °C 109 Auto-ignition temperature, °C 420 Relative density (water = 1) 1.4 Relative vapor density (air = 1) 6.8 Relative density at 20 °C of saturated mixture vapor/air (air = 1) 1.0 Vapor pressure, mm Hg at 20 °C 0.15 Solubility in water none Explosive limits, vol% in air 2.1-5.6 Relative molecular mass 195.5 Log P octanol/water 2.9	**COLORLESS TO BROWN FUMING LIQUID WITH PUNGENT ODOR** Vapor mixes readily with air. Decomposes when heated or burned, giving off corrosive vapors (→ *hydrogen chloride*). Reacts violently with light metals and amines, with risk of fire and explosion. Reacts slowly with water, forming → *hydrochloric acid* and → *benzoic acid*. Reacts with air to form hydrochloric acid fumes. Attacks some plastics.
	TLV-TWA not available
Gross formula: $C_7H_5Cl_3$	**Absorption route:** Can enter the body by inhalation or ingestion or through the skin. Harmful atmospheric concentrations build up fairly slowly on evaporation at 20°C, but harmful concentrations of airborne particles can build up much more rapidly. **Immediate effects:** Corrosive to the eyes, skin and respiratory tract. In substantial concentrations can impair consciousness. Inhalation of vapor/fumes can cause severe breathing difficulties (lung edema). In serious cases risk of unconsciousness and death. If liquid is swallowed, droplets can enter the lungs, with risk of pneumonia.

HAZARDS/SYMPTOMS	PREVENTIVE MEASURES	FIRE EXTINGUISHING/FIRST AID
Fire: combustible.	keep away from open flame, no smoking.	powder, water spray, foam, carbon dioxide, (halons).
	STRICT HYGIENE, AVOID ALL CONTACT	
Inhalation: *corrosive*; sore throat, cough, severe breathing difficulties.	ventilation, local exhaust or respiratory protection.	fresh air, rest, take to hospital.
Skin: *corrosive, is absorbed*; redness, pain, chemical burns.	gloves, protective clothing.	remove contaminated clothing, flush skin with water or shower, call a doctor if necessary.
Eyes: *corrosive*; redness, pain, impaired vision.	face shield.	flush with water, take to a doctor.
Ingestion: *corrosive*; sore throat, cough, abdominal pain, vomiting.		rinse mouth, DO NOT induce vomiting, take immediately to hospital.

SPILLAGE	STORAGE	LABELING / NFPA
collect leakage in sealable containers, soak up with sand or other inert absorbent and remove to safe place, render remainder harmless with caustic soda solution; (additional individual protection: CHEMICAL SUIT).	keep cool; ventilate at floor level.	 1 3 0

NOTES
Alcohol consumption increases toxic effects. Lung edema symptoms usually develop several hours later and are aggravated by physical exertion: rest and hospitalization essential. As first aid, a doctor or authorized person should consider administering a corticosteroid spray. Technical grade can contain hydrochloric acid and benzoic acid. Use airtight packaging.

Transport Emergency Card TEC(R)-833

HI: 80; UN-number: 2226

BENZOTRIFLUORIDE

PHYSICAL PROPERTIES	IMPORTANT CHARACTERISTICS
Boiling point, °C 102 Melting point, °C −29 Flash point, °C 12 Auto-ignition temperature, °C 620 Relative density (water = 1) 1.2 Relative vapor density (air = 1) 5.0 Relative density at 20 °C of saturated mixture vapor/air (air = 1) 1.2 Vapor pressure, mm Hg at 20 °C 31 Solubility in water none Relative molecular mass 146.1 Log P octanol/water 2.8	**COLORLESS LIQUID WITH PUNGENT ODOR** Vapor is heavier than air and spreads at ground level, with risk of ignition at a distance. Do not use compressed air when filling, emptying or processing. Toxic vapors of → *hydrogen fluoride* are a product of combustion. Reacts with hot water or steam, forming → *hydrogen fluoride* and → *benzoic acid*. Reacts violently with oxidants. Reacts with air to form corrosive vapors (→ *hydrogen fluoride*). TLV-TWA not available **Absorption route:** Can enter the body by inhalation or ingestion or through the skin. **Immediate effects:** Corrosive to the eyes, skin and respiratory tract. In substantial concentrations can impair consciousness. Inhalation of vapor/fumes can cause severe breathing difficulties (lung edema). In serious cases risk of death.
Gross formula: $C_7H_5F_3$	

HAZARDS/SYMPTOMS	PREVENTIVE MEASURES	FIRE EXTINGUISHING/FIRST AID
Fire: extremely flammable.	keep away from open flame and sparks, no smoking.	powder, carbon dioxide, (halons), DO NOT USE WATER-BASED EXTINGUISHERS.
Explosion: forms explosive air-vapor mixtures.	sealed machinery, ventilation, explosion-proof electrical equipment and lighting.	in case of fire: keep tanks/drums cool by spraying with water, but DO NOT spray substance with water.
Inhalation: *corrosive*; sore throat, cough, shortness of breath, severe breathing difficulties, dizziness.	ventilation, local exhaust or respiratory protection.	fresh air, rest, place in half-sitting position, take to hospital.
Skin: *corrosive, is absorbed*; redness, pain, serious burns.	gloves, protective clothing.	remove contaminated clothing, flush skin with water or shower, refer to a doctor.
Eyes: *corrosive*; redness, pain, impaired vision.	face shield, or combined eye and respiratory protection.	flush with water, take to a doctor.
Ingestion: *corrosive*; sore throat, abdominal pain, diarrhea, sleepiness.		rinse mouth, take immediately to hospital.

SPILLAGE	STORAGE	LABELING / NFPA
collect leakage in sealable containers, soak up with sand or other inert absorbent and remove to safe place, do not flush into sewer; (additional individual protection: CHEMICAL SUIT).	keep in dry, fireproof place, separate from oxidants.	R: 11 S: 16-23 🔥 Flammable ◇ 3 / 3 1

NOTES
Reacts slowly with water; does not dissolve. Lung edema symptoms usually develop several hours later and are aggravated by physical exertion: rest and hospitalization essential. As first aid, a doctor or authorized person should consider administering a corticosteroid spray. Use airtight packaging.

Transport Emergency Card TEC(R)-30G34

HI: 33; UN-number: 2338

CAS-No: [98-88-4]
benzenecarbonyl chloride
α-**chlorobenzaldehyde**

C_6H_5COCl

BENZOYL CHLORIDE

PHYSICAL PROPERTIES		IMPORTANT CHARACTERISTICS
Boiling point, °C	197	**COLORLESS LIQUID WITH PUNGENT ODOR**
Melting point, °C	−1	Vapor mixes readily with air. Decomposes when heated, giving off toxic and corrosive
Flash point, °C	72	→ *hydrochloric acid.* Reacts violently with water, amines, alcohols, DMSO and alkaline-earth and
Auto-ignition temperature, °C	600	alkali metals. Reacts violently with strong oxidants. Reacts with air to form corrosive fumes.
Relative density (water = 1)	1.2	Attacks many metals, giving off flammable gas (→ *hydrogen*).
Relative vapor density (air = 1)	4.9	

PHYSICAL PROPERTIES (cont.)		IMPORTANT CHARACTERISTICS (cont.)
Relative density at 20 °C of saturated mixture		**TLV-TWA** not available
vapor/air (air = 1)	1.00	
Vapor pressure, mm Hg at 20 °C	0.38	**Absorption route:** Can enter the body by inhalation or ingestion. Harmful atmospheric
Solubility in water	reaction	concentrations build up fairly slowly on evaporation, but much more rapidly in aerosol form.
Explosive limits, vol% in air	2.5-27	**Immediate effects:** Corrosive to the eyes, skin and respiratory tract. Inhalation of vapor/fumes
Relative molecular mass	140.6	can cause severe breathing difficulties (lung edema). In serious cases risk of death.

Gross formula:	C_7H_5ClO

HAZARDS/SYMPTOMS	PREVENTIVE MEASURES	FIRE EXTINGUISHING/FIRST AID
Fire: combustible, many chemical reactions can cause fire and explosion.	keep away from open flame, no smoking.	powder, carbon dioxide, (halons), DO NOT USE WATER-BASED EXTINGUISHERS.
Explosion: above 72°C: forms explosive air-vapor mixtures.	above 72°C: sealed machinery, ventilation.	
	STRICT HYGIENE	
Inhalation: *corrosive*; sore throat, cough, shortness of breath, severe breathing difficulties.	ventilation, local exhaust or respiratory protection.	fresh air, rest, place in half-sitting position, artificial respiration if necessary, take to hospital.
Skin: *corrosive*; redness, pain, burns.	gloves, protective clothing.	remove contaminated clothing, flush skin with water or shower, call a doctor.
Eyes: *corrosive*; redness, pain, impaired vision.	face shield.	flush with water, take to a doctor.
Ingestion: *corrosive*; sore throat, abdominal pain.		rinse mouth, immediately take to hospital.

SPILLAGE	STORAGE	LABELING / NFPA
flush away with water, collect leakage in sealable containers, flush away remainder with water; (additional individual protection: CHEMICAL SUIT).	keep dry, separate from oxidants, alcohols, amines and DMSO; ventilate at floor level.	R: 34 S: 26 Corrosive ◇ 2 / 3 2 / W

NOTES
Reacts violently with extinguishing agents such as water. Lung edema symptoms usually develop several hours later and are aggravated by physical exertion: rest and hospitalization essential. As first aid, a doctor or authorized person should consider administering a corticosteroid spray. Fight large-scale fires with large quantities of water.

Transport Emergency Card TEC(R)-64

HI: 80; UN-number: 1736

acetic acid, benzyl ester
acetoxymethylbenzene
α-acetoxytoluene
benzyl ethanoate
phenylmethyl acetate

$CH_3COOCH_2C_6H_5$

BENZYL ACETATE

PHYSICAL PROPERTIES	IMPORTANT CHARACTERISTICS	
Boiling point, °C 214 Melting point, °C −52 Flash point, °C 102 Auto-ignition temperature, °C 460 Relative density (water = 1) 1.1 Relative vapor density (air = 1) 5.2 Relative density at 20 °C of saturated mixture vapor/air (air = 1) 1.00 Vapor pressure, mm Hg at 45 °C 0.99 Solubility in water poor Relative molecular mass 150.2 Log P octanol/water 2.0 Gross formula: $C_9H_{10}O_2$	**COLORLESS LIQUID WITH CHARACTERISTIC ODOR** Vapor mixes readily with air. Reacts with strong oxidants. TLV-TWA not available **Absorption route:** Can enter the body by inhalation or ingestion or through the skin. Harmful atmospheric concentrations build up fairly slowly on evaporation, but much more rapidly in aerosol form. **Immediate effects:** Irritates the eyes, skin and respiratory tract. Liquid destroys the skin's natural oils. Affects the nervous system.	

HAZARDS/SYMPTOMS	PREVENTIVE MEASURES	FIRE EXTINGUISHING/FIRST AID
Fire: combustible.	keep away from open flame, no smoking.	powder, AFFF, foam, carbon dioxide, (halons).
Inhalation: sore throat, cough, drowsiness.	ventilation.	fresh air, rest.
Skin: *is absorbed*; redness.	gloves.	wash skin with soap and water.
Eyes: redness, pain.	safety glasses.	flush with water, take to a doctor if necessary.
Ingestion: abdominal pain, nausea.		rinse mouth, call a doctor.

SPILLAGE	STORAGE	LABELING / NFPA
collect leakage in sealable containers, soak up with sand or other inert absorbent, and remove to safe place.		 1 1 0

NOTES

CAS-No: [100-51-6]
α-**hydroxytoluene**
phenylcarbinol

$C_6H_5CH_2OH$

BENZYL ALCOHOL

PHYSICAL PROPERTIES		IMPORTANT CHARACTERISTICS	
Boiling point, °C	206	**COLORLESS LIQUID WITH CHARACTERISTIC ODOR**	
Melting point, °C	− 15	Vapor mixes readily with air. Reacts violently with strong oxidants. Attacks many plastics. On	
Flash point, °C	101	prolonged contact with air forms → *benzaldehyde*.	
Auto-ignition temperature, °C	435		
Relative density (water = 1)	1.04		
Relative vapor density (air = 1)	3.7	TLV-TWA	not available
Relative density at 20 °C of saturated mixture			
vapor/air (air = 1)	1.0	**Absorption route:** Can enter the body by inhalation or ingestion. Harmful atmospheric	
Vapor pressure, mm Hg at 20 °C	0.099	concentrations build up very slowly, if at all, on evaporation at approx. 20°C, but much more	
Solubility in water	poor	rapidly in aerosol form.	
Explosive limits, vol% in air	1.3-13	**Immediate effects:** Irritates the eyes, skin and respiratory tract. Liquid destroys the skin's natural	
Electrical conductivity, pS/m	2.7x10^7	oils. In substantial concentrations can impair consciousness. In serious cases risk of	
Relative molecular mass	108.1	unconsciousness.	
Log P octanol/water	1.1		
Gross formula:	C_7H_8O		

HAZARDS/SYMPTOMS	PREVENTIVE MEASURES	FIRE EXTINGUISHING/FIRST AID
Fire: combustible.	keep away from open flame, no smoking.	powder, AFFF, foam, carbon dioxide, (halons).
Inhalation: sore throat, cough, shortness of breath, dizziness, drowsiness.	ventilation.	fresh air, rest, place in half-sitting position, call a doctor.
Skin: redness, pain.	gloves.	remove contaminated clothing, flush skin with water or shower, call a doctor if necessary.
Eyes: impaired vision, redness, pain.	safety glasses.	flush with water, take to a doctor if necessary.
Ingestion: sore throat, cough, abdominal pain, vomiting; see also 'Inhalation'.		rinse mouth, immediately take to hospital.

SPILLAGE	STORAGE	LABELING / NFPA
collect leakage in sealable containers, soak up with sand or other inert absorbent and remove to safe place.	keep separate from oxidants.	R: 20/22 S: 26 ✖ Harmful

NOTES
Alcohol consumption increases toxic effects.

BENZYL BENZOATE

PHYSICAL PROPERTIES	IMPORTANT CHARACTERISTICS
Boiling point (decomposes), °C 323 Melting point, °C 21 Flash point, °C 148 Auto-ignition temperature, °C 480 Relative density (water = 1) 1.1 Relative vapor density (air = 1) 7.3 Vapor pressure, mm Hg at 44 °C 0.99 Solubility in water none Relative molecular mass 212.2 Log P octanol/water 4.0	**COLORLESS VISCOUS LIQUID OR WHITE SOLID IN VARIOUS FORMS WITH CHARACTERISTIC ODOR** Reacts with strong oxidants.
	TLV-TWA not available
	Absorption route: Can enter the body by inhalation or ingestion. Evaporation negligible at 20° C, but harmful concentrations of airborne particles can build up rapidly. **Immediate effects:** Irritates the eyes, skin and respiratory tract. **Effects of prolonged/repeated exposure:** Prolonged or repeated contact can cause skin disorders.
Gross formula: $C_{14}H_{12}O_2$	

HAZARDS/SYMPTOMS	PREVENTIVE MEASURES	FIRE EXTINGUISHING/FIRST AID
Fire: combustible.	keep away from open flame, no smoking.	powder, AFFF, foam, carbon dioxide, (halons).
Inhalation: sore throat, cough, shortness of breath.	local exhaust or respiratory protection.	fresh air, rest.
Skin: redness.	gloves.	remove contaminated clothing, flush skin with water or shower.
Eyes: redness, pain.	goggles.	flush with water, take to a doctor if necessary.
Ingestion: abdominal pain, nausea.		rinse mouth, call a doctor or take to hospital.

SPILLAGE	STORAGE	LABELING / NFPA
collect leakage in sealable containers or clean up spillage; (additional individual protection: P2 respirator).	keep separate from oxidants.	R: 22 S: 25 ✖ Harmful

NOTES

Transport Emergency Card TEC(R)-61G06

BENZYL BROMIDE

PHYSICAL PROPERTIES	IMPORTANT CHARACTERISTICS
Boiling point, °C 201 Melting point, °C −3 Flash point, °C 87 Relative density (water = 1) 1.4 Relative vapor density (air = 1) 5.8 Relative density at 20 °C of saturated mixture vapor/air (air = 1) 1.0 Vapor pressure, mm Hg at 20 °C 0.38 Solubility in water none Relative molecular mass 171.0 Log P octanol/water 2.9	**COLORLESS LIQUID** Vapor mixes readily with air. Corrosive vapors (→ *hydrogen bromide*) are a product of combustion. Decomposes slowly on contact with water or water vapor, forming → *benzyl alcohol* and corrosive → *hydrogen bromide*. Reacts violently with bases and strong oxidants. Reacts with air to form corrosive vapors (→ *hydrogen bromide*). Attacks many metals especially in the presence of moisture.
	TLV-TWA not available
	Absorption route: Can enter the body by inhalation or ingestion or through the skin. Harmful atmospheric concentrations can build up fairly rapidly on evaporation at approx. 20°C - even more rapidly in aerosol form. **Immediate effects:** Corrosive to the eyes, skin and respiratory tract. Affects the nervous system. Inhalation can cause lung edema. In serious cases risk of death.
Gross formula: C_7H_7Br	

HAZARDS/SYMPTOMS	PREVENTIVE MEASURES	FIRE EXTINGUISHING/FIRST AID
Fire: combustible.	keep away from open flame, no smoking.	powder, water spray, foam, carbon dioxide, (halons).
Explosion: above 87°C: forms explosive air-vapor mixtures.	above 87°C: sealed machinery, ventilation.	
	STRICT HYGIENE	
Inhalation: *corrosive*; cough, severe breathing difficulties, dizziness, drowsiness.	ventilation, local exhaust or respiratory protection.	fresh air, rest, place in half-sitting position, take to hospital.
Skin: *is absorbed, corrosive*; redness, pain, burns.	gloves, protective clothing.	remove contaminated clothing, flush skin with water or shower, refer to a doctor.
Eyes: *corrosive*; redness, pain, impaired vision.	face shield.	flush with water, take to a doctor.
Ingestion: *corrosive*; sore throat, abdominal pain.		rinse mouth, immediately take to hospital.

SPILLAGE	STORAGE	LABELING / NFPA
evacuate area, collect leakage in sealable containers, soak up with dry sand or other inert absorbent and remove to safe place; (additional individual protection: CHEMICAL SUIT).	keep dry, separate from bases; ventilate at floor level.	R: 36/37/38 S: 39 ✖ Irritant 2 / 2 / 0

NOTES
Lung edema symptoms usually develop several hours later and are aggravated by physical exertion: rest and hospitalization essential. As first aid, a doctor or authorized person should consider administering a corticosteroid spray. Under no circumstances use near flame or hot surface or when welding. Use airtight packaging.

Transport Emergency Card TEC(R)-61G06B

HI: 60; UN-number: 1737

BENZYL BUTYL ETHER

PHYSICAL PROPERTIES	IMPORTANT CHARACTERISTICS	
Boiling point, °C 223 Relative density (water = 1) 0.9 Relative vapor density (air = 1) 5.7 Relative density at 20 °C of saturated mixture vapor/air (air = 1) 1.0 Vapor pressure, mm Hg at 20 °C 0.11 Solubility in water none Relative molecular mass 164.2	**COLORLESS LIQUID WITH CHARACTERISTIC ODOR** Able to form peroxides. Reacts violently with oxidants.	
	TLV-TWA not available	
	Absorption route: Can enter the body by inhalation or ingestion. Harmful atmospheric concentrations build up very slowly, if at all, on evaporation at approx. 20°C, but much more rapidly in aerosol form. **Immediate effects:** Irritates the eyes and respiratory tract. Liquid destroys the skin's natural oils. Affects the nervous system.	
Gross formula: $C_{11}H_{16}O$		

HAZARDS/SYMPTOMS	PREVENTIVE MEASURES	FIRE EXTINGUISHING/FIRST AID
Fire: combustible.	keep away from open flame, no smoking.	AFFF, foam, powder, carbon dioxide, (halons).
Inhalation: drowsiness, sleepiness.	ventilation.	fresh air, rest.
Skin: redness.	gloves.	remove contaminated clothing, flush skin with water or shower.
Eyes: redness, pain.	safety glasses.	flush with water, take to a doctor if necessary.
Ingestion: drowsiness, sleepiness, nausea.		rinse mouth, call a doctor.

SPILLAGE	STORAGE	LABELING / NFPA
collect leakage in sealable containers, soak up with sand or other inert absorbent and remove to safe place.	keep in fireproof place, separate from oxidants.	

NOTES
Before distilling check for peroxides; if found, render harmless.

CAS-No: [100-44-7]
(chloromethyl)benzene
α-**chlorotoluene**

$C_6H_5CH_2Cl$

BENZYL CHLORIDE

PHYSICAL PROPERTIES	IMPORTANT CHARACTERISTICS	
Boiling point, °C 179	**COLORLESS LIQUID WITH PUNGENT ODOR**	
Melting point, °C −43	Vapor mixes readily with air. Decomposes when heated or burned or on contact with moisture,	
Flash point, °C 60	giving off toxic and corrosive vapors (→ *hydrochloric acid*). Reacts violently with various metals.	
Auto-ignition temperature, °C 585	Reacts with oxidants. Attacks some plastics.	
Relative density (water = 1) 1.1		
Relative vapor density (air = 1) 4.4	**TLV-TWA**	1 ppm	5.2 mg/m³
Relative density at 20 °C of saturated mixture		
vapor/air (air = 1) 1.0	**Absorption route:** Can enter the body by inhalation or ingestion or through the skin. Harmful	
Vapor pressure, mm Hg at 20 °C 0.91	atmospheric concentrations can build up fairly rapidly on evaporation at approx. 20°C - even	
Solubility in water, g/100 ml at 30 °C 0.05	more rapidly in aerosol form.	
Explosive limits, vol% in air 1.1-14	**Immediate effects:** Corrosive to the eyes, skin and respiratory tract. Affects the nervous system.	
Relative molecular mass 126.6	Inhalation can cause lung edema. In serious cases risk of death.	
Log P octanol/water 2.3	**Effects of prolonged/repeated exposure:** Can cause liver and kidney damage.	
Gross formula: C_7H_7Cl		

HAZARDS/SYMPTOMS	PREVENTIVE MEASURES	FIRE EXTINGUISHING/FIRST AID
Fire: combustible.	keep away from open flame, no smoking.	powder, AFFF, foam, carbon dioxide, (halons).
Explosion: above 60°C: forms explosive air-vapor mixtures.	above 60°C: sealed machinery, ventilation.	in case of fire: keep tanks/drums cool by spraying with water, but DO NOT spray substance with water.
Inhalation: *corrosive*; sore throat, cough, severe breathing difficulties, drowsiness.	ventilation, local exhaust or respiratory protection.	fresh air, rest, place in half-sitting position, take to hospital.
Skin: *corrosive, is absorbed*; redness, pain, serious burns.	gloves, protective clothing.	remove contaminated clothing, flush skin with water or shower, call a doctor.
Eyes: *corrosive*; redness, pain, impaired vision.	face shield.	flush with water, take to a doctor.
Ingestion: *corrosive*; abdominal pain, diarrhea, vomiting, drowsiness.		rinse mouth, take immediately to hospital.

SPILLAGE	STORAGE	LABELING / NFPA
collect leakage in sealable containers, soak up with sand or other inert absorbent and remove to safe place; (additional individual protection: breathing apparatus).	keep dry; ventilate at floor level.	R: 36/37/38 S: 39 ✖ Irritant

NOTES

Depending on the degree of exposure, regular medical checkups are advisable. Lung edema symptoms usually develop several hours later and are aggravated by physical exertion: rest and hospitalization essential. As first aid, a doctor or authorized person should consider administering a corticosteroid spray. Use airtight packaging made of special material.

Transport Emergency Card TEC(R)-772

HI: 68; UN-number: 1738

ω-**cyanotoluene**
phenylacetonitrile
α-**tolunitrile**

$C_6H_5CH_2C \equiv N$

BENZYLCYANIDE

PHYSICAL PROPERTIES		IMPORTANT CHARACTERISTICS
Boiling point, °C	234	**COLORLESS LIQUID WITH CHARACTERISTIC ODOR**
Melting point, °C	−24	Decomposes when heated, giving off → *hydrogen cyanide*. Reacts violently with acids, giving off
Flash point, °C	102	toxic gas (→ *hydrogen cyanide*).
Relative density (water = 1)	1.02	
Relative vapor density (air = 1)	4.1	TLV-TWA not available
Relative density at 20 °C of saturated mixture		
vapor/air (air = 1)	1.00	**Absorption route:** Can enter the body by inhalation or ingestion or through the skin. Insufficient
Vapor pressure, mm Hg at 20 °C	ca. 0.38	data on the rate at which harmful concentrations can build up.
Solubility in water	none	**Immediate effects:** Irritates the eyes, skin and respiratory tract. Impedes tissue respiration. In
Electrical conductivity, pS/m	2 x 10⁷	serious cases risk of unconsciousness.
Relative molecular mass	117.2	
Log P octanol/water	1.6	
Gross formula:	C_8H_7N	

HAZARDS/SYMPTOMS	PREVENTIVE MEASURES	FIRE EXTINGUISHING/FIRST AID
Fire: combustible.	keep away from open flame, no smoking.	powder, AFFF, foam, carbon dioxide, (halons).
	STRICT HYGIENE	
Inhalation: sore throat, severe breathing difficulties, headache, dizziness, nausea, drowsiness, unconsciousness, feeling of weakness.	ventilation, local exhaust or respiratory protection.	fresh air, rest, place in half-sitting position, artificial respiration if necessary, take to hospital.
Skin: *is absorbed;* redness; see also 'Inhalation'.	gloves, protective clothing.	remove contaminated clothing, wash skin with soap and water, call a doctor.
Eyes: redness, pain.	safety glasses.	flush with water, take to a doctor if necessary.
Ingestion: abdominal pain; see also 'Inhalation'.	do not eat, drink or smoke while working.	rinse mouth, immediately take to hospital.

SPILLAGE	STORAGE	LABELING / NFPA
collect leakage in sealable containers, render remainder harmless with bleaching lye; (additional individual protection: breathing apparatus).	keep in fireproof place, separate from acids.	1 2 0

NOTES
Special first aid required in the event of poisoning: antidotes must be available (with instructions).

Transport Emergency Card TEC(R)-852

HI: 60; UN-number: 2470

BENZYL ETHER

PHYSICAL PROPERTIES	IMPORTANT CHARACTERISTICS	
Boiling point, °C 298 Melting point, °C 4 Flash point, °C 135 Relative density (water = 1) 1.05 Relative vapor density (air = 1) 6.8 Relative density at 20 °C of saturated mixture vapor/air (air = 1) 1 Vapor pressure, mm Hg at 20 °C ca. 0.0015 Solubility in water none Relative molecular mass 198.3	**COLORLESS LIQUID** Flow, agitation etc. can cause build-up of electrostatic charge due to liquid's low conductivity. Able to form peroxides. Decomposes when heated above boiling point. Reacts violently with strong oxidants.	
	TLV-TWA not available	
	Absorption route: Can enter the body by inhalation or ingestion. Harmful atmospheric concentrations build up very slowly, if at all, on evaporation at approx. 20°C, but much more rapidly in aerosol form. **Immediate effects:** Irritates the eyes and respiratory tract. Liquid destroys the skin's natural oils. Affects the nervous system.	
Gross formula: $C_{14}H_{14}O$		

HAZARDS/SYMPTOMS	PREVENTIVE MEASURES	FIRE EXTINGUISHING/FIRST AID
Fire: combustible.	keep away from open flame, no smoking.	powder, AFFF, foam, carbon dioxide, (halons).
Explosion:	grounding.	
Inhalation: drowsiness, sleepiness.	ventilation, local exhaust or respiratory protection.	fresh air, rest.
Skin: redness.	gloves.	remove contaminated clothing, flush skin with water or shower.
Eyes: redness, pain.	safety glasses.	flush with water, take to a doctor if necessary.
Ingestion: nausea, drowsiness, sleepiness.		rinse mouth, call a doctor.

SPILLAGE	STORAGE	LABELING / NFPA
collect leakage in sealable containers, soak up with sand or other inert absorbent and remove to safe place.	keep separate from oxidants; ventilate at floor level.	

NOTES
Before distilling check for peroxides; if found, render harmless.

BENZYL ETHYL ETHER

PHYSICAL PROPERTIES	IMPORTANT CHARACTERISTICS
Boiling point, °C 185 Flash point, °C 51 Relative density (water = 1) 0.95 Relative vapor density (air = 1) 4.7 Relative density at 20 °C of saturated mixture vapor/air (air = 1) 1.0 Vapor pressure, mm Hg at 20 °C ca. 0.76 Explosive limits, vol% in air ?-? Relative molecular mass 136.2	**COLORLESS LIQUID WITH CHARACTERISTIC ODOR** Vapor mixes readily with air. Flow, agitation etc. can cause build-up of electrostatic charge due to liquid's low conductivity. Presumed to be able to form peroxides. Reacts violently with strong oxidants.
	TLV-TWA not available
	Absorption route: Can enter the body by inhalation or ingestion. Insufficient data on the rate at which harmful concentrations can build up. **Immediate effects:** Irritates the eyes, skin and respiratory tract. Liquid destroys the skin's natural oils.
Gross formula: $C_9H_{12}O$	

HAZARDS/SYMPTOMS	PREVENTIVE MEASURES	FIRE EXTINGUISHING/FIRST AID
Fire: flammable.	keep away from open flame and sparks, no smoking.	powder, AFFF, foam, carbon dioxide, (halons).
Explosion: above 51°C: forms explosive air-vapor mixtures.	above 51°C: sealed machinery, ventilation, explosion-proof electrical equipment, grounding.	in case of fire: keep tanks/drums cool by spraying with water.
Inhalation: drowsiness, sleepiness.	ventilation, local exhaust or respiratory protection.	fresh air, rest.
Skin: redness.	gloves.	remove contaminated clothing, flush skin with water or shower.
Eyes: redness, pain.	safety glasses.	flush with water, take to a doctor if necessary.
Ingestion: nausea, drowsiness, sleepiness.		rinse mouth, call a doctor.

SPILLAGE	STORAGE	LABELING / NFPA
collect leakage in sealable containers, soak up with sand or other inert absorbent and remove to safe place.	keep in fireproof place, separate from oxidants; add an inhibitor.	

NOTES
Before distilling check for peroxides; if found, render harmless.

Transport Emergency Card TEC(R)-30G08

BENZYL FORMATE

PHYSICAL PROPERTIES	IMPORTANT CHARACTERISTICS
Boiling point, °C 203 Melting point, °C 4 Flash point, °C 42 Relative density (water = 1) 1.1 Relative vapor density (air = 1) 4.7 Relative density at 20 °C of saturated mixture vapor/air (air = 1) 1.0 Vapor pressure, mm Hg at 20 °C 0.23 Solubility in water none Explosive limits, vol% in air ?-? Relative molecular mass 136.1 Gross formula: $C_8H_8O_2$	**COLORLESS LIQUID WITH CHARACTERISTIC ODOR** Vapor mixes readily with air. Reacts with strong oxidants. TLV-TWA not available **Absorption route:** Can enter the body by inhalation or ingestion. Insufficient data on the rate at which harmful concentrations can build up. **Immediate effects:** Irritates the eyes, skin and respiratory tract.

HAZARDS/SYMPTOMS	PREVENTIVE MEASURES	FIRE EXTINGUISHING/FIRST AID
Fire: flammable.	keep away from open flame and sparks, no smoking.	powder, AFFF, foam, carbon dioxide, (halons).
Explosion: above 42°C: forms explosive air-vapor mixtures.	above 42°C: sealed machinery, ventilation, explosion-proof electrical equipment.	in case of fire: keep tanks/drums cool by spraying with water.
Inhalation: sore throat, cough.	ventilation, local exhaust or respiratory protection.	fresh air, rest.
Skin: redness.	gloves.	remove contaminated clothing, flush skin with water or shower.
Eyes: redness, pain.	acid goggles.	flush with water, take to a doctor if necessary.
Ingestion: sore throat, abdominal pain.		rinse mouth, call a doctor.

SPILLAGE	STORAGE	LABELING / NFPA
collect leakage in sealable containers, soak up with sand or other inert absorbent and remove to safe place.	keep in fireproof place, separate from oxidants.	

NOTES

BERYLLIUM

PHYSICAL PROPERTIES	IMPORTANT CHARACTERISTICS	
Melting point, °C 1280 Relative density (water = 1) 1.85 Solubility in water none Relative atomic mass 9.0	**GRAY OR SILVERY SOLID IN VARIOUS FORMS** Reacts with acids, giving off flammable gas (→ *hydrogen*).	
	TLV-TWA	0.002 mg/m³ A2
	Absorption route: Can enter the body by inhalation or ingestion. **Immediate effects:** Irritates the eyes and respiratory tract. Inhalation can cause chronic lung disorders and inflammatory symptoms in many other organs. Where skin is broken can cause skin disorders which are slow to heal. **Effects of prolonged/repeated exposure:** Prolonged or repeated contact can cause skin disorders.	
Gross formula: Be		

HAZARDS/SYMPTOMS	PREVENTIVE MEASURES	FIRE EXTINGUISHING/FIRST AID
Fire: combustible.	keep away from open flame, no smoking.	special powder extinguisher, DO NOT USE WATER-BASED EXTINGUISHERS.
	STRICT HYGIENE, KEEP DUST UNDER CONTROL	IN ALL CASES CALL IN A DOCTOR
Inhalation: cough, shortness of breath.	local exhaust or respiratory protection.	fresh air, rest, take to hospital.
Skin: redness, pain.	gloves, protective clothing.	remove contaminated clothing, flush skin with water or shower, call a doctor.
Eyes: redness, pain.	face shield, or combined eye and respiratory protection.	flush with water, take to a doctor.
Ingestion: cough, shortness of breath.	do not eat, drink or smoke while working.	rinse mouth, call a doctor or take to hospital.

SPILLAGE	STORAGE	LABELING / NFPA
clean up spillage, carefully collect remainder; (additional individual protection: CHEMICAL SUIT).	keep separate from acids.	R: 26/27-37-39 S: 26-28-45 ☠ Toxic

NOTES
Do not take work clothing home. Reacts violently with extinguishing agents such as water, foam, carbon dioxide and halons. Depending on the degree of exposure, regular medical checkups are advisable. Unbreakable packaging preferred; if breakable, keep in unbreakable container. The warnings and measures on this card apply particularly to beryllium powder and processes in which beryllium metal can give off finely dispersed particles, fumes or vapor.

Transport Emergency Card TEC(R)-61G12   HI: 60; UN-number: 1567

BERYLLIUM OXIDE

PHYSICAL PROPERTIES		IMPORTANT CHARACTERISTICS	
Boiling point, °C	3900	**WHITE POWDER**	
Melting point, °C	2530	When heated above 1700°C gives off beryllium oxide vapor in concentrations dangerous to health. Such concentrations are reached at much lower temperatures in the presence of water vapor.	
Relative density (water = 1)	3.0		
Solubility in water	none		
Relative molecular mass	25.0		
		TLV-TWA	0.002 mg/m³ A2 ✳
		Absorption route: Can enter the body by inhalation or ingestion. Evaporation negligible at 20°C, but harmful concentrations of airborne particles can build up rapidly. **Immediate effects:** Irritates the eyes and respiratory tract. Inhalation can cause lung edema. Where skin is broken can cause skin disorders which are slow to heal. **Effects of prolonged/repeated exposure:** Prolonged or repeated contact can cause skin disorders. Can cause chronic lung disorders and inflammatory symptoms in many other organs. Has been found to cause a type of lung cancer in certain animal species under certain circumstances.	
Gross formula:	BeO		

HAZARDS/SYMPTOMS	PREVENTIVE MEASURES	FIRE EXTINGUISHING/FIRST AID
Fire: non-combustible.		in case of fire in immediate vicinity: preferably not water-based extinguishers.
	STRICT HYGIENE, KEEP DUST UNDER CONTROL	IN ALL CASES CALL IN A DOCTOR
Inhalation: cough, shortness of breath.	local exhaust or respiratory protection.	fresh air, rest, take to hospital.
Skin: redness, pain.	gloves, protective clothing.	remove contaminated clothing, flush skin with water or shower, call a doctor.
Eyes: redness, pain.	face shield, or combined eye and respiratory protection.	flush with water, take to a doctor.
Ingestion: cough, shortness of breath.	do not eat, drink or smoke while working.	rinse mouth, call a doctor or take to hospital.

SPILLAGE	STORAGE	LABELING / NFPA
evacuate area, call in an expert, clean up spillage, carefully collect remainder; (additional individual protection: CHEMICAL SUIT).		R: 26/27-37-39 S: 26-28-45 Note A Toxic

NOTES

✳ TLV equals that of beryllium. Do not take work clothing home. Depending on the degree of exposure, regular medical checkups are advisable. Lung edema symptoms usually develop several hours later and are aggravated by physical exertion: rest and hospitalization essential. As first aid, a doctor or authorized person should consider administering a corticosteroid spray. Unbreakable packaging preferred; if breakable, keep in unbreakable container.

Transport Emergency Card TEC(R)-61G11

UN-number: 1566

BICYCLOHEPTADIENE

PHYSICAL PROPERTIES	IMPORTANT CHARACTERISTICS
Boiling point, °C　　　　　　　　　　　90 Melting point, °C　　　　　　　　　　− 20 Flash point, °C　　　　　　　　　　　− 21 Auto-ignition temperature, °C　　　> 350 Relative density (water = 1)　　　　　0.9 Relative vapor density (air = 1)　　　3.14 Relative density at 20 °C of saturated mixture vapor/air (air = 1)　　　　　　　　1.148 Vapor pressure, mm Hg at 20 °C　　　52 Solubility in water　　　　　　　　none Explosive limits, vol% in air　　1.6 - 6.3 Electrical conductivity, pS/m　　　< 10⁴ Relative molecular mass　　　　　　92.1	**COLORLESS LIQUID WITH CHARACTERISTIC ODOR** Vapor is heavier than air and spreads at ground level, with risk of ignition at a distance. Flow, agitation etc. can cause build-up of electrostatic charge due to liquid's low conductivity. Do not use compressed air when filling, emptying or processing. Forms isomers when heated, incl. → *cycloheptatriene*. Reacts violently with strong oxidants and many other compounds.
	TLV-TWA　　　　　　　　not available
	Absorption route: Can enter the body by inhalation or ingestion. **Immediate effects:** Liquid destroys the skin's natural oils. In substantial concentrations can impair consciousness. If liquid is swallowed, droplets can enter the lungs, with risk of pneumonia. **Effects of prolonged/repeated exposure:** Can cause liver damage.
Gross formula:　　　　　　　C_7H_8	

HAZARDS/SYMPTOMS	PREVENTIVE MEASURES	FIRE EXTINGUISHING/FIRST AID
Fire: extremely flammable.	keep away from open flame and sparks, no smoking.	powder, AFFF, foam, carbon dioxide, (halons).
Explosion: forms explosive air-vapor mixtures.	sealed machinery, ventilation, explosion-proof electrical equipment and lighting, grounding.	in case of fire: keep tanks/drums cool by spraying with water.
Inhalation: cough, headache, drowsiness.	ventilation, local exhaust or respiratory protection.	fresh air, rest, artificial respiration if necessary, call a doctor.
Skin: redness, pain.	gloves.	remove contaminated clothing, flush skin with water or shower, refer to doctor if necessary.
Eyes: redness, pain.	safety glasses.	flush with water, take to a doctor.
Ingestion: nausea, cramps.	do not eat, drink or smoke while working.	rinse mouth, DO NOT induce vomiting, take immediately to hospital.

SPILLAGE	STORAGE	LABELING / NFPA
collect leakage in sealable containers, soak up with sand or other inert absorbent and remove to safe place; (additional individual protection: breathing apparatus).	keep in fireproof place, separate from oxidants; add a stabilizer.	

NOTES
Alcohol consumption increases toxic effects.

Transport Emergency Card TEC(R)-155　　　　　　　　　　　　　　　　　　　　　　　HI: 33; UN-number: 2251

BISMUTHTRIBROMIDE

PHYSICAL PROPERTIES	IMPORTANT CHARACTERISTICS	
Boiling point, °C 453 Melting point, °C 218 Relative density (water = 1) 5.7 Vapor pressure, mm Hg at 282 °C 9.9 Solubility in water reaction Relative molecular mass 448.7	**YELLOW HYGROSCOPIC POWDER WITH PUNGENT ODOR** Decomposes when heated above boiling point, giving off toxic vapors. Reacts with water or moist air, forming bismuthoxybromide and corrosive → *hydrogen bromide*. Attacks many metals, esp. in the presence of moisture.	
	TLV-TWA not available	
	Absorption route: Can enter the body by ingestion. Evaporation negligible at 20°C, but harmfull atmospheric concentrations build up fairly slowly on evaporation at 20°C, but harmful concentrations can build up rapidly on contact with moisture or in the form of airborne particles. **Immediate effects:** Corrosive to the eyes and skin.	
Gross formula: BiBr$_3$		

HAZARDS/SYMPTOMS	PREVENTIVE MEASURES	FIRE EXTINGUISHING/FIRST AID
Fire: non-combustible.		in case of fire in immediate vicinity: use any extinguishing agent.
Inhalation:	local exhaust or respiratory protection.	fresh air, rest, call a doctor.
Skin: *corrosive.*	gloves, protective clothing.	remove contaminated clothing, flush skin with water or shower.
Eyes: *corrosive.*	face shield.	flush with water, take to a doctor.
Ingestion: *corrosive.*		rinse mouth, call a doctor.

SPILLAGE	STORAGE	LABELING / NFPA
clean up spillage, carefully collect remainder; (additional individual protection: breathing apparatus).	keep dry.	

NOTES
Insufficient toxicological data on harmful effects to humans. Use airtight packaging.

BORAX

PHYSICAL PROPERTIES	IMPORTANT CHARACTERISTICS	
Boiling point, °C 320 Melting point, °C 75 Relative density (water = 1) 1.7 Solubility in water, g/100 ml 6 Relative molecular mass 381.4	**BLUEISH-GRAY OR GREEN CRYSTALS**	
	TLV-TWA	5 mg/m³
	Absorption route: Can enter the body by inhalation or ingestion. **Immediate effects:** Affects the nervous system. In serious cases risk of seizures, unconsciousness and death. **Effects of prolonged/repeated exposure:** Can cause kidney and brain damage.	
Gross formula: $B_4Na_2O_7.10H_2O$		

HAZARDS/SYMPTOMS	PREVENTIVE MEASURES	FIRE EXTINGUISHING/FIRST AID
Fire: non-combustible.		in case of fire in immediate vicinity: use any extinguishing agent.
Inhalation: cough, nausea, vomiting.	local exhaust or respiratory protection.	fresh air, rest, artificial respiration if necessary, take to hospital.
Skin:	gloves.	flush skin with water or shower.
Eyes: redness, pain.	goggles, or combined eye and respiratory protection.	flush with water, take to a doctor.
Ingestion: diarrhea, nausea, vomiting, drowsiness, muscle cramps.		rinse mouth, call a doctor or take to hospital.

SPILLAGE	STORAGE	LABELING / NFPA
clean up spillage, flush away remainder with water; (additional individual protection: P2 respirator).		

NOTES
From 60 °C to 200°C loss of crystal water. Apparent melting point due to loss of water of crystallization.

BORIC ACID

PHYSICAL PROPERTIES	IMPORTANT CHARACTERISTICS	
Melting point (decomposes), °C 169 Relative density (water = 1) 1.4 Vapor pressure, mm Hg at 20 °C <0.0076 Solubility in water, g/100 ml at 20 °C 5 Relative molecular mass 61.8	**COLORLESS CRYSTALS OR WHITE POWDER** Decomposes when heated above 100°C, forming water and boric anhydride.	
	TLV-TWA not available	
	Absorption route: Can enter the body by inhalation or ingestion. Evaporation negligible at 20°C, but harmful concentrations of airborne particles can build up rapidly. **Immediate effects:** In serious cases risk of seizures, unconsciousness and death. **Effects of prolonged/repeated exposure:** Can cause liver, kidney and brain damage.	
Gross formula: BH_3O_3		

HAZARDS/SYMPTOMS	PREVENTIVE MEASURES	FIRE EXTINGUISHING/FIRST AID
Fire: non-combustible.		in case of fire in immediate vicinity: use any extinguishing agent.
Inhalation: sore throat, cough.	local exhaust or respiratory protection.	fresh air, rest, consult a doctor if necessary.
Skin: redness.	gloves.	remove contaminated clothing, flush skin with water or shower, call a doctor.
Eyes: redness, pain.	goggles.	flush with water, take to a doctor if necessary.
Ingestion: diarrhea, cramps, vomiting.		rinse mouth, immediately take to hospital.

SPILLAGE	STORAGE	LABELING / NFPA
clean up spillage, flush away remainder with water; (additional individual protection: P2 respirator).		

NOTES

BORIUMTRIFLUORIDE
(cylinder)

PHYSICAL PROPERTIES	IMPORTANT CHARACTERISTICS
Boiling point, °C − 100 Melting point, °C − 127 Relative density (water = 1) 1.2 Relative vapor density (air = 1) 2.4 Solubility in water reaction Relative molecular mass 67.8	**COLORLESS COMPRESSED GAS WITH PUNGENT ODOR, GIVING OFF ON EXPOSURE TO AIR** Gas is heavier than air. Decomposes when heated and on contact with moist air and water, giving off toxic and corrosive fumes (→ *boric acid*, → *hydrogen fluoride*, → *fluoroboric acid*). Reacts violently with many substances (incl. metals), with risk of fire and explosion. Catalyzes many reactions such as polymerization, with risk of fire and explosion.
	TLV-TWA 1 ppm 2.8 mg/m³ C
	Absorption route: Can enter the body by inhalation. Harmful atmospheric concentrations can build up very rapidly if gas is released. **Immediate effects:** Corrosive to the eyes, skin and respiratory tract. Inhalation of vapor/fumes can cause lung edema. Can affect the upper respiratory tract, causing nose-bleed. Can cause death. Keep under medical observation. **Effects of prolonged/repeated exposure:** Prolonged or repeated contact to gas/fumes can cause lung disorders. Can affect the kidneys.
Gross formula: BF$_3$	

HAZARDS/SYMPTOMS	PREVENTIVE MEASURES	FIRE EXTINGUISHING/FIRST AID
Fire: non-combustible.		in case of fire in immediate vicinity: use any extinguishing agent.
Explosion:		in case of fire: keep undamaged cylinders cool by spraying with water, but DO NOT spray substance with water.
	STRICT HYGIENE	IN ALL CASES CALL IN A DOCTOR
Inhalation: *corrosive*, cought, sore throat, severe breathing difficulties.	ventilation, local exhaust or respiratory protection.	fresh air, rest, place in half-sitting position, artificial respiration if necessary, take immediately to hospital.
Skin: *corrosive*, redness, burning sensation, pain.	gloves, protective clothing.	remove contaminated clothing, flush skin with water or shower, refer to a doctor.
Eyes: *corrosive*; redness, pain, impaired vision.	face shield, or combined eye and respiratory protection.	flush thoroughly with water (remove contact lenses if easy), take to a doctor.

SPILLAGE	STORAGE	LABELING / NFPA
evacuate area, call in an expert, ventilate, combat gas cloud with water curtain; (additional individual protection: CHEMICAL SUIT).	if stored indoors: keep in cool, fireproof place.	R: 14-26-35 S: 9-26-28-36-45 ☣ Toxic NFPA: 4 / 0 / 1

NOTES
Boron trifluoride forms complexes with ether, acidic acid, phenol, etc. Forms monohydrate and/or dihydrate with substandard water. The dihydrate is a corrosive liquid with melting point at 6°C. TLV is maximum and must not be exceeded. Lung edema symptoms usually develop several hours later and are aggravated by physical exertion: rest and hospitalization essential. Do not spray leaking cylinder with water (to avoid corrosion). The boron trifluoride dihydrate has HI no. 80 and UN no. 2851.

Transport Emergency Card TEC(R)-20G02

HI: 286; UN-number: 1008

BORON TRICHLORIDE
(cylinder)

PHYSICAL PROPERTIES	IMPORTANT CHARACTERISTICS	
Boiling point, °C 12.5 Melting point, °C − 107 Relative density (water = 1) 1.4 Relative vapor density (air = 1) 4.1 Vapor pressure, mm Hg at 20 °C 988 Solubility in water reaction Relative molecular mass 117.2	**COLORLESS COMPRESSED LIQUEFIED GAS WITH PUNGENT ODOR** Decomposes in flame or when heated, giving off toxic gases (*chlorine*). Reacts violently with water, giving off → *boric acid* and → *hydrochloric acid*. Attacks most metals in the presence of moisture.	
	TLV-TWA	not available
	Absorption route: Can enter the body by inhalation or ingestion or through the skin. Harmful atmospheric concentrations can build up very rapidly if gas is released. **Immediate effects:** Corrosive to the eyes, skin and respiratory tract. Inhalation can cause lung edema. In serious cases risk of death. **Effects of prolonged/repeated exposure:** Can cause liver, kidney and brain damage.	
Gross formula: BCl_3		

HAZARDS/SYMPTOMS	PREVENTIVE MEASURES	FIRE EXTINGUISHING/FIRST AID
Fire: non-combustible.		in case of fire in immediate vicinity: preferably no water.
Explosion:		in case of fire: keep undamaged cylinders cool by spraying with water, but DO NOT spray substance with water.
Inhalation: *corrosive*; sore throat, cough, severe breathing difficulties, vomiting.	ventilation, local exhaust or respiratory protection.	fresh air, rest, place in half-sitting position, take to hospital.
Skin: *corrosive, is absorbed*; redness, pain, burns.	gloves, protective clothing.	remove contaminated clothing, flush skin with water or shower, call a doctor.
Eyes: *corrosive*; redness, pain, impaired vision.	face shield, or combined eye and respiratory protection.	flush with water, take to a doctor.

SPILLAGE	STORAGE	LABELING / NFPA
evacuate area, call in an expert, ventilate, under no circumstances spray liquid with water, combat gas cloud with water curtain; (additional individual protection: CHEMICAL SUIT).	if stored indoors, keep in cool, fireproof place.	◇

NOTES
Depending on the degree of exposure, regular medical checkups are advisable. Lung edema symptoms usually develop several hours later and are aggravated by physical exertion: rest and hospitalization essential. As first aid, a doctor or authorized person should consider administering a corticosteroid spray. Do not spray leaking cylinder with water (to avoid corrosion). Turn leaking cylinder so that leak is on top to prevent liquid boron trichloride escaping. Use cylinder with special fittings.

Transport Emergency Card TEC(R)-20G01

HI: 286; UN-number: 1741

BORON TRIFLUORIDE ETHERATE

PHYSICAL PROPERTIES	IMPORTANT CHARACTERISTICS
Boiling point, °C 126 Melting point, °C −60 Flash point, °C 64 Relative density (water = 1) 1.1 Relative vapor density (air = 1) 4.9 Relative density at 20 °C of saturated mixture vapor/air (air = 1) 1.04 Vapor pressure, mm Hg at 20 °C ca. 9.1 Solubility in water reaction Relative molecular mass 141.9	**COLORLESS FUMING LIQUID** Vapor mixes readily with air. Able to form peroxides. Decomposes when heated or burned, giving off toxic, corrosive vapors. Reacts violently with water and steam, giving off toxic, corrosive and flammable vapors, with risk of fire and explosion. Reacts violently with oxidants. Reacts with air to form corrosive vapors (→ *hydrogen fluoride*).
	TLV-TWA not available
	Absorption route: Can enter the body by inhalation or ingestion. Harmful atmospheric concentrations can build up very rapidly on evaporation at 20°C. **Immediate effects:** Corrosive to the eyes, skin and respiratory tract. Inhalation of vapor/fumes can cause severe breathing difficulties (lung edema). In serious cases risk of death. **Effects of prolonged/repeated exposure:** Can cause kidney damage.
Gross formula: $C_4H_{10}BF_3O$	

HAZARDS/SYMPTOMS	PREVENTIVE MEASURES	FIRE EXTINGUISHING/FIRST AID
Fire: combustible.	keep away from open flame, no smoking, avoid contact with water and steam.	powder, carbon dioxide, (halons), DO NOT USE WATER-BASED EXTINGUISHERS.
Explosion: above 64°C: forms explosive air-vapor mixtures, contact with water or steam can cause explosion.	above 64°C: sealed machinery, ventilation, explosion-proof electrical equipment.	in case of fire: keep tanks/drums cool by spraying with water, but DO NOT spray substance with water.
	STRICT HYGIENE	
Inhalation: *corrosive*; cough, severe breathing difficulties.	ventilation, local exhaust or respiratory protection.	fresh air, rest, place in half-sitting position, take to hospital.
Skin: *corrosive*; redness, pain, serious burns.	gloves, protective clothing.	remove contaminated clothing, flush skin with water or shower, call a doctor, take to hospital if necessary.
Eyes: *corrosive*; redness, pain, impaired vision.	face shield, or combined eye and respiratory protection.	flush with water, take to a doctor.
Ingestion: *corrosive*; sore throat, diarrhea, vomiting.	do not eat, drink or smoke while working.	rinse mouth, take immediately to hospital.

SPILLAGE	STORAGE	LABELING / NFPA
collect leakage in sealable containers, soak up with dry sand or other inert absorbent and remove to safe place; (additional individual protection: breathing apparatus).	keep in dry, fireproof place, separate from oxidants.	![NFPA diamond: 2 top, 3 left, 1 right, W bottom]

NOTES
Before distilling check for peroxides; if found, render harmless. TLV for *boron trifluoride* (1ppm C) applicable in most cases. Lung edema symptoms usually develop several hours later and are aggravated by physical exertion: rest and hospitalization essential. As first aid, a doctor or authorized person should consider administering a corticosteroid spray. Use airtight packaging. Unbreakable packaging preferred; if breakable, keep in unbreakable container.

Transport Emergency Card TEC(R)-80G20B

HI: 83; UN-number: 2604

BROMINE

PHYSICAL PROPERTIES	IMPORTANT CHARACTERISTICS	
Boiling point, °C 59	**FUMING REDDISH-BROWN LIQUID WITH PUNGENT ODOR**	
Melting point, °C −7	Vapor is heavier than air. Reacts violently (1) with many organic compounds, many metals and	
Relative density (water = 1) 3.1	phosphorus, and (2) when heated with hydrogen, with risk of fire and explosion in both cases.	
Relative vapor density (air = 1) 5.5	Attacks many materials. In aqueous solution forms → *hydrobromic acid* and *oxygen*.	
Relative density at 20 °C of saturated mixture		
vapor/air (air = 1) 2.0	TLV-TWA	0.1 ppm	0.66 mg/m³
Vapor pressure, mm Hg at 20 °C 16.7	STEL	0.3 ppm	2.0 mg/m³
Solubility in water, g/100 ml at 20 °C 4		
Relative molecular mass 159.8		
	Absorption route: Can enter the body by inhalation or ingestion or through the skin. Harmful atmospheric concentrations can build up very rapidly on evaporation at 20°C.	
	Immediate effects: Corrosive to the eyes, skin and respiratory tract. Inhalation of vapor/fumes can cause severe breathing difficulties (lung edema). In serious cases risk of death.	
	Effects of prolonged/repeated exposure: Prolonged or repeated contact can cause skin disorders.	
Gross formula: Br$_2$		

HAZARDS/SYMPTOMS	PREVENTIVE MEASURES	FIRE EXTINGUISHING/FIRST AID
Fire: non-combustible, many chemical reactions can cause fire and explosion.		in case of fire in immediate vicinity: use any extinguishing agent.
Explosion: many chemical reactions can cause explosion.		in case of fire: keep tanks/drums cool by spraying with water.
	STRICT HYGIENE	IN ALL CASES CALL A DOCTOR
Inhalation: *corrosive*; sore throat, cough, shortness of breath, severe breathing difficulties, dizziness.	ventilation, local exhaust or respiratory protection.	fresh air, rest, place in half-sitting position, take to hospital.
Skin: *is absorbed, corrosive*; redness, pain, serious burns.	gloves, protective clothing.	remove contaminated clothing, flush skin with water or·shower, take to a doctor or to hospital.
Eyes: *corrosive*; redness, pain, impaired vision.	face shield, or combined eye and respiratory protection.	flush with water, take to a doctor.
Ingestion: *corrosive*; sore throat, abdominal cramps, vomiting.	do not eat, drink or smoke while working.	rinse mouth, take immediately to hospital.

SPILLAGE	STORAGE	LABELING / NFPA
evacuate area, call in an expert, collect leakage in sealable containers, neutralize spillage with limewater and flush away; (additional individual protection: CHEMICAL SUIT).	keep cool and dry, separate from combustible substances and reducing agents; ventilate at floor level.	R: 26-35 S: 7/9-26 Corrosive

NOTES
Odor limit is above TLV. Lung edema symptoms usually develop several hours later and are aggravated by physical exertion: rest and hospitalization essential. As first aid, a doctor or authorized person should consider administering a corticosteroid spray. Use airtight packaging made of special material.

Transport Emergency Card TEC(R)-65

HI: 886; UN-number: 1744

BrCN
bromocyan
bromocyanide
bromocyanogen
cyanogen bromide

BROMINE CYANIDE

PHYSICAL PROPERTIES	IMPORTANT CHARACTERISTICS
Boiling point, °C 62 Melting point, °C 52 Relative density (water = 1) 2.0 Relative vapor density (air = 1) 3.6 Relative density at 20 °C of saturated mixture vapor/air (air = 1) 1.3 Vapor pressure, mm Hg at 20 °C 88 Solubility in water reaction Relative molecular mass 105.9	**COLORLESS CRYSTALS WITH PUNGENT ODOR** Vapor is heavier than air. Able to polymerize. In pure, dry state is fairly stable; in impure state decomposes rapidly, possibly explosively. Decomposes when heated or on contact with water, water vapor or alcohol, giving off toxic and corrosive gases (→ *hydrogen bromide* and → *hydrogen cyanide*). Reacts violently with acids, forming *bromine water* and → *hydrogen cyanide*. Reacts violently with oxidants, anhydrous ammonia, amines, phosgene, etc.
	TLV-TWA not available
	Absorption route: Can enter the body by inhalation or ingestion or through the skin. Harmful atmospheric concentrations can build up very rapidly on evaporation at 20°C. **Immediate effects:** Corrosive to the eyes, skin and respiratory tract. Impedes tissue respiration. Inhalation of vapor/fumes can cause severe breathing difficulties (lung edema). In serious cases risk of seizures, unconsciousness and death.
Gross formula: BrCN	

HAZARDS/SYMPTOMS	PREVENTIVE MEASURES	FIRE EXTINGUISHING/FIRST AID
Fire: non-combustible.		in case of fire in immediate vicinity: avoid direct contact with foam.
Explosion: decomposition or reaction with ammonia etc. can cause explosion.	sealed machinery.	
	STRICT HYGIENE, AVOID ALL CONTACT	IN ALL CASES CALL IN A DOCTOR
Inhalation: *corrosive*; sore throat, cough, severe breathing difficulties, headache, dizziness, nausea, unconsciousness.	ventilation, local exhaust or respiratory protection.	fresh air, rest, place in half-sitting position, artificial respiration if necessary, special treatment, take to hospital.
Skin: *corrosive, is absorbed*; redness, pain, burns, headache, dizziness.	gloves, protective clothing.	remove contaminated clothing, flush skin with water or shower, call a doctor or take directly to hospital.
Eyes: *corrosive*; redness, pain, impaired vision.	face shield, or combined eye and respiratory protection.	flush with water, take to a doctor if necessary.
Ingestion: *corrosive*; sore throat, abdominal pain, diarrhea, severe breathing difficulties, nausea, vomiting, unconsciousness.	do not eat, drink or smoke while working.	rinse mouth, rest, special treatment, take immediately to hospital.

SPILLAGE	STORAGE	LABELING / NFPA

evacuate area, call in an expert, render harmless with sodium hypochlorite solution or calcium hypochlorite (10%) and after verifying results flush away with water; (additional individual protection: CHEMICAL SUIT).	if stored indoors, keep in dry, fireproof place, separate from acids and as much as possible from other substances; ventilate.	
		
		         0
		   3 1

NOTES
Depending on the degree of exposure, regular medical checkups are advisable. Lung edema symptoms usually develop several hours later and are aggravated by physical exertion: rest and hospitalization essential. As first aid, a doctor or authorized person should consider administering a corticosteroid spray. Special first aid required in the event of poisoning: antidotes must be available (with instructions). Use airtight packaging. Unbreakable packaging preferred; if breakable, keep in unbreakable container.

Transport Emergency Card TEC(R)-61G09   **UN-number: 1889**

BROMO ACETONE

PHYSICAL PROPERTIES	IMPORTANT CHARACTERISTICS
Boiling point (decomposes), °C 137 Melting point, °C −37 Flash point, °C 45 Relative density (water = 1) 1.6 Relative vapor density (air = 1) 4.8 Relative density at 20 °C of saturated mixture vapor/air (air = 1) 1.04 Vapor pressure, mm Hg at 20 °C 9.1 Solubility in water poor Explosive limits, vol% in air ?-? Relative molecular mass 137.0	**COLORLESS LIQUID WITH PUNGENT ODOR, TURNING PURPLE ON EXPOSURE TO AIR** Vapor mixes readily with air. Corrosive and toxic vapor are a product of combustion (→ *hydrogen bromide*). Reacts violently with oxidants.
	TLV-TWA not available
	Absorption route: Can enter the body by inhalation or ingestion. Harmful atmospheric concentrations can build up fairly rapidly on evaporation at approx. 20°C - even more rapidly in aerosol form. **Immediate effects:** Corrosive to the eyes, skin and respiratory tract. Inhalation of vapor/fumes can cause severe breathing difficulties (lung edema). In serious cases risk of death. **Effects of prolonged/repeated exposure:** Prolonged or repeated contact can cause skin disorders.
Gross formula: C_3H_5BrO	

HAZARDS/SYMPTOMS	PREVENTIVE MEASURES	FIRE EXTINGUISHING/FIRST AID
Fire: flammable.	keep away from open flame and sparks, no smoking.	powder, water spray, foam, carbon dioxide, (halons).
Explosion: above 45°C: forms explosive air-vapor mixtures.	above 45°C: sealed machinery, ventilation, explosion-proof electrical equipment.	in case of fire: keep tanks/drums cool by spraying with water.
		IN ALL CASES CALL IN A DOCTOR
Inhalation: *corrosive*; sore throat, cough, shortness of breath, severe breathing difficulties, headache, dizziness.	ventilation, local exhaust or respiratory protection.	fresh air, rest, place in half-sitting position, take to hospital.
Skin: *corrosive*; redness, pain, serious burns.	gloves, protective clothing.	remove contaminated clothing, flush skin with water or shower, call a doctor or take to hospital.
Eyes: *corrosive*; redness, pain, impaired vision.	face shield, or combined eye and respiratory protection.	flush with water, take to a doctor.
Ingestion: *corrosive*; sore throat, abdominal pain, diarrhea.		rinse mouth, immediately take to hospital.

SPILLAGE	STORAGE	LABELING / NFPA
collect leakage in sealable containers, soak up with sand or other inert absorbent and remove to safe place; (additional individual protection: CHEMICAL SUIT).	keep in dark, fireproof place, separate from oxidants.	

NOTES
Lung edema symptoms usually develop several hours later and are aggravated by physical exertion: rest and hospitalization essential. As first aid, a doctor or authorized person should consider administering a corticosteroid spray. Bromo acetone is a tear gas.

Transport Emergency Card TEC(R)-61G07B

HI: 60; UN-number: 1569

o-, m-, p-BROMOANILINE

PHYSICAL PROPERTIES	IMPORTANT CHARACTERISTICS	
Boiling point, °C 230-250 Melting point, °C 20-66 Relative density (water = 1) 1.6 Relative vapor density (air = 1) 5.9 Vapor pressure, mm Hg at 120 °C 23 Solubility in water none Relative molecular mass 172.0 Log P octanol/water 2.2	**COLORLESS LIQUID OR COLORLESS CRYSTALS** Toxic gases (→ *hydrogen bromide* and → *nitrous vapors*) are a product of combustion. Decomposes when heated, giving off toxic vapors. Reacts with acids, forming toxic gases.	
	TLV-TWA not available	
	Absorption route: Can enter the body by inhalation or ingestion or through the skin. Harmful atmospheric concentrations build up very slowly, if at all, on evaporation at approx. 20°C, but much more rapidly in aerosol form. **Immediate effects:** In serious cases risk of unconsciousness and death. **Effects of prolonged/repeated exposure:** Can affect the blood.	
Gross formula: C_6H_6BrN		

HAZARDS/SYMPTOMS	PREVENTIVE MEASURES	FIRE EXTINGUISHING/FIRST AID
Fire: combustible.	keep away from open flame, no smoking.	powder, alcohol-resistant foam, water spray, carbon dioxide, (halons).
	STRICT HYGIENE, KEEP DUST UNDER CONTROL	
Inhalation: shortness of breath, headache, dizziness, blue skin, feeling of weakness.	local exhaust or respiratory protection.	fresh air, rest, call a doctor or take to hospital.
Skin: *is absorbed*; see also 'Inhalation'.	gloves, protective clothing.	remove contaminated clothing, wash skin with soap and water, call a doctor or take to hospital.
Eyes: redness, pain.	face shield.	flush with water, take to a doctor.
Ingestion: unconsciousness, see also 'Inhalation'.	do not eat, drink or smoke while working.	rinse mouth, immediately take to hospital.

SPILLAGE	STORAGE	LABELING / NFPA
clean up spillage, carefully collect remainder; (additional individual protection: P3 respirator).	keep separate from acids.	

NOTES
CAS Nos: o-bromoaniline 615-36-1; m-bromoaniline 591-19-5; p-bromoaniline 106-40-1. Depending on the degree of exposure, regular medical checkups are advisable. Effects on blood due to formation of methemoglobin. Special first aid required in the event of poisoning: antidotes must be available (with instructions).

Transport Emergency Card TEC(R)-61G11

CAS-No: [108-86-1]
monobromobenzene
phenyl bromide

C_6H_5Br

BROMOBENZENE

PHYSICAL PROPERTIES		IMPORTANT CHARACTERISTICS	
Boiling point, °C	156	**COLORLESS LIQUID WITH CHARACTERISTIC ODOR**	
Melting point, °C	−31	Vapor mixes readily with air. Flow, agitation etc. can cause build-up of electrostatic charge due to	
Flash point, °C	51	liquid's low conductivity. Toxic gases (→ *hydrogen bromide*) are a product of combustion.	
Auto-ignition temperature, °C	565	Reacts violently with strong oxidants and alkaline-earth and alkali metals.	
Relative density (water = 1)	1.5		
Relative vapor density (air = 1)	5.4	TLV-TWA	not available
Relative density at 20 °C of saturated mixture			
vapor/air (air = 1)	1.02	**Absorption route:** Can enter the body by inhalation or ingestion or through the skin. Harmful	
Vapor pressure, mm Hg at 20 °C	3.3	atmospheric concentrations build up fairly slowly on evaporation at 20°C, but much more rapidly	
Solubility in water, g/100 ml at 30 °C	0.05	in aerosol form.	
Explosive limits, vol% in air	0.5-2.5	**Immediate effects:** Irritates the eyes, skin and respiratory tract. Affects the nervous system. If	
Electrical conductivity, pS/m	1.2×10^3	liquid is swallowed, droplets can enter the lungs, with risk of pneumonia.	
Relative molecular mass	157.0	**Effects of prolonged/repeated exposure:** Can cause liver damage.	
Log P octanol/water	3.0		
Gross formula:	C_6H_5Br		

HAZARDS/SYMPTOMS	PREVENTIVE MEASURES	FIRE EXTINGUISHING/FIRST AID
Fire: flammable.	keep away from open flame and sparks, no smoking.	powder, water spray, foam, carbon dioxide, (halons).
Explosion: above 51°C: forms explosive air-vapor mixtures.	above 51°C: sealed machinery, ventilation, explosion-proof electrical equipment, grounding.	in case of fire: keep tanks/drums cool by spraying with water.
Inhalation: sore throat, cough, shortness of breath, heachache, drowsiness, sleepiness.	ventilation, local exhaust or respiratory protection.	fresh air, rest, call a doctor.
Skin: *is absorbed*; redness.	gloves.	remove contaminated clothing, wash skin with soap and water, call a doctor if necessary.
Eyes: redness, pain.	safety glasses.	flush with water, take to a doctor if necessary.
Ingestion: abdominal pain, nausea; see also 'Inhalation'.		rinse mouth, call a doctor or take to hospital.

SPILLAGE	STORAGE	LABELING / NFPA
collect leakage in sealable containers, soak up with sand or other inert absorbent and remove to safe place.	keep in fireproof place, separate from oxidants.	R: 10-38 ✖ Irritant

NOTES
Under no circumstances use near flame or hot surface or when welding.

Transport Emergency Card TEC(R)-30G36

HI: 30; UN-number: 2514

127

1-BROMO-3-CHLOROPROPANE

PHYSICAL PROPERTIES	IMPORTANT CHARACTERISTICS	
Boiling point, °C 144	**COLORLESS LIQUID**	
Melting point, °C −59	Vapor mixes readily with air. Decomposes in flame or on hot surface, giving off toxic and	
Relative density (water = 1) 1.6	corrosive vapors (→ *hydrogen bromide*, → *hydrogen chloride*). Decomposes when heated above	
Relative vapor density (air = 1) 5.4	boiling point, giving off corrosive vapors.	
Relative density at 20 °C of saturated mixture		
vapor/air (air = 1) 1.03	TLV-TWA not available	
Vapor pressure, mm Hg at 20 °C 5.0		
Solubility in water none		
Relative molecular mass 157.4	**Absorption route:** Can enter the body by inhalation or ingestion or through the skin. Harmful	
	atmospheric concentrations build up fairly slowly on evaporation at 20° C, but much more rapidly	
	in aerosol form.	
	Immediate effects: Irritates the eyes, skin and respiratory tract. Liquid destroys the skin's natural	
	oils. Affects the nervous system. In serious cases risk of unconsciousness.	
	Effects of prolonged/repeated exposure: Can cause liver and kidney damage.	
Gross formula: C_3H_6BrCl		

HAZARDS/SYMPTOMS	PREVENTIVE MEASURES	FIRE EXTINGUISHING/FIRST AID
Fire: non-combustible.		in case of fire in immediate vicinity: use any extinguishing agent.
Inhalation: sore throat, dizziness, drowsiness, unconsciousness.	ventilation, local exhaust or respiratory protection.	fresh air, rest, call a doctor or take to hospital.
Skin: *is absorbed*; redness, pain.	gloves.	remove contaminated clothing, flush skin with water or shower, call a doctor.
Eyes: redness, pain.	safety glasses.	flush with water, take to a doctor if necessary.
Ingestion: abdominal pain, nausea, dizziness, unconsciousness.		rinse mouth, take immediately to hospital.

SPILLAGE	STORAGE	LABELING / NFPA
collect leakage in sealable containers, soak up with sand or other inert absorbent and remove to safe place.	ventilate at floor level.	

NOTES
Alcohol consumption increases toxic effects. Depending on the degree of exposure, regular medical checkups are advisable. Under no circumstances use near flame or hot surface or when welding.

Transport Emergency Card TEC(R)-888

HI: 60; UN-number: 2688

CAS-No: [357-57-3]
2,3-dimethoxystrichnidin-10-one
brucine hydrate
2,3-dimethoxystrychnine

$C_{23}H_{26}O_4N_2 \cdot 4H_2O$

BRUCINE
(quarternary hydrate)

PHYSICAL PROPERTIES	IMPORTANT CHARACTERISTICS
Melting point, °C 105 (see notes) Relative density (water = 1) > 1 Solubility in water poor Relative molecular mass 466.5 Log P octanol/water 0.4	**COLORLESS TO WHITE CRYSTALS** Finely dispersed in air can cause dust explosion. In dry state can form electrostatic charge if stirred, transported pneumatically, poured etc. Decomposes when heated or burned, giving off toxic vapors (→ *nitrous vapors*). Reacts with strong oxidants.

TLV-TWA	not available

Absorption route: Can enter the body by inhalation or ingestion. Harmful atmospheric concentrations build up very slowly, if at all, on evaporation at approx. 20°C, but harmful concentrations of airborne particles can build up much more rapidly.
Immediate effects: Irritates the eyes and respiratory tract. Affects the nervous system and musculature, causing seizures and severe breathing difficulties. At high exposure levels can cause unconsciousness and death.

Gross formula: $C_{23}H_{26}N_2O_4 \cdot 4H_2O$

HAZARDS/SYMPTOMS	PREVENTIVE MEASURES	FIRE EXTINGUISHING/FIRST AID
Fire: combustible.	keep away from open flame, no smoking.	water spray, powder.
Explosion: finely dispersed particles form explosive mixtures on contact with air.	keep dust from accumulating, closed system, explosion-proof electrical equipment and lighting, prevent buildup of electrostatic charges by grounding, etc.	
	STRICT HYGIENE, AVOID ALL CONTACT	IN ALL CASES CALL IN A DOCTOR
Inhalation: severe breathing difficulties, disorientation, cramps, vomiting, seizures.	local exhaust or respiratory protection.	fresh air, rest, artificial respiration if necessary, take immediately to hospital.
Skin: redness.	gloves.	remove contaminated clothing, flush skin with water and wash with soap and water.
Eyes: redness, pain.	goggles, or combined eye and respiratory protection.	flush thoroughly with water (remove contact lenses if easy), take to a doctor.
Ingestion: abdominal pain, diarrhea; see also 'Inhalation'.		rinse mouth, take immediately to hospital.

SPILLAGE	STORAGE	LABELING / NFPA
clean up spillage, carefully collect remainder; (additional individual protection: respirator with A/P3 filter).	keep separate from oxidants; ventilate at floor level.	R: 26/28 S: 1-13-45 ☠ Toxic

NOTES
Apparent melting point due to loss of water of crystallization. Melting point of anhydrous brucine: 178°C. Log P octanol/water is estimated.

Transport Emergency Card TEC(R)-61G10

UN-number: 1570

biethylene
bivinyl
divinyl
vinylethylene

1,3-BUTADIENE
(cylinder)

PHYSICAL PROPERTIES	IMPORTANT CHARACTERISTICS
Boiling point, °C −5 Melting point, °C − 109 Flash point, °C flammable gas Auto-ignition temperature, °C 415 Relative density (water = 1) 0.6 Relative vapor density (air = 1) 1.9 Vapor pressure, mm Hg at 20 °C 1862 Solubility in water none Explosive limits, vol% in air 1.4-16.3 Minimum ignition energy, mJ 0.13 Relative molecular mass 54.1	**COLORLESS COMPRESSED LIQUEFIED GAS WITH CHARACTERISTIC ODOR** Gas is heavier than air and spreads at ground level, with risk of ignition at a distance, and can accumulate close to ground level, causing oxygen deficiency, with risk of unconsciousness. Readily able to form peroxides and polymerize violently, with risk of fire and explosion. Flow, agitation etc. can cause build-up of electrostatic charge due to liquid's low conductivity. At fast flow rates can ignite spontaneously on exposure to air. Reacts violently with oxidants and many other substances.

TLV-TWA	10 ppm	22 mg/m³ A2

Absorption route: Can enter the body by inhalation. Harmful atmospheric concentrations can build up very rapidly if gas is released.
Immediate effects: Irritates the eyes and respiratory tract. Can cause frostbite due to rapid evaporation. In substantial concentrations can impair consciousness.

Gross formula: C_4H_6

HAZARDS/SYMPTOMS	PREVENTIVE MEASURES	FIRE EXTINGUISHING/FIRST AID
Fire: extremely flammable.	keep away from open flame and sparks, no smoking.	shut off supply; if impossible and no danger to surrounding area, allow fire to burn itself out; otherwise extinguish with powder, carbon dioxide, (halons).
Explosion: forms explosive air-gas mixtures.	sealed machinery, ventilation, explosion-proof electrical equipment and lighting, grounding when pumping etc. in liquid form, non-sparking tools.	in case of fire: keep cylinders cool by spraying with water, fight fire from sheltered location.
Inhalation: cough, dizziness, drowsiness.	ventilation, local exhaust or respiratory protection.	fresh air, rest, call a doctor.
Skin: *in case of frostbite:* redness, pain, blisters.	insulating gloves.	*in case of frostbite:* DO NOT remove clothing, flush skin with water or shower, take to hospital.
Eyes: *in case of frostbite:* impaired vision, redness, pain.	face shield, or combined eye and respiratory protection.	flush with water, take to a doctor.

SPILLAGE	STORAGE	LABELING / NFPA
evacuate area, call in an expert, ventilate, under no circumstances spray liquid with water; (additional individual protection: breathing apparatus).	keep in cool, fireproof place; add an inhibitor.	 4 2 2

NOTES
Before distilling check for peroxides; if found, render harmless. On rendering explosive gas/vapor-air mixtures inert, see chapter 'Tables and formulas'. Alcohol consumption increases toxic (narcotic) effects. High atmospheric concentrations, e.g. in poorly ventilated spaces, can cause oxygen deficiency, with risk of unconsciousness. Turn leaking cylinder so that leak is on top to prevent liquid 1,3-butadiene escaping.

Transport Emergency Card TEC(R)-126   **HI: 239; UN-number: 1010**

CAS-No: [106-97-8]
n-butane
butyl hydride
methylethylmethane

C_4H_{10}

BUTANE
(cylinder)

PHYSICAL PROPERTIES		IMPORTANT CHARACTERISTICS	
Boiling point, °C	−0.5	**COLORLESS ODORLESS COMPRESSED LIQUEFIED GAS**	
Melting point, °C	−138	Gas is heavier than air and spreads at ground level, with risk of ignition at a distance. Flow, agitation etc. can cause build-up of electrostatic charge due to liquid's low conductivity.	
Flash point, °C	flammable gas		
Auto-ignition temperature, °C	365		
Relative density (water = 1)	0.58	TLV-TWA	800 ppm	1900 mg/m³
Relative vapor density (air = 1)	2.01		
Vapor pressure, mm Hg at 20 °C	1.6		
Solubility in water	none	**Absorption route:** Can enter the body by inhalation.	
Explosive limits, vol% in air	1.9-8.5	**Immediate effects:** Liquid can cause frostbite due to rapid evaporation. Can cause suffocation due to air saturation.	
Minimum ignition energy, mJ	0.25		
Electrical conductivity, pS/m	< 10⁴		
Relative molecular mass	58.1		
Gross formula:	C_4H_{10}		

HAZARDS/SYMPTOMS	PREVENTIVE MEASURES	FIRE EXTINGUISHING/FIRST AID
Fire: extremely flammable.	keep away from open flame and sparks, no smoking.	shut off supply; if impossible and no danger to surrounding area, allow fire to burn itself out; otherwise extinguish with powder, carbon dioxide, (halons).
Explosion: forms explosive air-gas mixtures.	sealed machinery, ventilation, explosion-proof electrical equipment and lighting, grounding, non-sparking tools.	in case of fire: keep cylinders cool by spraying with water, fight fire from sheltered location.
Inhalation: shortness of breath, headache, drowsiness, unconsciousness.	ventilation, local exhaust or respiratory protection.	fresh air, rest, artificial respiration if necessary, call a doctor.
Skin: *in case of frostbite:* redness, pain, blisters.	insulating gloves.	*in case of frostbite:* DO NOT remove clothing, flush skin with water or shower, take to a doctor.
Eyes: *in case of frostbite:* redness, pain, impaired vision.	face shield.	flush with water, take to a doctor if necessary.

SPILLAGE	STORAGE	LABELING / NFPA
evacuate area, call in an expert, ventilate, under no circumstances spray liquid with water; (additional individual protection: breathing apparatus).	keep in cool, fireproof place.	

NOTES

Use *propane* if ambient temperature drops below 5° C. Under no circumstances warm cylinder. High atmospheric concentrations, e.g. in poorly ventilated spaces, can cause oxygen deficiency, with risk of unconsciousness. The measures on this card also apply to neopentane (2,2-dimethylpropane C_5H_{12}). Turn leaking cylinder so that leak is on top to prevent liquid butane escaping.

Transport Emergency Card TEC(R)-27B

HI: 23; UN-number: 1011

1,2-BUTANEDIOL

PHYSICAL PROPERTIES	IMPORTANT CHARACTERISTICS	
Boiling point, °C 194 Melting point, °C −114 Flash point, °C 90 Auto-ignition temperature, °C ? Relative density (water = 1) 1.01 Relative vapor density (air = 1) 3.1 Relative density at 20 °C of saturated mixture vapor/air (air = 1) 1.0 Vapor pressure, mm Hg at 20 °C 0.076 Solubility in water ∞ Relative molecular mass 90.1	**COLORLESS VISCOUS LIQUID** Vapor mixes readily with air. Reacts with strong oxidants.	
	TLV-TWA not available	
	Absorption route: Can enter the body by inhalation or ingestion. Harmful atmospheric concentrations build up very slowly, if at all, on evaporation at approx. 20°C, but much more rapidly in aerosol form. **Immediate effects:** Irritates the eyes. In substantial concentrations can impair consciousness. **Effects of prolonged/repeated exposure:** Can cause kidney damage.	
Gross formula: $C_4H_{10}O_2$		

HAZARDS/SYMPTOMS	PREVENTIVE MEASURES	FIRE EXTINGUISHING/FIRST AID
Fire: combustible.	keep away from open flame, no smoking.	powder, alcohol-resistant foam, water spray, carbon dioxide, (halons).
Explosion: above 90°C: forms explosive air-vapor mixtures.	above 90°C: sealed machinery, ventilation.	
Inhalation: sore throat, cough, headache, dizziness, drowsiness.	ventilation.	fresh air, rest.
Skin: redness.	gloves.	remove contaminated clothing, flush skin with water or shower.
Eyes: redness, pain.	safety glasses.	flush with water, take to a doctor if necessary.
Ingestion: abdominal pain, headache, nausea, drowsiness.		rinse mouth, call a doctor.

SPILLAGE	STORAGE	LABELING / NFPA
collect leakage in sealable containers, flush away remainder with water.	keep separate from oxidants; ventilate at floor level.	(NFPA diamond: 2 top, 1 left, 0 right)

NOTES

1,3-BUTANEDIOL

PHYSICAL PROPERTIES		IMPORTANT CHARACTERISTICS	
Boiling point, °C	204	**COLORLESS VISCOUS HYGROSCOPIC LIQUID**	
Melting point, °C	0	Vapor mixes readily with air. Reacts with strong oxidants.	
Flash point, °C	109		
Auto-ignition temperature, °C	375	TLV-TWA　　　　　　　　not available	
Relative density (water = 1)	1.0		
Relative vapor density (air = 1)	3.1	**Absorption route:** Can enter the body by inhalation or ingestion. Harmful atmospheric concentrations build up very slowly, if at all, on evaporation at approx. 20°C, but much more rapidly in aerosol form.	
Relative density at 20 °C of saturated mixture vapor/air (air = 1)	1.0		
Vapor pressure, mm Hg at 20 °C	0.061	**Immediate effects:** Irritates the eyes.	
Solubility in water	∞		
Explosive limits, vol% in air	1.9-?		
Relative molecular mass	90.1		
Log P octanol/water	− 1.0		
Gross formula:	$C_4H_{10}O_2$		

HAZARDS/SYMPTOMS	PREVENTIVE MEASURES	FIRE EXTINGUISHING/FIRST AID
Fire: combustible.	keep away from open flame, no smoking.	powder, alcohol-resistant foam, water spray, carbon dioxide, (halons).
Inhalation: drowsiness.	ventilation.	fresh air, rest.
Skin: redness.	gloves.	remove contaminated clothing, flush skin with water or shower.
Eyes: redness.	safety glasses.	flush with water, take to a doctor if necessary.
Ingestion: drowsiness.		rinse mouth, call a doctor.

SPILLAGE	STORAGE	LABELING / NFPA
collect leakage in sealable containers, flush away remainder with water.	keep separate from oxidants.	1 / 1 / 0

NOTES
Log P octanol/water is estimated. Symptoms can occur only after swallowing large quantities of the liquid or inhaling high concentrations of the vapor.

1,4-BUTANEDIOL

PHYSICAL PROPERTIES		IMPORTANT CHARACTERISTICS	
Boiling point, °C	230	**COLORLESS HYGROSCOPIC VISCOUS LIQUID OR WHITE CRYSTALS**	
Melting point, °C	16	Vapor mixes readily with air. Reacts with strong oxidants.	
Flash point, °C	121		
Relative density (water = 1)	1.0	TLV-TWA — not available	
Relative vapor density (air = 1)	3.1		
Relative density at 20 °C of saturated mixture vapor/air (air = 1)	1.0	**Absorption route:** Can enter the body by inhalation or ingestion. Harmful atmospheric concentrations build up very slowly, if at all, on evaporation at approx. 20°C, but much more rapidly in aerosol form.	
Vapor pressure, mm Hg at 20 °C	< 0.076	**Immediate effects:** Irritates the eyes.	
Solubility in water	∞	**Effects of prolonged/repeated exposure:** Can cause kidney damage.	
Relative molecular mass	90.1		
Log P octanol/water	− 1.3		
Gross formula:	$C_4H_{10}O_2$		

HAZARDS/SYMPTOMS	PREVENTIVE MEASURES	FIRE EXTINGUISHING/FIRST AID
Fire: combustible.	keep away from open flame, no smoking.	powder, alcohol-resistant foam, water spray, carbon dioxide, (halons).
Inhalation: drowsiness.	ventilation.	fresh air, rest.
Skin: redness.	gloves.	remove contaminated clothing, flush skin with water and if possible wash with soap.
Eyes: redness, pain.	safety glasses.	flush with water, take to a doctor if necessary.
Ingestion: drowsiness.		rinse mouth, consult a doctor if necessary.

SPILLAGE	STORAGE	LABELING / NFPA
collect leakage in sealable containers or clean up spillage, flush away remainder with water.	keep separate from oxidants.	![NFPA diamond: health 1, flammability 1, reactivity 0]

NOTES
Log P octanol/water is estimated.

CAS-No: [513-85-9]
2,3-butylene glycol
2,3-dihydoxybutane

$CH_3CHOHCHOHCH_3$

2,3-BUTANEDIOL

PHYSICAL PROPERTIES		IMPORTANT CHARACTERISTICS	
Boiling point, °C	180	**COLORLESS LIQUID OR WHITE CRYSTALS**	
Melting point, °C	19	Vapor mixes readily with air. Reacts violently with strong oxidants.	
Flash point, °C	85		
Auto-ignition temperature, °C	402	TLV-TWA	not available
Relative density (water = 1)	1.01		
Relative density at 20 °C of saturated mixture		**Absorption route:** Can enter the body by inhalation or ingestion. Harmful atmospheric concentrations build up very slowly, if at all, on evaporation at approx. 20°C, but much more rapidly in aerosol form.	
vapor/air (air = 1)	1.0		
Vapor pressure, mm Hg at 20 °C	0.18		
Solubility in water	∞	**Immediate effects:** Irritates the eyes. Affects the nervous system.	
Relative molecular mass	90.1	**Effects of prolonged/repeated exposure:** Can cause kidney damage.	
Log P octanol/water	− 0.9		
Gross formula:	$C_4H_{10}O_2$		

HAZARDS/SYMPTOMS	PREVENTIVE MEASURES	FIRE EXTINGUISHING/FIRST AID
Fire: combustible.	keep away from open flame, no smoking.	powder, alcohol-resistant foam, water spray, carbon dioxide, (halons).
Explosion: above 85°C: forms explosive air-vapor mixtures.	above 85°C: sealed machinery, ventilation.	
Inhalation: sore throat, cough, headache, dizziness, drowsiness.	local exhaust or respiratory protection.	fresh air, rest, call a doctor.
Skin: redness.	gloves.	remove contaminated clothing, flush skin with water or shower.
Eyes: redness, pain.	safety glasses.	flush with water, take to a doctor.
Ingestion: abdominal pain, headache, nausea, drowsiness.		rinse mouth, call a doctor.

SPILLAGE	STORAGE	LABELING / NFPA
collect leakage in sealable containers, clean up spillage, flush away remainder with water.	keep dry, separate from oxidants.	1 1 0

NOTES

CAS-No: [78-92-2]
2-butanol
ethyl methyl carbinol
2-hydroxybutane

$CH_3CH_2CHOHCH_3$

sec-BUTANOL

PHYSICAL PROPERTIES		IMPORTANT CHARACTERISTICS	
Boiling point, °C	99	**COLORLESS LIQUID WITH CHARACTERISTIC ODOR**	
Melting point, °C	− 89	Reacts with alkali metals, giving off flammable gas (→ *hydrogen*). Reacts violently with strong	
Flash point, °C	24	oxidants. Attacks many plastics.	
Auto-ignition temperature, °C	390		
Relative density (water = 1)	0.8	TLV-TWA	100 ppm 303 mg/m³
Relative vapor density (air = 1)	2.6		
Relative density at 20 °C of saturated mixture		**Absorption route:** Can enter the body by inhalation or ingestion or through the skin. Harmful	
vapor/air (air = 1)	1.03	atmospheric concentrations build up fairly slowly on evaporation at 20°C, but much more rapidly	
Vapor pressure, mm Hg at 20°C	13	in aerosol form.	
Solubility in water, g/100 ml at 20 °C	12.5	**Immediate effects:** Irritates the eyes, skin and respiratory tract. Liquid destroys the skin's natural	
Explosive limits, vol% in air	1.7-9.8	oils. In substantial concentrations can impair consciousness.	
Electrical conductivity, pS/m	< 1x10⁷		
Relative molecular mass	74.1		
Log P octanol/water	0.6		
Gross formula:	$C_4H_{10}O$		

HAZARDS/SYMPTOMS	PREVENTIVE MEASURES	FIRE EXTINGUISHING/FIRST AID
Fire: flammable.	keep away from open flame and sparks, no smoking.	powder, AFFF, foam, carbon dioxide, (halons).
Explosion: above 24°C: forms explosive air-vapor mixtures.	above 24°C: sealed machinery, ventilation, explosion-proof electrical equipment.	in case of fire: keep tanks/drums cool by spraying with water.
Inhalation: sore throat, cough, shortness of breath, drowsiness.	ventilation, local exhaust or respiratory protection.	fresh air, rest, place in half-sitting position, call a doctor.
Skin: *is absorbed*; redness.	gloves.	remove contaminated clothing, flush skin with water or shower.
Eyes: redness, pain, impaired vision.	safety glasses.	flush with water, take to a doctor.
Ingestion: abdominal pain, diarrhea, vomiting.		rinse mouth, call a doctor or take to hospital.

SPILLAGE	STORAGE	LABELING / NFPA
collect leakage in sealable containers, soak up with sand or other inert absorbent and remove to safe place, flush away remainder with water; (additional individual protection: breathing apparatus).	keep in fireproof place, separate from oxidants and alkali metals.	R: 10-20 S: 16 Note C ✖ Harmful

NFPA diamond: 3 (left), 1 (bottom), 0 (right)

NOTES
Alcohol consumption increases toxic effects.

1-BUTENE
(cylinder)

PHYSICAL PROPERTIES	IMPORTANT CHARACTERISTICS
Boiling point, °C −6 Melting point, °C −185 Flash point, °C flammable gas Auto-ignition temperature, °C 384 Relative density (water = 1) 0.6 Relative vapor density (air = 1) 1.9 Vapor pressure, mm Hg at 20 °C 1900 Solubility in water none Explosive limits, vol% in air 1.6-10 Relative molecular mass 56.1 Gross formula: C_4H_8	**COLORLESS COMPRESSED LIQUEFIED GAS** Gas is heavier than air and spreads at ground level, with risk of ignition at a distance, and can accumulate close to ground level, causing oxygen deficiency, with risk of unconsciousness. Presumed to be able to form peroxides and thus to polymerize. Flow, agitation etc. can cause build-up of electrostatic charge due to liquid's low conductivity. Do not use compressed air when filling, emptying or processing. Reacts with strong acids, halogens and nitrogen oxides. Reacts violently with strong oxidants. TLV-TWA not available **Absorption route:** Can enter the body by inhalation. **Immediate effects:** Liquid can cause frostbite due to rapid evaporation. Can cause suffocation due to air saturation.

HAZARDS/SYMPTOMS	PREVENTIVE MEASURES	FIRE EXTINGUISHING/FIRST AID
Fire: extremely flammable.	keep away from open flame and sparks, no smoking.	shut off supply; if impossible and no danger to surrounding area, allow fire to burn itself out; otherwise extinguish with powder, carbon dioxide, (halons).
Explosion: forms explosive air-gas mixtures.	sealed machinery, ventilation, explosion-proof electrical equipment and lighting, grounding when pumping etc. in liquid form, non-sparking tools.	in case of fire: keep cylinders cool by spraying with water, fight fire from sheltered location.
Inhalation: cough, dizziness, shortness of breath, drowsiness, unconsciousness.	ventilation, local exhaust or respiratory protection.	fresh air, rest, call a doctor.
Skin: *in case of frostbite:* redness, pain, blisters.	insulating gloves.	*in case of frostbite:* DO NOT remove clothing, flush skin with water or shower.
Eyes: *in case of frostbite:* redness, pain, impaired vision.	face shield, or combined eye and respiratory protection.	flush with water, take to doctor.

SPILLAGE	STORAGE	LABELING / NFPA
evacuate area, call in an expert, ventilate, under no circumstances spray liquid with water; (additional individual protection: breathing apparatus).	if stored indoors, keep in cool, fireproof place, separate from oxidants.	◇ 4 / 1 0

NOTES
Check for peroxides; if found, render harmless. High atmospheric concentrations, e.g. in poorly ventilated spaces, can cause oxygen deficiency, with risk of unconsciousness. The measures on this card also apply to 2-butene (β-butylene). Physical properties of 2-butene: boiling point 4°C; melting point −139°C; auto-ignition temperature 313°C. Turn leaking cylinder so that leak is on top to prevent liquid 1-butene escaping.

Transport Emergency Card TEC(R)-500

HI: 23; UN-number: 1012

CAS-No: [111-76-2]
ethylene glycol n-butyl ether
ethylene glycol monobutyl ether

$C_4H_9OC_2H_4OH$

2-BUTOXYETHANOL

PHYSICAL PROPERTIES		IMPORTANT CHARACTERISTICS	
Boiling point, °C	171	**COLORLESS LIQUID WITH CHARACTERISTIC ODOR**	
Melting point, °C	−70	Vapor mixes readily with air. Able to form peroxides. Reacts violently with strong oxidants. Attacks light metals, giving off → *hydrogen*.	
Flash point, °C	61		
Auto-ignition temperature, °C	240		
Relative density (water = 1)	0.9	TLV-TWA (skin) 25 ppm 121 mg/m³	
Relative vapor density (air = 1)	4.1		
Relative density at 20 °C of saturated mixture vapor/air (air = 1)	1.0	**Absorption route:** Can enter the body by inhalation or ingestion or through the skin. Harmful atmospheric concentrations build up fairly slowly on evaporation at 20°C, but much more rapidly in aerosol form.	
Vapor pressure, mm Hg at 20 °C	0.76	**Immediate effects:** Irritates the eyes, skin and respiratory tract. In substantial concentrations can impair consciousness.	
Solubility in water	∞		
Explosive limits, vol% in air	1.1-10.6	**Effects of prolonged/repeated exposure:** Can cause liver and kidney damage. Can affect the blood.	
Relative molecular mass	118.2		
Gross formula: $C_6H_{14}O_2$			

HAZARDS/SYMPTOMS	PREVENTIVE MEASURES	FIRE EXTINGUISHING/FIRST AID
Fire: combustible.	keep away from open flame, no smoking.	powder, alcohol-resistant foam, water spray, carbon dioxide, (halons).
Explosion: above 61°C: forms explosive air-vapor mixtures.	above 61°C: sealed machinery, ventilation.	
Inhalation: sore throat, cough, headache, nausea.	ventilation, local exhaust or respiratory protection.	fresh air, rest.
Skin: *is absorbed*; redness.	gloves, protective clothing.	remove contaminated clothing, flush skin with water or shower.
Eyes: redness, pain.	face shield.	flush with water, take to a doctor if necessary.
Ingestion: abdominal pain, diarrhea, nausea.		rinse mouth, take immediately to hospital.

SPILLAGE	STORAGE	LABELING / NFPA
collect leakage in sealable containers, flush away remainder with water.	keep in dark place, separate from oxidants; ventilate at floor level.	R: 20/21/22-37 S: 24/25 ✖ Harmful [NFPA diamond: 2 / 2 / 0]

NOTES
Before distilling check for peroxides; if found, render harmless. Depending on the degree of exposure, regular medical checkups are advisable. Butyl Cellosolve and Butyl Ethoxol are trade names.

acetic acid, n-butyl ester
butyl acetate
1-butyl acetate
butyl ethanoate

$CH_3COOC_4H_9$

n-BUTYL ACETATE

PHYSICAL PROPERTIES	IMPORTANT CHARACTERISTICS
Boiling point, °C ... 127 Melting point, °C ... −77 Flash point, °C ... 22 Auto-ignition temperature, °C ... 370 Relative density (water = 1) ... 0.9 Relative vapor density (air = 1) ... 4.0 Relative density at 20 °C of saturated mixture vapor/air (air = 1) ... 1.03 Vapor pressure, mm Hg at 20 °C ... 8.1 Solubility in water, g/100 ml at 20 °C ... 0.7 Explosive limits, vol% in air ... 1.2-7.5 Relative molecular mass ... 116.2	**COLORLESS LIQUID WITH CHARACTERISTIC ODOR** Vapor mixes readily with air. Decomposes slowly when exposed to moisture or air, forming → *acetic acid* and → *n-butanol*. Reacts violently with strong oxidants, with risk of fire and explosion. Attacks many plastics.

PHYSICAL PROPERTIES (cont.)	IMPORTANT CHARACTERISTICS (cont.)		
	TLV-TWA	150 ppm	713 mg/m³
	STEL	200 ppm	950 mg/m³

Absorption route: Can enter the body by inhalation or ingestion. Harmful atmospheric concentrations build up fairly slowly on evaporation at 20°C, but much more rapidly in aerosol form.
Immediate effects: Irritates the eyes and respiratory tract. Liquid destroys the skin's natural oils. Vapor has narcotic effect in high concentrations.

Gross formula: $C_6H_{12}O_2$

HAZARDS/SYMPTOMS	PREVENTIVE MEASURES	FIRE EXTINGUISHING/FIRST AID
Fire: flammable.	keep away from open flame and sparks, no smoking.	powder, AFFF, foam, carbon dioxide, (halons).
Explosion: above 27°C: forms explosive air-vapor mixtures.	above 27°C: sealed machinery, ventilation, explosion-proof electrical equipment.	in case of fire: keep tanks/drums cool by spraying with water.
Inhalation: cough, shortness of breath, drowsiness.	ventilation, local exhaust or respiratory protection.	fresh air, rest, call a doctor.
Skin: redness.	gloves.	remove contaminated clothing, flush skin with water or shower.
Eyes: redness, pain.	safety glasses.	flush with water, take to a doctor.
Ingestion: sore throat, abdominal pain, nausea.		rinse mouth, call a doctor or take to hospital.

SPILLAGE	STORAGE	LABELING / NFPA
collect leakage in sealable containers, soak up with sand or other inert absorbent and remove to safe place.	keep in fireproof place, separate from oxidants.	R: 10 3 / 1 / 0

NOTES
Flash point highly dependent on presence of impurities (incl. alcohol content), and in some grades can be as much as 10-12°C higher than value given, with consequent substantial lowering of flammability at room temperature.

Transport Emergency Card TEC(R)-66

HI: 30; UN-number: 1123

sec-BUTYLACETATE

PHYSICAL PROPERTIES	IMPORTANT CHARACTERISTICS	
Boiling point, °C .... 112 Melting point, °C .... − 100 Flash point, °C .... 19 Auto-ignition temperature, °C .... ? Relative density (water = 1) .... 0.87 Relative vapor density (air = 1) ..... 4.0 Relative density at 20 °C of saturated mixture vapor/air (air = 1) .... 1.1 Vapor pressure, mm Hg at 20 °C ..... 19 Solubility in water, g/100 ml at 20 °C ... 3 Explosive limits, vol% in air ...... 1.7-9.8 Relative molecular mass 116.2	**COLORLESS LIQUID WITH CHARACTERISTIC ODOR** Vapor is heavier than air and spreads at ground level, with risk of ignition at a distance. Do not use compressed air when filling, emptying or processing. Decomposes slowly when exposed to light or on contact with water, forming → *acetic acid* and → *sec-butanol*. Reacts violently with strong oxidants. Attacks many plastics.	
	TLV-TWA 200 ppm 950 mg/m³	
	Absorption route: Can enter the body by inhalation or ingestion. Harmful atmospheric concentrations build up fairly slowly on evaporation at 20° C, but much more rapidly in aerosol form. **Immediate effects:** Irritates the eyes and respiratory tract. Liquid destroys the skin's natural oils. Vapor has narcotic effect in high concentrations.	
Gross formula: $C_6H_{12}O_2$		

HAZARDS/SYMPTOMS	PREVENTIVE MEASURES	FIRE EXTINGUISHING/FIRST AID
Fire: extremely flammable.	keep away from open flame and sparks, no smoking.	powder, AFFF, foam, carbon dioxide, (halons).
Explosion: forms explosive air-vapor mixtures.	sealed machinery, ventilation, explosion-proof electrical equipment and lighting.	in case of fire: keep tanks/drums cool by spraying with water.
Inhalation: cough, shortness of breath, drowsiness.	ventilation, local exhaust or respiratory protection.	fresh air, rest, call a doctor.
Skin: redness.	gloves.	remove contaminated clothing, flush skin with water or shower.
Eyes: redness.	safety glasses.	flush with water, take to a doctor.
Ingestion: sore throat, abdominal pain, nausea.		rinse mouth, call a doctor or take to hospital.

SPILLAGE	STORAGE	LABELING / NFPA
collect leakage in sealable containers, soak up with sand or other inert absorbent and remove to safe place.	keep in dry, dark, fireproof place, separate from oxidants.	R: 11 S: 16-23-29-33 Note C 🔥 Flammable

NOTES
Flash point highly dependent on presence of impurities (incl. alcohol content), and in some grades can be as much as 10-12° C higher than value given, with consequent substantial lowering of flammability at room temperature.

acrylic acid n-butyl ester
butyl 2-propenoate
2-propenoic acid butyl ester

$CH_2=CHCOOC_4H_9$

n-BUTYL ACRYLATE

PHYSICAL PROPERTIES		IMPORTANT CHARACTERISTICS
Boiling point, °C	146	**COLORLESS LIQUID WITH CHARACTERISTIC ODOR**
Melting point, °C	−65	Vapor mixes readily with air. Able to polymerize very readily, evolving large quantities of heat,
Flash point, °C	37	when moderately heated or exposed to light or on contact with peroxides. Reacts violently with
Auto-ignition temperature, °C	275	strong oxidants, with risk of fire and explosion.
Relative density (water = 1)	0.9	

TLV-TWA	10 ppm	52 mg/m³

Relative density at 20 °C of saturated mixture
vapor/air (air = 1) — 1.02

Vapor pressure, mm Hg at 20 °C — 3.8

Solubility in water, g/100 ml at 20 °C — 0.1

Explosive limits, vol% in air — 1.2-8.0

Relative molecular mass — 128.2

Absorption route: Can enter the body by inhalation or ingestion or through the skin. Harmful atmospheric concentrations can build up fairly rapidly on evaporation at approx. 20°C - even more rapidly in aerosol form.

Immediate effects: Corrosive to the eyes, skin and respiratory tract. Inhalation of vapor/fumes can cause severe breathing difficulties (lung edema). In serious cases risk of death.

Effects of prolonged/repeated exposure: Prolonged or repeated contact can cause skin disorders.

Gross formula: $C_7H_{12}O_2$

HAZARDS/SYMPTOMS	PREVENTIVE MEASURES	FIRE EXTINGUISHING/FIRST AID
Fire: flammable.	keep away from open flame and sparks, no smoking.	powder, AFFF, foam, carbon dioxide, (halons).
Explosion: above 37°C: forms explosive air-vapor mixtures.	above 37°C: sealed machinery, ventilation, explosion-proof electrical equipment.	in case of fire: keep tanks/drums cool by spraying with water.
	STRICT HYGIENE	IN ALL CASES CALL A DOCTOR
Inhalation: *corrosive*; sore throat, cough, shortness of breath, severe breathing difficulties.	ventilation, local exhaust or respiratory protection.	fresh air, rest, place in half-sitting position, take to hospital.
Skin: *is absorbed, corrosive*; redness, pain, burns.	gloves.	remove contaminated clothing, flush skin with water or shower, refer to a doctor.
Eyes: *corrosive*; redness, pain, impaired vision.	face shield.	flush with water, take to a doctor.
Ingestion: *corrosive*; sore throat, abdominal pain, vomiting.		rinse mouth, immediately take to hospital.

SPILLAGE	STORAGE	LABELING / NFPA
collect leakage in non-sealable containers, soak up with sand or other inert absorbent and remove to safe place; (additional individual protection: breathing apparatus).	keep in cool, dark, fireproof place; add an inhibitor.	R: 10-36/37/38-43 S: 9 Note D ✖ Irritant

NFPA diamond: 2 (top), 2 (left), 2 (right)

NOTES
Depending on the degree of exposure, regular medical checkups are advisable. Lung edema symptoms usually develop several hours later and are aggravated by physical exertion: rest and hospitalization essential. As first aid, a doctor or authorized person should consider administering a corticosteroid spray.

CH$_3$CH$_2$CH$_2$CH$_2$OH

n-butanol
1-butanol
butyric alcohol
1-hydroxybutane
propyl carbinol

n-BUTYL ALCOHOL

PHYSICAL PROPERTIES		IMPORTANT CHARACTERISTICS
Boiling point, °C	118	**COLORLESS LIQUID WITH CHARACTERISTIC ODOR**
Melting point, °C	−89	Reacts with strong oxidants and alkali metals, giving off flammable gas (→ *hydrogen*). Attacks
Flash point, °C	35	many plastics.
Auto-ignition temperature, °C	340	
Relative density (water = 1)	0.8	
Relative vapor density (air = 1)	2.6	TLV-TWA (Skin) 50 ppm 152 mg/m³ C
Relative density at 20 °C of saturated mixture		
vapor/air (air = 1)	1.01	**Absorption route:** Can enter the body by inhalation or ingestion or through the skin. Harmful
Vapor pressure, mm Hg at 20 °C	5.3	atmospheric concentrations can build up fairly rapidly on evaporation at approx. 20°C - even
Solubility in water, g/100 ml at 20 °C	8	more rapidly in aerosol form.
Explosive limits, vol% in air	1.4-11.3	**Immediate effects:** Irritates the eyes, skin and respiratory tract. Liquid destroys the skin's natural
Electrical conductivity, pS/m	9.1 x 10⁵	oils. In substantial concentrations can impair consciousness. In serious cases risk of
Relative molecular mass	74.1	unconsciousness.
Log P octanol/water	0.9	
Gross formula:	C$_4$H$_{10}$O	

HAZARDS/SYMPTOMS	PREVENTIVE MEASURES	FIRE EXTINGUISHING/FIRST AID
Fire: flammable.	keep away from open flame and sparks, no smoking.	powder, AFFF, foam, carbon dioxide, (halons).
Explosion: above 35°C: forms explosive air-vapor mixtures.	above 35°C: sealed machinery, ventilation, explosion-proof electrical equipment.	in case of fire: keep tanks/drums cool by spraying with water.
Inhalation: sore throat, cough, shortness of breath, drowsiness.	ventilation, local exhaust or respiratory protection.	fresh air, rest, place in half-sitting position, call a doctor.
Skin: *is absorbed*; redness.	gloves.	remove contaminated clothing, flush skin with water or shower.
Eyes: redness, pain, impaired vision.	safety glasses.	flush with water, take to a doctor.
Ingestion: abdominal pain, diarrhea, vomiting.		rinse mouth, call a doctor or take to hospital.

SPILLAGE	STORAGE	LABELING / NFPA
collect leakage in sealable containers, soak up with sand or other inert absorbent and remove to safe place; (additional individual protection: breathing apparatus).	keep separate from oxidants, in fireproof place.	R: 10-20 S: 16 Note C ✖ Harmful

NOTES
TLV is maximum and must not be exceeded. Alcohol consumption increases toxic effects.

Transport Emergency Card TEC(R)-583

HI: 30; UN-number: 1120

CAS-No: [75-65-0]
t-butanol
1,1-dimethylethanol
2-methyl-2-propanol
trimethyl carbinol

$(CH_3)_3COH$

tert-BUTYL ALCOHOL

PHYSICAL PROPERTIES		IMPORTANT CHARACTERISTICS	
Boiling point, °C	83	**COLORLESS LIQUID OR WHITE CRYSTALS WITH CHARACTERISTIC ODOR**	
Melting point, °C	25	Vapor mixes readily with air, forming explosive mixtures. Do not use compressed air when filling, emptying or processing. Reacts with alkaline-earth and alkali metals giving off flammable gas ($\rightarrow$ *hydrogen*). Reacts violently with oxidants, with risk of fire and explosion. Attacks many plastics.	
Flash point, °C	11		
Auto-ignition temperature, °C	470		
Relative density (water = 1)	0.8		
Relative vapor density (air = 1)	2.56		
Relative density at 20 °C of saturated mixture vapor/air (air = 1)	1.06	TLV-TWA 100 ppm 303 mg/m³	
Vapor pressure, mm Hg at 20 °C	3.0	STEL 150 ppm 455 mg/m³	
Solubility in water	∞		
Explosive limits, vol% in air	2.3-8	**Absorption route:** Can enter the body by inhalation or ingestion or through the skin. Harmful atmospheric concentrations can build up fairly rapidly on evaporation at approx. 20°C - even more rapidly in the form of airborne particles or aerosol.	
Electrical conductivity, pS/m	2.6 x 10⁶		
Relative molecular mass	74.1		
Log P octanol/water	0.4	**Immediate effects:** Irritates the eyes, skin and respiratory tract. Liquid destroys the skin's natural oils. In substantial concentrations can impair consciousness. In serious cases risk of unconsciousness.	
Gross formula:	$C_4H_{10}O$		

HAZARDS/SYMPTOMS	PREVENTIVE MEASURES	FIRE EXTINGUISHING/FIRST AID
Fire: extremely flammable.	keep away from open flame and sparks, no smoking.	powder, alcohol-resistant foam, large quantities of water, carbon dioxide, (halons).
Explosion: forms explosive air-vapor mixtures.	sealed machinery, ventilation, explosion-proof electrical equipment and lighting.	in case of fire: keep tanks/drums cool by spraying with water.
Inhalation: sore throat, cough, shortness of breath, drowsiness.	ventilation, local exhaust or respiratory protection.	fresh air, rest, place in half-sitting position, call a doctor.
Skin: *is absorbed*; redness.	gloves.	remove contaminated clothing, flush skin with water or shower, refer to a doctor if necessary.
Eyes: redness, pain, impaired vision.	safety glasses.	flush with water, take to a doctor.
Ingestion: abdominal pain, vomiting, drowsiness.		rinse mouth, call a doctor or take to hospital.

SPILLAGE	STORAGE	LABELING / NFPA
collect leakage in sealable containers or clean up spillage, flush away remainder with water; (additional individual protection: breathing apparatus).	keep in fireproof place, separate from oxidants and alkali metals.	R: 11-20 S: 9-16 🔥 Flammable ✖ Harmful NFPA: 3 1 0

NOTES
Alcohol consumption increases toxic effects.

Transport Emergency Card TEC(R)-164

HI: 30; UN-number: 1120

BUTYLAMINE

PHYSICAL PROPERTIES	IMPORTANT CHARACTERISTICS
Boiling point, °C 78 Melting point, °C −50 Flash point, °C −12 Auto-ignition temperature, °C 310 Relative density (water = 1) 0.7 Relative vapor density (air = 1) 2.5 Relative density at 20 °C of saturated mixture vapor/air (air = 1) 1.2 Vapor pressure, mm Hg at 20 °C 69 Solubility in water ∞ Explosive limits, vol% in air 1.7-10 Relative molecular mass 73.1 Log P octanol/water 0.8	**COLORLESS LIQUID WITH CHARACTERISTIC ODOR** Vapor is heavier than air and spreads at ground level, with risk of ignition at a distance. Do not use compressed air when filling, emptying or processing. Decomposes when heated or burned, giving off toxic gases (→ *nitrous vapors*). Reacts violently with oxidants and strong acids, with risk of fire and explosion. Attacks light metals and copper.
	TLV-TWA (skin) 5 ppm 15 mg/m³ C
	Absorption route: Can enter the body by inhalation or ingestion or through the skin. Harmful atmospheric concentrations can build up very rapidly on evaporation at 20°C. **Immediate effects:** Corrosive to the eyes, skin and respiratory tract. Inhalation of vapor/fumes can cause severe breathing difficulties (lung edema). In serious cases risk of death.
Gross formula: $C_4H_{11}N$	

HAZARDS/SYMPTOMS	PREVENTIVE MEASURES	FIRE EXTINGUISHING/FIRST AID
Fire: extremely flammable.	keep away from open flame and sparks, no smoking, avoid contact with strong oxidants.	powder, alcohol-resistant foam, large quantities of water, carbon dioxide, (halons).
Explosion: forms explosive air-vapor mixtures.	sealed machinery, ventilation, explosion-proof electrical equipment and lighting.	in case of fire: keep tanks/drums cool by spraying with water.
	STRICT HYGIENE	
Inhalation: *corrosive*; sore throat, cough, shortness of breath, severe breathing difficulties.	ventilation, local exhaust or respiratory protection.	fresh air, rest, place in half-sitting position, artificial respiration if necessary, take to hospital.
Skin: *corrosive, is absorbed*; redness, pain, burns.	gloves, protective clothing.	remove contaminated clothing, flush skin with water or shower, call a doctor.
Eyes: *corrosive*; redness, pain, impaired vision.	face shield, or combined eye and respiratory protection.	flush with water, take to a doctor.
Ingestion: *corrosive*; sore throat, abdominal pain, diarrhea, vomiting.		rinse mouth, immediately take to hospital.

SPILLAGE	STORAGE	LABELING / NFPA
evacuate area, call in an expert, collect leakage in sealable containers, soak up with sand or other inert absorbent and remove to safe place; (additional individual protection: breathing apparatus).	keep in fireproof place, separate from oxidants and strong acids.	R: 11-36/37/38 S: 16-26-29 🔥 Flammable ✖ Irritant NFPA 3 / 3 / 0

NOTES
TLV is maximum and must not be exceeded. Lung edema symptoms usually develop several hours later and are aggravated by physical exertion: rest and hospitalization essential. As first aid, a doctor or authorized person should consider administering a corticosteroid spray. Unbreakable packaging preferred; if breakable, keep in unbreakable container.

Transport Emergency Card TEC(R)-655

HI: 338; UN-number: 1125

CAS-No: [75-64-9]
2-aminoisobutane
2-amino-2-methylpropane
2-methyl-2-propanamine
trimethylaminomethane
trimethylcarbinylamine

$(CH_3)_3CNH_2$

tert-BUTYLAMINE

PHYSICAL PROPERTIES		IMPORTANT CHARACTERISTICS
Boiling point, °C	45	**COLORLESS LIQUID WITH CHARACTERISTIC ODOR**
Melting point, °C	−67	Vapor is heavier than air and spreads at ground level, with risk of ignition at a distance. Do not use
Flash point, °C	−9	compressed air when filling, emptying or processing. Toxic gases are a product of combustion
Auto-ignition temperature, °C	380	(→ *nitrous vapors*). Decomposes when heated, giving off combustible and toxic vapors
Relative density (water = 1)	0.7	(→ *nitrous vapors*). Reacts violently with strong oxidants, acids and anhydrides. Also reacts with
Relative vapor density (air = 1)	2.5	many other chemicals, e.g. alcohols, glycol ethers, nitriles, phenols and monomers.
Relative density at 20°C of saturated mixture		
vapor/air (air = 1)	1.6	TLV-TWA not available
Vapor pressure, mm Hg at 20°C	292	
Solubility in water	∞	**Absorption route:** Can enter the body by inhalation or ingestion or through the skin. Harmful
Explosive limits, vol% in air	1.7-8.9	atmospheric concentrations can build up very rapidly on evaporation at 20°C.
Relative molecular mass	73.1	**Immediate effects:** Corrosive to the eyes, skin and respiratory tract. Affects the nervous system.
Log P octanol/water	0.4	Inhalation of vapor/fumes can cause severe breathing difficulties (lung edema). In serious cases
		risk of death.
		Effects of prolonged/repeated exposure: Prolonged or repeated contact can cause skin
		disorders. Can cause kidney damage.
Gross formula:	$C_4H_{11}N$	

HAZARDS/SYMPTOMS	PREVENTIVE MEASURES	FIRE EXTINGUISHING/FIRST AID
Fire: extremely flammable.	keep away from open flame and sparks, no smoking.	powder, alcohol-resistant foam, large quantities of water, carbon dioxide, (halons).
Explosion: forms explosive air-vapor mixtures.	sealed machinery, ventilation, explosion-proof electrical equipment and lighting.	in case of fire: keep tanks/drums cool by spraying with water.
	STRICT HYGIENE	IN ALL CASES CALL IN A DOCTOR
Inhalation: *corrosive*; cough, severe breathing difficulties.	ventilation, local exhaust or respiratory protection.	fresh air, rest, place in half-sitting position, take to hospital.
Skin: *corrosive, is absorbed*; redness, pain, serious burns.	gloves, protective clothing.	remove contaminated clothing, flush skin with water or shower, call a doctor.
Eyes: *corrosive*; redness, pain, impaired vision.	face shield, or combined eye and respiratory protection.	flush with water, take to a doctor.
Ingestion: *corrosive*; diarrhea, abdominal cramps, vomiting.		rinse mouth, take immediately to hospital.

SPILLAGE	STORAGE	LABELING / NFPA
evacuate area, call in an expert, collect leakage in sealable containers, soak up with sand or other inert absorbent and remove to safe place; (additional individual protection: CHEMICAL SUIT).	keep in fireproof place, separate from strong oxidants, acids and alcohols.	◇ 4 2 0

NOTES
Depending on the degree of exposure, regular medical checkups are advisable. Lung edema symptoms usually develop several hours later and are aggravated by physical exertion: rest and hospitalization essential. As first aid, a doctor or authorized person should consider administering a corticosteroid spray. Unbreakable packaging preferred; if breakable, keep in unbreakable container.

Transport Emergency Card TEC(R)-30G23

n-BUTYL BENZOATE

PHYSICAL PROPERTIES	IMPORTANT CHARACTERISTICS
Boiling point, °C 250 Melting point, °C −22 Flash point, °C 107 Auto-ignition temperature, °C 481 Relative density (water = 1) 1.0 Relative vapor density (air = 1) 6.2 Relative density at 20 °C of saturated mixture vapor/air (air = 1) 1.0 Vapor pressure, mm Hg at 20 °C 0.015 Solubility in water none Relative molecular mass 178.2 Gross formula: $C_{11}H_{14}O_2$	**COLORLESS VISCOUS LIQUID WITH CHARACTERISTIC ODOR** Vapor mixes readily with air. Reacts violently with strong oxidants. TLV-TWA not available **Absorption route:** Can enter the body by inhalation or ingestion. Harmful atmospheric concentrations build up very slowly, if at all, on evaporation at approx. 20°C, but much more rapidly in aerosol form. **Immediate effects:** Irritates the eyes and skin.

HAZARDS/SYMPTOMS	PREVENTIVE MEASURES	FIRE EXTINGUISHING/FIRST AID
Fire: combustible.	keep away from open flame, no smoking.	powder, AFFF, foam, carbon dioxide, (halons).
Inhalation:	ventilation.	fresh air, rest, call a doctor if necessary.
Skin: redness, pain.	gloves.	remove contaminated clothing, flush skin with water or shower.
Eyes: redness, pain.	safety glasses.	flush with water, take to a doctor if necessary.
Ingestion:		rinse mouth, call a doctor if necessary.

SPILLAGE	STORAGE	LABELING / NFPA
collect leakage in sealable containers, soak up with sand or other inert absorbent and remove to safe place.	keep separate from oxidants.	1 1 0

NOTES
Insufficient toxicological data on harmful effects to humans.

CAS-No: [98-73-7]
p-t-butylbenzoic acid
4-tert-butylbenzoic acid
4-(1,1-dimethylethyl)benzoic acid
TBBA

$(CH_3)_3CC_6H_4COOH$

p-tert-BUTYLBENZOIC ACID

PHYSICAL PROPERTIES		IMPORTANT CHARACTERISTICS
Melting point, °C	163	**WHITE CRYSTALLINE POWDER**
Relative density (water = 1)	1.14	In dry state can form electrostatic charge if stirred, transported pneumatically, poured etc.
Vapor pressure, mm Hg at 20 °C	<0.076	Decomposes when heated above 80°C, giving off flammable vapors. Reacts with strong
Solubility in water	none	oxidants.
Relative molecular mass	178.1	

TLV-TWA	not available

Absorption route: Can enter the body by inhalation or ingestion. Evaporation negligible at 20°C, but harmful concentrations of airborne particles can build up rapidly.
Immediate effects: Irritates the eyes, skin and respiratory tract.

Gross formula: $C_{11}H_{14}O_2$

HAZARDS/SYMPTOMS	PREVENTIVE MEASURES	FIRE EXTINGUISHING/FIRST AID
Fire: combustible.	keep away from open flame, no smoking.	water spray, powder.
Explosion: finely dispersed particles form explosive mixtures on contact with air.	keep dust from accumulating; sealed machinery, explosion-proof electrical equipment and lighting, grounding.	
	KEEP DUST UNDER CONTROL	
Inhalation: sore throat, cough, shortness of breath.	local exhaust or respiratory protection.	fresh air, rest, call a doctor.
Skin: redness.	gloves.	remove contaminated clothing, flush skin with water or shower.
Eyes: redness, pain.	goggles.	flush with water, take to a doctor if necessary.
Ingestion: abdominal pain, nausea.		rinse mouth, call a doctor.

SPILLAGE	STORAGE	LABELING / NFPA
clean up spillage, flush away remainder with water; (additional individual protection: P2 respirator).	keep separate from oxidants.	

NOTES

n-BUTYLBUTYRATE

PHYSICAL PROPERTIES		IMPORTANT CHARACTERISTICS	
Boiling point, °C	166	**COLORLESS LIQUID WITH CHARACTERISTIC ODOR**	
Melting point, °C	−92	Vapor mixes readily with air. Reacts violently with strong oxidants. Attacks many plastics.	
Flash point, °C	50		
Auto-ignition temperature, °C	?	TLV-TWA not available	
Relative density (water = 1)	0.87		
Relative vapor density (air = 1)	5		
Relative density at 20 °C of saturated mixture		**Absorption route:** Can enter the body by inhalation or ingestion or through the skin. Insufficient	
vapor/air (air = 1)	1.01	data on the rate at which harmful concentrations can build up.	
Vapor pressure, mm Hg at 20 °C	1.4	**Immediate effects:** Irritates the eyes, skin and respiratory tract. Affects the nervous system.	
Solubility in water	none		
Explosive limits, vol% in air	?-?		
Relative molecular mass	144.2		
Gross formula: $C_8H_{16}O_2$			

HAZARDS/SYMPTOMS	PREVENTIVE MEASURES	FIRE EXTINGUISHING/FIRST AID
Fire: flammable.	keep away from open flame and sparks, no smoking.	powder, AFFF, foam, carbon dioxide, (halons).
Explosion: above 50°C: forms explosive air-vapor mixtures.	above 50°C: sealed machinery, ventilation, explosion-proof electrical equipment.	in case of fire: keep tanks/drums cool by spraying with water.
Inhalation: sore throat, cough, drowsiness.	ventilation, local exhaust or respiratory protection.	fresh air, rest, call a doctor.
Skin: *is absorbed*; redness.	gloves.	remove contaminated clothing, wash skin with water and soap, take to a doctor.
Eyes: redness, pain.	safety glasses.	flush with water, take to a doctor.
Ingestion: pain with swallowing, nausea.		rinse mouth, call a doctor.

SPILLAGE	STORAGE	LABELING / NFPA
collect leakage in sealable containers, soak up with sand or other inert absorbent and remove to safe place.	keep in fireproof place, separate from oxidants.	R: 10 Note C

NFPA diamond: blue 2, red 2, yellow 0

NOTES
Alcohol consumption increases toxic effects.

p-tert-BUTYLCATECHOL

PHYSICAL PROPERTIES	IMPORTANT CHARACTERISTICS
Boiling point, °C 285 Melting point, °C 56 Flash point, °C 129 Relative density (water = 1) 1.05 Relative vapor density (air = 1) 5.7 Solubility in water moderate Relative molecular mass 166.2	**WHITE CRYSTALS** Decomposes when heated, giving off toxic vapors. Reacts with strong oxidants.
	TLV-TWA not available
	Absorption route: Can enter the body by inhalation or ingestion. Harmful atmospheric concentrations build up very slowly, if at all, on evaporation at approx. 20°C, but harmful concentrations of airborne particles can build up much more rapidly. **Immediate effects:** Irritates the eyes, skin and respiratory tract. **Effects of prolonged/repeated exposure:** Prolonged or repeated contact can cause skin disorders (dipigmentation).
Gross formula: $C_{10}H_{14}O_2$	

HAZARDS/SYMPTOMS	PREVENTIVE MEASURES	FIRE EXTINGUISHING/FIRST AID
Fire: combustible.	keep away from open flame, no smoking.	powder, AFFF, foam, carbon dioxide, (halons); in case of fire in immediate vicinity: use any extinguishing agent.
	KEEP DUST UNDER CONTROL	
Inhalation: sore throat, cough, shortness of breath.	local exhaust or respiratory protection.	fresh air, rest, place in half-sitting position, call a doctor.
Skin: redness, pain.	gloves, protective clothing.	remove contaminated clothing, wash skin with soap and water, refer to a doctor.
Eyes: redness, pain.	goggles.	flush with water, take to a doctor.
Ingestion: abdominal cramps, vomiting.		rinse mouth, call a doctor or take to hospital.

SPILLAGE	STORAGE	LABELING / NFPA
clean up spillage, flush away remainder with water; (additional individual protection: P2 respirator).	keep separate from oxidants.	NFPA diamond: 1 (top), 2 (left), 0 (right)

NOTES
Depending on the degree of exposure, regular medical checkups are advisable. Some hours after intensive contact with the skin blister may form.

Transport Emergency Card TEC(R)-80G09

UN-number: 2921 (n.o.s.)

BUTYL CHLORIDE

PHYSICAL PROPERTIES	IMPORTANT CHARACTERISTICS
Boiling point, °C 78 Melting point, °C − 123 Flash point, °C − 12 Auto-ignition temperature, °C 240 Relative density (water = 1) 0.9 Relative vapor density (air = 1) 3.2 Relative density at 20 °C of saturated mixture vapor/air (air = 1) 1.2 Vapor pressure, mm Hg at 20 °C 81 Solubility in water, g/100 ml at 12 °C 0.7 Explosive limits, vol% in air 1.8-10.1 Electrical conductivity, pS/m 1×10^4 Relative molecular mass 92.6 Log P octanol/water 2.4 Gross formula: C_4H_9Cl	**COLORLESS LIQUID WITH PUNGENT ODOR** Vapor is heavier than air and spreads at ground level, with risk of ignition at a distance. Do not use compressed air when filling, emptying or processing. Decomposes when heated or burned, giving off corrosive vapors (→ *hydrochloric acid*). Decomposes slowly when exposed to moisture, giving off corrosive vapors (→ *hydrochloric acid*). Reacts violently with alkaline-earth and alkali metals, metal powders and oxidants, with risk of fire and explosion. Attacks aluminum and many plastics. TLV-TWA not available **Absorption route:** Can enter the body by inhalation or ingestion. Harmful atmospheric concentrations can build up fairly rapidly on evaporation at approx. 20°C - even more rapidly in aerosol form. **Immediate effects:** Irritates the eyes, skin and respiratory tract. Affects the nervous system.

HAZARDS/SYMPTOMS	PREVENTIVE MEASURES	FIRE EXTINGUISHING/FIRST AID
Fire: extremely flammable.	keep away from open flame and sparks, no smoking.	powder, AFFF, foam, carbon dioxide, (halons).
Explosion: forms explosive air-vapor mixtures.	sealed machinery, ventilation, explosion-proof electrical equipment and lighting.	in case of fire: keep tanks/drums cool by spraying with water.
Inhalation: sore throat, cough, dizziness, drowsiness.	ventilation, local exhaust or respiratory protection.	fresh air, rest, call a doctor.
Skin: redness.	gloves.	remove contaminated clothing, flush skin with water or shower.
Eyes: redness, pain.	acid goggles.	flush with water, take to a doctor if necessary.
Ingestion: abdominal cramps, nausea.		rinse mouth, call a doctor.

SPILLAGE	STORAGE	LABELING / NFPA
collect leakage in sealable containers, soak up with sand or other inert absorbent and remove to safe place; (additional individual protection: breathing apparatus).	keep in dry, fireproof place, separate from oxidants.	3 2 0

NOTES
If substance is likely to come in contact with water add inhibitor to prevent acidification.

tert-BUTYLCHLORIDE

PHYSICAL PROPERTIES	IMPORTANT CHARACTERISTICS
Boiling point, °C 51 Melting point, °C −25 Flash point, °C −25 Auto-ignition temperature, °C 602 Relative density (water = 1) 0.8 Relative vapor density (air = 1) 3.2 Relative density at 20 °C of saturated mixture vapor/air (air = 1) ca. 1.7 Vapor pressure, mm Hg at 32 °C 403 Solubility in water poor Explosive limits, vol% in air ?-? Relative molecular mass 92.6 Gross formula: C_4H_9Cl	**COLORLESS LIQUID WITH CHARACTERISTIC ODOR** Vapor is heavier than air and spreads at ground level, with risk of ignition at a distance. Decomposes when heated or burned, giving off corrosive vapors (→ *hydrochloric acid*). Decomposes slowly on contact with water and when exposed to light. Reacts violently with alkaline-earth, alkali metals and metal powders. Reacts violently with oxidants, with risk of fire and explosion. Attacks aluminum and many plastics.

TLV-TWA	not available

Absorption route: Can enter the body by inhalation or ingestion. Harmful atmospheric concentrations can build up fairly rapidly on evaporation at approx. 20°C - even more rapidly in aerosol form.
Immediate effects: Irritates the eyes, skin and respiratory tract. Affects the nervous system.

HAZARDS/SYMPTOMS	PREVENTIVE MEASURES	FIRE EXTINGUISHING/FIRST AID
Fire: extremely flammable.	keep away from open flame and sparks, no smoking.	powder, AFFF, foam, carbon dioxide, (halons).
Explosion: forms explosive air-vapor mixtures.	sealed machinery, ventilation, explosion-proof electrical equipment and lighting.	in case of fire: keep tanks/drums cool by spraying with water.
Inhalation: sore throat, cough, dizziness, drowsiness.	ventilation, local exhaust or respiratory protection.	fresh air, rest, call a doctor if necessary.
Skin: redness.	gloves.	remove contaminated clothing, flush skin with water or shower.
Eyes: redness, pain.	acid goggles.	flush with water, take to a doctor if necessary.
Ingestion: abdominal cramps, nausea.		rinse mouth, call a doctor.

SPILLAGE	STORAGE	LABELING / NFPA
collect leakage in sealable containers, soak up with sand or other inert absorbent and remove to safe place; (additional individual protection: breathing apparatus).	keep in dry, fireproof place, separate from oxidants.	3 / 2

NOTES

Insufficient toxicological data on harmful effects to humans. Under no circumstances use near flame or hot surface or when welding.

$$CH_2—CH—CH_2—CH_3$$

1,2-BUTYLENE OXIDE

PHYSICAL PROPERTIES		IMPORTANT CHARACTERISTICS
Boiling point, °C	63	**COLORLESS LIQUID WITH CHARACTERISTIC ODOR**
Melting point, °C	−150	Vapor is heavier than air and spreads at ground level, with risk of ignition at a distance. Potassium and sodium hydroxide, iron, tin and aluminum chlorides cause violent polymerization. Flow, agitation etc. can cause build-up of electrostatic charge due to liquid's low conductivity. Do not use compressed air when filling, emptying or processing. Reacts violently with strong oxidants, alcohols and acids.
Flash point, °C	−15	
Auto-ignition temperature, °C	439	
Relative density (water = 1)	0.83	
Relative vapor density (air = 1)	2.49	
Relative density at 20 °C of saturated mixture		
vapor/air (air = 1)	1.3	TLV-TWA not available
Vapor pressure, mm Hg at 20 °C	142	
Solubility in water	moderate	**Absorption route:** Can enter the body by inhalation or ingestion or through the skin. Harmful atmospheric concentrations can build up very rapidly on evaporation at 20°C.
Explosive limits, vol% in air	1.5-18.3	**Immediate effects:** Corrosive to the eyes, skin and respiratory tract. Affects the nervous system. Inhalation of vapor/fumes can cause severe breathing difficulties (lung edema). In serious cases risk of unconsciousness and death.
Relative molecular mass	72.1	
		Effects of prolonged/repeated exposure: Prolonged or repeated contact can cause skin disorders.
Gross formula:	C_4H_8O	

HAZARDS/SYMPTOMS	PREVENTIVE MEASURES	FIRE EXTINGUISHING/FIRST AID
Fire: extremely flammable.	keep away from open flame and sparks, no smoking.	powder, water spray, foam, carbon dioxide, (halons).
Explosion: forms explosive air-vapor mixtures.	sealed machinery, ventilation, explosion-proof electrical equipment and lighting, grounding.	in case of fire: keep tanks/drums cool by spraying with water.
Inhalation: *corrosieve*; sore throat, cough, shortness of breath, severe breathing difficulties, dizziness, drowsiness.	ventilation, local exhaust or respiratory protection.	fresh air, rest, place in half-sitting position, call a doctor and take to hospital.
Skin: *corrosive, is absorbed*; redness, pain, burns.	gloves, protective clothing.	remove contaminated clothing, flush skin with water or shower, refer to a doctor.
Eyes: *corrosive*; redness, pain, impaired vision.	face shield, or combined eye and respiratory protection.	flush with water, take to a doctor.
Ingestion: *corrosive*; sore throat, abdominal pain, diarrhea, vomiting.		rinse mouth, immediately take to hospital.

SPILLAGE	STORAGE	LABELING / NFPA
collect leakage in sealable containers, soak up with sand or other inert absorbent and remove to safe place; (additional individual protection: breathing apparatus).	keep in fireproof place, separate from other substances, add an inhibitor.	3 / 2 2

NOTES

Lung edema symptoms usually develop several hours later and are aggravated by physical exertion: rest and hospitalization essential. As first aid, a doctor or authorized person should consider administering a corticosteroid spray. The measures on this card also apply to 2,3-butylene oxide.

Transport Emergency Card TEC(R)-30G31   **HI: 339; UN-number: 3022**

BUTYL ETHER

PHYSICAL PROPERTIES	IMPORTANT CHARACTERISTICS
Boiling point, °C 142 Melting point, °C −95 Flash point, °C 25 Auto-ignition temperature, °C 175 Relative density (water = 1) 0.8 Relative vapor density (air = 1) 4.5 Relative density at 20 °C of saturated mixture vapor/air (air = 1) 1.02 Vapor pressure, mm Hg at 20 °C 4.9 Solubility in water none Explosive limits, vol% in air 0.9-8.5 Relative molecular mass 130.2 Gross formula: $C_8H_{18}O$	**COLORLESS LIQUID** Presumed to be able to form peroxides. Flow, agitation etc. can cause build-up of electrostatic charge due to liquid's low conductivity. Reacts violently with strong oxidants. TLV-TWA not available **Absorption route:** Can enter the body by inhalation or ingestion or through the skin. Harmful atmospheric concentrations build up fairly slowly on evaporation at 20° C, but much more rapidly in aerosol form. **Immediate effects:** Irritates the eyes, skin and respiratory tract. Affects the nervous system.

HAZARDS/SYMPTOMS	PREVENTIVE MEASURES	FIRE EXTINGUISHING/FIRST AID
Fire: flammable.	keep away from open flame and sparks, no smoking.	powder, AFFF, foam, carbon dioxide, (halons).
Explosion: above 25°C: forms explosive air-vapor mixtures.	above 25°C: sealed machinery, ventilation, explosion-proof electrical equipment, grounding.	in case of fire: keep tanks/drums cool by spraying with water.
Inhalation: cough, shortness of breath, drowsiness.	ventilation, local exhaust or respiratory protection.	fresh air, rest, call a doctor.
Skin: *is absorbed*; redness, pain.	gloves.	remove contaminated clothing, flush skin with water or shower.
Eyes: redness, pain.	face shield.	flush with water, take to a doctor.
Ingestion: sore throat, abdominal pain, nausea.		rinse mouth, call a doctor.

SPILLAGE	STORAGE	LABELING / NFPA
collect leakage in sealable containers, soak up with sand or other inert absorbent and remove to safe place.	keep in cool, dark, fireproof place, separate from oxidants; add an inhibitor.	R: 10-36/37/38 ✖ Irritant (NFPA diamond: 3 top, 2 left, 1 right)

NOTES
Before distilling check for peroxides; if found, render harmless. The measures on this card also apply to t-butyl ether. Use airtight packaging.

n-BUTYLFORMATE

PHYSICAL PROPERTIES	IMPORTANT CHARACTERISTICS	
Boiling point, °C 107 Melting point, °C −96 Flash point, °C 18 Auto-ignition temperature, °C 322 Relative density (water = 1) 0.9 Relative vapor density (air = 1) 3.5 Relative density at 20 °C of saturated mixture vapor/air (air = 1) 1.06 Vapor pressure, mm Hg at 20 °C 18 Solubility in water poor Explosive limits, vol% in air 1.7-8 Electrical conductivity, pS/m 109 Relative molecular mass 102.1 Gross formula: C$_5$H$_{10}$O$_2$	**COLORLESS LIQUID WITH PUNGENT ODOR** Vapor mixes readily with air, forming explosive mixtures. Flow, agitation etc. can cause build-up of electrostatic charge due to liquid's low conductivity. Do not use compressed air when filling, emptying or processing. Reacts violently with strong oxidants.	
	TLV-TWA not available	
	Absorption route: Can enter the body by inhalation or ingestion. Insufficient data on the rate at which harmful concentrations can build up. **Immediate effects:** Irritates the skin. Corrosive to the eyes and respiratory tract. Affects the nervous system. Inhalation can cause lung edema. In serious cases risk of death.	

HAZARDS/SYMPTOMS	PREVENTIVE MEASURES	FIRE EXTINGUISHING/FIRST AID
Fire: extremely flammable.	keep away from open flame and sparks, no smoking.	powder, AFFF, foam, carbon dioxide, (halons); in case of fire in immediate vicinity: use any extinguishing agent.
Explosion: forms explosive air-vapor mixtures.	sealed machinery, ventilation, explosion-proof electrical equipment and lighting.	in case of fire: keep tanks/drums cool by spraying with water.
Inhalation: *corrosive*; sore throat, cough, shortness of breath, severe breathing difficulties.	ventilation, local exhaust or respiratory protection.	fresh air, rest, place in half-sitting position, take to hospital.
Skin: redness, pain.	gloves.	remove contaminated clothing, flush skin with water or shower, refer to a doctor.
Eyes: *corrosive*; redness, pain, impaired vision.	face shield.	flush with water, take to a doctor.
Ingestion: *corrosive*; sore throat, abdominal pain, diarrhea.		rinse mouth, call a doctor or take to hospital.

SPILLAGE	STORAGE	LABELING / NFPA
collect leakage in sealable containers, soak up with sand or other inert absorbent and remove to safe place; (additional individual protection: breathing apparatus).	keep in fireproof place, separate from oxidants.	R: 11 S: 9-16-33 Note C 🔥 Flammable [3 / 2 / 0]

NOTES
Lung edema symptoms usually develop several hours later and are aggravated by physical exertion: rest and hospitalization essential. As first aid, a doctor or authorized person should consider administering a corticosteroid spray.

Transport Emergency Card TEC(R)-30G30

HI: 33; UN-number: 1128

CAS-No: [2426-08-6]
n-BGE
2,3-epoxypropyl butyl ether

$$CH_3(CH_2)_3—O—CH_2—CH—CH_2$$
(with epoxide O above CH—CH2)

n-BUTYL GLYCIDYL ETHER

PHYSICAL PROPERTIES		IMPORTANT CHARACTERISTICS

PHYSICAL PROPERTIES

Boiling point, °C	164
Melting point, °C	?
Flash point, °C	58
Relative density (water = 1)	0.9
Relative vapor density (air = 1)	4.5
Relative density at 20 °C of saturated mixture vapor/air (air = 1)	1.01
Vapor pressure, mm Hg at 20 °C	3.0
Solubility in water, g/100 ml at 20 °C	2
Relative molecular mass	130.2

Gross formula: $C_7H_{14}O_2$

IMPORTANT CHARACTERISTICS

COLORLESS LIQUID WITH CHARACTERISTIC ODOR
Vapor mixes readily with air. Presumed to be able to form peroxides. Reacts violently with oxidants.

TLV-TWA	25 ppm	133 mg/m³

Absorption route: Can enter the body by inhalation or ingestion or through the skin. Harmful atmospheric concentrations build up fairly slowly on evaporation at 20° C, but much more rapidly in aerosol form.
Immediate effects: Irritates the eyes, skin and respiratory tract. Affects the nervous system.
Effects of prolonged/repeated exposure: Prolonged or repeated contact can cause dermatitis; hypersensitivity can occur. Inhalation of high concentrations of fumes/vapor can cause lung disorders. Can cause liver damage.

HAZARDS/SYMPTOMS	PREVENTIVE MEASURES	FIRE EXTINGUISHING/FIRST AID
Fire: combustible.	keep away from open flame, no smoking.	powder, AFFF, foam, carbon dioxide, (halons).
Explosion: above 58°C: forms explosive air-vapor mixtures.	above 58°C: sealed machinery, ventilation.	
Inhalation: cough, shortness of breath, drowsiness, unconsciousness.	ventilation, local exhaust or respiratory protection.	fresh air, rest, call a doctor, take to hospital if necessary.
Skin: *is absorbed*; redness.	gloves, protective clothing.	remove contaminated clothing, flush skin with water or shower.
Eyes: redness, pain.	face shield.	flush with water, take to a doctor if necessary.
Ingestion: sore throat, abdominal pain, drowsiness.		rinse mouth, refer to a doctor.

SPILLAGE	STORAGE	LABELING / NFPA
collect leakage in sealable containers, soak up with sand or other inert absorbent and remove to safe place.	keep separate from oxidants; ventilate at floor level.	R: 20-43 S: 24/25 ✖ Harmful

NOTES
Before distilling check for peroxides; if found, render harmless.

BUTYL GLYCOLATE

PHYSICAL PROPERTIES	IMPORTANT CHARACTERISTICS	
Boiling point, °C 184 Melting point, °C − 26 Flash point, °C 68 Auto-ignition temperature, °C 404 Relative density (water = 1) 1.01 Relative vapor density (air = 1) 4.6 Relative density at 20 °C of saturated mixture vapor/air (air = 1) 1.01 Vapor pressure, mm Hg at 20 °C 1.0 Solubility in water, g/100 ml 7 Explosive limits, vol% in air 1.3-? Relative molecular mass 132.2	**COLORLESS LIQUID WITH CHARACTERISTIC ODOR** Vapor mixes readily with air. Reacts with strong oxidants.	
	TLV-TWA not available	
	Absorption route: Can enter the body by inhalation or ingestion. Harmful atmospheric concentrations build up very slowly, if at all, on evaporation at approx. 20°C, but much more rapidly in aerosol form. **Immediate effects:** Irritates the eyes, skin and respiratory tract.	
Gross formula: $C_6H_{12}O_3$		

HAZARDS/SYMPTOMS	PREVENTIVE MEASURES	FIRE EXTINGUISHING/FIRST AID
Fire: combustible.	keep away from open flame, no smoking.	powder, AFFF, foam, carbon dioxide, (halons).
Explosion: above 68°C: forms explosive air-vapor mixtures.	above 68°C: sealed machinery, ventilation.	
Inhalation: cough, headache, nausea.	ventilation, local exhaust or respiratory protection.	fresh air, rest, call a doctor.
Skin: redness.	gloves, protective clothing.	flush skin with water or shower.
Eyes: redness, pain, impaired vision.	safety glasses.	flush with water, take to a doctor if necessary.
Ingestion: vomiting; see also 'Inhalation'.		rinse mouth, call a doctor or take to hospital.

SPILLAGE	STORAGE	LABELING / NFPA
collect leakage in sealable containers, soak up with sand or other inert absorbent and remove to safe place.	keep separate from oxidants; ventilate at floor level.	

NOTES

Transport Emergency Card TEC(R)-30G15

n-BUTYLLITHIUM

PHYSICAL PROPERTIES	IMPORTANT CHARACTERISTICS
Decomposes below boiling point, °C 150 Flash point, °C see notes Relative density (water = 1) 0.76 Solubility in water reaction Explosive limits, vol% in air ?-? Relative molecular mass 63.9	**LIGHT YELLOW OILY LIQUID WITH PUNGENT ODOR** Ignites spontaneously on exposure to air. Lithium oxide vapors are a product of combustion. Decomposes when exposed to air or water vapor, giving off lithium oxide fumes. Strong reducing agent which reacts violently with oxidants. Reacts violently with water, alcohols, phenols and many other substances. Reacts with air to form irritating vapors.
	TLV-TWA not available
	Absorption route: Can enter the body by inhalation or ingestion. Harmful atmospheric concentrations can build up very rapidly on evaporation at 20°C. **Immediate effects:** Corrosive to the eyes, skin and respiratory tract. Inhalation of vapor/fumes can cause severe breathing difficulties (lung edema). In serious cases risk of death.
Gross formula: C_4H_9Li	

HAZARDS/SYMPTOMS	PREVENTIVE MEASURES	FIRE EXTINGUISHING/FIRST AID
Fire: extremely flammable, many chemical reactions can cause fire and explosion.	keep away from open flame and sparks, no smoking, avoid contact with water and many other substances.	special extinguishing powder, dry sand, NO other extinguishing agents, can reignite after being extinguished.
Explosion: contact with water and many other substances can cause explosion.		in case of fire: keep undamaged cylinders cool by spraying with water, but DO NOT spray substance with water.
	STRICT HYGIENE	
Inhalation: *corrosive*; sore throat, cough, shortness of breath, severe breathing difficulties.	ventilation, local exhaust or respiratory protection.	fresh air, rest, place in half-sitting position, take to hospital.
Skin: *corrosive*; redness, burns.	gloves, protective clothing.	remove contaminated clothing, flush skin with water or shower, call a doctor.
Eyes: *corrosive*; redness, pain, impaired vision.	face shield, or combined eye and respiratory protection.	flush with water, take to a doctor.
Ingestion: *corrosive*; sore throat, abdominal pain, diarrhea.		rinse mouth, take immediately to hospital.

SPILLAGE	STORAGE	LABELING / NFPA
evacuate area, call in an expert, dilute leakage with diesel/machine oil, soak up/cover with sand or other inert absorbent and remove to safe place, under no circumstances spray liquid with water; (additional individual protection: CHEMICAL SUIT).	keep in dry, fireproof place under dry nitrogen, separate from all other substances.	3 4 2 W

NOTES

Usually sold as 15-25% solution in hydrocarbons, which does not ignite spontaneously on exposure to air. For 15% solution in hexane see relevant entry. Combustible, but flash point not given in the literature. Reacts violently with extinguishing agents such as carbon dioxide, halons, water, foam and ordinary powder. Lung edema symptoms usually develop several hours later and are aggravated by physical exertion: rest and hospitalization essential. As first aid, a doctor or authorized person should consider administering a corticosteroid spray. Use cylinder with special fittings.

HI: X333; UN-number: 2445

n-BUTYLLITHIUM
(15% solution in hexane)

PHYSICAL PROPERTIES	IMPORTANT CHARACTERISTICS
Boiling point, °C — 69 Melting point, °C — −95 Flash point, °C — −22 Auto-ignition temperature, °C — 240 Relative density (water = 1) — 0.68 Relative vapor density (air = 1) — 3.0 Relative density at 20 °C of saturated mixture vapor/air (air = 1) — 1.3 Vapor pressure, mm Hg at 20 °C — 122 Solubility in water — reaction Explosive limits, vol% in air — 1.2-7.8 Minimum ignition energy, mJ — 0.24 Electrical conductivity, pS/m — 100 Gross formula: — C$_4$H$_9$Li	**CLEAR LIGHT YELLOW LIQUID WITH PUNGENT ODOR** Vapor is heavier than air and spreads at ground level, with risk of ignition at a distance. Flow, agitation etc. can cause build-up of electrostatic charge due to liquid's low conductivity. Do not use compressed air when filling, emptying or processing. Once solvent has evaporated (e.g. on clothing) can ignite spontaneously on exposure to air. Lithium oxide vapors are a product of combustion. Also decomposes in solution when exposed to air or water vapor, giving off → *lithium hydroxide* fumes. Strong reducing agent which reacts violently with oxidants. Reacts violently with water, alcohols, amines, phenols and many other substances, with risk of fire and explosion.

	TLV-TWA	✱

Absorption route: Can enter the body by inhalation or ingestion. Harmful atmospheric concentrations can build up fairly rapidly on evaporation at approx. 20°C - even more rapidly in aerosol form.
Immediate effects: Corrosive to the eyes, skin and respiratory tract. Inhalation of vapor/fumes can cause severe breathing difficulties (lung edema). In serious cases risk of death.

HAZARDS/SYMPTOMS	PREVENTIVE MEASURES	FIRE EXTINGUISHING/FIRST AID
Fire: extremely flammable, many chemical reactions can cause fire and explosion.	keep away from open flame and sparks, no smoking, avoid contact with other substances.	special extinguishing powder, dry sand, NO other extinguishing agents, can reignite after being extinguished.
Explosion: forms explosive air-vapor mixtures, many chemical reactions can cause explosion.	sealed machinery, ventilation, explosion-proof electrical equipment and lighting, grounding, non-sparking tools.	in case of fire: keep undamaged cylinders cool by spraying with water, but DO NOT spray substance with water.
	STRICT HYGIENE	
Inhalation: *corrosive*; sore throat, cough, shortness of breath, severe breathing difficulties.	ventilation, local exhaust or respiratory protection.	fresh air, rest, place in half-sitting position, take to hospital.
Skin: *corrosive*; redness, pain, burns.	gloves, protective clothing.	remove contaminated clothing, flush skin with water or shower.
Eyes: *corrosive*; redness, pain, impaired vision.	face shield, or combined eye and respiratory protection.	flush with water, take to a doctor.
Ingestion: *corrosive*; sore throat, abdominal pain, diarrhea.		rinse mouth, take immediately to hospital.

SPILLAGE	STORAGE	LABELING / NFPA
evacuate area, call in an expert, collect leakage in sealable containers, soak up remainder with sand or other inert absorbent and remove to safe place; (additional individual protection: CHEMICAL SUIT).	keep in dry, fireproof place under inert gas, separate from all other substances.	4 / 2 / 3 / W (NFPA diamond)

NOTES
✱ See TLV for hexane. Physical properties (boiling point, flash point, autoignition point, explosive limits, vapor pressure, relative density and conductivity) are based on hexane. For pure butyllithium see relevant entry. Reacts violently with extinguishing agents such as water, foam and ordinary powder. Remove contaminated clothing immediately (risk of spontaneous combustion). Lung edema symptoms usually develop several hours later and are aggravated by physical exertion: rest and hospitalization essential. As first aid, a doctor or authorized person should consider administering a corticosteroid spray. Use cylinder with special fittings. Pure butyllithium is not available commercially.

HI: X323; UN-number: 2445

$H_2C = C(CH_3)COO(CH_2)_3CH_3$

BUTYL METHACRYLATE

PHYSICAL PROPERTIES	IMPORTANT CHARACTERISTICS
Boiling point, °C 163 Melting point, °C −75 Flash point, °C 41 Auto-ignition temperature, °C 290 Relative density (water = 1) 0.9 Relative vapor density (air = 1) 4.9 Relative density at 20 °C of saturated mixture vapor/air (air = 1) 1.01 Vapor pressure, mm Hg at 20 °C 2.3 Solubility in water, g/100 ml at 20 °C 0.6 Explosive limits, vol% in air 2-8 Relative molecular mass 142,2 Gross formula: $C_8H_{14}O_2$	**COLORLESS LIQUID** Vapor mixes readily with air. Moderate heating, exposure to light, oxidants or contaminants can cause violent polymerization. Flow, agitation etc. can cause build-up of electrostatic charge due to liquid's low conductivity. TLV-TWA not available **Absorption route:** Can enter the body by inhalation or ingestion or through the skin. Harmful atmospheric concentrations build up fairly slowly on evaporation at 20°C, but much more rapidly in aerosol form. **Immediate effects:** Irritates the eyes, skin and respiratory tract. Inhalation of vapor/fumes can cause severe breathing difficulties (lung edema). **Effects of prolonged/repeated exposure:** Prolonged or repeated contact can cause eczema.

HAZARDS/SYMPTOMS	PREVENTIVE MEASURES	FIRE EXTINGUISHING/FIRST AID
Fire: flammable.	keep away from open flame and sparks, no smoking.	powder, AFFF, foam, carbon dioxide, (halons).
Explosion: above 41°C: forms explosive air-vapor mixtures.	above 41°C: sealed machinery, ventilation, explosion-proof electrical equipment, grounding.	in case of fire: keep tanks/drums cool by spraying with water.
Inhalation: sore throat, cough, shortness of breath, dizziness.	ventilation, local exhaust or respiratory protection.	fresh air, rest, place in half-sitting position, call a doctor.
Skin: *is absorbed*; redness, pain.	gloves.	remove contaminated clothing, flush skin with water or shower, refer to a doctor.
Eyes: redness, pain.	face shield.	flush with water, take to a doctor.
Ingestion: sore throat, abdominal pain.		rinse mouth, take immediately to hospital.

SPILLAGE	STORAGE	LABELING / NFPA
collect leakage in sealable containers, soak up with sand or other inert absorbent and remove to safe place.	keep in cool, fireproof place, separate from oxidants; add an inhibitor.	R: 10-36/37/38-43 Note D ✖ Irritant 2 / 2 / 0

NOTES
Lung edema symptoms usually develop several hours later and are aggravated by physical exertion: rest and hospitalization essential. As first aid, a doctor or authorized person should consider administering a corticosteroid spray.

n-BUTYL PROPIONATE

PHYSICAL PROPERTIES	IMPORTANT CHARACTERISTICS	
Boiling point, °C 146 Melting point, °C − 90 Flash point, °C 32 Auto-ignition temperature, °C 425 Relative density (water = 1) 0.87 Relative vapor density (air = 1) 4.5 Relative density at 20 °C of saturated mixture vapor/air (air = 1) 1.01 Vapor pressure, mm Hg at 20 °C 2.9 Solubility in water poor Explosive limits, vol% in air ?-? Relative molecular mass 130.2 Gross formula: $C_7H_{14}O_2$	**COLORLESS LIQUID WITH CHARACTERISTIC ODOR** Vapor mixes readily with air. Reacts with strong oxidants.	
	TLV-TWA not available	
	Absorption route: Can enter the body by inhalation or ingestion or through the skin. Harmful atmospheric concentrations build up very slowly, if at all, on evaporation at approx. 20°C, but much more rapidly in aerosol form. **Immediate effects:** Irritates the eyes, skin and respiratory tract.	

HAZARDS/SYMPTOMS	PREVENTIVE MEASURES	FIRE EXTINGUISHING/FIRST AID
Fire: flammable.	keep away from open flame and sparks, no smoking.	powder, AFFF, foam, carbon dioxide, (halons).
Explosion: above 32°C: forms explosive air-vapor mixtures.	above 32°C: sealed machinery, ventilation, explosion-proof electrical equipment.	in case of fire: keep tanks/drums cool by spraying with water.
Inhalation: sore throat, cough, drowsiness.	ventilation, local exhaust or respiratory protection.	fresh air, rest, call a doctor.
Skin: *is absorbed*; redness.	gloves.	remove contaminated clothing, flush skin with water or shower.
Eyes: redness, pain.	safety glasses.	flush with water, take to a doctor if necessary.
Ingestion: sore throat, abdominal pain, nausea.		rinse mouth, call a doctor.

SPILLAGE	STORAGE	LABELING / NFPA
collect leakage in sealable containers, soak up with sand or other inert absorbent and remove to safe place.	keep in fireproof place, separate from oxidants.	R: 10 Note C 3 2 0

NOTES

Transport Emergency Card TEC(R)-30G35

HI: 30; UN-number: 1914

CAS-No: [5593-70-4]
TBT
tetrabutyl titanate
titanium butylate

$Ti(C_4H_9O)_4$

BUTYL TITANATE

PHYSICAL PROPERTIES	IMPORTANT CHARACTERISTICS
Boiling point, °C 312 Melting point, °C −55 Flash point, °C 78 Relative density (water = 1) 0.99 Relative vapor density (air = 1) 11.5 Relative density at 20 °C of saturated mixture vapor/air (air = 1) 1.0 Vapor pressure, mm Hg at 20°C <0.0076 Solubility in water reaction Relative molecular mass 340.4	**COLORLESS TO LIGHT YELLOW LIQUID WITH CHARACTERISTIC ODOR** Decomposes on contact with water, evolving heat and giving off → *butanol* vapors. Reacts violently with oxidants. TLV-TWA not available **Absorption route:** Can enter the body by inhalation or ingestion. Harmful atmospheric concentrations build up very slowly, if at all, on evaporation at approx. 20°C, but much more rapidly in aerosol form. **Immediate effects:** Irritates the eyes and respiratory tract. Affects the nervous system. **Effects of prolonged/repeated exposure:** Prolonged or repeated contact can cause skin disorders.
Gross formula: $C_{16}H_{36}O_4Ti$	

HAZARDS/SYMPTOMS	PREVENTIVE MEASURES	FIRE EXTINGUISHING/FIRST AID
Fire: combustible.	keep away from open flame, no smoking.	powder, carbon dioxide, (halons), DO NOT USE WATER-BASED EXTINGUISHERS.
Explosion: above 78°C: forms explosive air-vapor mixtures.	above 78°C: sealed machinery, ventilation.	in case of fire: keep tanks/drums cool by spraying with water, but DO NOT spray substance with water.
Inhalation: sore throat, cough, headache, sleepiness.	ventilation, local exhaust or respiratory protection.	fresh air, rest, call a doctor.
Skin: redness.	gloves.	remove contaminated clothing, flush skin with water or shower.
Eyes: redness, pain.	safety glasses.	flush with water, take to a doctor if necessary.
Ingestion: sore throat, abdominal pain, nausea, sleepiness.		rinse mouth, call a doctor.

SPILLAGE	STORAGE	LABELING / NFPA
collect leakage in sealable containers, soak up with dry sand or other inert absorbent and remove to safe place.	keep dry, separate from oxidants; ventilate at floor level.	

NOTES
Use airtight packaging.

Transport Emergency Card TEC(R)-30G15

1-methyl-4-tert-butylbenzene
4-tert-butyltoluene
1-(1,1-dimethylethyl)-4-methylbenzene

$CH_3C_6H_4C(CH_3)_3$

p-tert-BUTYLTOLUENE

PHYSICAL PROPERTIES	IMPORTANT CHARACTERISTICS
Boiling point, °C 193 Melting point, °C −52 Flash point, °C 68 Relative density (water = 1) 0.9 Relative vapor density (air = 1) 5.1 Relative density at 20 °C of saturated mixture vapor/air (air = 1) 1.0 Vapor pressure, mm Hg at 20 °C 0.68 Solubility in water none Relative molecular mass 148.3 Gross formula: $C_{11}H_{16}$	**COLORLESS LIQUID WITH CHARACTERISTIC ODOR** Vapor mixes readily with air. Flow, agitation etc. can cause build-up of electrostatic charge due to liquid's low conductivity. Reacts with strong oxidants. TLV-TWA 10 ppm 61 mg/m³ STEL 20 ppm 121 mg/m³ **Absorption route:** Can enter the body by inhalation or ingestion or through the skin. Harmful atmospheric concentrations build up fairly slowly on evaporation, but much more rapidly in aerosol form. **Immediate effects:** Irritates the eyes, skin, respiratory tract and mucous membranes. Liquid destroys the skin's natural oils. Affects the nervous system and blood. **Effects of prolonged/repeated exposure:** Can cause liver and kidney damage.

HAZARDS/SYMPTOMS	PREVENTIVE MEASURES	FIRE EXTINGUISHING/FIRST AID
Fire: combustible.	keep away from open flame, no smoking.	powder, AFFF, foam, carbon dioxide, (halons).
Explosion: above 68°C: forms explosive air-vapor mixtures.	above 68°C: sealed machinery, ventilation, grounding.	
Inhalation: cough, shortness of breath, headache, sleepiness, feeling of weakness.	ventilation, local exhaust or respiratory protection.	fresh air, rest, call a doctor if necessary.
Skin: redness, pain.	gloves, protective clothing.	remove contaminated clothing, flush skin with water or shower.
Eyes: redness, pain.	safety glasses.	flush with water, take to a doctor.
Ingestion: sore throat, abdominal pain, shortness of breath, nausea, drowsiness, feeling of weakness.		rinse mouth, call a doctor or take to hospital.

SPILLAGE	STORAGE	LABELING / NFPA
collect leakage in sealable containers, soak up with sand or other inert absorbent and remove to safe place.	keep separate from oxidants; ventilate at floor level.	

NOTES
Alcohol consumption increases toxic effects. Depending on the degree of exposure, regular medical checkups are advisable.

Transport Emergency Card TEC(R)-153

HI: 30; UN-number: 2667

2-BUTYNE

PHYSICAL PROPERTIES	IMPORTANT CHARACTERISTICS
Boiling point, °C 27 Melting point, °C −32 Flash point, °C < −20 Auto-ignition temperature, °C ? Relative density (water = 1) 0.7 Relative vapor density (air = 1) 1.9 Relative density at 20 °C of saturated mixture vapor/air (air = 1) 1.7 Vapor pressure, mm Hg at 20 °C 597 Solubility in water none Explosive limits, vol% in air 1.4-? Relative molecular mass 54.1 Gross formula: C_4H_6	**COLORLESS LIQUID WITH CHARACTERISTIC ODOR** Vapor is heavier than air and spreads at ground level, with risk of ignition at a distance; can accumulate close to ground level, causing oxygen deficiency, with risk of unconsciousness. Flow, agitation etc. can cause build-up of electrostatic charge due to liquid's low conductivity. Do not use compressed air when filling, emptying or processing. Reacts with oxidants. **TLV-TWA** not available **Absorption route:** Can enter the body by inhalation or ingestion. **Immediate effects:** Liquid destroys the skin's natural oils. Can cause frostbite due to rapid evaporation.

HAZARDS/SYMPTOMS	PREVENTIVE MEASURES	FIRE EXTINGUISHING/FIRST AID
Fire: extremely flammable.	keep away from open flame and sparks, no smoking.	powder, AFFF, foam, carbon dioxide, (halons).
Explosion: forms explosive air-vapor mixtures.	sealed machinery, ventilation, explosion-proof electrical equipment and lighting, grounding.	in case of fire: keep tanks/drums cool by spraying with water.
Inhalation: headache, dizziness, drowsiness, unconsciousness. ✱	ventilation, local exhaust or respiratory protection.	fresh air, rest, artificial respiration if necessary, take to hospital.
Skin: *in case of frostbite*: redness, pain, sores.	insulating gloves.	*in case of frostbite*: DO NOT remove clothing, flush skin with water or shower, take to a doctor.
Eyes: *in case of frostbite*: redness, pain, impaired vision.	face shield.	flush with water, take to a doctor if necessary.
Ingestion: abdominal pain.		rinse mouth, call a doctor.

SPILLAGE	STORAGE	LABELING / NFPA
evacuate area, call in an expert, ventilate, collect leakage in sealable containers, soak up with sand or other inert absorbent and remove to safe place; (additional individual protection: breathing apparatus).	keep in cool, fireproof place, separate from oxidants.	(NFPA diamond, 4)

NOTES
✱ High atmospheric concentrations, e.g. in poorly ventilated spaces, can cause oxygen deficiency, with risk of unconsciousness.

BUTYRALDEHYDE

PHYSICAL PROPERTIES		IMPORTANT CHARACTERISTICS
Boiling point, °C	75	**COLORLESS LIQUID WITH PUNGENT ODOR**
Melting point, °C	− 100	Vapor is heavier than air and spreads at ground level, with risk of ignition at a distance. Do not use
Flash point, °C	−7	compressed air when filling, emptying or processing. Can ignite spontaneously when finely
Auto-ignition temperature, °C	230	dispersed in air. Reacts violently with strong oxidants and many other substances.
Relative density (water = 1)	0.8	
Relative vapor density (air = 1)	2.5	**TLV-TWA** not available
Relative density at 20 °C of saturated mixture		
vapor/air (air = 1)	1.18	**Absorption route:** Can enter the body by inhalation or ingestion. Harmful atmospheric
Vapor pressure, mm Hg at 20 °C	93.5	concentrations can build up fairly rapidly on evaporation at approx. 20°C - even more rapidly in
Solubility in water, g/100 ml at 20 °C	3.7	aerosol form.
Explosive limits, vol% in air	1.4-12.5	**Immediate effects:** Irritates the skin. Corrosive to the eyes and respiratory tract. Inhalation can
Relative molecular mass	72.1	cause lung edema. In serious cases risk of death.
Log P octanol/water	1.2	**Effects of prolonged/repeated exposure:** Prolonged or repeated contact can cause skin disorders.
Gross formula:	C_4H_8O	

HAZARDS/SYMPTOMS	PREVENTIVE MEASURES	FIRE EXTINGUISHING/FIRST AID
Fire: extremely flammable.	keep away from open flame and sparks, no smoking.	powder, AFFF, foam, carbon dioxide, (halons).
Explosion: forms explosive air-vapor mixtures.	sealed machinery, ventilation, explosion-proof electrical equipment and lighting.	in case of fire: keep tanks/drums cool by spraying with water.
Inhalation: *corrosive*; sore throat, cough, shortness of breath, severe breathing difficulties.	ventilation, local exhaust or respiratory protection.	fresh air, rest, place in half-sitting position, take to hospital.
Skin: redness, pain.	gloves.	remove contaminated clothing, flush skin with water or shower.
Eyes: *corrosive*; redness, pain, impaired vision.	face shield.	flush with water, take to a doctor if necessary.
Ingestion: *corrosive*; sore throat, abdominal pain.		rinse mouth, immediately take to hospital.

SPILLAGE	STORAGE	LABELING / NFPA
collect leakage in sealable containers, soak up with sand or other inert absorbent and remove to safe place, do not use sawdust or other combustible absorbents; (additional individual protection: breathing apparatus).	keep in dark, fireproof place, separate from oxidants.	R: 11 S: 9-29-33 🔥 Flammable 3 / 2 / 2

NOTES
Lung edema symptoms usually develop several hours later and are aggravated by physical exertion: rest and hospitalization essential. As first aid, a doctor or authorized person should consider administering a corticosteroid spray. The measures on this card also apply to isobutyraldehyde.

CAS-No: [107-92-6]
n-butanoic acid
n-butyric acid
ethylacetic acid
1-propanecarboxylic acid
propylformic acid

C_3H_7COOH

BUTYRIC ACID

PHYSICAL PROPERTIES		IMPORTANT CHARACTERISTICS
Boiling point, °C	164	**COLORLESS LIQUID WITH PUNGENT ODOR**
Melting point, °C	−4.5	Vapor mixes readily with air. Medium strong acid which reacts violently with bases and is
Flash point, °C	72	corrosive. Reacts violently with strong oxidants.
Auto-ignition temperature, °C	440	
Relative density (water = 1)	0.96	TLV-TWA not available
Relative vapor density (air = 1)	3.0	
Relative density at 20 °C of saturated mixture		**Absorption route:** Can enter the body by inhalation or ingestion. Harmful atmospheric
vapor/air (air = 1)	1.0	concentrations build up fairly slowly on evaporation, but much more rapidly in aerosol form.
Vapor pressure, mm Hg at 20 °C	0.76	**Immediate effects:** Irritates the eyes, skin and respiratory tract.
Solubility in water	∞	**Effects of prolonged/repeated exposure:** Can affect the blood. Can cause kidney damage.
Explosive limits, vol% in air	2-10	
Electrical conductivity, pS/m	< 1x10⁶	
Relative molecular mass	88.1	
Log P octanol/water	0.8	
Gross formula:	$C_4H_8O_2$	

HAZARDS/SYMPTOMS	PREVENTIVE MEASURES	FIRE EXTINGUISHING/FIRST AID
Fire: combustible.	keep away from open flame, no smoking.	AFFF, water spray, alcohol-resistant foam, carbon dioxide, (halons).
Explosion: above 72°C: forms explosive air-vapor mixtures.	above 72°C: sealed machinery, ventilation.	
Inhalation: sore throat, cough, shortness of breath.	ventilation, local exhaust or respiratory protection.	fresh air, rest, place in half-sitting position.
Skin: redness, pain.	gloves.	remove contaminated clothing, flush skin with water or shower, call a doctor.
Eyes: redness, pain, impaired vision.	safety glasses.	flush with water, take to a doctor if necessary.
Ingestion: sore throat, abdominal pain.		rinse mouth, call a doctor.

SPILLAGE	STORAGE	LABELING / NFPA
collect leakage in sealable containers, flush away remainder with water.	keep separate from oxidants and strong bases; ventilate at floor level.	R: 34 S: 26-36 Corrosive

NOTES
The measures on this card also apply to isobutyric acid.

Transport Emergency Card TEC(R)-80G20C

HI: 80; UN-number: 2820

CAS-No: [109-74-0]
butanenitrile
propyl cyanide

$CH_3CH_2CH_2C \equiv N$

N-BUTYRONITRILE

PHYSICAL PROPERTIES	IMPORTANT CHARACTERISTICS
Boiling point, °C 118 Melting point, °C −112 Flash point, °C 21 Auto-ignition temperature, °C 501 Relative density (water = 1) 0.8 Relative vapor density (air = 1) 2.4 Relative density at 20 °C of saturated mixture vapor/air (air = 1) 1.02 Vapor pressure, mm Hg at 20 °C 15.2 Solubility in water poor Explosive limits, vol% in air 1.65-? Relative molecular mass 69.1	**COLORLESS LIQUID WITH CHARACTERISTIC ODOR** Vapor mixes readily with air, forming explosive mixtures. Do not use compressed air when filling, emptying or processing. Toxic gases are a product of combustion (→ *nitrous vapors*). Decomposes on contact with steam or hot surface, giving off toxic gas (→ *hydrogen cyanide*). Reacts violently with acids, giving off toxic gas (→ *hydrogen cyanide*). Reacts violently with strong oxidants.
	TLV-TWA not available
	Absorption route: Can enter the body by inhalation or ingestion or through the skin. Harmful atmospheric concentrations can build up fairly rapidly on evaporation at approx. 20°C - even more rapidly in aerosol form. **Immediate effects:** Impedes tissue respiration.
Gross formula: C_4H_7N	

HAZARDS/SYMPTOMS	PREVENTIVE MEASURES	FIRE EXTINGUISHING/FIRST AID
Fire: extremely flammable.	keep away from open flame and sparks, no smoking.	powder, AFFF, foam, carbon dioxide, (halons).
Explosion: forms explosive air-vapor mixtures.	sealed machinery, ventilation, explosion-proof electrical equipment and lighting.	in case of fire: keep tanks/drums cool by spraying with water.
	STRICT HYGIENE	IN ALL CASES CALL IN A DOCTOR
Inhalation: severe breathing difficulties, drowsiness, feeling of weakness, unconsciousness.	ventilation, local exhaust or respiratory protection.	fresh air, rest, special treatment, take to hospital.
Skin: *is absorbed*; headache, severe breathing difficulties, feeling of weakness.	gloves, protective clothing.	remove contaminated clothing, flush skin with water or shower.
Eyes: redness.	face shield, or combined eye and respiratory protection.	flush with water, take to a doctor.
Ingestion: severe breathing difficulties, drowsiness, feeling of weakness, unconsciousness.	do not eat, drink or smoke while working.	fresh air, rest, special treatment, take immediately to hospital.

SPILLAGE	STORAGE	LABELING / NFPA
collect leakage in sealable containers, render remainder harmless with sodium hypochlorite solution; (additional individual protection: breathing apparatus).	keep in fireproof place, separate from oxidants and acids.	R: 10-23/24/25 S: 44 ☠ Toxic

NOTES
Special first aid required in the event of poisoning: antidotes must be available (with instructions).

NFPA diagram values: 3 (top), 3 (left), 0 (right)

Transport Emergency Card TEC(R)-30G45

HI: 336; UN-number: 2411

BUTYRYL CHLORIDE

PHYSICAL PROPERTIES		IMPORTANT CHARACTERISTICS
Boiling point, °C	101	**COLORLESS LIQUID WITH PUNGENT ODOR**
Melting point, °C	−89	Vapor is heavier than air and spreads at ground level, with risk of ignition at a distance. Do not use
Flash point, °C	<21	compressed air when filling, emptying or processing. Toxic and corrosive vapors are a product of
Relative density (water = 1)	1.03	combustion (→ *hydrochloric acid*, → *phosgene*). Reacts with water (slowly) and alcohols and
Relative vapor density (air = 1)	3.67	many other compounds, giving off corrosive vapors (→ *hydrochloric acid*). Reacts violently with
Relative density at 20 °C of saturated mixture		oxidants. Reacts with air to form corrosive vapors (→ *hydrochloric acid*).
vapor/air (air = 1)	1.10	
Vapor pressure, mm Hg at 20 °C	ca. 28	TLV-TWA not available
Solubility in water	reaction	
Explosive limits, vol% in air	?-?	**Absorption route:** Can enter the body by inhalation or ingestion.
Relative molecular mass	106.6	**Immediate effects:** Corrosive to the eyes, skin, mucous membranes and respiratory tract. Inhalation of vapor/fumes can cause severe breathing difficulties (lung edema). In serious cases risk of death.
Gross formula:	C_4H_7ClO	

HAZARDS/SYMPTOMS	PREVENTIVE MEASURES	FIRE EXTINGUISHING/FIRST AID
Fire: extremely flammable.	keep away from open flame and sparks, no smoking.	powder, AFFF, foam, carbon dioxide, (halons).
Explosion: forms explosive air-vapor mixtures.	sealed machinery, ventilation, explosion-proof electrical equipment and lighting.	in case of fire: keep tanks/drums cool by spraying with water.
	STRICT HYGIENE	IN ALL CASES CALL A DOCTOR
Inhalation: *corrosive*; sore throat, cough, shortness of breath, severe breathing difficulties.	ventilation, local exhaust or respiratory protection.	fresh air, rest, place in half-sitting position, artificial respiration, take to hospital.
Skin: *corrosive*; redness, pain, burns.	gloves, protective clothing.	remove contaminated clothing, flush skin with water or shower, refer to a doctor.
Eyes: *corrosive*; redness, pain, impaired vision.	face shield, or combined eye and respiratory protection.	flush with water, take to a doctor.
Ingestion: *corrosive*; sore throat, abdominal pain, diarrhea.		rinse mouth, call a doctor or take to hospital.

SPILLAGE	STORAGE	LABELING / NFPA
collect leakage in sealable containers, soak up with sand or other inert absorbent and remove to safe place; (additional individual protection: breathing apparatus).	keep in dry, fireproof place.	R: 11-34 S: 16-23-26-36 Flammable Corrosive

NOTES
Lung edema symptoms usually develop several hours later and are aggravated by physical exertion: rest and hospitalization essential. As first aid, a doctor or authorized person should consider administering a corticosteroid spray.

Transport Emergency Card TEC(R)-892

HI: 338; UN-number: 2353

CADMIUM

PHYSICAL PROPERTIES	IMPORTANT CHARACTERISTICS	
Boiling point, °C .... 765 Melting point, °C .... 321 Relative density (water = 1) .... 8.6 Solubility in water .... none Relative atomic mass 112.4	**SILVER-WHITE METAL OR GRAY POWDER** Can ignite spontaneously on exposure to air. Reacts violently with acids, giving off flammable gas (→ *hydrogen*). Cadmium powder reacts violently with strong oxidants.	
	TLV-TWA	0.05 mg/m³
	Absorption route: Can enter the body by inhalation or ingestion. **Immediate effects:** Irritates the eyes, skin and gastro-intestinal tract. Corrosive to the respiratory tract. Inhalation of vapor/fumes can cause lung edema. In serious cases risk of death. **Effects of prolonged/repeated exposure:** Can affect the blood. Can cause kidney damage. Can impair sense of smell.	
Gross formula: Cd		

HAZARDS/SYMPTOMS	PREVENTIVE MEASURES	FIRE EXTINGUISHING/FIRST AID
Fire: combustible.	keep away from open flame, no smoking.	special powder extinguisher, dry sand, NO other extinguishing agents.
Explosion: finely dispersed particles form explosive mixtures on contact with air.	keep dust from accumulating; sealed machinery, explosion-proof electrical equipment and lighting, grounding.	
	KEEP DUST UNDER CONTROL	
Inhalation: *corrosive*; sore throat, cough, shortness of breath, severe breathing difficulties.	local exhaust or respiratory protection.	fresh air, rest, place in half-sitting position, take to hospital.
Skin: redness.	gloves.	remove contaminated clothing, flush skin with water or shower.
Eyes: redness, pain.	goggles.	flush with water, take to a doctor.
Ingestion: abdominal pain, diarrhea, nausea, vomiting.	do not eat, drink or smoke while working.	rinse mouth, immediately take to hospital.

SPILLAGE	STORAGE	LABELING / NFPA
clean up spillage, carefully collect remainder; (additional individual protection: P3 respirator).	keep in fireproof place, separate from acids.	

NOTES
Do not take work clothing home. Reacts violently with extinguishing agents such as water, foam, carbon dioxide and halons. Lung edema symptoms usually develop several hours later and are aggravated by physical exertion: rest and hospitalization essential. As first aid, a doctor or authorized person should consider administering a corticosteroid spray. Unbreakable packaging preferred; if breakable, keep in unbreakable container.

CADMIUM ACETATE

PHYSICAL PROPERTIES	IMPORTANT CHARACTERISTICS	
Melting point, °C — 130 Relative density (water = 1) — 2.01 Solubility in water — good Relative molecular mass — 284.5	**COLORLESS CRYSTALS** Decomposes when heated, giving off toxic vapors.	
	TLV-TWA	0.02 mg/m³ ✱
	Absorption route: Can enter the body by inhalation or ingestion. **Immediate effects:** Irritates the eyes, skin, respiratory tract and gastro-intestinal tract. Inhalation of vapor/fumes can cause severe breathing difficulties (lung edema). In serious cases risk of death. **Effects of prolonged/repeated exposure:** Can affect the blood. Can cause kidney damage. Can impair sense of smell.	
Gross formula: $C_4H_6CdO_4 \cdot 3H_2O$		

HAZARDS/SYMPTOMS	PREVENTIVE MEASURES	FIRE EXTINGUISHING/FIRST AID
Fire: non-combustible.		in case of fire in immediate vicinity: use any extinguishing agent.
	KEEP DUST UNDER CONTROL	
Inhalation: sore throat, cough, shortness of breath, severe breathing difficulties.	local exhaust or respiratory protection.	fresh air, rest, place in half-sitting position, take to hospital.
Skin: redness.	gloves.	remove contaminated clothing, flush skin with water or shower.
Eyes: redness, pain.	goggles, or combined eye and respiratory protection.	flush with water, take to a doctor.
Ingestion: abdominal pain, diarrhea, nausea, vomiting.		rinse mouth, take immediately to hospital.

SPILLAGE	STORAGE	LABELING / NFPA
clean up spillage, carefully collect remainder; (additional individual protection: P3 respirator).		R: 20/21/22 S: 22 Note A ✖ Harmful

NOTES

Apparent melting point due to loss of water of crystallization. Melting point of anhydrous cadmium acetate 256° C, relative density 2.34, molecular weight 230.5. ✱ TLV equals that of cadmium. Depending on the degree of exposure, regular medical checkups are advisable. Lung edema symptoms usually develop several hours later and are aggravated by physical exertion: rest and hospitalization essential. As first aid, a doctor or authorized person should consider administering a corticosteroid spray.

Transport Emergency Card TEC(R)-61G11   **HI: 60; UN-number: 2570**

CADMIUM HYDROXIDE

PHYSICAL PROPERTIES	IMPORTANT CHARACTERISTICS	
Melting point (decomposes), °C 120 Solubility in water none Relative molecular mass 146.4	**WHITE POWDER** Decomposes when heated above 120°C, forming → *cadmium oxide*.	
	TLV-TWA	0.02 mg/m³ ✱
	Absorption route: Can enter the body by inhalation or ingestion. **Immediate effects:** Irritates the eyes, skin and gastro-intestinal tract. Corrosive to the respiratory tract. Inhalation can cause severe breathing difficulties (lung edema). In serious cases risk of death. **Effects of prolonged/repeated exposure:** Can impair sense of smell. Can affect the blood. Can cause kidney damage.	
Gross formula: CdH$_2$O$_2$		

HAZARDS/SYMPTOMS	PREVENTIVE MEASURES	FIRE EXTINGUISHING/FIRST AID
Fire: non-combustible.		in case of fire in immediate vicinity: use any extinguishing agent.
	KEEP DUST UNDER CONTROL	
Inhalation: *corrosive*; sore throat, cough, shortness of breath, severe breathing difficulties.	local exhaust or respiratory protection.	fresh air, rest, place in half-sitting position, take to hospital.
Skin: redness, pain.	gloves.	remove contaminated clothing, flush skin with water or shower.
Eyes: redness, pain, impaired vision.	goggles.	flush with water, take to a doctor.
Ingestion: abdominal pain, diarrhea, nausea, vomiting.		rinse mouth, take immediately to hospital.

SPILLAGE	STORAGE	LABELING / NFPA
clean up spillage, carefully collect remainder; (additional individual protection: P3 respirator).		R: 20/21/22 S: 22 Note A ✖ Harmful

NOTES
✱ TLV equals that of cadmium (total dust). Do not take work clothing home. Lung edema symptoms usually develop several hours later and are aggravated by physical exertion: rest and hospitalization essential. As first aid, a doctor or authorized person should consider administering a corticosteroid spray. Unbreakable packaging preferred; if breakable, keep in unbreakable container.

Transport Emergency Card TEC(R)-61G11   **UN-number: 2570**

CADMIUM OXIDE

PHYSICAL PROPERTIES	IMPORTANT CHARACTERISTICS	
Sublimation temperature, °C 1560 Relative density (water = 1) 7 à 8 Vapor pressure, mm Hg at 55 °C 10^{-6} Solubility in water, g/100 ml 0.005 Relative molecular mass 128.4	**BROWNISH-RED CRYSTALS OR YELLOW TO DARK BROWN POWDER** Decomposes slowly when heated above 700°C, giving off toxic → *cadmium* vapors.	
	TLV-TWA	0.01 mg/m³ A2 ✱
	Absorption route: Can enter the body by inhalation or ingestion. Evaporation negligible at 20°C, but harmful concentrations of airborne particles can build up rapidly. **Immediate effects:** Irritates the eyes, skin and gastro-intestinal tract. Corrosive to the respiratory tract. Inhalation can cause lung edema. In serious cases risk of death. **Effects of prolonged/repeated exposure:** Can affect the blood. Can cause kidney damage.	
Gross formula: CdO		

HAZARDS/SYMPTOMS	PREVENTIVE MEASURES	FIRE EXTINGUISHING/FIRST AID
Fire: non-combustible.		in case of fire in immediate vicinity: use any extinguishing agent.
	KEEP DUST UNDER CONTROL	
Inhalation: *corrosive*; sore throat, cough, shortness of breath, severe breathing difficulties.	local exhaust or respiratory protection.	fresh air, rest, place in half-sitting position, take to hospital.
Skin: redness.	gloves.	remove contaminated clothing, flush skin with water or shower, refer to a doctor.
Eyes: redness, pain.	goggles, or combined eye and respiratory protection.	flush with water, take to a doctor.
Ingestion: abdominal pain, diarrhea, nausea, vomiting.	do not eat, drink or smoke while working.	rinse mouth, call a doctor or take immediately to hospital.

SPILLAGE	STORAGE	LABELING / NFPA
clean up spillage, carefully collect remainder; (additional individual protection: P3 respirator).		R: 23/25-33-40 S: 22-44 ☠ Toxic

NOTES
Cadmium oxide exists in various crystal forms. At 700°C sublimation is already substantial. Do not take work clothing home. ✱ TLV equals that of cadmium (total dust); cadmium (respirable fraction) has TLV-TWA of 0.002mg/m³ (A2). Depending on the degree of exposure, regular medical checkups are advisable. Lung edema symptoms usually develop several hours later and are aggravated by physical exertion: rest and hospitalization essential. As first aid, a doctor or authorized person should consider administering a corticosteroid spray.

Transport Emergency Card TEC(R)-61G11

UN-number: 2570

CADMIUM SULFIDE

PHYSICAL PROPERTIES	IMPORTANT CHARACTERISTICS	
Sublimation temperature, °C 980 Relative density (water = 1) 4.8 Solubility in water none Relative molecular mass 144.5	**YELLOW OR ORANGE-BROWN POWDER** Decomposes on contact with acids, giving off toxic gas (→ *hydrogen sulfide*). Reacts with strong oxidants.	
	TLV-TWA 0.02 mg/m³ ✳	
	Absorption route: Can enter the body by inhalation or ingestion. **Immediate effects:** Irritates the eyes, skin and gastro-intestinal tract. Corrosive to the respiratory organs. Inhalation can cause lung edema. In serious cases risk of death. **Effects of prolonged/repeated exposure:** Can affect the blood. Can cause kidney damage. Can impair sense of smell.	
Gross formula: CdS		

HAZARDS/SYMPTOMS	PREVENTIVE MEASURES	FIRE EXTINGUISHING/FIRST AID
Fire: combustible.		water spray, powder.
	KEEP DUST UNDER CONTROL	
Inhalation: *corrosive*; sore throat, cough, shortness of breath, severe breathing difficulties.	local exhaust or respiratory protection.	fresh air, rest, place in half-sitting position, take to hospital.
Skin: redness.	gloves.	remove contaminated clothing, flush skin with water or shower.
Eyes: redness, pain.	goggles, or combined eye and respiratory protection.	flush with water, take to a doctor.
Ingestion: abdominal pain, diarrhea, nausea, vomiting.	do not eat, drink or smoke while working.	rinse mouth, take immediately to hospital.

SPILLAGE	STORAGE	LABELING / NFPA
clean up spillage, carefully collect remainder; (additional individual protection: P3 respirator).	keep separate from acids.	

NOTES
✳ TLV equals that of cadmium. Lung edema symptoms usually develop several hours later and are aggravated by physical exertion: rest and hospitalization essential. As first aid, a doctor or authorized person should consider administering a corticosteroid spray. Do not take work clothing home. Unbreakable packaging preferred; if breakable, keep in unbreakable container.

UN-number: 2570

CALCIUM

PHYSICAL PROPERTIES	IMPORTANT CHARACTERISTICS
Boiling point, °C 1484 Melting point, °C 850 Relative density (water = 1) 1.5 Vapor pressure, mm Hg at 983 °C 9.9 Solubility in water reaction Relative atomic mass 40.1	**SILVER-WHITE METALLIC-LIKE LUMPS OR POWDER** In finely dispersed form the substance can ignite spontaneously on exposure to air. Reacts with mixture of copper oxides and iron oxides to form shock-sensitive compounds. Strong reducing agent which reacts violently with oxidants. Reacts with cold water and reacts violently with warm water, acids, halogenated hydrocarbons, alkali hydroxides, carbonates and many other substances, with risk of fire and explosion. Reacts with air to form flammable gas (→ *hydrogen*).
	TLV-TWA not available
	Absorption route: Can enter the body by inhalation or ingestion. **Immediate effects:** Irritates the respiratory tract. Corrosive to the eyes and skin.
Gross formula: Ca	

HAZARDS/SYMPTOMS	PREVENTIVE MEASURES	FIRE EXTINGUISHING/FIRST AID
Fire: combustible, many chemical reactions can cause fire and explosion.	keep away from open flame, no smoking, avoid contact with water, acids and many other substances.	dry sand, special powder extinguisher, DO NOT USE WATER-BASED EXTINGUISHERS.
Explosion: reaction with water, acids and many other substances can cause explosion.		in case of fire: keep tanks/drums cool by spraying with water, but DO NOT spray substance with water.
	STRICT HYGIENE	
Inhalation: sore throat, cough, shortness of breath.	local exhaust or respiratory protection.	fresh air, rest, call a doctor.
Skin: *corrosive*; redness, pain, serious burns.	gloves, protective clothing.	remove contaminated clothing, first wipe skin clean, flush skin with water or shower, call a doctor, take to hospital if necessary.
Eyes: *corrosive*; redness, pain, impaired vision.	face shield.	flush with water, take to a doctor.
Ingestion: abdominal pain, vomiting.		rinse mouth, immediately take to hospital.

SPILLAGE	STORAGE	LABELING / NFPA
clean up spillage, carefully collect remainder, do not use sawdust or other combustible absorbents; (additional individual protection: P3 respirator).	keep in fireproof, dry place under petroleum, separate from many other substances.	R: 15 S: 8-24/25-43 🔥 Flammable

NOTES
Reacts violently with extinguishing agents such as water, foam, halons and carbon dioxide. Wash work clothing, do not dry clean. Use airtight packaging. Unbreakable packaging preferred; if breakable, keep in unbreakable container.

Transport Emergency Card TEC(R)-43G01   HI: X423; UN-number: 1401

CALCIUM BROMATE
(monohydrate)

PHYSICAL PROPERTIES	IMPORTANT CHARACTERISTICS
Melting point, °C — 180 Relative density (water = 1) — 3.3 Solubility in water — very good Relative molecular mass — 313.9 Gross formula: Br$_2$CaO$_6$.H$_2$O	**WHITE CRYSTALS OR POWDER** Decomposes when heated, giving off toxic and corrosive vapors and → *oxygen*, with increased risk of fire. Strong oxidant which reacts violently with combustible substances and reducing agents, with risk of fire and explosion. Reacts with strong acids, giving off corrosive vapors (→ *hydrogen bromide*). Reacts violently with textiles, oil, fat, sugar, sawdust, ammonium salts, carbon, sulfur, phosphorus, metal powder and sulfides, with risk of fire and explosion.

TLV-TWA	not available

Absorption route: Can enter the body by inhalation or ingestion. Evaporation negligible at 20°C, but harmful concentrations of airborne particles can build up rapidly.
Immediate effects: Irritates the eyes, skin and respiratory tract. Affects the nervous system.
Effects of prolonged/repeated exposure: Can affect the blood. Can cause kidney and brain damage. In serious cases risk of unconsciousness and death.

HAZARDS/SYMPTOMS	PREVENTIVE MEASURES	FIRE EXTINGUISHING/FIRST AID
Fire: not combustible but enhances combustion of other substances.	avoid contact with combustible substances.	in case of fire in immediate vicinity: use any extinguishing agent.
Explosion: contact/mixing with many substances can cause explosion.		in case of fire: keep tanks/drums cool by spraying with water.
	STRICT HYGIENE, KEEP DUST UNDER CONTROL	
Inhalation: cough, shortness of breath, headache, dizziness, blue skin.	local exhaust or respiratory protection.	fresh air, rest, take to hospital.
Skin: redness.	gloves.	remove contaminated clothing, flush skin with water or shower.
Eyes: redness, pain.	goggles.	flush with water, take to a doctor.
Ingestion: abdominal pain, dizziness, vomiting, blue skin.	do not eat, drink or smoke while working.	rinse mouth, immediately take to hospital.

SPILLAGE	STORAGE	LABELING / NFPA
clean up spillage, flush away remainder with water, do not use sawdust or other combustible absorbents; (additional individual protection: P2 respirator).	keep separate from combustible substances, reducing agents and strong acids.	◇

NOTES
Apparent melting point due to loss of water of crystallization. Flush contaminated clothing with water (fire hazard). Depending on the degree of exposure, regular medical checkups are advisable. Effects on blood due to formation of methemoglobin. Special first aid required in the event of poisoning: antidotes must be available (with instructions). Packaging: special material.

Transport Emergency Card TEC(R)-51G01   **UN-number: 1450**

calcium acetylide
calcium dicarbide
carbide

CaC$_2$

CALCIUM CARBIDE

PHYSICAL PROPERTIES	IMPORTANT CHARACTERISTICS
Boiling point, °C 2300 Melting point, °C > 447 Relative density (water = 1) 2.2 Solubility in water reaction Relative molecular mass 64.1	**GREY HYGROSCOPIC GRANULES OR POWDER** Decomposes on contact with water, giving off calcium hydroxide and → *acetylene* gas, with risk of fire and explosion. Resultant acetylene is usually contaminated with → *phosphine*. Reacts with moist air to form flammable → *acetylene* gas.
	TLV-TWA not available
	Absorption route: Can enter the body by inhalation or ingestion. **Immediate effects:** Corrosive to the eyes, skin and respiratory tract. Inhalation can cause severe breathing difficulties (lung edema).
Gross formula: C$_2$Ca	

HAZARDS/SYMPTOMS	PREVENTIVE MEASURES	FIRE EXTINGUISHING/FIRST AID
Fire: non-combustible, but reaction with water can cause fire and explosion.	avoid contact with water and moist air.	powder, dry sand, DO NOT USE WATER-BASED EXTINGUISHERS; in case of fire in immediate vicinity: do not use water or water-based extinguishers.
Inhalation: *corrosive*; sore throat, cough, shortness of breath, severe breathing difficulties.	local exhaust or respiratory protection.	fresh air, rest, place in half-sitting position, take to hospital.
Skin: *corrosive*; redness, pain, burns.	gloves, protective clothing.	remove contaminated clothing, flush skin with water or shower, refer to a doctor.
Eyes: *corrosive*; redness, pain, impaired vision.	goggles.	flush with water, take to a doctor.
Ingestion: *corrosive*; sore throat, cough, diarrhea, vomiting.		rinse mouth, take immediately to hospital.

SPILLAGE	STORAGE	LABELING / NFPA
clean up spillage, carefully collect remainder, avoid contact with water; (additional individual protection: P2 respirator).	keep dry; ventilate.	R: 15 S: 8-43 🔥 Flammable NFPA: 1 / 3 / 2 / W

NOTES
Lung edema symptoms usually develop several hours later and are aggravated by physical exertion: rest and hospitalization essential. As first aid, a doctor or authorized person should consider administering a corticosteroid spray. Use airtight packaging.

HI: 423; UN-number: 1402

aragonite
calcite
chalk
dolomite
marble

CALCIUM CARBONATE

PHYSICAL PROPERTIES		IMPORTANT CHARACTERISTICS	
Boiling point (decomposes), °C	825	**WHITE - SOMETIMES COLORED - LUMPS OR POWDER**	
Relative density (water = 1)	2.9	Decomposes when heated above 825°C, giving off → *carbon dioxide* and corrosive residue	
Solubility in water	none	(→ *calcium oxide*). Reacts violently with acids, giving off → *carbon dioxide* gas.	
Relative molecular mass	100.1		
		TLV-TWA	10 mg/m³
		Absorption route: Can enter the body by inhalation or ingestion. Evaporation negligible at 20°C, but unpleasant concentrations of airborne particles can build up rapidly. **Immediate effects:** Irritates the eyes.	
Gross formula:	CCaO$_3$		

HAZARDS/SYMPTOMS	PREVENTIVE MEASURES	FIRE EXTINGUISHING/FIRST AID
Fire: non-combustible.		in case of fire in immediate vicinity: use any extinguishing agent.
	KEEP DUST UNDER CONTROL	
Eyes: redness, pain.	goggles.	flush with water, take to a doctor if necessary.

SPILLAGE	STORAGE	LABELING / NFPA
clean up spillage, flush away remainder with water; (additional individual protection: P1 respirator).		

NOTES
Use P2 respirator when pulverizing, crushing or grinding. Calcium carbonate powder is a suitable absorbent for chemical spillages, esp. acids.

$Ca(ClO_3)_2 \cdot H_2O$

CALCIUM CHLORATE
(monohydrate)

PHYSICAL PROPERTIES	IMPORTANT CHARACTERISTICS	
Melting point, °C 100 Relative density (water = 1) 2.7 Solubility in water, g/100 ml at 8 °C 178 Relative molecular mass 243.0	**WHITE TO LIGHT YELLOW HYGROSCOPIC CRYSTALS** Decomposes when heated, giving off toxic vapor (→ *chlorine* and → *oxygen*), with increased risk of fire. Strong oxidant which reacts violently with combustible substances and reducing agents, with risk of fire and explosion. Reacts violently with aluminum, copper, carbon, sulfur, sulfides and ammonium salts, with risk of fire and explosion. Reacts with strong acids, giving off explosive and toxic gas (→ *chlorine dioxide*).	
	TLV-TWA not available	
	Absorption route: Can enter the body by inhalation or ingestion. Evaporation negligible at 20° C, but harmful concentrations of airborne particles can build up rapidly. **Immediate effects:** Irritates the eyes, skin and respiratory tract. **Effects of prolonged/repeated exposure:** Can affect the blood. Can cause kidney and heart damage. In serious cases risk of death.	
Gross formula: $CaCl_2O_6 \cdot H_2O$		

HAZARDS/SYMPTOMS	PREVENTIVE MEASURES	FIRE EXTINGUISHING/FIRST AID
Fire: not combustible but enhances combustion of other substances.		in case of fire in immediate vicinity: use any extinguishing agent.
Explosion: contact/mixing with many organic substances can cause explosion.		in case of fire: keep tanks/drums cool by spraying with water.
	STRICT HYGIENE	
Inhalation: sore throat, cough, headache, dizziness, blue skin, feeling of weakness.	local exhaust or respiratory protection.	fresh air, rest, call a doctor.
Skin: redness.	gloves.	remove contaminated clothing, flush skin with water or shower.
Eyes: redness, pain.	goggles.	flush with water, take to a doctor.
Ingestion: abdominal pain, nausea, blue skin.		rinse mouth, immediately take to hospital.

SPILLAGE	STORAGE	LABELING / NFPA
clean up spillage, flush away remainder with water, do not use sawdust or other combustible absorbents; (additional individual protection: P2 respirator).	keep dry, separate from combustible substances, reducing agents and strong acids.	 ◇ 0 1 ◇ 1 oxy

NOTES

Apparent melting point due to loss of water of crystallization. Flush contaminated clothing with water (fire hazard). Depending on the degree of exposure, regular medical checkups are advisable. Effects on blood due to formation of methemoglobin. Special first aid required in the event of poisoning: antidotes must be available (with instructions). Contact with wood, textiles, oil, fat, sugar, sawdust, ammonium salts, carbon, sulfur, phosphorus, metal powder, sulfides etc. can cause fire and explosion. Packaging: special material.

Transport Emergency Card TEC(R)-51G01

UN-number: 1452

CALCIUM CHLORIDE

PHYSICAL PROPERTIES	IMPORTANT CHARACTERISTICS	
Boiling point, °C **1600** Melting point, °C **772** Relative density (water = 1) **2.1** Solubility in water, g/100 ml **74.5** Relative molecular mass **111** Gross formula: **CaCl$_2$**	**WHITE HYGROSCOPIC CRYSTALS** When strongly heated gives off irritating vapors. Attacks many metals and building materials.	
	TLV-TWA not available	
	Absorption route: Can enter the body by inhalation or ingestion. Evaporation negligible at 20°C, but harmful concentrations of airborne particles can build up rapidly. **Immediate effects:** Irritates the eyes, skin and respiratory tract.	

HAZARDS/SYMPTOMS	PREVENTIVE MEASURES	FIRE EXTINGUISHING/FIRST AID
Fire: non-combustible.		in case of fire in immediate vicinity: use any extinguishing agent.
	KEEP DUST UNDER CONTROL	
Inhalation: cough, shortness of breath.	local exhaust or respiratory protection.	fresh air, rest, take to a doctor.
Skin: redness.	gloves.	remove contaminated clothing, flush skin with water or shower.
Eyes: redness, pain.	goggles.	flush with water, take to a doctor.
Ingestion: abdominal cramps, nausea.		rinse mouth, call a doctor.

SPILLAGE	STORAGE	LABELING / NFPA
clean up spillage, flush away remainder with water; (additional individual protection: P2 respirator).	keep dry.	R: 36 S: 22-24 ✖ Irritant

NOTES
Use airtight packaging.

UN-number: 1453

CALCIUM HYDRIDE

PHYSICAL PROPERTIES	IMPORTANT CHARACTERISTICS	
Melting point (decomposes), °C 675 Relative density (water = 1) 1.7 Solubility in water reaction Relative molecular mass 42.1	**COLORLESS CRYSTALS OR GRAY POWDER** Decomposes above 675°C, giving off → *hydrogen* gas, with risk of fire and explosion. Strong reducing agent which reacts violently with oxidants. Reacts violently with water, water vapor, alcohols and acids, giving off → *hydrogen* gas, with risk of fire and explosion.	
	TLV-TWA not available	
	Absorption route: Can enter the body by inhalation or ingestion. Evaporation negligible at 20°C, but harmful concentrations of airborne particles can build up rapidly. **Immediate effects:** Corrosive to the eyes, skin and respiratory tract. Inhalation can cause lung edema. In serious cases risk of death.	
Gross formula: CaH$_2$		

HAZARDS/SYMPTOMS	PREVENTIVE MEASURES	FIRE EXTINGUISHING/FIRST AID
Fire: extremely flammable, many chemical reactions can cause fire and explosion.	keep away from open flame and sparks, no smoking, avoid contact with hydrous substances and oxidants.	special powder extinguisher, dry sand, NO other extinguishing agents; in case of fire in immediate vicinity: do not use water.
Explosion:		in case of fire: keep tanks/drums cool by spraying with water, but DO NOT spray substance with water.
	STRICT HYGIENE	IN ALL CASES CALL A DOCTOR
Inhalation: *corrosive*; sore throat, cough, severe breathing difficulties.	local exhaust or respiratory protection.	fresh air, rest, place in half-sitting position, take to hospital.
Skin: *corrosive*; redness, pain, serious burns.	gloves, protective clothing.	remove contaminated clothing, flush skin with water or shower, call a doctor.
Eyes: *corrosive*; redness, pain, impaired vision.	goggles, or combined eye and respiratory protection.	flush with water, take to a doctor.
Ingestion: *corrosive*; abdominal pain, diarrhea, vomiting.		rinse mouth, take immediately to hospital.

SPILLAGE	STORAGE	LABELING / NFPA
evacuate area, call in an expert, clean up spillage, carefully collect remainder, do not use sawdust or other combustible absorbents, remove to safe place; (additional individual protection: P2 respirator).	keep in dry, fireproof place, separate from oxidants.	R: 15 S: 7/8-24/25-43 🔥 Flammable

NOTES
Melting point (measured in hydrogen) 816°C. Flush contaminated clothing with water (fire hazard). Reacts violently with extinguishing agents such as water and foam. Lung edema symptoms usually develop several hours later and are aggravated by physical exertion: rest and hospitalization essential. As first aid, a doctor or authorized person should consider administering a corticosteroid spray. Use airtight packaging.

Transport Emergency Card TEC(R)-43G04

HI: X423; UN-number: 1404

CALCIUM HYPOCHLORITE

PHYSICAL PROPERTIES	IMPORTANT CHARACTERISTICS	
Melting point (decomposes), °C 100 Relative density (water = 1) 2.35 Solubility in water reaction Relative molecular mass 143	**WHITE POWDER WITH PUNGENT ODOR** Decomposes when heated or exposed to sunlight, giving off → *oxygen*, with increased risk of fire, and corrosive, toxic → *chlorine*. Strong oxidant which reacts violently with combustible substances and reducing agents. Reacts with moisture and acids, giving off toxic and corrosive gas (→ *chlorine*).	
	TLV-TWA not available	
Gross formula: $CaCl_2O_2$	**Absorption route:** Can enter the body by inhalation or ingestion. Evaporation negligible at 20°C, but harmful concentrations of airborne particles can build up rapidly. **Immediate effects:** Corrosive to the eyes, skin and respiratory tract. Inhalation can cause lung edema. In serious cases risk of death. **Effects of prolonged/repeated exposure:** Prolonged or repeated contact can cause skin disorders.	

HAZARDS/SYMPTOMS	PREVENTIVE MEASURES	FIRE EXTINGUISHING/FIRST AID
Fire: not combustible but enhances combustion of other substances.	avoid contact with combustible substances.	large quantities of water.
Explosion: reactions with reducing agents can cause explosion.		
	KEEP DUST UNDER CONTROL	
Inhalation: *corrosive*; sore throat, cough, shortness of breath, severe breathing difficulties.	local exhaust or respiratory protection.	fresh air, rest, place in half-sitting position, artificial respiration if necessary, take to hospital.
Skin: *corrosive*; redness, pain, burns.	gloves.	remove contaminated clothing, flush skin with water or shower, take to a doctor.
Eyes: *corrosive*; redness, pain, impaired vision.	face shield, or combined eye and respiratory protection.	flush with water, take to a doctor.
Ingestion: *corrosive*; sore throat, abdominal pain, vomiting.		rinse mouth, rest, call a doctor or take to hospital.

SPILLAGE	STORAGE	LABELING / NFPA
clean up spillage, flush away remainder with water; (additional individual protection: breathing apparatus).	keep in dry, dark place, separate from combustible substances, reducing agents and acids.	R: 8-31-34 S: 2-26-43 Oxidant Corrosive

NOTES
Under no circumstances return leftover substance to stock. Addition of only small quantities of combustible materials starts a fire which is hard to extinguish. Lung edema symptoms usually develop several hours later and are aggravated by physical exertion: rest and hospitalization essential. As first aid, a doctor or authorized person should consider administering a corticosteroid spray.

UN-number: 1748

$Ca(NO_3)_2$

CALCIUM NITRATE

PHYSICAL PROPERTIES	IMPORTANT CHARACTERISTICS
Melting point, °C 561 Relative density (water = 1) 2.4 Solubility in water, g/100 ml at 18 °C 121 Relative molecular mass 164.1	**COLORLESS HYGROSCOPIC CRYSTALS** Reacts with organic contaminants to form shock-sensitive mixtures. Strong oxidant which reacts violently with combustible substances and reducing agents. Is corrosive and attacks many substances.

TLV-TWA	not available

Absorption route: Can enter the body by inhalation or ingestion. Evaporation negligible at 20°C, but harmful concentrations of airborne particles can build up rapidly.
Immediate effects: Irritates the eyes, skin and respiratory tract. In serious cases risk of unconsciousness.
Effects of prolonged/repeated exposure: Can affect the blood if ingested. Affects the blood vessels.

Gross formula: CaN_2O_6

HAZARDS/SYMPTOMS	PREVENTIVE MEASURES	FIRE EXTINGUISHING/FIRST AID
Fire: not combustible but enhances combustion of other substances.	avoid contact with combustible substances.	in case of fire in immediate vicinity: use any extinguishing agent.
Inhalation: sore throat, cough, shortness of breath, headache, dizziness.	local exhaust or respiratory protection.	fresh air, rest, call a doctor.
Skin: redness, pain.	gloves.	remove contaminated clothing, flush skin with water or shower.
Eyes: redness, pain.	face shield.	flush with water, take to a doctor.
Ingestion: abdominal cramps, nausea, blue skin, feeling of weakness.		rinse mouth, call a doctor or take to hospital.

SPILLAGE	STORAGE	LABELING / NFPA
clean up spillage, flush away remainder with water.	keep cool, separate from combustible substances and reducing agents.	

NOTES

Flush contaminated clothing with water (fire hazard). Depending on the degree of exposure, regular medical checkups are advisable. Effects on blood due to formation of methemoglobin. Special first aid required in the event of poisoning: antidotes must be available (with instructions). Use airtight packaging.

Transport Emergency Card TEC(R)-51G01

UN-number: 1454

$Ca(NO_2)_2.H_2O$

CALCIUM NITRITE

PHYSICAL PROPERTIES	IMPORTANT CHARACTERISTICS
Melting point, °C 100 Relative density (water = 1) 2.2 Solubility in water, g/100 ml at 6 °C 46 Relative molecular mass 150.1	**COLORLESS OR WHITE YELLOW HYGROSCOPIC CRYSTALS** Decomposes when heated and on contact with acids, giving off toxic vapors (→ *nitrous vapors*). Strong oxidant which reacts violently with combustible substances and reducing agents, with risk of fire and explosion. Reacts with many organic and inorganic substances, with ammonium salt and cyanides at higher temperature even explosively.

TLV-TWA	not available

Absorption route: Can enter the body by inhalation or ingestion. Evaporation negligible at 20°C, but harmful concentrations of airborne particles can build up rapidly.
Immediate effects: Can affect the smooth muscular tissue, causing low blood pressure. High concentrations can impair consciousness.
Effects of prolonged/repeated exposure: Can affect hemoglobin, causing formation of methemoglobin.

Gross formula: $CaN_2O_4.H_2O$

HAZARDS/SYMPTOMS	PREVENTIVE MEASURES	FIRE EXTINGUISHING/FIRST AID
Fire: not combustible but enhances combustion of other substances.	avoid contact with ammonium salts, cyanides and combustible substances and reducing agents.	in case of fire in immediate vicinity: use any extinguishing agent.
Explosion: strong heating can cause explosion.		in case of fire: keep tanks/drums cool by spraying with water.
	KEEP DUST UNDER CONTROL, STRICT HYGIENE	
Inhalation: nausea, vomiting, disorientation, feeling of weakness, unconsciousness.	local exhaust or respiratory protection.	fresh air, rest, call a doctor.
Skin:	gloves.	remove contaminated clothing, flush skin with water and wash thoroughly with soap and water.
Eyes: redness.	goggles.	flush with water, take to a doctor if necessary.
Ingestion: abdominal pain, blue lips or nails, see also: 'Inhalation'.	do not eat, drink or smoke while working.	rinse mouth, give no liquids, take immediately to hospital.

SPILLAGE	STORAGE	LABELING / NFPA
clean up spillage, flush away remainder with water; (additional individual protection: P2 respirator).	keep dry, separate form combustible substances, reducing agents, ammomium salts, cyanides and acids.	

NOTES
Apparent melting point due to loss of water of crystallization. Becomes shock-sensitive when contaminated with organic substances.

Transport Emergency Card TEC(R)-51G01

UN-number: 2627

CAS-No: [1305-78-8]   CaO
calx
lime
pebble lime
quicklime

CALCIUM OXIDE

PHYSICAL PROPERTIES	IMPORTANT CHARACTERISTICS	
Boiling point, °C 2850 Melting point, °C 2570 Relative density (water = 1) 3.4 Solubility in water reaction Relative molecular mass 56.0	**WHITE HYGROSCOPIC POWDER** Reacts violently with water, evolving large quantities of heat. Reacts violently with acids and light metals.	
	TLV-TWA	2 mg/m³
	Absorption route: Can enter the body by inhalation or ingestion. Evaporation negligible at 20°C, but harmful concentrations of airborne particles can build up rapidly. **Immediate effects:** Corrosive to the eyes, skin and respiratory tract. Inhalation can cause lung edema.	
Gross formula: CaO		

HAZARDS/SYMPTOMS	PREVENTIVE MEASURES	FIRE EXTINGUISHING/FIRST AID
Fire: non-combustible.		
Inhalation: *corrosive*; sore throat, cough, shortness of breath, severe breathing difficulties.	local exhaust or respiratory protection.	fresh air, rest, place in half-sitting position, take to hospital.
Skin: *corrosive*; redness, pain, burns.	gloves, protective clothing.	remove contaminated clothing, flush skin with water or shower, take to a doctor.
Eyes: *corrosive*; redness, pain, impaired vision.	goggles.	flush with water, take to a doctor if necessary.
Ingestion: *corrosive*; sore throat, abdominal pain, diarrhea.		rinse mouth, immediately take to hospital.

SPILLAGE	STORAGE	LABELING / NFPA
clean up spillage, flush away remainder with water; (additional individual protection: P2 respirator).	keep dry, separate from acids.	0 / 1 / 1

NOTES
Lung edema symptoms usually develop several hours later and are aggravated by physical exertion: rest and hospitalization essential. As first aid, a doctor or authorized person should consider administering a corticosteroid spray. Use airtight packaging.

Transport Emergency Card TEC(R)-80G13   **UN-number: 1910**

CALCIUM PHOSPHIDE

PHYSICAL PROPERTIES	IMPORTANT CHARACTERISTICS	
Melting point, °C 1600 Relative density (water = 1) 2.5 Solubility in water reaction Relative molecular mass 182.2	**GRAY LUMPS OR BROWNISH-RED CRYSTALS WITH CHARACTERISTIC ODOR** Decomposes when heated or burned, giving off toxic and corrosive vapors (→ *phosphorus oxides*). Strong reducing agent which reacts violently with oxidants. Reacts violently with water, water vapor and acids, giving off toxic and flammable vapors (→ *phosphine*), with risk of fire and explosion. Reacts violently with chlorine, chlorine monoxide, oxygen and sulfur, with risk of fire and explosion. Reacts with moisture in air to form → *phosphine*.	
	TLV-TWA	✱
	Absorption route: Can enter the body by inhalation or ingestion. Evaporation negligible at 20°C, but harmful concentrations of airborne particles can build up rapidly. **Immediate effects:** Irritates the eyes and respiratory tract. Affects the nervous system. In serious cases risk of death. Affects metabolism. **Effects of prolonged/repeated exposure:** Can cause heart, liver and kidney damage.	
Gross formula: Ca_3P_2		

HAZARDS/SYMPTOMS	PREVENTIVE MEASURES	FIRE EXTINGUISHING/FIRST AID
Fire: extremely flammable.	keep away from open flame and sparks, no smoking.	special powder extinguisher, DO NOT USE WATER-BASED EXTINGUISHERS.
Explosion: reaction with water or acids can cause explosion.		
	KEEP DUST UNDER CONTROL	
Inhalation: shortness of breath, headache, dizziness, vomiting, diarrhea, muscle cramps.	local exhaust or respiratory protection.	fresh air, rest, call a doctor and take to hospital.
Skin: redness.	gloves, protective clothing.	remove contaminated clothing, wash skin with soap and water.
Eyes: redness, pain.	face shield, or combined eye and respiratory protection.	flush with water, take to a doctor if necessary.
Ingestion: diarrhea, abdominal cramps, nausea.	do not eat, drink or smoke while working.	rinse mouth, take immediately to hospital.

SPILLAGE	STORAGE	LABELING / NFPA
evacuate area, call in an expert, clean up spillage, soak up with dry sand or other inert absorbent and remove to safe place; (additional individual protection: breathing apparatus).	keep in dry, fireproof place, separate from oxidants.	R: 15/29-28 S: 1/2-22-43-45 🔥 Flammable ☠ Toxic

NOTES
Reacts with moisture to form phosphine and phosphine dimer, which can ignite spontaneously on exposure to air: this is the main flash point determinant under these conditions. Reacts violently with water-based extinguishing agents. ✱ → *Phosphine* (TLV-TWA 0.3ppm and 0.42mg/m³, STEL 1ppm and 1.4mg/m³) is given off in acid/moist environment (stomach/lungs). Use airtight packaging.

Transport Emergency Card TEC(R)-42G01

HI: 43; UN-number: 1360

CALCIUM SULFIDE

PHYSICAL PROPERTIES	IMPORTANT CHARACTERISTICS	
Melting point (decomposes), °C 2000 Relative density (water = 1) 2.6 Solubility in water reaction Relative molecular mass 72.1	**YELLOW OR GRAY POWDER WITH CHARACTERISTIC ODOR** Decomposes on contact with water, water vapor and weak acids, giving off flammable and toxic gas (→ *hydrogen sulfide*). Reacts violently with lead dioxide, chlorates and nitrates and other strong oxidants, with risk of fire and explosion. Reacts with air to form toxic gases (→ *hydrogen sulfide*).	
	TLV-TWA not available	
	Absorption route: Can enter the body by inhalation or ingestion. **Immediate effects:** Irritates the respiratory tract. Corrosive to the eyes and skin. Affects the nervous system.	
Gross formula: CaS		

HAZARDS/SYMPTOMS	PREVENTIVE MEASURES	FIRE EXTINGUISHING/FIRST AID
Fire: combustible, many chemical reactions can cause fire and explosion.	keep away from open flame, no smoking, avoid contact with oxidants, water and acids.	powder, carbon dioxide, dry sand, (halons), DO NOT USE WATER-BASED EXTINGUISHERS.
Explosion: finely dispersed particles form explosive mixtures on contact with air.	keep dust from accumulating; sealed machinery, explosion-proof electrical equipment and lighting, grounding.	
	STRICT HYGIENE, KEEP DUST UNDER CONTROL	
Inhalation: sore throat, cough, shortness of breath, dizziness, drowsiness.	local exhaust or respiratory protection.	fresh air, rest, place in half-sitting position, call a doctor.
Skin: *corrosive*; redness, pain.	gloves, protective clothing.	remove contaminated clothing, first wipe skin clean, then flush skin with water or shower, take to a doctor if necessary.
Eyes: *corrosive*; redness, pain, impaired vision.	goggles, or combined eye and respiratory protection.	flush with water, take to a doctor.
Ingestion: diarrhea, vomiting, see also 'Inhalation'.		rinse mouth, rest, call a doctor or take immediately to hospital.

SPILLAGE	STORAGE	LABELING / NFPA
clean up spillage, flush away remainder with water; (additional individual protection: P2 respirator).	keep in fireproof, dry place, separate from oxidants and acids.	R: 31-36/37/38 S: 28 ✖ Irritant

NOTES

Transport Emergency Card TEC(R)-80G11

CALOMEL

PHYSICAL PROPERTIES	IMPORTANT CHARACTERISTICS	
Sublimation temperature, °C ca. 400 Relative density (water = 1) 7.2 Solubility in water, g/100 ml at 25 °C 0.0002 Relative molecular mass 472.1	**WHITE CRYSTALLINE POWDER, TURNING DARK ON EXPOSURE TO AIR** Decomposes slowly on exposure to light, forming → *mercury chloride* and → *mercury.*	
	TLV-TWA (skin) 0.05 mg/m³ ✱	
	Absorption route: Absorption via the intestines is small. In powder form calomel is not absorbed through the skin, however it has effects in oil form. Can enter the body by inhalation or ingestion. Evaporation negligible at 20°C, but harmful concentrations of airborne particles can build up rapidly. **Immediate effects:** Irritates the eyes, skin and respiratory tract. Affects the nervous system. **Effects of prolonged/repeated exposure:** Can cause liver and kidney damage.	
Gross formula: Cl_2Hg_2		

HAZARDS/SYMPTOMS	PREVENTIVE MEASURES	FIRE EXTINGUISHING/FIRST AID
Fire: non-combustible.		
Inhalation: sore throat, cough, shortness of breath, irritation of mucous mencous membrane.	local exhaust or respiratory protection.	fresh air, rest, call a doctor.
Skin: *is absorbed*; redness, pain.	gloves.	remove contaminated clothing, wash skin with soap and water, call a doctor.
Eyes: redness, pain.	goggles.	flush with water, take to a doctor.
Ingestion: metallic taste; abdominal pain, diarrhea, vomiting; low acute oral toxicity, but risk of mercury poisoning in the absence of diarrhea.		rinse mouth, immediately take to hospital.

SPILLAGE	STORAGE	LABELING / NFPA
clean up spillage, carefully collect remainder; (additional individual protection: P3 respirator).	keep in dark place.	R: 22 S: 2 ✖ Harmful

NOTES
✱ TLV equals that of lead. Do not take work clothing home. Depending on the degree of exposure, regular medical checkups are advisable. In rare cases fever and rash can develop about a week after exposure; this is not serious. Chronic exposure also causes mercury poisoning if the TLV value is exceeded for a long period. Nervous tissues in particular exhibit toxic effect. Unbreakable packaging preferred; if breakable, keep in unbreakable container.

CAS-No: [76-22-2]
2-bornanone
2-oxobornane
1,7,7-trimethylbicycol(2,2,1)heptanone-2

$C_{10}H_{16}O$

CAMPHOR

PHYSICAL PROPERTIES	IMPORTANT CHARACTERISTICS	
Boiling point, °C 204 Melting point, °C 176 Flash point, °C 66 Auto-ignition temperature, °C 460 Relative density (water = 1) 0.99 Relative vapor density (air = 1) 5.3 Relative density at 20 °C of saturated mixture vapor/air (air = 1) 1.00 Vapor pressure, mm Hg at 20 °C 2.3 Solubility in water, g/100 ml at 20 °C 0.1 Explosive limits, vol% in air 0.6-4.5 Relative molecular mass 152.2 Gross formula: $C_{10}H_{16}O$	**COLORLESS CRYSTALS OR WHITE LUMPS WITH CHARACTERISTIC ODOR** Sublimes even at room temperature. Vapor mixes readily with air. In dry state can form electrostatic charge if stirred, transported pneumatically, poured etc. Lots of soot during combustion. Reacts violently with strong oxidants.	

TLV-TWA	2 ppm	12 mg/m³ ✱
STEL	3 ppm	19 mg/m³

Absorption route: Can enter the body by inhalation or ingestion. Harmful atmospheric concentrations build up fairly slowly on evaporation at 20°C, but harmful concentrations of airborne particles can build up much more rapidly.
Immediate effects: Irritates the eyes, skin and respiratory tract. Affects the nervous system. In serious cases risk of seizures and death.

HAZARDS/SYMPTOMS	PREVENTIVE MEASURES	FIRE EXTINGUISHING/FIRST AID
Fire: combustible.	keep away from open flame, no smoking, avoid contact with strong oxidants.	water spray, powder.
Explosion: above 66°C: forms explosive air-vapor mixtures.	above 66°C: sealed machinery, ventilation.	
Inhalation: sore throat, shortness of breath, headache, running nose.	local exhaust or respiratory protection.	fresh air, rest, call a doctor if necessary.
Skin: redness.	gloves.	
Eyes: redness, pain.	safety glasses.	flush with water, take to a doctor if necessary.
Ingestion: nausea, vomiting, dizziness, sensation of heat, restlessness, cramps, seizures.	do not eat, drink or smoke while working.	rinse mouth, immediately take to hospital.

SPILLAGE	STORAGE	LABELING / NFPA
clean up spillage, carefully collect remainder; (addition individual protection: breathing apparatus; respirator with A/P2 filter)	keep separate from oxidants; ventilate at floor level.	 ⬥ 2 0 ◇ 0

NOTES
The melting synthetic camphor is 165°C. ✱ TLV as synthetic grade.

butylacetic acid
capronic acid
n-hexanoic acid
pentiformic acid
pentylformic acid

$C_5H_{11}COOH$

CAPROIC ACID

PHYSICAL PROPERTIES		IMPORTANT CHARACTERISTICS
Boiling point, °C	205	**COLORLESS VISCOUS LIQUID WITH CHARACTERISTIC ODOR**
Melting point, °C	−4	Vapor mixes readily with air. Reacts with strong oxidants.
Flash point, °C	102	
Auto-ignition temperature, °C	380	TLV-TWA not available
Relative density (water = 1)	0.9	
Relative vapor density (air = 1)	4.0	
Relative density at 20 °C of saturated mixture		**Absorption route:** Can enter the body by inhalation or ingestion. Harmful atmospheric
vapor/air (air = 1)	1.0	concentrations build up very slowly, if at all, on evaporation at approx. 20°C, but much more
Vapor pressure, mm Hg at 20 °C	7.6x10⁻⁴	rapidly in aerosol form.
Solubility in water, g/100 ml at 20 °C	1.1	**Immediate effects:** Irritates the eyes, skin and respiratory tract.
Explosive limits, vol% in air	1.3-9.3	
Relative molecular mass	116.2	
Log P octanol/water	1.9	
Gross formula:	$C_6H_{12}O_2$	

HAZARDS/SYMPTOMS	PREVENTIVE MEASURES	FIRE EXTINGUISHING/FIRST AID
Fire: combustible.	keep away from open flame, no smoking.	powder, AFFF, foam, carbon dioxide, (halons).
Inhalation: sore throat, cough.	ventilation.	fresh air, rest, refer to a doctor if necessary.
Skin: redness, pain.	gloves.	remove contaminated clothing, flush skin with water or shower.
Eyes: redness, pain.	face shield.	flush with water, take to a doctor.
Ingestion: sore throat, abdominal pain.		rinse mouth.

SPILLAGE	STORAGE	LABELING / NFPA
collect leakage in sealable containers, soak up with sand or other inert absorbent and remove to safe place.	keep separate from oxidants.	1 2 0

NOTES
Explosive limits apply to raised temperatures. Abnormal vapor pressure curve: saturation vapor pressure at 50°C: 0.084mm Hg.

HI: 80; UN-number: 2829

CAS-No: [105-60-2]
aminocaproic lactam
ε-caprolactam
2-oxohexamethyleneimine

CAPROLACTAM

PHYSICAL PROPERTIES		IMPORTANT CHARACTERISTICS	
Boiling point, °C	270	**WHITE CRYSTALS**	
Melting point, °C	69	Decomposes when heated, giving off toxic vapors (→ *nitrous vapors* and → *anhydrous*	
Flash point, °C	110	*ammonia*). Reacts violently with strong oxidants, giving off toxic vapors.	
Auto-ignition temperature, °C	375		
Relative density (water = 1)	1.02	TLV-TWA 4.3 ppm 20 mg/m³ ✱	
Relative vapor density (air = 1)	3.9	STEL 8.6 ppm 40 mg/m³	
Relative density at 20 °C of saturated mixture			
vapor/air (air = 1)	1.00	**Absorption route:** Can enter the body by inhalation or ingestion. Evaporation negligible at 20°C,	
Vapor pressure, mm Hg at 20 °C	0.0019	but harmful concentrations of airborne particles can build up rapidly.	
Solubility in water, g/100 ml	82	**Immediate effects:** Irritates the eyes, respiratory tract and central nervous system. Hot liquid can	
Explosive limits, vol% in air	1.4-8	cause burns.	
Relative molecular mass	113.2		
Gross formula:	$C_6H_{11}NO$		

HAZARDS/SYMPTOMS	PREVENTIVE MEASURES	FIRE EXTINGUISHING/FIRST AID
Fire: combustible.	keep away from open flame, no smoking, avoid contact with strong oxidants.	water spray, powder.
	KEEP DUST UNDER CONTROL	
Inhalation: vapor: cough, irritation of throat and lungs, stomach complaints, irritability, muscle cramps; dust: irritation of nasal mucous membrane and nosebleeds.	local exhaust or respiratory protection.	fresh air, rest, call a doctor if necessary.
Skin: infection and eczema.	gloves, protective clothing.	flush thoroughly with water.
Eyes: redness, pain.	face shield, or combined eye and respiratory protection.	flush with water, take to a doctor.
Ingestion: most commonly nausea and otherwise gastroenteritis.		rinse mouth, call a doctor.

SPILLAGE	STORAGE	LABELING / NFPA
allow warm spillage to set, clean up spillage, flush away remainder with water; (additional individual protection: P2 respirator).	keep separate from oxidants; ventilate at floor level.	

NOTES
✱ TLV equals that of caprolactam (vapor). TLV of caprolactam (dust): TLV-TWA 1mg/m³ and STEL 3mg/m³. Usually transported, stored and processed in molten state at approx. 80°C, in which form it is highly aggressive to the skin.

CARBON DIOXIDE
(cylinder)

PHYSICAL PROPERTIES	IMPORTANT CHARACTERISTICS	
Boiling point, °C −79 Relative density (water = 1) 0.8 Relative vapor density (air = 1) 1.5 Vapor pressure, mm Hg at 20 °C 43,776 Solubility in water, g/100 ml at 25 °C 0.16 Relative molecular mass 44.0	**COLORLESS, ODORLESS COMPRESSED LIQUEFIED GAS** Vapor is heavier than air and can accumulate close to ground level, causing oxygen deficiency, with risk of unconsciousness. At fast flow rates from cylinder can cause build-up of static electricity, which can ignite any explosive mixture present. Free-flowing liquid condenses to form → *dry ice*. At high temperatures reacts violently with anhydrous ammonia and various amines.	
	TLV-TWA 5000 ppm 9000 mg/m³ STEL 30,000 ppm 54,000 mg/m³	
	Absorption route: Can enter the body by inhalation. Can saturate the air if released, with risk of suffocation. **Immediate effects:** Can cause frostbite due to rapid evaporation. Inhalation can cause severe breathing difficulties. In serious cases risk of unconsciousness.	
Gross formula: CO_2		

HAZARDS/SYMPTOMS	PREVENTIVE MEASURES	FIRE EXTINGUISHING/FIRST AID
Fire: non-combustible.		in case of fire in immediate vicinity: use any extinguishing agent.
Explosion:		in case of fire: keep cylinders cool by spraying with water.
Inhalation: labored breathing, perspiration, severe breathing difficulties, headache, dizziness.	ventilation, local exhaust or respiratory protection.	fresh air, rest, artificial respiration if necessary, call a doctor or take to hospital.
Skin: *in case of frostbite*: redness, pain, serious burns.	insulating gloves.	*in case of frostbite*: DO NOT remove clothing, flush skin with water or shower, take to a doctor.
Eyes: *in case of frostbite*: redness, pain, impaired vision.	acid goggles.	flush with water, take to a doctor.

SPILLAGE	STORAGE	LABELING / NFPA
ventilate, under no circumstances spray liquid with water; (additional individual protection: breathing apparatus).	if stored indoors, keep in cool, fireproof place.	

NOTES
High atmospheric concentrations, e.g. in poorly ventilated spaces, can cause oxygen deficiency, with risk of unconsciousness. At atmospheric concentrations above 10% causes unconsciousness and death; given off in many fermentation processes (wine etc.) and is a major component of flue gas. Turn leaking cylinder so that leak is on top to prevent liquid carbon dioxide escaping.

Transport Emergency Card TEC(R)-11

HI: 20; UN-number: 1013

CARBON DIOXIDE
(solid)

PHYSICAL PROPERTIES		IMPORTANT CHARACTERISTICS
Sublimation temperature, °C	−79	**WHITE, EXTREMELY COLD CRYSTALS**
Relative density (water = 1)	1.6	Evaporates on exposure to air, giving off → *carbon dioxide* gas, and can accumulate close to ground level, causing oxygen deficiency, with risk of unconsciousness.
Relative vapor density (air = 1)	1.5	
Vapor pressure, mm Hg at 20 °C	43	
Solubility in water, g/100 ml at 20 °C	0.15	
Relative molecular mass	44	

TLV-TWA	5000 ppm	9000 mg/m³
STEL	30,000 ppm	54,000 mg/m³

Absorption route: Can enter the body by inhalation. Can saturate the air if released, with risk of suffocation.
Immediate effects: Irritates the eyes, skin and respiratory tract. Inhalation can cause severe breathing difficulties. In serious cases risk of unconsciousness. Can cause frostbite in skin and eyes.

Gross formula: CO_2

HAZARDS/SYMPTOMS	PREVENTIVE MEASURES	FIRE EXTINGUISHING/FIRST AID
Fire: non-combustible.		in case of fire in immediate vicinity: use any extinguishing agent.
Inhalation: severe breathing difficulties, headache, dizziness, unconsciousness.	ventilation, local exhaust or respiratory protection.	fresh air, rest, artificial respiration if necessary, take to hospital.
Skin: redness, pain, blisters.	insulating gloves.	*in case of frostbite*: DO NOT remove clothing, flush skin with water or shower, take to a doctor.
Eyes: *in case of frostbite*: redness, pain, impaired vision.	safety glasses.	flush with water, take to a doctor.

SPILLAGE	STORAGE	LABELING / NFPA
ventilate, clean up spillage (but do not put into sealed drums), under no circumstances spray solid with water; (additional individual protection: breathing apparatus).	keep cool; ventilate at floor level.	

NOTES
High atmospheric concentrations, e.g. in poorly ventilated spaces, can cause oxygen deficiency, with risk of unconsciousness. Above 12% atmospheric carbon dioxide causes unconsciousness and death.

Transport Emergency Card TEC(R)-11

HI: 22; UN-number: 1845

CARBON DISULFIDE

PHYSICAL PROPERTIES	IMPORTANT CHARACTERISTICS
Boiling point, °C — 46 Melting point, °C — − 112 Flash point, °C — − 40 Auto-ignition temperature, °C — 100 Relative density (water = 1) — 1.3 Relative vapor density (air = 1) — 2.6 Relative density at 20 °C of saturated mixture vapor/air (air = 1) — 1.6 Vapor pressure, mm Hg at 20 °C — 304 Solubility in water, g/100 ml at 20 °C — 0.2 Explosive limits, vol% in air — 1.0-60 Minimum ignition energy, mJ — 0.009 Relative molecular mass — 76.1 Log P octanol/water — 2.0	**COLORLESS LIQUID WITH CHARACTERISTIC ODOR** Vapor is heavier than air and spreads at ground level, with risk of ignition at a distance. Flow, agitation etc. can cause build-up of electrostatic charge due to liquid's low conductivity. Do not use compressed air when filling, emptying or processing. Toxic and corrosive vapors are a product of combustion (oxides of sulfur). Can decompose explosively if subjected to shock. Reacts violently with alkali metals and reducing agents, with risk of fire and explosion. Attacks many plastics and rubber.
	TLV-TWA (skin) 10 ppm 31 mg/m³
	Absorption route: Can enter the body by inhalation or ingestion or through the skin. Harmful atmospheric concentrations can build up very rapidly on evaporation at 20°C. **Immediate effects:** Irritates the eyes, skin and respiratory tract. Liquid destroys the skin's natural oils. Affects the nervous system. In serious cases risk of unconsciousness. **Effects of prolonged/repeated exposure:** Affects the vascular system.
Gross formula: CS_2	

HAZARDS/SYMPTOMS	PREVENTIVE MEASURES	FIRE EXTINGUISHING/FIRST AID
Fire: extremely flammable, many chemical reactions can cause fire and explosion.	keep away from open flame, sparks and hot surfaces (e.g. steam lines), no smoking.	powder, water spray, foam, carbon dioxide, (halons).
Explosion: forms explosive air-vapor mixtures.	sealed machinery, ventilation, explosion-proof electrical equipment and lighting, grounding, non-sparking tools, do not subject to shock or friction.	case of fire, keep tanks/drums cool by spraying with water.
Inhalation: sore throat, headache, dizziness, drowsiness, sleepiness.	ventilation, local exhaust or respiratory protection.	fresh air, rest, call a doctor.
Skin: redness, pain.	gloves, protective clothing.	remove contaminated clothing, wash skin with soap and water, call a doctor.
Eyes: redness, pain, impaired vision.	face shield, or combined eye and respiratory protection.	flush with water, take to a doctor if necessary.
Ingestion: sore throat, abdominal pain, diarrhea, headache, dizziness, drowsiness, sleepiness.		rinse mouth, take immediately to hospital.

SPILLAGE	STORAGE	LABELING / NFPA
evacuate area, call in an expert, cover liquid with water, collect leakage in sealable containers, soak up with sand or other inert absorbent and remove to safe place; (additional individual protection: breathing apparatus).	keep in fireproof place, separate from oxidants and combustible substances.	R: 12-26 S: 27-29-33-43-45 🔥 ☠️ Flammable Highly toxic 4 / 2 0 (NFPA diamond)

NOTES
Spontaneous combustion possible even on hot surfaces (lamps, central heating). Metal oxides can lower autoignition point. Log P octanol/water is estimated. Alcohol consumption increases toxic effects. Depending on the degree of exposure, regular medical checkups are advisable. Unbreakable packaging preferred; if breakable, keep in unbreakable container.

Transport Emergency Card TEC(R)-39

HI: 336; UN-number: 1131

carbonic oxide
carbon oxide
CO

CARBON MONOXIDE

(cylinder)

PHYSICAL PROPERTIES	IMPORTANT CHARACTERISTICS
Boiling point, °C − 191 Melting point, °C − 205 Flash point, °C flammable gas Auto-ignition temperature, °C 605 Relative vapor density (air = 1) 0.97 Vapor pressure, mm Hg at 20 °C 44,688 Solubility in water none Explosive limits, vol% in air 12-75 Minimum ignition energy, mJ 0.1 Relative molecular mass 28.0	**COLORLESS, ODORLESS COMPRESSED GAS** Mixes readily with air, forming explosive mixtures. TLV-TWA 50 ppm 57 mg/m³ STEL 400 ppm 458 mg/m³ **Absorption route:** Can enter the body by inhalation. Harmful atmospheric concentrations can build up very rapidly if gas is released. **Immediate effects:** Can affect the blood. In serious cases risk of breathing, heart rate and cardiovascular disorders, unconsciousness, seizures and death. Affects the nervous system. Can cause brain damage.
Gross formula: CO	

HAZARDS/SYMPTOMS	PREVENTIVE MEASURES	FIRE EXTINGUISHING/FIRST AID
Fire: extremely flammable.	keep away from open flame and sparks, no smoking.	shut off supply; if impossible and no danger to surrounding area, allow fire to burn itself out; otherwise extinguish with powder, carbon dioxide, (halons).
Explosion: forms explosive air-gas mixtures.	sealed machinery, ventilation, explosion-proof electrical equipment and lighting, non-sparking tools.	in case of fire: keep cylinders cool by spraying with water, fight fire from sheltered location.
Inhalation: headache, dizziness, unconsciousness.	ventilation, local exhaust or respiratory protection.	fresh air, rest, artificial respiration, administer oxygen and take immediately to hospital.

SPILLAGE	STORAGE	LABELING / NFPA
evacuate area, call in an expert, ventilate; (additional individual protection: breathing apparatus).	keep in cool, fireproof place.	4 / 3 / 0

NOTES
On rendering explosive gas/vapor-air mixtures inert, see chapter 'Tables and formulas'. Odorless, consequently no warning of exceeded TLV. Special first aid required in the event of poisoning: antidotes must be available (with instructions). Product of incomplete combustion of coal, oil, wood etc. Hospitalization essential if symptoms present.

Transport Emergency Card TEC(R)-20G05

HI: 236; UN-number: 1016

CARBON TETRACHLORIDE

PHYSICAL PROPERTIES		IMPORTANT CHARACTERISTICS	
Boiling point, °C	77	**COLORLESS LIQUID WITH CHARACTERISTIC ODOR**	
Melting point, °C	−23	Vapor is heavier than air. Flow, agitation etc. can cause build-up of electrostatic charge due to	
Relative density (water = 1)	1.6	liquid's low conductivity. Decomposes in flame or on hot surface, giving off toxic gas and	
Relative vapor density (air = 1)	5.3	corrosive vapor (→ *hydrochloric acid*). Decomposes on contact with light metals, evolving heat.	
Relative density at 20 °C of saturated mixture		Reacts violently with many compounds. Dissolves some plastics and rubber products.	
vapor/air (air = 1)	1.5			
Vapor pressure, mm Hg at 20 °C	91.2	TLV-TWA (skin) 5 ppm 31 mg/m³ A2	
Solubility in water, g/100 ml	0.1		
Electrical conductivity, pS/m	4x10^{-4}	**Absorption route:** Can enter the body by inhalation or ingestion or through the skin. Harmful	
Relative molecular mass	153.8	atmospheric concentrations can build up very rapidly on evaporation at 20°C.	
Log P octanol/water	2.6	**Immediate effects:** Irritates the eyes. Liquid destroys the skin's natural oils. Affects the nervous	
		system. In serious cases risk of unconsciousness.	
		Effects of prolonged/repeated exposure: Can cause liver and kidney damage.	
Gross formula:	CCl$_4$		

HAZARDS/SYMPTOMS	PREVENTIVE MEASURES	FIRE EXTINGUISHING/FIRST AID
Fire: non-combustible.		in case of fire in immediate vicinity: use any extinguishing agent.
	STRICT HYGIENE	
Inhalation: headache, dizziness, nausea, drowsiness, unconsciousness.	sealed machinery, ventilation, local exhaust or respiratory protection.	fresh air, rest, artificial respiration if necessary, call a doctor.
Skin: redness.	gloves, protective clothing.	remove contaminated clothing, flush skin with water or shower.
Eyes: redness, pain.	face shield, or combined eye and respiratory protection.	flush with water, take to a doctor.
Ingestion: abdominal pain, diarrhea, dizziness, unconsciousness.		rinse mouth, take immediately to hospital.

SPILLAGE	STORAGE	LABELING / NFPA
evacuate area, call in an expert, collect leakage in sealable containers, soak up with sand or other inert absorbent and remove to safe place; (additional individual protection: breathing apparatus).	ventilate at floor level.	R: 26/27 S: 2-38-45 ☠ Highly toxic NFPA: 3 / 0 / 0

NOTES

Log P octanol/water is estimated. Carbon tetrachloride (or products containing more than 1% by vol. carbon tetrachloride) should not be used as a solvent, cleaning agent or thinner, except in a closed system or under conditions offering at least equivalent protection against poisoning. Has been found to cause a type of liver cancer in certain animal species under certain circumstances. Odor limit is above TLV. Alcohol consumption increases toxic effects. Depending on the degree of exposure, regular medical checkups are advisable. Under no circumstances use near flame or hot surface or when welding. Not to be confused with *tetrachloroethene*.

Transport Emergency Card TEC(R)-102

HI: 60; UN-number: 1846

CARBORUNDUM

PHYSICAL PROPERTIES		IMPORTANT CHARACTERISTICS	
Sublimation temperature, °C	2700	**GREEN OR BLACK CRYSTALS**	
Relative density (water = 1)	3.2	Oxidizes slowly above 1000°C with oxygen from the air.	
Solubility in water	none		
Relative molecular mass	40.1	TLV-TWA	10 mg/m³
		Absorption route: Can enter the body by inhalation or ingestion. Evaporation negligible at 20°C, but unpleasant concentrations of airborne particles can build up. **Immediate effects:** High concentrations of carborundum dust irritate the eyes and throat.	
Gross formula:	CSi		

HAZARDS/SYMPTOMS	PREVENTIVE MEASURES	FIRE EXTINGUISHING/FIRST AID
Fire: non-combustible.		in case of fire in immediate vicinity: use any extinguishing agent.
	KEEP DUST UNDER CONTROL	
Inhalation: finely dispersed dust: sore throat, cough.	local exhaust or respiratory protection.	fresh air, rest.
Eyes: high concentrations of finely dispersed dust: redness, pain.	goggles.	flush with water, take to a doctor if necessary.
Ingestion:		rinse mouth.

SPILLAGE	STORAGE	LABELING / NFPA
clean up spillage; (additional individual protection: P1 respirator).		

NOTES
Insufficient medical data on the chronic harmful effects to humans. Carborundum always contains quartz. Exposure to quartz dust can cause lung disorders (→ *silicon dioxide*). It is unknown whether carborundum itself has these effects.

CAUSTIC SODA SOLUTION (33%)

PHYSICAL PROPERTIES		IMPORTANT CHARACTERISTICS	
Boiling point, °C	120	**COLORLESS VISCOUS AQUEOUS SOLUTION OF SODIUM HYDROXIDE**	
Melting point, °C	8	Strong base which reacts violently with acids and corrodes aluminum, zinc etc. Attacks lead and	
Relative density (water = 1)	1.3	alloys containing copper or silicon. Below 50°C does not attack steel, iron or tin.	
Solubility in water	∞		
Relative molecular mass	40	TLV-TWA	2 mg/m³ C ✷
		Absorption route: Can enter the body by inhalation or ingestion. Evaporation negligible at 20°C, but harmful atmospheric concentrations can build up rapidly in spray form. **Immediate effects:** Corrosive to the eyes, skin and respiratory tract. Inhalation of vapor/fumes can cause severe breathing difficulties (lung edema). In serious cases risk of unconsciousness and death.	
Gross formula:	HNaO		

HAZARDS/SYMPTOMS	PREVENTIVE MEASURES	FIRE EXTINGUISHING/FIRST AID
Fire: non-combustible.		in case of fire in immediate vicinity: use any extinguishing agent.
Inhalation: *corrosive*; sore throat, cough, shortness of breath.	ventilation.	fresh air, rest, place in half-sitting position, call a doctor.
Skin: *corrosive*; serious burns.	gloves, protective clothing.	remove contaminated clothing, flush skin with water or shower, refer to a doctor.
Eyes: *corrosive*; redness, pain, impaired vision.	face shield.	flush with water, take to a doctor.
Ingestion: *corrosive*; abdominal cramps, vomiting, diarrhea.	do not eat, drink or smoke while working.	rinse mouth, immediately take to hospital.

SPILLAGE	STORAGE	LABELING / NFPA
collect leakage in sealable containers, flush away remainder with water.	keep separate from acids.	R: 35 S: 2-26-27-37/39 Note B Corrosive NFPA: 3 0 1

NOTES
Melting point of 20% solution: −26°C; 50% solution: +12°C. ✷ TLV equals that of sodium hydroxide. TLV is maximum and must not be exceeded. Lung edema symptoms usually develop several hours later and are aggravated by physical exertion: rest and hospitalization essential. As first aid, a doctor or authorized person should consider administering a corticosteroid spray. Unbreakable packaging preferred; if breakable, keep in unbreakable container.

Transport Emergency Card TEC(R)-52

HI: 80; UN-number: 1824

leucoline
quinoline

CHINOLINE

PHYSICAL PROPERTIES	IMPORTANT CHARACTERISTICS
Boiling point, °C 238 Melting point, °C −15 Flash point, °C see notes Auto-ignition temperature, °C 480 Relative density (water = 1) 1.09 Relative vapor density (air = 1) 4.5 Relative density at 20 °C of saturated mixture vapor/air (air = 1) 1.0 Vapor pressure, mm Hg at 20 °C < 0.076 Solubility in water poor Explosive limits, vol% in air 1.2-7 Electrical conductivity, pS/m 2.2 x 10^6 Relative molecular mass 129.2 Log P octanol/water 2.0 Gross formula: C$_9$H$_7$N	**COLORLESS HYGROSCOPIC LIQUID WITH CHARACTERISTIC ODOR, TURNING BROWN ON EXPOSURE TO AIR** Vapor mixes readily with air. Toxic gases are a product of combustion (→ *nitrogendioxide*). Reacts violently with some strong oxidants. TLV-TWA not available **Absorption route:** Can enter the body by inhalation or ingestion or through the skin. Harmful atmospheric concentrations build up fairly slowly on evaporation at 20° C, but much more rapidly in aerosol form. **Immediate effects:** Irritates the eyes, skin and respiratory tract. Affects the nervous system. In serious cases risk of unconsciousness. **Effects of prolonged/repeated exposure:** Can cause liver and kidney damage.

HAZARDS/SYMPTOMS	PREVENTIVE MEASURES	FIRE EXTINGUISHING/FIRST AID
Fire: combustible.	keep away from open flame, no smoking.	powder, AFFF, foam, carbon dioxide, (halons).
Explosion: above 60°C: forms explosive air-vapor mixtures; see notes	above 60°C: sealed machinery, ventilation.	
Inhalation: cough, shortness of breath, dizziness, sleepiness.	ventilation, local exhaust or respiratory protection.	fresh air, rest, place in half-sitting position, call a doctor.
Skin: *is absorbed*; redness, dizziness, sleepiness.	gloves, protective clothing.	remove contaminated clothing, flush skin with water or shower, refer to a doctor.
Eyes: redness, pain, impaired vision.	face shield.	flush with water, take to a doctor.
Ingestion: abdominal pain, nausea, vomiting, sleepiness.		rinse mouth, call a doctor or take to hospital.

SPILLAGE	STORAGE	LABELING / NFPA
collect leakage in sealable containers, soak up with sand or other inert absorbent and remove to safe place; (additional individual protection: breathing apparatus).	keep in dark place, separate from strong oxidants, ventilate at floor level.	 2 — 1 — 0

NOTES
Flash point is between 60 and 110° C. Solubility in warm water is good. Reacts with some combinations of several substances, with risk of explosion. Alcohol consumption increases toxic effects.

Transport Emergency Card TEC(R)-61G06C

HI: 60; UN-number: 2656

CHLORAL

PHYSICAL PROPERTIES	IMPORTANT CHARACTERISTICS
Boiling point, °C 98 Melting point, °C −57 Relative density (water = 1) 1.5 Relative vapor density (air = 1) 5.1 Relative density at 20 °C of saturated mixture vapor/air (air = 1) 1.2 Vapor pressure, mm Hg at 20 °C 40 Solubility in water reaction Relative molecular mass 147.4	**COLORLESS LIQUID WITH CHARACTERISTIC ODOR** Vapor is heavier than air. Able to polymerize on exposure to light and on contact with → *sulfuric acid*. Decomposes in flame or on hot surface, giving off toxic and corrosive gases. Reacts with water, evolving large quantities of heat to form chloral hydrate. Reacts with concentrated bases (alkaline solutions), forming → *chloroform* and formiate. Attacks many materials.
	TLV-TWA not available
	Absorption route: Can enter the body by inhalation or ingestion. Harmful atmospheric concentrations can build up very rapidly on evaporation at 20°C. **Immediate effects:** Corrosive to the eyes, skin and respiratory tract. Affects the nervous system and heart. Inhalation can cause lung edema. In serious cases risk of unconsciousness and death.
Gross formula: C_2HCl_3O	

HAZARDS/SYMPTOMS	PREVENTIVE MEASURES	FIRE EXTINGUISHING/FIRST AID
Fire: non-combustible.		in case of fire in immediate vicinity: use any extinguishing agent.
Explosion:		in case of fire: keep tanks/drums cool by spraying with water, but DO NOT spray substance with water.
Inhalation: *corrosive*; sore throat, shortness of breath, sleepiness, unconsciousness.	ventilation, local exhaust or respiratory protection.	fresh air, rest, take to hospital.
Skin: redness, pain.	gloves.	remove contaminated clothing, flush skin with water or shower, call a doctor if necessary.
Eyes: *corrosive*; redness, pain, impaired vision.	face shield.	flush with water, take to a doctor if necessary.
Ingestion: *corrosive*, nauses, dizziness, sleepiness, unconsciousness.		rinse mouth, call a doctor or take to hospital.

SPILLAGE	STORAGE	LABELING / NFPA
collect leakage in sealable containers, soak up with sand or other inert absorbent and remove to safe place; (additional individual protection: breathing apparatus).	keep in dry, dark place, separate from strong bases, ventilate at floor level.	

NOTES
Addition of small amounts of dust or increase in oxygen content increase flammability. Alcohol consumption increases toxic effects. Lung edema symptoms usually develop several hours later and are aggravated by physical exertion: rest and hospitalization essential. As first aid, a doctor or authorized person should consider administering a corticosteroid spray. Under no circumstances use near flame or hot surface or when welding. Unbreakable packaging preferred; if breakable, keep in unbreakable container.

Transport Emergency Card TEC(R)-61G06B

HI: 60; UN-number: 2075 (hydrate)

CAS-No: [302-17-0]
trichloroethylidene glycol
trichloroacetaldehyde monohydrate

$CCl_3CH(OH)_2$

CHLORAL HYDRATE

PHYSICAL PROPERTIES	IMPORTANT CHARACTERISTICS
Boiling point (decomposes), °C 96 Melting point, °C 57 Relative density (water = 1) 1.9 Relative vapor density (air = 1) 5.7 Relative density at 20 °C of saturated mixture vapor/air (air = 1) 1.06 Vapor pressure, mm Hg at 20 °C 9.9 Solubility in water very good Relative molecular mass 165.4 Log P octanol/water 0.5	**TRANSPARENT COLORLESS CRYSTALS WITH CHARACTERISTIC ODOR** Decomposes in flame, giving off → *hydrochloric acid*. Reacts with strong bases, giving off → *chloroform* and formates. Reacts with strong oxidants. Attacks various metals.
	TLV-TWA not available
	Absorption route: Can enter the body by inhalation, through the skin or by ingestion. Harmful atmospheric concentrations can build up fairly rapidly on evaporation at approx. 20° C; harmful concentrations of airborne particles can build up even more rapidly. **Immediate effects:** Corrosive to the eyes, skin and respiratory tract. Inhalation of particles can cause lung edema. Can affect the central nervous system and heart circulation system, causing heart rate disorders and low blood pressure. Can impair consciousness. Can cause allergic skin reactions. Exposure to high concentrations can be fatal. Keep under medical observation. **Effects of prolonged/repeated exposure:** Can affect the liver and kidneys with resulting damage.
Gross formula: $C_2H_3ClO_2$	

HAZARDS/SYMPTOMS	PREVENTIVE MEASURES	FIRE EXTINGUISHING/FIRST AID
Fire: non-combustible.		in case of fire in immediate vicinity: use any extinguishing agent.
Explosion:		in case of fire: keep tanks/drums cool by spraying with water.
	STRICT HYGIENE	IN ALL CASES CALL IN A DOCTOR
Inhalation: *corrosive*; cough, nausea, sore throat, dizziness, unconsciousness.	local exhaust or respiratory protection.	fresh air, rest, place in half-sitting position, artificial respiration if necessary, take immediately to hospital.
Skin: *corrosive, is absorbed*; redness, burning sensation.	gloves.	remove contaminated clothing, flush skin with water or shower, refer to a doctor.
Eyes: *corrosive*; redness, pain, impaired vision.	face shield, or combined eye and respiratory protection.	flush thoroughly with water (remove contact lenses if easy), take to a doctor.
Ingestion: *corrosive*; abdominal cramps, disorientation, nausea, unconsciousness.	do not eat, drink or smoke while working.	rinse mouth, give no liquids, take immediately to hospital.

SPILLAGE	STORAGE	LABELING / NFPA
clean up spillage, carefully collect remainder; (additional individual protection: breathing apparatus).	keep in dark place, separate from oxidants and strong bases; ventilate at floor level.	R: 25-36/38 ☠ Toxic

NOTES
At boiling point decomposes by hydrolysis. Addition of small quantities of combustible substances or raising of oxygen content makes substance combustible. Alcohol consumption increases toxic (narcotic) effects. Lung edema symptoms usually develop several hours later and are aggravated by physical exertion: rest and hospitalization essential. As first aid, a doctor or authorized person should consider administering a suitable spray.

Transport Emergency Card TEC(R)-61G06B

HI: 60; UN-number: 2811

CHLORINATED PARAFFIN

PHYSICAL PROPERTIES	IMPORTANT CHARACTERISTICS	
Decomposes below boiling point, °C 205 Melting point, °C −30/+90 Relative density (water = 1) 1.0/1.7 Relative vapor density (air = 1) 19-26 Relative density at 20 °C of saturated mixture vapor/air (air = 1) 1 Vapor pressure, mm Hg at 20 °C <0.076 Solubility in water none Relative molecular mass 454-1097	**COLORLESS TO BROWN LIQUID OR SOLID** Decomposes when heated, giving off toxic and corrosive gas (→ *hydrogen chloride*).	
	TLV-TWA not available	
	Absorption route: Can enter the body by inhalation or ingestion or through the skin. Harmful atmospheric concentrations build up very slowly, if at all, on evaporation at approx. 20°C, but much more rapidly in aerosol form. **Immediate effects:** Affects the nervous system. **Effects of prolonged/repeated exposure:** Prolonged or repeated contact can cause skin disorders. Can cause liver, kidney and brain damage.	

HAZARDS/SYMPTOMS	PREVENTIVE MEASURES	FIRE EXTINGUISHING/FIRST AID
Fire: non-combustible.		in case of fire in immediate vicinity: use any extinguishing agent.
Inhalation: sore throat, cough, shortness of breath.	ventilation.	fresh air, rest, call a doctor.
Skin: *is absorbed*; redness.	gloves.	remove contaminated clothing, wash skin with soap and water, refer to a doctor.
Eyes: redness, pain.	safety glasses.	flush with water, take to a doctor.
Ingestion: abdominal pain, vomiting.		rinse mouth, call a doctor.

SPILLAGE	STORAGE	LABELING / NFPA
collect leakage in sealable containers, allow to set if possible, clean up spillage, carefully collect remainder.		

NOTES
Physical properties depend on composition of mixture (degree of chlorination). Cereclor is a trade name.

CHLORINE
(cylinder)

PHYSICAL PROPERTIES		IMPORTANT CHARACTERISTICS
Boiling point, °C	−34	**YELLOWISH-GREEN GAS WITH PUNGENT ODOR**
Melting point, °C	−102	Vapor is heavier than air. Local heating of steel apparatus could cause chlorine-iron fire. Forms explosive mixtures with hydrogen, acetylene or anhydrous ammonia which can be ignited even by strong sunlight. Corrosive when moist. Strong oxidant which reacts violently with combustible substances and reducing agents. Reacts violently with many organic compounds, phosphorus and metal powders. At room temperature attacks steel, copper, bronze etc. in moist state but not in dry state.
Relative density (water = 1)	1.4	
Relative vapor density (air = 1)	2.5	
Vapor pressure, mm Hg at 20 °C	5168	
Solubility in water, g/100 ml at 20 °C	0.7	
Relative molecular mass	70.9	

TLV-TWA	0.5 ppm	1.5 mg/m³ C
STEL	1.0 ppm	2.9 mg/m³

Absorption route: Can enter the body by inhalation. Harmful atmospheric concentrations can build up very rapidly if gas is released.
Immediate effects: Corrosive to the eyes, skin and respiratory tract. Can cause frostbite due to rapid evaporation. Inhalation of vapor/fumes can cause severe breathing difficulties (lung edema). In serious cases risk of death.

Gross formula: Cl$_2$

HAZARDS/SYMPTOMS	PREVENTIVE MEASURES	FIRE EXTINGUISHING/FIRST AID
Fire: non-combustible, many chemical reactions can cause fire and explosion.	avoid contact with combustible substances, hydrogen, acetylene, anhydrous ammonia, metal powders and phosphorus.	in case of fire in immediate vicinity: use any extinguishing agent.
Explosion: contact with hydrogen, acetylene, anhydrous ammonia, metal powders or phosphorus can cause explosion.	special equipment.	in case of fire: keep cylinder cool by spraying with water.
	STRICT HYGIENE	IN ALL CASES CALL A DOCTOR
Inhalation: *corrosive*; sore throat, cough, shortness of breath, severe breathing difficulties.	ventilation, local exhaust or respiratory protection.	fresh air, rest, place in half-sitting position, take to hospital.
Skin: *corrosive*; redness, pain, serious burns.	insulated gloves, protective clothing.	remove contaminated clothing, flush skin with water or shower, call a doctor.
Eyes: *corrosive*; redness, pain, impaired vision.	acid goggles, or combined eye and respiratory protection.	flush with water, take to a doctor.

SPILLAGE	STORAGE	LABELING / NFPA
evacuate area, call in an expert, render harmless with dilute caustic soda solution and flush away with water, under no circumstances spray liquid with water, combat gas cloud with water curtain; (additional individual protection: CHEMICAL SUIT); (additional individual protection at low concentrations: breathing apparatus).	if stored indoors, keep in cool, fireproof place.	0 / 3 / 0 / oxy

NOTES
Lung edema symptoms usually develop several hours later and are aggravated by physical exertion: rest and hospitalization essential. As first aid, a doctor or authorized person should consider administering a corticosteroid spray. Solubility in water applies to a solution above which chlorine gas has a vapor pressure of 760mm Hg. Do not spray leaking cylinder with water (to avoid corrosion). Turn leaking cylinder so that leak is on top to prevent liquid chlorine escaping. Do not use grease on faucets, stopcocks, etc. Use cylinder with special fittings.

Transport Emergency Card TEC(R)-2

HI: 266; UN-number: 1017

chlorine oxide
chlorine(IV) oxide
chlorine peroxide
ClO2

ClO$_2$

CHLORINE DIOXIDE

PHYSICAL PROPERTIES		IMPORTANT CHARACTERISTICS	
Boiling point, °C	11	**REDDISH-YELLOW TO YELLOWISH-GREEN GAS OR LIQUID WITH PUNGENT ODOR**	
Melting point, °C	−59	Gas is heavier than air and spreads at ground level. In concentrations as low as > 10% can	
Relative density (water = 1)	1.6	decompose explosively when heated, exposed to sunlight or subjected to shock or friction.	
Relative vapor density (air = 1)	2.3	Strong oxidant which reacts violently with combustible substances and reducing agents. Reacts	
Vapor pressure, mm Hg at 20 °C	1064	slowly with water to form → *perchloric acid*. Reacts violently with → *mercury*, → *carbon dioxide*	
Solubility in water, g/100 ml at 20 °C	0.3	and → *carbon monoxide*.	
Relative molecular mass	67.5		
		TLV-TWA	0.1 ppm	0.28 mg/m³
		STEL	0.3 ppm	0.83 mg/m³
		Absorption route: Can enter the body by inhalation. Harmful atmospheric concentrations can build up very rapidly if gas is released. **Immediate effects:** Corrosive to the eyes, skin and respiratory tract. Can cause frostbite due to rapid evaporation. Inhalation can cause lung edema. In serious cases risk of death.	
Gross formula:	ClO$_2$		

HAZARDS/SYMPTOMS	PREVENTIVE MEASURES	FIRE EXTINGUISHING/FIRST AID
Fire: non-combustible, many chemical reactions can cause fire and explosion.	avoid contact with combustible substances.	in case of fire in immediate vicinity: use any extinguishing agent.
Explosion: sunlight, shock or friction can cause explosion.	do not expose to sunlight or subject to shock or friction.	in case of fire: keep tanks/drums cool by spraying with water.
	STRICT HYGIENE	
Inhalation: *corrosive*; sore throat, cough, shortness of breath, severe breathing difficulties.	ventilation, local exhaust or respiratory protection.	fresh air, rest, place in half-sitting position, take to hospital.
Skin: *corrosive*; redness, pain, serious burns.	insulated gloves, protective clothing.	remove contaminated clothing, flush skin with water or shower, call a doctor.
Eyes: *corrosive*; redness, pain, impaired vision.	acid goggles, or combined eye and respiratory protection.	flush with water, take to a doctor.

SPILLAGE	STORAGE	LABELING / NFPA
evacuate area, call in an expert, ventilate, render harmless with dilute lye and then flush away with water; under no circumstances spray liquid with water, do not use sawdust or other combustible absorbents; (additional individual protection: CHEMICAL SUIT).	keep in cool, dark place under inert gas, separate from combustible substances and reducing agents; ventilate at floor level.	

NOTES
Flush contaminated clothing with water (fire hazard). Lung edema symptoms usually develop several hours later and are aggravated by physical exertion: rest and hospitalization essential. As first aid, a doctor or authorized person should consider administering a corticosteroid spray. Unbreakable packaging preferred; if breakable, keep in unbreakable container. Cannot be handled in pure state and is therefore prepared on location.

CAS-No: [78-95-5]
acetonyl chloride
1-chloro-2-ketopropane
chloromethyl methyl ketone
1-chloro-2-oxopropane
1-chloro-2-propanone

CH_3COCH_2Cl

CHLOROACETONE

PHYSICAL PROPERTIES		IMPORTANT CHARACTERISTICS	
Boiling point, °C	119	**COLORLESS LIQUID WITH PUNGENT ODOR**	
Melting point, °C	−45	Vapor mixes readily with air, forming explosive mixtures. Do not use compressed air when filling, emptying or processing. Decomposes when heated, giving off flammable, toxic and corrosive vapors. Reacts violently with strong oxidants, with risk of fire and explosion.	
Flash point, °C	7		
Relative density (water = 1)	1.16		
Relative vapor density (air = 1)	3.2		
Vapor pressure, mm Hg at 20 °C	21	TLV-TWA (skin)　　　　1 ppm　　　　3.8 mg/m³　C	
Solubility in water	good		
Relative molecular mass	92.5	**Absorption route:** Can enter the body by inhalation or ingestion. Harmful atmospheric concentrations can build up very rapidly on evaporation at 20°C.	
Log P octanol/water	0.3	**Immediate effects:** Corrosive to the eyes, skin and respiratory tract. Causes watering of the eyes. Inhalation can cause lung edema. In serious cases risk of death.	
Gross formula:	C_3H_5ClO		

HAZARDS/SYMPTOMS	PREVENTIVE MEASURES	FIRE EXTINGUISHING/FIRST AID
Fire: extremely flammable.	keep away from open flame and sparks, no smoking.	powder, AFFF, foam, carbon dioxide, (halons).
Explosion: forms explosive air-vapor mixtures.	sealed machinery, ventilation, explosion-proof electrical equipment and lighting.	in case of fire: keep tanks/drums cool by spraying with water.
	STRICT HYGIENE	
Inhalation: *corrosive*; sore throat, cough, shortness of breath, severe breathing difficulties.	ventilation, local exhaust or respiratory protection.	fresh air, rest, place in half-sitting position, take to hospital.
Skin: *corrosive*; redness, pain, serious burns.	gloves, protective clothing.	remove contaminated clothing, flush skin with water or shower, refer/take to a doctor.
Eyes: *corrosive*; redness, pain, impaired vision.	face shield.	flush with water, take to a doctor.
Ingestion: *corrosive*; sore throat, abdominal pain, vomiting.		rinse mouth, immediately take to hospital.

SPILLAGE	STORAGE	LABELING / NFPA
collect leakage in sealable containers, soak up with sand or other inert absorbent and remove to safe place; (additional individual protection: breathing apparatus).	keep in dark, fireproof place, separate from oxidants.	

NOTES

Log P octanol/water is estimated. Lung edema symptoms usually develop several hours later and are aggravated by physical exertion: rest and hospitalization essential. As first aid, a doctor or authorized person should consider administering a corticosteroid spray.

Transport Emergency Card TEC(R)-61G06B　　　　　　　　　　　　　　**HI: 60; UN-number: 1695**

CHLOROACETONITRILE

PHYSICAL PROPERTIES	IMPORTANT CHARACTERISTICS	
Boiling point, °C 126 Melting point, °C ? Flash point, °C 56 Relative density (water = 1) 1.19 Relative vapor density (air = 1) 2.61 Relative density at 20 °C of saturated mixture vapor/air (air = 1) 1 Vapor pressure, mm Hg at 20 °C 8.7 Solubility in water none Relative molecular mass 75.5	**COLORLESS LIQUID WITH PUNGENT ODOR** Vapor mixes readily with air. Toxic vapors are a product of combustion. Decomposes when heated, giving off toxic and flammable vapors (→ *hydrogen cyanide*). Reacts with steam and acids giving off toxic and flammable vapor (→ *hydrogen cyanide*).	
	TLV-TWA not available	
	Absorption route: Can enter the body by inhalation or ingestion. Harmful atmospheric concentrations can build up fairly rapidly on evaporation at approx. 20°C - even more rapidly in aerosol form.	
Gross formula: C_2H_2ClN		

HAZARDS/SYMPTOMS	PREVENTIVE MEASURES	FIRE EXTINGUISHING/FIRST AID
Fire: flammable.	keep away from open flame, no smoking.	powder, AFFF, foam, carbon dioxide, (halons).
Explosion: above 56°C: forms explosive air-vapor mixtures.	above 56°C: sealed machinery, ventilation.	
Inhalation: cough, anxiety.	ventilation, local exhaust or respiratory protection.	fresh air, rest, call a doctor.
Skin:	gloves.	remove contaminated clothing, flush skin with water or shower, call a doctor.
Eyes: redness, pain.	face shield, or combined eye and respiratory protection.	flush with water, take to a doctor.
Ingestion: abdominal pain, nausea, hydrocyanic acid can form.		rinse mouth, immediately take to hospital.

SPILLAGE	STORAGE	LABELING / NFPA
collect leakage in sealable containers, soak up with sand or other inert absorbent and remove to safe place; (additional individual protection: breathing apparatus).	keep in fireproof place.	R: 23/24/25 S: 44 ☠ Toxic NFPA: 2 / 3 / 0

NOTES
Insufficient toxicological data on harmful effects to humans.

chloracetyl chloride
chloroacetic acid chloride
monochloroacetyl chloride

ClCH₂COCl

CHLOROACETYL CHLORIDE

PHYSICAL PROPERTIES	IMPORTANT CHARACTERISTICS
Boiling point, °C 107 Melting point, °C − 22 Relative density (water = 1) 1.5 Relative vapor density (air = 1) 3.9 Vapor pressure, mm Hg at 20 °C 19 Solubility in water reaction Relative molecular mass 112.9	**COLORLESS TO LIGHT YELLOW LIQUID WITH PUNGENT (EYE-WATERING) ODOR** Decomposes in flame or on hot surface, giving off toxic gas. Decomposes on contact with water, forming → *hydrochloric acid* and → *chloroacetic acid*. Decomposes when heated, giving off toxic gas (→ *phosgene*). Reacts violently with water, alcohols and many organic compounds. Reacts violently with metal powders and sodium amide, with risk of fire and explosion. Reacts with air to form corrosive vapors (HCl).

| | TLV-TWA | 0.05 ppm | 0.23 mg/m³ |

Absorption route: Can enter the body by inhalation or ingestion or through the skin. Harmful atmospheric concentrations can build up very rapidly on evaporation at 20°C.
Immediate effects: Corrosive to the eyes, skin and respiratory tract. Inhalation of vapor/fumes can cause severe breathing difficulties (lung edema). In serious cases risk of death.

Gross formula: C₂H₂Cl₂O

HAZARDS/SYMPTOMS	PREVENTIVE MEASURES	FIRE EXTINGUISHING/FIRST AID
Fire: non-combustible, many chemical reactions can cause fire and explosion.	avoid contact with other substances.	in case of fire in immediate vicinity: DO NOT USE WATER-BASED EXTINGUISHERS.
Explosion:		in case of fire: keep tanks/drums cool by spraying with water, but DO NOT spray substance with water.
Inhalation: *corrosive*; sore throat, cough, shortness of breath, severe breathing difficulties.	ventilation, local exhaust or respiratory protection.	fresh air, rest, place in half-sitting position, take to hospital.
Skin: *corrosive*; redness, pain, serious burns.	gloves, protective clothing.	remove contaminated clothing, flush skin with water or shower, refer to a doctor if necessary.
Eyes: *corrosive*; redness, pain, impaired vision.	face shield, or combined eye and respiratory protection.	flush with water, take to a doctor.
Ingestion: *corrosive*; sore throat, abdominal pain, diarrhea, vomiting.		rinse mouth, immediately take to hospital.

SPILLAGE	STORAGE	LABELING / NFPA
call in an expert, collect leakage in sealable containers, soak up with quicklime, sodium/potassium chloride and remove to safe place; (additional individual protection: breathing apparatus).	keep dry, separate from all other substances.	R: 34-37 S: 9-26 Corrosive 3 / 0 / 1

NOTES
Reacts violently with extinguishing agents such as water and foam. Lung edema symptoms usually develop several hours later and are aggravated by physical exertion: rest and hospitalization essential. As first aid, a doctor or authorized person should consider administering a corticosteroid spray. Under no circumstances use near flame or hot surface or when welding. Use airtight packaging. Unbreakable packaging preferred; if breakable, keep in unbreakable container.

Transport Emergency Card TEC(R)-80G14

HI: X80; UN-number: 1752

CAS-No: [95-51-2]
1-amino-2-chlorobenzene
2-chloroaniline
2-chlorobenzenamine
2-chlorophenylamine

$ClC_6H_4NH_2$

o-CHLOROANILINE

PHYSICAL PROPERTIES	IMPORTANT CHARACTERISTICS	
Boiling point, °C 208	**YELLOW LIQUID WITH CHARACTERISTIC ODOR, TURNING BROWN ON EXPOSURE TO AIR**	
Melting point, °C − 14	Vapor mixes readily with air. Toxic gases are a product of combustion (→ *hydrochloric acid* and	
Flash point, °C 108	→ *nitrous vapors*). Reacts violently with strong oxidants.	
Auto-ignition temperature, °C >500		
Relative density (water = 1) 1.21	TLV-TWA not available	
Relative vapor density (air = 1) 4.4		
Relative density at 20 °C of saturated mixture vapor/air (air = 1) 1.0	**Absorption route:** Can enter the body by inhalation or ingestion or through the skin. Harmful atmospheric concentrations build up fairly slowly on evaporation at 20° C, but much more rapidly in aerosol form.	
Vapor pressure, mm Hg at 20° C 0.015	**Immediate effects:** Irritates the eyes, skin and respiratory tract.	
Solubility in water none	**Effects of prolonged/repeated exposure:** Prolonged or repeated contact can cause skin disorders. Can affect the blood. Can cause liver and kidney damage. In serious cases risk of death.	
Relative molecular mass 127.6		
Log P octanol/water 1.9		
Gross formula: C_6H_6ClN		

HAZARDS/SYMPTOMS	PREVENTIVE MEASURES	FIRE EXTINGUISHING/FIRST AID
Fire: combustible.	keep away from open flame, no smoking.	powder, water spray, foam, carbon dioxide, (halons).
Inhalation: shortness of breath, headache, dizziness, blue skin, feeling of weakness.	ventilation, local exhaust or respiratory protection.	fresh air, rest, take to hospital.
Skin: *is absorbed*; redness; see also 'Inhalation'.	gloves.	remove contaminated clothing, wash skin with soap and water, call a doctor.
Eyes: redness, pain.	face shield.	flush with water, take to a doctor if necessary.
Ingestion: abdominal pain, nausea; see also 'Inhalation'.	do not eat, drink or smoke while working.	rinse mouth, immediately take to hospital.

SPILLAGE	STORAGE	LABELING / NFPA
collect leakage in sealable containers, soak up with sand or other inert absorbent and remove to safe place.	keep separate from oxidants.	R: 23/24/25-33 S: 28-36/37-44 Note C ☠ Toxic

NOTES
Depending on the degree of exposure, regular medical checkups are advisable. Effects on blood due to formation of methemoglobin. Special first aid required in the event of poisoning: antidotes must be available (with instructions). The measures on this card also apply to m-chloroaniline or 3-chloroaniline: boiling point 230° C; melting point − 10° C; flash point 118° C; autoignition temperature > 540° C; vapor pressure 0.98mm Hg at 50° C.

Transport Emergency Card TEC(R)-773

HI: 60; UN-number: 2019

CAS-No: [106-47-8]
1-amino-4-chlorobenzene
p-chloroaminobenzene
4-chloro-1-aminobenzene
4-chlorobenzeneamine
4-chlorophenylamine

$ClC_6H_4NH_2$

p-CHLOROANILINE

PHYSICAL PROPERTIES	IMPORTANT CHARACTERISTICS
Boiling point, °C 232 Melting point, °C 70 Flash point, °C > 188 Relative density (water = 1) 1.4 Relative vapor density (air = 1) 4.4 Relative density at 20 °C of saturated mixture vapor/air (air = 1) 1.0 Vapor pressure, mm Hg at 20 °C 0.015 Solubility in water none Relative molecular mass 127.6 Log P octanol/water 1.8	**COLORLESS CRYSTALS** Dissolves in hot water. Decomposes when burned, forming → *hydrochloric acid* and → *nitrous vapors*. Reacts violently with oxidants.
	TLV-TWA not available
	Absorption route: Can enter the body by inhalation or ingestion or through the skin. Harmful atmospheric concentrations build up fairly slowly on evaporation at 20°C, but harmful concentrations of airborne particles can build up much more rapidly. **Immediate effects:** Irritates the eyes, skin and respiratory tract. **Effects of prolonged/repeated exposure:** Prolonged or repeated contact can cause skin disorders. Can affect the blood. Can cause liver and kidney damage. In serious cases risk of death.
Gross formula: C_6H_6ClN	

HAZARDS/SYMPTOMS	PREVENTIVE MEASURES	FIRE EXTINGUISHING/FIRST AID
Fire: combustible.	keep away from open flame, no smoking.	water spray, powder.
Inhalation: shortness of breath, headache, dizziness, blue skin, feeling of weakness.	local exhaust or respiratory protection.	fresh air, rest, take to hospital.
Skin: *is absorbed*; redness; see also 'Inhalation'.	gloves, protective clothing.	remove contaminated clothing, wash skin with soap and water, call a doctor.
Eyes: redness, pain.	close-fitting safety glasses.	flush with water, take to a doctor if necessary.
Ingestion: abdominal pain, nausea; see also 'Inhalation'.	do not eat, drink or smoke while working.	rinse mouth, immediately take to hospital.

SPILLAGE	STORAGE	LABELING / NFPA
clean up spillage, carefully collect remainder; (additional individual protection: P2 respirator).	keep separate from oxidants.	R: 23/24/25-33 S: 28-36/37-44 Note C ☠ Toxic ◇

NOTES
Depending on the degree of exposure, regular medical checkups are advisable. Effects on blood due to formation of methemoglobin. Special first aid required in the event of poisoning: antidotes must be available (with instructions).

$C_6H_3Cl(OH)(CH_3)$

4-chloro-3-methylphenol
4-chloro-m-cresol
p-chlorocresol
4-chloro-1-hydroxy-3-methylbenzene
2-chloro-5-hydroxytoluene

p-CHLORO-m-CRESOL

PHYSICAL PROPERTIES	IMPORTANT CHARACTERISTICS
Boiling point, °C 235 Melting point, °C 55 Flash point, °C 118 Auto-ignition temperature, °C 590 Relative density (water = 1) 1.4 Relative vapor density (air = 1) 49 Relative density at 20 °C of saturated mixture vapor/air (air = 1) 1.0 Vapor pressure, mm Hg at 20 °C 0.061 Solubility in water, g/100 ml at 20 °C 0.4 Relative molecular mass 142.6 Log P octanol/water 3.1	**COLORLESS TO LIGHT YELLOW CRYSTALLINE POWDER WITH CHARACTERISTIC ODOR** Vapor mixes readily with air. Decomposes when heated or burned, giving off toxic and corrosive gases.
	TLV-TWA not available
	Absorption route: Can enter the body by inhalation or ingestion or through the skin. Evaporation negligible at 20°C, but harmful concentrations of airborne particles can build up rapidly. **Immediate effects:** Corrosive to the eyes, skin and respiratory tract. Affects the nervous system. Inhalation can cause lung edema. In serious cases risk of death. **Effects of prolonged/repeated exposure:** Prolonged or repeated contact can cause eczema. Can cause liver, kidney and brain damage.
Gross formula: C_7H_7ClO	

HAZARDS/SYMPTOMS	PREVENTIVE MEASURES	FIRE EXTINGUISHING/FIRST AID
Fire: combustible.	keep away from open flame, no smoking.	water spray, powder.
	KEEP DUST UNDER CONTROL	
Inhalation: *corrosive*; severe breathing difficulties, headache, dizziness.	local exhaust or respiratory protection.	fresh air, rest, place in half-sitting position, take to hospital.
Skin: *corrosive*; redness, pain, burns, prolonged or repeated contact can cause eczema.	gloves, protective clothing.	remove contaminated clothing, flush skin with water or shower, call a doctor if necessary.
Eyes: *corrosive*; redness, pain, impaired vision.	face shield.	flush with water, take to doctor.
Ingestion: *corrosive*; abdominal pain, headache, dizziness, vomiting.		rinse mouth, take to doctor.

SPILLAGE	STORAGE	LABELING / NFPA
clean up spillage, carefully collect remainder; (additional individual protection: P2 respirator).		R: 2/22-38 S: 26-28 **✖** Harmful

NOTES
Dimorphous crystals, i.e. with 2 melting points: 55°C and 66°C. Technical grades of chlorocresol are a mixture of ortho- and para-isomers and 4,6-dichlorocresol. Lung edema symptoms usually develop several hours later and are aggravated by physical exertion: rest and hospitalization essential. As first aid, a doctor or authorized person should consider administering a corticosteroid spray.

Transport Emergency Card TEC(R)-61G12B

HI: 60; UN-number: 2669

2-chloroethanol
β-chloroethyl alcohol
ethylene chlorhydrine
ethylene chlorohydrin
2-monochloroethanol

CH_2ClCH_2OH

2-CHLOROETHYL ALCOHOL

PHYSICAL PROPERTIES	IMPORTANT CHARACTERISTICS
Boiling point, °C 128 Melting point, °C −68 Flash point, °C 55 Auto-ignition temperature, °C 425 Relative density (water = 1) 1.2 Relative density at 20 °C of saturated mixture vapor/air (air = 1) 1.01 Vapor pressure, mm Hg at 20 °C 5.6 Solubility in water ∞ Explosive limits, vol% in air 4.9-16 Relative molecular mass 80.5 Gross formula: C_2H_5ClO	**COLORLESS LIQUID WITH CHARACTERISTIC ODOR** Vapor mixes readily with air. Decomposes when heated, giving off toxic and corrosive gases. Reacts violently with oxidants. Reacts violently with hot, strong bases, e.g. caustic soda, giving off → ethylene oxide.

TLV-TWA (skin)	1 ppm	3.3 mg/m³ C

Absorption route: Can enter the body by inhalation or ingestion or through the skin. Harmful atmospheric concentrations can build up very rapidly on evaporation at 20°C.
Immediate effects: Irritates the eyes and respiratory tract. Inhalation of vapor/fumes can cause severe breathing difficulties (lung edema). In serious cases risk of death. Several hours can elapse between exposure and onset of symptoms.
Effects of prolonged/repeated exposure: Can cause liver, kidney and brain damage.

HAZARDS/SYMPTOMS	PREVENTIVE MEASURES	FIRE EXTINGUISHING/FIRST AID
Fire: combustible.	keep away from open flame, no smoking; avoid contact with hot, strong bases.	powder, alcohol-resistant foam, water spray, carbon dioxide, (halons).
Explosion: above 55°C: forms explosive air-vapor mixtures.	above 55°C: sealed machinery, ventilation.	in case of fire: keep tanks/drums cool by spraying with water.
	STRICT HYGIENE	
Inhalation: severe breathing difficulties, headache, dizziness, vomiting, lowered blood pressure, uncoordinated movements, delirium, shock, unconsciousness.	ventilation, local exhaust or respiratory protection.	fresh air, rest, place in half-sitting position, take to hospital.
Skin: *is absorbed rapidly*; irritation and redness appear sometime later.	gloves, protective clothing.	remove contaminated clothing, wash skin thoroughly with soap and water, call a doctor.
Eyes: redness, pain, impaired vision.	face shield.	flush with water, take to a doctor.
Ingestion: headache, severe breathing difficulties, dizziness, vomiting, lowered blood pressure, uncoordinated movements, delirium, shock, unconsciousness.		rinse mouth, take immediately to hospital.

SPILLAGE	STORAGE	LABELING / NFPA
call in an expert, collect leakage in sealable containers, soak up with sand or other inert absorbent and remove to safe place; (additional individual protection: breathing apparatus).	keep in dry, fireproof place, separate from oxidants.	R: 26/27/28 S: 7/9-28-45 ☠ Toxic

NFPA diagram: 2 (top), 3 (left), 0 (right)

NOTES
Depending on the degree of exposure, regular medical checkups are advisable. Lung edema symptoms usually develop several hours later and are aggravated by physical exertion: rest and hospitalization essential. As first aid, a doctor or authorized person should consider administering a corticosteroid spray.

Transport Emergency Card TEC(R)-729

HI: 60; UN-number: 1135

CAS-No: [67-66-3]
formyl trichloride
methane trichloride
methenyl chloride
trichloroform
trichloromethane

CHCl$_3$

CHLOROFORM

PHYSICAL PROPERTIES	IMPORTANT CHARACTERISTICS
Boiling point, °C 61 Melting point, °C −63 Relative density (water = 1) 1.5 Relative vapor density (air = 1) 4.1 Relative density at 20 °C of saturated mixture vapor/air (air = 1) 1.6 Vapor pressure, mm Hg at 20 °C 152 Solubility in water, g/100 ml at 20 °C 0.8 Electrical conductivity, pS/m < 10⁴ Relative molecular mass 119.4 Log P octanol/water 2.0	**COLORLESS LIQUID WITH CHARACTERISTIC ODOR** Vapor is heavier than air. Flow, agitation etc. can cause build-up of electrostatic charge due to liquid's low conductivity. Decomposes in flame or on hot surface, giving off corrosive vapors (→ *hydrochloric acid*). Reacts with light metals and strong bases, with risk of fire and explosion. Attacks many plastics.

PHYSICAL PROPERTIES cont.	TLV-TWA	10 ppm	49 mg/m³ A2

Absorption route: Can enter the body by inhalation or ingestion. Harmful atmospheric concentrations can build up very rapidly on evaporation at 20°C.
Immediate effects: Irritates the eyes, skin and respiratory tract. Liquid destroys the skin's natural oils. Affects the nervous system and heart. In serious cases risk of unconsciousness.
Effects of prolonged/repeated exposure: Can cause liver and kidney damage. Has been found to cause a type of liver cancer in certain animal species under certain circumstances.

Gross formula: CHCl$_3$

HAZARDS/SYMPTOMS	PREVENTIVE MEASURES	FIRE EXTINGUISHING/FIRST AID
Fire: non-combustible.		in case of fire in immediate vicinity: use any extinguishing agent.
Explosion:		in case of fire: keep tanks/drums cool by spraying with water.
Inhalation: cough, headache, nausea, drowsiness, unconsciousness.	ventilation, local exhaust or respiratory protection.	fresh air, rest, call a doctor.
Skin: redness, pain.	gloves.	remove contaminated clothing, flush skin with water or shower.
Eyes: redness, pain.	safety glasses, or combined eye and respiratory protection.	flush with water, take to a doctor if necessary.
Ingestion: headache, nausea, vomiting, drowsiness.		rinse mouth, DO NOT induce vomiting, take immediately to hospital.

SPILLAGE	STORAGE	LABELING / NFPA
collect leakage in sealable containers, soak up with sand or other inert absorbent and remove to safe place; (additional individual protection: breathing apparatus).	ventilate at floor level.	R: 20/22-38-40-48 S: 36/37 ✖ Harmful

NFPA diamond: 0 (top), 2 (left), 0 (right)

NOTES
Odor limit is above TLV. Alcohol consumption increases toxic effects. Depending on the degree of exposure, regular medical checkups are advisable. Under no circumstances use near flame or hot surface or when welding.

Transport Emergency Card TEC(R)-146

HI: 60; UN-number: 1888

CAS-No: [107-30-2]
chlorodimethyl ether
CMME
dimethylchloroether
methoxymethyl chloride
monochlorodimethyl ether

CH₃OCH₂Cl

CHLOROMETHYL METHYL ETHER

PHYSICAL PROPERTIES	IMPORTANT CHARACTERISTICS
Boiling point, °C 59 Melting point, °C −104 Flash point, °C −8 Relative density (water = 1) 1.1 Relative vapor density (air = 1) 2.8 Relative density at 20 °C of saturated mixture vapor/air (air = 1) 1.4 Vapor pressure, mm Hg at 20 °C 162 Solubility in water reaction Explosive limits, vol% in air ?-?	**COLORLESS LIQUID WITH CHARACTERISTIC ODOR** Vapor is heavier than air and spreads at ground level, with risk of ignition at a distance. Presumed to be able to form peroxides. Flow, agitation etc. can cause build-up of electrostatic charge due to liquid's low conductivity. Do not use compressed air when filling, emptying or processing. Toxic and corrosive vapors are a product of combustion (→ *hydrogen chloride*). Decomposes on contact with warm water, forming → *hydrochloric acid*.

TLV-TWA A2

Absorption route: Can enter the body by inhalation or ingestion or through the skin. Harmful atmospheric concentrations can build up very rapidly on evaporation at 20°C.
Immediate effects: Corrosive to the eyes, skin and respiratory tract. Affects the nervous system. Inhalation of vapor/fumes can cause severe breathing difficulties (lung edema). In serious cases risk of unconsciousness and death.
Effects of prolonged/repeated exposure: Can cause liver and kidney damage.

Gross formula: C₂H₅ClO

HAZARDS/SYMPTOMS	PREVENTIVE MEASURES	FIRE EXTINGUISHING/FIRST AID
Fire: extremely flammable.	keep away from open flame and sparks, no smoking.	powder, AFFF, foam, carbon dioxide, (halons), DO NOT USE WATER-BASED EXTINGUISHERS.
Explosion: forms explosive air-vapor mixtures.	sealed machinery, ventilation, explosion-proof electrical equipment and lighting, grounding.	in case of fire: keep tanks/drums cool by spraying with water, but DO NOT spray substance with water.
	STRICT HYGIENE, AVOID ALL CONTACT	
Inhalation: *corrosive*; sore throat, cough, severe breathing difficulties, headache, nausea.	ventilation, local exhaust or respiratory protection.	fresh air, rest, place in half-sitting position, take to hospital.
Skin: *corrosive, is absorbed*; redness, pain, burns.	gloves, protective clothing.	remove contaminated clothing, flush skin with water or shower, call a doctor.
Eyes: *corrosive*; redness, pain, impaired vision.	face shield, or combined eye and respiratory protection.	flush with water, take to a doctor.
Ingestion: *corrosive*; sore throat, abdominal pain, vomiting.		rinse mouth, immediately take to hospital.

SPILLAGE	STORAGE	LABELING / NFPA
evacuate area, call in an expert, collect leakage in sealable containers, soak up with sand or other inert absorbent and remove to safe place; (additional individual protection: CHEMICAL SUIT).	keep in dry, dark, fireproof place.	R: 45-11-20/21/22 S: 53-9-16-44 Note E Flammable Toxic 3 3 2

NOTES

Monochlorodimethyl ether can contain up to 7% bis(chloromethyl)ether, which can cause lung cancer in humans under certain circumstances. Monochlorodimethyl ether is itself a suspected carcinogen. Explosive limits not given in the literature. Before distilling check for peroxides; if found, render harmless. Lung edema symptoms usually develop several hours later and are aggravated by physical exertion: rest and hospitalization essential. As first aid, a doctor or authorized person should consider administering a corticosteroid spray. Under no circumstances use near flame or hot surface or when welding. Use airtight packaging.

Transport Emergency Card TEC(R)-30G32

HI: 336; UN-number: 1239

1-CHLORONAPHTALENE

PHYSICAL PROPERTIES		IMPORTANT CHARACTERISTICS
Boiling point, °C	259	**COLORLESS VISCOUS LIQUID WITH CHARACTERISTIC ODOR**
Melting point, °C	− 2.5	Vapor mixes readily with air. Flow, agitation etc. can cause build-up of electrostatic charge due to
Flash point, °C	121	liquid's low conductivity. Toxic and corrosive vapors (→ *hydrogen chloride*) *are a product of*
Auto-ignition temperature, °C	>558	*combustion.*
Relative density (water = 1)	1.2	
Relative vapor density (air = 1)	5.6	
Relative density at 20 °C of saturated mixture		TLV-TWA not available
vapor/air (air = 1)	1.00	
Vapor pressure, mm Hg at 20 °C	0.15	**Absorption route:** Can enter the body by inhalation or ingestion or through the skin. Harmful
Solubility in water	none	atmospheric concentrations build up very slowly, if at all, on evaporation at approx. 20°C, but
Electrical conductivity, pS/m	1800	much more rapidly in aerosol form.
Relative molecular mass	162.6	**Immediate effects:** Irritates the eyes, skin and respiratory tract.
		Effects of prolonged/repeated exposure: Can cause skin disorders and kidney damages.
Gross formula:	C$_{10}$H$_7$Cl	

HAZARDS/SYMPTOMS	PREVENTIVE MEASURES	FIRE EXTINGUISHING/FIRST AID
Fire: combustible.	keep away from open flame, no smoking.	powder, water spray, foam, carbon dioxide, (halons).
Explosion:	grounding.	
Inhalation: sore throat, cough, shortness of breath.	ventilation.	fresh air, rest, call a doctor.
Skin: *is absorbed*: redness, chronic exposure can cause acne.	gloves, protective clothing.	remove contaminated clothing, wash skin with soap and water.
Eyes: redness, pain.	face shield.	flush with water, take to a doctor if necessary.
Ingestion: sore throat, abdominal pain, nausea.		rinse mouth, call a doctor.

SPILLAGE	STORAGE	LABELING / NFPA
collect leakage in sealable containers, soak up with sand or other inert absorbent and remove to safe place.	if stored indoors, keep in fireproof place, ventilate at floor level.	

NFPA: 1 / 1 / 0

NOTES
The measures on this card also apply to 2-chloronaphtalene.

CHLORONITROANILINE

PHYSICAL PROPERTIES	IMPORTANT CHARACTERISTICS
Decomposes below boiling point, °C ca. 210 Melting point, °C 108 Flash point, °C 205 Auto-ignition temperature, °C 520 Relative density (water = 1) > 1 Relative vapor density (air = 1) 5.9 Relative density at 20 °C of saturated mixture vapor/air (air = 1) 1.0 Vapor pressure, mm Hg at 20 °C < 0.76 Solubility in water poor Relative molecular mass 172.6	**YELLOW NEEDLE-SHAPED CRYSTALS OR YELLOW POWDER** → *Hydrochloric acid* and → *nitrogen oxides* are a product of combustion. Decomposes when heated or on contact with strong acids, giving off toxic vapors (→ *hydrochloric acid* and → *nitrogen oxides*). Reacts with strong oxidants.
	TLV-TWA not available
	Absorption route: Can enter the body by inhalation or ingestion. Insufficient data on the rate at which harmful concentrations can build up.
Gross formula: $C_6H_5ClN_2O_2$	

HAZARDS/SYMPTOMS	PREVENTIVE MEASURES	FIRE EXTINGUISHING/FIRST AID
Fire: combustible.	keep away from open flame, no smoking.	water spray, powder.
Explosion: finely dispersed particles form explosive mixtures on contact with air.	keep dust from accumulating; sealed machinery, explosion-proof electrical equipment and lighting, grounding.	
Inhalation:	local exhaust or respiratory protection.	fresh air, rest, call a doctor.
Skin:	gloves.	remove contaminated clothing, flush skin with water or shower.
Eyes:	goggles, or combined eye and respiratory protection.	flush with water, take to a doctor if necessary.
Ingestion:	do not eat, drink or smoke while working.	rinse mouth, call a doctor.

SPILLAGE	STORAGE	LABELING / NFPA
clean up spillage, carefully collect remainder; (additional individual protection: respirator with A/P2 filter).	keep in fireproof place, keep separate from strong acids.	R: 26/27/28-33 S: 28-36/37-45 Note C ☣ ☠ Toxic

NOTES
Insufficient toxicological data on harmful effects to humans. The measures on this card also apply to p-chloro-o-nitroaniline (melting point: 116°C). Use airtight packaging.

Transport Emergency Card TEC(R)-874

HI: 60; UN-number: 2237

o-CHLORONITROBENZENE

PHYSICAL PROPERTIES	IMPORTANT CHARACTERISTICS	
Boiling point, °C 250 Melting point, °C 33 Flash point, °C 124 Relative density (water = 1) 1.3 Relative vapor density (air = 1) 5.4 Relative density at 20 °C of saturated mixture vapor/air (air = 1) 1.03 Vapor pressure, mm Hg at 20 °C < 0.076 Solubility in water none Relative molecular mass 157.6 Log P octanol/water 2.2	**YELLOWISH-GREEN CRYSTALS** Vapor mixes readily with air. Decomposes when heated or burned, giving off toxic and corrosive vapors (→ *nitrous vapors*, → *chlorine*, → *hydrochloric acid*, → *phosgene*). Reacts violently with reducing agents and many other substances, with risk of fire and explosion.	
	TLV-TWA (skin) 0.1 ppm 0.64 mg/m³	
Gross formula: $C_6H_4ClNO_2$	**Absorption route:** Can enter the body by inhalation or ingestion or through the skin. Harmful atmospheric concentrations can build up fairly rapidly on evaporation at approx. 20°C; harmful concentrations of airborne particles can build up even more rapidly. **Immediate effects:** Corrosive to the eyes, skin and respiratory tract. Inhalation of finely dispersed substance can cause severe breathing difficulties (lung edema). In serious cases risk of death. **Effects of prolonged/repeated exposure:** Prolonged or repeated contact can cause skin disorders. Can affect the blood. Can cause liver and kidney damage.	

HAZARDS/SYMPTOMS	PREVENTIVE MEASURES	FIRE EXTINGUISHING/FIRST AID
Fire: combustible, many chemical reactions can cause fire and explosion.	keep away from open flame, no smoking.	water spray, powder.
Explosion: finely dispersed particles form explosive mixtures on contact with air.	keep dust from accumulating; sealed machinery, explosion-proof electrical equipment and lighting, grounding.	in case of fire: keep tanks/drums cool by spraying with water.
	STRICT HYGIENE, KEEP DUST UNDER CONTROL	
Inhalation: headache, dizziness, nausea, vomiting, blue skin, feeling of weakness.	local exhaust or respiratory protection.	fresh air, rest, take to hospital.
Skin: *is absorbed*; redness; see also 'Inhalation'.	gloves, protective clothing.	remove contaminated clothing, wash skin with soap and water, call a doctor.
Eyes: redness, pain.	safety glasses, or combined eye and respiratory protection.	flush with water, take to a doctor.
Ingestion: headache, nausea, dizziness, vomiting, blue skin, feeling of weakness.	do not eat, drink or smoke while working.	rinse mouth, take to hospital.

SPILLAGE	STORAGE	LABELING / NFPA
clean up spillage, carefully collect remainder; (additional individual protection: respirator with A/P2 filter).	keep separate from combustible substances and reducing agents; ventilate at floor level.	NFPA diamond: blue 3, red 1, yellow 0

NOTES
The measures on this card also apply to m-chloronitrobenzene (synonym: m-nitrochlorobenzene), CAS-No 121-73-3, boiling point 236°C, melting point 46°C, vapor pressure at 20°C: < 0.076mm Hg. Addition of small quantities of combustible substances or raising of oxygen content makes substance combustible. Flush contaminated clothing with water. Depending on the degree of exposure, regular medical checkups are advisable. Lung edema symptoms usually develop several hours later and are aggravated by physical exertion: rest and hospitalization essential. As first aid, a doctor or authorized person should consider administering a corticosteroid spray. Special first aid required in the event of poisoning: antidotes must be available (with instructions). Effects on blood due to formation of methemoglobin.

Transport Emergency Card TEC(R)-877

HI: 60; UN-number: 1578

p-CHLORONITROBENZENE

PHYSICAL PROPERTIES	IMPORTANT CHARACTERISTICS
Boiling point, °C 242 Melting point, °C 83 Flash point, °C 127 Auto-ignition temperature, °C 510 Relative density (water = 1) 1.4 Relative vapor density (air = 1) 5.4 Relative density at 20 °C of saturated mixture vapor/air (air = 1) 1.03 Vapor pressure, mm Hg at 20 °C <0.076 Solubility in water none Relative molecular mass 157.6 Log P octanol/water 2.4	**YELLOWISH-GREEN POWDER OR CRYSTALS** Vapor mixes readily with air. Decomposes when heated, giving off toxic gases (→ *nitrogen oxides*, → *hydrochloric acid*, → *phosgene* and → *chlorine*). Strong oxidant which reacts violently with combustible substances and reducing agents. Reacts with many substances, with risk of fire and explosion.

	TLV-TWA (skin)	0.1 ppm	0.64 mg/m³

Absorption route: Can enter the body by inhalation or ingestion or through the skin. Harmful atmospheric concentrations can build up fairly rapidly on evaporation at approx. 20°C; harmful concentrations of airborne particles can build up even more rapidly.
Effects of prolonged/repeated exposure: Prolonged or repeated contact can cause skin disorders. Can affect the blood. Can cause liver and kidney damage. In serious cases risk of death.

Gross formula:	$C_6H_4ClNO_2$

HAZARDS/SYMPTOMS	PREVENTIVE MEASURES	FIRE EXTINGUISHING/FIRST AID
Fire: combustible, many chemical reactions can cause fire and explosion.	keep away from open flame, no smoking.	water spray, powder.
Explosion: finely dispersed particles form explosive mixtures on contact with air.	keep dust from accumulating, sealed machinery, explosion-proof electrical equipment and lighting, grounding.	
	STRICT HYGIENE, KEEP DUST UNDER CONTROL	
Inhalation: headache, dizziness, nausea, vomiting, blue skin, feeling of weakness.	local exhaust or respiratory protection.	fresh air, rest, take to hospital.
Skin: *is absorbed*; redness; see also 'Inhalation'.	gloves, protective clothing.	remove contaminated clothing, wash skin with soap and water, call a doctor.
Eyes: redness, pain.	safety glasses, or combined eye and respiratory protection.	flush with water, take to a doctor.
Ingestion: headache, dizziness, nausea, vomiting, blue skin, feeling of weakness.	do not eat, drink or smoke while working.	rinse mouth, call a doctor or take to hospital.

SPILLAGE	STORAGE	LABELING / NFPA
clean up spillage, carefully collect remainder; (additional individual protection: breathing apparatus).	keep separate from combustible substances and reducing agents; ventilate at floor level.	R: 23/24/25-33 S: 28-37-44 ☠ Toxic	3 1 0

NOTES
Depending on the degree of exposure, regular medical checkups are advisable. Effects on blood due to formation of methemoglobin. Special first aid required in the event of poisoning: antidotes must be available (with instructions).

Transport Emergency Card TEC(R)-877

HI: 60; UN-number: 1578

1,1-CHLORONITROETHANE

PHYSICAL PROPERTIES	IMPORTANT CHARACTERISTICS
Boiling point, °C 124 Melting point, °C 1 Flash point, °C 56 Relative density (water = 1) 1.3 Relative vapor density (air = 1) 3.8 Relative density at 20 °C of saturated mixture vapor/air (air = 1) 1.04 Vapor pressure, mm Hg at 20 °C ca. 9.9 Solubility in water moderate Relative molecular mass 109.5	**YELLOW LIQUID** Vapor mixes readily with air. Corrosive and toxic vapors are a product of combustion (→ *hydrochloric acid* and → *nitrous vapors*). Can deflagrate (burn explosively) when heated. The substance is shock-sensitive. Reacts violently with strong oxidants.
	TLV-TWA not available
	Absorption route: Can enter the body by inhalation or ingestion. Harmful atmospheric concentrations can build up fairly rapidly on evaporation at approx. 20°C - even more rapidly in aerosol form. **Immediate effects:** Corrosive to the eyes, skin and respiratory tract. Affects the nervous system. Inhalation can cause lung edema. In serious cases risk of unconsciousness and death. **Effects of prolonged/repeated exposure:** Can cause liver and kidney damage.
Gross formula: $C_2H_4ClNO_2$	

HAZARDS/SYMPTOMS	PREVENTIVE MEASURES	FIRE EXTINGUISHING/FIRST AID
Fire: flammable.	keep away from open flame, no smoking.	powder, water spray, foam, carbon dioxide, (halons).
Explosion: above 56°C: forms explosive air-vapor mixtures; elevated temperature, friction or shock can cause explosion.	above 56°C: sealed machinery, ventilation, explosion-proof electrical equipment, do not subject to shock or friction.	in case of fire: keep tanks/drums cool by spraying with water, fight fire from sheltered location.
Inhalation: *corrosive*; sore throat, cough, shortness of breath, severe breathing difficulties, headache, drowsiness.	ventilation, local exhaust or respiratory protection.	fresh air, rest, place in half-sitting position, take to hospital.
Skin: *corrosive*; redness, pain, burns.	gloves, protective clothing.	remove contaminated clothing, flush skin with water or shower, refer to a doctor.
Eyes: *corrosive*; redness, pain, impaired vision.	face shield, or combined eye and respiratory protection.	flush with water, take to a doctor if necessary.
Ingestion: *corrosive*; diarrhea, abdominal cramps, vomiting, drowsiness.		rinse mouth, immediately take to hospital.

SPILLAGE	STORAGE	LABELING / NFPA
collect leakage in sealable containers, soak up with sand or other inert absorbent and remove to safe place; (additional individual protection: breathing apparatus).	keep in fireproof place, separate from oxidants.	2 3

NOTES
Depending on the degree of exposure, regular medical checkups are advisable. Lung edema symptoms usually develop several hours later and are aggravated by physical exertion: rest and hospitalization essential. As first aid, a doctor or authorized person should consider administering a corticosteroid spray.

Transport Emergency Card TEC(R)-61G05

CAS-No: [106-48-9]
4-chloro-1-hydroxybenzene
4-chlorophenol
4-hydroxychlorobenzene

C_6H_4ClOH

p-CHLOROPHENOL

PHYSICAL PROPERTIES	IMPORTANT CHARACTERISTICS
Boiling point, °C 220 Melting point, °C 43 Flash point, °C 121 Relative density (water = 1) 1.3 Relative vapor density (air = 1) 4.4 Relative density at 20 °C of saturated mixture vapor/air (air = 1) 1.0 Vapor pressure, mm Hg at 20 °C 0.099 Solubility in water, g/100 ml at 20 °C 2.7 Electrical conductivity, pS/m 9.5×10^6 Relative molecular mass 128.6 Log P octanol/water 2.4 Gross formula: C_6H_5ClO	**WHITE OR LIGHT YELLOW CRYSTALS WITH CHARACTERISTIC ODOR** Vapor mixes readily with air. Decomposes when heated or burned, giving off toxic and corrosive vapors, (→ *hydrochloric acid*). Reacts with oxidants. Attacks aluminum and copper. TLV-TWA not available **Absorption route:** Can enter the body by inhalation or ingestion or through the skin. Harmful atmospheric concentrations build up fairly slowly on evaporation at 20°C, but harmful concentrations of airborne particles can build up much more rapidly. **Immediate effects:** Irritates the eyes, skin and respiratory tract. Affects the nervous system. In substantial concentrations can cause agitation, trembling, seizures etc. In serious cases risk of unconsciousness and death. **Effects of prolonged/repeated exposure:** Can cause liver and kidney damage.

HAZARDS/SYMPTOMS	PREVENTIVE MEASURES	FIRE EXTINGUISHING/FIRST AID
Fire: combustible.	keep away from open flame, no smoking.	powder, water spray, foam, carbon dioxide, (halons).
	AVOID ALL CONTACT WITH SKIN	
Inhalation: cough, shortness of breath, headache, dizziness.	local exhaust or respiratory protection.	fresh air, rest, place in half-sitting position, call a doctor.
Skin: *is absorbed*; redness, headache, dizziness.	gloves, protective clothing.	remove contaminated clothing, wash skin with soap and water, consult a doctor if necessary.
Eyes: redness, pain.	face shield.	flush with water, take to a doctor.
Ingestion: abdominal pain, vomiting.		rinse mouth, immediately take to hospital.

SPILLAGE	STORAGE	LABELING / NFPA
clean up spillage, carefully collect remainder; (additional individual protection: breathing apparatus).	ventilate at floor level.	R: 20/21/22 S: 2-28 Note C ✖ Harmful

NFPA diagram: 3 (left) / 1 (top) / 0 (right)

NOTES
Depending on the degree of exposure, regular medical checkups are advisable.

Transport Emergency Card TEC(R)-61G12C

HI: 60; UN-number: 2020

CHLOROPICRIN

PHYSICAL PROPERTIES		IMPORTANT CHARACTERISTICS	
Boiling point, °C	112	**COLORLESS OILY LIQUID WITH PUNGENT ODOR**	
Melting point, °C	−64	Vapor is heavier than air. Decomposes when heated or exposed to light, giving off corrosive and toxic gases (→ *hydrochloric acid* and → *nitrous vapors*). Can decompose explosively if heated rapidly or subjected to shock. Strong oxidant which reacts violently with combustible substances and reducing agents. Reacts violently with alkaline-earth and alkali metals and metal powder, with risk of fire and explosion.	
Relative density (water = 1)	1.7		
Relative vapor density (air = 1)	5.7		
Relative density at 20 °C of saturated mixture			
vapor/air (air = 1)	1.11		
Vapor pressure, mm Hg at 20 °C	19		
Solubility in water, g/100 ml	0.02	TLV-TWA	0.1 ppm	0.67 mg/m³
Relative molecular mass	164.4		
Log P octanol/water	2.4	**Absorption route:** Can enter the body by inhalation or ingestion or through the skin. Harmful atmospheric concentrations can build up very rapidly on evaporation at 20°C.	
		Immediate effects: Corrosive to the eyes, skin and respiratory tract. Inhalation of vapor/fumes can cause severe breathing difficulties (lung edema). In serious cases risk of death.	
		Effects of prolonged/repeated exposure: Can affect the blood. Can damage the cardiac muscles.	
Gross formula:	CCl_3NO_2		

(TLV-TWA row: 0.1 ppm and 0.67 mg/m³)

HAZARDS/SYMPTOMS	PREVENTIVE MEASURES	FIRE EXTINGUISHING/FIRST AID
Fire: non-combustible, many chemical reactions can cause fire and explosion.	avoid contact with reducing agents and combustible substances.	in case of fire in immediate vicinity: use any extinguishing agent.
Explosion: many chemical reactions or rapid heating can cause explosion.	do not subject to shock or friction.	in case of fire: keep tanks/drums cool by spraying with water, fight fire from sheltered location.
	STRICT HYGIENE	
Inhalation: *corrosive*; sore throat, cough, severe breathing difficulties, dizziness, nausea, vomiting, blue skin, feeling of weakness.	ventilation, local exhaust or respiratory protection.	fresh air, rest, place in half-sitting position, take to hospital.
Skin: *corrosive, is absorbed*; redness, pain, burns.	gloves, protective clothing.	remove contaminated clothing, flush skin with water or shower, take to hospital.
Eyes: *corrosive*; redness, pain, impaired vision.	face shield, or combined eye and respiratory protection.	flush with water, take to doctor if necessary.
Ingestion: *corrosive*; sore throat, cough, severe breathing difficulties, nausea, dizziness, vomiting, blue skin, feeling of weakness.	do not eat, drink or smoke while working.	rinse mouth, take immediately to hospital.

SPILLAGE	STORAGE	LABELING / NFPA
evacuate area, call in an expert, collect leakage in sealable containers, soak up with sand or other inert absorbent and remove to safe place, do not use sawdust or other combustible absorbents; (additional individual protection: CHEMICAL SUIT).	keep in dark place, separate from combustible substances and reducing agents; ventilate at floor level.	R: 26/27/28-36/37/38 S: 26-36-45 Toxic NFPA: 0 / 4 / 3

NOTES
Log P octanol/water is estimated. Flush contaminated clothing with water (fire hazard). Odor limit is above TLV. Depending on the degree of exposure, regular medical checkups are advisable. Lung edema symptoms usually develop several hours later and are aggravated by physical exertion: rest and hospitalization essential. As first aid, a doctor or authorized person should consider administering a corticosteroid spray. Effects on blood due to formation of methemoglobin. Special first aid required in the event of poisoning: antidotes must be available (with instructions). Under no circumstances use near flame or hot surface or when welding. Use airtight packaging. Unbreakable packaging preferred; if breakable, keep in unbreakable container.

Transport Emergency Card TEC(R)-162

HI: 66; UN-number: 1580

CAS-No: [126-99-8]
2-chloro-1,3 butadiene
chlorobutadiene
2-chloro-1,3-diene
chloroprene
β-chloroprene

$CH_2 = CH\text{-}CCl = CH_2$

2-CHLOROPRENE

PHYSICAL PROPERTIES	IMPORTANT CHARACTERISTICS
Boiling point, °C · · · 59 Melting point, °C · · · −130 Flash point, °C · · · −20 Relative density (water = 1) · · · 0.96 Relative vapor density (air = 1) · · · 3.0 Relative density at 20 °C of saturated mixture vapor/air (air = 1) · · · 1.5 Vapor pressure, mm Hg at 20 °C · · · 203 Solubility in water · · · poor Explosive limits, vol% in air · · · 2.5-20 Relative molecular mass · · · 88.5 Gross formula: · · · C_4H_5Cl	**COLORLESS LIQUID WITH CHARACTERISTIC ODOR** Vapor is heavier than air and spreads at ground level, with risk of ignition at a distance. Readily able to form peroxides and thus to polymerize. Flow, agitation etc. can cause build-up of electrostatic charge due to liquid's low conductivity. Do not use compressed air when filling, emptying or processing. Corrosive vapors (→ *hydrochloric acid*) and toxic vapors (→ *phosgene*) are a product of combustion. Reacts violently with alkaline-earth and alkali metals and various metal powders. Reacts with oxidants. Attacks some plastics. TLV-TWA (skin) 10 ppm 36 mg/m³ **Absorption route:** Can enter the body by inhalation or ingestion or through the skin. Harmful atmospheric concentrations can build up very rapidly on evaporation at approx. 20°C. **Immediate effects:** Irritates the eyes, skin and respiratory tract. Inhalation of vapor/fumes can cause severe breathing difficulties (lung edema). In serious cases risk of death. **Effects of prolonged/repeated exposure:** Prolonged or repeated contact can cause skin disorders. Prolonged contact can cause hair loss. Can cause liver and kidney damage.

HAZARDS/SYMPTOMS	PREVENTIVE MEASURES	FIRE EXTINGUISHING/FIRST AID
Fire: extremely flammable.	keep away from open flame and sparks, no smoking.	powder, AFFF, foam, carbon dioxide, (halons).
Explosion: forms explosive air-vapor mixtures.	sealed machinery, ventilation, explosion-proof electrical equipment and lighting, grounding.	in case of fire: keep tanks/drums cool by spraying with water.
	STRICT HYGIENE	
Inhalation: cough, shortness of breath, severe breathing difficulties.	ventilation, local exhaust or respiratory protection.	fresh air, rest, place in half-sitting position, call a doctor.
Skin: irritation, redness, pain.	gloves, protective clothing.	remove contaminated clothing, flush skin with water or shower, refer to a doctor.
Eyes: redness, pain, impaired vision.	face shield, or combined eye and respiratory protection.	flush with water, take to a doctor.
Ingestion: diarrhea, abdominal cramps, vomiting, unconsciousness.		take immediately to hospital.

SPILLAGE	STORAGE	LABELING / NFPA
evacuate area, call in an expert, collect leakage in non-sealable containers, soak up with sand or other inert absorbent and remove to safe place; (additional individual protection: breathing apparatus).	keep in cool, dark fireproof place under inert gas, separate from oxidants; add an inhibitor.	R: 12-20 S: 9-16-29-23 Note D 🔥 Flammable ✖ Harmful ◇ 3 2 0

NOTES
Without inhibitor can polymerize at room temperature; with inhibitor when heated above 40°C. Before distilling check for peroxides; if found, render harmless. Odor limit is above TLV. Depending on the degree of exposure, regular medical checkups are advisable. Lung edema symptoms usually develop several hours later and are aggravated by physical exertion: rest and hospitalization essential. As first aid, a doctor or authorized person should consider administering a corticosteroid spray. Special first aid required in the event of poisoning: antidotes must be available (with instructions). Use airtight packaging. Unbreakable packaging preferred; if breakable, keep in unbreakable container.

Transport Emergency Card TEC(R)-700

HI: 336; UN-number: 1991

α-CHLOROPROPIONIC ACID

PHYSICAL PROPERTIES	IMPORTANT CHARACTERISTICS
Boiling point, °C 186 Melting point, °C −12 Flash point, °C 107 Auto-ignition temperature, °C 550 Relative density (water = 1) 1.3 Relative vapor density (air = 1) 3.8 Relative density at 20 °C of saturated mixture vapor/air (air = 1) 1.0 Vapor pressure, mm Hg at 20 °C ca. 0.23 Solubility in water ∞ Relative molecular mass 108.5 Log P octanol/water 1.0	**COLORLESS TO LIGHT YELLOW VISCOUS LIQUID** Vapor mixes readily with air. Corrosive fumes (→ *hydrochloric acid*) are a product of combustion. In aqueous solution is a medium strong acid. Reacts violently with strong bases. Reacts with strong oxidants.
	TLV-TWA not available
	Absorption route: Can enter the body by inhalation or ingestion. Harmful atmospheric concentrations build up fairly slowly on evaporation, but much more rapidly in aerosol form. **Immediate effects:** Corrosive to the eyes, skin and respiratory tract, even in highly dilute solution. Inhalation can cause lung edema. In serious cases risk of death. **Effects of prolonged/repeated exposure:** Repeated contact can cause lung disorders.
Gross formula: C$_3$H$_5$ClO$_2$	

HAZARDS/SYMPTOMS	PREVENTIVE MEASURES	FIRE EXTINGUISHING/FIRST AID
Fire: combustible.	keep away from open flame, no smoking.	AFFF, water spray, carbon dioxide, (halons).
Inhalation: *corrosive*; sore throat, cough, shortness of breath, severe breathing difficulties.	ventilation, local exhaust or respiratory protection.	fresh air, rest, place in half-sitting position, take to hospital.
Skin: *corrosive*; redness, pain, blisters.	gloves, protective clothing.	remove contaminated clothing, flush skin with water or shower, call a doctor if necessary.
Eyes: *corrosive*; redness, pain, impaired vision.	face shield.	flush with water, take to a doctor.
Ingestion: *corrosive*; sore throat, abdominal pain.		rinse mouth, take to hospital.

SPILLAGE	STORAGE	LABELING / NFPA
collect leakage in sealable containers, soak up with sand or other inert absorbent and remove to safe place, neutralize with 5% (aqueous) solution of sodium carbonate; (additional individual protection: breathing apparatus).	keep separate from oxidants and strong bases.	R: 22-35 S: 23-26-28-36 Corrosive

NOTES
Log P octanol/water is estimated. Lung edema symptoms usually develop several hours later and are aggravated by physical exertion: rest and hospitalization essential. As first aid, a doctor or authorized person should consider administering a corticosteroid spray.

CHLOROSULFONIC ACID

PHYSICAL PROPERTIES	IMPORTANT CHARACTERISTICS
Boiling point, °C 151 Melting point, °C −80 Relative density (water = 1) 1.8 Relative vapor density (air = 1) 4.0 Relative density at 20 °C of saturated mixture vapor/air (air = 1) 1.0 Vapor pressure, mm Hg at 20 °C 0.38 Solubility in water reaction Relative molecular mass 116.5	**COLORLESS TO LIGHT YELLOW LIQUID WITH PUNGENT ODOR** Vapor mixes readily with air. Decomposes violently on contact with water or water vapor, giving off fumes of → *hydrochloric acid* and → *sulfuric acid*. Strong oxidant which reacts violently with combustible substances and reducing agents. Strong acid which reacts violently with bases and is corrosive. Reacts violently with phosphorus, with risk of fire and explosion. Reacts with air to form corrosive fumes. Attacks many metals, giving off flammable gas (→ *hydrogen*).
	TLV-TWA not available
	Absorption route: Can enter the body by inhalation or ingestion. Harmful atmospheric concentrations can build up very rapidly on evaporation at 20°C. **Immediate effects:** Corrosive to the eyes, skin and respiratory tract. Inhalation can cause lung edema. In serious cases risk of death.
Gross formula: ClHO$_3$S	

HAZARDS/SYMPTOMS	PREVENTIVE MEASURES	FIRE EXTINGUISHING/FIRST AID
Fire: non-combustible, many chemical reactions can cause fire and explosion.	avoid contact with combustible substances.	in case of fire in immediate vicinity: preferably not water.
Inhalation: *corrosive*; sore throat, cough, shortness of breath, severe breathing difficulties.	ventilation, local exhaust or respiratory protection.	fresh air, rest, place in half-sitting position, take to hospital.
Skin: *corrosive*; redness, pain, burns.	gloves, protective clothing.	remove contaminated clothing, flush skin with water or shower, refer to a doctor.
Eyes: *corrosive*; redness, pain, impaired vision.	face shield, or combined eye and respiratory protection.	flush with water, take to a doctor.
Ingestion: *corrosive*; sore throat, abdominal pain, diarrhea, vomiting.		rinse mouth, take immediately to hospital.

SPILLAGE	STORAGE	LABELING / NFPA
collect leakage in sealable containers, soak up with dry sand or other inert absorbent and remove to safe place, combat gas cloud with water curtain; (additional individual protection: CHEMICAL SUIT).	keep dry, separate from combustible substances, reducing agents and strong bases; ventilate at floor level.	R: 14-35-37 S: 26 Corrosive 0 / 3 / 2 / W

NOTES
Lung edema symptoms usually develop several hours later and are aggravated by physical exertion: rest and hospitalization essential. As first aid, a doctor or authorized person should consider administering a corticosteroid spray. Persons with lung conditions should take precautions against inhalation. Unbreakable packaging preferred; if breakable, keep in unbreakable container.

Transport Emergency Card TEC(R)-51

HI: 88; UN-number: 1754

CHROMIUM TRIOXIDE

PHYSICAL PROPERTIES		IMPORTANT CHARACTERISTICS	
Decomposes below boiling point, °C	198	**PURPLISH-RED CRYSTALS**	
Melting point, °C	196	Decomposes when heated above 200°C, forming chromium (III) oxide (Cr_2O_3) and → *oxygen*, with increased risk of fire. Strong oxidant which reacts violently with combustible substances and reducing agents, with risk of fire and explosion. In aqueous solution is a strong acid which reacts violently with bases and is corrosive.	
Relative density (water = 1)	2.7		
Solubility in water, g/100 ml	61		
Relative molecular mass	100.0		
		TLV-TWA	0.05 mg/m³ ✷
		Absorption route: Can enter the body by inhalation or ingestion. Evaporation negligible at 20°C, but harmful concentrations of airborne particles can build up rapidly. **Immediate effects:** Corrosive to the eyes, skin and respiratory tract. Can cause kidney damage. Inhalation can cause lung edema. **Effects of prolonged/repeated exposure:** Prolonged or repeated contact can cause skin disorders.	
Gross formula:	CrO_3		

HAZARDS/SYMPTOMS	PREVENTIVE MEASURES	FIRE EXTINGUISHING/FIRST AID
Fire: non-combustible, but makes other substances more combustible.	avoid contact with combustible substances.	in case of fire in immediate vicinity: use any extinguishing agent.
Inhalation: *corrosive*; sore throat, cough, shortness of breath, severe breathing difficulties.	local exhaust or respiratory protection.	fresh air, rest, place in half-sitting position, take to hospital.
Skin: *corrosive*; redness, pain, burns.	gloves, protective clothing.	remove contaminated clothing, flush skin with water or shower, refer to a doctor.
Eyes: *corrosive*; redness, pain, impaired vision.	face shield, or combined eye and respiratory protection.	flush with water, take to a doctor.
Ingestion: *corrosive*; sore throat, diarrhea, abdominal cramps, vomiting.	do not eat, drink or smoke while working.	rinse mouth, take immediately to hospital.

SPILLAGE	STORAGE	LABELING / NFPA
clean up spillage, carefully collect remainder; (additional individual protection: P3 respirator).	keep dry, separate from combustible substances, reducing agents and bases.	R: 8-35-43 S: 28 Oxidant Corrosive

NOTES

✷ TLV equals that of chromium(VI)compounds (water soluble). Depending on the degree of exposure, regular medical checkups are advisable. Lung edema symptoms usually develop several hours later and are aggravated by physical exertion: rest and hospitalization essential. As first aid, a doctor or authorized person should consider administering a corticosteroid spray. Packaging: special material.

Transport Emergency Card TEC(R)-866

HI: 50; UN-number: 1463

CAS-No: [104-55-2]
3-phenylpropenal
cinnamyl aldehyde

$C_6H_5CH=CHCHO$

CINNAMIC ALDEHYDE

PHYSICAL PROPERTIES		IMPORTANT CHARACTERISTICS	
Boiling point, °C	249	**LIGHT YELLOW VISCOUS LIQUID WITH CHARACTERISTIC ODOR**	
Melting point, °C	−8	Vapor mixes readily with air. Presumed to be able to form peroxides. Strong reducing agent which reacts violently with oxidants, with risk of fire and explosion. Reacts violently with alkalis and acids when moderately heated. Attacks aluminum, iron and some plastics.	
Flash point, °C	111		
Relative density (water = 1)	1.05		
Relative vapor density (air = 1)	4.6		
Relative density at 20 °C of saturated mixture vapor/air (air = 1)	1.0	TLV-TWA	not available
Vapor pressure, mm Hg at 120 °C	9.9	**Absorption route:** Can enter the body by inhalation or ingestion. Harmful atmospheric concentrations build up very slowly, if at all, on evaporation at approx. 20°C, but much more rapidly in aerosol form.	
Solubility in water	none	**Immediate effects:** Irritates the eyes, skin and respiratory tract.	
Relative molecular mass	132.2	**Effects of prolonged/repeated exposure:** Prolonged or repeated contact can cause eczema.	
Gross formula:	C_9H_8O		

HAZARDS/SYMPTOMS	PREVENTIVE MEASURES	FIRE EXTINGUISHING/FIRST AID
Fire: combustible.	keep away from open flame, no smoking.	powder, AFFF, foam, carbon dioxide, (halons).
Explosion: contact with oxidants can cause explosion.		
Inhalation: sore throat, cough.	ventilation.	fresh air, rest.
Skin: redness.	gloves.	remove contaminated clothing, flush skin with water or shower.
Eyes: redness, pain.	safety glasses.	flush with water, take to a doctor if necessary.
Ingestion: abdominal pain.		rinse mouth.

SPILLAGE	STORAGE	LABELING / NFPA
collect leakage in sealable containers, soak up with sand or other inert absorbent and remove to safe place, do not use sawdust or other combustible absorbents.	keep in dark place, separate from oxidants; add an inhibitor.	

NOTES
Before distilling check for peroxides; if found, render harmless. Packaging: special material.

CAS-No: [77-92-9]
β-hydroxytricarballylic acid
β-hydroxytricarboxylic acid
2-hydroxy-1,2,3-propanetricarboxylic acid

$(HOOC)CH_2C(OH)(COOH)CH_2(COOH)$

CITRIC ACID

PHYSICAL PROPERTIES	IMPORTANT CHARACTERISTICS	
Decomposes below boiling point, °C ca. 200 Melting point, °C 153 Auto-ignition temperature, °C 1000 Relative density (water = 1) 1.54 Vapor pressure, mm Hg at 20 °C <0.076 Solubility in water, g/100 ml at 20 °C 59.2 Relative molecular mass 192.1 Log P octanol/water − 1.7 Gross formula: $C_6H_8O_7$	**COLORLESS CRYSTALS** In aqueous solution is a medium strong acid. Reacts with bases and is corrosive. Reacts with strong oxidants.	
	TLV-TWA not available	
	Absorption route: Can enter the body by inhalation or ingestion. Evaporation negligible at 20°C, but harmful concentrations of airborne particles can build up rapidly. **Immediate effects:** Irritates the eyes, skin and respiratory tract.	

HAZARDS/SYMPTOMS	PREVENTIVE MEASURES	FIRE EXTINGUISHING/FIRST AID
Fire: combustible.	keep away from open flame, no smoking.	water spray, powder.
Inhalation: sore throat, cough, shortness of breath.	local exhaust or respiratory protection.	fresh air, rest, call a doctor.
Skin: redness, pain.	gloves.	remove contaminated clothing, flush skin with water or shower.
Eyes: redness, pain.	goggles.	flush with water, take to a doctor if necessary.
Ingestion: sore throat, abdominal pain.		rinse mouth, call a doctor.

SPILLAGE	STORAGE	LABELING / NFPA
clean up spillage, flush away remainder with water; (additional individual protection: P2 respirator).	keep separate from strong bases.	

NOTES
Apparent melting point of citric acid monohydrate: 100°C.

CITRONELLA
(mixture of essential oils)

PHYSICAL PROPERTIES		IMPORTANT CHARACTERISTICS
Boiling point, °C	207	**LIGHT YELLOW LIQUID WITH CHARACTERISTIC ODOR, TURNING BROWN ON EXPOSURE TO AIR**
Flash point, °C	74	Vapor mixes readily with air. Reacts with strong oxidants.
Relative density (water = 1)	0.9	
Relative vapor density (air = 1)	5.4	
Vapor pressure, mm Hg at 88 °C	11	TLV-TWA not available
Solubility in water	poor	
Relative molecular mass	154.3	**Absorption route:** Can enter the body by inhalation or ingestion. Harmful atmospheric concentrations build up very slowly, if at all, on evaporation at approx. 20°C, but much more rapidly in aerosol form.
		Immediate effects: Irritates the eyes, skin and respiratory tract.
		Effects of prolonged/repeated exposure: Prolonged or repeated contact can cause skin disorders.
Gross formula:	$C_{10}H_{18}O$	

HAZARDS/SYMPTOMS	PREVENTIVE MEASURES	FIRE EXTINGUISHING/FIRST AID
Fire: combustible.	keep away from open flame, no smoking.	powder, AFFF, foam, carbon dioxide, (halons).
Inhalation: sore throat, cough.	ventilation.	fresh air, rest, call a doctor.
Skin: redness.	gloves.	remove contaminated clothing, wash skin with soap and water.
Eyes: redness.	safety glasses.	flush with water, take to a doctor if necessary.
Ingestion: abdominal cramps, nausea.		rinse mouth, call a doctor.

SPILLAGE	STORAGE	LABELING / NFPA
collect leakage in sealable containers, soak up with sand or other inert absorbent and remove to safe place.	keep in dry, dark place.	

NOTES
Textiles impregnated with citronella can be self-heating. Citronella is the main component of citronella oil. Physical properties of citronella are given. CAS No. for citronella (3,7-dimethyl-6-octenal): 141-26-4.

COBALT
(powder)

PHYSICAL PROPERTIES	IMPORTANT CHARACTERISTICS	
Boiling point, °C ca. 3000 Melting point, °C 1495 Relative density (water = 1) 8.9 Solubility in water none Relative molecular mass 58.9	**GREY POWDER** Can ignite spontaneously on exposure to air (pyrophoric). Can promote decomposition of various organic substances.	
	TLV-TWA	0,05 mg/m³ ✱
	Absorption route: Can enter the body by inhalation or ingestion. Evaporation negligible at 20°C, but harmful concentrations of airborne particles can build up rapidly. **Immediate effects:** Irritates the eyes, skin and respiratory tract. **Effects of prolonged/repeated exposure:** Prolonged or repeated contact can cause eczema. Can affect the blood. Can cause heart and thyroid damage. Prolonged exposure to particles can cause lung disorders.	
Gross formula: Co		

HAZARDS/SYMPTOMS	PREVENTIVE MEASURES	FIRE EXTINGUISHING/FIRST AID
Fire: extremely flammable.	keep away from open flame and sparks, no smoking.	special powder extinguisher, dry sand, NO other extinguishing agents.
Explosion: finely dispersed particles can explode on contact with air.	keep dust from accumulating; sealed machinery, explosion-proof electrical equipment and lighting, grounding.	
Inhalation: cough, shortness of breath.	local exhaust or respiratory protection.	fresh air, rest, call a doctor.
Skin: redness.	gloves, protective clothing.	wash skin with soap and water, call a doctor if necessary.
Eyes: redness.	goggles, or combined eye and respiratory protection.	flush with water, take to a doctor if necessary.
Ingestion: abdominal pain, nausea.	do not eat, drink or smoke while working.	rinse mouth, call a doctor.

SPILLAGE	STORAGE	LABELING / NFPA
clean up spillage, carefully collect remainder; (additional individual protection: P2 respirator).	keep in fireproof place.	

NOTES
✱ TLV equals that of cobalt (metal dust and fume). Do not take work clothing home. Unbreakable packaging preferred; if breakable, keep in unbreakable container.

Transport Emergency Card TEC(R)-42G03

bis(acetato)cobalt
cobalt acetate
cobalt diacetate
cobaltous acetate

$Co(C_2H_3O_2)_2.4H_2O$

COBALT (II) ACETATE

PHYSICAL PROPERTIES	IMPORTANT CHARACTERISTICS
Melting point, °C 140 Relative density (water = 1) 1.7 Solubility in water ∞ Relative molecular mass 249.1	**RED CRYSTALS** TLV-TWA not available **Absorption route:** Can enter the body by inhalation or ingestion. Evaporation negligible at 20°C, but harmful concentrations of airborne particles (powder) can build up rapidly. **Immediate effects:** Irritates the eyes, skin and respiratory tract. **Effects of prolonged/repeated exposure:** Prolonged or repeated contact can cause skin disorders (eczema). Can affect the blood. Can cause liver and kidney damage. Prolonged exposure to particles can cause lung disorders.
Gross formula: $C_4H_6CoO_4.4H_2O$	

HAZARDS/SYMPTOMS	PREVENTIVE MEASURES	FIRE EXTINGUISHING/FIRST AID
Fire: non-combustible.		
Inhalation: sore throat, cough, shortness of breath, dizziness.	local exhaust or respiratory protection.	fresh air, rest, call a doctor.
Skin: redness, pain.	gloves.	remove contaminated clothing, wash skin with soap and water, refer to a doctor.
Eyes: redness, pain.	goggles.	flush with water, take to a doctor if necessary.
Ingestion: diarrhea, abdominal cramps, vomiting, heat sensation.	do not eat, drink or smoke while working.	rinse mouth, call a doctor or take to hospital.

SPILLAGE	STORAGE	LABELING / NFPA
clean up spillage, carefully collect remainder; (additional individual protection: P2 respirator).		

NOTES
Apparent melting point due to loss of water of crystallization.

CAS-No: [7791-13-1]
cobalt muriate
cobaltous chloride

$CoCl_2.6H_2O$

COBALT (II) CHLORIDE

PHYSICAL PROPERTIES		IMPORTANT CHARACTERISTICS
Boiling point, °C	110	**DARK RED CRYSTALS OR BLUE POWDER**
Boiling point (decomposes), °C	400	Decomposes in flame or on hot surface, giving off toxic vapors. Decomposes when heated above 120°C in water vapor and anhydrous cobalt chloride.
Melting point, °C	86	
Relative density (water = 1)	1.92	
Solubility in water, g/100 ml at 0 °C	76	**TLV-TWA** not available
Relative molecular mass	237.9	

Absorption route: Can enter the body by inhalation or ingestion. Evaporation negligible at 20°C, but harmful concentrations of airborne particles can build up rapidly.
Immediate effects: Irritates the eyes, skin and respiratory tract.
Effects of prolonged/repeated exposure: Can affect the blood and the thyroid. Prolonged or repeated contact can cause skin disorders. Can cause heart disorders and kidney damage.

Gross formula: $Cl_2Co.6H_2O$

HAZARDS/SYMPTOMS	PREVENTIVE MEASURES	FIRE EXTINGUISHING/FIRST AID
Fire: non-combustible.		in case of fire in immediate vicinity: use any extinguishing agent.
Inhalation: cough, nausea, red complexion, ringing in the ears.	local exhaust or respiratory protection.	fresh air, rest, call a doctor.
Skin: redness.	gloves, protective clothing.	remove contaminated clothing, flush skin and wash thoroughly wish water and soap.
Eyes: redness, pain.	goggles.	flush with water, take to a doctor.
Ingestion: abdominal pain, diarrhea, vomiting; see also 'Inhalation'.	do not eat, drink or smoke while working.	rinse mouth, call a doctor.

SPILLAGE	STORAGE	LABELING / NFPA
clean up spillage, carefully collect remainder; (additional individual protection: P2 respirator).		

NOTES
Boiling point of anhydrous cobalt chloride: 1049°C. Melting point of anhydrous cobalt chloride: 724°C. Apparent melting point due to loss of water of crystallization. Depending on the degree of exposure, regular medical checkups are advisable.

Transport Emergency Card TEC(R)-61G11

COBALT (II) NITRATE

PHYSICAL PROPERTIES	IMPORTANT CHARACTERISTICS
Boiling point (decomposes), °C 75 Melting point, °C 55 Relative density (water = 1) 1.9 Solubility in water, g/100 ml 134 Relative molecular mass 291.1	**RED CRYSTALS** Decomposes when heated above 75°C, forming cobaltoxide and → *nitrous vapors*. Strong oxidant which reacts violently with combustible substances and reducing agents.
	TLV-TWA not available
	Absorption route: Can enter the body by inhalation or ingestion. Evaporation negligible at 20°C, but harmful concentrations of airborne particles can build up rapidly. **Immediate effects:** Irritates the eyes, skin and respiratory tract. Inhalation can cause severe breathing difficulties. **Effects of prolonged/repeated exposure:** Can affect the thyroid. Can affect the blood. Can cause kidney damage.
Gross formula: $CoN_2O_6.6H_2O$	

HAZARDS/SYMPTOMS	PREVENTIVE MEASURES	FIRE EXTINGUISHING/FIRST AID
Fire: not combustible but enhances combustion of other substances.	avoid contact with combustible substances.	in case of fire in immediate vicinity: use any extinguishing agent.
	KEEP DUST UNDER CONTROL	
Inhalation: cough, severe breathing difficulties, nausea.	local exhaust or respiratory protection.	fresh air, rest, place in half-sitting position, call a doctor.
Skin: redness.	gloves, protective clothing.	remove contaminated clothing, flush skin with water or shower, refer to a doctor.
Eyes: redness, pain.	face shield, or combined eye and respiratory protection.	flush with water, take to a doctor.
Ingestion: abdominal pain, diarrhea, vomiting.		rinse mouth, call a doctor.

SPILLAGE	STORAGE	LABELING / NFPA
clean up spillage, carefully collect remainder; (additional individual protection: P2 respirator).	keep separate from combustible substances and reducing agents.	

NOTES
Apparent melting point due to loss of water of crystallization. Flush contaminated clothing with water (fire hazard). Depending on the degree of exposure, regular medical checkups are advisable.

COBALT (II) SULFATE

PHYSICAL PROPERTIES	IMPORTANT CHARACTERISTICS	
Melting point (decomposes), °C 735 Relative density (water = 1) 3.71 Solubility in water, g/100 ml at 20 °C 36.2 Relative molecular mass 155	**DARK BLUE CRYSTALS** Toxic cobaltoxides are a product of combustion.	
	TLV-TWA not available	
Gross formula: CoO$_4$S	**Absorption route:** Can enter the body by inhalation or ingestion. Evaporation negligible at 20° C, but harmful concentrations of airborne particles can build up rapidly. **Immediate effects:** Irritates the eyes, skin and respiratory tract. **Effects of prolonged/repeated exposure:** Prolonged or repeated contact can cause skin disorders. Can affect the thyroid. Can affect the blood. Can cause kidney damage.	

HAZARDS/SYMPTOMS	PREVENTIVE MEASURES	FIRE EXTINGUISHING/FIRST AID
Fire: non-combustible.		in case of fire in immediate vicinity: use any extinguishing agent.
	KEEP DUST UNDER CONTROL	
Inhalation: cough, shortness of breath, nausea.	local exhaust or respiratory protection.	fresh air, rest, call a doctor.
Skin: redness.	gloves, protective clothing.	remove contaminated clothing, flush skin with water or shower, refer to a doctor if necessary.
Eyes: redness, pain.	face shield, or combined eye and respiratory protection.	flush with water, take to a doctor.
Ingestion: abdominal pain, vomiting.	do not eat, drink or smoke while working.	rinse mouth, call a doctor.

SPILLAGE	STORAGE	LABELING / NFPA
clean up spillage, carefully collect remainder; (additional individual protection: P2 respirator).		

NOTES
Depending on the degree of exposure, regular medical checkups are advisable.

COLCHICINE

PHYSICAL PROPERTIES	IMPORTANT CHARACTERISTICS	
Boiling point, °C ? Melting point, °C 150 Solubility in water, g/100 ml 4.5	**LIGHT YELLOW POWDER, DISCOLORING ON EXPOSURE TO LIGHT** Decomposes when heated or burned, giving off toxic vapors. Reacts with strong oxidants.	
	TLV-TWA not available	
	Absorption route: Can enter the body by inhalation or ingestion. **Effects of prolonged/repeated exposure:** Can affect the blood. Can cause liver, kidney and brain damage.	
Gross formula: $C_{22}H_{25}NO_6$		

HAZARDS/SYMPTOMS	PREVENTIVE MEASURES	FIRE EXTINGUISHING/FIRST AID
Fire: combustible.	keep away from open flame, no smoking.	water spray, powder.
	STRICT HYGIENE, KEEP DUST UNDER CONTROL	
Inhalation: nausea, vomiting, diarrhea, abdominal cramps.	local exhaust or respiratory protection.	fresh air, rest, take to hospital.
Skin: redness.	gloves.	remove contaminated clothing, flush skin with water or shower.
Eyes: redness, pain, impaired vision.	goggles, or combined eye and respiratory protection.	flush with water, take to a doctor if necessary.
Ingestion: sore throat, abdominal pain; see also 'Inhalation'.	do not eat, drink or smoke while working.	rinse mouth, take immediately to hospital.

SPILLAGE	STORAGE	LABELING / NFPA
clean up spillage, carefully collect remainder; (additional individual protection: P3 respirator).		R: 26/28 S: 1-13-45 ☠ Toxic

NOTES
Do not take work clothing home. Depending on the degree of exposure, regular medical checkups are advisable. Immediate treatment of symptoms necessary in the event of poisoning.

Transport Emergency Card TEC(R)-61G12

COLLODION

(nitrocellulose solution in ether)

PHYSICAL PROPERTIES	IMPORTANT CHARACTERISTICS
Boiling point, °C ca. 35 1) Flash point, °C −18 Auto-ignition temperature, °C 170 Relative density (water = 1) 0.71 1) Relative vapor density (air = 1) 2.6 Relative density at 20 °C of saturated mixture vapor/air (air = 1) 1.9 Vapor pressure, mm Hg at 20 °C ca. 448 Solubility in water moderate Explosive limits, vol% in air 1.7-48	**CLEAR VISCOUS SOLUTION OF NITROCELLULOSE ETHYL ALCOHOL AND DIETHYL ETHER** Vapor is heavier than air and spreads at ground level, with risk of ignition at a distance. Once dry due to evaporation can deflagrate (burn explosively), giving off → *nitrous vapors*, and is shock-sensitive. Decomposes when moderately heated or on contact with air, giving off → *carbon monoxide* and → *nitrous vapors*.
	TLV-TWA not available
	Absorption route: Solvent can enter the body by inhalation or ingestion. **Immediate effects:** Solvent irritates the eyes and respiratory tract and affects the central nervous system.

HAZARDS/SYMPTOMS	PREVENTIVE MEASURES	FIRE EXTINGUISHING/FIRST AID
Fire: extremely flammable.	keep away from open flame and sparks, no smoking.	large quantities of water, water spray.
Explosion: forms explosive air-vapor mixtures, heating the dry substance can cause explosion.	sealed machinery, ventilation, explosion-proof electrical equipment and lighting.	in case of fire: keep tanks/drums cool by spraying with water.
Inhalation: sore throat, headache, dizziness, drowsiness.	ventilation, local exhaust or respiratory protection.	fresh air, rest, call a doctor.
Skin: redness.		remove contaminated clothing, wash skin with soap and water.
Eyes: redness, pain.	safety glasses, or combined eye and respiratory protection.	flush with water, take to a doctor.
Ingestion: sore throat, abdominal pain, vomiting.	do not eat, drink or smoke while working.	rinse mouth, call a doctor.

SPILLAGE	STORAGE	LABELING / NFPA
evacuate area, call in an expert, ventilate, collect leakage in sealable containers, soak up with sand or other inert absorbent and remove to safe place; (additional individual protection: breathing apparatus).	keep in cool, dark, fireproof place; prevent evaporation.	 4 1 0

NOTES
1)Physical properties of ether are given. Do not take work clothing home. Evaporation of solvent leaves nitrocellulose residue (guncotton) which is highly flammable and explosive. Other solvents are also used, e.g. various alcohols. Guncotton can be self-igniting under certain circumstances. Alcohol consumption increases toxic effects of solvent.

Transport Emergency Card TEC(R)-30G01

HI: 33; UN-number: 2059

COPPER
(powder)

PHYSICAL PROPERTIES	IMPORTANT CHARACTERISTICS	
Boiling point, °C 2567 Melting point, °C 1083 Relative density (water = 1) 8.9 Vapor pressure, mm Hg at 1628 °C 0.99 Solubility in water none Relative atomic mass 63.5	**RED POWDER, TURNING GREEN ON CONTACT WITH MOIST AIR** Reacts with → *acetylene*, other alkynes and azides to form shock-sensitive compounds. Mixtures of finely dispersed powder with various strong oxidants can explode if heated or subjected to shock or friction. Toxic vapors/fumes are given off when heated. Attacked by many acids and → *anhydrous ammonia* and dissolves readily in dilute nitric acid or warm concentrated → *sulfuric acid*.	
	TLV-TWA 1 mg/m³	
Gross formula: Cu	**Absorption route:** Can enter the body by inhalation of dust or fumes or by ingestion. Evaporation negligible at 20°C, but harmful concentrations of airborne particles can build up rapidly. **Immediate effects:** Irritates the eyes, skin and respiratory tract. Inhalation of powder/fumes can cause metal fume fever. **Effects of prolonged/repeated exposure:** Prolonged or repeated contact can cause dermatitis. Inhalation of high concentrations of vapor can cause lung disorders.	

HAZARDS/SYMPTOMS	PREVENTIVE MEASURES	FIRE EXTINGUISHING/FIRST AID
Fire: flammable (in powder form).	keep away from open flame, no smoking.	special powder extinguisher, dry sand, NO other extinguishing agents.
	KEEP DUST UNDER CONTROL	
Inhalation: burning sensation, cough, headache, sore throat.	local exhaust or respiratory protection.	fresh air, rest, refer to a doctor.
Skin:	gloves.	
Eyes: redness.	goggles.	
Ingestion: abdominal pain, diarrhea, nausea.		rinse mouth, give no liquids, refer to a doctor.

SPILLAGE	STORAGE	LABELING / NFPA
clean up spillage, carefully collect remainder; (additional individual protection: P2 respirator).	keep separate from strong acids and ammonia.	

NOTES
TLV-TWA for fumes 0.2mg/m³.

COPPER (II) ACETATE

PHYSICAL PROPERTIES	IMPORTANT CHARACTERISTICS
Boiling point (decomposes), °C 240 Melting point, °C 115 Relative density (water = 1) 1.9 Solubility in water, g/100 ml 7.2 Relative molecular mass 199.6	**DARK GREEN POWDER** Decomposes when heated above 240°C, giving off irritating and flammable vapor (→ *acetic acid*).
	TLV-TWA not available
	Absorption route: Can enter the body by inhalation or ingestion. Evaporation negligible at 20°C, but harmful concentrations of airborne particles can build up rapidly. **Immediate effects:** Irritates the skin and respiratory tract. Corrosive to the eyes. **Effects of prolonged/repeated exposure:** Prolonged or repeated contact can cause skin disorders. Can affect the blood. Can cause liver and kidney damage.
Gross formula: $C_4H_6CuO_4 \cdot H_2O$	

HAZARDS/SYMPTOMS	PREVENTIVE MEASURES	FIRE EXTINGUISHING/FIRST AID
Fire: non-combustible.		in case of fire in immediate vicinity: use any extinguishing agent.
	KEEP DUST UNDER CONTROL	
Inhalation: sore throat, cough, shortness of breath, affects the digestive tract, metal fume fever.	local exhaust or respiratory protection.	fresh air, rest, call a doctor, take to hospital if necessary.
Skin: redness, pain.	gloves.	remove contaminated clothing, flush skin with water or shower, refer to a doctor.
Eyes: *corrosive*; redness, pain, impaired vision.	goggles.	flush with water, take to a doctor.
Ingestion: *corrosive*; sore throat, abdominal pain, diarrhea, vomiting.		rinse mouth, take immediately to hospital.

SPILLAGE	STORAGE	LABELING / NFPA
clean up spillage, carefully collect remainder; (additional individual protection: P2 respirator).		

NOTES

Transport Emergency Card TEC(R)-80G09

COPPER (II) CARBONATE
(basic)

PHYSICAL PROPERTIES	IMPORTANT CHARACTERISTICS
Melting point (decomposes), °C 200 Relative density (water = 1) 4.0 Solubility in water none Relative molecular mass 221.1	**GREEN POWDER**
	TLV-TWA not available
	Absorption route: Can enter the body by inhalation or ingestion. Evaporation negligible at 20°C, but harmful concentrations of airborne particles can build up rapidly. **Immediate effects:** Irritates the skin and respiratory tract. Corrosive to the eyes. **Effects of prolonged/repeated exposure:** Prolonged or repeated contact can cause skin disorders. Can affect the blood. Can cause liver and kidney damage.
Gross formula: $CH_2Cu_2O_5$	

HAZARDS/SYMPTOMS	PREVENTIVE MEASURES	FIRE EXTINGUISHING/FIRST AID
Fire: non-combustible.		
	KEEP DUST UNDER CONTROL	
Inhalation: sore throat, cough, shortness of breath, headache.	local exhaust or respiratory protection.	fresh air, rest, take to hospital.
Skin: redness, pain.	gloves.	remove contaminated clothing, flush skin with water or shower, refer to a doctor.
Eyes: *corrosive*; redness, pain, impaired vision.	goggles.	flush with water, take to a doctor.
Ingestion: sore throat, abdominal pain, diarrhea, vomiting.	do not eat, drink or smoke while working.	rinse mouth, take immediately to hospital.

SPILLAGE	STORAGE	LABELING / NFPA
clean up spillage, carefully collect remainder: (additional individual protection: P2 respirator).		

NOTES

COPPER (I) CHLORIDE

PHYSICAL PROPERTIES		IMPORTANT CHARACTERISTICS	
Decomposes below boiling point, °C	1490	**WHITE CRYSTALS, TURNING GREEN ON EXPOSURE TO AIR**	
Melting point, °C	430		
Relative density (water = 1)	4.1	TLV-TWA	not available
Vapor pressure, mm Hg at 546 °C	0.99		
Solubility in water	poor	**Absorption route:** Can enter the body by inhalation or ingestion. Evaporation negligible at 20° C, but harmful concentrations of airborne particles can build up rapidly. **Immediate effects:** Irritates the eyes, skin, respiratory tract and gastro-intestinal tract. **Effects of prolonged/repeated exposure:** In large quanties can affect the blood. Can cause liver and kidney damage.	
Relative molecular mass	99		
Gross formula:	ClCu		

HAZARDS/SYMPTOMS	PREVENTIVE MEASURES	FIRE EXTINGUISHING/FIRST AID
Fire: non-combustible.		in case of fire in immediate vicinity: use any extinguishing agent.
Inhalation: sore throat, cough, shortness of breath.	local exhaust or respiratory protection.	fresh air, rest, place in half-sitting position, call a doctor.
Skin: redness, pain.	gloves.	remove contaminated clothing, flush skin with water or shower.
Eyes: redness, pain.	goggles.	flush with water, take to a doctor.
Ingestion: abdominal pain, diarrhea, vomiting.	do not eat, drink or smoke while working.	rinse mouth, rest, call a doctor.

SPILLAGE	STORAGE	LABELING / NFPA
clean up spillage, carefully collect remainder; (additional individual protection: P2 respirator).		R: 22 S: 22 ✖ Harmful

NOTES

COPPER (II) CHLORIDE

PHYSICAL PROPERTIES	IMPORTANT CHARACTERISTICS	
Decomposes below boiling point, °C 993 Melting point, °C 498 Relative density (water = 1) 3.4 Solubility in water good Relative molecular mass 134.4	**YELLOWISH-BROWN HYGROSCOPIC POWDER** Reacts with sodium or potassium to form shock-sensitive mixtures. Decomposes when heated, forming monovalent copper compounds and → *chlorine*. Reacts with moist air to form dihydrate.	
	TLV-TWA not available	
	Absorption route: Can enter the body by inhalation or ingestion. Evaporation negligible at 20°C, but harmful concentrations of airborne particles can build up rapidly. **Immediate effects:** Irritates the skin and respiratory tract. Corrosive to the eyes. **Effects of prolonged/repeated exposure:** Prolonged or repeated contact can cause skin disorders. Can affect the blood. Can cause liver and kidney damage.	
Gross formula: Cl$_2$Cu		

HAZARDS/SYMPTOMS	PREVENTIVE MEASURES	FIRE EXTINGUISHING/FIRST AID
Fire: non-combustible.		in case of fire in immediate vicinity: use any extinguishing agent.
	KEEP DUST UNDER CONTROL	
Inhalation: sore throat, cough, shortness of breath.	local exhaust or respiratory protection.	fresh air, rest, call a doctor and take to hospital.
Skin: redness, pain.	gloves.	remove contaminated clothing, flush skin with water or shower, refer to a doctor.
Eyes: *corrosive*; redness, pain, impaired vision.	goggles.	flush with water, take to a doctor.
Ingestion: *corrosive*; abdominal pain, diarrhea, vomiting.	do not eat, drink or smoke while working.	rinse mouth, take immediately to hospital.

SPILLAGE	STORAGE	LABELING / NFPA
clean up spillage, carefully collect remainder; (additional individual protection: P2 respirator).	keep dry.	

NOTES
Above 300°C some decomposition already occurs; melting point given is of mixture of monovalent and divalent copper salts formed.

HI: 60; UN-number: 2802

COPPER (II) HYDROXIDE

PHYSICAL PROPERTIES	IMPORTANT CHARACTERISTICS	
Melting point (decomposes), °C ? Relative density (water = 1) 3.4 Solubility in water none Relative molecular mass 97.6	**BLUE TO BLUEISH-GREEN, YELLOW OR LIGHT BLUE CRYSTALS OR LIGHT BLUE POWDER** Decomposes when moderately heated, forming black cupric oxide.	
	TLV-TWA not available	
	Absorption route: Can enter the body by inhalation or ingestion. Evaporation negligible at 20°C, but harmful concentrations of airborne particles can build up rapidly. **Immediate effects:** Irritates the skin and respiratory tract. Corrosive to the eyes. **Effects of prolonged/repeated exposure:** Prolonged or repeated contact can cause skin disorders. Can affect the blood. Can cause liver and kidney damage.	
Gross formula: CuH$_2$O$_2$		

HAZARDS/SYMPTOMS	PREVENTIVE MEASURES	FIRE EXTINGUISHING/FIRST AID
Fire: non-combustible.		in case of fire in immediate vicinity: use any extinguishing agent.
Inhalation: sore throat, cough, shortness of breath, headache.	local exhaust or respiratory protection.	fresh air, rest, call a doctor, take to hospital if necessary.
Skin: redness, pain.	gloves.	remove contaminated clothing, flush skin with water or shower, refer to a doctor.
Eyes: *corrosive*; redness, pain, impaired vision.	goggles, or combined eye and respiratory protection.	flush with water, take to a doctor.
Ingestion: sore throat, abdominal pain, diarrhea, vomiting.	do not eat, drink or smoke while working.	rinse mouth, take immediately to hospital.

SPILLAGE	STORAGE	LABELING / NFPA
clean up spillage, carefully collect remainder; (additional individual protection: P2 respirator).		

NOTES

COPPER (II) NITRATE

PHYSICAL PROPERTIES	IMPORTANT CHARACTERISTICS	
Decomposes below boiling point, °C 170 Melting point, °C 114.5 Relative density (water = 1) 2.3 Solubility in water, g/100 ml at 0 °C 138 Relative molecular mass 241.6	**BLUE HYGROSCOPIC CRYSTALS** Risk of spontaneous combustion with wood, paper, textiles and other combustible substances. Decomposes when heated above 170°C, giving off toxic gases (→ *nitrous vapors*). Strong oxidant which reacts violently with combustible substances and reducing agents, with risk of fire and explosion. In aqueous solution is a medium strong acid which reacts violently with strong bases. Reacts violently with ether, ferrocyanide and tin. Attacks many metals.	
	TLV-TWA not available	
	Absorption route: Can enter the body by inhalation or ingestion. Evaporation negligible at 20°C, but harmful concentrations of airborne particles can build up rapidly. **Immediate effects:** Irritates the eyes, skin and respiratory tract. In serious cases risk of unconsciousness. **Effects of prolonged/repeated exposure:** Can affect the blood if swallowed.	
Gross formula: $CuN_2O_6.3H_2O$		

HAZARDS/SYMPTOMS	PREVENTIVE MEASURES	FIRE EXTINGUISHING/FIRST AID
Fire: non-combustible, but makes other substances more combustible.	avoid contact with combustible substances.	in case of fire in immediate vicinity: use any extinguishing agent.
Explosion: mixing with strong reducing agents can cause explosion.		
Inhalation: sore throat, cough, shortness of breath.	local exhaust or respiratory protection.	fresh air, rest, call a doctor.
Skin: redness, pain.	gloves, protective clothing.	remove contaminated clothing, flush skin with water or shower.
Eyes: redness, pain.	face shield.	flush with water, take to a doctor.
Ingestion: abdominal cramps, nausea, dizziness, blue skin, feeling of weakness.		rinse mouth, call a doctor or take to hospital.

SPILLAGE	STORAGE	LABELING / NFPA
clean up spillage, carefully collect remainder; (additional individual protection: P2 respirator).	keep dry, separate from combustible substances and reducing agents; ventilate at floor level.	

NOTES
Flush contaminated clothing with water (fire hazard). Effects on blood due to formation of methemoglobin. Special first aid required in the event of poisoning: antidotes must be available (with instructions).

COPPER (II) SULFATE

PHYSICAL PROPERTIES	IMPORTANT CHARACTERISTICS	
Melting point (decomposes), °C 110 Relative density (water = 1) 2.3 Solubility in water, g/100 ml at 0 °C 32 Relative molecular mass 249.7	**BLUE CRYSTALS, BLUE CRYSTALLINE GRANULES OR BLUE POWDER** Acidic in aqueous solution.	
	TLV-TWA	1 mg/m³ ✱
	Absorption route: Can enter the body by inhalation or ingestion. **Immediate effects:** Irritates the skin and respiratory tract. Corrosive to the eyes. **Effects of prolonged/repeated exposure:** Prolonged or repeated contact can cause skin disorders. Can affect the blood. Can cause liver and kidney damage.	
Gross formula: $CuO_4.S.5H_2O$		

HAZARDS/SYMPTOMS	PREVENTIVE MEASURES	FIRE EXTINGUISHING/FIRST AID
Fire: non-combustible.		in case of fire in immediate vicinity: use any extinguishing agent.
	KEEP DUST UNDER CONTROL	
Inhalation: sore throat, cough, shortness of breath.	local exhaust or respiratory protection.	fresh air, rest, call a doctor and take to hospital.
Skin: redness, pain.	gloves.	remove contaminated clothing, flush skin with water or shower, refer to a doctor.
Eyes: *corrosive*; redness, pain, impaired vision.	goggles.	flush with water, take to a doctor.
Ingestion: *corrosive*; sore throat, abdominal pain, diarrhea, vomiting.	do not eat, drink or smoke while working.	rinse mouth, take immediately to hospital.

SPILLAGE	STORAGE	LABELING / NFPA
clean up spillage, carefully collect remainder; (additional individual protection: P2 respirator).		

NOTES
✱ TLV equals that of copper (dusts). Anhydrous product is white. Apparent melting point due to loss of water of crystallization. Other apparent melting points: 150 < C.

HI: 60; UN-number: 1645

CAS-No: [95-48-7]
o-cresylic acid
1-hydroxy-2-methylbenzene
2-hydroxytoluene
2-methylphenol

$CH_3C_6H_4OH$

o-CRESOL

PHYSICAL PROPERTIES	IMPORTANT CHARACTERISTICS
Boiling point, °C 191 Melting point, °C 31 Flash point, °C 81 Auto-ignition temperature, °C 555 Relative density (water = 1) 1.05 Relative vapor density (air = 1) 3.7 Relative density at 20 °C of saturated mixture vapor/air (air = 1) 1.0 Vapor pressure, mm Hg at 20 °C 0.23 Solubility in water, g/100 ml 2 Explosive limits, vol% in air 1-? Electrical conductivity, pS/m 3.7×10^5 Relative molecular mass 108.1 Log P octanol/water 2.0 Gross formula: C_7H_8O	**COLORLESS TO YELLOWISH-BROWN LIQUID OR CRYSTALS WITH CHARACTERISTIC ODOR** Vapor mixes readily with air. Strong reducing agent which reacts violently with oxidants. TLV-TWA (skin) 5 ppm 22 mg/m³ **Absorption route:** Can enter the body by inhalation or ingestion or through the skin. Harmful atmospheric concentrations build up fairly slowly on evaporation at 20° C, but much more rapidly in aerosol form. **Immediate effects:** Corrosive to the eyes, skin and respiratory tract. Risk of burns, esp. when working in direct sunlight. Affects the nervous system. Inhalation can cause severe breathing difficulties. In serious cases risk of unconsciousness and death. **Effects of prolonged/repeated exposure:** Can cause kidney damage.

HAZARDS/SYMPTOMS	PREVENTIVE MEASURES	FIRE EXTINGUISHING/FIRST AID
Fire: combustible.	keep away from open flame, no smoking.	powder, AFFF, foam, carbon dioxide, (halons).
Explosion: above 81°C: forms explosive air-vapor mixtures.	above 81°C: sealed machinery, ventilation, explosion-proof electrical equipment.	
	STRICT HYGIENE	IN ALL CASES CALL A DOCTOR
Inhalation: *corrosive*; sore throat, cough, shortness of breath, severe breathing difficulties, headache, dizziness, drowsiness, unconsciousness.	ventilation, local exhaust or respiratory protection.	fresh air, rest, place in half-sitting position, take to hospital.
Skin: *corrosive, is absorbed*; redness, pain, serious burns; see also 'Inhalation'.	gloves, protective clothing.	*quickly* remove contaminated clothing, flush skin thoroughly with water and *rush to hospital*.
Eyes: *corrosive*; redness, pain, impaired vision.	face shield.	flush with water, take to a doctor.
Ingestion: *corrosive*; abdominal pain, vomiting, diarrhea, chest pain; see also 'Inhalation'.	do not eat, drink or smoke while working.	rinse mouth, take immediately to hospital.

SPILLAGE	STORAGE	LABELING / NFPA
collect leakage in sealable containers, soak up with sand or other inert absorbent and remove to safe place, clean up spillage; (additional individual protection: CHEMICAL SUIT).	keep separate from oxidants.	R: 24/25-34 S: 2-28-44 Note C ☠ Toxic

NOTES
Depending on the degree of exposure, regular medical checkups are advisable. Skin absorbs even diluted cresol. DO NOT touch contaminated clothing or skin with bare hands. Following thorough washing with water, any cresol remaining can be removed from skin with a mixture of a polyethene glycol and alcohol (70:30), which must be available (with instructions). The measures on this card also apply to meta-cresol and para-cresol. Flash point of technical-grade cresol mixtures ranges upwards from 40° C. Corrosive crystals, formed by sublimation, spread readily. Melting point of meta-cresol 11°C; colorless to yellowish-brown liquid. Para-cresol: finely dispersed particles explosive on contact with air; prevention: keep dust from accumulating.

Transport Emergency Card TEC(R)-37A

HI: 60; UN-number: 2076

$CH_3CH=CHCHO$

CROTONALDEHYDE

PHYSICAL PROPERTIES	IMPORTANT CHARACTERISTICS
Boiling point, °C 102 Melting point, °C −74 Flash point, °C 13 Auto-ignition temperature, °C 230 Relative density (water = 1) 0.9 Relative vapor density (air = 1) 2.4 Relative density at 20°C of saturated mixture vapor/air (air = 1) 1.06 Vapor pressure, mm Hg at 20°C 30 Solubility in water, g/100 ml at 20°C 18 Explosive limits, vol% in air 3.1-15.5 Relative molecular mass 70.9	**COLORLESS LIQUID WITH PUNGENT ODOR** Vapor mixes readily with air, forming explosive mixtures. Presumed to be able to form peroxides. Readily able to polymerize when heated or on contact with many other substances, with risk of fire and explosion. Do not use compressed air when filling, emptying or processing. Strong reducing agent which reacts violently with oxidants. Reacts violently with strong oxidants and many other compounds, with risk of fire and explosion. Attacks plastics.

	TLV-TWA	2 ppm	6 mg/m³

Absorption route: Can enter the body by inhalation or ingestion. Harmful atmospheric concentrations can build up very rapidly on evaporation at 20°C.
Immediate effects: Corrosive to the eyes, skin and respiratory tract. Affects the nervous system. Inhalation of vapor/fumes can cause severe breathing difficulties (lung edema). In serious cases risk of death.

Gross formula: C_4H_6O

HAZARDS/SYMPTOMS	PREVENTIVE MEASURES	FIRE EXTINGUISHING/FIRST AID
Fire: extremely flammable, many chemical reactions can cause fire and explosion.	keep away from open flame and sparks, no smoking, avoid contact with strong oxidants.	powder, AFFF, foam, carbon dioxide, (halons).
Explosion: forms explosive air-vapor mixtures.	sealed machinery, ventilation, explosion-proof electrical equipment and lighting, grounding.	in case of fire: keep tanks/drums cool by spraying with water.
	STRICT HYGIENE	IN ALL CASES CALL IN A DOCTOR
Inhalation: *corrosive*; sore throat, cough, shortness of breath, severe breathing difficulties.	ventilation, local exhaust or respiratory protection.	fresh air, rest, place in half-sitting position, take to hospital.
Skin: *corrosive*; redness, pain, burns.	gloves, protective clothing.	remove contaminated clothing, flush skin with water or shower, refer to a doctor.
Eyes: *corrosive*; redness, pain, impaired vision.	face shield, or combined eye and respiratory protection.	flush with water, take to a doctor.
Ingestion: sore throat, abdominal pain, vomiting.		rinse mouth, immediately take to hospital.

SPILLAGE	STORAGE	LABELING / NFPA
evacuate area, call in an expert, collect leakage in sealable containers, soak up with sand or other inert absorbent and remove to safe place, do not use sawdust or other combustible absorbents; (additional individual protection: breathing apparatus).	keep in cool, fireproof place, separate from oxidants and strong bases; add an inhibitor.	R: 11-23-36/37/38 S: 29-33-44 Flammable Toxic 3 / 3 / 2

NOTES
Technical grades contain approx. 7% water, which has a stabilizing effect. Melting point of technical grades: −5°C. Lung edema symptoms usually develop several hours later and are aggravated by physical exertion: rest and hospitalization essential. As first aid, a doctor or authorized person should consider administering a corticosteroid spray. Unbreakable packaging preferred; if breakable, keep in unbreakable container.

Transport Emergency Card TEC(R)-30G30

HI: 33; UN-number: 1143

CAS-No: [3724-65-0]
trans-2-butenoic acid
β-methacrylic acid
2-propane carbonic acid

$CH_3CHCHCOOH$

CROTONIC ACID

PHYSICAL PROPERTIES		IMPORTANT CHARACTERISTICS
Boiling point, °C	185	**COLORLESS NEEDLES OR WHITE CRYSTALS WITH CHARACTERISTIC ODOR**
Melting point, °C	72	Vapor mixes readily with air. Able to polymerize violently when exposed to UV radiation and/or
Flash point, °C	88	moisture, evolving considerable quantities of heat. In dry state can form electrostatic charge if
Auto-ignition temperature, °C	396	stirred, transported pneumatically, poured etc. Decomposes when heated, giving off toxic gases.
Relative density (water = 1)	1.02	Strong reducing agent which reacts violently with oxidants and peroxides, with risk of fire or
Relative vapor density (air = 1)	3.0	explosion. Reacts violently with strong bases.
Relative density at 20 °C of saturated mixture		

IMPORTANT CHARACTERISTICS

COLORLESS NEEDLES OR WHITE CRYSTALS WITH CHARACTERISTIC ODOR
Vapor mixes readily with air. Able to polymerize violently when exposed to UV radiation and/or moisture, evolving considerable quantities of heat. In dry state can form electrostatic charge if stirred, transported pneumatically, poured etc. Decomposes when heated, giving off toxic gases. Strong reducing agent which reacts violently with oxidants and peroxides, with risk of fire or explosion. Reacts violently with strong bases.

PHYSICAL PROPERTIES	
Boiling point, °C	185
Melting point, °C	72
Flash point, °C	88
Auto-ignition temperature, °C	396
Relative density (water = 1)	1.02
Relative vapor density (air = 1)	3.0
Relative density at 20 °C of saturated mixture vapor/air (air = 1)	1.0
Vapor pressure, mm Hg at 20 °C	0.19
Solubility in water, g/100 ml at 15 °C	8.3
Relative molecular mass	86.1
Log P octanol/water	0.7

TLV-TWA	not available

Absorption route: Can enter the body by inhalation or ingestion. Harmful atmospheric concentrations build up fairly slowly on evaporation at 20°C, but harmful concentrations of airborne particles can build up much more rapidly.
Immediate effects: Corrosive to the eyes, skin and respiratory tract.

Gross formula: $C_4H_6O_2$

HAZARDS/SYMPTOMS	PREVENTIVE MEASURES	FIRE EXTINGUISHING/FIRST AID
Fire: combustible.	keep away from open flame, no smoking.	water spray, powder.
Explosion: above 88°C: forms explosive air-vapor mixtures.	above 88°C: sealed machinery, ventilation.	
	KEEP DUST UNDER CONTROL	
Inhalation: *corrosive*; sore throat, cough, shortness of breath.	local exhaust or respiratory protection.	fresh air, rest, place in half-sitting position, call a doctor.
Skin: *corrosive*; redness, pain, burns.	gloves.	remove contaminated clothing, wash skin with soap and water, refer to a doctor.
Eyes: *corrosive*; redness, pain, impaired vision.	face shield.	flush with water, take to a doctor.
Ingestion: *corrosive*; sore throat, abdominal pain, vomiting.		rinse mouth, call a doctor or take to hospital.

SPILLAGE	STORAGE	LABELING / NFPA
clean up spillage, flush away remainder with water; (additional individual protection: P2 respirator).	keep in dry, dark place, separate from oxidants and strong bases; ventilate at floor level.	R: 34 S: 15-26 Note D Corrosive

NOTES

Use airtight packaging.

CUMARINE

PHYSICAL PROPERTIES		IMPORTANT CHARACTERISTICS	
Boiling point, °C	302	**COLORLESS CRYSTALS, FLAKES OR POWDER WITH CHARACTERISTIC ODOR**	
Melting point, °C	71		
Relative density (water = 1)	0.93	TLV-TWA	not available
Vapor pressure, mm Hg at 20°C	<0.076		
Solubility in water, g/100 ml at 20°C	0.25	**Absorption route:** Can enter the body by inhalation or ingestion. Evaporation negligible at 20°C, but harmful concentrations of airborne particles can build up rapidly.	
Relative molecular mass	146.1		
Log P octanol/water	1.4	**Effects of prolonged/repeated exposure:** Can cause liver damage.	
Gross formula:	$C_9H_6O_2$		

HAZARDS/SYMPTOMS	PREVENTIVE MEASURES	FIRE EXTINGUISHING/FIRST AID
Fire: combustible.	keep away from open flame, no smoking.	water spray, powder.
Explosion: finely dispersed particles form explosive mixtures on contact with air.	keep dust from accumulating; sealed machinery, explosion-proof electrical equipment and lighting, grounding.	
	KEEP DUST UNDER CONTROL	
Inhalation:	local exhaust or respiratory protection.	fresh air, rest, call a doctor.
Skin:	gloves.	remove contaminated clothing, flush skin with water or shower, call a doctor.
Eyes: redness, pain.	goggles.	flush with water, take to a doctor if necessary.
Ingestion:		rinse mouth, call a doctor if necessary.

SPILLAGE	STORAGE	LABELING / NFPA
clean up spillage, carefully collect remainder; (additional individual protection: P2 respirator).		

NOTES

Insufficient toxicological data on harmful effects to humans. In contrast to cumarine compounds, cumarine has no influence on the blood coagulation.

CAS-No: [98-82-8]
cumol
isopropyl benzene
1-methylethyl benzene
2-phenylpropane

$C_6H_5CH(CH_3)_2$

CUMENE

PHYSICAL PROPERTIES		IMPORTANT CHARACTERISTICS	
Boiling point, °C	152	**COLORLESS LIQUID WITH CHARACTERISTIC ODOR**	
Melting point, °C	−96	Vapor mixes readily with air. Able to form peroxides. Flow, agitation etc. can cause build-up of electrostatic charge due to liquid's low conductivity. Reacts violently with strong oxidants.	
Flash point, °C	31		
Auto-ignition temperature, °C	420		
Relative density (water = 1)	0.9	TLV-TWA (skin) — 50 ppm — 246 mg/m³	
Relative vapor density (air = 1)	4.1		
Relative density at 20 °C of saturated mixture vapor/air (air = 1)	1.01	**Absorption route:** Can enter the body by inhalation or ingestion or through the skin. Harmful atmospheric concentrations build up fairly slowly on evaporation at 20°C, but much more rapidly in aerosol form.	
Vapor pressure, mm Hg at 20 °C	3.3	**Immediate effects:** Irritates the eyes, skin and respiratory tract. Liquid destroys the skin's natural oils. Affects the nervous system. In serious cases risk of unconsciousness. If liquid is swallowed, droplets can enter the lungs, with risk of pneumonia.	
Solubility in water	none		
Explosive limits, vol% in air	0.8-6.0		
Relative molecular mass	120.2		
Log P octanol/water	3.7		
Gross formula:	C_9H_{12}		

HAZARDS/SYMPTOMS	PREVENTIVE MEASURES	FIRE EXTINGUISHING/FIRST AID
Fire: flammable.	keep away from open flame and sparks, no smoking.	powder, AFFF, foam, carbon dioxide, (halons).
Explosion: above 31°C: forms explosive air-vapor mixtures.	above 31°C: sealed machinery, ventilation, explosion-proof electrical equipment, grounding.	in case of fire: keep tanks/drums cool by spraying with water.
Inhalation: sore throat, cough, shortness of breath, headache, nausea, drowsiness, unconsciousness.	ventilation, local exhaust or respiratory protection.	fresh air, rest, take to hospital if necessary.
Skin: redness.	gloves, protective clothing.	remove contaminated clothing, flush skin with water or shower.
Eyes: redness, pain.	safety glasses.	flush with water, take to a doctor if necessary.
Ingestion: abdominal pain, vomiting; see also 'Inhalation'.		rinse mouth, DO NOT induce vomiting, take to hospital if necessary.

SPILLAGE	STORAGE	LABELING / NFPA
collect leakage in sealable containers, soak up with sand or other inert absorbent and remove to safe place.	keep in cool, dark, fireproof place, separate from oxidants; add an inhibitor.	R: 10-37 ✖ Irritant

NOTES
Before distilling check for peroxides; if found, render harmless. Alcohol consumption increases toxic effects. Absorbed slowly through the skin.

Transport Emergency Card TEC(R)-594

HI: 30; UN-number: 1918

CAS-No: [80-15-9]
α,α-dimethylbenzyl hydroperoxide
α-cumenyl hydroperoxide

$C_6H_5C(CH_3)_2OOH$

CUMENE HYDROPEROXIDE
(70-80%)

PHYSICAL PROPERTIES		IMPORTANT CHARACTERISTICS
Boiling point (decomposes), °C	153	**COLORLESS TO LIGHT YELLOW LIQUID WITH CHARACTERISTIC ODOR**
Melting point, °C	− 10	Flow, agitation etc. can cause build-up of electrostatic charge due to liquid's low conductivity.
Flash point, °C	79	Can produce thermal explosion when heated. Decomposes when heated above 50°C, evolving
Relative density (water = 1)	1.01	heat. Strong oxidant which reacts violently with combustible substances and reducing agents.
Relative vapor density (air = 1)	5.3	Contact with rust, organic contaminants, salts of metals and tertiary amines can lead to violent
Solubility in water, g/100 ml	1.5	explosion. Reacts violently with inorganic acids and bases, with risk of fire and explosion.
Electrical conductivity, pS/m	< 10⁴	
Relative molecular mass	152.2	TLV-TWA not available

Absorption route: Can enter the body by inhalation or ingestion. Harmful atmospheric concentrations build up very slowly, if at all, on evaporation at approx. 20°C, but much more rapidly in aerosol form.
Immediate effects: Corrosive to the eyes, skin and respiratory tract. Inhalation of vapor/fumes can cause severe breathing difficulties (lung edema).

Gross formula: $C_9H_{12}O_2$

HAZARDS/SYMPTOMS	PREVENTIVE MEASURES	FIRE EXTINGUISHING/FIRST AID
Fire: combustible, many chemical reactions can cause fire and explosion.	keep away from open flame, no smoking, avoid contact with many substances (see Important characteristics).	water spray, powder, carbon dioxide, (halons).
Explosion: heating and contamination in pressure resistant containers can cause explosion.	do not store in pressure resistant containers, avoid contact with many substances (see Important characteristics).	in case of fire: keep tanks/drums cool by spraying with water for a long time after the fire.
Inhalation: *corrosive*; sore throat, cough, shortness of breath, severe breathing difficulties.	ventilation, local exhaust or respiratory protection.	fresh air, rest, place in half-sitting position, take to hospital.
Skin: *corrosive*; redness, pain, burns.	gloves.	remove contaminated clothing, flush skin with water or shower, call a doctor if necessary.
Eyes: *corrosive*; redness, pain, impaired vision.	face shield.	flush with water, take to a doctor.
Ingestion: *corrosive*; abdominal pain, diarrhea, vomiting.		rinse mouth, immediately take to hospital.

SPILLAGE	STORAGE	LABELING / NFPA
call in an expert, collect leakage in non-sealable containers of suitable material, soak up with sand or other inert absorbent and remove to safe place.	keep cool, separate from other substances than organic peroxides.	R: 11-35 S: 3/7/9-14-27-37/39 Oxidant Corrosive

NOTES
Most technical grades are dissolved in a cumene-containing solvent. Store in original packaging; do not return product which has been taken out! Flush contaminated clothing with water (fire hazard). Lung edema symptoms usually develop several hours later and are aggravated by physical exertion: rest and hospitalization essential. As first aid, a doctor or authorized person should consider administering a corticosteroid spray.

Transport Emergency Card TEC(R)-649

HI: 539; UN-number: 2116

chlorine cyanide
chlorocyan
chlorocyanide
chlorocyanogen
CICN

CYANOGEN CHLORIDE
(cylinder)

PHYSICAL PROPERTIES		IMPORTANT CHARACTERISTICS
Boiling point, °C	13	**COMPRESSED LIQUEFIED GAS WITH PUNGENT ODOR**
Melting point, °C	−6	Gas is heavier than air. Moisture or chlorine can cause violent polymerization. Decomposes when
Relative density (water = 1)	1.2	heated or burned, giving off toxic and corrosive vapors (→ *hydrogen cyanide* and → *hydrochloric*
Relative vapor density (air = 1)	2.2	*acid*). Reacts with alkalis, olefins and amines, and slowly with moisture or water. Attacks copper
Vapor pressure, mm Hg at 20 °C	988	and brass.
Solubility in water, g/100 ml at 20 °C	6.5	
Relative molecular mass	61.5	
Log P octanol/water	0.6	

TLV-TWA	0.3 ppm	0.75 mg/m³ C

Absorption route: Can enter the body by inhalation or ingestion or through the skin. Harmful atmospheric concentrations can build up very rapidly if gas is released.
Immediate effects: Corrosive to the eyes, skin and respiratory tract. Can cause frostbite due to rapid evaporation. Inhalation of vapor/fumes can cause severe breathing difficulties (lung edema). In serious cases risk of seizures, unconsciousness and death. Impedes tissue respiration.

Gross formula: CCIN

HAZARDS/SYMPTOMS	PREVENTIVE MEASURES	FIRE EXTINGUISHING/FIRST AID
Fire: non-combustible.		in case of fire in immediate vicinity: use any extinguishing agent.
Explosion:		in case of fire: keep cylinders cool by spraying with water.
	STRICT HYGIENE	
Inhalation: *corrosive*; sore throat, cough, severe breathing difficulties, headache, dizziness, nausea, feeling of weakness.	ventilation, local exhaust or respiratory protection.	fresh air, rest, place in half-sitting position, artificial respiration if necessary, take to hospital.
Skin: *corrosive*; redness, pain, sores.	insulated gloves, protective clothing.	remove contaminated clothing; *in case of frostbite*: DO NOT remove clothing, flush skin with water or shower, call a doctor.
Eyes: *corrosive*; redness, pain, impaired vision.	face shield, or combined eye and respiratory protection.	flush with water, take to a doctor.
Ingestion: *corrosive*; sore throat, abdominal pain, diarrhea, severe breathing difficulties, nausea.		rinse mouth, immediately take to hospital.

SPILLAGE	STORAGE	LABELING / NFPA
evacuate area, call in an expert, ventilate, under no circumstances spray liquid with water, combat gas cloud with water curtain; (additional individual protection: CHEMICAL SUIT).	if stored indoors, keep in cool, fireproof place; ventilate at floor level.	

NOTES
Log P octanol/water is estimated. TLV is maximum and must not be exceeded. Lung edema symptoms usually develop several hours later and are aggravated by physical exertion: rest and hospitalization essential. As first aid, a doctor or authorized person should consider administering a corticosteroid spray. Special first aid required in the event of poisoning: antidotes must be available (with instructions). Do not use materials containing copper. Do not spray leaking cylinder with water (to avoid corrosion). Turn leaking cylinder so that leak is on top to prevent liquid cyanogen chloride escaping.

Transport Emergency Card TEC(R)-20G15

HI: 236; UN-number: 1589

CAS-No: [108-77-0]
2,4,6-trichloro-1,3,5-triazine
cyanuric acid chloride
cyanuryl chloride
trichlorocyanidine

CYANURIC CHLORIDE

PHYSICAL PROPERTIES	IMPORTANT CHARACTERISTICS
Boiling point, °C 194 Melting point, °C 146 Relative density (water = 1) 1.3 Relative vapor density (air = 1) 6.4 Relative density at 20 °C of saturated mixture vapor/air (air = 1) 1.0 Vapor pressure, mm Hg at 20 °C ca. 0.38 Solubility in water reaction Relative molecular mass 184.4	**COLORLESS CRYSTALS WITH PUNGENT ODOR** Vapor mixes readily with air. Toxic and corrosive gases are a product of combustion (→ *hydrochloric acid*, → *chlorine*, → *nitrous vapors*). Decomposes on contact with moisture in the presence of → *cyanic acid* or → *hydrogen chloride*. Reacts violently with ethanol, evolving heat, with risk of fire and explosion. Reacts with air to form → *hydrochloric acid* fumes.
	TLV-TWA not available
	Absorption route: Can enter the body by inhalation or ingestion. Harmful atmospheric concentrations build up fairly slowly on evaporation at 20°C, but harmful concentrations can build up much more rapidly on contact with moisture or in the form of airborne particles. **Immediate effects:** Corrosive to the eyes, skin and respiratory tract. Inhalation of vapor/fumes can cause severe breathing difficulties (lung edema). In serious cases risk of death.
Gross formula: $C_3Cl_3N_3$	

HAZARDS/SYMPTOMS	PREVENTIVE MEASURES	FIRE EXTINGUISHING/FIRST AID
Fire: non-combustible.		in case of fire in immediate vicinity: do not use water-based extinguishers.
	STRICT HYGIENE	
Inhalation: *corrosive*; sore throat, cough, shortness of breath, severe breathing difficulties.	ventilation, local exhaust or respiratory protection.	fresh air, rest, place in half-sitting position, take to hospital.
Skin: *corrosive*; redness, pain, burns.	gloves.	remove contaminated clothing, flush skin with water or shower, take to a doctor.
Eyes: *corrosive*; redness, pain, impaired vision.	face shield, or combined eye and respiratory protection.	flush with water, take to a doctor.
Ingestion: *corrosive*; sore throat, abdominal pain, vomiting.		rinse mouth, immediately take to hospital.

SPILLAGE	STORAGE	LABELING / NFPA
clean up spillage, carefully collect remainder; (additional individual protection: breathing apparatus).	keep dry, separate from acids and alcohols.	R: 36/37/38 S: 28 **✖** Irritant

NOTES

Lung edema symptoms usually develop several hours later and are aggravated by physical exertion: rest and hospitalization essential. As first aid, a doctor or authorized person should consider administering a corticosteroid spray. Use airtight packaging.

Transport Emergency Card TEC(R)-80G20C

HI: 80; UN-number: 2670

CYCLOHEPTATRIENE

PHYSICAL PROPERTIES	IMPORTANT CHARACTERISTICS	
Boiling point, °C 117 Melting point, °C −80 Flash point, °C 4 Auto-ignition temperature, °C ? Relative density (water = 1) 0.9 Relative vapor density (air = 1) 3.2 Relative density at 20 °C of saturated mixture vapor/air (air = 1) 1.05 Vapor pressure, mm Hg at 20 °C 17 Solubility in water none Explosive limits, vol% in air ?-? Relative molecular mass 92.1 Gross formula: C_7H_8	**COLORLESS TO DARK YELLOW LIQUID WITH CHARACTERISTIC ODOR, TURNING YELLOW ON EXPOSURE TO AIR** Vapor mixes readily with air, forming explosive mixtures. Flow, agitation etc. can cause build-up of electrostatic charge due to liquid's low conductivity. Do not use compressed air when filling, emptying or processing. Reacts with many compounds. Reacts violently with strong oxidants, with risk of fire and explosion.	
	TLV-TWA not available	
	Absorption route: Can enter the body by inhalation or ingestion or through the skin. Insufficient data on the rate at which harmful concentrations can build up. **Immediate effects:** Irritates the eyes, skin and respiratory tract. Liquid destroys the skin's natural oils. Affects the nervous system. If liquid is swallowed, droplets can enter the lungs, with risk of pneumonia.	

HAZARDS/SYMPTOMS	PREVENTIVE MEASURES	FIRE EXTINGUISHING/FIRST AID
Fire: extremely flammable.	keep away from open flame and sparks, no smoking.	powder, AFFF, foam, carbon dioxide, (halons).
Explosion: forms explosive air-vapor mixtures.	sealed machinery, ventilation, explosion-proof electrical equipment and lighting, grounding.	in case of fire: keep tanks/drums cool by spraying with water.
Inhalation: sore throat, cough, headache, sleepiness.	ventilation, local exhaust or respiratory protection.	fresh air, rest, call a doctor.
Skin: redness.	gloves.	remove contaminated clothing, flush skin with water or shower.
Eyes: redness, pain.	safety glasses.	flush with water, take to a doctor if necessary.
Ingestion: abdominal pain, headache, nausea.		rinse mouth, DO NOT induce vomiting, call a doctor or take to hospital.

SPILLAGE	STORAGE	LABELING / NFPA
collect leakage in sealable containers, soak up with sand or other inert absorbent and remove to safe place; (additional individual protection: breathing apparatus).	keep in fireproof place, separate from oxidants.	

NOTES

Transport Emergency Card TEC(R)-30G01   **UN-number: 2603**

CAS-No: [110-82-7]

benzene hexahydride
hexahydrobenzene
hexamethylene
hexanaphthene

C_6H_{12}

CYCLOHEXANE

PHYSICAL PROPERTIES		IMPORTANT CHARACTERISTICS	
Boiling point, °C	81	**COLORLESS LIQUID WITH CHARACTERISTIC ODOR**	
Melting point, °C	7	Vapor is heavier than air and spreads at ground level, with risk of ignition at a distance. Flow, agitation etc. can cause build-up of electrostatic charge due to liquid's low conductivity. Do not use compressed air when filling, emptying or processing.	
Flash point, °C	−18		
Auto-ignition temperature, °C	260		
Relative density (water = 1)	0.8		
Relative vapor density (air = 1)	2.9		
Relative density at 20 °C of saturated mixture vapor/air (air = 1)	1.2	**TLV-TWA** 300 ppm 1030 mg/m³	
Vapor pressure, mm Hg at 20 °C	79		
Solubility in water	none	**Absorption route:** Can enter the body by inhalation or ingestion. Harmful atmospheric concentrations can build up fairly rapidly on evaporation at approx. 20°C - even more rapidly in aerosol form.	
Explosive limits, vol% in air	1.2-8.4		
Minimum ignition energy, mJ	0.22	**Immediate effects:** Liquid destroys the skin's natural oils. Affects the nervous system. In serious cases risk of unconsciousness.	
Electrical conductivity, pS/m	1.9		
Relative molecular mass	84.2		
Gross formula:	C_6H_{12}		

HAZARDS/SYMPTOMS	PREVENTIVE MEASURES	FIRE EXTINGUISHING/FIRST AID
Fire: extremely flammable.	keep away from open flame and sparks, no smoking.	powder, AFFF, foam, carbon dioxide, (halons).
Explosion: forms explosive air-vapor mixtures.	sealed machinery, ventilation, explosion-proof electrical equipment and lighting, grounding, non-sparking tools.	in case of fire: keep tanks/drums cool by spraying with water.
Inhalation: headache, drowsiness, unconsciousness.	ventilation, local exhaust or respiratory protection.	fresh air, rest, call a doctor.
Skin: redness.	gloves.	remove contaminated clothing, wash skin with soap and water.
Eyes: redness.	safety glasses.	flush with water, take to a doctor if necessary.
Ingestion: sore throat, diarrhea, nausea.		rinse mouth, DO NOT induce vomiting, call a doctor or take to hospital.

SPILLAGE	STORAGE	LABELING / NFPA
collect leakage in sealable containers, soak up with sand or other inert absorbent and remove to safe place; (additional individual protection: breathing apparatus).	keep in fireproof place.	R: 11 S: 9-16-33 🔥 Flammable

NFPA diamond: 3 / 1 / 0

NOTES
Alcohol consumption increases toxic effects.

Transport Emergency Card TEC(R)-103

HI: 33; UN-number: 1145

CYCLOHEXANOL

PHYSICAL PROPERTIES	IMPORTANT CHARACTERISTICS	
Boiling point, °C 161	**COLORLESS HYGROSCOPIC VISCOUS LIQUID OR WHITE CRYSTALS WITH CHARACTERISTIC ODOR**	
Melting point, °C 24	Vapor mixes readily with air. Reacts violently with strong oxidants. Attacks some plastics.	
Flash point, °C 68		
Auto-ignition temperature, °C 300	TLV-TWA (skin) 50 ppm 206 mg/m³	
Relative density (water = 1) 0.95		
Relative vapor density (air = 1) 3.5	**Absorption route:** Can enter the body by inhalation or ingestion or through the skin. Harmful atmospheric concentrations build up very slowly, if at all, on evaporation at approx. 20°C, but much more rapidly in aerosol form.	
Relative density at 20 °C of saturated mixture vapor/air (air = 1) 1.0		
Vapor pressure, mm Hg at 20 °C 0.99	**Immediate effects:** Irritates the eyes, skin and respiratory tract. In substantial concentrations can impair consciousness. Can cause liver, kidney and lung damage.	
Solubility in water, g/100 ml at 20 °C 5.7		
Explosive limits, vol% in air see notes		
Relative molecular mass 100.2		
Log P octanol/water 1.2		
Gross formula: $C_6H_{12}O$		

HAZARDS/SYMPTOMS	PREVENTIVE MEASURES	FIRE EXTINGUISHING/FIRST AID
Fire: combustible.	keep away from open flame, no smoking.	powder, AFFF, foam, carbon dioxide, (halons).
Explosion: above 68°C: forms explosive air-vapor mixtures.	above 68°C: sealed machinery, ventilation.	in case of fire: keep tanks/drums cool by spraying with water.
Inhalation: sore throat, cough.	ventilation.	fresh air, rest, call a doctor if necessary.
Skin: *is absorbed*; redness.	gloves.	remove contaminated clothing, flush skin with water or shower.
Eyes: redness, pain.	safety glasses.	flush with water, take to a doctor.
Ingestion: cough, abdominal pain, diarrhea.		rinse mouth, call a doctor.

SPILLAGE	STORAGE	LABELING / NFPA
collect leakage in sealable containers, soak up with sand or other inert absorbent and remove to safe place.	keep separate from oxidants; ventilate at floor level.	R: 20/22-37/38 S: 24/25 ✖ Harmful

NOTES
Estimated lower explosive limit: 1.2% by vol.

Transport Emergency Card TEC(R)-30G15

CYCLOHEXANONE

PHYSICAL PROPERTIES	IMPORTANT CHARACTERISTICS
Boiling point, °C 156 Melting point, °C −32 Flash point, °C 43 Auto-ignition temperature, °C 420 Relative density (water = 1) 0.95 Relative vapor density (air = 1) 3.4 Relative density at 20 °C of saturated mixture vapor/air (air = 1) 1.01 Vapor pressure, mm Hg at 20 °C 3.6 Solubility in water moderate Explosive limits, vol% in air 1.1-9.4 Electrical conductivity, pS/m 5 × 10⁴ Relative molecular mass 98.1 Log P octanol/water 0.8	**COLORLESS LIQUID WITH CHARACTERISTIC ODOR** Vapor mixes readily with air. Presumed to be able to form peroxides. Flow, agitation etc. can cause build-up of electrostatic charge due to liquid's low conductivity. Reacts violently with strong oxidants. Do not use copper, brass, bronze or lead fittings. Attacks many plastics. TLV-TWA (skin) 25 ppm 100 mg/m³ **Absorption route:** Can enter the body by inhalation or ingestion. Harmful atmospheric concentrations build up fairly slowly on evaporation at 20°C, but much more rapidly in aerosol form. **Immediate effects:** Irritates the eyes, skin and respiratory tract. Liquid destroys the skin's natural oils. Affects the nervous system.
Gross formula: $C_6H_{10}O$	

HAZARDS/SYMPTOMS	PREVENTIVE MEASURES	FIRE EXTINGUISHING/FIRST AID
Fire: flammable.	keep away from open flame and sparks, no smoking.	powder, AFFF, foam, carbon dioxide, (halons).
Explosion: above 43°C: forms explosive air-vapor mixtures.	above 43°C: sealed machinery, ventilation, explosion-proof electrical equipment, grounding.	in case of fire: keep tanks/drums cool by spraying with water.
Inhalation: cough, drowsiness.	ventilation, local exhaust or respiratory protection.	fresh air, rest, call a doctor if necessary.
Skin: redness.	gloves.	remove contaminated clothing, flush skin with water or shower, refer to a doctor if necessary.
Eyes: *irritation*; redness, pain.	safety glasses.	flush with water, take to a doctor if necessary.
Ingestion: abdominal cramps; see also 'Inhalation'.		rinse mouth, DO NOT induce vomiting, call a doctor or take to hospital.

SPILLAGE	STORAGE	LABELING / NFPA
collect leakage in sealable containers, soak up with sand or other inert absorbent and remove to safe place.	keep in fireproof place, separate from oxidants.	R: 10-20 S: 25 ✖ Harmful

NOTES
Before distilling check for peroxides; if found, render harmless. Alcohol consumption increases toxic effects.

Transport Emergency Card TEC(R)-108

HI: 30; UN-number: 1915

$$H_2C \overset{\displaystyle \overset{H_2}{C} - \overset{H_2}{C}}{\underset{\displaystyle \underset{H_2}{C} - \underset{H_2}{C}}{}} C = N - OH$$

CYCLOHEXANON OXIME

PHYSICAL PROPERTIES	IMPORTANT CHARACTERISTICS
Boiling point, °C — 208 Melting point, °C — 90 Flash point, °C — 103 Auto-ignition temperature, °C — 265 Relative density (water = 1) — 0.98 Relative vapor density (air = 1) — 3.9 Relative density at 20 °C of saturated mixture vapor/air (air = 1) — 1.0 Vapor pressure, mm Hg at 20 °C — 0.15 Solubility in water, g/100 ml at 20 °C — 1.5 Explosive limits, vol% in air — 1.3-? Relative molecular mass — 113.2	**WHITE CRYSTALS WITH CHARACTERISTIC ODOR** Finely dispersed in air can cause dust explosion. In dry state can form electrostatic charge if stirred, transported pneumatically, poured etc. Can decompose explosively if heated, esp. in the presence of acids, giving off nitrous vapors. Reacts with strong oxidants.
	TLV-TWA not available
	Absorption route: Can enter the body by inhalation or ingestion. Harmful atmospheric concentrations build up fairly slowly on evaporation at approx. 20°C, but harmful concentrations of airborne particles can build up much more rapidly. **Immediate effects:** Can affect the heart, causing low blood pressure. Exposure to high concentrations can cause unconsciousness. **Effects of prolonged/repeated exposure:** Can affect the red blood cells, causing anemia. Can cause liver and kidney damage.
Gross formula: $C_6H_{11}NO$	

HAZARDS/SYMPTOMS	PREVENTIVE MEASURES	FIRE EXTINGUISHING/FIRST AID
Fire: combustible.	keep away from open flame, no smoking.	water spray, powder.
Explosion: finely dispersed particles form explosive mixtures on contact with air.	keep dust from accumulating, closed system, explosion-proof electrical equipment and lighting, prevent buildup of electrostatic charges by grounding, etc.	
	KEEP DUST UNDER CONTROL	
Inhalation: cough, dizziness, feeling of weakness.	local exhaust or respiratory protection.	fresh air, rest, consult a doctor.
Skin:	gloves.	
Eyes: irritation, redness, pain.	goggles.	flush thoroughly with water (remove contact lenses if easy), take to a doctor.
Ingestion: dizziness, nausea, unconsciousness.		rinse mouth, give no liquids, take immediately to hospital.

SPILLAGE	STORAGE	LABELING / NFPA
clean up spillage, carefully collect remainder; (additional individual protection: respirator with A/P2 filter).	keep separate from oxidants; ventilate at floor level.	

NOTES

CYCLOHEXENE

PHYSICAL PROPERTIES	IMPORTANT CHARACTERISTICS
Boiling point, °C — 83 Melting point, °C — −104 Flash point, °C — < −7 Auto-ignition temperature, °C — >244 Relative density (water = 1) — 0.8 Relative vapor density (air = 1) — 2.8 Relative density at 20 °C of saturated mixture vapor/air (air = 1) — 1.2 Vapor pressure, mm Hg at 20 °C — 71 Solubility in water — none Relative molecular mass — 82.1	**COLORLESS LIQUID** Vapor is heavier than air and spreads at ground level, with risk of ignition at a distance. Able to form peroxides and polymerize. Flow, agitation etc. can cause build-up of electrostatic charge due to liquid's low conductivity. Do not use compressed air when filling, emptying or processing. Reacts violently with many compounds.
	TLV-TWA 300 ppm 1010 mg/m³
Gross formula: C_6H_{10}	**Absorption route:** Can enter the body by inhalation or ingestion. Harmful atmospheric concentrations build up fairly slowly on evaporation at 20°C, but much more rapidly in aerosol form. **Immediate effects:** Irritates the eyes, skin and respiratory tract. Affects the nervous system. If liquid is swallowed, droplets can enter the lungs, with risk of pneumonia.

HAZARDS/SYMPTOMS	PREVENTIVE MEASURES	FIRE EXTINGUISHING/FIRST AID
Fire: extremely flammable.	keep away from open flame and sparks, no smoking.	powder, AFFF, foam, carbon dioxide, (halons).
Explosion: forms explosive air-vapor mixtures.	sealed machinery, ventilation, explosion-proof electrical equipment and lighting, grounding.	in case of fire: keep tanks/drums cool by spraying with water.
Inhalation: severe breathing difficulties, headache, dizziness, drowsiness.	ventilation, local exhaust or respiratory protection.	fresh air, rest, artificial respiration if necessary, call a doctor.
Skin: redness, burning sensation.	gloves.	remove contaminated clothing, flush skin with water or shower.
Eyes: redness.	safety glasses.	flush with water, take to a doctor if necessary.
Ingestion: severe breathing difficulties, nausea.		rinse mouth, DO NOT induce vomiting, take immediately to hospital.

SPILLAGE	STORAGE	LABELING / NFPA
collect leakage in sealable containers, soak up with sand or other inert absorbent and remove to safe place.	keep in fireproof place; add an inhibitor.	3 1 ◇ 0

NOTES
Before distilling check for peroxides; if found, render harmless. High atmospheric concentrations, e.g. in poorly ventilated spaces, can cause oxygen deficiency, with risk of unconsciousness.

Transport Emergency Card TEC(R)-508

HI: 33; UN-number: 2256

CAS-No: [108-91-8]
aminocyclohexane
aminohexahydrobenzene
CHA
cyclohexanamine

$C_6H_{11}NH_2$

CYCLOHEXYLAMINE

PHYSICAL PROPERTIES		IMPORTANT CHARACTERISTICS
Boiling point, °C	134	**COLORLESS LIQUID WITH CHARACTERISTIC ODOR, TURNING DARK ON EXPOSURE TO AIR**
Melting point, °C	−18	Vapor mixes readily with air. Decomposes when heated or burned, giving off toxic gases. Reacts violently with strong acids. Reacts violently with strong oxidants, with risk of fire and explosion. Attacks copper and its alloys.
Flash point, °C	26	
Auto-ignition temperature, °C	290	
Relative density (water = 1)	0.9	
Relative vapor density (air = 1)	3.4	
Relative density at 20 °C of saturated mixture vapor/air (air = 1)	1.03	TLV-TWA (skin) 10 ppm 41 mg/m³
Vapor pressure, mm Hg at 20 °C	10.6	**Absorption route:** Can enter the body by inhalation or ingestion or through the skin. Harmful atmospheric concentrations can build up fairly rapidly on evaporation at approx. 20°C - even more rapidly in aerosol form.
Solubility in water	∞	**Immediate effects:** Corrosive to the eyes, skin and respiratory tract. Affects the nervous system. Inhalation of vapor/fumes can cause severe breathing difficulties (lung edema). In serious cases risk of death.
Explosive limits, vol% in air	1.5-9.4	
Relative molecular mass	99.2	
Gross formula:	$C_6H_{13}N$	

HAZARDS/SYMPTOMS	PREVENTIVE MEASURES	FIRE EXTINGUISHING/FIRST AID
Fire: flammable.	keep away from open flame and sparks, no smoking.	powder, alcohol-resistant foam, large quantities of water, carbon dioxide, (halons).
Explosion: above 26°C: forms explosive air-vapor mixtures.	above 26°C: sealed machinery, ventilation, explosion-proof electrical equipment.	in case of fire: keep tanks/drums cool by spraying with water.
		IN ALL CASES CALL IN A DOCTOR
Inhalation: *corrosive*; cough, shortness of breath, dizziness, nausea.	ventilation, local exhaust or respiratory protection.	fresh air, rest, place in half-sitting position, take to hospital.
Skin: *corrosive*; redness, pain, serious burns.	gloves, protective clothing.	remove contaminated clothing, flush skin with water or shower, refer to a doctor.
Eyes: *corrosive*; redness, pain, impaired vision.	face shield.	flush with water, take to a doctor.
Ingestion: *corrosive*; abdominal cramps, nausea.		rinse mouth, immediately take to hospital.

SPILLAGE	STORAGE	LABELING / NFPA
collect leakage in sealable containers, flush away remainder with water; (additional individual protection: breathing apparatus).	keep in fireproof place, separate from oxidants and strong acids.	R: 10-21/22-34 S: 36/37/39 Corrosive

NOTES
Lung edema symptoms usually develop several hours later and are aggravated by physical exertion: rest and hospitalization essential. As first aid, a doctor or authorized person should consider administering a corticosteroid spray.

Transport Emergency Card TEC(R)-71

HI: 83; UN-number: 2357

255

1,3-CYCLOPENTADIENE

PHYSICAL PROPERTIES	IMPORTANT CHARACTERISTICS
Boiling point, °C 40 Melting point, °C −97 Flash point, °C < −50 Auto-ignition temperature, °C 640 Relative density (water = 1) 0.8 Relative vapor density (air = 1) 2.3 Relative density at 20 °C of saturated mixture vapor/air (air = 1) 1.6 Vapor pressure, mm Hg at 20 °C 369 Solubility in water none Explosive limits, vol% in air - Minimum ignition energy, mJ 0.67 Relative molecular mass 66.1 Gross formula: C_5H_6	**COLORLESS LIQUID** Vapor is heavier than air and spreads at ground level, with risk of ignition at a distance. Able to polymerize when moderately heated or - in the absence of inhibitor - on contact with peroxide or trichloroacetic acid. Flow, agitation etc. can cause build-up of electrostatic charge due to liquid's low conductivity. Reacts violently with oxidants, sulfuric acid, nitric acid, magnesium and alkaline-earth metals, with risk of fire and explosion. TLV-TWA 75 ppm 203 mg/m³ **Absorption route:** Can enter the body by inhalation or ingestion. Harmful atmospheric concentrations can build up very rapidly on evaporation at 20°C. **Immediate effects:** Liquid destroys the skin's natural oils. Affects the nervous system. If liquid is swallowed, droplets can enter the lungs, with risk of pneumonia.

HAZARDS/SYMPTOMS	PREVENTIVE MEASURES	FIRE EXTINGUISHING/FIRST AID
Fire: extremely flammable.	keep away from open flame and sparks, no smoking.	powder, AFFF, foam, carbon dioxide, (halons).
Explosion: forms explosive air-vapor mixtures.	sealed machinery, ventilation, explosion-proof electrical equipment and lighting, grounding.	in case of fire: keep tanks/drums cool by spraying with water.
Inhalation: headache, dizziness, drowsiness, unconsciousness.	ventilation, local exhaust or respiratory protection.	fresh air, rest, call a doctor.
Skin: redness.	gloves.	remove contaminated clothing, flush skin with water or shower.
Eyes: redness, pain.	safety glasses.	flush with water, take to a doctor if necessary.
Ingestion: abdominal pain, nausea.		rinse mouth, DO NOT induce vomiting, take immediately to hospital.

SPILLAGE	STORAGE	LABELING / NFPA
collect leakage in sealable containers, soak up with sand or other inert absorbent and remove to safe place; (additional individual protection: breathing apparatus).	keep in fireproof place, separate from oxidants and strong acids; add an inhibitor.	

NOTES
Explosive limits not given in the literature. Before distilling check for peroxides; if found, render harmless. Alcohol consumption increases toxic effects. High atmospheric concentrations, e.g. in poorly ventilated spaces, can cause oxygen deficiency, with risk of unconsciousness. Dimerization occurs spontaneously on contact with oxygen. Monomer is perishable above −80°C. Use airtight packaging.

Transport Emergency Card TEC(R)-30G03

CYCLOPENTANE

PHYSICAL PROPERTIES		IMPORTANT CHARACTERISTICS	
Boiling point, °C	49	**COLORLESS LIQUID WITH CHARACTERISTIC ODOR**	
Melting point, °C	−93	Vapor is heavier than air and spreads at ground level, with risk of ignition at a distance. Flow, agitation etc. can cause build-up of electrostatic charge due to liquid's low conductivity. Do not use compressed air when filling, emptying or processing. Reacts violently with strong oxidants, with risk of fire and explosion.	
Flash point, °C	< −20		
Auto-ignition temperature, °C	380		
Relative density (water = 1)	0.75		
Relative vapor density (air = 1)	2.42		
Relative density at 20 °C of saturated mixture vapor/air (air = 1)	1.47	TLV-TWA	600 ppm	1720 mg/m³
Vapor pressure, mm Hg at 20 °C	255		
Solubility in water	none	**Absorption route:** Can enter the body by inhalation or ingestion. Harmful atmospheric concentrations can build up fairly rapidly on evaporation at approx. 20°C - even more rapidly in aerosol form.	
Explosive limits, vol% in air	1.5-8.7		
Minimum ignition energy, mJ	0.54		
Electrical conductivity, pS/m	< 10⁴	**Immediate effects:** Liquid destroys the skin's natural oils. Affects the nervous system. In serious cases risk of unconsciousness.	
Relative molecular mass	70		
Gross formula:	C_5H_{10}		

HAZARDS/SYMPTOMS	PREVENTIVE MEASURES	FIRE EXTINGUISHING/FIRST AID
Fire: extremely flammable.	keep away from open flame and sparks, no smoking.	powder, AFFF, foam, carbon dioxide, (halons).
Explosion: forms explosive air-vapor mixtures.	sealed machinery, ventilation, explosion-proof electrical equipment and lighting, grounding, non-sparking tools.	in case of fire: keep tanks/drums cool by spraying with water.
Inhalation: headache, drowsiness, unconsciousness.	ventilation, local exhaust or respiratory protection.	fresh air, rest, call a doctor.
Skin: redness.	gloves.	remove contaminated clothing, wash skin with soap and water.
Eyes: redness.	safety glasses.	flush with water, take to a doctor if necessary.
Ingestion: sore throat, diarrhea, nausea.		rinse mouth, DO NOT induce vomiting, call a doctor.

SPILLAGE	STORAGE	LABELING / NFPA
collect leakage in sealable containers, soak up with sand or other inert absorbent and remove to safe place; (additional individual protection: breathing apparatus).	keep in fireproof place, keep separate from strong oxidants.	R: 11 S: 9-16-29-33 🔥 Flammable

NFPA diagram: 3 (top), 1 (left), 0 (right)

NOTES
Alcohol consumption increases toxic (narcotic) effects.

C_5H_8

CYCLOPENTENE

PHYSICAL PROPERTIES		IMPORTANT CHARACTERISTICS
Boiling point, °C	46	**COLORLESS LIQUID WITH CHARACTERISTIC ODOR**
Melting point, °C	−93	Vapor is heavier than air and spreads at ground level, with risk of ignition at a distance. Presumed
Flash point, °C	−30	to be able to form peroxides. Able to polymerize. Flow, agitation etc. can cause build-up of
Auto-ignition temperature, °C	395	electrostatic charge due to liquid's low conductivity. Do not use compressed air when filling,
Relative density (water = 1)	0.77	emptying or processing. Reacts violently with many compounds.
Relative vapor density (air = 1)	2.35	
Relative density at 20 °C of saturated mixture vapor/air (air = 1)	1.6	TLV-TWA not available
Vapor pressure, mm Hg at 20 °C	318	**Absorption route:** Can enter the body by inhalation. Harmful atmospheric concentrations can
Solubility in water	none	build up fairly rapidly on evaporation at approx. 20°C - even more rapidly in aerosol form.
Explosive limits, vol% in air	?-?	**Immediate effects:** Irritates the eyes, skin and respiratory tract. If liquid is swallowed, droplets
Electrical conductivity, pS/m	< 10^4	can enter the lungs, with risk of pneumonia. Can cause narcosis.
Relative molecular mass	68.1	
Gross formula:	C_5H_8	

HAZARDS/SYMPTOMS	PREVENTIVE MEASURES	FIRE EXTINGUISHING/FIRST AID
Fire: extremely flammable.	keep away from open flame and sparks, no smoking.	powder, AFFF, foam, carbon dioxide, (halons).
Explosion: forms explosive air-vapor mixtures.	sealed machinery, ventilation, explosion-proof electrical equipment and lighting, grounding.	in case of fire: keep tanks/drums cool by spraying with water.
Inhalation: severe breathing difficulties, headache, dizziness, drowsiness.	ventilation, local exhaust or respiratory protection.	fresh air, rest, artificial respiration if necessary.
Skin: redness, burning sensation.	gloves.	remove contaminated clothing, flush skin with water or shower.
Eyes: redness.	safety glasses.	flush with water, take to a doctor if necessary.
Ingestion: severe breathing difficulties, nausea.		rinse mouth, DO NOT induce vomiting, take immediately to hospital.

SPILLAGE	STORAGE	LABELING / NFPA
collect leakage in non-sealable containers, soak up with sand or other inert absorbent and remove to safe place; (additional individual protection: breathing apparatus).	keep in fireproof place; add an inhibitor.	3 / 1 / 1

NOTES
Explosive limits not given in the literature. High atmospheric concentrations, e.g. in poorly ventilated spaces, can cause oxygen deficiency, with risk of unconsciousness.

Transport Emergency Card TEC(R)-533

HI: 33; UN-number: 2246

CYCLOPROPANE
(cylinder)

PHYSICAL PROPERTIES	IMPORTANT CHARACTERISTICS
Boiling point, °C −33 Melting point, °C −127 Flash point, °C flammable gas Auto-ignition temperature, °C 498 Relative density (water = 1) 0.6 Relative vapor density (air = 1) 1.5 Vapor pressure, mm Hg at 20 °C 4864 Solubility in water moderate Explosive limits, vol% in air 2.4-10.4 Minimum ignition energy, mJ 0.17 Electrical conductivity, pS/m < 10⁴ Relative molecular mass 42.1	**COLORLESS COMPRESSED LIQUEFIED GAS WITH CHARACTERISTIC ODOR** Gas is heavier than air and spreads at ground level, with risk of ignition at a distance. Flow, agitation etc. can cause build-up of electrostatic charge due to liquid's low conductivity. Do not use compressed air when filling, emptying or processing. TLV-TWA not available **Absorption route:** Can enter the body by inhalation. **Immediate effects:** Liquid can cause frostbite due to rapid evaporation. Affects the nervous system. Can cause heart rate disorders. In serious cases risk of unconsciousness.
Gross formula: C_3H_6	

HAZARDS/SYMPTOMS	PREVENTIVE MEASURES	FIRE EXTINGUISHING/FIRST AID
Fire: extremely flammable.	keep away from open flame and sparks, no smoking.	shut off supply; if impossible and no danger to surrounding area, allow fire to burn itself out; otherwise extinguish with powder, carbon dioxide, (halons).
Explosion: forms explosive air-gas mixtures.	sealed machinery, ventilation, explosion-proof electrical equipment and lighting, grounding, non-sparking tools.	in case of fire: keep cylinders cool by spraying with water, fight fire from sheltered location.
Inhalation: headache, dizziness, nausea, drowsiness, sleepiness, unconsciousness.	ventilation, local exhaust or respiratory protection.	fresh air, rest, take to hospital.
Skin: *in case of frostbite*: redness, pain, sores.	insulating gloves.	*in case of frosbite*: DO NOT remove clothing, flush skin and clothing with water.
Eyes: *in case of frostbite*: redness, pain, impaired vision.	face shield.	flush with water, take to a doctor.

SPILLAGE	STORAGE	LABELING / NFPA
evacuate area, call in an expert, ventilate; (additional individual protection: breathing apparatus).	keep in cool, fireproof place.	 4 1 0

NOTES
High atmospheric concentrations, e.g. in poorly ventilated spaces, can cause oxygen deficiency, with risk of unconsciousness. Turn leaking cylinder so that leak is on top to prevent liquid cyclopropane escaping. Close valve when not in use; check hoses and lines regularly; test for leaking connections with soap solution.

Transport Emergency Card TEC(R)-662

HI: 23; UN-number: 1027

$CH_3C_6H_4CH(CH_3)_2$

p-CYMENE

PHYSICAL PROPERTIES		IMPORTANT CHARACTERISTICS
Boiling point, °C	177	**COLORLESS LIQUID WITH CHARACTERISTIC ODOR** Vapor mixes readily with air. Flow, agitation etc. can cause build-up of electrostatic charge due to liquid's low conductivity. Reacts with strong oxidants. Light-sensitive.
Melting point, °C	−68	
Flash point, °C	47	
Auto-ignition temperature, °C	435	
Relative density (water = 1)	0.9	TLV-TWA not available
Relative vapor density (air = 1)	4.6	
Relative density at 20 °C of saturated mixture vapor/air (air = 1)	1.01	**Absorption route:** Can enter the body by inhalation or ingestion. Harmful atmospheric concentrations build up fairly slowly on evaporation at 20°C, but much more rapidly in aerosol form.
Vapor pressure, mm Hg at 20 °C	1.5	**Immediate effects:** Irritates the eyes, skin and respiratory tract. Liquid destroys the skin's natural oils. Affects the nervous system. If liquid is swallowed, droplets can enter the lungs, with risk of pneumonia.
Solubility in water	none	
Explosive limits, vol% in air	0.7-5.6	
Relative molecular mass	134.2	
Gross formula:	$C_{10}H_{14}$	

HAZARDS/SYMPTOMS	PREVENTIVE MEASURES	FIRE EXTINGUISHING/FIRST AID
Fire: flammable.	keep away from open flame and sparks, no smoking.	powder, AFFF, foam, carbon dioxide, (halons).
Explosion: above 47°C: forms explosive air-vapor mixtures.	above 47°C: sealed machinery, ventilation, explosion-proof electrical equipment, grounding.	in case of fire: keep tanks/drums cool by spraying with water.
Inhalation: sore throat, headache, dizziness, drowsiness.	ventilation, local exhaust or respiratory protection.	fresh air, rest, call a doctor.
Skin: redness.	gloves, protective clothing.	remove contaminated clothing, flush skin with water or shower.
Eyes: redness, pain.	safety glasses.	flush with water, take to a doctor if necessary.
Ingestion: abdominal pain, severe breathing difficulties, nausea; see also 'Inhalation'.		rinse mouth, DO NOT induce vomiting, take immediately to hospital.

SPILLAGE	STORAGE	LABELING / NFPA
collect leakage in sealable containers, soak up with sand or other inert absorbent and remove to safe place.	keep in dark, fireproof place, separate from oxidants.	2 / 2 / 0

NOTES
Alcohol consumption increases toxic effects. The measures on this card also apply to ortho and meta-cymene and technical mixtures of isomers.

Transport Emergency Card TEC(R)-30G35

HI: 30; UN-number: 2046

CAS-No: [91-17-8]
bicyclo(4,4,0)decane
decalin

$C_{10}H_{18}$

DECAHYDRONAPHTHALENE (cis/trans)

PHYSICAL PROPERTIES	IMPORTANT CHARACTERISTICS
Boiling point, °C ca. 190 Melting point, °C ca. −35 Flash point, °C 58 Auto-ignition temperature, °C 250 Relative density (water = 1) 0.9 Relative vapor density (air = 1) 4.8 Relative density at 20 °C of saturated mixture vapor/air (air = 1) 1.01 Vapor pressure, mm Hg at 20 °C 0.76 à 3.0 Solubility in water none Explosive limits, vol% in air 0.7-5 Relative molecular mass 138.2	**COLORLESS LIQUID WITH CHARACTERISTIC ODOR** Vapor mixes readily with air. Flow, agitation etc. can cause build-up of electrostatic charge due to liquid's low conductivity. Reacts violently with strong oxidants. TLV-TWA　　　　　　　　not available **Absorption route:** Can enter the body by inhalation or ingestion. Harmful atmospheric concentrations can build up fairly rapidly on evaporation at approx. 20°C - even more rapidly in aerosol form. **Immediate effects:** Irritates the eyes, skin and respiratory tract. Liquid destroys the skin's natural oils. Affects the nervous system. If liquid is swallowed, droplets can enter the lungs, with risk of pneumonia.
Gross formula:　　　　　　$C_{10}H_{18}$	

HAZARDS/SYMPTOMS	PREVENTIVE MEASURES	FIRE EXTINGUISHING/FIRST AID
Fire: combustible.	keep away from open flame, no smoking.	powder, AFFF, foam, carbon dioxide, (halons).
Explosion: above 58°C: forms explosive air-vapor mixtures.	above 58°C: sealed machinery, ventilation, grounding.	in case of fire: keep tanks/drums cool by spraying with water.
Inhalation: cough, shortness of breath, dizziness, vomiting.	ventilation, local exhaust or respiratory protection.	fresh air, rest, call a doctor.
Skin: redness.	gloves.	remove contaminated clothing, wash skin with soap and water.
Eyes: redness, pain.	face shield.	flush with water, take to a doctor if necessary.
Ingestion: abdominal pain, nausea, dizziness.		rinse mouth, DO NOT induce vomiting, call a doctor.

SPILLAGE	STORAGE	LABELING / NFPA
collect leakage in sealable containers, soak up with sand or other inert absorbent and remove to safe place; (additional individual protection: breathing apparatus).	keep in fireproof place, separate from oxidants.	 2 2　　0

NOTES
Decaline (trans): boiling point: 187° C; melting point: −30° C; flash point: 50° C; vapor pressure at 20° C: 3.0 mm Hg; explosive limits: 0.7-5.4%. Decaline (cis) boiling point: 195° C; melting point: 43° C; flash point: 61° C; vapor pressure explosive limits: 0.7-4.9%. Not to be confused with napthalene ($C_{10}H_8$).

Transport Emergency Card TEC(R)-558　　　　　　　　　　　　　　　　　　　　　　　　**HI: 30; UN-number: 1147**

DAA
dimethylacetonylcarbinol
4-hydroxyl-2-keto-4-methylpentane
4-hydroxy-4-methylpentan-2-one
4-hydroxy-4-methyl-2-pentanone

$CH_3COCH_2C(CH_3)_2OH$

DIACETONE ALCOHOL

PHYSICAL PROPERTIES		IMPORTANT CHARACTERISTICS	
Boiling point, °C	166	**COLORLESS LIQUID**	
Melting point, °C	− 47	Vapor mixes readily with air. Decomposes when heated or on contact with acids or bases, forming → *acetone* and → *mesityl oxide*. Reacts violently with strong oxidants. Reacts with alkali metals, giving off → *hydrogen*.	
Flash point, °C	58		
Auto-ignition temperature, °C	640		
Relative density (water = 1)	0.9		
Relative vapor density (air = 1)	4.0		
Relative density at 20 °C of saturated mixture			
vapor/air (air = 1)	1.0		
Vapor pressure, mm Hg at 20 °C	0.84		
Solubility in water	∞		
Explosive limits, vol% in air	1.8-6.9		
Relative molecular mass	116.2		

| | | TLV-TWA | 50 ppm | 238 mg/m³ |

Absorption route: Can enter the body by inhalation or ingestion. Harmful atmospheric concentrations build up fairly slowly on evaporation at 20°C, but much more rapidly in aerosol form.
Immediate effects: Irritates the eyes, skin and respiratory tract. Liquid destroys the skin's natural oils. Affects the nervous system (narcotic effect). Inhalation can cause severe breathing difficulties. In serious cases risk of unconsciousness.
Effects of prolonged/repeated exposure: Prolonged or repeated contact can cause skin disorders. Can cause liver and kidney damage.

Gross formula: $C_6H_{12}O_2$

HAZARDS/SYMPTOMS	PREVENTIVE MEASURES	FIRE EXTINGUISHING/FIRST AID
Fire: combustible.	keep away from open flame, no smoking.	powder, alcohol-resistant foam, water spray, carbon dioxide, (halons).
Explosion: above 58°C: forms explosive air-vapor mixtures.	above 58°C: sealed machinery, ventilation.	in case of fire: keep tanks/drums cool by spraying with water.
Inhalation: headache, drowsiness, unconsciousness.	ventilation, local exhaust or respiratory protection.	fresh air, rest, call a doctor.
Skin: redness, pain.	gloves.	remove contaminated clothing, flush skin with water or shower, refer to a doctor.
Eyes: redness, pain.	face shield.	flush with water, take to a doctor.
Ingestion: abdominal pain, vomiting.		rinse mouth, take to hospital.

SPILLAGE	STORAGE	LABELING / NFPA
collect leakage in sealable containers, soak up with sand or other inert absorbent and remove to safe place, flush away remainder with water.	keep in fireproof place, separate from oxidants, strong acids and strong bases.	R: 36 S: 24/25 ✖ Irritant

NOTES
Technical grades contain acetone and thus have a much lower flash point (starting at − 18°C).

DIACETYL

PHYSICAL PROPERTIES		IMPORTANT CHARACTERISTICS	
Boiling point, °C	88	**GREEN TO YELLOW LIQUID WITH CHARACTERISTIC ODOR**	
Melting point, °C	−3	Vapor is heavier than air and spreads at ground level, with risk of ignition at a distance.	
Flash point, °C	6		
Auto-ignition temperature, °C	365	TLV-TWA	not available
Relative density (water = 1)	0.99		
Relative vapor density (air = 1)	2.97		
Relative density at 20 °C of saturated mixture vapor/air (air = 1)	1.1	**Absorption route:** Can enter the body by inhalation or ingestion.	
Vapor pressure, mm Hg at 20 °C	49	**Immediate effects:** Irritates the eyes, skin and respiratory tract. Affects the nervous system (narcotic effect).	
Solubility in water	good		
Relative molecular mass	86.1		
Gross formula:	$C_4H_6O_2$		

HAZARDS/SYMPTOMS	PREVENTIVE MEASURES	FIRE EXTINGUISHING/FIRST AID
Fire: extremely flammable.	keep away from open flame and sparks, no smoking.	powder, alcohol-resistant foam, large quantities of water, carbon dioxide, (halons).
Explosion: forms explosive air-vapor mixtures.	sealed machinery, ventilation, explosion-proof electrical equipment and lighting.	in case of fire: keep tanks/drums cool by spraying with water.
Inhalation: sore throat, cough, drowsiness, unconsciousness.	ventilation, local exhaust or respiratory protection.	fresh air, rest.
Skin: redness.	gloves.	remove contaminated clothing, flush skin with water or shower.
Eyes: redness, pain.	face shield.	flush with water, take to a doctor if necessary.
Ingestion: sore throat, abdominal pain.		rinse mouth, call a doctor.

SPILLAGE	STORAGE	LABELING / NFPA
collect leakage in sealable containers, flush away remainder with water.	keep in fireproof place.	3 1 0

NOTES

$(CH_2=CHCH_2)_2NH$

DIALLYLAMINE

PHYSICAL PROPERTIES	IMPORTANT CHARACTERISTICS	
Boiling point, °C 111	**COLORLESS LIQUID WITH CHARACTERISTIC ODOR**	
Melting point, °C −100	Vapor mixes readily with air, forming explosive mixtures. Do not use compressed air when filling, emptying or processing. Decomposes when heated or burned giving off toxic gases, giving off toxic vapors (incl. → *nitrous vapors*). Reacts violently with oxidants and strong acids. Attacks copper, tin, aluminum and zinc.	
Flash point, °C 21		
Relative density (water = 1) 0.8		
Relative vapor density (air = 1) 3.4		
Relative density at 20 °C of saturated mixture vapor/air (air = 1) 1.06	**TLV-TWA** not available	
Vapor pressure, mm Hg at 20 °C ca. 18		
Solubility in water, g/100 ml at 20 °C 9	**Absorption route:** Can enter the body by inhalation or ingestion. Harmful atmospheric concentrations can build up very rapidly on evaporation at 20°C.	
Explosive limits, vol% in air ?-?	**Immediate effects:** Corrosive to the eyes, skin and respiratory tract. Inhalation of vapor/fumes can cause severe breathing difficulties (lung edema). In serious cases risk of death.	
Relative molecular mass 97.2		
Gross formula: $C_6H_{11}N$		

HAZARDS/SYMPTOMS	PREVENTIVE MEASURES	FIRE EXTINGUISHING/FIRST AID
Fire: extremely flammable.	keep away from open flame and sparks, no smoking.	powder, AFFF, foam, carbon dioxide, (halons).
Explosion: forms explosive air-vapor mixtures.	sealed machinery, ventilation, explosion-proof electrical equipment and lighting.	in case of fire: keep tanks/drums cool by spraying with water.
Inhalation: *corrosive*; sore throat, cough, shortness of breath, severe breathing difficulties.	ventilation, local exhaust or respiratory protection.	fresh air, rest, place in half-sitting position, artificial respiration if necessary, take to hospital.
Skin: *corrosive*; redness, pain, burns.	gloves, protective clothing.	remove contaminated clothing, flush skin with water or shower, call a doctor.
Eyes: *corrosive*; redness, pain, impaired vision.	face shield, or combined eye and respiratory protection.	flush with water, take to a doctor.
Ingestion: *corrosive*; sore throat, abdominal pain, diarrhea, vomiting.		rinse mouth, call a doctor or take to hospital.

SPILLAGE	STORAGE	LABELING / NFPA
collect leakage in sealable containers, soak up with sand or other inert absorbent and remove to safe place; (additional individual protection: breathing apparatus).	keep in fireproof place, separate from oxidants and strong acids.	3 / 3 1

NOTES
Explosive limits not given in the literature. Lung edema symptoms usually develop several hours later and are aggravated by physical exertion: rest and hospitalization essential. As first aid, a doctor or authorized person should consider administering a corticosteroid spray. Unbreakable packaging preferred; if breakable, keep in unbreakable container.

Transport Emergency Card TEC(R)-30G31

HI: 338; UN-number: 2359

$C_6H_4(COOCH_2CH=CH_2)_2$

DIALLYL PHTALATE

PHYSICAL PROPERTIES		IMPORTANT CHARACTERISTICS
Boiling point, °C	290	**COLORLESS VISCOUS LIQUID WITH CHARACTERISTIC ODOR**
Melting point, °C	−70	Vapor mixes readily with air. Polymerizes to a solid when heated or on exposure to peroxides.
Flash point, °C	166	Decomposes when heated above boiling point, giving off toxic vapors. Reacts violently with
Auto-ignition temperature, °C	435	strong oxidants, with risk of fire and explosion.
Relative density (water = 1)	1.1	
Relative vapor density (air = 1)	8.5	

TLV-TWA	5 mg/m³

Absorption route: Can enter the body by inhalation or ingestion. Harmful atmospheric concentrations build up fairly slowly on evaporation at 20°C, but much more rapidly in aerosol form.
Immediate effects: Irritates the eyes, skin and respiratory tract.
Effects of prolonged/repeated exposure: Prolonged or repeated contact can cause skin disorders (eczema).

PHYSICAL PROPERTIES (cont.)	
Relative density at 20°C of saturated mixture	
vapor/air (air = 1)	1.00
Vapor pressure, mm Hg at 20°C	1.5×10^{-3}
Solubility in water	none
Relative molecular mass	246.3
Gross formula:	$C_{14}H_{14}O_4$

HAZARDS/SYMPTOMS	PREVENTIVE MEASURES	FIRE EXTINGUISHING/FIRST AID
Fire: combustible.	keep away from open flame, no smoking.	powder, AFFF, foam, carbon dioxide, (halons).
Inhalation: sore throat, cough, headache.	ventilation.	fresh air, rest, call a doctor if necessary.
Skin: redness.	gloves.	remove contaminated clothing, flush skin with water or shower.
Eyes: redness, pain.	safety glasses.	flush with water, take to a doctor if necessary.
Ingestion: abdominal pain.		rinse mouth, call a doctor or take to hospital.

SPILLAGE	STORAGE	LABELING / NFPA
collect leakage in sealable containers, soak up with sand or other inert absorbent.	keep separate from oxidants; ventilate at floor level.	R: 22 S: 24/25 ✖ Harmful

NOTES

1,3-DIAMINOBUTANE

PHYSICAL PROPERTIES	IMPORTANT CHARACTERISTICS	
Boiling point, °C 143 Flash point, °C 52 Relative density (water = 1) 0.85 Relative vapor density (air = 1) 3.0 Relative density at 20 °C of saturated mixture vapor/air (air = 1) 1.01 Vapor pressure, mm Hg at 20 °C ca. 4.5 Solubility in water ∞ Explosive limits, vol% in air ?-? Relative molecular mass 88.2	**COLORLESS LIQUID WITH CHARACTERISTIC ODOR** Vapor mixes readily with air. → *Nitrous vapors* are a product of combustion. In aqueous solution is a strong base which reacts violently with acids and corrodes aluminum, zinc etc. Reacts violently with strong oxidants.	
	TLV-TWA not available	
	Absorption route: Can enter the body by inhalation or ingestion or through the skin. Harmful atmospheric concentrations can build up fairly rapidly on evaporation at approx. 20° C - even more rapidly in aerosol form. **Immediate effects:** Corrosive to the eyes, skin and respiratory tract. Inhalation can cause lung edema. In serious cases risk of death.	
Gross formula: $C_4H_{12}N_2$		

HAZARDS/SYMPTOMS	PREVENTIVE MEASURES	FIRE EXTINGUISHING/FIRST AID
Fire: flammable.	keep away from open flame and sparks, no smoking.	powder, alcohol-resistant foam, large quantities of water, carbon dioxide, (halons).
Explosion: above 52°C: forms explosive air-vapor mixtures.	above 52°C: sealed machinery, ventilation, explosion-proof electrical equipment.	in case of fire: keep tanks/drums cool by spraying with water.
Inhalation: *corrosive*; sore throat, cough, shortness of breath, severe breathing difficulties.	ventilation, local exhaust or respiratory protection.	fresh air, rest, place in half-sitting position, take to hospital.
Skin: *corrosive, is absorbed*; redness, pain, burns.	gloves, protective clothing.	remove contaminated clothing, flush skin with water or shower, refer to a doctor.
Eyes: *corrosive*; redness, pain, impaired vision.	face shield.	flush with water, take to a doctor.
Ingestion: *corrosive*; sore throat, abdominal pain, diarrhea.		rinse mouth, take to hospital.

SPILLAGE	STORAGE	LABELING / NFPA
collect leakage in sealable containers, flush away remainder with water; (additional individual protection: breathing apparatus).	keep in fireproof place, separate from oxidants and acids.	

NOTES
Explosive limits not given in the literature. Lung edema symptoms usually develop several hours later and are aggravated by physical exertion: rest and hospitalization essential. As first aid, a doctor or authorized person should consider administering a corticosteroid spray. The measures on this card also apply to 1,4-diaminobutane and tetramethylenediamine.

Transport Emergency Card TEC(R)-80G15

CAS-No: [462-94-2]
cadaverine
pentamethylenediamine
pentane-1,5-diamine

$NH_2(CH_2)_5NH_2$

1,5-DIAMINOPENTANE

PHYSICAL PROPERTIES	IMPORTANT CHARACTERISTICS
Boiling point, °C 179 Melting point, °C 9 Flash point, °C 62 Relative density (water = 1) 0.87 Relative vapor density (air = 1) 3.5 Relative density at 20 °C of saturated mixture vapor/air (air = 1) 1.0 Vapor pressure, mm Hg at 20 °C ca. 0.68 Solubility in water ∞ Relative molecular mass 102.2 Log P octanol/water −0.8 Gross formula: $C_5H_{14}N_2$	**COLORLESS VISCOUS HYGROSCOPIC LIQUID WITH CHARACTERISTIC ODOR** Vapor mixes readily with air. Decomposes when heated, giving off flammable and toxic gases (→ *nitrous vapors*). In aqueous solution is a medium strong base which reacts violently with strong acids. Reacts violently with strong oxidants. Attacks copper, nickel, steel and many plastics. Reacts with mercury to form explosive compounds.

PHYSICAL PROPERTIES (cont.)	
	TLV-TWA not available
	Absorption route: Can enter the body by inhalation or ingestion or through the skin. Harmful atmospheric concentrations build up fairly slowly on evaporation at 20°C, but much more rapidly in aerosol form. **Immediate effects:** Corrosive to the eyes, skin and respiratory tract. Inhalation of vapor/fumes can cause severe breathing difficulties (lung edema). In serious cases risk of death. **Effects of prolonged/repeated exposure:** Prolonged or repeated contact can cause (allergic) skin disorders.

HAZARDS/SYMPTOMS	PREVENTIVE MEASURES	FIRE EXTINGUISHING/FIRST AID
Fire: combustible.	keep away from open flame, no smoking, avoid contact with oxidants.	powder, alcohol-resistant foam, water spray, carbon dioxide, (halons).
Explosion: above 62°C: forms explosive air-vapor mixtures.	above 62°C: sealed machinery, ventilation.	
Inhalation: *corrosive*; sore throat, cough, shortness of breath.	ventilation, local exhaust or respiratory protection.	fresh air, rest, artificial respiration if necessary, take to hospital.
Skin: *corrosive, is absorbed*; redness, pain, burns.	gloves, protective clothing.	remove contaminated clothing, flush skin with water or shower, refer to a doctor.
Eyes: *corrosive*; redness, pain, impaired vision.	face shield.	flush with water, take to a doctor.
Ingestion: *corrosive*; sore throat, abdominal pain, diarrhea, vomiting.		rinse mouth, immediately take to hospital.

SPILLAGE	STORAGE	LABELING / NFPA
collect leakage in sealable containers, neutralize remainder and flush away with water.	keep separate from oxidants and strong acids; ventilate at floor level.	

NOTES
Log P octanol/water is estimated. Lung edema symptoms usually develop several hours later and are aggravated by physical exertion: rest and hospitalization essential. As first aid, a doctor or authorized person should consider administering a corticosteroid spray. Packaging: special material.

Transport Emergency Card TEC(R)-80G15

2,4-DIAMINOPHENOL DIHYDROCHLORIDE

PHYSICAL PROPERTIES	IMPORTANT CHARACTERISTICS	
Decomposes below boiling point, °C 230-240 Solubility in water, g/100 ml at 15 °C 27 Relative molecular mass 197.1	**WHITE CRYSTALS** Decomposes, giving off toxic and corrosive vapors (→ *hydrochloric acid* and → *nitrous vapors*). Reacts violently with strong oxidants.	
	TLV-TWA not available	
	Absorption route: Can enter the body by inhalation or ingestion or through the skin. Evaporation negligible at 20°C, but harmful concentrations of airborne particles can build up rapidly. **Immediate effects:** Irritates the eyes, skin and respiratory tract. At high exposure levels can cause high blood pressure, abdominal disorders, dizziness and convulsions. **Effects of prolonged/repeated exposure:** Prolonged or repeated contact can cause (allergic) skin disorders.	
Gross formula: $C_6H_{10}ClN_2O$		

HAZARDS/SYMPTOMS	PREVENTIVE MEASURES	FIRE EXTINGUISHING/FIRST AID
Fire: non-combustible.		in case of fire in immediate vicinity: use any extinguishing agent.
	KEEP DUST UNDER CONTROL	
Inhalation: sore throat, cough.	local exhaust or respiratory protection.	fresh air, rest, place in half-sitting position, call a doctor.
Skin: *is absorbed*; redness.	gloves.	remove contaminated clothing, flush skin with water or shower, refer to a doctor.
Eyes: redness, pain, serious damage can occur rapidly.	goggles.	flush with water, take to a doctor if necessary.
Ingestion: abdominal cramps, nausea.		rinse mouth, call a doctor.

SPILLAGE	STORAGE	LABELING / NFPA
clean up spillage, flush away remainder with water; (additional individual protection: P2 respirator).		

NOTES

1,2-DIAMINOPROPANE

PHYSICAL PROPERTIES	IMPORTANT CHARACTERISTICS
Boiling point, °C — 119 Flash point, °C — 24 Auto-ignition temperature, °C — 360-416 Relative density (water = 1) — 0.9 Relative vapor density (air = 1) — 2.6 Relative density at 20 °C of saturated mixture vapor/air (air = 1) — 1.02 Vapor pressure, mm Hg at 20 °C — ca. 7.6 Solubility in water — ∞ Explosive limits, vol% in air — 2.2-11.1 Relative molecular mass — 74.1 Log P octanol/water — −1.8	**COLORLESS HYGROSCOPIC LIQUID WITH CHARACTERISTIC ODOR** Vapor mixes readily with air. Reacts with mercury to form shock-sensitive compounds. Decomposes when heated or burned, giving off toxic gases (→ *nitrous vapors*). Reacts violently with oxidants, with risk of fire and explosion. Attacks copper and its alloys.
	TLV-TWA not available
	Absorption route: Can enter the body by inhalation or ingestion or through the skin. Harmful atmospheric concentrations can build up fairly rapidly on evaporation at approx. 20°C - even more rapidly in aerosol form. **Immediate effects:** Corrosive to the eyes, skin and respiratory tract. Inhalation of vapor/fumes can cause severe breathing difficulties (lung edema). In serious cases risk of death.
Gross formula: $C_3H_{10}N_2$	

HAZARDS/SYMPTOMS	PREVENTIVE MEASURES	FIRE EXTINGUISHING/FIRST AID
Fire: flammable.	keep away from open flame and sparks, no smoking.	powder, alcohol-resistant foam, large quantities of water, carbon dioxide, (halons).
Explosion: above 24°C: forms explosive air-vapor mixtures.	above 24°C: sealed machinery, ventilation, explosion-proof electrical equipment.	in case of fire: keep tanks/drums cool by spraying with water.
	STRICT HYGIENE	
Inhalation: *corrosive*; sore throat, cough, severe breathing difficulties, nausea.	ventilation, local exhaust or respiratory protection.	fresh air, rest, artificial respiration if necessary, take to hospital.
Skin: *corrosive, is absorbed*; redness, pain, burns.	gloves, protective clothing.	remove contaminated clothing, flush skin with water or shower, call a doctor.
Eyes: *corrosive*; redness, pain, impaired vision.	face shield.	flush with water, take to a doctor.
Ingestion: *corrosive*; sore throat, abdominal pain, nausea, vomiting.		rinse mouth, take to hospital.

SPILLAGE	STORAGE	LABELING / NFPA
collect leakage in sealable containers, flush away remainder with water; (additional individual protection: breathing apparatus).	keep in dry, fireproof place, separate from strong acids.	3 / 2 0

NOTES
Log P octanol/water is estimated. Lung edema symptoms usually develop several hours later and are aggravated by physical exertion: rest and hospitalization essential. As first aid, a doctor or authorized person should consider administering a corticosteroid spray.

Transport Emergency Card TEC(R)-670

HI: 83; UN-number: 2258

1,3-DIAMINOPROPANE

PHYSICAL PROPERTIES	IMPORTANT CHARACTERISTICS
Boiling point, °C 139 Melting point, °C −12 Flash point, °C 24 Relative density (water = 1) 0.88 Relative vapor density (air = 1) 2.5 Relative density at 20 °C of saturated mixture vapor/air (air = 1) 1.01 Vapor pressure, mm Hg at 20 °C ca. 5.3 Solubility in water ∞ Explosive limits, vol% in air ?-? Relative molecular mass 74.1	**COLORLESS LIQUID WITH CHARACTERISTIC ODOR** Vapor mixes readily with air. → *Nitrous gases* are a product of combustion. Reacts violently with strong acids and strong oxidants.
	TLV-TWA not available
	Absorption route: Can enter the body by inhalation or ingestion or through the skin. Harmful atmospheric concentrations can build up fairly rapidly on evaporation at approx. 20°C - even more rapidly in aerosol form. **Immediate effects:** Corrosive to the eyes, skin and respiratory tract. Inhalation can cause lung edema. **Effects of prolonged/repeated exposure:** Prolonged or repeated contact can cause skin disorders.
Gross formula: $C_3H_{10}N_2$	

HAZARDS/SYMPTOMS	PREVENTIVE MEASURES	FIRE EXTINGUISHING/FIRST AID
Fire: flammable.	keep away from open flame and sparks, no smoking.	powder, alcohol-resistant foam, large quantities of water, carbon dioxide, (halons).
Explosion: above 24°C: forms explosive air-vapor mixtures.	above 24°C: sealed machinery, ventilation, explosion-proof electrical equipment.	in case of fire: keep tanks/drums cool by spraying with water.
Inhalation: *corrosive*; sore throat, cough, shortness of breath, severe breathing difficulties.	ventilation, local exhaust or respiratory protection.	fresh air, rest, place in half-sitting position, take to hospital.
Skin: *corrosive, is absorbed*; redness, pain, burns.	gloves, protective clothing.	remove contaminated clothing, flush skin with water or shower, call a doctor.
Eyes: *corrosive*; redness, pain, impaired vision.	face shield.	flush with water, take to a doctor.
Ingestion: *corrosive*; sore throat, abdominal pain, diarrhea.		rinse mouth, call a doctor or take to hospital.

SPILLAGE	STORAGE	LABELING / NFPA
collect leakage in sealable containers, flush away remainder with water; (additional individual protection: breathing apparatus).	keep in fireproof place, separate from oxidants and acids.	

NOTES
Explosive limits not given in the literature. Lung edema symptoms usually develop several hours later and are aggravated by physical exertion: rest and hospitalization essential. As first aid, a doctor or authorized person should consider administering a corticosteroid spray.

Transport Emergency Card TEC(R)-80G15

1-DIAZO-2-NAPHTOL-4-SULFONIC ACID

PHYSICAL PROPERTIES	IMPORTANT CHARACTERISTICS	
Decomposes below melting point, °C — 100 Solubility in water — poor Relative molecular mass — 250.2	**YELLOW NEEDLE-SHAPED CRYSTALS OR PASTE** Decomposes explosively when heated above 100°C, giving off toxic and corrosive vapors (→ *sulfur oxides* and → *nitrous vapors*).	
	TLV-TWA	not available
	Absorption route: Can enter the body by inhalation or ingestion. **Immediate effects:** Irritates the eyes, skin and respiratory tract.	
Gross formula: $C_{10}H_6N_2O_4S$		

HAZARDS/SYMPTOMS	PREVENTIVE MEASURES	FIRE EXTINGUISHING/FIRST AID
Fire: flammable.	keep away from open flame and sparks, no smoking, avoid contact with hot surfaces (e.g. steam lines).	large quantities of water.
Explosion: heating above 100°C can cause explosion.		in case of fire: keep tanks/drums cool by spraying with water, fight fire from sheltered location.
Inhalation: sore throat, cough.	local exhaust or respiratory protection.	fresh air, rest, call a doctor.
Skin: redness, pain.	gloves.	remove contaminated clothing, flush skin with water or shower.
Eyes: redness, pain.	acid goggles.	flush with water, take to a doctor if necessary.
Ingestion: abdominal pain, nausea.	do not eat, drink or smoke while working.	rinse mouth, call a doctor.

SPILLAGE	STORAGE	LABELING / NFPA
clean up spillage, carefully collect remainder; (additional individual protection: P2 respirator).	keep in fireproof place, keep separate from strong bases.	

NOTES

CAS-No: [94-36-0]
benzoic acid peroxide
benzoyl peroxide
benzoyl superoxide

$(C_6H_5CO)_2O_2$

DIBENZOYL PEROXIDE
(50%)

PHYSICAL PROPERTIES	IMPORTANT CHARACTERISTICS	
Decomposes below melting point, °C 80 Flash point, °C see notes Solubility in water none Relative molecular mass 242.2	**WHITE PASTE OR POWDER** Strong oxidant which reacts violently with combustible substances and reducing agents. Reacts with acids, bases, amines and many other substances.	
	TLV-TWA	5 mg/m³
	Absorption route: Can enter the body by inhalation or ingestion. Evaporation negligible at 20°C, but harmful atmospheric concentrations can build up rapidly in aerosol form. **Immediate effects:** Irritates the eyes, skin and respiratory tract. Dermatitis and allergic reactions (eczema and asthma) are common. Headache, drowsiness, lowering of pulse rate and body temperature.	
Gross formula: $C_{14}H_{10}O_4$		

HAZARDS/SYMPTOMS	PREVENTIVE MEASURES	FIRE EXTINGUISHING/FIRST AID
Fire: combustible.	keep away from open flame, no smoking, avoid contamination.	large quantities of water, NOT FOAM EXTINGUISHER.
Explosion: contamination or heating can cause violent decomposition.	do not heat above 80°C.	in case of fire: keep tanks/drums cool by spraying with water, fight fire from sheltered location.
Inhalation: sore throat, cough, (due to products of decomposition).	local exhaust or respiratory protection.	fresh air, rest, place in half-sitting position, call a doctor.
Skin: redness, pain.	gloves.	remove contaminated clothing, flush skin with water or shower.
Eyes: redness, pain, impaired vision.	goggles, or combined eye and respiratory protection.	flush with water, take to doctor.
Ingestion: abdominal pain, nausea, vomiting.		rinse mouth, call a doctor.

SPILLAGE	STORAGE	LABELING / NFPA
render harmless with large quantites of 10% caustic soda solution, flush away remainder with water; (additional individual protection: P2 respirator).	keep cool, separate from other substances.	R: 3-36/37/38 S: 3/7/9-14-27-34-37/39 Explosive Irritant

NOTES
Under no circumstances mix directly with accelerator. Combustible, but flash point not given in the literature. Flush contaminated clothing with water (fire hazard). Store in original packaging; do not replace once removed.

Transport Emergency Card TEC(R)-52G01

HI: 11; UN-number: 2087 (2089)

DIBORANE
(cylinder)

PHYSICAL PROPERTIES	IMPORTANT CHARACTERISTICS
Boiling point, °C −93 Melting point, °C −165 Flash point, °C flammable gas Auto-ignition temperature, °C 38 Relative density (water = 1) 0.46 Relative vapor density (air = 1) 0.96 Vapor pressure, mm Hg at 20 °C 31,920 Solubility in water reaction Explosive limits, vol% in air 0.8-90 Relative molecular mass 27.7	**COLORLESS COMPRESSED GAS OR REFRIGERATED COMPRESSED LIQUEFIED GAS WITH CHARACTERISTIC ODOR** Mixes readily with air, forming explosive mixtures. Decomposes slowly at room temperature. Strong reducing agent which reacts violently with oxidants. Reacts explosively with chlorine and violently with halogen-containing compounds. Reacts with water and water vapor also in the air, giving off → *hydrogen* gas. Reacts with lithium and aluminum, forming hydrides which ignite spontaneously on exposure to air.

	PHYSICAL PROPERTIES

TLV-TWA	0.1 ppm	0.11 mg/m³

Absorption route: Can enter the body by inhalation or through the skin. Harmful atmospheric concentrations can build up very rapidly if gas is released.
Immediate effects: Corrosive to the eyes and respiratory tract. Inhalation of vapor/fumes can cause severe breathing difficulties (lung edema). In serious cases risk of death.

Gross formula: B_2H_6

HAZARDS/SYMPTOMS	PREVENTIVE MEASURES	FIRE EXTINGUISHiNG/FIRST AID
Fire: extremely flammable.	keep away from open flame and sparks, no smoking.	shut off supply; if impossible and no danger to surrounding area, allow fire to burn itself out; otherwise extinguish with special powder extinguisher, DO NOT USE WATER-BASED EXTINGUISHERS, DO NOT USE HALONS.
Explosion: forms explosive air-gas mixtures.	sealed machinery, ventilation, explosion-proof electrical equipment and lighting.	in case of fire: keep undamaged cylinders cool by spraying with water, but DO NOT spray substance with water, fight fire from sheltered location.
	STRICT HYGIENE	IN ALL CASES CALL IN A DOCTOR
Inhalation: *corrosive*, sore throat, cough, fever, (can seem to be like metal fume fever).	ventilation, local exhaust or respiratory protection.	fresh air, rest, place in half-sitting position, artificial respiration if necessary, take to hospital.
Skin: *is absorbed*; redness, pain, burns.	gloves.	flush clothing with water before removing, then flush skin with water or shower, call a doctor.
Eyes: *corrosive*; redness, pain, impaired vision.	face shield, or combined eye and respiratory protection.	flush with water, take to a doctor.
Ingestion: nausea, vomiting.		rinse mouth, call a doctor.

SPILLAGE	STORAGE	LABELING / NFPA
evacuate area, call in an expert, remove leaking cylinder to safe place, under no circumstances spray liquid with water, combat gas cloud with water curtain; (additional individual protection: CHEMICAL SUIT).	keep in fireproof, cool place, separate from oxidants and halogen-containing substances, ventilation, or under refrigeration.	4 / 3 / 3 / W

NOTES
Diborane is supplied in refrigerated cylinders (<0° C) or mixed with other gases in small cylinders at normal temperature. Reacts violently with extinguishing agents containing halogens. Odor limit is above TLV. Odor is not detectable after prolonged exposure. Lung edema symptoms usually develop several hours later and are aggravated by physical exertion: rest and hospitalization essential. As first aid, a doctor or authorized person should consider administering a corticosteroid spray. Turn leaking cylinder so that leak is on top to prevent liquid diborane escaping. Packaging: special material. Use Teflon or Viton packaging material.

Transport Emergency Card TEC(R)-20G21

HI: 236

BrCH$_2$CHBrCH$_2$Cl

1,2-DIBROMO-3-CHLOROPROPANE

PHYSICAL PROPERTIES	IMPORTANT CHARACTERISTICS
Boiling point, °C ... 196 Sublimation temperature, °C ... 7 Flash point, °C ... 77 Relative density (water = 1) ... 2.1 Relative vapor density (air = 1) ... 8.2 Relative density at 20°C of saturated mixture vapor/air (air = 1) ... 1.01 Vapor pressure, mm Hg at 20°C ... 0.76 Solubility in water, g/100 ml at 20°C ... 0.1 Relative molecular mass ... 236.4 Gross formula: C$_3$H$_5$Br$_2$Cl	**COLORLESS TO BROWN LIQUID WITH PUNGENT ODOR** Vapor mixes readily with air. Decomposes when heated or burned, giving off corrosive vapors (→ *hydrobromic acid* and → *hydrochloric acid*). Reacts violently with light metals (aluminum and magnesium). Attacks many metals when moist. TLV-TWA not available **Absorption route:** Can enter the body by inhalation or ingestion or through the skin. Harmful atmospheric concentrations can build up very rapidly on evaporation at 20°C. **Immediate effects:** Corrosive to the eyes, skin and respiratory tract. Inhalation of vapor/fumes can cause severe breathing difficulties (lung edema). Can cause damage to the gonads, halting or reducing production of sperm cells. **Effects of prolonged/repeated exposure:** Can cause liver and kidney damage. Has been found to cause a type of abdominal and breast cancer in certain animal species under certain circumstances.

HAZARDS/SYMPTOMS	PREVENTIVE MEASURES	FIRE EXTINGUISHING/FIRST AID
Fire: combustible.	keep away from open flame, no smoking.	powder, water spray, foam, carbon dioxide, (halons).
Explosion: above 77°C: forms explosive air-vapor mixtures.	above 77°C: sealed machinery, ventilation.	
Inhalation: *corrosive*; sore throat, cough, shortness of breath, nausea.	ventilation, local exhaust or respiratory protection.	fresh air, rest, take to hospital.
Skin: *is absorbed*; redness, pain.	gloves, protective clothing.	remove contaminated clothing, flush skin with water and wash thoroughly with soap and water.
Eyes: *corrosive*; redness, pain, impaired vision.	face shield, or combined eye and respiratory protection.	flush with water, take to a doctor if necessary.
Ingestion: sore throat, nausea.		rinse mouth, take immediately to hospital.

SPILLAGE	STORAGE	LABELING / NFPA
collect leakage in sealable containers, soak up with dry sand or other inert absorbent and remove to safe place; (additional individual protection: CHEMICAL SUIT).		R: 45-46-20/21-25-48 S: 53-44 Note E ☠ Toxic

NOTES
Depending on the degree of exposure, regular medical checkups are advisable. Lung edema symptoms usually develop several hours later and are aggravated by physical exertion: rest and hospitalization essential. As first aid, a doctor or authorized person should consider administering a corticosteroid spray. Dibromochloropropane is an insecticide.

Transport Emergency Card TEC(R)-61G06C

HI: 60; UN-number: 2872

CAS-No: [111-92-2]
N-butyl-1-butanamine
DBA
N-dibutylamine
di-N-butylamine
DNBA

$(C_4H_9)NH$

DIBUTYLAMINE

PHYSICAL PROPERTIES		IMPORTANT CHARACTERISTICS	
Boiling point, °C	159	**COLORLESS LIQUID WITH CHARACTERISTIC ODOR**	
Melting point, °C	−59	Vapor mixes readily with air. Decomposes when heated or burned, giving off toxic vapors	
Flash point, °C	42	(→ *nitrous vapors*). In aqueous solution is a strong base which reacts violently with acids and	
Auto-ignition temperature, °C	260	corrodes aluminum, zinc etc. Reacts violently with strong oxidants. Also reacts with many	
Relative density (water = 1)	0.8	organic compounds. Attacks copper, tin and their alloys.	
Relative vapor density (air = 1)	4.5		
Relative density at 20 °C of saturated mixture vapor/air (air = 1)	1.00	**TLV-TWA** not available	
Vapor pressure, mm Hg at 20 °C	2.0		
Solubility in water	poor	**Absorption route:** Can enter the body by inhalation or ingestion. Harmful atmospheric	
Explosive limits, vol% in air	1.1-?	concentrations can build up fairly rapidly on evaporation at approx. 20°C - even more rapidly in	
Relative molecular mass	129.3	aerosol form.	
Log P octanol/water	2.7	**Immediate effects:** Corrosive to the eyes, skin and respiratory tract. Inhalation can cause lung	
		edema. In serious cases risk of death.	
Gross formula:	$C_8H_{19}N$		

HAZARDS/SYMPTOMS	PREVENTIVE MEASURES	FIRE EXTINGUISHING/FIRST AID
Fire: flammable.	keep away from open flame and sparks, no smoking.	powder, AFFF, foam, carbon dioxide, (halons).
Explosion: above 42°C: forms explosive air-vapor mixtures.	above 42°C: sealed machinery, ventilation, explosion-proof electrical equipment.	in case of fire: keep tanks/drums cool by spraying with water.
	STRICT HYGIENE	
Inhalation: *corrosive*; sore throat, cough, shortness of breath, severe breathing difficulties.	ventilation, local exhaust or respiratory protection.	fresh air, rest, place in half-sitting position, take to hospital.
Skin: *corrosive*; redness, pain, serious burns.	gloves, protective clothing.	remove contaminated clothing, flush skin with water or shower, call a doctor or take directly to hospital.
Eyes: *corrosive*; redness, pain, impaired vision.	face shield.	flush with water, take to a doctor.
Ingestion: *corrosive*; sore throat, abdominal pain, diarrhea, nausea.		rinse mouth, take immediately to hospital.

SPILLAGE	STORAGE	LABELING / NFPA
collect leakage in sealable containers, soak up with sand or other inert absorbent and remove to safe place; (additional individual protection: breathing apparatus).	keep in fireproof place, separate from oxidants and strong acids.	R: 10-20/21/22 ✖ Harmful

NFPA: 3 / 2 / 0

NOTES
Lung edema symptoms usually develop several hours later and are aggravated by physical exertion: rest and hospitalization essential. As first aid, a doctor or authorized person should consider administering a corticosteroid spray.

Transport Emergency Card TEC(R)-80G15

HI: 83; UN-number: 2248

$C_6H_5N(C_4H_9)_2$

N,N-DIBUTYLANILINE

PHYSICAL PROPERTIES	IMPORTANT CHARACTERISTICS	
Boiling point, °C 275 Melting point, °C −32 Flash point, °C 110 Relative density (water = 1) 0.9 Relative vapor density (air = 1) 7.1 Relative density at 20°C of saturated mixture vapor/air (air = 1) 1.0 Vapor pressure, mm Hg at 30°C 0.076 Solubility in water poor Relative molecular mass 205.3 Gross formula: $C_{14}H_{23}N$	**LIGHT YELLOW VISCOUS LIQUID WITH CHARACTERISTIC ODOR** Vapor mixes readily with air. → *Nitrous gases* are a product of combustion. Decomposes when heated and on contact with acids, forming → *aniline* and → *nitrous vapors*. Reacts violently with strong oxidants.	
	TLV-TWA not available	
	Absorption route: Can enter the body by inhalation or ingestion or through the skin. Harmful atmospheric concentrations build up very slowly, if at all, on evaporation at approx. 20°C, but much more rapidly in aerosol form. **Immediate effects:** Irritates the eyes and respiratory tract. **Effects of prolonged/repeated exposure:** Prolonged or repeated contact can cause skin disorders. Can affect the blood. Can cause liver and kidney damage.	

HAZARDS/SYMPTOMS	PREVENTIVE MEASURES	FIRE EXTINGUISHING/FIRST AID
Fire: combustible.	keep away from open flame, no smoking.	powder, AFFF, foam, carbon dioxide, (halons).
	STRICT HYGIENE	
Inhalation: shortness of breath, headache, dizziness, blue skin, feeling of weakness, muscle weakness.	ventilation, local exhaust or respiratory protection.	fresh air, rest, place in half-sitting position, call a doctor.
Skin: *is absorbed*; redness; see also 'Inhalation'.	gloves, protective clothing.	remove contaminated clothing, flush skin with water or shower, call a doctor.
Eyes: redness, pain.	face shield.	flush with water, take to a doctor.
Ingestion: abdominal pain, headache, nausea, blue skin, muscle weakness.	do not eat, drink or smoke while working.	rinse mouth, immediately take to hospital.

SPILLAGE	STORAGE	LABELING / NFPA
collect leakage in sealable containers, soak up with sand or other inert absorbent and remove to safe place.	keep separate from oxidants and acids; ventilate at floor level.	1 / 3 / 0

NOTES

Depending on the degree of exposure, regular medical checkups are advisable. Effects on blood due to formation of methemoglobin. Special first aid required in the event of poisoning: antidotes must be available (with instructions).

Transport Emergency Card TEC(R)-61G06

butylhydroxytoluene
2,6-di-tert-butyl-p-cresol
2,6-di-tert-butyl-4-methylphenol

$C_6H_2(C_3H_9)_2(CH_3)OH$

DI-tert-BUTYL-4-METHYLPHENOL

PHYSICAL PROPERTIES		IMPORTANT CHARACTERISTICS	
Boiling point, °C	265	**WHITE CRYSTALS**	
Melting point, °C	70	Vapor mixes readily with air. Decomposes when heated, giving off toxic vapors. Reacts with strong oxidants.	
Flash point, °C	127		
Relative density (water = 1)	1.05		
Relative vapor density (air = 1)	7.6	TLV-TWA	not available
Relative density at 20 °C of saturated mixture			
vapor/air (air = 1)	1.0	**Absorption route:** Can enter the body by inhalation or ingestion. Harmful atmospheric concentrations build up very slowly, if at all, on evaporation at approx. 20° C, but harmful concentrations of airborne particles can build up much more rapidly.	
Vapor pressure, mm Hg at 20 °C	0.015		
Solubility in water	none		
Relative molecular mass	220.3		
Gross formula:	$C_{13}H_{24}O$		

HAZARDS/SYMPTOMS	PREVENTIVE MEASURES	FIRE EXTINGUISHING/FIRST AID
Fire: combustible.	keep away from open flame, no smoking.	water spray, powder; in case of fire in immediate vicinity: use any extinguishing agent.
Inhalation:	local exhaust or respiratory protection.	
Skin:	gloves.	remove contaminated clothing, flush skin with water or shower, call a doctor if necessary.
Eyes:	goggles.	flush with water, take to a doctor if necessary.
Ingestion:		rinse mouth, call a doctor.

SPILLAGE	STORAGE	LABELING / NFPA
clean up spillage, flush away remainder with water; (additional individual protection: P2 respirator).	keep separate from oxidants.	

NOTES
Insufficient toxicological data on harmful effects to humans. Topanol OC is a trade name.

bis(tert-butyl)peroxide
tert-butyl peroxide
bis(1,1-dimethylethyl)peroxide
DTBP

$(CH_3)_3COOC(CH_3)_3$

DI-tert-BUTYL PEROXIDE

PHYSICAL PROPERTIES	IMPORTANT CHARACTERISTICS
Boiling point, °C 111 Melting point, °C −40 Flash point, °C 12 Relative density (water = 1) 0.8 Relative vapor density (air = 1) 5.0 Relative density at 20 °C of saturated mixture vapor/air (air = 1) 1.1 Vapor pressure, mm Hg at 20 °C 20 Solubility in water poor Explosive limits, vol% in air ?-? Minimum ignition energy, mJ 0.5 Relative molecular mass 146.2 Gross formula: $C_8H_{18}O_2$	**COLORLESS LIQUID WITH CHARACTERISTIC ODOR** Vapor is heavier than air and spreads at ground level, with risk of ignition at a distance. Do not use compressed air when filling, emptying or processing. Decomposes when heated or on contact with other substances, with risk of fire and explosion. Strong oxidant which reacts violently with combustible substances and reducing agents. TLV-TWA not available **Absorption route:** Can enter the body by inhalation or ingestion. Harmful atmospheric concentrations can build up fairly rapidly on evaporation at approx. 20°C - even more rapidly in aerosol form. **Immediate effects:** Irritates the eyes, skin and respiratory tract.

HAZARDS/SYMPTOMS	PREVENTIVE MEASURES	FIRE EXTINGUISHING/FIRST AID
Fire: extremely flammable.	keep away from open flame and sparks, no smoking; avoid contact with hot surfaces (steam lines), prevent contamination.	large quantities of water, powder, AFFF, foam, carbon dioxide, (halons).
Explosion: forms explosive vapor-air mixtures, heating in sealed machinery can cause explosion.	ventilation, explosion-proof electrical equipment and lighting, non-sparking tools.	in case of fire: keep tanks/drums cool by spraying with water, fight fire from sheltered location.
Inhalation: sore throat, cough, shortness of breath.	ventilation, local exhaust or respiratory protection.	fresh air, rest, place in half-sitting position, call a doctor.
Skin: redness, pain.	gloves.	remove contaminated clothing, flush skin with water or shower.
Eyes: redness, pain.	face shield.	flush with water, take to a doctor.
Ingestion: abdominal cramps, vomiting.		rinse mouth, call a doctor.

SPILLAGE	STORAGE	LABELING / NFPA
collect leakage in non-sealable containers, soak up with sand or other inert absorbent and remove to safe place, do not use sawdust or other combustible absorbents, do not flush into sewer.	keep in cool, dark, fireproof place, separate from other substances than organic peroxides.	R: 11-37/38 S: 3/7/9-14-27-37/39 Oxidant Irritant 3 / 2 / 4 / oxy

NOTES
Explosive limits not given in the literature. Flush contaminated clothing with water (fire hazard). Packaging: special material. Store in original packaging; do not return product which has been taken out.

Transport Emergency Card TEC(R)-52G01

HI: 539; UN-number: 2102

CAS-No: [84-74-2]
DBP
dibutyl-1,2-benzenedicarboxylate
di-n-butyl phthalate

$C_6H_4(COOC_4H_9)_2$

DIBUTYL PHTHALATE

PHYSICAL PROPERTIES		IMPORTANT CHARACTERISTICS	
Boiling point, °C	340	**COLORLESS ODORLESS VISCOUS LIQUID**	
Melting point, °C	−35	Vapor mixes readily with air. Reacts violently with oxidants.	
Flash point, °C	157		
Auto-ignition temperature, °C	400	TLV-TWA	5 mg/m³
Relative density (water = 1)	1.05		
Relative vapor density (air = 1)	9.6	**Absorption route:** Can enter the body by inhalation or ingestion. Evaporation negligible at 20°C, but harmful atmospheric concentrations can build up rapidly in aerosol form.	
Relative density at 20°C of saturated mixture vapor/air (air = 1)	1.0		
Vapor pressure, mm Hg at 20°C	1.52×10^{-5}	**Immediate effects:** Irritates the eyes, skin and respiratory tract. Affects the nervous system.	
Solubility in water	none		
Explosive limits, vol% in air	see notes		
Relative molecular mass	278.3		
Gross formula:	$C_{16}H_{22}O_4$		

HAZARDS/SYMPTOMS	PREVENTIVE MEASURES	FIRE EXTINGUISHING/FIRST AID
Fire: combustible.	keep away from open flame, no smoking.	powder, AFFF, foam, carbon dioxide, (halons).
Inhalation: sore throat, cough, dizziness.	ventilation.	fresh air, rest, call a doctor if necessary.
Skin: redness, pain.	gloves.	remove contaminated clothing, flush skin with water or shower.
Eyes: redness, pain.	safety glasses.	flush with water, take to a doctor if necessary.
Ingestion: abdominal pain, nausea, dizziness.		rinse mouth, call a doctor.

SPILLAGE	STORAGE	LABELING / NFPA
collect leakage in sealable containers, soak up with sand or other inert absorbent and remove to safe place.	keep separate from oxidants.	

NOTES
Literature gives a lower explosive limit of 0.5%, but this is not reached until 235°C.

$(C_4H_9)_2Sn(OOCC_{11}H_{23})_2$

DIBUTYLTIN LAURATE

PHYSICAL PROPERTIES	IMPORTANT CHARACTERISTICS	
Melting point, °C 23 Flash point, °C 226 Relative density (water = 1) 1.05 Solubility in water none Relative molecular mass 631.6	**LIGHT YELLOW LIQUID OR COLORLESS CRYSTALS** Decomposes when heated or burned, giving off toxic vapors. Reacts with oxidants.	
	TLV-TWA (skin)	0.1 mg/m³ ✳
	Absorption route: Can enter the body by inhalation or ingestion or through the skin. Evaporation negligible at 20°C, but harmful atmospheric concentrations can build up rapidly in aerosol form. **Immediate effects:** Corrosive to the eyes, skin and respiratory tract. Affects the nervous system. Inhalation can cause lung edema. In serious cases risk of unconsciousness and death. **Effects of prolonged/repeated exposure:** Prolonged or repeated contact can cause skin disorders. Can cause liver damage.	
Gross formula: $C_{32}H_{64}O_4Sn$		

HAZARDS/SYMPTOMS	PREVENTIVE MEASURES	FIRE EXTINGUISHING/FIRST AID
Fire: combustible.	keep away from open flame, no smoking.	powder, AFFF, foam, carbon dioxide, (halons).
	STRICT HYGIENE	
Inhalation: *corrosive*; cough, shortness of breath, severe breathing difficulties, nausea.	ventilation, local exhaust or respiratory protection.	fresh air, rest, take to hospital.
Skin: *corrosive*; redness, pain.	gloves, protective clothing.	remove contaminated clothing, flush skin with water or shower, call a doctor.
Eyes: *corrosive*; redness, pain, impaired vision.	face shield.	flush with water, take to a doctor if necessary.
Ingestion: *corrosive*; abdominal pain, diarrhea, nausea, vomiting.	do not eat, drink or smoke while working.	rinse mouth, call a doctor or take to hospital.

SPILLAGE	STORAGE	LABELING / NFPA
collect leakage in sealable containers, soak up with sand or other inert absorbent and remove to safe place, carefully collect remainder; (additional individual protection: breathing apparatus).	ventilate at floor level.	

NOTES
Freezing point is 6°C (transition from liquid to solid phase by cooling). ✳ TLV equals that of tin. Lung edema symptoms usually develop several hours later and are aggravated by physical exertion: rest and hospitalization essential. As first aid, a doctor or authorized person should consider administering a corticosteroid spray.

DI-n-BUTYLTIN OXIDE

PHYSICAL PROPERTIES	IMPORTANT CHARACTERISTICS
Decomposes below melting point, °C ? Auto-ignition temperature, °C 279 Relative density (water = 1) 1.6 Solubility in water none Relative molecular mass see notes	**WHITE POWDER** Finely dispersed in air can cause dust explosion. In dry state can form electrostatic charge if stirred, transported pneumatically, poured etc. Decomposes when heated or burned, giving off toxic vapors of → tin/tin oxides. Reacts with oxidants.

TLV-TWA (skin)	0.1 mg/m³ ✶
STEL	0.2 mg/m³

Absorption route: Can enter the body by inhalation, through the skin or by ingestion. Evaporation negligible at 20°C, but harmful concentrations of airborne particles can build up rapidly.
Immediate effects: Corrosive to the eyes, skin and respiratory tract. Inhalation can cause lung edema. Can cause death. Keep under medical observation.
Effects of prolonged/repeated exposure: Prolonged or repeated contact can cause dermatitis. Can affect the central and peripheral nervous system, causing neurological disorders. Can cause paralysis, visual disorders and EEG disorders.

HAZARDS/SYMPTOMS	PREVENTIVE MEASURES	FIRE EXTINGUISHING/FIRST AID
Fire: combustible.	keep away from open flame, no smoking.	water spray, powder.
Explosion: finely dispersed particles form explosive mixtures on contact with air.	keep dust from accumulating, closed system, explosion-proof electrical equipment and lighting, prevent buildup of electrostatic charges by grounding, etc.	
	KEEP DUST UNDER CONTROL, STRICT HYGIENE	IN ALL CASES CALL IN A DOCTOR
Inhalation: *corrosive*; burning sensation, cough, headache, difficulty breathing, sore throat.	local exhaust or respiratory protection.	fresh air, rest, place in half-sitting position, take immediately to hospital.
Skin: *corrosive, is absorbed*; redness, burning sensation, pain.	gloves.	remove contaminated clothing, flush skin with water or shower, consult a doctor.
Eyes: *corrosive*; redness, pain, impaired vision.	face shield, or combined eye and respiratory protection.	flush thoroughly with water (remove contact lenses if easy), take to a doctor.
Ingestion: *corrosive*; abdominal cramps, burning sensation, nausea, sore throat.	do not eat, drink or smoke while working.	rinse mouth, immediately take to hospital.

SPILLAGE	STORAGE	LABELING / NFPA
clean up spillage, carefully collect remainder; (additional individual protection: P3 respirator).	keep separate from oxidants.	

NOTES
✶ TLV equals that of tin (organic compounds). Compound with quadrivalent tin occurs in semi-polymeric state, thus relative molecular mass is variable. This also determines stability of product. Decomposition temperature not given in the literature. Other data indicate loss of stability below 300°C. Unbreakable packaging preferred; if breakable, keep in unbreakable container. Lung edema symptoms usually develop several hours later and are aggravated by physical exertion: rest and hospitalization essential. As first aid, a doctor or authorized person should consider administering a suitable spray.

UN-number: 2788

bichloracetic acid
DCA
dichloracetic acid
dichlorethanoic acid
dichloroethanoic acid

$CHCl_2COOH$

DICHLOROACETIC ACID

PHYSICAL PROPERTIES	IMPORTANT CHARACTERISTICS
Decomposes below boiling point, °C 194 Melting point, °C 10 Relative density (water = 1) 1.6 Relative vapor density (air = 1) 4.4 Relative density at 20 °C of saturated mixture vapor/air (air = 1) 1.0 Vapor pressure, mm Hg at 44 °C 0.99 Solubility in water very good Relative molecular mass 129.0 Log P octanol/water 0.7	**COLORLESS LIQUID WITH PUNGENT ODOR** Vapor mixes readily with air. Decomposes when heated or burned, giving off toxic → *phosgene* and corrosive vapors (→ *hydrochloric acid*). Strong acid which reacts violently with bases and is corrosive.
	TLV-TWA not available
	Absorption route: Can enter the body by inhalation or ingestion or through the skin. Insufficient data on the rate at which harmful concentrations can build up. **Immediate effects:** Corrosive to the eyes, skin and respiratory tract. Inhalation of vapor/fumes can cause severe breathing difficulties (lung edema). In serious cases risk of death.
Gross formula: $C_2H_2Cl_2O_2$	

HAZARDS/SYMPTOMS	PREVENTIVE MEASURES	FIRE EXTINGUISHING/FIRST AID
Fire: non-combustible.		in case of fire in immediate vicinity: use any extinguishing agent.
Inhalation: *corrosive*; sore throat, cough, shortness of breath, severe breathing difficulties.	ventilation, local exhaust or respiratory protection.	fresh air, rest, place in half-sitting position, take to hospital.
Skin: *corrosive*; redness, pain, serious burns.	gloves, protective clothing.	remove contaminated clothing, flush skin with water or shower, call a doctor if necessary.
Eyes: *corrosive*; redness, pain, impaired vision.	face shield.	flush with water, take to a doctor.
Ingestion: *corrosive*; diarrhea, nausea, vomiting.		rinse mouth, take immediately to hospital.

SPILLAGE	STORAGE	LABELING / NFPA
collect leakage in sealable containers, render remainder harmless with soda and remove to safe place.	ventilate at floor level.	R: 35 S: 26 Corrosive

NOTES
Log P octanol/water is estimated. Lung edema symptoms usually develop several hours later and are aggravated by physical exertion: rest and hospitalization essential. As first aid, a doctor or authorized person should consider administering a corticosteroid spray. UNDER NO CIRCUMSTANCES add water to acid; when diluting ALWAYS add acid to water. Unbreakable packaging preferred; if breakable, keep in unbreakable container.

Transport Emergency Card TEC(R)-80G20B

HI: 80; UN-number: 1764

2,5-DICHLOROANILINE

PHYSICAL PROPERTIES	IMPORTANT CHARACTERISTICS	
Boiling point, °C .... 251 Melting point, °C .... 50 Flash point, °C .... 166 Relative vapor density (air = 1) .... 5.6 Relative density at 20 °C of saturated mixture vapor/air (air = 1) .... 1.0 Vapor pressure, mm Hg at 20 °C .... <0.076 Solubility in water .... poor Relative molecular mass 162.0	**COLORLESS CRYSTALLINE NEEDLES** Vapor mixes readily with air. Decomposes when heated or burned, giving off corrosive and toxic vapors (→ *hydrochloric acid* and → *nitrous vapors*). Reacts with strong oxidants.	
	TLV-TWA not available	
	Absorption route: Can enter the body by inhalation or ingestion or through the skin. Harmful atmospheric concentrations build up very slowly, if at all, on evaporation at approx. 20°C, but harmful concentrations of airborne particles can build up much more rapidly. **Immediate effects:** In serious cases risk of death. **Effects of prolonged/repeated exposure:** Prolonged or repeated contact can cause skin disorders. Can affect the blood. Can cause liver and kidney damage.	
Gross formula: $C_6H_5Cl_2N$		

HAZARDS/SYMPTOMS	PREVENTIVE MEASURES	FIRE EXTINGUISHING/FIRST AID
Fire: combustible.	keep away from open flame, no smoking.	water spray, powder.
	KEEP DUST UNDER CONTROL	
Inhalation: shortness of breath, headache, dizziness, blue skin, feeling of weakness.	local exhaust or respiratory protection.	fresh air, rest, call a doctor.
Skin: *is absorbed*; redness; see also 'Inhalation'.	gloves, protective clothing.	remove contaminated clothing, wash skin with soap and water, call a doctor.
Eyes: redness, pain.	safety glasses.	flush with water, take to a doctor if necessary.
Ingestion: abdominal pain, nausea; see also 'Inhalation'.	do not eat, drink or smoke while working.	rinse mouth, take immediately to hospital.

SPILLAGE	STORAGE	LABELING / NFPA
clean up spillage, carefully collect remainder; (additional individual protection: P2 respirator).	ventilate at floor level.	

NOTES
Depending on the degree of exposure, regular medical checkups are advisable. Effects on blood due to formation of methemoglobin. Special first aid required in the event of poisoning: antidotes must be available (with instructions).

3,4-DICHLOROANILINE

PHYSICAL PROPERTIES	IMPORTANT CHARACTERISTICS
Boiling point, °C 272 Melting point, °C 72 Flash point, °C 166 Relative vapor density (air = 1) 5.6 Relative density at 20 °C of saturated mixture vapor/air (air = 1) 1.0 Vapor pressure, mm Hg at 20 °C ca. 0.076 Solubility in water poor Relative molecular mass 162.0 Log P octanol/water 2.7	**COLORLESS CRYSTALS** Vapor mixes readily with air. Decomposes when heated, giving off toxic vapors. TLV-TWA not available **Absorption route:** Can enter the body by inhalation or ingestion or through the skin. Insufficient data on the rate at which harmful concentrations can build up. **Immediate effects:** In serious cases risk of death. **Effects of prolonged/repeated exposure:** Prolonged or repeated contact can cause skin disorders. Can affect the blood. Can cause liver and kidney damage.
Gross formula: $C_6H_5Cl_2N$	

HAZARDS/SYMPTOMS	PREVENTIVE MEASURES	FIRE EXTINGUISHING/FIRST AID
Fire: combustible.	keep away from open flame, no smoking.	AFFF, water spray, powder.
	KEEP DUST UNDER CONTROL	
Inhalation: shortness of breath, headache, dizziness, blue skin, feeling of weakness.	local exhaust or respiratory protection.	fresh air, rest, call a doctor.
Skin: *is absorbed*; redness; see also 'Inhalation'.	gloves, protective clothing.	remove contaminated clothing, wash skin with water and soap, call a doctor, take to hospital.
Eyes: redness, pain.	safety glasses.	flush with water, take to a doctor if necessary.
Ingestion: abdominal pain, nausea; see also 'Inhalation'.	do not eat, drink or smoke while working.	rinse mouth, immediately take to hospital.

SPILLAGE	STORAGE	LABELING / NFPA
clean up spillage, carefully collect remainder; (additional individual protection: P2 respirator).	ventilate at floor level.	◇ 1 / 3 / 0

NOTES
Depending on the degree of exposure, regular medical checkups are advisable. Special first aid required in the event of poisoning: antidotes must be available (with instructions). Effects on blood due to formation of methemoglobin.

Transport Emergency Card TEC(R)-837

HI: 60; UN-number: 1590

p-DICHLOROBENZENE

PHYSICAL PROPERTIES		IMPORTANT CHARACTERISTICS	
Boiling point, °C	174	**WHITE CRYSTALS WITH CHARACTERISTIC ODOR**	
Melting point, °C	53	Vapor mixes readily with air. Decomposes when heated, giving off toxic and corrosive gases.	
Flash point, °C	66	Reacts violently with alkali metals and metal powders.	
Auto-ignition temperature, °C	>500		
Relative density (water = 1)	1.3		
Relative vapor density (air = 1)	5.1		
Relative density at 20 °C of saturated mixture			
vapor/air (air = 1)	1.01		
Vapor pressure, mm Hg at 20 °C	1.3		
Solubility in water	none		
Relative molecular mass	147.0		
Log P octanol/water	3.4		

	TLV-TWA	75 ppm	451 mg/m³
	STEL	110 ppm	661 mg/m³

Absorption route: Can enter the body by inhalation or ingestion. Harmful atmospheric concentrations build up fairly slowly on evaporation, but harmful concentrations of airborne particles can build up much more rapidly.
Immediate effects: Irritates the eyes, skin and respiratory tract.
Effects of prolonged/repeated exposure: Can cause liver and kidney damage. Destroys the skin's natural oils.

Gross formula: $C_6H_4Cl_2$

HAZARDS/SYMPTOMS	PREVENTIVE MEASURES	FIRE EXTINGUISHING/FIRST AID
Fire: combustible.	keep away from open flame, no smoking.	powder, water spray, foam, carbon dioxide, (halons).
Explosion: above 66°C: forms explosive air-vapor mixtures.	above 66°C: sealed machinery, ventilation.	
	KEEP DUST UNDER CONTROL	
Inhalation: cough, restlessness, feeling of weakness, trembling.	ventilation, local exhaust or respiratory protection.	fresh air, rest, call a doctor.
Skin: redness, pain.	gloves.	remove contaminated clothing, flush skin with water or shower.
Eyes: redness, burning sensation.	safety glasses.	flush with water, take to doctor.
Ingestion: abdominal pain, vomiting; see also 'Inhalation'.		rinse mouth, take to hospital.

SPILLAGE	STORAGE	LABELING / NFPA
clean up spillage, carefully collect remainder; (additional individual protection: respirator with A/P2 filter).	ventilate at floor level.	R: 22 S: 2-24/25 ✖ Harmful

NOTES

Depending on the degree of exposure, regular medical checkups are advisable.

HI: 60; UN-number: 1592

CAS-No: [75-71-8]
fluorocarbon 12
freon 12
propellant 12
refrigerant 12
R 12

CCl_2F_2

DICHLORODIFLUROMETHANE
(cylinder)

PHYSICAL PROPERTIES	IMPORTANT CHARACTERISTICS
Boiling point, °C − 30 Melting point, °C − 158 Relative density (water = 1) 1.5 Vapor pressure, mm Hg at 20 °C 4332 Solubility in water, g/100 ml at 20 °C 0.03 Relative molecular mass 120.9 Gross formula: CCl_2F_2	**COLORLESS COMPRESSED LIQUEFIED GAS** Gas is heavier than air and can accumulate close to ground level, causing oxygen deficiency, with risk of unconsciousness. Decomposes in flame or on hot surface, giving off toxic and corrosive gas. Decomposes slowly on contact with water. Reacts violently with molten aluminum and with magnesium, with risk of fire and explosion. Attacks zinc and magnesium and its alloys. TLV-TWA 1000 ppm 4950 mg/m³ **Absorption route:** Can enter the body by inhalation. Harmful atmospheric concentrations can build up very rapidly if gas is released. **Immediate effects:** Liquid can cause frostbite due to rapid evaporation. Affects the nervous system.

HAZARDS/SYMPTOMS	PREVENTIVE MEASURES	FIRE EXTINGUISHING/FIRST AID
Fire: non-combustible.		in case of fire in immediate vicinity: use any extinguishing agent.
Explosion:		in case of fire: keep cylinders cool by spraying with water.
Inhalation: sleepiness, ringing in the ears.	ventilation, local exhaust or respiratory protection.	fresh air, rest.
Skin: *in case of frostbite*: redness, pain, sores.	insulating gloves.	*in case of frostbite*: DO NOT remove clothing, flush skin with water or shower, refer to a doctor.
Eyes: *in case of frostbite*: redness, pain, impaired vision.	acid goggles, or combined eye and respiratory protection.	*in case of frostbite*: flush with water, take to a doctor.

SPILLAGE	STORAGE	LABELING / NFPA
ventilate, under no circumstances spray liquid with water; (additional individual protection: breathing apparatus).	if stored indoors, keep in cool, fireproof place.	

NOTES
High atmospheric concentrations, e.g. in poorly ventilated spaces, can cause oxygen deficiency, with risk of unconsciousness. Freon 12, Frigen 12 and Arcton 12 are trade names. Under no circumstances use near flame or hot surface or when welding. Turn leaking cylinder so that leak is on top to prevent liquid dichlorodifluoromethane escaping.

Transport Emergency Card TEC(R)-20G08   **HI: 20; UN-number: 1028**

CAS-No: [542-88-1]
BCME
bis-chloromethyl ether
chloro(chloromethoxy)methane
chloromethyl ether

$(CH_2Cl)_2O$

DICHLORODIMETHYL ETHER

PHYSICAL PROPERTIES		IMPORTANT CHARACTERISTICS	
Boiling point, °C	104	**COLORLESS LIQUID WITH PUNGENT ODOR**	
Melting point, °C	−42	Vapor mixes readily with air, forming explosive mixtures. Presumed to be able to form peroxides. Do not use compressed air when filling, emptying or processing. Decomposes when heated or burned and on contact with water or moist air, giving off corrosive vapors (→ *hydrochloric acid*). Attacks many plastics.	
Flash point, °C	< 19		
Relative density (water = 1)	1.3		
Relative vapor density (air = 1)	4.0		
Relative density at 20 °C of saturated mixture vapor/air (air = 1)	1.0	TLV-TWA	0.001 ppm
Vapor pressure, mm Hg at 20 °C	0.11	**Absorption route:** Can enter the body by inhalation or ingestion or through the skin. Harmful atmospheric concentrations can build up very rapidly on evaporation at 20°C. **Immediate effects:** Corrosive to the eyes, skin and respiratory tract. Affects the nervous system. Inhalation of vapor/fumes can cause severe breathing difficulties (lung edema). In serious cases risk of unconsciousness and death. **Effects of prolonged/repeated exposure:** Can cause liver and kidney damage. Has been found to cause a type of lung cancer in humans under certain circumstances.	
Solubility in water	reaction		
Explosive limits, vol% in air	see notes		
Relative molecular mass	115		
Gross formula:	$C_2H_4Cl_2O$		

HAZARDS/SYMPTOMS	PREVENTIVE MEASURES	FIRE EXTINGUISHING/FIRST AID
Fire: extremely flammable.	keep away from open flame and sparks, no smoking.	powder, water spray, foam, carbon dioxide, (halons).
Explosion: forms explosive air-vapor mixtures.	sealed machinery, ventilation, explosion-proof electrical equipment and lighting.	in case of fire: keep tanks/drums cool by spraying with water, but DO NOT spray substance with water.
	STRICT HYGIENE, AVOID ALL CONTACT	IN ALL CASES CALL IN A DOCTOR
Inhalation: *corrosive*; sore throat, headache, dizziness, nausea, severe breathing difficulties.	local exhaust or respiratory protection.	fresh air, rest, place in half-sitting position, take to hospital.
Skin: *corrosive*; redness, pain.	gloves, protective clothing.	remove contaminated clothing, flush skin with water or shower, refer to a doctor.
Eyes: *corrosive*; redness, pain, impaired vision.	face shield, or combined eye and respiratory protection.	flush with water, take to a doctor.
Ingestion: *corrosive*; sore throat, abdominal pain, vomiting.		rinse mouth, immediately take to hospital.

SPILLAGE	STORAGE	LABELING / NFPA
evacuate area, call in an expert, collect leakage in sealable containers, soak up with sand or other inert absorbent and remove to safe place; (additional individual protection: CHEMICAL SUIT).	keep in fireproof, dry and dark place.	R: 45-10-22-24-26 S: 53/45 Note E ☠ Highly toxic

NOTES
Explosive limits not given in the literature. Before distilling check for peroxides; if found, render harmless. Odor limit is above TLV. Lung edema symptoms usually develop several hours later and are aggravated by physical exertion: rest and hospitalization essential. As first aid, a doctor or authorized person should consider administering a corticosteroid spray. Use airtight packaging. Unbreakable packaging preferred; if breakable, keep in unbreakable container.

UN-number: 2249

CAS-No: [75-34-3]
as-dichloroethane
ethylidene chloride
ethylidene dichloride
1,1-ethylidene dichloride

CH_3CHCl_2

1,1-DICHLOROETHANE

PHYSICAL PROPERTIES		IMPORTANT CHARACTERISTICS	
Boiling point, °C	57	**COLORLESS LIQUID WITH CHARACTERISTIC ODOR**	
Melting point, °C	−97	Vapor is heavier than air and spreads at ground level, with risk of ignition at a distance.	
Flash point, °C	−9	Decomposes when heated or burned, giving off corrosive vapors (→ *hydrochloric acid*). Reacts	
Auto-ignition temperature, °C	660	violently with alkaline-earth and alkali metals, with risk of fire and explosion. Reacts violently with	
Relative density (water = 1)	1.2	strong oxidants. Attacks many plastics.	
Relative vapor density (air = 1)	3.4		
Relative density at 20 °C of saturated mixture		TLV-TWA 200 ppm 810 mg/m³	
vapor/air (air = 1)	1.6	STEL 250 ppm 1010 mg/m³	
Vapor pressure, mm Hg at 20 °C	185		
Solubility in water, g/100 ml	0.5	**Absorption route:** Can enter the body by inhalation. Harmful atmospheric concentrations can	
Explosive limits, vol% in air	5.6-16	build up fairly rapidly on evaporation at approx. 20°C - even more rapidly in aerosol form.	
Electrical conductivity, pS/m	2x10⁵	**Immediate effects:** The vapor of the substance irritates the eyes and respiratory tract. Can affect	
Relative molecular mass	99.0	the central nervous system. Exposure to high concentrations can cause unconsciousness.	
		Effects of prolonged/repeated exposure: Prolonged or repeated contact can cause dermatitis.	
		Can affect the liver and kidneys.	
Gross formula:	$C_2H_4Cl_2$		

HAZARDS/SYMPTOMS	PREVENTIVE MEASURES	FIRE EXTINGUISHING/FIRST AID
Fire: extremely flammable.	keep away from open flame and sparks, no smoking.	powder, water spray, foam, carbon dioxide, (halons).
Explosion: forms explosive air-vapor mixtures.	sealed machinery, ventilation, explosion-proof electrical equipment and lighting; do not use compressed air when filling, emptying or processing.	in case of fire: keep tanks/drums cool by spraying with water.
	PREVENT MIST FORMATION	
Inhalation: irritation, dizziness, sleepiness, drowsiness, unconsciousness.	ventilation, local exhaust or respiratory protection.	fresh air, rest, place in half-sitting position, refer to a doctor.
Skin: *is absorbed*; redness.	gloves.	remove contaminated clothing, flush skin with water and wash with soap and water.
Eyes: irritation, redness, impaired vision.	safety glasses.	flush thoroughly with water (remove contact lenses if easy), take to a doctor.
Ingestion: burning sensation. cough, dizziness, nausea, unconsciousness.		rinse mouth, call a doctor.

SPILLAGE	STORAGE	LABELING / NFPA
collect leakage in sealable containers, soak up with sand or other inert absorbent and remove to safe place; (additional individual protection: breathing apparatus).	keep in fireproof place, separate from oxidants.	Highly flammable Harmful

NOTES
Odor gives inadequate warning of exceeded TLV. Under no circumstances use near flame or hot surface or when welding. Ethylene dichloride and ethylene chloride are synonyms for 1,2-dichloroethane!

Transport Emergency Card TEC(R)-30G34

HI: 336; UN-number: 2362

ethane dichloride
ethylene chloride
ethylene dichloride
1,2-ethylene dichloride

$ClCH_2CH_2Cl$

1,2-DICHLOROETHANE

PHYSICAL PROPERTIES		IMPORTANT CHARACTERISTICS	
Boiling point, °C	84	**COLORLESS LIQUID WITH CHARACTERISTIC ODOR**	
Melting point, °C	−36	Vapor is heavier than air and spreads at ground level, with risk of ignition at a distance. Flow, agitation etc. can cause build-up of electrostatic charge due to liquid's low conductivity. Do not use compressed air when filling,emptying or processing. Toxic and corrosive vapors are a product of combustion (→ *phosgene* and → *hydrochloric acid*). Decomposes on contact with steam, forming → *hydrochloric acid*. Reacts violently with alkali amides and various metal powders with risk of fire and explosion. Reacts violently with strong oxidants. Attacks many plastics.	
Flash point, °C	13		
Auto-ignition temperature, °C	413		
Relative density (water = 1)	1.3		
Relative vapor density (air = 1)	3.4		
Relative density at 20 °C of saturated mixture vapor/air (air = 1)	1.2		
Vapor pressure, mm Hg at 20 °C	66.1		
Solubility in water, g/100 ml at 20 °C	0.8	TLV-TWA	10 ppm	40 mg/m³
Explosive limits, vol% in air	6.2-16		
Electrical conductivity, pS/m	4x10³	**Absorption route:** Can enter the body by inhalation or ingestion or through the skin. Harmful atmospheric concentrations can build up fairly rapidly on evaporation at approx. 20°C - even more rapidly in aerosol form.	
Relative molecular mass	99		
		Immediate effects: Vapors irritate the respiratory tract and skin. Liquid destroys the skin's natural oils. Liquid is corrosive to the eyes and irritates the skin and respiratory tract. Affects the nervous system. In serious cases risk of unconsciousness and death.	
Gross formula:	C₂H₄Cl₂	**Effects of prolonged/repeated exposure:** Can cause liver and kidney damage. Has been found to cause a type of cancer in certain animal species under certain circumstances.	

HAZARDS/SYMPTOMS	PREVENTIVE MEASURES	FIRE EXTINGUISHING/FIRST AID
Fire: extremely flammable.	keep away from open flame and sparks, no smoking.	powder, water spray, foam, carbon dioxide, (halons).
Explosion: forms explosive air-vapor mixtures.	sealed machinery, ventilation, explosion-proof electrical equipment and lighting, grounding.	in case of fire: keep tanks/drums cool by spraying with water.
	STRICT HYGIENE	
Inhalation: irritation, dizziness, nausea, drowsiness.	ventilation, local exhaust or respiratory protection.	fresh air, rest, refer to a doctor.
Skin: *is absorbed*; redness.	gloves, protective clothing.	remove contaminated clothing, wash skin with soap and water, refer to a doctor.
Eyes: redness, pain, impaired vision.	acid goggles.	flush with water, take to a doctor.
Ingestion: abdominal cramps, dizziness, vomiting.		rinse mouth, take immediately to hospital.

SPILLAGE	STORAGE	LABELING / NFPA
collect leakage in sealable containers, soak up with sand or other inert absorbent and remove to safe place; (additional individual protection: breathing apparatus).	keep in fireproof place, separate from oxidants.	R: 45-11-22-36/37/38 S: 53-16-29-44 Note E 🔥 Flammable ☠ Toxic NFPA: 2 3 0

NOTES
Alcohol consumption increases toxic effects. Depending on the degree of exposure, regular medical checkups are advisable. Vapor can attack the cornea.

Transport Emergency Card TEC(R)-605

HI: 33; UN-number: 1184

CAS-No: [75-35-4]
1,1-DCE
1,1-dichloroethylene
vinylidene chloride
vinylidene dichloride
vinylidine chloride

$CH_2 = CCl_2$

1,1-DICHLOROETHENE

PHYSICAL PROPERTIES		IMPORTANT CHARACTERISTICS
Boiling point, °C	32	**COLORLESS HIGHLY VOLATILE LIQUID WITH CHARACTERISTIC ODOR**
Melting point, °C	−122	Vapor is heavier than air and spreads at ground level, with risk of ignition at a distance. Readily
Flash point, °C	−28	able to form peroxides. Able to polymerize when moderately heated. Toxic (→ *phosgene*) and
Auto-ignition temperature, °C	530	corrosive vapors (→ *hydrochloric acid*) are a product of combustion. Reacts violently with
Relative density (water = 1)	1.2	alkaline-earth and alkali metals and metal powders, with risk of fire and explosion. Reacts
Relative vapor density (air = 1)	3.35	violently with strong oxidants.

Relative density at 20 °C of saturated mixture			
vapor/air (air = 1)	2.5	TLV-TWA	5 ppm	20 mg/m³
Vapor pressure, mm Hg at 20 °C	505	STEL	20 ppm	79 mg/m³

Solubility in water, g/100 ml at 25 °C	0.25	**Absorption route:** Can enter the body by inhalation or ingestion. Harmful atmospheric
Explosive limits, vol% in air	6.5-15.5	concentrations can build up very rapidly on evaporation at 20°C.
Relative molecular mass	96.9	**Immediate effects:** Irritates the eyes, skin and respiratory tract. Affects the nervous system.
		Inhalation of vapor/fumes can cause severe breathing difficulties (lung edema).
		Effects of prolonged/repeated exposure: Can cause liver and kidney damage.

Gross formula:	$C_2H_2Cl_2$

HAZARDS/SYMPTOMS	PREVENTIVE MEASURES	FIRE EXTINGUISHING/FIRST AID
Fire: extremely flammable.	keep away from open flame and sparks, no smoking.	powder, water spray, foam, carbon dioxide, (halons).
Explosion: forms explosive air-vapor mixtures.	sealed machinery, ventilation, explosion-proof electrical equipment and lighting.	in case of fire: keep tanks/drums cool by spraying with water.
	STRICT HYGIENE	IN ALL CASES CALL IN A DOCTOR
Inhalation: cough, dizziness, drowsiness, unconsciousness.	ventilation, local exhaust or respiratory protection.	fresh air, rest, place in half-sitting position, call a doctor.
Skin: redness, pain.	gloves.	remove contaminated clothing, flush skin with water or shower.
Eyes: redness, pain.	safety glasses.	flush with water, take to a doctor.
Ingestion: cough, dizziness, drowsiness, unconsciousness.		rinse mouth, take immediately to hospital.

SPILLAGE	STORAGE	LABELING / NFPA
evacuate area, call in an expert, soak up with sand or other inert absorbent and remove to safe place; (additional individual protection: breathing apparatus).	keep in cool, dark, fireproof place, separate from oxidants; add an inhibitor.	R: 12-20-40 S: 7-16-29 Note D Flammable Harmful

NOTES

Before distilling check for peroxides; if found, render harmless. Stabilized by addition of inhibitor; even moderate heating can render inhibitor ineffective, with risk of polymerization. Sometimes referred to incorrectly as 1,1-dichloroethane. Odor limit is above TLV. Alcohol consumption increases toxic effects. Depending on the degree of exposure, regular medical checkups are advisable. Lung edema symptoms usually develop several hours later and are aggravated by physical exertion: rest and hospitalization essential. As first aid, a doctor or authorized person should consider administering a corticosteroid spray. Unbreakable packaging preferred; if breakable, keep in unbreakable container.

Transport Emergency Card TEC(R)-641

HI: 339; UN-number: 1303

CAS-No: [111-44-4]
bis(2-chloroethyl)ether
2,2'-dichlorodiethyl ether
dichloroethyl oxide

$(H_2ClCCH_2)_2O$

sym-DICHLOROETHYL ETHER

PHYSICAL PROPERTIES		IMPORTANT CHARACTERISTICS	
Boiling point, °C	178	**COLORLESS LIQUID WITH PUNGENT ODOR**	
Melting point, °C	−52	Vapor mixes readily with air. Presumed to be able to form peroxides. Decomposes when heated	
Flash point, °C	55	or on contact with water or steam, giving off corrosive vapors (→ *hydrochloric acid*). Reacts	
Auto-ignition temperature, °C	365	violently with various metal powders and oxidants, with risk of fire and explosion. Attacks	
Relative density (water = 1)	1.2	aluminum.	
Relative vapor density (air = 1)	4.9		
Relative density at 20 °C of saturated mixture		TLV-TWA (skin)	5 ppm	29 mg/m³
vapor/air (air = 1)	1.0	STEL	10 ppm	58 mg/m³
Vapor pressure, mm Hg at 20 °C	0.84		
Solubility in water, g/100 ml	1.1	**Absorption route:** Can enter the body by inhalation or ingestion or through the skin. Harmful	
Explosive limits, vol% in air	0.8-?	atmospheric concentrations build up fairly slowly on evaporation, but much more rapidly in	
Relative molecular mass	143.0	aerosol form.	
		Immediate effects: Irritates the skin and respiratory tract and corrosive to the eyes. Liquid	
		destroys the skin's natural oils. Affects the nervous system.	
		Effects of prolonged/repeated exposure: Can cause liver and kidney damage.	
Gross formula:	$C_4H_8Cl_2O$		

HAZARDS/SYMPTOMS	PREVENTIVE MEASURES	FIRE EXTINGUISHING/FIRST AID
Fire: combustible.	keep away from open flame, no smoking.	powder, water spray, foam, carbon dioxide, (halons).
Explosion: above 55°C: forms explosive air-vapor mixtures.	above 55°C: sealed machinery, ventilation.	
Inhalation: cough, headache, drowsiness, sleepiness.	ventilation, local exhaust or respiratory protection.	fresh air, rest, call a doctor.
Skin: *is absorbed*; redness.	gloves, protective clothing.	remove contaminated clothing, flush skin with water or shower, call a doctor.
Eyes: redness, pain, impaired vision.	face shield.	flush with water, take to a doctor.
Ingestion: headache, vomiting, drowsiness, stomachache.		rinse mouth, immediately take to hospital.

SPILLAGE	STORAGE	LABELING / NFPA
collect leakage in sealable containers, soak up with sand or other inert absorbent and remove to safe place.	keep in fireproof place, separate from oxidants.	R: 10-26/27/28-40 S: 7/9-27-38-45 Toxic

NOTES
Before distilling check for peroxides; if found, render harmless. Alcohol consumption increases toxic effects. Depending on the degree of exposure, regular medical checkups are advisable. Has been found to cause a type of lung cancer in mice under certain circumstances. The measures on this card also apply to dichloropropyl ether (flash point 85°C). 2,2'-dichloroethyl ether is an insecticide.

Transport Emergency Card TEC(R)-61G05

HI: 63; UN-number: 1916

DICHLOROMONOFLUOROMETHANE
(cylinder)

PHYSICAL PROPERTIES	IMPORTANT CHARACTERISTICS
Boiling point, °C 9 Melting point, °C − 135 Auto-ignition temperature, °C 550 Relative density (water = 1) 1.4 Relative vapor density (air = 1) 3.6 Vapor pressure, mm Hg at 20 °C 1140 Solubility in water, g/100 ml at 25 °C 0.95 Relative molecular mass 102.9	**COLORLESS COMPRESSED LIQUEFIED GAS** Gas is heavier than air. Decomposes in flame or on hot surface, giving off toxic and corrosive vapors. Decomposes very slowly on contact with water. Reacts violently with liquid aluminum, with risk of fire and explosion. Attacks zinc, magnesium and its alloys.

TLV-TWA	10 ppm	42 mg/m³

Absorption route: Can enter the body by inhalation. Harmful atmospheric concentrations can build up very rapidly if gas is released.
Immediate effects: Liquid can cause frostbite due to rapid evaporation.

Gross formula: CHCl$_2$F

HAZARDS/SYMPTOMS	PREVENTIVE MEASURES	FIRE EXTINGUISHING/FIRST AID
Fire: non-combustible.		in case of fire in immediate vicinity: use any extinguishing agent.
Explosion:		in case of fire: keep cylinders cool by spraying with water.
Inhalation: sleepiness, ringing in the ears.	ventilation, local exhaust or respiratory protection.	fresh air, rest.
Skin: *in case of frostbite*: redness, pain, sores.	insulating gloves.	*in case of frostbite*: DO NOT remove clothing, flush skin with water or shower, refer to a doctor.
Eyes: *in case of frostbite*: redness, pain, impaired vision.	acid goggles, or combined eye and respiratory protection.	*in case of frostbite*: flush with water, take to a doctor.

SPILLAGE	STORAGE	LABELING / NFPA
ventilate, under no circumstances spray liquid with water; (additional individual protection: breathing apparatus).	if stored indoors, keep in cool, fireproof place.	

NOTES

Under certain circumstances can form combustible air-vapor mixtures, which are difficult to ignite. High atmospheric concentrations, e.g. in poorly ventilated spaces, can cause oxygen deficiency, with risk of unconsciousness. Freon 21, Frigen 21, Arcton 21, etc., are trade names. Under no circumstances use near flame or hot surface or when welding. Turn leaking cylinder so that leak is on top to prevent liquid dichloromonofluoromethane escaping.

Transport Emergency Card TEC(R)-20G08

HI: 20; UN-number: 1029

2,4-DICHLOROPHENOL

PHYSICAL PROPERTIES		IMPORTANT CHARACTERISTICS
Boiling point, °C	210	**COLORLESS TO LIGHT YELLOW CRYSTALS**
Melting point, °C	45	Vapor mixes readily with air. Decomposes when heated, giving off toxic vapors. Reacts violently
Flash point, °C	114	with oxidants and strong acids.
Relative density (water = 1)	1.4	
Relative vapor density (air = 1)	5.6	TLV-TWA not available
Vapor pressure, mm Hg at 20 °C	0.076	
Solubility in water	poor	**Absorption route:** Can enter the body by inhalation or ingestion. Harmful atmospheric
Relative molecular mass	163	concentrations build up very slowly, if at all, on evaporation at approx. 20°C, but much more
		rapidly in aerosol form.
		Immediate effects: Irritates the eyes, skin and respiratory tract. Liquid destroys the skin's natural
		oils. Affects the nervous system.
		Effects of prolonged/repeated exposure: Prolonged or repeated contact can cause skin
		disorders. Can cause liver and kidney damage.
Gross formula: $C_6H_4Cl_2O$		

HAZARDS/SYMPTOMS	PREVENTIVE MEASURES	FIRE EXTINGUISHING/FIRST AID
Fire: combustible.	keep away from open flame, no smoking.	powder, water spray, foam, carbon dioxide, (halons).
Explosion: finely dispersed particles form explosive mixtures on contact with air.		
	KEEP DUST UNDER CONTROL	
Inhalation: sore throat, cough, tiredness, feeling of weakness, sweating, thirst, muscle cramps, raised temperature.	local exhaust or respiratory protection.	fresh air, rest, cool down, take to hospital.
Skin: redness, pain.	gloves.	remove contaminated clothing, wash skin with soap and water, refer to a doctor.
Eyes: redness, pain.	face shield.	flush with water, take to a doctor if necessary.
Ingestion: abdominal pain, diarrhea, vomiting; see also 'Inhalation'.		rinse mouth, take immediately to hospital.

SPILLAGE	STORAGE	LABELING / NFPA
clean up spillage, carefully collect remainder; (additional individual protection: P2 respirator).		R: 22-36/38 S: 26-28 ✖ Harmful

NOTES
Alcohol consumption increases toxic effects.

$Cl_2C_6H_3OCH_2COOH$

2,4-DICHLOROPHENOXYACETIC ACID

PHYSICAL PROPERTIES	IMPORTANT CHARACTERISTICS
Decomposes below boiling point, °C 180 Melting point, °C 138 Relative density (water = 1) 1.56 Relative vapor density (air = 1) 7.6 Relative density at 20 °C of saturated mixture vapor/air (air = 1) 1.0 Vapor pressure, mm Hg at 20 °C <0.76 Solubility in water none Relative molecular mass 221.0 Log P octanol/water 2.8	**WHITE TO YELLOW POWDER** Corrosive gases (→ *hydrochloric acid*) are a product of combustion. Decomposes when heated above 180°C, giving off toxic and corrosive gases. Can decompose slowly when exposed to direct sunlight. Reacts with strong oxidants.
	TLV-TWA 10 mg/m³
	Absorption route: Can enter the body by inhalation or ingestion. Evaporation negligible at 20°C, but harmful concentrations of airborne particles can build up rapidly. **Immediate effects:** Irritates the eyes, skin and respiratory tract. Affects carbohydrate metabolism and nervous system. Can cause heart rate disorders. **Effects of prolonged/repeated exposure:** Prolonged or repeated contact can cause skin disorders. Can cause liver and kidney damage.
Gross formula: $C_8H_6Cl_2O_3$	

HAZARDS/SYMPTOMS	PREVENTIVE MEASURES	FIRE EXTINGUISHING/FIRST AID
Fire: hardly combustible.		in case of fire in immediate vicinity: use any extinguishing agent.
Explosion:		in case of fire: keep tanks/drums cool by spraying with water.
	KEEP DUST UNDER CONTROL	
Inhalation: sore throat, cough, drowsiness, muscle weakness, muscle cramps.	local exhaust or respiratory protection.	fresh air, rest, call a doctor.
Skin: redness.	gloves, protective clothing.	remove contaminated clothing, flush skin with water or shower.
Eyes: redness, pain.	goggles, or combined eye and respiratory protection.	flush with water, take to a doctor.
Ingestion: abdominal pain, nausea; see also 'Inhalation'.		rinse mouth, call a doctor or take directly to hospital.

SPILLAGE	STORAGE	LABELING / NFPA
clean up spillage, carefully collect remainder; (additional individual protection: P2 respirator).	keep in dark place.	R: 20/21/22 S: 2-13 ✖ Harmful

NOTES
Esters and salts of 2,4-D are used as insecticides and can have higher volatility (vapor pressure); toxicity similar to that of 2,4-D (or 2,4-DP). Solvent contained in commercial products often presents a greater danger than active substance. Alcohol consumption increases toxic effects. The measures on this card also apply to 2,4-DP: melting point: 118°C.

Transport Emergency Card TEC(R)-61G53

HI: 60; UN-number: 1609

CAS-No: [78-87-5]
α,β-**dichloropropane**
α,β-**propylene dichloride**
propylene chloride
propylene dichloride

$CH_2ClCHClCH_3$

1,2-DICHLOROPROPANE

PHYSICAL PROPERTIES	IMPORTANT CHARACTERISTICS
Boiling point, °C 96 Melting point, °C −80 Flash point, °C 15 Auto-ignition temperature, °C 555 Relative density (water = 1) 1.2 Relative vapor density (air = 1) 3.9 Relative density at 20 °C of saturated mixture vapor/air (air = 1) 1.2 Vapor pressure, mm Hg at 20 °C 43 Explosive limits, vol% in air 3.4-14.5 Relative molecular mass 113	**COLORLESS LIQUID WITH CHARACTERISTIC ODOR** Vapor (gas) is heavier than air and spreads at ground level, with risk of ignition at a distance. Do not use compressed air when filling, emptying or processing. Decomposes when heated or burned, giving off corrosive vapors (→ *hydrochloric acid*). Reacts with oxidants, violently with light metals, with risk of fire and explosion. Attacks aluminum and many plastics.

TLV-TWA	75 ppm	347 mg/m³
STEL	110 ppm	508 mg/m³

Absorption route: Can enter the body by inhalation or ingestion. Harmful atmospheric concentrations can build up fairly rapidly on evaporation at approx. 20°C - even more rapidly in aerosol form.
Immediate effects: Irritates the eyes, skin and respiratory tract. Liquid destroys the skin's natural oils. Affects the nervous system.
Effects of prolonged/repeated exposure: Can cause liver, kidney or heart damage.

Gross formula: $C_3H_6Cl_2$

HAZARDS/SYMPTOMS	PREVENTIVE MEASURES	FIRE EXTINGUISHING/FIRST AID
Fire: extremely flammable.	keep away from open flame and sparks, no smoking.	powder, AFFF, foam, carbon dioxide, (halons).
Explosion: forms explosive air-vapor mixtures.	sealed machinery, ventilation, explosion-proof electrical equipment and lighting.	in case of fire: keep tanks/drums cool by spraying with water.
Inhalation: sore throat, cough, shortness of breath, headache.	ventilation, local exhaust or respiratory protection.	fresh air, rest, call a doctor.
Skin: redness, pain.	gloves.	remove contaminated clothing, flush skin with water or shower.
Eyes: redness, pain.	safety glasses.	flush with water, take to a doctor if necessary.
Ingestion: abdominal pain, headache, vomiting, sleepiness.		rinse mouth, call a doctor or take to hospital.

SPILLAGE	STORAGE	LABELING / NFPA
collect leakage in sealable containers, soak up with sand or other inert absorbent and remove to safe place; (additional individual protection: breathing apparatus).	keep in fireproof place, separate from oxidants.	R: 11-20 S: 9-16-29-33 Note C 🔥 Flammable ✖ Harmful NFPA 2-3-0

NOTES
Alcohol consumption increases toxic effects. Depending on the degree of exposure, regular medical checkups are advisable. The measures on this card also apply to 1,3-dichloropropane and 2,2-dichloropropane.

Transport Emergency Card TEC(R)-617

HI: 33; UN-number: 1279

CAS-No: [542-75-6]

3-chloroallyl chloride
3-chloropropenyl chloride
dichloropropene
1,3-dichloro-1-propene
1,3-dichloropropylene

$CICH = CHCH_2Cl$

1,3-DICHLOROPROPENE

PHYSICAL PROPERTIES	IMPORTANT CHARACTERISTICS
Boiling point, °C 108 Flash point, °C 35 Relative density (water = 1) 1.2 Relative vapor density (air = 1) 3.8 Relative density at 20 °C of saturated mixture vapor/air (air = 1) 1.11 Vapor pressure, mm Hg at 20 °C 30 Solubility in water none Explosive limits, vol% in air 5.3-14.5 Relative molecular mass 111.0	**COLORLESS TO LIGHT YELLOW LIQUID WITH CHARACTERISTIC ODOR** Flow, agitation etc. can cause build-up of electrostatic charge due to liquid's low conductivity. Toxic vapors are a product of combustion (→ *phosgene*, → *hydrochloric acid*). Reacts with light metals, evolving heat. Reacts violently with strong oxidants. Attacks steel at high temperatures.
	TLV-TWA (skin) 1 ppm 4.5 mg/m³
	Absorption route: Can enter the body by inhalation or ingestion or through the skin. Harmful atmospheric concentrations can build up very rapidly on evaporation at 20°C. **Immediate effects:** Irritates the eyes, skin and respiratory tract. Liquid destroys the skin's natural oils. Affects the nervous system. **Effects of prolonged/repeated exposure:** Can cause liver and kidney damage.
Gross formula: $C_3H_4Cl_2$	

HAZARDS/SYMPTOMS	PREVENTIVE MEASURES	FIRE EXTINGUISHING/FIRST AID
Fire: flammable.	keep away from open flame and sparks, no smoking.	powder, water spray, foam, carbon dioxide, (halons).
Explosion: above 35°C: forms explosive air-vapor mixtures.	above 35°C: sealed machinery, ventilation, explosion-proof electrical equipment, grounding.	in case of fire: keep tanks/drums cool by spraying with water.
Inhalation: sore throat, cough, shortness of breath, headache, drowsiness.	ventilation, local exhaust or respiratory protection.	fresh air, rest, call a doctor.
Skin: *is absorbed*; redness, pain.	gloves, protective clothing.	remove contaminated clothing, flush skin with water or shower, refer to a doctor.
Eyes: redness, pain.	safety glasses, or combined eye and respiratory protection.	flush with water, take to a doctor if necessary.
Ingestion: abdominal pain, nausea.		rinse mouth, call a doctor or take to hospital.

SPILLAGE	STORAGE	LABELING / NFPA
collect leakage in sealable containers, soak up with sand or other inert absorbent and remove to safe place; (additional individual protection: breathing apparatus).	keep in fireproof place, separate from oxidants.	R: 11-22 S: 9-16-29-33 Note C 🔥 ✖ Flammable Harmful NFPA: 3 / 3 / 0

NOTES
Alcohol consumption increases toxic effects. Depending on the degree of exposure, regular medical checkups are advisable. 1,3-dichloropropene is an insecticide. N.B. propylene dichloride = dichloropropane.

Transport Emergency Card TEC(R)-664

HI: 30; UN-number: 2047

DICHROMATE/SULFURIC ACID SOLUTION
(Sol. of potassium/sodium dichromate in sulfuric acid of various concentrations)

PHYSICAL PROPERTIES	IMPORTANT CHARACTERISTICS	
Boiling point, °C ca. 340 Relative density (water = 1) ca. 1.8 Solubility in water ∞	**CLEAR ORANGE LIQUID** Wood, paper and textiles can ignite spontaneously on contact with solution. Strong acid which reacts violently with bases and is corrosive. Reacts violently with organic solvents, with risk of fire and explosion. Reacts violently with hydrochloric acid and chlorides, giving off toxic gas (→ *chlorine*).	
	TLV-TWA	✱
	Absorption route: Can enter the body by inhalation or ingestion or through the skin. Harmful atmospheric concentrations build up very slowly, if at all, on evaporation at approx. 20°C, but much more rapidly in aerosol form. **Immediate effects:** Corrosive to the eyes, skin and respiratory tract. Inhalation of vapor/fumes can cause severe breathing difficulties (lung edema). In serious cases risk of death. **Effects of prolonged/repeated exposure:** Prolonged or repeated contact can cause skin disorders.	

HAZARDS/SYMPTOMS	PREVENTIVE MEASURES	FIRE EXTINGUISHING/FIRST AID
Fire: not combustible but enhances combustion of other substances.	avoid contact with combustible substances.	
Inhalation: *corrosive*; sore throat, cough, shortness of breath, severe breathing difficulties.	ventilation.	fresh air, rest, place in half-sitting position, take to hospital.
Skin: *corrosive*; redness, pain, burns.	gloves, protective clothing.	remove contaminated clothing, flush skin with water or shower, refer to a doctor.
Eyes: *corrosive*; redness, pain, impaired vision.	face shield.	flush with water, take to a doctor.
Ingestion: *corrosive*; diarrhea, abdominal cramps, vomiting.		rinse mouth, immediately take to hospital.

SPILLAGE	STORAGE	LABELING / NFPA
collect leakage in sealable containers, soak up with sand or other inert absorbent and remove to safe place, carefully collect remainder.	keep separate from combustible substances, reducing agents and bases.	

NOTES
Chromic acid is actually chromium trioxide and its aqueous solution. Flush contaminated clothing with water (fire hazard). ✱ No TLV for dichromate/sulfuric acid solution; TLVs of components (chromium trioxide and sulfuric acid) are 0.05mg/m³ and 1mg/m³ (STEL 3mg/m³) respectively. Lung edema symptoms usually develop several hours later and are aggravated by physical exertion: rest and hospitalization essential. As first aid, a doctor or authorized person should consider administering a corticosteroid spray. UNDER NO CIRCUMSTANCES add water to acid; when diluting ALWAYS add acid to water.

Transport Emergency Card TEC(R)-80G02

HI: 88; UN-number: 2240

CAS-No: [461-58-5]
cyanoguanidine

$$H_2N-\overset{\overset{N}{\|}}{C}-\overset{\overset{H}{}}{N}-\overset{H}{}C\equiv N$$

DICYANDIAMIDE

PHYSICAL PROPERTIES	IMPORTANT CHARACTERISTICS	
Melting point, °C 211 Relative density (water = 1) 1.4 Vapor pressure, mm Hg at 20 °C <0.076 Solubility in water, g/100 ml at 20 °C 3.2 Relative molecular mass 84.1 Log P octanol/water −1.5	**WHITE CRYSTALLINE POWDER** Decomposes when heated above melting point or when aqueous solution is heated above 80°C, giving off → *anhydrous ammonia*. Reacts with strong acids, giving off toxic gas (→ *hydrogen cyanide*). Forms explosive mixtures with strong oxidants.	
	TLV-TWA not available	
	Absorption route: Can enter the body by inhalation or ingestion. Evaporation negligible at 20°C, but unpleasant concentrations of airborne particles can build up.	
Gross formula: $C_2H_2N_4$		

HAZARDS/SYMPTOMS	PREVENTIVE MEASURES	FIRE EXTINGUISHING/FIRST AID
Fire: non-combustible.		in case of fire in immediate vicinity: use any extinguishing agent.
	KEEP DUST UNDER CONTROL	
Inhalation:	local exhaust or respiratory protection.	fresh air, rest, call a doctor if necessary.
Skin:	gloves.	remove contaminated clothing, flush skin with water or shower.
Eyes:	goggles.	flush with water, take to a doctor if necessary.
Ingestion:		rinse mouth, call a doctor.

SPILLAGE	STORAGE	LABELING / NFPA
clean up spillage, flush away remainder with water; (additional individual protection: P2 respirator).	keep separate from oxidants and strong acids.	

NOTES
Log P octanol/water is estimated. Insufficient toxicological data on harmful effects to humans.

CAS-No: [77-73-6]
bicyclopentadiene
biscyclopentadiene
1,3-cyclopentadiene, dimer
DCPD

$C_{10}H_{12}$

DICYCLOPENTADIENE

PHYSICAL PROPERTIES	IMPORTANT CHARACTERISTICS	
Boiling point, °C 170	**COLORLESS CRYSTALS WITH CHARACTERISTIC ODOR**	
Melting point, °C 33	Vapor mixes readily with air. In dry state can form electrostatic charge if stirred, transported pneumatically, poured etc. Reacts violently with oxidants. Able to form peroxides.	
Flash point, °C 32		
Auto-ignition temperature, °C 503		
Relative density (water = 1) 1.0	TLV-TWA	5 ppm	27 mg/m³
Relative vapor density (air = 1) 4.55		
Relative density at 20 °C of saturated mixture vapor/air (air = 1) 1.01	**Absorption route:** Can enter the body by inhalation or ingestion. Harmful atmospheric concentrations build up fairly slowly on evaporation at 20°C, but harmful concentrations of airborne particles can build up much more rapidly.	
Vapor pressure, mm Hg at 20 °C 1.4		
Solubility in water none		
Explosive limits, vol% in air ?-?	**Immediate effects:** Irritates the eyes, skin and respiratory tract.	
Relative molecular mass 132.2		
Gross formula: $C_{10}H_{12}$		

HAZARDS/SYMPTOMS	PREVENTIVE MEASURES	FIRE EXTINGUISHING/FIRST AID
Fire: flammable.	keep away from open flame and sparks, no smoking.	water spray, powder.
Explosion: above 32°C: forms explosive air-vapor mixtures.	above 32°C: sealed machinery, ventilation, explosion-proof electrical equipment, grounding.	in case of fire: keep tanks/drums cool by spraying with water.
Inhalation: cough, headache, dizziness, sleepiness.	local exhaust or respiratory protection.	fresh air, rest, call a doctor.
Skin: redness.	gloves.	remove contaminated clothing, flush skin with water or shower.
Eyes: redness, pain.	goggles.	flush with water, take to a doctor if necessary.
Ingestion: abdominal pain, nausea.		rinse mouth, DO NOT induce vomiting, call a doctor.

SPILLAGE	STORAGE	LABELING / NFPA
clean up spillage, carefully collect remainder; (additional individual protection: respirator with A/P2 filter or breathing apparatus).	keep in fireproof place, separate from oxidants.	3 / 1 / 1

NOTES
Explosive limits not given in the literature. Before distilling check for peroxides; if found, render harmless.

Transport Emergency Card TEC(R)-689

HI: 30; UN-number: 2048

$(CH_2CH_2OH)_2NH$

DEA
diethylolamine
di(2-hydroxyethyl)amine
diolamine
2,2'-iminobisethanol

DIETHANOLAMINE

PHYSICAL PROPERTIES		IMPORTANT CHARACTERISTICS
Boiling point (decomposes), °C	263	**VISCOUS HYGROSCOPIC LIQUID OR WHITE CRYSTALS**
Melting point, °C	28	Vapor mixes readily with air. Reacts violently with strong oxidants. Attacks copper, zinc, aluminum and their alloys.
Flash point, °C	138	
Auto-ignition temperature, °C	660	
Relative density (water = 1)	1.1	
Relative vapor density (air = 1)	3.7	
Relative density at 20 °C of saturated mixture vapor/air (air = 1)	1.0	
Vapor pressure, mm Hg at 20 °C	< 0.076	**Absorption route:** Can enter the body by inhalation or ingestion. Harmful atmospheric concentrations build up very slowly, if at all, on evaporation at approx. 20°C, but much more rapidly in aerosol form.
Solubility in water	∞	
Relative molecular mass	105.1	
Log P octanol/water	− 1.4	

		TLV-TWA	3 ppm	13 mg/m³

Immediate effects: Irritates the eyes, skin and respiratory tract.

Gross formula: $C_4H_{11}NO_2$

HAZARDS/SYMPTOMS	PREVENTIVE MEASURES	FIRE EXTINGUISHING/FIRST AID
Fire: combustible.	keep away from open flame, no smoking.	powder, water spray, foam, carbon dioxide, (halons).
Inhalation: sore throat, cough.	local exhaust or respiratory protection.	fresh air, rest.
Skin: redness.	gloves.	remove contaminated clothing, wash skin with soap and water or shower, refer to a doctor.
Eyes: redness, pain.	face shield.	flush with water, take to a doctor if necessary.
Ingestion: abdominal pain, diarrhea, vomiting.		rinse mouth, call a doctor, take to hospital.

SPILLAGE	STORAGE	LABELING / NFPA
clean up spillage, flush away remainder with water.	keep dry, separate from oxidants.	R: 36/38 S: 26 ✖ Irritant

NFPA: 1 / 1 / 0

NOTES

DIETHYLACETAL

PHYSICAL PROPERTIES	IMPORTANT CHARACTERISTICS
Boiling point, °C 103 Melting point, °C −100 Flash point, °C −21 Auto-ignition temperature, °C 230 Relative density (water = 1) 0.8 Relative vapor density (air = 1) 4.1 Relative density at 20 °C of saturated mixture vapor/air (air = 1) 1.11 Vapor pressure, mm Hg at 20 °C 27 Solubility in water, g/100 ml at 25 °C 4.6 Explosive limits, vol% in air 1.6-10.4 Relative molecular mass 118.2 Log P octanol/water 0.8 Gross formula: $C_6H_{14}O_2$	**COLORLESS LIQUID WITH CHARACTERISTIC ODOR** Vapor is heavier than air and spreads at ground level, with risk of ignition at a distance. Presumed to be able to form peroxides when exposed to light. Flow, agitation etc. can cause build-up of electrostatic charge due to liquid's low conductivity. Do not use compressed air when filling, emptying or processing. Reacts violently with strong oxidants. TLV-TWA not available **Absorption route:** Can enter the body by inhalation or ingestion. Harmful atmospheric concentrations can build up fairly rapidly on evaporation at approx. 20°C - even more rapidly in aerosol form. **Immediate effects:** Irritates the eyes, skin and respiratory tract. Affects the nervous system.

HAZARDS/SYMPTOMS	PREVENTIVE MEASURES	FIRE EXTINGUISHING/FIRST AID
Fire: extremely flammable.	keep away from open flame and sparks, no smoking.	powder, AFFF, foam, carbon dioxide, (halons).
Explosion: forms explosive air-vapor mixtures.	sealed machinery, ventilation, explosion-proof electrical equipment and lighting, grounding.	in case of fire: keep tanks/drums cool by spraying with water.
Inhalation: sore throat, cough, headache, dizziness.	ventilation, local exhaust or respiratory protection.	fresh air, rest, call a doctor.
Skin: redness, pain.	gloves.	remove contaminated clothing, flush skin with water or shower.
Eyes: redness, pain.	safety glasses.	flush with water, take to a doctor if necessary.
Ingestion: abdominal pain, nausea, sleepiness.		rinse mouth, call a doctor.

SPILLAGE	STORAGE	LABELING / NFPA
collect leakage in sealable containers, soak up with sand or other inert absorbent and remove to safe place; (additional individual protection: breathing apparatus).	keep in dark, fireproof place, separate from oxidants.	R: 11-36/38 S: 9-16-33 🔥 ✖ Flammable Irritant NFPA: 3 2 0

NOTES
Log P octanol/water is estimated. Before distilling check for peroxides; if found, render harmless.

$(C_2H_5)_2AlCl$

DIETHYLALUMINUM CHLORIDE

PHYSICAL PROPERTIES	IMPORTANT CHARACTERISTICS
Boiling point, °C 208 Melting point, °C −74 Relative density (water = 1) 0.97 Relative vapor density (air = 1) 4.1 Vapor pressure, mm Hg at 20 °C 0.23 Solubility in water reaction Relative molecular mass 120.5	**COLORLESS FUMING LIQUID** Aluminum oxide fumes and → *hydrochloric acid* fumes are a product of combustion. Can ignite spontaneously on exposure to air. Strong reducing agent which reacts violently with oxidants. Reacts violently with water, alcohols, phenols, amines, carbon dioxide, oxides of sulfur and nitrogen oxides, with risk of fire and explosion. Reacts with air to form aluminum oxide fumes and → *hydrochloric acid* fumes.
	TLV-TWA 2 mg/m³
	Absorption route: Can enter the body by inhalation or ingestion. Harmful atmospheric concentrations can build up fairly rapidly on evaporation at approx. 20°C - even more rapidly in aerosol form. **Immediate effects:** Corrosive to the eyes, skin and respiratory tract. Inhalation of vapor/fumes can cause severe breathing difficulties (lung edema). In serious cases risk of death.
Gross formula: $C_4H_{10}AlCl$	

HAZARDS/SYMPTOMS	PREVENTIVE MEASURES	FIRE EXTINGUISHING/FIRST AID
Fire: extremely flammable, many chemical reactions can cause fire and explosion.	keep away from open flame and sparks, no smoking, avoid contact with air, water and many other substances.	special powder extinguisher, dry sand, NO other extinguishing agents.
Explosion: contact with water and many other substances can cause explosion.		in case of fire: keep undamaged cylinders cool by spraying with water, but DO NOT spray substance with water.
	STRICT HYGIENE	
Inhalation: *corrosive*; sore throat, cough, severe breathing difficulties.	completely sealed machinery, ventilation, local exhaust or respiratory protection.	fresh air, rest, place in half-sitting position, take to hospital.
Skin: *corrosive*; serious burns, slow healing.	gloves, protective clothing.	remove contaminated clothing, flush skin with water or shower, refer to a doctor.
Eyes: *corrosive*; redness, pain, impaired vision.	face shield, or combined eye and respiratory protection.	flush with water, take to a doctor.
Ingestion: *corrosive*; sore throat, abdominal pain, diarrhea.		rinse mouth, take immediately to hospital.

SPILLAGE	STORAGE	LABELING / NFPA
evacuate area, call in an expert, dilute leakage with diesel/machine oil, soak up/cover with dry sand/table salt and remove to safe place; (additional individual protection: CHEMICAL SUIT).	keep in fireproof place under dry nitrogen, separate from all other substances.	NFPA diamond: 4 (top), 3 (left), 3 (right), W (bottom)

NOTES
Usually sold as 15-30% solution in hydrocarbons: flash point and explosive limits depend on solvent used. For 15-30% solution see relevant entry. Reacts violently with extinguishing agents such as water, foam, carbon dioxide and ordinary powder. Lung edema symptoms usually develop several hours later and are aggravated by physical exertion: rest and hospitalization essential. As first aid, a doctor or authorized person should consider administering a corticosteroid spray. Use cylinder with special fittings.

Transport Emergency Card TEC(R)-42G02

HI: X333; UN-number: 2221

DIETHYLALUMINUMCHLORIDE
(15% solution in hexane)

PHYSICAL PROPERTIES	IMPORTANT CHARACTERISTICS
Boiling point, °C — 69 Melting point, °C — −95 Flash point, °C — −26 Auto-ignition temperature, °C — 240 Relative density (water = 1) — 0.74 Relative vapor density (air = 1) — 3.0 Vapor pressure, mm Hg at 20 °C — 122 Solubility in water — reaction Explosive limits, vol% in air — 1.1-7.5 Minimum ignition energy, mJ — 0.24 Electrical conductivity, pS/m — 100	**COLORLESS LIQUID WITH PUNGENT ODOR** Vapor is heavier than air and spreads at ground level, with risk of ignition at a distance. Flow, agitation etc. can cause build-up of electrostatic charge due to liquid's low conductivity. Once solvent has evaporated (e.g. on clothing) can ignite spontaneously. Aluminum oxide fumes and → *hydrochloric acid* fumes are a product of combustion. Also decomposes in solution when exposed to air or water vapor, giving off → *hydrochloric* fumes. Strong reducing agent which reacts violently with oxidants. Reacts violently with water, alcohols, phenols, amines, ketones, carbon dioxide, sulfur oxides and nitrogen oxides, with risk of fire and explosion. Reacts with air to form → *hydrochloric acid* fumes.

TLV-TWA	2 mg/m³ ✷

Absorption route: Can enter the body by inhalation or ingestion. Harmful atmospheric concentrations can build up very rapidly on evaporation at 20°C. **Immediate effects:** Corrosive to the eyes, skin and respiratory tract. Inhalation of vapor/fumes can cause severe breathing difficulties (lung edema). In serious cases risk of death.

Gross formula: $C_4H_{10}AlCl$

HAZARDS/SYMPTOMS	PREVENTIVE MEASURES	FIRE EXTINGUISHING/FIRST AID
Fire: extremely flammable, many chemical reactions can cause fire and explosion.	keep away from open flame and sparks, no smoking; avoid contact with water, alcohols, oxidants and many other substances.	special extinguishing powder, dry sand, NO other extinguishing agents, can reignite after being extinguished.
Explosion: forms explosive air-vapor mixtures; contact with water, alcohols and many other substances can cause explosion.	sealed machinery, ventilation, explosion-proof electrical equipment and lighting, grounding.	in case of fire: keep undamaged cylinders cool by spraying with water, but DO NOT spray substance with water.
	STRICT HYGIENE	
Inhalation: *corrosive*; sore throat, cough, severe breathing difficulties.	completely sealed machinery, ventilation.	fresh air, rest, place in half-sitting position, take to hospital.
Skin: *corrosive*; redness, serious burns, slow healing.	gloves, protective clothing.	remove contaminated clothing, flush skin with water or shower, refer to a doctor.
Eyes: *corrosive*; redness, pain, impaired vision.	face shield, or combined eye and respiratory protection.	flush with water, take to a doctor.
Ingestion: *corrosive*; sore throat, abdominal pain, diarrhea.		rinse mouth, immediately take to hospital.

SPILLAGE	STORAGE	LABELING / NFPA
evacuate area, call in an expert, collect leakage in sealable containers, soak up remainder with dry sand/table salt, under no circumstances spray liquid with water; (additional individual protection: CHEMICAL SUIT).	keep in dry, fireproof place under inert gas, separate from all other substances.	 4 3 3 W

NOTES
Physical properties (boiling point, flash point, autoignition point, explosive limits, vapor pressure, relative density and conductivity) are based on hexane. For pure diethylaluminumchloride see relevant card. Remove contaminated clothing immediately and flush with water (risk of spontaneous combustion). Lung edema symptoms usually develop several hours later and are aggravated by physical exertion: rest and hospitalization essential. ✷ For TLV of → *hexane* see relevant card. Use cylinder with special fittings.

Transport Emergency Card TEC(R)-42G02

HI: X333; UN-number: 2220

DIETHYLALUMINUM HYDRIDE

PHYSICAL PROPERTIES	IMPORTANT CHARACTERISTICS	
Decomposes below boiling point, °C ... 40 Melting point, °C ... −59 Relative density (water = 1) ... 0.79 Relative vapor density (air = 1) ... 2.9 Vapor pressure, mm Hg at 20 °C ... <0.76 Solubility in water ... reaction Relative molecular mass ... 86.1	**COLORLESS FUMING LIQUID** Decomposes spontaneously on exposure to air. Aluminum oxide vapors are a product of combustion. Strong reducing agent which reacts violently with oxidants. Reacts violently with water, alcohols, phenols, amines, carbon dioxide, oxides of sulfur, nitrogen oxides, halogens, halogenated hydrocarbons and many other substances, with risk of fire and explosion. Reacts with air to form irritating vapors.	
	TLV-TWA	2 mg/m³
Gross formula:　　　　　　$C_4H_{11}Al$	**Absorption route:** Can enter the body by inhalation or ingestion. Harmful atmospheric concentrations can build up fairly rapidly on evaporation at approx. 20°C - even more rapidly in aerosol form. **Immediate effects:** Corrosive to the eyes, skin and respiratory tract. Inhalation of vapor/fumes can cause severe breathing difficulties (lung edema). In serious cases risk of death.	

HAZARDS/SYMPTOMS	PREVENTIVE MEASURES	FIRE EXTINGUISHING/FIRST AID
Fire:	keep away from open flame and sparks, no smoking.	special extinguishing powder, dry sand, NO other extinguishing agents, can reignite after being extinguished.
Explosion: contact with water and many other substances can cause explosion.		in case of fire: keep undamaged cylinders cool by spraying with water, but DO NOT spray substance with water.
	STRICT HYGIENE	
Inhalation: *corrosive*; sore throat, cough, severe breathing difficulties.	completely sealed machinery, ventilation, local exhaust or respiratory protection.	fresh air, rest, place in half-sitting position, take to hospital.
Skin: *corrosive*; serious burns, slow healing.	gloves, protective clothing.	remove contaminated clothing, flush skin with water or shower, refer to a doctor.
Eyes: *corrosive*; redness, pain, impaired vision.	face shield, or combined eye and respiratory protection.	flush with water, take to a doctor.
Ingestion: *corrosive*; sore throat, abdominal pain, diarrhea.		rinse mouth, take immediately to hospital.

SPILLAGE	STORAGE	LABELING / NFPA
evacuate area, call in an expert, dilute leakage with diesel/machine oil, soak up/cover with dry sand/table salt and remove to safe place, under no circumstances spray liquid with water; (additional individual protection: CHEMICAL SUIT).	keep in fireproof place under dry nitrogen, separate from all other substances.	(NFPA diamond: 3 / 3 / W)

NOTES
Usually sold as 15-30% solution in hydrocarbons, which does not ignite spontaneously on exposure to air. For 15% solution, see relevant entry. Reacts violently with extinguishing agents such as carbon dioxide, halons, water, foam and ordinary powder. Lung edema symptoms usually develop several hours later and are aggravated by physical exertion: rest and hospitalization essential. As first aid, a doctor or authorized person should consider administering a corticosteroid spray. Use cylinder with special fittings.

Transport Emergency Card TEC(R)-42G02　　　　　　　　　　　　　　　　　**HI: X323; UN-number: 2220**

CAS-No: [87-27-2]
DEAH

$(C_2H_5)_2AlH$

DIETHYLALUMINUM HYDRIDE
(15% solution in hexane)

PHYSICAL PROPERTIES	IMPORTANT CHARACTERISTICS
Boiling point, °C ... 69 Melting point, °C ... −95 Flash point, °C ... −26 Auto-ignition temperature, °C ... 240 Relative density (water = 1) ... 0.72 Relative vapor density (air = 1) ... 3.0 Relative density at 20 °C of saturated mixture vapor/air (air = 1) ... 1.3 Vapor pressure, mm Hg at 20 °C ... 121.6 Solubility in water ... reaction Explosive limits, vol% in air ... 1.1-7.5 Minimum ignition energy, mJ ... 0.24 Electrical conductivity, pS/m ... 100 Gross formula: $C_4H_{11}Al$	**COLORLESS FUMING LIQUID** Vapor is heavier than air and spreads at ground level, with risk of ignition at a distance. Flow, agitation etc. can cause build-up of electrostatic charge due to liquid's low conductivity. Do not use compressed air when filling, emptying or processing. Once solvent has evaporated (e.g. on clothing) can ignite spontaneously on exposure to air. Aluminum oxide vapors are a product of combustion. Strong reducing agent which reacts violently with oxidants. Reacts violently with water, alcohols, phenols, halogenated hydrocarbons and many other substances, with risk of fire and explosion.

TLV-TWA	2 mg/m³ ✱

Absorption route: Can enter the body by inhalation or ingestion. Harmful atmospheric concentrations can build up fairly rapidly on evaporation at approx. 20°C - even more rapidly in aerosol form.
Immediate effects: Corrosive to the eyes, skin and respiratory tract. Inhalation of vapor/fumes can cause severe breathing difficulties (lung edema). In serious cases risk of death.

HAZARDS/SYMPTOMS	PREVENTIVE MEASURES	FIRE EXTINGUISHING/FIRST AID
Fire: extremely flammable, many chemical reactions can cause fire and explosion.	keep away from open flame and sparks, no smoking, avoid contact with water, alcohols, phenols and many other substances.	special extinguishing powder, dry sand, NO other extinguishing agents, can reignite after being extinguished.
Explosion: forms explosive air-vapor mixtures; contact with water, phenols, alcohols and many other substances can cause explosion.	sealed machinery, ventilation, explosion-proof electrical equipment and lighting, grounding.	in case of fire: keep undamaged cylinders cool by spraying with water, but DO NOT spray substance with water.
	STRICT HYGIENE	
Inhalation: *corrosive*; sore throat, cough, severe breathing difficulties.	sealed machinery, ventilation, local exhaust or respiratory protection.	fresh air, rest, place in half-sitting position, take to hospital.
Skin: *corrosive*; redness, pain, serious burns, slow healing.	gloves, protective clothing.	remove contaminated clothing, flush skin with water or shower, refer to a doctor.
Eyes: *corrosive*; redness, pain, impaired vision.	face shield, or combined eye and respiratory protection.	flush with water, take to a doctor.
Ingestion: *corrosive*; sore throat, abdominal pain, diarrhea.		rinse mouth, immediately take to hospital.

SPILLAGE	STORAGE	LABELING / NFPA
evacuate area, call in an expert, collect leakage in sealable containers, soak up remainder with dry sand/table salt and remove to safe place, under no circumstances spray liquid with water; (additional individual protection: CHEMICAL SUIT).	keep in dry, fireproof place under inert gas, separate from all other substances.	3 / 3 / W

NOTES
Physical properties (boiling point, flash point, autoignition point, explosive limits, vapor pressure, relative density, conductivity and ignition energy) are based on hexane. For pure diethyl aluminum hydride see relevant sheet. Reacts violently with extinguishing agents such as water, foam, halons, carbon dioxide and ordinary powder. Remove contaminated clothing immediately and flush with water (risk of spontaneous combustion). Lung edema symptoms usually develop several hours later and are aggravated by physical exertion: rest and hospitalization essential. As first aid, a doctor or authorized person should consider administering a corticosteroid spray. ✱ For TLV of *hexane* see relevant card. Use cylinder with special fittings.

Transport Emergency Card TEC(R)-41G02

HI: X333; UN-number: 2220

DIETHYLAMINE

PHYSICAL PROPERTIES	IMPORTANT CHARACTERISTICS
Boiling point, °C 56 Melting point, °C −50 Flash point, °C < −26 Auto-ignition temperature, °C 310 Relative density (water = 1) 0.7 Relative vapor density (air = 1) 2.5 Relative density at 20 °C of saturated mixture vapor/air (air = 1) 1.4 Vapor pressure, mm Hg at 20 °C 190 Solubility in water ∞ Explosive limits, vol% in air 1.7-10.1 Relative molecular mass 73.1 Log P octanol/water 0.5	**COLORLESS LIQUID WITH PUNGENT ODOR** Vapor (gas) is heavier than air and spreads at ground level, with risk of ignition at a distance. Do not use compressed air when filling, emptying or processing. Decomposes when heated or burned, giving off toxic vapors (→ *nitrous vapors*). Strong base which reacts violently with acids and corrodes aluminum, zinc etc. Reacts violently with many organic substances and mercury. Reacts violently with strong oxidants, with risk of fire and explosion. Attacks aluminum, copper, lead, tin, zinc and alloys.

TLV-TWA	10 ppm	30 mg/m³	
STEL	25 ppm	75 mg/m³	

Absorption route: Can enter the body by inhalation or ingestion or through the skin. Harmful atmospheric concentrations can build up very rapidly on evaporation at 20° C.
Immediate effects: Corrosive to the eyes, skin and respiratory tract. Inhalation can cause lung edema. In serious cases risk of death.

Gross formula: $C_4H_{11}N$

HAZARDS/SYMPTOMS	PREVENTIVE MEASURES	FIRE EXTINGUISHING/FIRST AID
Fire: extremely flammable.	keep away from open flame and sparks, no smoking.	powder, alcohol-resistant foam, large quantities of water, carbon dioxide, (halons).
Explosion: forms explosive air-vapor mixtures.	sealed machinery, ventilation, explosion-proof electrical equipment and lighting.	in case of fire: keep tanks/drums cool by spraying with water.
	STRICT HYGIENE	
Inhalation: *corrosive*; sore throat, cough, shortness of breath, severe breathing difficulties.	ventilation, local exhaust or respiratory protection.	fresh air, rest, place in half-sitting position, take to hospital.
Skin: *corrosive*; redness, pain, serious burns.	gloves, protective clothing.	remove contaminated clothing, flush skin with water or shower, call a doctor.
Eyes: *corrosive*; redness, pain, impaired vision.	face shield, or combined eye and respiratory protection.	flush with water, take to a doctor.
Ingestion: *corrosive*; sore throat, abdominal pain.		rinse mouth, call a doctor or take to hospital.

SPILLAGE	STORAGE	LABELING / NFPA
collect leakage in sealable containers, soak up with sand or other inert absorbent and remove to safe place; (additional individual protection: CHEMICAL SUIT).	keep in fireproof place, separate from acids and oxidants.	R: 11-36/37 S: 16-26-29 🔥 Flammable ✖ Irritant NFPA 3-3-0

NOTES
Forms a hydrate with melting point − 19° C. Lung edema symptoms usually develop several hours later and are aggravated by physical exertion: rest and hospitalization essential. As first aid, a doctor or authorized person should consider administering a corticosteroid spray. Unbreakable packaging preferred; if breakable, keep in unbreakable container.

Transport Emergency Card TEC(R)-635

HI: 338; UN-number: 1154

CAS-No: [100-37-8]
N,N-diethylethanolamine

$(C_2H_5)_2NCH_2CH_2OH$

DIETHYLAMINOETHANOL

PHYSICAL PROPERTIES		IMPORTANT CHARACTERISTICS
Boiling point, °C	161	**COLORLESS HYGROSCOPIC LIQUID WITH CHARACTERISTIC ODOR**
Melting point, °C	−70	Vapor mixes readily with air. Decomposes when heated or burned, giving off toxic vapors
Flash point, °C	60	(→ *nitrous vapors*). Reacts violently with acids and acid anhydrides. Reacts violently with
Auto-ignition temperature, °C	260	oxidants, with risk of fire and explosion. Attacks light metals.
Relative density (water = 1)	0.9	
Relative vapor density (air = 1)	4.0	
Relative density at 20 °C of saturated mixture		
vapor/air (air = 1)	1.01	
Vapor pressure, mm Hg at 20 °C	1.4	
Solubility in water	∞	
Explosive limits, vol% in air	1.8-28.0	
Relative molecular mass	117.2	
Log P octanol/water	0.4	

TLV-TWA (skin) 10 ppm 48 mg/m³

Absorption route: Can enter the body by inhalation or ingestion or through the skin. Harmful atmospheric concentrations build up fairly slowly on evaporation at 20°C, but much more rapidly in aerosol form.
Immediate effects: Corrosive to the eyes, skin and respiratory tract. Affects the nervous system. Inhalation of vapor/fumes can cause severe breathing difficulties (lung edema). In serious cases risk of death.

Gross formula: $C_6H_{15}NO$

HAZARDS/SYMPTOMS	PREVENTIVE MEASURES	FIRE EXTINGUISHING/FIRST AID
Fire: combustible.	keep away from open flame, no smoking.	powder, alcohol-resistant foam, water spray, carbon dioxide, (halons).
Explosion: above 60°C: forms explosive air-vapor mixtures.	above 60°C: sealed machinery, ventilation.	
Inhalation: *corrosive*; sore throat, cough, shortness of breath, severe breathing difficulties.	ventilation, local exhaust or respiratory protection.	fresh air, rest, place in half-sitting position, take to hospital.
Skin: *corrosive*; redness, pain, serious burns.	gloves, protective clothing.	remove contaminated clothing, flush skin with water or shower, refer to a doctor.
Eyes: *corrosive*; redness, pain, impaired vision.	face shield.	flush with water, take to a doctor.
Ingestion: *corrosive*; sore throat, abdominal pain, diarrhea, vomiting.		rinse mouth, take immediately to hospital.

SPILLAGE	STORAGE	LABELING / NFPA
collect leakage in sealable containers, flush away remainder with water.	keep separate from oxidants and strong acids; ventilate at floor level.	R: 36/37/38 S: 28 ✖ Irritant

NFPA diamond: 2 (top), 3 (left), 0 (right)

NOTES

Log P octanol/water is estimated. Lung edema symptoms usually develop several hours later and are aggravated by physical exertion: rest and hospitalization essential. As first aid, a doctor or authorized person should consider administering a corticosteroid spray.

Transport Emergency Card TEC(R)-30G35

HI: 30; UN-number: 2686

N,N-DIETHYLANILINE

PHYSICAL PROPERTIES	IMPORTANT CHARACTERISTICS
Boiling point, °C 216 Melting point, °C −38 Flash point, °C 88 Auto-ignition temperature, °C 332 Relative density (water = 1) 0.94 Relative vapor density (air = 1) 5.2 Relative density at 20 °C of saturated mixture vapor/air (air = 1) 1.0 Vapor pressure, mm Hg at 20 °C 0.099 Solubility in water, g/100 ml at 12 °C 1.4 Relative molecular mass 149.2 Gross formula: $C_{10}H_{15}N$	**COLORLESS TO YELLOW LIQUID WITH CHARACTERISTIC ODOR** Vapor mixes readily with air. → *Nitrous vapors* are a product of combustion. Decomposes when heated or burned, giving off toxic gases (→ *nitrous vapors*). Reacts violently with strong oxidants. TLV-TWA not available **Absorption route:** Can enter the body by inhalation or ingestion or through the skin. Harmful atmospheric concentrations build up fairly slowly on evaporation at 20° C, but much more rapidly in aerosol form. **Immediate effects:** Irritates the skin. In serious cases risk of unconsciousness. **Effects of prolonged/repeated exposure:** Can affect the blood. Can cause liver and kidney damage.

HAZARDS/SYMPTOMS	PREVENTIVE MEASURES	FIRE EXTINGUISHING/FIRST AID
Fire: combustible.	keep away from open flame, no smoking.	powder, water spray, foam, carbon dioxide, (halons).
Explosion: above 88°C: forms explosive air-vapor mixtures.	above 88°C: sealed machinery, ventilation.	
	STRICT HYGIENE	IN ALL CASES CALL IN A DOCTOR
Inhalation: shortness of breath, dizziness, blue skin, feeling of weakness.	ventilation, local exhaust or respiratory protection.	fresh air, rest, artificial respiration if necessary, take to hospital.
Skin: *is absorbed*; see also 'Inhalation'.	gloves, protective clothing.	remove contaminated clothing, wash skin with soap and water.
Eyes: redness, pain.	face shield.	flush with water, take to a doctor if necessary.
Ingestion: see also 'Inhalation'.	do not eat, drink or smoke while working.	rinse mouth, immediately take to hospital.

SPILLAGE	STORAGE	LABELING / NFPA
collect leakage in sealable containers, soak up with sand or other inert absorbent and remove to safe place; (additional individual protection: breathing apparatus).	keep in dark place, separate from oxidants.	R: 23/24/25-33 S: 28-37-44 ☠ Toxic	3 2 0

NOTES
Depending on the degree of exposure, regular medical checkups are advisable. Effects on blood due to formation of methemoglobin. Special first aid required in the event of poisoning: antidotes must be available (with instructions).

Transport Emergency Card TEC(R)-61G06C

HI: 60; UN-number: 2432

DIETHYL CARBONATE

PHYSICAL PROPERTIES	IMPORTANT CHARACTERISTICS
Boiling point, °C 126 Melting point, °C −43 Flash point, °C 25 Relative density (water = 1) 0.98 Relative vapor density (air = 1) 4.1 Relative density at 20 °C of saturated mixture vapor/air (air = 1) 1.04 Vapor pressure, mm Hg at 20 °C 9.9 Explosive limits, vol% in air ?-? Electrical conductivity, pS/m 9.1×10^4 Relative molecular mass 118.1 Gross formula: $C_5H_{10}O_3$	**COLORLESS LIQUID WITH CHARACTERISTIC ODOR** Vapor mixes readily with air. Reacts violently with oxidants, with risk of fire and explosion. TLV-TWA not available **Absorption route:** Can enter the body by inhalation or ingestion. Insufficient data on the rate at which harmful concentrations can build up. **Immediate effects:** Irritates the eyes and respiratory tract.

HAZARDS/SYMPTOMS	PREVENTIVE MEASURES	FIRE EXTINGUISHING/FIRST AID
Fire: flammable.	keep away from open flame and sparks, no smoking.	powder, AFFF, foam, carbon dioxide, (halons).
Explosion: above 25°C: forms explosive air-vapor mixtures.	above 25°C: sealed machinery, ventilation, explosion-proof electrical equipment.	in case of fire: keep tanks/drums cool by spraying with water.
Inhalation: sore throat, cough, shortness of breath.	ventilation, local exhaust or respiratory protection.	fresh air, rest, call a doctor.
Skin: redness.	gloves.	remove contaminated clothing, flush skin with water or shower.
Eyes: redness, pain.	safety glasses.	flush with water, take to doctor if necessary.
Ingestion: abdominal pain, nausea.		rinse mouth, call a doctor.

SPILLAGE	STORAGE	LABELING / NFPA
collect leakage in sealable containers, soak up with sand or other inert absorbent and remove to safe place; (additional individual protection: breathing apparatus).	keep in fireproof place, separate from oxidants.	

NOTES
Explosive limits not given in the literature.

CAS-No: [111-46-6]
diglycol
ethylene diglycol
glycol ether
2,2'-oxibisethanol
2,2'-oxydiethanol

$O(CH_2CH_2OH)_2$

DIETHYLENE GLYCOL

PHYSICAL PROPERTIES		IMPORTANT CHARACTERISTICS	
Boiling point, °C	244	**COLORLESS, ODORLESS, VISCOUS HYGROSCOPIC LIQUID**	
Melting point, °C	−6	Reacts violently with strong oxidants.	
Flash point, °C	124		
Auto-ignition temperature, °C	224	TLV-TWA not available	
Relative density (water = 1)	1.1		
Relative vapor density (air = 1)	3.7		
Relative density at 20 °C of saturated mixture		**Absorption route:** Can enter the body by inhalation or ingestion. Harmful atmospheric	
vapor/air (air = 1)	1.0	concentrations build up very slowly, if at all, on evaporation at approx. 20°C, but much more	
Vapor pressure, mm Hg at 30 °C	0.0076	rapidly in aerosol form.	
Solubility in water	∞	**Immediate effects:** Irritates the eyes and respiratory tract. Liquid destroys the skin's natural oils.	
Explosive limits, vol% in air	1.8-12.2	Affects the nervous system.	
Electrical conductivity, pS/m	5.86 x 10⁷	**Effects of prolonged/repeated exposure:** Can cause liver and kidney damage.	
Relative molecular mass	106.1		
Log P octanol/water	−2.0		
Gross formula: $C_4H_{10}O_3$			

HAZARDS/SYMPTOMS	PREVENTIVE MEASURES	FIRE EXTINGUISHING/FIRST AID
Fire: combustible.	keep away from open flame, no smoking.	powder, alcohol-resistant foam, water spray, carbon dioxide, (halons).
Inhalation: sore throat, dizziness, drowsiness.	ventilation.	fresh air, rest, call a doctor if necessary.
Skin: redness.	gloves.	remove contaminated clothing, flush skin with water or shower.
Eyes: redness, pain.	acid goggles.	flush with water, take to a doctor if necessary.
Ingestion: abdominal pain, headache, nausea, vomiting, drowsiness.		rinse mouth, call a doctor or take to hospital.

SPILLAGE	STORAGE	LABELING / NFPA
collect leakage in sealable containers, flush away remainder with water.	keep dry, separate from oxidants; ventilate at floor level.	◇ 1 / 1 0

NOTES
Log P octanol/water is estimated. Depending on the degree of exposure, regular medical checkups are advisable.

CAS-No: [112-36-7]
bis(2-ethoxyethyl)ether
diethyl carbitol
2-ethoxyethyl ether

$(C_2H_5OCH_2CH_2)_2O$

DIETHYLENE GLYCOL DIETHYL ETHER

PHYSICAL PROPERTIES		IMPORTANT CHARACTERISTICS	
Boiling point, °C	189	**COLORLESS VISCOUS LIQUID**	
Melting point, °C	−44	Vapor mixes readily with air. Able to form peroxides. Irritating fumes are a product of combustion.	
Flash point, °C	82		
Auto-ignition temperature, °C	205	TLV-TWA not available	
Relative density (water = 1)	0.9		
Relative vapor density (air = 1)	5.6		
Relative density at 20 °C of saturated mixture		**Absorption route:** Can enter the body by inhalation or ingestion or through the skin. Harmful	
vapor/air (air = 1)	1.0	atmospheric concentrations build up very slowly, if at all, on evaporation at approx. 20°C, but	
Vapor pressure, mm Hg at 20 °C	ca. 0.59	much more rapidly in aerosol form.	
Solubility in water	∞	**Immediate effects:** Irritates the eyes, skin and respiratory tract.	
Relative molecular mass	162.2	**Effects of prolonged/repeated exposure:** Can cause kidney damage.	
Gross formula:	$C_8H_{18}O_3$		

HAZARDS/SYMPTOMS	PREVENTIVE MEASURES	FIRE EXTINGUISHING/FIRST AID
Fire: combustible.	keep away from open flame, no smoking.	powder, alcohol-resistant foam, water spray, carbon dioxide, (halons).
Explosion: above 82°C: forms explosive air-vapor mixtures.	above 82°C: sealed machinery, ventilation.	
Inhalation: cough, shortness of breath.	ventilation.	fresh air, rest.
Skin: *is absorbed*; redness.	gloves.	remove contaminated clothing, wash skin with soap and water.
Eyes: redness.	face shield.	flush with water, take to a doctor if necessary.
Ingestion: headache, nausea, vomiting.		rinse mouth, call a doctor or take to hospital.

SPILLAGE	STORAGE	LABELING / NFPA
collect leakage in sealable containers, flush away remainder with water.		

NOTES
Before distilling check for peroxides; if found, render harmless. Depending on the degree of exposure, regular medical checkups are advisable.

CAS-No: [112-34-5]
2-(2-butoxyethoxy)ethanol
butyl carbitol
diethylene glycol butyl ether
diethylene glycol n-butyl ether

$C_4H_9(OC_2H_4)_2OH$

DIETHYLENE GLYCOL MONOBUTYL ETHER

PHYSICAL PROPERTIES		IMPORTANT CHARACTERISTICS
Boiling point, °C	230	**COLORLESS, HYGROSCOPIC VISCOUS LIQUID WITH CHARACTERISTIC ODOR**
Melting point, °C	−68	Vapor mixes readily with air. Able to form peroxides. Reacts violently with strong bases. Reacts
Flash point, °C	78	violently with strong oxidants. Attacks light metals (incl. aluminum), giving off → *hydrogen*.
Auto-ignition temperature, °C	225	
Relative density (water = 1)	0.95	TLV-TWA not available
Relative vapor density (air = 1)	5.6	
Relative density at 20 °C of saturated mixture		**Absorption route:** Can enter the body by inhalation or ingestion or through the skin. Harmful
vapor/air (air = 1)	1.0	atmospheric concentrations build up very slowly, if at all, on evaporation at approx. 20°C, but
Vapor pressure, mm Hg at 20 °C	0.023	much more rapidly in aerosol form.
Solubility in water	∞	**Immediate effects:** Irritates the eyes, skin and respiratory tract. Liquid destroys the skin's natural
Explosive limits, vol% in air	0.8-9.4	oils.
Relative molecular mass	162.2	**Effects of prolonged/repeated exposure:** Causes kidney damage.
Log P octanol/water	0.3	
Gross formula:	$C_8H_{18}O_3$	

HAZARDS/SYMPTOMS	PREVENTIVE MEASURES	FIRE EXTINGUISHING/FIRST AID
Fire: combustible.	keep away from open flame, no smoking, avoid contact with strong oxidants.	powder, alcohol-resistant foam, water spray, carbon dioxide, (halons).
Explosion: above 78°C: forms explosive air-vapor mixtures.	above 78°C: sealed machinery, ventilation.	
Inhalation: sore throat, cough.	ventilation.	fresh air, rest, call a doctor if necessary.
Skin: *is absorbed*; redness, rawness.	gloves.	remove contaminated clothing, flush skin with water or shower.
Eyes: redness, pain, impaired vision.	face shield.	flush with water, take to a doctor if necessary.
Ingestion: abdominal pain, headache, nausea.		rinse mouth, call a doctor or take to hospital.

SPILLAGE	STORAGE	LABELING / NFPA
collect leakage in sealable containers, flush away remainder with water.	keep in dry, dark place, separate from strong oxidants and strong bases; ventilate at floor level.	

NOTES
Log P octanol/water is estimated. Before distilling check for peroxides; if found, render harmless. Butyl Dioxitol and Butyl Carbitol are trade names. Packaging: special material.

CAS-No: [111-90-0]
carbitol
diethylene glycol ethyl ether
ethoxydiglycol
2-(2-ethoxyethoxy) ethanol

$C_2H_5(OC_2H_4)_2OH$

DIETHYLENE GLYCOL MONOETHYL ETHER

PHYSICAL PROPERTIES		IMPORTANT CHARACTERISTICS	
Boiling point, °C	202	**COLORLESS VISCOUS LIQUID**	
Melting point, °C	−80	Vapor mixes readily with air. Presumed to be able to form peroxides on exposure to air. Reacts violently with strong oxidants.	
Flash point, °C	94		
Auto-ignition temperature, °C	204		
Relative density (water = 1)	0.99	**TLV-TWA** not available	
Relative vapor density (air = 1)	4.6		
Relative density at 20 °C of saturated mixture			
vapor/air (air = 1)	1.0	**Absorption route:** Can enter the body by inhalation or ingestion or through the skin. Harmful atmospheric concentrations build up very slowly, if at all, on evaporation at approx. 20°C, but much more rapidly in aerosol form.	
Vapor pressure, mm Hg at 20 °C	0.99		
Solubility in water	∞		
Explosive limits, vol% in air	1.2-8.5	**Immediate effects:** Irritates the eyes. Can cause kidney damage.	
Relative molecular mass	134.2		
Log P octanol/water	0.9		
Gross formula: $C_6H_{14}O_3$			

HAZARDS/SYMPTOMS	PREVENTIVE MEASURES	FIRE EXTINGUISHING/FIRST AID
Fire: combustible.	keep away from open flame, no smoking.	powder, water spray, foam, carbon dioxide, (halons).
Explosion: above 94°C: forms explosive air-vapor mixtures.	above 94°C: sealed machinery, ventilation.	
Inhalation: cough.	ventilation.	fresh air, rest.
Skin: redness.	gloves.	remove contaminated clothing, wash skin with soap and water.
Eyes: redness.	face shield.	flush with water, take to a doctor if necessary.
Ingestion: abdominal pain, diarrhea.		rinse mouth, take to hospital if necessary.

SPILLAGE	STORAGE	LABELING / NFPA
collect leakage in sealable containers, flush away remainder with water.		◇ 1 / 1 0

NOTES
The measures on this card also apply to diethylene glycol monomethyl ether. Dioxitol and Carbitol are trade names. Log P octanol/water is estimated.

CAS-No: [111-40-0]
aminoethylethandiamine
3-azapentane-1,5-diamine
DETA
2,2'-diaminodiethyl amine
2,2'-iminobisethylamine

$(NH_2C_2H_4)_2NH$

DIETHYLENETRIAMINE

PHYSICAL PROPERTIES	IMPORTANT CHARACTERISTICS
Boiling point, °C 207 Melting point, °C −39 Flash point, °C 102 Auto-ignition temperature, °C 395 Relative density (water = 1) 0.95 Relative vapor density (air = 1) 3.5 Relative density at 20°C of saturated mixture vapor/air (air = 1) 1.0 Vapor pressure, mm Hg at 20°C 0.23 Solubility in water ∞ Relative molecular mass 103.2	**COLORLESS LIQUID WITH CHARACTERISTIC ODOR** Vapor mixes readily with air. Toxic and corrosive vapors are a product of combustion (→ *nitrous vapors* and → *hydrogen cyanide*). Medium strong base. Reacts with chlorinated hydrocarbons. Reacts with strong oxidants. Promotes spontaneous decomposition of nitrogen compounds. Attacks copper, nickel and cobalt.
	TLV-TWA (skin) 1 ppm 4.2 mg/m³
	Absorption route: Can enter the body by inhalation or ingestion or through the skin. Harmful atmospheric concentrations build up fairly slowly on evaporation at 20°C, but much more rapidly in aerosol form. **Immediate effects:** Corrosive to the eyes, skin and respiratory tract. Can damage the liver. Inhalation of vapor/fumes can cause severe breathing difficulties (lung edema). **Effects of prolonged/repeated exposure:** Repeated slight contact with skin frequently causes hypersensitivity (eczema). Prolonged or repeated exposure to vapor can cause lung disorders (asthma).
Gross formula: $C_4H_{13}N_3$	

HAZARDS/SYMPTOMS	PREVENTIVE MEASURES	FIRE EXTINGUISHING/FIRST AID
Fire: combustible.	keep away from open flame, no smoking.	water spray, alcohol-resistant foam, powder, carbon dioxide, NO halons.
	STRICT HYGIENE	IN ALL CASES CALL IN A DOCTOR
Inhalation: *corrosive*; cough, shortness of breath, severe breathing difficulties.	ventilation, local exhaust or respiratory protection.	fresh air, rest, place in half-sitting position, take to hospital.
Skin: *corrosive*; redness, pain, burns.	gloves, protective clothing.	remove contaminated clothing, flush skin with water or shower, call a doctor.
Eyes: *corrosive*; redness, pain, impaired vision.	face shield.	flush with water, take to doctor.
Ingestion: *corrosive*; sore throat, abdominal pain, diarrhea, vomiting.		rinse mouth, take immediately to hospital.

SPILLAGE	STORAGE	LABELING / NFPA
collect leakage in sealable containers, flush away remainder with water.	keep separate from strong acids and nitrogen compounds; ventilate at floor level.	R: 21/22-34-43 S: 26-36/37/39 Corrosive 1 / 3 0

NOTES
Depending on the degree of exposure, regular medical checkups are advisable. Lung edema symptoms usually develop several hours later and are aggravated by physical exertion: rest and hospitalization essential. As first aid, a doctor or authorized person should consider administering a corticosteroid spray. Unbreakable packaging preferred; if breakable, keep in unbreakable container.

Transport Emergency Card TEC(R)-80G16

HI: 80; UN-number: 2079

1,2-diethoxy ethane
ethylene glycol diethyl ether

$C_2H_5OC_2H_4OC_2H_5$

DIETHYL GLYCOL

PHYSICAL PROPERTIES	IMPORTANT CHARACTERISTICS
Boiling point, °C 121	**COLORLESS LIQUID WITH CHARACTERISTIC ODOR**
Melting point, °C .... −74	Vapor mixes readily with air. Presumed to be able to form peroxides on contact with oxidants. Reacts violently with strong oxidants.
Flash point, °C 35	
Auto-ignition temperature, °C 207	
Relative density (water = 1) 0.85	
Relative vapor density (air = 1) 4.1	**TLV-TWA** not available
Relative density at 20 °C of saturated mixture vapor/air (air = 1) 1.04	**Absorption route:** Can enter the body by inhalation or ingestion or through the skin. Harmful atmospheric concentrations build up fairly slowly on evaporation, but much more rapidly in aerosol form.
Vapor pressure, mm Hg at 20 °C 9.1	
Solubility in water .... none	
Explosive limits, vol% in air .... ?-?	**Immediate effects:** Irritates the eyes, skin and respiratory tract. Affects the nervous system.
Relative molecular mass 118.2	**Effects of prolonged/repeated exposure:** Prolonged contact can cause skin disorders (burns). Can cause liver and kidney damage.
Gross formula: $C_6H_{14}O_2$	

HAZARDS/SYMPTOMS	PREVENTIVE MEASURES	FIRE EXTINGUISHING/FIRST AID
Fire: flammable.	keep away from open flame and sparks, no smoking.	powder, AFFF, foam, carbon dioxide, (halons).
Explosion: above 35°C: forms explosive air-vapor mixtures.	above 35°C: sealed machinery, ventilation, explosion-proof electrical equipment.	in case of fire: keep tanks/drums cool by spraying with water.
Inhalation: sore throat, cough, drowsiness, sleepiness.	ventilation, local exhaust or respiratory protection.	fresh air, rest.
Skin: *is absorbed*; redness.	gloves.	remove contaminated clothing, flush skin with water or shower.
Eyes: redness, pain.	face shield.	flush with water, take to a doctor if necessary.
Ingestion: abdominal pain, nausea, vomiting.		rinse mouth, call a doctor or take to hospital.

SPILLAGE	STORAGE	LABELING / NFPA
collect leakage in sealable containers, soak up with sand or other inert absorbent and remove to safe place.	keep in fireproof place, separate from oxidants.	3 / 0

NOTES
Explosive limits not given in the literature. Before distilling check for peroxides; if found, render harmless.

Transport Emergency Card TEC(R)-30G08

HI: 30; UN-number: 1153

metacetone
3-pentanone
propione

$C_2H_5COC_2H_5$

DIETHYL KETONE

PHYSICAL PROPERTIES	IMPORTANT CHARACTERISTICS
Boiling point, °C 102 Melting point, °C −40 Flash point, °C 12 Auto-ignition temperature, °C 445 Relative density (water = 1) 0.8 Relative vapor density (air = 1) 3.0 Relative density at 20°C of saturated mixture vapor/air (air = 1) 1.04 Vapor pressure, mm Hg at 20°C 15 Solubility in water, g/100 ml at 20°C 4.7 Explosive limits, vol% in air 1.6-? Electrical conductivity, pS/m $> 10^4$ Relative molecular mass 86.1 Gross formula: $C_5H_{10}O$	**COLORLESS LIQUID WITH CHARACTERISTIC ODOR** Vapor mixes readily with air, forming explosive mixtures. Do not use compressed air when filling, emptying or processing. Reacts violently with oxidants, with risk of fire and explosion. Attacks many plastics. TLV-TWA 200 ppm 705 mg/m³ **Absorption route:** Can enter the body by inhalation or ingestion or through the skin. **Immediate effects:** Irritates the eyes, skin and respiratory tract. Liquid destroys the skin's natural oils. Affects the nervous system.

HAZARDS/SYMPTOMS	PREVENTIVE MEASURES	FIRE EXTINGUISHING/FIRST AID
Fire: extremely flammable.	keep away from open flame and sparks, no smoking.	powder, AFFF, foam, carbon dioxide, (halons).
Explosion: forms explosive air-vapor mixtures.	sealed machinery, ventilation, explosion-proof electrical equipment and lighting.	in case of fire: keep tanks/drums cool by spraying with water.
Inhalation: sore throat, cough, headache, dizziness, drowsiness, sleepiness.	ventilation, local exhaust or respiratory protection.	fresh air, rest, place in half-sitting position, call a doctor.
Skin: *is absorbed*; redness, pain.	gloves.	remove contaminated clothing, flush skin with water or shower, refer to a doctor.
Eyes: redness, pain.	safety glasses.	flush with water, take to a doctor.
Ingestion: sore throat, abdominal pain, nausea.		rinse mouth, call a doctor or take to hospital.

SPILLAGE	STORAGE	LABELING / NFPA
collect leakage in sealable containers, soak up with sand or other inert absorbent and remove to safe place, do not flush into sewer.	keep in fireproof place, separate from oxidants.	R: 11 S: 9-16-33 🔥 Flammable

NFPA:
```
      3
    1   0
```

NOTES

DIETHYL OXALATE

PHYSICAL PROPERTIES	IMPORTANT CHARACTERISTICS	
Boiling point, °C 185 Melting point, °C −41 Flash point, °C 76 Relative density (water = 1) 1.08 Relative vapor density (air = 1) 5.0 Relative density at 20 °C of saturated mixture vapor/air (air = 1) 1.0 Vapor pressure, mm Hg at 20 °C ca. 0.38 Solubility in water poor Electrical conductivity, pS/m 7.12x10⁷ Relative molecular mass 146.1	**COLORLESS VISCOUS LIQUID WITH CHARACTERISTIC ODOR** Vapor mixes readily with air. Decomposes slowly on contact with water forming → *oxalic acid*. Reacts with strong oxidants.	
	TLV-TWA not available	
	Absorption route: Can enter the body by inhalation or ingestion or through the skin. Harmful atmospheric concentrations build up fairly slowly on evaporation at 20°C, but much more rapidly in aerosol form. **Immediate effects:** Irritates the eyes, skin and respiratory tract. Liquid destroys the skin's natural oils. Affects the nervous system. If liquid is swallowed, droplets can enter the lungs, with risk of pneumonia. **Effects of prolonged/repeated exposure:** Can cause kidney damage.	
Gross formula: C₆H₁₀O₄		

HAZARDS/SYMPTOMS	PREVENTIVE MEASURES	FIRE EXTINGUISHING/FIRST AID
Fire: combustible.	keep away from open flame, no smoking.	powder, AFFF, foam, carbon dioxide, (halons).
Explosion: above 76°C: forms explosive air-vapor mixtures.	above 76°C: sealed machinery, ventilation.	
Inhalation: headache, nausea.	ventilation.	fresh air, rest.
Skin: redness.	gloves.	remove contaminated clothing, flush skin with water or shower, absorb as much of the substance with an absorbent and wash until there is no longer a sour reaction.
Eyes: redness, pain.	face shield.	flush with water, take to a doctor if necessary.
Ingestion: headache, abdominal pain, nausea.		rinse mouth, call a doctor or take to hospital.

SPILLAGE	STORAGE	LABELING / NFPA
collect leakage in sealable containers, soak up with sand or other inert absorbent and remove to safe place.	keep dry, separate from oxidants; ventilate at floor level.	R: 22/36 S: 23 ✖ Harmful

NOTES
Alcohol consumption increases toxic effects.

DEP
ethyl phthalate
phthalic acid diethyl ester

$C_6H_4(COOC_2H_5)_2$

p-DIETHYL PHTHALATE

PHYSICAL PROPERTIES		IMPORTANT CHARACTERISTICS	
Boiling point, °C	302	**WHITE POWDER**	
Melting point, °C	44	Vapor mixes readily with air. Reacts with strong oxidants.	
Flash point, °C	117		
Auto-ignition temperature, °C	?	TLV-TWA	5 mg/m³
Relative density (water = 1)	1.1		
Relative vapor density (air = 1)	7.7		
Relative density at 20 °C of saturated mixture		**Absorption route:** Can enter the body by inhalation or ingestion. Harmful atmospheric	
vapor/air (air = 1)	1.0	concentrations build up very slowly, if at all, on evaporation at approx. 20°C, but harmful	
Vapor pressure, mm Hg at 20 °C	ca. 0.076	concentrations of airborne particles can build up much more rapidly.	
Solubility in water	none	**Immediate effects:** Irritates the eyes, skin and respiratory tract. In high concentrations affects	
Relative molecular mass	222.2	the nervous system.	
Gross formula:	$C_{12}H_{14}O_4$		

HAZARDS/SYMPTOMS	PREVENTIVE MEASURES	FIRE EXTINGUISHING/FIRST AID
Fire: combustible.	keep away from open flame, no smoking.	water spray, powder.
Inhalation: dizziness, drowsiness.	local exhaust or respiratory protection.	fresh air, rest.
Skin: redness.	gloves.	remove contaminated clothing, flush skin with water or shower.
Eyes: redness, pain.	goggles.	flush with water, take to a doctor if necessary.
Ingestion: abdominal pain, nausea, dizziness.		rinse mouth, call a doctor.

SPILLAGE	STORAGE	LABELING / NFPA
clean up spillage, flush away remainder with water; (additional individual protection: P2 respirator).		

NOTES

DIETHYL SUCCINATE

PHYSICAL PROPERTIES	IMPORTANT CHARACTERISTICS
Boiling point, °C 217 Melting point, °C −21 Flash point, °C 90 Relative density (water = 1) 1.04 Relative vapor density (air = 1) 6.0 Relative density at 20 °C of saturated mixture vapor/air (air = 1) 1.0 Vapor pressure, mm Hg at 55 °C 0.99 Solubility in water poor Relative molecular mass 174.2	**COLORLESS LIQUID WITH CHARACTERISTIC ODOR** Vapor mixes readily with air. Reacts with strong oxidants. TLV-TWA not available **Absorption route:** Can enter the body by inhalation or ingestion. Harmful atmospheric concentrations build up very slowly, if at all, on evaporation at approx. 20°C, but much more rapidly in aerosol form. **Immediate effects:** Irritates the eyes, skin and respiratory tract.
Gross formula: $C_8H_{14}O_4$	

HAZARDS/SYMPTOMS	PREVENTIVE MEASURES	FIRE EXTINGUISHING/FIRST AID
Fire: combustible.	keep away from open flame, no smoking.	powder, AFFF, foam, carbon dioxide, (halons).
Explosion: above 90°C: forms explosive air-vapor mixtures.	above 90°C: sealed machinery, ventilation.	
Inhalation: sore throat, cough.	ventilation.	fresh air, rest, call a doctor.
Skin: redness.	gloves.	remove contaminated clothing, wash skin with soap and water, take to a doctor if necessary.
Eyes: redness, pain.	safety glasses.	flush with water, take to a doctor if necessary.
Ingestion: sore throat, abdominal pain.		rinse mouth, call a doctor.

SPILLAGE	STORAGE	LABELING / NFPA
collect leakage in sealable containers, flush away remainder with water.	keep separate from strong oxidants.	1 / 1 / 0

NOTES
Insufficient toxicological data on harmful effects to humans.

Transport Emergency Card TEC(R)-30G15

CAS-No: [64-67-5]
diethyl sulphate
ethyl sulfate
sulfuric acid diethyl ester

$(C_2H_5)_2SO_4$

DIETHYLSULFATE

PHYSICAL PROPERTIES	IMPORTANT CHARACTERISTICS
Decomposes below boiling point, °C　208 Melting point, °C　−25 Flash point, °C　104 Auto-ignition temperature, °C　436 Relative density (water = 1)　1.2 Relative vapor density (air = 1)　5.3 Relative density at 20°C of saturated mixture vapor/air (air = 1)　1.0 Vapor pressure, mm Hg at 20°C　0.19 Solubility in water　reaction Relative molecular mass　154.2	**COLORLESS OILY LIQUID WITH CHARACTERISTIC ODOR, TURNING BROWN ON EXPOSURE TO AIR** Vapor mixes readily with air. Decomposes when heated, giving off combustible and toxic vapors (→ *ether* and → *sulfur dioxide*). Reacts with oxidants. Reacts violently with warm water and water vapor, forming → *monoethylsulfate* and → *ethanol*.
	TLV-TWA　　　　　　　not available
	Absorption route: Can enter the body by inhalation or ingestion. Harmful atmospheric concentrations build up fairly slowly on evaporation at 20°C, but much more rapidly in aerosol form. **Immediate effects:** Corrosive to the eyes, skin and respiratory tract. Inhalation of vapor/fumes can cause severe breathing difficulties (lung edema).
Gross formula:　　　　　$C_4H_{10}O_4S$	

HAZARDS/SYMPTOMS	PREVENTIVE MEASURES	FIRE EXTINGUISHING/FIRST AID
Fire: combustible.	keep away from open flame, no smoking.	powder, water spray, foam, carbon dioxide, (halons).
Explosion:		in case of fire: keep tanks/drums cool by spraying with water.
Inhalation: *corrosive*; sore throat, cough, shortness of breath, severe breathing difficulties.	ventilation, local exhaust or respiratory protection.	fresh air, rest, place in half-sitting position, take to hospital.
Skin: *corrosive*; redness, pain, burns.	gloves, protective clothing.	remove contaminated clothing, dab skin with tissues, then flush with water because of violent reaction, refer to a doctor.
Eyes: *corrosive*; redness, pain, impaired vision.	face shield, or combined eye and respiratory protection.	flush with water, take to a doctor.
Ingestion: *corrosive*; abdominal pain, diarrhea, vomiting.		rinse mouth, immediately take to hospital.

SPILLAGE	STORAGE	LABELING / NFPA
collect leakage in sealable containers, soak up with sand or other inert absorbent and remove to safe place; (additional individual protection: breathing apparatus).	keep in dry, fireproof place.	R: 45-46-20/21/22-34 S: 53-26-44 Note E ☠ Toxic

NFPA diagram: 3 (left), 1 (top), 1 (right)

NOTES
Lung edema symptoms usually develop several hours later and are aggravated by physical exertion: rest and hospitalization essential. As first aid, a doctor or authorized person should consider administering a corticosteroid spray. A study from 1979 indicates a possible increased risk of cancer of the upper respiratory tubes in humans. Is carcinogenic in animals. Use airtight packaging. Unbreakable packaging preferred; if breakable, keep in unbreakable container.

Transport Emergency Card TEC(R)-61G08B　　　　　　　　　　　　　　　　**HI: 60; UN-number: 1594**

CAS-No: [352-93-2]
ethyl sulfide
ethylthioethane
3-thiapentane
1,1'-thiobisethane
thioethylether

$(C_2H_5)_2S$

DIETHYL SULFIDE

PHYSICAL PROPERTIES		IMPORTANT CHARACTERISTICS
Boiling point, °C	92	**COLORLESS LIQUID WITH CHARACTERISTIC ODOR**
Melting point, °C	− 104	Vapor is heavier than air and spreads at ground level, with risk of ignition at a distance. Do not use compressed air when filling, emptying or processing. Toxic and corrosive gases are a product of combustion (→ *sulfur dioxide*). Decomposes when heated or on contact with acids or water, giving off toxic gas (→ *hydrogen sulfide*). Reacts with strong oxidants.
Flash point, °C	− 10	
Relative density (water = 1)	0.8	
Relative vapor density (air = 1)	3.1	
Relative density at 20 °C of saturated mixture vapor/air (air = 1)	1.13	
Vapor pressure, mm Hg at 20 °C	50	TLV-TWA not available
Solubility in water	poor	
Explosive limits, vol% in air	?-?	**Absorption route:** Can enter the body by inhalation or ingestion. Insufficient data available on the rate at which harmful concentrations can build up.
Relative molecular mass	90.2	
Log P octanol/water	2.0	
Gross formula:	$C_4H_{10}S$	

HAZARDS/SYMPTOMS	PREVENTIVE MEASURES	FIRE EXTINGUISHING/FIRST AID
Fire: extremely flammable.	keep away from open flame and sparks, no smoking.	powder, AFFF, foam, carbon dioxide, (halons).
Explosion: forms explosive air-vapor mixtures.	sealed machinery, ventilation, explosion-proof electrical equipment and lighting.	in case of fire: keep tanks/drums cool by spraying with water, but DO NOT spray substance with water.
Inhalation:	ventilation, local exhaust or respiratory protection.	fresh air, rest, call a doctor.
Skin:	gloves.	remove contaminated clothing, flush skin with water or shower.
Eyes:	face shield.	flush with water, take to a doctor if necessary.
Ingestion:		rinse mouth, call a doctor.

SPILLAGE	STORAGE	LABELING / NFPA
collect leakage in sealable containers, soak up with sand or other inert absorbent and remove to safe place; (additional individual protection: breathing apparatus).	keep in dry, fireproof place, separate from strong acids.	

NOTES
Explosive limits not given in the literature. Insufficient medical data on effects on human health. Since it decomposes on contact with water, it could cause symptoms of H_2S-poisoning if inhaled or ingested. Use airtight packaging.

Transport Emergency Card TEC(R)-30G32

HI: 336; UN-number: 2375

difluoroethane
ethylene fluoride
ethylidene difluoride
ethylidene fluoride

1,1-DIFLUOROETHANE
(cylinder)

PHYSICAL PROPERTIES	IMPORTANT CHARACTERISTICS
Boiling point, °C — 25 Melting point, °C — 117 Flash point, °C flammable gas Relative density (water = 1) 0.9 Relative vapor density (air = 1) 2.3 Vapor pressure, mm Hg at 20 °C 4.3 Solubility in water poor Explosive limits, vol% in air 3.7-18 Relative molecular mass 66.1	**COLORLESS COMPRESSED LIQUEFIED GAS WITH CHARACTERISTIC ODOR** Gas is heavier than air and spreads at ground level, with risk of ignition at a distance. Flow, agitation etc. can cause build-up of electrostatic charge due to liquid's low conductivity. Decomposes when heated or burned, giving off toxic gases (→ *hydrogen fluoride*). Reacts violently with strong oxidants. Attacks metals in the presence of moisture.
	TLV-TWA not available
Gross formula: C$_2$H$_4$F$_2$	**Absorption route:** Can enter the body by inhalation. Harmful atmospheric concentrations can build up very rapidly if gas is released. **Immediate effects:** Irritates the eyes. Cold liquid can cause frostbite. Affects the nervous system. In serious cases risk of unconsciousness.

HAZARDS/SYMPTOMS	PREVENTIVE MEASURES	FIRE EXTINGUISHING/FIRST AID
Fire: extremely flammable.	keep away from open flame and sparks, no smoking.	shut off supply; if impossible and no danger to surrounding area, allow fire to burn itself out; otherwise extinguish with powder, carbon dioxide, (halons).
Explosion: forms explosive air-gas mixtures.	sealed machinery, ventilation, explosion-proof electrical equipment and lighting, grounding when pumping etc. in liquid form.	in case of fire: keep cylinders cool by spraying with water.
Inhalation: cough, dizziness, drowsiness.	ventilation, local exhaust or respiratory protection.	fresh air, rest, call a doctor.
Skin: *in case of frostbite*: redness, pain, sores.	insulated gloves.	*in case of frostbite*: DO NOT remove clothing, flush skin with water or shower, call a doctor if necessary.
Eyes: *in case of frostbite*: redness, pain, impaired vision.	acid goggles, or combined eye and respiratory protection.	flush with water, take to a doctor.

SPILLAGE	STORAGE	LABELING / NFPA
evacuate area, call in an expert, ventilate, under no circumstances spray liquid with water; (additional individual protection: breathing apparatus).	if stored indoors, keep in cool, fireproof place.	

NOTES
Alcohol consumption increases toxic effects. High atmospheric concentrations, e.g. in poorly ventilated spaces, can cause oxygen deficiency, with risk of unconsciousness. Do not spray leaking cylinder with water (to avoid corrosion). Turn leaking cylinder so that leak is on top to prevent liquid difluorethane escaping.

Transport Emergency Card TEC(R)-20G08

HI: 23; UN-number: 1030

1,1-difluoroethene
difluoro-1,1 ethylene
vinylidene difluoride
vinylidene fluoride

$CH_2=CF_2$

1,1-DIFLUOROETHYLENE
(cylinder)

PHYSICAL PROPERTIES	IMPORTANT CHARACTERISTICS
Boiling point, °C −82 Melting point, °C −144 Flash point, °C flammable gas Auto-ignition temperature, °C 480 Relative density (water = 1) 0.62 Relative vapor density (air = 1) 2.2 Vapor pressure, mm Hg at 21 °C 28 Solubility in water poor Explosive limits, vol% in air 5.5-21.3 Relative molecular mass 64.0	**COLORLESS COMPRESSED LIQUEFIED GAS WITH CHARACTERISTIC ODOR** Gas is heavier than air and spreads at ground level, with risk of ignition at a distance. Able to form peroxides. Under certain conditions can polymerize violently, evolving large quantities of heat. Flow, agitation etc. can cause build-up of electrostatic charge due to liquid's low conductivity. Decomposes when heated or burned, giving off toxic gases (→ *hydrogen fluoride*). Reacts violently with oxidants.
	TLV-TWA not available
	Absorption route: Can enter the body by inhalation. Harmful atmospheric concentrations can build up very rapidly if gas is released. **Immediate effects:** Irritates the eyes and respiratory tract. Liquid can cause frostbite due to rapid evaporation. In serious cases risk of unconsciousness.
Gross formula: $C_2H_2F_2$	

HAZARDS/SYMPTOMS	PREVENTIVE MEASURES	FIRE EXTINGUISHING/FIRST AID
Fire: extremely flammable.	keep away from open flame and sparks, no smoking.	shut off supply; if impossible and no danger to surrounding area, allow fire to burn itself out; otherwise extinguish with powder, carbon dioxide, (halons).
Explosion: forms explosive air-gas mixtures.	sealed machinery, ventilation, explosion-proof electrical equipment and lighting, grounding when pumping etc. in liquid form.	in case of fire: keep cylinders cool by spraying with water.
Inhalation: sore throat, cough, shortness of breath, nausea.	ventilation, local exhaust or respiratory protection.	fresh air, rest.
Skin: *in case of frostbite:* redness, pain, sores.	insulating gloves.	*in case of frostbite:* DO NOT remove clothing, flush skin with water or shower, call a doctor if necessary.
Eyes: *in case of frostbite:* redness, pain, impaired vision.	acid goggles, or combined eye and respiratory protection.	flush with water, take to a doctor.

SPILLAGE	STORAGE	LABELING / NFPA
evacuate area, call in an expert, ventilate, under no circumstances spray liquid with water; (additional individual protection: breathing apparatus).	if stored indoors, keep in cool, fireproof place, separate from oxidants.	 4 1 2

NOTES
Before distilling check for peroxides; if found, render harmless. High atmospheric concentrations, e.g. in poorly ventilated spaces, can cause oxygen deficiency, with risk of unconsciousness. Do not spray leaking cylinder with water (to avoid corrosion). Turn leaking cylinder so that leak is on top to prevent liquid difluoroethylene escaping.

Transport Emergency Card TEC(R)-683

HI: 239; UN-number: 1959

CAS-No: [2238-07-5]
di-(2,3-epoxypropyl)ether
DGE

$$CH_2-CH-CH_2-O-CH_2-CH-CH_2$$

DIGLYCIDYL ETHER

PHYSICAL PROPERTIES	IMPORTANT CHARACTERISTICS	
Boiling point, °C 260 Flash point, °C see notes Relative density (water = 1) 1.3 Relative vapor density (air = 1) 4.5 Relative density at 20 °C of saturated mixture vapor/air (air = 1) 1.00 Vapor pressure, mm Hg at 20 °C 0.091 Relative molecular mass 130.1	**COLORLESS LIQUID** Vapor mixes readily with air. Presumed to be able to form peroxides. Reacts with strong oxidants.	
	TLV-TWA 0.1 ppm 0.5 mg/m³	
	Absorption route: Can enter the body by inhalation or ingestion or through the skin. Harmful atmospheric concentrations can build up fairly rapidly on evaporation at approx. 20°C - even more rapidly in aerosol form. **Immediate effects:** Corrosive to the eyes, skin and respiratory tract. Affects the nervous system, blood and internal tract. Inhalation of vapor/fumes can cause severe breathing difficulties (lung edema). **Effects of prolonged/repeated exposure:** Prolonged or repeated contact can cause eczema or other skin disorders. Can affect the blood. Can cause liver and kidney damage.	
Gross formula: $C_6H_{10}O_3$		

HAZARDS/SYMPTOMS	PREVENTIVE MEASURES	FIRE EXTINGUISHING/FIRST AID
Fire: combustible.	keep away from open flame, no smoking.	powder, water spray, foam, carbon dioxide, (halons).
Inhalation: sore throat, cough, shortness of breath, severe breathing difficulties.	ventilation, local exhaust or respiratory protection.	fresh air, rest, place in half-sitting position, take to hospital.
Skin: redness, pain, serious burns.	gloves, protective clothing.	remove contaminated clothing, flush skin with water or shower, take to hospital.
Eyes: redness, pain, impaired vision.	face shield.	flush with water, take to a doctor.

SPILLAGE	STORAGE	LABELING / NFPA
collect leakage in sealable containers, soak up with sand or other inert absorbent and remove to safe place; (additional individual protection: breathing apparatus).	ventilate at floor level.	

NOTES
Combustible, but flash point and explosive limits not given in the literature. Before distilling check for peroxides; if found, render harmless. Lung edema symptoms usually develop several hours later and are aggravated by physical exertion: rest and hospitalization essential. As first aid, a doctor or authorized person should consider administering a corticosteroid spray.

2,3-DIHYDROPYRAN

PHYSICAL PROPERTIES	IMPORTANT CHARACTERISTICS
Boiling point, °C 84 Melting point, °C −70 Flash point, °C −18 Relative density (water = 1) 0.9 Relative vapor density (air = 1) 2.9 Relative density at 20 °C of saturated mixture vapor/air (air = 1) 1.16 Vapor pressure, mm Hg at 20 °C ca. 65 Solubility in water, g/100 ml at 25 °C 1.6 Explosive limits, vol% in air ?-? Minimum ignition energy, mJ 0.36 Relative molecular mass 84.1 Gross formula: C_5H_8O	**COLORLESS LIQUID WITH CHARACTERISTIC ODOR** Vapor is heavier than air and spreads at ground level, with risk of ignition at a distance. Presumed to be able to form peroxides. Flow, agitation etc. can cause build-up of electrostatic charge due to liquid's low conductivity. Do not use compressed air when filling, emptying or processing. Reacts violently with strong oxidants, with risk of fire and explosion. TLV-TWA not available **Absorption route:** Can enter the body by inhalation or ingestion. **Immediate effects:** Irritates the eyes, skin and respiratory tract.

HAZARDS/SYMPTOMS	PREVENTIVE MEASURES	FIRE EXTINGUISHING/FIRST AID
Fire: extremely flammable.	keep away from open flame and sparks, no smoking.	powder, AFFF, foam, carbon dioxide, (halons).
Explosion: forms explosive air-vapor mixtures.	sealed machinery, ventilation, explosion-proof electrical equipment and lighting, grounding, non-sparking tools.	in case of fire: keep tanks/drums cool by spraying with water.
Inhalation: sore throat, cough, shortness of breath.	ventilation, local exhaust or respiratory protection.	fresh air, rest, call a doctor.
Skin: redness, pain.	gloves.	remove contaminated clothing, flush skin with water or shower.
Eyes: redness, pain, impaired vision.	safety glasses.	flush with water, take to a doctor if necessary.
Ingestion: sore throat, abdominal pain, nausea.		rinse mouth, call a doctor.

SPILLAGE	STORAGE	LABELING / NFPA
collect leakage in sealable containers, soak up with sand or other inert absorbent and remove to safe place, do not flush into sewer; (additional individual protection: breathing apparatus).	keep in fireproof place, separate from oxidants.	 3 / 2 / 0

NOTES
Explosive limits not given in the literature. Before distilling check for peroxides; if found, render harmless. Insufficient toxicological data on harmful effects to humans.

Transport Emergency Card TEC(R)-30G30

HI: 33; UN-number: 2376

DIISOBUTYL ALUMINUM CHLORIDE

PHYSICAL PROPERTIES	IMPORTANT CHARACTERISTICS
Decomposes below boiling point, °C 138 Melting point, °C −39.5 Relative density (water = 1) 0.9 Relative vapor density (air = 1) 6.0 Vapor pressure, mm Hg at 60 °C 0.030 Solubility in water reaction Relative molecular mass 176.5	**COLORLESS FUMING LIQUID** Ignites spontaneously on exposure to air. Aluminum oxide fumes and → *hydrochloric acid* fumes are a product of combustion. Strong reducing agent which reacts violently with oxidants. Reacts violently with water, alcohols, phenols, amines, carbon dioxide, oxides of sulfur and nitrogen oxides, with risk of fire and explosion. Reacts with air to form aluminum oxide fumes and → *hydrochloric acid* fumes.

| | TLV-TWA | 2 mg/m³ |

| | **Absorption route:** Can enter the body by inhalation or ingestion. Harmful atmospheric concentrations can build up fairly rapidly on evaporation at approx. 20° C - even more rapidly in aerosol form.
Immediate effects: Corrosive to the eyes, skin and respiratory tract. Inhalation of vapor/fumes can cause severe breathing difficulties (lung edema). In serious cases risk of death. |

Gross formula: $C_8H_{18}AlCl$

HAZARDS/SYMPTOMS	PREVENTIVE MEASURES	FIRE EXTINGUISHING/FIRST AID
Fire: extremely flammable, many chemical reactions can cause fire and explosion.	keep away from open flame and sparks, no smoking, avoid contact with air, water and many other substances.	special extinguishing powder, dry sand, NO other extinguishing agents, can reignite after being extinguished.
Explosion: contact with water and many other substances can cause explosion.		in case of fire: keep undamaged cylinders cool by spraying with water, but DO NOT spray substance with water.
	STRICT HYGIENE	
Inhalation: *corrosive*; sore throat, cough, severe breathing difficulties.	completely sealed machinery, ventilation, local exhaust or respiratory protection.	fresh air, rest, place in half-sitting position, take to hospital.
Skin: *corrosive*; serious burns, slow healing.	gloves, protective clothing.	remove contaminated clothing, flush skin with water or shower, refer to a doctor.
Eyes: *corrosive*; redness, pain, impaired vision.	face shield, or combined eye and respiratory protection.	flush with water, take to a doctor.
Ingestion: *corrosive*; sore throat, abdominal pain, diarrhea.		rinse mouth, take immediately to hospital.

SPILLAGE	STORAGE	LABELING / NFPA
evacuate area, call in an expert, dilute leakage with diesel/machine oil, soak up/cover with dry sand/table salt and remove to safe place; (additional individual protection: CHEMICAL SUIT).	keep in fireproof place under dry nitrogen, separate from all other substances.	

NOTES
Often sold as 15-30% solution in hydrocarbons: flash point and explosive limits depend on solvent used. 15-30% solution does not ignite spontaneously on exposure to air (see relevant entry). Reacts violently with extinguishing agents such as carbon dioxide, water, foam and ordinary powder. Lung edema symptoms usually develop several hours later and are aggravated by physical exertion: rest and hospitalization essential. As first aid, a doctor or authorized person should consider administering a corticosteroid spray. Use cylinder with special fittings.

Transport Emergency Card TEC(R)-42G02

HI: X333; UN-number: 2221

DIISOBUTYL ALUMINUM CHLORIDE

(15% solution in hexane)

PHYSICAL PROPERTIES		IMPORTANT CHARACTERISTICS
Boiling point, °C	69	**COLORLESS FUMING LIQUID**
Melting point, °C	− 95	Vapor is heavier than air and spreads at ground level, with risk of ignition at a distance. Aluminum oxide fumes and → *hydrochloric acid* fumes are a product of combustion. Can ignite spontaneously on exposure to air once solvent has evaporated. Strong reducing agent which reacts violently with oxidants. Reacts violently with water, alcohols, phenols, amines, carbon dioxide, oxides of sulfur, nitrogen oxides and many other substances, with risk of fire and explosion.
Flash point, °C	− 26	
Auto-ignition temperature, °C	240	
Relative density (water = 1)	0.73	
Relative vapor density (air = 1)	3.0	
Relative density at 20 °C of saturated mixture vapor/air (air = 1)	1.3	
Vapor pressure, mm Hg at 20 °C	122	
Solubility in water	reaction	**TLV-TWA** 2 mg/m³ ✷
Explosive limits, vol% in air	1.1-7.5	
Minimum ignition energy, mJ	0.24	**Absorption route:** Can enter the body by inhalation or ingestion. Harmful atmospheric concentrations can build up fairly rapidly on evaporation at approx. 20°C - even more rapidly in aerosol form.
Electrical conductivity, pS/m	100	
		Immediate effects: Corrosive to the eyes, skin and respiratory tract. Inhalation of vapor/fumes can cause severe breathing difficulties (lung edema). In serious cases risk of death.
Gross formula:	$C_8H_{18}AlCl$	

HAZARDS/SYMPTOMS	PREVENTIVE MEASURES	FIRE EXTINGUISHING/FIRST AID
Fire: extremely flammable, many chemical reactions can cause fire and explosion.	keep away from open flame and sparks, no smoking, avoid contact with water, alcohols and many other substances.	special extinguishing powder, dry sand, NO other extinguishing agents, can reignite after being extinguished.
Explosion: forms explosive air-vapor mixtures, contact with water and many other substances can cause explosion.	sealed machinery, ventilation, explosion-proof electrical equipment and lighting, grounding.	in case of fire: keep undamaged cylinders cool by spraying with water, but DO NOT spray substance with water.
	STRICT HYGIENE	
Inhalation: *corrosive*; sore throat, cough, severe breathing difficulties.	sealed machinery, ventilation, local exhaust or respiratory protection.	fresh air, rest, place in half-sitting position, take to hospital.
Skin: *corrosive*; redness, serious burns, slow healing.	gloves, protective clothing.	remove contaminated clothing, flush skin with water or shower, refer to a doctor.
Eyes: *corrosive*; redness, pain, impaired vision.	face shield, or combined eye and respiratory protection.	flush with water, take to a doctor.
Ingestion: *corrosive*; sore throat, abdominal pain, diarrhea.		rinse mouth, take immediately to hospital.

SPILLAGE	STORAGE	LABELING / NFPA
evacuate area, call in an expert, collect leakage in sealable containers, soak up remainder with dry sand/table salt, under no circumstances spray liquid with water; (additional individual protection: CHEMICAL SUIT).	keep in dry, fireproof place under inert gas, separate from all other substances.	

NOTES
Physical properties (boiling point, flash point, autoignition point, explosive limits, vapor pressure, relative density, conductivity and ignition energy) are based on hexane. For pure diisobutyl aluminum chloride, see relevant entry. ✷ For TLV of *hexane*, see relevant entry. Reacts violently with extinguishing agents such as water, foam, halons, carbon dioxide and ordinary powder. Remove contaminated clothing immediately and flush with water (risk of spontaneous combustion). Lung edema symptoms usually develop several hours later and are aggravated by physical exertion: rest and hospitalization essential. As first aid, a doctor or authorized person should consider administering a corticosteroid spray. Use cylinder with special fittings.

Transport Emergency Card TEC(R)-42G02

HI: X333; UN-number: 2220

DIISOBUTYL ALUMINUM HYDRIDE

PHYSICAL PROPERTIES	IMPORTANT CHARACTERISTICS	
Boiling point, °C 114(0.76mm Hg) Melting point, °C −70 Relative density (water = 1) 0.79 Relative vapor density (air = 1) 4.1 Vapor pressure, mm Hg at 60 °C 0.076 Solubility in water reaction Relative molecular mass 142.2	**COLORLESS FUMING LIQUID** Ignites spontaneously on exposure to air. Aluminum oxide vapors are a product of combustion. Strong reducing agent which reacts violently with oxidants. Reacts violently with water, phenols, alcohols, amines, carbon dioxide, oxides of sulfur, nitrogen oxides, halogens, halogenated hydrocarbons and many other substances, with risk of fire and explosion. Reacts with air to form irritating vapors.	
	TLV-TWA	2 mg/m³
	Absorption route: Can enter the body by inhalation or ingestion. Harmful atmospheric concentrations can build up fairly rapidly on evaporation at approx. 20°C - even more rapidly in aerosol form. **Immediate effects:** Corrosive to the eyes, skin and respiratory tract. Inhalation of vapor/fumes can cause severe breathing difficulties (lung edema). In serious cases risk of death.	
Gross formula: $C_8H_{19}Al$		

HAZARDS/SYMPTOMS	PREVENTIVE MEASURES	FIRE EXTINGUISHING/FIRST AID
Fire: extremely flammable, many chemical reactions can cause fire and explosion.	keep away from open flame and sparks, no smoking, avoid contact with air, water and many other substances.	special extinguishing powder, dry sand, NO other extinguishing agents, can reignite after being extinguished.
Explosion: contact with water and many other substances can cause explosion.		in case of fire: keep undamaged cylinders cool by spraying with water, but DO NOT spray substance with water.
	STRICT HYGIENE	
Inhalation: *corrosive*; sore throat, cough, severe breathing difficulties.	completely sealed machinery, ventilation, local exhaust or respiratory protection.	fresh air, rest, place in half-sitting position, take to hospital.
Skin: *corrosive*; serious burns, slow healing.	gloves, protective clothing.	remove contaminated clothing, flush skin with water or shower, refer to a doctor.
Eyes: *corrosive*; redness, pain, impaired vision.	face shield, or combined eye and respiratory protection.	flush with water, take to a doctor.
Ingestion: *corrosive*; sore throat, abdominal pain, diarrhea.		rinse mouth, take immediately to hospital.

SPILLAGE	STORAGE	LABELING / NFPA
evacuate area, call in an expert, dilute leakage with diesel/machine oil, soak up/cover with dry sand/table salt and remove to safe place, under no circumstances spray liquid with water; (additional individual protection: CHEMICAL SUIT).	keep in fireproof place under dry nitrogen, separate from all other substances.	 3 3 W

NOTES
Usually sold as 15-30% solution in hydrocarbons, which does not ignite spontaneously on exposure to air. For 15% solution see relevant entry. Reacts violently with extinguishing agents such as carbon dioxide, halons, foam, water and ordinary powder. Lung edema symptoms usually develop several hours later and are aggravated by physical exertion: rest and hospitalization essential. As first aid, a doctor or authorized person should consider administering a corticosteroid spray. Use cylinder with special fittings.

Transport Emergency Card TEC(R)-42G02

HI: X333; UN-number: 3051

DIISOBUTYL ALUMINUM HYDRIDE
(15% solution in hexane)

PHYSICAL PROPERTIES	IMPORTANT CHARACTERISTICS
Boiling point, °C — 69 Melting point, °C — −95 Flash point, °C — −26 Auto-ignition temperature, °C — 260 Relative density (water = 1) — 0.73 Relative vapor density (air = 1) — 3.0 Relative density at 20°C of saturated mixture vapor/air (air = 1) — 1.3 Vapor pressure, mm Hg at 20°C — 121.6 Solubility in water — reaction Explosive limits, vol% in air — 1.1-7.5 Minimum ignition energy, mJ — 0.24 Electrical conductivity, pS/m — 100	**COLORLESS FUMING LIQUID** Vapor is heavier than air and spreads at ground level, with risk of ignition at a distance. Flow, agitation etc. can cause build-up of electrostatic charge due to liquid's low conductivity. Do not use compressed air when filling, emptying or processing. Once solvent has evaporated (e.g. on clothing) ignites spontaneously on exposure to air. Aluminum oxide vapors are a product of combustion. Strong reducing agent which reacts violently with oxidants. Reacts violently with water, alcohols, phenols, amines, halogenated hydrocarbons, carbon dioxide, nitrogen oxides, oxides of sulfur and many other substances, with risk of fire and explosion.
	TLV-TWA 2 mg/m³ ✱
Gross formula: $C_8H_{19}Al$	**Absorption route:** Can enter the body by inhalation or ingestion. Harmful atmospheric concentrations can build up fairly rapidly on evaporation at approx. 20°C - even more rapidly in aerosol form. **Immediate effects:** Corrosive to the eyes, skin and respiratory tract. Inhalation of vapor/fumes can cause severe breathing difficulties (lung edema). In serious cases risk of death.

HAZARDS/SYMPTOMS	PREVENTIVE MEASURES	FIRE EXTINGUISHING/FIRST AID
Fire: extremely flammable, many chemical reactions can cause fire and explosion.	keep away from open flame and sparks, no smoking, avoid contact with water, alcohols and many other substances.	special extinguishing powder, dry sand, NO other extinguishing agents, can reignite after being extinguished.
Explosion: forms explosive air-vapor mixtures; contact with water, alcohols, phenols and many other substances can cause explosion.	sealed machinery, ventilation, explosion-proof electrical equipment and lighting, grounding.	in case of fire: keep undamaged cylinders cool by spraying with water, but DO NOT spray substance with water.
	STRICT HYGIENE	
Inhalation: *corrosive*; sore throat, cough, severe breathing difficulties.	sealed machinery, ventilation, local exhaust or respiratory protection.	fresh air, rest, place in half-sitting position, take to hospital.
Skin: *corrosive*; redness, pain, serious burns, slow healing.	gloves, protective clothing.	remove contaminated clothing, flush skin with water or shower, call a doctor if necessary.
Eyes: *corrosive*; redness, pain, impaired vision.	face shield, or combined eye and respiratory protection.	flush with water, take to a doctor.
Ingestion: *corrosive*; sore throat, abdominal pain, diarrhea.		rinse mouth, immediately take to hospital.

SPILLAGE	STORAGE	LABELING / NFPA
evacuate area, call in an expert, collect leakage in sealable containers, soak up remainder with dry sand/table salt and remove to safe place, under no circumstances spray liquid with water; (additional individual protection: CHEMICAL SUIT).	keep in dry, fireproof place under inert gas, separate from all other substances.	 3 / 3 / W

NOTES

Physical properties (boiling point, flash point, autoignition point, explosive limits, vapor pressure, relative density, conductivity and ignition temperature) are based on hexane. For pure diisobutyl aluminum hydride see relevant card. ✱ For TLV of *hexane* see relevant sheet. Reacts violently with extinguishing agents such as water, foam, halons, carbon dioxide and ordinary powder. Remove contaminated clothing immediately and flush with water (risk of spontaneous combustion). Lung edema symptoms usually develop several hours later and are aggravated by physical exertion: rest and hospitalization essential. As first aid, a doctor or authorized person should consider administering a corticosteroid spray. Use airtight packaging. Use cylinder with special fittings.

Transport Emergency Card TEC(R)-42G02

HI: X333; UN-number: 2220

DIISOBUTYLCARBINOL

PHYSICAL PROPERTIES	IMPORTANT CHARACTERISTICS	
Boiling point, °C 178 Melting point, °C −65 Flash point, °C 74 Auto-ignition temperature, °C 426 Relative density (water = 1) 0.8 Relative vapor density (air = 1) 5.0 Relative density at 20 °C of saturated mixture vapor/air (air = 1) 1.0 Vapor pressure, mm Hg at 20 °C 0.23 Solubility in water, g/100 ml at 20 °C 0.06 Explosive limits, vol% in air see notes Relative molecular mass 144.3 Gross formula: $C_9H_{20}O$	**COLORLESS LIQUID WITH CAMPHOR ODOR** Vapor mixes readily with air. Reacts with oxidants, with risk of fire and explosion. TLV-TWA not available **Absorption route:** Can enter the body by ingestion. Insufficient data on the rate at which harmful concentrations can build up.	

HAZARDS/SYMPTOMS	PREVENTIVE MEASURES	FIRE EXTINGUISHING/FIRST AID
Fire: combustible.	keep away from open flame, no smoking.	powder, AFFF, foam, carbon dioxide, (halons).
Explosion: above 74°C: forms explosive air-vapor mixtures.	above 74°C: sealed machinery, ventilation.	
Inhalation:	ventilation, local exhaust or respiratory protection.	fresh air, rest, call a doctor if necessary.
Skin:	gloves.	remove contaminated clothing, flush skin with water or shower, call a doctor if necessary.
Eyes:	safety glasses.	flush with water, take to a doctor if necessary.
Ingestion:		rinse mouth, call a doctor if necessary.

SPILLAGE	STORAGE	LABELING / NFPA
collect leakage in sealable containers, soak up with sand or other inert absorbent and remove to safe place.	keep separate from oxidants; ventilate at floor level.	2 1 0

NOTES
At 100°C explosion limits are between 0.8 and 6.1% by vol. Insufficient toxicological data on harmful effects to humans.

Transport Emergency Card TEC(R)-30G37

HI: 30; UN-number: 1993

CAS-No: [25167-70-8]
2,4,4-trimethylpentene-1

$(CH_3)_3\text{-C-CH}_2\text{-C-}(CH_3)=CH_2$

DIISOBUTYLENE

PHYSICAL PROPERTIES		IMPORTANT CHARACTERISTICS
Boiling point, °C	101	**COLORLESS LIQUID WITH CHARACTERISTIC ODOR**
Melting point, °C	− 101	Vapor (gas) is heavier than air and spreads at ground level, with risk of ignition at a distance.
Flash point, °C	− 7	Presumed to be able to form peroxides. Able to polymerize, with risk of fire and explosion. Flow,
Auto-ignition temperature, °C	415	agitation etc. can cause build-up of electrostatic charge due to liquid's low conductivity. Do not
Relative density (water = 1)	0.72	use compressed air when filling, emptying or processing. Reacts violently with strong oxidants.
Relative vapor density (air = 1)	3.97	
Relative density at 20 °C of saturated mixture vapor/air (air = 1)	1.12	TLV-TWA not available
Vapor pressure, mm Hg at 20 °C	31	
Solubility in water	poor	**Absorption route:** Can enter the body by inhalation or ingestion.
Explosive limits, vol% in air	0.7-5.8	**Immediate effects:** Irritates the eyes, skin and respiratory tract. If liquid is swallowed, droplets
Minimum ignition energy, mJ	0.96	can enter the lungs, with risk of pneumonia.
Electrical conductivity, pS/m	< 10⁴	
Relative molecular mass	112.2	
Gross formula:	C_8H_{16}	

HAZARDS/SYMPTOMS	PREVENTIVE MEASURES	FIRE EXTINGUISHING/FIRST AID
Fire: extremely flammable.	keep away from open flame and sparks, no smoking.	powder, AFFF, foam, carbon dioxide, (halons).
Explosion: forms explosive air-vapor mixtures.	sealed machinery, ventilation, explosion-proof electrical equipment and lighting, grounding.	in case of fire: keep tanks/drums cool by spraying with water.
Inhalation: cough, headache, dizziness, drowsiness.	ventilation, local exhaust or respiratory protection.	fresh air, rest, call a doctor.
Skin: redness.	gloves.	remove contaminated clothing, wash skin with soap and water.
Eyes: redness, pain.	safety glasses.	flush with water, take to a doctor if necessary.
Ingestion: abdominal pain, headache, nausea, dizziness.		rinse mouth, DO NOT induce vomiting, call a doctor or take to hospital.

SPILLAGE	STORAGE	LABELING / NFPA
collect leakage in sealable containers, soak up with sand or other inert absorbent and remove to safe place, do not flush into sewer.	keep in fireproof place, separate from oxidants.	R: 11 S: 9-16-29-33 🔥 Flammable NFPA: 3 / 2 / 0

NOTES
Mixture of α and β-diisobutylene isomers. The measures on this card also apply to 2,4,4-trimethylpentene-2 (flash point 2°C, autoignition temperature 308°C).

2,6-dimethyl-4-heptanone
DIBK
isobutyl ketone
isovalerone

$(CH_3CH(CH_3)CH_2)_2C=O$

DIISOBUTYL KETONE

PHYSICAL PROPERTIES		IMPORTANT CHARACTERISTICS	
Boiling point, °C	168	**COLORLESS LIQUID WITH CHARACTERISTIC ODOR**	
Melting point, °C	−46	Vapor mixes readily with air. Flow, agitation etc. can cause build-up of electrostatic charge due to	
Flash point, °C	49	liquid's low conductivity. Reacts with strong oxidants. Attacks many plastics.	
Auto-ignition temperature, °C	345		
Relative density (water = 1)	0.8	TLV-TWA 25 ppm 145 mg/m³	
Relative vapor density (air = 1)	4.9		
Relative density at 20 °C of saturated mixture		**Absorption route:** Can enter the body by inhalation or ingestion. Harmful atmospheric	
vapor/air (air = 1)	1.01	concentrations build up fairly slowly on evaporation at 20°C, but much more rapidly in aerosol	
Vapor pressure, mm Hg at 20 °C	2.0	form.	
Solubility in water, g/100 ml	0.05	**Immediate effects:** Irritates the eyes, skin and respiratory tract.	
Explosive limits, vol% in air	0.8-7.1	**Effects of prolonged/repeated exposure:** Prolonged or repeated contact can cause skin	
Relative molecular mass	142.2	disorders.	
Gross formula:	$C_9H_{18}O$		

HAZARDS/SYMPTOMS	PREVENTIVE MEASURES	FIRE EXTINGUISHING/FIRST AID
Fire: flammable.	keep away from open flame and sparks, no smoking.	powder, AFFF, foam, carbon dioxide, (halons).
Explosion: above 49°C: forms explosive air-vapor mixtures.	above 49°C: sealed machinery, ventilation, explosion-proof electrical equipment, grounding.	in case of fire: keep tanks/drums cool by spraying with water.
Inhalation: sore throat, cough, headache.	ventilation, local exhaust or respiratory protection.	fresh air, rest.
Skin: *is absorbed*; redness.	gloves.	remove contaminated clothing, flush skin with water or shower.
Eyes: redness.	safety glasses.	flush with water, take to a doctor.
Ingestion: vomiting, abdominal pain, cramps, nausea.		DO NOT induce vomiting, call a doctor.

SPILLAGE	STORAGE	LABELING / NFPA
collect leakage in sealable containers, soak up with sand or other inert absorbent and remove to safe place.	keep in fireproof place.	R: 10-37 S: 24 ✖ Irritant

NOTES
Odor limit is above TLV.

DIISOOCTYL PHTHALATE

PHYSICAL PROPERTIES	IMPORTANT CHARACTERISTICS
Boiling point, °C 370 Melting point, °C −43 Flash point, °C 232 Relative density (water = 1) 0.99 Relative vapor density (air = 1) 13.5 Relative density at 20 °C of saturated mixture vapor/air (air = 1) 1.0 Vapor pressure, mm Hg at 20 °C < 0.076 Solubility in water none Relative molecular mass 390.5	**COLORLESS VISCOUS LIQUID WITH CHARACTERISTIC ODOR** Vapor mixes readily with air. Flow, agitation etc. can cause build-up of electrostatic charge due to liquid's low conductivity. Reacts with strong oxidants.
	TLV-TWA not available
	Absorption route: Can enter the body by ingestion. Evaporation negligible at 20°C, but harmful atmospheric concentrations can build up rapidly in aerosol form. **Immediate effects:** Irritates the eyes, skin and respiratory tract. **Effects of prolonged/repeated exposure:** Prolonged or repeated contact can cause skin disorders.
Gross formula: $C_{24}H_{38}O_4$	

HAZARDS/SYMPTOMS	PREVENTIVE MEASURES	FIRE EXTINGUISHING/FIRST AID
Fire: combustible.	keep away from open flame, no smoking.	powder, AFFF, foam, carbon dioxide, (halons).
Inhalation: sore throat, cough.	ventilation.	fresh air, rest.
Skin: redness.	gloves.	remove contaminated clothing, flush skin with water or shower.
Eyes: redness, pain.	safety glasses.	flush with water, take to a doctor if necessary.
Ingestion: abdominal pain, nausea.		rinse mouth, call a doctor.

SPILLAGE	STORAGE	LABELING / NFPA
collect leakage in sealable containers, soak up with sand or other inert absorbent and remove to safe place.		

NOTES
Hexaplas DIOP is a trade name.

DIISOPROPYL CARBINOL

PHYSICAL PROPERTIES	IMPORTANT CHARACTERISTICS	
Boiling point, °C — 140 Melting point, °C — −70 Flash point, °C — 49 Relative density (water = 1) — 0.8 Relative vapor density (air = 1) — 4.0 Relative density at 20 °C of saturated mixture vapor/air (air = 1) — 1.0 Vapor pressure, mm Hg at 20 °C — 1.9 Relative molecular mass — 116.2	**COLORLESS LIQUID WITH CHARACTERISTIC ODOR** Vapor mixes readily with air. Reacts with strong oxidants.	
	TLV-TWA — not available	
	Absorption route: Can enter the body by inhalation or ingestion. Harmful atmospheric concentrations build up fairly slowly on evaporation at 20° C, but much more rapidly in aerosol form. **Immediate effects:** Irritates the eyes and skin.	
Gross formula: — $C_7H_{16}O$		

HAZARDS/SYMPTOMS	PREVENTIVE MEASURES	FIRE EXTINGUISHING/FIRST AID
Fire: flammable.	keep away from open flame and sparks, no smoking.	powder, AFFF, foam, carbon dioxide, (halons).
Explosion: above 49°C: forms explosive air-vapor mixtures.	above 49°C: sealed machinery, ventilation, explosion-proof electrical equipment and lighting.	in case of fire: keep tanks/drums cool by spraying with water.
Inhalation: sore throat, cough, dizziness, sleepiness.	ventilation, local exhaust or respiratory protection.	fresh air, rest, call a doctor.
Skin: *is absorbed*; redness, pain; see also 'Inhalation'.	gloves.	remove contaminated clothing, wash skin with soap and water.
Eyes: redness, pain.	face shield.	flush with water, take to a doctor if necessary.
Ingestion: nausea, drowsiness.		rinse mouth, call a doctor.

SPILLAGE	STORAGE	LABELING / NFPA
collect leakage in sealable containers, soak up with sand or other inert absorbent and remove to safe place.	keep in fireproof place.	

NOTES
Alcohol consumption increases toxic effects.

Transport Emergency Card TEC(R)-30G08

DIKETENE

PHYSICAL PROPERTIES	IMPORTANT CHARACTERISTICS
Boiling point, °C .... 127 Melting point, °C .... −7 Flash point, °C .... 33 Auto-ignition temperature, °C .... 275 Relative density (water = 1) .... 1.09 Relative vapor density (air = 1) .... 2.9 Relative density at 20 °C of saturated mixture vapor/air (air = 1) .... 1.02 Vapor pressure, mm Hg at 20 °C .... 7.0 Solubility in water .... reaction Explosive limits, vol% in air .... 0.2-11 Relative molecular mass 84	**COLORLESS LIQUID WITH PUNGENT ODOR** Vapor mixes readily with air. Can polymerize when moderately heated. Reacts violently with water, with risk of fire and explosion. Forms explosive vapor mixtures on contact with bases, alcohols and amines.
	TLV-TWA not available
	Absorption route: Can enter the body by inhalation or ingestion or through the skin. Harmful atmospheric concentrations can build up fairly rapidly on evaporation at approx. 20°C - even more rapidly in aerosol form. **Immediate effects:** Corrosive to the eyes, skin and respiratory tract. Inhalation of vapor/fumes can cause severe breathing difficulties (lung edema). In serious cases risk of death.
Gross formula: $C_4H_4O_2$	

HAZARDS/SYMPTOMS	PREVENTIVE MEASURES	FIRE EXTINGUISHING/FIRST AID
Fire: flammable, many chemical reactions can cause fire and explosion.	keep away from open flame and sparks, no smoking, avoid contact with other substances.	carbon dioxide, dry sand, special powder extinguisher, DO NOT USE WATER-BASED EXTINGUISHERS.
Explosion: above 33°C: forms explosive air-vapor mixtures, many chemical reactions can cause explosion.	above 33°C: sealed machinery, ventilation, explosion-proof electrical equipment.	in case of fire: keep tanks/drums cool by spraying with water, but DO NOT spray substance with water, fight fire from protected location.
Inhalation: *corrosive*; sore throat, cough, shortness of breath, severe breathing difficulties.	ventilation, local exhaust or respiratory protection.	fresh air, rest, place in half-sitting position, take to hospital.
Skin: *is absorbed*, *corrosive*; redness, pain, serious burns.	gloves, protective clothing.	remove contaminated clothing, flush skin with water or shower, refer to a doctor.
Eyes: *corrosive*; redness, pain, impaired vision.	face shield, or combined eye and respiratory protection.	flush with water, take to a doctor.
Ingestion: *corrosive*; sore throat, abdominal pain, diarrhea, vomiting.		rinse mouth, immediately take to hospital.

SPILLAGE	STORAGE	LABELING / NFPA
evacuate area, call in an expert, collect leakage in non-sealable containers, soak up with dry sand or other inert absorbent and remove to safe place (additional individual protection: breathing apparatus).	keep in dry, fireproof place under inert gas, separate from other chemicals; add an inhibitor.	R: 10-20 S: 3 Note D ✖ Harmful

NOTES
Reacts violently with extinguishing agents such as bicarbonate-based powders, halons and water. Special precautions also needed for transport. Lung edema symptoms usually develop several hours later and are aggravated by physical exertion: rest and hospitalization essential. As first aid, a doctor or authorized person should consider administering a corticosteroid spray. Unbreakable packaging preferred; if breakable, keep in unbreakable container.

DIMETHYLACETAL

PHYSICAL PROPERTIES	IMPORTANT CHARACTERISTICS
Boiling point, °C .... 64 Melting point, °C .... − 113 Flash point, °C .... − 28 Relative density (water = 1) .... 0.85 Relative vapor density (air = 1) .... 3.1 Relative density at 20 °C of saturated mixture vapor/air (air = 1) .... 1.4 Vapor pressure, mm Hg at 20 °C .... ca. 144 Solubility in water .... ∞ Relative molecular mass 90.1	**COLORLESS LIQUID WITH CHARACTERISTIC ODOR** Substance is heavier than air and spreads at ground level, with risk of ignition at a distance. Able to form peroxides. Do not use compressed air when filling, emptying or processing. Reacts with strong acids and with oxidants.
	TLV-TWA not available
	Absorption route: Can enter the body by inhalation or ingestion. Harmful atmospheric concentrations can build up fairly rapidly on evaporation at approx. 20°C - even more rapidly in aerosol form. **Immediate effects:** Irritates the eyes, skin and respiratory tract. Affects the nervous system.
Gross formula: $C_4H_{10}O_2$	

HAZARDS/SYMPTOMS	PREVENTIVE MEASURES	FIRE EXTINGUISHING/FIRST AID
Fire: extremely flammable.	keep away from open flame and sparks, no smoking.	powder, alcohol-resistant foam, large quantities of water, carbon dioxide, (halons).
Explosion: forms explosive air-vapor mixtures.	sealed machinery, ventilation, explosion-proof electrical equipment and lighting.	in case of fire: keep tanks/drums cool by spraying with water.
Inhalation: sore throat, cough, headache, dizziness, drowsiness.	ventilation, local exhaust or respiratory protection.	fresh air, rest, call a doctor.
Skin: redness, pain.	gloves.	remove contaminated clothing, flush skin with water or shower.
Eyes: redness, pain.	face shield.	flush with water, take to a doctor.
Ingestion: abdominal pain, nausea, sleepiness.		rinse mouth, call a doctor.

SPILLAGE	STORAGE	LABELING / NFPA
collect leakage in sealable containers, soak up with sand or other inert absorbent and remove to safe place; (additional individual protection: breathing apparatus).	keep in fireproof, dark place, separate from oxidants and strong acids, add an inhibitor.	R: 11 S: 9-16-33 🔥 Flammable

NOTES
Before distilling check for peroxides; if found, render harmless. Alcohol consumption increases toxic effects.

Transport Emergency Card TEC(R)-30G30

HI: 33; UN-number: 2377

CAS-No: [127-19-5]
acetic acid dimethylamide
acetyldimethylamine
dimethylacetamide
DMAC

$CH_3CON(CH_3)_2$

N,N-DIMETHYL ACETAMIDE

PHYSICAL PROPERTIES	IMPORTANT CHARACTERISTICS
Boiling point, °C 164 Melting point, °C −20 Flash point, °C 66 Auto-ignition temperature, °C ? Relative density (water = 1) 0.9 Relative vapor density (air = 1) 3.0 Relative density at 20 °C of saturated mixture vapor/air (air = 1) 1.0 Vapor pressure, mm Hg at 20 °C 1.3 Solubility in water ∞ Explosive limits, vol% in air 2-11.5 Relative molecular mass 87.1 Log P octanol/water 0.8	**COLORLESS LIQUID WITH CHARACTERISTIC ODOR** Vapor mixes readily with air. → *Nitrous vapors* are a product of combustion. Decomposes when heated, giving off toxic and corrosive vapors. Reacts violently with chlorinated solvents and strong oxidants. Attacks many plastics.

TLV-TWA (skin)	10 ppm	36 mg/m³

Absorption route: Can enter the body by inhalation or ingestion or through the skin. Harmful atmospheric concentrations build up fairly slowly on evaporation at 20° C, but much more rapidly in aerosol form.
Immediate effects: Irritates the eyes, skin and respiratory tract. Liquid destroys the skin's natural oils. Affects the nervous system. In substantial concentrations can impair consciousness.
Effects of prolonged/repeated exposure: Can cause liver and kidney damage.

Gross formula: C_4H_9NO

HAZARDS/SYMPTOMS	PREVENTIVE MEASURES	FIRE EXTINGUISHING/FIRST AID
Fire: combustible.	keep away from open flame, no smoking.	water spray, alcohol-resistant foam, powder, carbon dioxide, DO NOT USE HALONS.
Explosion: above 66°C: forms explosive air-vapor mixtures.	above 66°C: sealed machinery, ventilation.	
Inhalation: headache, nausea, drowsiness, impaired sight/hearing, intolerance for alcohol.	ventilation, local exhaust or respiratory protection.	fresh air, rest, call a doctor.
Skin: *is absorbed*; redness; see also 'Inhalation'.	gloves, protective clothing.	remove contaminated clothing, wash skin with soap and water, refer/take to a doctor.
Eyes: redness, pain.	face shield.	flush with water, take to a doctor if necessary.
Ingestion: diarrhea, abdominal cramps; see also 'Inhalation'.	do not eat, drink or smoke while working.	rinse mouth, call a doctor or take immediately to hospital.

SPILLAGE	STORAGE	LABELING / NFPA
collect leakage in sealable containers, soak up with sand or other inert absorbent and remove to safe place.	keep separate from chlorinated solvents and strong oxidants; ventilate at floor level.	R: 20/21-36 S: 26-28-36 ✖ Harmful

NOTES
Most types of gloves are to a certain extent permeable to dimethyl acetamide. Alcohol consumption inadvisable after exposure. Depending on the degree of exposure, regular medical checkups are advisable. Absorbed very rapidly through the skin.

Transport Emergency Card TEC(R)-30G23

DIMETHYLAMINE
(40% aqueous solution)

PHYSICAL PROPERTIES	IMPORTANT CHARACTERISTICS
Boiling point, °C — 54 Melting point, °C — −37 Flash point, °C — −18 Auto-ignition temperature, °C — 390 Relative density (water = 1) — 0.9 Vapor pressure, mm Hg at 20 °C — 218 Solubility in water — ∞ Explosive limits, vol% in air — 2.6-12.3 Relative molecular mass — 45.1 Log P octanol/water — −0.2	**COLORLESS LIQUID WITH PUNGENT ODOR** Vapor (gas) is heavier than air and spreads at ground level, with risk of ignition at a distance. Do not use compressed air when filling, emptying or processing. Toxic gases are a product of combustion (→ *nitrous vapors*). Strong base which reacts violently with acids and corrodes aluminum, zinc etc. Reacts violently with mercury and strong oxidants, with risk of fire and explosion. Attacks copper, zinc, aluminum and their alloys.

	TLV-TWA	10 ppm	18 mg/m³

Absorption route: Can enter the body by inhalation or ingestion. Harmful atmospheric concentrations can build up very rapidly on evaporation at 20°C.
Immediate effects: Corrosive to the eyes, skin and respiratory tract. Inhalation can cause lung edema. In serious cases risk of death.
Effects of prolonged/repeated exposure: Prolonged or repeated contact can cause skin disorders.

Gross formula: C₂H₇N

HAZARDS/SYMPTOMS	PREVENTIVE MEASURES	FIRE EXTINGUISHING/FIRST AID
Fire: extremely flammable.	keep away from open flame and sparks, no smoking.	powder, alcohol-resistant foam, large quantities of water, carbon dioxide, (halons).
Explosion: forms explosive air-vapor mixtures.	sealed machinery, ventilation, explosion-proof electrical equipment and lighting.	in case of fire: keep tanks/drums cool by spraying with water.
Inhalation: *corrosive*; sore throat, cough, shortness of breath, severe breathing difficulties.	ventilation, local exhaust or respiratory protection.	fresh air, rest, place in half-sitting position, take to hospital.
Skin: *corrosive*; redness, pain, serious burns.		remove contaminated clothing, flush skin with water or shower, take to a doctor.
Eyes: *corrosive*; redness, pain, impaired vision.	face shield, or combined eye and respiratory protection.	flush with water, take to a doctor.
Ingestion: *corrosive*; sore throat, abdominal pain, vomiting.		rinse mouth, immediately take to hospital.

SPILLAGE	STORAGE	LABELING / NFPA
evacuate area, call in an expert, collect leakage in sealable containers, flush away remainder with water; if necessary, render harmless with sodium sulfate; (additional individual protection: CHEMICAL SUIT).	keep in cool, fireproof place, separate from oxidants and acids.	R: 13-36/37 S: 16-26-29 Note C Flammable Irritant

NOTES
Log P octanol/water is estimated. Lung edema symptoms usually develop several hours later and are aggravated by physical exertion: rest and hospitalization essential. As first aid, a doctor or authorized person should consider administering a corticosteroid spray.

Transport Emergency Card TEC(R)-30G33

HI: 338; UN-number: 1160

CAS-No: [108-01-0]
deanol
dimethylaminoethanol
dimethylethanolamine
N,N-dimethyl-2-hydroxyethylamine
β-hydroxyethyldimethylamine

$(CH_3)_2NCH_2CH_2OH$

2-DIMETHYLAMINOETHANOL

PHYSICAL PROPERTIES		IMPORTANT CHARACTERISTICS
Boiling point, °C	134	**COLORLESS LIQUID WITH PUNGENT ODOR**
Melting point, °C	−59	Vapor mixes readily with air. → *Nitrous vapors* are a product of combustion. Reacts violently with oxidants. Attacks copper and its alloys.
Flash point, °C	31	
Auto-ignition temperature, °C	220	
Relative density (water = 1)	0.9	
Relative vapor density (air = 1)	3.0	TLV-TWA not available
Relative density at 20 °C of saturated mixture vapor/air (air = 1)	1.01	**Absorption route:** Can enter the body by inhalation or ingestion or through the skin.
Vapor pressure, mm Hg at 20 °C	4.3	**Immediate effects:** Irritates the eyes, skin and respiratory tract. Affects the nervous system. Inhalation of vapor/fumes can cause severe breathing difficulties (lung edema). In serious cases risk of death.
Solubility in water	∞	
Explosive limits, vol% in air	?-?	
Relative molecular mass	89.1	
Gross formula: $C_4H_{11}NO$		

HAZARDS/SYMPTOMS	PREVENTIVE MEASURES	FIRE EXTINGUISHING/FIRST AID
Fire: flammable.	keep away from open flame and sparks, no smoking.	powder, alcohol-resistant foam, water spray, carbon dioxide, (halons).
Explosion: above 31°C: forms explosive air-vapor mixtures.	above 31°C: sealed machinery, ventilation, explosion-proof electrical equipment.	in case of fire: keep tanks/drums cool by spraying with water.
Inhalation: *corrosive*; sore throat, cough, shortness of breath, severe breathing difficulties.	ventilation, local exhaust or respiratory protection.	fresh air, rest, place in half-sitting position, take to hospital.
Skin: *is absorbed, corrosive*; redness, pain, serious burns.	gloves, protective clothing.	remove contaminated clothing, flush skin with water or shower, refer to a doctor.
Eyes: *corrosive*; redness, pain, impaired vision.	face shield.	flush with water, take to a doctor.
Ingestion: *corrosive*; sore throat, abdominal pain, muscle cramps, vomiting, diarrhea.		rinse mouth, immediately take to hospital.

SPILLAGE	STORAGE	LABELING / NFPA
collect leakage in sealable containers, soak up with sand or other inert absorbent and remove to safe place; (additional individual protection: breathing apparatus).	keep in fireproof place, separate from oxidants.	R: 10 - 36/37/38 S: 28 ✖ Irritant 2 / 2 0

NOTES
Explosive limits not given in the literature. Lung edema symptoms usually develop several hours later and are aggravated by physical exertion: rest and hospitalization essential. As first aid, a doctor or authorized person should consider administering a corticosteroid spray.

Transport Emergency Card TEC(R)-30G35

HI: 30; UN-number: 2051

$(CH_3)_2N(CH_2)_3NH_2$

3-DIMETHYLAMINOPROPYLAMINE

PHYSICAL PROPERTIES	IMPORTANT CHARACTERISTICS
Boiling point, °C 123 Melting point, °C <70 Flash point, °C 35 Relative density (water = 1) 0.8 Relative vapor density (air = 1) 3.5 Relative density at 20 °C of saturated mixture vapor/air (air = 1) 1.03 Vapor pressure, mm Hg at 20 °C ca. 10 Solubility in water ∞ Explosive limits, vol% in air ?-? Relative molecular mass 102.2	**COLORLESS LIQUID WITH CHARACTERISTIC ODOR** Vapor mixes readily with air. Toxic vapors (→ *nitrous vapors*) are a product of combustion. Decomposes when heated, giving off combustible and toxic vapors. Reacts violently with strong oxidants and strong acids.
	TLV-TWA not available
	Absorption route: Can enter the body by inhalation or ingestion or through the skin. Harmful atmospheric concentrations can build up fairly rapidly on evaporation at approx. 20°C - even more rapidly in aerosol form. **Immediate effects:** Corrosive to the eyes, skin and respiratory tract. Inhalation of vapor/fumes can cause severe breathing difficulties (lung edema). In serious cases risk of unconsciousness. **Effects of prolonged/repeated exposure:** Prolonged or repeated contact can cause skin disorders. Prolonged or repeated contact with vapor can cause lung disorders. Can cause liver and kidney damage.
Gross formula: $C_5H_{14}N_2$	

HAZARDS/SYMPTOMS	PREVENTIVE MEASURES	FIRE EXTINGUISHING/FIRST AID
Fire: flammable.	keep away from open flame and sparks, no smoking.	powder, alcohol-resistant foam, water spray, carbon dioxide, (halons).
Explosion: above 35°C: forms explosive air-vapor mixtures.	above 35°C: sealed machinery, ventilation, explosion-proof electrical equipment.	in case of fire: keep tanks/drums cool by spraying with water.
Inhalation: *corrosive*; sore throat, cough, severe breathing difficulties, nausea.	ventilation, local exhaust or respiratory protection.	fresh air, rest, place in half-sitting position, take to hospital.
Skin: *corrosive, is absorbed*; redness, pain, serious burns.	gloves, protective clothing.	remove contaminated clothing, flush skin with water or shower, call a doctor if necessary.
Eyes: *corrosive*; redness, pain, impaired vision.	face shield, or combined eye and respiratory protection.	flush with water, take to a doctor.
Ingestion: *corrosive*; abdominal pain, diarrhea, vomiting.	do not eat, drink or smoke while working.	rinse mouth, take immediately to hospital.

SPILLAGE	STORAGE	LABELING / NFPA
collect leakage in sealable containers, flush away remainder with water; (additional individual protection: breathing apparatus).	keep in fireproof place, separate from oxidants.	R: 10-22-34-43 S: 26-36/37/39 Corrosive NFPA: 3 - 2 - 0

NOTES
Explosive limits not given in the literature. Depending on the degree of exposure, regular medical checkups are advisable. Lung edema symptoms usually develop several hours later and are aggravated by physical exertion: rest and hospitalization essential. As first aid, a doctor or authorized person should consider administering a corticosteroid spray.

Transport Emergency Card TEC(R)-80G15

DIMETHYLAMMONIUMCHLORIDE

PHYSICAL PROPERTIES	IMPORTANT CHARACTERISTICS	
Melting point, °C 171 Relative density (water = 1) 0.85 Solubility in water ∞ Relative molecular mass 81.5	**WHITE POWDER** Corrosive vapors (→ *hydrochloric acid*) are a product of combustion. Decomposes when heated, giving off flammable and corrosive vapors.	
	TLV-TWA not available	
	Absorption route: Can enter the body by inhalation or ingestion. Evaporation negligible at 20°C, but harmful concentrations of airborne particles can build up rapidly. **Immediate effects:** Irritates the eyes, skin and respiratory tract. **Effects of prolonged/repeated exposure:** Prolonged or repeated contact can cause skin disorders.	
Gross formula: C_2H_8ClN		

HAZARDS/SYMPTOMS	PREVENTIVE MEASURES	FIRE EXTINGUISHING/FIRST AID
Fire: combustible.	keep away from open flame, no smoking.	water spray, powder; in case of fire in immediate vicinity: use any extinguishing agent.
Inhalation: cough.	local exhaust or respiratory protection.	fresh air, rest.
Skin: redness.	gloves.	remove contaminated clothing, flush skin with water or shower.
Eyes: redness, pain.	goggles.	flush with water, take to a doctor if necessary.
Ingestion: sore throat, abdominal pain.	do not eat, drink or smoke while working.	rinse mouth, call a doctor if necessary.

SPILLAGE	STORAGE	LABELING / NFPA
clean up spillage, flush away remainder with water; (additional individual protection: P2 respirator).		

NOTES

CAS-No: [121-69-7]
N,N-dimethylbenzeneamine
dimethylamino benzene
dimethylaniline
dimethylphenylamine

$C_6H_5N(CH_3)_2$

N,N-DIMETHYLANILINE

PHYSICAL PROPERTIES		IMPORTANT CHARACTERISTICS
Boiling point, °C	193	**YELLOW TO YELLOWISH-BROWN OILY LIQUID WITH CHARACTERISTIC ODOR, TURNING BROWN ON EXPOSURE TO AIR**
Melting point, °C	2	Vapor mixes readily with air. Toxic gases are a product of combustion (→ *nitrous vapors*).
Flash point, °C	63	Decomposes when heated, giving off toxic gases (→ *aniline*). Reacts violently with strong
Auto-ignition temperature, °C	370	oxidants.
Relative density (water = 1)	0.96	
Relative vapor density (air = 1)	4.2	

TLV-TWA (skin)		5 ppm	25 mg/m³
STEL		10 ppm	50 mg/m³

PHYSICAL PROPERTIES (cont.)	
Relative density at 20 °C of saturated mixture vapor/air (air = 1)	1.00
Vapor pressure, mm Hg at 20°C	0.53
Solubility in water	none
Explosive limits, vol% in air	1-7
Relative molecular mass	121.2
Log P octanol/water	2.5

Absorption route: Can enter the body by inhalation or ingestion or through the skin. Harmful atmospheric concentrations build up fairly slowly on evaporation at 20°C, but much more rapidly in aerosol form.
Immediate effects: Acute exposure to high concentrations causes immediate unconsciousness, which can last many hours, followed by severe abdominal pain and impairment of vision.
Effects of prolonged/repeated exposure: Can affect the blood.

Gross formula: $C_8H_{11}N$

HAZARDS/SYMPTOMS	PREVENTIVE MEASURES	FIRE EXTINGUISHING/FIRST AID
Fire: combustible.	keep away from open flame, no smoking, avoid contact with strong oxidants.	powder, AFFF, foam, carbon dioxide, (halons).
Explosion: above 63°C: forms explosive air-vapor mixtures.	above 63°C: sealed machinery, ventilation.	in case of fire: keep tanks/drums cool by spraying with water.
	STRICT HYGIENE	IN ALL CASES CALL A DOCTOR
Inhalation: severe breathing difficulties, headache, blue skin, feeling of weakness.	ventilation, local exhaust or respiratory protection.	fresh air, rest, artificial respiration if necessary, take to hospital.
Skin: *is absorbed*; see also 'Inhalation'.	gloves, protective clothing.	remove contaminated clothing, wash skin with soap and water, call a doctor.
Eyes: redness, pain.	face shield.	flush with water, take to doctor.
Ingestion: headache, severe breathing difficulties, feeling of weakness, blue skin.	do not eat, drink or smoke while working.	rinse mouth, take immediately to hospital.

SPILLAGE	STORAGE	LABELING / NFPA
collect leakage in sealable containers, soak up with sand or other inert absorbent and remove to safe place.	keep in fireproof place, separate from oxidants.	R: 23/24/25-33 S: 28-37-44 ☠ Toxic

NFPA diamond: 3 (left), 2 (top), 0 (right)

NOTES

Depending on the degree of exposure, regular medical checkups are advisable. Effects on blood due to formation of methemoglobin. Special first aid required in the event of poisoning: antidotes must be available (with instructions).

DIMETHYL CARBONATE

PHYSICAL PROPERTIES	IMPORTANT CHARACTERISTICS
Boiling point, °C 90 Melting point, °C 2 Flash point, °C 18 Relative density (water = 1) 1.07 Relative vapor density (air = 1) 3.1 Relative density at 20 °C of saturated mixture vapor/air (air = 1) 1.1 Vapor pressure, mm Hg at 20 °C ca. 40 Solubility in water none Explosive limits, vol% in air ?-? Relative molecular mass 90.1 Gross formula: C$_3$H$_6$O$_3$	**COLORLESS LIQUID WITH CHARACTERISTIC ODOR** Vapor is heavier than air and spreads at ground level, with risk of ignition at a distance. Do not use compressed air when filling, emptying or processing. Reacts violently with strong oxidants. TLV-TWA not available **Absorption route:** Can enter the body by inhalation or ingestion. Harmful atmospheric concentrations can build up fairly rapidly on evaporation at approx. 20°C - even more rapidly in aerosol form. **Immediate effects:** Corrosive to the eyes, skin and respiratory tract. Affects the nervous system. Inhalation of vapor/fumes can cause severe breathing difficulties (lung edema). In serious cases risk of death.

HAZARDS/SYMPTOMS	PREVENTIVE MEASURES	FIRE EXTINGUISHING/FIRST AID
Fire: extremely flammable.	keep away from open flame and sparks, no smoking.	powder, AFFF, foam, carbon dioxide, (halons).
Explosion: forms explosive air-vapor mixtures.	sealed machinery, ventilation, explosion-proof electrical equipment and lighting.	in case of fire: keep tanks/drums cool by spraying with water.
Inhalation: *corrosive*, sore throat, cough, severe breathing difficulties, dizziness.	ventilation, local exhaust or respiratory protection.	fresh air, rest, place in half-sitting position, take to hospital.
Skin: *corrosive*; redness, pain, burns.	gloves.	remove contaminated clothing, flush skin with water or shower, take to hospital.
Eyes: *corrosive*; redness, pain, impaired vision.	face shield.	flush with water, take to a doctor.
Ingestion: *corrosive*; sore throat, nausea, abdominal pain, diarrhea.		rinse mouth, take immediately to hospital.

SPILLAGE	STORAGE	LABELING / NFPA
collect leakage in sealable containers, soak up with sand or other inert absorbent and remove to safe place; (additional individual protection: breathing apparatus).	keep in fireproof place, separate from oxidants.	R: 11-20/21/22 S: 9-29 Flammable Harmful NFPA: 3 / 2 / 1

NOTES
Explosive limits not given in the literature. Lung edema symptoms usually develop several hours later and are aggravated by physical exertion: rest and hospitalization essential. As first aid, a doctor or authorized person should consider administering a corticosteroid spray. Symptoms of poisoning depend partly on the presence of contaminants and can thus vary.

Transport Emergency Card TEC(R)-673

HI: 33; UN-number: 1161

1,4-DIMETHYLCYCLOHEXANE- trans

PHYSICAL PROPERTIES		IMPORTANT CHARACTERISTICS
Boiling point, °C	119	**COLORLESS LIQUID WITH CHARACTERISTIC ODOR**
Melting point, °C	−37	Vapor mixes readily with air, forming explosive mixtures. Flow, agitation etc. can cause build-up
Flash point, °C	11	of electrostatic charge due to liquid's low conductivity. Do not use compressed air when filling,
Auto-ignition temperature, °C	304	emptying or processing. Reacts violently with strong oxidants.
Relative density (water = 1)	0.8	
Relative vapor density (air = 1)	3.9	TLV-TWA — not available
Relative density at 20 °C of saturated mixture		
vapor/air (air = 1)	1.06	**Absorption route:** Can enter the body by inhalation or ingestion. Harmful atmospheric
Vapor pressure, mm Hg at 20 °C	17.5	concentrations build up fairly slowly on evaporation at 20°C, but much more rapidly in aerosol
Solubility in water	none	form.
Explosive limits, vol% in air	?-?	**Immediate effects:** Irritates the eyes. Affects the nervous system. In serious cases risk of
Electrical conductivity, pS/m	< 10⁴	unconsciousness.
Relative molecular mass	112.2	
Gross formula:	C_8H_{16}	

HAZARDS/SYMPTOMS	PREVENTIVE MEASURES	FIRE EXTINGUISHING/FIRST AID
Fire: extremely flammable.	keep away from open flame and sparks, no smoking.	powder, AFFF, foam, carbon dioxide, (halons).
Explosion: forms explosive air-vapor mixtures.	sealed machinery, ventilation, explosion-proof electrical equipment and lighting, grounding.	in case of fire: keep tanks/drums cool by spraying with water.
Inhalation: dizziness, drowsiness, unconsciousness.	ventilation, local exhaust or respiratory protection.	fresh air, rest, place in half-sitting position, call a doctor.
Skin: redness.	gloves.	remove contaminated clothing, flush skin with water or shower.
Eyes: redness, pain.	safety glasses.	flush with water, take to a doctor.
Ingestion: abdominal pain, vomiting; see also 'Inhalation'.		rinse mouth, call a doctor or take to hospital.

SPILLAGE	STORAGE	LABELING / NFPA
collect leakage in sealable containers, soak up with sand or other inert absorbent and remove to safe place, do not flush into sewer.	keep in fireproof place, separate from oxidants.	R: 11 S: 9-16-33 🔥 Flammable 3 / 1 / 0

NOTES
Explosive limits not given in the literature.

$(CH_3)_2SiCl_2$

DIMETHYLDICHLOROSILANE

PHYSICAL PROPERTIES		IMPORTANT CHARACTERISTICS
Boiling point, °C	70	**COLORLESS FUMING LIQUID WITH PUNGENT ODOR**
Melting point, °C	−76	Vapor is heavier than air and spreads at ground level, with risk of ignition at a distance. Do not use
Flash point, °C	−9	compressed air when filling, emptying or processing. → *Hydrochloric acid* fumes are a product of
Auto-ignition temperature, °C	380	combustion. Reacts violently with water (vapor), alcohols, acetone, amines, anhydrous ammonia
Relative density (water = 1)	1.1	and many other substances, giving off corrosive vapors (→ *hydrochloric acid*). Reacts with air to
Relative vapor density (air = 1)	4.4	form corrosive vapors (→ *hydrochloric acid*).
Relative density at 20 °C of saturated mixture		
vapor/air (air = 1)	1.5	TLV-TWA not available
Vapor pressure, mm Hg at 20 °C	110	
Solubility in water	reaction	
Explosive limits, vol% in air	3.4-9.5	**Absorption route:** Can enter the body by inhalation or ingestion. Harmful atmospheric
Minimum ignition energy, mJ	2.0-2.4	concentrations can build up very rapidly on evaporation at 20°C.
Relative molecular mass	129.1	**Immediate effects:** Corrosive to the eyes, skin and respiratory tract. Inhalation of vapor/fumes
		can cause severe breathing difficulties (lung edema). In serious cases risk of death.
Gross formula:	$C_2H_6Cl_2Si$	

HAZARDS/SYMPTOMS	PREVENTIVE MEASURES	FIRE EXTINGUISHING/FIRST AID
Fire: extremely flammable.	keep away from open flame and sparks, no smoking.	large quantities AFFF, powder, carbon dioxide, (halons); DO NOT USE OTHER WATER-BASED EXTINGUISHERS.
Explosion: forms explosive air-vapor mixtures.	sealed machinery, ventilation, explosion-proof electrical equipment and lighting.	in case of fire: keep tanks/drums cool by spraying with water, but DO NOT spray substance with water.
Inhalation: *corrosive*; sore throat, cough, shortness of breath, severe breathing difficulties.	ventilation, local exhaust or respiratory protection.	fresh air, rest, place in half-sitting position, take to hospital.
Skin: *corrosive*; redness, pain, burns.	gloves, protective clothing.	remove contaminated clothing, flush skin with water or shower, refer to a doctor.
Eyes: *corrosive*; redness, pain, impaired vision.	face shield, or combined eye and respiratory protection.	flush with water, take to a doctor if necessary.
Ingestion: *corrosive*; sore throat, abdominal pain, diarrhea, nausea.		rinse mouth, immediately take to hospital.

SPILLAGE	STORAGE	LABELING / NFPA
collect leakage in sealable containers, soak up with sand or other inert absorbent and remove to safe place, sprinkle remainder thoroughly with soda and flush away with water after waiting 10 minutes; (additional individual protection: CHEMICAL SUIT).	keep in dry, fireproof place, separate from other substances.	R: 11-36/37/38 🔥 Flammable ✖ Irritant (NFPA diamond: 3 top, 3 left, 1 right)

NOTES
Reacts violently with extinguishing agents such as water. Lung edema symptoms usually develop several hours later and are aggravated by physical exertion: rest and hospitalization essential. As first aid, a doctor or authorized person should consider administering a corticosteroid spray. Use airtight packaging.

Transport Emergency Card TEC(R)-30G40

HI: X338; UN-number: 1162

DIMETHYL ETHER
(cylinder)

PHYSICAL PROPERTIES	IMPORTANT CHARACTERISTICS
Boiling point, °C − 25 Melting point, °C − 141 Flash point, °C flammable gas Auto-ignition temperature, °C 235 Relative density (water = 1) 0.7 Relative vapor density (air = 1) 1.6 Vapor pressure, mm Hg at 20 °C 4028 Solubility in water, g/100 ml at 18 °C 7.6 Explosive limits, vol% in air 3.0-18.6 Minimum ignition energy, mJ 0.29 Relative molecular mass 46.1 Gross formula: C_2H_6O	**COLORLESS COMPRESSED LIQUEFIED GAS** Gas is heavier than air and spreads at ground level, with risk of ignition at a distance. Presumed to be able to form peroxides. TLV-TWA not available **Absorption route:** Can enter the body by inhalation. Harmful atmospheric concentrations can build up very rapidly if gas is released. **Immediate effects:** Irritates the eyes and respiratory tract. Liquid can cause frostbite due to rapid evaporation. Affects the nervous system.

HAZARDS/SYMPTOMS	PREVENTIVE MEASURES	FIRE EXTINGUISHING/FIRST AID
Fire: extremely flammable.	keep away from open flame and sparks, no smoking.	shut off supply; if impossible and no danger to surrounding area, allow fire to burn itself out; otherwise extinguish with powder, carbon dioxide, (halons).
Explosion: forms explosive air-gas mixtures.	sealed machinery, ventilation, explosion-proof electrical equipment and lighting, non-sparking tools.	in case of fire: keep cylinders cool by spraying with water.
Inhalation: cough, shortness of breath, headache, sleepiness.	ventilation, local exhaust or respiratory protection.	fresh air, rest, artificial respiration if necessary, call a doctor.
Skin: *in case of frostbite*: redness, pain, sores.	insulating gloves.	*in case of frostbite*: DO NOT remove clothing, flush skin with water or shower, take to hospital if necessary.
Eyes: *in case of frostbite*: redness, pain, impaired vision.	face shield, or combined eye and respiratory protection.	flush with water, take to a doctor.

SPILLAGE	STORAGE	LABELING / NFPA
evacuate area, call in an expert, ventilate, under no circumstances spray liquid with water; (additional individual protection: breathing apparatus).	keep in cool, fireproof place.	4 2 1

NOTES
Before distilling check for peroxides; if found, render harmless. High atmospheric concentrations, e.g. in poorly ventilated spaces, can cause oxygen deficiency, with risk of unconsciousness. Symptoms of effects on the nervous system occur only at high concentrations. Pure DME, unlike technical grades, does not form peroxides. Turn leaking cylinder so that leak is on top to prevent liquid dimethyl ether escaping.

Transport Emergency Card TEC(R)-676

HI: 23; UN-number: 1033

CAS-No: [68-12-2]
N,N-dimethylformamide
DMF
DMFA
N-formyldimethylamine

$(CH_3)_2NCHO$

DIMETHYLFORMAMIDE

PHYSICAL PROPERTIES		IMPORTANT CHARACTERISTICS
Boiling point, °C	153	**COLORLESS LIQUID WITH CHARACTERISTIC ODOR**
Melting point, °C	−61	Vapor mixes readily with air. Decomposes when heated above 350°C or burned, giving off toxic
Flash point, °C	58	vapors. Reacts violently with nitrates, carbon tetrachloride and hexachlorocyclohexane, with risk
Auto-ignition temperature, °C	445	of fire and explosion. Attacks some plastics.
Relative density (water = 1)	0.95	

COLORLESS LIQUID WITH CHARACTERISTIC ODOR
Vapor mixes readily with air. Decomposes when heated above 350°C or burned, giving off toxic vapors. Reacts violently with nitrates, carbon tetrachloride and hexachlorocyclohexane, with risk of fire and explosion. Attacks some plastics.

PHYSICAL PROPERTIES			IMPORTANT CHARACTERISTICS		
Boiling point, °C		153			
Melting point, °C		−61			
Flash point, °C		58			
Auto-ignition temperature, °C		445			
Relative density (water = 1)		0.95			
Relative vapor density (air = 1)		2.5	**TLV-TWA (skin)**	10 ppm	30 mg/m³
Relative density at 20°C of saturated mixture					
vapor/air (air = 1)		1.01			
Vapor pressure, mm Hg at 20°C		3.04			
Solubility in water		∞			
Explosive limits, vol% in air		2.2-16			
Electrical conductivity, pS/m		6x10⁶			
Relative molecular mass		73.1			
Log P octanol/water		−0.7			

Absorption route: Can enter the body by inhalation or ingestion or - above all - through the skin. Harmful atmospheric concentrations can build up fairly rapidly on evaporation at approx. 20°C - even more rapidly in aerosol form.
Immediate effects: Irritates the eyes, skin and respiratory tract.
Effects of prolonged/repeated exposure: Prolonged or repeated contact can cause skin disorders. Can cause liver and kidney damage.

Gross formula: C_2H_7NO

HAZARDS/SYMPTOMS	PREVENTIVE MEASURES	FIRE EXTINGUISHING/FIRST AID
Fire: combustible.	keep away from open flame, no smoking.	powder, alcohol-resistant foam, water spray, carbon dioxide, (halons).
Explosion: above 58°C: forms explosive air-vapor mixtures.	above 58°C: sealed machinery, ventilation.	
	AVOID EXPOSURE OF PREGNANT WOMEN AS FAR AS POSSIBLE	
Inhalation: headache, dizziness, vomiting, diarrhea, abdominal cramps.	ventilation, local exhaust or respiratory protection.	fresh air, rest, call a doctor.
Skin: *is absorbed*; redness, pain.	gloves, protective clothing.	remove contaminated clothing, wash skin with soap and water, refer to a doctor.
Eyes: redness, pain, impaired vision.	face shield.	flush with water, take to a doctor.
Ingestion: diarrhea, abdominal cramps, headache, dizziness, vomiting.		rinse mouth, call a doctor or take to hospital.

SPILLAGE	STORAGE	LABELING / NFPA
collect leakage in sealable containers, soak up with sand or other inert absorbent and remove to safe place; (additional individual protection: breathing apparatus).	keep separate from oxidants; ventilate at floor level.	R: 20/21-36 S: 26-28-36 ✖ Harmful 2 / 1 / 0

NOTES
Log P octanol/water is estimated. Alcohol consumption increases toxic effects. Depending on the degree of exposure, regular medical checkups are advisable.

CAS-No: [57-14-7]
asym-dimethylhydrazine
dimazine
dimethylhydrazine
N,N-dimethylhydrazine
UDMH

$(CH_3)_2N-NH_2$

1,1-DIMETHYLHYDRAZINE

PHYSICAL PROPERTIES	IMPORTANT CHARACTERISTICS
Boiling point, °C 63 Melting point, °C −58 Flash point, °C −18 Auto-ignition temperature, °C 240 Relative density (water = 1) 0.8 Relative vapor density (air = 1) 2.1 Relative density at 20 °C of saturated mixture vapor/air (air = 1) 1.2 Vapor pressure, mm Hg at 20 °C 110 Solubility in water good Explosive limits, vol% in air 2.4-95 Relative molecular mass 60.1	**COLORLESS LIQUID WITH CHARACTERISTIC ODOR** Vapor is heavier than air and spreads at ground level, with risk of ignition at a distance. Do not use compressed air when filling, emptying or processing. → *Nitrous vapors* are a product of combustion. Strong reducing agent which reacts violently with oxidants such as nitrogen tetroxide, hydrogen peroxide and nitric acid, with risk of fire and explosion. Reacts violently with strong acids and oxidants.
	TLV-TWA (skin) 0.5 ppm 1.2 mg/m³ A2
	Absorption route: Can enter the body by inhalation or ingestion or through the skin. **Immediate effects:** Corrosive to the eyes, skin and respiratory tract. In substantial concentrations can cause seizures. Inhalation of vapor/fumes can cause severe breathing difficulties (lung edema). In serious cases risk of death. **Effects of prolonged/repeated exposure:** Can affect the blood. Can cause liver and kidney damage. Has been found to cause a type of lung and liver cancer in certain animal species under certain circumstances.
Gross formula: $C_2H_8N_2$	

HAZARDS/SYMPTOMS	PREVENTIVE MEASURES	FIRE EXTINGUISHING/FIRST AID
Fire: extremely flammable.	keep away from open flame and sparks, no smoking.	water spray, foam, powder, carbon dioxide.
Explosion: forms explosive air-vapor mixtures.	sealed machinery, ventilation, explosion-proof electrical equipment and lighting.	in case of fire: keep tanks/drums cool by spraying with water.
	STRICT HYGIENE	IN ALL CASES CALL IN A DOCTOR
Inhalation: *corrosive*; sore throat, cough, severe breathing difficulties, vomiting, cramps.	ventilation, local exhaust or respiratory protection.	fresh air, rest, call a doctor or take to hospital.
Skin: *corrosive, is absorbed*; redness, pain.	gloves, protective clothing.	remove contaminated clothing, flush skin with water or shower, call a doctor.
Eyes: *corrosive*; redness, pain, impaired vision.	face shield, or combined eye and respiratory protection.	flush with water, take to a doctor.
Ingestion: *corrosive*; abdominal pain, vomiting, pain with swallowing.		rinse mouth, immediately take to hospital.

SPILLAGE	STORAGE	LABELING / NFPA
evacuate area, call in an expert, collect leakage in sealable containers, soak up with sand or other inert absorbent and remove to safe place, do not use sawdust or other combustible substances; (additional individual protection: CHEMICAL SUIT).	keep in fireproof place, separate from oxidants and acids.	R: 45-11-23/25-34 S: 53-16-33-34 Note E Flammable Toxic 3 / 3 1

NOTES
Odor limit is above TLV. Depending on the degree of exposure, regular medical checkups are advisable. Lung edema symptoms usually develop several hours later and are aggravated by physical exertion: rest and hospitalization essential. As first aid, a doctor or authorized person should consider administering a corticosteroid spray. Use airtight packaging. Unbreakable packaging preferred; if breakable, keep in unbreakable container.

Transport Emergency Card TEC(R)-30G31

HI: 338; UN-number: 1163

CAS-No: [131-11-3]
1,2-benzenedicarboxylic acid dimethyl ester
dimethyl benzeneorthodicarboxylate
DMP
methyl phthalate
phthalic acid dimethyl ester

$C_6H_4(COOCH_3)_2$

DIMETHYL PHTHALATE

PHYSICAL PROPERTIES		IMPORTANT CHARACTERISTICS	
Boiling point, °C	284	**COLORLESS ODORLESS VISCOUS LIQUID**	
Melting point, °C	2	Vapor mixes readily with air. Reacts with strong oxidants.	
Flash point, °C	146		
Auto-ignition temperature, °C	490	TLV-TWA	5 mg/m³
Relative density (water = 1)	1.2		
Relative vapor density (air = 1)	6.7	**Absorption route:** Can enter the body by inhalation or ingestion. Evaporation negligible at 20°C, but harmful atmospheric concentrations can build up rapidly in aerosol form.	
Relative density at 20 °C of saturated mixture vapor/air (air = 1)	1.0	**Immediate effects:** Irritates the eyes, skin and respiratory tract, and stomach if swallowed. In substantial concentrations can impair consciousness.	
Vapor pressure, mm Hg at 20 °C	9.9×10^{-3}		
Solubility in water, g/100 ml at 20 °C	0.43		
Explosive limits, vol% in air	see notes		
Electrical conductivity, pS/m	8.1×10^5		
Relative molecular mass	194.2		
Gross formula:	$C_{10}H_{10}O_4$		

HAZARDS/SYMPTOMS	PREVENTIVE MEASURES	FIRE EXTINGUISHING/FIRST AID
Fire: combustible.	keep away from open flame, no smoking.	powder, water spray, foam, carbon dioxide, (halons).
Inhalation: sore throat, cough, dizziness.	ventilation.	fresh air, rest, call a doctor if necessary.
Skin: redness, pain.	gloves.	remove contaminated clothing, flush skin with water or shower.
Eyes: redness, pain.	safety glasses.	flush with water, take to doctor if necessary.
Ingestion: abdominal pain, nausea, dizziness.		rinse mouth, call a doctor.

SPILLAGE	STORAGE	LABELING / NFPA
collect leakage in sealable containers, soak up with sand or other inert absorbent and remove to safe place.	ventilate at floor level.	

NOTES
Lower explosive limit of 0.9% given in the literature, but this is only reached at a temperature of 180°C.

CAS-No: [77-78-1]
dimethyl sulphate
DMS
sulfuric acid dimethyl ester

$(CH_3)_2SO_4$

DIMETHYL SULFATE

PHYSICAL PROPERTIES	IMPORTANT CHARACTERISTICS
Boiling point, °C — 188 Melting point, °C — −32 Flash point, °C — 83 Auto-ignition temperature, °C — 470 Relative density (water = 1) — 1.3 Relative vapor density (air = 1) — 4.4 Relative density at 20 °C of saturated mixture vapor/air (air = 1) — 1.0 Vapor pressure, mm Hg at 20°C — 4.6 Solubility in water, g/100 ml at 18 °C — 2.8 Explosive limits, vol% in air — 3.6-23.2 Relative molecular mass — 126.1	**CLEAR COLORLESS VISCOUS LIQUID** Vapor mixes readily with air. Decomposes when heated, giving off → *sulfur dioxide* and flammable vapors. Reacts violently with strong oxidants, bases and acids, with risk of fire and explosion.

PHYSICAL PROPERTIES	IMPORTANT CHARACTERISTICS		
	TLV-TWA (skin)	0.1 ppm	0.52 mg/m³ C A2

PHYSICAL PROPERTIES	IMPORTANT CHARACTERISTICS
	Absorption route: Can enter the body by inhalation or ingestion or through the skin. Harmful atmospheric concentrations can build up very rapidly on evaporation at 20°C. **Immediate effects:** Corrosive to the eyes, skin and respiratory tract. In high concentrations can cause seizures and impairment of consciousness. Inhalation of vapor/fumes can cause severe breathing difficulties (lung edema). In serious cases risk of death. **Effects of prolonged/repeated exposure:** Can cause liver and kidney damage. Has been found to cause a type of cancer in certain animal species under certain circumstances.
Gross formula: $C_2H_6O_4S$	

HAZARDS/SYMPTOMS	PREVENTIVE MEASURES	FIRE EXTINGUISHING/FIRST AID
Fire: combustible.	keep away from open flame, no smoking.	powder, water spray, foam, carbon dioxide, (halons).
Explosion: above 83°C: forms explosive air-vapor mixtures.	above 83°C: sealed machinery, ventilation.	
	STRICT HYGIENE, AVOID ALL CONTACT	IN ALL CASES CALL A DOCTOR
Inhalation: *corrosive*; sore throat, cough, shortness of breath.	ventilation, local exhaust or respiratory protection.	fresh air, rest, artificial respiration if necessary, take to hospital.
Skin: *corrosive, is absorbed*; redness, pain, serious burns.	gloves, protective clothing.	remove contaminated clothing, flush skin with water or shower, refer to a doctor.
Eyes: *corrosive*; redness, pain, impaired vision.	face shield, or combined eye and respiratory protection.	flush with water, take to a doctor.
Ingestion: sore throat, abdominal pain, vomiting.	do not eat, drink or smoke while working.	rinse mouth, take immediately to hospital.

SPILLAGE	STORAGE	LABELING / NFPA
evacuate area, call in an expert, collect leakage in sealable containers, soak up with sand or other inert absorbent and remove to safe place; (additional individual protection: CHEMICAL SUIT).	keep separate from oxidants, strong acids and strong bases; ventilate at floor level.	R: 45-25-26-34 S: 53-26-27-45 Note E ☠ Highly toxic

NOTES
TLV is maximum and must not be exceeded. Lung edema symptoms usually develop several hours later and are aggravated by physical exertion: rest and hospitalization essential. As first aid, a doctor or authorized person should consider administering a corticosteroid spray. Unbreakable packaging preferred; if breakable, keep in unbreakable container.

Transport Emergency Card TEC(R)-54

HI: 66; UN-number: 1595

CAS-No: [67-68-5]
sulficyl bis(methane)
DMSO

$(CH_3)_2SO$

DIMETHYL SULFOXIDE

PHYSICAL PROPERTIES		IMPORTANT CHARACTERISTICS	
Boiling point, °C	189	**COLORLESS VISCOUS HYGROSCOPIC LIQUID**	
Melting point, °C	8	Vapor mixes readily with air. Corrosive gases are a product of combustion ($\rightarrow$ *sulfur dioxide*). Decomposes when heated above 100°C, giving off flammable gases and corrosive vapors ($\rightarrow$ *sulfur dioxide*). Reacts violently with acetyl chloride, cyanuric chloride, benzene sulfonyl chloride, sulfuric and phosphoric haloids, with risk of fire and explosion. Reacts violently with strong oxidants. Attacks many plastics.	
Flash point, °C	95		
Auto-ignition temperature, °C	270		
Relative density (water = 1)	1.1		
Relative vapor density (air = 1)	2.7		
Relative density at 20 °C of saturated mixture			
vapor/air (air = 1)	1.0	TLV-TWA not available	
Vapor pressure, mm Hg at 20 °C	0.38		
Solubility in water	∞	**Absorption route:** Can enter the body by inhalation or ingestion or - very rapidly - through the skin. Harmful atmospheric concentrations build up fairly slowly on evaporation at 20°C, but much more rapidly in aerosol form. **Immediate effects:** Irritates the eyes, skin and respiratory tract.	
Electrical conductivity, pS/m	2×10^5		
Relative molecular mass	78.1		
Log P octanol/water	−2.0		
Gross formula: C_2H_6OS			

HAZARDS/SYMPTOMS	PREVENTIVE MEASURES	FIRE EXTINGUISHING/FIRST AID
Fire: combustible, many chemical reactions can cause fire and explosion.	keep away from open flame, no smoking.	powder, alcohol-resistant foam, water spray, carbon dioxide, (halons).
Explosion: above 95°C: forms explosive air-vapor mixtures.	above 95°C: sealed machinery, ventilation.	
Inhalation: cough.	ventilation, local exhaust or respiratory protection.	fresh air, rest.
Skin: *is absorbed*; redness, nausea, sleepiness.	gloves.	remove contaminated clothing, wash skin with soap and water, call a doctor.
Eyes: redness, pain, impaired vision.	face shield.	flush with water, take to a doctor if necessary.
Ingestion: abdominal pain, diarrhea, vomiting.		rinse mouth, call a doctor or take to hospital.

SPILLAGE	STORAGE	LABELING / NFPA
collect leakage in sealable containers, soak up with sand or other inert absorbent and remove to safe place, flush away remainder with water; (additional individual protection: breathing apparatus).	keep separate from oxidants.	

NOTES
Usually solidifies at a temperature below melting point.

$C_6H_4(COOCH_3)_2$

DIMETHYL TEREPHTHALATE

PHYSICAL PROPERTIES		IMPORTANT CHARACTERISTICS
Boiling point, °C	285	**COLORLESS CRYSTALS OR WHITE POWDER**
Melting point, °C	140	In dry state can form electrostatic charge if stirred, transported pneumatically, poured etc.
Flash point, °C	141	Reacts with strong oxidants.
Auto-ignition temperature, °C	518	
Relative density (water = 1)	1.04	TLV-TWA not available
Relative vapor density (air = 1)	6.7	
Vapor pressure, mm Hg at 150 °C	13	**Absorption route:** Can enter the body by inhalation or ingestion. Evaporation negligible at 20°C, but harmful concentrations of airborne particles can build up rapidly.
Solubility in water	none	
Explosive limits, vol% in air	0.8-11.8	**Immediate effects:** Irritates the eyes and respiratory tract.
Relative molecular mass	194.2	
Gross formula:	$C_{10}H_{10}O_4$	

HAZARDS/SYMPTOMS	PREVENTIVE MEASURES	FIRE EXTINGUISHING/FIRST AID
Fire: combustible.	keep away from open flame, no smoking.	water spray, powder.
Explosion: finely dispersed particles form explosive mixtures on contact with air.	keep dust from accumulating; sealed machinery, explosion-proof electrical equipment and lighting, grounding.	
Inhalation: tickling of the throat.	local exhaust or respiratory protection.	fresh air, rest.
Eyes: redness.	goggles.	flush with water, take to a doctor if necessary.
Ingestion: abdominal pain.		rinse mouth, call a doctor if necessary.

SPILLAGE	STORAGE	LABELING / NFPA
clean up spillage, carefully collect remainder; (additional individual protection: P2 respirator).	ventilate at floor level.	

NOTES
Processing, transshipping and transporting in molten form can cause explosive gas-air mixtures.

CAS-No: [97-02-9]
1-amino-2,4-dinitrobenzene
dinitro-2,4 aniline
2,4-dinitrobenzeneamine
2,4-dinitrophenylamine

$(NO_2)_2C_6H_3NH_2$

2,4-DINITROANILINE

PHYSICAL PROPERTIES	IMPORTANT CHARACTERISTICS
Melting point, °C 188 Flash point, °C 224 Relative density (water = 1) 1.6 Relative vapor density (air = 1) 6.3 Relative density at 20 °C of saturated mixture vapor/air (air = 1) 1.07 Vapor pressure, mm Hg at 20 °C 9.8 Solubility in water none Relative molecular mass 183.1	**YELLOW CRYSTALS WITH CHARACTERISTIC ODOR** Vapor mixes readily with air. → *Nitrous vapors* are a product of combustion. Decomposes when moderately heated, giving off toxic vapors which form an explosive mixture with air. Reacts violently with oxidants.
	TLV-TWA　　　　　　　　　　not available
	Absorption route: Can enter the body by inhalation or ingestion or through the skin. Harmful atmospheric concentrations can build up very rapidly on evaporation at 20°C. **Immediate effects:** Irritates the eyes, skin and respiratory tract. **Effects of prolonged/repeated exposure:** Can affect the blood. Can cause liver and kidney damage.
Gross formula: $C_6H_5N_3O_4$	

HAZARDS/SYMPTOMS	PREVENTIVE MEASURES	FIRE EXTINGUISHING/FIRST AID
Fire: combustible.	keep away from open flame, no smoking.	water spray, powder.
Explosion: heating can cause explosion.		in case of fire: keep tanks/drums cool by spraying with water, fight fire from sheltered location.
	STRICT HYGIENE, KEEP DUST UNDER CONTROL	
Inhalation: shortness of breath, headache, dizziness, blue skin, feeling of weakness.	local exhaust or respiratory protection.	fresh air, rest, place in half-sitting position, call a doctor.
Skin: *is absorbed*; redness; see also 'Inhalation'.	gloves, protective clothing.	remove contaminated clothing, flush skin with water or shower, call a doctor.
Eyes: redness, pain.	safety glasses, or combined eye and respiratory protection.	flush with water, take to a doctor if necessary.
Ingestion: abdominal pain.	do not eat, drink or smoke while working.	rinse mouth, take to hospital.

SPILLAGE	STORAGE	LABELING / NFPA
clean up spillage, carefully collect remainder; (additional individual protection: breathing apparatus).	keep separate from oxidants.	R: 26/27/28-33 S: 28-36/37-45 ☠ Toxic

NFPA diamond: blue 3, red 1, yellow 3

NOTES
In confined spaces combustion can cause explosion. Fight fires in advanced stage from explosion-proof shelter. Depending on the degree of exposure, regular medical checkups are advisable. Effects on blood due to formation of methemoglobin. Special first aid required in the event of poisoning: antidotes must be available (with instructions).

Transport Emergency Card TEC(R)-875

HI: 60; UN-number: 1596

3,5-DINITROBENZOYLCHLORIDE

PHYSICAL PROPERTIES	IMPORTANT CHARACTERISTICS
Melting point, °C 69 Flash point, °C see notes Vapor pressure, mm Hg at 196 °C 11 Solubility in water reaction Relative molecular mass 230.6	**YELLOW CRYSTALS WITH PUNGENT ODOR** Can deflagrate (burn explosively) when heated. Decomposes when heated or burned or on contact with moist air, giving off toxic and corrosive vapors.
	TLV-TWA not available
	Absorption route: Can enter the body by inhalation or ingestion. Evaporation negligible at 20°C, but harmful concentrations of airborne particles can build up rapidly. **Immediate effects:** Irritates the eyes, skin and respiratory tract.
Gross formula: $C_7H_3ClN_2O_5$	

HAZARDS/SYMPTOMS	PREVENTIVE MEASURES	FIRE EXTINGUISHING/FIRST AID
Fire: combustible.	keep away from open flame, no smoking.	water spray, powder.
Explosion: subjection to friction/shock can cause explosion.		in case of fire: keep tanks/drums cool by spraying with water, fight fire from sheltered location.
Inhalation: sore throat, cough, shortness of breath.	local exhaust or respiratory protection.	fresh air, rest, call a doctor.
Skin: redness, pain.	gloves.	remove contaminated clothing, flush skin with water or shower.
Eyes: redness, pain.	goggles, or combined eye and respiratory protection.	flush with water, take to a doctor if necessary.
Ingestion: sore throat, abdominal pain, nausea, vomiting.		rinse mouth, call a doctor or take to hospital.

SPILLAGE	STORAGE	LABELING / NFPA
evacuate area, call in an expert, clean up spillagen, carefully collect remainder, do not use sawdust or other combustible absorbents; (additional individual protection: respirator with A/P2 filter).	keep in fireproof place, separate from combustible substances and reducing agents.	

NOTES
Combustible, but flash point and explosive limits not given in the literature. In confined spaces combustion can cause explosion. Use airtight packaging.

CAS-No: [97-00-7]
1-chloro-2,4-dinitrobenzene
1,3-dinitro-4-chlorobenzene

$ClC_6H_3(NO_2)_2$

DINITROCHLOROBENZENE

PHYSICAL PROPERTIES		IMPORTANT CHARACTERISTICS
Boiling point, °C	315	**LIGHT YELLOW CRYSTALS**
Melting point, °C	52	Toxic vapors are a product of combustion (incl. → *nitrous vapors*). Can deflagrate (burn
Flash point, °C	194	explosively) when heated. Can decompose explosively if subjected to shock. Strong oxidant
Auto-ignition temperature, °C	432	which reacts violently with combustible substances and reducing agents.
Relative density (water = 1)	1.7	
Relative vapor density (air = 1)	7.0	TLV-TWA not available
Relative density at 20 °C of saturated mixture		
vapor/air (air = 1)	1.00	**Absorption route:** Can enter the body by inhalation or ingestion or through the skin. Evaporation
Vapor pressure, mm Hg at 20 °C	< 0.076	negligible at 20°C, but harmful concentrations of airborne particles can build up rapidly.
Solubility in water	none	**Immediate effects:** Irritates the eyes, skin and respiratory tract. In serious cases risk of death.
Explosive limits, vol% in air	1.9-22	**Effects of prolonged/repeated exposure:** Prolonged or repeated contact can cause skin
Relative molecular mass	202.6	disorders. Can affect the blood. Can cause liver and kidney damage.
Gross formula:	$C_6H_3ClN_2O_4$	

HAZARDS/SYMPTOMS	PREVENTIVE MEASURES	FIRE EXTINGUISHING/FIRST AID
Fire: combustible, many chemical reactions can cause fire and explosion.	keep away from open flame, no smoking.	large quantities of water.
Explosion: heating in confined space can cause explosion.	do not subject to shocks or friction.	in case of fire: keep tanks/drums cool by spraying with water, fight fire from sheltered location.
Inhalation: shortness of breath, headache, dizziness, blue skin, feeling of weakness.	local exhaust or respiratory protection.	fresh air, rest, take to hospital.
Skin: *is absorbed*; redness, pain; see also 'Inhalation'.	gloves, protective clothing.	remove contaminated clothing, wash skin with soap and water, call a doctor.
Eyes: redness, pain.	vizor.	flush with water, take to a doctor if necessary.
Ingestion: abdominal pain, nausea; see also 'Inhalation'.	do not eat, drink or smoke while working.	rinse mouth, take immediately to hospital.

SPILLAGE	STORAGE	LABELING / NFPA
clean up spillage, carefully collect remainder and remove to safe place, do not use sawdust or other combustible absorbents; (additional individual protection: P3 respirator).	keep in fireproof place, separate from combustible substances and reducing agents.	R: 23/24/25-33 S: 28-37-44 Note C ☠ Toxic

NFPA diamond: blue 3, yellow 1, red 4

NOTES
The measures on this sheet also apply to all modifications. Is absorbed by rubber, neoprene and leather: use suitable gloves (plastic). Melting points for α-isomer: 53° C; β-isomer: 43° C; γ-isomer: 27° C. In confined spaces combustion can cause explosion. Depending on the degree of exposure, regular medical checkups are advisable. Effects on blood due to formation of methemoglobin. Special first aid required in the event of poisoning: antidotes must be available (with instructions).

Transport Emergency Card TEC(R)-94A

HI: 60; UN-number: 1577

$(HO)C_6H_3(NO_2)_2$

2,4-DINITROPHENOL

PHYSICAL PROPERTIES	IMPORTANT CHARACTERISTICS
Boiling point, °C sublimes Melting point, °C 115 Relative density (water = 1) 1.7 Solubility in water, g/100 ml at 18 °C 0.6 Relative molecular mass 184.1 Log P octanol/water 1.5	**LIGHT YELLOW POWDER WITH CHARACTERISTIC ODOR** In dry state can form electrostatic charge if stirred, transported pneumatically, poured etc. Burns explosively when heated or subjected to shock or friction. Reacts violently with combustible substances and reducing agents. Forms shock-sensitive compounds with bases and anhydrous ammonia.

TLV-TWA	not available

Absorption route: Can enter the body by inhalation or ingestion or through the skin. Evaporation negligible at 20°C, but harmful concentrations of airborne particles can build up rapidly.
Immediate effects: Prolonged or repeated contact causes yellowing of the skin. At high exposure levels can affect various organs. In serious cases risk of death.

Gross formula: $C_6H_4N_2O_5$

HAZARDS/SYMPTOMS	PREVENTIVE MEASURES	FIRE EXTINGUISHING/FIRST AID
Fire: combustible.	keep away from open flame and sparks, no smoking, avoid contact with combustible substances.	large quantities of water.
Explosion: shock or friction can cause explosion.	sealed machinery, ventilation, explosion-proof electrical equipment and lighting.	in case of fire: keep tanks/drums cool by spraying with water.
Inhalation: nausea, abdominal cramps, fever, thirst, sweating, rapid pulse.	local exhaust or respiratory protection.	fresh air, rest in a cool place, take to hospital.
Skin: *is absorbed.*	gloves, protective clothing.	remove contaminated clothing, wash skin with soap and water, take to hospital.
Eyes: redness, pain.	face shield.	flush with water, take to a doctor.
Ingestion: abdominal pain, abdominal cramps, nausea, fever, thirst, sweating, rapid pulse.	do not eat, drink or smoke while working.	rinse mouth, take immediately to hospital.

SPILLAGE	STORAGE	LABELING / NFPA
evacuate area, call in an expert, dampen spillage with water and clean up, soak up remainder with sand or other inert absorbent and remove to safe place; (additional individual protection: breathing apparatus).	keep in cool, dark, fireproof place, separate from combustible substances and reducing agents; add 15% water.	R: 23/24/25-33 S: 28-37-44 Note C ☠ Toxic

NOTES

Because of explosive properties is used in the form of a water paste. The measures listed apply to the pure form. Flush contaminated clothing with water (fire hazard). Depending on the degree of exposure, regular medical checkups are advisable.

UN-number: 76

$(C_9H_{19})_2C_6H_3(OH)$

DINONYLPHENOL

PHYSICAL PROPERTIES		IMPORTANT CHARACTERISTICS	
Boiling point, °C	?	**COLORLESS LIQUID**	
Melting point, °C	?	Decomposes when heated, giving off toxic vapors. Reacts with strong oxidants.	
Flash point, °C	see notes		
Relative density (water = 1)	0.9	TLV-TWA	not available
Relative vapor density (air = 1)	12		
Vapor pressure, mm Hg at 180°C	9.9	**Absorption route:** Can enter the body by inhalation or ingestion or through the skin. Harmful atmospheric concentrations build up very slowly, if at all, on evaporation at approx. 20°C, but much more rapidly in aerosol form.	
Solubility in water	none		
Relative molecular mass	346.6		
		Immediate effects: Corrosive to the eyes, skin and respiratory tract. Inhalation of vapor/fumes can cause severe breathing difficulties (lung edema). In serious cases risk of death.	
Gross formula:	$C_{24}H_{42}O$		

HAZARDS/SYMPTOMS	PREVENTIVE MEASURES	FIRE EXTINGUISHING/FIRST AID
Fire: combustible.	keep away from open flame, no smoking.	powder, AFFF, foam, carbon dioxide, (halons); in case of fire in immediate vicinity: use any extinguishing agent.
Inhalation: *corrosive*; sore throat, cough, severe breathing difficulties, unconsciousness.	ventilation, local exhaust or respiratory protection.	fresh air, rest, place in half-sitting position, take to hospital.
Skin: *corrosive*; redness, pain, serious burns.	gloves, protective clothing.	remove contaminated clothing, flush skin with water or shower, call a doctor.
Eyes: *corrosive*; redness, pain, impaired vision.	face shield.	flush with water, take to a doctor.
Ingestion: *corrosive*, abdominal pain, pain with swallowing.	do not eat, drink or smoke while working.	rinse mouth, immediately take to hospital.

SPILLAGE	STORAGE	LABELING / NFPA
collect leakage in sealable containers, soak up with sand or other inert absorbent and remove to safe place.		

NOTES
Combustible, but flash point and explosive limits not given in the literature. Lung edema symptoms usually develop several hours later and are aggravated by physical exertion: rest and hospitalization essential. As first aid, a doctor or authorized person should consider administering a corticosteroid spray.

Transport Emergency Card TEC(R)-80G08

CAS-No: [117-81-7]
bis(2-ethylhexyl) phthalate
di(2-ethylhexyl) phthalate
di-sec-octyl phthalate
DOP

$C_6H_4(COOCH_2CH(C_2H_5)C_4H_9)_2$

DIOCTYL PHTHALATE

PHYSICAL PROPERTIES		IMPORTANT CHARACTERISTICS
Boiling point, °C	385	**LIGHT-COLORED VISCOUS LIQUID WITH CHARACTERISTIC ODOR**
Melting point, °C	−55	Decomposes slowly when heated above 170°C, giving off toxic gases.
Flash point, °C	199	
Auto-ignition temperature, °C	410	TLV-TWA not available
Relative density (water = 1)	0.99	
Relative vapor density (air = 1)	13.5	**Absorption route:** Can enter the body by ingestion.
Relative density at 20 °C of saturated mixture vapor/air (air = 1)	1.0	**Immediate effects:** Irritates the eyes, skin and respiratory tract.
Vapor pressure, mm Hg at 150 °C	0.068	**Effects of prolonged/repeated exposure:** Prolonged or repeated contact can cause skin
Solubility in water	none	disorders. Has been found to cause a type of liver cancer in certain animal species under certain
Relative molecular mass	390.5	circumstances.
Gross formula: $C_{24}H_{38}O_4$		

HAZARDS/SYMPTOMS	PREVENTIVE MEASURES	FIRE EXTINGUISHING/FIRST AID
Fire: combustible.	keep away from open flame, no smoking.	powder, AFFF, foam, carbon dioxide, (halons).
Inhalation: sore throat, cough.	ventilation.	fresh air, rest.
Skin: redness.	gloves.	remove contaminated clothing, flush skin with water or shower.
Eyes: redness, pain.	safety glasses.	flush with water, take to a doctor if necessary.
Ingestion: abdominal pain, nausea.		rinse mouth, call a doctor.

SPILLAGE	STORAGE	LABELING / NFPA
collect leakage in sealable containers, soak up with sand or an other inert absorbent and remove to safe place.		

NOTES
Diisooctyl phthalate is a mixture of phthalic esters of octylalcohol isomers; physical properties (boiling point, melting point and flash point) differ from those given above for DOP.

CAS-No: [123-91-1]
1,4-diethylene dioxide
diethylene ether
dioxan
dioxyethylene ether
tetrahydro-p-dioxin

1,4-DIOXANE

PHYSICAL PROPERTIES		IMPORTANT CHARACTERISTICS
Boiling point, °C	101	**COLORLESS LIQUID WITH CHARACTERISTIC ODOR**
Melting point, °C	12	Vapor mixes readily with air, forming explosive mixtures. Able to form peroxides. Flow, agitation
Flash point, °C	11	etc. can cause build-up of electrostatic charge due to liquid's low conductivity. Do not use
Auto-ignition temperature, °C	375	compressed air when filling, emptying or processing. Reacts violently with concentrated strong
Relative density (water = 1)	1.03	acids and strong oxidants. Attacks many plastics.
Relative vapor density (air = 1)	3.0	
Relative density at 20 °C of saturated mixture vapor/air (air = 1)	1.08	

TLV-TWA (skin)	50 ppm	180 mg/m³	

Vapor pressure, mm Hg at 20 °C	31.2
Solubility in water	∞
Explosive limits, vol% in air	1.9-22.5
Minimum ignition energy, mJ	< 0.3
Electrical conductivity, pS/m	0.5
Relative molecular mass	88.1
Log P octanol/water	− 0.4

Absorption route: Can enter the body by inhalation or ingestion or through the skin. Harmful atmospheric concentrations can build up fairly rapidly on evaporation at 20° C - even more rapidly in aerosol form.
Immediate effects: Irritates the eyes, skin and respiratory tract. Affects the nervous system. In serious cases risk of unconsciousness. If liquid is swallowed, droplets can enter the lungs, with risk of pneumonia.
Effects of prolonged/repeated exposure: Can affect the blood. Can cause liver and kidney damage.

Gross formula: $C_4H_8O_2$

HAZARDS/SYMPTOMS	PREVENTIVE MEASURES	FIRE EXTINGUISHING/FIRST AID
Fire: extremely flammable.	keep away from open flame and sparks, no smoking.	powder, alcohol-resistant foam, large quantities of water, carbon dioxide, (halons).
Explosion: forms explosive air-vapor mixtures.	sealed machinery, ventilation, explosion-proof electrical equipment and lighting, grounding, non-sparking hand tools.	in case of fire: keep tanks/drums cool by spraying with water.
Inhalation: sore throat, headache, dizziness, drowsiness.	ventilation, local exhaust or respiratory protection.	fresh air, rest, call a doctor.
Skin: *is absorbed*; redness.	gloves, protective clothing.	remove contaminated clothing, wash skin with soap and water.
Eyes: redness, pain.	face shield.	flush with water, take to a doctor.
Ingestion: sore throat, abdominal pain, nausea.		rinse mouth, DO NOT induce vomiting, take immediately to hospital.

SPILLAGE	STORAGE	LABELING / NFPA
collect leakage in sealable containers, flush away remainder with water; (additional individual protection: breathing apparatus).	keep in cool, dark, fireproof place, separate from oxidants and strong acids; add an inhibitor.	R: 11-36/37-40 S: 16-36/37 🔥 Flammable ✖ Harmful NFPA: 2 3 1

NOTES
Reacts with 15% water to form an azeotrope which boils at 88° C. Before distilling check for peroxides; if found, render harmless. Hydroquinone is one of the inhibitors added to prevent peroxide-formation. Odor limit is above TLV. Alcohol consumption increases toxic effects. Depending on the degree of exposure, regular medical checkups are advisable. Use airtight packaging.

Transport Emergency Card TEC(R)-546

HI: 33; UN-number: 1165

biphenyl
1,1'-biphenyl
dibenzene
phenylbenzene

DIPHENYL

PHYSICAL PROPERTIES		IMPORTANT CHARACTERISTICS	
Boiling point, °C	255	**WHITE CRYSTALS OR FLAKES WITH CHARACTERISTIC ODOR**	
Melting point, °C	70	Vapor mixes readily with air. Reacts violently with strong oxidants.	
Flash point, °C	113		
Auto-ignition temperature, °C	540	TLV-TWA 0.2 ppm 1.3 mg/m³	
Relative density (water = 1)	1.15		
Relative vapor density (air = 1)	5.3	**Absorption route:** Can enter the body by inhalation or ingestion or through the skin. Harmful	
Relative density at 20 °C of saturated mixture		atmospheric concentrations build up fairly slowly on evaporation at 20°C, but much more rapidly	
vapor/air (air = 1)	1.0	at higher temperatures or in the form of airborne particles.	
Vapor pressure, mm Hg at 20 °C	5.3×10^{-3}	**Immediate effects:** Irritates the eyes, skin and respiratory tract.	
Solubility in water	none	**Effects of prolonged/repeated exposure:** Prolonged or repeated contact can cause eczema.	
Explosive limits, vol% in air	0.6-5.8	Can cause liver and kidney damage. Also disorders of the central and peripheral nervous system	
Relative molecular mass	154.2	such as headache, convulsions, numbness and pain in limbs, paralysis.	
Log P octanol/water	3.8		
Gross formula:	$C_{12}H_{10}$		

HAZARDS/SYMPTOMS	PREVENTIVE MEASURES	FIRE EXTINGUISHING/FIRST AID
Fire: combustible.	keep away from open flame, no smoking.	powder, AFFF, foam, carbon dioxide, (halons).
Explosion: finely dispersed particles form explosive mixtures on contact with air.	keep dust from accumulating; sealed machinery, explosion-proof electrical equipment and lighting, grounding.	
Inhalation: cough, shortness of breath, headache.	local exhaust or respiratory protection.	fresh air, rest, call a doctor.
Skin: redness, pain.	gloves.	remove contaminated clothing, flush skin with water or shower.
Eyes: redness, pain.	safety glasses.	flush with water, take to a doctor if necessary.
Ingestion: abdominal pain, diarrhea, nausea, vomiting.	do not eat, drink or smoke while working.	rinse mouth, take to hospital.

SPILLAGE	STORAGE	LABELING / NFPA
clean up spillage, carefully collect remainder; (additional individual protection: respirator with A/P2 filter).	keep separate from oxidants.	1 / 2 / 0

NOTES
Component of many heat transfer media, often mixed with → *diphenyl oxide*. Depending on the degree of exposure, regular medical checkups are advisable. Dowtherm A is a trade name. Expands greatly on melting; relative density at 80°C: 0.98.

CAS-No: [101-68-8]
4-4'-diphenylmethane diisocyanate
methylene bisphenyl isocyanate
4,4'-methylenediphenyl diisocyanate

$OCNC_6H_4CH_2C_6H_4NCO$

DIPHENYLMETHANE-4,4'-DIISOCYANATE

PHYSICAL PROPERTIES		IMPORTANT CHARACTERISTICS	
Boiling point, °C	314	**WHITE TO LIGHT YELLOW STICKY FLAKES**	
Melting point, °C	37	Decomposes when heated or burned, giving off toxic vapors (→ *hydrogen cyanide*, → *nitrous*	
Flash point, °C	196	*vapors* and → *carbon monoxide*). Reacts violently with acids, bases, alcohols and amines, with	
Relative density (water = 1)	1.2	risk of fire and explosion. Reacts slowly with cold water, giving off → *carbon monoxide* etc.	
Vapor pressure, mm Hg at 25 °C	1.4×10^{-4}		
Solubility in water	reaction	TLV-TWA	0.005 ppm	0.051 mg/m³ C
Relative molecular mass	250.3		

Absorption route: Can enter the body by inhalation or ingestion or through the skin. Harmful atmospheric concentrations build up very slowly, if at all, on evaporation at approx. 20°C, but harmful concentrations of airborne particles can build up much more rapidly.
Immediate effects: Irritates the eyes, skin and respiratory tract. Inhalation can cause severe breathing difficulties (asthma).

Gross formula: $C_{15}H_{10}N_2O_2$

HAZARDS/SYMPTOMS	PREVENTIVE MEASURES	FIRE EXTINGUISHING/FIRST AID
Fire: combustible, many chemical reactions can cause fire and explosion.	keep away from open flame, no smoking, avoid contact with acids, bases, alcohols and amines.	powder, carbon dioxide, (halons), large quantities of water, NO OTHER WATER-BASED EXTINGUISHERS.
Explosion: violent reactions can cause explosions.		in case of fire: keep tanks/drums cool by spraying with water, but do direct contact of substance with water.
	STRICT HYGIENE, KEEP DUST UNDER CONTROL	
Inhalation: sore throat, cough, severe breathing difficulties, tightness of the chest.	local exhaust or respiratory protection.	fresh air, rest, place in half-sitting position, call a doctor.
Skin: *is absorbed*; redness, pain.	gloves, protective clothing.	remove contaminated clothing, flush skin with water or shower, refer to a doctor.
Eyes: redness, pain, impaired vision.	safety glasses, or combined eye and respiratory protection.	flush with water, take to a doctor.
Ingestion: abdominal cramps, nausea, vomiting.	do not eat, drink or smoke while working.	take immediately to hospital.

SPILLAGE	STORAGE	LABELING / NFPA
clean up spillage, carefully collect remainder, render harmless with aqueous solution of 5% ammonia and 10% isopropanol; (additional individual protection: breathing apparatus).	keep dry, separate from acids, bases, amines and alcohols; ventilate at floor level.	R: 20-36/37/38-42 S: 26-28-38-45 ✖ Harmful

NOTES

Physical properties apply to 100%-pure substance. Effectiveness of preventive measures depends largely on working conditions: consult an expert before using. Do not take work clothing home. TLV is maximum and must not be exceeded. In case of fire, water should be used for extinguishing only IN EXTREMELY LARGE QUANTITIES. Depending on the degree of exposure, regular medical checkups are advisable. Asthma symptoms usually develop several hours later and are aggravated by physical exertion: rest and hospitalization essential. Persons with history of asthma symptoms should under no circumstances come in contact again with substance. Symptoms listed are acute; the following can additionally develop after some 30 mins-10 hours: fever, joint and muscle pains, perspiration, shivering, edema, general feeling of malaise. Later still hypersensitivity can develop: eczema and esp. asthma. Use airtight packaging made of special material.

Transport Emergency Card TEC(R)-634

HI: 60; UN-number: 2489

biphenyl oxide
diphenyl ether
1,1'-oxybisbenzene
phenyl ether

$C_6H_5OC_6H_5$

DIPHENYL OXIDE

PHYSICAL PROPERTIES		IMPORTANT CHARACTERISTICS
Boiling point, °C	258	**COLORLESS CRYSTALS OR LIQUID WITH CHARACTERISTIC ODOR**
Melting point, °C	27	Vapor mixes readily with air. Flow, agitation etc. can cause build-up of electrostatic charge due to
Flash point, °C	115	liquid's low conductivity.
Auto-ignition temperature, °C	618	
Relative density (water = 1)	1.08	
Relative vapor density (air = 1)	5.9	

TLV-TWA: 1 ppm — 7 mg/m³ ✱
STEL: 2 ppm — 14 mg/m³

Relative density at 20 °C of saturated mixture vapor/air (air = 1): 1.0
Vapor pressure, mm Hg at 20 °C: 0.061
Solubility in water: none
Explosive limits, vol% in air: 0.8-15
Electrical conductivity, pS/m: 30
Relative molecular mass: 170.2
Log P octanol/water: 4.3

Absorption route: Can enter the body by inhalation or ingestion. Harmful atmospheric concentrations build up fairly slowly on evaporation at 20°C, but much more rapidly in aerosol form.
Immediate effects: Irritates the eyes, skin and respiratory tract.
Effects of prolonged/repeated exposure: Can damage the liver.

Gross formula: $C_{12}H_{10}O$

HAZARDS/SYMPTOMS	PREVENTIVE MEASURES	FIRE EXTINGUISHING/FIRST AID
Fire: combustible.	keep away from open flame, no smoking.	water spray, powder.
Explosion:	keep dust from accumulating, sealed machinery, explosion-proof electrical equipment and lighting, grounding when pumping etc. in liquid form.	
Inhalation: sore throat, cough, shortness of breath.	ventilation, local exhaust or respiratory protection.	fresh air, rest, call a doctor.
Skin: redness, pain.	gloves.	remove contaminated clothing, flush skin with water or shower, refer to a doctor.
Eyes: redness, pain.	safety glasses.	flush with water, take to a doctor if necessary.
Ingestion: sore throat, abdominal pain, diarrhea, vomiting.		rinse mouth, call a doctor or take to hospital.

SPILLAGE	STORAGE	LABELING / NFPA
clean up spillage, carefully collect remainder and remove to safe place; (additional individual protection: P2 respirator).	keep in fireproof place.	1 / 1 / 0

NOTES

Component of many heat transfer media, often mixed with → *diphenyl*. Dowtherm is a trade name. ✱ TLV equals that of diphenyl oxide (vapor).

DIPHENYLSULFONIC ACID

PHYSICAL PROPERTIES	IMPORTANT CHARACTERISTICS	
Boiling point, °C decomposes Melting point, °C 138 Solubility in water good Relative molecular mass 253.1	**WHITE HYGROSCOPIC CRYSTALS** Decomposes when heated, giving off toxic (→ *diphenyl*) and corrosive (→ *sulfuric acid*) vapors. In aqueous solution is a strong acid which reacts violently with bases and is corrosive. Attacks many metals, giving off flammable gas (→ *hydrogen*).	

TLV-TWA	not available

Absorption route: Can enter the body by inhalation or ingestion. Evaporation negligible at 20°C, but harmful concentrations of airborne particles can build up rapidly.
Immediate effects: Corrosive to the eyes, skin and respiratory tract. Inhalation of vapor/fumes can cause severe breathing difficulties (lung edema).

Gross formula: $C_{12}H_{10}O_3S.H_2O$

HAZARDS/SYMPTOMS	PREVENTIVE MEASURES	FIRE EXTINGUISHING/FIRST AID
Fire: non-combustible.		in case of fire in immediate vicinity: use any extinguishing agent.
Inhalation: *corrosive*; shortness of breath, severe breathing difficulties.	local exhaust or respiratory protection.	fresh air, rest, place in half-sitting position, take to hospital.
Skin: *corrosive*; redness, pain, serious burns.	gloves.	remove contaminated clothing, flush skin with water or shower, call a doctor.
Eyes: *corrosive*; redness, pain, impaired vision.	goggles.	flush with water, take to a doctor.
Ingestion: *corrosive*; sore throat, abdominal pain, diarrhea.	do not eat, drink or smoke while working.	rinse mouth, immediately take to hospital.

SPILLAGE	STORAGE	LABELING / NFPA
clean up spillage, soak up remainder with dry sand and remove to safe place; (additional individual protection: P2 respirator).	keep dry, separate from strong bases.	

NOTES
Lung edema symptoms usually develop several hours later and are aggravated by physical exertion: rest and hospitalization essential. As first aid, a doctor or authorized person should consider administering a corticosteroid spray.

DI-n-PROPYLALUMINUM HYDRIDE

PHYSICAL PROPERTIES		IMPORTANT CHARACTERISTICS
Boiling point, °C	?	**COLORLESS FUMING LIQUID**
Melting point, °C	−63	Ignites spontaneously on exposure to air. Aluminum oxide vapors are a product of combustion.
Relative density (water = 1)	0.8	Strong reducing agent which reacts violently with oxidants. Reacts violently with water, phenols,
Relative vapor density (air = 1)	3.9	alcohols, amines, carbon dioxide, nitrogen oxides, oxides of sulfur, halogens, halogenated
Vapor pressure, mm Hg at 20 °C	?	hydrocarbons and many other substances, with risk of fire and explosion. Reacts with air to form
Solubility in water	reaction	irritating vapors.
Relative molecular mass	114	

	TLV-TWA	2 mg/m³

Absorption route: Can enter the body by inhalation or ingestion. Harmful atmospheric concentrations can build up fairly rapidly on evaporation at approx. 20°C - even more rapidly in aerosol form.
Immediate effects: Corrosive to the eyes, skin and respiratory tract. Inhalation of vapor/fumes can cause severe breathing difficulties (lung edema). In serious cases risk of death.

Gross formula: $C_6H_{15}Al$

HAZARDS/SYMPTOMS	PREVENTIVE MEASURES	FIRE EXTINGUISHING/FIRST AID
Fire: extremely flammable, many chemical reactions can cause fire and explosion.	keep away from open flame and sparks, no smoking, avoid contact with air, water and many other substances.	special extinguishing powder, dry sand, NO other extinguishing agents, can reignite after being extinguished.
Explosion: contact with water and many other substances can cause explosion.		in case of fire: keep undamaged cylinders cool by spraying with water, but DO NOT spray substance with water.
	STRICT HYGIENE	
Inhalation: *corrosive*; sore throat, cough, severe breathing difficulties.	completely sealed machinery, ventilation, local exhaust or respiratory protection.	fresh air, rest, place in half-sitting position, take to hospital.
Skin: *corrosive*; serious burns, slow healing.	gloves, protective clothing.	remove contaminated clothing, flush skin with water or shower, refer to a doctor.
Eyes: *corrosive*; redness, pain, impaired vision.	face shield, or combined eye and respiratory protection.	flush with water, take to a doctor.
Ingestion: *corrosive*; sore throat, abdominal pain, diarrhea.		rinse mouth, take immediately to hospital.

SPILLAGE	STORAGE	LABELING / NFPA
evacuate area, call in an expert, dilute leakage with diesel/machine oil, soak up/cover with dry sand/table salt and remove to safe place; (additional individual protection: CHEMICAL SUIT).	keep in fireproof place under dry nitrogen, separate from all other substances.	

NOTES

Usually sold as 15-30% solution in hydrocarbons, which does not ignite spontaneously on exposure to air. For 15% solution see relevant entry. Reacts violently with extinguishing agents such as carbon dioxide, halons, foam, water and ordinary powder. Lung edema symptoms usually develop several hours later and are aggravated by physical exertion: rest and hospitalization essential. As first aid, a doctor or authorized person should consider administering a corticosteroid spray. Use cylinder with special fittings.

Transport Emergency Card TEC(R)-42G02

HI: X333; UN-number: 2221

DI-n-PROPYLALUMINUM HYDRIDE
(15 % solution in hexane)

PHYSICAL PROPERTIES		IMPORTANT CHARACTERISTICS
Boiling point, °C	69	**COLORLESS FUMING LIQUID**
Melting point, °C	−95	Vapor is heavier than air and spreads at ground level, with risk of ignition at a distance. Flow,
Flash point, °C	−26	agitation etc. can cause build-up of electrostatic charge due to liquid's low conductivity. Do not
Auto-ignition temperature, °C	240	use compressed air when filling, emptying or processing. Once solvent has evaporated (e.g. on
Relative density (water = 1)	0.74	clothing) ignites spontaneously on exposure to air. Aluminum oxide vapors are a product of
Relative vapor density (air = 1)	3.0	combustion. Strong reducing agent which reacts violently with oxidants. Reacts violently with
Relative density at 20 °C of saturated mixture		alcohols, phenols, halogenated hydrocarbons, carbon dioxide, nitrogen oxides, oxides of sulfur
vapor/air (air = 1)	1.3	and many other compounds, with risk of fire and explosion.
Vapor pressure, mm Hg at 20 °C	122	

Solubility in water	reaction	**TLV-TWA** 2 mg/m³ ✱
Explosive limits, vol% in air	1.1-7.5	
Minimum ignition energy, mJ	0.24	**Absorption route:** Can enter the body by inhalation or ingestion. Harmful atmospheric
Electrical conductivity, pS/m	100	concentrations can build up fairly rapidly on evaporation at approx. 20°C - even more rapidly in
		aerosol form.
		Immediate effects: Corrosive to the eyes, skin and respiratory tract. Inhalation of vapor/fumes
		can cause severe breathing difficulties (lung edema). In serious cases risk of death.
Gross formula:	C_9H_{10}	

HAZARDS/SYMPTOMS	PREVENTIVE MEASURES	FIRE EXTINGUISHING/FIRST AID
Fire: extremely flammable, many chemical reactions can cause fire and explosion.	keep away from open flame and sparks, no smoking, avoid contact with water, alcohols, phenols and many other substances.	special extinguishing powder, dry sand, NO other extinguishing agents, can reignite after being extinguished.
Explosion: forms explosive air-vapor mixtures; contact with water, phenols, alcohols and many other substances can cause explosion.	sealed machinery, ventilation, explosion-proof electrical equipment and lighting, grounding, non-sparking hand tools.	in case of fire: keep tanks/drums cool by spraying with water, but DO NOT spray substance with water.
	STRICT HYGIENE	
Inhalation: *corrosive*; sore throat, cough, severe breathing difficulties.	sealed machinery, ventilation, local exhaust or respiratory protection.	fresh air, rest, place in half-sitting position, take to hospital.
Skin: *corrosive*; redness, serious burns, slow healing.	gloves, protective clothing.	remove contaminated clothing, flush skin with water or shower, call a doctor if necessary.
Eyes: *corrosive*; redness, pain, impaired vision.	face shield.	flush with water, take to a doctor.
Ingestion: *corrosive*; sore throat, abdominal pain, diarrhea.		rinse mouth, immediately take to hospital.

SPILLAGE	STORAGE	LABELING / NFPA
evacuate area, call in an expert, collect leakage in sealable containers, soak up with dry sand or dry table salt and remove to safe place, under no circumstances spray liquid with water; (additional individual protection: CHEMICAL SUIT).	keep in dry, fireproof place under inert gas, separate from all other substances.	3 / 3 / W

NOTES
Physical properties (boiling point, flash point, autoignition point, explosive limits, vapor pressure, relative density, conductivity and ignition energy) are based on hexane. For pure substance: → *di-n-propylaluminum hydride*. Reacts violently with extinguishing agents such as water, foam, halons, carbon dioxide and ordinary powder. Remove contaminated clothing immediately (risk of spontaneous combustion). Lung edema symptoms usually develop several hours later and are aggravated by physical exertion: rest and hospitalization essential. As first aid, a doctor or authorized person should consider administering a corticosteroid spray. As first aid, a doctor or authorized person should consider administering a corticosteroid spray. ✱ For TLV of *hexane* see relevant sheet.

Transport Emergency Card TEC(R)-42G02

HI: X333; UN-number: 2220

CAS-No: [110-98-5]
bis(2-hydroxypropyl) ether
2,2'-dihydroxydipropyl ether
1,1'-dimethyldiethylene glycol
1,1'-oxydi-2-propanol

$(CH_3CHOHCH_2)_2O$

DIPROPYLENE GLYCOL

PHYSICAL PROPERTIES		IMPORTANT CHARACTERISTICS	
Boiling point, °C	232	**COLORLESS, ODORLESS VISCOUS LIQUID**	
Melting point, °C	−40	Vapor mixes readily with air. Oxidizes at high temperatures. Reacts with strong oxidants.	
Flash point, °C	137		
Auto-ignition temperature, °C	310	TLV-TWA not available	
Relative density (water = 1)	1.02		
Relative vapor density (air = 1)	4.6		
Relative density at 20 °C of saturated mixture		**Absorption route:** Can enter the body by inhalation or ingestion. Evaporation negligible at 20°C, but harmful atmospheric concentrations can build up rapidly in aerosol form.	
vapor/air (air = 1)	1.0	**Immediate effects:** Irritates the eyes, skin and respiratory tract.	
Vapor pressure, mm Hg at 20 °C	0.0076		
Solubility in water	∞		
Explosive limits, vol% in air	2.9-12.6		
Relative molecular mass	134.2		
Log P octanol/water	−1.2		
Gross formula: $C_6H_{14}O$			

HAZARDS/SYMPTOMS	PREVENTIVE MEASURES	FIRE EXTINGUISHING/FIRST AID
Fire: combustible.	keep away from open flame, no smoking.	powder, AFFF, foam, carbon dioxide, (halons).
Inhalation: sore throat, cough, headache.	ventilation.	fresh air, rest.
Skin: redness, pain.	gloves.	remove contaminated clothing, flush skin with water or shower.
Eyes: redness, pain.	acid goggles.	flush with water, take to a doctor if necessary.
Ingestion: abdominal pain, nausea.		rinse mouth, call a doctor.

SPILLAGE	STORAGE	LABELING / NFPA
collect leakage in sealable containers, flush away remainder with water.	keep separate from oxidants.	

NOTES
Log P octanol/water is estimated.

3,3'-diaminodipropylamine
3,3'-iminobispropylamine

HN(C$_3$H$_6$NH$_2$)$_2$

DIPROPYLENE TRIAMINE

PHYSICAL PROPERTIES	IMPORTANT CHARACTERISTICS
Boiling point, °C 241 Melting point, °C −6 Flash point, °C 80 Relative density (water = 1) 0.9 Relative vapor density (air = 1) 4.5 Relative density at 20 °C of saturated mixture vapor/air (air = 1) 1.00 Vapor pressure, mm Hg at 20 °C <0.076 Solubility in water ∞ Relative molecular mass 131.2	**COLORLESS LIQUID WITH CHARACTERISTIC ODOR** Vapor mixes readily with air. Decomposes when heated or burned, giving off toxic vapors (incl. → *nitrous vapors*). Strong base which reacts violently with acids and corrodes aluminum, zinc etc. Reacts violently with strong oxidants. Attacks copper and its alloys.
	TLV-TWA not available
	Absorption route: Can enter the body by inhalation or ingestion or through the skin. Harmful atmospheric concentrations build up fairly slowly on evaporation at 20°C, but much more rapidly in aerosol form. **Immediate effects:** Corrosive to the eyes, skin and respiratory tract. Inhalation of vapor/fumes can cause severe breathing difficulties (lung edema). **Effects of prolonged/repeated exposure:** Prolonged or repeated contact can cause skin disorders. Repeated contact can cause asthma. Can cause liver damage.
Gross formula: C$_6$H$_{17}$N$_3$	

HAZARDS/SYMPTOMS	PREVENTIVE MEASURES	FIRE EXTINGUISHING/FIRST AID
Fire: combustible.	keep away from open flame, no smoking.	powder, alcohol-resistant foam, water spray, carbon dioxide, (halons).
Explosion: above 80°C: forms explosive air-vapor mixtures.	above 80°C: sealed machinery, ventilation.	
Inhalation: *corrosive*; sore throat, cough, shortness of breath, severe breathing difficulties.	ventilation, local exhaust or respiratory protection.	fresh air, rest, place in half-sitting position, take to hospital.
Skin: *corrosive, is absorbed*; redness, pain, burns.	gloves, protective clothing.	remove contaminated clothing, flush skin with water or shower, refer to a doctor.
Eyes: *corrosive*; redness, pain, impaired vision.	face shield.	flush with water, take to a doctor.
Ingestion: *corrosive*; sore throat, abdominal pain, vomiting.		rinse mouth, call a doctor or take to hospital.

SPILLAGE	STORAGE	LABELING / NFPA
collect leakage in sealable containers, flush away remainder with water.	keep separate from oxidants and acids; ventilate at floor level.	R: 21/22-34-43 S: 26-36/37/39 Corrosive

NOTES
Lung edema symptoms usually develop several hours later and are aggravated by physical exertion: rest and hospitalization essential. As first aid, a doctor or authorized person should consider administering a corticosteroid spray.

Transport Emergency Card TEC(R)-80G15

HI: 80; UN-number: 2269

CAS-No: [7558-79-4]
disodium acid phosphate
disodium hydrogen phosphate
disodium orthophosphate

$Na_2HPO_4.12H_2O$

DISODIUMPHOSPHATE

PHYSICAL PROPERTIES	IMPORTANT CHARACTERISTICS	
Melting point, °C 35 Relative density (water = 1) 1.5 Solubility in water, g/100 ml at 20 °C 7.7 Relative molecular mass 358.1	**COLORLESS CRYSTALS** In aqueous solution is a weak base.	
	TLV-TWA not available	
	Absorption route: Can enter the body by inhalation or ingestion. Evaporation negligible at 20°C, but harmful concentrations of airborne particles can build up rapidly. **Immediate effects:** Irritates the eyes, skin and respiratory tract.	
Gross formula: $HNa_2O_4P.12H_2O$		

HAZARDS/SYMPTOMS	PREVENTIVE MEASURES	FIRE EXTINGUISHING/FIRST AID
Fire: non-combustible.		in case of fire in immediate vicinity: use any extinguishing agent.
Inhalation: sore throat, cough.	local exhaust or respiratory protection.	fresh air, rest.
Skin: redness.	gloves.	remove contaminated clothing, flush skin with water or shower, refer to a doctor.
Eyes: redness, pain.	goggles.	flush with water, take to a doctor.
Ingestion: abdominal pain.		rinse mouth, call a doctor or take to hospital.

SPILLAGE	STORAGE	LABELING / NFPA
clean up spillage, flush away remainder with water; (additional individual protection: P2 respirator).		

NOTES
Apparent melting point due to loss of water of crystallization.

CAS-No: [1321-74-0]
DVB
vinylstyrene

$C_6H_4(CHCH_2)_2$

1,4-DIVINYLBENZENE
(inhibited)

PHYSICAL PROPERTIES	IMPORTANT CHARACTERISTICS
Boiling point, °C 200 Melting point, °C −67 Flash point, °C 74 Relative density (water = 1) 0.9 Relative vapor density (air = 1) 4.5 Relative density at 20 °C of saturated mixture vapor/air (air = 1) 1.0 Vapor pressure, mm Hg at 20 °C 0.34 Solubility in water none Explosive limits, vol% in air 1.1-6.2 Relative molecular mass 130.1	**COLORLESS LIQUID WITH CHARACTERISTIC ODOR** Vapor mixes readily with air. Can polymerize above 0°C, with risk of fire and explosion. Flow, agitation etc. can cause build-up of electrostatic charge due to liquid's low conductivity. Reacts violently with oxidants, with risk of fire and explosion.
	TLV-TWA not available
	Absorption route: Can enter the body by inhalation or ingestion or through the skin. Evaporation negligible at 20°C, but harmful atmospheric concentrations can build up rapidly in aerosol form. **Immediate effects:** Irritates the eyes and respiratory tract and, to a lesser extent, the skin. Liquid destroys the skin's natural oils. Affects the nervous system. In serious cases risk of unconsciousness. **Effects of prolonged/repeated exposure:** Can cause heart, kidney and liver damage.
Gross formula: $C_{10}H_{10}$	

HAZARDS/SYMPTOMS	PREVENTIVE MEASURES	FIRE EXTINGUISHING/FIRST AID
Fire: combustible.	keep away from open flame, no smoking.	powder, AFFF, foam, carbon dioxide, (halons).
Explosion: above 74°C: forms explosive air-vapor mixtures.	above 74°C: sealed machinery, ventilation, grounding.	
Inhalation: sore throat, cough, shortness of breath, dizziness, feeling of weakness.	ventilation, local exhaust or respiratory protection.	fresh air, rest, place in half-sitting position, call a doctor.
Skin: *is absorbed*; redness, pain; see also 'Inhalation'.	gloves, protective clothing.	remove contaminated clothing, flush skin with water or shower, refer to a doctor.
Eyes: redness, pain.	face shield.	flush with water, take to a doctor.
Ingestion: abdominal pain, vomiting; see also 'Inhalation'.		rinse mouth, take immediately to hospital.

SPILLAGE	STORAGE	LABELING / NFPA
collect leakage in sealable containers, soak up with sand or other inert absorbent and remove to safe place.	keep in cool, dark place under inert gas, separate from oxidants; add an inhibitor.	 2 2 2

NOTES
Alcohol consumption increases toxic effects. Depending on the degree of exposure, regular medical checkups are advisable. Divinyl benzene without inhibitor must be kept in cold storage. The measures on this sheet also apply to the isomers of 1,4-divinyl benzene.

$C_{10}H_{21}CH=CH_2$

1-DODECANE

PHYSICAL PROPERTIES		IMPORTANT CHARACTERISTICS	
Boiling point, °C	213	**COLORLESS LIQUID WITH CHARACTERISTIC ODOR**	
Melting point, °C	−35	Vapor mixes readily with air. Presumed to be able to form peroxides. Flow, agitation etc. can cause build-up of electrostatic charge due to liquid's low conductivity. Reacts with oxidants, inorganic acids and halogenated compounds and molten sulfur.	
Flash point, °C	58		
Auto-ignition temperature, °C	255		
Relative density (water = 1)	0.76		
Relative vapor density (air = 1)	5.8		
Relative density at 20 °C of saturated mixture vapor/air (air = 1)	1	TLV-TWA not available	
Vapor pressure, mm Hg at 20 °C	0.17	**Absorption route:** Can enter the body by inhalation or ingestion. Harmful atmospheric concentrations build up very slowly, if at all, on evaporation at approx. 20°C, but much more rapidly in aerosol form.	
Solubility in water	none		
Electrical conductivity, pS/m	< 10⁴		
Relative molecular mass	168.3		
		Immediate effects: Irritates the eyes, skin and respiratory tract.	
Gross formula:	$C_{12}H_{24}$		

HAZARDS/SYMPTOMS	PREVENTIVE MEASURES	FIRE EXTINGUISHING/FIRST AID
Fire: combustible.	keep away from open flame, no smoking.	powder, AFFF, foam, carbon dioxide, (halons).
Explosion: above 58°C: forms explosive air-vapor mixtures.	above 58°C: sealed machinery, ventilation, grounding.	
Inhalation: sore throat, cough.	ventilation, local exhaust or respiratory protection.	fresh air, rest, place in half-sitting position, call a doctor.
Skin: redness.	gloves.	remove contaminated clothing, flush skin with water or shower, call a doctor.
Eyes: redness, pain.	safety glasses.	flush with water, take to a doctor.
Ingestion: abdominal pain, nausea.		rinse mouth, call a doctor.

SPILLAGE	STORAGE	LABELING / NFPA
collect leakage in sealable containers, soak up with sand or other inert absorbent and remove to safe place.	keep separate from oxidants, inorganic acids and halogenated compounds; ventilate at floor level.	

NOTES

Transport Emergency Card TEC(R)-30G15

$$CH_3 (CH_2)_{10}C-O-O-C (CH_2)_{10}CH_3$$
$$\overset{||}{O} \qquad\qquad \overset{||}{C}$$

DODECANOYL PEROXIDE

PHYSICAL PROPERTIES	IMPORTANT CHARACTERISTICS
Melting point, °C 54 Decomposes below melting point, °C 50 Flash point, °C see notes Auto-ignition temperature, °C 112 Solubility in water none Relative molecular mass 398.6	**WHITE COARSE POWDER** Can deflagrate (burn explosively) when heated. Strong oxidant which reacts violently with combustible substances and reducing agents, with risk of fire and explosion.
	TLV-TWA not available
	Absorption route: No data on absorption of particles into the body. Evaporation negligible at 20°C, but harmful atmospheric concentrations can build up rapidly in aerosol form. **Immediate effects:** In high concentrations is corrosive to the eyes and respiratory tract. Inhalations of high concentrations of particles can cause lung edema. Can affect the respiratory tract. Exposure to high concentrations can be fatal. Keep under medical observation.
Gross formula: $C_{24}H_{46}O_4$	

HAZARDS/SYMPTOMS	PREVENTIVE MEASURES	FIRE EXTINGUISHING/FIRST AID
Fire: flammable.	avoid contact with combustible substances and hot surfaces (steam lines).	large quantities of water, water spray; in case of fire in immediate vicinity: use any extinguishing agent.
Explosion: contamination or heating can cause explosion.	do not heat above 25°C.	in case of fire: keep tanks/drums cool by spraying with water, fight fire from sheltered location.
	KEEP DUST UNDER CONTROL	
Inhalation: *corrosive*; cough, headache, difficulty breathing, sore throat.	local exhaust or respiratory protection.	fresh air, rest, place in half-sitting position, take immediately to hospital.
Skin: redness, burning sensation.	gloves, protective clothing.	remove contaminated clothing, flush skin with water and wash with soap and water.
Eyes: *corrosive*; redness, pain, impaired vision.	goggles.	flush thoroughly with water (remove contact lenses if easy), take to a doctor.
Ingestion: *corrosive*; abdominal pain, burning sensation, nausea, sore throat, vomiting.		rinse mouth, immediately take to hospital.

SPILLAGE	STORAGE	LABELING / NFPA
clean up spillage, soak up remainder with sand or other inert absorbent and remove to safe place, do not use sawdust or other combustible absorbents; (additional individual protection: P2 respirator).	keep in cool, fireproof place, separate from other substances.	◇

NOTES
Flash point presumably the same as melting point. Flush contaminated clothing with water (fire hazard). Lung edema symptoms usually develop several hours later and are aggravated by physical exertion: rest and hospitalization essential. As first aid, a doctor or authorized person should consider administering a suitable spray. Insufficient data on harmful effects to humans: exercise great caution.

Transport Emergency Card TEC(R)-52G01

UN-number: 2124

DODECYLBENZENE

PHYSICAL PROPERTIES		IMPORTANT CHARACTERISTICS	
Boiling point, °C	230-300	**COLORLESS LIQUID WITH CHARACTERISTIC ODOR**	
Flash point, °C	135	Vapor mixes readily with air.	
Relative density (water = 1)	0.86		
Relative vapor density (air = 1)	8.4	TLV-TWA	not available
Relative density at 20 °C of saturated mixture vapor/air (air = 1)	1.0	**Absorption route:** Can enter the body by inhalation or ingestion or through the skin. Harmful atmospheric concentrations build up very slowly, if at all, on evaporation at approx. 20°C, but much more rapidly in aerosol form.	
Vapor pressure, mm Hg at 20 °C	<0.076		
Solubility in water	none		
Relative molecular mass	246.4	**Immediate effects:** Irritates the eyes, skin and respiratory tract.	
Gross formula:	$C_{18}H_{30}$		

HAZARDS/SYMPTOMS	PREVENTIVE MEASURES	FIRE EXTINGUISHING/FIRST AID
Fire: combustible.	keep away from open flame, no smoking.	powder, AFFF, foam, carbon dioxide, (halons).
Inhalation: sore throat, cough; see also 'Ingestion'.	ventilation.	fresh air, rest, call a doctor.
Skin: *is absorbed*; redness.	gloves.	remove contaminated clothing, flush skin with water or shower, call a doctor.
Eyes: redness, pain.	safety glasses.	flush with water, take to a doctor if necessary.
Ingestion: abdominal pain, headache, nausea, dizziness, drowsiness.		rinse mouth, call a doctor.

SPILLAGE	STORAGE	LABELING / NFPA
collect leakage in sealable containers, soak up with sand or other inert absorbent and remove to safe place.	ventilate at floor level.	

NOTES
Technical product is a mixture of isomers; boiling point depends on composition.

1-dodecanethiol
dodecyl mercaptan
lauryl mercaptan
1-mercaptododecane

$CH_3(CH_2)_{11}SH$

n-DODECYL MERCAPTAN

PHYSICAL PROPERTIES	IMPORTANT CHARACTERISTICS	
Boiling point, °C 270-285 Melting point, °C −7 Flash point, °C 88 Relative density (water = 1) 0.8 Relative vapor density (air = 1) 7.0 Relative density at 20 °C of saturated mixture vapor/air (air = 1) 1.0 Solubility in water none Relative molecular mass 202.2	**COLORLESS TO LIGHT YELLOW LIQUID WITH CHARACTERISTIC ODOR** Vapor mixes readily with air. Toxic vapors are a product of combustion (→ *sulfur dioxide*). Decomposes when heated, giving off toxic and flammable gas (→ *hydrogen sulfide*). Reacts violently with strong oxidants.	
	TLV-TWA not available	
	Absorption route: Can enter the body by inhalation or ingestion. Harmful atmospheric concentrations build up very slowly, if at all, on evaporation at approx. 20°C, but much more rapidly in aerosol form. **Immediate effects:** Irritates the eyes, skin and respiratory tract. Affects the nervous system.	
Gross formula: $C_{12}H_{26}S$		

HAZARDS/SYMPTOMS	PREVENTIVE MEASURES	FIRE EXTINGUISHING/FIRST AID
Fire: combustible.	keep away from open flame, no smoking.	powder, AFFF, foam, carbon dioxide, (halons).
Inhalation: sore throat, shortness of breath, headache, nausea.	ventilation.	fresh air, rest, call a doctor if necessary.
Skin: redness.	gloves.	remove contaminated clothing, flush skin with water or shower.
Eyes: redness, pain.	safety glasses.	flush with water, take to a doctor if necessary.
Ingestion: abdominal pain, headache, nausea, vomiting, drowsiness.		rinse mouth, call a doctor.

SPILLAGE	STORAGE	LABELING / NFPA
collect leakage in sealable containers, soak up with sand or other inert absorbent and remove to safe place, to avoid stench do not flush into sewer.	keep separate from oxidants; ventilate at floor level.	 1 2 0

NOTES

HI: 30; UN-number: 1228

tert-DODECYL MERCAPTAN

PHYSICAL PROPERTIES	IMPORTANT CHARACTERISTICS
Boiling point, °C 227-248 Melting point, °C −45 Flash point, °C 82 Auto-ignition temperature, °C 230 Relative density (water = 1) 0.9 Relative vapor density (air = 1) 7.0 Relative density at 20 °C of saturated mixture vapor/air (air = 1) 1.0 Vapor pressure, mm Hg at 20 °C 0.076 Solubility in water none Relative molecular mass 202.4 Gross formula: $C_{12}H_{26}S$	**COLORLESS TO LIGHT YELLOW LIQUID WITH CHARACTERISTIC ODOR** Toxic gases are a product of combustion (→ *sulfur dioxide*). Reacts violently with oxidants. TLV-TWA not available **Absorption route:** Can enter the body by inhalation or ingestion. Harmful atmospheric concentrations build up fairly slowly on evaporation at approx. 20°C, but much more rapidly in aerosol form. **Immediate effects:** Irritates the eyes, skin and respiratory tract. Affects the nervous system. In serious cases risk of seizures, unconsciousness and death.

HAZARDS/SYMPTOMS	PREVENTIVE MEASURES	FIRE EXTINGUISHING/FIRST AID
Fire: combustible.	keep away from open flame, no smoking.	powder, AFFF, foam, carbon dioxide, (halons).
Explosion: above 82°C: forms explosive air-vapor mixtures.	above 82°C: sealed machinery, ventilation.	in case of fire: keep tanks/drums cool by spraying with water.
Inhalation: sore throat, shortness of breath, headache, nausea, drowsiness, unconsciousness, cyanosis.	ventilation, local exhaust or respiratory protection.	fresh air, rest, call a doctor.
Skin: redness.	gloves.	remove contaminated clothing, flush skin with water or shower, refer to a doctor if necessary.
Eyes: redness, pain.	safety glasses.	flush with water, take to a doctor if necessary.
Ingestion: abdominal pain, headache, nausea, drowsiness, unconsciousness.		rinse mouth, take to hospital.

SPILLAGE	STORAGE	LABELING / NFPA
collect leakage in sealable containers, soak up with sand or other inert absorbent and remove to safe place; to avoid excessive stench, keep liquid out of sewer, basements and excavation sites; (additional individual protection: breathing apparatus).	keep separate from oxidants; ventilate at floor level.	1 / 2 / 0

NOTES
Liquid gives off a very unpleasant, penetrating odor. Use airtight packaging. Unbreakable packaging preferred; if breakable, keep in unbreakable container.

Transport Emergency Card TEC(R)-30G17

HI: 30; UN-number: 1228

EDTA

PHYSICAL PROPERTIES	IMPORTANT CHARACTERISTICS	
Decomposes below boiling point, °C 150 Relative density (water = 1) 0.86 Solubility in water, g/100 ml at 20 °C 0.05 Relative molecular mass 292.2 Gross formula: $C_{10}H_{16}N_2O_8$	**WHITE POWDER** Decomposes when heated above 150°C, giving off toxic gases ($\rightarrow$ *nitrous vapors*).	
	TLV-TWA not available	
	Absorption route: Can enter the body by inhalation or ingestion. Evaporation negligible at 20°C, but unpleasant concentrations of airborne particles can build up. **Immediate effects:** Irritates the eyes, skin and respiratory tract.	

HAZARDS/SYMPTOMS	PREVENTIVE MEASURES	FIRE EXTINGUISHING/FIRST AID
Fire: not combustible but enhances combustion of other substances.		in case of fire in immediate vicinity: use any extinguishing agent.
Inhalation: sore throat, cough.	local exhaust or respiratory protection.	fresh air, rest, call a doctor.
Skin: redness, pain.	gloves.	remove contaminated clothing, flush skin with water or shower.
Eyes: redness, pain.	goggles.	flush with water, take to a doctor if necessary.
Ingestion: sore throat.		rinse mouth, call a doctor.

SPILLAGE	STORAGE	LABELING / NFPA
clean up spillage, flush away remainder with water; (additional individual protection: P1 respirator).		

NOTES

CAS-No: [106-89-8]
chloromethyloxirane
1-chloro-2,3-epoxypropane
chloropropylene oxide
3-chloro-1,2-propylene oxide
1,2-epoxy-3-chloropropane

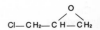

EPICHLOROHYDRIN

PHYSICAL PROPERTIES		IMPORTANT CHARACTERISTICS
Boiling point, °C	116	**COLORLESS LIQUID WITH CHARACTERISTIC ODOR**
Melting point, °C	−48	Vapor mixes readily with air. Moderate heating or heating with acid or lye can cause
Flash point, °C	28	polymerization. Corrosive vapors are a product of combustion (→ *hydrochloric acid*). Reacts
Auto-ignition temperature, °C	385	violently with alkaline-earth and alkali metals, various metal powders and strong oxidants, with
Relative density (water = 1)	1.2	risk of fire and explosion.
Relative vapor density (air = 1)	3.2	
Relative density at 20 °C of saturated mixture		
vapor/air (air = 1)	1.05	TLV-TWA (skin)　　　　0.1 ppm　　　　0.38 mg/m³　A2
Vapor pressure, mm Hg at 20 °C	13	
Solubility in water, g/100 ml at 10 °C	6	**Absorption route:** Can enter the body by inhalation or ingestion or through the skin. Harmful
Explosive limits, vol% in air	2.3-34.4	atmospheric concentrations can build up fairly rapidly on evaporation at approx. 20°C - even
Electrical conductivity, pS/m	3.4 x 10⁶	more rapidly in aerosol form.
Relative molecular mass	92.5	**Immediate effects:** Corrosive to the eyes, skin and respiratory tract. Inhalation can cause lung
		edema. In serious cases risk of death.
		Effects of prolonged/repeated exposure: Prolonged or repeated contact can cause skin
		disorders. Can cause temporary sterility and liver, kidney and adrenal gland damage. Has been
		found to cause a type of nose and throat cancer in certain animal species under certain
		circumstances.
Gross formula:	C_3H_5ClO	

HAZARDS/SYMPTOMS	PREVENTIVE MEASURES	FIRE EXTINGUISHING/FIRST AID
Fire: flammable.	keep away from open flame and sparks, no smoking.	powder, water spray, foam, carbon dioxide (halons).
Explosion: above 28°C: forms explosive air-vapor mixtures.	above 28°C: sealed machinery, ventilation, explosion-proof electrical equipment.	in case of fire: keep tanks/drums cool by spraying with water.
	STRICT HYGIENE	
Inhalation: *corrosive*; sore throat, cough, shortness of breath, severe breathing difficulties, headache.	ventilation, local exhaust or respiratory protection.	fresh air, rest, place in half-sitting position, take to hospital.
Skin: *corrosive*; redness, pain, burns.	gloves, protective clothing.	remove contaminated clothing, flush skin with water or shower, take to hospital.
Eyes: *corrosive*; redness, pain, impaired vision.	face shield, or combined eye and respiratory protection.	flush with water, take to doctor.
Ingestion: *corrosive*; abdominal pain, vomiting.		rinse mouth, take immediately to hospital.

SPILLAGE	STORAGE	LABELING / NFPA
evacuate area, call in an expert, collect leakage in sealable containers, soak up with sand or other inert absorbent and remove to safe place; (additional individual protection: CHEMICAL SUIT).	keep in fireproof place, separate from oxidants, strong acids and strong bases.	R: 45-10-23/24/25-34-43 S: 53-9-44 Note E

Toxic

NFPA: 3 / 3 / 2

NOTES

Odor limit is above TLV. Depending on the degree of exposure, regular medical checkups are advisable. Lung edema symptoms usually develop several hours later and are aggravated by physical exertion: rest and hospitalization essential. As first aid, a doctor or authorized person should consider administering a corticosteroid spray.

Transport Emergency Card TEC(R)-6

HI: 63; UN-number: 2023

CAS-No: [74-84-0]
bimethyl
dimethyl
ethyl hydride
methylmethane

C_2H_6

ETHANE
(cylinder)

PHYSICAL PROPERTIES	IMPORTANT CHARACTERISTICS
Boiling point, °C −89 Melting point, °C −183 Flash point, °C flammable gas Auto-ignition temperature, °C 515 Relative density (water = 1) 0.4 Relative vapor density (air = 1) 1.04 Vapor pressure, mm Hg at 20 °C 29 Solubility in water none Explosive limits, vol% in air 3.0-12.5 Minimum ignition energy, mJ 0.24 Electrical conductivity, pS/m $< 10^4$ Relative molecular mass 30.1	**COLORLESS, ODORLESS COMPRESSED LIQUEFIED GAS** Mixes readily with air, forming explosive mixtures. Flow, agitation etc. can cause build-up of electrostatic charge due to liquid's low conductivity. Do not use compressed air when filling, emptying or processing.
	TLV-TWA not available
	Absorption route: Can saturate the air if released, with risk of suffocation. **Immediate effects:** Can cause frostbite due to rapid evaporation. Inhalation can cause severe breathing difficulties.
Gross formula: C_2H_6	

HAZARDS/SYMPTOMS	PREVENTIVE MEASURES	FIRE EXTINGUISHING/FIRST AID
Fire: extremely flammable.	keep away from open flame and sparks, no smoking.	shut off supply; if impossible and no danger to surrounding area, allow fire to burn itself out; otherwise extinguish with powder, carbon dioxide, (halons).
Explosion: forms explosive air-gas mixtures.	sealed machinery, ventilation, explosion-proof electrical equipment and lighting, grounding, non-sparking tools.	in case of fire: keep cylinders cool by spraying with water, fight fire from sheltered location.
		IN ALL CASES CALL A DOCTOR
Inhalation: shortness of breath, headache, drowsiness, unconsciousness.	ventilation, local exhaust or respiratory protection.	fresh air, rest, artificial respiration if necessary, call a doctor.
Skin: *is absorbed*; redness, pain, blisters.	insulated gloves.	*in case of frostbite*: DO NOT remove clothing, flush skin with water or shower, refer to a doctor.
Eyes: *in case of frostbite*: redness, pain, impaired vision.	face shield.	flush with water, take to a doctor if necessary.
Ingestion: abdominal pain, nausea; see also 'Inhalation'.		rinse mouth, take immediately to hospital.

SPILLAGE	STORAGE	LABELING / NFPA
evacuate area, call in an expert, ventilate, under no circumstances spray liquid with water; (additional individual protection: breathing apparatus).	keep in cool, fireproof place.	

NOTES
High atmospheric concentrations can cause oxygen deficiency, with risk of unconsciousness and death. Turn leaking cylinder so that leak is on top to prevent liquid ethane escaping.

Transport Emergency Card TEC(R)-868

HI: 23; UN-number: 1035

ETHANOL

PHYSICAL PROPERTIES	IMPORTANT CHARACTERISTICS
Boiling point, °C 78 Melting point, °C − 117 Flash point, °C 12 Auto-ignition temperature, °C 370 Relative density (water = 1) 0.8 Relative vapor density (air = 1) 1.6 Relative density at 20 °C of saturated mixture vapor/air (air = 1) 1.04 Vapor pressure, mm Hg at 20 °C 45 Solubility in water ∞ Explosive limits, vol% in air 3.4-19 Electrical conductivity, pS/m 1.3×10^5 Relative molecular mass 46.1 Log P octanol/water − 0.3	**COLORLESS LIQUID WITH CHARACTERISTIC ODOR** Vapor mixes readily with air, forming explosive mixtures. Do not use compressed air when filling, emptying or processing. Reacts violently with strong oxidants, with risk of fire and explosion. TLV-TWA 1000 ppm 1880 mg/m³ **Absorption route:** Can enter the body by inhalation or ingestion. Harmful atmospheric concentrations build up fairly slowly on evaporation at approx. 20°C, but much more rapidly in aerosol form. **Immediate effects:** Irritates the eyes, skin and respiratory tract. Liquid destroys the skin's natural oils. In high concentrations or if swallowed can cause agitation and/or impairment of consciousness. **Effects of prolonged/repeated exposure:** Oral ingestion can cause liver damage.
Gross formula: C₂H₆O	

HAZARDS/SYMPTOMS	PREVENTIVE MEASURES	FIRE EXTINGUISHING/FIRST AID
Fire: extremely flammable.	keep away from open flame and sparks, no smoking.	powder, alcohol-resistant foam, large quantities of water, carbon dioxide, (halons); in case of fire in immediate vicinity: use any extinguishing agent.
Explosion: forms explosive air-vapor mixtures.	sealed machinery, ventilation, explosion-proof electrical equipment and lighting.	in case of fire: keep tanks/drums cool by spraying with water.
Inhalation: cough, headache, dizziness, drowsiness.	ventilation, local exhaust or respiratory protection.	fresh air, rest, call a doctor if necessary.
Skin: redness.	gloves.	remove contaminated clothing, flush skin with water or shower.
Eyes: redness, pain.	safety glasses.	flush with water, take to a doctor if necessary.
Ingestion: headache, dizziness, drowsiness.		rinse mouth, take to a doctor if necessary.

SPILLAGE	STORAGE	LABELING / NFPA
collect leakage in sealable containers, flush away remainder with water.	keep in fireproof place, separate from oxidants.	R: 11 S: 7-16 🔥 Flammable NFPA: 3 / 0 / 0

NOTES

Alcohol consumption increases narcotic effects. Flash point of solutions: 95 vol.%: 14°C; 80 vol.%: 20°C; 70 vol.%: 21°C; 60 vol.%: 22°C; 50 vol.%: 24°C; 40 vol.%: 26°C; 30 vol.%: 29°C; 20 vol.%: 36°C; 10 vol.%: 49°C; 5 vol.%: 62°C.

Transport Emergency Card TEC(R)-32

HI: 33; UN-number: 1170

CAS-No: [141-43-5]
aminoethyl alcohol
2-aminoethyl alcohol
ethylolamine
glycinol
olamine

$H_2NC_2H_4OH$

ETHANOLAMINE

PHYSICAL PROPERTIES		IMPORTANT CHARACTERISTICS		
Boiling point, °C	171	**COLORLESS LIQUID WITH CHARACTERISTIC ODOR**		
Melting point, °C	10	Vapor mixes readily with air. Strong base which reacts violently with acids and corrodes aluminum, zinc etc. Decomposes when heated or burned, giving off toxic gases. Reacts violently with oxidants. Above 60°C reacts with aluminum, giving off → *hydrogen*. Attacks copper, copper alloys and rubber.		
Flash point, °C	85			
Auto-ignition temperature, °C	410			
Relative density (water = 1)	1.0			
Relative vapor density (air = 1)	2.1			
Relative density at 20 °C of saturated mixture				
vapor/air (air = 1)	1.0	TLV-TWA	3 ppm	7.5 mg/m³
Vapor pressure, mm Hg at 20°C	0.46	STEL	6 ppm	15 mg/m³
Solubility in water	∞			
Relative molecular mass	61.08	**Absorption route:** Can enter the body by inhalation or ingestion. Harmful atmospheric concentrations build up fairly slowly on evaporation at 20°C, but much more rapidly in aerosol form.		
Log P octanol/water	−1.3			
		Immediate effects: Irritates the eyes, skin and respiratory tract. Affects the nervous system. In substantial concentrations can impair consciousness.		
		Effects of prolonged/repeated exposure: Prolonged or repeated contact can cause skin disorders.		
Gross formula:	C_2H_7NO			

HAZARDS/SYMPTOMS	PREVENTIVE MEASURES	FIRE EXTINGUISHING/FIRST AID
Fire: combustible.	keep away from open flame, no smoking.	powder, alcohol-resistant foam, water spray, carbon dioxide, (halons).
Explosion: above 85°C: forms explosive air-vapor mixtures.	above 85°C: sealed machinery, ventilation.	
Inhalation: sore throat, cough, shortness of breath, headache.	ventilation, local exhaust or respiratory protection.	fresh air, rest, place in half-sitting position, call a doctor.
Skin: redness, pain.	gloves.	remove contaminated clothing, flush skin with water or shower, refer to a doctor.
Eyes: redness, pain, impaired vision.	face shield.	flush with water, take to a doctor.
Ingestion: abdominal pain, nausea, vomiting.		rinse mouth, call a doctor or take to hospital.

SPILLAGE	STORAGE	LABELING / NFPA
collect leakage in sealable containers, flush away remainder with water.	keep separate from oxidants and strong acids; ventilate at floor level.	R: 20-36/37/38 ✖ Irritant

NFPA: 2 / 2 / 0

NOTES
Odor limit is above TLV. Alcohol consumption increases toxic effects.

Transport Emergency Card TEC(R)-80G16

HI: 80; UN-number: 2491

CAS-No: [74-85-1]
bicarburetted hydrogen
ethylene
olefiant gas

$CH_2=CH_2$

ETHENE
(cylinder)

PHYSICAL PROPERTIES	IMPORTANT CHARACTERISTICS
Boiling point, °C −104 Melting point, °C −169 Flash point, °C flammable gas Auto-ignition temperature, °C 425 Relative density (water = 1) see notes Relative vapor density (air = 1) 0.98 Vapor pressure, mm Hg at 10°C 39,520 Solubility in water poor Explosive limits, vol% in air 2.7-34 Minimum ignition energy, mJ 0.07 Electrical conductivity, pS/m <10⁴ Relative molecular mass 28.1 Gross formula: C₂H₄	**COLORLESS COMPRESSED GAS WITH CHARACTERISTIC ODOR** Mixes readily with air, forming explosive mixtures. Many substances, e.g. oxidants and chlorine compounds, can cause polymerization, with risk of fire and explosion. Flow, agitation etc. can cause build-up of electrostatic charge due to liquid's low conductivity. Reacts violently with strong oxidants. Reacts violently with hydrogen bromide, chlorine, hydrogen chloride and nitrogen oxides, with risk of fire and explosion. Attacks cast iron. TLV-TWA not available **Absorption route:** Can saturate the air if released, with risk of suffocation. **Immediate effects:** In substantial concentrations can impair consciousness.

HAZARDS/SYMPTOMS	PREVENTIVE MEASURES	FIRE EXTINGUISHING/FIRST AID
Fire: extremely flammable.	keep away from open flame and sparks, no smoking.	shut off supply; if impossible and no danger to surrounding area, allow fire to burn itself out; otherwise extinguish with powder, carbon dioxide, (halons).
Explosion: forms explosive air-gas mixtures.	sealed machinery, ventilation, explosion-proof electrical equipment and lighting, grounding, non-sparking tools.	in case of fire: keep cylinders cool by spraying with water, fight fire from sheltered location.
Inhalation: severe breathing difficulties, drowsiness.	ventilation, local exhaust or respiratory protection.	fresh air, rest, artificial respiration if necessary, call a doctor.

SPILLAGE	STORAGE	LABELING / NFPA
evacuate area, call in an expert, ventilate; (additional individual protection: breathing apparatus).	keep in cool, fireproof place, separate from oxidants and halogen acids.	 4 1 ◇ 2

NOTES
At an ambient temperature of >10°C occurs only as gas in cylinder. Density at critical temperature (10°C): 0.22kg/l. High atmospheric concentrations, e.g. in poorly ventilated spaces, can cause oxygen deficiency, with risk of unconsciousness.

Transport Emergency Card TEC(R)-136

HI: 23; UN-number: 1962

CAS-No: [60-29-7]
diethyl oxide
diethyl ether
oxybis-1,1'-ethane
ethyl ether
ethyl oxide

$C_2H_5OC_2H_5$

ETHER

PHYSICAL PROPERTIES		IMPORTANT CHARACTERISTICS
Boiling point, °C	35	**COLORLESS LIQUID WITH CHARACTERISTIC ODOR**
Melting point, °C	− 116	Vapor is heavier than air and spreads at ground level, with risk of ignition at a distance. Able to
Flash point, °C	− 45	form peroxides. Flow, agitation etc. can cause build-up of electrostatic charge due to liquid's low
Auto-ignition temperature, °C	160	conductivity. Do not use compressed air when filling, emptying or processing. Decomposes
Relative density (water = 1)	0.71	when heated above 550°C, forming aldehydes. Reacts violently with strong oxidants, with risk of
Relative vapor density (air = 1)	2.6	fire and explosion. Attacks many plastics and zinc.
Relative density at 20 °C of saturated mixture		
vapor/air (air = 1)	1.9	
Vapor pressure, mm Hg at 20 °C	448	
Solubility in water, g/100 ml at 20 °C	6.9	
Explosive limits, vol% in air	1.7-48	**Absorption route:** Can enter the body by inhalation or ingestion. Harmful atmospheric
Minimum ignition energy, mJ	0.19	concentrations can build up fairly rapidly on evaporation at approx. 20°C - even more rapidly in
Electrical conductivity, pS/m	10	aerosol form.
Relative molecular mass	74.1	**Immediate effects:** Irritates the eyes, skin and respiratory tract. Liquid destroys the skin's natural
Log P octanol/water	0.8	oils. In substantial concentrations can impair consciousness.
Gross formula:	$C_4H_{10}O$	

Additional data row for TLV:

TLV-TWA	400 ppm	1210 mg/m³
STEL	500 ppm	1520 mg/m³

HAZARDS/SYMPTOMS	PREVENTIVE MEASURES	FIRE EXTINGUISHING/FIRST AID
Fire: extremely flammable.	keep away from open flame and sparks, no smoking.	powder, AFFF, foam, carbon dioxide, (halons).
Explosion: forms explosive air-vapor mixtures.	sealed machinery, ventilation, explosion-proof electrical equipment and lighting, grounding, non-sparking tools.	in case of fire: keep tanks/drums cool by spraying with water.
Inhalation: dizziness, sleepiness, unconsciousness.	ventilation, local exhaust or respiratory protection.	fresh air, rest, artificial respiration if necessary.
Skin: redness, in some cases frostbite.	gloves.	remove contaminated clothing; *in case of frostbite*: DO NOT remove clothing, flush skin with water or shower.
Eyes: redness.	safety glasses.	
Ingestion: dizziness, vomiting, sleepiness.		rinse mouth, DO NOT induce vomiting, call a doctor or take to hospital.

SPILLAGE	STORAGE	LABELING / NFPA
evacuate area, call in an expert, collect leakage in sealable containers, soak up with sand or other inert absorbent and remove to safe place; (additional individual protection: breathing apparatus).	keep in cool, dark, fireproof place, separate from oxidants; add an inhibitor.	R: 12-19 S: 9-16-29-33 ⬦ Flammable NFPA: 4 (top), 2 (left), 1 (right)

NOTES

Before distilling check for peroxides; if found, render harmless. Alcohol consumption increases toxic effects. Unbreakable packaging preferred; if breakable, keep in unbreakable container.

Transport Emergency Card TEC(R)-72

HI: 33; UN-number: 1155

2-ETHOXYETHANOL

PHYSICAL PROPERTIES	IMPORTANT CHARACTERISTICS
Boiling point, °C 135 Melting point, °C − 100 Flash point, °C 40 Auto-ignition temperature, °C 235 Relative density (water = 1) 0.9 Relative vapor density (air = 1) 3.1 Relative density at 20 °C of saturated mixture vapor/air (air = 1) 1.01 Vapor pressure, mm Hg at 20 °C 3.8 Solubility in water ∞ Explosive limits, vol% in air 1.8 - 15.7 Relative molecular mass 90.1 Log P octanol/water − 0.5 Gross formula: $C_4H_{10}O_2$	**COLORLESS VISCOUS LIQUID** Vapor mixes readily with air. Reacts with oxidants. Attacks rubber. TLV-TWA (skin)　　　5 ppm　　　18 mg/m³ **Absorption route:** Can enter the body by inhalation or ingestion or through the skin. **Immediate effects:** Irritates the eyes, skin and respiratory tract. **Effects of prolonged/repeated exposure:** Can affect the blood. Can cause liver and kidney damage.

HAZARDS/SYMPTOMS	PREVENTIVE MEASURES	FIRE EXTINGUISHING/FIRST AID
Fire: flammable.	keep away from open flame and sparks, no smoking.	powder, alcohol-resistant foam, water spray, carbon dioxide, (halons).
Explosion: above 40°C: forms explosive air-vapor mixtures.	above 40°C: sealed machinery, ventilation, explosion-proof electrical equipment.	in case of fire: keep tanks/drums cool by spraying with water.
Inhalation: sore throat, cough.		fresh air, rest, call a doctor.
Skin: *is absorbed*; redness, pain.	gloves, protective clothing.	remove contaminated clothing, flush skin with water or shower, call a doctor.
Eyes: redness, pain, impaired vision.	face shield.	flush with water, take to a doctor if necessary.
Ingestion: abdominal pain, nausea.		rinse mouth, call a doctor or take to hospital.

SPILLAGE	STORAGE	LABELING / NFPA
collect leakage in sealable containers, flush away remainder with water.	keep in fireproof place, separate from oxidants.	R: 10-36 S: 24 ❎ Irritant NFPA: 2 (top) 2 (left) 0 (right)

NOTES
Alcohol consumption increases toxic effects. Depending on the degree of exposure, regular medical checkups are advisable. Cellosolve is a trade name.

acetic acid 2-ethoxyethyl ester
ethylene glycol ethyl ether acetate

$CH_3COOC_2H_4OC_2H_5$

2-ETHOXYETHYL ACETATE

PHYSICAL PROPERTIES		IMPORTANT CHARACTERISTICS	
Boiling point, °C	156	**COLORLESS LIQUID WITH CHARACTERISTIC ODOR**	
Melting point, °C	−62	Vapor mixes readily with air. Able to form peroxides. Reacts violently with oxidants. Attacks rubber.	
Flash point, °C	49		
Auto-ignition temperature, °C	380		
Relative density (water = 1)	0.97	TLV-TWA (skin)	5 ppm 27 mg/m³
Relative vapor density (air = 1)	4.6		
Relative density at 20 °C of saturated mixture vapor/air (air = 1)	1.01	**Absorption route:** Can enter the body by inhalation or ingestion.	
Vapor pressure, mm Hg at 20 °C	2.0	**Immediate effects:** Irritates the eyes and respiratory tract. Liquid destroys the skin's natural oils. In high concentrations can cause muscle paralysis and tingling in face, arms and legs.	
Solubility in water, g/100 ml	30	**Effects of prolonged/repeated exposure:** Can cause kidney damage.	
Explosive limits, vol% in air	1.7 - 8.3		
Relative molecular mass	132.2		
Gross formula:	$C_6H_{12}O_3$		

HAZARDS/SYMPTOMS	PREVENTIVE MEASURES	FIRE EXTINGUISHING/FIRST AID
Fire: flammable.	keep away from open flame and sparks, no smoking.	powder, alcohol-resistant foam, water spray, carbon dioxide, (halons).
Explosion: above 49°C: forms explosive air-vapor mixtures.	above 49°C: sealed machinery, ventilation, explosion-proof electrical equipment.	in case of fire: keep tanks/drums cool by spraying with water.
Inhalation: sore throat, cough, drowsiness.	ventilation, local exhaust or respiratory protection.	fresh air, rest.
Skin: *is absorbed*; redness.	gloves.	remove contaminated clothing, flush skin with water or shower.
Eyes: redness, pain.	safety glasses.	flush with water, take to a doctor.
Ingestion: headache, vomiting, drowsiness.		rinse mouth, take immediately to hospital.

SPILLAGE	STORAGE	LABELING / NFPA
collect leakage in sealable containers, flush away remainder with water.	keep in fireproof place, separate from oxidants.	R: 10-20/21 S: 24 ☒ Harmful

NOTES
Alcohol consumption increases toxic effects. Depending on the degree of exposure, regular medical checkups are advisable. Cellosolve acetate is a trade name.

Transport Emergency Card TEC(R)-549

HI: 30; UN-number: 1172

acetic acid ethyl ester
acetic ether
acetoxyethane
ethyl acetic ester
ethyl ethanoate

$CH_3COOC_2H_5$

ETHYL ACETATE

PHYSICAL PROPERTIES		IMPORTANT CHARACTERISTICS		
Boiling point, °C	77	**COLORLESS LIQUID WITH CHARACTERISTIC ODOR**		
Melting point, °C	− 84	Vapor is heavier than air and spreads at ground level, with risk of ignition at a distance. Do not use compressed air when filling, emptying or processing. Reacts violently with some strong oxidants.		
Flash point, °C	− 4			
Auto-ignition temperature, °C	427			
Relative density (water = 1)	0.9	TLV-TWA	400 ppm	1440 mg/m³
Relative vapor density (air = 1)	3.0			
Relative density at 20 °C of saturated mixture vapor/air (air = 1)	1.2	**Absorption route:** Can enter the body by inhalation or ingestion. Harmful atmospheric concentrations build up fairly slowly on evaporation at 20°C, but much more rapidly in aerosol form.		
Vapor pressure, mm Hg at 20 °C	73.7			
Solubility in water, g/100 ml at 20 °C	8.5	**Immediate effects:** Irritates the eyes, skin and respiratory tract. Liquid destroys the skin's natural oils. In substantial concentrations can impair consciousness.		
Explosive limits, vol% in air	2-12			
Minimum ignition energy, mJ	0.46			
Electrical conductivity, pS/m	> 10⁵			
Relative molecular mass	88.1			
Log P octanol/water	0.7			
Gross formula:	$C_4H_8O_2$			

HAZARDS/SYMPTOMS	PREVENTIVE MEASURES	FIRE EXTINGUISHING/FIRST AID
Fire: extremely flammable.	keep away from open flame and sparks, no smoking.	powder, AFFF, foam, carbon dioxide, (halons).
Explosion: forms explosive air-vapor mixtures.	sealed machinery, ventilation, explosion-proof electrical equipment and lighting, grounding.	in case of fire: keep tanks/drums cool by spraying with water.
Inhalation: cough, shortness of breath, dizziness.	ventilation, local exhaust or respiratory protection.	fresh air, rest, call a doctor.
Skin: redness.	gloves.	remove contaminated clothing, flush skin with water or shower.
Eyes: redness, pain.	face shield.	flush with water, take to a doctor if necessary.
Ingestion: sore throat, abdominal pain, diarrhea.		rinse mouth, call a doctor or take to hospital.

SPILLAGE	STORAGE	LABELING / NFPA
collect leakage in sealable containers, soak up with sand or other inert absorbent and remove to safe place; (additional individual protection: breathing apparatus).	keep in fireproof place, separate from oxidants.	R: 11 S: 16-23-29-33 🔥 Flammable 3 / 1 / 0

NOTES
Alcohol consumption increases toxic effects.

CAS-No: [141-97-9]
diacetic ester
ethyl acetylacetate
ethyl 3-oxobutanoate
3-oxobutanoic acid ethyl ester

$CH_3COCH_2COOC_2H_5$

ETHYL ACETOACETATE

PHYSICAL PROPERTIES		IMPORTANT CHARACTERISTICS	
Boiling point, °C	180	**COLORLESS LIQUID WITH CHARACTERISTIC ODOR**	
Melting point, °C	−39	Vapor mixes readily with air. Reacts with strong oxidants.	
Flash point, °C	84		
Auto-ignition temperature, °C	295	TLV-TWA	not available
Relative density (water = 1)	1.03		
Relative vapor density (air = 1)	4.5	**Absorption route:** Can enter the body by inhalation or ingestion.	
Relative density at 20 °C of saturated mixture		**Immediate effects:** Irritates the eyes, skin and respiratory tract. In serious cases risk of	
vapor/air (air = 1)	1.0	unconsciousness.	
Vapor pressure, mm Hg at 29 °C	0.99		
Solubility in water, g/100 ml	11.6		
Relative molecular mass	130.1		
Log P octanol/water	1.2		
Gross formula:	$C_6H_{10}O_3$		

HAZARDS/SYMPTOMS	PREVENTIVE MEASURES	FIRE EXTINGUISHING/FIRST AID
Fire: combustible.	keep away from open flame, no smoking.	powder, AFFF, foam, carbon dioxide, (halons).
Explosion: above 84°C: forms explosive air-vapor mixtures.	above 84°C: sealed machinery, ventilation.	
Inhalation: cough, dizziness.	ventilation, local exhaust or respiratory protection.	fresh air, rest, call a doctor.
Skin: redness, pain.	gloves.	remove contaminated clothing, flush skin with water or shower.
Eyes: redness, pain.	safety glasses.	flush with water, take to a doctor.
Ingestion: sore throat, abdominal pain, dizziness.		rinse mouth, call a doctor.

SPILLAGE	STORAGE	LABELING / NFPA
collect leakage in sealable containers, soak up with sand or other inert absorbent and remove to safe place, flush away remainder with water.		2 2 0

NOTES
Log P octanol/water is estimated.

Transport Emergency Card TEC(R)-30G37

CAS-No: [140-88-5]
acrylic acid, ethyl ester
ethoxy carbonyl ethylene
ethyl propenoate
ethyl 2-propenoate
2-propenoic acid, ethyl ester

$CH_2CHCOOC_2H_5$

ETHYL ACRYLATE

PHYSICAL PROPERTIES	IMPORTANT CHARACTERISTICS
	COLORLESS LIQUID WITH PUNGENT ODOR

Boiling point, °C	100	
Melting point, °C	−72	
Flash point, °C	9	
Auto-ignition temperature, °C	350	
Relative density (water = 1)	0.9	
Relative vapor density (air = 1)	3.5	
Relative density at 20 °C of saturated mixture vapor/air (air = 1)	1.09	
Vapor pressure, mm Hg at 20 °C	30	
Solubility in water, g/100 ml at 20 °C	1.5	
Explosive limits, vol% in air	1.7-13	
Relative molecular mass	100.1	

Vapor is heavier than air and spreads at ground level, with risk of ignition at a distance. Moderate heating or exposure to light or peroxides readily causes polymerization, evolving heat. Do not use compressed air when filling, emptying or processing. Reacts violently with oxidants, with risk of fire and explosion.

TLV-TWA (skin)	5 ppm	20 mg/m³ A2
STEL	15 ppm	61 mg/m³

Absorption route: Can enter the body by inhalation or ingestion or through the skin. Harmful atmospheric concentrations can build up fairly rapidly on evaporation at approx. 20°C - even more rapidly in aerosol form.
Immediate effects: Corrosive to the eyes, skin and respiratory tract. Inhalation of vapor/fumes can cause severe breathing difficulties (lung edema). In serious cases risk of death.
Effects of prolonged/repeated exposure: Prolonged or repeated contact can cause skin disorders. Can cause liver and kidney damage.

Gross formula: $C_5H_8O_2$

HAZARDS/SYMPTOMS	PREVENTIVE MEASURES	FIRE EXTINGUISHING/FIRST AID
Fire: extremely flammable.	keep away from open flame and sparks, no smoking.	powder, AFFF, foam, carbon dioxide, (halons).
Explosion: forms explosive air-vapor mixtures.	sealed machinery, ventilation, explosion-proof electrical equipment and lighting.	in case of fire: keep tanks/drums cool by spraying with water.
	STRICT HYGIENE	IN ALL CASES CALL A DOCTOR
Inhalation: *corrosive*; sore throat, cough, shortness of breath, severe breathing difficulties, headache.	ventilation, local exhaust or respiratory protection.	fresh air, rest, place in half-sitting position, take to hospital.
Skin: *corrosive*; redness, pain, chemical burns.	gloves, protective clothing.	remove contaminated clothing, flush skin with water or shower, refer to a doctor.
Eyes: *corrosive*; redness, pain, impaired vision.	face shield, or combined eye and respiratory protection.	flush with water, take to doctor.
Ingestion: *corrosive*; sore throat, abdominal pain, vomiting.		rinse mouth, take immediately to hospital.

SPILLAGE	STORAGE	LABELING / NFPA
collect leakage in sealable containers, soak up with sand or other inert absorbent and remove to safe place; (additional individual protection: breathing apparatus).	keep in cool, dark, fireproof place, separate from oxidants; add an inhibitor.	R: 11-36/37/38-20/22-43 S: 9-16-33 🔥 ✖ Flammable Irritant 3 / 2 2

NOTES
Depending on the degree of exposure, regular medical checkups are advisable. Lung edema symptoms usually develop several hours later and are aggravated by physical exertion: rest and hospitalization essential. As first aid, a doctor or authorized person should consider administering a corticosteroid spray.

Transport Emergency Card TEC(R)-671

HI: 339; UN-number: 1917

$(C_2H_5)AlCl_2$

ETHYL ALUMINUM DICHLORIDE

PHYSICAL PROPERTIES	IMPORTANT CHARACTERISTICS
Boiling point, °C 194 Melting point, °C − 32 Relative density (water = 1) 1.2 Relative vapor density (air = 1) 4.6 Vapor pressure, mm Hg at 60 °C 4.9 Solubility in water reaction Relative molecular mass 134	**COLORLESS FUMING LIQUID** Aluminum oxide fumes and → *hydrochloric acid* fumes are a product of combustion. Can ignite spontaneously on exposure to air. Strong reducing agent which reacts violently with oxidants. Reacts violently with water, alcohols, phenols, amines, carbon dioxide, oxides of sulfur and nitrogen oxides, with risk of fire and explosion. Reacts with air to form aluminum oxide fumes and → *hydrochloric acid* fumes.

	IMPORTANT CHARACTERISTICS
	TLV-TWA 2 mg/m³
Gross formula: $C_2H_5AlCl_2$	**Absorption route:** Can enter the body by inhalation or ingestion. Harmful atmospheric concentrations can build up fairly rapidly on evaporation at approx. 20°C - even more rapidly in aerosol form. **Immediate effects:** Corrosive to the eyes, skin and respiratory tract. Inhalation of vapor/fumes can cause severe breathing difficulties (lung edema). In serious cases risk of death.

HAZARDS/SYMPTOMS	PREVENTIVE MEASURES	FIRE EXTINGUISHING/FIRST AID
Fire: extremely flammable, many chemical reactions can cause fire and explosion.	keep away from open flame and sparks, no smoking, avoid contact with air, water and many other substances.	special extinguishing powder, dry sand, NO other extinguishing agents, can reignite after being extinguished.
Explosion: contact with water and many other substances can cause explosion.		in case of fire: keep undamaged cylinders cool by spraying with water, but DO NOT spray substance with water.
	STRICT HYGIENE	
Inhalation: *corrosive*; sore throat, cough, severe breathing difficulties.	completely sealed machinery, ventilation, local exhaust or respiratory protection.	fresh air, rest, place in half-sitting position, take to hospital.
Skin: *corrosive*; serious burns, slow healing.	gloves, protective clothing.	remove contaminated clothing, flush skin with water or shower, refer to a doctor.
Eyes: *corrosive*; redness, pain, impaired vision.	face shield, or combined eye and respiratory protection.	flush with water, take to a doctor.
Ingestion: *corrosive*; sore throat, abdominal pain, diarrhea.		rinse mouth, take immediately to hospital.

SPILLAGE	STORAGE	LABELING / NFPA
evacuate area, call in an expert, dilute leakage with diesel/machine oil, soak up/cover with dry sand/table salt and remove to safe place; (additional individual protection: CHEMICAL SUIT).	keep in fireproof place under dry nitrogen, separate from all other substances.	

NOTES
Often sold as 15-30% solution in hydrocarbons: flash point and explosive limits depend on solvent used. For 15% solution see relevant sheet. Reacts violently with extinguishing agents such as carbon dioxide, water, foam and ordinary powder. Lung edema symptoms usually develop several hours later and are aggravated by physical exertion: rest and hospitalization essential. As first aid, a doctor or authorized person should consider administering a corticosteroid spray. Use cylinder with special fittings.

Transport Emergency Card TEC(R)-42G02

HI: X333; UN-number: 2220

$(C_2H_5)AlCl_2$

ETHYL ALUMINUM DICHLORIDE
(15% solution in hexane)

PHYSICAL PROPERTIES	IMPORTANT CHARACTERISTICS
Boiling point, °C — 69 Melting point, °C — −95 Flash point, °C — −26 Auto-ignition temperature, °C — 240 Relative density (water = 1) — 0.75 Relative vapor density (air = 1) — 3.0 Relative density at 20 °C of saturated mixture vapor/air (air = 1) — 1.3 Vapor pressure, mm Hg at 20 °C — 122 Solubility in water — reaction Explosive limits, vol% in air — 1.1-7.5 Minimum ignition energy, mJ — 0.24 Electrical conductivity, pS/m — 100 Gross formula: $C_2H_5AlCl_2$	**COLORLESS FUMING LIQUID** Vapor is heavier than air and spreads at ground level, with risk of ignition at a distance. Flow, agitation etc. can cause build-up of electrostatic charge due to liquid's low conductivity. Aluminum oxide fumes and → *hydrochloric acid* fumes are a product of combustion. Can ignite spontaneously on exposure to air once solvent has evaporated. Strong reducing agent which reacts violently with oxidants. Reacts violently with water, alcohols, phenols, amines, carbon dioxide, sulfur oxides, nitrogen oxides, and many other substances, with risk of fire and explosion.

	IMPORTANT CHARACTERISTICS (cont.)
TLV-TWA	2 mg/m³

Absorption route: Can enter the body by inhalation or ingestion. Harmful atmospheric concentrations can build up fairly rapidly on evaporation at approx. 20°C - even more rapidly in aerosol form.
Immediate effects: Corrosive to the eyes, skin and respiratory tract. Inhalation of vapor/fumes can cause severe breathing difficulties (lung edema).

HAZARDS/SYMPTOMS	PREVENTIVE MEASURES	FIRE EXTINGUISHING/FIRST AID
Fire: extremely flammable, many chemical reactions can cause fire and explosion.	keep away from open flame and sparks, no smoking, avoid contact with water, alcohols and many other substances.	special extinguishing powder, dry sand, NO other extinguishing agents, can reignite after being extinguished.
Explosion: forms explosive air-vapor mixtures; contact with water, alcohols and many other substances can cause explosion.	sealed machinery, ventilation, explosion-proof electrical equipment and lighting, grounding.	in case of fire: keep undamaged cylinders cool by spraying with water, but DO NOT spray substance with water.
	STRICT HYGIENE	
Inhalation: *corrosive*; sore throat, cough, severe breathing difficulties.	sealed machinery, local exhaust or respiratory protection.	fresh air, rest, place in half-sitting position, take to hospital.
Skin: *corrosive*; redness, serious burns, slow healing.	gloves, protective clothing.	remove contaminated clothing, shower, call a doctor if necessary.
Eyes: *corrosive*; redness, pain, impaired vision.	face shield, or combined eye and respiratory protection.	flush with water, take to a doctor.
Ingestion: *corrosive*; sore throat, abdominal pain, diarrhea.		rinse mouth, immediately take to hospital.

SPILLAGE	STORAGE	LABELING / NFPA
evacuate area, call in an expert, collect leakage in sealable containers, soak up remainder with dry sand/table salt, under no circumstances spray liquid with water; (additional individual protection: CHEMICAL SUIT).	keep in fireproof place under inert gas, separate from all other substances.	

NOTES

Physical properties (boiling point, flash point, autoignition point, explosive limits, vapor pressure, relative density, conductivity and ignition energy) are based on hexane. For pure ethylaluminumdichloride see relevant sheet. For TLV of → *hexane* see relevant sheet. Reacts violently with extinguishing agents such as water, foam, carbon dioxide and ordinary powder. Remove contaminated clothing immediately and flush with water (risk of spontaneous combustion). As first aid, a doctor or authorized person should consider administering a corticosteroid spray. Use cylinder with special fittings.

Transport Emergency Card TEC(R)-42G02

HI: X333; UN-number: 2221

ETHYLALUMINUM SESQUICHLORIDE

PHYSICAL PROPERTIES	IMPORTANT CHARACTERISTICS
Boiling point, °C 204 Melting point, °C −20 Relative density (water = 1) 1.1 Relative vapor density (air = 1) 8.53 Vapor pressure, mm Hg at 20 °C 0.28 Solubility in water reaction Relative molecular mass 247.5	**COLORLESS FUMING LIQUID WITH PUNGENT ODOR** Can ignite spontaneously on exposure to air. Aluminum fumes and → *hydrochloric acid* fumes are a product of combustion. Strong reducing agent which reacts violently with oxidants. Reacts violently with water, alcohols, phenols, amines, carbon dioxide, oxides of sulfur, nitrogen oxides and many other substances, with risk of fire and explosion.

	TLV-TWA	2 mg/m³

Absorption route: Can enter the body by inhalation or ingestion. Harmful atmospheric concentrations can build up fairly rapidly on evaporation at approx. 20°C - even more rapidly in aerosol form.
Immediate effects: Corrosive to the eyes, skin and respiratory tract. Inhalation of vapor/fumes can cause severe breathing difficulties (lung edema). In serious cases risk of death.

Gross formula: $C_6H_{15}Al_2Cl_3$

HAZARDS/SYMPTOMS	PREVENTIVE MEASURES	FIRE EXTINGUISHING/FIRST AID
Fire: extremely flammable, many chemical reactions can cause fire and explosion.	keep away from open flame and sparks, no smoking, avoid contact with air, water and many other substances.	special extinguishing powder, dry sand, NO other extinguishing agents, can reignite after being extinguished.
Explosion: contact with water and many other substances can cause explosion.		in case of fire: keep undamaged cylinders cool by spraying with water, but DO NOT spray substance with water.
	STRICT HYGIENE	
Inhalation: *corrosive*; sore throat, cough, severe breathing difficulties.	sealed machinery, ventilation, local exhaust or respiratory protection.	fresh air, place in half-sitting position, take to hospital.
Skin: *corrosive*; serious burns, slow healing.	gloves, protective clothing.	remove contaminated clothing, flush skin with water or shower, refer to a doctor.
Eyes: *corrosive*; redness, pain, impaired vision.	face shield, or combined eye and respiratory protection.	flush with water, take to a doctor.
Ingestion: *corrosive*; sore throat, abdominal pain, diarrhea.		rinse mouth, take immediately to hospital.

SPILLAGE	STORAGE	LABELING / NFPA
evacuate area, call in an expert, dilute leakage with diesel/machine oil, soak up/cover with dry sand/table salt and remove to safe place, under no circumstances spray liquid with water; (additional individual protection: CHEMICAL SUIT).	keep in dry, fireproof place under inert gas, separate from all other substances.	 3 / 3 / W

NOTES
Usually sold as 15-30% solution in hydrocarbons, which does not ignite spontaneously on exposure to air. For 15% solution see relevant sheet. Lung edema symptoms usually develop several hours later and are aggravated by physical exertion: rest and hospitalization essential. As first aid, a doctor or authorized person should consider administering a corticosteroid spray. Use cylinder with special fittings.

ETHYL ALUMINUM SESQUICHLORIDE

(15% solution in hexane)

PHYSICAL PROPERTIES	IMPORTANT CHARACTERISTICS
Boiling point, °C — 69 Melting point, °C — −95 Flash point, °C — −26 Auto-ignition temperature, °C — 240 Relative density (water = 1) — 0.8 Relative vapor density (air = 1) — 3.0 Relative density at 20 °C of saturated mixture vapor/air (air = 1) — 1.3 Vapor pressure, mm Hg at 20 °C — 122 Solubility in water — reaction Explosive limits, vol% in air — 1.1-7.5 Minimum ignition energy, mJ — 0.24 Electrical conductivity, pS/m — 100 Gross formula: $C_6H_{15}Al_2Cl_3$	**COLORLESS FUMING LIQUID WITH PUNGENT ODOR** Vapor is heavier than air and spreads at ground level, with risk of ignition at a distance. Flow, agitation etc. can cause build-up of electrostatic charge due to liquid's low conductivity. Do not use compressed air when filling, emptying or processing. Once solvent has evaporated (e.g. on clothing) can ignite spontaneously on exposure to air. Aluminum fumes and → *hydrochloric acid* fumes are a product of combustion. Strong reducing agent which reacts violently with oxidants. Reacts violently with water, alcohols, phenols and many other substances, with risk of fire and explosion.

TLV-TWA	2 mg/m³

Absorption route: Can enter the body by inhalation or ingestion. Harmful atmospheric concentrations can build up fairly rapidly on evaporation at approx. 20°C - even more rapidly in aerosol form.
Immediate effects: Corrosive to the eyes, skin and respiratory tract. Inhalation of vapor/fumes can cause severe breathing difficulties (lung edema). In serious cases risk of death.

HAZARDS/SYMPTOMS	PREVENTIVE MEASURES	FIRE EXTINGUISHING/FIRST AID
Fire: extremely flammable, many chemical reactions can cause fire and explosion.	keep away from open flame and sparks, no smoking, avoid contact with water and many other substances.	special extinguishing powder, dry sand, NO other extinguishing agents, can reignite after being extinguished.
Explosion: forms explosive air-vapor mixtures, contact with water and many other substances can cause explosion.	sealed machinery, ventilation, explosion-proof electrical equipment and lighting, grounding.	in case of fire: keep undamaged cylinders cool by spraying with water, but DO NOT spray substance with water.
	STRICT HYGIENE	
Inhalation: *corrosive*; sore throat, cough, severe breathing difficulties.	sealed machinery, ventilation, local exhaust or respiratory protection.	fresh air, rest, place in half-sitting position, take to hospital.
Skin: *corrosive*; redness, serious burns, slow healing.	gloves, protective clothing.	remove contaminated clothing, flush skin with water or shower, call a doctor if necessary.
Eyes: *corrosive*; redness, pain, impaired vision.	face shield, or combined eye and respiratory protection.	flush with water, take to a doctor.
Ingestion: *corrosive*; sore throat, abdominal pain, diarrhea.		rinse mouth, take immediately to hospital.

SPILLAGE	STORAGE	LABELING / NFPA
evacuate area, call in an expert, collect leakage in sealable containers, soak up remainder with dry sand/table salt and remove to safe place; (additional individual protection: CHEMICAL SUIT).	keep in dry, fireproof place under inert gas, separate from all other substances.	

NOTES
Physical properties (boiling point, flash point, autoignition point, explosive limits, vapor pressure, relative density, conductivity and ignition energy) are based on hexane. For pure ethyl aluminum sesquichloride, see relevant sheet. For TLV of *hexane*, see relevant sheet. Reacts violently with extinguishing agents such as water, foam, carbon dioxide and ordinary powder. Remove contaminated clothing immediately and flush with water (risk of spontaneous combustion). Lung edema symptoms usually develop several hours later and are aggravated by physical exertion: rest and hospitalization essential. As first aid, a doctor or authorized person should consider administering a corticosteroid spray. Use cylinder with special fittings.

Transport Emergency Card TEC(R)–42G02

HI: X333; UN-number: 2220

C₂H₅NH₂

aminoethane
1-aminoethane
EA
ethanamine
monoethylamine

ETHYLAMINE
(cylinder)

PHYSICAL PROPERTIES		IMPORTANT CHARACTERISTICS
Boiling point, °C	17	**COLORLESS COMPRESSED LIQUEFIED GAS WITH PUNGENT ODOR**
Melting point, °C	−81	Gas is heavier than air and spreads at ground level, with risk of ignition at a distance. Do not use
Flash point, °C	−49	compressed air when filling, emptying or processing. Toxic gases (→ *nitrous vapors*) are a
Auto-ignition temperature, °C	380	product of combustion. Medium strong base which reacts violently with acids and corrodes
Relative density (water = 1)	0.7	aluminum, zinc etc. Reacts violently with strong oxidants. Reacts violently with many organic
Relative vapor density (air = 1)	1.6	compounds, with risk of fire and explosion.
Vapor pressure, mm Hg at 20 °C	912	
Solubility in water	∞	
Explosive limits, vol% in air	3.5-14	
Minimum ignition energy, mJ	2.4	**Absorption route:** Can enter the body by inhalation. Harmful atmospheric concentrations can
Electrical conductivity, pS/m	7 x 10⁵	build up very rapidly if gas is released.
Relative molecular mass	45.1	**Immediate effects:** Corrosive to the eyes, skin and respiratory tract. Can cause frostbite due to
Log P octanol/water	−0.2	rapid evaporation. Inhalation of vapor/fumes can cause severe breathing difficulties (lung

TLV-TWA: 10 ppm, 18 mg/m³ (placed in Important Characteristics section)

Gross formula: C₂H₇N

TLV-TWA — 10 ppm — 18 mg/m³

Absorption route: Can enter the body by inhalation. Harmful atmospheric concentrations can build up very rapidly if gas is released.
Immediate effects: Corrosive to the eyes, skin and respiratory tract. Can cause frostbite due to rapid evaporation. Inhalation of vapor/fumes can cause severe breathing difficulties (lung edema). In serious cases risk of death.
Effects of prolonged/repeated exposure: Prolonged or repeated exposure can cause nerve tissue disorders.

HAZARDS/SYMPTOMS	PREVENTIVE MEASURES	FIRE EXTINGUISHING/FIRST AID
Fire: extremely flammable.	keep away from open flame and sparks, no smoking.	shut off supply; if impossible and no danger to surrounding area, allow fire to burn itself out; otherwise extinguish with powder, carbon dioxide, (halons).
Explosion: forms explosive air-gas mixtures.	sealed machinery, ventilation, explosion-proof electrical equipment and lighting.	in case of fire: keep cylinders cool by spraying with water, fight fire from sheltered location.
Inhalation: *corrosive*; sore throat, cough, shortness of breath, severe breathing difficulties.	ventilation, local exhaust or respiratory protection.	fresh air, rest, place in half-sitting position, take to hospital.
Skin: *corrosive*; *in case of frostbite*: redness, pain, serious burns.	gloves, protective clothing.	*in case of frostbite*: DO NOT remove clothing, flush skin with water or shower, call a doctor if necessary.
Eyes: *corrosive*, *in case of frostbite*: redness, pain, impaired vision.	face shield, or combined eye and respiratory protection.	flush with water, take to a doctor.

SPILLAGE	STORAGE	LABELING / NFPA
evacuate area, ventilate, combat gas cloud with water curtain, remove cylinder to safe place if possible; (additional individual protection: CHEMICAL SUIT).	if stored indoors, keep in cool, fireproof place, separate from oxidants and acids.	◇ 4 / 3 0

NOTES

Also supplied commercially in the form of 70% aqueous solution. Log P octanol/water is estimated. Lung edema symptoms usually develop several hours later and are aggravated by physical exertion: rest and hospitalization essential. As first aid, a doctor or authorized person should consider administering a corticosteroid spray. Turn leaking cylinder so that leak is on top to prevent liquid ethylamine escaping.Use cylinder with special fittings.

Transport Emergency Card TEC(R)-20G12

HI: 236; UN-number: 1036

amyl ethyl ketone
EAK
5-methyl-3-heptanone

$C_2H_5COCH_2CH(CH_3)C_2H_5$

ETHYL AMYL KETONE

PHYSICAL PROPERTIES	IMPORTANT CHARACTERISTICS
Boiling point, °C 160 Melting point, °C −57 Flash point, °C 59 Auto-ignition temperature, °C ? Relative density (water = 1) 0.8 Relative vapor density (air = 1) 4.4 Relative density at 20 °C of saturated mixture vapor/air (air = 1) 1.0 Vapor pressure, mm Hg at 25 °C 2.1 Solubility in water, g/100 ml at 20 °C 0.3 Relative molecular mass 128.2 Gross formula: $C_8H_{16}O$	**COLORLESS LIQUID WITH PUNGENT ODOR** Vapor mixes readily with air. Reacts violently with oxidants. TLV-TWA 25 ppm 131 mg/m³ **Absorption route:** Can enter the body by inhalation or ingestion. Harmful atmospheric concentrations build up fairly slowly on evaporation, but much more rapidly in aerosol form. **Immediate effects:** Irritates the eyes and respiratory tract. Liquid destroys the skin's natural oils. Affects the nervous system. In serious cases risk of unconsciousness.

HAZARDS/SYMPTOMS	PREVENTIVE MEASURES	FIRE EXTINGUISHING/FIRST AID
Fire: combustible.	keep away from open flame, no smoking.	powder, AFFF, foam, carbon dioxide, (halons).
Explosion: above 59°C: forms explosive air-vapor mixtures.	above 59°C: sealed machinery, ventilation, explosion-proof electrical equipment.	
Inhalation: cough, shortness of breath, headache, nausea.	ventilation, local exhaust or respiratory protection.	fresh air, rest, call a doctor if necessary.
Skin: redness.	gloves.	remove contaminated clothing, flush skin with water or shower.
Eyes: redness, pain.	face shield.	flush with water, take to a doctor if necessary.
Ingestion: nausea, drowsiness.		rinse mouth, call a doctor or take to hospital.

SPILLAGE	STORAGE	LABELING / NFPA
collect leakage in sealable containers, soak up with sand or other inert absorbent and remove to safe place.	keep separate from oxidants.	R: 10-36/37 S: 23 ❎ Irritant

NOTES
Alcohol consumption increases toxic effects. Depending on the degree of exposure, regular medical checkups are advisable.

Transport Emergency Card TEC(R)-30G35

HI: 30; UN-number: 2271

N-ETHYLANILINE

PHYSICAL PROPERTIES	IMPORTANT CHARACTERISTICS
Boiling point, °C 205 Melting point, °C −63 Flash point, °C 85 Auto-ignition temperature, °C 480 Relative density (water = 1) 0.96 Relative density at 20 °C of saturated mixture vapor/air (air = 1) 1.0 Vapor pressure, mm Hg at 20 °C 3.0 Solubility in water none Relative molecular mass 121.2 Log P octanol/water 2.3 Gross formula: $C_8H_{11}N$	**COLORLESS VISCOUS LIQUID WITH CHARACTERISTIC ODOR, TURNING BROWN ON EXPOSURE TO AIR** Vapor mixes readily with air. → *Nitrous vapors* are a product of combustion. Reacts violently with strong oxidants and concentrated strong acids, giving off toxic vapors (→ *nitrous vapors* and → *aniline*), with risk of fire and explosion. Reacts violently with concentrated nitric acid, with risk of fire and explosion. TLV-TWA not available **Absorption route:** Can enter the body by inhalation or ingestion or through the skin. Evaporation negligible at 20°C, but harmful atmospheric concentrations can build up rapidly in aerosol form. **Immediate effects:** In serious cases risk of death. **Effects of prolonged/repeated exposure:** Prolonged or repeated contact can cause skin disorders. Can affect the blood. Can cause liver and kidney damage.

HAZARDS/SYMPTOMS	PREVENTIVE MEASURES	FIRE EXTINGUISHING/FIRST AID
Fire: combustible.	keep away from open flame, no smoking, avoid contact with strong oxidants and nitric acid.	powder, AFFF, foam, carbon dioxide, (halons).
Explosion: above 85°C: forms explosive air-vapor mixtures.	above 85°C: sealed machinery, ventilation.	
Inhalation: shortness of breath, headache, dizziness, see also: 'Ingestion'.	ventilation, local exhaust or respiratory protection.	fresh air, rest, call a doctor.
Skin: *is absorbed*; redness; see also 'Inhalation'.	gloves, protective clothing.	remove contaminated clothing, wash skin with soap and water, call a doctor if necessary.
Eyes: redness, pain.	face shield.	flush with water, take to a doctor if necessary.
Ingestion: abdominal pain, nausea, vomiting, unconsciousness, dizziness, blue skin, feeling of weakness.	do not eat, drink or smoke while working.	rinse mouth, immediately take to hospital.

SPILLAGE	STORAGE	LABELING / NFPA
collect leakage in sealable containers, soak up with sand or other inert absorbent and remove to safe place.	keep in dark place, separate from oxidants and strong acids, ventilate at floor level.	R: 23/24/25-33 S: 28-37-44 ☠ Toxic NFPA: 3 / 2 / 0

NOTES
Depending on the degree of exposure, regular medical checkups are advisable. Effects on blood due to formation of methemoglobin. Special first aid required in the event of poisoning: antidotes must be available (with instructions).

Transport Emergency Card TEC(R)-61G06C

HI: 60; UN-number: 2272

EB
ethylbenzol
phenylethane

ETHYLBENZENE

PHYSICAL PROPERTIES		IMPORTANT CHARACTERISTICS	
Boiling point, °C	136	**COLORLESS LIQUID WITH CHARACTERISTIC ODOR**	
Melting point, °C	−95	Vapor mixes readily with air. Flow, agitation etc. can cause build-up of electrostatic charge due to	
Flash point, °C	23	liquid's low conductivity. Reacts violently with strong oxidants, with risk of fire and explosion.	
Auto-ignition temperature, °C	430		
Relative density (water = 1)	0.9		
Relative vapor density (air = 1)	3.7	TLV-TWA 100 ppm 434 mg/m³	
Relative density at 20 °C of saturated mixture		STEL 125 ppm 543 mg/m³	
vapor/air (air = 1)	1.02		
Vapor pressure, mm Hg at 20 °C	7.2	**Absorption route:** Can enter the body by inhalation or ingestion. Harmful atmospheric	
Solubility in water	none	concentrations build up fairly slowly on evaporation at 20°C, but much more rapidly in aerosol	
Explosive limits, vol% in air	1.0-7.8	form.	
Electrical conductivity, pS/m	<10⁴	**Immediate effects:** Irritates the eyes, skin and respiratory tract. In substantial concentrations	
Relative molecular mass	106.2	can impair consciousness. In serious cases risk of unconsciousness. If liquid is swallowed,	
Log P octanol/water	3.2	droplets can enter the lungs, with risk of pneumonia.	
		Effects of prolonged/repeated exposure: Prolonged or repeated contact can cause skin	
		disorders.	
Gross formula:	C_8H_{10}		

HAZARDS/SYMPTOMS	PREVENTIVE MEASURES	FIRE EXTINGUISHING/FIRST AID
Fire: flammable.	keep away from open flame and sparks, no smoking.	powder, AFFF, foam, carbon dioxide, (halons).
Explosion: above 23°C: forms explosive air-vapor mixtures.	above 23°C: sealed machinery, ventilation, explosion-proof electrical equipment, grounding.	in case of fire: keep tanks/drums cool by spraying with water.
Inhalation: cough, severe breathing difficulties, headache, dizziness, tiredness, sleepiness, unconsciousness, respiratory tract irritation (mucous membrane).	ventilation, local exhaust or respiratory protection.	fresh air, rest, place in half-sitting position, call a doctor.
Skin: redness, pain.	gloves.	remove contaminated clothing, flush skin with water or shower.
Eyes: redness, pain, impaired vision, watering of the eyes.	face shield.	flush with water, take to a doctor.
Ingestion: abdominal pain, nausea, vomiting.		rinse mouth, DO NOT induce vomiting, call a doctor.

SPILLAGE	STORAGE	LABELING / NFPA
collect leakage in sealable containers, soak up with sand or other inert absorbent and remove to safe place.	keep in fireproof place, separate from oxidants.	R: 11-20 S: 16-24/25-29 Flammable Irritant

NOTES
Absorbed into body fat and detectable in the blood.

Transport Emergency Card TEC(R)-522 **HI: 33; UN-number: 1175**

ETHYL BENZOATE

PHYSICAL PROPERTIES	IMPORTANT CHARACTERISTICS
Boiling point, °C .... 213 Melting point, °C .... −35 Flash point, °C .... 88 Auto-ignition temperature, °C .... 490 Relative density (water = 1) .... 1.05 Relative vapor density (air = 1) .... 5.2 Relative density at 20 °C of saturated mixture vapor/air (air = 1) .... 1.0 Vapor pressure, mm Hg at 20 °C .... ca. 0.12 Solubility in water .... none Relative molecular mass 150.2	**COLORLESS LIQUID WITH CHARACTERISTIC ODOR** Vapor mixes readily with air. Flow, agitation etc. can cause build-up of electrostatic charge due to liquid's low conductivity. Reacts violently with strong oxidants.
	TLV-TWA not available
	Absorption route: Can enter the body by inhalation or ingestion. Harmful atmospheric concentrations build up very slowly, if at all, on evaporation at approx. 20°C, but much more rapidly in aerosol form. **Immediate effects:** Irritates the eyes, skin and respiratory tract.
Gross formula: $C_9H_{10}O_2$	

HAZARDS/SYMPTOMS	PREVENTIVE MEASURES	FIRE EXTINGUISHING/FIRST AID
Fire: combustible.	keep away from open flame, no smoking.	powder, AFFF, foam, carbon dioxide, (halons).
Explosion: above 88°C: forms explosive air-vapor mixtures.	above 88°C: sealed machinery, ventilation, grounding.	in case of fire: keep tanks/drums cool by spraying with water.
Inhalation: sore throat, cough, shortness of breath.	ventilation.	fresh air, rest, call a doctor.
Skin: redness.	gloves.	remove contaminated clothing, flush skin with water or shower.
Eyes: redness, pain.	safety glasses.	flush with water, take to a doctor if necessary.
Ingestion: abdominal pain, diarrhea, nausea.		rinse mouth, call a doctor or take to hospital.

SPILLAGE	STORAGE	LABELING / NFPA
collect leakage in sealable containers, soak up with sand or other inert absorbent and remove to safe place.	keep separate from oxidants; ventilate at floor level.	1 / 1 / 0

NOTES

Transport Emergency Card TEC(R)-30G15

ETHYL BROMIDE

PHYSICAL PROPERTIES	IMPORTANT CHARACTERISTICS	
Boiling point, °C 38 Melting point, °C −119 Flash point, °C ? Auto-ignition temperature, °C 510 Relative density (water = 1) 1.46 Relative vapor density (air = 1) 3.8 Relative density at 20 °C of saturated mixture vapor/air (air = 1) 2.4 Vapor pressure, mm Hg at 20 °C 388 Solubility in water, g/100 ml 0.9 Explosive limits, vol% in air 6.7-11.3 Electrical conductivity, pS/m < 2 x 10⁸ Relative molecular mass 109.0 Log P octanol/water 1.7 Gross formula: C_2H_5Br	**COLORLESS LIQUID WITH CHARACTERISTIC ODOR, TURNING YELLOW ON EXPOSURE TO AIR** Vapor is heavier than air and spreads at ground level, with risk of ignition at a distance. Decomposes in flame or on hot surface, giving off toxic and corrosive vapors (→ *hydrogen bromide*). Decomposes on contact with steam or hot water, giving off → *hydrogen bromide*. Reacts violently with strong oxidants, metal powders and alkaline-earth and alkali metals, with risk of fire and explosion. Attacks plastics.	
	TLV-TWA 200 ppm 891 mg/m³ STEL 250 ppm 1110 mg/m³	
	Absorption route: Can enter the body by inhalation or ingestion or through the skin. Harmful atmospheric concentrations can build up fairly rapidly on evaporation at approx. 20°C - even more rapidly in aerosol form. **Immediate effects:** Corrosive to the eyes, skin and respiratory tract. High concentrations can impair consciousness. Inhalation of vapor/fumes can cause severe breathing difficulties (lung edema). In serious cases risk of unconsciousness and death. **Effects of prolonged/repeated exposure:** Can cause liver and kidney damage.	

HAZARDS/SYMPTOMS	PREVENTIVE MEASURES	FIRE EXTINGUISHING/FIRST AID
Fire: combustible only at temperatures above 500°C; non-combustible under normal conditions.		powder, water spray, foam, carbon dioxide, (halons).
Explosion:		in case of fire: keep tanks/drums cool by spraying with water.
	STRICT HYGIENE	IN ALL CASES CALL IN A DOCTOR
Inhalation: *corrosive*; cough, shortness of breath, dizziness, drowsiness.	ventilation, local exhaust or respiratory protection.	fresh air, rest, place in half-sitting position, take to hospital.
Skin: *corrosive, is absorbed*; redness, pain, serious burns.	gloves, protective clothing.	remove contaminated clothing, flush skin with water or shower, take to hospital.
Eyes: *corrosive*; redness, pain, impaired vision.	face shield.	flush with water, take to a doctor.
Ingestion: *corrosive*; diarrhea, vomiting, unconsciousness.		rinse mouth, immediately take to hospital.

SPILLAGE	STORAGE	LABELING / NFPA
collect leakage in sealable containers, soak up with sand or other inert absorbent and remove to safe place; (additional individual protection: breathing apparatus).	keep in cool, fireproof place, separate from oxidants.	 1 2 0

NOTES
Log P octanol/water is estimated. Lung edema symptoms usually develop several hours later and are aggravated by physical exertion: rest and hospitalization essential. As first aid, a doctor or authorized person should consider administering a corticosteroid spray. Under certain circumstances can form combustible vapor-air mixtures which are difficult to ignite. Addition of small quantities of combustible substances or raising of oxygen content increases flammability considerably. Alcohol consumption increases toxic effects. Under no circumstances use near flame or hot surface or when welding.

Transport Emergency Card TEC(R)-701

HI: 60; UN-number: 1891

CAS-No: [628-81-9]
n-butyl ethyl ether
2-ethoxybutane

$C_2H_5OC_4H_9$

ETHYL-n-BUTYL ETHER

PHYSICAL PROPERTIES		IMPORTANT CHARACTERISTICS	
Boiling point, °C	92	**COLORLESS LIQUID WITH CHARACTERISTIC ODOR**	
Melting point, °C	− 124	Vapor is heavier than air and spreads at ground level, with risk of ignition at a distance. Able to	
Flash point, °C	− 1	form peroxides. Flow, agitation etc. can cause build-up of electrostatic charge due to liquid's low	
Auto-ignition temperature, °C	?	conductivity. Do not use compressed air when filling, emptying or processing. Reacts violently	
Relative density (water = 1)	0.8	with oxidants. Attacks many plastics.	
Relative vapor density (air = 1)	3.5		
Relative density at 20 °C of saturated mixture		**TLV-TWA**	not available
vapor/air (air = 1)	1.1		
Vapor pressure, mm Hg at 20 °C	44	**Absorption route:** Can enter the body by inhalation or ingestion.	
Solubility in water	none	**Immediate effects:** Irritates the eyes and respiratory tract. In substantial concentrations can	
Explosive limits, vol% in air	?-?	impair consciousness.	
Relative molecular mass	102		
Log P octanol/water	2.0		
Gross formula:	$C_6H_{14}O$		

HAZARDS/SYMPTOMS	PREVENTIVE MEASURES	FIRE EXTINGUISHING/FIRST AID
Fire: extremely flammable.	keep away from open flame and sparks, no smoking.	powder, AFFF, foam, carbon dioxide, (halons).
Explosion: forms explosive air-vapor mixtures.	sealed machinery, ventilation, explosion-proof equipment and lighting, grounding.	in case of fire: keep tanks/drums cool by spraying with water.
Inhalation: dizziness, nausea, drowsiness, sleepiness.	ventilation, local exhaust or respiratory protection.	fresh air, rest, call a doctor.
Skin: redness.	gloves.	remove contaminated clothing, flush skin with water or shower.
Eyes: redness, pain.	safety glasses.	flush with water, take to a doctor.

SPILLAGE	STORAGE	LABELING / NFPA
collect leakage in sealable containers, soak up with sand or other inert absorbent and remove to safe place, do not flush into sewer; (additional individual protection: breathing apparatus).	keep in cool, dark, fireproof place, separate from oxidants; add an inhibitor.	NFPA diamond: 3 (top), 2 (left), 0 (right)

NOTES
Explosive limits not given in the literature. Before distilling check for peroxides; if found, render harmless. Alcohol consumption increases toxic effects.

Transport Emergency Card TEC(R)-30G30

HI: 33; UN-number: 1179

chloroethane
hydrochloric ether
monochloroethane
muriatic ether

CH_3CH_2Cl

ETHYL CHLORIDE
(cylinder)

PHYSICAL PROPERTIES	IMPORTANT CHARACTERISTICS
Boiling point, °C — 12 Melting point, °C — −138 Flash point, °C — flammable gas Auto-ignition temperature, °C — 510 Relative density (water = 1) — 0.9 Relative vapor density (air = 1) — 2.2 Vapor pressure, mm Hg at 20 °C — 1064 Solubility in water, g/100 ml — 0.6 Explosive limits, vol% in air — 3.6-15.4 Electrical conductivity, pS/m — <3x10⁵ Relative molecular mass — 64.5 Log P octanol/water — 1.5 Gross formula: C_2H_5Cl	**COMPRESSED LIQUEFIED GAS WITH CHARACTERISTIC ODOR** Gas is heavier than air and spreads at ground level, with risk of ignition at a distance. Flow, agitation etc. can cause build-up of electrostatic charge due to liquid's low conductivity. Corrosive vapors are a product of combustion (→ *hydrochloric acid*). Reacts with water and steam, giving off corrosive vapors (→ *hydrochloric acid*). Reacts violently with alkali metals, aluminum, zinc and magnesium. **TLV-TWA** 1000 ppm 2640 mg/m³ **Absorption route:** Can enter the body by inhalation or ingestion or through the skin. Harmful atmospheric concentrations can build up very rapidly if gas is released. **Immediate effects:** Irritates the eyes, skin and respiratory tract. Can cause frostbite due to rapid evaporation. Affects the nervous system. In serious cases risk of unconsciousness. **Effects of prolonged/repeated exposure:** Can cause liver and kidney damage.

HAZARDS/SYMPTOMS	PREVENTIVE MEASURES	FIRE EXTINGUISHING/FIRST AID
Fire: extremely flammable.	keep away from open flame and sparks, no smoking.	shut off supply; if impossible and no danger to surrounding area, allow fire to burn itself out; otherwise extinguish with powder, carbon dioxide, (halons).
Explosion: forms explosive air-gas mixtures.	sealed machinery, ventilation, explosion-proof electrical equipment and lighting, grounding when pumping etc. in liquid form.	in case of fire: keep intact cylinders cool by spraying with water, but DO NOT spray substance with water.
Inhalation: headache, dizziness, drowsiness.		fresh air, rest, place in half-sitting position, call a doctor.
Skin: *is absorbed; in case of frostbite:* redness, pain.	insulated gloves, protective clothing.	*in case of frostbite:* DO NOT remove clothing, flush skin with water or shower, take to a doctor.
Eyes: redness, pain, impaired vision, blisters.	face shield, or combined eye and respiratory protection.	flush with water, take to a doctor.
Ingestion: sore throat, abdominal cramps, headache; see also 'Inhalation'.		rinse mouth, take immediately to hospital.

SPILLAGE	STORAGE	LABELING / NFPA
evacuate area, call in an expert, ventilate, under no circumstances spray liquid with water; (additional individual protection: breathing apparatus).	keep in cool, fireproof place.	NFPA diamond: (top) 4, (left) 2, (right) 0

NOTES
Log P octanol/water is estimated. Alcohol consumption increases toxic effects. Depending on the degree of exposure, regular medical checkups are advisable. Under no circumstances use near flame or hot surface or when welding. Turn leaking cylinder so that leak is on top to prevent liquid ethyl chloride escaping. High atmospheric concentrations can cause oxygen deficiency, with risk of unconsciousness.

Transport Emergency Card TEC(R)-616

HI: 236; UN-number: 1037

CAS-No: [105-39-5]
chloroacetic acid, ethyl ester
ethyl α-chloroacetate
ethyl 2-chloroacetate
ethyl chloroethanoate
ethyl monochloroacetate

$CH_2ClCOOC_2H_5$

ETHYL CHLOROACETATE

PHYSICAL PROPERTIES		IMPORTANT CHARACTERISTICS	
Boiling point, °C	145	**COLORLESS LIQUID WITH PUNGENT ODOR**	
Melting point, °C	− 26	Vapor mixes readily with air. Corrosive fumes are a product of combustion (→ *hydrochloric acid*). Decomposes in hot water, forming → *acetic acid*. Reacts violently with alkaline-earth and alkali metals, with risk of fire and explosion.	
Flash point, °C	54		
Relative density (water = 1)	1.15		
Relative vapor density (air = 1)	4.2		
Relative density at 20 °C of saturated mixture vapor/air (air = 1)	1.01	TLV-TWA	not available
Vapor pressure, mm Hg at 20 °C	3.7	**Absorption route:** Can enter the body by inhalation or ingestion or through the skin. Harmful atmospheric concentrations can build up fairly rapidly on evaporation at approx. 20°C - even more rapidly in aerosol form.	
Solubility in water	none		
Explosive limits, vol% in air	?-?		
Relative molecular mass	122.6	**Immediate effects:** Corrosive to the eyes, skin and respiratory tract. Affects the nervous system. Inhalation can cause lung edema. In serious cases risk of death.	
Gross formula:	$C_4H_7ClO_2$		

HAZARDS/SYMPTOMS	PREVENTIVE MEASURES	FIRE EXTINGUISHING/FIRST AID
Fire: flammable.	keep away from open flame and sparks, no smoking.	powder, water spray, foam, carbon dioxide, (halons).
Explosion: above 54°C: forms explosive air-vapor mixtures.	above 54°C: sealed machinery, ventilation, explosion-proof electrical equipment.	in case of fire: keep tanks/drums cool by spraying with water.
Inhalation: *corrosive*; sore throat, severe breathing difficulties, dizziness.	ventilation, local exhaust or respiratory protection.	fresh air, rest, place in half-sitting position, take to hospital.
Skin: *corrosive*; redness, pain, burns.	gloves, protective clothing.	remove contaminated clothing, flush skin with water or shower, refer to a doctor.
Eyes: *corrosive*; redness, pain, impaired vision.	face shield.	flush with water, take to a doctor.
Ingestion: *corrosive*; abdominal pain, nausea.		rinse mouth, take to hospital.

SPILLAGE	STORAGE	LABELING / NFPA
collect leakage in sealable containers, soak up with sand or other inert absorbent and remove to safe place; (additional individual protection: breathing apparatus).	keep in fireproof place.	R: 23/24/25 S: 7/9-44 ☠️ Toxic

NOTES
Explosive limits not given in the literature. Lung edema symptoms usually develop several hours later and are aggravated by physical exertion: rest and hospitalization essential. As first aid, a doctor or authorized person should consider administering a corticosteroid spray. The measures on this card also apply to methyl chloroacetate: flash point: 47°C.

Transport Emergency Card TEC(R)-904

HI: 63; UN-number: 1181

ETHYL CHLOROFORMATE

PHYSICAL PROPERTIES	IMPORTANT CHARACTERISTICS
Boiling point, °C 95 Melting point, °C −81 Flash point, °C 16 Auto-ignition temperature, °C 500 Relative density (water = 1) 1.14 Relative vapor density (air = 1) 3.7 Relative density at 20 °C of saturated mixture vapor/air (air = 1) 1.15 Vapor pressure, mm Hg at 20 °C 42 Solubility in water none Relative molecular mass 108.5	**COLORLESS LIQUID WITH PUNGENT ODOR** Vapor is heavier than air and spreads at ground level, with risk of ignition at a distance. Do not use compressed air when filling, emptying or processing. Corrosive vapors are a product of combustion (→ *hydrogen chloride*). Decomposes slowly on contact with water, forming → *hydrochloric acid*, → *carbon dioxide* and → *ethanol*. Reacts violently with strong oxidants. Reacts with air to form corrosive vapors (→ *hydrochloric acid*). Attacks many metals, esp. in the presence of moisture.
	TLV-TWA not available
	Absorption route: Can enter the body by inhalation or ingestion. Harmful atmospheric concentrations can build up very rapidly on evaporation at 20°C. **Immediate effects:** Corrosive to the eyes, skin and respiratory tract. Inhalation of vapor/fumes can cause severe breathing difficulties (lung edema). In serious cases risk of death.
Gross formula: C$_3$H$_5$ClO$_2$	

HAZARDS/SYMPTOMS	PREVENTIVE MEASURES	FIRE EXTINGUISHING/FIRST AID
Fire: extremely flammable.	keep away from open flame and sparks, no smoking.	powder, water spray, foam, carbon dioxide, (halons).
Explosion: forms explosive air-vapor mixtures.	sealed machinery, ventilation, explosion-proof electrical equipment and lighting.	in case of fire: keep tanks/drums cool by spraying with water, but DO NOT spray substance with water.
	STRICT HYGIENE	
Inhalation: *corrosive*; sore throat, cough, severe breathing difficulties, nausea.	ventilation, local exhaust or respiratory protection.	fresh air, rest, place in half-sitting position, take to hospital.
Skin: *corrosive*; redness, pain, burns.	gloves, protective clothing.	remove contaminated clothing, flush skin with water or shower, call a doctor if necessary.
Eyes: *corrosive*; redness, pain, impaired vision.	face shield, or combined eye and respiratory protection.	flush with water, take to a doctor.
Ingestion: *corrosive*; sore throat, abdominal pain, diarrhea, vomiting.		rinse mouth, take immediately to hospital.

SPILLAGE	STORAGE	LABELING / NFPA
collect leakage in sealable containers, soak up with chalk, sand or other inert absorbent and remove to safe place or render harmless with bicarbonate solution, flush away remainder with water; (additional individual protection: breathing apparatus).	keep in dry, fireproof place, separate from oxidants.	R: 11-23-36/37/38 S: 9-16-38-44 Flammable Toxic 3 / 1

NOTES

Sometimes contaminated with small quantities of → *phosgene*. Lung edema symptoms usually develop several hours later and are aggravated by physical exertion: rest and hospitalization essential. As first aid, a doctor or authorized person should consider administering a corticosteroid spray. Use airtight packaging. Unbreakable packaging preferred; if breakable, keep in unbreakable container.

Transport Emergency Card TEC(R)-30G33

HI: 336; UN-number: 1182

ETHYLENE
(liquid, refrigerated)

PHYSICAL PROPERTIES	IMPORTANT CHARACTERISTICS
Boiling point, °C — 104 Melting point, °C — 169 Flash point, °C flammable gas Auto-ignition temperature, °C 425 Relative density (water = 1) 0.6 Relative vapor density (air = 1) 0.98 Vapor pressure, mm Hg at 10 °C 39,520 Solubility in water poor Explosive limits, vol% in air 2.7-34 Minimum ignition energy, mJ 0.07 Relative molecular mass 28.1 Gross formula: C_2H_4	**COLORLESS, EXTREMELY COLD LIQUID** Cold gas is heavier than air. Flow, agitation etc. can cause build-up of electrostatic charge due to liquid's low conductivity. Reacts violently with oxidants and halogen acids, with risk of fire and explosion. TLV-TWA not available **Absorption route:** Can enter the body by inhalation. Can saturate the air if released, with risk of suffocation. **Immediate effects:** Cold liquid can cause frostbite. In substantial concentrations can impair consciousness.

HAZARDS/SYMPTOMS	PREVENTIVE MEASURES	FIRE EXTINGUISHING/FIRST AID
Fire: extremely flammable.	keep away from open flame and sparks, no smoking.	shut off supply; if impossible and no danger to surrounding area, allow fire to burn itself out; otherwise extinguish with powder, carbon dioxide, (halons).
Explosion: forms explosive air-gas mixtures.	sealed machinery, ventilation, explosion-proof electrical equipment and lighting, grounding when pumping etc. in liquid form, non-sparking tools.	
Inhalation: severe breathing difficulties, drowsiness.	ventilation, local exhaust or respiratory protection.	fresh air, rest, call a doctor.
Skin: *in case of frostbite:* redness, pain, blisters.	insulating gloves.	*in case of frostbite:* DO NOT remove clothing, flush skin with water or shower, refer to a doctor if necessary.
Eyes: *in case of frostbite:* redness, pain, impaired vision.	face shield.	flush with water, take to a doctor.

SPILLAGE	STORAGE	LABELING / NFPA
evacuate area, call in an expert, ventilate, under no circumstances spray liquid with water; (additional individual protection: breathing apparatus).	keep in fireproof place, under refrigeration.	R: 13 S: 9-16-33 Flammable

NOTES

High atmospheric concentrations, e.g. in poorly ventilated spaces, can cause oxygen deficiency, with risk of unconsciousness. Use special insulated pressure vessel.

CAS-No: [96-49-1]
carbonic acid, cyclic ethylene ester
dioxolone-2
cyclic ethylene carbonate
ethylene glycol carbonate
glycol carbonate

OCOCH$_2$CH$_2$O

ETHYLENE CARBONATE

PHYSICAL PROPERTIES		IMPORTANT CHARACTERISTICS	
Boiling point, °C	244	**COLORLESS LIQUID**	
Melting point, °C	36	Reacts with strong oxidants.	
Flash point, °C	143		
Relative density (water = 1)	1.3	TLV-TWA	not available
Relative vapor density (air = 1)	3.0		
Relative density at 20 °C of saturated mixture vapor/air (air = 1)	1.0	**Absorption route:** Can enter the body by inhalation or ingestion. Harmful atmospheric concentrations build up very slowly, if at all, on evaporation at approx. 20°C, but much more rapidly in aerosol form.	
Vapor pressure, mm Hg at 20 °C	<0.076	**Immediate effects:** Irritates the eyes, skin and respiratory tract.	
Solubility in water	good	**Effects of prolonged/repeated exposure:** Ingestion of large quantities can cause kidney damage.	
Relative molecular mass	88.0		
Gross formula:	C$_3$H$_4$O$_3$		

HAZARDS/SYMPTOMS	PREVENTIVE MEASURES	FIRE EXTINGUISHING/FIRST AID
Fire: combustible.	keep away from open flame, no smoking.	powder, water spray, foam, carbon dioxide, (halons).
Inhalation: sore throat, cough.	local exhaust or respiratory protection.	fresh air, rest.
Skin: redness, pain.	gloves.	remove contaminated clothing, flush skin with water or shower.
Eyes: redness, pain.	safety glasses.	flush with water, take to a doctor if necessary.
Ingestion: abdominal pain, nausea.		rinse mouth, call a doctor or take to hospital.

SPILLAGE	STORAGE	LABELING / NFPA
collect leakage in sealable containers or clean up spillage, flush away with water.		1 / 2 1

NOTES

Transport Emergency Card TEC(R)-590

CAS-No: [109-78-4]
hydracryonitrile
3-hydroxypropanenitrile
β-hydroxypropionitrile

$HOCH_2CH_2CN$

ETHYLENE CYANOHYDRIN

PHYSICAL PROPERTIES		IMPORTANT CHARACTERISTICS
Boiling point (decomposes), °C	229	**COLORLESS LIQUID WITH CHARACTERISTIC ODOR, TURNING YELLOW ON EXPOSURE TO LIGHT**
Melting point, °C	−46	Vapor mixes readily with air. Amines or inorganic bases can cause polymerization. Decomposes
Flash point, °C	>21	when heated or on contact with acids, giving off toxic and flammable vapors (→ *hydrogen*
Auto-ignition temperature, °C	?	*cyanide*). Strong reducing agent which reacts violently with oxidants. Reacts violently with
Relative density (water = 1)	1.05	sodium hydroxide. Reacts violently with hot water and steam, giving off flammable toxic vapors
Relative vapor density (air = 1)	2.5	(→ *hydrogen cyanide*). Attacks copper.
Relative density at 20 °C of saturated mixture		
vapor/air (air = 1)	1.0	TLV-TWA not available
Vapor pressure, mm Hg at 20 °C	0.076	
Solubility in water	∞	**Absorption route:** Can enter the body by inhalation or ingestion. Harmful atmospheric
Explosive limits, vol% in air	?-?	concentrations build up fairly slowly on evaporation at 20°C, but much more rapidly in aerosol
Relative molecular mass	71.1	form or due to chemical reaction.
		Immediate effects: Irritates the eyes, skin and respiratory tract.
Gross formula:	C_3H_5NO	

HAZARDS/SYMPTOMS	PREVENTIVE MEASURES	FIRE EXTINGUISHING/FIRST AID
Fire: flammable.	keep away from open flame and sparks, no smoking, avoid contact with oxidants, acids and strong bases.	powder, alcohol-resistant foam, large quantities of water, carbon dioxide, (halons).
Explosion: above 21°C: forms explosive air-vapor mixtures.	above 21°C: sealed machinery, ventilation, explosion-proof electrical equipment.	in case of fire: keep tanks/drums cool by spraying with water.
Inhalation: sore throat, cough.	ventilation, local exhaust or respiratory protection.	fresh air, rest, call a doctor.
Skin: redness, pain.	gloves.	remove contaminated clothing, flush skin with water or shower.
Eyes: redness, pain.	face shield.	flush with water, take to a doctor if necessary.
Ingestion: abdominal pain, nausea.		rinse mouth, call a doctor.

SPILLAGE	STORAGE	LABELING / NFPA
collect leakage in sealable containers, soak up with sand or other inert absorbent and remove to safe place.	keep in cool, dark, fireproof place, separate from oxidants, acids and bases.	(NFPA diamond: blue 2, red 1, yellow 2)

NOTES
The commercial product with 22% water has a flash point of 129°C. Explosive limits not given in the literature.

Transport Emergency Card TEC(R)-61G07

CAS-No: [107-15-3]
1,2-diaminoethane
dimethylenediamine
1,2-ethanediamine
ethylene diamine
1,2-ethylenediamine

$H_2NCH_2CH_2NH_2$

ETHYLENEDIAMINE

PHYSICAL PROPERTIES		IMPORTANT CHARACTERISTICS	
Boiling point, °C	117	**COLORLESS TO LIGHT YELLOW HYGROSCOPIC LIQUID WITH CHARACTERISTIC ODOR**	
Melting point, °C	11	Vapor mixes readily with air. Toxic and corrosive vapors are a product of combustion (→ *nitrous*	
Flash point, °C	34	*vapors* and → *hydrogen cyanide*). Medium strong base. Reacts with chlorinated hydrocarbons.	
Auto-ignition temperature, °C	385	Reacts with strong oxidants. Promotes spontaneous decomposition of nitrogen compounds.	
Relative density (water = 1)	0.9	Attacks copper, nickel and cobalt.	
Relative vapor density (air = 1)	2.1		
Relative density at 20 °C of saturated mixture			
vapor/air (air = 1)	1.01	**TLV-TWA**	10 ppm	25 mg/m³
Vapor pressure, mm Hg at 20 °C	10		
Solubility in water	∞	**Absorption route:** Can enter the body by inhalation or ingestion. Harmful atmospheric	
Explosive limits, vol% in air	2.7-16.6	concentrations can build up fairly rapidly on evaporation at approx. 20°C - even more rapidly in	
Electrical conductivity, pS/m	9x10⁸	aerosol form.	
Relative molecular mass	60.1	**Immediate effects:** Corrosive to the eyes, skin and respiratory tract. Inhalation of vapor/fumes	
		can cause severe breathing difficulties (lung edema). In serious cases risk of death.	
		Effects of prolonged/repeated exposure: Prolonged or repeated contact can cause skin	
		disorders.	
Gross formula:	$C_2H_8N_2$		

HAZARDS/SYMPTOMS	PREVENTIVE MEASURES	FIRE EXTINGUISHING/FIRST AID
Fire: flammable.	keep away from open flame and sparks, no smoking.	water spray, alcohol-resistant foam, powder, carbon dioxide, DO NOT USE HALONS.
Explosion: above 34°C: forms explosive air-vapor mixtures.	above 34°C: sealed machinery, ventilation, explosion-proof electrical equipment.	in case of fire: keep tanks/drums cool by spraying with water.
Inhalation: *corrosive*; sore throat, cough, shortness of breath, severe breathing difficulties.	ventilation, local exhaust or respiratory protection.	fresh air, rest, artificial respiration if necessary, take to hospital.
Skin: redness, pain, burns.	gloves, protective clothing.	remove contaminated clothing, flush skin with water or shower, call a doctor.
Eyes: *corrosive*; redness, pain, impaired vision.	face shield.	flush with water, take to a doctor.
Ingestion: *corrosive*; sore throat, abdominal pain, vomiting.		rinse mouth, immediately take to hospital.

SPILLAGE	STORAGE	LABELING / NFPA
collect leakage in sealable containers, flush away remainder with water; (additional individual protection: breathing apparatus).	keep in fireproof place, separate from oxidants, strong acids and nitrogen compounds.	R: 10-21/22-34-43 S: 9-26-36/37/39 Corrosive 3 / 3 / 0

NOTES
Lung edema symptoms usually develop several hours later and are aggravated by physical exertion: rest and hospitalization essential. As first aid, a doctor or authorized person should consider administering a corticosteroid spray. Unbreakable packaging preferred; if breakable, keep in unbreakable container.

Transport Emergency Card TEC(R)-77

HI: 83; UN-number: 1604

1,2-dibromoethane
sym-dibromoethane
EDB
ethylene bromide
1,2-ethylene dibromide

ETHYLENE DIBROMIDE

PHYSICAL PROPERTIES		IMPORTANT CHARACTERISTICS	
Boiling point, °C	132	**COLORLESS LIQUID WITH CHARACTERISTIC ODOR**	
Melting point, °C	10	Vapor mixes readily with air. Flow, agitation etc. can cause build-up of electrostatic charge due to liquid's low conductivity. Decomposes in flame or on hot surface, giving off toxic vapors ($\rightarrow$ *hydrogen bromide*). Decomposes slowly on exposure to light. Reacts with various metal powders, with risk of fire and explosion. Reacts with bases and strong oxidants. Attacks some plastics.	
Relative density (water = 1)	2.2		
Relative vapor density (air = 1)	6.5		
Relative density at 20 °C of saturated mixture			
vapor/air (air = 1)	1.08		
Vapor pressure, mm Hg at 20 °C	11		
Solubility in water, g/100 ml	0.4	TLV-TWA (skin)	A2
Electrical conductivity, pS/m	<2 x 10⁴		
Relative molecular mass	187.9		
		Absorption route: Can enter the body by inhalation or ingestion or through the skin. Harmful atmospheric concentrations can build up fairly rapidly on evaporation at approx. 20°C - even more rapidly in aerosol form. **Immediate effects:** Irritates the eyes, skin and respiratory tract. Affects the nervous system. **Effects of prolonged/repeated exposure:** Can damage the liver.	
Gross formula:	C$_2$H$_4$Br$_2$		

HAZARDS/SYMPTOMS	PREVENTIVE MEASURES	FIRE EXTINGUISHING/FIRST AID
Fire: non-combustible.		in case of fire in immediate vicinity: use any extinguishing agent.
	STRICT HYGIENE	IN ALL CASES CALL IN A DOCTOR
Inhalation: sore throat, cough, shortness of breath.	ventilation, local exhaust or respiratory protection.	fresh air, rest, place in half-sitting position, call a doctor.
Skin: redness, pain.	gloves, protective clothing.	remove contaminated clothing, flush skin with water or shower.
Eyes: redness, pain.	face shield, or combined eye and respiratory protection.	flush with water, take to a doctor.
Ingestion: sore throat, abdominal cramps.		rinse mouth, take immediately to hospital.

SPILLAGE	STORAGE	LABELING / NFPA
collect leakage in sealable containers, soak up with sand or other inert absorbent and remove to safe place; (additional individual protection: breathing apparatus).	keep in dark place, separate from oxidants and strong bases; ventilate at floor level.	R: 45-23/24/25-36/37/38 S: 53-44 Note E Toxic

NOTES
Odor limit is above TLV. Under no circumstances use near flame or hot surface or when welding.

CAS-No: [107-21-1]
1,2-dihydroxyethane
1,2-ethanediol
ethylene alcohol
glycol
monoethylene glycol

$HOCH_2CH_2OH$

ETHYLENE GLYCOL

PHYSICAL PROPERTIES		IMPORTANT CHARACTERISTICS	
Boiling point, °C	197	**ODORLESS LIGHT VISCOUS LIQUID**	
Melting point, °C	− 13	Vapor mixes readily with air. Reacts violently with strong oxidants and oxidizing acids.	
Flash point, °C	111		
Auto-ignition temperature, °C	410	TLV-TWA	50 ppm 127 mg/m³ C ✱
Relative density (water = 1)	1.1		
Relative vapor density (air = 1)	2.1	**Absorption route:** Can enter the body by inhalation or ingestion. Harmful atmospheric concentrations build up very slowly, if at all, on evaporation at approx. 20°C, but much more rapidly in aerosol form.	
Vapor pressure, mm Hg at 20 °C	0.091		
Solubility in water	∞		
Explosive limits, vol% in air	3.2-53	**Immediate effects:** Irritates the eyes and skin. Affects the nervous system. Can cause liver, kidney and brain damage. In serious cases risk of unconsciousness.	
Electrical conductivity, pS/m	1.16x10⁸		
Relative molecular mass	62.1		
Log P octanol/water	− 1.9		
Gross formula:	$C_2H_6O_2$		

HAZARDS/SYMPTOMS	PREVENTIVE MEASURES	FIRE EXTINGUISHING/FIRST AID
Fire: combustible.	keep away from open flame, no smoking.	powder, alcohol-resistant foam, water spray, carbon dioxide, (halons).
Inhalation: headache, dizziness, sleepiness.	ventilation.	fresh air, rest.
Skin: redness.	gloves.	remove contaminated clothing, flush skin with water or shower.
Eyes: redness.	acid goggles.	flush with water, take to a doctor if necessary.
Ingestion: diarrhea, abdominal cramps, headache, dizziness, vomiting, sleepiness.		rinse mouth, immediately take to hospital.

SPILLAGE	STORAGE	LABELING / NFPA
collect leakage in sealable containers, flush away remainder with water.	keep dry, separate from oxidants and oxidizing acids; ventilate at floor level.	R: 22 S: 2 ✖ Harmful

NOTES
✱ TLV equals that of ethylene glycol (vapor and mist). TLV is maximum and must not be exceeded. Depending on the degree of exposure, regular medical checkups are advisable.

CAS-No: [112-48-1]
1,2-dibutoxyethane
1,1'-(1,2-ethanediylbis(oxy))bisbutane

$C_4H_9OC_2H_4OC_4H_9$

ETHYLENE GLYCOL DIBUTYL ETHER

PHYSICAL PROPERTIES		IMPORTANT CHARACTERISTICS
Boiling point, °C	203	**COLORLESS LIQUID WITH CHARACTERISTIC ODOR**
Melting point, °C	−69	Vapor mixes readily with air. Reacts with strong oxidants.
Flash point, °C	85	
Relative density (water = 1)	0.8	TLV-TWA not available
Relative vapor density (air = 1)	6.0	
Relative density at 20 °C of saturated mixture vapor/air (air = 1)	1.0	**Absorption route:** Can enter the body by inhalation or ingestion or through the skin. Evaporation negligible at 20°C, but harmful atmospheric concentrations can build up rapidly in aerosol form.
Vapor pressure, mm Hg at 20 °C	0.091	**Immediate effects:** Irritates the eyes, skin and respiratory tract. Affects the nervous system.
Solubility in water	poor	**Effects of prolonged/repeated exposure:** Can affect the blood. Can cause liver and kidney damage.
Relative molecular mass	174.0	
Gross formula:	$C_{10}H_{22}O_2$	

HAZARDS/SYMPTOMS	PREVENTIVE MEASURES	FIRE EXTINGUISHING/FIRST AID
Fire: combustible.	keep away from open flame, no smoking.	powder, AFFF, foam, carbon dioxide, (halons).
Explosion: above 85°C: forms explosive air-vapor mixtures.	above 85°C: sealed machinery, ventilation.	
Inhalation: sore throat, cough, shortness of breath.	ventilation.	fresh air, rest, place in half-sitting position, call a doctor.
Skin: *is absorbed*; redness.	gloves.	remove contaminated clothing, flush skin with water or shower, call a doctor.
Eyes: redness, pain.	face shield.	flush with water, take to a doctor.
Ingestion: abdominal pain, vomiting, drowsiness, sleepiness, unconsciousness.		rinse mouth, take immediately to hospital.

SPILLAGE	STORAGE	LABELING / NFPA
collect leakage in sealable containers, soak up with sand or other inert absorbent and remove to safe place.	keep separate from oxidants.	2 / 1 / 0

NOTES
Depending on the degree of exposure, regular medical checkups are advisable. Dibutyl Cellosolve is a trade name.

Transport Emergency Card TEC(R)-30G115

ETHYLENE GLYCOL DIMETHYL ETHER

PHYSICAL PROPERTIES	IMPORTANT CHARACTERISTICS
Boiling point, °C 83 Melting point, °C −58 Flash point, °C <21 Auto-ignition temperature, °C 230 Relative density (water = 1) 0.9 Relative vapor density (air = 1) 3.1 Relative density at 20 °C of saturated mixture vapor/air (air = 1) 1.13 Vapor pressure, mm Hg at 20 °C 4.9 Solubility in water good Explosive limits, vol% in air 1.6-10.4 Relative molecular mass 90.1 Gross formula: $C_4H_{10}O_2$	**COLORLESS LIQUID WITH CHARACTERISTIC ODOR** Vapor is heavier than air and spreads at ground level, with risk of ignition at a distance. Able to form peroxides. Do not use compressed air when filling, emptying or processing. Reacts violently with strong oxidants. TLV-TWA not available **Absorption route:** Can enter the body by inhalation or ingestion or through the skin. Harmful atmospheric concentrations can build up fairly rapidly on evaporation at approx. 20°C - even more rapidly in aerosol form. **Immediate effects:** Irritates the eyes and respiratory tract. Affects the nervous system. **Effects of prolonged/repeated exposure:** Can affect the blood. Can cause liver and kidney damage.

HAZARDS/SYMPTOMS	PREVENTIVE MEASURES	FIRE EXTINGUISHING/FIRST AID
Fire: extremely flammable.	keep away from open flame and sparks, no smoking.	powder, AFFF, foam, carbon dioxide, (halons).
Explosion: forms explosive air-vapor mixtures.	sealed machinery, ventilation, explosion-proof electrical equipment and lighting.	in case of fire: keep tanks/drums cool by spraying with water.
Inhalation: sore throat, cough, shortness of breath.	ventilation, local exhaust or respiratory protection.	fresh air, rest, call a doctor.
Skin: *is absorbed*; redness.	gloves.	remove contaminated clothing, flush skin with water or shower, call a doctor.
Eyes: redness, pain.	safety glasses.	flush with water, take to a doctor.
Ingestion: abdominal pain, vomiting, drowsiness, sleepiness, unconsciousness.		rinse mouth, take immediately to hospital.

SPILLAGE	STORAGE	LABELING / NFPA
collect leakage in sealable containers, soak up with sand or other inert absorbent and remove to safe place; (additional individual protection: breathing apparatus).	keep in fireproof place, separate from oxidants.	R: 10-20-19 S: 24/25 ❌ Harmful 2 / 2 / 0

NOTES
Before distilling check for peroxides; if found, render harmless. Depending on the degree of exposure, regular medical checkups are advisable. Special first aid required in the event of poisoning: antidotes must be available (with instructions).

Transport Emergency Card TEC(R)-30G01

CAS-No: [112-07-2]
acetic acid, 2-butoxyethyl ester
2-butoxyethanol acetate
2-butoxyethyl-acetate

$C_4H_9OCH_2CH_2OOCCH_3$

ETHYLENE GLYCOL MONOBUTYL ETHER ACETATE

PHYSICAL PROPERTIES	IMPORTANT CHARACTERISTICS
Boiling point, °C 192 Melting point, °C −63 Flash point, °C 71 Auto-ignition temperature, °C 340 Relative density (water = 1) 0.94 Relative vapor density (air = 1) 5.5 Relative density at 20 °C of saturated mixture vapor/air (air = 1) 1.0 Vapor pressure, mm Hg at 20 °C 0.30 Solubility in water, g/100 ml 1.7 Explosive limits, vol% in air 0.9-8.5 Relative molecular mass 160.2 Gross formula: $C_8H_{16}O_3$	**COLORLESS LIQUID WITH CHARACTERISTIC ODOR** Vapor mixes readily with air. Able to form peroxides. Reacts with strong oxidants.
	TLV-TWA not available
	Absorption route: Can enter the body by inhalation or ingestion or through the skin. Harmful atmospheric concentrations build up very slowly, if at all, on evaporation at approx. 20°C, but much more rapidly in aerosol form. **Immediate effects:** Irritates the eyes and respiratory tract. Liquid destroys the skin's natural oils. Exposure to high concentrations can cause kidney damage. **Effects of prolonged/repeated exposure:** Can affect the kidneys.

HAZARDS/SYMPTOMS	PREVENTIVE MEASURES	FIRE EXTINGUISHING/FIRST AID
Fire: combustible.	keep away from open flame, no smoking.	powder, AFFF, foam, carbon dioxide, (halons).
Explosion: above 71°C: forms explosive air-vapor mixtures.	above 71°C: sealed machinery, ventilation.	
Inhalation: sore throat, cough, headache, drowsiness.	ventilation, local exhaust or respiratory protection.	fresh air, rest.
Skin: *is absorbed*; redness.	gloves.	remove contaminated clothing, wash skin with soap and water.
Eyes: redness, pain.	face shield.	flush with water, take to a doctor if necessary.
Ingestion: sore throat, abdominal pain, headache, vomiting, drowsiness.		rinse mouth, call a doctor or take to hospital.

SPILLAGE	STORAGE	LABELING / NFPA
collect leakage in sealable containers, soak up with sand or other inert absorbent and remove to safe place.	keep separate from oxidants.	R: 20/21 S: 24 ✖ Harmful

NOTES
Alcohol consumption increases toxic effects. Depending on the degree of exposure, regular medical checkups are advisable. Butyl Cellosolve actetate is a trade name. Before distilling check for peroxides; if found, render harmless.

Transport Emergency Card TEC(R)-61G06C

409

ETHYLENE GLYCOL MONOPHENYL ETHER

PHYSICAL PROPERTIES	IMPORTANT CHARACTERISTICS	
Boiling point, °C 242 Melting point, °C 14 Flash point, °C 121 Auto-ignition temperature, °C ? Relative density (water = 1) 1.1 Relative vapor density (air = 1) 4.77 Relative density at 20 °C of saturated mixture vapor/air (air = 1) 1.0 Vapor pressure, mm Hg at 20 °C < 0.076 Solubility in water, g/100 ml at 20 °C 2.6 Relative molecular mass 138.2 Log P octanol/water 1.2	**COLORLESS LIQUID WITH CHARACTERISTIC ODOR** Vapor mixes readily with air. Reacts with strong oxidants.	
	TLV-TWA not available	
	Absorption route: Can enter the body by inhalation or ingestion. Harmful atmospheric concentrations build up very slowly, if at all, on evaporation at approx. 20°C, but much more rapidly in aerosol form. **Immediate effects:** Irritates the eyes, skin and respiratory tract.	
Gross formula: $C_8H_{10}O_2$		

HAZARDS/SYMPTOMS	PREVENTIVE MEASURES	FIRE EXTINGUISHING/FIRST AID
Fire: combustible.	keep away from open flame, no smoking.	powder, alcohol-resistant foam, water spray, carbon dioxide, (halons).
Inhalation: sore throat, cough, headache.	ventilation.	fresh air, rest.
Skin: redness.	gloves.	remove contaminated clothing, flush skin with water or shower.
Eyes: redness, pain.	face shield.	flush with water, take to a doctor if necessary.
Ingestion: sore throat, abdominal pain, nausea.		rinse mouth, call a doctor.

SPILLAGE	STORAGE	LABELING / NFPA
collect leakage in sealable containers, soak up with sand or other inert absorbent and remove to safe place.	keep separate from oxidants.	

NOTES

ETHYLENEIMINE

PHYSICAL PROPERTIES		IMPORTANT CHARACTERISTICS
Boiling point, °C	55	**COLORLESS LIQUID WITH CHARACTERISTIC ODOR**
Melting point, °C	−71	Vapor is heavier than air and spreads at ground level, with risk of ignition at a distance. Can
Flash point, °C	−13	polymerize spontaneously when heated (even moderately) on contact with acids or atmospheric
Auto-ignition temperature, °C	320	carbon dioxide, with risk of fire and explosion. Do not use compressed air when filling, emptying
Relative density (water = 1)	0.8	or processing. Reacts with silver and silver alloys to form shock-sensitive compounds.
Relative vapor density (air = 1)	1.5	Decomposes when heated or burned, giving off toxic vapors (→ *nitrous vapors*). In aqueous
Relative density at 20 °C of saturated mixture		solution is a strong base which reacts violently with acids and corrodes aluminum, zinc etc.
vapor/air (air = 1)	1.11	Attacks rubber and plastics.
Vapor pressure, mm Hg at 20 °C	163	
Solubility in water	∞	
Minimum ignition energy, mJ	0.48	
Electrical conductivity, pS/m	8 x 10⁸	
Relative molecular mass	43.1	

		TLV-TWA (skin)	0.5 ppm	0.88 mg/m³

Absorption route: Can enter the body by inhalation or ingestion. Harmful atmospheric concentrations can build up very rapidly on evaporation at 20°C.
Immediate effects: Corrosive to the eyes, skin and respiratory tract. Inhalation of vapor/fumes can cause severe breathing difficulties (lung edema).
Effects of prolonged/repeated exposure: Prolonged or repeated contact can cause skin disorders. Can cause kidney damage. Has been found to cause a type of cancer in rats under certain circumstances.

Gross formula:	C_2H_5N

HAZARDS/SYMPTOMS	PREVENTIVE MEASURES	FIRE EXTINGUISHING/FIRST AID
Fire: extremely flammable.	keep away from open flame and sparks, no smoking; avoid contact with acids and carbon dioxide.	alcohol-resistant foam, powder, (halons), large quantities of water, water spray, DO NOT use carbon dioxide.
Explosion: forms explosive air-vapor mixtures; contact with acids and carbon dioxide can cause explosion.	sealed machinery, ventilation, explosion-proof electrical equipment and lighting, non-sparking tools.	in case of fire: keep tanks/drums cool by spraying with water, fight fire from sheltered location.
	STRICT HYGIENE	
Inhalation: *corrosive*; sore throat, cough, shortness of breath, severe breathing difficulties.	ventilation, local exhaust or respiratory protection.	fresh air, rest, place in half-sitting position, take to hospital.
Skin: *corrosive*; redness, pain, burns.	gloves, protective clothing.	remove contaminated clothing, flush skin with water or shower, call a doctor.
Eyes: *corrosive*; redness, pain, impaired vision.	face shield, or combined eye and respiratory protection.	flush with water, take to a doctor.
Ingestion: *corrosive*; sore throat, diarrhea, abdominal cramps, vomiting.		rinse mouth, call a doctor or take to hospital.

SPILLAGE	STORAGE	LABELING / NFPA
evacuate area, call in an expert, collect leakage in sealable containers, soak up with sand or other inert absorbent and remove to safe place, do not use sawdust or other combustible absorbents; (additional individual protection: breathing apparatus).	keep in fireproof place, separate form acids and combustible substances; add an inhibitor.	R: 11-26/27/28-40 S: 9-29-36-45 Note D Flammable Toxic 3 / 3 3

NOTES

Vapor contains no inhibitor; condensate can therefore polymerize in exhaust or ventilation facilities, with risk of breakdown. Flush contaminated clothing with water (fire hazard). Depending on the degree of exposure, regular medical checkups are advisable. Lung edema symptoms usually develop several hours later and are aggravated by physical exertion: rest and hospitalization essential. As first aid, a doctor or authorized person should consider administering a corticosteroid spray. Use airtight packaging. Unbreakable packaging preferred; if breakable, keep in unbreakable container.

Transport Emergency Card TEC(R)-30G32

HI: 336; UN-number: 1185

CAS-No: [75-21-8]
dihydrooxirene
1,2-epoxyethane
oxacyclopropane
oxane
oxirane

ETHYLENE OXIDE
(cylinder)

PHYSICAL PROPERTIES	IMPORTANT CHARACTERISTICS
Boiling point, °C 11 Melting point, °C −112 Flash point, °C flammable gas Auto-ignition temperature, °C 429 Relative density (water = 1) 0.9 Relative vapor density (air = 1) 1.5 Vapor pressure, mm Hg at 20 °C 1140 Solubility in water good Explosive limits, vol% in air 2.6-100 Minimum ignition energy, mJ 0.06 Relative molecular mass 44.1	**COLORLESS COMPRESSED LIQUEFIED GAS WITH CHARACTERISTIC ODOR** Gas is heavier than air and spreads at ground level, with risk of ignition at a distance. Readily able to polymerize when heated or on contact with metal salts, with risk of fire and explosion. Flow, agitation etc. can cause build-up of electrostatic charge due to liquid's low conductivity. Can deflagrate (burn explosively) when heated. Decomposes on contact with copper, aluminum, mercury, magnesium and alloys, with risk of fire and explosion. Reacts violently with amines, alcohols, oxidants and many other compounds, with risk of fire and explosion.
	TLV-TWA 1 ppm 1.8 mg/m³ A2
	Absorption route: Can enter the body by inhalation. Harmful atmospheric concentrations can build up very rapidly if gas is released. **Immediate effects:** Irritates the eyes, skin and respiratory tract. Liquid can cause frostbite due to rapid evaporation. Affects the nervous system.
Gross formula: C_2H_4O	

HAZARDS/SYMPTOMS	PREVENTIVE MEASURES	FIRE EXTINGUISHING/FIRST AID
Fire: extremely flammable, many chemical reactions can cause fire and explosion.	keep away from open flame and sparks, no smoking.	shut off supply; if impossible and no danger to surrounding area, allow fire to burn itself out; otherwise extinguish with powder, carbon dioxide, (halons).
Explosion: forms explosive air-gas mixtures; heating can cause explosion, even in absence of air.	sealed machinery, ventilation, explosion-proof electrical equipment and lighting, grounding when pumping etc. in liquid form, non-sparking tools.	in case of fire: keep cylinders cool by spraying with water, fight fire from sheltered location.
Inhalation: cough, headache, dizziness, nausea, vomiting, diarrhea.	ventilation, local exhaust or respiratory protection.	fresh air, rest, call a doctor.
Skin: *in case of frostbite*: redness, pain, burns.	insulating gloves.	*in case of frostbite*: DO NOT remove clothing, flush skin with water or shower, refer to a doctor.
Eyes: *in case of frostbite*: redness, pain, impaired vision.	acid goggles, or combined eye and respiratory protection.	flush with water, take to a doctor if necessary.
Ingestion: sore throat, abdominal pain, chest pain; see also 'Inhalation'.		rinse mouth, call a doctor or take to hospital.

SPILLAGE	STORAGE	LABELING / NFPA
evacuate area, call in an expert, ventilate, under no circumstances spray liquid with water, combat gas cloud with water curtain; (additional individual protection: breathing apparatus).	keep in cool, fireproof place under inert gas, separate from other chemicals.	4 / 2 3

NOTES
Can react with water to form a hydrate with a melting point of 125° C. In confined spaces combustion can cause explosion. Odor limit is above TLV. Alcohol consumption increases toxic effects. Turn leaking cylinder so that leak is on top to prevent liquid ethylene oxide escaping. Use cylinder with special fittings.

Transport Emergency Card TEC(R)-16 (20G15)

HI: 236; UN-number: 1040

ETHYL FORMATE

PHYSICAL PROPERTIES		IMPORTANT CHARACTERISTICS	
Boiling point, °C	54	**COLORLESS LIQUID WITH CHARACTERISTIC ODOR**	
Melting point, °C	−80	Vapor is heavier than air and spreads at ground level, with risk of ignition at a distance. Do not use compressed air when filling, emptying or processing. Decomposes slowly in water, forming → *formic acid* and → *ethanol*. Reacts violently with oxidants.	
Flash point, °C	−20		
Auto-ignition temperature, °C	440		
Relative density (water = 1)	0.9		
Relative vapor density (air = 1)	2.6		
Relative density at 20 °C of saturated mixture		TLV-TWA	100 ppm	303 mg/m³
vapor/air (air = 1)	1.4		
Vapor pressure, mm Hg at 20 °C	195	**Absorption route:** Can enter the body by inhalation or ingestion or through the skin. Harmful atmospheric concentrations can build up fairly rapidly on evaporation at approx. 20°C - even more rapidly in aerosol form.	
Solubility in water, g/100 ml at 20 °C	12		
Explosive limits, vol% in air	2.7-16.5		
Electrical conductivity, pS/m	1.45x10⁵	**Immediate effects:** Irritates the eyes, skin and respiratory tract. In substantial concentrations can impair consciousness.	
Relative molecular mass	74.1		
Gross formula:	C$_3$H$_6$O$_2$		

HAZARDS/SYMPTOMS	PREVENTIVE MEASURES	FIRE EXTINGUISHING/FIRST AID
Fire: extremely flammable.	keep away from open flame and sparks, no smoking.	powder, AFFF, foam, carbon dioxide, (halons).
Explosion: forms explosive air-vapor mixtures.	sealed machinery, ventilation, explosion-proof electrical equipment and lighting.	in case of fire: keep tanks/drums cool by spraying with water.
Inhalation: sore throat, cough, shortness of breath, headache.	ventilation, local exhaust or respiratory protection.	fresh air, rest, call a doctor.
Skin: *is absorbed*; redness.	gloves.	remove contaminated clothing, flush skin with water or shower.
Eyes: redness, pain.	face shield.	flush with water, take to a doctor.
Ingestion: sore throat, abdominal pain, nausea.		rinse mouth, call a doctor.

SPILLAGE	STORAGE	LABELING / NFPA
collect leakage in sealable containers, soak up with sand or other inert absorbent and remove to safe place; (additional individual protection: breathing apparatus).	keep in dry, fireproof place, separate from oxidants.	R: 11 S: 9-16-33 Flammable 2 / 3 / 0

NOTES

Transport Emergency Card TEC(R)-537

HI: 33; UN-number: 1190

413

$C_4H_9CH(C_2H_5)CHO$

2-ETHYLHEXANAL

PHYSICAL PROPERTIES	IMPORTANT CHARACTERISTICS	
Boiling point, °C 163 Melting point, °C −76 Flash point, °C 52 Relative density (water = 1) 0.9 Relative vapor density (air = 1) 4.5 Relative density at 20 °C of saturated mixture vapor/air (air = 1) 1.0 Vapor pressure, mm Hg at 20 °C 0.99 Solubility in water poor Explosive limits, vol% in air see notes Relative molecular mass 128.2 Gross formula: $C_8H_{16}O$	**YELLOW LIQUID WITH PUNGENT ODOR** Vapor mixes readily with air. Able to form peroxides on exposure to air. Reacts violently with oxidants. Can ignite spontaneously on contact with porous combustible substances.	
	TLV-TWA not available	
	Absorption route: Can enter the body by inhalation or ingestion. Insufficient data on the rate at which harmful concentrations can build up. **Immediate effects:** Irritates the eyes, skin and respiratory tract.	

HAZARDS/SYMPTOMS	PREVENTIVE MEASURES	FIRE EXTINGUISHING/FIRST AID
Fire: flammable.	keep away from open flame and sparks, no smoking.	powder, AFFF, foam, carbon dioxide, (halons).
Explosion: above 52°C: forms explosive air-vapor mixtures.	above 52°C: sealed machinery, ventilation, explosion-proof electrical equipment.	in case of fire: keep tanks/drums cool by spraying with water.
Inhalation: sore throat, cough, headache.	ventilation, local exhaust or respiratory protection.	fresh air, rest, call a doctor.
Skin: redness, pain.	gloves.	remove contaminated clothing, flush skin with water or shower, refer to a doctor.
Eyes: redness, pain.	face shield.	flush with water, take to a doctor.
Ingestion: nausea, vomiting.		rinse mouth, call a doctor.

SPILLAGE	STORAGE	LABELING / NFPA
collect leakage in sealable containers, soak up remainder with sand or other inert absorbent and remove to safe place.	keep in fireproof place, separate from oxidants.	2 2 X 1

NOTES
Explosive limits at 90°C and 135°C respectively: 0.8 and 7.2% by vol. Flush contaminated clothing with water (fire hazard). Under certain circumstances can ignite spontaneously on exposure to air.

CAS-No: [149-57-5]
2-butylbutanoic acid
butylethylacetic acid
2-ethylcaproic acid
3-heptanecarboxylic acid

$C_4H_9CH(C_2H_5)COOH$

2-ETHYLHEXOIC ACID

PHYSICAL PROPERTIES		IMPORTANT CHARACTERISTICS
Boiling point, °C	228	**COLORLESS LIQUID WITH CHARACTERISTIC ODOR**
Melting point, °C	−83	Vapor mixes readily with air. Flow, agitation etc. can cause build-up of electrostatic charge due to liquid's low conductivity. Reacts with strong oxidants.
Flash point, °C	118	
Auto-ignition temperature, °C	371	
Relative density (water = 1)	0.9	TLV-TWA not available
Relative density at 20 °C of saturated mixture vapor/air (air = 1)	1.0	**Absorption route:** Can enter the body by inhalation or ingestion or through the skin. Harmful atmospheric concentrations build up very slowly, if at all, on evaporation at approx. 20°C, but much more rapidly in aerosol form.
Vapor pressure, mm Hg at 20 °C	0.030	
Solubility in water	poor	
Explosive limits, vol% in air	0.8-6.0	**Immediate effects:** Irritates the eyes, skin and respiratory tract.
Relative molecular mass	144.2	
Gross formula:	$C_8H_{16}O_2$	

HAZARDS/SYMPTOMS	PREVENTIVE MEASURES	FIRE EXTINGUISHING/FIRST AID
Fire: combustible.	keep away from open flame, no smoking.	powder, AFFF, foam, carbon dioxide, (halons).
Inhalation: sore throat, cough.	ventilation.	fresh air, rest.
Skin: *is absorbed*; redness.	gloves.	remove contaminated clothing, flush skin with water or shower.
Eyes: redness, pain.	face shield.	flush with water, take to a doctor if necessary.
Ingestion: sore throat, abdominal pain.		rinse mouth, call a doctor if necessary.

SPILLAGE	STORAGE	LABELING / NFPA
collect leakage in sealable containers, soak up with sand or other inert absorbent and remove to safe place.	keep separate from oxidants.	1 1 0

NOTES

415

$$CH_2 = CHCOOC_8H_{17}$$

2-ETHYLHEXYL ACRYLATE

PHYSICAL PROPERTIES	IMPORTANT CHARACTERISTICS	
Boiling point, °C 229 Melting point, °C −90 Flash point, °C 82 Auto-ignition temperature, °C 245 Relative density (water = 1) 0.9 Relative vapor density (air = 1) 6.3 Relative density at 20 °C of saturated mixture vapor/air (air = 1) 1.0 Vapor pressure, mm Hg at 20 °C 0.099 Solubility in water none Explosive limits, vol% in air 0.8-? Relative molecular mass 184.3	**COLORLESS LIQUID WITH CHARACTERISTIC ODOR** Vapor mixes readily with air. Exposure to light, moderate heating, contact with peroxides or reducing agents, or contaminants readily cause polymerization. Reacts violently with strong oxidants.	
	TLV-TWA not available	
Gross formula: $C_{11}H_{20}O_2$	**Absorption route:** Can enter the body by inhalation or ingestion or through the skin. Harmful atmospheric concentrations can build up fairly rapidly on evaporation at approx. 20°C - even more rapidly in aerosol form. **Immediate effects:** Irritates the skin. Corrosive to the eyes and respiratory tract. Inhalation can cause lung edema.	

HAZARDS/SYMPTOMS	PREVENTIVE MEASURES	FIRE EXTINGUISHING/FIRST AID
Fire: combustible.	keep away from open flame, no smoking.	powder, AFFF, foam, carbon dioxide, (halons).
Explosion: above 82°C: forms explosive air-vapor mixtures.	above 82°C: sealed machinery, ventilation.	in case of fire: keep tanks/drums cool by spraying with water.
Inhalation: *corrosive*; sore throat, cough, shortness of breath.	ventilation, local exhaust or respiratory protection.	fresh air, rest, place in half-sitting position, take to a doctor.
Skin: redness, pain.	gloves.	remove contaminated clothing, flush skin with water or shower, refer to a doctor.
Eyes: *corrosive*; redness, pain, impaired vision.	face shield, or combined eye and respiratory protection.	flush with water, take to a doctor.
Ingestion: abdominal pain, vomiting.		rinse mouth, immediately take to hospital.

SPILLAGE	STORAGE	LABELING / NFPA
collect leakage in sealable containers, soak up with sand or other inert absorbent and remove to safe place; (additional individual protection: breathing apparatus).	keep cool; add an inhibitor.	R: 37/38-43 Note D ✖ Irritant

NOTES

Inhibitor added to prevent spontaneous polymerization is highly irritating to the eyes, skin and respiratory tract. Depending on the degree of exposure, regular medical checkups are advisable. Lung edema symptoms usually develop several hours later and are aggravated by physical exertion: rest and hospitalization essential. Use aluminum or stainless steel containers. Do not keep under inert gas.

Transport Emergency Card TEC(R)-30G15

$C_4H_9CH(C_2H_5)CH_2NH_2$

2-ETHYLHEXYLAMINE-1

PHYSICAL PROPERTIES	IMPORTANT CHARACTERISTICS
Boiling point, °C — 169 Melting point, °C — < −76 Flash point, °C — 50 Auto-ignition temperature, °C — 265 Relative density (water = 1) — 0.8 Relative density at 20 °C of saturated mixture vapor/air (air = 1) — 1.0 Vapor pressure, mm Hg at 20 °C — 0.21 Solubility in water, g/100 ml — 0.25 Explosive limits, vol% in air — ?-? Relative molecular mass — 129.2	**COLORLESS LIQUID WITH PUNGENT ODOR** Vapor mixes readily with air. Decomposes when heated or burned, giving off toxic and corrosive vapors. Strong base which reacts violently with acids and corrodes aluminum, zinc etc. Reacts violently with strong oxidants and organic oxygen compounds. Attacks copper and its alloys. TLV-TWA not available **Absorption route:** Can enter the body by inhalation or ingestion or through the skin. **Immediate effects:** Corrosive to the eyes, skin and respiratory tract. Inhalation of vapor/fumes can cause severe breathing difficulties (lung edema). **Effects of prolonged/repeated exposure:** Prolonged or repeated contact can cause skin disorders. Can cause liver and kidney damage.
Gross formula: $C_8H_{19}N$	

HAZARDS/SYMPTOMS	PREVENTIVE MEASURES	FIRE EXTINGUISHING/FIRST AID
Fire: flammable.	keep away from open flame and sparks, no smoking.	powder, AFFF, foam, carbon dioxide, (halons).
Explosion: above 50°C: forms explosive air-vapor mixtures.	above 50°C: sealed machinery, ventilation, explosion-proof electrical equipment.	in case of fire: keep tanks/drums cool by spraying with water.
Inhalation: *corrosive*; sore throat, cough, shortness of breath, severe breathing difficulties.	ventilation, local exhaust or respiratory protection.	fresh air, rest, place in half-sitting position, take to hospital.
Skin: *corrosive, is absorbed*; redness, pain, serious burns.	gloves, protective clothing.	remove contaminated clothing, flush skin with water or shower, refer to a doctor.
Eyes: *corrosive*; redness, pain, impaired vision.	face shield, combined with respiratory protection.	flush with water, take to a doctor.
Ingestion: *corrosive*; sore throat, abdominal pain, nausea.	do not eat, drink or smoke while working.	rinse mouth, take immediately to hospital.

SPILLAGE	STORAGE	LABELING / NFPA
collect leakage in sealable containers, soak up with sand or other inert absorbent and remove to safe place.	keep in fireproof place, separate from oxidants and strong acids.	NFPA diamond: Health 2, Flammability 2, Reactivity 0

NOTES
Explosive limits not given in the literature. Lung edema symptoms usually develop several hours later and are aggravated by physical exertion: rest and hospitalization essential. As first aid, a doctor or authorized person should consider administering a corticosteroid spray.

Transport Emergency Card TEC(R)-80G15

HI: 83; UN-number: 2276

$$CH_3-(CH_2)_3-CH(C_2H_5)-CH_2-O-CH_2-\overset{\displaystyle O}{CH}-CH_2$$

ETHYLHEXYL GLYCIDYL ETHER

PHYSICAL PROPERTIES	IMPORTANT CHARACTERISTICS	
Boiling point (decomposes), °C 150 Melting point, °C ? Flash point, °C 95 Relative density (water = 1) 0.9 Relative vapor density (air = 1) 6.4 Relative density at 20 °C of saturated mixture vapor/air (air = 1) > 1.05 Solubility in water poor Relative molecular mass 186.3	**COLORLESS LIQUID** Vapor mixes readily with air. Able to form peroxides. Reacts with many substances, e.g. amines, carbonic acids, aldehydes, alcohols and halogen acids.	
	TLV-TWA not available	
	Absorption route: Can enter the body by inhalation or ingestion or through the skin. Insufficient data on the rate at which harmful concentrations can build up. **Immediate effects:** Irritates the eyes, skin and respiratory tract. Liquid destroys the skin's natural oils. Affects the nervous system. **Effects of prolonged/repeated exposure:** Prolonged or repeated contact can cause skin disorders (eczema).	
Gross formula: $C_{11}H_{22}O_2$		

HAZARDS/SYMPTOMS	PREVENTIVE MEASURES	FIRE EXTINGUISHING/FIRST AID
Fire: combustible.	keep away from open flame, no smoking.	powder, AFFF, foam, carbon dioxide, (halons).
Explosion: above 95°C: forms explosive air-vapor mixtures.	above 95°C: sealed machinery, ventilation.	
Inhalation: sore throat, cough, shortness of breath, drowsiness, restlessness.	ventilation, local exhaust or respiratory protection.	fresh air, rest, call a doctor.
Skin: *is absorbed*; redness.	gloves.	remove contaminated clothing, flush skin with water or shower.
Eyes: redness, pain.	face shield.	flush with water, take to a doctor.
Ingestion: abdominal pain, nausea, vomiting, drowsiness.		rinse mouth, call a doctor or take to hospital.

SPILLAGE	STORAGE	LABELING / NFPA
collect leakage in sealable containers, soak up with sand or other inert absorbent and remove to safe place.	ventilate at floor level.	

NOTES
Before distilling check for peroxides; if found, render harmless. Very few data available.

5-ETHYLIDENE-2-NORBORNENE

PHYSICAL PROPERTIES		IMPORTANT CHARACTERISTICS	
Boiling point, °C	148	**COLORLESS LIQUID WITH CHARACTERISTIC ODOR**	
Melting point, °C	−80	Vapor mixes readily with air. Able to polymerize. Flow, agitation etc. can cause build-up of electrostatic charge due to liquid's low conductivity. Reacts violently with strong oxidants.	
Flash point, °C	38		
Auto-ignition temperature, °C	?		
Relative density (water = 1)	0.9		
Relative vapor density (air = 1)	4.2		
Relative density at 20 °C of saturated mixture vapor/air (air = 1)	1.02	TLV-TWA 5 ppm 25 mg/m³ C	
Vapor pressure, mm Hg at 20 °C	4.3	**Absorption route:** Can enter the body by inhalation or ingestion. Harmful atmospheric concentrations can build up fairly rapidly on evaporation at approx. 20°C - even more rapidly in aerosol form.	
Solubility in water	none		
Explosive limits, vol% in air	?-?		
Electrical conductivity, pS/m	< 10⁴	**Immediate effects:** Irritates the eyes and respiratory tract.	
Relative molecular mass	120.2		
Gross formula:	C_9H_{12}		

HAZARDS/SYMPTOMS	PREVENTIVE MEASURES	FIRE EXTINGUISHING/FIRST AID
Fire: flammable.	keep away from open flame and sparks, no smoking.	powder, AFFF, foam, carbon dioxide, (halons).
Explosion: above 38°C: forms explosive air-vapor mixtures.	above 38°C: sealed machinery, ventilation, explosion-proof electrical equipment, grounding.	in case of fire: keep tanks/drums cool by spraying with water.
Inhalation: sore throat, cough, shortness of breath.	ventilation, local exhaust or respiratory protection.	fresh air, rest, call a doctor if necessary.
Skin:	gloves.	remove contaminated clothing, flush skin with water or shower.
Eyes: redness, pain.	safety glasses.	flush with water, take to a doctor if necessary.
Ingestion:		rinse mouth, call a doctor if necessary.

SPILLAGE	STORAGE	LABELING / NFPA
(additional individual protection: breathing apparatus).	keep in fireproof place under inert gas, separate from oxidants, add an inhibitor.	

NOTES
Explosive limits not given in the literature. TLV is maximum and must not be exceeded. Insufficient toxicological data on harmful effects to humans.

ETHYL IODIDE

PHYSICAL PROPERTIES	IMPORTANT CHARACTERISTICS
Boiling point, °C 72 Melting point, °C − 108 Flash point, °C see notes Relative density (water = 1) 1.9 Relative vapor density (air = 1) 5.4 Relative density at 20 °C of saturated mixture vapor/air (air = 1) 1.6 Vapor pressure, mm Hg at 20 °C 106 Solubility in water poor Relative molecular mass 156.0 Log P octanol/water 2.0	**COLORLESS LIQUID WITH CHARACTERISTIC ODOR, TURNING RED ON EXPOSURE TO AIR** Vapor is heavier than air. Decomposes on contact with air or moisture when exposed to light, forming → *iodine*. Reacts violently with strong oxidants, giving off corrosive vapors.
	TLV-TWA not available
	Absorption route: Can enter the body by inhalation or ingestion or through the skin. **Immediate effects:** Corrosive to the eyes, skin and respiratory tract. In substantial concentrations can cause agitation, impairment of consciousness and loss of balance. Inhalation can cause lung edema. In serious cases risk of death. **Effects of prolonged/repeated exposure:** Prolonged or repeated contact can cause skin disorders.
Gross formula: C_2H_5I	

HAZARDS/SYMPTOMS	PREVENTIVE MEASURES	FIRE EXTINGUISHING/FIRST AID
Fire: combustible.	keep away from open flame, no smoking.	powder, water spray, foam, carbon dioxide, (halons).
Explosion:		in case of fire: keep tanks/drums cool by spraying with water.
Inhalation: *corrosive*; sore throat, cough, severe breathing difficulties, dizziness, sleepiness.	ventilation, local exhaust or respiratory protection.	fresh air, rest, place in half-sitting position, take to hospital.
Skin: *corrosive, is absorbed*; redness, pain, burns.	gloves, protective clothing.	remove contaminated clothing, flush skin with water or shower, take to a doctor.
Eyes: *corrosive*; redness, pain, impaired vision.	face shield, or combined eye and respiratory protection.	flush with water, take to a doctor.
Ingestion: *corrosive*; abdominal pain, nausea, dizziness, sleepiness, unconsciousness.	do not eat, drink or smoke while working.	rinse mouth, take immediately to hospital.

SPILLAGE	STORAGE	LABELING / NFPA
collect leakage in sealable containers, soak up with sand or other inert absorbent and remove to safe place; (additional individual protection: breathing apparatus).	keep in cool, dry, dark place, separate from oxidants; ventilate at floor level.	

NOTES
Combustible, but flash point not given in the literature. Lung edema symptoms usually develop several hours later and are aggravated by physical exertion: rest and hospitalization essential. As first aid, a doctor or authorized person should consider administering a corticosteroid spray. Use airtight packaging.

Transport Emergency Card TEC(R)-61G08

ETHYL ISOCYANATE

PHYSICAL PROPERTIES	IMPORTANT CHARACTERISTICS
Boiling point, °C 60 Melting point, °C < −50 Flash point, °C −6 Auto-ignition temperature, °C ? Relative density (water = 1) 0.9 Relative vapor density (air = 1) 2.5 Relative density at 20 °C of saturated mixture vapor/air (air = 1) 1.3 Vapor pressure, mm Hg at 20 °C ca. 160 Solubility in water reaction Explosive limits, vol% in air ?-? Relative molecular mass 71.1	**COLORLESS LIQUID WITH PUNGENT ODOR** Vapor is heavier than air and spreads at ground level, with risk of ignition at a distance. Do not use compressed air when filling, emptying or processing. Toxic vapors are a product of combustion (→ *nitrous vapors*). Reacts violently with water, acids, bases, alcohols and amines, with risk of fire and explosion. Reacts violently with strong oxidants. Reaction products are → *carbon monoxide*, → *carbon dioxide*, → *hydrocyanic acid*, → *nitrous vapors*.
	TLV-TWA not available
	Absorption route: Can enter the body by inhalation or ingestion or through the skin. Harmful atmospheric concentrations can build up very rapidly on evaporation at 20°C. **Immediate effects:** Corrosive to the eyes, skin and respiratory tract. Inhalation can cause severe breathing difficulties. **Effects of prolonged/repeated exposure:** Prolonged or repeated contact can cause skin disorders. Prolonged, repeated intensive contact with the vapor can cause lung disorders (asthma).
Gross formula: C_3H_5NO	

HAZARDS/SYMPTOMS	PREVENTIVE MEASURES	FIRE EXTINGUISHING/FIRST AID
Fire: extremely flammable.	keep away from open flame and sparks, no smoking.	powder, carbon dioxide, (halons), DO NOT USE WATER-BASED EXTINGUISHERS.
Explosion: forms explosive air-vapor mixtures.	sealed machinery, ventilation, explosion-proof electrical equipment and lighting.	in case of fire: keep tanks/drums cool by spraying with water, but DO NOT spray substance with water.
	STRICT HYGIENE	
Inhalation: *corrosive*; sore throat, cough, shortness of breath, severe breathing difficulties.	ventilation, local exhaust or respiratory protection.	fresh air, rest, place in half-sitting position, take to hospital.
Skin: *corrosive, is absorbed*; redness, pain, burns.	gloves, protective clothing.	remove contaminated clothing, flush skin with water or shower, take to a doctor.
Eyes: *corrosive*; redness, pain, impaired vision.	face shield, or combined eye and respiratory protection.	flush with water, take to a doctor.
Ingestion: *corrosive*; abdominal pain, diarrhea, vomiting.		rinse mouth, immediately take to hospital.

SPILLAGE	STORAGE	LABELING / NFPA
evacuate area, call in an expert, collect leakage in sealable containers, soak up with dry sand or other inert absorbent and remove to safe place; (additional individual protection: CHEMICAL SUIT).	keep in cool, dry, fireproof place, separate from all other substances.	

NOTES
Explosive limits and autoignition temperature not given in the literature. Reacts violently with extinguishing agents such as water and foam. Lung edema symptoms usually develop several hours later and are aggravated by physical exertion: rest and hospitalization essential. As first aid, a doctor or authorized person should consider administering a corticosteroid spray. Unbreakable packaging preferred; if breakable, keep in unbreakable container.

Transport Emergency Card TEC(R)-61G03

HI: 336; UN-number: 2481

diethyl malonate
malonic ester

$CH_2(COOC_2H_5)_2$

ETHYL MALONATE

PHYSICAL PROPERTIES		IMPORTANT CHARACTERISTICS	
Boiling point, °C	199	**COLORLESS LIQUID WITH CHARACTERISTIC ODOR** Vapor mixes readily with air. Reacts with strong oxidants.	
Melting point, °C	−50		
Flash point, °C	84		
Auto-ignition temperature, °C	424	TLV-TWA not available	
Relative density (water = 1)	1.06		
Relative vapor density (air = 1)	5.5	**Absorption route:** Can enter the body by inhalation or ingestion. Harmful atmospheric concentrations build up very slowly, if at all, on evaporation at approx. 20°C, but much more rapidly in aerosol form. **Immediate effects:** Irritates the eyes, skin and respiratory tract. **Effects of prolonged/repeated exposure:** Prolonged or repeated contact can cause skin disorders.	
Relative density at 20 °C of saturated mixture vapor/air (air = 1)	1.00		
Vapor pressure, mm Hg at 20 °C	0.23		
Solubility in water, g/100 ml	2		
Relative molecular mass	160.2		
Gross formula:	$C_7H_{12}O_4$		

HAZARDS/SYMPTOMS	PREVENTIVE MEASURES	FIRE EXTINGUISHING/FIRST AID
Fire: combustible.	keep away from open flame, no smoking.	powder, AFFF, foam, carbon dioxide, (halons).
Explosion: above 84°C: forms explosive air-vapor mixtures.	above 84°C: sealed machinery, ventilation.	
Inhalation: sore throat, cough.	ventilation.	fresh air, rest, call a doctor.
Skin: redness, pain.	gloves.	remove contaminated clothing, flush skin with water or shower.
Eyes: redness, pain.	safety glasses.	flush with water, take to a doctor if necessary.
Ingestion: sore throat, abdominal pain, diarrhea, feeling of weakness.		rinse mouth, call a doctor.

SPILLAGE	STORAGE	LABELING / NFPA
collect leakage in sealable containers, soak up with sand or other inert absorbent and remove to safe place.	ventilate at floor level.	

NOTES

Transport Emergency Card TEC(R)-30G15

CAS-No: [75-08-1]
ethanethiol
ethyl hydrosulfide
ethyl sulfhydrate
mercaptoethane
thioethyl alcohol

C_2H_5SH

ETHYLMERCAPTAN

PHYSICAL PROPERTIES		IMPORTANT CHARACTERISTICS	
Boiling point, °C	35	**COLORLESS LIQUID WITH CHARACTERISTIC ODOR**	
Melting point, °C	− 148	Vapor is heavier than air and spreads at ground level, with risk of ignition at a distance. Do not use compressed air when filling, emptying or processing. Toxic gases are a product of combustion. Reacts violently with strong oxidants, with risk of fire and explosion. Reacts with acids giving off flammable and toxic gas (→ *hydrogen sulfide*).	
Flash point, °C	< 0		
Auto-ignition temperature, °C	295		
Relative density (water = 1)	0.8		
Relative vapor density (air = 1)	2.1		
Relative density at 20 °C of saturated mixture vapor/air (air = 1)	1.09	TLV-TWA	0.5 ppm	1.3 mg/m³
Vapor pressure, mm Hg at 20 °C	448		
Explosive limits, vol% in air	2.8-18.2	**Absorption route:** Can enter the body by inhalation or ingestion. Harmful atmospheric concentrations can build up very rapidly on evaporation at 20°C.	
Relative molecular mass	62.1	**Immediate effects:** Irritates the eyes, skin and respiratory tract. In substantial concentrations can impair consciousness. In serious cases risk of unconsciousness.	
		Effects of prolonged/repeated exposure: Can cause liver damage.	
Gross formula:	C_2H_6S		

HAZARDS/SYMPTOMS	PREVENTIVE MEASURES	FIRE EXTINGUISHING/FIRST AID
Fire: extremely flammable.	keep away from open flame and sparks, no smoking.	powder, AFFF, foam, carbon dioxide, (halons); in case of fire in immediate vicinity: use any extinguishing agent.
Explosion: forms explosive air-vapor mixtures.	sealed machinery, ventilation, explosion-proof electrical equipment and lighting.	in case of fire: keep tanks/drums cool by spraying with water.
Inhalation: sore throat, shortness of breath, unconsciousness, blue skin.	ventilation, local exhaust or respiratory protection.	fresh air, rest, call a doctor.
Skin: redness, pain.	gloves.	remove contaminated clothing, flush skin with water or shower, refer to a doctor.
Eyes: redness, pain.	face shield, or combined eye and respiratory protection.	flush with water, take to a doctor.
Ingestion: sore throat, abdominal pain; see also 'Inhalation'.		rinse mouth, call a doctor.

SPILLAGE	STORAGE	LABELING / NFPA
evacuate area, call in an expert, collect leakage in sealable containers, soak up with sand or other inert absorbent and remove to safe place; (additional individual protection: breathing apparatus).	keep in fireproof, cool place, separate from oxidants and strong acids.	R: 11-20 S: 16-25 🔥 Flammable ✖ Harmful NFPA: 4 2 0

NOTES
Also commercially available in cylinders.

Transport Emergency Card TEC(R)-688

HI: 336; UN-number: 2363

CAS-No: [105-37-3]
ethylpropanoate
propanoic acid ethyl ester
propionic ether

$C_2H_5COOC_2H_5$

ETHYL PROPIONATE

PHYSICAL PROPERTIES	IMPORTANT CHARACTERISTICS
Boiling point, °C 99 Melting point, °C −74 Flash point, °C 12 Auto-ignition temperature, °C 475 Relative density (water = 1) 0.89 Relative vapor density (air = 1) 3.5 Relative density at 20 °C of saturated mixture vapor/air (air = 1) 1.09 Vapor pressure, mm Hg at 20 °C 27 Solubility in water, g/100 ml at 15 °C 2.5 Explosive limits, vol% in air 1.8-11 Electrical conductivity, pS/m 8.3×10^{10} Relative molecular mass 102.1 Log P octanol/water 1.2 Gross formula: $C_5H_{10}O_2$	**COLORLESS LIQUID WITH CHARACTERISTIC ODOR** Vapor is heavier than air and spreads at ground level, with risk of ignition at a distance. Do not use compressed air when filling, emptying or processing. Reacts with strong oxidants. TLV-TWA not available **Absorption route:** Can enter the body by inhalation or ingestion. Harmful atmospheric concentrations build up fairly slowly on evaporation at 20°C, but much more rapidly in aerosol form. **Immediate effects:** Irritates the eyes and respiratory tract.

HAZARDS/SYMPTOMS	PREVENTIVE MEASURES	FIRE EXTINGUISHING/FIRST AID
Fire: extremely flammable.	keep away from open flame and sparks, no smoking.	powder, AFFF, foam, carbon dioxide, (halons).
Explosion: forms explosive air-vapor mixtures.	sealed machinery, ventilation, explosion-proof electrical equipment and lighting.	in case of fire: keep tanks/drums cool by spraying with water.
Inhalation: cough, shortness of breath.	ventilation, local exhaust or respiratory protection.	fresh air, rest, call a doctor if necessary.
Skin: redness.	gloves.	remove contaminated clothing, flush skin with water or shower.
Eyes: redness, pain.	safety glasses.	flush with water, take to a doctor if necessary.
Ingestion: abdominal pain, vomiting.		rinse mouth, call a doctor or take to hospital.

SPILLAGE	STORAGE	LABELING / NFPA
collect leakage in sealable containers, soak up with sand or other inert absorbent and remove to safe place; (additional individual protection: breathing apparatus).	keep in fireproof place, separate from oxidants.	R: 11 S: 16-23-29-33 🔥 Flammable

NOTES

CAS-No: [78-10-4]
ethyl orthosilicate
silicic acid tetraethyl ester
tetraethoxysilane
tetraethyl orthosilicate
tetraethyl silicate

$Si(OC_2H_5)_4$

ETHYL SILICATE

PHYSICAL PROPERTIES	IMPORTANT CHARACTERISTICS
Boiling point, °C — 166 Melting point, °C — −82 Flash point, °C — 37 Relative density (water = 1) — 0.9 Relative vapor density (air = 1) — 7.2 Relative density at 20 °C of saturated mixture vapor/air (air = 1) — 1.01 Vapor pressure, mm Hg at 20 °C — 13 Solubility in water — none Explosive limits, vol% in air — 1.3-23 Electrical conductivity, pS/m — <3x10⁶ Relative molecular mass — 208.3	**COLORLESS LIQUID WITH CHARACTERISTIC ODOR** Vapor mixes readily with air. Decomposes when heated, giving off combustible and toxic vapors. Decomposes slowly on contact with water. Reacts violently with strong oxidants and strong acids.

TLV-TWA	10 ppm	85 mg/m³

Absorption route: Can enter the body by inhalation or ingestion. Harmful atmospheric concentrations build up fairly slowly on evaporation at 20°C, but much more rapidly in aerosol form.
Immediate effects: Corrosive to the eyes, skin and respiratory tract. Liquid destroys the skin's natural oils. Inhalation of vapor/fumes can cause severe breathing difficulties (lung edema).
Effects of prolonged/repeated exposure: Can cause liver and kidney damage.

Gross formula: $C_8H_{20}O_4Si$

HAZARDS/SYMPTOMS	PREVENTIVE MEASURES	FIRE EXTINGUISHING/FIRST AID
Fire: flammable.	keep away from open flame and sparks, no smoking.	powder, AFFF, foam, carbon dioxide, (halons).
Explosion: above 37°C: forms explosive air-vapor mixtures.	above 37°C: sealed machinery, ventilation, explosion-proof electrical equipment.	in case of fire: keep tanks/drums cool by spraying with water.
Inhalation: *corrosive*; sore throat, cough, shortness of breath, severe breathing difficulties.	ventilation, local exhaust or respiratory protection.	fresh air, rest, place in half-sitting position, take to hospital.
Skin: *corrosive*; redness, pain, burns.	gloves.	remove contaminated clothing, flush skin with water or shower, refer to a doctor.
Eyes: *corrosive*; redness, pain, impaired vision.	face shield.	flush with water, take to a doctor.
Ingestion: *corrosive*, abdominal pain, diarrhea, pain with swallowing.		rinse mouth, immediately take to hospital.

SPILLAGE	STORAGE	LABELING / NFPA
collect leakage in sealable containers, soak up with sand or other inert absorbent and remove to safe place.	keep in fireproof place.	R: 10-20-36/37 ✖ Harmful

NFPA: 2 / 2 / 0

NOTES

Lung edema symptoms usually develop several hours later and are aggravated by physical exertion: rest and hospitalization essential. As first aid, a doctor or authorized person should consider administering a corticosteroid spray.

Transport Emergency Card TEC(R)-556

HI: 30; UN-number: 1292

FERROUS (III) CHLORIDE

PHYSICAL PROPERTIES		IMPORTANT CHARACTERISTICS	
Decomposes below boiling point, °C	324	**BROWN HYGROSCOPIC CRYSTALS WITH CHARACTERISTIC ODOR**	
Melting point, °C	306	Decomposes when heated, giving off toxic → *chlorine* gas. In aqueous solution is a medium	
Relative density (water = 1)	2.8	strong acid which attacks many metals, giving off → *hydrogen* gas. Reacts violently with strong	
Solubility in water, g/100 ml at 20 °C	92	bases.	
Relative molecular mass	162.2		
		TLV-TWA	1 mg/m³ ✳
		Absorption route: Can enter the body by ingestion. Evaporation negligible at 20°C, but harmful concentrations of airborne particles can build up rapidly. **Immediate effects:** Corrosive to the eyes and gastro-intestinal mucous membranes; irritates the skin.	
Gross formula:	Cl$_3$Fe		

HAZARDS/SYMPTOMS	PREVENTIVE MEASURES	FIRE EXTINGUISHING/FIRST AID
Fire: non-combustible.		in case of fire in immediate vicinity: use any extinguishing agent.
Inhalation:	local exhaust or respiratory protection.	
Skin: redness, pain.	gloves, protective clothing.	remove contaminated clothing, flush skin with water or shower.
Eyes: *corrosive*; redness, pain, impaired vision.	face shield.	flush with water, take to a doctor.
Ingestion: *corrosive*; sore throat, abdominal pain, nausea.		rinse mouth, immediately take to hospital.

SPILLAGE	STORAGE	LABELING / NFPA
clean up spillage, flush away remainder with water.	keep dry, separate from strong bases.	

NOTES
Apparent melting point of iron chloride hydrate (6H$_2$O) due to loss of water of crystallization: 37°C. ✳ TLV equals that of iron salts (soluble). Use airtight packaging.

iron sulfate
iron vitriol

FERROUS (II) SULFATE

PHYSICAL PROPERTIES	IMPORTANT CHARACTERISTICS	
Melting point (decomposes), °C 64 Relative density (water = 1) 1.9 Solubility in water, g/100 ml at 20 °C 25.6 Relative molecular mass 278	**GREEN CRYSTALS, TURNING BROWN ON EXPOSURE TO AIR** Decomposes when heated above 400° C, giving off toxic and corrosive vapors (→ *sulfur dioxide*). In aqueous solution is a medium strong acid. Reacts with oxygen in air to form ferric sulfate.	
	TLV-TWA	1 mg/m³ ✳
	Absorption route: Can enter the body by inhalation or ingestion. Evaporation negligible at 20° C, but harmful concentrations of airborne particles can build up rapidly.	
Gross formula: FeO$_4$S.7H$_2$O		

HAZARDS/SYMPTOMS	PREVENTIVE MEASURES	FIRE EXTINGUISHING/FIRST AID
Fire: non-combustible.		in case of fire in immediate vicinity: use any extinguishing agent.
Inhalation: tickling of the throat.		fresh air, rest.
Skin:	gloves.	
Eyes: redness, pain.	goggles, or combined eye and respiratory protection.	flush with water, take to a doctor if necessary.
Ingestion: abdominal pain, vomiting, constipation.		rinse mouth, call a doctor.

SPILLAGE	STORAGE	LABELING / NFPA
clean up spillage, flush away remainder with water; (additional individual protection: P2 respirator).	keep separate from oxidants and strong bases; keep in tightly-closed packaging.	◇

NOTES
Apparent melting point due to loss of water of crystallization. ✳ TLV equals that of iron.

FLUOBORIC ACID
(25-78% solution)

PHYSICAL PROPERTIES	IMPORTANT CHARACTERISTICS
Boiling point (decomposes), °C 130 Relative density (water = 1) 1.8 Relative vapor density (air = 1) 3.0 Relative density at 20 °C of saturated mixture vapor/air (air = 1) < 1.1 Vapor pressure, mm Hg at 20 °C < < 7.6 Solubility in water ∞ Relative molecular mass 87.8	**COLORLESS LIQUID WITH PUNGENT ODOR** Vapor mixes readily with air. Decomposes when heated above 130°C, giving off toxic and corrosive vapors (incl. → *hydrogen fluoride*). Strong acid which reacts violently with bases and is corrosive. Attacks many metals, giving off flammable gas (→ *hydrogen*).
	TLV-TWA not available
	Absorption route: Can enter the body by inhalation or ingestion. Harmful atmospheric concentrations can build up fairly rapidly on evaporation at approx. 20°C - even more rapidly in aerosol form. **Immediate effects:** Corrosive to the eyes, skin and respiratory tract. Inhalation of vapor/fumes can cause severe breathing difficulties (lung edema). In serious cases risk of death.
Gross formula: BF$_4$H	

HAZARDS/SYMPTOMS	PREVENTIVE MEASURES	FIRE EXTINGUISHING/FIRST AID
Fire: non-combustible.		in case of fire in immediate vicinity: use any extinguishing agent.
	STRICT HYGIENE	IN ALL CASES CALL A DOCTOR
Inhalation: *corrosive*; sore throat, cough, severe breathing difficulties.	ventilation, local exhaust or respiratory protection.	fresh air, rest, place in half-sitting position, take to hospital.
Skin: *corrosive*; redness, pain, serious burns.	gloves, protective clothing.	remove contaminated clothing, flush skin with water or shower, call a doctor.
Eyes: *corrosive*; redness, pain, impaired vision.	face shield.	flush with water, take to a doctor.
Ingestion: *corrosive*; abdominal pain, diarrhea.		rinse mouth, take immediately to hospital.

SPILLAGE	STORAGE	LABELING / NFPA
collect leakage in sealable containers, absorb spillage with chalk (CaCO₃); (additional individual protection: CHEMICAL SUIT).✶)	keep separate from bases; ventilate at floor level.	0 3 0

NOTES
Physical properties given relate to 78% solution. ✶) For under-40% commercial solution breathing apparatus is sufficient. Lung edema symptoms generally take a few hours to become apparent and are aggravated by physical exertion: rest and hospitalization essential. As first aid, a doctor or authorized person should consider administering a corticosteroid spray. Use airtight packaging made of special material.

Transport Emergency Card TEC(R)-732

HI: 80; UN-number: 1775

FLUORINE
(cylinder)

PHYSICAL PROPERTIES	IMPORTANT CHARACTERISTICS	
Boiling point, °C −188 Melting point, °C −219 Relative vapor density (air = 1) 1.3 Solubility in water reaction Relative molecular mass 38.0	**LIGHT GREENISH-YELLOW COMPRESSED GAS WITH PUNGENT ODOR** Gas is heavier than air. Strong oxidant which reacts violently with combustible substances and reducing agents, with risk of fire and explosion. Reacts violently with very many substances, with risk of fire and explosion. Reacts violently with water, giving off corrosive vapors (→ *ozone* and → *hydrogen fluoride*). Attacks many metals, esp. in the presence of water or moisture.	
	TLV-TWA 1 ppm 1.6 mg/m³ STEL 2 ppm 3.1 mg/m³	
	Absorption route: Can enter the body by inhalation. Harmful atmospheric concentrations can build up very rapidly if gas is released. **Immediate effects:** Corrosive to the eyes, skin and respiratory tract. Can cause frostbite due to rapid evaporation. Inhalation of vapor/fumes can cause severe breathing difficulties (lung edema). In serious cases risk of death.	
Gross formula: F_2		

HAZARDS/SYMPTOMS	PREVENTIVE MEASURES	FIRE EXTINGUISHING/FIRST AID
Fire: non-combustible, many chemical reactions can cause fire and explosion.	avoid contact with other substances.	
Explosion:		in case of fire: keep intact cylinders cool by spraying with water, but DO NOT spray substance with water.
	STRICT HYGIENE	
Inhalation: *corrosive*; sore throat, cough, shortness of breath, severe breathing difficulties.	ventilation, local exhaust or respiratory protection.	fresh air, rest, place in half-sitting position, take to hospital.
Skin: *corrosive*; *in case of frostbite*: redness, pain, severe sores.	insulated gloves, protective clothing.	*in case of frostbite*: DO NOT remove clothing, flush skin with large quantities of water, take to hospital.
Eyes: *corrosive*; redness, pain, impaired vision.	face shield, or combined eye and respiratory protection.	flush with water, take to doctor.

SPILLAGE	STORAGE	LABELING / NFPA
evacuate area, call in an expert, ventilate, combat gas cloud with water curtain; (additional individual protection: CHEMICAL SUIT).	if stored indoors, keep in cool, dry, fireproof place, separate from other chemicals; ventilate at floor level.	◇ 0 / 4 / 4 / W

NOTES

Lung edema symptoms usually develop several hours later and are aggravated by physical exertion: rest and hospitalization essential. As first aid, a doctor or authorized person should consider administering a corticosteroid spray. Do not spray leaking cylinder with water (to avoid corrosion). Turn leaking cylinder so that leak is on top to prevent liquid fluorine escaping. Use cylinder with special fittings.

Transport Emergency Card TEC(R)-20G03

UN-number: 1045

1-FLUORO-2,4-DINITROBENZENE

PHYSICAL PROPERTIES	IMPORTANT CHARACTERISTICS	
Boiling point, °C 296 Melting point, °C 26 Flash point, °C see notes Relative density (water = 1) 1.5 Relative density at 20 °C of saturated mixture vapor/air (air = 1) 1.0 Vapor pressure, mm Hg at 20 °C 0.76 Solubility in water none Relative molecular mass 186.1	**LIQUID OR CRYSTALS** Decomposes when heated or burned, giving off → *nitrous vapors* and → *hydrogen fluoride.* Reacts violently with many substances.	
	TLV-TWA not available	
	Absorption route: Can enter the body by inhalation or ingestion. Harmful atmospheric concentrations build up very slowly, if at all, on evaporation at approx. 20°C, but much more rapidly in aerosol form. **Immediate effects:** Corrosive to the eyes, skin and respiratory tract. **Effects of prolonged/repeated exposure:** Prolonged or repeated contact can cause skin disorders (eczema).	
Gross formula: $C_6H_3FN_2O_4$		

HAZARDS/SYMPTOMS	PREVENTIVE MEASURES	FIRE EXTINGUISHING/FIRST AID
Fire: combustible.	keep away from open flame, no smoking.	powder, water spray, foam, carbon dioxide, (halons).
Inhalation: *corrosive*; sore throat, cough, shortness of breath, headache.	local exhaust or respiratory protection.	fresh air, rest, call a doctor.
Skin: *corrosive*; redness, pain.	gloves.	remove contaminated clothing, flush skin with water or shower, call a doctor.
Eyes: *corrosive*; redness, pain.	face shield, or combined eye and respiratory protection.	flush with water, take to a doctor.
Ingestion: *corrosive*; sore throat, abdominal pain, nausea, vomiting.	do not eat, drink or smoke while working.	rinse mouth, call a doctor.

SPILLAGE	STORAGE	LABELING / NFPA
collect leakage in sealable container, soak up with sand or other inert absorbent and remove to save place, carefully collect remainder; (additional individual protection: P3 respirator).	keep in fireproof place.	

NOTES
Combustible, but flash point and explosive limits not given in the literature.

Transport Emergency Card TEC(R)-61G06A/12

CAS-No: [50-00-0]
formalin
methanal
methyl aldehyde
methylene oxide
oxomethane

FORMALDEHYDE

(37% solution in water with 10% methanol)

PHYSICAL PROPERTIES		IMPORTANT CHARACTERISTICS
Boiling point, °C	96	**COLORLESS SOLUTION OF METHANAL WITH PUNGENT ODOR**
Melting point, °C	− 15	Technical grades usually contain methanol as inhibitor. Vapor mixes readily with air. More or less
Flash point, °C	60	readily able to polymerize, depending on methanol content. Decomposes when heated, giving off
Auto-ignition temperature, °C	ca. 400	→ *formic acid*. Strong reducing agent which reacts violently with oxidants and various organic
Relative density (water = 1)	1.1	substances. Reacts with hydrochloric acid, giving off highly toxic → *bis(chloromethyl)ether*.
Relative vapor density (air = 1)	1.04	
Vapor pressure, mm Hg at 20 °C	ca. 1.52	

	TLV-TWA	1 ppm	1.2 mg/m³ A2
	STEL	2 ppm	2.5 mg/m³

Solubility in water ∞
Explosive limits, vol% in air 7-73
Relative molecular mass 30.0
Log P octanol/water 0

Absorption route: Can enter the body by inhalation or ingestion. Harmful atmospheric concentrations can build up fairly rapidly on evaporation at approx. 20°C - even more rapidly in aerosol form.
Immediate effects: Corrosive to the eyes, skin and respiratory tract. Inhalation can cause lung edema. Can cause liver and kidney damage if swallowed.
Effects of prolonged/repeated exposure: Prolonged or repeated contact can cause skin disorders.

Gross formula: CH_2O

HAZARDS/SYMPTOMS	PREVENTIVE MEASURES	FIRE EXTINGUISHING/FIRST AID
Fire: combustible.	keep away from open flame, no smoking.	powder, alcohol-resistant foam, water spray, carbon dioxide, (halons).
Explosion: above 60°C: forms explosive air-vapor mixtures.	above 60°C: sealed machinery, ventilation.	in case of fire: keep tanks/drums cool by spraying with water.
Inhalation: *corrosive*; sore throat, cough, shortness of breath, severe breathing difficulties.	ventilation, local exhaust or respiratory protection.	fresh air, rest, place in half-sitting position, take to hospital.
Skin: *corrosive*; redness, pain, burns.	gloves, protective clothing.	remove contaminated clothing, flush skin with water or shower, refer to a doctor.
Eyes: *corrosive*; redness, pain, impaired vision.	face shield, or combined eye and respiratory protection.	flush with water, take to a doctor.
Ingestion: *corrosive*; sore throat, abdominal pain, diarrhea.		rinse mouth, take immediately to hospital.

SPILLAGE	STORAGE	LABELING / NFPA
collect leakage in sealable containers, flush away remainder with water; (additional individual protection: breathing apparatus).	keep cool, separate from oxidants; ventilate at floor level.	R: 23/24/25-34-40-43 S: 26-36/37-44-51 Note B + D Toxic

NFPA: 3 / 2 / 0

NOTES

Flash point and autoignition point depend on methanol content. Flash point: methanol-free: 80°C; 10% methanol: 60°C; 15% methanol: 50°C. Explosive limits are for methanol-free. Log P octanol/water is estimated. Minimum storage temperature (to avoid polymerization): 1% methanol: 21°C; 7% methanol: 16°C; 15% methanol: 5°C. Lung edema symptoms usually develop several hours later and are aggravated by physical exertion: rest and hospitalization essential. As first aid, a doctor or authorized person should consider administering a corticosteroid spray. Special first aid and treatment are required if formaldehyde has been ingested.

Transport Emergency Card TEC(R)-80

HI: 80; UN-number: 2209

carbamaldehyde
methanamide

OCHNH$_2$

FORMAMIDE

PHYSICAL PROPERTIES		IMPORTANT CHARACTERISTICS	
Boiling point (decomposes), °C	210	**COLORLESS TO LIGHT YELLOW VISCOUS LIQUID WITH CHARACTERISTIC ODOR**	
Melting point, °C	2.6	Vapor mixes readily with air. Toxic gases are a product of combustion (→ *nitrous vapors*).	
Flash point, °C	154	Decomposes when heated above 180°C, forming water, → *anhydrous ammonia*, → *carbon*	
Auto-ignition temperature, °C	>500	*dioxide* and → *prussic acid*. In aqueous solution, esp. in alkaline or acidic environment,	
Relative density (water = 1)	1.1	hydrolyzes slowly, forming ammonium formate. Reacts with hydrolyzing agents to form	
Relative vapor density (air = 1)	1.6	→ *prussic acid*.	
Relative density at 20 °C of saturated mixture			
vapor/air (air = 1)	1.0	TLV-TWA (skin) 10 ppm 18 mg/m³	
Vapor pressure, mm Hg at 20 °C	0.023		
Solubility in water	∞	**Absorption route:** Can enter the body by inhalation or ingestion or through the skin. Harmful	
Electrical conductivity, pS/m	2 x 10⁷	atmospheric concentrations build up very slowly, if at all, on evaporation at approx. 20°C, but	
Relative molecular mass	45.0	much more rapidly in aerosol form.	
Log P octanol/water	−1.6	**Immediate effects:** Irritates the eyes, skin and respiratory tract.	
		Effects of prolonged/repeated exposure: Can damage the liver.	
Gross formula:	CH$_3$NO		

HAZARDS/SYMPTOMS	PREVENTIVE MEASURES	FIRE EXTINGUISHING/FIRST AID
Fire: combustible.	keep away from open flame, no smoking.	powder, alcohol-resistant foam, water spray, carbon dioxide, (halons).
	AVOID EXPOSURE OF PREGNANT WOMEN AS FAR AS POSSIBLE	
Inhalation: headache, dizziness, nausea, vomiting, abdominal cramps.	ventilation, local exhaust or respiratory protection.	fresh air, rest, call a doctor.
Skin: *is absorbed*; redness, pain.	gloves.	remove contaminated clothing, flush skin with water or shower, refer to a doctor.
Eyes: redness, pain.	face shield.	flush with water, take to a doctor.
Ingestion: abdominal cramps, headache, nausea, vomiting, dizziness.		rinse mouth, take immediately to hospital.

SPILLAGE	STORAGE	LABELING / NFPA
collect leakage in sealable containers, soak up with sand or other inert absorbent en remove so safe place.		(NFPA diamond: blue 2, red 1)

NOTES
Log P octanol/water is estimated. Alcohol consumption increases toxic effects. Depending on the degree of exposure, regular medical checkups are advisable.

aminic acid
formylic acid
hydrogen carboxylic acid
methanoic acid

FORMIC ACID

PHYSICAL PROPERTIES		IMPORTANT CHARACTERISTICS	
Boiling point, °C	101	**COLORLESS FUMING LIQUID WITH PUNGENT ODOR**	
Melting point, °C	8	Vapor mixes readily with air. Decomposes when heated, giving off toxic and flammable gas	
Flash point, °C	69	(→ *carbon monoxide* and/or → *hydrogen*). Acids promote decomposition. Strong reducing	
Auto-ignition temperature, °C	520	agent which reacts violently with oxidants. Reacts violently with bases and is corrosive.	
Relative density (water = 1)	1.2		
Relative vapor density (air = 1)	1.6	TLV-TWA	5 ppm	9.4 mg/m³
Relative density at 20 °C of saturated mixture			
vapor/air (air = 1)	1.03	**Absorption route:** Can enter the body by inhalation or ingestion. Harmful atmospheric	
Vapor pressure, mm Hg at 20 °C	32.3	concentrations can build up fairly rapidly on evaporation at approx. 20°C - even more rapidly in	
Solubility in water	∞	aerosol form.	
Explosive limits, vol% in air	14-33	**Immediate effects:** Corrosive to the eyes, skin and respiratory tract. Inhalation can cause lung	
Electrical conductivity, pS/m	5.8x10⁹	edema.	
Relative molecular mass	46.0		
Log P octanol/water	− 0.5		
Gross formula:	CH₂O₂		

HAZARDS/SYMPTOMS	PREVENTIVE MEASURES	FIRE EXTINGUISHING/FIRST AID
Fire: combustible.	keep away from open flame, no smoking.	powder, alcohol-resistant foam, water spray, carbon dioxide, (halons).
Explosion: above 69°C: forms explosive air-vapor mixtures.	above 69°C: sealed machinery, ventilation.	
Inhalation: *corrosive*; sore throat, cough, shortness of breath, severe breathing difficulties.	ventilation, local exhaust or respiratory protection.	fresh air, rest, place in half-sitting position, take to hospital.
Skin: *corrosive*; redness, pain, burns.	gloves, protective clothing.	remove contaminated clothing, flush skin with water or shower, refer to a doctor if necessary.
Eyes: *corrosive*; redness, pain, impaired vision.	face shield.	flush with water, take to a doctor if necessary.
Ingestion: *corrosive*; sore throat, abdominal pain, diarrhea.		rinse mouth, take immediately to hospital.

SPILLAGE	STORAGE	LABELING / NFPA
collect leakage in sealable containers, flush away remainder with water; (additional individual protection: CHEMICAL SUIT).	keep separate from oxidants and strong bases; ventilate at floor level.	R: 35 S: 2-23-26 Note B Corrosive (NFPA: 3 / 2 / 0)

NOTES
Odor limit is above TLV. Explosive limits of 90% aqueous solution: 18-57% by vol. Lung edema symptoms usually develop several hours later and are aggravated by physical exertion: rest and hospitalization essential. As first aid, a doctor or authorized person should consider administering a corticosteroid spray.

Transport Emergency Card TEC(R)-89

HI: 80; UN-number: 1779

FUEL OIL

PHYSICAL PROPERTIES	IMPORTANT CHARACTERISTICS
Boiling point, °C 200-400 Melting point, °C −10-+40 Flash point, °C >55 Auto-ignition temperature, °C 220-300 Relative density (water = 1) 0.9-1.1 Relative vapor density (air = 1) ca. 4 Relative density at 20 °C of saturated mixture vapor/air (air = 1) 1.0 Vapor pressure, mm Hg at 50 °C ca. 0.76 Solubility in water none Explosive limits, vol% in air 1.5-? Electrical conductivity, pS/m $< 10^4$	**BLACKISH-BROWN VISCOUS MASS, DARK LIQUID ABOVE 50 °C, WITH CHARACTERISTIC ODOR** Mixture of mainly higher hydrocarbons. Vapor mixes readily with air. Flow, agitation etc. can cause build-up of electrostatic charge due to liquid's low conductivity. TLV-TWA ✱ **Absorption route:** Can enter the body by ingestion. Evaporation negligible at 20°C, but harmful atmospheric concentrations can build up rapidly in aerosol form. See 'Notes'. **Immediate effects:** Liquid destroys the skin's natural oils. Hot liquid can cause burns. If liquid is swallowed, droplets can enter the lungs, with risk of pneumonia.

HAZARDS/SYMPTOMS	PREVENTIVE MEASURES	FIRE EXTINGUISHING/FIRST AID
Fire: combustible.	keep away from open flame, no smoking.	powder, AFFF, foam, carbon dioxide, (halons).
Explosion: above 55°C: forms explosive air-vapor mixtures.	above 55°C: sealed machinery, ventilation, grounding.	
Inhalation: only at very high temperatures.		
Skin: redness.	heat-insulated gloves, protective clothing.	remove contaminated clothing, wash skin with soap and water.
Eyes: redness, pain.	face shield, or combined eye and respiratory protection.	flush with water, take to a doctor if necessary.
Ingestion: irritation of mucous membranes, burning sensation in mouth and throat, cough, nausea, stomachache, abdominal pain, vomiting, diarrhea, drowsiness.		rinse mouth, DO NOT induce vomiting, take to hospital.

SPILLAGE	STORAGE	LABELING / NFPA
collect leakage in sealable containers, allow remainder to set, clean up spillage.		

NOTES
Protection is for processing of warm/hot liquid. ✱ No TLV due to varying composition. Can contain harmful components which readily evaporate from liquid above 50°C.

Transport Emergency Card TEC(R)-752   **HI: 30; UN-number: 1202**

FUEL OIL
(heating oil)

PHYSICAL PROPERTIES	IMPORTANT CHARACTERISTICS
Boiling point, °C 155-390 Melting point, °C < − 10 Flash point, °C > 55 Auto-ignition temperature, °C > 220 Relative density (water = 1) 0.8-0.9 Relative density at 20 °C of saturated mixture vapor/air (air = 1) 1.0 Vapor pressure, mm Hg at 20 °C 3.04 Solubility in water none Explosive limits, vol% in air 0.6-6.5 Electrical conductivity, pS/m < 10⁴	**COLORED LIQUID WITH CHARACTERISTIC ODOR** Vapor mixes readily with air. Flow, agitation etc. can cause build-up of electrostatic charge due to liquid's low conductivity. Mixture of hydrocarbons; physical properties vary, depending on composition. TLV-TWA ✱ **Absorption route:** Can enter the body by inhalation or ingestion. **Immediate effects:** Irritates the eyes. Liquid destroys the skin's natural oils. If liquid is swallowed, droplets can enter the lungs, with risk of pneumonia. **Effects of prolonged/repeated exposure:** At high exposure levels can attack the peripheral nervous system, possibly due to the presence of hexane.

HAZARDS/SYMPTOMS	PREVENTIVE MEASURES	FIRE EXTINGUISHING/FIRST AID
Fire: combustible.	keep away from open flame, no smoking.	powder, AFFF, foam, carbon dioxide, (halons).
Explosion: above 55°C: forms explosive air-vapor mixtures.	above 55°C: sealed machinery, ventilation, grounding.	
Inhalation: sore throat, dizziness, nausea, vomiting, drowsiness, unconsciousness, impaired vision.	ventilation.	fresh air, rest.
Skin: redness.	gloves.	remove contaminated clothing, wash skin with soap and water.
Eyes: redness, pain.	safety glasses.	flush with water, take to a doctor if necessary.
Ingestion: sore throat, abdominal pain, diarrhea, severe breathing difficulties, nausea, vomiting, dizziness, chest pain.		rinse mouth, DO NOT induce vomiting, take immediately to hospital.

SPILLAGE	STORAGE	LABELING / NFPA
collect leakage in sealable containers, soak up with sand or other inert absorbent and remove to safe place.		2 0 0

NOTES
✱ No TLV can be given due to varying composition of domestic fuel oil.

Transport Emergency Card TEC(R)-30G15

HI: 30; UN-number: 1202

FUMARIC ACID

PHYSICAL PROPERTIES	IMPORTANT CHARACTERISTICS	
Sublimation temperature, °C — see notes Melting point, °C — 287 Flash point, °C — 230 Auto-ignition temperature, °C — 375 Relative density (water = 1) — 1.6 Vapor pressure, mm Hg at 165 °C — 1.7 Solubility in water, g/100 ml at 25 °C — 0.6 Relative molecular mass — 116.0 Log P octanol/water — 0.3	**WHITE CRYSTALS** Decomposes when heated above 350°C, giving off corrosive vapors (→ *maleic anhydride*). Reacts with strong oxidants.	
	TLV-TWA not available	
Gross formula: $C_4H_4O_4$	**Absorption route:** Can enter the body by inhalation or ingestion. Evaporation negligible at 20°C, but harmful concentrations of airborne particles can build up rapidly. **Immediate effects:** Irritates the eyes, skin and respiratory tract.	

HAZARDS/SYMPTOMS	PREVENTIVE MEASURES	FIRE EXTINGUISHING/FIRST AID
Fire: combustible.	keep away from open flame, no smoking.	water spray, powder.
Explosion: finely dispersed particles form explosive mixtures on contact with air.	keep dust from accumulating; sealed machinery, explosion-proof electrical equipment and lighting, grounding.	
Inhalation: sore throat, cough.	local exhaust or respiratory protection.	fresh air, rest, call a doctor.
Skin: redness.	gloves.	remove contaminated clothing, flush skin with water or shower.
Eyes: redness, pain.	goggles.	flush with water, take to a doctor if necessary.
Ingestion: sore throat, abdominal pain, vomiting.		rinse mouth, call a doctor.

SPILLAGE	STORAGE	LABELING / NFPA
clean up spillage, flush away remainder with water; (additional individual protection: P2 respirator).		R: 36 S: 26 ☒ Irritant

NOTES
Sublimates when heated above 200°C. Ignition temperature of dust cloud 740°C. Log P octanol/water is estimated.

CAS-No: [98-01-1]
fural
2-furaldehyde
2-furancarboxaldehyde
furfuraldehyde

FURFURAL

<table>
<tr><td colspan="2">PHYSICAL PROPERTIES</td><td colspan="2">IMPORTANT CHARACTERISTICS</td></tr>
<tr>
<td>

Boiling point, °C   162
Melting point, °C   −37
Flash point, °C   60
Auto-ignition temperature, °C   315
Relative density (water = 1)   1.2
Relative vapor density (air = 1)   3.3
Relative density at 20 °C of saturated mixture
vapor/air (air = 1)   1.00
Vapor pressure, mm Hg at 20 °C   1.1
Solubility in water, g/100 ml   8.3
Explosive limits, vol% in air   2.1-19.3
Relative molecular mass   96.1
</td>
<td colspan="3">

COLORLESS TO LIGHT YELLOW LIQUID WITH CHARACTERISTIC ODOR, TURNING BROWN ON EXPOSURE TO AIR
Vapor mixes readily with air. Reacts violently with oxidants. Reacts violently with strong acids and bases, with risk of fire and explosion (due to resin formation). Attacks many plastics.

| TLV-TWA (skin) | 2 ppm | 7.9 mg/m³ |

Absorption route: Can enter the body by inhalation or ingestion. Harmful atmospheric concentrations can build up fairly rapidly on evaporation at approx. 20°C - even more rapidly in aerosol form.
Immediate effects: Corrosive to the eyes, skin and respiratory tract. Inhalation of vapor/fumes can cause severe breathing difficulties (lung edema). In serious cases risk of death. Can affect the central nervous system, causing weakness, coma.
Effects of prolonged/repeated exposure: Prolonged or repeated contact can cause dermatitis.
</td>
</tr>
<tr>
<td>Gross formula: $C_5H_4O_2$</td>
<td colspan="3"></td>
</tr>
<tr><td>HAZARDS/SYMPTOMS</td><td>PREVENTIVE MEASURES</td><td colspan="2">FIRE EXTINGUISHING/FIRST AID</td></tr>
<tr>
<td>**Fire:** combustible.</td>
<td>keep away from open flame, no smoking.</td>
<td colspan="2">powder, water spray, foam, carbon dioxide, (halons).</td>
</tr>
<tr>
<td>**Explosion:** above 60°C: forms explosive air-vapor mixtures.</td>
<td>above 60°C: sealed machinery, ventilation.</td>
<td colspan="2"></td>
</tr>
<tr>
<td>**Inhalation:** *corrosive*; sore throat, cough, shortness of breath, severe breathing difficulties.</td>
<td>ventilation, local exhaust or respiratory protection.</td>
<td colspan="2">fresh air, rest, place in half-sitting position, take to hospital.</td>
</tr>
<tr>
<td>**Skin:** *corrosive*; redness, pain, burns.</td>
<td>gloves, protective clothing.</td>
<td colspan="2">remove contaminated clothing, flush skin with water or shower, refer to a doctor.</td>
</tr>
<tr>
<td>**Eyes:** *corrosive*; redness, pain, impaired vision.</td>
<td>face shield.</td>
<td colspan="2">flush with water, take to a doctor.</td>
</tr>
<tr>
<td>**Ingestion:** *corrosive*; sore throat, abdominal pain, vomiting.</td>
<td></td>
<td colspan="2">rinse mouth, immediately take to hospital.</td>
</tr>
<tr><td>SPILLAGE</td><td>STORAGE</td><td colspan="2">LABELING / NFPA</td></tr>
<tr>
<td>collect leakage in sealable containers, soak up with sand or other inert absorbent and remove to safe place; (additional individual protection: breathing apparatus).</td>
<td>keep in dark place, separate from oxidants, strong acids and strong bases; ventilate at floor level.</td>
<td colspan="2">

R: 23/25
S: 24/25-44

Toxic

(NFPA diamond: 2 / 2 / 0)
</td>
</tr>
<tr><td colspan="4">NOTES</td></tr>
<tr><td colspan="4">Lung edema symptoms usually develop several hours later and are aggravated by physical exertion: rest and hospitalization essential. As first aid, a doctor or authorized person should consider administering a corticosteroid spray.</td></tr>
</table>

Transport Emergency Card TEC(R)-84

HI: 30; UN-number: 1199

CAS-No: [98-00-0]
2-furanmethanol
2-furancarbinol
furfuralalcohol
2-furyl carbinol
2-hydroxymethylfuran

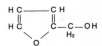

FURFURYL ALCOHOL

PHYSICAL PROPERTIES		IMPORTANT CHARACTERISTICS	
Boiling point, °C	171	**COLORLESS LIQUID, TURNING YELLOW ON EXPOSURE TO AIR**	
Melting point, °C	−31	Vapor mixes readily with air. Reacts violently with strong acids and oxidants, with risk of fire and explosion.	
Flash point, °C	75		
Auto-ignition temperature, °C	390		
Relative density (water = 1)	1.1	TLV-TWA (skin) 10 ppm 40 mg/m³	
Relative density at 20 °C of saturated mixture vapor/air (air = 1)	1.0	STEL 15 ppm 60 mg/m³	
Vapor pressure, mm Hg at 20 °C	0.38	**Absorption route:** Can enter the body by inhalation or ingestion or through the skin. Harmful atmospheric concentrations build up fairly slowly on evaporation at 20°C, but much more rapidly in aerosol form.	
Solubility in water	∞		
Explosive limits, vol% in air	1.8-16.3		
Relative molecular mass	98.1	**Immediate effects:** Irritates the skin and respiratory tract. Corrosive to the eyes. Liquid destroys the skin's natural oils. Affects the nervous system. In serious cases risk of unconsciousness and death.	
Gross formula:	$C_5H_6O_2$		

HAZARDS/SYMPTOMS	PREVENTIVE MEASURES	FIRE EXTINGUISHING/FIRST AID
Fire: combustible.	keep away from open flame, no smoking.	powder, alcohol-resistant foam, water spray, carbon dioxide, (halons).
Explosion: above 75°C: forms explosive air-vapor mixtures.	above 75°C: sealed machinery, ventilation.	
Inhalation: headache, dizziness, vomiting, unconsciousness.	ventilation, local exhaust or respiratory protection.	fresh air, rest, take to hospital.
Skin: *is absorbed*; redness, pain.	gloves.	remove contaminated clothing, flush skin with water or shower, refer to a doctor.
Eyes: *corrosive*; redness, pain, impaired vision.	face shield.	flush with water, take to a doctor.
Ingestion: diarrhea, nausea, dizziness.		rinse mouth, call a doctor or take to hospital.

SPILLAGE	STORAGE	LABELING / NFPA
collect leakage in sealable containers, flush away remainder with water.	keep separate from oxidants and strong acids.	R: 20/21/22 ✖ Harmful

NFPA: 1 / 2 / 1

NOTES
Odor limit is above TLV. Alcohol consumption increases toxic effects.

Transport Emergency Card TEC(R)-61G06C

HI: 60; UN-number: 2874

GALLIC ACID

PHYSICAL PROPERTIES	IMPORTANT CHARACTERISTICS	
Melting point (decomposes), °C 240 Solubility in water, g/100 ml at 20 °C 1 Relative molecular mass 188.1 Log P octanol/water −0.2	**WHITE GRAYISH CRYSTALS OR POWDER** Reacts violently with some oxidants.	
	TLV-TWA not available	
	Absorption route: Can enter the body by inhalation or ingestion. **Immediate effects:** Irritates the eyes, skin and respiratory tract.	
Gross formula: $C_7H_6O_5.H_2O$		

HAZARDS/SYMPTOMS	PREVENTIVE MEASURES	FIRE EXTINGUISHING/FIRST AID
Fire: combustible.	keep away from open flame, no smoking.	water spray, powder.
Inhalation: sore throat, cough.	local exhaust or respiratory protection.	fresh air, rest, call a doctor if necessary.
Skin: redness.	gloves.	remove contaminated clothing, flush skin with water or shower.
Eyes: redness, pain.	safety glasses.	flush with water, take to a doctor if necessary.
Ingestion: sore throat, abdominal pain.		rinse mouth, call a doctor.

SPILLAGE	STORAGE	LABELING / NFPA
clean up spillage, flush away remainder with water; (additional individual protection: P2 respirator).	keep separate from oxidants.	

NOTES
Loss of crystalwater when heated above 100°C. Log P octanol/water is estimated.

GAS OIL

PHYSICAL PROPERTIES		IMPORTANT CHARACTERISTICS	
Boiling point, °C	180-370	**COLORLESS OR COLORED LIQUID WITH CHARACTERISTIC ODOR**	
Melting point, °C	<0[1]	Mixture of hydrocarbons; physical properties vary, depending on the composition. Vapor mixes readily with air. Flow, agitation etc. can cause build-up of electrostatic charge due to liquid's low conductivity.	
Flash point, °C	>55		
Auto-ignition temperature, °C	>220		
Relative density (water = 1)	0.8-0.9		
Relative vapor density (air = 1)	7		
Relative density at 20 °C of saturated mixture vapor/air (air = 1)	1	TLV-TWA	not available
Vapor pressure, mm Hg at 20 °C	<0.76	**Absorption route:** Can enter the body by inhalation or ingestion. Harmful atmospheric concentrations build up very slowly, if at all, on evaporation at approx. 20°C, but much more rapidly in aerosol form.	
Solubility in water	none		
Explosive limits, vol% in air	0.6-6.5		
Electrical conductivity, pS/m	<10[4]	**Immediate effects:** Liquid destroys the skin's natural oils. If liquid is swallowed, droplets can enter the lungs, with risk of pneumonia.	
Relative molecular mass	ca. 170		

HAZARDS/SYMPTOMS	PREVENTIVE MEASURES	FIRE EXTINGUISHING/FIRST AID
Fire: combustible.	keep away from open flame, no smoking.	powder, AFFF, foam, carbon dioxide, (halons).
Explosion: above 55°C: forms explosive air-vapor mixtures.	above 55°C: sealed machinery, ventilation, grounding.	
Inhalation: headache.	ventilation.	fresh air, rest.
Skin: redness.	gloves.	remove contaminated clothing, wash skin with soap and water.
Eyes: redness, pain.	safety glasses.	flush with water, take to a doctor if necessary.
Ingestion: abdominal pain, nausea, severe breathing difficulties.		rinse mouth, DO NOT induce vomiting, take immediately to hospital.

SPILLAGE	STORAGE	LABELING / NFPA
collect leakage in sealable containers, soak up with sand or other inert absorbent and remove to safe place.		 2 0 0

NOTES
The measures on this card also apply to diesel oil. Gas oil is used as fuel for small boilers, furnaces and engines (not in road vehicles); diesel oil is used as fuel for road vehicle engines. Dyes are usually added to gas oil to comply with customs regulations. 1)Melting point of gas oil is artificially lowered in winter.

Transport Emergency Card TEC(R)--26

HI: 30; UN-number: 1202

GASOLINE

PHYSICAL PROPERTIES	IMPORTANT CHARACTERISTICS	
Boiling point, °C · 38-205 Melting point, °C · < -20 Flash point, °C · < -20 Auto-ignition temperature, °C · > 220 Relative density (water = 1) · 0.7-0.8 Relative vapor density (air = 1) · ca. 4 Relative density at 20 °C of saturated mixture vapor/air (air = 1) · ca. 1.15 Vapor pressure, mm Hg at 20 °C · 38-304 Solubility in water · none Explosive limits, vol% in air · 0.6-8 Minimum ignition energy, mJ · > 60 Electrical conductivity, pS/m · 0.1 (pure) Relative molecular mass · ca. 115	**COLORED LIQUID WITH CHARACTERISTIC ODOR** Vapor is heavier than air and spreads at ground level, with risk of ignition at a distance. Flow, agitation etc. can cause build-up of electrostatic charge due to liquid's low conductivity. Do not use compressed air when filling, emptying or processing.	
	TLV-TWA · 300 ppm · 890 mg/m³ ✱ STEL · 500 ppm · 1480 mg/m³	
	Absorption route: Can enter the body by inhalation or ingestion. Harmful atmospheric concentrations can build up fairly rapidly on evaporation at approx. 20°C - even more rapidly in aerosol form. **Immediate effects:** Liquid destroys the skin's natural oils. Affects the nervous system. If liquid is swallowed, droplets can enter the lungs, with risk of pneumonia.	

HAZARDS/SYMPTOMS	PREVENTIVE MEASURES	FIRE EXTINGUISHING/FIRST AID
Fire: extremely flammable.	keep away from open flame and sparks, no smoking.	powder, AFFF, foam, carbon dioxide, (halons).
Explosion: forms explosive air-vapor mixtures.	sealed machinery, ventilation, explosion-proof electrical equipment and lighting, grounding.	in case of fire: keep tanks/drums cool by spraying with water.
Inhalation: headache, dizziness, unconsciousness.	ventilation, local exhaust or respiratory protection.	fresh air, rest, artificial respiration if necessary, call a doctor.
Skin: redness.	gloves.	remove contaminated clothing, wash skin with soap and water.
Eyes: redness, pain.	safety glasses.	flush with water, take to a doctor.
Ingestion: nausea, vomiting, dizziness, chest pain.		rinse mouth, DO NOT induce vomiting, call a doctor or take to hospital.

SPILLAGE	STORAGE	LABELING / NFPA
collect leakage in sealable containers, soak up with dry sand or other inert absorbent and remove to safe place, do not flush into sewer; (additional individual protection: breathing apparatus).	keep in fireproof place.	R: 11 S: 9-16-29-33 🔥 Flammable · 1 - 3 / 0

NOTES
Referred to as petrol in British English. ✱ Vehicle gasoline can contain up to 7.5% benzene and up to 0.4kg/m³ (as lead) lead alkyles (TEL, TML). The presence of these components affects the TLV. Risk of electrostatic charges forming in end product is slight due to addition of dopes. Alcohol consumption increases toxic effects.

Transport Emergency Card TEC(R)-530

HI: 33; UN-number: 1203

1,3-diformal propane
glutaral
glutaric dialdehyde
pentanedial

OHC(CH$_2$)$_3$CHO

GLUTARALDEHYDE

PHYSICAL PROPERTIES	IMPORTANT CHARACTERISTICS	
Boiling point (decomposes), °C 188 Melting point, °C ca. − 10 Relative density (water = 1) 0.98 Relative vapor density (air = 1) 3.5 Relative density at 20 °C of saturated mixture vapor/air (air = 1) 1.0 Vapor pressure, mm Hg at 70 °C 9.9 Solubility in water reaction Relative molecular mass 100.1	**OILY LIQUID WITH PUNGENT ODOR** Vapor mixes readily with air. Decomposes when heated above 188°C, giving off acid vapors. Strong reducing agent which reacts violently with oxidants. Reacts with water, forming soluble polymers.	
	TLV-TWA	0.2 ppm	0.82 mg/m³ C
	Absorption route: Can enter the body by inhalation or ingestion or through the skin. Harmful atmospheric concentrations can build up fairly rapidly on evaporation at approx. 20°C - even more rapidly in aerosol form. **Immediate effects:** Irritates the eyes, skin and respiratory tract. Affects the nervous system. In serious cases risk of unconsciousness. **Effects of prolonged/repeated exposure:** Prolonged or repeated contact can cause eczema.	
Gross formula: C$_5$H$_8$O$_2$		

HAZARDS/SYMPTOMS	PREVENTIVE MEASURES	FIRE EXTINGUISHING/FIRST AID
Fire: hardly combustible.		in case of fire in immediate vicinity: use any extinguishing agent.
Inhalation: cough, shortness of breath, dizziness, drowsiness.	ventilation, local exhaust or respiratory protection.	fresh air, rest, call a doctor.
Skin: *is absorbed*; redness, dizziness, drowsiness.	gloves.	remove contaminated clothing, flush skin with water or shower, call/take to a doctor.
Eyes: redness, pain.	face shield.	flush with water, take to a doctor if necessary.
Ingestion: abdominal pain, nausea.		rinse mouth, take immediately to hospital.

SPILLAGE	STORAGE	LABELING / NFPA
collect leakage in sealable containers, flush away remainder with water; (additional individual protection: breathing apparatus).	keep separate from oxidants; ventilate at floor level.	

NOTES
Usually sold as 50% aqueous solution. TLV is maximum and must not be exceeded.

1,3-propane dicarboxylic acid
pentanedioic acid

$HOOC(CH_2)_3COOH$

GLUTARIC ACID

PHYSICAL PROPERTIES	IMPORTANT CHARACTERISTICS
Boiling point (decomposes), °C 302 Melting point, °C 97 Relative density (water = 1) 1.4 Relative vapor density (air = 1) 4.5 Relative density at 20 °C of saturated mixture vapor/air (air = 1) 1.0 Solubility in water, g/100 ml at 20°C 64 Relative molecular mass 132.1 Log P octanol/water −0.3	**WHITE CRYSTALS** Decomposes slowly when heated to boiling point. In aqueous solution is a medium strong acid. Reacts with strong oxidants and bases.

	TLV-TWA not available
	Absorption route: Can enter the body by inhalation or ingestion. Evaporation negligible at 20°C, but harmful concentrations of airborne particles can build up rapidly. **Immediate effects:** Irritates the eyes, skin and respiratory tract.

Gross formula: $C_5H_8O_4$

HAZARDS/SYMPTOMS	PREVENTIVE MEASURES	FIRE EXTINGUISHING/FIRST AID
Fire: combustible.	keep away from open flame, no smoking.	water spray, powder.
Inhalation: sore throat, cough.	local exhaust or respiratory protection.	fresh air, rest, place in half-sitting position, call a doctor if necessary.
Skin: redness, pain.	gloves.	remove contaminated clothing, flush skin with water or shower.
Eyes: redness, pain.	face shield.	flush with water, take to a doctor.
Ingestion: sore throat, abdominal pain, vomiting.		rinse mouth, consult a doctor.

SPILLAGE	STORAGE	LABELING / NFPA
clean up spillage, flush away remainder with water; (additional individual protection: P2 respirator).	keep separate from oxidants and strong bases.	

NOTES
Log P octanol/water is estimated.

CAS-No: [102-76-1]
**1,2,3-propane triol triacetate
triacetine**

$C_3H_5(OOCCH_3)_3$

GLYCERIN TRIACETATE

PHYSICAL PROPERTIES	IMPORTANT CHARACTERISTICS
Boiling point, °C 258 Melting point, °C 4 Flash point, °C 138 Auto-ignition temperature, °C 433 Relative density (water = 1) 1.16 Relative vapor density (air = 1) 7.5 Relative density at 20 °C of saturated mixture vapor/air (air = 1) 1.0 Vapor pressure, mm Hg at 20 °C ca. 0.076 Solubility in water, g/100 ml at 20 °C 7 Explosive limits, vol% in air 1.0-? Relative molecular mass 218.2 Log P octanol/water 0.1 Gross formula: $C_9H_{14}O_6$	**COLORLESS VISCOUS LIQUID WITH CHARACTERISTIC ODOR** Reacts with strong oxidants. TLV-TWA not available **Absorption route:** Can enter the body by inhalation or ingestion. Evaporation negligible at 20°C, but unpleasant concentrations can build up in aerosol form. **Immediate effects:** Does not affect the eyes, skin and respiratory tract.

HAZARDS/SYMPTOMS	PREVENTIVE MEASURES	FIRE EXTINGUISHING/FIRST AID
Fire: combustible.	keep away from open flame, no smoking.	powder, water spray, foam, carbon dioxide, (halons).
Inhalation:	ventilation.	
Eyes:	safety glasses.	

SPILLAGE	STORAGE	LABELING / NFPA
collect leakage in sealable containers, flush away remainder with water.		 ◇ 1 / 1 / 0

NOTES
When cooled at −37°C the substance goes by the solid state. Log P octanol/water is estimated.

GLYCEROL

PHYSICAL PROPERTIES	IMPORTANT CHARACTERISTICS
Boiling point, °C 290 Melting point, °C 18 Flash point, °C 160 Auto-ignition temperature, °C 400 Relative density (water = 1) 1.3 Relative vapor density (air = 1) 3.2 Relative density at 20 °C of saturated mixture vapor/air (air = 1) 1.0 Vapor pressure, mm Hg at 50 °C 0.0023 Solubility in water ∞ Electrical conductivity, pS/m 6.4×10^6 Relative molecular mass 92.1 Log P octanol/water −2.6	**COLORLESS ODORLESS HYGROSCOPIC VISCOUS LIQUID** Vapor mixes readily with air. Able to polymerize above 149°C. Decomposes when heated above 290°C, giving off corrosive gas (→ *acrolein*). Reacts violently with strong oxidants, with risk of fire and explosion.
	TLV-TWA 10 mg/m³ ✳
Gross formula: $C_3H_8O_3$	**Absorption route:** Can enter the body by inhalation or ingestion. Harmful atmospheric concentrations build up very slowly, if at all, on evaporation at approx. 20°C, but much more rapidly in aerosol form. **Immediate effects:** Irritates the eyes and skin. **Effects of prolonged/repeated exposure:** Can cause kidney damage.

HAZARDS/SYMPTOMS	PREVENTIVE MEASURES	FIRE EXTINGUISHING/FIRST AID
Fire: combustible.	keep away from open flame, no smoking.	powder, alcohol-resistant foam, water spray, carbon dioxide, (halons).
Skin: redness.	gloves.	remove contaminated clothing, wash skin with soap and water.
Eyes: redness, pain.	safety glasses.	flush with water, take to a doctor if necessary.
Ingestion: diarrhea, headache, nausea.		rinse mouth, call a doctor or take immediately to hospital.

SPILLAGE	STORAGE	LABELING / NFPA
collect leakage in sealable containers, flush away remainder with water.	keep dry, separate from oxidants.	◇ 1 / 1 0

NOTES
Does not solidify until well below melting point. Log P octanol/water is estimated. ✳ TLV equals that of glycerin mist.

GLYCEROLTRIBUTYRATE

PHYSICAL PROPERTIES		IMPORTANT CHARACTERISTICS	
Boiling point, °C	315	**COLORLESS VISCOUS LIQUID WITH CHARACTERISTIC ODOR**	
Melting point, °C	−75	Vapor mixes readily with air. Reacts with strong oxidants.	
Flash point, °C	180		
Auto-ignition temperature, °C	407	TLV-TWA not available	
Relative density (water = 1)	1.03		
Relative vapor density (air = 1)	10.5		
Relative density at 20 °C of saturated mixture		**Absorption route:** Can enter the body by inhalation or ingestion. Evaporation negligible at 20°C, but harmful atmospheric concentrations can build up rapidly in aerosol form.	
vapor/air (air = 1)	1.0		
Vapor pressure, mm Hg at 20 °C	<0.076		
Solubility in water	none		
Explosive limits, vol% in air	0.5-?		
Relative molecular mass	302.4		
Log P octanol/water	2.5		
Gross formula: $C_{15}H_{26}O_6$			

HAZARDS/SYMPTOMS	PREVENTIVE MEASURES	FIRE EXTINGUISHING/FIRST AID
Fire: combustible.	keep away from open flame, no smoking.	powder, AFFF, foam, carbon dioxide, (halons).
Inhalation:	ventilation.	fresh air, rest.
Skin:	gloves.	remove contaminated clothing, flush skin with water or shower.
Eyes:		flush with water, take to a doctor.
Ingestion:		rinse mouth, call a doctor if necessary.

SPILLAGE	STORAGE	LABELING / NFPA
collect leakage in sealable containers, soak up with sand or other inert absorbent and remove to safe place.	keep separate from oxidants.	

NOTES
Explosive limit is definite at 208°C. Insufficient toxicological data on harmful effects to humans. Few indications of toxicity found in animal experiments.

CAS-No: [556-52-5]
2,3-epoxy-1-propanol
epihydrin alcohol
glycidyl alcohol
3-hydroxy-1,2-epoxypropane
3-hydroxypropylene oxide

$$HO-CH_2-CH-CH_2$$
with epoxide O bridging the CH and CH$_2$

GLYCIDOL

PHYSICAL PROPERTIES		IMPORTANT CHARACTERISTICS
Boiling point (decomposes), °C	162	**COLORLESS LIQUID**
Melting point, °C	−70	Vapor mixes readily with air. Polymerizes on contact with strong acids and bases, active metals and certain metal salts. Reacts violently with strong oxidants.
Flash point, °C	71	
Auto-ignition temperature, °C	415	
Relative density (water = 1)	1.1	TLV-TWA 25 ppm 76 mg/m³
Relative vapor density (air = 1)	2.6	
Relative density at 20 °C of saturated mixture vapor/air (air = 1)	1.0	**Absorption route:** Can enter the body by inhalation or ingestion or through the skin.
Vapor pressure, mm Hg at 25 °C	0.84	**Immediate effects:** Irritates the eyes, skin and respiratory tract. Affects the nervous system. In serious cases risk of unconsciousness.
Solubility in water	∞	**Effects of prolonged/repeated exposure:** Prolonged or repeated contact can cause skin disorders.
Relative molecular mass	74.1	
Gross formula:	C$_3$H$_6$O$_2$	

HAZARDS/SYMPTOMS	PREVENTIVE MEASURES	FIRE EXTINGUISHING/FIRST AID
Fire: combustible.	keep away from open flame, no smoking.	water spray, alcohol-resistant foam, powder, carbon dioxide, (halons).
Explosion: above 71°C: forms explosive air-vapor mixtures.	above 71°C: sealed machinery, ventilation.	
Inhalation: cough, shortness of breath, unconsciousness.	ventilation, local exhaust or respiratory protection.	fresh air, rest, place in half-siting position, call a doctor, take to hospital if necessary.
Skin: *is absorbed*; redness, pain.	gloves, protective clothing.	remove contaminated clothing, flush skin with water or shower, refer to a doctor.
Eyes: redness, pain, impaired vision.	face shield.	flush with water, take to a doctor.
Ingestion: abdominal pain, shortness of breath, nausea, vomiting, unconsciousness, seizures.		rinse mouth, give plenty of water to drink, DO NOT INDUCE VOMITING, take immediately to hospital.

SPILLAGE	STORAGE	LABELING / NFPA
collect leakage in non-sealable containers, flush away remainder with water.	keep separate from oxidants, strong acids and strong bases; ventilate at floor level.	R: 23-21/22-36/37/38-42/43 S: 44 Toxic

NOTES

Transport Emergency Card TEC(R)-61G07

OHCH₂COOH

hydroxyacetic acid
hydroxyethanoic acid

GLYCOLLIC ACID

PHYSICAL PROPERTIES	IMPORTANT CHARACTERISTICS
Decomposes below boiling point, °C 100 Melting point, °C 80 Relative density (water = 1) 1.27 Solubility in water very good Relative molecular mass 76.0 Log P octanol/water − 1.1	**COLORLESS HYGROSCOPIC CRYSTALS WITH CHARACTERISTIC ODOR** Decomposes when heated above 100°C, giving off flammable vapors. In aqueous solution is a medium strong acid.
	TLV-TWA not available
	Absorption route: Can enter the body by inhalation or ingestion. Evaporation negligible at 20°C, but harmful concentrations of airborne particles can build up rapidly. **Immediate effects:** Corrosive to the eyes, skin and respiratory tract.
Gross formula: $C_2H_4O_3$	

HAZARDS/SYMPTOMS	PREVENTIVE MEASURES	FIRE EXTINGUISHING/FIRST AID
Fire: combustible.	keep away from open flame, no smoking.	powder, water spray, foam, carbon dioxide, (halons).
Inhalation: *corrosive*; sore throat, cough, shortness of breath.	local exhaust or respiratory protection.	fresh air, rest, call a doctor.
Skin: *corrosive*; redness, pain, burns.	gloves.	remove contaminated clothing, flush skin with water or shower, refer to a doctor.
Eyes: *corrosive*; redness, pain, impaired vision.	face shield.	flush with water, take to a doctor.
Ingestion: *corrosive*; sore throat, cough.		rinse mouth, call a doctor or take to hospital.

SPILLAGE	STORAGE	LABELING / NFPA
clean up spillage, flush away remainder with water; (additional individual protection: P2 respirator).		

NOTES
Often sold in 70% aqueous solution.

CAS-No: [107-22-2]
diformyl
oxalic aldehyde

OCHCHO

GLYOXAL

PHYSICAL PROPERTIES		IMPORTANT CHARACTERISTICS
Boiling point, °C	50	**LIGHT YELLOW LIQUID OR YELLOW CRYSTALS WITH CHARACTERISTIC ODOR**
Melting point, °C	15	Vapor is heavier than air and spreads at ground level, with risk of ignition at a distance. Able to
Flash point, °C	?	polymerize violently when heated above 50°C and/or on contact with bases, water or moist air,
Relative density (water = 1)	1.2	with risk of fire and explosion. Strong reducing agent which reacts violently with water, strong
Relative vapor density (air = 1)	2.0	acids, bases and many other substances. Attacks steel, aluminum and copper.
Vapor pressure, mm Hg at 20 °C	ca. 236	
Solubility in water	reaction	TLV-TWA not available
Explosive limits, vol% in air	?-?	
Relative molecular mass	58.0	
		Absorption route: Can enter the body by inhalation or ingestion or through the skin. Insufficient data on the rate at which harmful concentrations can build up.
		Immediate effects: Irritates the eyes, skin and respiratory tract.
Gross formula:	$C_2H_2O_2$	

HAZARDS/SYMPTOMS	PREVENTIVE MEASURES	FIRE EXTINGUISHING/FIRST AID
Fire: flammable.	keep away from open flame and sparks, no smoking.	powder, carbon dioxide, (halons), DO NOT USE WATER-BASED EXTINGUISHERS.
Explosion: forms explosive air-vapor mixtures, polymerization reactions can cause explosions.	sealed machinery, ventilation, explosion-proof electrical equipment and lighting.	in case of fire: keep tanks/drums cool by spraying with water, but DO NOT spray substance with water.
Inhalation: sore throat, cough, shortness of breath.	ventilation, local exhaust or respiratory protection.	fresh air, rest, call a doctor.
Skin: *is absorbed*; redness, pain.	gloves.	remove contaminated clothing, flush skin with water or shower.
Eyes: redness, pain.	face shield.	flush with water, take to a doctor if necessary.
Ingestion: abdominal pain, pain during swallowing and vomiting.		rinse mouth, call a doctor or take to hospital.

SPILLAGE	STORAGE	LABELING / NFPA
call in an expert, collect leakage in non-sealable containers, soak up with sand or other inert absorbent and remove to safe place, do not use sawdust or other combustible absorbents; (additional individual protection: breathing apparatus).	keep in cool, dry, fireproof place, separate from strong acids and bases, add an inhibitor.	R: 36/38 S: 26-28 Note B ✖ Irritant

NOTES
Combustible, but flash point and explosive limits not given in the literature. Explosive limits not given in the literature. Reacts violently with extinguishing agents such as water or foam. Due to its high reactivity (e.g. polymerization) glyoxal is not obtainable as pure substance. An inhibited 40% solution in water is much more available. Use airtight packaging.

o-hydroxyanisole
1-hydroxy-2-methoxybenzene
2-methoxyphenol
methylcatechol
pyrocatechol monomethyl ether

$HOC_6H_4OCH_3$

GUAIACOL

PHYSICAL PROPERTIES	IMPORTANT CHARACTERISTICS
Boiling point, °C 205 Melting point, °C 28 Flash point, °C 82 Auto-ignition temperature, °C 385 Relative density (water = 1) 1.12 Relative vapor density (air = 1) 4.3 Relative density at 20 °C of saturated mixture vapor/air (air = 1) 1.00 Vapor pressure, mm Hg at 20 °C 0.15 Solubility in water, g/100 ml at 15 °C 1.7 Relative molecular mass 124.1 Log P octanol/water 1.9 Gross formula: $C_7H_8O_2$	**LIGHT YELLOW CRYSTALS WITH CHARACTERISTIC ODOR, TURNING BROWN ON EXPOSURE TO AIR** Vapor mixes readily with air. Reacts with oxidants. TLV-TWA not available **Absorption route:** Can enter the body by inhalation or ingestion or through the skin. Evaporation negligible at 20°C, but harmful atmospheric concentrations can build up rapidly in aerosol form. **Immediate effects:** Irritates the skin and respiratory tract. Corrosive to the eyes. Affects the nervous system.

HAZARDS/SYMPTOMS	PREVENTIVE MEASURES	FIRE EXTINGUISHING/FIRST AID
Fire: combustible.	keep away from open flame, no smoking.	powder, water spray, foam, carbon dioxide, (halons).
Explosion: above 82°C: forms explosive air-vapor mixtures.	above 82°C: sealed machinery, ventilation.	
	KEEP DUST UNDER CONTROL	
Inhalation: sore throat, cough.	local exhaust or respiratory protection.	fresh air, rest, call a doctor.
Skin: *is absorbed*; see also: 'Ingestion'.	gloves, protective clothing.	remove contaminated clothing, flush skin with water or shower, call a doctor.
Eyes: *corrosive*; redness, pain.	face shield.	flush with water, take to a doctor.
Ingestion: cough, abdominal pain, diarrhea, burning in mouth and throat, drowsiness, feeling of weakness.	do not eat, drink or smoke while working.	rinse mouth, immediately take to hospital.

SPILLAGE	STORAGE	LABELING / NFPA
clean up spillage, carefully collect remainder; (additional individual protection: P2 respirator).	ventilate at floor level.	

NOTES
At 20°C the substance can still be in liquid form because guaiacol solidifies very slowly. Log P octanol/water is estimated.

2-bromo-2-chloro-1,1,1-trifluoroethane

C(HBrCl)C(F$_3$)

HALOTHANE

PHYSICAL PROPERTIES	IMPORTANT CHARACTERISTICS	
Boiling point, °C .... 50 Melting point, °C .... − 118 Relative density (water = 1) .... 1.87 Relative vapor density (air = 1) .... 6.9 Relative density at 20 °C of saturated mixture vapor/air (air = 1) .... 2.9 Vapor pressure, mm Hg at 20 °C ..... 246 Solubility in water, g/100 ml at 20 °C 0.45 Relative molecular mass 197.3	**COLORLESS LIQUID WITH CHARACTERISTIC ODOR** Vapor is heavier than air and can accumulate close to ground level, causing oxygen deficiency, with risk of unconsciousness. Toxic and corrosive gases are a product of combustion. Decomposes when exposed to light.	
	TLV-TWA	50 ppm	404 mg/m³
	Absorption route: Can enter the body by inhalation, through the skin and by ingestion. Harmful atmospheric concentrations can build up very rapidly on evaporation at 20°C. **Immediate effects:** Irritates the eyes, skin and respiratory tract. Can affect the central nervous system, heart and liver. Can impair consciousness. Exposure to high concentrations can cause heart rate disorders, severe breathing difficulties and unconsciousness. Keep under medical observation. **Effects of prolonged/repeated exposure:** Can affect the liver, nervous system and kidneys. Can cause birth defects.	
Gross formula: C$_2$HBrClF$_3$		

HAZARDS/SYMPTOMS	PREVENTIVE MEASURES	FIRE EXTINGUISHING/FIRST AID
Fire: non-combustible.		in case of fire in immediate vicinity: use any extinguishing agent.
	STRICT HYGIENE, AVOID EXPOSURE OF PREGNANT WOMEN	
Inhalation: dizziness, sleepiness, drowsiness, nausea.	ventilation, local exhaust or respiratory protection.	fresh air, rest, artificial respiration if necessary, take immediately to hospital.
Skin: *is absorbed.*	gloves.	remove contaminated clothing, flush skin with water and wash with soap and water, consult a doctor.
Eyes: redness.	face shield, or combined eye and respiratory protection.	flush thoroughly with water (remove contact lenses if easy), take to a doctor.
Ingestion: disorientation, dizziness, sleepiness, drowsiness, nausea.		

SPILLAGE	STORAGE	LABELING / NFPA
call in an expert, collect leakage in sealable containers, soak up remainder with sand or other inert absorbent and remove to safe place; (additional individual protection: breathing apparatus).	keep in cool, dark place; ventilate at floor level.	 ⬦

NOTES

HELIUM
(cylinder)

PHYSICAL PROPERTIES		IMPORTANT CHARACTERISTICS	
Boiling point, °C	− 269	**COLORLESS, ODORLESS COMPRESSED GAS**	
Relative vapor density (air = 1)	0.14	Gas is lighter than air.	
Solubility in water	none		
Relative atomic mass	4.0	TLV-TWA	not available
Log P octanol/water	0.7		
		Absorption route: Can saturate the air if released, with risk of suffocation. **Immediate effects:** Inhalation can cause severe breathing difficulties. In serious cases risk of unconsciousness.	
Gross formula:	He		

HAZARDS/SYMPTOMS	PREVENTIVE MEASURES	FIRE EXTINGUISHING/FIRST AID
Fire: non-combustible.		in case of fire in immediate vicinity: use any extinguishing agent.
Explosion:		in case of fire: keep cylinders cool by spraying with water.
Inhalation: severe breathing difficulties, headache, dizziness, unconsciousness.	ventilation, local exhaust or respiratory protection.	fresh air, rest, artificial respiration if necessary, take to hospital.

SPILLAGE	STORAGE	LABELING / NFPA
ventilate; (additional individual protection: breathing apparatus).	if stored indoors, keep in cool, fireproof place; ventilate.	

NOTES
Log P octanol/water is estimated. High atmospheric concentrations, e.g. in poorly ventilated spaces, can cause oxygen deficiency, with risk of unconsciousness.

Transport Emergency Card TEC(R)-20G01

HI: 20; UN-number: 1046

HELIUM
(liquid, refrigerated)

PHYSICAL PROPERTIES	IMPORTANT CHARACTERISTICS	
Boiling point, °C − 268.9 Relative density (water = 1) see notes Relative vapor density (air = 1) 0.14 Vapor pressure, mm Hg at − 269 °C 760 Solubility in water none Relative atomic mass 4.0 Log P octanol/water 0.7	**COLORLESS, ODORLESS, EXTREMELY COLD LIQUID** *Cold* vapor is heavier than air.	
	TLV-TWA not available	
	Absorption route: If liquid is released, air can become saturated due to evaporation, with risk of suffocation. **Immediate effects:** Cold liquid can cause frostbite.	
Gross formula: He		

HAZARDS/SYMPTOMS	PREVENTIVE MEASURES	FIRE EXTINGUISHING/FIRST AID
Fire: non-combustible.		in case of fire in immediate vicinity: use any extinguishing agent, but do not spray liquid with water.
Inhalation: see Notes.	ventilation, local exhaust or respiratory protection.	fresh air, rest, artificial respiration if necessary, call a doctor, take to hospital if necessary.
Skin: *in case of frostbite*: redness, pain, blisters, sores.	insulating gloves.	*in case of frostbite*: DO NOT remove clothing, flush skin with water or shower, take to a doctor.
Eyes: *in case of frostbite*: redness, pain, impaired vision.	face shield, or combined eye and respiratory protection.	flush with water, take to a doctor.

SPILLAGE	STORAGE	LABELING / NFPA
ventilate, under no circumstances spray liquid with water; (additional individual protection: breathing apparatus).	ventilate, refrigerate.	◇

NOTES

Density of liquid at boiling point 0.147kg/l. Critical temperature − 267.9° C. Log P octanol/water is estimated. High atmospheric concentrations, e.g. in poorly ventilated spaces, can cause oxygen deficiency, with risk of unconsciousness. Use special insulated container.

Transport Emergency Card TEC(R)-829   **HI: 22; UN-number: 1963**

dipropylmethane
n-heptane
heptyl hydride

HEPTANE

PHYSICAL PROPERTIES	IMPORTANT CHARACTERISTICS
Boiling point, °C — 98 Melting point, °C — −91 Flash point, °C — −4 Auto-ignition temperature, °C — 215 Relative density (water = 1) — 0.68 Relative vapor density (air = 1) — 3.45 Relative density at 20 °C of saturated mixture vapor/air (air = 1) — 1.12 Vapor pressure, mm Hg at 20 °C — 37 Solubility in water, g/100 ml — 0.05 Explosive limits, vol% in air — 1.1-7 Minimum ignition energy, mJ — 0.24 Electrical conductivity, pS/m — 6.6 Relative molecular mass — 100.2 Gross formula: C₇H₁₆	**COLORLESS LIQUID WITH CHARACTERISTIC ODOR** Vapor is heavier than air and spreads at ground level, with risk of ignition at a distance. Flow, agitation etc. can cause build-up of electrostatic charge due to liquid's low conductivity. Do not use compressed air when filling, emptying or processing. Usually contains isomers and other hydrocarbons.

TLV-TWA	400 ppm	1640 mg/m³
STEL	500 ppm	2050 mg/m³

Absorption route: Can enter the body by inhalation or ingestion. Harmful atmospheric concentrations build up fairly slowly on evaporation at 20°C, but much more rapidly in aerosol form.
Immediate effects: Irritates the eyes. High concentrations can affect the central nervous system. Liquid destroys the skin's natural oils. If liquid is swallowed, droplets can enter the lungs, with risk of pneumonia.

HAZARDS/SYMPTOMS	PREVENTIVE MEASURES	FIRE EXTINGUISHING/FIRST AID
Fire: extremely flammable.	keep away from open flame and sparks, no smoking.	powder, AFFF, foam, carbon dioxide, (halons).
Explosion: forms explosive air-vapor mixtures.	sealed machinery, ventilation, explosion-proof electrical equipment and lighting, grounding, non-sparking tools.	in case of fire: keep tanks/drums cool by spraying with water.
Inhalation: severe breathing difficulties, dizziness, nausea, drowsiness, sleepiness, unconsciousness, light in the head, disorientation, giggly, drunk-like walk.	ventilation, local exhaust or respiratory protection.	fresh air, rest, place in half-sitting position, artifical respiration if necessary, call a doctor.
Skin: redness.	gloves.	remove contaminated clothing, flush skin with water or shower.
Eyes: redness, pain.	safety glasses.	flush with water, take to a doctor if necessary.
Ingestion: abdominal pain, nausea; see also 'Inhalation'.		rinse mouth, DO NOT induce vomiting, call a doctor or take to hospital.

SPILLAGE	STORAGE	LABELING / NFPA
collect leakage in sealable containers, soak up with sand or other inert absorbent and remove to safe place.	keep in fireproof place.	R: 11 S: 9-16-23-29-33 Note C 🔥 Flammable

NFPA: 3 / 1 / 0

NOTES

Isomers of heptane have boiling points between 79°C and 98°C. 5000ppm for four minutes produces all the symptoms stated under 'Inhalation', but not unconsciousness and symptoms to the eyes and mucosa. In case of chronically high contact with heptane, there are no long-term damages, in contrast to hexane, but hexane is often present as contamination or admixture.

Transport Emergency Card TEC(R)-504

HI: 33; UN-number: 1206

amylmethyl carbinol
heptanol-2
sec-heptyl alcohol
methyl-n-amyl carbinol

$CH_3(CH_2)_4CHOHCH_3$

2-HEPTANOL

PHYSICAL PROPERTIES		IMPORTANT CHARACTERISTICS
Boiling point, °C	160	**COLORLESS LIQUID WITH CHARACTERISTIC ODOR**
Melting point, °C	?	Vapor mixes readily with air. Reacts with strong oxidants.
Flash point, °C	71	
Relative density (water = 1)	0.8	TLV-TWA not available
Relative vapor density (air = 1)	4.0	
Relative density at 20 °C of saturated mixture vapor/air (air = 1)	1.0	**Absorption route:** Can enter the body by inhalation or ingestion or through the skin. Harmful atmospheric concentrations build up fairly slowly on evaporation at 20°C, but much more rapidly in aerosol form.
Vapor pressure, mm Hg at 20 °C	0.99	
Solubility in water, g/100 ml	0.35	
Relative molecular mass	116.2	**Immediate effects:** Irritates the eyes, skin and respiratory tract. Affects the nervous system.
Gross formula:	$C_7H_{16}O$	

HAZARDS/SYMPTOMS	PREVENTIVE MEASURES	FIRE EXTINGUISHING/FIRST AID
Fire: combustible.	keep away from open flame, no smoking.	powder, AFFF, foam, carbon dioxide, (halons).
Explosion: above 71°C: forms explosive air-vapor mixtures.	above 71°C: sealed machinery, ventilation.	
Inhalation: sore throat, cough, dizziness, sleepiness.	ventilation, local exhaust or respiratory protection.	fresh air, rest, call a doctor.
Skin: *is absorbed*; redness.	gloves.	remove contaminated clothing, flush skin with water or shower.
Eyes: redness, pain.	acid goggles.	flush with water, take to a doctor if necessary.
Ingestion: headache, nausea, dizziness, sleepiness.		rinse mouth, call a doctor.

SPILLAGE	STORAGE	LABELING / NFPA
collect leakage in sealable containers, soak up with sand or other inert absorbent and remove to safe place.	keep separate from strong oxidants; ventilate at floor level.	◇ 2 / 0 0

NOTES
Alcohol consumption increases toxic effects. The measures on this card also apply to 1-heptanol and 3-heptanol. 1-heptanol: boiling point: 176°C; melting point: −35°C. 3-heptanol: boiling point: 156°C; melting point: −70°C; flash point: 60°C.

Transport Emergency Card TEC(R)-30G37

HI: 30; UN-number: 1987

enanthic acid
n-heptanoic acid
n-heptylic acid
1-hexanecarcoxylic acid

$CH_3(CH_2)_5COOH$

n-HEPTOIC ACID

PHYSICAL PROPERTIES	IMPORTANT CHARACTERISTICS
Boiling point, °C 223 Melting point, °C −7.5 Flash point, °C > 107 Auto-ignition temperature, °C 289 Relative density (water = 1) 0.9 Relative vapor density (air = 1) 4.5 Relative density at 20 °C of saturated mixture vapor/air (air = 1) 1.0 Vapor pressure, mm Hg at 20 °C < 0.076 Solubility in water, g/100 ml at 15 °C 0.24 Relative molecular mass 130.1 Log P octanol/water 2.7 Gross formula: $C_7H_{14}O_2$	**COLORLESS VISCOUS LIQUID WITH CHARACTERISTIC ODOR** Vapor mixes readily with air. TLV-TWA not available **Absorption route:** Can enter the body by inhalation or ingestion. Harmful atmospheric concentrations build up very slowly, if at all, on evaporation at approx. 20°C, but much more rapidly in aerosol form. **Immediate effects:** Irritates the eyes, skin and respiratory tract.

HAZARDS/SYMPTOMS	PREVENTIVE MEASURES	FIRE EXTINGUISHING/FIRST AID
Fire: combustible.	keep away from open flame, no smoking.	powder, AFFF, foam, carbon dioxide, (halons).
Inhalation: sore throat, cough.	ventilation.	fresh air, rest, call a doctor.
Skin: redness.	gloves.	remove contaminated clothing, flush skin with water or shower, call a doctor if necessary.
Eyes: redness, pain.	safety glasses.	flush with water, take to a doctor if necessary.
Ingestion: sore throat, abdominal pain.		rinse mouth, call a doctor or take to hospital.

SPILLAGE	STORAGE	LABELING / NFPA
collect leakage in sealable containers, soak up with sand or other inert absorbent and remove to safe place.	ventilate at floor level.	◇

NOTES
Log P octanol/water is estimated.

HEXACHLOROBENZENE

PHYSICAL PROPERTIES		IMPORTANT CHARACTERISTICS
Boiling point, °C	326	**WHITE CRYSTALS WITH FAINT ODOR**
Melting point, °C	230	Decomposes when heated or burned, giving off corrosive (→ *hydrochloric acid*) and highly toxic
Flash point, °C	242	vapors. Reacts violently with dimethylformamide.
Relative density (water = 1)	2.0	
Vapor pressure, mm Hg at 114 °C	0.76	TLV-TWA not available
Solubility in water	none	
Relative molecular mass	284.8	**Absorption route:** Can enter the body by inhalation or ingestion. Evaporation negligible at 20°C,
Log P octanol/water	6.2	but harmful atmospheric concentrations can build up rapidly if heated or in aerosol form.
		Inhalation of aerosol can cause toxic reactions.
		Immediate effects: Irritates the eyes, skin and respiratory tract.
		Effects of prolonged/repeated exposure: Metabolic disorders in red blood-cell colorant
		(porphyrin) can be detected by urine test. Hypersensitivity to light. Inhalation/ingestion of large
		quantities can cause nerve tissue and liver disorders.
Gross formula: C_6Cl_6		

HAZARDS/SYMPTOMS	PREVENTIVE MEASURES	FIRE EXTINGUISHING/FIRST AID
Fire: combustible.	keep away from open flame, no smoking.	water spray, powder.
Inhalation: irritation, cough.	ventilation, local exhaust or respiratory protection.	fresh air, rest, place in half-sitting position, take to hospital.
Skin: redness, irritation from prolonged contact.	gloves, protective clothing.	remove contaminated clothing, flush skin with water or shower, call a doctor.
Eyes: irritation, redness, pain.	face shield.	flush with water, take to a doctor.
Ingestion: irritation, sore throat, abdominal pain, diarrhea, vomiting, porphyria (see above), light-sensitivity.	do not eat, drink or smoke while working.	rinse mouth, take immediately to hospital.

SPILLAGE	STORAGE	LABELING / NFPA
if fluid allow to set, clean up spillage, carefully collect remainder; (additional individual protection: respirator with A/P2 filter).	keep separate from dimethylformamide; ventilate at floor level.	

NOTES
Depending on the degree of exposure, regular medical checkups are advisable.

perchlorobutadiene

$Cl_2C = CClCCl = CCl_2$

HEXACHLOROBUTADIENE

PHYSICAL PROPERTIES		IMPORTANT CHARACTERISTICS	
Boiling point, °C	212	**COLORLESS LIQUID**	
Melting point, °C	−21	Vapor mixes readily with air. Decomposes when heated above 600°C or burned, giving off toxic	
Flash point, °C	90	gas (→ *phosgene*) and corrosive vapors (→ *hydrochloric acid*). Reacts with aluminum, evolving	
Auto-ignition temperature, °C	610	heat. Attacks rubber and some plastics.	
Relative density (water = 1)	1.7		
Relative vapor density (air = 1)	9.0	TLV-TWA (skin) 0.02 ppm 0.21 mg/m³ A2	
Relative density at 20 °C of saturated mixture			
vapor/air (air = 1)	1.0	**Absorption route:** Can enter the body by inhalation or ingestion or through the skin. Harmful	
Vapor pressure, mm Hg at 20 °C	0.38	atmospheric concentrations can build up very rapidly on evaporation at 20°C.	
Solubility in water, g/100 ml at 20 °C	0.05	**Immediate effects:** Irritates the eyes, skin and respiratory tract.	
Relative molecular mass	260.8	**Effects of prolonged/repeated exposure:** Causes kidney damage, and to a lesser extent liver	
		damage and central nervous system disorders. Has been found to cause a type of kidney cancer	
		in some animal species under certain circumstances.	
Gross formula:	C_4Cl_6		

HAZARDS/SYMPTOMS	PREVENTIVE MEASURES	FIRE EXTINGUISHING/FIRST AID
Fire: combustible.	keep away from open flame, no smoking.	powder, water spray, foam, carbon dioxide, (halons).
Explosion: above 90°C: forms explosive air-vapor mixtures.	above 90°C: sealed machinery, ventilation.	
Inhalation: cough, shortness of breath.	ventilation, local exhaust or respiratory protection.	fresh air, rest, call a doctor.
Skin: *is absorbed*; redness.	gloves, protective clothing.	remove contaminated clothing, flush skin with water or shower, refer to a doctor.
Eyes: redness, pain.	face shield, or combined eye and respiratory protection.	flush with water, take to a doctor if necessary.
Ingestion: sore throat, abdominal pain, stomachache, nausea.		rinse mouth, call a doctor or take to hospital.

SPILLAGE	STORAGE	LABELING / NFPA
evacuate area, call in an expert, collect leakage in sealable containers, soak up with sand or other inert absorbent and remove to safe place; (additional individual protection: breathing apparatus).	ventilate at floor level.	1 / 2 / 1

NOTES
Before distilling check for peroxides; if found, render harmless. Depending on the degree of exposure, regular medical checkups are advisable. Under no circumstances use near flame or hot surface or when welding.

Transport Emergency Card TEC(R)-613

HI: 60; UN-number: 2279

CAS-No: [58-89-9]
γ-benzene hexachloride
γ-hexachlorane
γ-hexachlorocyclohexane
lindane

$C_6H_6Cl_6$

γ-1,2,3,4,5,6-HEXACHLOROCYCLOHEXANE

PHYSICAL PROPERTIES	IMPORTANT CHARACTERISTICS	
Boiling point (decomposes), °C 288 Melting point, °C 113 Relative density (water = 1) 1.9 Relative density at 20 °C of saturated mixture vapor/air (air = 1) 1.0 Vapor pressure, mm Hg at 20 °C 0.038 Solubility in water, g/100 ml at 20 °C 0.001 Relative molecular mass 290.8	**WHITE POWDER** Decomposes when heated or burned, giving off toxic and corrosive vapors (→ *phosgene* and → *hydrochloric acid*).	
	TLV-TWA (skin)	0.5 mg/m³
	Absorption route: Can enter the body by inhalation or ingestion or through the skin. Harmful atmospheric concentrations can build up fairly rapidly on evaporation at approx. 20°C; harmful concentrations of airborne particles can build up even more rapidly. **Immediate effects:** Irritates the eyes, skin and respiratory tract. **Effects of prolonged/repeated exposure:** Prolonged intensive contact can affect the blood.	
Gross formula: $C_6H_6Cl_6$		

HAZARDS/SYMPTOMS	PREVENTIVE MEASURES	FIRE EXTINGUISHING/FIRST AID
Fire: non-combustible.		in case of fire in immediate vicinity: use any extinguishing agent.
	STRICT HYGIENE, KEEP DUST UNDER CONTROL	
Inhalation: headache, muscle ache, trembling, shaking, nausea, vomiting, unconsciousness, seizures.	local exhaust or respiratory protection.	fresh air, rest, place in half-sitting position, artificial respiration if necessary.
Skin: *is absorbed*; redness; see also 'Inhalation'.	gloves.	remove contaminated clothing, wash skin with soap and water, call a doctor.
Eyes: redness, pain.	safety glasses, or combined eye and respiratory protection.	flush with water, take to a doctor if necessary.
Ingestion: abdominal pain, diarrhea, vomiting; see also 'Inhalation'.	do not eat, drink or smoke while working.	rinse mouth, immediately take to hospital.

SPILLAGE	STORAGE	LABELING / NFPA
clean up spillage, carefully collect remainder; (additional individual protection: breathing apparatus).	ventilate at floor level.	R: 23/24/25-36/38 S: 2-13-44 ☠ Toxic

NOTES
At 20°C saturated vapor/air mixture contains 560mg HCH/m³. Remarkably volatile, despite being a solid. The substance is usually used in dissolved form in a flammable solvent an thus can present fire and explosion hazards. Under no circumstances use near flame or hot surface or when welding.

Transport Emergency Card TEC(R)-61G53

HI: 60; UN-number: 2761

CAS-No: [67-72-1]
carbon hexachloride
ethylene hexachloride
1,1,1,2,2,2-hexachloroethane
perchloroethane

C_2Cl_6

HEXACHLOROETHANE

PHYSICAL PROPERTIES		IMPORTANT CHARACTERISTICS	
Sublimation temperature, °C	186	**COLORLESS TO WHITE CRYSTALS WITH CAMPHORIC ODOR**	
Relative density (water = 1)	2.1	Vapor mixes readily with air. Decomposes when heated above 300°C, giving off toxic gas ($\rightarrow$ *phosgene*). Reacts with zinc, aluminium and iron, forming $\rightarrow$ *perchloroethane*. Reacts with alkalis, giving off flammable gas.	
Relative vapor density (air = 1)	8.2		
Relative density at 20°C of saturated mixture vapor/air (air = 1)	1.0		
Vapor pressure, mm Hg at 33°C	0.99		
Solubility in water	none	TLV-TWA (skin) 1 ppm 10 mg/m³	
Relative molecular mass	236.7		
		Absorption route: Can enter the body by inhalation or ingestion or through the skin. **Immediate effects:** Irritates the eyes, skin and respiratory tract. Affects the nervous system and causes feeling of weakness, difficulty in walking and muscle tremors. In serious cases risk of unconsciousness. **Effects of prolonged/repeated exposure:** Can cause liver and kidney damage.	
Gross formula:	C_2Cl_6		

HAZARDS/SYMPTOMS	PREVENTIVE MEASURES	FIRE EXTINGUISHING/FIRST AID
Fire: non-combustible.		in case of fire in immediate vicinity: use any extinguishing agent.
	KEEP DUST UNDER CONTROL	
Inhalation: sore throat, cough, headache, dizziness, sleepiness.	ventilation, local exhaust or respiratory protection.	fresh air, rest, refer to a doctor.
Skin: *is absorbed*; redness, pain.	gloves.	remove contaminated clothing, wash skin with soap and water, refer to a doctor if necessary.
Eyes: redness, pain, impaired vision, profuse watering of the eyes, light-sensitivity, cramping of the eyelids.	goggles.	flush with water, take to a doctor if necessary.
Ingestion: sore throat, abdominal pain, stomachache, headache, dizziness, sleepiness.		rinse mouth, consult a doctor.

SPILLAGE	STORAGE	LABELING / NFPA
clean up spillage, carefully collect remainder; (additional individual protection: breathing apparatus).	ventilate at floor level.	

NOTES
Under no circumstances use near flame or hot surface or when welding.

HEXACHLORONAPHTHALENE

PHYSICAL PROPERTIES		IMPORTANT CHARACTERISTICS	
Boiling point, °C	ca. 350	**WHITE OR YELLOW WAX-LIKE SUBSTANCE WITH CHARACTERISTIC ODOR**	
Melting point, °C	137	Decomposes when heated or burned, giving off corrosive ($\rightarrow$ *hydrochloric acid*) and highly toxic vapors.	
Solubility in water	none		
Relative molecular mass	334.8		
		TLV-TWA (skin)	0.2 mg/m³
		Absorption route: Can enter the body by inhalation or ingestion or through the skin. Evaporation negligible at 20°C, but when heated moderately or in aerosol form harmful atmospheric concentrations of airborne particles can build up rapidly. **Effects of prolonged/repeated exposure:** Prolonged or repeated contact can cause skin disorders (chlorine acne). Skin disorders and liver damage can occur.	
Gross formula:	$C_{10}H_2Cl_6$		

HAZARDS/SYMPTOMS	PREVENTIVE MEASURES	FIRE EXTINGUISHING/FIRST AID
Fire: combustible.	keep away from open flame, no smoking.	water spray, powder.
Inhalation: nausea, vomiting, jaundice, dark-colored urine, fever, disorientation, unconsciousness.	local exhaust or respiratory protection.	fresh air, rest, take immediately to hospital.
Skin: *is absorbed*; redness, pimples.	gloves, protective clothing.	remove contaminated clothing, wash skin with soap and water, refer to a doctor.
Eyes: redness, pain.	face shield.	flush with water, take to a doctor.
Ingestion: nausea, vomiting, jaundice, dark-colored urine, fever, disorientation, unconsciousness.	do not eat, drink or smoke while working.	rinse mouth, immediately take to hospital.

SPILLAGE	STORAGE	LABELING / NFPA
if warm allow to set, clean up spillage, carefully collect remainder; (additional individual protection: respirator with A/P2 filter).	ventilate at floor level.	

NOTES
Concentrations of approximately 1-2 mg/m³ can be fatal due to liver poisoning. Strict regulations must be observed when working with this substance. Emergency measures must be prepared so that provisions can be laid down.

$H_2N(CH_2)_6NH_2$

HEXAMETHYLENEDIAMINE

PHYSICAL PROPERTIES		IMPORTANT CHARACTERISTICS
Boiling point, °C	204	**WHITE HYGROSCOPIC CRYSTALS OR SHINY LEAFLETS**
Melting point, °C	41	Toxic vapors (→ *nitrous vapors*) are a product of combustion. Decomposes when heated, giving
Flash point, °C	80	off combustible and toxic vapors (→ *anhydrous ammonia*). Strong reducing agent which reacts
Auto-ignition temperature, °C	310	violently with oxidants. In aqueous solution is a strong base which reacts violently with acids and
Relative density (water = 1)	0.8¹)	corrodes aluminum, zinc etc.
Vapor pressure, mm Hg at 50 °C	1.5	
Solubility in water, g/100 ml	∞	TLV-TWA not available
Relative molecular mass	116.2	
		Absorption route: Can enter the body by inhalation or ingestion. Evaporation negligible at 20°C, but harmful concentrations of airborne particles can build up rapidly.
		Immediate effects: Corrosive to the eyes, skin and respiratory tract. Inhalation can cause lung edema. In serious cases risk of death.
		Effects of prolonged/repeated exposure: Repeated contact can cause allergic skin disorders (hypersensitivity). Can affect the blood. Can cause liver, kidney and heart damage.
Gross formula:	$C_6H_{16}N_2$	

HAZARDS/SYMPTOMS	PREVENTIVE MEASURES	FIRE EXTINGUISHING/FIRST AID
Fire: combustible.	keep away from open flame, no smoking.	alcohol-resistant foam, powder, (halons); avoid contact with water and carbon dioxide if possible.
Explosion: above 80°C: forms explosive air-vapor mixtures.	above 80°C: sealed machinery, ventilation.	
	KEEP DUST UNDER CONTROL	
Inhalation: *corrosive*; sore throat, cough, shortness of breath, severe breathing difficulties, headache, dizziness, fainting.	local exhaust or respiratory protection.	fresh air, rest, place in half-sitting position, take to hospital.
Skin: *corrosive*; redness, pain, burns.	heat-insulated gloves.	remove contaminated clothing, flush skin with water or shower, refer to a doctor.
Eyes: *corrosive*; redness, pain, impaired vision.	goggles, face shield or combined eye and respiratory protection.	flush with water, take to a doctor.
Ingestion: *corrosive*; sore throat, abdominal pain, diarrhea, vomiting.	do not eat, drink or smoke while working.	rinse mouth, take to hospital.

SPILLAGE	STORAGE	LABELING / NFPA
collect leakage in sealable containers (or scoop up what has already set), clean up spillage, flush away remainder with water; (additional individual protection: breathing apparatus; P2 respirator).	keep dry, separate from oxidants and acids.	

NOTES
1) Often transported and stored in liquid state at 55°C. Depending on the degree of exposure, regular medical checkups are advisable. Lung edema symptoms usually develop several hours later and are aggravated by physical exertion: rest and hospitalization essential. As first aid, a doctor or authorized person should consider administering a corticosteroid spray. Use airtight packaging.

Transport Emergency Card TEC(R)-80G16

HI: 80; UN-number: 2280 (solid)

HEXAMETHYLENETETRAMINE

PHYSICAL PROPERTIES		IMPORTANT CHARACTERISTICS
Sublimation temperature, °C	260	**WHITE VITREOUS GRANULES**
Melting point, °C	280	Decomposes when heated, giving off → *nitrous vapors*. Reacts violently with sodium peroxide.
Flash point, °C	250	
Relative density (water = 1)	1.27	TLV-TWA not available
Relative vapor density (air = 1)	4.9	
Solubility in water, g/100 ml at 20 °C	150	**Absorption route:** Can enter the body by ingestion.
Relative molecular mass	140.2	**Immediate effects:** Irritates the eyes, skin and respiratory tract. Vapor given off on heating is
Log P octanol/water	−2.2	corrosive to the eyes, skin and respiratory tract and can cause lung edema if inhaled.
Gross formula:	$C_6H_{12}N_4$	

HAZARDS/SYMPTOMS	PREVENTIVE MEASURES	FIRE EXTINGUISHING/FIRST AID
Fire: combustible.	keep away from open flame, no smoking.	water spray, powder.
Explosion: finely dispersed particles form explosive mixtures on contact with air.	keep dust from accumulating; sealed machinery, explosion-proof electrical equipment and lighting, grounding.	
Inhalation: vapor (from heating): *corrosive*; cough, shortness of breath.	local exhaust or respiratory protection.	fresh air, rest, place in half-sitting position, call a doctor.
Skin: irritation, redness, pain, burns.	gloves.	remove contaminated clothing, flush skin with water or shower, refer to a doctor.
Eyes: irritation, redness, pain.	goggles.	flush with water, take to a doctor if necessary.
Ingestion: irritation, sore throat, abdominal pain, nausea, bladder irritation, rash.	do not eat, drink or smoke while working.	rinse mouth, take to hospital.

SPILLAGE	STORAGE	LABELING / NFPA
clean up spillage, flush away remainder with water; (additional individual protection: P1 respirator).		

NOTES

Log P octanol/water is estimated. Lung edema symptoms usually develop several hours later and are aggravated by physical exertion: rest and hospitalization essential. As first aid, a doctor or authorized person should consider administering a corticosteroid spray. Technical grades can contain formaldehyde and/or ammonia and are thus more aggressive than pure substance.

UN-number: 1328

n-HEXANE

PHYSICAL PROPERTIES	IMPORTANT CHARACTERISTICS	
Boiling point, °C 69	**COLORLESS LIQUID WITH CHARACTERISTIC ODOR**	
Melting point, °C −95	Vapor (gas) is heavier than air and spreads at ground level, with risk of ignition at a distance. Flow,	
Flash point, °C −22	agitation etc. can cause build-up of electrostatic charge due to liquid's low conductivity. Do not	
Auto-ignition temperature, °C 240	use compressed air when filling, emptying or processing. Liquid can contain isomers.	
Relative density (water = 1) 0.66		
Relative vapor density (air = 1) 2.98	**TLV-TWA**	50 ppm	176 mg/m³
Relative density at 20 °C of saturated mixture		
vapor/air (air = 1) 1.31	**Absorption route:** Can enter the body by inhalation or ingestion. Harmful atmospheric	
Vapor pressure, mm Hg at 20 °C 122	concentrations can build up fairly rapidly on evaporation at approx. 20°C - even more rapidly in	
Solubility in water none	aerosol form.	
Explosive limits, vol% in air 1.2-7.8	**Immediate effects:** Irritates the eyes. Liquid destroys the skin's natural oils. If liquid is swallowed,	
Minimum ignition energy, mJ 0.24	droplets can enter the lungs, with risk of pneumonia.	
Electrical conductivity, pS/m 100	**Effects of prolonged/repeated exposure:** Prolonged exposure to high concentrations can	
Relative molecular mass 86.2	cause disorders of the peripheral and central nervous system.	
Gross formula: C_6H_{14}		

HAZARDS/SYMPTOMS	PREVENTIVE MEASURES	FIRE EXTINGUISHING/FIRST AID
Fire: extremely flammable.	keep away from open flame and sparks, no smoking.	powder, AFFF, foam, carbon dioxide, (halons).
Explosion: forms explosive air-vapor mixtures.	sealed machinery, ventilation, explosion-proof electrical equipment and lighting, grounding, non-sparking tools.	in case of fire: keep tanks/drums cool by spraying with water.
Inhalation: nausea, drowsiness, sleepiness; in high concentrations: severe breathing difficulties.	ventilation, local exhaust or respiratory protection.	fresh air, rest, place in half-sitting position, artificial respiration if necessary, call a doctor.
Skin: redness.	gloves.	remove contaminated clothing, flush skin with water or shower.
Eyes: redness, pain.	safety glasses.	flush with water, take to a doctor if necessary.
Ingestion: abdominal pain, nausea; irritation of the mouth, throat, esophagus and stomach.		rinse mouth, DO NOT induce vomiting, call a doctor or take to hospital.

SPILLAGE	STORAGE	LABELING / NFPA
collect leakage in sealable containers, soak up with sand or other inert absorbent and remove to safe place; (additional individual protection: breathing apparatus).	keep in fireproof place.	R: 11-20-48 S: 9-16-24/25-29-51 Flammable Harmful

NOTES
Depending on the degree of exposure, regular medical checkups are advisable.

2,5-dihydroxyhexane
2,5-hexylene glycol

$CH_3CHOH(CH_2)_2CHOHCH_3$

2,5-HEXANEDIOL

PHYSICAL PROPERTIES	IMPORTANT CHARACTERISTICS
Boiling point, °C 220 Flash point, °C 104 Relative density (water = 1) 0.96 Relative vapor density (air = 1) 4.07 Solubility in water very good Relative molecular mass 118.2	**COLORLESS LIQUID** Reacts with strong oxidants. TLV-TWA not available **Absorption route:** Can enter the body by inhalation or ingestion. Evaporation negligible at 20° C, but harmful atmospheric concentrations can build up rapidly in aerosol form. **Immediate effects:** Irritates the eyes, skin and respiratory tract.
Gross formula: $C_6H_{14}O_2$	

HAZARDS/SYMPTOMS	PREVENTIVE MEASURES	FIRE EXTINGUISHING/FIRST AID
Fire: combustible.	keep away from open flame, no smoking, avoid contact with strong oxidants.	water spray, powder; in case of fire in immediate vicinity: use any extinguishing agent.
Inhalation: cough, dizziness, headache.		fresh air, rest, consult a doctor.
Skin:	gloves.	flush skin with water and wash with soap and water.
Eyes: redness.	face shield.	
Ingestion: abdominal pain, headache, nausea, vomiting.		rinse mouth, refer to a doctor.

SPILLAGE	STORAGE	LABELING / NFPA
collect leakage in sealable containers, flush away remainder with water.	keep separate from strong oxidants.	◇ 1 / 2 0

NOTES
Insufficient data on harmful effects to humans: exercise caution.

$C_6H_{13}OH$

amyl carbinol
caproyl alcohol
hexyl alcohol
pentyl carbinol

HEXANOL

PHYSICAL PROPERTIES	IMPORTANT CHARACTERISTICS
Boiling point, °C 157 Melting point, °C − 45 Flash point, °C 63 Auto-ignition temperature, °C 290 Relative density (water = 1) 0.8 Relative vapor density (air = 1) 3.5 Relative density at 20 °C of saturated mixture vapor/air (air = 1) 1.0 Vapor pressure, mm Hg at 20 °C 0.99 Solubility in water poor Explosive limits, vol% in air 1.3-7.7 Electrical conductivity, pS/m 2×10^7 Relative molecular mass 102.2 Log P octanol/water 2.0 Gross formula: $C_6H_{14}O$	**COLORLESS LIQUID WITH CHARACTERISTIC ODOR** Vapor mixes readily with air. Reacts with strong oxidants. TLV-TWA not available **Absorption route:** Can enter the body by inhalation or ingestion or through the skin. Harmful atmospheric concentrations build up fairly slowly on evaporation at 20° C, but much more rapidly in aerosol form. **Immediate effects:** Irritates the eyes, skin and respiratory tract. Affects the nervous system.

HAZARDS/SYMPTOMS	PREVENTIVE MEASURES	FIRE EXTINGUISHING/FIRST AID
Fire: combustible.	keep away from open flame, no smoking.	powder, AFFF, foam, carbon dioxide, (halons).
Explosion: above 63°C: forms explosive air-vapor mixtures.	above 63°C: sealed machinery, ventilation.	
Inhalation: sore throat, cough, dizziness, sleepiness.	ventilation, local exhaust or respiratory protection.	fresh air, rest, call a doctor.
Skin: *is absorbed*; redness.	gloves.	remove contaminated clothing, flush skin with water or shower.
Eyes: redness, pain.	face shield.	flush with water, take to a doctor if necessary.
Ingestion: headache, nausea, dizziness, sleepiness.		rinse mouth, call a doctor.

SPILLAGE	STORAGE	LABELING / NFPA
collect leakage in sealable containers, soak up with sand or other inert absorbent and remove to safe place.		R: 22 S: 24/25 ✖ Harmful (NFPA diamond: 2 top, 1 left, 0 right)

NOTES
Alcohol consumption increases toxic effects. The measures on this card also apply to most isomers and mixtures, except 2-2-dimethyl butanol and 4-methyl-2-pentanol.

Transport Emergency Card TEC(R)-30G37

HI: 30; UN-number: 1987

CAS-No: [107-41-5]
2-methyl-2,4-pentanediol
2,4-dihydroxy-2-methylpentane
2-methylpentane-2,4-diol
α,α,α'-trimethyltrimethylene glyco

$(CH_3)_2COHCH_2CHOHCH_3$

HEXYLENE GLYCOL

PHYSICAL PROPERTIES		IMPORTANT CHARACTERISTICS	
Boiling point, °C	196	**COLORLESS VISCOUS LIQUID**	
Melting point, °C	−100	Vapor mixes readily with air.	
Flash point, °C	96		
Relative density (water = 1)	0.9	TLV-TWA 25 ppm 121 mg/m³ C	
Relative vapor density (air = 1)	4.1		
Relative density at 20 °C of saturated mixture		**Absorption route:** Can enter the body by inhalation or ingestion. Harmful atmospheric concentrations build up very slowly, if at all, on evaporation at approx. 20°C, but much more rapidly in aerosol form.	
vapor/air (air = 1)	1.0		
Vapor pressure, mm Hg at 20 °C	0.053	**Immediate effects:** Corrosive to the eyes, skin and respiratory tract. Affects the nervous system.	
Solubility in water	∞		
Relative molecular mass	118.2		
Log P octanol/water	−0.1		
Gross formula: $C_6H_{14}O_2$			

HAZARDS/SYMPTOMS	PREVENTIVE MEASURES	FIRE EXTINGUISHING/FIRST AID
Fire: combustible.	keep away from open flame, no smoking.	powder, alcohol-resistant foam, water spray, carbon dioxide, (halons).
Explosion: above 96°C: forms explosive air-vapor mixtures.	above 96°C: sealed machinery, ventilation.	
Inhalation: cough, headache, agitation followed by drowsiness, unconsciousness.	ventilation, local exhaust or respiratory protection.	fresh air, rest, call a doctor.
Skin: redness.	gloves.	remove contaminated clothing, flush skin with water or shower.
Eyes: redness, pain.	safety glasses.	flush with water, take to an eye doctor.
Ingestion: nausea, vomiting, agitation followed by drowsiness, unconsciousness.		rinse mouth, call a doctor or take to hospital.

SPILLAGE	STORAGE	LABELING / NFPA
collect leakage in sealable containers, flush away remainder with water.		R: 36/38 ✖ Irritant

NOTES
Log P octanol/water is estimated. TLV is maximum and must not be exceeded because of eye irritation, esp. corneal erosion, which is slow to heal.

CAS-No: [302-01-2]
diamide
diamine

H_2N-NH_2

HYDRAZINE
(100%)

PHYSICAL PROPERTIES		IMPORTANT CHARACTERISTICS
Boiling point, °C	113	**COLORLESS FUMING HYGROSCOPIC LIQUID WITH PUNGENT ODOR**
Melting point, °C	1.5	Vapor mixes readily with air. Toxic gases are a product of combustion (→ *nitrous vapors*). Can
Flash point, °C	52	ignite spontaneously on exposure to air if absorbed into porous materials. Can decompose
Auto-ignition temperature, °C	270	explosively when heated. Decomposes on contact with many metals and metal oxides, with risk
Relative density (water = 1)	1.01	of fire and explosion. Strong reducing agent which reacts violently with oxidants, with risk of fire
Relative vapor density (air = 1)	1.05	or explosion. Strong base which reacts violently with acids and corrodes aluminum, zinc etc.
Relative density at 20 °C of saturated mixture		Reacts violently with halogens, with risk of fire and explosion.
vapor/air (air = 1)	1.0	
Vapor pressure, mm Hg at 20 °C	16	TLV-TWA (skin) 0.01 ppm A2 0.013 mg/m³ A2
Solubility in water	∞	
Explosive limits, vol% in air	4.7-100	**Absorption route:** Can enter the body by inhalation or ingestion or through the skin. Harmful
Relative molecular mass	32.1	atmospheric concentrations can build up very rapidly on evaporation at 20°C.
Log P octanol/water	−1.1	**Immediate effects:** Corrosive to the eyes, skin and respiratory tract. Affects the nervous system.
		Inhalation of vapor/fumes can cause severe breathing difficulties (lung edema). In serious cases
		risk of death.
		Effects of prolonged/repeated exposure: Prolonged or repeated contact can cause skin
		disorders. Can affect the blood. Can cause liver and kidney damage. Repeated exposure to high
		concentrations may cause cancer in humans.
Gross formula:	H_4N_2	

HAZARDS/SYMPTOMS	PREVENTIVE MEASURES	FIRE EXTINGUISHING/FIRST AID
Fire: flammable, many chemical reactions can cause fire and explosion.	keep away from open flame and sparks, no smoking, avoid contact with oxidants.	powder, alcohol-resistant foam, water spray, carbon dioxide, (halons).
Explosion: above 52°C: forms explosive air-vapor mixtures.	above 52°C: sealed machinery, ventilation, explosion-proof electrical equipment, non-sparking tools.	in case of fire: keep tanks/drums cool by spraying with water, fight fire from sheltered location.
	STRICT HYGIENE	IN ALL CASES CALL IN A DOCTOR
Inhalation: *corrosive*; sore throat, cough, shortness of breath, severe breathing difficulties, dizziness, vomiting.	ventilation, local exhaust or respiratory protection.	fresh air, rest, place in half-sitting position, take to hospital.
Skin: *corrosive, is absorbed*; redness, pain, burns.	gloves, protective clothing.	remove contaminated clothing, flush skin with water or shower, refer to a doctor.
Eyes: *corrosive*; redness, pain, impaired vision.	face shield, or combined eye and respiratory protection.	flush with water, take to a doctor if necessary.
Ingestion: *corrosive*; abdominal pain, dizziness, vomiting.		rinse mouth, immediately take to hospital.

SPILLAGE	STORAGE	LABELING / NFPA
evacuate area, call in an expert, collect leakage in sealable containers, flush away remainder with water, do not use sawdust or other combustible absorbents; (additional individual protection: CHEMICAL SUIT).	keep in fireproof place, separate from oxidants and strong acids.	R: 10-26/27/28-34-40 S: 36/37/39-45 ☠ Toxic

NOTES

Log P octanol/water is estimated. Combustible material impregnated with hydrazine solution can ignite spontaneously once dry. Flush contaminated clothing with water (fire hazard). Odor limit is above TLV. Depending on the degree of exposure, regular medical checkups are advisable. Lung edema symptoms usually develop several hours later and are aggravated by physical exertion: rest and hospitalization essential. As first aid, a doctor or authorized person should consider administering a corticosteroid spray. Use airtight packaging. Unbreakable packaging preferred; if breakable, keep in unbreakable container.

Transport Emergency Card TEC(R)-85

HI: 886; UN-number: 2029

HYDRAZINESOLUTION
(15%)

PHYSICAL PROPERTIES		IMPORTANT CHARACTERISTICS	
Boiling point, °C	102	**COLORLESS TO YELLOW AQUEOUS SOLUTION OF HYDRAZINE WITH CHARACTERISTIC ODOR**	
Melting point, °C	−14	Vapor mixes readily with air. Strong reducing agent which reacts violently with oxidants. Strong base which reacts violently with acids and corrodes aluminum, zinc etc. Attacks glass.	
Relative density (water = 1)	1.0		
Solubility in water	∞		
Relative molecular mass	50.1		
Log P octanol/water	−1.1		
		TLV-TWA (skin)　　0.01 ppm A2　　0.013 mg/m³ A2 ✶	
		Absorption route: Can enter the body by inhalation or ingestion or through the skin. **Immediate effects:** Corrosive to the eyes, skin and respiratory tract. Affects the nervous system. Inhalation of vapor/fumes can cause severe breathing difficulties (lung edema). In serious cases risk of death. **Effects of prolonged/repeated exposure:** Prolonged or repeated contact can cause eczema. Can cause liver and kidney damage. Repeated exposure to high concentrations may cause cancer in humans.	
Gross formula:	$H_4N_2.H_2O$		

HAZARDS/SYMPTOMS	PREVENTIVE MEASURES	FIRE EXTINGUISHING/FIRST AID
Fire: non-combustible.		
	STRICT HYGIENE	IN ALL CASES CALL IN A DOCTOR
Inhalation: *corrosive*; sore throat, cough, severe breathing difficulties, dizziness, nausea, vomiting.	ventilation, local exhaust or respiratory protection.	fresh air, rest, place in half-sitting position, take to hospital.
Skin: *corrosive*; redness, pain, burns.	gloves.	remove contaminated clothing, flush skin with water or shower, call a doctor.
Eyes: *corrosive*; redness, pain, impaired vision.	face shield.	flush with water, take to a doctor.
Ingestion: *corrosive*; abdominal pain, diarrhea, nausea, cramps, disturbances to heartbeat, lowered blood sugar, dizziness, vomiting.		rinse mouth, immediately take to hospital.

SPILLAGE	STORAGE	LABELING / NFPA
collect leakage in sealable containers, flush away remainder with water, do not use sawdust or other combustible absorbents; (additional individual protection: breathing apparatus).	keep cool, separate from oxidants and strong acids; ventilate at floor level.	R: 24/25-34-40 S: 36/37/39 Note B 　Corrosive

NOTES
Log P octanol/water is estimated. Flush contaminated clothing with water (fire hazard). Odor limit is above TLV. N_2H_4 level above an open container of 15% hydrazine solution at 20° C in a well-ventilated room is below TLV. Depending on the degree of exposure, regular medical checkups are advisable. Lung edema symptoms usually develop several hours later and are aggravated by physical exertion: rest and hospitalization essential. As first aid, a doctor or authorized person should consider administering a corticosteroid spray. Unbreakable packaging preferred; if breakable, keep in unbreakable container. ✶ TLV equals that of hydrazine (N_2H_4).

Transport Emergency Card TEC(R)-85A (C 40%)

HI: 86; UN-number: 2030

HYDROCARBON-SOLVENTS

(aliphatic - initial boiling point > 180°C)

PHYSICAL PROPERTIES	IMPORTANT CHARACTERISTICS
Boiling point, °C — > 180 Melting point, °C — < − 20 Flash point, °C — 55-100 Auto-ignition temperature, °C — 210 Relative density (water = 1) — 0.73-0.87 Relative vapor density (air = 1) — 4.5 Vapor pressure, mm Hg at 20 °C — < 0.38 Solubility in water — none Explosive limits, vol% in air — 0.6-8.0	**COLORLESS LIQUID WITH CHARACTERISTIC ODOR** Mixture of aliphatic hydrocarbons with lower boiling point above 180°C. The lower the boiling point, the higher the vapor pressure at 20°C and the lower the flash point. Flow, agitation etc. can cause build-up of electrostatic charge due to liquid's low conductivity. TLV-TWA not available **Absorption route:** Can enter the body by inhalation or ingestion. Harmful atmospheric concentrations build up very slowly, if at all, on evaporation at approx. 20°C, but much more rapidly in aerosol form. **Immediate effects:** Liquid destroys the skin's natural oils. Affects the nervous system. If liquid is swallowed, droplets can enter the lungs, with risk of pneumonia.

HAZARDS/SYMPTOMS	PREVENTIVE MEASURES	FIRE EXTINGUISHING/FIRST AID
Fire: combustible.	keep away from open flame, no smoking.	powder, AFFF, foam, carbon dioxide, (halons).
Explosion: above 55°C: forms explosive air-vapor mixtures.	above 55°C: sealed machinery, ventilation, explosion-proof electrical equipment, grounding.	
Inhalation: headache, dizziness, nausea, drowsiness.	ventilation, local exhaust or respiratory protection.	fresh air, rest, call a doctor.
Skin: redness.	gloves.	remove contaminated clothing, wash skin with soap and water.
Eyes: redness, pain.	safety glasses.	flush with water, take to a doctor if necessary.
Ingestion: vomiting, diarrhea; see also 'Inhalation'.		rinse mouth, DO NOT induce vomiting, take to hospital.

SPILLAGE	STORAGE	LABELING / NFPA
collect leakage in sealable containers, soak up with sand or other inert absorbent and remove to safe place.	keep in fireproof place.	

NOTES
Alcohol consumption increases toxic effects. Shell Sol is a trade name.

Transport Emergency Card TEC(R)-752

HI: 30; UN-number: 1993

HYDROCARBON-SOLVENTS

(aromatic - initial boiling point > 180°C)

PHYSICAL PROPERTIES

Boiling point, °C	> 180
Melting point, °C	< −20
Flash point, °C	55-100
Auto-ignition temperature, °C	475
Relative density (water = 1)	0.73-0.87
Relative vapor density (air = 1)	ca. 4.5
Vapor pressure, mm Hg at 20 °C	< 0.38
Solubility in water	none
Explosive limits, vol% in air	1.0-6.0
Electrical conductivity, pS/m	10^{-1}

IMPORTANT CHARACTERISTICS

COLORLESS LIQUID WITH CHARACTERISTIC ODOR
Mixture of aromatic hydrocarbons with lower boiling point above 180°C. Flow, agitation etc. can cause build-up of electrostatic charge due to liquid's low conductivity. Carbon monoxide, carbon dioxide and water are products of combustion.

TLV-TWA	not available

Absorption route: Can enter the body by inhalation or ingestion. Harmful atmospheric concentrations build up fairly slowly on evaporation, but much more rapidly in aerosol form.
Immediate effects: Liquid destroys the skin's natural oils. Affects the nervous system. If liquid is swallowed, droplets can enter the lungs, with risk of pneumonia.

HAZARDS/SYMPTOMS | PREVENTIVE MEASURES | FIRE EXTINGUISHING/FIRST AID

HAZARDS/SYMPTOMS	PREVENTIVE MEASURES	FIRE EXTINGUISHING/FIRST AID
Fire: combustible.	keep away from open flame, no smoking.	powder, AFFF, foam, carbon dioxide, (halons).
Explosion: above 55°C: forms explosive air-vapor mixtures.	above 55°C: sealed machinery, ventilation, grounding.	
Inhalation: headache, dizziness, drowsiness.	ventilation, local exhaust or respiratory protection.	fresh air, rest, artificial respiration if necessary, call a doctor.
Skin: redness, rawness, burning sensation.	gloves.	remove contaminated clothing, flush skin with water or shower.
Eyes: redness, burning sensation.	safety glasses.	flush with water, take to a doctor.
Ingestion: severe breathing difficulties, nausea.		DO NOT induce vomiting, take to hospital.

SPILLAGE | STORAGE | LABELING / NFPA

SPILLAGE	STORAGE	LABELING / NFPA
collect leakage in sealable containers, soak up with sand or other inert absorbent and remove to safe place.	keep in fireproof place.	

NOTES

Alcohol consumption increases toxic effects. Can contain polycyclic aromatics, some of which are known to cause a type of cancer in humans under certain circumstances. Shell Sol is a trade name.

Transport Emergency Card TEC(R)-752

HI: 30; UN-number: 1202

471

HYDROCHLORIC ACID
(ca. 36%)

PHYSICAL PROPERTIES	IMPORTANT CHARACTERISTICS
Boiling point, °C 57 Melting point, °C −35 Relative density (water = 1) 1.2 Relative vapor density (air = 1) 1.3 Relative density at 20 °C of saturated mixture vapor/air (air = 1) 1.04 Vapor pressure, mm Hg at 20 °C 95 Solubility in water ∞ Relative molecular mass 36.5 Log P octanol/water 0.3	**COLORLESS, FUMING AQUEOUS SOLUTION OF HYDROGEN CHLORIDE WITH PUNGENT ODOR** Reacts with air to form corrosive acid fumes which are heavier than air and spread at ground level. Strong acid which reacts violently with bases and is corrosive. Reacts violently with strong oxidants, giving off toxic gas (→ *chlorine*). Attacks many metals, giving off flammable gas (→ *hydrogen*).
	TLV-TWA 5 ppm 7.5 mg/m³ C
	Absorption route: Can enter the body by inhalation or ingestion. Harmful atmospheric concentrations can build up very rapidly on evaporation at 20°C. **Immediate effects:** Corrosive to the eyes, skin and respiratory tract. Inhalation of vapor/fumes can cause severe breathing difficulties (lung edema).
Gross formula: ClH	

HAZARDS/SYMPTOMS	PREVENTIVE MEASURES	FIRE EXTINGUISHING/FIRST AID
Fire: non-combustible.		in case of fire in immediate vicinity: use any extinguishing agent.
Inhalation: *corrosive*; sore throat, cough, shortness of breath, severe breathing difficulties.	ventilation, local exhaust or respiratory protection.	fresh air, rest, place in half-sitting position, take to hospital.
Skin: *corrosive*; redness, pain, burns.	gloves, protective clothing.	remove contaminated clothing, flush skin with water or shower, call a doctor.
Eyes: *corrosive*; redness, pain, impaired vision.	face shield, or combined eye and respiratory protection.	flush with water, take to a doctor.
Ingestion: *corrosive*; severe breathing difficulties, vomiting, blisters on lips and tongue; burning pain in mouth and throat, esophagus and stomach.		rinse mouth, immediately take to hospital.

SPILLAGE	STORAGE	LABELING / NFPA
render harmless with a weak solution of sodium bicarbonate, collect leakage in sealable containers, flush away remainder with water, combat gas cloud with water curtain; (additional individual protection: CHEMICAL SUIT).	keep cool, separate from oxidants and strong bases.	R: 34-37 S: 2-26 Note B Corrosive 0 / 3 / 0

NOTES
Log P octanol/water is estimated. Can react with formaldehyde, forming highly toxic → *dichloromethyl ether*. TLV is maximum and must not be exceeded. Lung edema symptoms usually develop several hours later and are aggravated by physical exertion: rest and hospitalization essential. As first aid, a doctor or authorized person should consider administering a corticosteroid spray. Unbreakable packaging preferred; if breakable, keep in unbreakable container.

Transport Emergency Card TEC(R)-50B

HI: 80; UN-number: 1789

CAS-No: [74-90-8]
formonitrile
HCN
hydrogen cyanide
Prussic acid

H−C≡N

HYDROCYANIC ACID
(20% solution in water)

PHYSICAL PROPERTIES	IMPORTANT CHARACTERISTICS
Boiling point, °C 43 Melting point, °C −11 Flash point, °C −8 Auto-ignition temperature, °C 535 (HCN) Relative density (water = 1) 0.96 Relative vapor density (air = 1) 0.9 (HCN) Relative density at 20 °C of saturated mixture vapor/air (air = 1) ca. 1.0 Vapor pressure, mm Hg at 20 °C 374 Solubility in water ∞ Explosive limits, vol% in air 5.4-46.6 (HCN) Relative molecular mass 27.0 (HCN) Log P octanol/water 0.7 Gross formula: CHN	**COLORLESS LIQUID WITH CHARACTERISTIC ODOR** Vapor mixes readily with air, forming explosive mixtures. Readily able to polymerize when moderately heated and on contact with alkaline substances. TLV-TWA (skin) 10 ppm C 11 mg/m³ C ✱ **Absorption route:** Can enter the body by inhalation or ingestion or through the skin. Harmful atmospheric concentrations can build up very rapidly on evaporation at 20°C. **Immediate effects:** Irritates the respiratory tract. Affects the respiratory enzymes. In serious cases risk of seizures, unconsciousness and death.

HAZARDS/SYMPTOMS	PREVENTIVE MEASURES	FIRE EXTINGUISHING/FIRST AID
Fire: extremely flammable.	keep away from open flame and sparks, no smoking.	powder, alcohol-resistant foam, large quantities of water, carbon dioxide, (halons).
Explosion: forms explosive air-vapor mixtures.	sealed machinery, ventilation, explosion-proof electrical equipment and lighting.	in case of fire: keep tanks/drums cool by spraying with water.
	STRICT HYGIENE	
Inhalation: sore throat, severe breathing difficulties, headache, nausea, unconsciousness.	ventilation, local exhaust or respiratory protection.	fresh air, rest, special treatment, artificial respiration if necessary, take to hospital.
Skin: *is absorbed*; redness.	gloves, protective clothing.	remove contaminated clothing, flush skin with water or shower.
Eyes: redness, pain.	face shield, or combined eye and respiratory protection.	flush with water, take to a doctor if necessary.
Ingestion: sore throat, headache, severe breathing difficulties, nausea, unconsciousness.		rinse mouth, special treatment, call a doctor or take to hospital.

SPILLAGE	STORAGE	LABELING / NFPA
evacuate area, call in an expert, render harmless with sodium hypochlorite solution; (additional individual protection: CHEMICAL SUIT).	keep in cool, fireproof place; add an inhibitor.	R: 12-26/27/28 S: 7/9-13-16-45 Flammable Toxic 4 4 2

NOTES
✱ TLV equals that of hydrogen cyanide. 10% hydrocyanic acid solution has flash point of 2°C and a vapor pressure of 228mm Hg at 20°C. Log P octanol/water is estimated. Special first aid required in the event of poisoning: antidotes must be available (with instructions). Unbreakable packaging preferred; if breakable, keep in unbreakable container.

Transport Emergency Card TEC(R)-698

HI: 663; UN-number: 1613

CAS-No: [1333-74-0]
hydrogen, compressed
hydrogen gas
molecular hydrogen

H_2

HYDROGEN
(cylinder)

PHYSICAL PROPERTIES	IMPORTANT CHARACTERISTICS	
Boiling point, °C −253 Flash point, °C flammable gas Relative vapor density (air = 1) 0.07 Solubility in water none Explosive limits, vol% in air 4-76 Minimum ignition energy, mJ 0.01 Relative molecular mass 2.0	**COLORLESS, ODORLESS COMPRESSED GAS** Gas is lighter than air. Reacts with chlorine to form hydrogen chloride. This reaction can be initiated e.g. by UV radiation. Reacts violently with acetylene, nitrous oxide, nitric oxide and fluorine, with risk of fire and explosion. Reacts with oxygen or air to form oxyhydrogen.	
	TLV-TWA not available	
	Absorption route: Can saturate the air if released, with risk of suffocation.	
Gross formula: H_2		

HAZARDS/SYMPTOMS	PREVENTIVE MEASURES	FIRE EXTINGUISHING/FIRST AID
Fire: extremely flammable.	keep away from open flame and sparks, no smoking.	shut off supply; if impossible and no danger to surrounding area, allow fire to burn itself out; otherwise extinguish with powderx, carbon dioxide, (halons).
Explosion: forms explosive air-gas mixtures.	sealed machinery, ventilation, explosion-proof electrical equipment and lighting, non-sparking hand tools.	in case of fire: keep cylinders cool by spraying with water.
Inhalation: severe breathing difficulties, headache, dizziness, unconsciousness.	ventilation.	fresh air, rest, artificial respiration if necessary, take to hospital.

SPILLAGE	STORAGE	LABELING / NFPA
ventilate.	in stored indoors, keep in fireproof place; ventilate at ceiling level.	(NFPA diamond: blue 0, red 4, yellow 0)

NOTES
Can ignite spontaneously and explosively in air at fast flow rates from cylinder or in the event of gas leak. High atmospheric concentrations, e.g. in poorly ventilated spaces, can cause oxygen deficiency, with risk of unconsciousness. Use special hydrogen detector. Ventilate storage area at highest level.

Transport Emergency Card TEC(R)-20G04

HI: 23; UN-number: 1049

HYDROGEN BROMIDE
(cylinder)

PHYSICAL PROPERTIES	IMPORTANT CHARACTERISTICS	
Boiling point, °C −67 Melting point, °C −87 Relative density (water = 1) 1.8 Relative vapor density (air = 1) 2.8 Vapor pressure, mm Hg at 20 °C 16 Solubility in water, g/100 ml at 25 °C 193 Relative molecular mass 80.9	**COLORLESS COMPRESSED LIQUEFIED GAS WITH PUNGENT ODOR** Gas is heavier than air. In aqueous solution is a strong acid which reacts violently with bases and is corrosive; not corrosive in dry state. Reacts violently with many organic compounds. Reacts with air to form corrosive vapors (which spread at ground level). Attacks base metals, giving off flammable gas (→ *hydrogen*).	
	TLV-TWA 3 ppm 9.9 mg/m³ C	
	Absorption route: Can enter the body by inhalation. Harmful atmospheric concentrations can build up very rapidly if gas is released. **Immediate effects:** Corrosive to the eyes, skin and respiratory tract. Can cause frostbite due to rapid evaporation. Inhalation can cause lung edema. In serious cases risk of death.	
Gross formula: BrH		

HAZARDS/SYMPTOMS	PREVENTIVE MEASURES	FIRE EXTINGUISHING/FIRST AID
Fire: non-combustible.		in case of fire in immediate vicinity: use any extinguishing agent.
Explosion:		in case of fire: keep cylinders cool by spraying with water.
	STRICT HYGIENE	
Inhalation: *corrosive*; sore throat, cough, severe breathing difficulties.	ventilation, local exhaust or respiratory protection.	fresh air, rest, place in half-sitting position, take to hospital.
Skin: *in case of frostbite*: redness, pain, severe sores.	insulated gloves, protective clothing.	*in case of frostbite*: DO NOT remove clothing, flush skin with water or shower, take to hospital.
Eyes: *corrosive*; redness, pain, impaired vision.	face shield, or combined eye and respiratory protection.	flush with water, take to doctor.

SPILLAGE	STORAGE	LABELING / NFPA
evacuate area, call in an expert, ventilate, under no circumstances spray liquid with water, combat gas cloud with water curtain; (additional individual protection: breathing apparatus).	if stored indoors, keep in cool, fireproof place, separate from bases; ventilate at floor level.	NFPA diamond: 0 (top), 3 (left), 0 (right)

NOTES
Lung edema symptoms usually develop several hours later and are aggravated by physical exertion: rest and hospitalization essential. As first aid, a doctor or authorized person should consider administering a corticosteroid spray. Do not spray leaking cylinder with water (to avoid corrosion). Turn leaking cylinder so that leak is on top to prevent liquid hydrogen bromide escaping. Use cylinder with special fittings.

Transport Emergency Card TEC(R)-668

HI: 286; UN-number: 1048

HYDROGEN BROMIDE
(47% solution in water)

PHYSICAL PROPERTIES	IMPORTANT CHARACTERISTICS	
Boiling point, °C 126 Melting point, °C − 11 Relative density (water = 1) 1.5 Relative vapor density (air = 1) 1.7 Relative density at 20 °C of saturated mixture vapor/air (air = 1) < 1.1 Solubility in water ∞ Relative molecular mass 80.9	**COLORLESS TO LIGHT YELLOW LIQUID WITH PUNGENT ODOR, TURNING DARK ON EXPOSURE TO AIR OR LIGHT** Vapor mixes readily with air. Strong acid which reacts violently with bases and is corrosive. Reacts with strong oxidants, giving off → *bromine*. Reacts with air to form corrosive vapors. Attacks many metals, giving off flammable gas (→ *hydrogen*).	
	TLV-TWA 3 ppm 9.9 mg/m³ C ✷	
	Absorption route: Can enter the body by inhalation or ingestion. Harmful atmospheric concentrations can build up fairly rapidly on evaporation at approx. 20°C - even more rapidly in aerosol form. **Immediate effects:** Corrosive to the eyes, skin and respiratory tract. Inhalation of vapor/fumes can cause severe breathing difficulties (lung edema). In serious cases risk of death.	
Gross formula: BrH		

HAZARDS/SYMPTOMS	PREVENTIVE MEASURES	FIRE EXTINGUISHING/FIRST AID
Fire: non-combustible.		in case of fire in immediate vicinity: use any extinguishing agent.
	STRICT HYGIENE	
Inhalation: *corrosive*; sore throat, cough, shortness of breath, severe breathing difficulties.	ventilation, local exhaust or respiratory protection.	fresh air, rest, place in half-sitting position, take to hospital.
Skin: *corrosive*; redness, pain, serious burns.	gloves, protective clothing.	remove contaminated clothing, flush skin with water or shower, call a doctor.
Eyes: *corrosive*; redness, pain, impaired vision.	face shield, or combined eye and respiratory protection.	flush with water, take to a doctor.
Ingestion: *corrosive*; sore throat, abdominal pain, diarrhea, nausea, vomiting.	do not eat, drink or smoke while working.	rinse mouth, take immediately to hospital.

SPILLAGE	STORAGE	LABELING / NFPA
collect leakage in sealable containers, neutralize spillage with sodium/calcium carbonate powder and then flush away with water; (additional individual protection: CHEMICAL SUIT).	keep in dark place, separate from oxidants and strong bases; ventilate at floor level.	R: 34-37 S: 7/9-26 Note B Corrosive 3 ◇ 0 / 0

NOTES
47% aqueous solution is an azeotropic mixture. ✷ TLV equals that of hydrogen bromide. Lung edema symptoms usually develop several hours later and are aggravated by physical exertion: rest and hospitalization essential. As first aid, a doctor or authorized person should consider administering a corticosteroid spray. Concentrated solutions give off corrosive fumes on exposure to air. Unbreakable packaging preferred; if breakable, keep in unbreakable container.

Transport Emergency Card TEC(R)-80G01 (80G06)

HI: 80; UN-number: 1788

HYDROGEN CHLORIDE
(cylinder)

PHYSICAL PROPERTIES		IMPORTANT CHARACTERISTICS	
Boiling point, °C	−85	**COLORLESS COMPRESSED LIQUEFIED GAS WITH PUNGENT ODOR**	
Melting point, °C	−114	Gas is heavier than air. In aqueous solution is a strong acid which reacts violently with bases and	
Relative density (water = 1)	0.86	is corrosive. Reacts violently with strong oxidants, giving off toxic gas (→ *chlorine*). Reacts with	
Relative vapor density (air = 1)	1.3	air to form corrosive mixtures (→ *hydrochloric acid*). Attacks many metals, giving off flammable	
Vapor pressure, mm Hg at 20 °C	31.9	gas (→ *hydrogen*).	
Solubility in water, g/100 ml at 20 °C	72		
Relative molecular mass	36.5		
Log P octanol/water	0.3	TLV-TWA	5 ppm	7 mg/m³ C

Absorption route: Can enter the body by inhalation. Harmful atmospheric concentrations can build up very rapidly if gas is released.
Immediate effects: Corrosive to the eyes, skin and respiratory tract. Can cause frostbite due to rapid evaporation. Inhalation can cause lung edema.

Gross formula: ClH

HAZARDS/SYMPTOMS	PREVENTIVE MEASURES	FIRE EXTINGUISHING/FIRST AID
Fire: non-combustible.		in case of fire in immediate vicinity: use any extinguishing agent.
Explosion:		in case of fire: keep cylinder cool by spraying with water.
	STRICT HYGIENE	IN ALL CASES CALL A DOCTOR
Inhalation: *corrosive*; sore throat, cough, shortness of breath, severe breathing difficulties, headache.	ventilation, local exhaust or respiratory protection.	fresh air, rest, place in half-sitting position, take to hospital.
Skin: *corrosive*; redness, pain, serious burns.	insulated gloves, protective clothing.	remove contaminated clothing; *in case of frostbite*: DO NOT remove clothing, flush skin with water or shower, refer to a doctor.
Eyes: *corrosive*; redness, pain, impaired vision.	face shield, or combined eye and respiratory protection.	flush with water, take to a doctor.

SPILLAGE	STORAGE	LABELING / NFPA
evacuate area, call in an expert, ventilate, under no circumstances spray liquid with water, combat gas cloud with water curtain; (additional individual protection: CHEMICAL SUIT).	if stored indoors, keep in cool, fireproof place, separate from oxidants.	 0 3 0 0

NOTES
Log P octanol/water is estimated. TLV is maximum and must not be exceeded. Lung edema symptoms usually develop several hours later and are aggravated by physical exertion: rest and hospitalization essential. As first aid, a doctor or authorized person should consider administering a corticosteroid spray. Do not spray leaking cylinder with water (to avoid corrosion). Turn leaking cylinder so that leak is on top to prevent liquid hydrogen chloride escaping. Use cylinder with special fittings.

CAS-No: [74-90-8]
formonitrile
HCN
hydrocyanic acid
Prussic acid

HCN

HYDROGEN CYANIDE

(cylinder)

PHYSICAL PROPERTIES		IMPORTANT CHARACTERISTICS
Boiling point, °C	26	**COLORLESS HIGHLY VOLATILE LIQUID WITH CHARACTERISTIC ODOR**
Melting point, °C	−13	Vapor mixes readily with air, forming explosive mixtures. Moderate heating or small quantities of
Flash point, °C	−18	water (>2%) and/or traces of alkali can cause explosive polymerization, with risk of fire and
Auto-ignition temperature, °C	535	explosion. Do not use compressed air when filling, emptying or processing. Toxic gases are a
Relative density (water = 1)	0.7	product of combustion (→ *nitrous vapors*). Reacts with oxidants, with risk of fire and explosion.
Relative vapor density (air = 1)	0.93	

Relative density at 20 °C of saturated mixture vapor/air (air = 1)	0.94	TLV-TWA (skin) 10 ppm 11 mg/m³ C
Vapor pressure, mm Hg at 20 °C	631	
Solubility in water	very good	**Absorption route:** Can enter the body by inhalation or ingestion or through the skin. Harmful
Explosive limits, vol% in air	5.4-46.6	atmospheric concentrations can build up very rapidly on evaporation at 20°C.
Electrical conductivity, pS/m	10⁷	**Immediate effects:** Irritates the eyes, skin and respiratory tract. Impedes tissue respiration. In
Relative molecular mass	27.0	serious cases risk of seizures, unconsciousness and death.
Log P octanol/water	0.7	

Gross formula: CHN

HAZARDS/SYMPTOMS	PREVENTIVE MEASURES	FIRE EXTINGUISHING/FIRST AID
Fire: extremely flammable.	keep away from open flame and sparks, no smoking.	shut off supply; if impossible and no danger to surrounding area, allow fire to burn itself out; otherwise extinguish with powder, carbon dioxide, alcohol-resistant foam, (halons).
Explosion: forms explosive air-vapor mixtures.	sealed machinery, ventilation, explosion-proof electrical equipment and lighting.	in case of fire: keep cylinder cool by spraying with water.
Inhalation: sore throat, severe breathing difficulties, headache, dizziness, nausea, unconsciousness.	ventilation, local exhaust or respiratory protection.	fresh air, rest, place in half-sitting position; if possible special treatment (see Notes), otherwise administer 100% oxygen and take to hospital; avoid mouth-to-mouth resuscitation if possible (risk to person assisting).
Skin: *is absorbed*; redness, pain; see also 'Inhalation'.	gloves, protective clothing.	remove contaminated clothing, wash skin with soap and water, call a doctor.
Eyes: redness, pain.	face shield, or combined eye and respiratory protection.	flush with water, take to a doctor if necessary.

SPILLAGE	STORAGE	LABELING / NFPA
evacuate area, call in an expert, collect leakage in sealable containers, first dilute spillage and then render harmless with 15% → *sodium hypochlorite* solution, flush away with water; (additional individual protection: CHEMICAL SUIT).	keep in cool, fireproof place; add an inhibitor.	4 / 2 / 4

NOTES
Log P octanol/water is estimated. TLV is maximum and must not be exceeded. Special first aid required in the event of poisoning: antidotes must be available (with instructions). Limited storage life. Cylinder must be returned to supplier within one year. Turn leaking cylinder so that leak is on top to prevent liquid hydrogen cyanide escaping. Use cylinder with special fittings.

Transport Emergency Card TEC(R)-61G02   HI: 663; UN-number: 1051

HYDROGEN FLUORIDE
(hydrogen fluoride solution 30-80%)

PHYSICAL PROPERTIES	IMPORTANT CHARACTERISTICS
Boiling point, °C 113-47 Melting point, °C <35 Relative density (water = 1) 1.1-1.3 Relative vapor density (air = 1) 0.7 (HF) Relative density at 20 °C of saturated mixture vapor/air (air = 1) 1.0-1.4 Vapor pressure, mm Hg at 20 °C 1.8-158 Solubility in water ∞ Relative molecular mass 20.0 (HF) Log P octanol/water −0.9	**COLORLESS LIQUID WITH PUNGENT ODOR** Vapor is heavier than air. Strong acid which reacts violently with many metals, giving off → *hydrogen*. Reacts with air to form corrosive vapors which are heavier than air and spread at ground level. Attacks glass and other siliceous materials.
	TLV-TWA 3 ppm 2.6 mg/m³ C ✳
	Absorption route: Can enter the body by inhalation or ingestion or through the skin. Harmful atmospheric concentrations can build up very rapidly if gas is released. **Immediate effects:** Corrosive to the eyes, skin and respiratory tract. Inhalation of vapor/fumes can cause severe breathing difficulties (lung edema). In serious cases risk of death.
Gross formula: FH	

HAZARDS/SYMPTOMS	PREVENTIVE MEASURES	FIRE EXTINGUISHING/FIRST AID
Fire: non-combustible.		in case of fire in immediate vicinity: use any extinguishing agent.
Explosion:		in case of fire: keep tanks/drums cool by spraying with water.
Inhalation: *corrosive*; sore throat, cough, severe breathing difficulties.	ventilation, local exhaust or respiratory protection.	fresh air, rest, place in half-sitting position, take to hospital.
Skin: *corrosive*; redness, pain, serious burns.	gloves, protective clothing.	remove contaminated clothing, flush skin with water or shower, take immediately to hospital.
Eyes: *corrosive*; redness, pain, impaired vision.	face shield, or combined eye and respiratory protection.	flush with water, take to doctor.
Ingestion: *corrosive*; sore throat, abdominal pain, diarrhea, vomiting.		rinse mouth, take immediately to hospital.

SPILLAGE	STORAGE	LABELING / NFPA
evacuate area, call in an expert, collect leakage in sealable containers, soak up with sand or other inert absorbent and remove (not using metal containers) to safe place; (additional individual protection: CHEMICAL SUIT).	keep cool, separate from strong bases; ventilate at floor level.	R: 26/27/28-35 S: 7/9-26-36/37-45 Note B Toxic Corrosive 0 / 4 / 1

NOTES
Log P octanol/water is estimated. ✳ TLV equals that of fluorine. Lung edema symptoms usually develop several hours later and are aggravated by physical exertion: rest and hospitalization essential. As first aid, a doctor or authorized person should consider administering a corticosteroid spray. Skin conditions develop only after some time; sores are very painful and do not heal easily, requiring special medical treatment. Inhalation/ingestion of substantial quantities or burns covering large areas of skin can cause symptoms requiring special treatment, due to decreased levels of calcium and magnesium in the blood. Use airtight packaging made of special material.

Transport Emergency Card TEC(R)-703/704

HI: 886; UN-number: 1790

CAS-No: [7664-39-3]
fluorohydric acid
HF
anhydrous hydrofluoric acid
anhydrous hydrogen fluoride

HYDROGEN FLUORIDE
(cylinder)

PHYSICAL PROPERTIES	IMPORTANT CHARACTERISTICS	
Boiling point, °C 20 Melting point, °C −83 Relative density (water = 1) 0.95 Relative vapor density (air = 1) 2.5 Vapor pressure, mm Hg at 20°C 0.76 Solubility in water ∞ Relative molecular mass 20.0 Log P octanol/water −0.9	**COLORLESS GAS OR HYGROSCOPIC LIQUID WITH PUNGENT ODOR** Vapor is heavier than air and reacts with water vapor to form fumes which spread at ground level. In aqueous solution gives off large quantities of heat. Reacts with alcohols and unsaturated compounds. Attacks glass and other silicatious materials. Can cause polyethene to become brittle.	
	TLV-TWA 3 ppm 2.6 mg/m³ C ✳	
	Absorption route: Can enter the body by inhalation or ingestion or through the skin. Harmful atmospheric concentrations can build up very rapidly on evaporation at 20°C. **Immediate effects:** Corrosive to the eyes, skin and respiratory tract. Inhalation can cause lung edema. In serious cases risk of death.	
Gross formula: FH		

HAZARDS/SYMPTOMS	PREVENTIVE MEASURES	FIRE EXTINGUISHING/FIRST AID
Fire: non-combustible.		in case of fire in immediate vicinity: use any extinguishing agent.
Explosion:		in case of fire: keep cylinder cool by spraying with water.
	STRICT HYGIENE	
Inhalation: *corrosive*; sore throat, cough, severe breathing difficulties.	ventilation, local exhaust or respiratory protection.	fresh air, rest, place in half-sitting position, take to hospital.
Skin: *corrosive*; redness, pain, serious burns.	gloves, protective clothing.	remove contaminated clothing, flush skin with water or shower, take to hospital.
Eyes: *corrosive*; redness, pain, impaired vision.	acid goggles, or combined eye and respiratory protection.	flush with water, take to a doctor.

SPILLAGE	STORAGE	LABELING / NFPA
evacuate area, call in an expert, ventilate, combat gas cloud with water curtain; (additional individual protection: CHEMICAL SUIT).	if stored indoors, keep in cool, fireproof place.	0 / 4 / 1

NOTES

Molecules of hydrogen fluoride associate in both liquid and gas phase, depending on temperature; at 20°C the average association is 3.7HF per molecule, resulting in a molecular mass for the gas of 74, on which the relative density is based. Log P octanol/water is estimated. ✳ TLV equals that of fluorine. Lung edema symptoms usually develop several hours later and are aggravated by physical exertion: rest and hospitalization essential. As first aid, a doctor or authorized person should consider administering a corticosteroid spray. Inhalation/ingestion of substantial quantities, or burns covering large areas of skin, can cause symptoms requiring special treatment, due to decreased levels of calcium and magnesium in the blood. Do not spray leaking cylinder with water (to avoid corrosion). Turn leaking cylinder so that leak is on top to prevent liquid hydrogen fluoride escaping. Use cylinder with special fittings.

Transport Emergency Card TEC(R)-78

HI: 886; UN-number: 1052

HYDROGEN PEROXIDE
(10%)

PHYSICAL PROPERTIES	IMPORTANT CHARACTERISTICS
Boiling point, °C ca. 100 Melting point, °C ca. 0 Relative density (water = 1) 1.0 Relative density at 20 °C of saturated mixture vapor/air (air = 1) 1.0 Vapor pressure, mm Hg at 30 °C 0.076 Solubility in water ∞ Relative molecular mass 34.0 Log P octanol/water − 1.1	**COLORLESS LIQUID** Vapor mixes readily with air. Decomposes when heated, on contact with rough surface or due to contaminants, giving off oxygen, with increased risk of fire. At pH values above 7 (in basic environment) solution can decompose violently, esp. in the presence of metal ions. Strong oxidant which reacts violently with combustible substances and reducing agents. Attacks many substances, esp. textiles and paper.

	TLV-TWA	1 ppm	1.4 mg/m³ ✶

Absorption route: Can enter the body by inhalation or ingestion. Harmful atmospheric concentrations build up fairly slowly on evaporation, but much more rapidly in aerosol form.
Immediate effects: Irritates the skin. Corrosive to the eyes and respiratory tract.

Gross formula: H_2O_2

HAZARDS/SYMPTOMS	PREVENTIVE MEASURES	FIRE EXTINGUISHING/FIRST AID
Fire: not combustible but enhances combustion of other substances.	do not heat.	in case of fire in immediate vicinity: use any extinguishing agent.
Explosion: violent decomposition (especially due to mixing with other substances) can cause explosion.		in case of fire: keep tanks/drums cool by spraying with water.
Inhalation: *corrosive*; sore throat, cough, shortness of breath.	ventilation, local exhaust or respiratory protection.	fresh air, rest, place in half-sitting position, call a doctor.
Skin: redness, pain.	gloves.	remove contaminated clothing, flush skin with water or shower.
Eyes: *corrosive*; redness, pain, impaired vision.	face shield.	flush with water, take to a doctor.
Ingestion: *corrosive*; sore throat, abdominal pain, vomiting.		rinse mouth, call a doctor or take to hospital.

SPILLAGE	STORAGE	LABELING / NFPA
flush away with water.	keep in cool, dark place, separate from combustible substances, reducing agents and strong bases.	◇ 0 / 2 1 / oxy

NOTES
Log P octanol/water is estimated. ✶ TLV equals that of hydrogen peroxide: 90%.

HYDROGEN PEROXIDE
(ca. 35%)

PHYSICAL PROPERTIES	IMPORTANT CHARACTERISTICS
Boiling point, °C — 108 Melting point, °C — -33 Relative density (water = 1) — 1.1 Relative vapor density (air = 1) — 1.2 Relative density at 20 °C of saturated mixture vapor/air (air = 1) — 1.0 Vapor pressure, mm Hg at 30 °C — 0.30 Solubility in water — ∞ Relative molecular mass — 34.0 Log P octanol/water — -1.1	**COLORLESS LIQUID** Vapor mixes readily with air. Decomposes when moderately heated, on contact with rough surface or due to contaminants, giving off oxygen. At pH values above 7 (in basic environment) solution can decompose violently, esp. in the presence of metal ions. Strong oxidant which reacts violently with combustible substances and reducing agents, with risk of fire and explosion. Attacks many organic substances, esp. paper and textiles.
	TLV-TWA 1 ppm 1.4 mg/m³ ✱
	Absorption route: Can enter the body by inhalation or ingestion. Harmful atmospheric concentrations build up fairly slowly on evaporation at 20°C, but much more rapidly in aerosol form. **Immediate effects:** Irritates the skin. Corrosive to the eyes and respiratory tract. Inhalation can cause lung edema.
Gross formula: H₂O₂	

HAZARDS/SYMPTOMS	PREVENTIVE MEASURES	FIRE EXTINGUISHING/FIRST AID
Fire: non-combustible, but makes other substances more combustible; many chemical reactions can cause fire and explosion.	avoid contact with combustible substances, do not heat.	in case of fire in immediate vicinity: use any extinguishing agent.
Explosion: decomposition and mixing with other substances can cause explosion, do not heat solution, avoid contamination.		in case of fire: keep tanks/drums cool by spraying with water.
Inhalation: *corrosive*; sore throat, cough, shortness of breath, severe breathing difficulties.	ventilation, local exhaust or respiratory protection.	fresh air, rest, place in half-sitting position, take to hospital.
Skin: redness, pain.	gloves, protective clothing.	remove contaminated clothing, flush skin with water or shower, call a doctor.
Eyes: *corrosive*; redness, pain, impaired vision.	face shield.	flush with water, take to a doctor if necessary.
Ingestion: *corrosive*; sore throat, abdominal pain, diarrhea, vomiting.		rinse mouth, immediately take to hospital.

SPILLAGE	STORAGE	LABELING / NFPA
flush away with water.	keep in cool, dark place, separate from combustible substances, reducing agents and strong bases.	R: 34 S: 28-39 Note B 🛠 Corrosive

NFPA diamond: 2 (left) / 1 (right) / 0 (top) / oxy (bottom)

NOTES
Log P octanol/water is estimated. ✱ TLV as H₂O₂: 90%. Flush contaminated clothing with water (fire hazard). If a container of hydrogen peroxide becomes warm, remove to safe place and empty, then flush with water. Lung edema symptoms usually develop several hours later and are aggravated by physical exertion: rest and hospitalization essential. As first aid, a doctor or authorized person should consider administering a corticosteroid spray. Packaging: special material, non-airtight with safety vent.

HYDROGEN PEROXIDE
(50%)

PHYSICAL PROPERTIES		IMPORTANT CHARACTERISTICS	
Boiling point, °C	114	**COLORLESS LIQUID**	
Melting point, °C	−50	Vapor mixes readily with air. Decomposes violently when heated, on contact with rough surface	
Relative density (water = 1)	1.2	or due to contaminants, giving off → *oxygen*, with increased risk of fire. At pH values above 7 (in	
Relative vapor density (air = 1)	1.2	basic environment) solution can decompose violently, esp. in the presence of metal ions. Strong	
Relative density at 20 °C of saturated mixture		oxidant which reacts violently with combustible substances and reducing agents, with risk of fire	
vapor/air (air = 1)	1.0	and explosion. Attacks many organic substances, esp. paper and textiles.	
Vapor pressure, mm Hg at 30 °C	0.76		
Solubility in water	∞	TLV-TWA	1 ppm	1.4 mg/m³ ✱
Relative molecular mass	34.0			
Log P octanol/water	−1.1	**Absorption route:** Can enter the body by inhalation or ingestion. Harmful atmospheric	
		concentrations can build up fairly rapidly on evaporation at approx. 20°C - even more rapidly in	
		aerosol form.	
		Immediate effects: Corrosive to the eyes, skin and respiratory tract. Inhalation of vapor/fumes	
		can cause severe breathing difficulties (lung edema).	
Gross formula:	H_2O_2		

HAZARDS/SYMPTOMS	PREVENTIVE MEASURES	FIRE EXTINGUISHING/FIRST AID
Fire: non-combustible, but makes other substances more combustible: many chemical reactions can cause fire and explosion.	do not heat.	in case of fire in immediate vicinity: use any extinguishing agent.
Explosion: decomposition or mixing with other substances can cause explosion.	do not heat solution, avoid contamination.	in case of fire: keep tanks/drums cool by spraying with water.
Inhalation: *corrosive*; sore throat, cough, shortness of breath, severe breathing difficulties.	ventilation, local exhaust or respiratory protection.	fresh air, rest, place in half-sitting position, take to hospital.
Skin: *corrosive*; redness, pain, burns.	gloves, protective clothing.	remove contaminated clothing, flush skin with water or shower, call a doctor.
Eyes: *corrosive*; redness, pain, impaired vision.	face shield.	flush with water, take to a doctor.
Ingestion: *corrosive*; sore throat, abdominal pain, vomiting, chest pain.		rinse mouth, take to hospital.

SPILLAGE	STORAGE	LABELING / NFPA
flush away with water; (additional individual protection: breathing apparatus).	keep in cool, dark place, separate from combustible substances, reducing agents and strong bases; add an inhibitor.	R: 34 S: 28-39 Note B Corrosive

NOTES

Log P octanol/water is estimated. ✱ TLV as H_2O_2: 90%. Flush contaminated clothing with water (fire hazard). If container becomes warm, remove to safe place and drain, flush away with water. Lung edema symptoms usually develop several hours later and are aggravated by physical exertion: rest and hospitalization essential. As first aid, a doctor or authorized person should consider administering a corticosteroid spray. Packaging: special material, non-airtight with safety vent.

Transport Emergency Card TEC(R)-43

HI: 85; UN-number: 2014

HYDROGEN SULFIDE
(cylinder)

PHYSICAL PROPERTIES	IMPORTANT CHARACTERISTICS
Boiling point, °C −60 Melting point, °C −86 Flash point, °C flammable gas Auto-ignition temperature, °C 260 Relative density (water = 1) 0.8 Relative vapor density (air = 1) 1.2 Vapor pressure, mm Hg at 20 °C 13,680 Solubility in water, g/100 ml at 10 °C 0.6 Explosive limits, vol% in air 4.0-46 Minimum ignition energy, mJ 0.07 Electrical conductivity, pS/m 1000 Relative molecular mass 34.1 Log P octanol/water 1.2 Gross formula: H₂S	**COLORLESS COMPRESSED LIQUEFIED GAS WITH CHARACTERISTIC ODOR** Gas is heavier than air and spreads at ground level, with risk of ignition at a distance. Flow, agitation etc. can cause build-up of electrostatic charge due to liquid's low conductivity. Toxic corrosive vapors (→ *sulfur dioxides*) are a product of combustion. Reacts violently with oxidants, with risk of fire and explosion. Attacks many metals. TLV-TWA 10 ppm 14 mg/m³ STEL 15 ppm 21 mg/m³ **Absorption route:** Can enter the body by inhalation. Harmful atmospheric concentrations can build up very rapidly if gas is released. **Immediate effects:** Irritates the eyes and respiratory tract. Liquid can cause frostbite due to rapid evaporation. Affects the nervous system. Inhalation of high concentrations of gas can cause lung edema. In serious cases risk of rapid unconsciousness and death.

HAZARDS/SYMPTOMS	PREVENTIVE MEASURES	FIRE EXTINGUISHING/FIRST AID
Fire: extremely flammable.	keep away from open flame and sparks, no smoking.	shut off supply; if impossible and no danger to surrounding area, allow fire to burn itself out; otherwise extinguish with powder, carbon dioxide, (halons).
Explosion: forms explosive air-gas mixtures.	sealed machinery, ventilation, explosion-proof electrical equipment and lighting, grounding when pumping etc. in liquid form, non-sparking hand tools.	in case of fire: keep cylinders cool by spraying with water, fight fire from sheltered location.
Inhalation: sore throat, cough, severe breathing difficulties, headache, dizziness, unconsciousness.	ventilation, local exhaust or respiratory protection.	fresh air, rest, place in half-sitting position, take to hospital.
Skin: *in case of frostbite*; redness, pain, serious burns.	insulating gloves.	*in case of frostbite*: DO NOT remove clothing, flush skin with water or shower, take to hospital.
Eyes: *in case of frostbite*: redness, pain, impaired vision.	acid goggles, or combined eye and respiratory protection.	flush with water, take to a doctor if necessary.

SPILLAGE	STORAGE	LABELING / NFPA
evacuate area, call in an expert, ventilate, under no circumstances spray liquid with water; (additional individual protection: breathing apparatus).	keep in cool, fireproof place.	4 3 0

NOTES
Log P octanol/water is estimated. Depending on the degree of exposure, regular medical checkups are advisable. Lung edema symptoms usually develop several hours later and are aggravated by physical exertion: rest and hospitalization essential. As first aid, a doctor or authorized person should consider administering a corticosteroid spray. Turn leaking cylinder so that leak is on top to prevent liquid hydrogen sulfide escaping. Use cylinder with special fittings.

Transport Emergency Card TEC(R)-826   **HI: 236; UN-number: 1053**

CAS-No: [123-31-9]
1,4-benzenediol
benzohydroquinone
p-dihydroxybenzene
hydroquinol
p-hydroxyphenol

$C_6H_4(OH)_2$

HYDROQUINONE

PHYSICAL PROPERTIES	IMPORTANT CHARACTERISTICS
Boiling point, °C 285 Sublimation temperature, °C 169 Flash point, °C 165 Auto-ignition temperature, °C 515 Relative density (water = 1) 1.33 Relative vapor density (air = 1) 3.81 Relative density at 20 °C of saturated mixture vapor/air (air = 1) 1.0 Vapor pressure, mm Hg at 20 °C 9.9×10^{-4} Solubility in water, g/100 ml at 15 °C 5.9 Relative molecular mass 110.1 Log P octanol/water 0.5	**COLORLESS CRYSTALS** Vapor mixes readily with air. TLV-TWA 2 mg/m³ **Absorption route:** Can enter the body by inhalation or ingestion or, to a limited extent, through the skin. Evaporation negligible at 20°C, but harmful concentrations of airborne particles can build up rapidly. **Immediate effects:** Corrosive to the eyes, skin and respiratory tract. Inhalation can cause lung edema. In serious cases risk of death. **Effects of prolonged/repeated exposure:** Can affect the blood. Prolonged exposure to fumes, dust or vapor can cause lung disorders.
Gross formula: $C_6H_6O_2$	

HAZARDS/SYMPTOMS	PREVENTIVE MEASURES	FIRE EXTINGUISHING/FIRST AID
Fire: combustible.	keep away from open flame, no smoking.	water spray, powder.
	KEEP DUST UNDER CONTROL	
Inhalation: *corrosive*; sore throat, cough, shortness of breath, severe breathing difficulties, headache, blue skin, feeling of weakness, tinnitus, dizziness, nausea, diarrhea, muscle spasms (delirium, unconsciousness).	local exhaust or respiratory protection.	fresh air, rest, place in half-sitting position, take to hospital.
Skin: *corrosive*; redness, pain, serious burns.	gloves.	remove contaminated clothing, flush skin with water or shower, refer to a doctor.
Eyes: *corrosive*; redness, pain, impaired vision.	face shield, or combined eye and respiratory protection.	flush with water, take to an eye doctor.
Ingestion: *corrosive*; vomiting, unconsciousness, hemolytic anemia and jaundice, liver problems; see also 'Inhalation'.	do not eat, drink or smoke while working.	rinse mouth, take to hospital.

SPILLAGE	STORAGE	LABELING / NFPA
clean up spillage, flush away remainder with water; (additional individual protection: P2 respirator).		R: 20/22 S: 2-24/25-39 ✖ Harmful

NFPA diamond: 1 (top), 2 (left), 0 (right)

NOTES

Depending on the degree of exposure, regular medical checkups are advisable. Lung edema symptoms usually develop several hours later and are aggravated by physical exertion: rest and hospitalization essential. As first aid, a doctor or authorized person should consider administering a corticosteroid spray. Effects on blood due to formation of methemoglobin. Special first aid required in the event of poisoning: antidotes must be available (with instructions). At room temperature, in the presence of moisture, oxidizes to quinone, which causes much worse eye irritation than hydroquinone, with consequent risk of conjunctivitis and corneal erosion.

Transport Emergency Card TEC(R)-61G12C

HI: 60; UN-number: 2662

CAS-No: [107-89-1]
Aldol
β-hydroxybutyraldehyde

$CH_3CH(OH)CH_2CHO$

3-HYDROXYBUTANAL

PHYSICAL PROPERTIES		IMPORTANT CHARACTERISTICS	
Boiling point (decomposes), °C	193	**COLORLESS TO LIGHT YELLOW LIQUID WITH CHARACTERISTIC ODOR**	
Melting point, °C	−88	Vapor mixes readily with air. Decomposes when heated above 85°C, forming → *crotonaldehyde.*	
Flash point, °C	83	Reacts violently with strong oxidants.	
Auto-ignition temperature, °C	245		
Relative density (water = 1)	1.1	TLV-TWA not available	
Relative vapor density (air = 1)	3.0		
Relative density at 20 °C of saturated mixture vapor/air (air = 1)	1.0	**Absorption route:** Can enter the body by inhalation or ingestion. Harmful atmospheric concentrations build up very slowly, if at all, on evaporation at approx. 20°C, but much more rapidly in aerosol form.	
Vapor pressure, mm Hg at 20 °C	<0.076		
Solubility in water	good		
Relative molecular mass	88.1	**Immediate effects:** Irritates the eyes, skin and respiratory tract.	
Gross formula:	$C_4H_8O_2$		

HAZARDS/SYMPTOMS	PREVENTIVE MEASURES	FIRE EXTINGUISHING/FIRST AID
Fire: combustible.	keep away from open flame, no smoking.	powder, alcohol-resistant foam, water spray, carbon dioxide, (halons).
Explosion: above 83°C: forms explosive air-vapor mixtures, due to formation of crotonaldehyde vapor.	above 83°C: sealed machinery, ventilation.	in case of fire: keep tanks/drums cool by spraying with water.
Inhalation: sore throat, cough.	ventilation, local exhaust or respiratory protection.	fresh air, rest.
Skin: redness.	gloves.	remove contaminated clothing, flush skin with water or shower.
Eyes: redness.	face shield.	flush with water, take to a doctor if necessary.
Ingestion: sore throat, abdominal pain, stomachache.		rinse mouth, call a doctor.

SPILLAGE	STORAGE	LABELING / NFPA
collect leakage in sealable containers, soak up with sand or other inert absorbent and remove to safe place, flush away remainder with water.	keep separate from oxidants; ventilate at floor level.	2 / 3 / 2

NOTES

Transport Emergency Card TEC(R)-61G06B

HI: 60; UN-number: 2839

CAS-No: [111-41-1]
2-((2-aminoethyl)amino)ethanol
N-aminoethylethanol amine
1-(2-hydroxyethylamino)-2-aminoethane
2-(2-hydroxyethylamino)ethylamine

$NH_2CH_2CH_2NHCH_2CH_2OH$

HYDROXYETHYLETHYLENEDIAMINE

PHYSICAL PROPERTIES		IMPORTANT CHARACTERISTICS
Boiling point, °C	238	**COLORLESS LIQUID WITH CHARACTERISTIC ODOR**
Melting point, °C	?	Vapor mixes readily with air. Decomposes when heated or burned, giving off toxic vapors. Reacts with oxidants.
Flash point, °C	135	
Auto-ignition temperature, °C	368	
Relative density (water = 1)	1.03	TLV-TWA not available
Relative vapor density (air = 1)	3.6	
Relative density at 20 °C of saturated mixture vapor/air (air = 1)	1.0	**Absorption route:** Can enter the body by inhalation or ingestion. Harmful atmospheric concentrations build up very slowly, if at all, on evaporation at approx. 20°C, but much more rapidly in aerosol form.
Vapor pressure, mm Hg at 20 °C	0.0099	**Immediate effects:** Irritates the skin. Corrosive to the eyes and respiratory tract. Inhalation of vapor/fumes can cause severe breathing difficulties (lung edema).
Solubility in water	∞	**Effects of prolonged/repeated exposure:** Prolonged or repeated contact can cause skin disorders.
Relative molecular mass	104.2	
Gross formula: $C_4H_{12}N_2O$		

HAZARDS/SYMPTOMS	PREVENTIVE MEASURES	FIRE EXTINGUISHING/FIRST AID
Fire: combustible.	keep away from open flame, no smoking.	powder, alcohol-resistant foam, water spray, carbon dioxide, (halons).
Inhalation: *corrosive*; cough, severe breathing difficulties.	ventilation.	fresh air, rest, place in half-sitting position, call a doctor.
Skin: redness, pain.	gloves.	remove contaminated clothing, flush skin with water or shower, call a doctor if necessary.
Eyes: *corrosive*; redness, pain, impaired vision.	safety glasses.	flush with water, take to a doctor.
Ingestion: sore throat, abdominal cramps, diarrhea, chest pain, nausea, vomiting.		rinse mouth, call a doctor or take to hospital.

SPILLAGE	STORAGE	LABELING / NFPA
collect leakage in sealable containers, flush away remainder with water.	ventilate at floor level.	<div style="text-align:center">1 1 0</div>

NOTES
Lung edema symptoms usually develop several hours later and are aggravated by physical exertion: rest and hospitalization essential. As first aid, a doctor or authorized person should consider administering a corticosteroid spray.

HO(CH$_2$)$_2$NHNH$_2$

β-HYDROXYETHYLHYDRAZINE

PHYSICAL PROPERTIES	IMPORTANT CHARACTERISTICS	
Boiling point (decomposes), °C — 203 Melting point, °C — −70 Flash point, °C — 85 Relative density (water = 1) — 1.1 Relative vapor density (air = 1) — 2.6 Relative density at 20 °C of saturated mixture vapor/air (air = 1) — 1.0 Vapor pressure, mm Hg at 145 °C — 4.6 Solubility in water — ∞ Relative molecular mass — 76.1	**COLORLESS LIGHT VISCOUS LIQUID** Vapor mixes readily with air. Toxic gases are a product of combustion (→ *nitrous vapors*). Decomposes when heated above 203°C, giving off toxic gases (→ *nitrous vapors*). Reducing agent. Reacts with oxidants.	
	TLV-TWA not available	
	Absorption route: Can enter the body by inhalation or ingestion or through the skin. **Immediate effects:** Irritates the eyes, skin and respiratory tract. Affects the nervous system. Inhalation of vapor/fumes can cause severe breathing difficulties (lung edema). In serious cases risk of seizures and death. **Effects of prolonged/repeated exposure:** Heart rate disorders, anemia, low blood-sugar level. Prolonged or repeated contact can cause eczema. Can affect the blood. Can cause liver damage.	
Gross formula: C$_2$H$_8$N$_2$O		

HAZARDS/SYMPTOMS	PREVENTIVE MEASURES	FIRE EXTINGUISHING/FIRST AID
Fire: combustible.	keep away from open flame, no smoking.	powder, alcohol-resistant foam, water spray, carbon dioxide, (halons).
Explosion: above 85°C: forms explosive air-vapor mixtures.	above 85°C: sealed machinery, ventilation.	
	STRICT HYGIENE	IN ALL CASES CALL IN A DOCTOR
Inhalation: *corrosive*; cough, shortness of breath, severe breathing difficulties, vomiting, jaundice.	ventilation, local exhaust or respiratory protection.	fresh air, rest, place in half-sitting position, artificial respiration if necessary, take to hospital.
Skin: *corrosive, is absorbed*; redness, pain.	gloves, protective clothing.	remove contaminated clothing, flush skin with water or shower, call a doctor if necessary.
Eyes: *corrosive*; redness, pain, impaired vision.	face shield.	flush with water, take to a doctor.
Ingestion: cough, shortness of breath, severe breathing difficulties, cramps, vomiting, jaundice, abdominal pain.	do not eat, drink or smoke while working.	rinse mouth, immediately take to hospital.

SPILLAGE	STORAGE	LABELING / NFPA
collect leakage in sealable containers, soak up with sand or other inert absorbent and remove to safe place, do not use sawdust or other combustible absorbents; (additional individual protection: breathing apparatus).	keep separate from combustible substances and oxidants; ventilate at floor level.	

NOTES
Flush contaminated clothing with water (fire hazard). Depending on the degree of exposure, regular medical checkups are advisable. Lung edema symptoms usually develop several hours later and are aggravated by physical exertion: rest and hospitalization essential. As first aid, a doctor or authorized person should consider administering a corticosteroid spray.

Transport Emergency Card TEC(R)-80G12

CAS-No: [80-05-7]
4,4'-(1-methyl ethylidene)bisphenol
2,2-bis(4-hydroxyfenyl)propane
difenylol propane

$(CH_3)_2C(C_6H_4OH)_2$

2,2 BIS-(4-HYDROXYFENYL)PROPANE

PHYSICAL PROPERTIES	IMPORTANT CHARACTERISTICS	
Boiling point, °C (4.0mm Hg) 220 Melting point, °C 153 Flash point, °C 207 Auto-ignition temperature, °C 600 Relative density (water = 1) 1.2 Vapor pressure, mm Hg at 190°C 0.65 Solubility in water none Relative molecular mass 228.3	**WHITE CRYSTALS WITH CHARACTERISTIC ODOR** Reacts violently with strong oxidants, with risk of fire and explosion.	
	TLV-TWA not available	
	Absorption route: Can enter the body by inhalation or ingestion. Evaporation negligible at 20°C, but harmful atmospheric concentrations can build up rapidly in aerosol form. **Immediate effects:** Irritates the eyes, skin and respiratory tract. **Effects of prolonged/repeated exposure:** Prolonged or repeated contact can cause skin disorders.	
Gross formula: $C_{15}H_{16}O_2$		

HAZARDS/SYMPTOMS	PREVENTIVE MEASURES	FIRE EXTINGUISHING/FIRST AID
Fire: combustible.	keep away from open flame, no smoking.	water spray, powder.
Explosion: finely dispersed particles form explosive mixtures on contact with air.	keep dust from accumulating; sealed machinery, explosion-proof electrical equipment and lighting, grounding.	
Inhalation: sore throat, cough.	local exhaust or respiratory protection.	fresh air, rest.
Skin: redness.	gloves.	remove contaminated clothing, flush skin with water or shower.
Eyes: redness, pain.	goggles.	flush with water, take to a doctor.
Ingestion: nausea, vomiting.		rinse mouth, call a doctor.

SPILLAGE	STORAGE	LABELING / NFPA
clean up spillage, carefully collect remainder; (additional individual protection: P2 respirator).	separate from strong oxidants, bulk storage under inert gas.	

NOTES
Bisfenol A is a trade name.

HYDROXYLAMINE

PHYSICAL PROPERTIES	IMPORTANT CHARACTERISTICS
Decomposes below boiling point, °C <70 Melting point, °C 33 Flash point, °C see notes Relative density (water = 1) 1.2 Vapor pressure, mm Hg at 47 °C 9.9 Solubility in water ∞ Relative molecular mass 33.0 Log P octanol/water −1.5	**COLORLESS HYGROSCOPIC NEEDLES OR FLAKES** Decomposes at ambient temperature on contact with atmospheric moisture and carbon dioxide. Decomposes explosively when heated above 70° C and/or due to contaminants. Reacts violently with oxidants and many other substances, with risk of fire and explosion. Attacks aluminum, copper, tin and zinc.
	TLV-TWA not available
	Absorption route: Can enter the body by inhalation or ingestion or through the skin. **Immediate effects:** Irritates the eyes, skin and respiratory tract. **Effects of prolonged/repeated exposure:** Prolonged or repeated contact can cause skin disorders. Can affect the blood.
Gross formula: H_3NO	

HAZARDS/SYMPTOMS	PREVENTIVE MEASURES	FIRE EXTINGUISHING/FIRST AID
Fire: flammable.	keep away from open flame and sparks, no smoking, avoid contact with strong oxidants.	large quantities of water, powder, carbon dioxide, (halons).
Explosion: heating and/or catalytic decomposition can cause explosion.		in case of fire: keep tanks/drums cool by spraying with water, fight fire from sheltered location.
Inhalation: sore throat, shortness of breath, headache, dizziness, blue skin, feeling of weakness.	local exhaust or respiratory protection.	fresh air, rest, call a doctor.
Skin: *is absorbed*; redness, pain.	gloves, protective clothing.	remove contaminated clothing, wash skin with soap and water, call a doctor.
Eyes: redness, pain.	face shield.	flush with water, take to a doctor if necessary.
Ingestion: sore throat, nausea, vomiting, shortness of breath, headache, dizziness, feeling of weakness, blue skin.	do not eat, drink or smoke while working.	rinse mouth, immediately take to hospital.

SPILLAGE	STORAGE	LABELING / NFPA
clean up spillage, flush away remainder with water, do not use sawdust or other combustible absorbents; (additional individual protection: P2 respirator).	keep in cool, dry, fireproof place, separate from oxidants.	0 2 3

NOTES
Combustible, but flash point and explosive limits not given in the literature. Log P octanol/water is estimated. Depending on the degree of exposure, regular medical checkups are advisable. Effects on blood due to formation of methemoglobin. Special first aid required in the event of poisoning: antidotes must be available (with instructions). Use airtight packaging made of special material.

Transport Emergency Card TEC(R)-80G15

$NH_2OH.H_2SO_4$

HYDROXYLAMINE HYDROSULFATE

PHYSICAL PROPERTIES	IMPORTANT CHARACTERISTICS	
Decomposes below boiling point, °C 57 Relative density (water = 1) 1.9 Solubility in water good Relative molecular mass 131.1	**WHITE TO BROWN HYGROSCOPIC CRYSTALS** Decomposes when moderately heated, giving off corrosive vapors. Strong reducing agent which reacts violently with oxidants. In aqueous solution is a strong acid which reacts violently with bases and is corrosive. Attacks many metals, giving off flammable gas (→ *hydrogen*).	
	TLV-TWA not available	
	Absorption route: Can enter the body by ingestion or through the skin. **Immediate effects:** Corrosive to the eyes, skin and mucous membranes. Vapor given off on heating/decomposition is corrosive to the respiratory tract if inhaled and can cause lung edema. **Effects of prolonged/repeated exposure:** Can affect the blood.	
Gross formula: H_5NO_5S		

HAZARDS/SYMPTOMS	PREVENTIVE MEASURES	FIRE EXTINGUISHING/FIRST AID
Fire: non-combustible.		in case of fire in immediate vicinity: use any extinguishing agent.
Explosion:		in case of fire: keep tanks/drums cool by spraying with water.
	STRICT HYGIENE	
Inhalation: vapor (from decompositon): *corrosive*; sore throat, severe breathing difficulties, dizziness, blue skin, headache.	local exhaust or respiratory protection.	fresh air, rest, place in half-sitting position, take to hospital.
Skin: *corrosive*; redness, pain, serious burns.	gloves.	remove contaminated clothing, wash skin with soap and water, take to hospital.
Eyes: *corrosive*; redness, pain, impaired vision.	face shield.	flush with water, take to a doctor.
Ingestion: *corrosive*; sore throat, nausea, stomachache, abdominal pain, shortness of breath, feeling of weakness, headache, blue skin.		rinse mouth, immediately take to hospital.

SPILLAGE	STORAGE	LABELING / NFPA
clean up spillage, neutralize remainder with soda and flush away with water; (additional individual protection: P2 respirator).	if stored indoors, keep in dry, fireproof place, separate from oxidants and strong bases; ventilate at floor level.	

NOTES
Not to be confused with hydroxylamine sulfate. Depending on the degree of exposure, regular medical checkups are advisable. Lung edema symptoms usually develop several hours later and are aggravated by physical exertion: rest and hospitalization essential. As first aid, a doctor or authorized person should consider administering a corticosteroid spray. Effects on blood due to formation of methemoglobin. Special first aid required in the event of poisoning: antidotes must be available (with instructions).

Transport Emergency Card TEC(R)-80G19

$(NH_2OH)_2.H_2SO_4$

HYDROXYLAMINE SULFATE

PHYSICAL PROPERTIES	IMPORTANT CHARACTERISTICS
Melting point (decomposes), °C 120 Relative density (water = 1) 1.9 Solubility in water, g/100 ml at 20°C 68.5 Relative molecular mass 164.1	**WHITE HYGROSCOPIC CRYSTALS OR POWDER WITH CHARACTERISTIC ODOR** Can produce thermal explosion when heated. Decomposes when heated - explosively when subjected to friction - giving off corrosive vapors (→ *oxides of sulfur*, → *nitrous vapors*). With 8% water can decompose explosively at temperatures as low as 90°C. Strong oxidant which reacts violently with combustible substances and reducing agents. Reacts with bases, forming → *hydroxylamine*.
	TLV-TWA not available
	Absorption route: Can enter the body by ingestion, and presumed to be able to enter through the skin. **Immediate effects:** Corrosive to the eyes, skin and mucous membranes. Vapors given off on heating/decomposition attack the respiratory tract and can cause lung edema. **Effects of prolonged/repeated exposure:** Can affect the blood.
Gross formula: $H_8N_2O_6S$	

HAZARDS/SYMPTOMS	PREVENTIVE MEASURES	FIRE EXTINGUISHING/FIRST AID
Fire: combustible.	keep away from open flame, no smoking.	water spray, powder.
Explosion: risk of explosion if subjected to friction.	do not subject to shocks or friction.	
	STRICT HYGIENE	
Inhalation: *corrosive*; sore throat, cough, severe breathing difficulties, headache, dizziness, blue skin, feeling of weakness.	local exhaust or respiratory protection.	fresh air, rest, place in half-sitting position, take to hospital.
Skin: *corrosive*; redness, pain, burns.	gloves.	remove contaminated clothing, flush skin with water or shower, call a doctor.
Eyes: *corrosive*; redness, pain, impaired vision.	goggles.	flush with water, take to a doctor.
Ingestion: *corrosive*; sore throat, abdominal pain, diarrhea, stomachache, headache, shortness of breath, nausea, blue skin, feeling of weakness.	do not eat, drink or smoke while working.	rinse mouth, take immediately to hospital.

SPILLAGE	STORAGE	LABELING / NFPA
clean up spillage, flush away remainder with water; (additional individual protection: P2 respirator).	keep dry, separate from oxidants.	

NOTES
Flush contaminated clothing with water (fire hazard). Depending on the degree of exposure, regular medical checkups are advisable. Lung edema symptoms usually develop several hours later and are aggravated by physical exertion: rest and hospitalization essential. As first aid, a doctor or authorized person should consider administering a corticosteroid spray. Effects on blood due to formation of methemoglobin. Special first aid required in the event of poisoning: antidotes must be available (with instructions).

Transport Emergency Card TEC(R)-80G21C

HI: 80; UN-number: 2865

IODINE

PHYSICAL PROPERTIES		IMPORTANT CHARACTERISTICS	
Boiling point, °C	183	**DARK PURPLE CRYSTALS OR BLACK-GRAY FLAKES WITH PUNGENT ODOR**	
Melting point, °C	114	Vapor mixes readily with air. Reacts with anhydrous ammonia to form shock-sensitive	
Relative density (water = 1)	4.93	compounds. Reacts violently with reducing agents, sulfur, antimony, alkali metals, metal	
Relative vapor density (air = 1)	8.8	powders and phosphorus, with risk of fire and explosion.	
Vapor pressure, mm Hg at 20 °C	0.21		
Solubility in water, g/100 ml	0.03	TLV-TWA	0.1 ppm	1.0 mg/m³ C
Relative molecular mass	253.8		
		Absorption route: Can enter the body by inhalation or ingestion. Harmful atmospheric concentrations build up fairly slowly on evaporation, but much more rapidly in aerosol form. **Immediate effects:** Irritates the skin. Corrosive to the eyes, mucous membranes and respiratory tract. Inhalation of vapor/fumes can cause severe breathing difficulties (lung edema). In serious cases risk of death. **Effects of prolonged/repeated exposure:** Inhalation or other contact can cause hypersensitivity of the skin, mucous membranes and respiratory tract. Prolonged exposure can cause increased thyroid activity.	
Gross formula:	I_2		

HAZARDS/SYMPTOMS	PREVENTIVE MEASURES	FIRE EXTINGUISHING/FIRST AID
Fire: non-combustible, many chemical reactions can cause fire and explosion.	avoid contact with reducing agents, sulfur, antimony, alkali metals, metal powders, phosphorus and ammonia.	in case of fire in immediate vicinity: use any extinguishing agent.
	KEEP DUST UNDER CONTROL	
Inhalation: *corrosive*; sore throat, cough, severe breathing difficulties, fever.	local exhaust or respiratory protection.	fresh air, rest, place in half-sitting position, call a doctor and take to hospital.
Skin: irritation, redness, pain, burns, depigmentation.	gloves.	remove contaminated clothing, flush skin with water or shower, call a doctor.
Eyes: *corrosive*; redness, pain, impaired vision.	face shield, or combined eye and respiratory protection.	flush with water, take to a doctor.
Ingestion: *corrosive*; sore throat, pain in esophagus, nausea, vomiting, stomachache, diarrhea, fever.	do not eat, drink or smoke while working.	rinse mouth, call a doctor and take immediately to hospital.

SPILLAGE	STORAGE	LABELING / NFPA
clean up spillage, dissolve remainder in sodium thiosulfate solution or sodium (bi)sulfite solution and flush away with water; (additional individual protection (against dust): P2 respirator).	keep separate from combustible substances, reducing agents, metal powders, alkali metals, antimony, sulfur and phosphorus.	R: 20/21 S: 23-25 ☒ Harmful

NOTES
TLV is maximum and must not be exceeded. Lung edema symptoms usually develop several hours later and are aggravated by physical exertion: rest and hospitalization essential. As first aid, a doctor or authorized person should consider administering a corticosteroid spray. Can cause iodism in sensitive subjects, with symptoms of mucous membrane irritation, cough, bronchitis, asthma, headache, cold sores. Subjects with hypersensitivity to iodine should avoid all contact with substance, since reactions can be violent or fatal.

CAS-No: [123-92-2]
acetic acid, isopentyl ester
isoamyl ethanoate
isopentyl acetate
3-methyl-1-butyl acetate

$CH_3COOC_5H_{11}$

ISOAMYL ACETATE

PHYSICAL PROPERTIES		IMPORTANT CHARACTERISTICS	
Boiling point, °C	142	**COLORLESS LIQUID WITH CHARACTERISTIC ODOR**	
Melting point, °C	−78	Vapor mixes readily with air. Reacts violently with strong oxidants.	
Flash point, °C	25		
Auto-ignition temperature, °C	380	TLV-TWA 100 ppm 532 mg/m³	
Relative density (water = 1)	0.87		
Relative vapor density (air = 1)	4.5	**Absorption route:** Can enter the body by inhalation or ingestion. Harmful atmospheric	
Relative density at 20 °C of saturated mixture		concentrations build up fairly slowly on evaporation at 20°C, but much more rapidly in aerosol	
vapor/air (air = 1)	1.02	form.	
Vapor pressure, mm Hg at 20 °C	5.3	**Immediate effects:** Irritates the eyes, skin and respiratory tract. Liquid destroys the skin's natural	
Solubility in water, g/100 ml	0.25	oils.	
Explosive limits, vol% in air	1-10		
Electrical conductivity, pS/m	1.6x10⁵		
Relative molecular mass	130.2		
Gross formula:	$C_7H_{14}O_2$		

HAZARDS/SYMPTOMS	PREVENTIVE MEASURES	FIRE EXTINGUISHING/FIRST AID
Fire: flammable.	keep away from open flame and sparks, no smoking.	powder, AFFF, foam, carbon dioxide, (halons).
Explosion: above 25°C: forms explosive air-vapor mixtures.	above 25°C: sealed machinery, ventilation, explosion-proof electrical equipment.	in case of fire: keep tanks/drums cool by spraying with water.
Inhalation: sore throat, shortness of breath, headache, dizziness, sleepiness, increased pulse rate, tiredness.	ventilation, local exhaust or respiratory protection.	fresh air, rest, call a doctor.
Skin: redness, pain.	gloves.	remove contaminated clothing, flush skin with water or shower, call a doctor if necessary.
Eyes: redness, pain.	safety glasses.	flush with water, take to a doctor if necessary.
Ingestion: sore throat, abdominal pain, nausea; see also 'Inhalation'.		rinse mouth, call a doctor or take to hospital.

SPILLAGE	STORAGE	LABELING / NFPA
collect leakage in sealable containers, soak up with sand or other inert absorbent and remove to safe place.	keep in fireproof place, separate from oxidants.	⬥ 3 1 0

NOTES
Alcohol consumption increases toxic effects.

Transport Emergency Card TEC(R)-30G08

HI: 30; UN-number: 1104

CAS-No: [123-51-3]
isoamylol
isopentyl alcohol
3-methyl-1-butanol

$(CH_3)_2CHCH_2CH_2OH$

ISOAMYL ALCOHOL (PRIMARY)

PHYSICAL PROPERTIES		IMPORTANT CHARACTERISTICS	
Boiling point, °C	131	**COLORLESS LIQUID WITH CHARACTERISTIC ODOR**	
Melting point, °C	− 117	Vapor mixes readily with air. Reacts violently with strong oxidants.	
Flash point, °C	43		
Auto-ignition temperature, °C	340	TLV-TWA	100 ppm	361 mg/m³
Relative density (water = 1)	0.8	STEL	125 ppm	452 mg/m³
Relative vapor density (air = 1)	3.0		
Relative density at 20 °C of saturated mixture vapor/air (air = 1)	1.01	**Absorption route:** Can enter the body by inhalation or ingestion or through the skin. Evaporation negligible at 20°C, but harmful atmospheric concentrations can build up rapidly in aerosol form.	
Vapor pressure, mm Hg at 20 °C	2.3	**Immediate effects:** Irritates the eyes, skin and respiratory tract.	
Solubility in water, g/100 ml	2.67		
Explosive limits, vol% in air	1.2 - 8		
Electrical conductivity, pS/m	1.4 x 10⁵		
Relative molecular mass	88.2		
Log P octanol/water	1.2		
Gross formula:	$C_5H_{12}O$		

HAZARDS/SYMPTOMS	PREVENTIVE MEASURES	FIRE EXTINGUISHING/FIRST AID
Fire: flammable.	keep away from open flame and sparks, no smoking.	powder, AFFF, foam, carbon dioxide, (halons).
Explosion: above 43°C: forms explosive air-vapor mixtures.	above 43°C: sealed machinery, ventilation, explosion-proof electrical equipment.	in case of fire: keep tanks/drums cool by spraying with water.
Inhalation: cough, shortness of breath, headache, dizziness, nausea, vomiting, diarrhea.	ventilation, local exhaust or respiratory protection.	fresh air, rest, call a doctor.
Skin: redness, pain.	gloves.	remove contaminated clothing, flush skin with water or shower.
Eyes: redness, pain.	safety glasses.	flush with water, take to a doctor.
Ingestion: abdominal pain, diarrhea, headache, shortness of breath, nausea, dizziness, vomiting.		rinse mouth, call a doctor.

SPILLAGE	STORAGE	LABELING / NFPA
collect leakage in sealable containers, soak up with sand or other inert absorbent and remove to safe place.	keep in fireproof place, separate from oxidants.	R: 10-20 S: 24/25 Note C ✖ Harmful

NFPA: 2 / 1 / 0

NOTES

Alcohol consumption increases toxic effects. The measures on this card also apply to isoamyl alcohol (secundary). The boiling point is 113°C and the flash point is 30°C: UN number 1105.

ISOAMYL FORMATE

PHYSICAL PROPERTIES	IMPORTANT CHARACTERISTICS	
Boiling point, °C — 124 Melting point, °C — −74 Flash point, °C — 22 Auto-ignition temperature, °C — 320 Relative density (water = 1) — 0.9 Relative vapor density (air = 1) — 4.0 Relative density at 20 °C of saturated mixture vapor/air (air = 1) — 1.04 Vapor pressure, mm Hg at 17 °C — 9.9 Solubility in water, g/100 ml at 20 °C — 0.3 Explosive limits, vol% in air — 1.7-10 Relative molecular mass — 116.1 Gross formula: $C_6H_{12}O_2$	**COLORLESS LIQUID WITH CHARACTERISTIC ODOR** Vapor mixes readily with air, forming explosive mixtures. Do not use compressed air when filling, emptying or processing. Reacts violently with strong oxidants, with risk of fire and explosion. TLV-TWA not available **Absorption route:** Can enter the body by inhalation or ingestion. Harmful atmospheric concentrations can build up fairly rapidly on evaporation at approx. 20°C - even more rapidly in aerosol form. **Immediate effects:** Irritates the eyes, skin and respiratory tract. In substantial concentrations can cause unconsciousness.	

HAZARDS/SYMPTOMS	PREVENTIVE MEASURES	FIRE EXTINGUISHING/FIRST AID
Fire: flammable.	keep away from open flame and sparks, no smoking.	powder, AFFF, foam, carbon dioxide, (halons).
Explosion: above 22°C: forms explosive air-vapor mixtures.	above 22°C: sealed machinery, ventilation, explosion-proof electrical equipment.	in case of fire: keep tanks/drums cool by spraying with water.
Inhalation: sore throat, cough, shortness of breath, headache.	ventilation, local exhaust or respiratory protection.	fresh air, rest, call a doctor.
Skin: redness.	gloves.	remove contaminated clothing, flush skin with water or shower.
Eyes: redness, pain.	safety glasses.	flush with water, take to a doctor if necessary.
Ingestion: sore throat, abdominal pain, stomachache, headache, shortness of breath.		rinse mouth, call a doctor.

SPILLAGE	STORAGE	LABELING / NFPA
collect leakage in sealable containers, soak up with sand or other inert absorbent and remove to safe place.	keep in fireproof place, separate from oxidants.	R: 10 Note C

NOTES

CAS-No: [75-28-5]
1,1-dimethylethane
2-methylpropane
trimethylmethane

C_4H_{10}

ISOBUTANE
(cylinder)

PHYSICAL PROPERTIES	IMPORTANT CHARACTERISTICS
Boiling point, °C −12 Melting point, °C −160 Flash point, °C flammable gas Auto-ignition temperature, °C 460 Relative density (water = 1) 0.6 Relative vapor density (air = 1) 2.1 Vapor pressure, mm Hg at 20 °C 2280 Solubility in water none Explosive limits, vol% in air 1.8-8.4 Electrical conductivity, pS/m < 10⁴ Relative molecular mass 58.1 Gross formula: C_4H_{10}	**COLORLESS COMPRESSED LIQUEFIED GAS** Vapor (gas) is heavier than air and spreads at ground level, with risk of ignition at a distance. Flow, agitation etc. can cause build-up of electrostatic charge due to liquid's low conductivity. Do not use compressed air when filling, emptying or processing. TLV-TWA not available **Absorption route:** Can enter the body by inhalation. **Immediate effects:** Liquid can cause frostbite due to rapid evaporation. Can cause suffocation due to air saturation.

HAZARDS/SYMPTOMS	PREVENTIVE MEASURES	FIRE EXTINGUISHING/FIRST AID
Fire: extremely flammable.	keep away from open flame and sparks, no smoking.	shut off supply; if impossible and no danger to surrounding area, allow fire to burn itself out; otherwise extinguish with powder, carbon dioxide, (halons).
Explosion: forms explosive air-gas mixtures.	sealed machinery, ventilation, explosion-proof electrical equipment and lighting, grounding, non-sparking tools.	in case of fire: keep cylinders cool by spraying with water, fight fire from sheltered location.
Inhalation: shortness of breath, headache, drowsiness, unconsciousness, heart complaints.	ventilation, local exhaust or respiratory protection.	fresh air, rest, artificial respiration if necessary, call a doctor.
Skin: contact with the liquid: frostbite, redness, pain, serious burns, blisters.	insulating gloves.	*in case of frostbite:* DO NOT remove clothing, flush skin with water or shower, take to a doctor.
Eyes: spattering of the liquid: redness, pain, impaired vision.	face shield.	flush with water, take to an eye doctor.

SPILLAGE	STORAGE	LABELING / NFPA
evacuate area, call in an expert, ventilate; (additional individual protection: breathing apparatus).	keep in cool, fireproof place.	

NOTES
High atmospheric concentrations, e.g. in poorly ventilated spaces, can cause oxygen deficiency, with risk of unconsciousness. The measures on this card also apply to neopentane (2,2-dimethylpropane, C_5H_{12}). Use *propane* if ambient temperature drops below 5° C. Under no circumstances warm cylinder. Turn leaking cylinder so that leak is on top to prevent liquid isobutane escaping.

Transport Emergency Card TEC(R)-501

HI: 23; UN-number: 1969

CAS-No: [78-83-1]
isobutylalcohol
isopropylcarbinol
2-methyl-1-propanol

$(CH_3)_2CHCH_2OH$

ISOBUTANOL

PHYSICAL PROPERTIES	IMPORTANT CHARACTERISTICS	
Boiling point, °C 108 Melting point, °C − 108 Flash point, °C 27 Auto-ignition temperature, °C 430 Relative density (water = 1) 0.8 Relative vapor density (air = 1) 2.6 Relative density at 20 °C of saturated mixture vapor/air (air = 1) 1.02 Vapor pressure, mm Hg at 20 °C 9.1 Solubility in water, g/100 ml 9.5 Explosive limits, vol% in air 1.2 - 10.9 Electrical conductivity, pS/m 1.6x10⁶ Relative molecular mass 74.1 Log P octanol/water 0.7	**COLORLESS LIQUID WITH CHARACTERISTIC ODOR** Vapor mixes readily with air. Reacts with strong oxidants and with alkali metals, giving off flammable gas (→ *hydrogen*).	
	TLV-TWA 50 ppm 152 mg/m³	
	Absorption route: Can enter the body by inhalation or ingestion or through the skin. **Immediate effects:** Irritates the eyes, skin and respiratory tract. Liquid destroys the skin's natural oils. Affects the nervous system.	
Gross formula: $C_4H_{10}O$		

HAZARDS/SYMPTOMS	PREVENTIVE MEASURES	FIRE EXTINGUISHING/FIRST AID
Fire: flammable.	keep away from open flame and sparks, no smoking.	powder, AFFF, foam, carbon dioxide, (halons).
Explosion: above 27°C: forms explosive air-vapor mixtures.	above 27°C: sealed machinery, ventilation, explosion-proof electrical equipment.	in case of fire: keep tanks/drums cool by spraying with water.
Inhalation: sore throat, cough, shortness of breath, headache, drowsiness.	ventilation, local exhaust or respiratory protection.	fresh air, rest, place in half-sitting position, call a doctor.
Skin: *is absorbed*; redness.	gloves.	remove contaminated clothing, flush skin with water or shower.
Eyes: redness, pain, impaired vision.	safety glasses.	flush with water, take to a doctor.
Ingestion: abdominal pain, vomiting, drowsiness.		rinse mouth, call a doctor, take to hospital.

SPILLAGE	STORAGE	LABELING / NFPA
collect leakage in sealable containers, flush away remainder with water.	keep in fireproof place, keep separate from strong oxidants.	R: 10-20 S: 16 Note C ❌ Harmful

NOTES
Odor limit is above TLV. Alcohol consumption increases toxic effects. In case of chronic exposure: loss of apetite, loss of eye sight, corneal damage, disturbances of the liver function.

Transport Emergency Card TEC(R)-583

HI: 30; UN-number: 1212

2-amino-2,2-dimethylethanol
2-aminoisobutanol
2-amino-2-methyl-1-propanol
2-hydroxymethyl-2-propylamine
hydroxy-tert-butylamine

$(CH_3)_2C(NH_2)CH_2OH$

ISOBUTANOLAMINE

PHYSICAL PROPERTIES		IMPORTANT CHARACTERISTICS
Boiling point, °C	165	**COLORLESS VISCOUS LIQUID OR CRYSTALLINE MASS**
Melting point, °C	ca. 26	Toxic gases are a product of combustion ($\rightarrow$ *nitrous vapors*). In aqueous solution is a medium strong base which reacts with acids. Reacts with strong oxidants, with risk of fire and explosion.
Flash point, °C	67	
Relative density (water = 1)	0.9	
Relative vapor density (air = 1)	3.1	TLV-TWA not available
Relative density at 20 °C of saturated mixture vapor/air (air = 1)	1.0	**Absorption route:** Can enter the body by inhalation or ingestion. Harmful atmospheric concentrations build up fairly slowly on evaporation at 20°C, but much more rapidly in aerosol form.
Vapor pressure, mm Hg at 20 °C	ca. 0.76	**Immediate effects:** Corrosive to the eyes, skin and respiratory tract. Also corrosive if swallowed. Inhalation of powder can cause lung edema. If liquid is swallowed, droplets can enter the lungs, with risk of pneumonia. Exposure to high concentrations can be fatal. Keep under medical observation.
Solubility in water	∞	
Relative molecular mass	89.1	
Gross formula:	$C_4H_{11}NO$	

HAZARDS/SYMPTOMS	PREVENTIVE MEASURES	FIRE EXTINGUISHING/FIRST AID
Fire: combustible.	keep away from open flame, no smoking.	powder, alcohol-resistant foam, water spray, carbon dioxide, (halons).
Explosion: above 67°C: forms explosive air-vapor mixtures.	above 67°C: sealed machinery, ventilation.	
	KEEP DUST UNDER CONTROL, STRICT HYGIENE	
Inhalation: *corrosive*; burning sensation, cough, headache, difficulty breathing, nausea, sore throat.	ventilation, local exhaust or respiratory protection.	fresh air, rest, place in half-sitting position, artificial respiration if necessary, take immediately to hospital.
Skin: *corrosive*; redness, burning sensation, pain.	gloves.	remove contaminated clothing, flush skin with water or shower, consult a doctor.
Eyes: *corrosive*; redness, pain, serious burns.	face shield.	flush thoroughly with water (remove contact lenses if easy), take to a doctor.
Ingestion: *corrosive*; abdominal cramps, burning sensation, cough, nausea, sore throat.		rinse mouth, DO NOT induce vomiting, take immediately to hospital.

SPILLAGE	STORAGE	LABELING / NFPA
collect leakage in sealable containers, flush away remainder with water.	keep separate from oxidants and strong acids; ventilate at floor level.	R: 36/38 ✖ Irritant

NOTES

Lung edema symptoms usually develop several hours later and are aggravated by physical exertion: rest and hospitalization essential. As first aid, a doctor or authorized person should consider administering a suitable spray.

acetic acid, iso-butyl ester
acetic acid, 2-methylpropyl ester
2-methylpropyl acetate
β-methylpropyl ethanoate

$CH_3COOCH_2CH(CH_3)CH_3$

ISOBUTYLACETATE

PHYSICAL PROPERTIES	IMPORTANT CHARACTERISTICS
Boiling point, °C 117 Melting point, °C −99 Flash point, °C 18 Auto-ignition temperature, °C 420 Relative density (water = 1) 0.9 Relative vapor density (air = 1) 4.0 Relative density at 20°C of saturated mixture vapor/air (air = 1) 1.05 Vapor pressure, mm Hg at 20°C 15 Solubility in water, g/100 ml at 20°C 0.7 Explosive limits, vol% in air 2.4-10.5 Electrical conductivity, pS/m 2.5×10^{10} Relative molecular mass 116.2	**COLORLESS LIQUID WITH CHARACTERISTIC ODOR** Vapor mixes readily with air, forming explosive mixtures. Do not use compressed air when filling, emptying or processing. Decomposes slowly when exposed to light or on contact with water, forming → *acetic acid* and → *isobutanol*. Reacts violently with strong oxidants.
	TLV-TWA 150 ppm 713 mg/m³
	Absorption route: Can enter the body by inhalation or ingestion. Harmful atmospheric concentrations build up fairly slowly on evaporation at 20°C, but much more rapidly in aerosol form. **Immediate effects:** Irritates the eyes and respiratory tract. Liquid destroys the skin's natural oils. Vapor has narcotic effect in high concentrations.
Gross formula: $C_6H_{12}O_2$	

HAZARDS/SYMPTOMS	PREVENTIVE MEASURES	FIRE EXTINGUISHING/FIRST AID
Fire: extremely flammable.	keep away from open flame and sparks, no smoking.	powder, AFFF, foam, carbon dioxide, (halons).
Explosion: forms explosive air-vapor mixtures.	sealed machinery, ventilation, explosion-proof electrical equipment and lighting.	in case of fire: keep tanks/drums cool by spraying with water.
Inhalation: cough, shortness of breath, drowsiness.	ventilation, local exhaust or respiratory protection.	fresh air, rest, call a doctor.
Skin: redness.	gloves.	remove contaminated clothing, flush skin with water or shower.
Eyes: redness, pain.	safety glasses.	flush with water, take to a doctor.
Ingestion: sore throat, abdominal pain, nausea.		rinse mouth, call a doctor or take to hospital.

SPILLAGE	STORAGE	LABELING / NFPA
collect leakage in sealable containers, soak up with sand or other inert absorbent and remove to safe place.	keep in dry, dark, fireproof place, separate from oxidants.	R: 11 S: 16-23-29-33 Note C 🔥 Flammable 3 / 1 / 0

NOTES
Flash point highly dependent on presence of impurities (incl. alcohol content), and in some grades can be as much as 10-12°C higher than value given, with consequent substantial lowering of flammability at room temperature.

Transport Emergency Card TEC(R)-507

HI: 33; UN-number: 1213

ISOBUTYL ALUMINUM DICHLORIDE

PHYSICAL PROPERTIES	IMPORTANT CHARACTERISTICS	
Decomposes below boiling point, °C 50 Melting point, °C −30 Relative density (water = 1) 1.12 Relative vapor density (air = 1) 5.4 Solubility in water reaction Relative molecular mass 155.0	**COLORLESS FUMING LIQUID WITH PUNGENT ODOR** Ignites spontaneously on exposure to air. Aluminum oxide fumes and → *hydrochloric acid* fumes are a product of combustion. Strong reducing agent which reacts violently with oxidants. Reacts violently with water, alcohols, amines, phenols, carbon dioxide, oxides of sulfur, nitrogen oxides, halogens and halogenated hydrocarbons.	
	TLV-TWA	2 mg/m^3
	Absorption route: Can enter the body by inhalation or ingestion. Harmful atmospheric concentrations can build up fairly rapidly on evaporation at approx. 20°C - even more rapidly in aerosol form. **Immediate effects:** Corrosive to the eyes, skin and respiratory tract. Inhalation of vapor/fumes can cause severe breathing difficulties (lung edema). In serious cases risk of death.	
Gross formula: $C_4H_9AlCl_2$		

HAZARDS/SYMPTOMS	PREVENTIVE MEASURES	FIRE EXTINGUISHING/FIRST AID
Fire: extremely flammable, many chemical reactions can cause fire and explosion.	keep away from open flame and sparks, no smoking, avoid contact with water, air and many other substances.	special extinguishing powder, dry sand, NO other extinguishing agents, can reignite after being extinguished.
Explosion: contact with water and many other substances can cause explosion.		in case of fire: keep undamaged cylinders cool by spraying with water, but DO NOT spray substance with water.
Inhalation: *corrosive*; sore throat, cough, severe breathing difficulties.	completely sealed machinery, ventilation, local exhaust or respiratory protection.	fresh air, rest, place in half-sitting position, take to hospital.
Skin: *corrosive*; serious burns, slow healing.	gloves, protective clothing.	remove contaminated clothing, refer to a doctor.
Eyes: *corrosive*; redness, pain, impaired vision.	face shield, or combined eye and respiratory protection.	flush with water, take to a doctor.
Ingestion: *corrosive*; sore throat, abdominal pain, diarrhea.		rinse mouth, take immediately to hospital.

SPILLAGE	STORAGE	LABELING / NFPA
evacuate area, call in an expert, dilute leakage with diesel/machine oil, soak up/cover with dry sand/table salt and remove to safe place; (additional individual protection: CHEMICAL SUIT).	keep in fireproof place under dry nitrogen, separate from all other substances.	

NOTES
Usually sold as 15-30% solution in hydrocarbons. Flash point and explosive limits depend on solvent used. 15% solution does not ignite spontaneously on exposure to air. For 15% solution see relevant entry. Reacts violently with extinguishing agents such as water, foam, carbon dioxide and ordinary powder. Lung edema symptoms usually develop several hours later and are aggravated by physical exertion: rest and hospitalization essential. As first aid, a doctor or authorized person should consider administering a corticosteroid spray. Use cylinder with special fittings.

Transport Emergency Card TEC(R)-42G02

HI: X333; UN-number: 2221

ISOBUTYL ALUMINUM DICHLORIDE

(15% solution in hexane)

PHYSICAL PROPERTIES	IMPORTANT CHARACTERISTICS
Boiling point, °C 69 Melting point, °C − 95 Flash point, °C − 26 Auto-ignition temperature, °C 240 Relative density (water = 1) 0.79 Relative vapor density (air = 1) 3.0 Relative density at 20 °C of saturated mixture vapor/air (air = 1) 1.3 Vapor pressure, mm Hg at 20 °C 122 Solubility in water reaction Explosive limits, vol% in air 1.1-7.5 Minimum ignition energy, mJ 0.24 Electrical conductivity, pS/m 100	**COLORLESS FUMING LIQUID WITH PUNGENT ODOR** Vapor is heavier than air and spreads at ground level, with risk of ignition at a distance. Flow, agitation etc. can cause build-up of electrostatic charge due to liquid's low conductivity. Do not use compressed air when filling, emptying or processing. Once solvent has evaporated (e.g. on clothing) can ignite spontaneously on exposure to air. Aluminum oxide fumes and → *hydrochloric acid* fumes are a product of combustion. Strong reducing agent which reacts violently with oxidants. Reacts violently with water, alcohols, phenols, amines, oxides of carbon, nitrogen and sulfur and many other substances.

	TLV-TWA 2 mg/m³ ✳

Gross formula: $C_4H_9AlCl_2$	**Absorption route:** Can enter the body by inhalation or ingestion. Harmful atmospheric concentrations can build up fairly rapidly on evaporation at approx. 20° C - even more rapidly in aerosol form. **Immediate effects:** Corrosive to the eyes, skin and respiratory tract. Inhalation of vapor/fumes can cause severe breathing difficulties (lung edema). In serious cases risk of death.

HAZARDS/SYMPTOMS	PREVENTIVE MEASURES	FIRE EXTINGUISHING/FIRST AID
Fire: extremely flammable, many chemical reactions can cause fire and explosion.	keep away from open flame and sparks, no smoking, avoid contact with water, alcohols, phenols and many other substances.	special extinguishing powder, dry sand, NO other extinguishing agents, can reignite after being extinguished.
Explosion: forms explosive air-vapor mixtures; contact with water, phenols, alcohols and many other substances can cause explosion.	sealed machinery, ventilation, explosion-proof electrical equipment and lighting, grounding.	in case of fire: keep undamaged cylinders cool by spraying with water, but DO NOT spray substance with water.
	STRICT HYGIENE	
Inhalation: *corrosive*; sore throat, cough, severe breathing difficulties.	sealed machinery, ventilation, local exhaust or respiratory protection.	fresh air, rest, place in half-sitting position, take to hospital.
Skin: serious burns, slow healing.	gloves, protective clothing.	remove contaminated clothing, flush skin with water or shower, refer to a doctor.
Eyes: *corrosive*; redness, pain, impaired vision.	face shield, or combined eye and respiratory protection.	flush with water, take to a doctor.
Ingestion: *corrosive*; sore throat, abdominal pain, diarrhea.		rinse mouth, take immediately to hospital.

SPILLAGE	STORAGE	LABELING / NFPA
evacuate area, call in an expert, collect leakage in sealable containers, soak up remainder with dry sand/table salt and remove to safe place; (additional individual protection: CHEMICAL SUIT).	keep in dry, fireproof place under inert gas, separate from all other substances.	

NOTES
Physical properties (boiling point, flash point, autoignition point, explosive limits, vapor pressure, relative density, conductivity and ignition energy) are based on hexane. For pure isobutyl aluminum dichloride see relevant entry. ✳ For TLV of *hexane* see relevant entry. Remove contaminated clothing immediately (risk of spontaneous combustion). Lung edema symptoms usually develop several hours later and are aggravated by physical exertion: rest and hospitalization essential. As first aid, a doctor or authorized person should consider administering a corticosteroid spray. Use cylinder with special fittings.

Transport Emergency Card TEC(R)-42G02

HI: X333; UN-number: 2220

CAS-No: [513-36-0]
1-chloro-2-methylpropane

$CH_2ClCH(CH_3)_2$

ISOBUTYL CHLORIDE

PHYSICAL PROPERTIES	IMPORTANT CHARACTERISTICS
Boiling point, °C 68 Melting point, °C −130 Flash point, °C ca. −18 Relative density (water = 1) 0.9 Relative vapor density (air = 1) 3.2 Relative density at 20 °C of saturated mixture vapor/air (air = 1) 1.3 Vapor pressure, mm Hg at 20 °C 119 Solubility in water none Explosive limits, vol% in air 2.0-8.8 Electrical conductivity, pS/m 10^4 Relative molecular mass 92.6	**COLORLESS LIQUID WITH CHARACTERISTIC ODOR** Vapor is heavier than air and spreads at ground level, with risk of ignition at a distance. Flow, agitation etc. can cause build-up of electrostatic charge due to liquid's low conductivity. Decomposes when heated or burned, giving off corrosive vapors (→ *hydrochloric acid*). Reacts violently with strong oxidants and metal powders, with risk of fire and explosion.
	TLV-TWA not available
	Absorption route: Can enter the body by inhalation of vapor or ingestion. Harmful atmospheric concentrations can build up fairly rapidly on evaporation at approx. 20°C - even more rapidly in aerosol form. **Immediate effects:** Irritates the eyes, skin and respiratory tract. Can affect the nervous system.
Gross formula: C_4H_9Cl	

HAZARDS/SYMPTOMS	PREVENTIVE MEASURES	FIRE EXTINGUISHING/FIRST AID
Fire: extremely flammable.	keep away from open flame and sparks, no smoking.	powder, AFFF, foam, carbon dioxide, (halons).
Explosion: forms explosive air-vapor mixtures.	sealed machinery, ventilation, explosion-proof electrical equipment and lighting, grounding; do not use compressed air when filling, emptying or processing.	in case of fire: keep tanks/drums cool by spraying with water.
Inhalation: cough, dizziness, sore throat.	ventilation, local exhaust or respiratory protection.	fresh air, rest, refer to a doctor.
Skin: redness.	gloves.	remove contaminated clothing, flush skin with water and wash with soap and water.
Eyes: redness, pain.	safety glasses.	flush thoroughly with water (remove contact lenses if easy), take to a doctor.
Ingestion: abdominal cramps, dizziness, nausea.		rinse mouth, give no liquids, consult a doctor.

SPILLAGE	STORAGE	LABELING / NFPA
collect leakage in sealable containers, soak up with sand or other inert absorbent and remove to safe place.	keep in fireproof place, separate from oxidants.	3 / 2 / 0

NOTES

Insufficient data on harmful effects to humans: exercise great caution. Flash point (−18°C) estimated from vapor pressure curve and lower explosive limit.

Transport Emergency Card TEC(R)-30G32

HI: 33; UN-number: 1127

ISOBUTYLENE

(cylinder)

PHYSICAL PROPERTIES	IMPORTANT CHARACTERISTICS
Boiling point, °C −7 Melting point, °C −140 Flash point, °C flammable gas Auto-ignition temperature, °C 465 Relative density (water = 1) 0.6 Relative vapor density (air = 1) 1.99 Vapor pressure, mm Hg at 20 °C 1976 Solubility in water <0.5 Explosive limits, vol% in air 1.8-9.6 Electrical conductivity, pS/m <0.6x10⁴ Relative molecular mass 56.1	**COLORLESS COMPRESSED LIQUEFIED GAS WITH CHARACTERISTIC ODOR** Gas is heavier than air and spreads at ground level, with risk of ignition at a distance, and can accumulate close to ground level, causing oxygen deficiency, with risk of unconsciousness. Presumed to be able to form peroxides. Able to polymerize, with risk of fire and explosion. Flow, agitation etc. can cause build-up of electrostatic charge due to liquid's low conductivity. Reacts violently with strong oxidants. Reacts violently with hydrogen bromide, hydrogen chloride, chlorine, fluorine and nitrogen oxides.
	TLV-TWA not available
Gross formula: C_4H_8	**Absorption route:** Can enter the body by inhalation. **Immediate effects:** Liquid can cause frostbite due to rapid evaporation. Can cause suffocation due to air saturation.

HAZARDS/SYMPTOMS	PREVENTIVE MEASURES	FIRE EXTINGUISHING/FIRST AID
Fire: extremely flammable.	keep away from open flame and sparks, no smoking.	shut off supply; if impossible and no danger to surrounding area, allow fire to burn itself out; otherwise extinguish with powder, carbon dioxide, (halons).
Explosion: forms explosive air-gas mixtures.	sealed machinery, ventilation, explosion-proof electrical equipment and lighting, grounding, non-sparking tools.	in case of fire: keep cylinder cool by spraying with water, fight fire from sheltered location.
Inhalation: shortness of breath, headache, drowsiness, unconsciousness.	ventilation, local exhaust or respiratory protection.	fresh air, rest, artificial respiration if necessary, call a doctor.
Skin: *in case of frostbite:* redness, pain, blisters.	insulating gloves.	*in case of frostbite:* DO NOT remove clothing, call a doctor.
Eyes: *in case of frostbite:* redness, pain, impaired vision.	face shield.	flush with water, take to a doctor.

SPILLAGE	STORAGE	LABELING / NFPA
evacuate area, call in an expert, ventilate, under no circumstances spray liquid with water; (additional individual protection: breathing apparatus).	keep in cool, fireproof place; add an inhibitor.	4 / 1 0

NOTES
High atmospheric concentrations, e.g. in poorly ventilated spaces, can cause oxygen deficiency, with risk of unconsciousness. Turn leaking cylinder so that leak is on top to prevent liquid isobutylene escaping.

CAS-No: [108-80-5]
cyanuric acid
trihydroxycyanidine
2,4,6-trihydroxy-1,3,5-triazine

ISOCYANURIC ACID

PHYSICAL PROPERTIES		IMPORTANT CHARACTERISTICS	
Melting point (decomposes), °C	320	**WHITE CRYSTALS**	
Relative density (water = 1)	2.5	→ *Nitrous vapors* are a product of combustion. Decomposes when heated above 320°C, giving off corrosive and explosive → *cyanic acid*.	
Vapor pressure, mm Hg at 20 °C	?		
Solubility in water, g/100 ml at 20 °C	0.5		
Relative molecular mass	129.1	**TLV-TWA** not available	
		Absorption route: Can enter the body by inhalation or ingestion. Evaporation negligible at 20°C, but harmful concentrations of airborne particles can build up rapidly. **Immediate effects:** Irritates the eyes and respiratory tract.	
Gross formula:	$C_3H_3N_3O_3$		

HAZARDS/SYMPTOMS	PREVENTIVE MEASURES	FIRE EXTINGUISHING/FIRST AID
Fire: hardly combustible.		in case of fire in immediate vicinity: use any extinguishing agent.
Explosion:		in case of fire: keep tanks/drums cool by spraying with water.
Inhalation: sore throat, cough.	local exhaust or respiratory protection.	fresh air, rest, place in half-sitting position, take to hospital.
Skin:	gloves.	
Eyes: redness, pain.	goggles.	flush with water, take to a doctor if necessary.
Ingestion: abdominal pain.		rinse mouth, call a doctor or take immediately to hospital.

SPILLAGE	STORAGE	LABELING / NFPA
clean up spillage, carefully collect remainder; (additional individual protection: P2 respirator).		

NOTES
Insufficient toxicological data on harmful effects to humans. Poses little danger in industrial applications, but under certain temperature and pressure conditions can decompose, forming highly toxic cyanic acid.

ISODECYLALCOHOL

PHYSICAL PROPERTIES	IMPORTANT CHARACTERISTICS	
Boiling point, °C 212-224 Melting point, °C 7 Flash point, °C 96 Auto-ignition temperature, °C 266 Relative density (water = 1) 0.8 Relative vapor density (air = 1) 5.5 Relative density at 20 °C of saturated mixture vapor/air (air = 1) 1 Vapor pressure, mm Hg at 20°C 7.6×10^{-3} Solubility in water, g/100 ml 2.5 Explosive limits, vol% in air 0.8-? Relative molecular mass 158	**COLORLESS LIGHT VISCOUS LIQUID WITH CHARACTERISTIC ODOR** Mixture of isomers. Vapor mixes readily with air. Reacts with strong oxidants.	
	TLV-TWA not available	
	Absorption route: san enter the body by inhalation or ingestion or through the skin. Harmful atmospheric concentrations build up very slowly, if at all, on evaporation at approx. 20°C, but much more rapidly in aerosol form. **Immediate effects:** Irritates the eyes, skin and respiratory tract. Liquid destroys the skin's natural oils.	
Gross formula: $C_{10}H_{22}O$		

HAZARDS/SYMPTOMS	PREVENTIVE MEASURES	FIRE EXTINGUISHING/FIRST AID
Fire: combustible.	keep away from open flame, no smoking.	powder, AFFF, foam, carbon dioxide, (halons).
Explosion: above 96°C: forms explosive air-vapor mixtures.	above 96°C: sealed machinery, ventilation.	
Inhalation: sore throat, headache, dizziness, nausea, drowsiness.	ventilation.	fresh air, rest, call a doctor.
Skin: redness.	gloves.	remove contaminated clothing, wash skin with soap and water.
Eyes: redness, pain.	safety glasses.	flush with water, take to a doctor if necessary.
Ingestion: sore throat, abdominal pain, stomachache, see also 'Inhalation'.		rinse mouth, call a doctor or take to hospital.

SPILLAGE	STORAGE	LABELING / NFPA
collect leakage in sealable containers, soak up with sand or other inert absorbent and remove to safe place.	keep separate from oxidants.	

NOTES

Transport Emergency Card TEC(R)-30G15

CAS-No: [540-84-1]
2,2,4-trimethylpentane

$(CH_3)_3CCH_2CH(CH_3)_2$

ISOOCTANE

PHYSICAL PROPERTIES		IMPORTANT CHARACTERISTICS	
Boiling point, °C	99	**COLORLESS LIQUID WITH CHARACTERISTIC ODOR**	
Melting point, °C	−116	Vapor is heavier than air and spreads at ground level, with risk of ignition at a distance. Flow, agitation etc. can cause build-up of electrostatic charge due to liquid's low conductivity. Do not use compressed air when filling, emptying or processing.	
Flash point, °C	−12		
Auto-ignition temperature, °C	418		
Relative density (water = 1)	0.7		
Relative vapor density (air = 1)	3.9	TLV-TWA not available	
Relative density at 20 °C of saturated mixture vapor/air (air = 1)	1.15	**Absorption route:** Can enter the body by inhalation or ingestion. Harmful atmospheric concentrations can build up fairly rapidly on evaporation at 20°C - even more rapidly in aerosol form.	
Vapor pressure, mm Hg at 20 °C	40		
Solubility in water	none		
Explosive limits, vol% in air	1.1-6	**Immediate effects:** Liquid destroys the skin's natural oils. Affects the nervous system. If liquid is swallowed, droplets can enter the lungs, with risk of pneumonia.	
Minimum ignition energy, mJ	1.35		
Electrical conductivity, pS/m	< 10⁴		
Relative molecular mass	114.2		
Gross formula:	C_8H_{18}		

HAZARDS/SYMPTOMS	PREVENTIVE MEASURES	FIRE EXTINGUISHING/FIRST AID
Fire: extremely flammable.	keep away from open flame and sparks, no smoking.	powder, AFFF, foam, carbon dioxide, (halons).
Explosion: forms explosive air-vapor mixtures.	sealed machinery, ventilation, explosion-proof electrical equipment and lighting, grounding, non-sparking tools.	in case of fire: keep tanks/drums cool by spraying with water.
Inhalation: severe breathing difficulties, drowsiness, sleepiness.	ventilation, local exhaust or respiratory protection.	fresh air, rest, artificial respiration if necessary, call a doctor.
Skin: redness.	gloves.	remove contaminated clothing, flush skin with water or shower.
Eyes: redness, pain.	safety glasses.	flush with water, take to a doctor if necessary.
Ingestion: abdominal pain, nausea.		rinse mouth, DO NOT induce vomiting, call a doctor or take to hospital.

SPILLAGE	STORAGE	LABELING / NFPA
collect leakage in sealable containers, soak up with sand or other inert absorbent and remove to safe place, do not flush into sewer; (additional individual protection: breathing apparatus).	keep in fireproof place.	R: 11 S: 9-16-29-33 Note C 🔥 Flammable

NFPA diamond: blue 0 (left), red 3 (top), yellow 0 (right)

NOTES
High atmospheric concentrations, e.g. in poorly ventilated spaces, can cause oxygen deficiency, with risk of unconsciousness.

Transport Emergency Card TEC(R)-95

HI: 33; UN-number: 1262

ISOPENTANE

PHYSICAL PROPERTIES	IMPORTANT CHARACTERISTICS
Boiling point, °C 28 Melting point, °C − 160 Flash point, °C > − 51 Auto-ignition temperature, °C 420 Relative density (water = 1) 0.6 Relative vapor density (air = 1) 2.5 Relative density at 20 °C of saturated mixture vapor/air (air = 1) 2.45 Vapor pressure, mm Hg at 20 °C 752 Solubility in water none Explosive limits, vol% in air 1.4 - 7.6 Minimum ignition energy, mJ 0.7 Relative molecular mass 72.2	**COLORLESS LIQUID WITH CHARACTERISTIC ODOR** Vapor is heavier than air and spreads at ground level, with risk of ignition at a distance. Flow, agitation etc. can cause build-up of electrostatic charge due to liquid's low conductivity. Do not use compressed air when filling, emptying or processing. TLV-TWA not available **Absorption route:** Can enter the body by inhalation or ingestion. **Immediate effects:** Irritates the eyes, skin and respiratory tract. Liquid destroys the skin's natural oils. If liquid is swallowed, droplets can enter the lungs, with risk of pneumonia.
Gross formula: C_5H_{12}	

HAZARDS/SYMPTOMS	PREVENTIVE MEASURES	FIRE EXTINGUISHING/FIRST AID
Fire: extremely flammable.	keep away from open flame and sparks, no smoking.	powder, AFFF, foam, carbon dioxide, (halons).
Explosion: forms explosive air-vapor mixtures.	sealed machinery, ventilation, explosion-proof electrical equipment and lighting, grounding.	in case of fire: keep tanks/drums cool by spraying with water.
Inhalation: sore throat, shortness of breath, dizziness, drowsiness.	ventilation, local exhaust or respiratory protection.	fresh air, rest, call a doctor.
Skin: redness.	gloves.	remove contaminated clothing, flush skin with water or shower.
Eyes: redness, pain.	safety glasses.	flush with water, take to a doctor if necessary.
Ingestion: sore throat, abdominal pain, nausea, stomachache, drowsiness.		rinse mouth, call a doctor and take to hospital if necessary.

SPILLAGE	STORAGE	LABELING / NFPA
evacuate area, call in an expert, collect leakage in sealable containers, soak up with sand or other inert absorbent and remove to safe place; (additional individual protection: breathing apparatus).	keep in fireproof place.	R: 11 S: 9-16-29-33 🔥 Flammable NFPA: 1 / 4 / 0

NOTES
Alcohol consumption increases toxic effects. High atmospheric concentrations, e.g. in poorly ventilated spaces, can cause oxygen deficiency, with risk of unconsciousness.

Transport Emergency Card TEC(R)-592

HI: 33; UN-number: 1265

CAS-No: [78-79-5]
2-methyl-1,3-butadiene
3-methyl-1,3-butadiene

$CH_2 = C(CH_3)-C = CH_2$

ISOPRENE

PHYSICAL PROPERTIES		IMPORTANT CHARACTERISTICS
Boiling point, °C	34	**COLORLESS LIQUID WITH CHARACTERISTIC ODOR**
Melting point, °C	− 146	Vapor is heavier than air and spreads at ground level, with risk of ignition at a distance. Readily
Flash point, °C	< − 20	able to form peroxides and thus to polymerize. Heating or peroxides cause polymerization, with
Auto-ignition temperature, °C	220	risk of fire and explosion. Flow, agitation etc. can cause build-up of electrostatic charge due to
Relative density (water = 1)	0.7	liquid's low conductivity. Do not use compressed air when filling, emptying or processing. Reacts
Relative vapor density (air = 1)	2.35	violently with reducing agents, oxidants, sulfuric acid, nitric acid, chlorosulfonic acid and acetic
Relative density at 20 °C of saturated mixture		acid, with risk of fire and explosion. Attacks some plastics.
vapor/air (air = 1)	1.8	
Vapor pressure, mm Hg at 20 °C	ca. 456	TLV-TWA not available
Solubility in water	none	
Explosive limits, vol% in air	1.0-9.7	**Absorption route:** Can enter the body by inhalation or ingestion.
Relative molecular mass	68.1	**Immediate effects:** Irritates the eyes, skin and respiratory tract. Affects the nervous system. In serious cases risk of unconsciousness.
Gross formula:	C_5H_8	

HAZARDS/SYMPTOMS	PREVENTIVE MEASURES	FIRE EXTINGUISHING/FIRST AID
Fire: extremely flammable.	keep away from open flame and sparks, no smoking; avoid contact with oxidants, sulfuric acid and nitric acid.	powder, AFFF, foam, carbon dioxide, (halons).
Explosion: forms explosive air-vapor mixtures.	sealed machinery, ventilation, explosion-proof electrical equipment and lighting, grounding.	in case of fire: keep tanks/drums cool by spraying with water.
Inhalation: sore throat, cough, dizziness, drowsiness.	ventilation, local exhaust or respiratory protection.	fresh air, rest, call a doctor.
Skin: redness.	gloves.	remove contaminated clothing, flush skin with water or shower.
Eyes: redness, pain.	safety glasses.	flush with water, take to a doctor if necessary.
Ingestion: sore throat, abdominal pain, stomachache.		rinse mouth, call a doctor or take to hospital.

SPILLAGE	STORAGE	LABELING / NFPA
evacuate area, collect leakage in sealable containers, soak up with sand or other inert absorbent and remove to safe place; (additional individual protection: breathing apparatus).	keep in cool, dark, fireproof place, separate from oxidants and strong acids; add an inhibitor.	R: 12 S: 9-16-29-33 Note D Flammable

NOTES

Before distilling check for peroxides; if found, render harmless. Alcohol consumption increases toxic effects. High atmospheric concentrations, e.g. in poorly ventilated spaces, can cause oxygen deficiency, with risk of unconsciousness.

Transport Emergency Card TEC(R)-531

HI: 339; UN-number: 1218

CAS-No: [78-96-6]
1-amino-2-hydroxypropane
1-amino-propanol-2
2-hydroxypropylamine
monoisopropanolamine

$CH_3CHOHCH_2NH_2$

ISOPROPANOLAMINE

PHYSICAL PROPERTIES		IMPORTANT CHARACTERISTICS
Boiling point, °C	160	**COLORLESS LIQUID WITH CHARACTERISTIC ODOR**
Melting point, °C	−2	Vapor mixes readily with air. Toxic vapors are a product of combustion. Reacts with strong
Flash point, °C	71	oxidants.
Auto-ignition temperature, °C	335	
Relative density (water = 1)	0.96	
Relative vapor density (air = 1)	2.6	TLV-TWA not available
Relative density at 20 °C of saturated mixture		
vapor/air (air = 1)	1.0	**Absorption route:** Can enter the body by inhalation or ingestion or through the skin. Harmful
Vapor pressure, mm Hg at 20 °C	1.4	atmospheric concentrations can build up fairly rapidly on evaporation at approx. 20°C - even
Solubility in water	∞	more rapidly in aerosol form.
Relative molecular mass	75.1	**Immediate effects:** Irritates the eyes, skin and respiratory tract.
Log P octanol/water	−1.0	**Effects of prolonged/repeated exposure:** Prolonged or repeated contact can cause skin disorders.
Gross formula: C_3H_9NO		

HAZARDS/SYMPTOMS	PREVENTIVE MEASURES	FIRE EXTINGUISHING/FIRST AID
Fire: combustible.	keep away from open flame, no smoking.	powder, alcohol-resistant foam, water spray, carbon dioxide, (halons).
Explosion: above 71°C: forms explosive air-vapor mixtures.	above 71°C: sealed machinery, ventilation.	
Inhalation: sore throat, cough, shortness of breath.	ventilation, local exhaust or respiratory protection.	fresh air, rest.
Skin: *is absorbed*; redness.	gloves.	remove contaminated clothing, wash skin with soap and water.
Eyes: redness, pain.	face shield.	flush with water, take to a doctor if necessary.
Ingestion: sore throat, abdominal pain.		rinse mouth, call a doctor.

SPILLAGE	STORAGE	LABELING / NFPA
collect leakage in sealable containers, flush away remainder with water.	keep separate from strong oxidants.	R: 34 S: 23-26-36 Corrosive 2 2 0

NOTES

510

CAS-No: [108-21-4]
acetic acid, 1-methylethyl ester
2-acetoxypropane
1-methylethyl acetate
2-propyl acetate

$CH_3COOCH(CH_3)_2$

ISOPROPYL ACETATE

PHYSICAL PROPERTIES		IMPORTANT CHARACTERISTICS
Boiling point, °C	89	**COLORLESS LIQUID WITH CHARACTERISTIC ODOR**
Melting point, °C	−73	Vapor mixes readily with air, forming explosive mixtures. Do not use compressed air when filling,
Flash point, °C	4	emptying or processing. Decomposes slowly on contact with steel when exposed to air, forming
Auto-ignition temperature, °C	460	→ *acetic acid* and → *isopropyl alcohol.* Reacts violently with strong oxidants. Liquid dissolves
Relative density (water = 1)	0.9	many plastics.
Relative vapor density (air = 1)	3.5	
Relative density at 20 °C of saturated mixture		
vapor/air (air = 1)	1.02	
Vapor pressure, mm Hg at 20 °C	46.4	
Solubility in water, g/100 ml at 20 °C	3.1	
Explosive limits, vol% in air	1.8-8	
Relative molecular mass	102.1	

| | TLV-TWA | 250 ppm | 1040 mg/m³ |
| | STEL | 310 ppm | 1290 mg/m³ |

Absorption route: Can enter the body by inhalation or ingestion. Harmful atmospheric concentrations build up fairly slowly on evaporation at approx. 20°C, but much more rapidly in aerosol form.
Immediate effects: Irritates the eyes, skin and respiratory tract. In serious cases risk of unconsciousness.

Gross formula: $C_5H_{10}O_2$

HAZARDS/SYMPTOMS	PREVENTIVE MEASURES	FIRE EXTINGUISHING/FIRST AID
Fire: extremely flammable.	keep away from open flame and sparks, no smoking.	powder, AFFF, foam, carbon dioxide, (halons).
Explosion: forms explosive air-vapor mixtures.	sealed machinery, ventilation, explosion-proof electrical equipment and lighting.	in case of fire: keep tanks/drums cool by spraying with water, but DO NOT spray substance with water.
Inhalation: cough, shortness of breath, dizziness.	ventilation, local exhaust or respiratory protection.	fresh air, rest.
Skin: irritation, redness.	gloves.	remove contaminated clothing, flush skin with water or shower.
Eyes: redness, pain.	safety glasses.	flush with water, take to a doctor.
Ingestion: sore throat, abdominal pain, diarrhea.		rinse mouth, call a doctor or take to hospital.

SPILLAGE	STORAGE	LABELING / NFPA
collect leakage in sealable containers, soak up with sand or other inert absorbent and remove to safe place; (additional individual protection: breathing apparatus).	keep in fireproof place, separate from oxidants.	R: 11 S: 16-23-29-33 Note C Flammable

NOTES

Flush contaminated clothing with water (fire hazard). Packaging: special material. Do not use iron containers.

Transport Emergency Card TEC(R)-524

HI: 33; UN-number: 1220

dimethylcarbinol
isopropanol
2-propanol
sec-propyl alcohol

CH$_3$CHOHCH$_3$

ISOPROPYL ALCOHOL

PHYSICAL PROPERTIES	IMPORTANT CHARACTERISTICS
Boiling point, °C 82 Melting point, °C −90 Flash point, °C 12 Auto-ignition temperature, °C 425 Relative density (water = 1) 0.8 Relative vapor density (air = 1) 2.1 Relative density at 20 °C of saturated mixture vapor/air (air = 1) 1.05 Vapor pressure, mm Hg at 20 °C 32.68 Solubility in water ∞ Explosive limits, vol% in air 2 - 12 Minimum ignition energy, mJ 0.65 Electrical conductivity, pS/m 5.8x10^6 Relative molecular mass 60.1 Log P octanol/water 0.1 Gross formula: C$_3$H$_8$O	**COLORLESS LIQUID** Vapor mixes readily with air, forming explosive mixtures. Do not use compressed air when filling, emptying or processing. Reacts violently with oxidants. Reacts violently with oleum, phosgene, aluminum, crotonaldehyde and alkaline-earth and alkali metals, with risk of fire and explosion.

	TLV-TWA	400 ppm	983 mg/m³
	STEL	200 ppm	1230 mg/m³

Absorption route: Can enter the body by inhalation or ingestion. Harmful atmospheric concentrations build up fairly slowly on evaporation at 20° C, but much more rapidly in aerosol form.
Immediate effects: Irritates the skin and respiratory tract. Corrosive to the eyes. Liquid destroys the skin's natural oils and is absorbed in small quantities. In substantial concentrations can impair consciousness.

HAZARDS/SYMPTOMS	PREVENTIVE MEASURES	FIRE EXTINGUISHING/FIRST AID
Fire: extremely flammable.	keep away from open flame and sparks, no smoking.	powder, alcohol-resistant foam, large quantities of water, carbon dioxide, (halons).
Explosion: forms explosive air-vapor mixtures.	sealed machinery, ventilation, explosion-proof electrical equipment and lighting.	in case of fire: keep tanks/drums cool by spraying with water.
Inhalation: sore throat, cough, shortness of breath, headache.	ventilation, local exhaust or respiratory protection.	fresh air, rest, call a doctor if necessary.
Skin: redness, pain.	gloves.	remove contaminated clothing, flush skin with water or shower, call a doctor if necessary.
Eyes: *corrosive*; redness, pain, impaired vision.	face shield.	flush with water, take to a doctor.
Ingestion: sore throat, cough, abdominal pain, diarrhea, abdominal cramps, stomachache, headache, vomiting.		rinse mouth, take to hospital if necessary.

SPILLAGE	STORAGE	LABELING / NFPA
collect leakage in sealable containers, flush away remainder with water; (additional individual protection: breathing apparatus).	keep in fireproof place, separate from oxidants.	R: 11 S: 7-16 Note C 🔥 Flammable

NFPA diamond: 3 (top), 1 (left), 0 (right)

NOTES
Alcohol consumption increases toxic effects. Log P octanol/water is estimated.

Transport Emergency Card TEC(R)-544

HI: 33; UN-number: 1219

CAS-No: [75-31-0]
2-aminopropane
1-methylethylamine
MIPA
2-propylamine

$(CH_3)_2CHNH_2$

ISOPROPYLAMINE

PHYSICAL PROPERTIES		IMPORTANT CHARACTERISTICS	
Boiling point, °C	32	**COLORLESS LIQUID WITH CHARACTERISTIC ODOR**	
Melting point, °C	−95	Vapor is heavier than air and spreads at ground level, with risk of ignition at a distance. Do not use compressed air when filling, emptying or processing. Toxic gases are a product of combustion (→ nitrous vapors). Strong base which reacts violently with acids and corrodes aluminum, zinc etc. Reacts violently with nitroparaffins, halogenated hydrocarbons, oxidants and many other substances. Reacts with mercury. Attacks copper and copper compounds.	
Flash point, °C	< −20		
Auto-ignition temperature, °C	400		
Relative density (water = 1)	0.69		
Relative vapor density (air = 1)	2.0		
Relative density at 20 °C of saturated mixture			

vapor/air (air = 1)	1.62	TLV-TWA	5 ppm	12 mg/m³
Vapor pressure, mm Hg at 20 °C	483	STEL	10 ppm	24 mg/m³
Solubility in water	∞		
Explosive limits, vol% in air	2.3-10.4	**Absorption route:** Can enter the body by inhalation or ingestion or through the skin. Harmful atmospheric concentrations can build up very rapidly on evaporation at 20°C.
Minimum ignition energy, mJ	2	
Relative molecular mass	59.1	**Immediate effects:** Corrosive to the eyes, skin and respiratory tract. Inhalation of vapor/fumes can cause severe breathing difficulties (lung edema). In serious cases risk of death.
Log P octanol/water	0	

Effects of prolonged/repeated exposure: Prolonged or repeated contact can cause skin disorders.

Gross formula: C_3H_9N

HAZARDS/SYMPTOMS	PREVENTIVE MEASURES	FIRE EXTINGUISHING/FIRST AID
Fire: extremely flammable.	keep away from open flame and sparks, no smoking.	powder, alcohol-resistant foam, large quantities of water, carbon dioxide, (halons).
Explosion: forms explosive air-vapor mixtures.	sealed machinery, ventilation, explosion-proof electrical equipment and lighting.	in case of fire: keep tanks/drums cool by spraying with water.
Inhalation: *corrosive*; sore throat, cough, shortness of breath, severe breathing difficulties.	ventilation, local exhaust or respiratory protection.	fresh air, rest, take to hospital.
Skin: *is absorbed, corrosive*; redness, pain, serious burns.	gloves, protective clothing.	remove contaminated clothing, flush skin with water or shower, call a doctor.
Eyes: *corrosive*; redness, pain, impaired vision.	face shield, or combined eye and respiratory protection.	flush with water, take to an eye doctor.
Ingestion: *corrosive*; sore throat, diarrhea, abdominal cramps, vomiting, see also: 'Inhalation'.		rinse mouth, immediately take to hospital.

SPILLAGE	STORAGE	LABELING / NFPA
evacuate area, call in an expert, collect leakage in sealable containers, soak up with with sand or other inert absorbent and remove to safe place; (additional individual protection: CHEMICAL SUIT).	keep in fireproof place, separate from oxidants, strong acids and halogenated hydrocarbons.	R: 12-36/37/38 S: 16-26-29 Flammable Irritant NFPA 3 4 0

NOTES

Lung edema symptoms usually develop several hours later and are aggravated by physical exertion: rest and hospitalization essential. As first aid, a doctor or authorized person should consider administering a corticosteroid spray. At 5ppm the ammonia air is clearly recognizable. Unbreakable packaging preferred; if breakable, keep in unbreakable container.

Transport Emergency Card TEC(R)-656

HI: 338; UN-number: 1221

ISOPROPYL CHLORIDE

PHYSICAL PROPERTIES	IMPORTANT CHARACTERISTICS
Boiling point, °C 35 Melting point, °C −117 Flash point, °C −32 Auto-ignition temperature, °C 590 Relative density (water = 1) 0.86 Relative vapor density (air = 1) 2.71 Relative density at 20 °C of saturated mixture vapor/air (air = 1) 1.95 Vapor pressure, mm Hg at 20 °C 429 Solubility in water, g/100 ml at 20 °C 0.31 Explosive limits, vol% in air 2.8-10.7 Relative molecular mass 78.5	**COLORLESS LIQUID WITH CHARACTERISTIC ODOR** Vapor is heavier than air and spreads at ground level, with risk of ignition at a distance. Flow, agitation etc. can cause build-up of electrostatic charge due to liquid's low conductivity. Do not use compressed air when filling, emptying or processing. Corrosive gases are a product of combustion (→ *hydrochloric acid*). Reacts violently with strong oxidants, zinc/aluminum powder and alkaline-earth metals, with risk of fire and explosion.
	TLV-TWA not available
	Absorption route: Can enter the body by inhalation or ingestion. Harmful atmospheric concentrations can build up very rapidly on evaporation at 20°C. **Immediate effects:** Inhalation of large quantities causes unconsciousness, with risk of heart rate, liver and kidney disorders.
Gross formula: C_3H_7Cl	

HAZARDS/SYMPTOMS	PREVENTIVE MEASURES	FIRE EXTINGUISHING/FIRST AID
Fire: extremely flammable.	keep away from open flame and sparks, no smoking.	powder, AFFF, foam, carbon dioxide, (halons).
Explosion: forms explosive air-vapor mixtures.	sealed machinery, ventilation, explosion-proof electrical equipment and lighting, grounding.	in case of fire: keep tanks/drums cool by spraying with water.
Inhalation: nausea, vomiting, drowsiness, sleepiness.	ventilation, local exhaust or respiratory protection.	fresh air, rest, call a doctor.
Skin: redness, irritation.	gloves.	remove contaminated clothing, wash skin with soap and water.
Eyes: redness, pain.	safety glasses, or combined eye and respiratory protection.	flush with water, take to a doctor if necessary.
Ingestion: sore throat, cough, nausea, vomiting, drowsiness.		rinse mouth, take to hospital.

SPILLAGE	STORAGE	LABELING / NFPA
collect leakage in sealable containers, soak up with sand or other inert absorbent and remove to safe place; (additional individual protection: breathing apparatus).	keep in cool, fireproof place, separate from oxidants.	R: 11-20/21/22 S: 9-29 🔥 Flammable ✖ Harmful (NFPA diamond: 4 top, 2 left, 0 right)

NOTES
Since boiling point is virtually equal to human body temperature, ingestion entails inhalation, with risk of unconsciousness.

Transport Emergency Card TEC(R)-30G34

HI: 33; UN-number: 2356

CAS-No: [108-20-3]
diisopropyl oxide
diisopropyl ether
di-(1-methylethyl)ether
2-isopropoxypropane
2,2'-oxybispropane

$(CH_3)_2\ CHOCH(CH_3)_2$

ISOPROPYL ETHER

PHYSICAL PROPERTIES

Boiling point, °C	68
Melting point, °C	−86
Flash point, °C	−28
Auto-ignition temperature, °C	405
Relative density (water = 1)	0.7
Relative vapor density (air = 1)	3.5
Relative density at 20 °C of saturated mixture vapor/air (air = 1)	1.4
Vapor pressure, mm Hg at 20 °C	133
Solubility in water, g/100 ml at 20 °C	0.9
Explosive limits, vol% in air	1.0-21
Relative molecular mass	102.2

Gross formula: $C_6H_{14}O$

IMPORTANT CHARACTERISTICS

COLORLESS LIQUID WITH CHARACTERISTIC ODOR
Readily able to form peroxides. Flow, agitation etc. can cause build-up of electrostatic charge due to liquid's low conductivity. Do not use compressed air when filling, emptying or processing. Reacts violently with oxidants, with risk of fire and explosion.

TLV-TWA	250 ppm	1040 mg/m³
STEL	310 ppm	1300 mg/m³

Absorption route: Can enter the body by inhalation or ingestion. Harmful atmospheric concentrations can build up fairly rapidly on evaporation at approx. 20°C - even more rapidly in aerosol form.
Immediate effects: Irritates the eyes and respiratory tract. Liquid destroys the skin's natural oils. In substantial concentrations can cause unconsciousness.

HAZARDS/SYMPTOMS	PREVENTIVE MEASURES	FIRE EXTINGUISHING/FIRST AID
Fire: extremely flammable.	keep away from open flame and sparks, no smoking.	powder, AFFF, foam, carbon dioxide, (halons); in case of fire in immediate vicinity: use any extinguishing agent.
Explosion: forms explosive air-vapor mixtures.	sealed machinery, ventilation, explosion-proof electrical equipment and lighting, grounding.	in case of fire: keep tanks/drums cool by spraying with water.
Inhalation: sore throat, cough, shortness of breath, headache.	ventilation, local exhaust or respiratory protection.	fresh air, rest, call a doctor.
Skin: redness.	gloves.	remove contaminated clothing, flush skin with water or shower.
Eyes: redness, pain.	safety glasses.	flush with water, take to a doctor if necessary.
Ingestion: sore throat, cough, shortness of breath, abdominal pain, nausea.		rinse mouth, call a doctor.

SPILLAGE	STORAGE	LABELING / NFPA
collect leakage in sealable containers, soak up with sand or other inert absorbent and remove to safe place; (additional individual protection: breathing apparatus).	keep in cool, dark, fireproof place, separate from oxidants; add an inhibitor.	R: 11-19 S: 9-16-33 Flammable

NOTES

Before distilling check for peroxides; if found, render harmless. Use airtight packaging.

Transport Emergency Card TEC(R)-128

HI: 33; UN-number: 1159

CAS-No: [109-59-1]
2-isopropoxyethanol
ethylene glycol monoisopropyl ether

$(CH_3)_2CHOCH_2CH_2OH$

ISOPROPYL GLYCOL

PHYSICAL PROPERTIES		IMPORTANT CHARACTERISTICS	
		COLORLESS LIQUID WITH CHARACTERISTIC ODOR Vapor mixes readily with air. Reacts with oxidants.	
Boiling point, °C	143		
Melting point, °C	< −60		
Flash point, °C	54	**TLV-TWA** 25 ppm 106 mg/m³	
Auto-ignition temperature, °C	345		
Relative density (water = 1)	0.9	**Absorption route:** Can enter the body by inhalation or ingestion or through the skin. Harmful atmospheric concentrations build up fairly slowly on evaporation at 20°C, but much more rapidly in aerosol form.	
Relative vapor density (air = 1)	3.6		
Relative density at 20°C of saturated mixture vapor/air (air = 1)	1.01		
Vapor pressure, mm Hg at 20°C	2.7	**Immediate effects:** Irritates the skin and respiratory tract. Corrosive to the eyes. Liquid destroys the skin's natural oils.	
Solubility in water	∞		
Explosive limits, vol% in air	1.6 - 13.0	**Effects of prolonged/repeated exposure:** Hemoglobin in urine and anemia are characteristic symptoms. Prolonged or high-level exposure can cause kidney, liver and spleen damage.	
Relative molecular mass	104.2		
Gross formula:	$C_5H_{12}O_2$		

HAZARDS/SYMPTOMS	PREVENTIVE MEASURES	FIRE EXTINGUISHING/FIRST AID
Fire: flammable.	keep away from open flame and sparks, no smoking.	powder, alcohol-resistant foam, water spray, carbon dioxide, (halons).
Explosion: above 54°C: forms explosive air-vapor mixtures.	above 54°C: sealed machinery, ventilation, explosion-proof electrical equipment.	in case of fire: keep tanks/drums cool by spraying with water.
Inhalation: cough, headache, dizziness, nausea, hemoglobin in the urine.	ventilation, local exhaust or respiratory protection.	fresh air, rest, call a doctor.
Skin: *is absorbed*; redness; see also 'Inhalation'.	gloves, protective clothing.	remove contaminated clothing, flush skin with water or shower, call a doctor.
Eyes: *corrosive*; redness, pain, impaired vision.	face shield.	flush with water, take to a doctor.
Ingestion: sore throat, cough, abdominal pain, headache, dizziness, vomiting, hemoglobin in the urine.		rinse mouth, rest, take to hospital if necessary.

SPILLAGE	STORAGE	LABELING / NFPA
collect leakage in sealable containers, flush away remainder with water.	keep in fireproof place.	R: 20/21-36 S: 24/25 ✖ Harmful

NOTES
Alcohol consumption increases toxic effects. Depending on the degree of exposure, regular medical checkups are advisable.

KRYPTON
(cylinder)

PHYSICAL PROPERTIES	IMPORTANT CHARACTERISTICS
Boiling point, °C −153 Melting point, °C −157 Relative vapor density (air = 1) 2.9 Solubility in water none Relative atomic mass 83.8 Log P octanol/water 1.2	**COLORLESS, ODORLESS COMPRESSED GAS** Gas is heavier than air and can accumulate close to ground level, causing oxygen deficiency, with risk of unconsciousness.
	TLV-TWA not available
	Absorption route: Can saturate the air if released, with risk of suffocation. **Immediate effects:** Inhalation can cause severe breathing difficulties.
Gross formula: Kr	

HAZARDS/SYMPTOMS	PREVENTIVE MEASURES	FIRE EXTINGUISHING/FIRST AID
Fire: non-combustible.		in case of fire in immediate vicinity: use any extinguishing agent.
Explosion:		in case of fire: keep cylinder cool by spraying with water, fight fire from sheltered location.
Inhalation: severe breathing difficulties, headache, dizziness, unconsciousness.	ventilation, local exhaust or respiratory protection.	fresh air, rest, artificial respiration if necessary, take to hospital.

SPILLAGE	STORAGE	LABELING / NFPA
ventilate; (additional individual protection: breathing apparatus).	if stored indoors, keep in fireproof place.	

NOTES
Critical temperature −64°C. Log P octanol/water is estimated. High atmospheric concentrations, e.g. in poorly ventilated spaces, can cause oxygen deficiency, with risk of unconsciousness.

Transport Emergency Card TEC(R)-20G01

HI: 20; UN-number: 1056

KRYPTON
(liquid, cooled)

PHYSICAL PROPERTIES	IMPORTANT CHARACTERISTICS
Boiling point, °C — 153 Melting point, °C — 157 Relative density (water = 1) see notes Relative vapor density (air = 1) 2.9 Solubility in water none Relative atomic mass 83.8 Log P octanol/water 1.2	**COLORLESS, ODORLESS, EXTREMELY COLD LIQUID** Vapor is heavier than air and can accumulate close to ground level, causing oxygen deficiency, with risk of unconsciousness.
	TLV-TWA not available
	Absorption route: If liquid is released, air can become saturated due to evaporation, with risk of suffocation. **Immediate effects:** Cold liquid can cause frostbite.
Gross formula: Kr	

HAZARDS/SYMPTOMS	PREVENTIVE MEASURES	FIRE EXTINGUISHING/FIRST AID
Fire: non-combustible.		in case of fire in immediate vicinity: use any extinguishing agent.
Explosion:		in case of fire: keep tanks/drums cool by spraying with water, but DO NOT spray substance with water.
Inhalation: see 'Notes'.	ventilation, local exhaust or respiratory protection.	fresh air, rest, call a doctor, take to hospital if necessary.
Skin: *in case of frostbite*: redness, pain, blisters, sores.	insulated gloves.	*in case of frostbite*: DO NOT remove clothing, flush skin with water or shower, call a doctor.
Eyes: *in case of frostbite*: redness, pain, impaired vision.	face shield, or combined eye and respiratory protection.	flush with water, take to a doctor.

SPILLAGE	STORAGE	LABELING / NFPA
ventilate, under no circumstances spray liquid with water.	ventilate at floor level, keep refrigerated.	

NOTES
Density of liquid at boiling point 2.4kg/l. Critical temperature − 64°C. Log P octanol/water is estimated. High atmospheric concentrations, e.g. in poorly ventilated spaces, can cause oxygen deficiency, with risk of unconsciousness. Use special insulated drums.

Transport Emergency Card TEC(R)-20G22   **HI: 22; UN-number: 1970**

CAS-No: [598-82-3]
2-hydroxypropionic acid
α-hydroxypropionic acid
milk acid

$CH_3CHOHCOOH$

LD-LACTIC ACID

PHYSICAL PROPERTIES		IMPORTANT CHARACTERISTICS	
Boiling point, °C	?	**COLORLESS VISCOUS LIQUID OR HYGROSCOPIC CRYSTALS**	
Melting point, °C	18	In aqueous solution is a corrosive acid.	
Flash point, °C	>74		
Relative density (water = 1)	1.2	TLV-TWA not available	
Solubility in water	∞		
Log P octanol/water	−0.6		
		Absorption route: Can enter the body by inhalation or ingestion.	
		Immediate effects: Irritates the skin and respiratory tract. Corrosive to the eyes.	
Gross formula:	$C_3H_6O_3$		

HAZARDS/SYMPTOMS	PREVENTIVE MEASURES	FIRE EXTINGUISHING/FIRST AID
Fire: combustible.	keep away from open flame, no smoking.	powder, alcohol-resistant foam, water spray, carbon dioxide, (halons).
Inhalation: sore throat, cough, shortness of breath.	local exhaust or respiratory protection.	fresh air, rest, call a doctor.
Skin: redness, pain.	gloves.	remove contaminated clothing, flush skin with water or shower.
Eyes: *corrosive*; redness, pain, impaired vision.	face shield.	flush with water, take to a doctor.
Ingestion: sore throat, abdominal pain.		rinse mouth, call a doctor.

SPILLAGE	STORAGE	LABELING / NFPA
collect leakage in sealable containers, clean up spillage, flush away remainder with water.	keep dry.	

NOTES
Melting point of L(+) and D(-) lactic acid: 53° C. Commercial product is a racemic mixture.

Transport Emergency Card TEC(R)-80G08

LANOLIN

PHYSICAL PROPERTIES		IMPORTANT CHARACTERISTICS	
Melting point, °C	37	**WHITE TO YELLOW PASTE**	
Flash point, °C	238	Consists of cholesterol ester of higher fatty acids. Can contain 20-30% water.	
Auto-ignition temperature, °C	445		
Solubility in water	none	TLV-TWA	not available
Relative molecular mass	?		
		Absorption route: Can enter the body by ingestion.	

HAZARDS/SYMPTOMS	PREVENTIVE MEASURES	FIRE EXTINGUISHING/FIRST AID
Fire: combustible.	keep away from open flame, no smoking.	water spray, powder.
Ingestion: in large quantities: abdominal pain, diarrhea.		DO NOT INDUCE VOMITING, call a doctor.

SPILLAGE	STORAGE	LABELING / NFPA
clean up spillage.		

NOTES

LAUGHING GAS
(cylinder)

PHYSICAL PROPERTIES		IMPORTANT CHARACTERISTICS	
Boiling point, °C	− 89	**COLORLESS COMPRESSED LIQUEFIED GAS WITH CHARACTERISTIC ODOR**	
Melting point, °C	− 91	Gas is heavier than air. When heated above 650°C in the presence of air gives off toxic gases	
Relative density (water = 1)	0.8	(→ *nitrous vapors*). Above 300°C is a strong oxidant. Can form explosive gas mixtures with	
Relative vapor density (air = 1)	1.5	anhydrous ammonia, carbon monoxide, hydrogen, hydrogen sulfide, phosphine and many other	
Vapor pressure, mm Hg at 20 °C	38.8	flammable vapors or gases.	
Solubility in water, g/100 ml at 20 °C	0.12		
Relative molecular mass	44.0		
		TLV-TWA 50 ppm 90 mg/m³	
		Absorption route: Can enter the body by inhalation. Harmful atmospheric concentrations can build up very rapidly if gas is released. **Immediate effects:** Affects the nervous system.	
Gross formula:	N$_2$O		

HAZARDS/SYMPTOMS	PREVENTIVE MEASURES	FIRE EXTINGUISHING/FIRST AID
Fire: not combustible but enhances combustion of other substances.	keep away from open flame and sparks, no smoking.	in case of fire in immediate vicinity: use any extinguishing agent.
Explosion: mixtures with ammonia, carbon monoxide, hydrogen, hydrogen sulfide, phosphine and other flammable vapors are explosive.		in case of fire: keep cylinders cool by spraying with water.
Inhalation: headache, sleepiness, unconsciousness.	ventilation, local exhaust or respiratory protection.	fresh air, rest, call a doctor.

SPILLAGE	STORAGE	LABELING / NFPA
ventilate; (additional individual protection: breathing apparatus).	if stored indoors, keep in cool, fireproof place, separate from combustible substances and reducing agents.	

NOTES
High atmospheric concentrations, e.g. in poorly ventilated spaces, can cause oxygen deficiency, with risk of unconsciousness. Solubility in water at 0°C 130g/100ml. Used as anesthetic. Turn leaking cylinder so that leak is on top to prevent liquid laughing gas escaping.

Transport Emergency Card TEC(R)-20G22

HI: 225; UN-number: 1070

LAURIC ACID

PHYSICAL PROPERTIES	IMPORTANT CHARACTERISTICS
Melting point, °C 44 Flash point, °C see notes Relative density (water = 1) 0.9 Vapor pressure, mm Hg at 131 °C 0.99 Solubility in water none Relative molecular mass 200.4 Log P octanol/water 4.2	**WHITE CRYSTALS** Decomposes when heated, giving off acid vapors. Reacts with oxidants. TLV-TWA not available **Absorption route:** Can enter the body by inhalation or ingestion. Evaporation negligible at 20°C, but unpleasant concentrations of airborne particles can build up. **Immediate effects:** Irritates the eyes and respiratory tract.
Gross formula: $C_{12}H_{24}O_2$	

HAZARDS/SYMPTOMS	PREVENTIVE MEASURES	FIRE EXTINGUISHING/FIRST AID
Fire: combustible.	keep away from open flame, no smoking.	powder, AFFF, foam, carbon dioxide, (halons).
Inhalation: sore throat, cough.	local exhaust or respiratory protection.	fresh air, rest.
Skin:	gloves.	
Eyes: redness, pain.	goggles.	flush with water, take to a doctor if necessary.
Ingestion: sore throat.		rinse mouth.

SPILLAGE	STORAGE	LABELING / NFPA
clean up spillage.		

NOTES
Combustible, but flash point and explosive limits not given in the literature.

LEAD
(powder)

PHYSICAL PROPERTIES	IMPORTANT CHARACTERISTICS	
Boiling point, °C — 1740 Melting point, °C — 327 Relative density (water = 1) — 11.3 Solubility in water — none Relative atomic mass — 207.2	**DARK GRAY POWDER** Gives off harmful lead vapors when heated above melting point. Dissolves in nitric acid but not in sulfuric or hydrochloric acid.	
	TLV-TWA	0.15 mg/m³ ✱
	Absorption route: Evaporation negligible at 20°C, but harmful concentrations of airborne particles can build up rapidly. **Effects of prolonged/repeated exposure:** Can cause kidney damage. Can affect the blood.	
Gross formula: Pb		

HAZARDS/SYMPTOMS	PREVENTIVE MEASURES	FIRE EXTINGUISHING/FIRST AID
Fire: non-combustible.		in case of fire in immediate vicinity: use any extinguishing agent.
	STRICT HYGIENE, KEEP DUST UNDER CONTROL	
Inhalation: headache, nausea, abdominal cramps, constipation.	local exhaust or respiratory protection.	fresh air, rest, call a doctor.
Skin:	gloves.	
Eyes:	goggles.	
Ingestion: abdominal cramps, headache, nausea, constipation.	do not eat, drink or smoke while working.	rinse mouth, call a doctor or take to hospital.

SPILLAGE	STORAGE	LABELING / NFPA
clean up spillage, carefully collect remainder; (additional individual protection: P2 respirator).		

NOTES
Lead level in blood is highly recommended as a yardstick for exposure. Do not take work clothing home. Depending on the degree of exposure, regular medical checkups are advisable. ✱ TLV equals that of lead (inorganic fumes and dust).

LEAD (II) ACETATE

PHYSICAL PROPERTIES	IMPORTANT CHARACTERISTICS	
Boiling point (decomposes), °C 200 Melting point, °C 75 Relative density (water = 1) 2.6 Solubility in water, g/100 ml at 20 °C 62.5 Relative molecular mass 379.3	**COLORLESS CRYSTALS OR WHITE POWDER** Decomposes when heated and on contact with acids, forming → *acetic acid*. Forms explosive compounds with lead bromate.	
	TLV-TWA	0.15 mg/m³ ✱
	Absorption route: Can enter the body by inhalation or ingestion. Evaporation negligible at 20°C, but harmful concentrations of airborne particles can build up rapidly. **Immediate effects:** Irritates the eyes, skin and respiratory tract. **Effects of prolonged/repeated exposure:** Can affect the blood. Can cause kidney damage.	
Gross formula: $C_4H_6O_4Pb \cdot 3H_2O$		

HAZARDS/SYMPTOMS	PREVENTIVE MEASURES	FIRE EXTINGUISHING/FIRST AID
Fire: non-combustible.		in case of fire in immediate vicinity: use any extinguishing agent.
	STRICT HYGIENE, KEEP DUST UNDER CONTROL	
Inhalation: sore throat, cough, headache, nausea, constipation, abdominal cramps.	local exhaust or respiratory protection.	fresh air, rest, call a doctor.
Skin: redness.	gloves.	remove contaminated clothing, flush skin with water or shower.
Eyes: redness.	goggles.	flush with water, take to a doctor.
Ingestion: headache, nausea, constipation, abdominal cramps.	do not eat, drink or smoke while working.	rinse mouth, call a doctor or take to hospital.

SPILLAGE	STORAGE	LABELING / NFPA
clean up spillage, carefully collect remainder; (additional individual protection: P2 respirator).	keep separate from lead bromate.	R: 20/22-33 S: 13-20/21 Note A ✖ Harmful

NOTES
Apparent melting point due to loss of water of crystallization. ✱ TLV equals that of lead. Do not take work clothing home. Depending on the degree of exposure, regular medical checkups are advisable. Lead level in blood is highly recommended as a yardstick for exposure.

LEAD (II) CARBONATE

PHYSICAL PROPERTIES	IMPORTANT CHARACTERISTICS	
Melting point (decomposes), °C　　315 Relative density (water = 1)　　6.6 Solubility in water　　none Relative molecular mass　　267.2	**WHITE CRYSTALLINE POWDER** Decomposes in hot water, in lead oxide on contact with acids (forming → *carbon dioxide*).	
	TLV-TWA	0.15 mg/m³ ✳
	Absorption route: Can enter the body by inhalation or ingestion. Evaporation negligible at 20°C, but harmful concentrations of airborne particles can build up rapidly. **Immediate effects:** Irritates the eyes and respiratory tract. **Effects of prolonged/repeated exposure:** Can affect the blood. Can cause kidney damage.	
Gross formula:　　CO₃Pb		

HAZARDS/SYMPTOMS	PREVENTIVE MEASURES	FIRE EXTINGUISHING/FIRST AID
Fire: non-combustible.		in case of fire in immediate vicinity: use any extinguishing agent.
	STRICT HYGIENE, KEEP DUST UNDER CONTROL	
Inhalation: sore throat, cough, headache, dizziness, constipation, abdominal cramps.	local exhaust or respiratory protection.	fresh air, rest, call a doctor.
Skin: redness.	gloves.	remove contaminated clothing, flush skin with water or shower.
Eyes: redness, pain.	goggles.	flush with water, take to a doctor if necessary.
Ingestion: abdominal cramps,	do not eat, drink or smoke while working.	rinse mouth, take to hospital.

SPILLAGE	STORAGE	LABELING / NFPA
clean up spillage, carefully collect remainder; (additional individual protection: P2 respirator).		R: 20/22-33 S: 13-20/21 Note A ✖ Harmful

NOTES
✳ TLV equals that of lead. Do not take work clothing home. Depending on the degree of exposure, regular medical checkups are advisable. Lead level in blood is highly recommended as a yardstick for exposure.

LEAD (II) CHROMATE

PHYSICAL PROPERTIES	IMPORTANT CHARACTERISTICS	
Melting point (decomposes), °C 844 Relative density (water = 1) 6.1 Vapor pressure, mm Hg at 20 °C 0 Solubility in water none Relative molecular mass 323.2	**ORANGE-YELLOW POWDER** Can ignite when mixed with combustible substances. Decomposes at melting point, giving off oxygen etc., with increased risk of fire.	
	TLV-TWA	✱
	Absorption route: Can enter the body by inhalation or ingestion. Evaporation negligible at 20° C, but harmful atmospheric concentrations can build up rapidly in aerosol form. **Immediate effects:** Corrosive to the eyes, skin and respiratory tract. Inhalation can cause lung edema. **Effects of prolonged/repeated exposure:** Can affect the blood. Can cause kidney damage. Has been found to cause a type of cancer in humans under certain circumstances.	
Gross formula: CrO$_4$Pb		

HAZARDS/SYMPTOMS	PREVENTIVE MEASURES	FIRE EXTINGUISHING/FIRST AID
Fire: non-combustible.		in case of fire in immediate vicinity: use any extinguishing agent.
	STRICT HYGIENE, KEEP DUST UNDER CONTROL	
Inhalation: *corrosive*; sore throat, cough, nausea, metallic taste.	local exhaust or respiratory protection.	fresh air, rest, call a doctor or take directly to hospital.
Skin: irritation.	gloves.	remove contaminated clothing, call a doctor.
Eyes: *corrosive*; redness, pain.	goggles, or combined eye and respiratory protection.	flush with water, take to doctor.
Ingestion: *corrosive*; sore throat, abdominal cramps, nausea, vomiting, metallic taste, constipation.	do not eat, drink or smoke while working.	rinse mouth, take immediately to hospital.

SPILLAGE	STORAGE	LABELING / NFPA
first dampen spillage, then clean up, carefully collect remainder; (additional individual protection: P3 respirator)		R: 33-40 S: 22 ✖ Harmful

NOTES
Welding, cutting and heating of materials treated with lead chromate can cause release of toxic lead oxide fumes. Depending on the degree of exposure, regular medical checkups are advisable. Lung edema symptoms usually develop several hours later and are aggravated by physical exertion: rest and hospitalization essential. As first aid, a doctor or authorized person should consider administering a corticosteroid spray. Unbreakable packaging preferred; if breakable, keep in unbreakable container. ✱ TLV-TWA of lead (II) chromate as chromium: 0.012 mg/m³ (A2); TLV-TWA as lead: 0.05 mg/m³ (A2).

Transport Emergency Card TEC(R)-221

LEAD (II) NAPHTHENATE

PHYSICAL PROPERTIES		IMPORTANT CHARACTERISTICS	
Melting point, °C	100	**YELLOW PASTE**	
Relative density (water = 1)	> 1	Contains 16 to 37% lead. Toxic vapors are a product of combustion.	
Solubility in water	none		
		TLV-TWA	0.15 mg/m³ ✳
		Absorption route: Can enter the body by ingestion or through the skin. Harmful atmospheric concentrations build up very slowly, if at all, on evaporation at approx. 20°C, but much more rapidly in aerosol form. **Immediate effects:** Irritates the eyes and skin. **Effects of prolonged/repeated exposure:** Can affect the blood. Can cause kidney damage.	

HAZARDS/SYMPTOMS	PREVENTIVE MEASURES	FIRE EXTINGUISHING/FIRST AID
Fire: combustible.	keep away from open flame, no smoking.	water spray, powder.
Inhalation:	ventilation.	
Skin: *is absorbed*; redness.	gloves.	remove contaminated clothing, wash skin with soap and water, call a doctor.
Eyes: redness.	safety glasses.	flush with water, take to a doctor if necessary.
Ingestion: sore throat, abdominal pain, vomiting.		rinse mouth, call a doctor.

SPILLAGE	STORAGE	LABELING / NFPA
clean up spillage, carefully collect remainder.		R: 20/22-33 S: 13-20/21 Note A ✖ Harmful

NOTES
✳ TLV equals that of lead. Lead level in blood is highly recommended as a yardstick for exposure.

Transport Emergency Card TEC(R)-221

LEAD (II) NITRATE

PHYSICAL PROPERTIES	IMPORTANT CHARACTERISTICS
Decomposes below melting point, °C 290 Relative density (water = 1) 4.6 Solubility in water, g/100 ml at 20 °C 52 Relative molecular mass 331.2	**WHITE CRYSTALS OR WHITE POWDER** Decomposes when heated above 200°C, giving off toxic → *nitrous vapors* and → *oxygen*, with increased risk of fire. Strong oxidant which reacts violently with combustible substances and reducing agents. Reacts violently with ammoniumthiocyanate, carbon, potassiumacetate, lead hypophosphite and many other substances.

	TLV-TWA	0.15 mg/m³ ✳

Absorption route: Can enter the body by inhalation or ingestion. Evaporation negligible at 20°C, but harmful concentrations of airborne particles can build up rapidly.
Immediate effects: Irritates the skin. Corrosive to the eyes and respiratory tract. Inhalation can cause lung edema. In serious cases risk of death.
Effects of prolonged/repeated exposure: Can affect the blood and can cause ringing and paralysis of the limbs. Can cause liver and kidney damage.

Gross formula: PbN$_2$O$_6$

HAZARDS/SYMPTOMS	PREVENTIVE MEASURES	FIRE EXTINGUISHING/FIRST AID
Fire: not combustible but enhances combustion of other substances.	avoid contact with combustible substances.	in case of fire in immediate vicinity: water spray preferred extinguishing agent.
Explosion:		in case of fire: keep tanks/drums cool by spraying with water.
	STRICT HYGIENE, KEEP DUST UNDER CONTROL	
Inhalation: *corrosive*; sore throat, cough, shortness of breath, severe breathing difficulties, headache.	local exhaust or respiratory protection.	fresh air, rest, place in half-sitting position, take to hospital.
Skin: redness, pain.	gloves.	remove contaminated clothing, flush skin with water or shower, refer to a doctor.
Eyes: *corrosive*; redness, pain, impaired vision.	goggles, or combined eye and respiratory protection.	flush with water, take to a doctor.
Ingestion: *corrosive*; abdominal pain, nausea.	do not eat, drink or smoke while working.	rinse mouth, immediately take to hospital.

SPILLAGE	STORAGE	LABELING / NFPA
clean up spillage, carefully collect remainder; (additional individual protection: P2 respirator).	keep separate from combustible substances and reducing agents.	R: 20/22-33 S: 13-20/21 Note A ✖ Harmful

NOTES

✳ TLV equals that of lead. Flush contaminated clothing with water (fire hazard). Do not take work clothing home. Lung edema symptoms usually develop several hours later and are aggravated by physical exertion: rest and hospitalization essential. As first aid, a doctor or authorized person should consider administering a corticosteroid spray. Special first aid required in the event of poisoning: antidotes must be available (with instructions). Depending on the degree of exposure, regular medical checkups are advisable. Lead level in blood is highly recommended as a yardstick for exposure.

LEADPEROXIDE

PHYSICAL PROPERTIES	IMPORTANT CHARACTERISTICS	
Decomposes below melting point, °C 290 Relative density (water = 1) 9.4 Solubility in water none Relative molecular mass 239.2	**BROWN CRYSTALLINE POWDER** Decomposes when heated, giving off → *lead* vapors and → *oxygen*, with increased risk of fire. Strong oxidant which reacts violently with combustible substances and reducing agents. Reacts with → *hydrochloric acid*, forming → *chlorine*.	
	TLV-TWA	0.15 mg/m³ ✳
	Absorption route: Can enter the body by inhalation or ingestion. Evaporation negligible at 20°C, but harmful concentrations of airborne particles can build up rapidly. **Immediate effects:** Irritates the eyes, skin and respiratory tract. In substantial concentrations can cause tingling and paralysis of the limbs. **Effects of prolonged/repeated exposure:** Can affect the blood. Can cause kidney damage.	
Gross formula: O$_2$Pb		

HAZARDS/SYMPTOMS	PREVENTIVE MEASURES	FIRE EXTINGUISHING/FIRST AID
Fire: not combustible but enhances combustion of other substances.	avoid contact with combustible substances.	in case of fire in immediate vicinity: use any extinguishing agent.
Explosion:		in case of fire: keep tanks/drums cool by spraying with water.
	STRICT HYGIENE, KEEP DUST UNDER CONTROL	
Inhalation: sore throat, cough, headache, dizziness, drowsiness.	local exhaust or respiratory protection.	fresh air, rest, call a doctor.
Skin: redness.	gloves.	remove contaminated clothing, flush skin with water or shower.
Eyes: redness, pain.	goggles, or combined eye and respiratory protection.	flush with water, take to a doctor.
Ingestion: abdominal cramps, headache, nausea, constipation.		rinse mouth, call a doctor or take to hospital.

SPILLAGE	STORAGE	LABELING / NFPA
clean up spillage, carefully collect remainder; (additional individual protection: P2 respirator).	keep separate from combustible substances, reducing agents and hydrochloric acid.	R: 20/22-23 S: 13-20/21 Note A ✖ Harmful

NOTES
✳ TLV equals that of lead. Do not take work clothing home. Depending on the degree of exposure, regular medical checkups are advisable. Lead level in blood is highly recommended as a yardstick for exposure.

CAS-No: [138-86-3]
4-isopropenyl-1-methyl-1-cyclohexene
cajeputene
dipentene
dl-p-mentha-1,8-diene

$CH_3C_6H_8C_3H_5$

LIMONENE

PHYSICAL PROPERTIES		IMPORTANT CHARACTERISTICS
Boiling point, °C	176	**COLORLESS LIQUID WITH CHARACTERISTIC ODOR**
Melting point, °C	−96	Vapor mixes readily with air. Presumed to be able to form peroxides and polymerize. Flow,
Flash point, °C	45	agitation etc. can cause build-up of electrostatic charge due to liquid's low conductivity. Reacts
Auto-ignition temperature, °C	255	violently with strong oxidants.
Relative density (water = 1)	0.8	
Relative vapor density (air = 1)	4.7	TLV-TWA not available
Relative density at 20 °C of saturated mixture		
vapor/air (air = 1)	1.0	**Absorption route:** Can enter the body by inhalation or ingestion. Harmful atmospheric
Vapor pressure, mm Hg at 20 °C	1.6	concentrations build up very slowly, if at all, on evaporation at approx. 20°C, but much more
Solubility in water	none	rapidly in aerosol form.
Explosive limits, vol% in air	0.7-6.1	**Immediate effects:** Irritates the eyes, skin and respiratory tract.
Electrical conductivity, pS/m	<10⁴	**Effects of prolonged/repeated exposure:** Prolonged or repeated contact can cause skin
Relative molecular mass	136	disorders.
Gross formula:	$C_{10}H_{16}$	

HAZARDS/SYMPTOMS	PREVENTIVE MEASURES	FIRE EXTINGUISHING/FIRST AID
Fire: flammable.	keep away from open flame and sparks, no smoking.	powder, AFFF, foam, carbon dioxide, (halons).
Explosion: above 45°C: forms explosive air-vapor mixtures.	above 45°C: sealed machinery, ventilation, explosion-proof electrical equipment, grounding.	in case of fire: keep tanks/drums cool by spraying with water.
Inhalation: sore throat, cough, shortness of breath, dizziness, nausea.	ventilation.	fresh air, rest, call a doctor.
Skin: redness, pain.	gloves.	remove contaminated clothing, wash skin with soap and water, consult a doctor if necessary.
Eyes: redness, pain.	safety glasses.	flush with water, take to a doctor if necessary.
Ingestion: abdominal pain, diarrhea, dizziness, nausea, vomiting.		rinse mouth, call a doctor.

SPILLAGE	STORAGE	LABELING / NFPA
collect leakage in non-sealable containers, soak up with sand or other inert absorbent and remove to safe place.	keep in cool, fireproof place, separate from oxidants; add an inhibitor.	R: 10-38 S: 28 ✖ Irritant

NOTES

LITHIUM

PHYSICAL PROPERTIES	IMPORTANT CHARACTERISTICS
Boiling point, °C 1347 Melting point, °C 180 Relative density (water = 1) 0.53 Vapor pressure, mm Hg at 723 °C 0.99 Solubility in water reaction Relative atomic mass 6.9	**SOFT SILVER-WHITE METAL** In finely dispersed form, can ignite spontaneously. Reacts violently with water, giving off → *hydrogen* and corrosive fumes. Reacts violently with oxidizing substances, acids and many other compounds, with risk of fire and explosion. Reacts with air to form lithium hydroxide fumes.
	TLV-TWA not available
	Absorption route: Can enter the body by inhalation or ingestion. **Immediate effects:** Corrosive to the eyes, skin and respiratory tract. Inhalation of vapor/fumes can cause severe breathing difficulties (lung edema). In serious cases risk of death. **Effects of prolonged/repeated exposure:** Can cause kidney damage.
Gross formula: Li	

HAZARDS/SYMPTOMS	PREVENTIVE MEASURES	FIRE EXTINGUISHING/FIRST AID
Fire: extremely flammable, many chemical reactions can cause fire and explosion.	keep away from open flame and sparks, no smoking; avoid contact with water.	dry sand, special powder extinguisher.
Explosion: finely dispersed particles form explosive mixtures on contact with air.		
	STRICT HYGIENE	
Inhalation: *corrosive*; sore throat, cough, shortness of breath, severe breathing difficulties.	local exhaust or respiratory protection.	fresh air, rest, place in half-sitting position, take to hospital.
Skin: *corrosive*; redness, pain, burns.	gloves.	remove contaminated clothing, flush skin with water or shower, refer to a doctor.
Eyes: *corrosive*; redness, pain.	close-fitting safety glasses, or combined eye and respiratory protection.	flush with water, take to a doctor.
Ingestion: *corrosive*; sore throat, abdominal pain, vomiting.	do not eat, drink or smoke while working.	rinse mouth, immediately take to hospital.

SPILLAGE	STORAGE	LABELING / NFPA
clean up spillage, soak up remainder with dry sand or other inert absorbent and remove to safe place.	keep in fireproof, dry place; under kerosene.	R: 14/15-34 S: 8-43 Flammable Corrosive

NOTES
Flush contaminated clothing with water (fire hazard). Reacts violently with extinguishing agents such as carbon dioxide, bicarbonate powder and halons. Depending on the degree of exposure, regular medical checkups are advisable. Lung edema symptoms usually develop several hours later and are aggravated by physical exertion: rest and hospitalization essential. As first aid, a doctor or authorized person should consider administering a corticosteroid spray.

Transport Emergency Card TEC(R)-43G01

UN-number: 1415

aluminum lithium hydride
LAH
lithium aluminohydride
lithium tetrahydroaluminate

LITHIUM ALUMINUM HYDRIDE

PHYSICAL PROPERTIES	IMPORTANT CHARACTERISTICS
Melting point (decomposes), °C 125 Relative density (water = 1) 0.9 Solubility in water reaction Relative molecular mass 37.9	**WHITE POWDER OR WHITE CRYSTALS** Can ignite spontaneously on exposure to air due to friction etc. Decomposes when heated to 125°C, forming aluminum, lithium hydride and flammable gas (→ *hydrogen*). Reacts violently with water, acids, alcohols, ethers and many other compounds, with risk of fire and explosion.

TLV-TWA	not available

Absorption route: Can enter the body by inhalation or ingestion.
Immediate effects: Corrosive to the eyes, skin and respiratory tract. Affects the nervous system. Inhalation can cause lung edema. In serious cases risk of death.

Gross formula: AlH₄Li

HAZARDS/SYMPTOMS	PREVENTIVE MEASURES	FIRE EXTINGUISHING/FIRST AID
Fire: extremely flammable, many chemical reactions can cause fire and explosion.	keep away from open flame and sparks, no smoking.	dry sand, special powder extinguisher, DO NOT USE WATER-BASED EXTINGUISHERS.
Explosion: finely dispersed particles form explosive mixtures on contact with air.	keep dust from accumulating, do not subject to shocks or friction.	
	KEEP DUST UNDER CONTROL	
Inhalation: *corrosive*; sore throat, cough, shortness of breath.	local exhaust or respiratory protection.	fresh air, rest, call a doctor.
Skin: *corrosive*; redness, pain, serious burns.	gloves, protective clothing.	remove contaminated clothing, flush skin with water or shower, call a doctor if necessary.
Eyes: *corrosive*; redness, pain, impaired vision.	close-fitting safety glasses.	flush with water, take to a doctor.
Ingestion: *corrosive*; sore throat, abdominal pain, vomiting.	do not eat, drink or smoke while working.	rinse mouth, take immediately to hospital.

SPILLAGE	STORAGE	LABELING / NFPA
evacuate area, call in an expert, clean up spillage and remove to safe place in sealed containers; (additional individual protection: P3 respirator).	keep in fireproof place, separate from all other substances.	R: 15 S: 7/8-24/25-43 🔥 Flammable

NFPA diamond: 1 (top), 2 (right), 3 (left), W (bottom)

NOTES
TLV for lithium hydride: 0.025mg/m³. Lung edema symptoms usually develop several hours later and are aggravated by physical exertion: rest and hospitalization essential. As first aid, a doctor or authorized person should consider administering a corticosteroid spray.

LITHIUMCARBONATE

PHYSICAL PROPERTIES	IMPORTANT CHARACTERISTICS
Decomposes below boiling point, °C 1200 Melting point, °C 723 Relative density (water = 1) 2.1 Solubility in water, g/100 ml 1.3 Relative molecular mass 73.9	**WHITE CRYSTALS** In aqueous solution is a strong base which reacts violently with acids and corrodes aluminum, zinc etc.
	TLV-TWA not available
	Absorption route: Can enter the body by inhalation or ingestion. Evaporation negligible at 20°C, but harmful concentrations of airborne particles can build up rapidly. **Immediate effects:** Corrosive to the eyes and respiratory tract. In substantial concentrations can impair consciousness. In serious cases risk of seizures and unconsciousness. **Effects of prolonged/repeated exposure:** Can cause kidney damage, heart rate disorders and disorders of the central nervous system, due to disturbances of the sodium/potassium balance.
Gross formula: CLi$_2$O$_3$	

HAZARDS/SYMPTOMS	PREVENTIVE MEASURES	FIRE EXTINGUISHING/FIRST AID
Fire: non-combustible.		in case of fire in immediate vicinity: preferably not water.
	KEEP DUST UNDER CONTROL	
Inhalation: sore throat, cough.	local exhaust or respiratory protection.	fresh air, rest, call a doctor.
Skin: redness.	gloves.	remove contaminated clothing, flush skin with water or shower.
Eyes: pain, impaired vision.	goggles.	flush with water, take to a doctor.
Ingestion: muscular spasms, disorientation, vomiting, unconsciousness.		rinse mouth, take to hospital.

SPILLAGE	STORAGE	LABELING / NFPA
clean up spillage (additional individual protection: P2 respirator).	keep separate from acids.	

NOTES
Depending on the degree of exposure, regular medical checkups are advisable.

LITHIUM HYDROXIDE

PHYSICAL PROPERTIES	IMPORTANT CHARACTERISTICS
Boiling point (decomposes), °C 924 Melting point, °C 450 Relative density (water = 1) 1.5 Solubility in water, g/100 ml at 20 °C 12 Relative molecular mass 24.0	**COLORLESS HYGROSCOPIC CRYSTALS** In aqueous solution is a strong base which reacts violently with acids and corrodes aluminum, zinc etc.
	TLV-TWA not available
	Absorption route: Can enter the body by inhalation or ingestion. Evaporation negligible at 20°C, but harmful concentrations of airborne particles can build up rapidly. **Immediate effects:** Corrosive to the eyes, skin and respiratory tract. Inhalation can cause lung edema. In serious cases risk of unconsciousness. **Effects of prolonged/repeated exposure:** Kidney damage and disorders of the heart and central nervous system can result from impairment of the sodium/potassium balance in the body.
Gross formula: HLiO	

HAZARDS/SYMPTOMS	PREVENTIVE MEASURES	FIRE EXTINGUISHING/FIRST AID
Fire: non-combustible.		in case of fire in immediate vicinity: use any extinguishing agent.
	STRICT HYGIENE, KEEP DUST UNDER CONTROL	
Inhalation: *corrosive*; sore throat, cough, shortness of breath, severe breathing difficulties; see also 'ingestion'.	local exhaust or respiratory protection.	fresh air, rest, place in half-sitting position, take to hospital.
Skin: *corrosive*; redness, pain, burns.	gloves, protective clothing.	remove contaminated clothing, flush skin with water or shower, call a doctor if necessary.
Eyes: *corrosive*; redness, pain, impaired vision.	face shield.	flush with water, take to a doctor.
Ingestion: *corrosive*; abdominal pain, diarrhea, vomiting, unconsciousness, muscle weakness, disorientation.		rinse mouth, take immediately to hospital.

SPILLAGE	STORAGE	LABELING / NFPA
clean up spillage, flush away remainder with water; (additional individual protection: P1 spirator).	keep dry, separate from acids.	0 / 3 / 0

NOTES
Depending on the degree of exposure, regular medical checkups are advisable. Lung edema symptoms generally take a few hours to become apparent and are aggravated by physical exertion: rest and hospitalization essential. As first aid, a doctor or authorized person should consider administering a corticosteroid spray.

Transport Emergency Card TEC(R)-751

HI: 80; UN-number: 2680

LPG
(cylinder)

PHYSICAL PROPERTIES	IMPORTANT CHARACTERISTICS
Boiling point, °C ca. −20 Melting point, °C ca. −160 Flash point, °C flammable gas Auto-ignition temperature, °C ca. 400 Relative density (water = 1) ca. 0.6 Relative vapor density (air = 1) ca. 1.8 Vapor pressure, mm Hg at 20°C 3040-6080 Solubility in water none Explosive limits, vol% in air 1.5-10 Minimum ignition energy, mJ 0.25 Electrical conductivity, pS/m < 10⁴ Relative molecular mass ca. 50	**COLORLESS COMPRESSED LIQUEFIED GAS** Gas is heavier than air and spreads at ground level, with risk of ignition at a distance. Flow, agitation etc. can cause build-up of electrostatic charge due to liquid's low conductivity. Do not use compressed air when filling, emptying or processing.

TLV-TWA	1000 ppm	1800 mg/m³

Absorption route: Can enter the body by inhalation.
Immediate effects: Liquid can cause frostbite due to rapid evaporation. In substantial concentrations can impair consciousness.

HAZARDS/SYMPTOMS	PREVENTIVE MEASURES	FIRE EXTINGUISHING/FIRST AID
Fire: extremely flammable.	keep away from open flame and sparks, no smoking.	shut off supply; if impossible and no danger to surrounding area, allow fire to burn itself out; otherwise extinguish with powder, carbon dioxide, (halons).
Explosion: forms explosive air-gas mixtures.	sealed machinery, ventilation, explosion-proof electrical equipment and lighting, grounding, non-sparking tools.	in case of fire: keep cylinders cool by spraying with water, fight fire from sheltered location.
Inhalation: shortness of breath, headache, drowsiness.	ventilation, local exhaust or respiratory protection.	fresh air, rest, artificial respiration if necessary, call a doctor.
Skin: *in case of frostbite*: redness, pain, blisters, sores.	insulating gloves.	*in case of frostbite*: DO NOT remove clothing, flush skin with water or shower, take to a doctor.
Eyes: *in case of frostbite*: redness, pain, impaired vision.	face shield.	flush with water, take to a doctor if necessary.

SPILLAGE	STORAGE	LABELING / NFPA
evacuate area, call in an expert, ventilate, under no circumstances spray liquid with water; (additional individual protection: breathing apparatus).	keep in cool, fireproof place.	NFPA diamond: 4 (top), 1 (left), 0 (right)

NOTES

A mercaptan is added to many commercial products as odorizer. High atmospheric concentrations, e.g. in poorly ventilated spaces, can cause oxygen deficiency, with risk of unconsciousness. Turn leaking cylinder so that leak is on top to prevent LPG escaping.

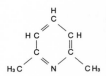

2,6-LUTIDINE

PHYSICAL PROPERTIES		IMPORTANT CHARACTERISTICS
Boiling point, °C	144	**COLORLESS OILY LIQUID WITH CHARACTERISTIC ODOR**
Melting point, °C	−6	Vapor mixes readily with air. Decomposes when heated or burned, giving off toxic → *nitrous*
Flash point, °C	38	*vapors*. Reacts with strong oxidants.
Relative density (water = 1)	0.93	
Relative vapor density (air = 1)	3.7	TLV-TWA not available
Vapor pressure, mm Hg at 20 °C	ca. 3.6	
Solubility in water, g/100 ml at 20 °C	40	**Absorption route:** Can enter the body by inhalation or ingestion or through the skin. Insufficient
Explosive limits, vol% in air	?-?	data on the rate at which harmful concentrations can build up.
Relative molecular mass	107.2	**Immediate effects:** Irritates the eyes, skin and respiratory tract. In substantial concentrations
		can impair consciousness.
		Effects of prolonged/repeated exposure: Can cause liver and kidney damage.
Gross formula:	C_7H_9N	

HAZARDS/SYMPTOMS	PREVENTIVE MEASURES	FIRE EXTINGUISHING/FIRST AID
Fire: flammable.	keep away from open flame and sparks, no smoking.	powder, alcohol-resistant foam, water spray, carbon dioxide, (halons).
Explosion: above 38°C: forms explosive air-vapor mixtures.	above 38°C: sealed machinery, ventilation, explosion-proof electrical equipment.	in case of fire: keep tanks/drums cool by spraying with water.
Inhalation: headache, dizziness, nausea, vomiting, drowsiness.	ventilation, local exhaust or respiratory protection.	fresh air, rest, call a doctor.
Skin: *is absorbed*; redness, pain; see also 'Inhalation'.	gloves.	remove contaminated clothing, flush skin and wash thoroughly with soap and water, call a doctor.
Eyes: redness, pain.	safety glasses.	flush with water, take to a doctor.
Ingestion: sore throat, abdominal pain, diarrhea.	do not eat, drink or smoke while working.	rinse mouth, call a doctor or take to hospital.

SPILLAGE	STORAGE	LABELING / NFPA
collect leakage in sealable containers, soak up with sand or other inert absorbent and remove to safe place; (additional individual protection: breathing apparatus).	keep in fireproof place; keep separate from oxidants.	

NOTES
Solubility in water decreases with increasing temperature. Explosive limits not given in the literature. Depending on the degree of exposure, regular medical checkups are advisable.

Transport Emergency Card TEC(R)-61G04

MAGNESIUM

PHYSICAL PROPERTIES	IMPORTANT CHARACTERISTICS
Boiling point, °C 1100 Melting point, °C 650 Auto-ignition temperature, °C 550 Relative vapor density (air = 1) 1.74 Vapor pressure, mm Hg at 621 °C 0.76 Solubility in water reaction Relative molecular mass 24.3	**SILVER-WHITE SHINING METAL OR METAL POWDER** Magnesium particles can be charged electrostatically and can ignite due to friction. Burns with an intense flame. Strong reducing agent which reacts violently with oxidants. Reacts violently with many substances, with risk of fire and explosion. Reacts with acids, giving off flammable → hydrogen.
	TLV-TWA not available
	Absorption route: Can enter the body by inhalation or ingestion. **Immediate effects:** Finely dispersed substance irritates the eyes and respiratory tract.
Gross formula: Mg	

HAZARDS/SYMPTOMS	PREVENTIVE MEASURES	FIRE EXTINGUISHING/FIRST AID
Fire: combustible, in powder form extremely flammable.	keep away from open flame and sparks, no smoking; avoid contact with water, acids, halogens and various other substances.	special powder extinguisher, dry sand, NO other extinguishing agents, DO NOT USE WATER-BASED EXTINGUISHERS.
Explosion: finely dispersed particles form explosive mixtures on contact with air.	do not subject to shocks or friction.	
	KEEP DUST UNDER CONTROL	
Inhalation: sore throat, cough.	local exhaust or respiratory protection.	fresh air, rest.
Skin: 1).		flush wound immediately with water, then remove contaminated clothing, take to a doctor.
Eyes: redness, pain.	goggles.	flush with water, take to a doctor.
Ingestion: abdominal pain, diarrhea.		rinse mouth, call a doctor.

SPILLAGE	STORAGE	LABELING / NFPA
clean up spillage, carefully collect remainder.	keep in dry, fireproof place; keep separate from oxidants and strong acids.	R: 11-15 (see 'Notes') S: 7/8-43 🔥 Flammable

NOTES

1)Wounds contaminated with magnesium take a long time to heal. Reacts violently with extinguishing agents such as water, halons, carbon dioxide and powder. R: 15 and 17 need to be mentioned on the label of unstabilized magnesium powder.

HI: 43; UN-number: 1418 (1869)

MAGNESIUM CHLORIDE

PHYSICAL PROPERTIES		IMPORTANT CHARACTERISTICS
Boiling point, °C	1412	**WHITE HYGROSCOPIC CRYSTALS**
Melting point, °C	708	Decomposes when heated slowly above 300°C, giving off toxic → *chlorine* gas. A considerable
Relative density (water = 1)	2.3	amount of heat is evolved when dissolved in water.
Solubility in water, g/100 ml at 20 °C	54	
Relative molecular mass	95.2	TLV-TWA not available
		Absorption route: Can enter the body by inhalation or ingestion.
		Immediate effects: Irritates the eyes and respiratory tract.
Gross formula:	Cl$_2$Mg	

HAZARDS/SYMPTOMS	PREVENTIVE MEASURES	FIRE EXTINGUISHING/FIRST AID
Fire: non-combustible.		in case of fire in immediate vicinity: preferably not water.
Inhalation: cough.	local exhaust or respiratory protection.	fresh air, rest.
Skin:	gloves.	remove contaminated clothing, flush skin with water or shower.
Eyes: redness, pain.	safety glasses.	flush with water, take to a doctor if necessary.
Ingestion: diarrhea, vomiting.		rinse mouth, call a doctor.

SPILLAGE	STORAGE	LABELING / NFPA
clean up spillage, flush away remainder with water.	keep dry.	

NOTES

MAGNESIUMHYDROXIDE

PHYSICAL PROPERTIES	IMPORTANT CHARACTERISTICS	
Melting point (decomposes), °C 350 Relative density (water = 1) 2.4 Solubility in water none Relative molecular mass 58.3	**WHITE POWDER** Decomposes when heated, forming → *magnesium oxide* and water vapor. Reacts with acids, evolving heat.	
	TLV-TWA not available	
	Absorption route: Can enter the body by inhalation or ingestion. Evaporation negligible at 20°C, but unpleasant atmospheric concentrations can build up rapidly in aerosol form. **Immediate effects:** Irritates the eyes and respiratory tract. Inhalation can cause metal fume fever.	
Gross formula: H_2MgO_2		

HAZARDS/SYMPTOMS	PREVENTIVE MEASURES	FIRE EXTINGUISHING/FIRST AID
Fire: non-combustible.		in case of fire in immediate vicinity: use any extinguishing agent.
Inhalation: sore throat, cough, headache, nausea, fever.	local exhaust or respiratory protection.	fresh air, rest, call a doctor.
Skin: redness.	gloves.	remove contaminated clothing, flush skin with water or shower.
Eyes: redness, pain.	goggles.	flush with water, take to a doctor if necessary.
Ingestion: abdominal pain.		rinse mouth, call a doctor.

SPILLAGE	STORAGE	LABELING / NFPA
clean up spillage, flush away remainder with water; (additional individual protection: P2 respirator).	keep separate from acids.	

NOTES

MAGNESIUM NITRATE

PHYSICAL PROPERTIES	IMPORTANT CHARACTERISTICS
Boiling point (decomposes), °C — 330 Melting point, °C — 89 Relative density (water = 1) — 1.5 Solubility in water, g/100 ml — 125 Relative molecular mass — 256.4	**WHITE DELIQUESCENT CRYSTALS** Decomposes when heated, giving off → *oxygen*, with increased risk of fire, and when heated above 330°C, giving off toxic and corrosive → *nitrous vapors*. Strong oxidant which reacts violently with combustible substances and reducing agents. Reacts violently with dimethylformamide, with risk of fire and explosion.

	TLV-TWA	not available

Absorption route: Can enter the body by inhalation or ingestion.
Immediate effects: Irritates the eyes, skin and respiratory tract. In serious cases risk of seizures.
Effects of prolonged/repeated exposure: Can affect the blood if ingested.

Gross formula: $MgN_2O_6.6H_2O$

HAZARDS/SYMPTOMS	PREVENTIVE MEASURES	FIRE EXTINGUISHING/FIRST AID
Fire: not combustible but enhances combustion of other substances.	avoid contact with combustible substances.	in case of fire in immediate vicinity: use any extinguishing agent.
Explosion: mixing with strong reducing agents can cause explosion.		
	KEEP DUST UNDER CONTROL	
Inhalation: cough, shortness of breath.	local exhaust or respiratory protection.	fresh air, rest, place in half-sitting position, call a doctor.
Skin: redness, pain.	gloves, protective clothing.	remove contaminated clothing, flush skin with water or shower.
Eyes: redness, pain.	face shield.	flush with water, take to a doctor.
Ingestion: abdominal cramps, feeling of weakness, muscle cramps, blue skin.		rinse mouth, call a doctor or take to hospital.

SPILLAGE	STORAGE	LABELING / NFPA
clean up spillage, flush away remainder with water.	keep dry, separate from combustible substances and reducing agents.	

NOTES

Apparent melting point due to loss of water of crystallization. Flush contaminated clothing with water (fire hazard). Effects on blood due to formation of methemoglobin. Special first aid required in the event of poisoning: antidotes must be available (with instructions).

UN-number: 1474

MAGNESIUM OXIDE

PHYSICAL PROPERTIES		IMPORTANT CHARACTERISTICS	
Boiling point, °C	3600	**WHITE POWDER**	
Melting point, °C	2852	Reacts with strong acids.	
Relative density (water = 1)	3.6		
Solubility in water	none	TLV-TWA	10 mg/m³ ✱
Relative molecular mass	40.3		
		Absorption route: Can enter the body by inhalation or ingestion. Evaporation negligible at 20°C, but unpleasant concentrations of airborne particles can build up. **Immediate effects:** Irritates the eyes and respiratory tract.	
Gross formula:	MgO		

HAZARDS/SYMPTOMS	PREVENTIVE MEASURES	FIRE EXTINGUISHING/FIRST AID
Fire: non-combustible.		in case of fire in immediate vicinity: use any extinguishing agent.
Inhalation: sore throat, cough, pain behind breast bone, fever.	local exhaust or respiratory protection.	
Skin:		remove contaminated clothing, flush skin with water or shower.
Eyes: redness, pain.	goggles.	flush with water, take to a doctor if necessary.
Ingestion: diarrhea.		rinse mouth, call a doctor.

SPILLAGE	STORAGE	LABELING / NFPA
clean up spillage, flush away remainder with water; (additional individual protection: P2 respirator).		

NOTES
Excessive exposure to MgO vapor produced can cause metal fume fever. ✱ TLV equals that of magnesium oxide fume.

MAGNESIUM SULFATE

PHYSICAL PROPERTIES	IMPORTANT CHARACTERISTICS	
Melting point (decomposes), °C 1124 Relative density (water = 1) 2.7 Solubility in water, g/100 ml at 0 °C 26 Relative molecular mass 120.4	**WHITE POWDER** Decomposes when heated above 1100°C, giving off toxic and corrosive vapors (→ *sulfur dioxide*).	
	TLV-TWA not available	
	Absorption route: Can enter the body by inhalation or ingestion. Evaporation negligible at 20°C, but harmful concentrations of airborne particles can build up rapidly. **Immediate effects:** Irritates the eyes and respiratory tract.	
Gross formula: MgO₄S		

HAZARDS/SYMPTOMS	PREVENTIVE MEASURES	FIRE EXTINGUISHING/FIRST AID
Fire: non-combustible.		in case of fire in immediate vicinity: use any extinguishing agent.
Inhalation: sore throat, cough.	local exhaust or respiratory protection.	fresh air, rest.
Skin:	gloves.	
Eyes: redness, pain.	safety glasses.	flush with water, take to a doctor if necessary.
Ingestion: abdominal pain, diarrhea.		rinse mouth, call a doctor if necessary.

SPILLAGE	STORAGE	LABELING / NFPA
clean up spillage, flush away remainder with water; (additional individual protection: P1 respirator).		

NOTES

CAS-No: [141-82-2]
cis-butanedioic acid
maleinic acid

$CH_2(COOH)_2$

MALEIC ACID

PHYSICAL PROPERTIES		IMPORTANT CHARACTERISTICS	
Melting point (decomposes), °C	135	**COLORLESS CRYSTALS**	
Relative density (water = 1)	1.6	Decomposes when heated, giving off toxic gases. In aqueous solution is a strong acid which reacts violently with bases and is corrosive.	
Vapor pressure, mm Hg at 20 °C	?		
Solubility in water, g/100 ml at 20 °C	154		
Relative molecular mass	104.1	**TLV-TWA**	not available
Log P octanol/water	−0.7		
		Absorption route: Can enter the body by inhalation or ingestion. Evaporation negligible at 20°C, but harmful concentrations of airborne particles can build up rapidly. **Immediate effects:** Corrosive to the eyes, skin and respiratory tract.	
Gross formula:	$C_3H_4O_4$		

HAZARDS/SYMPTOMS	PREVENTIVE MEASURES	FIRE EXTINGUISHING/FIRST AID
Fire: combustible.	keep away from open flame, no smoking.	water spray, powder.
	KEEP DUST UNDER CONTROL	
Inhalation: *corrosive*; sore throat, cough, shortness of breath.	local exhaust or respiratory protection.	fresh air, rest, place in half-sitting position, call a doctor.
Skin: *corrosive*; redness, pain.	gloves.	remove contaminated clothing, flush skin with water or shower, call a doctor if necessary.
Eyes: *corrosive*; redness, pain, impaired vision.	goggles.	flush with water, take to a doctor.
Ingestion: *corrosive*; sore throat, abdominal pain, vomiting.		rinse mouth, take to hospital.

SPILLAGE	STORAGE	LABELING / NFPA
clean up spillage, flush away remainder with water; (additional individual protection: P2 respirator).	keep separate from strong bases.	

NOTES
Log P octanol/water is estimated.

CAS-No: [108-31-6]
cis-butenedioic anhydride
dihydro-2,5-dioxofuran
2,5-furandione
MAA

MALEIC ANHYDRIDE

PHYSICAL PROPERTIES		IMPORTANT CHARACTERISTICS	
Boiling point, °C	202	**COLORLESS CRYSTALLINE NEEDLES WITH PUNGENT ODOR**	
Melting point, °C	53	Sublimates readily. In aqueous solution is a strong acid which reacts violently with bases and is	
Flash point, °C	102	corrosive. Reacts slowly with water, giving off → maleinic acid. Reacts with strong oxidants.	
Auto-ignition temperature, °C	380		
Relative density (water = 1)	1.4	TLV-TWA 0.25 ppm 1 mg/m³	
Relative density at 20 °C of saturated mixture			
vapor/air (air = 1)	1.0	**Absorption route:** Can enter the body by inhalation or ingestion. Harmful atmospheric	
Vapor pressure, mm Hg at 20 °C	4.5×10^{-5}	concentrations build up very slowly, if at all, on evaporation at approx. 20°C - much more rapidly if	
Solubility in water, g/100 ml at 25 °C	79	moderately heated or if concentrations of airborne particles are allowed to build up.	
Relative molecular mass	98.1	**Immediate effects:** Corrosive to the eyes, skin and respiratory tract. Inhalation can cause lung	
		edema. In serious cases risk of death.	
		Effects of prolonged/repeated exposure: Prolonged or repeated contact can cause skin	
		disorders.	
Gross formula:	$C_4H_2O_3$		

HAZARDS/SYMPTOMS	PREVENTIVE MEASURES	FIRE EXTINGUISHING/FIRST AID
Fire: combustible.	keep away from open flame, no smoking.	powder, alcohol-resistant foam, water spray, carbon dioxide, (halons).
Explosion: finely dispersed particles form explosive mixtures on contact with air.	keep dust from accumulating, sealed machinery, explosion-proof electrical equipment and lighting, grounding.	
	STRICT HYGIENE	IN ALL CASES CALL IN A DOCTOR
Inhalation: *corrosive*; sore throat, cough, severe breathing difficulties.	ventilation, local exhaust or respiratory protection.	fresh air, rest, place in half-sitting position, take to hospital.
Skin: *corrosive*; redness, pain, serious burns.	gloves.	remove contaminated clothing, flush skin with water or shower, refer to a doctor.
Eyes: *corrosive*; redness, pain, impaired vision.	face shield.	flush with water, take to a doctor.
Ingestion: *corrosive*; sore throat, abdominal pain, vomiting.		rinse mouth, take immediately to hospital.

SPILLAGE	STORAGE	LABELING / NFPA
clean up spillage, flush away remainder with water; (additional individual protection: P2 respirator).	keep dry, separate from strong bases; ventilate at floor level.	R: 22-36/37/38-42 S: 22-28-39 ✖ Irritant

NOTES

Odor limit is above TLV. Lung edema symptoms usually develop several hours later and are aggravated by physical exertion: rest and hospitalization essential. As first aid, a doctor or authorized person should consider administering a corticosteroid spray. Use airtight packaging. Also supplied in molten form. Saturated vapor pressure rises sharply, to 1.5mm Hg at 50°C.

Transport Emergency Card TEC(R)-824

HI: 80; UN-number: 2215

CAS-No: [110-16-7]
cis-butenedioic acid
maleic acid
malenic acid
toxilic acid

HOOCHC = CHCOOH

MALEINIC ACID

PHYSICAL PROPERTIES	IMPORTANT CHARACTERISTICS
Decomposes below boiling point, °C 135 Melting point, °C 130 Flash point, °C ca. 100 Relative density (water = 1) 1.6 Relative vapor density (air = 1) 4.0 Vapor pressure, mm Hg at 20 °C <0.076 Solubility in water, g/100 ml at 20 °C 78 Relative molecular mass 116.1 Log P octanol/water −0.5	**WHITE CRYSTALS OR WHITE POWDER WITH CHARACTERISTIC ODOR** In dry state can form electrostatic charge if stirred, transported pneumatically, poured etc. In aqueous solution is a medium strong acid. Reacts violently with strong oxidants and strong bases.
	TLV-TWA not available
	Absorption route: Can enter the body by inhalation or ingestion. Evaporation negligible at 20°C, but harmful concentrations of airborne particles can build up rapidly. **Immediate effects:** Irritates the eyes, skin and respiratory tract.
Gross formula: $C_4H_4O_4$	

HAZARDS/SYMPTOMS	PREVENTIVE MEASURES	FIRE EXTINGUISHING/FIRST AID
Fire: combustible.	keep away from open flame, no smoking.	water spray, powder.
Explosion:	keep dust from accumulating, sealed machinery, explosion-proof electrical equipment and lighting, grounding.	
	KEEP DUST UNDER CONTROL	
Inhalation: sore throat, cough, shortness of breath.	local exhaust or respiratory protection.	fresh air, rest, call a doctor if necessary.
Skin: redness, pain.	gloves.	remove contaminated clothing, flush skin with water or shower, refer to a doctor.
Eyes: redness, pain.	face shield.	flush with water, take to a doctor.
Ingestion: sore throat, abdominal pain, vomiting.		rinse mouth, call a doctor.

SPILLAGE	STORAGE	LABELING / NFPA
clean up spillage, flush away remainder with water; (additional individual protection: P2 respirator).	keep separate from oxidants and strong bases.	R: 22-36/37/38 S: 26-28-37 ✖ Irritant

NOTES
When heated above 100°C is partially converted to transisomer → *fumaric acid*. Log P octanol/water is estimated.

CAS-No: [90-64-2]
amygdalic acid
benzoglycolic acid
α-hydroxy-α-toluic acid
α-phenylhydroxyacetic acid
uromaline

$C_6H_5CHOHCOOH$

MANDELIC ACID

PHYSICAL PROPERTIES	IMPORTANT CHARACTERISTICS
Decomposes below boiling point, °C ? Melting point, °C 121 Relative density (water = 1) 1.3 Vapor pressure, mm Hg at 20 °C <0.076 Solubility in water, g/100 ml at 20 °C 16 Relative molecular mass 152.1 Log P octanol/water 0.2 Gross formula: $C_8H_8O_3$	**WHITE CRYSTALS OR POWDER WITH CHARACTERISTIC ODOR, TURNING BROWN ON EXPOSURE TO AIR** Decomposes when heated above melting point, giving off harmful vapors. In aqueous solution is a medium strong acid. Reacts with strong bases, evolving heat. Reacts with strong oxidants. TLV-TWA not available **Absorption route:** Can enter the body by inhalation or ingestion. Evaporation negligible at 20°C, but harmful concentrations of airborne particles can build up rapidly. **Effects of prolonged/repeated exposure:** In large quantities can cause kidney damage.

HAZARDS/SYMPTOMS	PREVENTIVE MEASURES	FIRE EXTINGUISHING/FIRST AID
Fire: combustible.	keep away from open flame, no smoking.	water spray, powder.
Inhalation:	local exhaust or respiratory protection.	fresh air, rest.
Skin:	gloves.	
Eyes: redness.	goggles.	flush with water, take to a doctor if necessary.
Ingestion: diarrhea, nausea.		rinse mouth, call a doctor.

SPILLAGE	STORAGE	LABELING / NFPA
clean up spillage, flush away remainder with water; (additional individual protection: P2 respirator).	keep separate from oxidants and strong bases; ventilate at floor level.	

NOTES
Log P octanol/water is estimated.

MANGANESE ACETATE

PHYSICAL PROPERTIES	IMPORTANT CHARACTERISTICS	
Boiling point (decomposes), °C 323 Melting point, °C 80 Relative density (water = 1) 1.6 Solubility in water good Relative molecular mass 245	**LIGHT RED CRYSTALS** Decomposes when heated, giving off toxic and acid vapors.	
	TLV-TWA	5 mg/m³ ✷
	Absorption route: Can enter the body by inhalation or ingestion. Evaporation negligible at 20° C, but harmful concentrations of airborne particles can build up rapidly. **Immediate effects:** Irritates the eyes and respiratory tract. Affects the nervous system. Inhalation can cause metal fume fever. Contact with vapor or inhalation of particles can cause pneumonia. **Effects of prolonged/repeated exposure:** Can cause brain damage.	
Gross formula: $C_4H_6MnO_4.4H_2O$		

HAZARDS/SYMPTOMS	PREVENTIVE MEASURES	FIRE EXTINGUISHING/FIRST AID
Fire: non-combustible.		in case of fire in immediate vicinity: use any extinguishing agent.
	KEEP DUST UNDER CONTROL	
Inhalation: cough, shortness of breath, nausea, fever.	local exhaust or respiratory protection.	fresh air, rest, call a doctor.
Skin: redness.	gloves.	remove contaminated clothing, flush skin with water or shower.
Eyes: redness, pain.	goggles, or combined eye and respiratory protection.	flush with water, take to a doctor if necessary.
Ingestion: abdominal pain, nausea.	do not eat, drink or smoke while working.	rinse mouth, take immediately to hospital.

SPILLAGE	STORAGE	LABELING / NFPA
clean up spillage, carefully collect remainder; (additional individual protection: P2 respirator).		

NOTES
Apparent melting point due to loss of water of crystallization. ✷ TLV equals that of manganese (dust and compounds). TLV for manganese (fume): TLV-TWA 1mg/m³ and STEL 3mg/m³.

manganese binoxide
manganese black

MANGANESE DIOXIDE

PHYSICAL PROPERTIES		IMPORTANT CHARACTERISTICS	
Melting point (decomposes), °C	535	**DARK BROWN POWDER**	
Relative density (water = 1)	5.0	Decomposes when heated above 535°C, giving off manganese (III) oxide and → *oxygen*, with increased risk of fire. Strong oxidant which reacts violently with combustible substances and reducing agents; can explode when mixed with organic solids. Reacts with hydrochloric acid, giving off toxic gas (→ *chlorine*).	
Solubility in water	none		
Relative molecular mass	86.9		
		TLV-TWA	1 mg/m³ ✱
		STEL	3 mg/m³
		Absorption route: Can enter the body by inhalation or ingestion. Evaporation negligible at 20°C, but harmful concentrations of airborne particles can build up rapidly. **Immediate effects:** Irritates the eyes and respiratory tract. Can cause brain damage. Inhalation of particles can cause pneumonia.	
Gross formula:	MnO₂		

HAZARDS/SYMPTOMS	PREVENTIVE MEASURES	FIRE EXTINGUISHING/FIRST AID
Fire: not combustible but enhances combustion of other substances, many chemical reactions can cause fire and explosion.	avoid contact with combustible substances.	in case of fire in immediate vicinity: use any extinguishing agent.
	KEEP DUST UNDER CONTROL	
Inhalation: cough.	local exhaust or respiratory protection.	fresh air, rest.
Skin: redness.	gloves.	remove contaminated clothing, flush skin with water or shower.
Eyes: redness.	goggles, or combined eye and respiratory protection.	flush with water, take to a doctor if necessary.
Ingestion: abdominal pain, nausea.		rinse mouth, call a doctor or take to hospital.

SPILLAGE	STORAGE	LABELING / NFPA
clean up spillage, flush away remainder with water, do not use sawdust or other combustible absorbents; (additional individual protection: P2 respirator).	keep separate from combustible substances, reducing agents and hydrochloric acid.	R: 20/22 S: 25 ✖ Harmful

NOTES
✱ TLV equals that of manganese (fume). Flush contaminated clothing with water (fire hazard). Most common industrial hazard is chronic intoxication, with symptoms of central nervous system disorders.

MANGANOUS SULFATE
(monohydrate)

PHYSICAL PROPERTIES	IMPORTANT CHARACTERISTICS	
Decomposes below boiling point, °C 850 Melting point, °C 57-117 Relative density (water = 1) 3.0 Solubility in water, g/100 ml 100 Relative molecular mass 169.1	**PINK HYGROSCOPIC CRYSTALS** Decomposes when heated above 850°C, giving off corrosive and toxic vapors (→ *sulfur dioxide*, → *sulfur trioxide*, manganese oxides).	
	TLV-TWA	5 mg/m³ ✹
	Absorption route: Can enter the body by inhalation, through the skin or by ingestion. Evaporation negligible at 20°C, but harmful atmospheric concentrations can build up rapidly in aerosol form. **Immediate effects:** Irritates the eyes, skin and respiratory tract. Inhalation of powder can cause bronchitis, pharyngitis and pneumonia. Can affect the central nervous system, liver, kidneys and gonads. Can cause listlessness, sleepiness, speech difficulties, stiffness, tremors (Parkinson's disease). Can result in long-term invalidity. **Effects of prolonged/repeated exposure:** Inhalation of high concentrations of powder can cause lung disorders.	
Gross formula: $MnO_4S.H_2O$		

HAZARDS/SYMPTOMS	PREVENTIVE MEASURES	FIRE EXTINGUISHING/FIRST AID
Fire: non-combustible.		in case of fire in immediate vicinity: use any extinguishing agent.
	KEEP DUST UNDER CONTROL	IN ALL CASES CALL IN A DOCTOR
Inhalation: burning sensation, cough, difficulty breathing.	local exhaust or respiratory protection.	fresh air, rest, place in half-sitting position, consult a doctor.
Skin: *is absorbed*; redness, burning sensation.	gloves.	remove contaminated clothing, flush skin with water and wash with soap and water.
Eyes: redness, pain, impaired vision.	face shield.	flush thoroughly with water (remove contact lenses if easy), take to a doctor.
Ingestion: abdominal cramps, nausea, sore throat.		rinse mouth, immediately take to hospital.

SPILLAGE	STORAGE	LABELING / NFPA
clean up spillage, flush away remainder with water; (additional individual protection: P2 respirator).	keep dry.	

NOTES
Apparent melting point due to loss of water of crystallization. Other apparent melting points: $MnSO_4.4H_2O$ 30°C. Sulfate loses all its water of crystallization at 400-450°C. Melting point of anhydrous manganese sulfate approx. 700°C. ✹ TLV equals that of manganese (dust and compounds).

CAS-No: [108-78-1]
1,3,5-triazine-2,4,6-triamine

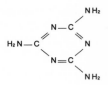

MELAMINE

PHYSICAL PROPERTIES	IMPORTANT CHARACTERISTICS
Melting point (decomposes), °C　345 Auto-ignition temperature, °C　>500 Relative density (water = 1)　1.6 Vapor pressure, mm Hg at 315 °C　51 Solubility in water, g/100 ml at 20 °C　0.3 Relative molecular mass　126.1 Gross formula:　$C_3H_6N_6$	**COLORLESS TO WHITE CRYSTALS** Sublimates. Toxic gases are a product of combustion. Decomposes when heated at a temperature as low as 270°C, giving off toxic vapors such as → *nitrous vapors*, → *hydrocyanic acid* and → *ammonia*. TLV-TWA　　　　　not available **Absorption route:** Can enter the body by inhalation or ingestion. Evaporation negligible at 20°C, but harmful concentrations of airborne particles can build up rapidly. **Immediate effects:** Irritates the eyes, skin and respiratory tract. **Effects of prolonged/repeated exposure:** Prolonged or repeated contact can cause skin disorders. Can cause kidney damage.

HAZARDS/SYMPTOMS	PREVENTIVE MEASURES	FIRE EXTINGUISHING/FIRST AID
Fire: hardly combustible.		in case of fire in immediate vicinity: use any extinguishing agent.
Explosion: finely dispersed particles form explosive mixtures on contact with air.	keep dust from accumulating; sealed machinery, explosion-proof electrical equipment and lighting, grounding.	
Inhalation: sore throat.	local exhaust or respiratory protection.	fresh air, rest.
Skin: redness.	gloves.	remove contaminated clothing, wash skin with soap and water.
Eyes: redness, pain.	goggles.	flush with water, take to a doctor if necessary.
Ingestion: abdominal pain.		rinse mouth, call a doctor.

SPILLAGE	STORAGE	LABELING / NFPA
clean up spillage, carefully collect remainder; (additional individual protection: P2 respirator).		

NOTES

2-MERCAPTOBENZOTHIAZOLE

PHYSICAL PROPERTIES	IMPORTANT CHARACTERISTICS
Boiling point (decomposes), °C — see notes Melting point, °C — 180 Auto-ignition temperature, °C — 628 Relative density (water = 1) — 1.4 Solubility in water — none Explosive limits, vol% in air — 15–? Relative molecular mass — 167.3	**LIGHT YELLOW POWDER OR CRYSTALS WITH CHARACTERISTIC ODOR** In dry state can form electrostatic charge if stirred, transported pneumatically, poured etc. Toxic corrosive gases are a product of combustion (→ *sulfur dioxide*, → *nitrous vapors*). Decomposes when heated or on contact with acids, giving off toxic gases (incl. → *hydrogen sulfide*). Reacts with oxidants.
	TLV-TWA not available
	Absorption route: Can enter the body by inhalation or ingestion. Evaporation negligible at 20°C, but harmful concentrations of airborne particles can build up rapidly. **Immediate effects:** Corrosive to the eyes, skin and respiratory tract. Inhalation of vapor/fumes can cause severe breathing difficulties (lung edema). **Effects of prolonged/repeated exposure:** Prolonged or repeated contact can cause skin disorders.
Gross formula: $C_7H_5NS_2$	

HAZARDS/SYMPTOMS	PREVENTIVE MEASURES	FIRE EXTINGUISHING/FIRST AID
Fire: combustible.	keep away from open flame, no smoking.	water spray, foam.
Explosion: finely dispersed particles form explosive mixtures on contact with air.	keep dust from accumulating, sealed machinery, explosion-proof electrical equipment and lighting, grounding.	
Inhalation: *corrosive*; sore throat, cough, severe breathing difficulties.	local exhaust or respiratory protection.	fresh air, rest, place in half-sitting position, take to hospital.
Skin: *corrosive*; redness, pain.	gloves.	remove contaminated clothing, flush skin with water or shower, take to a doctor.
Eyes: *corrosive*; redness, pain, impaired vision.	goggles, or combined eye and respiratory protection.	flush with water, take to a doctor.
Ingestion: *corrosive*; abdominal pain, diarrhea, vomiting.	do not eat, drink or smoke while working.	rinse mouth, call a doctor or take to hospital.

SPILLAGE	STORAGE	LABELING / NFPA
clean up spillage, carefully collect remainder; (additional individual protection: P2 respirator).	keep separate from acids.	

NOTES
Decomposition temperature not given in the literature. Lung edema symptoms usually develop several hours later and are aggravated by physical exertion: rest and hospitalization essential. As first aid, a doctor or authorized person should consider administering a corticosteroid spray.

Transport Emergency Card TEC(R)-80G09

CAS-No: [60-24-2]
1-ethanol-2-thiol
2-hydroxyethyl mercaptan
2-thioethanol
thioethylene glycol
thioglycol

$HSCH_2CH_2OH$

2-MERCAPTOETHANOL

PHYSICAL PROPERTIES	IMPORTANT CHARACTERISTICS	
Boiling point (decomposes), °C 157 Melting point, °C < −100 Flash point, °C 74 Relative density (water = 1) 1.1 Relative vapor density (air = 1) 2.7 Relative density at 20 °C of saturated mixture vapor/air (air = 1) 1.0 Vapor pressure, mm Hg at 20 °C 0.99 Solubility in water ∞ Relative molecular mass 78.1 Gross formula: C_2H_6OS	**COLORLESS LIQUID WITH CHARACTERISTIC ODOR** Vapor mixes readily with air. Decomposes when heated or burned, giving off toxic gases (→ *oxides of sulfur*). Reacts with oxidants. TLV-TWA not available **Absorption route:** Can enter the body by inhalation or ingestion. Insufficient data on the rate at which harmful concentrations can build up. **Immediate effects:** Irritates the eyes, skin and respiratory tract.	

HAZARDS/SYMPTOMS	PREVENTIVE MEASURES	FIRE EXTINGUISHING/FIRST AID
Fire: combustible.	keep away from open flame, no smoking.	powder, alcohol-resistant foam, water spray, carbon dioxide, (halons).
Explosion: above 74°C: forms explosive air-vapor mixtures.	above 74°C: sealed machinery, ventilation.	
Inhalation: sore throat, cough, shortness of breath.	ventilation, local exhaust or respiratory protection.	fresh air, rest, call a doctor.
Skin: redness, pain.	gloves.	remove contaminated clothing, flush skin with water or shower.
Eyes: redness, pain.	safety glasses.	flush with water, take to a doctor if necessary.
Ingestion: sore throat, abdominal pain, nausea.		rinse mouth, call a doctor.

SPILLAGE	STORAGE	LABELING / NFPA
collect leakage in sealable containers, soak up with sand or other inert absorbent and remove to safe place; (additional individual protection: breathing apparatus).	keep separate from oxidants; ventilate at floor level.	(NFPA diamond: blue 2, red 2)

NOTES

Transport Emergency Card TEC(R)-60G06B   **HI: 60; UN-number: 2966**

MERCURIC (II) OXIDE
(yellow and red)

PHYSICAL PROPERTIES	IMPORTANT CHARACTERISTICS	
Decomposes below melting point, °C 500 Relative density (water = 1) 11.1 Solubility in water, g/100 ml 0.005 Relative molecular mass 216.6	**RED OR YELLOW CRYSTALLINE POWDER** Decomposes when heated above 500°C or exposed to light, giving off toxic → *mercury* vapors and → *oxygen*, with increased risk of fire. Strong oxidant which reacts violently with combustible substances and reducing agents, with risk of explosion.	
	TLV-TWA (skin)	0.05 mg/m³ ✳
	Absorption route: Can enter the body by inhalation or ingestion or through the skin. Evaporation negligible at 20°C, but harmful concentrations of airborne particles can build up rapidly. **Immediate effects:** Irritates the eyes, skin and respiratory tract. Can cause gastric disorders. **Effects of prolonged/repeated exposure:** Prolonged or repeated contact can cause skin disorders. Can cause kidney damage.	
Gross formula: HgO		

HAZARDS/SYMPTOMS	PREVENTIVE MEASURES	FIRE EXTINGUISHING/FIRST AID
Fire: not combustible but enhances combustion of other substances.	avoid contact with reducing agents.	in case of fire in immediate vicinity: use any extinguishing agent.
	STRICT HYGIENE, KEEP DUST UNDER CONTROL	
Inhalation: sore throat, cough.	local exhaust or respiratory protection.	fresh air, rest, call a doctor.
Skin: *is absorbed*; irritation, redness.	gloves.	remove contaminated clothing, wash skin with soap and water, call a doctor if necessary.
Eyes: redness, pain.	goggles, or combined eye and respiratory protection.	flush with water, take to a doctor.
Ingestion: *corrosive*; abdominal pain, diarrhea, vomiting, feeling of weakness.	do not eat, drink or smoke while working.	rinse mouth, immediately take to hospital.

SPILLAGE	STORAGE	LABELING / NFPA
clean up spillage, carefully collect remainder, do not use sawdust or other combustible absorbents; (additional individual protection: P3 respirator).	keep in dark place, separate from combustible substances and reducing agents.	R: 26/27/28-33 S: 1/2-13-28-45 Note A ☠ Toxic

NOTES

✳ TLV equals that of mercury. Do not take work clothing home. Depending on the degree of exposure, regular medical checkups are advisable.

Transport Emergency Card TEC(R)-61G11

HI: 60; UN-number: 1645

MERCURY

PHYSICAL PROPERTIES	IMPORTANT CHARACTERISTICS
Boiling point, °C 357 Melting point, °C −39 Relative density (water = 1) 13.6 Relative vapor density (air = 1) 6.9 Relative density at 20 °C of saturated mixture vapor/air (air = 1) 1.0 Vapor pressure, mm Hg at 20 °C 1.5x10⁻³ Solubility in water none Relative atomic mass 200.6	**SILVERY, HEAVY LIQUID** Vapor mixes readily with air. If spilled forms very fine droplets which spread widely and readily adhere to clothing and skin. Reacts with many metals (except iron) to form amalgams. Reacts violently with strong nitric acid, acetylene, ammonia, chlorine etc. to form shock-sensitive compounds.

| | TLV-TWA (skin) | 0.05 mg/m³ |

Relative density, physical properties left column ends with:

Gross formula: Hg

Absorption route: Can enter the body by inhalation or ingestion.
Immediate effects: Irritates the eyes and skin. In substantial concentrations can cause impairment of consciousness, muscle weakness and tingling in arms and legs.
Effects of prolonged/repeated exposure: Can cause kidney, brain and nerve damage.

HAZARDS/SYMPTOMS	PREVENTIVE MEASURES	FIRE EXTINGUISHING/FIRST AID
Fire: non-combustible.		
	GUARD AGAINST SPILLAGE	
Inhalation: sore throat, headache, nausea.	ventilation, local exhaust or respiratory protection, apparatus in sump, ventilate at floor level, smooth and seamless floor.	fresh air, rest, call a doctor.
Skin: redness.	do not handle mercury if skin is broken, wear work clothes and gloves.	remove contaminated clothing, flush skin with water or shower, wash skin with soap and water, call a doctor if necessary.
Eyes: irritation, redness.	safety glasses.	flush with water, take to doctor if necessary.
Ingestion: vomiting, diarrhea.	do not eat, drink or smoke while working.	rinse mouth, take immediately to hospital.

SPILLAGE	STORAGE	LABELING / NFPA
clean up spillage, vacuum remainder with special equipment or wet-vac; cover remainder with (1) zinc or copper powder, or (2) special absorbents, and then carefully collect; measure mercury vapor concentration.	keep cool; ventilate at floor level.	R: 23-33 S: 7-44 ☠ Toxic

NOTES

Do not take work clothing home. Vapor saturation level at room temperature (14mg/m³) is far above TLV: risk of chronic poisoning. Regular checks with special apparatus of areas where mercury is used are recommended (check also instruments containing mercury). Covering spilled mercury with sulfur is not recommended since this is effective only as long as mercury remains covered. Use airtight packaging. Unbreakable packaging preferred; if breakable, keep in unbreakable container. TLV-TWA for women of childbearing age: 0.025mg/m³ over 8-hour period.

UN-number: 2809

MERCURY (II) ACETATE

PHYSICAL PROPERTIES	IMPORTANT CHARACTERISTICS	
Melting point (decomposes), °C 178 Solubility in water, g/100 ml at 10 °C 25 Relative molecular mass 318.7	**WHITE CRYSTALS OR WHITE POWDER** Decomposes when heated, giving off → *mercury* vapor and flammable vapors. Decomposes when exposed to light.	
	TLV-TWA (skin)	0.05 mg/m³ ✳
	Absorption route: Can enter the body by inhalation. Evaporation negligible at 20°C, but harmful concentrations of airborne particles can build up rapidly. **Immediate effects:** Corrosive to the eyes, skin and respiratory tract. Inhalation can cause lung edema. In serious cases risk of death. **Effects of prolonged/repeated exposure:** Prolonged or repeated contact can cause skin disorders. Can cause kidney damage.	
Gross formula: $C_4H_6HgO_4$		

HAZARDS/SYMPTOMS	PREVENTIVE MEASURES	FIRE EXTINGUISHING/FIRST AID
Fire: non-combustible.		in case of fire in immediate vicinity: use any extinguishing agent.
	STRICT HYGIENE, KEEP DUST UNDER CONTROL	
Inhalation: *corrosive*; sore throat, cough, shortness of breath, severe breathing difficulties, headache.	local exhaust or respiratory protection.	fresh air, rest, take to hospital.
Skin: *corrosive, is absorbed*; redness, pain, burns.	gloves, protective clothing.	remove contaminated clothing, flush skin with water or shower, call a doctor if necessary.
Eyes: *corrosive*; redness, pain, impaired vision.	goggles, or combined eye and respiratory protection.	flush with water, take to a doctor.
Ingestion: *corrosive*; metallic taste, abdominal pain, diarrhea, vomiting.	do not eat, drink or smoke while working.	rinse mouth, immediately take to hospital.

SPILLAGE	STORAGE	LABELING / NFPA
clean up spillage, carefully collect remainder; (additional individual protection: P3 respirator).	keep in dark place.	R: 26/27/28-33 S: 1/2-13-28-45 Note A ☠ Toxic

NOTES
Mercury acetate belongs to inorganic mercury salts. ✳ TLV equals that of mercury. Depending on the degree of exposure, regular medical checkups are advisable. Lung edema symptoms usually develop several hours later and are aggravated by physical exertion: rest and hospitalization essential. As first aid, a doctor or authorized person should consider administering a corticosteroid spray. Use airtight packaging.

MERCURY (II) BROMIDE

PHYSICAL PROPERTIES	IMPORTANT CHARACTERISTICS
Boiling point (decomposes), °C — 322 Melting point, °C — 236 Relative density (water = 1) — 6.1 Vapor pressure, mm Hg at 136.5 °C — 0.99 Solubility in water, g/100 ml at 25 °C — 0.6 Relative molecular mass — 360.4	**WHITE CRYSTALS OR WHITE POWDER** Reacts with sodium or potassium to form shock-sensitive mixtures. Decomposes when heated and when exposed to light, giving off toxic vapors (→ bromine). Reacts violently with indium at 350°C. The solution in water is corrosive.

	TLV-TWA (skin)	0.05 mg/m³ ✱

Absorption route: Can enter the body by inhalation. Evaporation negligible at 20°C, but harmful concentrations of airborne particles can build up rapidly.
Immediate effects: Corrosive to the eyes, skin and respiratory tract. Inhalation can cause lung edema. In serious cases risk of death. Can cause gastro-intestinal disorders.
Effects of prolonged/repeated exposure: Can cause liver and kidney damage.

Gross formula: Br$_2$Hg

HAZARDS/SYMPTOMS	PREVENTIVE MEASURES	FIRE EXTINGUISHING/FIRST AID
Fire: non-combustible.		in case of fire in immediate vicinity: use any extinguishing agent.
	STRICT HYGIENE, KEEP DUST UNDER CONTROL	
Inhalation: *corrosive*; sore throat, cough, shortness of breath, severe breathing difficulties, headache.	local exhaust or respiratory protection.	fresh air, rest, take to hospital.
Skin: *corrosive, is absorbed*; redness, pain, burns.	gloves, protective clothing.	remove contaminated clothing, flush skin with water or shower, call a doctor if necessary.
Eyes: *corrosive*; redness, pain, impaired vision.	goggles, or combined eye and respiratory protection.	flush with water, take to a doctor.
Ingestion: *corrosive*; abdominal pain, diarrhea, vomiting, metallic taste.	do not eat, drink or smoke while working.	rinse mouth, immediately take to hospital.

SPILLAGE	STORAGE	LABELING / NFPA
clean up spillage, carefully collect remainder; (additional individual protection: P3 respirator).	keep in dark place.	R: 26/27/28-33 S: 1/2-13-28-45 Note A ☠ Toxic

NOTES

The decomposition product sublimes. ✱ TLV equals that of mercury. Depending on the degree of exposure, regular medical checkups are advisable. Lung edema symptoms usually develop several hours later and are aggravated by physical exertion: rest and hospitalization essential. As first aid, a doctor or authorized person should consider administering a corticosteroid spray. The measures on this card also apply to mercury iodide. Use airtight packaging.

MERCURY (II) CHLORIDE

PHYSICAL PROPERTIES	IMPORTANT CHARACTERISTICS	
Boiling point, °C 303 Melting point, °C 277 Relative density (water = 1) 5.4 Vapor pressure, mm Hg at 20 °C 7.6x10^{-4} Solubility in water, g/100 ml at 20 °C 7.4 Relative molecular mass 271.5 Log P octanol/water 0.1	**WHITE CRYSTALS OR POWDER** Decomposes when heated (e.g. due to fire), giving off toxic vapors ($\rightarrow$ *mercury* and $\rightarrow$ *chlorine*). Reacts with light metals.	
	TLV-TWA (skin) 0.05 mg/m³ ✹	
	Absorption route: Can enter the body by inhalation or ingestion or through the skin (in solution). Evaporation negligible at 20°C, but harmful concentrations of airborne particles can build up rapidly. **Immediate effects:** Irritates the eyes and skin. In substantial concentrations can cause impairment of consciousness, muscle weakness and tingling in arms and legs. **Effects of prolonged/repeated exposure:** Can cause gastro-intestinal disorders, brain and kidney damage.	
Gross formula: Cl$_2$Hg		

HAZARDS/SYMPTOMS	PREVENTIVE MEASURES	FIRE EXTINGUISHING/FIRST AID
Fire: non-combustible.		in case of fire in immediate vicinity: use any extinguishing agent.
	STRICT HYGIENE, KEEP DUST UNDER CONTROL	
Inhalation: sore throat, cough.	local exhaust or respiratory protection.	fresh air, rest, call a doctor.
Skin: redness, pain, blisters.	gloves, protective clothing.	remove contaminated clothing, flush skin with water or shower, call a doctor if necessary.
Eyes: *irritation*; redness, pain, impaired vision.	goggles, or combined eye and respiratory protection.	flush with water, take to a doctor if necessary.
Ingestion: metallic taste, abdominal pain, diarrhea, vomiting.	do not eat, drink or smoke while working.	rinse mouth, take immediately to hospital.

SPILLAGE	STORAGE	LABELING / NFPA
clean up spillage, carefully collect remainder; (additional individual protection: P3 respirator).		R: 26/27/28-33 S: 1/2-13-28-45 Note A ☠ Toxic

NOTES
Solubility in water at 100°C: 55g/100ml. Log P octanol/water is estimated. ✹ TLV equals that of mercury. Do not take work clothing home. Depending on the degree of exposure, regular medical checkups are advisable. When strongly heated, e.g. due to fire, can produce high concentrations of mercury vapors. Unbreakable packaging preferred; if breakable, keep in unbreakable container.

Transport Emergency Card TEC(R)-61G11B

HI: 60; UN-number: 1624

MERCURY (II) NITRATE

PHYSICAL PROPERTIES		IMPORTANT CHARACTERISTICS
Decomposes below boiling point, °C	?	**YELLOW HYGROSCOPIC CRYSTALS OR YELLOW POWDER**
Melting point, °C	79	Reacts with acetylene, ethanol, phosphine and sulfur to form shock-sensitive compounds.
Relative density (water = 1)	4.39	Decomposes when heated, giving off toxic vapors (→ *mercury* vapor and → *nitrogen oxides*).
Solubility in water	good	Strong oxidant which reacts violently with combustible substances and reducing agents.
Relative molecular mass	333.6	Decomposes when exposed to light. The solution in water is corrosive.

TLV-TWA (skin)		0.05 mg/m³ ✶

Absorption route: Can enter the body by inhalation or ingestion or through the skin. Evaporation negligible at 20°C, but harmful concentrations of airborne particles can build up rapidly.
Immediate effects: Corrosive to the eyes, skin and respiratory tract. Inhalation can cause lung edema. In serious cases risk of seizures and death.
Effects of prolonged/repeated exposure: Can cause liver and kidney damage.

Gross formula: HgN_2O_6

HAZARDS/SYMPTOMS	PREVENTIVE MEASURES	FIRE EXTINGUISHING/FIRST AID
Fire: not combustible but enhances combustion of other substances.	avoid contact with combustible substances.	in case of fire in immediate vicinity: use any extinguishing agent.
Explosion: reacts with reducing agents, with risk of explosion.		
	STRICT HYGIENE, KEEP DUST UNDER CONTROL	
Inhalation: *corrosive*; sore throat, cough, shortness of breath, severe breathing difficulties, headache.	local exhaust or respiratory protection.	fresh air, rest, take to hospital.
Skin: *corrosive, is absorbed*; redness, burns.	gloves, protective clothing.	remove contaminated clothing, flush skin with water or shower, call a doctor if necessary.
Eyes: *corrosive*; redness, pain, impaired vision.	goggles, or combined eye and respiratory protection.	flush with water, take to a doctor.
Ingestion: *corrosive*; sore throat, abdominal pain, diarrhea, vomiting, metallic taste.	do not eat, drink or smoke while working.	rinse mouth, immediately take to hospital.

SPILLAGE	STORAGE	LABELING / NFPA
clean up spillage, carefully collect remainder, do not use sawdust or other combustible absorbents; (additional individual protection: P3 respirator).	keep in dark, dry place, separate from combustible substances and reducing agents.	R: 26/27/28-33 S: 1/2-13-28-45 Note A ☠ Toxic

NOTES
✶ TLV equals that of mercury. Flush contaminated clothing with water (fire hazard). Depending on the degree of exposure, regular medical checkups are advisable. Lung edema symptoms usually develop several hours later and are aggravated by physical exertion: rest and hospitalization essential. As first aid, a doctor or authorized person should consider administering a corticosteroid spray. Use airtight packaging.

Transport Emergency Card TEC(R)-61G36/37

UN-number: 1625

MERCURY (II) SULFATE

PHYSICAL PROPERTIES	IMPORTANT CHARACTERISTICS	
Decomposes below melting point, °C ? Relative density (water = 1) 6.5 Solubility in water reaction Relative molecular mass 296.7	**WHITE CRYSTALLINE POWDER** Decomposes when heated, giving off toxic vapors. Decomposes when exposed to light. Reacts with water, forming insoluble basic mercury sulfur and → *sulfuric acid*.	
	TLV-TWA (skin)	0.05 mg/m³ ✳
	Absorption route: Can enter the body by inhalation. Evaporation negligible at 20°C, but harmful concentrations of airborne particles can build up rapidly. **Immediate effects:** Irritates the skin. Corrosive to the eyes and respiratory tract. Inhalation can cause lung edema. In serious cases risk of death. Can cause gastro-intestinal disorders. **Effects of prolonged/repeated exposure:** Prolonged or repeated contact can cause skin disorders. Can cause kidney damage.	
Gross formula: HgO_4S		

HAZARDS/SYMPTOMS	PREVENTIVE MEASURES	FIRE EXTINGUISHING/FIRST AID
Fire: non-combustible.	avoid contact with reducing agents.	in case of fire in immediate vicinity: use any extinguishing agent.
	STRICT HYGIENE, KEEP DUST UNDER CONTROL	
Inhalation: *corrosive*; sore throat, cough, shortness of breath, severe breathing difficulties, headache.	local exhaust or respiratory protection.	fresh air, rest, take to hospital.
Skin: *is absorbed*; redness, pain.	gloves.	remove contaminated clothing, flush skin with water or shower, refer to a doctor.
Eyes: *corrosive*; redness, pain, impaired vision.	goggles, or combined eye and respiratory protection.	flush with water, take to a doctor.
Ingestion: *corrosive*; sore throat, abdominal pain, diarrhea, metallic taste, vomiting.	do not eat, drink or smoke while working.	rinse mouth, immediately take to hospital.

SPILLAGE	STORAGE	LABELING / NFPA
clean up spillage, carefully collect remainder; (additional individual protection: P3 respirator).	keep in dark, dry place.	R: 26/27/28-33 S: 1/2-13-28-45 Note A ☠ Toxic

NOTES
Decomposition temperature not given in the literature. Depending on the degree of exposure, regular medical checkups are advisable. Lung edema symptoms usually develop several hours later and are aggravated by physical exertion: rest and hospitalization essential. As first aid, a doctor or authorized person should consider administering a corticosteroid spray. Use airtight packaging. ✳TLV equals that of mercury.

Transport Emergency Card TEC(R)-61G11

UN-number: 1645

MERCURY (II) THIOCYANATE

PHYSICAL PROPERTIES	IMPORTANT CHARACTERISTICS	
Boiling point, °C 165 Solubility in water, g/100 ml at 25°C 0.07 Relative molecular mass 316.7	**WHITE POWDER** Decomposes when heated, giving off toxic vapors.	
	TLV-TWA (skin)	0.05 mg/m³ ✷
	Absorption route: Can enter the body by inhalation or ingestion or through the skin. Evaporation negligible at 20°C, but harmful concentrations of airborne particles can build up rapidly. **Immediate effects:** Irritates the skin. Corrosive to the eyes and respiratory tract. Can cause gastro-intestinal disorders. Inhalation can cause lung edema. In serious cases risk of death. **Effects of prolonged/repeated exposure:** Can cause liver and kidney damage.	
Gross formula: $C_2HgN_2S_2$		

HAZARDS/SYMPTOMS	PREVENTIVE MEASURES	FIRE EXTINGUISHING/FIRST AID
Fire: non-combustible.		in case of fire in immediate vicinity: use any extinguishing agent.
	STRICT HYGIENE, KEEP DUST UNDER CONTROL	
Inhalation: *corrosive*; sore throat, cough, shortness of breath, severe breathing difficulties, headache.	local exhaust or respiratory protection.	fresh air, rest, take to hospital.
Skin: *is absorbed*; redness, pain.	gloves, protective clothing.	remove contaminated clothing, flush skin with water or shower, refer to a doctor.
Eyes: *corrosive*; redness, pain, impaired vision.	goggles, or combined eye and respiratory protection.	flush with water, take to a doctor.
Ingestion: *corrosive*; metallic taste, sore throat, abdominal pain, diarrhea.	do not eat, drink or smoke while working.	rinse mouth, take immediately to hospital.

SPILLAGE	STORAGE	LABELING / NFPA
clean up spillage, carefully collect remainder; (additional individual protection: P2 respirator).	keep in dark place.	R: 26/27/28-33 S: 1/2-13-28-45 Note A ☠ Toxic

NOTES
✷ TLV equals that of mercury vapor (all forms except alkyl). Depending on the degree of exposure, regular medical checkups are advisable. Lung edema symptoms usually develop several hours later and are aggravated by physical exertion: rest and hospitalization essential. As first aid, a doctor or authorized person should consider administering a corticosteroid spray. Use airtight packaging.

CAS-No: [108-67-8]
1,3,5-trimethylbenzene
trimethylbenzol

$C_6H_3(CH_3)_3$

MESITYLENE

PHYSICAL PROPERTIES		IMPORTANT CHARACTERISTICS		
Boiling point, °C	165	**COLORLESS LIQUID WITH CHARACTERISTIC ODOR**		
Melting point, °C	−45	Vapor mixes readily with air. Flow, agitation etc. can cause build-up of electrostatic charge due to liquid's low conductivity. Reacts violently with strong oxidants, with risk of fire and explosion.		
Flash point, °C	47			
Auto-ignition temperature, °C	550			
Relative density (water = 1)	0.86	TLV-TWA	25 ppm	125 mg/m³
Relative vapor density (air = 1)	4.1			
Relative density at 20 °C of saturated mixture vapor/air (air = 1)	1.01	**Absorption route:** Can enter the body by inhalation or ingestion or through the skin. Harmful atmospheric concentrations build up fairly slowly on evaporation at 20°C, but much more rapidly in aerosol form.		
Vapor pressure, mm Hg at 20 °C	1.0			
Solubility in water	none			
Explosive limits, vol% in air	?-?	**Immediate effects:** Irritates the eyes, skin and respiratory tract. Liquid destroys the skin's natural oils. Affects the nervous system. Can cause liver damage. If liquid is swallowed, droplets can enter the lungs, with risk of pneumonia.		
Relative molecular mass	120.2			
Gross formula:	C_9H_{12}			

HAZARDS/SYMPTOMS	PREVENTIVE MEASURES	FIRE EXTINGUISHING/FIRST AID
Fire: flammable.	keep away from open flame and sparks, no smoking.	powder, AFFF, foam, carbon dioxide, (halons).
Explosion: above 47°C: forms explosive air-vapor mixtures.	above 47°C: sealed machinery, ventilation, explosion-proof electrical equipment, grounding.	in case of fire: keep tanks/drums cool by spraying with water.
Inhalation: sore throat, cough, shortness of breath, headache, nausea.	ventilation, local exhaust or respiratory protection.	fresh air, rest, take to hospital if necessary.
Skin: *is absorbed*; redness.	gloves.	remove contaminated clothing, flush skin with water or shower.
Eyes: redness, pain.	face shield.	flush with water, take to a doctor if necessary.
Ingestion: sore throat, cough, abdominal pain, headache, vomiting.		rinse mouth, DO NOT induce vomiting, take to hospital if necessary.

SPILLAGE	STORAGE	LABELING / NFPA
collect leakage in sealable containers, soak up with sand or other inert absorbent and remove to safe place.	keep in fireproof place, separate from oxidants.	R: 10-37 ⊠ Irritant

NOTES

Explosive limits not given in the literature. Alcohol consumption increases toxic effects. Absorbed slowly through the skin.

CAS-No: [141-79-7]
isopropylideneacetone
methyl isobutenyl ketone
4-methyl-3-penten-2-one

$(CH_3)_2C = CHCOCH_3$

MESITYL OXIDE

PHYSICAL PROPERTIES		IMPORTANT CHARACTERISTICS	
Boiling point, °C	130	**COLORLESS VISCOUS LIQUID WITH CHARACTERISTIC ODOR**	
Melting point, °C	−59	Vapor mixes readily with air. Able to form peroxides. Reacts violently with strong oxidants and	
Flash point, °C	31	many other substances. Attacks many plastics and copper.	
Auto-ignition temperature, °C	340		
Relative density (water = 1)	0.85	TLV-TWA	15 ppm	60 mg/m³
Relative vapor density (air = 1)	3.4	STEL	25 ppm	100 mg/m³
Relative density at 20 °C of saturated mixture			
vapor/air (air = 1)	1.02	**Absorption route:** Can enter the body by inhalation or ingestion or through the skin. Harmful	
Vapor pressure, mm Hg at 20 °C	7.6	atmospheric concentrations can build up fairly rapidly on evaporation at approx. 20°C - even	
Solubility in water, g/100 ml at 20 °C	3	more rapidly in aerosol form.	
Explosive limits, vol% in air	1.4-7.2	**Immediate effects:** Irritates the eyes, skin and respiratory tract. Liquid destroys the skin's natural	
Relative molecular mass	98.1	oils. In substantial concentrations can impair consciousness.	
Log P octanol/water	1.2		
Gross formula:	$C_6H_{10}O$		

HAZARDS/SYMPTOMS	PREVENTIVE MEASURES	FIRE EXTINGUISHING/FIRST AID
Fire: flammable.	keep away from open flame and sparks, no smoking.	powder, AFFF, foam, carbon dioxide, (halons).
Explosion: above 31°C: forms explosive air-vapor mixtures.	above 31°C: sealed machinery, ventilation, explosion-proof electrical equipment.	in case of fire: keep tanks/drums cool by spraying with water.
	STRICT HYGIENE	
Inhalation: cough, headache, drowsiness.	ventilation, local exhaust or respiratory protection.	fresh air, rest, call a doctor if necessary.
Skin: redness.	gloves.	remove contaminated clothing, flush skin with water or shower, refer to doctor if necessary.
Eyes: redness, pain.	face shield.	flush with water, take to a doctor.
Ingestion: abdominal cramps; see also 'Inhalation'.		rinse mouth, DO NOT induce vomiting, call a doctor or take to hospital.

SPILLAGE	STORAGE	LABELING / NFPA
collect leakage in sealable containers, soak up with sand or other inert absorbent and remove to safe place; (additional individual protection: breathing apparatus).	keep in fireproof place, separate from oxidants; avoid contact with copper.	R: 10-20/21/22 S: 25 ✖ Harmful

NOTES

Before distilling check for peroxides; if found, render harmless. Alcohol consumption increases toxic effects. Depending on the degree of exposure, regular medical checkups are advisable.

Transport Emergency Card TEC(R)-130

HI: 30; UN-number: 1229

MESYL CHLORIDE

PHYSICAL PROPERTIES	IMPORTANT CHARACTERISTICS
Boiling point, °C 164 Melting point, °C − 32 Flash point, °C > 100 Relative density (water = 1) 1.5 Relative vapor density (air = 1) 4.0 Relative density at 20 °C of saturated mixture vapor/air (air = 1) 1.01 Vapor pressure, mm Hg at 20 °C 2.0 Solubility in water reaction Relative molecular mass 114.6	**LIGHT YELLOW LIQUID WITH PUNGENT ODOR** Vapor mixes readily with air. Decomposes when heated, giving off flammable vapors (→ *methane*) and toxic vapors (→ *sulfur dioxide* and → *hydrochloric acid*). Reacts violently with bases, ammonia and many other substances, with risk of fire and explosion. Reacts with water, giving off corrosive vapors.
	TLV-TWA not available
Gross formula: CH₃ClO₂S	**Absorption route:** Can enter the body by inhalation or ingestion. Harmful atmospheric concentrations can build up very rapidly on evaporation at 20°C. **Immediate effects:** Corrosive to the eyes, skin and respiratory tract. Inhalation of vapor/fumes can cause lung edema. In serious cases risk of death.

HAZARDS/SYMPTOMS	PREVENTIVE MEASURES	FIRE EXTINGUISHING/FIRST AID
Fire: combustible.	keep away from open flame, no smoking.	AFFF, powder, carbon dioxide, (halons); in case of fire in immediate vicinity: preferably not water.
	STRICT HYGIENE	
Inhalation: *corrosive*; sore throat, cough, shortness of breath, severe breathing difficulties.	ventilation, local exhaust or respiratory protection.	fresh air, rest, place in half-sitting position, take to hospital.
Skin: *corrosive*; redness, pain, burns.	gloves, protective clothing.	remove contaminated clothing, flush skin with water or shower, take to a doctor.
Eyes: *corrosive*; redness, pain, impaired vision.	face shield, or combined eye and respiratory protection.	flush with water, take to a doctor.
Ingestion: *corrosive*; abdominal pain, diarrhea.		rinse mouth, take immediately to hospital.

SPILLAGE	STORAGE	LABELING / NFPA
collect leakage in sealable containers, neutralize spillage with bicarbonate or soda solution and flush away with water.	keep dry, separate from bases and ammonia.	

NOTES
Lung edema symptoms usually develop several hours later and are aggravated by physical exertion: rest and hospitalization essential. As first aid, a doctor or authorized person should consider administering a corticosteroid spray. Use airtight packaging.

Transport Emergency Card TEC(R)-80G10

METALDEHYDE

PHYSICAL PROPERTIES	IMPORTANT CHARACTERISTICS
Sublimation temperature, °C 112 Melting point, °C 246 Flash point, °C 36 Relative density (water = 1) ? Relative vapor density (air = 1) 6.1 Relative density at 20 °C of saturated mixture vapor/air (air = 1) 1.00 Vapor pressure, mm Hg at 20 °C 0.23 Solubility in water none Relative molecular mass 176.2	**WHITE SOLID IN VARIOUS FORMS** Vapor mixes readily with air. In dry state can form electrostatic charge if stirred, transported pneumatically, poured etc. Decomposes when heated above 80°C, giving off flammable and harmful vapors (→ *acetaldehyde*). Reacts with strong oxidants.
	TLV-TWA not available
	Absorption route: Can enter the body by inhalation or ingestion. Harmful atmospheric concentrations build up very slowly, if at all, on evaporation at approx. 20°C, but harmful concentrations of airborne particles can build up much more rapidly. **Immediate effects:** Irritates the eyes and skin. Affects the nervous system. Ingestion can cause unconsciousness, seizures and death in serious cases. **Effects of prolonged/repeated exposure:** Can cause brain, liver and kidney damage.
Gross formula: $C_8H_{16}O_4$	

HAZARDS/SYMPTOMS	PREVENTIVE MEASURES	FIRE EXTINGUISHING/FIRST AID
Fire: flammable.	keep away from open flame and sparks, no smoking; avoid contact with hot surfaces (steam lines) because of acetaldehyde formation.	water spray, powder.
Explosion: above 36°C: forms explosive air-vapor mixtures.	above 36°C: sealed machinery, ventilation, explosion-proof electrical equipment.	in case of fire: keep tanks/drums cool by spraying with water.
Inhalation: sore throat, cough.	local exhaust or respiratory protection.	fresh air, rest, refer to a doctor if necessary.
Skin: redness, pain.	gloves.	remove contaminated clothing, flush skin with water or shower.
Eyes: redness, pain.	safety glasses.	flush with water, take to a doctor if necessary.
Ingestion: abdominal pain, cramps, vomiting, unconsciousness.		rinse mouth, take immediately to hospital.

SPILLAGE	STORAGE	LABELING / NFPA
clean up spillage, carefully collect remainder; (additional individual protection: P2 respirator).	keep in fireproof place, separate from oxidants.	R: 10-20/22 S: 2-24/25 ✖ Harmful

NFPA diamond: blue 1, red 3, yellow 1

NOTES
Metaldehyde is a polymer of acetaldehyde: both tetramer $(C_2H_4O)_4$ and hexamer $(C_2H_4O)_6$ can occur. Melting point is measured in pressure vessel to prevent sublimation.

Transport Emergency Card TEC(R)-41G01

UN-number: 1332

$(HPO_3)n$

METAPHOSPHORIC ACID

PHYSICAL PROPERTIES	IMPORTANT CHARACTERISTICS
Sublimation temperature, °C ? Relative density (water = 1) 2.4 Solubility in water poor	**COLORLESS TRANSPARENT HYGROSCOPIC SOLID** Decomposes when heated, giving off toxic → *phosphorus oxide* vapors. In aqueous solution slowly transforms into a solution of → *phosphorus acid*, which is a medium strong acid and reacts with bases and is corrosive.
	TLV-TWA not available
	Absorption route: Can enter the body by inhalation or ingestion. Evaporation negligible at 20°C, but harmful concentrations of airborne particles can build up rapidly. **Immediate effects:** Corrosive to the eyes, skin and respiratory tract. Inhalation of vapor/fumes can cause severe breathing difficulties (lung edema). In serious cases risk of death.
Gross formula: HO_3P	

HAZARDS/SYMPTOMS	PREVENTIVE MEASURES	FIRE EXTINGUISHING/FIRST AID
Fire: non-combustible.		in case of fire in immediate vicinity: use any extinguishing agent.
Inhalation: *corrosive*; sore throat, severe breathing difficulties, vomiting.	local exhaust or respiratory protection.	fresh air, rest, place in half-sitting position, take to hospital.
Skin: *corrosive*; redness, pain, serious burns.	gloves.	remove contaminated clothing, flush skin with water or shower, call a doctor.
Eyes: *corrosive*; redness, pain, impaired vision.	face shield.	flush with water, take to a doctor.
Ingestion: *corrosive*; abdominal pain, vomiting, gray discoloration in mouth-throat cavity.		rinse mouth, immediately take to hospital.

SPILLAGE	STORAGE	LABELING / NFPA
clean up spillage, flush away remainder with water; (additional individual protection: P2 respirator).	keep dry, separate from strong bases.	

NOTES
TLV for → *phosphoric acid* can usually be applied. Lung edema symptoms usually develop several hours later and are aggravated by physical exertion: rest and hospitalization essential. As first aid, a doctor or authorized person should consider administering a corticosteroid spray.

Transport Emergency Card TEC(R)-80G09

CAS-No: [126-98-7]
2-cyano-1-propene
isopropene cyanide
isopropenylnitril
2-methylacrylonitrile
2-methyl-2-propenenitrile

$CH_2=C(CH_3)\text{-}CN$

METHACRYLONITRILE

PHYSICAL PROPERTIES		IMPORTANT CHARACTERISTICS
Boiling point, °C	90	**COLORLESS LIQUID WITH CHARACTERISTIC ODOR**
Melting point, °C	−36	Vapor is heavier than air and spreads at ground level, with risk of ignition at a distance. Able to polymerize, esp. in the presence of strong acids or bases or when exposed to light. Decomposes when heated or burned, giving off toxic gases (→ *nitrous vapors*). Reacts violently with strong oxidants.
Flash point, °C	13	
Relative density (water = 1)	0.8	
Relative vapor density (air = 1)	2.3	
Relative density at 20 °C of saturated mixture vapor/air (air = 1)	1.1	

TLV-TWA (skin) 1 ppm 2.7 mg/m³

Absorption route: Can enter the body by inhalation or ingestion or through the skin. Harmful atmospheric concentrations can build up very rapidly on evaporation at 20°C.
Immediate effects: Irritates the eyes. Impedes tissue respiration. In serious cases risk of death.

Vapor pressure, mm Hg at 20 °C	65
Solubility in water	poor
Explosive limits, vol% in air	2-6.8
Relative molecular mass	67.1

Gross formula: C_4H_5N

HAZARDS/SYMPTOMS	PREVENTIVE MEASURES	FIRE EXTINGUISHING/FIRST AID
Fire: extremely flammable.	keep away from open flame and sparks, no smoking.	powder, AFFF, foam, carbon dioxide, (halons).
Explosion: forms explosive air-vapor mixtures.	sealed machinery, ventilation, explosion-proof electrical equipment and lighting.	in case of fire: keep tanks/drums cool by spraying with water.
	STRICT HYGIENE	IN ALL CASES CALL IN A DOCTOR
Inhalation: sore throat, dizziness, nausea, feeling of weakness, severe breathing difficulties.	ventilation, local exhaust or respiratory protection.	special anti-acrylonitrile treatment, take to hospital.
Skin: *is absorbed*; redness; see also 'Inhalation'.	gloves, protective clothing.	remove contaminated clothing, flush skin with water or shower, refer to a doctor for special anti-acrylonitrile treatment.
Eyes: redness, pain.	face shield, or combined eye and respiratory protection.	flush with water, take to a doctor.
Ingestion: sore throat, severe breathing difficulties, nausea, dizziness, feeling of weakness.		rinse mouth, special anti-acrylonitrile treatment, take immediately to hospital.

SPILLAGE	STORAGE	LABELING / NFPA
evacuate area, call in an expert, collect leakage in sealable containers, soak up with sand or other inert absorbent and remove to safe place, render remainder harmless with sodium hypochlorite solution; (additional individual protection: breathing apparatus).	keep in dark, fireproof place, separate from oxidants, strong acids and strong bases.	R: 11-23/24/25-43 S: 9-16-18-29-45 Note D Flammable Toxic 3 2 2

NOTES
Odor limit is above TLV. Special first aid required in the event of poisoning: antidotes must be available (with instructions). Administer oxygen (100%) if necessary, avoid mouth-to-mouth resuscitation if possible (risk to person assisting).

METHANE

(cylinder)

PHYSICAL PROPERTIES		IMPORTANT CHARACTERISTICS	
Boiling point, °C	− 162	**COLORLESS, ODORLESS COMPRESSED GAS**	
Melting point, °C	− 182	Lighter than air.	
Flash point, °C	flammable gas		
Auto-ignition temperature, °C	537	TLV-TWA not available	
Relative vapor density (air = 1)	0.6		
Solubility in water	none	**Absorption route:** Can enter the body by inhalation. Can saturate the air if released, with risk of suffocation.	
Explosive limits, vol% in air	5-16		
Minimum ignition energy, mJ	0.28		
Relative molecular mass	16		
Gross formula:	CH$_4$		

HAZARDS/SYMPTOMS	PREVENTIVE MEASURES	FIRE EXTINGUISHING/FIRST AID
Fire: extremely flammable.	keep away from open flame and sparks, no smoking.	shut off supply; if impossible and no danger to surrounding area, allow fire to burn itself out; otherwise extinguish with powder, carbon dioxide, (halons).
Explosion: forms explosive air-gas mixtures.	sealed machinery, ventilation, explosion-proof electrical equipment and lighting, grounding, non-sparking tools.	in case of fire: keep cylinders cool by spraying with water, fight fire from sheltered location.
Inhalation: severe breathing difficulties, headache, drowsiness, unconsciousness.	ventilation, local exhaust or respiratory protection.	fresh air, rest, artificial respiration if necessary, take to hospital.
Eyes:	face shield.	

SPILLAGE	STORAGE	LABELING / NFPA
ventilate; (additional individual protection: breathing apparatus).	keep in cool, fireproof place; ventilate.	4 / 0 / 0

NOTES
High atmospheric concentrations, e.g. in poorly ventilated spaces, can cause oxygen deficiency, with risk of unconsciousness. The measures on this card also apply to natural gas (in cylinders). Close valve when not in use; check hoses and lines regularly; test for leaking connections with soap solution. These measures apply in all cases where gas could escape and when storing. Use cylinder with special fittings.

Transport Emergency Card TEC(R)-20G04

HI: 23; UN-number: 1971

CH₃OH equivalently CH_3OH

carbinol
methyl alcohol
methyl hydrate
methylol
wood alcohol

METHANOL

PHYSICAL PROPERTIES	IMPORTANT CHARACTERISTICS		
Boiling point, °C — 65	**COLORLESS LIQUID WITH CHARACTERISTIC ODOR**		
Melting point, °C — −98	Vapor mixes readily with air, readily forming explosive mixtures. Do not use compressed air when filling, emptying or processing. Reacts violently with strong oxidants.		
Flash point, °C — 11			
Auto-ignition temperature, °C — 455			
Relative density (water = 1) — 0.8	TLV-TWA (skin)	200 ppm	262 mg/m³
Relative vapor density (air = 1) — 1.1	STEL	250 ppm	328 mg/m³
Relative density at 20 °C of saturated mixture vapor/air (air = 1) — 1.01	**Absorption route:** Can enter the body by inhalation or ingestion or through the skin.		
Vapor pressure, mm Hg at 20 °C — 96.5	**Immediate effects:** Irritates the eyes, skin and respiratory tract. Liquid destroys the skin's natural oils. In substantial concentrations can impair consciousness and vision. In serious cases risk of unconsciousness and death.		
Solubility in water — ∞			
Explosive limits, vol% in air — 5.5 - 36.5			
Minimum ignition energy, mJ — 0.14			
Relative molecular mass — 32			
Log P octanol/water — −0.7			
Gross formula: CH₄O			

HAZARDS/SYMPTOMS	PREVENTIVE MEASURES	FIRE EXTINGUISHING/FIRST AID
Fire: extremely flammable.	keep away from open flame and sparks, no smoking.	powder, alcohol-resistant foam, large quantities of water, carbon dioxide, (halons).
Explosion: forms explosive air-vapor mixtures.	sealed machinery, ventilation, explosion-proof electrical equipment and lighting, non-sparking tools.	in case of fire: keep tanks/drums cool by spraying with water.
Inhalation: cough, headache, dizziness, drowsiness, unconsciousness.	ventilation, local exhaust or respiratory protection.	fresh air, rest, call a doctor.
Skin: redness.	gloves.	remove contaminated clothing, flush skin with water or shower, refer to a doctor if necessary.
Eyes: redness, impaired vision.	safety glasses.	flush with water, take to a doctor.
Ingestion: cough, diarrhea, abdominal cramps, headache, drowsiness.		take immediately to hospital.

SPILLAGE	STORAGE	LABELING / NFPA
collect leakage in sealable containers, flush away remainder with water; (additional individual protection: breathing apparatus).	keep in fireproof place, separate from oxidants.	R: 11-23/25 S: 2-7-16-24 Flammable Toxic [NFPA diamond: 1 / 3 / 0]

NOTES
Odor limit is above TLV. Depending on the degree of exposure, regular medical checkups are advisable. Special first aid and treatment are necessary.

Transport Emergency Card TEC(R)-36

HI: 336; UN-number: 1230

METHOXYETHANE

(cylinder)

PHYSICAL PROPERTIES	IMPORTANT CHARACTERISTICS
Boiling point, °C 11 Flash point, °C flammable gas Auto-ignition temperature, °C 190 Relative density (water = 1) 0.7 Relative vapor density (air = 1) 2.1 Vapor pressure, mm Hg at 20 °C 1262 Solubility in water ∞ Explosive limits, vol% in air 2.0 - 10.1 Relative molecular mass 60.1 Gross formula: C_3H_8O	**COLORLESS COMPRESSED LIQUEFIED GAS** Gas is heavier than air and spreads at ground level, with risk of ignition at a distance. Presumed to be able to form peroxides. Reacts violently with oxidants, with risk of fire and explosion. Attacks many plastics. TLV-TWA not available **Absorption route:** Can enter the body by inhalation. Harmful atmospheric concentrations can build up very rapidly if gas is released. **Immediate effects:** Irritates the eyes and respiratory tract. Can cause frostbite due to rapid evaporation. In serious cases risk of unconsciousness.

HAZARDS/SYMPTOMS	PREVENTIVE MEASURES	FIRE EXTINGUISHING/FIRST AID
Fire: extremely flammable.	keep away from open flame and sparks, no smoking, avoid contact with hot surfaces (e.g. steam lines).	shut off supply; if impossible and no danger to surrounding area, allow fire to burn itself out; otherwise extinguish with powder, carbon dioxide, (halons).
Explosion: forms explosive air-gas mixtures.	sealed machinery, ventilation, explosion-proof electrical equipment and lighting, non-sparking tools.	in case of fire: keep cylinders cool by spraying with water, fight fire from sheltered location.
Inhalation: cough, shortness of breath, headache, sleepiness, unconsciousness.	ventilation, local exhaust or respiratory protection.	fresh air, rest, call a doctor if necessary.
Skin: *in case of frostbite:* redness, pain, sores.	insulating gloves.	*in case of frostbite:* DO NOT remove clothing, flush skin with water or shower, call a doctor if necessary.
Eyes: *in case of frostbite:* redness, pain, impaired vision.	face shield, or combined eye and respiratory protection.	flush with water, take to a doctor.

SPILLAGE	STORAGE	LABELING / NFPA
evacuate area, call in an expert, ventilate, under no circumstances spray liquid with water; (additional individual protection: breathing apparatus).	keep in cool, fireproof place.	4 / 2 / 1

NOTES
Before distilling check for peroxides; if found, render harmless. High atmospheric concentrations, e.g. in poorly ventilated spaces, can cause oxygen deficiency, with risk of unconsciousness. Turn leaking cylinder so that leak is on top to prevent methoxyethane escaping.

Transport Emergency Card TEC(R)-30G01

UN-number: 1039

CAS-No: [79-41-4]
α-methacrylic acid
2-methylpropenoic acid

$CH_2 = C(CH_3)COOH$

METHYACRYLIC ACID

PHYSICAL PROPERTIES	IMPORTANT CHARACTERISTICS
Boiling point, °C 161 Melting point, °C 15 Flash point, °C 77 Relative density (water = 1) 1.01 Relative vapor density (air = 1) 2.97 Relative density at 20 °C of saturated mixture vapor/air (air = 1) 1.0 Vapor pressure, mm Hg at 20 °C 0.68 Solubility in water, g/100 ml at 20 °C 9.7 Relative molecular mass 86.1	**COLORLESS LIQUID OR SOLID WITH PUNGENT ODOR** Vapor mixes readily with air. Moderate heating, exposure to light, peroxides or traces of hydrochloric acid can cause violent polymerization, with risk of fire and explosion. Reducing agent which reacts violently with oxidants.

	TLV-TWA	20 ppm	70 mg/m³

Absorption route: Can enter the body by inhalation or ingestion. Harmful atmospheric concentrations build up fairly slowly on evaporation, but much more rapidly in aerosol form.
Immediate effects: Corrosive to the eyes, skin and respiratory tract. Inhalation of vapor/fumes can cause severe breathing difficulties (lung edema).

Gross formula: $C_4H_6O_2$

HAZARDS/SYMPTOMS	PREVENTIVE MEASURES	FIRE EXTINGUISHING/FIRST AID
Fire: combustible.	keep away from open flame, no smoking.	powder, AFFF, foam, carbon dioxide, (halons).
Explosion: above 77°C: forms explosive air-vapor mixtures.	above 77°C: sealed machinery, ventilation.	in case of fire: keep tanks/drums cool by spraying with water.
Inhalation: *corrosive*; sore throat, cough, shortness of breath, severe breathing difficulties.	ventilation, local exhaust or respiratory protection.	fresh air, rest, place in half-sitting position, artificial respiration if necessary, take to hospital.
Skin: *corrosive*; redness, pain, burns.	gloves.	remove contaminated clothing, flush skin with water or shower, refer to a doctor.
Eyes: *corrosive*; redness, pain, impaired vision.	face shield.	flush with water, take to a doctor if necessary.
Ingestion: *corrosive*; sore throat, abdominal pain.		rinse mouth, call a doctor or take to hospital.

SPILLAGE	STORAGE	LABELING / NFPA
collect leakage in sealable containers, collect solids, soak up with sand or other inert absorbent and remove to safe place.	keep in cool, dark place, separate from oxidants; add an inhibitor.	R: 34 S: 15-26 Note D Corrosive 3 ◆ 2 / 2

NOTES
Lung edema symptoms usually develop several hours later and are aggravated by physical exertion: rest and hospitalization essential. As first aid, a doctor or authorized person should consider administering a corticosteroid spray.

Transport Emergency Card TEC(R)-195

HI: 89; UN-number: 2531

CAS-No: [79-20-9]
acetic acid methyl ester
methyl ethanoate

CH_3COOCH_3

METHYL ACETATE

PHYSICAL PROPERTIES		IMPORTANT CHARACTERISTICS		
Boiling point, °C	57	**COLORLESS LIQUID WITH CHARACTERISTIC ODOR**		
Melting point, °C	−98	Vapor (gas) is heavier than air and spreads at ground level, with risk of ignition at a distance. Do not use compressed air when filling, emptying or processing. Decomposes on contact with acids or bases, giving off → *methanol*. Reacts violently with strong oxidants, with risk of fire and explosion. Attacks various plastics.		
Flash point, °C	−10			
Auto-ignition temperature, °C	454			
Relative density (water = 1)	0.9			
Relative vapor density (air = 1)	2.6			
Relative density at 20 °C of saturated mixture vapor/air (air = 1)	1.34	TLV-TWA	200 ppm	606 mg/m³
		STEL	250 ppm	757 mg/m³
Vapor pressure, mm Hg at 20 °C	167			
Solubility in water, g/100 ml at 20 °C	30	**Absorption route:** Can enter the body by inhalation or ingestion or through the skin. Harmful atmospheric concentrations can build up fairly rapidly on evaporation at approx. 20°C - even more rapidly in aerosol form.		
Explosive limits, vol% in air	3.1-16			
Electrical conductivity, pS/m	3.4 x 10⁸			
Relative molecular mass	74.1	**Immediate effects:** Irritates the eyes, skin and respiratory tract. Liquid destroys the skin's natural oils. Affects the nervous system.		
Log P octanol/water	0.2			
Gross formula:	$C_3H_6O_2$			

HAZARDS/SYMPTOMS	PREVENTIVE MEASURES	FIRE EXTINGUISHING/FIRST AID
Fire: extremely flammable.	keep away from open flame and sparks, no smoking.	powder, alcohol-resistant foam, large quantities of water, carbon dioxide, (halons).
Explosion: forms explosive air-vapor mixtures.	sealed machinery, ventilation, explosion-proof electrical equipment and lighting, non-sparking tools.	in case of fire: keep tanks/drums cool by spraying with water.
Inhalation: sore throat, cough, headache, dizziness, drowsiness.	ventilation, local exhaust or respiratory protection.	fresh air, rest, place in half-sitting position, call a doctor, take to hospital if necessary.
Skin: redness, pain.	gloves.	remove contaminated clothing, flush skin with water or shower.
Eyes: redness, pain, impaired vision.	safety glasses.	flush with water, take to doctor if necessary.
Ingestion: abdominal pain, chest pain, vomiting, drowsiness.		rinse mouth, call a doctor or take to hospital.

SPILLAGE	STORAGE	LABELING / NFPA
collect leakage in sealable containers, soak up with sand or other inert absorbent and remove to safe place; (additional individual protection: breathing apparatus).	keep in fireproof place, separate from oxidants.	R: 11 S: 16-23-29-33 🔥 Flammable NFPA: 3 / 1 / 0

NOTES
Alcohol consumption increases toxic effects.

CAS-No: [105-45-3]
1-methoxybutane-1,3-dione
methyl acetylacetate
methyl 3-oxobutanoate
methyl 3-oxobutyrate

$CH_3COCH_2COOCH_3$

METHYL ACETOACETATE

PHYSICAL PROPERTIES	IMPORTANT CHARACTERISTICS
Boiling point, °C 170 Melting point, °C 28 Flash point, °C 67 Auto-ignition temperature, °C 280 Relative density (water = 1) 1.08 Relative vapor density (air = 1) 4.0 Relative density at 20 °C of saturated mixture vapor/air (air = 1) 1.0 Vapor pressure, mm Hg at 20 °C 0.99 Solubility in water, g/100 ml at 20 °C 38 Relative molecular mass 116 Gross formula: $C_5H_8O_3$	**COLORLESS LIQUID OR COLORLESS CRYSTALS WITH CHARACTERISTIC ODOR** Reacts with strong oxidants. TLV-TWA not available **Absorption route:** Can enter the body by inhalation or ingestion. Harmful atmospheric concentrations build up very slowly, if at all, on evaporation at approx. 20°C, but much more rapidly in aerosol form. **Immediate effects:** Irritates the eyes, skin and respiratory tract. In serious cases risk of unconsciousness.

HAZARDS/SYMPTOMS	PREVENTIVE MEASURES	FIRE EXTINGUISHING/FIRST AID
Fire: combustible.	keep away from open flame, no smoking.	powder, alcohol-resistant foam, water spray, carbon dioxide, (halons).
Explosion: above 67°C: forms explosive air-vapor mixtures.	above 67°C: sealed machinery, ventilation.	
Inhalation: cough, dizziness, heart palpitations.	ventilation, local exhaust or respiratory protection.	fresh air, rest, call a doctor.
Skin: redness, pain.	gloves.	remove contaminated clothing, flush skin with water or shower.
Eyes: redness, pain.	safety glasses.	flush with water, take to a doctor.
Ingestion: sore throat, abdominal pain, dizziness.		rinse mouth, call a doctor.

SPILLAGE	STORAGE	LABELING / NFPA
clean up spillage, flush away remainder with water.		2 2 0

NOTES

Transport Emergency Card TEC(R)-30G37

allylene
propyne
1-propyne

$CH_3C \equiv CH$

METHYL ACETYLENE
(cylinder)

PHYSICAL PROPERTIES		IMPORTANT CHARACTERISTICS	
Boiling point, °C	− 23	**COLORLESS COMPRESSED LIQUEFIED GAS**	
Melting point, °C	− 103	Vapor (gas) is heavier than air and spreads at ground level, with risk of ignition at a distance. Able to form peroxides. Flow, agitation etc. can cause build-up of electrostatic charge due to liquid's low conductivity. Reacts with copper, silver, magnesium and alloys to form shock-sensitive compounds. Reacts violently with oxidants.	
Flash point, °C	flammable gas		
Auto-ignition temperature, °C	?		
Relative density (water = 1)	0.6		
Relative vapor density (air = 1)	1.4		
Vapor pressure, mm Hg at 21 °C	3952	TLV-TWA 1000 ppm 1640 mg/m³	
Explosive limits, vol% in air	1.7 -?		
Minimum ignition energy, mJ	0.11	**Absorption route:** Can enter the body by inhalation.	
Relative molecular mass	40.1	**Immediate effects:** Affects the nervous system.	
Gross formula:	C_3H_4		

HAZARDS/SYMPTOMS	PREVENTIVE MEASURES	FIRE EXTINGUISHING/FIRST AID
Fire: extremely flammable.	keep away from open flame and sparks, no smoking.	shut off supply; if impossible and no danger to surrounding area, allow fire to burn itself out; otherwise extinguish with powder, carbon dioxide, (halons).
Explosion: forms explosive air-gas mixtures.	sealed machinery, ventilation, explosion-proof electrical equipment and lighting, grounding, non-sparking tools.	in case of fire: keep cylinders cool by spraying with water, fight fire from sheltered location.
Inhalation: headache, severe breathing difficulties, drowsiness, unconsciousness.	ventilation, local exhaust or respiratory protection.	fresh air, rest, artificial respiration if necessary, call a doctor.
Skin: rapid evaporation can cause frostbite.	insulating gloves.	*in case of frostbite*: DO NOT remove clothing, just flush with water and call a doctor.
Eyes: *in case of frostbite*: redness, pain, impaired vision.	safety glasses, or combined eye and respiratory protection.	flush with water, take to a doctor.

SPILLAGE	STORAGE	LABELING / NFPA
evacuate area, call in an expert, ventilate, under no circumstances spray liquid with water; (additional individual protection: breathing apparatus).	keep in cool, fireproof place.	 4 2 2

NOTES
Material used for lines must not exceed 63% copper content. High atmospheric concentrations, e.g. in poorly ventilated spaces, can cause oxygen deficiency, with risk of unconsciousness. Turn leaking cylinder so that leak is on top to prevent liquid methyl acetylene escaping.

Transport Emergency Card TEC(R)-20G13

HI: 239; UN-number: 1060

METHYL ACETYLENE-PROPADIENE MIXTURE
(cylinder)

PHYSICAL PROPERTIES	IMPORTANT CHARACTERISTICS
Boiling point, °C −48/−4 Melting point, °C < −100 Flash point, °C flammable gas Auto-ignition temperature, °C 450 Relative density (water = 1) 0.5 Relative vapor density (air = 1) 1.5 Vapor pressure, mm Hg at 20 °C 1.1 - 7.3 Solubility in water poor Explosive limits, vol% in air 2 - 12 Minimum ignition energy, mJ 0.11 Relative molecular mass ca. 43	**COLORLESS COMPRESSED LIQUEFIED GAS WITH CHARACTERISTIC ADDED ODOR** Vapor (gas) is heavier than air and spreads at ground level, with risk of ignition at a distance. Reacts with copper to form shock-sensitive compounds.
	TLV-TWA 1000 ppm 1800 mg/m³
	Absorption route: Can enter the body by inhalation. **Immediate effects:** Liquid can cause frostbite due to rapid evaporation. Affects the nervous system.
Gross formula: C_3H_4	

HAZARDS/SYMPTOMS	PREVENTIVE MEASURES	FIRE EXTINGUISHING/FIRST AID
Fire: extremely flammable.	keep away from open flame and sparks, no smoking.	shut off supply; if impossible and no danger to surrounding area, allow fire to burn itself out; otherwise extinguish with powder, carbon dioxide, (halons).
Explosion: forms explosive air-gas mixtures.	sealed machinery, ventilation, explosion-proof electrical equipment and lighting, grounding, non-sparking tools.	in case of fire: keep cylinders cool by spraying with water, fight fire from sheltered location.
Inhalation: headache, drowsiness, shortness of breath, unconsciousness.	ventilation, local exhaust or respiratory protection.	fresh air, rest, artificial respiration if necessary, call a doctor.
Skin: *in case of frostbite:* redness, pain, blisters, sores.	insulating gloves.	*in case of frostbite:* DO NOT remove clothing, flush skin with water or shower, take to a doctor.
Eyes: *in case of frostbite:* impaired vision, redness, pain.	safety glasses, or combined eye and respiratory protection.	flush with water, take to a doctor.

SPILLAGE	STORAGE	LABELING / NFPA
evacuate area, call in an expert, ventilate, under no circumstances spray liquid with water; (additional individual protection: breathing apparatus).	keep in cool, fireproof place.	

NOTES

Material used for lines must not exceed 63% copper content. Turn leaking cylinder so that leak is on top to prevent liquid methyl acetylene-propadiene mixture escaping. Close valve when not in use; check hoses and lines regularly; test for leaking connections with soap solution. Preventive measures apply to production, filling and storage. Use cylinder with special fittings.

Transport Emergency Card TEC(R)-20G13

HI: 239; UN-number: 1060

CAS-No: [96-33-3]
acrylic acid, methyl ester
methyl propenoate
methyl-2-propenoate
2-propenoic acid, methyl ester

$CH_2 = CHCOOCH_3$

METHYL ACRYLATE

PHYSICAL PROPERTIES		IMPORTANT CHARACTERISTICS
Boiling point, °C	80	**COLORLESS LIQUID WITH PUNGENT ODOR**
Melting point, °C	−75	Vapor is heavier than air and spreads at ground level, with risk of ignition at a distance. Without
Flash point, °C	−3	inhibitor polymerizes readily when exposed to light and heat; with inhibitor polymerizes when
Auto-ignition temperature, °C	390	moderately heated; in both cases evolves heat. Do not use compressed air when filling, emptying
Relative density (water = 1)	0.95	or processing. Reacts violently with oxidants.
Relative vapor density (air = 1)	3.0	

Relative density at 20 °C of saturated mixture vapor/air (air = 1)	1.17	
Vapor pressure, mm Hg at 20 °C	71	TLV-TWA (skin) 10 ppm 35 mg/m³

Absorption route: Can enter the body by inhalation or ingestion or through the skin.
Immediate effects: Corrosive to the eyes, skin and respiratory tract. Inhalation of vapor/fumes can cause severe breathing difficulties (lung edema). In serious cases risk of death.
Effects of prolonged/repeated exposure: Prolonged or repeated contact with skin can cause hypersensitivity. Can cause liver and kidney damage.

Solubility in water, g/100 ml at 20 °C	6
Explosive limits, vol% in air	2.8-25
Relative molecular mass	86.1

Gross formula:	$C_4H_6O_2$

HAZARDS/SYMPTOMS	PREVENTIVE MEASURES	FIRE EXTINGUISHING/FIRST AID
Fire: extremely flammable.	keep away from open flame and sparks, no smoking.	powder, AFFF, foam, carbon dioxide, (halons).
Explosion: forms explosive air-vapor mixtures.	sealed machinery, ventilation, explosion-proof electrical equipment and lighting.	in case of fire: keep tanks/drums cool by spraying with water.
	STRICT HYGIENE	IN ALL CASES CALL A DOCTOR
Inhalation: *corrosive*; sore throat, cough, shortness of breath, severe breathing difficulties.	ventilation, local exhaust or respiratory protection.	fresh air, rest, place in half-sitting position, take to hospital.
Skin: *corrosive*; redness, pain, chemical burns.	gloves, protective clothing.	remove contaminated clothing, flush skin with water or shower, refer to a doctor.
Eyes: *corrosive*; redness, pain, impaired vision.	face shield.	flush with water, take to doctor.
Ingestion: *corrosive*; sore throat, abdominal pain, vomiting.		rinse mouth, take immediately to hospital.

SPILLAGE	STORAGE	LABELING / NFPA
collect leakage in sealable containers, soak up with sand or other inert absorbent and remove to safe place, do not flush into sewer; (additional individual protection: breathing apparatus).	keep refrigerated in dark, fireproof place, separate from oxidants; add an inhibitor.	R: 11-36/37/38-20/22-43 S: 9-16-29-33 🔥 Flammable ✖ Irritant ◇ 3 2 2

NOTES
Depending on the degree of exposure, regular medical checkups are advisable. Lung edema symptoms usually develop several hours later and are aggravated by physical exertion: rest and hospitalization essential. As first aid, a doctor or authorized person should consider administering a corticosteroid spray. Explosive limits apply to methyl acrylate stabilized with 0.5% hydroquinone (by weight). Use airtight packaging.

Transport Emergency Card TEC(R)-177

HI: 339; UN-number: 1919

dimethoxymethane
formal
formaldehyde dimethylacetal
methylene dimethyl ether

$CH_2(OCH_3)_2$

METHYLAL

PHYSICAL PROPERTIES		IMPORTANT CHARACTERISTICS	
Boiling point, °C	45	**COLORLESS LIQUID WITH CHARACTERISTIC ODOR**	
Melting point, °C	−105	Vapor is heavier than air and spreads at ground level, with risk of ignition at a distance. Presumed to be able to form peroxides. Do not use compressed air when filling, emptying or processing. Reacts violently with strong oxidants.	
Flash point, °C	−18		
Auto-ignition temperature, °C	235		
Relative density (water = 1)	0.9		
Relative density at 20 °C of saturated mixture vapor/air (air = 1)	1.7	TLV-TWA 1000 ppm 3110 mg/m³	
Vapor pressure, mm Hg at 20 °C	334	**Absorption route:** Can enter the body by inhalation or ingestion. Harmful atmospheric concentrations can build up very rapidly on evaporation at 20°C.	
Solubility in water, g/100 ml at 20 °C	32		
Explosive limits, vol% in air	?-?	**Immediate effects:** Irritates the eyes, skin and respiratory tract. Liquid destroys the skin's natural oils. In substantial concentrations can impair consciousness. In serious cases risk of unconsciousness.	
Relative molecular mass	76.1		
Log P octanol/water	0		
		Effects of prolonged/repeated exposure: Can cause liver and kidney damage.	
Gross formula:	$C_3H_8O_2$		

HAZARDS/SYMPTOMS	PREVENTIVE MEASURES	FIRE EXTINGUISHING/FIRST AID
Fire: extremely flammable.	keep away from open flame and sparks, no smoking.	powder, alcohol-resistant foam, large quantities of water, carbon dioxide, (halons).
Explosion: forms explosive air-vapor mixtures.	sealed machinery, ventilation, explosion-proof electrical equipment and lighting.	in case of fire: keep tanks/drums cool by spraying with water.
Inhalation: irritation, sore throat, cough, headache, dizziness, drowsiness.	ventilation, local exhaust or respiratory protection.	fresh air, rest, call a doctor.
Skin: redness, pain.	gloves.	remove contaminated clothing, flush skin with water or shower.
Eyes: redness, pain.	safety glasses.	flush with water, take to a doctor if necessary.
Ingestion: abdominal pain, nausea, sleepiness.		rinse mouth, call a doctor.

SPILLAGE	STORAGE	LABELING / NFPA
collect leakage in sealable containers, soak up with sand or other inert absorbent and remove to safe place; (additional individual protection: breathing apparatus).	keep in dark, fireproof place, separate from oxidants	 3 2 ◇ 2

NOTES
Explosive limits not given in the literature. Before distilling check for peroxides; if found, render harmless.

Transport Emergency Card TEC(R)-674

HI: 33; UN-number: 1234

CAS-No: [917-65-7]
MADC

$(CH_3)AlCl_2$

METHYLALUMINUM DICHLORIDE

PHYSICAL PROPERTIES		IMPORTANT CHARACTERISTICS	
Boiling point, °C	152	**COLORLESS FUMING LIQUID WITH PUNGENT ODOR**	
Melting point, °C	−73	Aluminum oxide fumes and → *hydrochloric acid* fumes are a product of combustion. Can ignite spontaneously on exposure to air. Strong reducing agent which reacts violently with oxidants. Reacts violently with water, alcohols, phenols, amines, carbon dioxide, oxides of sulfur and nitrogen oxides, with risk of fire and explosion.	
Relative density (water = 1)	1.36		
Relative vapor density (air = 1)	3.9		
Vapor pressure, mm Hg at 20 °C	3.5		
Solubility in water	reaction		
Relative molecular mass	112.9	TLV-TWA	2 mg/m³
		Absorption route: Can enter the body by inhalation or ingestion. Harmful atmospheric concentrations can build up fairly rapidly on evaporation at approx. 20° C - even more rapidly in aerosol form. **Immediate effects:** Corrosive to the eyes, skin and respiratory tract. Inhalation of vapor/fumes can cause severe breathing difficulties (lung edema). In serious cases risk of death.	
Gross formula:	CH_3AlCl_2		

HAZARDS/SYMPTOMS	PREVENTIVE MEASURES	FIRE EXTINGUISHING/FIRST AID
Fire: extremely flammable, many chemical reactions can cause fire and explosion.	keep away from open flame and sparks, no smoking, avoid contact with air, water and many other substances.	special extinguishing powder, dry sand, NO other extinguishing agents, can reignite after being extinguished.
Explosion: contact with water and many other substances can cause explosion.		in case of fire: keep undamaged cylinders cool by spraying with water, but DO NOT spray substance with water.
	STRICT HYGIENE	
Inhalation: *corrosive*; sore throat, cough, severe breathing difficulties.		fresh air, rest, place in half-sitting position, take to hospital.
Skin: *corrosive*; serious burns, slow healing.	gloves, protective clothing.	remove contaminated clothing, flush skin with water or shower, refer to a doctor.
Eyes: *corrosive*; redness, pain, impaired vision.	face shield, or combined eye and respiratory protection.	flush with water, take to a doctor.
Ingestion: *corrosive*; sore throat, abdominal pain, diarrhea.		rinse mouth, take immediately to hospital.

SPILLAGE	STORAGE	LABELING / NFPA
evacuate area, call in an expert, dilute leakage with diesel/machine oil, soak up/cover with dry sand/table salt and remove to safe place, under no circumstances spray liquid with water; (additional individual protection: CHEMICAL SUIT).	keep in dry, fireproof place under dry nitrogen, separate from all other substances.	

NOTES
Usually sold as 15-30% solution in hydrocarbons, which does not ignite spontaneously on exposure to air. For 15% solution see relevant entry. Reacts violently with extinguishing agents such as halons, carbon dioxide, water, foam and ordinary powder. Lung edema symptoms usually develop several hours later and are aggravated by physical exertion: rest and hospitalization essential. As first aid, a doctor or authorized person should consider administering a corticosteroid spray. Use cylinder with special fittings.

Transport Emergency Card TEC(R)-42G02

HI: X333; UN-number: 2221

METHYLALUMINUM DICHLORIDE

(15% solution in hexane)

PHYSICAL PROPERTIES	IMPORTANT CHARACTERISTICS
Boiling point, °C 69 Melting point, °C −95 Flash point, °C −26 Auto-ignition temperature, °C 240 Relative density (water = 1) 0.8 Relative vapor density (air = 1) 3.0 Relative density at 20 °C of saturated mixture vapor/air (air = 1) 1.3 Vapor pressure, mm Hg at 20 °C 122 Solubility in water reaction Explosive limits, vol% in air 1.1-7.5 Minimum ignition energy, mJ 0.24 Electrical conductivity, pS/m 100	**COLORLESS FUMING LIQUID WITH PUNGENT ODOR** Vapor is heavier than air and spreads at ground level, with risk of ignition at a distance. Flow, agitation etc. can cause build-up of electrostatic charge due to liquid's low conductivity. Do not use compressed air when filling, emptying or processing. Once solvent has evaporated (e.g. on clothing) can ignite spontaneously on exposure to air. Aluminum oxide fumes and → *hydrochloric acid* fumes are a product of combustion. Strong reducing agent which reacts violently with oxidants. Reacts violently with water, alcohols, phenols, amines, halogenated hydrocarbons, carbon dioxide, oxides of sulfur, nitrogen oxides and many other substances, with risk of fire and explosion.

	TLV-TWA	2 mg/m³ ✱

Absorption route: Can enter the body by inhalation or ingestion. Harmful atmospheric concentrations can build up fairly rapidly on evaporation at approx. 20°C - even more rapidly in aerosol form.
Immediate effects: Corrosive to the eyes, skin and respiratory tract. Inhalation of vapor/fumes can cause severe breathing difficulties (lung edema). In serious cases risk of death.
Effects of prolonged/repeated exposure: Can cause liver and kidney damage.

Gross formula: CH₃AlCl₂

HAZARDS/SYMPTOMS	PREVENTIVE MEASURES	FIRE EXTINGUISHING/FIRST AID
Fire: extremely flammable, many chemical reactions can cause fire and explosion.	keep away from open flame and sparks, no smoking, avoid contact with water, alcohols, phenols and many other substances.	special extinguishing powder, dry sand, NO other extinguishing agents, can reignite after being extinguished.
Explosion: forms explosive air-vapor mixtures; contact with water, alcohols, phenols and many other substances can cause explosion.	sealed machinery, ventilation, explosion-proof electrical equipment and lighting, grounding, non-sparking tools.	in case of fire: keep undamaged cylinders cool by spraying with water, but DO NOT spray substance with water.
	STRICT HYGIENE	
Inhalation: *corrosive*; sore throat, cough, severe breathing difficulties.	sealed machinery.	fresh air, rest, place in half-sitting position, take to hospital.
Skin: *corrosive*; redness, serious burns, slow healing.	gloves, protective clothing.	remove contaminated clothing, flush skin with water or shower, call a doctor if necessary.
Eyes: *corrosive*; redness, pain, impaired vision.	face shield, or combined eye and respiratory protection.	flush with water, take to a doctor.
Ingestion: *corrosive*; sore throat, abdominal pain, diarrhea.		rinse mouth, immediately take to hospital.

SPILLAGE	STORAGE	LABELING / NFPA
evacuate area, call in an expert, collect leakage in sealable containers, soak up remainder with dry sand/table salt and remove to safe place, under no circumstances spray liquid with water; (additional individual protection: CHEMICAL SUIT).	keep in dry, fireproof place under inert gas, separate from all other substances.	

NOTES

Physical properties (boiling point, flash point, autoignition point, explosive limits, vapor pressure, relative density, conductivity and ignition energy) are based on hexane. For pure methyl aluminum dichloride see relevant card. ✱ For TLV of *hexane* see relevant card. Reacts violently with extinguishing agents such as water, foam, halons, carbon dioxide and ordinary powder. Remove contaminated clothing immediately and flush with water (risk of spontaneous combustion). Lung edema symptoms usually develop several hours later and are aggravated by physical exertion: rest and hospitalization essential. As first aid, a doctor or authorized person should consider administering a corticosteroid spray. Use cylinder with special fittings.

Transport Emergency Card TEC(R)-42G02

HI: X333; UN-number: 2220

CAS-No: [12542-85-7]
MASC

$(CH_3)_3Al_2Cl_3$

METHYLALUMINUM SESQUICHLORIDE

PHYSICAL PROPERTIES		IMPORTANT CHARACTERISTICS
Boiling point, °C	143	**COLORLESS FUMING LIQUID**
Melting point, °C	23	Aluminum oxide fumes and → *hydrochloric acid* fumes are a product of combustion. Can ignite
Relative density (water = 1)	1.6	spontaneously on exposure to air. Strong reducing agent which reacts violently with oxidants.
Relative vapor density (air = 1)	7.0	Reacts violently with water, alcohols, phenols, amines, carbon dioxide, oxides of sulfur and
Vapor pressure, mm Hg at 20 °C	5.2	nitrogen oxides, with risk of fire and explosion. Reacts with air to form aluminum oxide fumes and
Solubility in water	reaction	→ *hydrochloric acid* fumes.
Relative molecular mass	203.5	

TLV-TWA: 2 mg/m³

Absorption route: Can enter the body by inhalation or ingestion. Harmful atmospheric concentrations can build up fairly rapidly on evaporation at approx. 20°C - even more rapidly in aerosol form.
Immediate effects: Corrosive to the eyes, skin and respiratory tract. Inhalation of vapor/fumes can cause severe breathing difficulties (lung edema). In serious cases risk of death.

Gross formula: $C_3H_9Al_2Cl_3$

HAZARDS/SYMPTOMS	PREVENTIVE MEASURES	FIRE EXTINGUISHING/FIRST AID
Fire: extremely flammable, many chemical reactions can cause fire and explosion.	keep away from open flame and sparks, no smoking, avoid contact with air, water and many other substances.	special extinguishing powder, dry sand, NO other extinguishing agents, can reignite after being extinguished.
Explosion: contact with water and many other substances can cause explosion.		in case of fire: keep tanks/drums cool by spraying with water, but NO direct contact of substance with water.
	STRICT HYGIENE	
Inhalation: *corrosive*; sore throat, cough, severe breathing difficulties.	completely sealed machinery, ventilation, local exhaust or respiratory protection.	fresh air, rest, place in half-sitting position, take to hospital.
Skin: *corrosive*; serious burns, slow healing.	gloves, protective clothing.	remove contaminated clothing, flush skin with water or shower, refer to a doctor.
Eyes: *corrosive*; redness, pain, impaired vision.	face shield, or combined eye and respiratory protection.	flush with water, take to a doctor.
Ingestion: *corrosive*; sore throat, abdominal pain, diarrhea.		rinse mouth, take immediately to hospital.

SPILLAGE	STORAGE	LABELING / NFPA
evacuate area, call in an expert, dilute leakage with diesel/machine oil, soak up/cover with dry sand/table salt and remove to safe place; (additional individual protection: CHEMICAL SUIT).	keep in fireproof place under dry nitrogen, separate from all other substances.	3 / 3 / W

NOTES

Sometimes sold as 15-30% solution in hydrocarbons, which does not ignite spontaneously on exposure to air. For 15% solution see relevant entry. Reacts violently with extinguishing agents such as carbon dioxide and halons. Lung edema symptoms usually develop several hours later and are aggravated by physical exertion: rest and hospitalization essential. As first aid, a doctor or authorized person should consider administering a corticosteroid spray. Use cylinder with special fittings.

Transport Emergency Card TEC(R)-42G02

HI: X333; UN-number: 2220

METHYLALUMINUM SESQUICHLORIDE

(15% solution in hexane)

PHYSICAL PROPERTIES	IMPORTANT CHARACTERISTICS
Boiling point, °C — 69 Melting point, °C — −95 Flash point, °C — −26 Auto-ignition temperature, °C — 240 Relative density (water = 1) — 0.8 Relative vapor density (air = 1) — 3.0 Relative density at 20 °C of saturated mixture vapor/air (air = 1) — 1.3 Vapor pressure, mm Hg at 20 °C — 121.6 Solubility in water — reaction Explosive limits, vol% in air — 1.1-7.5 Minimum ignition energy, mJ — 0.24 Electrical conductivity, pS/m — 100	**COLORLESS FUMING LIQUID** Vapor is heavier than air and spreads at ground level, with risk of ignition at a distance. Flow, agitation etc. can cause build-up of electrostatic charge due to liquid's low conductivity. Once solvent has evaporated (e.g. on clothing) can ignite spontaneously on exposure to air. Aluminum oxide fumes and → *hydrochloric acid* fumes are a product of combustion. Strong reducing agent which reacts violently with oxidants. Reacts violently with water, alcohols, phenols, amines, carbon dioxide, nitrogen oxides and many other substances, with risk of fire and explosion.

	TLV-TWA	2 mg/m³ ✶

Absorption route: Can enter the body by inhalation or ingestion. Harmful atmospheric concentrations can build up fairly rapidly on evaporation at approx. 20°C - even more rapidly in aerosol form.
Immediate effects: Corrosive to the eyes, skin and respiratory tract. Inhalation of vapor/fumes can cause severe breathing difficulties (lung edema). In serious cases risk of death.
Effects of prolonged/repeated exposure: Can cause liver and kidney damage.

Gross formula: $C_3H_9Al_2Cl_3$

HAZARDS/SYMPTOMS	PREVENTIVE MEASURES	FIRE EXTINGUISHING/FIRST AID
Fire: extremely flammable, many chemical reactions can cause fire and explosion.	keep away from open flame and sparks, no smoking, avoid contact with water, alcohols, phenols and many other substances.	special extinguishing powder, dry sand, NO other extinguishing agents, can reignite after being extinguished.
Explosion: forms explosive air-vapor mixtures, contact with water and many other substances can cause explosion.	sealed machinery, ventilation, explosion-proof electrical equipment and lighting, grounding.	in case of fire: keep undamaged cylinders cool by spraying with water, but DO NOT spray substance with water.
	STRICT HYGIENE	
Inhalation: *corrosive*; sore throat, cough, severe breathing difficulties.	sealed machinery, local exhaust or respiratory protection.	fresh air, rest, place in half-sitting position, take to hospital.
Skin: *corrosive*; redness, pain, serious burns, slow healing.	gloves, protective clothing.	remove contaminated clothing, flush skin with water or shower, call a doctor if necessary.
Eyes: *corrosive*; redness, pain, impaired vision.	face shield, or combined eye and respiratory protection.	flush with water, take to a doctor.
Ingestion: *corrosive*; sore throat, abdominal pain, diarrhea.		rinse mouth, immediately take to hospital.

SPILLAGE	STORAGE	LABELING / NFPA
evacuate area, call in an expert, collect leakage in sealable containers, soak up with dry sand/table salt, under no circumstances spray liquid with water; (additional individual protection: P2 respirator).	keep in dry, fireproof place under inert gas, separate from all other substances.	3 / 3 / W

NOTES

Physical properties (boiling point, melting point, autoignition point, explosive limits, vapor pressure, relative density, conductivity and ignition energy) are based on hexane. For pure methyl aluminum sesquichloride see relevant card. ✶ For TLV of *hexane* see relevant card. Reacts violently with extinguishing agents such as water, foam, halons, carbon dioxide and ordinary powder. Remove contaminated clothing immediately and flush with water (risk of spontaneous combustion). Lung edema symptoms usually develop several hours later and are aggravated by physical exertion: rest and hospitalization essential. As first aid, a doctor or authorized person should consider administering a corticosteroid spray. Use cylinder with special fittings.

Transport Emergency Card TEC(R)-42G02

HI: X333; UN-number: 2221

CAS-No: [74-89-5]
aminomethane
methanamine
monomethylamine

CH_3NH_2

METHYLAMINE
(in water, 40%)

PHYSICAL PROPERTIES	IMPORTANT CHARACTERISTICS
	COLORLESS LIQUID WITH PUNGENT ODOR

PHYSICAL PROPERTIES	
Boiling point, °C	48
Melting point, °C	−39
Flash point, °C	−10
Auto-ignition temperature, °C	430
Relative density (water = 1)	0.85
Relative vapor density (air = 1)	1.08
Relative density at 20 °C of saturated mixture vapor/air (air = 1)	1.02
Vapor pressure, mm Hg at 20 °C	236
Solubility in water	∞
Explosive limits, vol% in air	4.9-20.8
Relative molecular mass	31.1
Log P octanol/water	−0.6

Gross formula: CH_5N

COLORLESS LIQUID WITH PUNGENT ODOR
Vapor mixes readily with air, forming explosive mixtures. Do not use compressed air when filling, emptying or processing. Reacts with mercury to form shock-sensitive compounds. In aqueous solution is a strong base which reacts violently with acids and corrodes aluminum, zinc etc. Attacks many metals.

TLV-TWA	10 ppm	13 mg/m³

Absorption route: Can enter the body by inhalation or ingestion or through the skin. Harmful atmospheric concentrations can build up very rapidly on evaporation at 20°C.
Immediate effects: Corrosive to the eyes, skin and respiratory tract. Inhalation of vapor/fumes can cause severe breathing difficulties (lung edema). In serious cases risk of death.

HAZARDS/SYMPTOMS	PREVENTIVE MEASURES	FIRE EXTINGUISHING/FIRST AID
Fire: extremely flammable.	keep away from open flame and sparks, no smoking.	powder, alcohol-resistant foam, water spray, carbon dioxide, (halons).
Explosion:	sealed machinery, ventilation, explosion-proof electrical equipment and lighting.	
Inhalation: *corrosive*; sore throat, cough, shortness of breath, severe breathing difficulties, vomiting, diarrhea, abdominal cramps.	ventilation, local exhaust or respiratory protection.	fresh air, rest, place in half-sitting position, take to hospital.
Skin: *corrosive*; redness, pain.	gloves, protective clothing.	remove contaminated clothing, flush skin with water or shower, refer to a doctor.
Eyes: *corrosive*; redness, pain, impaired vision.	face shield, or combined eye and respiratory protection.	flush with water, take to a doctor.
Ingestion: *corrosive*; diarrhea, abdominal cramps, vomiting.		rinse mouth, take immediately to hospital.

SPILLAGE	STORAGE	LABELING / NFPA
collect leakage in sealable containers, flush away remainder with water, combat gas cloud with water curtain; (additional individual protection: breathing apparatus).	keep in cool, fireproof place, separate from acids and mercury.	R: 13-36/37 S: 16-26-29 Note C Flammable Irritant 3 4 0

NOTES
Lung edema symptoms usually develop several hours later and are aggravated by physical exertion: rest and hospitalization essential. As first aid, a doctor or authorized person should consider administering a corticosteroid spray. Unbreakable packaging preferred; if breakable, keep in unbreakable container.

Transport Emergency Card TEC(R)-72825

HI: 338; UN-number: 1235

CAS-No: [100-61-8]
anilinomethane
N-methylaminobenzene
N-methylbenzeneamine
methylphenylamine
N-methylphenylamine

$C_6H_5\text{-NHCH}_3$

N-METHYLANILINE

PHYSICAL PROPERTIES		IMPORTANT CHARACTERISTICS	
Boiling point, °C	196	**COLORLESS TO YELLOW VISCOUS LIQUID WITH CHARACTERISTIC ODOR, TURNING BROWN ON EXPOSURE TO AIR**	
Melting point, °C	−57	Vapor mixes readily with air. Toxic vapors ($\rightarrow$ *nitrous vapors*) are a product of combustion. Decomposes when heated, giving off toxic vapors. Reacts violently with concentrated strong acids and strong oxidants. Attacks plastics.	
Flash point, °C	80		
Relative density (water = 1)	0.99		
Relative vapor density (air = 1)	3.7		
Relative density at 20 °C of saturated mixture		TLV-TWA (skin) 0.5 ppm 2.2 mg/m³	
vapor/air (air = 1)	1.0		
Vapor pressure, mm Hg at 20 °C	0.30	**Absorption route:** Can enter the body by inhalation or ingestion or through the skin.	
Solubility in water	poor	**Immediate effects:** In serious cases risk of unconsciousness.	
Relative molecular mass	107.2	**Effects of prolonged/repeated exposure:** Can affect the blood. Can cause liver and kidney damage.	
Log P octanol/water	1.7		
Gross formula: C_7H_9N			

HAZARDS/SYMPTOMS	PREVENTIVE MEASURES	FIRE EXTINGUISHING/FIRST AID
Fire: combustible.	keep away from open flame, no smoking.	powder, AFFF, foam, carbon dioxide, (halons).
Explosion: above 80°C: forms explosive air-vapor mixtures.	above 80°C: sealed machinery, ventilation, grounding.	in case of fire: keep tanks/drums cool by spraying with water.
	STRICT HYGIENE	IN ALL CASES CALL IN A DOCTOR
Inhalation: shortness of breath, dizziness, blue skin, feeling of weakness.	ventilation, local exhaust or respiratory protection.	fresh air, rest, take to hospital.
Skin: *is absorbed*; see also 'Inhalation'.	gloves, protective clothing.	remove contaminated clothing, wash skin with soap and water, call a doctor.
Eyes: redness, pain.	face shield, or combined eye and respiratory protection.	flush with water, take to a doctor if necessary.
Ingestion: see also 'Inhalation'.	do not eat, drink or smoke while working.	rinse mouth, immediately take to hospital.

SPILLAGE	STORAGE	LABELING / NFPA
collect leakage in sealable containers, soak up with sand or other inert absorbent and remove to safe place; (additional individual protection: CHEMICAL SUIT).	keep separate from oxidants and acids; ventilate at floor level.	R: 23/24/25-33 S: 28-37-44 ☠ Toxic NFPA: 2 / 3 / 0

NOTES
Depending on the degree of exposure, regular medical checkups are advisable. Effects on blood due to formation of methemoglobin. Special first aid required in the event of poisoning: antidotes must be available (with instructions).

Transport Emergency Card TEC(R)-846

HI: 60; UN-number: 2294

METHYLATED SPIRIT

(80-90% ethanol)

PHYSICAL PROPERTIES		IMPORTANT CHARACTERISTICS	
Boiling point, °C	77	**LIGHT BLUE LIQUID WITH CHARACTERISTIC ODOR**	
Melting point, °C	−80	Vapor mixes readily with air, forming explosive mixtures. Do not use compressed air when filling, emptying or processing. Reacts violently with strong oxidants. Commercial product made up of 80-90% ethanol denatured with approx. 3% methanol and 2-7% aldehydes, ketones and esters.	
Flash point, °C	12		
Auto-ignition temperature, °C	425		
Relative density (water = 1)	0.8		
Relative vapor density (air = 1)	1.5	TLV-TWA	1000 ppm	1880 mg/m³
Relative density at 20 °C of saturated mixture vapor/air (air = 1)	ca. 1.03	**Absorption route:** Can enter the body by inhalation or ingestion. Harmful atmospheric concentrations build up fairly slowly on evaporation at 20°C, but much more rapidly in aerosol form.	
Vapor pressure, mm Hg at 28 °C	ca. 45		
Solubility in water	∞		
Explosive limits, vol% in air	3-15	**Immediate effects:** Irritates the eyes and respiratory tract. Liquid destroys the skin's natural oils. Affects the nervous system. If swallowed can impair the vision.	
Log P octanol/water	−0.3		

HAZARDS/SYMPTOMS	PREVENTIVE MEASURES	FIRE EXTINGUISHING/FIRST AID
Fire: extremely flammable.	keep away from open flame and sparks, no smoking.	powder, alcohol-resistant foam, large quantities of water, carbon dioxide, (halons).
Explosion: forms explosive air-vapor mixtures.	sealed machinery, ventilation, explosion-proof electrical equipment and lighting.	in case of fire: keep tanks/drums cool by spraying with water.
Inhalation: headache, dizziness, drowsiness.	ventilation, local exhaust or respiratory protection.	fresh air, rest, call a doctor.
Skin: redness.	gloves.	remove contaminated clothing, flush skin with water or shower.
Eyes: redness, pain.	safety glasses.	flush with water, take to a doctor if necessary.
Ingestion: headache, dizziness, drowsiness.		rinse mouth, call a doctor or take to hospital.

SPILLAGE	STORAGE	LABELING / NFPA
collect leakage in sealable containers, flush away remainder with water.	keep in fireproof place, separate from oxidants.	R: 11 S: 7-16 🔥 Flammable

NOTES
Alcohol consumption increases toxic effects.

METHYL BENZOATE

PHYSICAL PROPERTIES	IMPORTANT CHARACTERISTICS
Boiling point, °C 199 Melting point, °C − 12 Flash point, °C 83 Relative density (water = 1) 1.1 Relative vapor density (air = 1) 4.7 Relative density at 20 °C of saturated mixture vapor/air (air = 1) 1.0 Vapor pressure, mm Hg at 20 °C 0.19 Solubility in water none Electrical conductivity, pS/m 1.4×10^9 Relative molecular mass 136.1 Log P octanol/water 2.1 Gross formula: $C_8H_8O_2$	**COLORLESS LIQUID WITH CHARACTERISTIC ODOR** Vapor mixes readily with air. TLV-TWA not available **Absorption route:** Can enter the body by inhalation or ingestion. Insufficient data on the rate at which harmful concentrations can build up. **Immediate effects:** Irritates the eyes, skin and respiratory tract.

HAZARDS/SYMPTOMS	PREVENTIVE MEASURES	FIRE EXTINGUISHING/FIRST AID
Fire: combustible.	keep away from open flame, no smoking.	powder, AFFF, foam, carbon dioxide, (halons).
Explosion: above 83°C: forms explosive air-vapor mixtures.	above 83°C: sealed machinery, ventilation.	
Inhalation: sore throat, cough, shortness of breath.	ventilation, local exhaust or respiratory protection.	fresh air, rest, call a doctor.
Skin: redness, pain.	gloves.	remove contaminated clothing, flush skin with water or shower.
Eyes: redness, pain.	acid goggles.	flush with water, take to a doctor.
Ingestion: abdominal pain, diarrhea, vomiting.		rinse mouth, call a doctor or take to hospital.

SPILLAGE	STORAGE	LABELING / NFPA
collect leakage in sealable containers, soak up with sand or other inert absorbent and remove to safe place.	ventilate at floor level.	◇ 2 / 0 0

NOTES

Transport Emergency Card TEC(R)-30G15

UN-number: 2938

CAS-No: [98-85-1]
(1-hydroxyethyl)benzene
methylphenylcarbinol
1-phenethyl alcohol
styrallyl alcohol

$C_6H_5CH(CH_3)OH$

α-METHYLBENZYL ALCOHOL

PHYSICAL PROPERTIES	IMPORTANT CHARACTERISTICS
Boiling point, °C 204 Melting point, °C 21 Flash point, °C 96 Relative density (water = 1) 1.0 Relative vapor density (air = 1) 4.21 Relative density at 20 °C of saturated mixture vapor/air (air = 1) 1.0 Vapor pressure, mm Hg at 20 °C 0.099 Solubility in water poor Relative molecular mass 122.2 Gross formula: $C_8H_{10}O$	**COLORLESS LIQUID WITH CHARACTERISTIC ODOR** Vapor mixes readily with air. Reacts with strong oxidants. TLV-TWA not available **Absorption route:** Can enter the body by inhalation or ingestion. Harmful atmospheric concentrations build up very slowly, if at all, on evaporation at approx. 20°C, but much more rapidly in aerosol form. **Immediate effects:** Irritates the eyes, upper respiratory tract and skin.

HAZARDS/SYMPTOMS	PREVENTIVE MEASURES	FIRE EXTINGUISHING/FIRST AID
Fire: combustible.	keep away from open flame, no smoking.	powder, AFFF, foam, carbon dioxide, (halons).
Explosion: above 96°C: forms explosive air-vapor mixtures.	above 96°C: sealed machinery, ventilation.	
Inhalation: *irritation.*	ventilation.	fresh air, rest, call a doctor if necessary.
Skin: *irritation.*	gloves.	remove contaminated clothing, flush skin with water or shower.
Eyes: *irritation;* redness, pain.	safety glasses.	flush with water, take to a doctor if necessary.
Ingestion:		rinse mouth, call a doctor.

SPILLAGE	STORAGE	LABELING / NFPA
evacuate area, call in an expert, ventilate, under no circumstances spray liquid with water; (additional individual protection: GAS SUIT).		

NOTES
Insufficient toxicological data on harmful effects to humans.

Transport Emergency Card TEC(R)-30G15

585

METHYL BROMIDE
(cylinder)

PHYSICAL PROPERTIES	IMPORTANT CHARACTERISTICS
Boiling point, °C .... 4 Melting point, °C .... − 93 Auto-ignition temperature, °C ... 535 Relative density (water = 1) ..... 1.7 Relative vapor density (air = 1) .. 3.3 Vapor pressure, mm Hg at 20 °C .. 684 Solubility in water, g/100 ml at 20 °C .. 1.5 Relative molecular mass .... 95.0 Gross formula: CH$_3$Br	**COMPRESSED LIQUEFIED GAS** Gas is heavier than air. Decomposes in flame or on hot surface, giving off toxic and corrosive vapors. Reacts with water to form a voluminous hydrate. Reacts violently with aluminum, with risk of fire and explosion. Attacks zinc, magnesium and alkali metals.

TLV-TWA (skin)	5 ppm	19 mg/m^3

Absorption route: Can enter the body by inhalation or through the skin. Harmful atmospheric concentrations can build up very rapidly if gas is released.
Immediate effects: Corrosive to the respiratory tract. Can cause frostbite due to rapid evaporation. Affects the nervous system. In substantial concentrations can cause shaking, seizures etc. Can impair the vision. Inhalation can cause lung edema. In serious cases risk of death. Can cause kidney damage.
Effects of prolonged/repeated exposure: Can affect the central nervous system, causing feeling of weakness, muscle pain, confusion. Can cause kidney damage.

HAZARDS/SYMPTOMS	PREVENTIVE MEASURES	FIRE EXTINGUISHING/FIRST AID
Fire: non-combustible.	avoid contact with aluminum.	in case of fire in immediate vicinity: use any extinguishing agent.
Explosion:		in case of fire: keep cylinder cool by spraying with water.
	STRICT HYGIENE	
Inhalation: *corrosive*; sore throat, severe breathing difficulties, headache, nausea, vomiting, unconsciousness.	ventilation, local exhaust or respiratory protection.	fresh air, rest, place in half-sitting position, artificial respiration if necessary, take to hospital.
Skin: *is absorbed*; *in case of frostbite:* redness, pain, serious burns.	insulated gloves.	*in case of frostbite:* DO NOT remove clothing, flush skin with water or shower, refer to a doctor.
Eyes: redness, pain.	face shield, or combined eye and respiratory protection.	flush with water, take to a doctor.

SPILLAGE	STORAGE	LABELING / NFPA
evacuate area, call in an expert, ventilate, under no circumstances spray liquid with water; (additional individual protection: CHEMICAL SUIT).	if stored indoors, keep in cool, fireproof place; ventilate at floor level.	◇ 3 / 1 / 0

NOTES

Under certain circumstances can give off combustible vapor-air mixtures (8.6-20% by vol.) which are difficult to ignite. Addition of small quantities of dust or raising of oxygen content make substance flammable. Odor limit is above TLV. Bromide level in blood is one recommended yardstick for exposure. Depending on the degree of exposure, regular medical checkups are advisable. Lung edema symptoms usually develop several hours later and are aggravated by physical exertion: rest and hospitalization essential. As first aid, a doctor or authorized person should consider administering a corticosteroid spray. Under no circumstances use near flame or hot surface or when welding. Turn leaking cylinder so that leak is on top to prevent liquid methyl bromide escaping. Use cylinder with special fittings.

Transport Emergency Card TEC(R)-111

HI: 26; UN-number: 1062

CAS-No: [1634-04-4]
tert-butyl methyl ether
2-methoxy-2-methyl propane
methyl 1,1-dimethylethyl ether
MTBE

$(CH_3)_3COCH_3$

METHYL-tert-BUTYLETHER

PHYSICAL PROPERTIES	IMPORTANT CHARACTERISTICS
Boiling point, °C 55 Melting point, °C − 109 Flash point, °C − 28 Auto-ignition temperature, °C 460 Relative density (water = 1) 0.75 Relative vapor density (air = 1) 3.1 Relative density at 20 °C of saturated mixture vapor/air (air = 1) 1.5 Vapor pressure, mm Hg at 25 °C 249 Solubility in water, g/100 ml at 20 °C 5.1 Explosive limits, vol% in air 1-8 Electrical conductivity, pS/m 1.6×10^4 Relative molecular mass 88.2	**COLORLESS LIQUID WITH CHARACTERISTIC ODOR** Vapor is heavier than air and spreads at ground level, with risk of ignition at a distance. Presumed to be able to form peroxides. Do not use compressed air when filling, emptying or processing. Reacts with strong oxidants.
	TLV-TWA not available
	Absorption route: Can enter the body by inhalation or ingestion or through the skin. Harmful atmospheric concentrations can build up very rapidly on evaporation at approx. 20°C. **Immediate effects:** Irritates the eyes, skin and respiratory tract. Liquid destroys the skin's natural oils. Affects the nervous system. In substantial concentrations can impair consciousness. In serious cases risk of unconsciousness.
Gross formula: $C_5H_{12}O$	

HAZARDS/SYMPTOMS	PREVENTIVE MEASURES	FIRE EXTINGUISHING/FIRST AID
Fire: extremely flammable.	keep away from open flame and sparks, no smoking.	powder, water spray, foam, carbon dioxide, (halons).
Explosion: forms explosive air-vapor mixtures.	sealed machinery, ventilation, explosion-proof electrical equipment and lighting.	in case of fire: keep tanks/drums cool by spraying with water.
Inhalation: dizziness, sleepiness, unconsciousness.	ventilation, local exhaust or respiratory protection.	fresh air, rest, artificial respiration if necessary.
Skin: *is absorbed*; irritation, redness.	gloves.	remove contaminated clothing, flush skin with water or shower.
Eyes: redness.	face shield.	flush with water, take to a doctor.
Ingestion: vomiting, sleepiness.		rinse mouth, DO NOT induce vomiting, call a doctor or take to hospital.

SPILLAGE	STORAGE	LABELING / NFPA
collect leakage in sealable containers, soak up with sand or other inert absorbent and remove to safe place; (additional individual protection: breathing apparatus).	keep in cool, fireproof place, separate from oxidants.	R: 11-36/38 S: 24/25 ✖ Irritant

NOTES
Before distilling check for peroxides; if found, render harmless. Alcohol consumption increases toxic effects.

Transport Emergency Card TEC(R)-30G30

HI: 33; UN-number: 2398

CAS-No: [591-78-6] $C_4H_9COCH_3$
2-hexanone
MBK
methyl-n-butyl ketone

METHYL BUTYL KETONE

PHYSICAL PROPERTIES	IMPORTANT CHARACTERISTICS
Boiling point, °C 127 Melting point, °C −57 Flash point, °C 23 Auto-ignition temperature, °C 423 Relative density (water = 1) 0.8 Relative vapor density (air = 1) 3.5 Relative density at 20 °C of saturated mixture vapor/air (air = 1) 1.01 Vapor pressure, mm Hg at 20 °C 2.7 Solubility in water, g/100 ml at 20 °C 2 Explosive limits, vol% in air 1.2-8.0 Relative molecular mass 100.2	**COLORLESS LIQUID WITH CHARACTERISTIC ODOR** Vapor mixes readily with air. Reacts violently with light metals and strong oxidants. Attacks many plastics. TLV-TWA (skin) 5 ppm 20 mg/m³ **Absorption route:** Can enter the body by inhalation or ingestion or through the skin. Harmful atmospheric concentrations can build up fairly rapidly on evaporation at approx. 20°C - even more rapidly in aerosol form. **Immediate effects:** Irritates the eyes, skin and respiratory tract. Liquid destroys the skin's natural oils. In substantial concentrations can cause paralysis, tingling in arms and legs and impairment of consciousness.
Gross formula: $C_6H_{12}O$	

HAZARDS/SYMPTOMS	PREVENTIVE MEASURES	FIRE EXTINGUISHING/FIRST AID
Fire: flammable.	keep away from open flame and sparks, no smoking.	powder, AFFF, foam, carbon dioxide, (halons).
Explosion: above 23°C: forms explosive air-vapor mixtures.	above 23°C: sealed machinery, ventilation, explosion-proof electrical equipment.	in case of fire: keep tanks/drums cool by spraying with water.
Inhalation: cough, headache, drowsiness.	ventilation, local exhaust or respiratory protection.	fresh air, rest, call a doctor.
Skin: *is absorbed*; redness.	gloves, protective clothing.	remove contaminated clothing, flush skin with water or shower, call a doctor.
Eyes: irritation, redness, pain.	face shield.	flush with water, take to doctor if necessary.
Ingestion: sore throat, abdominal pain, diarrhea, nausea.		rinse mouth, DO NOT induce vomiting, call a doctor or take to hospital.

SPILLAGE	STORAGE	LABELING / NFPA
collect leakage in sealable containers, soak up with sand or other inert absorbent and remove to safe place; (additional individual protection: breathing apparatus).	keep in fireproof place, separate from oxidants.	R: 11-23-48 S: 9-16-29-44-51 Toxic Flammable 3 2 0

NOTES
Alcohol consumption increases toxic effects. Depending on the degree of exposure, regular medical checkups are advisable.

Transport Emergency Card TEC(R)-30G08

METHYL CHLORIDE
(cylinder)

PHYSICAL PROPERTIES	IMPORTANT CHARACTERISTICS

PHYSICAL PROPERTIES

Boiling point, °C	−24
Melting point, °C	−98
Flash point, °C	flammable gas
Auto-ignition temperature, °C	630
Relative density (water = 1)	0.9
Relative vapor density (air = 1)	1.8
Vapor pressure, mm Hg at 20 °C	3800
Solubility in water	poor
Explosive limits, vol% in air	7.1-18.5
Relative molecular mass	50.5

Gross formula: CH₃Cl

IMPORTANT CHARACTERISTICS

COLORLESS COMPRESSED LIQUEFIED GAS
Gas is heavier than air and spreads at ground level, with risk of ignition at a distance. Toxic and corrosive vapors are a product of combustion (→ *hydrochloric acid*). Reacts violently with aluminum, with risk of fire and explosion. Reacts with magnesium, zinc, anhydrous ammonia and amines. Can react with acetylene and fluorine.

TLV-TWA (skin)	50 ppm	103 mg/m³
STEL	100 ppm	207 mg/m³

Absorption route: Can enter the body by inhalation. Harmful atmospheric concentrations can build up very rapidly if gas is released.
Immediate effects: Can cause frostbite due to rapid evaporation. In substantial concentrations can cause shaking, seizures etc. In serious cases risk of unconsciousness.
Effects of prolonged/repeated exposure: Can cause liver and kidney damage.

HAZARDS/SYMPTOMS	PREVENTIVE MEASURES	FIRE EXTINGUISHING/FIRST AID
Fire: extremely flammable.	keep away from open flame and sparks, no smoking.	shut off supply; if impossible and no danger to surrounding area, allow fire to burn itself out; otherwise extinguish with powder, carbon dioxide, (halons).
Explosion: forms explosive air-gas mixtures.	sealed machinery, ventilation, explosion-proof electrical equipment and lighting.	in case of fire: keep cylinder cool by spraying with water.
Inhalation: headache, dizziness, nausea, sleepiness.	ventilation, local exhaust or respiratory protection.	fresh air, rest, call a doctor.
Skin: *in case of frostbite:* redness, pain, serious burns.	insulated gloves.	*in case of frostbite:* DO NOT remove clothing, flush skin with water or shower, call a doctor.
Eyes: redness, pain.	face shield, or combined eye and respiratory protection.	flush with water, take to a doctor if necessary.

SPILLAGE	STORAGE	LABELING / NFPA
evacuate area, call in an expert, ventilate, under no circumstances spray liquid with water; (additional individual protection: breathing apparatus).	keep in cool, fireproof place.	4 / 2 / 0

NOTES

Alcohol consumption increases toxic effects. Turn leaking cylinder so that leak is on top to prevent liquid methyl chloride escaping.

METHYL CHLOROFORM

PHYSICAL PROPERTIES		IMPORTANT CHARACTERISTICS	
Boiling point, °C	74	**COLORLESS LIQUID WITH CHARACTERISTIC ODOR**	
Melting point, °C	−32	Decomposes in flame or on hot surface, giving off toxic (→ *chlorine*, → *phosgene*) and corrosive	
Auto-ignition temperature, °C	537	(→ *hydrochloric acid*) vapors. Decomposes when heated above 177°C. Reacts violently with	
Relative density (water = 1)	1.3	alkaline-earth and alkali metals, various metal powders and amides.	
Relative vapor density (air = 1)	4.6		
Relative density at 20 °C of saturated mixture		TLV-TWA	350 ppm	1910 mg/m³
vapor/air (air = 1)	1.5	STEL	450 ppm	2460 mg/m³
Vapor pressure, mm Hg at 20 °C	101		
Solubility in water, g/100 ml	0.05	**Absorption route:** Can enter the body by inhalation or ingestion. Harmful atmospheric	
Electrical conductivity, pS/m	10³	concentrations can build up fairly rapidly on evaporation at approx. 20°C - even more rapidly in	
Relative molecular mass	133.4	aerosol form.	
		Immediate effects: Irritates the eyes, skin and respiratory tract. Liquid destroys the skin's natural	
		oils. Affects the nervous system.	
Gross formula:	$C_2H_3Cl_3$		

HAZARDS/SYMPTOMS	PREVENTIVE MEASURES	FIRE EXTINGUISHING/FIRST AID
Fire: non-combustible.		in case of fire in immediate vicinity: use any extinguishing agent.
Explosion: explosive air-vapor mixtures are difficult to ignite.		in case of fire: keep tanks/drums cool by spraying with water.
Inhalation: sore throat, headache, dizziness, drowsiness.	ventilation, local exhaust or respiratory protection.	fresh air, rest, call a doctor.
Skin: redness, pain.	gloves.	remove contaminated clothing, flush skin with water or shower, call a doctor if necessary.
Eyes: redness, pain.	safety glasses.	flush with water, take to a doctor if necessary.
Ingestion: sore throat, abdominal pain, headache, nausea, drowsiness.		rinse mouth, call a doctor or take to hospital.

SPILLAGE	STORAGE	LABELING / NFPA
collect leakage in sealable containers, soak up with sand or other inert absorbent and remove to safe place; (additional individual protection: breathing apparatus).	ventilate at floor level.	R: 20 S: 24/25 Note F ❌ Harmful

NFPA diamond: 1 (top), 2 (left), 0 (right)

NOTES
Under certain conditions can form combustible vapor-air mixtures (8-15.5% by vol.), which are difficult to ignite. Addition of small quantities of combustible substances or raising of oxygen content increases flammability considerably. Residues of evaporation can be combustible. Do not use aluminum packaging for unstabilized methyl chloroform. Alcohol consumption increases toxic effects. Chlorothene and Chlorothane are trade names. Under no circumstances use near flame or hot surface or when welding. Use airtight packaging.

Transport Emergency Card TEC(R)-721

HI: 60; UN-number: 2831

METHYL CHLOROFORMATE

PHYSICAL PROPERTIES	IMPORTANT CHARACTERISTICS
Boiling point, °C 71 Melting point, °C −61 Flash point, °C 5 Auto-ignition temperature, °C 485 Relative density (water = 1) 1.2 Relative vapor density (air = 1) 3.2 Relative density at 20 °C of saturated mixture vapor/air (air = 1) 1.3 Vapor pressure, mm Hg at 20 °C 102 Solubility in water poor Explosive limits, vol% in air 10.8-? Relative molecular mass 94.5	**COLORLESS LIQUID WITH PUNGENT ODOR** Vapor is heavier than air and spreads at ground level, with risk of ignition at a distance. Do not use compressed air when filling, emptying or processing. Toxic and corrosive vapors are a product of combustion. Decomposes when heated, giving off toxic vapors (→ *phosgene*). Reacts violently with strong oxidants. Reacts slowly with water, giving off → *methanol*, → *carbon dioxide* and → *hydrochloric acid*. Reacts with air to form corrosive vapors (→ *hydrochloric acid*). Attacks many metals, esp. in the presence of moisture.
	TLV-TWA not available
	Absorption route: Can enter the body by inhalation or ingestion or through the skin. **Immediate effects:** Corrosive to the eyes, skin and respiratory tract. Inhalation of vapor/fumes can cause severe breathing difficulties (lung edema). In serious cases risk of death.
Gross formula: C$_2$H$_3$ClO$_2$	

HAZARDS/SYMPTOMS	PREVENTIVE MEASURES	FIRE EXTINGUISHING/FIRST AID
Fire: extremely flammable.	keep away from open flame and sparks, no smoking.	powder, water spray, foam, carbon dioxide, (halons).
Explosion: forms explosive air-vapor mixtures.	sealed machinery, ventilation, explosion-proof electrical equipment and lighting.	in case of fire: keep tanks/drums cool by spraying with water.
Inhalation: *corrosive*; sore throat, cough, severe breathing difficulties, nausea.	ventilation, local exhaust or respiratory protection.	fresh air, rest, place in half-sitting position, take to hospital.
Skin: *is absorbed, corrosive*; redness, pain, serious burns.	gloves, protective clothing.	remove contaminated clothing, flush skin with water or shower, call a doctor if necessary.
Eyes: *corrosive*; redness, pain, impaired vision.	face shield, or combined eye and respiratory protection.	flush with water, take to a doctor.
Ingestion: *corrosive*; abdominal pain, diarrhea, vomiting.		rinse mouth, take immediately to hospital.

SPILLAGE	STORAGE	LABELING / NFPA
evacuate area, render harmless with ammonia water, collect leakage in sealable containers, soak up with sand or other inert absorbent and remove to safe place, flush away remainder with water; (additional individual protection: CHEMICAL SUIT).	keep in cool, dry, fireproof place, separate from oxidants; ventilate.	R: 11-23-36/37/38 S: 9-16-33-44 Flammable Toxic

NOTES
Lung edema symptoms usually develop several hours later and are aggravated by physical exertion: rest and hospitalization essential. As first aid, a doctor or authorized person should consider administering a corticosteroid spray. Use airtight packaging.

Transport Emergency Card TEC(R)-30G33

HI: 336; UN-number: 1238

4-chloro-2-methylphenoxyacetic acid
4-chloro-o-cresoxyacetic acid
4-chloro-o-toloxyacetic acid
MCP
MCPA

$HOOCCH_2OC_6H_3CH_3Cl$

2-METHYL-4-CHLOROPHENOXYACETIC ACID

PHYSICAL PROPERTIES	IMPORTANT CHARACTERISTICS
Boiling point, °C ... ? Melting point, °C ... 118 Vapor pressure, mm Hg at °C ... ? Solubility in water, g/100 ml at 20 °C ... 0.1 Relative molecular mass ... 200.6 Gross formula: $C_9H_9ClO_3$	**WHITE TO YELLOW POWDER WITH CHARACTERISTIC ODOR** Decomposes when heated, giving off toxic and corrosive vapors. TLV-TWA not available **Absorption route:** Can enter the body by inhalation or ingestion. Evaporation negligible at 20°C, but harmful concentrations of airborne particles can build up rapidly. **Immediate effects:** Irritates the eyes, skin and respiratory tract. Affects metabolism. In serious cases risk of unconsciousness. **Effects of prolonged/repeated exposure:** Prolonged or repeated contact can cause skin disorders. Can cause liver damage.

HAZARDS/SYMPTOMS	PREVENTIVE MEASURES	FIRE EXTINGUISHING/FIRST AID
Fire: combustible.	keep away from open flame, no smoking.	water spray, powder.
	KEEP DUST UNDER CONTROL	
Inhalation: cough, drowsiness.	local exhaust or respiratory protection.	fresh air, rest, call a doctor.
Skin: redness, pain.	gloves.	remove contaminated clothing, flush skin with water or shower.
Eyes: redness, pain.	goggles.	
Ingestion: vomiting, drowsiness.		rinse mouth, call a doctor.

SPILLAGE	STORAGE	LABELING / NFPA
clean up spillage, carefully collect remainder; (additional individual protection: P2 respirator).		R: 20/21/22 S: 2-13 Note A ✖ Harmful

NOTES
The measures on this card also apply to 2-(2-methyl-4-chlorophenoxy)propionic acid (MCPP). Solvent contained in commercial products is often more toxic than active substance.

CAS-No: [108-87-2]
cyclohexylmethane
heptanaphthene
hexahydrotoluene
toluene hexahydride

$C_6H_{11}CH_3$

METHYLCYCLOHEXANE

PHYSICAL PROPERTIES	IMPORTANT CHARACTERISTICS
Boiling point, °C 101 Melting point, °C − 126 Flash point, °C − 4 Auto-ignition temperature, °C 258 Relative density (water = 1) 0.8 Relative vapor density (air = 1) 3.4 Relative density at 20 °C of saturated mixture vapor/air (air = 1) 1.1 Vapor pressure, mm Hg at 20 °C 36 Solubility in water none Explosive limits, vol% in air 1.1-6.7 Minimum ignition energy, mJ 0.27 Electrical conductivity, pS/m < 10⁴ Relative molecular mass 98.2	**COLORLESS LIQUID WITH CHARACTERISTIC ODOR** Vapor is heavier than air and spreads at ground level, with risk of ignition at a distance. Flow, agitation etc. can cause build-up of electrostatic charge due to liquid's low conductivity. Do not use compressed air when filling, emptying or processing. Reacts violently with strong oxidants.
	TLV-TWA 400 ppm 1610 mg/m³
	Absorption route: Can enter the body by inhalation or ingestion. Harmful atmospheric concentrations build up fairly slowly on evaporation at 20°C, but much more rapidly in aerosol form. **Immediate effects:** Irritates the eyes and respiratory tract. Liquid destroys the skin's natural oils. In substantial concentrations can impair consciousness.
Gross formula: C_7H_{14}	

HAZARDS/SYMPTOMS	PREVENTIVE MEASURES	FIRE EXTINGUISHING/FIRST AID
Fire: extremely flammable.	keep away from open flame and sparks, no smoking.	powder, AFFF, foam, carbon dioxide, (halons).
Explosion: forms explosive air-vapor mixtures.	sealed machinery, ventilation, explosion-proof electrical equipment and lighting, grounding, non-sparking tools.	in case of fire: keep tanks/drums cool by spraying with water.
Inhalation: headache, dizziness, nausea, drowsiness, unconsciousness.	ventilation, local exhaust or respiratory protection.	fresh air, rest, call a doctor.
Skin: irritation, redness.	gloves.	remove contaminated clothing, flush skin with water or shower.
Eyes: redness, pain.	safety glasses.	flush with water, take to a doctor.
Ingestion: abdominal pain; see also 'Inhalation'.		rinse mouth, DO NOT induce vomiting, call a doctor or take to hospital.

SPILLAGE	STORAGE	LABELING / NFPA
collect leakage in sealable containers, soak up with sand or other inert absorbent and remove to safe place, do not flush into sewer.	keep in fireproof place, separate from oxidants.	R: 11 S: 9-16-33 🔥 Flammable 2 3 0

NOTES

METHYLENE CHLORIDE

PHYSICAL PROPERTIES		IMPORTANT CHARACTERISTICS
Boiling point, °C	40	**COLORLESS LIQUID WITH CHARACTERISTIC ODOR**
Melting point, °C	− 96	Vapor is heavier than air. Flow, agitation etc. can cause build-up of electrostatic charge due to
Flash point, °C	none	liquid's low conductivity. Decomposes in flame or on hot surface, giving off toxic and corrosive
Auto-ignition temperature, °C	605	vapors (→ *hydrochloric acid*). Reacts violently with nitric acid. Reacts with alkaline-earth and
Relative density (water = 1)	1.3	alkali metals. Attacks many plastics.
Relative density at 20 °C of saturated mixture		
vapor/air (air = 1)	1.9	
Vapor pressure, mm Hg at 20 °C	357	
Solubility in water, g/100 ml at 20 °C	2	**Absorption route:** Can enter the body by inhalation or ingestion. Harmful atmospheric
Electrical conductivity, pS/m	4.3 x 10³	concentrations can build up fairly rapidly on evaporation at approx. 20°C - even more rapidly in
Relative molecular mass	84.9	aerosol form.
		Immediate effects: Irritates the eyes, skin and respiratory tract. Liquid destroys the skin's natural
		oils. Can cause formation of carboxyhemoglobin.
		Effects of prolonged/repeated exposure: Can affect the blood. Can damage the liver. In serious
		cases risk of unconsciousness.
Gross formula:	CH_2Cl_2	

Note: TLV-TWA row between characteristics: TLV-TWA | 50 ppm | 174 mg/m³ A2

HAZARDS/SYMPTOMS	PREVENTIVE MEASURES	FIRE EXTINGUISHING/FIRST AID
Fire: non-combustible.		in case of fire in immediate vicinity: use any extinguishing agent.
Explosion:		in case of fire: keep tanks/drums cool by spraying with water.
Inhalation: headache, dizziness, nausea, drowsiness, unconsciousness.	ventilation, local exhaust or respiratory protection.	fresh air, rest, call a doctor.
Skin: redness.	gloves.	remove contaminated clothing, wash skin with soap and water.
Eyes: redness, pain.	safety glasses.	flush with water, take to doctor if necessary.
Ingestion: abdominal pain, chest pain; see also 'Inhalation'.		rinse mouth, call a doctor or take to hospital.

SPILLAGE	STORAGE	LABELING / NFPA
collect leakage in sealable containers, soak up with sand or other inert absorbent and remove to safe place; (additional individual protection: breathing apparatus).	ventilate at floor level.	R: 20 S: 24 ✖ Harmful

NFPA diamond: 1 (top), 2 (left), 0 (right)

NOTES

Combustible air-vapor mixtures (which are difficult to ignite) can form under certain circumstances (13-22 vol.%). Combustibility can be greatly increased by (1) addition of small quantities of combustible substances or (2) raising the oxygen content. Odor limit is above TLV. Alcohol consumption increases toxic effects. High atmospheric concentrations, e.g. in poorly ventilated areas, can cause oxygen deficiency, with risk of unconsciousness. Under no circumstances use near flame or hot surface or when welding.

Transport Emergency Card TEC(R)-720

HI: 60; UN-number: 1593

CAS-No: [109-83-1]
(2-hydroxyethyl)methylamine
N-methylaminoethanol
methylethanolamine
monomethylethanolamine

$H_3CNHCH_2CH_2OH$

N-METHYLETHANOLAMINE

PHYSICAL PROPERTIES

Boiling point, °C	156
Melting point, °C	−5
Flash point, °C	74
Auto-ignition temperature, °C	350
Relative density (water = 1)	0.9
Relative vapor density (air = 1)	2.6
Relative density at 20 °C of saturated mixture vapor/air (air = 1)	1.00
Vapor pressure, mm Hg at 20 °C	0.71
Solubility in water	∞
Explosive limits, vol% in air	1.6-19.8
Relative molecular mass	75.1

Gross formula: C_3H_9NO

IMPORTANT CHARACTERISTICS

COLORLESS VISCOUS LIQUID WITH CHARACTERISTIC ODOR
Vapor mixes readily with air. Decomposes when heated or burned, giving off toxic gases (→ *nitrous vapors*). Strong base which reacts violently with acids and corrodes aluminum, zinc etc. Reacts with strong oxidants. Attacks copper and its alloys.

TLV-TWA	not available

Absorption route: Can enter the body by inhalation or ingestion. Harmful atmospheric concentrations can build up fairly rapidly on evaporation at approx. 20°C - even more rapidly in aerosol form.
Immediate effects: Corrosive to the eyes, skin and respiratory tract. Inhalation of vapor/fumes can cause severe breathing difficulties (lung edema). In serious cases risk of death.
Effects of prolonged/repeated exposure: Prolonged or repeated contact can cause skin disorders. Can cause liver and kidney damage.

HAZARDS/SYMPTOMS	PREVENTIVE MEASURES	FIRE EXTINGUISHING/FIRST AID
Fire: combustible.	keep away from open flame, no smoking.	powder, alcohol-resistant foam, water spray, carbon dioxide, (halons).
Explosion: above 74°C: forms explosive air-vapor mixtures.	above 74°C: sealed machinery, ventilation.	
Inhalation: *corrosive*; sore throat, cough, severe breathing difficulties.	ventilation, local exhaust or respiratory protection.	fresh air, rest, place in half-sitting position, take to hospital.
Skin: *corrosive*; redness, pain, serious burns.	gloves, protective clothing.	remove contaminated clothing, flush skin with water or shower, refer to a doctor.
Eyes: *corrosive*; redness, pain, impaired vision.	face shield.	flush with water, take to a doctor.
Ingestion: *corrosive*; sore throat, abdominal pain, diarrhea, vomiting.	do not eat, drink or smoke while working.	rinse mouth, take immediately to hospital.

SPILLAGE	STORAGE	LABELING / NFPA
collect leakage in sealable containers, flush away remainder with water.	keep separate from oxidants and acids; ventilate at floor level.	R: 34 S: 23/26/36 Corrosive

NOTES

Lung edema symptoms usually develop several hours later and are aggravated by physical exertion: rest and hospitalization essential. As first aid, a doctor or authorized person should consider administering a corticosteroid spray.

Transport Emergency Card TEC(R)-80G15

595

2-butanone
ethylmethylketone
MEK
methyl acetone
methylethylketone

$CH_3COC_2H_5$

METHYL ETHYL KETONE

PHYSICAL PROPERTIES		IMPORTANT CHARACTERISTICS
Boiling point, °C	80	**COLORLESS LIQUID WITH CHARACTERISTIC ODOR**
Melting point, °C	−86	Vapor is heavier than air and spreads at ground level, with risk of ignition at a distance. Presumed
Flash point, °C	−6	to be able to form peroxides. Do not use compressed air when filling, emptying or processing.
Auto-ignition temperature, °C	505	Reacts with hydrogen peroxide-nitric acid combination to form shock-sensitive compounds.
Relative density (water = 1)	0.8	Reacts violently with strong oxidants, with risk of fire and explosion. Attacks copper in the
Relative vapor density (air = 1)	2.5	presence of acetylene.

Relative density at 20 °C of saturated mixture			
vapor/air (air = 1)	1.15	TLV-TWA	200 ppm	590 mg/m³
Vapor pressure, mm Hg at 20 °C	80	STEL	300 ppm	885 mg/m³

Solubility in water, g/100 ml at 20 °C	29
Explosive limits, vol% in air	1.8-11.5
Minimum ignition energy, mJ	0.53
Electrical conductivity, pS/m	3.6 x 10⁵
Relative molecular mass	72.1
Log P octanol/water	0.3

Absorption route: Can enter the body by inhalation or ingestion or through the skin. Harmful atmospheric concentrations can build up fairly rapidly on evaporation at approx. 20°C - even more rapidly in aerosol form.
Immediate effects: Irritates the eyes, skin and respiratory tract. Liquid destroys the skin's natural oils. In substantial concentrations can impair consciousness.

Gross formula: C_4H_8O

HAZARDS/SYMPTOMS	PREVENTIVE MEASURES	FIRE EXTINGUISHING/FIRST AID
Fire: extremely flammable.	keep away from open flame and sparks, no smoking.	powder, alcohol-resistant foam, large quantities of water, carbon dioxide, (halons).
Explosion: forms explosive air-vapor mixtures.	sealed machinery, ventilation, explosion-proof electrical equipment and lighting, non-sparking tools.	in case of fire: keep tanks/drums cool by spraying with water.
Inhalation: cough, headache, drowsiness.	ventilation, local exhaust or respiratory protection.	fresh air, rest, call a doctor if necessary.
Skin: *is absorbed*; redness.	gloves.	remove contaminated clothing, flush skin with water or shower, refer to a doctor if necessary.
Eyes: *irritation*; redness, pain.	safety glasses.	flush with water, take to a doctor.
Ingestion: abdominal cramps; see also 'Inhalation'.		rinse mouth, DO NOT induce vomiting, call a doctor or take to hospital.

SPILLAGE	STORAGE	LABELING / NFPA
collect leakage in sealable containers, soak up with sand or other inert absorbent and remove to safe place; (additional individual protection: breathing apparatus).	keep in fireproof place, separate from oxidants.	R: 11-36/37 S: 9-16-25-33 🔥 Flammable ✖ Irritant (NFPA: 3 / 1 / 0)

NOTES
Before distilling check for peroxides; if found, render harmless.

CAS-No: [1338-23-4]
2-butanoneperoxide
ethyl methyl ketone peroxide
MEK peroxide
MEKP

$C_8H_{16}O_4$

METHYL ETHYL KETONE PEROXIDE
(solution with approx. 10 % active oxygen content)

PHYSICAL PROPERTIES	IMPORTANT CHARACTERISTICS
Boiling point (decomposes), °C · · · · · 80 Flash point, °C · · · · · ca. 100 Relative density (water = 1) · · · · · 1.2 Solubility in water · · · · · none Relative molecular mass · · · · · 176.2	**COLORLESS SOLUTION OF METHYLETHYLKETONEPEROXIDE IN DIMETHYLPHTHALATE** Flow, agitation etc. can cause build-up of electrostatic charge due to liquid's low conductivity. Decomposes if heated rapidly above 80°C, with risk of fire and explosion. Strong oxidant which reacts violently with combustible substances and reducing agents. Reacts violently with salts and oxides of heavy metals, tertiary amines, strong acids and bases.

	TLV-TWA	0.2 ppm	1.5 mg/m³ C

Absorption route: Can enter the body by inhalation or ingestion. Insufficient data on the rate at which harmful concentrations can build up.
Immediate effects: Irritates the skin and respiratory tract. Corrosive to the eyes. Inhalation of vapor/fumes can cause severe breathing difficulties (lung edema). In large quantities can cause liver and kidney damage.

Gross formula: $C_8H_{16}O_4$

HAZARDS/SYMPTOMS	PREVENTIVE MEASURES	FIRE EXTINGUISHING/FIRST AID
Fire: combustible.	keep away from open flame, no smoking.	large quantities of water, water spray, powder, carbon dioxide, (halons), DO NOT USE FOAM.
Explosion: heating or contact or reactions with other substances can cause explosion.	grounding.	in case of fire: keep tanks/drums cool by spraying with water, fight fire from sheltered location.
Inhalation: sore throat, cough, shortness of breath, severe breathing difficulties.	ventilation, local exhaust or respiratory protection.	fresh air, rest, call a doctor.
Skin: redness, pain.	gloves, protective clothing.	remove contaminated clothing, flush skin with water or shower.
Eyes: *corrosive*; redness, pain, impaired vision.	face shield, or combined eye and respiratory protection.	flush with water, take to a doctor if necessary.
Ingestion: *corrosive*; abdominal pain, diarrhea, vomiting.		rinse mouth, immediately take to hospital.

SPILLAGE	STORAGE	LABELING / NFPA
evacuate area, call in an expert, collect leakage in non-sealable containers, soak up with sand or other inert absorbent and remove to safe place, do not use sawdust or other combustible absorbents.	keep in fireproof and cool place, separate from all other substances.	

NOTES

The physical properties depend on the dilution agent used. This sheet applies for MEKP in dimethylphtalate. Flush contaminated clothing with water (fire hazard). Under no circumstances mix directly with accelerator. Store in original packaging; do not replace once removed. TLV is maximum and must not be exceeded. Lung edema symptoms usually develop several hours later and are aggravated by physical exertion: rest and hospitalization essential. Butanox is a trade name. Packaging: special material.

Transport Emergency Card TEC(R)-51G01

HI: 539; UN-number: 2250, 2563

2-butanone oxime
ethyl methyl ketoxime
MEK-oxime

$CH_3CH_2C=NOHCH_3$

METHYL ETHYL KETONOXIME

PHYSICAL PROPERTIES	IMPORTANT CHARACTERISTICS
Boiling point, °C 152 Melting point, °C −30 Flash point, °C 62 Auto-ignition temperature, °C see notes Relative density (water = 1) 0.9 Relative vapor density (air = 1) 3.0 Relative density at 20 °C of saturated mixture vapor/air (air = 1) 1.01 Vapor pressure, mm Hg at 20 °C 2.7 Solubility in water good Relative molecular mass 87.1	**COLORLESS TO LIGHT YELLOW LIQUID** Vapor mixes readily with air. Toxic vapors are a product of combustion (→ *nitrogen dioxide*). Can produce thermal explosion when heated. Reacts violently with strong oxidants, with risk of fire and explosion.
	TLV-TWA not available
	Absorption route: Can enter the body by inhalation or ingestion. Insufficient data on the rate at which harmful concentrations can build up. **Immediate effects:** Irritates the eyes, skin and respiratory tract. In serious cases risk of death.
Gross formula: C_4H_9NO	

HAZARDS/SYMPTOMS	PREVENTIVE MEASURES	FIRE EXTINGUISHING/FIRST AID
Fire: combustible.	keep away from open flame, no smoking.	powder, alcohol-resistant foam, water spray, carbon dioxide, (halons).
Explosion: above 62°C: forms explosive air-vapor mixtures.	above 62°C: sealed machinery, ventilation.	in case of fire: keep tanks/drums cool by spraying with water.
Inhalation: sore throat, cough, drowsiness, sleepiness.	ventilation, local exhaust or respiratory protection.	fresh air, rest, call a doctor.
Skin: redness, pain.	gloves.	remove contaminated clothing, flush skin with water or shower, refer to a doctor if necessary.
Eyes: redness, pain, impaired vision.	face shield.	flush with water, take to a doctor if necessary.
Ingestion: abdominal pain, diarrhea, vomiting.		rinse mouth, take immediately to hospital.

SPILLAGE	STORAGE	LABELING / NFPA
collect leakage in sealable containers, soak up with sand or other inert absorbent and remove to safe place; (additional individual protection: breathing apparatus).	keep separate from oxidants; ventilate at floor level.	R: 36-43 S: 23-24 ✖ Irritant

NOTES
Acidic contaminants promote thermal decomposition. Depending on the degree of exposure, regular medical checkups are advisable.

Transport Emergency Card TEC(R)-30G15

METHYL FORMATE

PHYSICAL PROPERTIES	IMPORTANT CHARACTERISTICS
Boiling point, °C 32 Melting point, °C − 100 Flash point, °C < − 20 Auto-ignition temperature, °C 450 Relative density (water = 1) 0.97 Relative vapor density (air = 1) 2.1 Relative density at 20 °C of saturated mixture vapor/air (air = 1) 1.7 Vapor pressure, mm Hg at 20 °C 486 Solubility in water, g/100 ml at 20 °C 30 Explosive limits, vol% in air 5-23 Electrical conductivity, pS/m 1.92x10⁹ Relative molecular mass 60.1 Gross formula: $C_2H_4O_2$	**COLORLESS LIQUID WITH CHARACTERISTIC ODOR** Vapor (gas) is heavier than air and spreads at ground level, with risk of ignition at a distance. Do not use compressed air when filling, emptying or processing. Decomposes on contact with acids, bases or water. Reacts violently with strong oxidants, with risk of fire and explosion. Attacks some plastics.

TLV-TWA	100 ppm	246 mg/m³
STEL	150 ppm	368 mg/m³

Absorption route: Can enter the body by inhalation or ingestion. Harmful atmospheric concentrations can build up very rapidly on evaporation at 20°C.
Immediate effects: Irritates the eyes, skin and respiratory tract. Affects the nervous system. Inhalation can cause optic nerve damage. Inhalation of vapor/fumes can cause severe breathing difficulties (lung edema).

HAZARDS/SYMPTOMS	PREVENTIVE MEASURES	FIRE EXTINGUISHING/FIRST AID
Fire: extremely flammable.	keep away from open flame and sparks, no smoking.	powder, alcohol-resistant foam, large quantities of water, carbon dioxide, (halons); in case of fire in immediate vicinity: use any extinguishing agent.
Explosion: forms explosive air-vapor mixtures.	sealed machinery, ventilation, explosion-proof electrical equipment and lighting.	in case of fire: keep tanks/drums cool by spraying with water.
Inhalation: sore throat, cough, headache, dizziness.	ventilation, local exhaust or respiratory protection.	fresh air, rest, place in half-sitting position, take to hospital.
Skin: redness, pain.	gloves.	remove contaminated clothing, flush skin with water or shower, call a doctor if necessary.
Eyes: redness, pain.	face shield.	flush with water, take to a doctor if necessary.
Ingestion: abdominal pain, chest pain, vomiting.		rinse mouth, call a doctor or take to hospital.

SPILLAGE	STORAGE	LABELING / NFPA
collect leakage in sealable containers, soak up with sand or other inert absorbent and remove to safe place; (additional individual protection: breathing apparatus).	keep in cool, fireproof place, separate from oxidants.	R: 12 S: 9-16-33 🔥 Flammable

NFPA diamond: 4 (top), 2 (left), 0 (right)

NOTES
Lung edema symptoms usually develop several hours later and are aggravated by physical exertion: rest and hospitalization essential. As first aid, a doctor or authorized person should consider administering a corticosteroid spray.

EGME
ethylene glycol monomethyl ether
2-methoxyethanol

$CH_3OCH_2CH_2OH$

METHYL GLYCOL

PHYSICAL PROPERTIES	IMPORTANT CHARACTERISTICS
Boiling point, °C — 124 Melting point, °C — −86 Flash point, °C — 39 Auto-ignition temperature, °C — 285 Relative density (water = 1) — 0.97 Relative vapor density (air = 1) — 2.6 Relative density at 20 °C of saturated mixture vapor/air (air = 1) — 1.02 Vapor pressure, mm Hg at 20 °C — 8.4 Solubility in water — ∞ Explosive limits, vol% in air — 2.5-20 Relative molecular mass — 76.1 Log P octanol/water — −0.7 Gross formula: $C_3H_8O_2$	**COLORLESS HYGROSCOPIC LIQUID WITH CHARACTERISTIC ODOR** Able to form peroxides. Reacts with strong bases. Reacts violently with strong oxidants, with risk of fire and explosion. Attacks many plastics. Attacks many metals, giving off flammable gas (→ *hydrogen*). TLV-TWA (skin) 5 ppm 16 mg/m³ **Absorption route:** Can enter the body by inhalation or ingestion or through the skin. Harmful atmospheric concentrations can build up fairly rapidly on evaporation at approx. 20°C - even more rapidly in aerosol form. **Immediate effects:** Irritates the eyes and respiratory tract. Liquid destroys the skin's natural oils. Affects the nervous system. **Effects of prolonged/repeated exposure:** Can affect the blood. Has been found to affect reproduction and fetuses in laboratory animals under certain circumstances.

HAZARDS/SYMPTOMS	PREVENTIVE MEASURES	FIRE EXTINGUISHING/FIRST AID
Fire: flammable.	keep away from open flame and sparks, no smoking.	powder, alcohol-resistant foam, water spray, carbon dioxide, (halons).
Explosion: above 39°C: forms explosive air-vapor mixtures.	above 39°C: sealed machinery, ventilation, explosion-proof electrical equipment.	in case of fire: keep tanks/drums cool by spraying with water.
Inhalation: cough, headache, nausea, drowsiness.	ventilation, local exhaust or respiratory protection.	fresh air, rest, call a doctor.
Skin: *is absorbed*; redness; see also 'Inhalation'.	gloves, protective clothing.	remove contaminated clothing, flush skin with water or shower, call a doctor.
Eyes: redness, pain, impaired vision.	safety glasses.	flush with water, take to a doctor.
Ingestion: abdominal cramps, vomiting, drowsiness.		rinse mouth, DO NOT induce vomiting, take immediately to hospital.

SPILLAGE	STORAGE	LABELING / NFPA
collect leakage in sealable containers, flush away remainder with water; (additional individual protection: breathing apparatus).	keep in dry, dark, fireproof place, separate from oxidants and strong bases.	R: 10-37-20/21/22 S: 24/25 ✖ Harmful 2 / 2 / 0

NOTES
Log P octanol/water is estimated. Before distilling check for peroxides; if found, render harmless. Alcohol consumption increases toxic effects. Depending on the degree of exposure, regular medical checkups are advisable. Methyl Oxitol and Methyl Cellosolve are trade names.

HI: 30; UN-number: 1188

CAS-No: [110-49-6]
EGMEA
ethylene glycol monomethyl ether acetate
2-methoxyethyl acetate
methyl glycol monoacetate

$CH_3COOC_2H_4OCH_3$

METHYL GLYCOL ACETATE

PHYSICAL PROPERTIES		IMPORTANT CHARACTERISTICS		
Boiling point, °C	145	**COLORLESS LIQUID WITH CHARACTERISTIC ODOR**		
Melting point, °C	−65	Vapor mixes readily with air. Able to form peroxides on exposure to air. Reacts violently with strong oxidants.		
Flash point, °C	47			
Auto-ignition temperature, °C	380			
Relative density (water = 1)	1.0	TLV-TWA (skin)	5 ppm	24 mg/m³
Relative vapor density (air = 1)	4.1			
Relative density at 20 °C of saturated mixture		**Absorption route:** Can enter the body by inhalation or ingestion or through the skin. Harmful atmospheric concentrations can build up fairly rapidly on evaporation at approx. 20°C - even more rapidly in aerosol form.		
vapor/air (air = 1)	1.03			
Vapor pressure, mm Hg at 20 °C	7.1	**Immediate effects:** Irritates the eyes, skin and respiratory tract. Liquid destroys the skin's natural oils. Affects the nervous system.		
Solubility in water	∞			
Explosive limits, vol% in air	1.7-8.2	**Effects of prolonged/repeated exposure:** Can cause kidney damage. Has been found to affect reproduction and fetuses in laboratory animals under certain circumstances.		
Relative molecular mass	118.1			
Gross formula:	$C_5H_{10}O_3$			

HAZARDS/SYMPTOMS	PREVENTIVE MEASURES	FIRE EXTINGUISHING/FIRST AID
Fire: flammable.	keep away from open flame and sparks, no smoking.	powder, water spray, foam, carbon dioxide, (halons).
Explosion: above 47°C: forms explosive air-vapor mixtures.	above 47°C: sealed machinery, ventilation, explosion-proof electrical equipment.	in case of fire: keep tanks/drums cool by spraying with water.
Inhalation: sore throat, cough.	ventilation, local exhaust or respiratory protection.	fresh air, rest.
Skin: *is absorbed*; redness.	gloves, protective clothing.	remove contaminated clothing, wash skin with soap and water.
Eyes: redness, pain.	face shield.	flush with water, take to a doctor if necessary.
Ingestion: abdominal pain, headache, vomiting.		rinse mouth, call a doctor or take to hospital.

SPILLAGE	STORAGE	LABELING / NFPA
collect leakage in sealable containers, flush away remainder with water; (additional individual protection: breathing apparatus).	keep in fireproof place.	R: 10-20/21 S: 24 ✖ Harmful

NOTES
Alcohol consumption increases toxic effects. Depending on the degree of exposure, regular medical checkups are advisable. Methyl Cellosolve acetate and Methyl Oxitol acetate are trade names.

Transport Emergency Card TEC(R)-30G35

HI: 30; UN-number: 1189

CAS-No: [60-34-4]
hydrazomethane
N-methylhydrazine
1-methylhydrazine
MMH

CH_3NHNH_2

METHYL HYDRAZINE

PHYSICAL PROPERTIES	IMPORTANT CHARACTERISTICS	
Boiling point, °C .... 88 Melting point, °C .... −52 Flash point, °C .... −8 Auto-ignition temperature, °C ...... 194 Relative density (water = 1) .... 0.9 Relative vapor density (air = 1) ...... 1.6 Relative density at 20°C of saturated mixture vapor/air (air = 1) .... 1.03 Vapor pressure, mm Hg at 20°C ...... 37 Solubility in water .... ∞ Explosive limits, vol% in air ..... 2.5-97 Relative molecular mass 46.1	**COLORLESS HYGROSCOPIC LIQUID WITH CHARACTERISTIC ODOR** Vapor mixes readily with air, forming explosive mixtures. Can ignite spontaneously on exposure to air if finely dispersed or absorbed into porous matererial, e.g. earth, wood or textiles. Strong reducing agent which reacts violently with oxidants, e.g. as chlorine, hydrogen peroxide and concentrated nitric acid, with risk of fire and explosion. Medium strong base which reacts violently with strong acids.	
	TLV-TWA (skin)	0,2 ppm	0,38 mg/m³ C A2
Gross formula: CH_6N_2	**Absorption route:** Can enter the body by inhalation or ingestion or through the skin. Harmful atmospheric concentrations can build up very rapidly on evaporation at 20°C. **Immediate effects:** Corrosive to the eyes, skin and respiratory tract. Affects the nervous system. In serious cases risk of seizures. **Effects of prolonged/repeated exposure:** Prolonged or repeated contact can cause skin disorders. Can cause liver, kidney and brain damage, esp. with prolonged exposure. Can affect the blood.	

HAZARDS/SYMPTOMS	PREVENTIVE MEASURES	FIRE EXTINGUISHING/FIRST AID
Fire: extremely flammable, many chemical reactions can cause fire and explosion.	keep away from open flame and sparks, no smoking, avoid contact with strong oxidants and hot surfaces (e.g. steam lines).	large quantities of water, alcohol-resistant foam, powder, carbon dioxide, (halons).
Explosion: forms explosive air-vapor mixtures, contact with metal oxides can cause explosion.	sealed machinery, ventilation, explosion-proof electrical equipment and lighting.	in case of fire: keep tanks/drums cool by spraying with water.
	STRICT HYGIENE	IN ALL CASES CALL A DOCTOR
Inhalation: *corrosive*; cough, shortness of breath, severe breathing difficulties, vomiting, diarrhea, seizures.	ventilation, local exhaust or respiratory protection.	fresh air, rest, take to hospital.
Skin: *corrosive*; redness, pain, serious burns, sensitization.	gloves, protective clothing.	remove contaminated clothing, flush skin with water or shower, call a doctor.
Eyes: *corrosive*; redness, pain, impaired vision.	face shield, or combined eye and respiratory protection.	flush with water, take to a doctor.
Ingestion:		rinse mouth, take immediately to hospital.

SPILLAGE	STORAGE	LABELING / NFPA
evacuate area, call in an expert, collect leakage in sealable containers, thin spillage with 2-3 parts water and neutralize with 20% sulfuric acid, flush area with water; (additional individual protection: breathing apparatus).	keep in fireproof place under inert gas, separate from oxidants and strong bases.	3 / 3 2

NOTES
Flush contaminated clothing with water (fire hazard). TLV is maximum and must not be exceeded. Depending on the degree of exposure, regular medical checkups are advisable. Unbreakable packaging preferred; if breakable, keep in unbreakable container.

METHYL IODIDE

PHYSICAL PROPERTIES		IMPORTANT CHARACTERISTICS	
Boiling point, °C	42	**COLORLESS LIQUID, TURNING BROWN ON EXPOSURE TO AIR**	
Melting point, °C	−66	Vapor is heavier than air. Decomposes when heated above 270°C, giving off corrosive and toxic	
Relative density (water = 1)	2.3	gases (→ *hydrogen iodide*). Reacts slowly with hot water, giving off → *hydrogen iodide*.	
Relative vapor density (air = 1)	4.9		
Relative density at 20°C of saturated mixture			
vapor/air (air = 1)	2.7	TLV-TWA (skin) 2 ppm 12 mg/m³	
Vapor pressure, mm Hg at 20°C	333		
Solubility in water, g/100 ml	2	**Absorption route:** Can enter the body by inhalation or ingestion or through the skin. Harmful atmospheric concentrations can build up very rapidly on evaporation at 20°C.	
Electrical conductivity, pS/m	<2 x 10⁶	**Immediate effects:** Corrosive to the eyes, skin and respiratory tract. Affects the nervous system. Inhalation can cause lung edema. In serious cases risk of seizures and death.	
Relative molecular mass	141.9	**Effects of prolonged/repeated exposure:** Prolonged or repeated contact can cause skin disorders. Can cause brain damage.	
Log P octanol/water	1.7		
Gross formula: CH₃I			

HAZARDS/SYMPTOMS	PREVENTIVE MEASURES	FIRE EXTINGUISHING/FIRST AID
Fire: non-combustible.		in case of fire in immediate vicinity: use any extinguishing agent.
Explosion:		in case of fire: keep tanks/drums cool by spraying with water.
Inhalation: *corrosive*; sore throat, cough, severe breathing difficulties, dizziness, sleepiness.	ventilation, local exhaust or respiratory protection.	fresh air, rest, place in half-sitting position, take to hospital.
Skin: *corrosive, is absorbed*; redness, pain, burns.	gloves, protective clothing.	remove contaminated clothing, flush skin with water or shower, call a doctor.
Eyes: *corrosive*; redness, pain, impaired vision.	face shield, or combined eye and respiratory protection.	flush with water, take to a doctor.
Ingestion: *corrosive*; abdominal pain, nausea, dizziness, sleepiness, unconsciousness.		rinse mouth, take immediately to hospital.

SPILLAGE	STORAGE	LABELING / NFPA
evacuate area, call in an expert, collect leakage in sealable containers, soak up with sand or other inert absorbent and remove to safe place; (additional individual protection: CHEMICAL SUIT).	keep in dark place; ventilate at floor level.	R: 21-23/25-37/38-40 S: 36/37-38-44 ☠ Toxic

NOTES

Lung edema symptoms usually develop several hours later and are aggravated by physical exertion: rest and hospitalization essential. As first aid, a doctor or authorized person should consider administering a corticosteroid spray. Under no circumstances use near flame or hot surface or when welding. Use airtight packaging. Unbreakable packaging preferred; if breakable, keep in unbreakable container.

Transport Emergency Card TEC(R)-61G06B

HI: 60; UN-number: 2644

1,3-dimethylbutanol
methylamyl alcohol
4-methyl-2-pentanol
4-methyl-2-pentyl alcohol
MIBC

$(CH_3)_2CHCH_2CHOHCH_3$

METHYLISOBUTYL CARBINOL

PHYSICAL PROPERTIES	IMPORTANT CHARACTERISTICS	
Boiling point, °C 132	**COLORLESS LIQUID**	
Melting point, °C −60	Vapor mixes readily with air. Reacts violently with strong oxidants. Reacts with alkali metals, giving off flammable gas (→ *hydrogen*).	
Flash point, °C 41		
Auto-ignition temperature, °C ?		
Relative density (water = 1) 0.8	TLV-TWA (skin) 25 ppm 104 mg/m³	
Relative vapor density (air = 1) 3.5	STEL 40 ppm 167 mg/m³	
Relative density at 20°C of saturated mixture vapor/air (air = 1) 1.02	**Absorption route:** Can enter the body by inhalation or ingestion or through the skin. Harmful atmospheric concentrations can build up fairly rapidly on evaporation at approx. 20°C - even more rapidly in aerosol form.	
Vapor pressure, mm Hg at 20°C 5.3		
Solubility in water, g/100 ml 2		
Explosive limits, vol% in air 1.0-5.5	**Immediate effects:** Irritates the eyes, skin and respiratory tract. Liquid destroys the skin's natural oils. Affects the nervous system.	
Relative molecular mass 102.2		
Gross formula: $C_6H_{14}O$		

HAZARDS/SYMPTOMS	PREVENTIVE MEASURES	FIRE EXTINGUISHING/FIRST AID
Fire: flammable.	keep away from open flame and sparks, no smoking.	powder, AFFF, foam, carbon dioxide, (halons).
Explosion: above 41°C: forms explosive air-vapor mixtures.	above 41°C: sealed machinery, ventilation, explosion-proof electrical equipment.	in case of fire: keep tanks/drums cool by spraying with water.
Inhalation: sore throat, cough, shortness of breath, headache, dizziness, drowsiness.	ventilation, local exhaust or respiratory protection.	fresh air, rest, place in half-sitting position, call a doctor.
Skin: *is absorbed*; redness, pain.	gloves, protective clothing.	remove contaminated clothing, wash skin with soap and water, call a doctor.
Eyes: redness, pain, impaired vision.	face shield.	flush with water, take to a doctor.
Ingestion: abdominal pain, headache, vomiting, drowsiness.		rinse mouth, take to hospital.

SPILLAGE	STORAGE	LABELING / NFPA
collect leakage in sealable containers, soak up with sand or other inert absorbent and remove to safe place; (additional individual protection: breathing apparatus).	keep in fireproof place, separate from oxidants.	R: 10-37 S: 24/25 ✖ Irritant 2 / 2 / 0

NOTES

CAS-No: [108-10-1]
hexone
iso-butyl methyl ketone
isopropylacetone
4-methyl-2-pentanone
MIBK

$(CH_3)_2CHCH_2COCH_3$

METHYL ISOBUTYL KETONE

PHYSICAL PROPERTIES

Boiling point, °C	116
Melting point, °C	−80
Flash point, °C	14
Auto-ignition temperature, °C	475
Relative density (water = 1)	0.8
Relative vapor density (air = 1)	3.5
Relative density at 20 °C of saturated mixture vapor/air (air = 1)	1.02
Vapor pressure, mm Hg at 20 °C	5.3
Solubility in water, g/100 ml at 20 °C	2
Explosive limits, vol% in air	1.2-8
Electrical conductivity, pS/m	5.2×10^6
Relative molecular mass	100.2

Gross formula: $C_6H_{12}O$

IMPORTANT CHARACTERISTICS

COLORLESS LIQUID WITH CHARACTERISTIC ODOR
Vapor mixes readily with air, forming explosive mixtures. Do not use compressed air when filling, emptying or processing. Reacts violently with strong oxidants. Attacks many plastics.

TLV-TWA	50 ppm	205 mg/m³
STEL	75 ppm	307 mg/m³

Absorption route: Can enter the body by inhalation or ingestion. Harmful atmospheric concentrations build up fairly slowly on evaporation at 20°C, but much more rapidly in aerosol form.
Immediate effects: Irritates the eyes, skin and respiratory tract. Liquid destroys the skin's natural oils. In substantial concentrations can impair consciousness.

HAZARDS/SYMPTOMS	PREVENTIVE MEASURES	FIRE EXTINGUISHING/FIRST AID
Fire: extremely flammable.	keep away from open flame and sparks, no smoking.	powder, AFFF, foam, carbon dioxide, (halons).
Explosion: forms explosive air-vapor mixtures.	sealed machinery, ventilation, explosion-proof electrical equipment and lighting.	in case of fire: keep tanks/drums cool by spraying with water.
Inhalation: cough, headache, drowsiness.	ventilation, local exhaust or respiratory protection.	fresh air, rest, call a doctor if necessary.
Skin: redness.	gloves.	remove contaminated clothing, flush skin with water or shower, refer to a doctor if necessary.
Eyes: *irritation*; redness, pain.	safety glasses.	flush with water, take to doctor.
Ingestion: abdominal cramps; see also 'Inhalation'.		rinse mouth, call a doctor or take to hospital.

SPILLAGE	STORAGE	LABELING / NFPA
collect leakage in sealable containers, soak up with sand or other inert absorbent and remove to safe place.	keep in fireproof place, separate from oxidants.	R: 11 S: 9-16-23-33 Flammable

NOTES

Alcohol consumption increases toxic effects.

CAS-No: [624-83-9]
isocyanatomethane
isocyanic acid methyl ester
methylcarbylamine
MIC

CH_3NCO

METHYL ISOCYANATE

PHYSICAL PROPERTIES	IMPORTANT CHARACTERISTICS	
Boiling point, °C 38	**COLORLESS LIQUID WITH PUNGENT ODOR**	
Melting point, °C < −20	Vapor is heavier than air and spreads at ground level, with risk of ignition at a distance. Do not use	
Flash point, °C −7	compressed air when filling, emptying or processing. Decomposes when heated above boiling	
Auto-ignition temperature, °C 534	point or burned, giving off toxic gases (→ *nitrous vapors*). Reacts slowly with moisture and cold	
Relative density (water = 1) 0.96	water, violently with warm water, evolving heat and giving off carbon dioxide. Reacts violently with	
Relative vapor density (air = 1) 2.0	amines, alcohols, acids and strong oxidants. Attacks steel, copper and copper alloys.	
Relative density at 20 °C of saturated mixture		
vapor/air (air = 1) 1.4	TLV-TWA (skin) 0.02 ppm 0.047 mg/m³	
Vapor pressure, mm Hg at 20 °C 352		
Solubility in water reaction	**Absorption route:** Can enter the body by inhalation or ingestion or through the skin.	
Explosive limits, vol% in air 5.3-26	**Immediate effects:** Corrosive to the eyes, skin and respiratory tract. Inhalation can cause lung	
Relative molecular mass 57.1	edema. Inhalation can cause severe breathing difficulties (asthma).	
	Effects of prolonged/repeated exposure: Prolonged or repeated contact can cause skin	
	disorders.	
Gross formula: C_2H_3NO		

HAZARDS/SYMPTOMS	PREVENTIVE MEASURES	FIRE EXTINGUISHING/FIRST AID
Fire: extremely flammable, many chemical reactions can cause fire and explosion.	keep away from open flame and sparks, no smoking, avoid contact with water, acids, alcohols and amines.	powder, carbon dioxide, (halons), DO NOT USE WATER-BASED EXTINGUISHERS.
Explosion: forms explosive air-vapor mixtures.	sealed machinery, ventilation, explosion-proof electrical equipment and lighting.	in case of fire: keep tanks/drums cool by spraying with water, but DO NOT spray substance with water, fight fire from protected location.
	STRICT HYGIENE	IN ALL CASES CALL IN A DOCTOR
Inhalation: *corrosive*; sore throat, cough, shortness of breath, severe breathing difficulties.	ventilation, local exhaust or respiratory protection.	fresh air, rest, artificial respiration if necessary, take to hospital.
Skin: *corrosive, is absorbed*; redness, pain, burns.	gloves, protective clothing.	remove contaminated clothing, flush skin with water or shower, take to hospital.
Eyes: *corrosive*; redness, pain, impaired vision.	face shield, or combined eye and respiratory protection.	flush with water, take to a doctor.
Ingestion: *corrosive*; diarrhea, abdominal cramps, vomiting.		rinse mouth, take to a doctor.

SPILLAGE	STORAGE	LABELING / NFPA
evacuate area, call in an expert, collect leakage in sealable containers, soak up with sand or other inert absorbent and remove to safe place, do not flush into sewer; (additional individual protection: CHEMICAL SUIT).	keep in cool, dry, fireproof place, separate from all other substances.	R: 12-23/24/25-36/37/38 S: 9-30-43-44 Flammable Toxic NFPA: 4 3 2

NOTES
Reacts violently with extinguishing agents such as water and foam. Odor limit is above TLV. Depending on the degree of exposure, regular medical checkups are advisable. Lung edema symptoms usually develop several hours later and are aggravated by physical exertion: rest and hospitalization essential. As first aid, a doctor or authorized person should consider administering a corticosteroid spray. Packaging: special material. Can be rendered harmless with dilute spirits of ammonia (50% water, 45% ethyl alcohol, 5% ammonia).

Transport Emergency Card TEC(R)-61C03

HI: 336; UN-number: 2480

CAS-No: [74-93-1]
mercaptomethane
methanethiol
methyl sulfhydrate
thiomethanol

CH_3SH

METHYL MERCAPTANE
(cylinder)

PHYSICAL PROPERTIES		IMPORTANT CHARACTERISTICS

PHYSICAL PROPERTIES

Boiling point, °C	6
Melting point, °C	− 123
Flash point, °C	flammable gas
Auto-ignition temperature, °C	?
Relative density (water = 1)	0.87
Relative vapor density (air = 1)	1.7
Vapor pressure, mm Hg at 20 °C	1292
Solubility in water, g/100 ml at 20 °C	2.3
Explosive limits, vol% in air	3.9-21.8
Relative molecular mass	48.1

Gross formula: CH_4S

IMPORTANT CHARACTERISTICS

COLORLESS COMPRESSED LIQUEFIED GAS WITH CHARACTERISTIC ODOR
Gas is heavier than air and spreads at ground level, with risk of ignition at a distance. Do not use compressed air when filling, emptying or processing. Toxic gases are a product of combustion (→ *sulfur dioxide*). Reacts violently with strong oxidants, with risk of fire and explosion. Reacts with acids, giving off flammable and toxic → *hydrogen sulfide*. Reacts with light metals.

TLV-TWA	0.5 ppm	0.98 mg/m³

Absorption route: Can enter the body by inhalation. Harmful atmospheric concentrations can build up very rapidly if gas is released.
Immediate effects: Irritates the eyes, skin and respiratory tract. Liquid can cause frostbite due to rapid evaporation. In serious cases risk of unconsciousness.
Effects of prolonged/repeated exposure: Can cause liver damage.

HAZARDS/SYMPTOMS	PREVENTIVE MEASURES	FIRE EXTINGUISHING/FIRST AID
Fire: extremely flammable.	keep away from open flame and sparks, no smoking.	shut off supply; if impossible and no danger to surrounding area, allow fire to burn itself out; otherwise extinguish with powder, carbon dioxide, (halons); in case of fire in immediate vicinity: use any extinguishing agent.
Explosion: forms explosive air-gas mixtures.	sealed machinery, ventilation, explosion-proof electrical equipment and lighting.	in case of fire: keep cylinders cool by spraying with water.
Inhalation: sore throat, cough, shortness of breath, headache, nausea, unconsciousness.	ventilation, local exhaust or respiratory protection.	fresh air, rest, call a doctor.
Skin: *in case of frostbite*: redness, pain, sores.	gloves.	*in case of frostbite*: DO NOT remove clothing, flush skin with water or shower, refer to a doctor.
Eyes: *in case of frostbite*: redness, pain, impaired vision.	face shield, or combined eye and respiratory protection.	flush with water, take to a doctor.

SPILLAGE	STORAGE	LABELING / NFPA
evacuate area, call in an expert, ventilate, remove leaking cylinder to safe place; (additional individual protection: breathing apparatus).	keep in fireproof, cool place; keep separate from oxidants and strong acids.	R: 13-20 S: 16-25 Flammable Harmful

NOTES

Turn leaking cylinder so that leak is on top to prevent liquid methyl mercaptan escaping.

CAS-No: [80-62-6]
methacrylic acid, methyl ester
methyl-α-methylacrylate
methyl-2-methyl-2-propenoate
methyl 2-methyl propanoate
2-methyl propenoic acid, methyl ester

$CH_2 = C(CH_3)COOCH_3$

METHYL METHACRYLATE

PHYSICAL PROPERTIES	IMPORTANT CHARACTERISTICS
Boiling point, °C 101 Melting point, °C −48 Flash point, °C 10 Auto-ignition temperature, °C 430 Relative density (water = 1) 0.9 Relative vapor density (air = 1) 3.5 Relative density at 20 °C of saturated mixture vapor/air (air = 1) 1.1 Vapor pressure, mm Hg at 20 °C 36 Solubility in water, g/100 ml at 20 °C 1.5 Explosive limits, vol% in air 2.1-12.5 Relative molecular mass 100.1 Gross formula: $C_5H_8O_2$	**COLORLESS LIQUID WITH CHARACTERISTIC ODOR** Vapor (gas) is heavier than air and spreads at ground level, with risk of ignition at a distance. Readily able to polymerize when moderately heated or exposed to light or due to contaminants. Do not use compressed air when filling, emptying or processing. Reacts with many substances. Reacts violently with strong oxidants, with risk of fire and explosion. TLV-TWA 100 ppm 410 mg/m³ **Absorption route:** Can enter the body by inhalation or ingestion or through the skin. Harmful atmospheric concentrations can build up fairly rapidly on evaporation at approx. 20° C - even more rapidly in aerosol form. **Immediate effects:** Irritates the eyes, skin and respiratory tract. Inhalation of vapor/fumes can cause severe breathing difficulties (lung edema). **Effects of prolonged/repeated exposure:** Prolonged or repeated contact can cause eczema.

HAZARDS/SYMPTOMS	PREVENTIVE MEASURES	FIRE EXTINGUISHING/FIRST AID
Fire: extremely flammable.	keep away from open flame and sparks, no smoking.	powder, AFFF, foam, carbon dioxide, (halons).
Explosion: forms explosive air-vapor mixtures.	sealed machinery, ventilation, explosion-proof electrical equipment and lighting.	in case of fire: keep tanks/drums cool by spraying with water.
Inhalation: sore throat, cough, shortness of breath, dizziness.	ventilation, local exhaust or respiratory protection.	fresh air, rest, place in half-sitting position, call a doctor.
Skin: redness, pain.	gloves.	remove contaminated clothing, flush skin with water or shower, refer to a doctor.
Eyes: redness, pain.	face shield.	flush with water, take to a doctor.
Ingestion: sore throat, abdominal pain, vomiting.		rinse mouth, immediately take to hospital.

SPILLAGE	STORAGE	LABELING / NFPA
collect leakage in non-sealable containers, soak up with sand or other inert absorbent and remove to safe place; (additional individual protection: breathing apparatus).	keep in cool, dark, fireproof place, separate from oxidants; add an inhibitor.	R: 11-36/37/38-43 S: 9-16-29-33 Note D 🔥 Flammable ✖ Irritant NFPA: 3 2 2

NOTES
Lung edema symptoms usually develop several hours later and are aggravated by physical exertion: rest and hospitalization essential. As first aid, a doctor or authorized person should consider administering a corticosteroid spray.

Transport Emergency Card TEC(R)-196

HI: 339; UN-number: 1247

CAS-No: [109-02-4]
N-methyl morpholine
4-methyl morpholine

N-METHYL MORPHOLINE

PHYSICAL PROPERTIES	IMPORTANT CHARACTERISTICS
Boiling point, °C 115	**COLORLESS LIQUID WITH CHARACTERISTIC ODOR**
Melting point, °C .... −66	Vapor mixes readily with air. Decomposes when heated or burned, giving off toxic vapors
Flash point, °C 24	(→ *nitrous vapors*). Reacts with oxidants.
Auto-ignition temperature, °C 190	
Relative density (water = 1) 0.9	TLV-TWA not available
Relative vapor density (air = 1) 3.5	
Relative density at 20 °C of saturated mixture	**Absorption route:** Can enter the body by inhalation or ingestion or through the skin. Insufficient
vapor/air (air = 1) 1.05	data on the rate at which harmful concentrations can build up.
Vapor pressure, mm Hg at 20 °C 15	**Immediate effects:** Corrosive to the eyes, skin and respiratory tract. Inhalation of vapor/fumes
Solubility in water .... very good	can cause severe breathing difficulties (lung edema). In serious cases risk of death.
Explosive limits, vol% in air 2-10.6	**Effects of prolonged/repeated exposure:** Can cause liver and kidney damage.
Relative molecular mass 101.2	
Log P octanol/water .... −0.3	
Gross formula: $C_5H_{11}NO$	

HAZARDS/SYMPTOMS	PREVENTIVE MEASURES	FIRE EXTINGUISHING/FIRST AID
Fire: flammable.	keep away from open flame and sparks, no smoking, avoid contact with hot surfaces (e.g. steam lines).	powder, alcohol-resistant foam, large quantities of water, carbon dioxide, (halons).
Explosion: above 24°C: forms explosive air-vapor mixtures.	above 24°C: sealed machinery, ventilation, explosion-proof electrical equipment.	in case of fire: keep tanks/drums cool by spraying with water.
Inhalation: *corrosive*; sore throat, cough, shortness of breath, severe breathing difficulties.	ventilation, local exhaust or respiratory protection.	fresh air, rest, place in half-sitting position, take to hospital.
Skin: *is absorbed, corrosive*; redness, pain, serious burns.	gloves, protective clothing.	remove contaminated clothing, flush skin with water or shower, call a doctor if necessary.
Eyes: *corrosive*; redness, pain, impaired vision.	face shield.	flush with water, take to a doctor.
Ingestion: *corrosive*; sore throat, abdominal pain, nausea.		rinse mouth, call a doctor or take to hospital.

SPILLAGE	STORAGE	LABELING / NFPA
collect leakage in sealable containers, soak up with sand or other inert absorbent and remove to safe place; (additional individual protection: breathing apparatus).	keep in fireproof place, separate from oxidants.	3 / 2 / 0

NOTES
Lung edema symptoms usually develop several hours later and are aggravated by physical exertion: rest and hospitalization essential. As first aid, a doctor or authorized person should consider administering a corticosteroid spray.

Transport Emergency Card TEC(R)-80G15

HI: 83; UN-number: 2535

$C_{10}H_7OCONHCH_3$

carbaryl
1-naphthalenol methyl carbamate
1-naphthyl-N-methylcarbamate

N-METHYL-1-NAPHTHYLCARBAMATE

PHYSICAL PROPERTIES	IMPORTANT CHARACTERISTICS	
Melting point, °C 142 Relative density (water = 1) 1.2 Solubility in water, g/100 ml at 30 °C 0.012 Relative molecular mass 201.0	**WHITE CRYSTALS OR WHITE POWDER WITH FAINT ODOR** Toxic gases are a product of combustion. Decomposes when heated, giving off toxic gases.	
	TLV-TWA	5 mg/m³
Gross formula: $C_{12}H_{11}NO_2$	**Absorption route:** Can enter the body by inhalation or ingestion or through the skin or eyes. Evaporation negligible at 20°C, but harmful concentrations of airborne particles can build up rapidly. **Immediate effects:** Irritates the eyes, skin and respiratory tract. Affects the nervous system (by blocking cholinesterase). In substantial concentrations can cause seizures.	

HAZARDS/SYMPTOMS	PREVENTIVE MEASURES	FIRE EXTINGUISHING/FIRST AID
Fire: combustible.	keep away from open flame, no smoking.	foam, water spray, powder.
Explosion: finely dispersed particles can explode on contact with air.	keep dust from accumulating; sealed machinery, explosion-proof equipment and lighting, grounding.	
	KEEP DUST UNDER CONTROL	
Inhalation: sore throat, severe breathing difficulties, headache, nausea, feeling of weakness.	local exhaust or respiratory protection.	fresh air, rest, call a doctor.
Skin: *is absorbed*; see also 'Inhalation'.	gloves, protective clothing.	remove contaminated clothing, wash skin with soap and water, call a doctor, do not touch with bare hands.
Eyes: redness, pain, impaired vision.	goggles, or combined eye and respiratory protection.	flush with water, take to a doctor if necessary.
Ingestion: abdominal cramps, nausea, vomiting; see also 'Inhalation'.		rinse mouth, take immediately to hospital.

SPILLAGE	STORAGE	LABELING / NFPA
clean up spillage, carefully collect remainder; (additional individual protection: P2 respirator).	ventilate.	R: 20/22-37 S: 2-13 ✖ Harmful

NOTES
Inflammability of commercial product depends on medium. Depending on the degree of exposure, regular medical checkups are advisable. Special first aid required in the event of poisoning: antidotes must be available (with instructions). Carbaryl is an insecticide.

Transport Emergency Card TEC(R)-61G47

METHYLPROPIONATE

PHYSICAL PROPERTIES	IMPORTANT CHARACTERISTICS
Boiling point, °C 80 Melting point, °C −88 Flash point, °C −2 Auto-ignition temperature, °C 465 Relative density (water = 1) 0.9 Relative vapor density (air = 1) 3.0 Relative density at 20 °C of saturated mixture vapor/air (air = 1) 1.2 Vapor pressure, mm Hg at 20 °C 65 Solubility in water moderate Explosive limits, vol% in air 2.4-13 Relative molecular mass 88.1 Gross formula: $C_4H_8O_2$	**COLORLESS LIQUID WITH CHARACTERISTIC ODOR** Vapor is heavier than air and spreads at ground level, with risk of ignition at a distance. Do not use compressed air when filling, emptying or processing. Reacts violently with strong oxidants. TLV-TWA not available **Absorption route:** Can enter the body by inhalation or ingestion. Harmful atmospheric concentrations can build up fairly rapidly on evaporation at approx. 20°C - even more rapidly in aerosol form. **Immediate effects:** Irritates the eyes and respiratory tract.

HAZARDS/SYMPTOMS	PREVENTIVE MEASURES	FIRE EXTINGUISHING/FIRST AID
Fire: extremely flammable.	keep away from open flame and sparks, no smoking.	powder, AFFF, foam, carbon dioxide, (halons).
Explosion: forms explosive air-vapor mixtures.	sealed machinery, ventilation, explosion-proof electrical equipment and lighting.	in case of fire: keep tanks/drums cool by spraying with water.
Inhalation: sore throat, cough.	ventilation, local exhaust or respiratory protection.	fresh air, rest, call a doctor if necessary.
Skin: redness.	gloves.	remove contaminated clothing, flush skin with water or shower.
Eyes: redness, pain.	safety glasses.	flush with water, take to a doctor if necessary.
Ingestion: abdominal pain, vomiting.		rinse mouth, call a doctor or take to hospital.

SPILLAGE	STORAGE	LABELING / NFPA
collect leakage in sealable containers, soak up with sand or other inert absorbent and remove to safe place; (additional individual protection: breathing apparatus).	keep in fireproof place, separate from oxidants.	R: 11 S: 16-23-29-33 🔥 Flammable (NFPA diamond: 3 / 1 / 0)

NOTES

METHYL PROPYL ETHER

PHYSICAL PROPERTIES	IMPORTANT CHARACTERISTICS
Boiling point, °C 39 Melting point, °C − 69 Flash point, °C < − 20 Relative density (water = 1) 0.7 Relative vapor density (air = 1) 2.6 Relative density at 20 °C of saturated mixture vapor/air (air = 1) 1.7 Vapor pressure, mm Hg at 20 °C ca. 380 Solubility in water, g/100 ml 4 Explosive limits, vol% in air 2-? Relative molecular mass 74.1	**COLORLESS HIGHLY VOLATILE LIQUID WITH CHARACTERISTIC ODOR** Vapor is heavier than air and spreads at ground level, with risk of ignition at a distance. Able to form peroxides. Flow, agitation etc. can cause build-up of electrostatic charge due to liquid's low conductivity. Do not use compressed air when filling, emptying or processing. Reacts violently with oxidants. TLV-TWA not available **Absorption route:** Can enter the body by inhalation or ingestion. Harmful atmospheric concentrations can build up fairly rapidly on evaporation at approx. 20°C - even more rapidly in aerosol form. **Immediate effects:** Irritates the eyes and respiratory tract. In substantial concentrations can impair consciousness.
Gross formula: $C_4H_{10}O$	

HAZARDS/SYMPTOMS	PREVENTIVE MEASURES	FIRE EXTINGUISHING/FIRST AID
Fire: extremely flammable.	keep away from open flame and sparks, no smoking, avoid contact with hot surfaces (e.g. steam lines).	powder, AFFF, foam, carbon dioxide, (halons).
Explosion: forms explosive air-vapor mixtures.	sealed machinery, ventilation, explosion-proof electrical equipment and lighting, grounding, non-sparking tools.	in case of fire: keep tanks/drums cool by spraying with water.
Inhalation: dizziness, nausea, drowsiness, sleepiness.	ventilation, local exhaust or respiratory protection.	fresh air, rest, call a doctor.
Skin: redness.	gloves.	remove contaminated clothing, flush skin with water or shower.
Eyes: redness, pain.	acid goggles.	flush with water, take to a doctor.
Ingestion: nausea, dizziness, drowsiness, sleepiness.		rinse mouth, call a doctor.

SPILLAGE	STORAGE	LABELING / NFPA
collect leakage in sealable containers, soak up with sand or other inert absorbent and remove to safe place; (additional individual protection: breathing apparatus).	keep in cool, dark, fireproof place, separate from oxidants; add an inhibitor or keep only small quantities on hand.	3 / 0 / 0

NOTES
Before distilling check for peroxides; if found, render harmless. Alcohol consumption increases toxic (narcotic) effects.

Transport Emergency Card TEC(R)-30G30

HI: 33; UN-number: 2612

METHYLPROPYLKETONE

PHYSICAL PROPERTIES		IMPORTANT CHARACTERISTICS	
Boiling point, °C	102	**COLORLESS LIQUID WITH CHARACTERISTIC ODOR**	
Melting point, °C	− 78	Vapor mixes readily with air, forming explosive mixtures. Flow, agitation etc. can cause build-up of electrostatic charge due to liquid's low conductivity. Do not use compressed air when filling, emptying or processing. Reacts violently with strong oxidants, with risk of fire and explosion. Reacts violently with bromotrifluoride, with risk of fire and explosion.	
Flash point, °C	7		
Auto-ignition temperature, °C	505		
Relative density (water = 1)	0.8		
Relative vapor density (air = 1)	2.98		
Relative density at 20 °C of saturated mixture vapor/air (air = 1)	1.03	TLV-TWA	200 ppm	705 mg/m³
Vapor pressure, mm Hg at 20 °C	12	STEL	250 ppm	881 mg/m³
Solubility in water, g/100 ml at 20 °C	4		
Explosive limits, vol% in air	1.5-8.2	**Absorption route:** Can enter the body by inhalation or ingestion. Harmful atmospheric concentrations build up fairly slowly on evaporation at 20°C, but much more rapidly in aerosol form. **Immediate effects:** Irritates the eyes, skin and respiratory tract. Liquid destroys the skin's natural oils. In substantial concentrations can impair consciousness. **Effects of prolonged/repeated exposure:** Prolonged or repeated contact can cause skin disorders.	
Electrical conductivity, pS/m	< 10⁴		
Relative molecular mass	86.1		
Gross formula:	$C_5H_{10}O$		

HAZARDS/SYMPTOMS	PREVENTIVE MEASURES	FIRE EXTINGUISHING/FIRST AID
Fire: extremely flammable.	keep away from open flame and sparks, no smoking, avoid contact with strong oxidants.	powder, AFFF, foam, carbon dioxide, (halons).
Explosion: forms explosive air-vapor mixtures.	sealed machinery, ventilation, explosion-proof electrical equipment and lighting, grounding.	in case of fire: keep tanks/drums cool by spraying with water.
Inhalation: sore throat, cough, drowsiness, sleepiness.	ventilation, local exhaust or respiratory protection.	fresh air, rest, call a doctor.
Skin: redness.	gloves.	remove contaminated clothing, flush skin with water or shower.
Eyes: redness, pain.	safety glasses.	flush with water, take to a doctor.
Ingestion: abdominal pain, nausea, drowsiness.		rinse mouth, call a doctor.

SPILLAGE	STORAGE	LABELING / NFPA
collect leakage in sealable containers, soak up with sand or other inert absorbent and remove to safe place.	keep in fireproof place, keep separate from strong oxidants.	3 2 0

NOTES
Alcohol consumption increases toxic (narcotic) effects.

Transport Emergency Card TEC(R)-30G30

HI: 33; UN-number: 1249

CAS-No: [872-50-4]
N-methylpyrollidone
1-methylpyrollidone
1-methyl-2-pyrollidinone
NMP

C_5H_9NO

1-METHYL-2-PYROLLIDINONE

PHYSICAL PROPERTIES		IMPORTANT CHARACTERISTICS	
Boiling point, °C	202	**COLORLESS TO LIGHT YELLOW HYGROSCOPIC LIQUID WITH CHARACTERISTIC ODOR** Vapor mixes readily with air. Decomposes when heated or burned, giving off toxic vapors ($\rightarrow$ *nitrous vapors*). Reacts with strong acids. Attacks aluminum and light metals.	
Melting point, °C	−24		
Flash point, °C	96		
Auto-ignition temperature, °C	346		
Relative density (water = 1)	1.03	TLV-TWA 100 ppm 400 mg/m³	
Relative vapor density (air = 1)	3.42		
Relative density at 20 °C of saturated mixture vapor/air (air = 1)	1.0	**Absorption route:** Can enter the body by ingestion. Harmful atmospheric concentrations build up very slowly, if at all, on evaporation at approx. 20°C, but much more rapidly in aerosol form.	
Vapor pressure, mm Hg at 25 °C	0.53		
Solubility in water	∞		
Explosive limits, vol% in air	0.9-?		
Electrical conductivity, pS/m	2×10^6		
Relative molecular mass	99.1		
Gross formula:	C_5H_9NO		

HAZARDS/SYMPTOMS	PREVENTIVE MEASURES	FIRE EXTINGUISHING/FIRST AID
Fire: combustible.	keep away from open flame, no smoking.	powder, alcohol-resistant foam, water spray, carbon dioxide, (halons).
Explosion: above 96°C: forms explosive air-vapor mixtures.	above 96°C: sealed machinery, ventilation.	
Inhalation:	ventilation.	
Skin:	gloves.	remove contaminated clothing, flush skin with water or shower.
Eyes:	safety glasses.	
Ingestion:		call a doctor if necessary.

SPILLAGE	STORAGE	LABELING / NFPA
collect leakage in sealable containers, flush away remainder with water.	keep separate from strong acids.	R: 36/38 S: 41 ✖ Irritant [NFPA diamond: 2 / 1 / 0]

NOTES
Insufficient toxicological data on harmful effects to humans. Few indications of toxicity found in animal experiments.

METHYL SALICYLATE

PHYSICAL PROPERTIES	IMPORTANT CHARACTERISTICS
Boiling point, °C 223 Melting point, °C −8.6 Flash point, °C 101 Auto-ignition temperature, °C 450 Relative density (water = 1) 1.18 Relative vapor density (air = 1) 5.2 Relative density at 20 °C of saturated mixture vapor/air (air = 1) 1.0 Vapor pressure, mm Hg at 20 °C ca. 0.076 Solubility in water, g/100 ml at 20 °C 0.07 Relative molecular mass 152.1 Gross formula: $C_8H_8O_3$	**COLORLESS TO YELLOW VISCOUS LIQUID WITH CHARACTERISTIC ODOR** Vapor mixes readily with air. Reacts with strong oxidants.
	TLV-TWA not available
	Absorption route: Can enter the body by ingestion. Harmful atmospheric concentrations build up very slowly, if at all, on evaporation at approx. 20° C, but much more rapidly in aerosol form. **Immediate effects:** Irritates the eyes and skin. Affects the nervous system. In serious cases risk of seizures, unconsciousness and death.

HAZARDS/SYMPTOMS	PREVENTIVE MEASURES	FIRE EXTINGUISHING/FIRST AID
Fire: combustible.	keep away from open flame, no smoking.	powder, AFFF, foam, carbon dioxide, (halons); in case of fire in immediate vicinity: use any extinguishing agent.
Inhalation:	ventilation.	take to hospital.
Skin: redness, pain.	gloves.	remove contaminated clothing, flush skin with water or shower.
Eyes: redness, pain, impaired vision.	acid goggles.	flush with water, take to a doctor.
Ingestion: diarrhea, headache, dizziness.	do not eat, drink or smoke while working.	take to hospital.

SPILLAGE	STORAGE	LABELING / NFPA
collect leakage in sealable containers, soak up with sand or other inert absorbent and remove to safe place.	keep separate from oxidants.	

NOTES
Also referred to as sweet-birch oil.

CAS-No: [681-84-5]
methyl orthosilicate
silicic acid tetramethyl ester
tetramethoxysilane
tetramethyl orthosilicate

$Si(OCH_3)_4$

METHYL SILICATE

PHYSICAL PROPERTIES	IMPORTANT CHARACTERISTICS	
Boiling point, °C 121 Melting point, °C −8 Flash point, °C <21 Relative density (water = 1) 1.03 Relative vapor density (air = 1) 5.3 Relative density at 20 °C of saturated mixture vapor/air (air = 1) 1.07 Vapor pressure, mm Hg at 20 °C 12 Solubility in water reaction Relative molecular mass 152.2	**COLORLESS LIQUID WITH PUNGENT ODOR** Vapor mixes readily with air, forming explosive mixtures. Decomposes on contact with water, forming → *methanol* and → *silicic acid*.	
	TLV-TWA 1 ppm 6 mg/m³	
	Absorption route: Can enter the body by inhalation or ingestion. Harmful atmospheric concentrations can build up very rapidly on evaporation at 20°C. **Immediate effects:** Irritates the skin. Corrosive to the eyes and respiratory tract. Inhalation of vapor/fumes can cause severe breathing difficulties (lung edema). **Effects of prolonged/repeated exposure:** Can cause kidney damage and serious eye damage.	
Gross formula: $C_4H_{12}O_4Si$		

HAZARDS/SYMPTOMS	PREVENTIVE MEASURES	FIRE EXTINGUISHING/FIRST AID
Fire: extremely flammable.	keep away from open flame and sparks, no smoking.	AFFF, powder, carbon dioxide, (halons), NO OTHER WATER-BASED EXTINGUISHERS, DO NOT USE WATER.
Explosion: forms explosive air-vapor mixtures.	sealed machinery, ventilation, explosion-proof electrical equipment and lighting.	in case of fire: keep tanks/drums cool by spraying with water, but DO NOT spray substance with water.
	STRICT HYGIENE	
Inhalation: *corrosive*; sore throat, cough, severe breathing difficulties.	ventilation, local exhaust or respiratory protection.	fresh air, rest, place in half-sitting position, take to hospital.
Skin: redness, pain.	gloves, protective clothing.	remove contaminated clothing, flush skin with water or shower, refer to a doctor.
Eyes: *corrosive*; redness, pain, impaired vision.	face shield, or combined eye and respiratory protection.	flush with water and take to hospital (eye doctor).
Ingestion: *corrosive*; abdominal cramps, pain with swallowing.		rinse mouth, immediately take to hospital.

SPILLAGE	STORAGE	LABELING / NFPA
evacuate area, call in an expert, collect leakage in sealable containers, soak up with sand or other inert absorbent and remove to safe place; (additional individual protection: CHEMICAL SUIT).	keep in dry, fireproof place.	 3 / 3 / 1

NOTES
Lung edema symptoms usually develop several hours later and are aggravated by physical exertion: rest and hospitalization essential. As first aid, a doctor or authorized person should consider administering a corticosteroid spray. Use airtight packaging. Unbreakable packaging preferred; if breakable, keep in unbreakable container.

Transport Emergency Card TEC(R)-30G32

HI: 336; UN-number: 2606

CAS-No: [98-83-9]
isopropenylbenzene
(1-methylethenyl)benzene
1-methyl-1-phenylethene
2-phenylpropylene

$C_6H_5C(CH_3) = CH_2$

α-METHYLSTYRENE

PHYSICAL PROPERTIES		IMPORTANT CHARACTERISTICS	
Boiling point, °C	165	**COLORLESS LIQUID WITH CHARACTERISTIC ODOR**	
Melting point, °C	− 23	Vapor mixes readily with air. Able to form peroxides and can polymerize when heated or on contact with catalysts. Reacts with strong oxidants.	
Flash point, °C	54		
Auto-ignition temperature, °C	445		
Relative density (water = 1)	0.9	TLV-TWA 50 ppm 242 mg/m³	
Relative vapor density (air = 1)	4.1	STEL 100 ppm 483 mg/m³	
Relative density at 20 °C of saturated mixture vapor/air (air = 1)	1.01	**Absorption route:** Can enter the body by inhalation or ingestion. Harmful atmospheric concentrations build up fairly slowly on evaporation at 20°C, but much more rapidly in aerosol form.	
Vapor pressure, mm Hg at 20 °C	2.3		
Solubility in water	none	**Immediate effects:** Irritates the eyes, skin and respiratory tract. In substantial concentrations can impair consciousness.	
Explosive limits, vol% in air	0.9-6.6		
Relative molecular mass	118.2	**Effects of prolonged/repeated exposure:** Prolonged or repeated contact can cause skin disorders. Can cause liver and kidney damage.	
Gross formula:	C_9H_{10}		

HAZARDS/SYMPTOMS	PREVENTIVE MEASURES	FIRE EXTINGUISHING/FIRST AID
Fire: flammable.	keep away from open flame and sparks, no smoking.	powder, AFFF, foam, carbon dioxide, (halons).
Explosion: above 54°C: forms explosive air-vapor mixtures.	above 54°C: sealed machinery, ventilation, explosion-proof electrical equipment.	in case of fire: keep tanks/drums cool by spraying with water.
Inhalation: cough, shortness of breath.	ventilation, local exhaust or respiratory protection.	fresh air, rest, call a doctor.
Skin: redness, pain.	gloves.	remove contaminated clothing, flush skin with water or shower, call a doctor.
Eyes: redness, pain, impaired vision.	face shield.	flush with water, take to a doctor.
Ingestion: abdominal cramps, vomiting.		rinse mouth, call a doctor or take to hospital.

SPILLAGE	STORAGE	LABELING / NFPA
collect leakage in sealable containers, soak up with sand or other inert absorbent and remove to safe place.	keep in fireproof place, separate from oxidants; add an inhibitor.	R: 10-36/37 ✖ Irritant

NFPA diamond: 2 (top) / 2 (left) / 1 (right)

NOTES
Before distilling check for peroxides; if found, render harmless.

Transport Emergency Card TEC(R)-883

HI: 30; UN-number: 2303

METHYLTRICHLOROSILANE

PHYSICAL PROPERTIES	IMPORTANT CHARACTERISTICS
Boiling point, °C 66 Melting point, °C −78 Flash point, °C −9 Auto-ignition temperature, °C 404 Relative density (water = 1) 1.3 Relative vapor density (air = 1) 5.2 Relative density at 20 °C of saturated mixture vapor/air (air = 1) 1.7 Vapor pressure, mm Hg at 20 °C 136.8 Solubility in water reaction Explosive limits, vol% in air 5.1-? Relative molecular mass 149.5 Gross formula: CH₃Cl₃Si	**COLORLESS FUMING LIQUID WITH PUNGENT ODOR** Vapor is heavier than air and spreads at ground level, with risk of ignition at a distance. Corrosive fumes (→ *hydrochloric acid*) are a product of combustion. Reacts violently with water (vapor) and many other substances, usually giving off corrosive fumes (→ *hydrochloric acid*). Reacts violently with strong oxidants, with risk of fire and explosion. Reacts with air to form corrosive fumes (→ *hydrochloric acid*), which spread at ground level. Attacks many metals. TLV-TWA not available **Absorption route:** Can enter the body by inhalation of vapor, through the skin or by ingestion. Harmful atmospheric concentrations can build up very rapidly on evaporation at 20°C. **Immediate effects:** Corrosive to the eyes, skin and respiratory tract. Inhalation of vapor can cause lung edema. If liquid is swallowed, droplets can enter the lungs, with risk of pneumonia. Can cause death. Keep under medical observation.

HAZARDS/SYMPTOMS	PREVENTIVE MEASURES	FIRE EXTINGUISHING/FIRST AID
Fire: extremely flammable.	keep away from open flame and sparks, no smoking.	powder, carbon dioxide, DO NOT USE WATER-BASED EXTINGUISHERS.
Explosion: forms explosive air-vapor mixtures.	sealed machinery, ventilation, explosion-proof electrical equipment and lighting; do not use compressed air when filling, emptying or processing.	in case of fire: keep tanks/drums cool by spraying with water, but NO direct contact of substance with water.
	KEEP DUST UNDER CONTROL/PREVENT MIST FORMATION, STRICT HYGIENE	IN ALL CASES CALL IN A DOCTOR
Inhalation: *corrosive*; burning sensation, cough, difficulty breathing, sore throat.	sealed machinery; ventilation.	fresh air, rest, place in half-sitting position, take immediately to hospital.
Skin: *corrosive, is absorbed*; redness, burning sensation, pain.	gloves, protective clothing.	remove contaminated clothing, flush skin with water or shower, consult a doctor.
Eyes: *corrosive*; redness, pain, impaired vision.	face shield, or combined eye and respiratory protection.	flush thoroughly with water (remove contact lenses if easy), take to a doctor.
Ingestion: *corrosive*; abdominal cramps, burning sensation, nausea, sore throat.		rinse mouth, give no liquids, take immediately to hospital.

SPILLAGE	STORAGE	LABELING / NFPA
evacuate area, call in an expert, ventilate, collect leakage in sealable containers, soak up with dry sand or other inert absorbent mixed with sodium carbonate or soda and remove to safe place; (additional individual protection: CHEMICAL SUIT).	keep in dry, fireproof place, separate from all other substances.	R: 11-14-36/37/38 S: 26-39 Flammable Irritant 3 3 2 W

NOTES
Reacts violently with extinguishing agents such as water and foam. Lung edema symptoms usually develop several hours later and are aggravated by physical exertion: rest and hospitalization essential. As first aid, a doctor or authorized person should consider administering a suitable spray. Use airtight packaging. Unbreakable packaging preferred; if breakable, keep in unbreakable container.

Transport Emergency Card TEC(R)-30G40   **HI: X338; UN-number: 1250**

CAS-No: [78-94-4]
3-buten-2-one
acetyl ethylene
butenone
2-butenone
methylene acetone

$CH_3COCH=CH_2$

METHYL VINYL KETONE

PHYSICAL PROPERTIES	IMPORTANT CHARACTERISTICS
Boiling point, °C 81 Flash point, °C −7 Auto-ignition temperature, °C 491 Relative density (water = 1) 0.86 Relative vapor density (air = 1) 2.4 Relative density at 20 °C of saturated mixture vapor/air (air = 1) 1.12 Vapor pressure, mm Hg at 20 °C 76 Solubility in water good Explosive limits, vol% in air 2.1-15.6 Relative molecular mass 70.1	**COLORLESS LIQUID WITH PUNGENT ODOR** Vapor is heavier than air and spreads at ground level, with risk of ignition at a distance. Presumed to be able to form peroxides. Able to polymerize violently when moderately heated, with risk of fire and explosion. Do not use compressed air when filling, emptying or processing. Reacts violently with strong oxidants.
	TLV-TWA not available
	Absorption route: Can enter the body by inhalation or ingestion. Harmful atmospheric concentrations can build up fairly rapidly on evaporation at approx. 20°C - even more rapidly in aerosol form. **Immediate effects:** Corrosive to the eyes, skin and respiratory tract. Inhalation of vapor/fumes can cause severe breathing difficulties (lung edema). In serious cases risk of death.
Gross formula: C_4H_6O	

HAZARDS/SYMPTOMS	PREVENTIVE MEASURES	FIRE EXTINGUISHING/FIRST AID
Fire: extremely flammable.	keep away from open flame and sparks, no smoking.	powder, alcohol-resistant foam, large quantities of water, carbon dioxide, (halons).
Explosion: forms explosive air-vapor mixtures.	sealed machinery, ventilation, explosion-proof electrical equipment and lighting.	in case of fire: keep tanks/drums cool by spraying with water.
Inhalation: *corrosive*; sore throat, cough, shortness of breath.	ventilation, local exhaust or respiratory protection.	fresh air, rest, place in half-sitting position, take to hospital.
Skin: *corrosive*; redness, pain, burns.	gloves.	remove contaminated clothing, flush skin with water or shower, refer to a doctor.
Eyes: *corrosive*; redness, pain, impaired vision, watering of the eyes.	face shield.	flush with water, take to a doctor.
Ingestion: *corrosive*; sore throat, abdominal pain, diarrhea, vomiting.		rinse mouth, call a doctor or take to hospital.

SPILLAGE	STORAGE	LABELING / NFPA
collect leakage in sealable containers, soak up with sand or other inert absorbent and remove to safe place; (additional individual protection: breathing apparatus).	keep in cool, fireproof place, separate from oxidants; add an inhibitor.	3 / 3 / 2

NOTES
Before distilling check for peroxides; if found, render harmless. Lung edema symptoms usually develop several hours later and are aggravated by physical exertion: rest and hospitalization essential. As first aid, a doctor or authorized person should consider administering a corticosteroid spray.

Transport Emergency Card TEC(R)-598

HI: 339; UN-number: 1251

CAS-No: [79-11-8]
chloroacetic acid
α-chloroacetic acid
chloroethanoic acid
MCA
monochloroethanoic acid

CH$_2$ClCOOH

MONOCHLOROACETIC ACID

PHYSICAL PROPERTIES		IMPORTANT CHARACTERISTICS
Boiling point, °C	189	**WHITE HYGROSCOPIC CRYSTALS OR WHITE POWDER WITH PUNGENT ODOR**
Melting point, °C	ca. 62	Vapor mixes readily with air. Decomposes when heated or burned, giving off toxic gas
Flash point, °C	126	(→ *phosgene*) and corrosive vapors (→ *hydrochloric acid*). Acidic in aqueous solution, reacts
Relative density (water = 1)	1.58	with bases and is corrosive.
Relative vapor density (air = 1)	3.3	
Relative density at 20 °C of saturated mixture		TLV-TWA not available
vapor/air (air = 1)	1.0	
Vapor pressure, mm Hg at 20 °C	0.17	**Absorption route:** Can enter the body by inhalation or ingestion. Insufficient data on the rate at
Solubility in water	∞	which harmful concentrations can build up.
Explosive limits, vol% in air	8-?	**Immediate effects:** Corrosive to the eyes, skin and respiratory tract. Inhalation of vapor/fumes
Relative molecular mass	94.5	can cause severe breathing difficulties (lung edema). In serious cases risk of death.
Log P octanol/water	−0.1	
Gross formula:	C$_2$H$_3$ClO$_2$	

HAZARDS/SYMPTOMS	PREVENTIVE MEASURES	FIRE EXTINGUISHING/FIRST AID
Fire: combustible.	keep away from open flame, no smoking.	powder, alcohol-resistant foam, water spray, carbon dioxide, (halons).
	KEEP DUST UNDER CONTROL	
Inhalation: *corrosive*; sore throat, cough, severe breathing difficulties.	local exhaust or respiratory protection.	fresh air, rest, place in half-sitting position, take to hospital.
Skin: *corrosive*; redness, pain, serious burns.	gloves, protective clothing.	remove contaminated clothing, flush skin with water or shower, call a doctor.
Eyes: *corrosive*; redness, pain, impaired vision.	face shield.	flush with water, take to a doctor.
Ingestion: *corrosive*; sore throat, abdominal pain.	do not eat, drink or smoke while working.	rinse mouth, immediately take to hospital.

SPILLAGE	STORAGE	LABELING / NFPA
clean up spillage, render remainder harmless with soda, then flush away with water; (additional individual protection: respirator with A/P2 filter).	keep dry, separate from strong bases; ventilate at floor level.	R: 23/24/25-35 S: 22-36/37/39 Toxic NFPA: 1 / 3 / 0

NOTES
Log P octanol/water is estimated. After flushing skin with water, neutralize with 5% aqueous solution of sodium carbonate. Melting point of α-monochloroacetic acid: 63° C; β: 56° C; γ: 53° C. Lung edema symptoms usually develop several hours later and are aggravated by physical exertion: rest and hospitalization essential. As first aid, a doctor or authorized person should consider administering a corticosteroid spray.

Transport Emergency Card TEC(R)-57

HI: 80; UN-number: 1751

benzene chloride
chlorobenzene
phenyl chloride

MONOCHLOROBENZENE

PHYSICAL PROPERTIES

Boiling point, °C	132
Melting point, °C	−45
Flash point, °C	28
Auto-ignition temperature, °C	590
Relative density (water = 1)	1.1
Relative vapor density (air = 1)	3.9
Relative density at 20 °C of saturated mixture vapor/air (air = 1)	1.03
Vapor pressure, mm Hg at 20 °C	9.1
Solubility in water, g/100 ml at 20 °C	0.04
Explosive limits, vol% in air	1.3-11
Electrical conductivity, pS/m	7x10^3
Relative molecular mass	112.6
Log P octanol/water	2.8

Gross formula: C$_6$H$_5$Cl

IMPORTANT CHARACTERISTICS

COLORLESS LIQUID WITH CHARACTERISTIC ODOR
Vapor mixes readily with air. Flow, agitation etc. can cause build-up of electrostatic charge due to liquid's low conductivity. Decomposes when heated or burned, giving off toxic gas (→ *phosgene*) and corrosive vapors (→ *hydrochloric acid*). Reacts violently with alkali metals and strong oxidants.

TLV-TWA	75 ppm	345 mg/m^3

Absorption route: Can enter the body by inhalation or ingestion. Harmful atmospheric concentrations build up fairly slowly on evaporation at 20°C, but much more rapidly in aerosol form.
Immediate effects: Irritates the eyes, skin and respiratory tract. Liquid destroys the skin's natural oils. In substantial concentrations can impair consciousness. In serious cases risk of unconsciousness.
Effects of prolonged/repeated exposure: Can cause liver and kidney damage.

HAZARDS/SYMPTOMS	PREVENTIVE MEASURES	FIRE EXTINGUISHING/FIRST AID
Fire: flammable.	keep away from open flame and sparks, no smoking.	
Explosion: above 28°C: forms explosive air-vapor mixtures.	above 28°C: sealed machinery, ventilation, explosion-proof electrical equipment, grounding.	in case of fire: keep tanks/drums cool by spraying with water.
Inhalation: headache, dizziness, sleepiness, unconsciousness.	ventilation, local exhaust or respiratory protection.	fresh air, rest, call a doctor.
Skin: redness.	gloves.	remove contaminated clothing, wash skin with soap and water.
Eyes: redness, pain.	acid goggles.	flush with water, take to a doctor if necessary.
Ingestion: abdominal pain, headache, nausea, dizziness, sleepiness, unconsciousness.		rinse mouth, take immediately to hospital.

SPILLAGE	STORAGE	LABELING / NFPA
collect leakage in sealable containers, soak up with sand or other inert absorbent and remove to safe place.	keep in fireproof place, separate from oxidants.	R: 10-20 S: 24/25 ✖ Harmful

LABELING / NFPA diamond: 3 (top), 2 (left), 0 (right)

NOTES

Alcohol consumption increases toxic effects.

MONOCHLORODIFLUOROMETHANE
(cylinder)

PHYSICAL PROPERTIES		IMPORTANT CHARACTERISTICS
Boiling point, °C	−41	**COLORLESS, ODORLESS COMPRESSED LIQUEFIED GAS**
Melting point, °C	−146	Gas is heavier than air and can accumulate close to ground level, causing oxygen deficiency, with risk of unconsciousness. Flow, agitation etc. can cause build-up of electrostatic charge due to liquid's low conductivity. Decomposes in flame or on hot surface, giving off corrosive and toxic vapors. Reacts violently with alkali metals. Attacks aluminum, magnesium, zinc and alloys.
Relative density (water = 1)	1.2	
Relative vapor density (air = 1)	3.1	
Vapor pressure, mm Hg at 20 °C	6992	
Solubility in water, g/100 ml at 21 °C	0.30	
Relative molecular mass	86.5	

		TLV-TWA	1000 ppm	3540 mg/m³

Absorption route: Can enter the body by inhalation. Can saturate the air if released, with risk of suffocation.
Immediate effects: Liquid destroys the skin's natural oils. Liquid can cause frostbite due to rapid evaporation. Affects the nervous system.

Gross formula: CHCIF$_2$

HAZARDS/SYMPTOMS	PREVENTIVE MEASURES	FIRE EXTINGUISHING/FIRST AID
Fire: non-combustible.		in case of fire in immediate vicinity: use any extinguishing agent.
Explosion:		in case of fire: keep cylinders cool by spraying with water.
Inhalation: sleepiness, tinnitus.	ventilation, local exhaust or respiratory protection.	fresh air, rest, call a doctor if necessary.
Skin: *in case of frostbite*: redness, pain, sores.	insulating gloves.	*in case of frostbite*: DO NOT remove clothing, flush skin with water or shower, refer to a doctor.
Eyes: *in case of frostbite*: redness, pain, impaired vision.	acid goggles, or combined eye and respiratory protection.	*in case of frostbite*: flush with water, take to a doctor.

SPILLAGE	STORAGE	LABELING / NFPA
ventilate, under no circumstances spray liquid with water; (additional individual protection: breathing apparatus).	if stored indoors, keep in cool, fireproof place; ventilate at floor level.	

NOTES

High atmospheric concentrations, e.g. in poorly ventilated spaces, can cause oxygen deficiency, with risk of unconsciousness. Freon 22 and Frigen 22 (and many others) are trade names. Under no circumstances use near flame or hot surface or when welding. Turn leaking cylinder so that leak is on top to prevent liquid monochlorodifluoromethane escaping.

Transport Emergency Card TEC(R)-217

HI: 20; UN-number: 1018

CAS-No: [95-57-8]
1-chloro-2-hydroxybenzene
2-chloro-1-hydroxybenzene
o-chlorophenol
2-chlorophenol
2-hydroxychlorobenzene

C_6H_4ClOH

o-MONOCHLOROPHENOL

PHYSICAL PROPERTIES		IMPORTANT CHARACTERISTICS
Boiling point, °C	175	**COLORLESS LIQUID WITH CHARACTERISTIC ODOR**
Melting point, °C	7	Vapor mixes readily with air. Decomposes when heated or burned, giving off toxic and corrosive
Flash point, °C	64	vapors, ($\rightarrow$ *hydrochloric acid*). Reacts with oxidants. Attacks aluminum and copper.
Auto-ignition temperature, °C	> 550	
Relative density (water = 1)	1.3	TLV-TWA not available
Relative vapor density (air = 1)	4.4	
Relative density at 20 °C of saturated mixture		
vapor/air (air = 1)	1.01	**Absorption route:** Can enter the body by inhalation or ingestion or through the skin. Harmful
Vapor pressure, mm Hg at 12 °C	0.99	atmospheric concentrations can build up fairly rapidly on evaporation at approx. 20°C - even
Solubility in water, g/100 ml	2.9	more rapidly in aerosol form.
Electrical conductivity, pS/m	3.5 x 10^5	**Immediate effects:** Irritates the eyes, skin and respiratory tract. In substantial concentrations
Relative molecular mass	128.6	can cause agitation, trembling, seizures etc. In serious cases risk of unconsciousness and death.
Log P octanol/water	2.2	**Effects of prolonged/repeated exposure:** Can cause liver and kidney damage.
Gross formula: C_6H_5ClO		

HAZARDS/SYMPTOMS	PREVENTIVE MEASURES	FIRE EXTINGUISHING/FIRST AID
Fire: combustible.	keep away from open flame, no smoking.	powder, water spray, foam, carbon dioxide, (halons).
Explosion: above 64°C: forms explosive air-vapor mixtures.	above 64°C: sealed machinery, ventilation.	
	AVOID ALL CONTACT WITH SKIN	
Inhalation: cough, shortness of breath, headache, dizziness.	ventilation, local exhaust or respiratory protection.	fresh air, rest, place in half-sitting position, call a doctor.
Skin: *is absorbed*; redness, headache, dizziness.	gloves, protective clothing.	remove contaminated clothing, flush skin with water or shower, call a doctor if necessary.
Eyes: redness, pain.	face shield.	flush with water, take to a doctor.
Ingestion: abdominal pain, vomiting.		rinse mouth, take immediately to hospital.

SPILLAGE	STORAGE	LABELING / NFPA
collect leakage in sealable containers, soak up with sand or other inert absorbent and remove to safe place; (additional individual protection: breathing apparatus).	ventilate at floor level.	R: 20/21/22 S: 2 - 28 ✖ Harmful

NOTES

Depending on the degree of exposure, regular medical checkups are advisable.

Transport Emergency Card TEC(R)-799

HI: 68; UN-number: 2021

CAS-No: [75-72-9]
chlorotrifluoromethane
R13

CF_3Cl

MONOCHLOROTRIFLUOROMETHANE
(cylinder)

PHYSICAL PROPERTIES		IMPORTANT CHARACTERISTICS	
Boiling point, °C	−81	**COLORLESS, ODORLESS COMPRESSED LIQUEFIED GAS**	
Melting point, °C	−181	Gas is heavier than air and can accumulate close to ground level, causing oxygen deficiency, with	
Relative density (water = 1)	1.3	risk of unconsciousness. Decomposes in flame or on hot surface, giving off toxic and corrosive	
Relative vapor density (air = 1)	3.6	gases. Reacts with aluminum when subjected to rise in termperatur. Attacks magnesium (alloys)	
Vapor pressure, mm Hg at 20 °C	24,320	and zinc.	
Solubility in water	none		
Relative molecular mass	104.5	TLV-TWA	not available
		Absorption route: Can enter the body by inhalation. Can saturate the air if released, with risk of suffocation. **Immediate effects:** Liquid can cause frostbite due to rapid evaporation. Destroys the skin's natural oils. Affects the nervous system.	
Gross formula:	$CCIF_3$		

HAZARDS/SYMPTOMS	PREVENTIVE MEASURES	FIRE EXTINGUISHING/FIRST AID
Fire: non-combustible.		in case of fire in immediate vicinity: use any extinguishing agent.
Explosion:		in case of fire: keep cylinders cool by spraying with water.
Inhalation: headache, dizziness, sleepiness, unconsciousness.	ventilation, local exhaust or respiratory protection.	fresh air, rest, call a doctor if necessary.
Skin: *in case of frostbite:* redness, burns.	insulating gloves.	*in case of frostbite:* DO NOT remove clothing, flush skin with water or shower, refer to a doctor.
Eyes: *in case of frostbite:* redness, pain, impaired vision.	face shield, or combined eye and respiratory protection.	flush with water, take to a doctor.

SPILLAGE	STORAGE	LABELING / NFPA
ventilate; (additional individual protection: breathing apparatus).	if stored indoors, keep in cool, fireproof place.	

NOTES
High atmospheric concentrations, e.g. in poorly ventilated spaces, can cause oxygen deficiency, with risk of unconsciousness. Under no circumstances use near flame or hot surface or when welding. Turn leaking cylinder so that leak is on top to prevent liquid monochlorotrifluoromethane escaping. Used as aerosol propellant, extinguishing agent, etc. After use as extinguishing agent, provide good ventilation.

Transport Emergency Card TEC(R)-644

HI: 20; UN-number: 1022

MONOMETHYLFORMAMIDE

PHYSICAL PROPERTIES	IMPORTANT CHARACTERISTICS
Boiling point, °C 180-185 Melting point, °C − 40 Flash point, °C 98 Relative density (water = 1) 1.0 Relative vapor density (air = 1) 2.04 Relative density at 20 °C of saturated mixture vapor/air (air = 1) 1.0 Vapor pressure, mm Hg at 20 °C ca. 0.61 Solubility in water good Electrical conductivity, pS/m 10^8 Relative molecular mass 59.1	**COLORLESS TO LIGHT YELLOW LIQUID WITH CHARACTERISTIC ODOR** Vapor mixes readily with air. Decomposes when heated, giving off toxic gases. Reacts with oxidants. TLV-TWA not available **Absorption route:** Can enter the body by inhalation or ingestion or through the skin. Harmful atmospheric concentrations build up fairly slowly on evaporation at 20° C, but much more rapidly in aerosol form. **Immediate effects:** Irritates the eyes, skin and respiratory tract. **Effects of prolonged/repeated exposure:** Can cause liver and kidney damage.
Gross formula: C$_2$H$_5$NO	

HAZARDS/SYMPTOMS	PREVENTIVE MEASURES	FIRE EXTINGUISHING/FIRST AID
Fire: combustible.	keep away from open flame, no smoking.	powder, alcohol-resistant foam, water spray, carbon dioxide, (halons).
Explosion: above 98°C: forms explosive air-vapor mixtures.	above 98°C: sealed machinery, ventilation.	
	AVOID EXPOSURE OF PREGNANT WOMEN AS FAR AS POSSIBLE	
Inhalation: headache, dizziness, nausea, vomiting, abdominal cramps.	ventilation, local exhaust or respiratory protection.	fresh air, rest, call a doctor.
Skin: *is absorbed*; redness, pain.	gloves, protective clothing.	remove contaminated clothing, flush skin with water or shower, refer to a doctor.
Eyes: redness, pain.	safety glasses.	flush with water, take to a doctor.
Ingestion: abdominal cramps, headache, nausea, dizziness, vomiting.		rinse mouth, call a doctor or take to hospital.

SPILLAGE	STORAGE	LABELING / NFPA
collect leakage in sealable containers, soak up with sand or other inert absorbent and remove to safe place.		

NOTES
Alcohol consumption increases toxic effects. Depending on the degree of exposure, regular medical checkups are advisable. Women of childbearing age should avoid contact with substance to avoid damage to unborn children.

Transport Emergency Card TEC(R)-30G15

CAS-No: [110-91-8]
diethylene imidoxide
diethylene oximide
diethylenimide oxide
1-oxa-4-azacyclohexane
tetrahydro-1,4-oxazine

MORPHOLINE

PHYSICAL PROPERTIES		IMPORTANT CHARACTERISTICS		
Boiling point, °C	129	**COLORLESS HYGROSCOPIC LIQUID WITH CHARACTERISTIC ODOR**		
Melting point, °C	−5	Vapor mixes readily with air. Decomposes when burned, giving off toxic gases (→ *nitrous*		
Flash point, °C	38	*vapors*). Strong base which reacts violently with acids and corrodes aluminum, zinc etc. Reacts		
Auto-ignition temperature, °C	310	violently with strong oxidants, with risk of fire and explosion. Attacks copper and its alloys.		
Relative density (water = 1)	1.01			
Relative vapor density (air = 1)	3.0	TLV-TWA (skin)	20 ppm	71 mg/m³
Relative density at 20 °C of saturated mixture				
vapor/air (air = 1)	1.02	**Absorption route:** Can enter the body by inhalation or ingestion or through the skin. Harmful		
Vapor pressure, mm Hg at 20 °C	8.1	atmospheric concentrations can build up fairly rapidly on evaporation at approx. 20°C - even		
Solubility in water	∞	more rapidly in aerosol form.		
Explosive limits, vol% in air	2-11.2	**Immediate effects:** Corrosive to the eyes, skin and respiratory tract. Inhalation can cause lung		
Relative molecular mass	87.1	edema.		
Log P octanol/water	−1.1	**Effects of prolonged/repeated exposure:** Can cause liver and kidney damage.		
Gross formula:	C_4H_9NO			

HAZARDS/SYMPTOMS	PREVENTIVE MEASURES	FIRE EXTINGUISHING/FIRST AID
Fire: flammable.	keep away from open flame and sparks, no smoking.	powder, alcohol-resistant foam, water spray, carbon dioxide, (halons).
Explosion: above 38°C: forms explosive air-vapor mixtures.	above 38°C: sealed machinery, ventilation, explosion-proof electrical equipment.	in case of fire: keep tanks/drums cool by spraying with water.
	STRICT HYGIENE	
Inhalation: *corrosive*; sore throat, cough, shortness of breath, severe breathing difficulties.	ventilation, local exhaust or respiratory protection.	fresh air, rest, place in half-sitting position, take to hospital.
Skin: *corrosive, is absorbed*; redness, pain, burns.	gloves, protective clothing.	remove contaminated clothing, flush skin with water or shower, take to a doctor.
Eyes: *corrosive*; redness, pain, impaired vision.	face shield.	flush with water, take to a doctor.
Ingestion: *corrosive*; abdominal pain, diarrhea, vomiting.		rinse mouth, immediately take to hospital.

SPILLAGE	STORAGE	LABELING / NFPA
collect leakage in sealable containers, flush away remainder with water; (additional individual protection: breathing apparatus).	keep in dry, fireproof place, separate from oxidants and strong acids.	R: 10-20/21/22-34 S: 23-36 ![corrosive symbol] Corrosive ◇ 3 / 2 / 0

NOTES
Depending on the degree of exposure, regular medical checkups are advisable. Lung edema symptoms usually develop several hours later and are aggravated by physical exertion: rest and hospitalization essential. As first aid, a doctor or authorized person should consider administering a corticosteroid spray. Packaging: special material.

Transport Emergency Card TEC(R)-697

HI: 30; UN-number: 2054

CAS-No: [91-20-3]
naphthalin
naphthene
tar camphor
white tar

$C_{10}H_8$

NAPHTHALENE

PHYSICAL PROPERTIES		IMPORTANT CHARACTERISTICS		
Boiling point, °C	218	**WHITE CRYSTALS WITH CHARACTERISTIC ODOR**		
Melting point, °C	80	Vapor mixes readily with air. In dry state can form electrostatic charge if stirred, transported		
Flash point, °C	79	pneumatically, poured etc. In molten state can also form electrostatic charge. Reacts with strong		
Auto-ignition temperature, °C	526	oxidants, particularly violently with chromium trioxide (CrO_3).		
Relative density (water = 1)	1.2			
Relative vapor density (air = 1)	4.4	TLV-TWA	10 ppm	52 mg/m³
Relative density at 20 °C of saturated mixture		STEL	15 ppm	79 mg/m³
vapor/air (air = 1)	1.00			
Vapor pressure, mm Hg at 20 °C	0.030	**Absorption route:** Can enter the body by inhalation or ingestion or through the skin. Harmful		
Solubility in water	none	atmospheric concentrations build up very slowly, if at all, on evaporation at approx. 20°C, but		
Explosive limits, vol% in air	0.9-5.9	harmful concentrations of airborne particles can build up much more rapidly.		
Relative molecular mass	128.2	**Immediate effects:** Irritates the eyes and skin. In serious cases risk of seizures and death.		
Log P octanol/water	3.3	**Effects of prolonged/repeated exposure:** Prolonged or repeated contact can cause skin		
		disorders. Causes kidney damage. Can affect the blood. Can damage the lens of the eye.		
Gross formula:	$C_{10}H_8$			

HAZARDS/SYMPTOMS	PREVENTIVE MEASURES	FIRE EXTINGUISHING/FIRST AID
Fire: combustible.	keep away from open flame, no smoking.	water spray, powder.
Explosion: above 80°C: forms explosive air-vapor mixtures; finely dispersed particles form explosive mixtures on contact with air.	above 80°C: sealed machinery, ventilation, explosion-proof electrical equipment, grounding.	
Inhalation: sore throat, cough, headache, nausea.	local exhaust or respiratory protection.	fresh air, rest, call a doctor if necessary.
Skin: redness, pain.	gloves.	remove contaminated clothing, flush skin with water or shower, call a doctor if necessary.
Eyes: redness, pain, impaired vision.	goggles.	flush with water, take to a doctor.
Ingestion: nausea, vomiting; see also 'Inhalation'.		rinse mouth, immediately take to hospital.

SPILLAGE	STORAGE	LABELING / NFPA
clean up spillage, carefully collect remainder; (additional individual protection: P2 respirator).	keep separate from oxidants; ventilate at floor level.	2 / 2 / 0

NOTES
Log P octanol/water is estimated. Depending on the degree of exposure, regular medical checkups are advisable.

Transport Emergency Card TEC(R)-41G01

HI: 40; UN-number: 1334

NATURAL GAS

(cylinder)

PHYSICAL PROPERTIES	IMPORTANT CHARACTERISTICS
Boiling point, °C − 161 Melting point, °C − 183 Flash point, °C flammable gas Auto-ignition temperature, °C 670 Relative density (water = 1) 0.5 Relative vapor density (air = 1) 0.6 Solubility in water none Explosive limits, vol% in air 5-15.8 Relative molecular mass ~ 17.4	**COLORLESS GAS WITH CHARACTERISTIC ODOR** Mixes readily with air, forming explosive mixtures. Gas is lighter than air. Identifiable odor is added to household fuel gas. TLV-TWA not available **Absorption route:** Can enter the body by inhalation.
Gross formula: CH_4	

HAZARDS/SYMPTOMS	PREVENTIVE MEASURES	FIRE EXTINGUISHING/FIRST AID
Fire: extremely flammable.	keep away from open flame and sparks, no smoking.	shut off supply; if impossible and no danger to surrounding area, allow fire to burn itself out; otherwise extinguish with powder, carbon dioxide, (halons).
Explosion: forms explosive air-gas mixtures.	sealed machinery, ventilation, explosion-proof electrical equipment and lighting, grounding, non-sparking tools.	
Inhalation: headache, drowsiness, severe breathing difficulties, unconsciousness.	ventilation, local exhaust or respiratory protection.	fresh air, rest, take to hospital if necessary.

SPILLAGE	STORAGE	LABELING / NFPA
evacuate area, call in an expert, ventilate; (additional individual protection: breathing apparatus).		

NOTES
High atmospheric concentrations, e.g. in poorly ventilated spaces, can cause oxygen deficiency, with risk of unconsciousness.

Transport Emergency Card TEC(R)-622

HI: 223; UN-number: 1972

NATURAL GAS
(liquid, refrigerated)

PHYSICAL PROPERTIES	IMPORTANT CHARACTERISTICS
Boiling point, °C − 162 Melting point, °C − 182 Flash point, °C flammable gas Auto-ignition temperature, °C 670 Relative density (water = 1) see notes Relative vapor density (air = 1) 0.6 Solubility in water none Explosive limits, vol% in air 5-16 Minimum ignition energy, mJ 0.28 Electrical conductivity, pS/m < 10⁴ Relative molecular mass 17.4 Gross formula: CH₄	**COLORLESS, EXTREMELY COLD LIQUID** Cold gas is heavier than air; otherwise lighter than air. Identifiable odor is added to household fuel gas. Flow, agitation etc. can cause build-up of electrostatic charge due to liquid's low conductivity. Do not use compressed air when filling, emptying or processing. TLV-TWA not available **Absorption route:** Can enter the body by inhalation. Can saturate the air if released, with risk of suffocation. **Immediate effects:** Cold liquid can cause frostbite.

HAZARDS/SYMPTOMS	PREVENTIVE MEASURES	FIRE EXTINGUISHING/FIRST AID
Fire: extremely flammable.	keep away from open flame and sparks, no smoking.	shut off supply; if impossible and no danger to surrounding area, allow fire to burn itself out; otherwise extinguish with powder, carbon dioxide, (halons).
Explosion: forms explosive air-gas mixtures.	sealed machinery, ventilation, explosion-proof electrical equipment and lighting, grounding, non-sparking tools.	in case of fire: keep tanks/drums cool by spraying with water, but DO NOT spray substance with water.
Inhalation: severe breathing difficulties, headache, drowsiness, unconsciousness.	ventilation, local exhaust or respiratory protection.	fresh air, rest, artificial respiration if necessary, take to hospital.
Skin: *in case of frostbite*: redness, pain, blisters.	insulating gloves.	*in case of frostbite*: DO NOT remove clothing, flush skin with water or shower, take to a doctor.
Eyes: *in case of frostbite*: redness, pain, impaired vision.	face shield, or combined eye and respiratory protection.	flush with water, take to a doctor.

SPILLAGE	STORAGE	LABELING / NFPA
evacuate area, ventilate, under no circumstances spray liquid with water; (additional individual protection: breathing apparatus).	keep in fireproof place, under refrigeration; ventilate.	4 3 1

NOTES
Density at boiling point: 0.45kg/l. High atmospheric concentrations, e.g. in poorly ventilated spaces, can cause oxygen deficiency, with risk of unconsciousness. The measures on this card also apply to liquefied methane.

CAS-No: [7440-01-9]
neon, compressed
neon gas

NEON
(cylinder)

PHYSICAL PROPERTIES	IMPORTANT CHARACTERISTICS
Boiling point, °C −246 Melting point, °C −249 Relative vapor density (air = 1) 0.70 Solubility in water, g/100 ml at 0 °C 1.3 x 10⁻³ Relative atomic mass 20.2	**COLORLESS, ODORLESS COMPRESSED GAS** Gas is lighter than air.

Boiling point, °C −246
Melting point, °C −249
Relative vapor density (air = 1) 0.70
Solubility in water, g/100 ml at 0 °C 1.3×10^{-3}
Relative atomic mass 20.2

COLORLESS, ODORLESS COMPRESSED GAS
Gas is lighter than air.

TLV-TWA not available

Absorption route: If liquid is released, air can become saturated due to evaporation, with risk of suffocation.
Immediate effects: Cold liquid can cause frostbite.

Gross formula: Ne

HAZARDS/SYMPTOMS	PREVENTIVE MEASURES	FIRE EXTINGUISHING/FIRST AID
Fire: non-combustible.		in case of fire in immediate vicinity: use any extinguishing agent.
Explosion:		in case of fire: keep cylinders cool by spraying with water.
Inhalation: severe breathing difficulties, headache, dizziness, unconsciousness.	ventilation, local exhaust or respiratory protection.	fresh air, rest, artificial respiration if necessary.

SPILLAGE	STORAGE	LABELING / NFPA
ventilate; (additional individual protection: breathing apparatus).	if stored indoors, keep in cool, fireproof place; ventilate.	

NOTES
Critical temperature −229° C. High atmospheric concentrations, e.g. in poorly ventilated spaces, can cause oxygen deficiency, with risk of unconsciousness.

Transport Emergency Card TEC(R)-20G22

HI: 22; UN-number: 1913

NEON
(liquid, refrigerated)

PHYSICAL PROPERTIES	IMPORTANT CHARACTERISTICS
Boiling point, °C − 246 Melting point, °C − 249 Relative density (water = 1) see notes Relative vapor density (air = 1) 0.70 Vapor pressure, mm Hg at − 246 °C 760 Solubility in water, g/100 ml at 0 °C 1.3×10^{-3} Relative atomic mass 20.2	**COLORLESS, ODORLESS, EXTREMELY COLD LIQUID** *Cold* vapor is heavier than air. TLV-TWA not available **Absorption route:** Can saturate the air if released, with risk of suffocation. **Immediate effects:** Inhalation can cause severe breathing difficulties. In serious cases risk of unconsciousness.
Gross formula: Ne	

HAZARDS/SYMPTOMS	PREVENTIVE MEASURES	FIRE EXTINGUISHING/FIRST AID
Fire: non-combustible.		in case of fire in immediate vicinity: use any extinguishing agent, but do not spray liquid with water.
Inhalation: see Notes.	ventilation, local exhaust or respiratory protection.	fresh air, rest, artificial respiration if necessary, call a doctor, take to hospital if necessary.
Skin: *in case of frostbite*: redness, pain, blisters, sores.	insulating gloves.	*in case of frostbite*: DO NOT remove clothing, flush skin with water or shower, take to a doctor.
Eyes: *in case of frostbite*: redness, pain, impaired vision.	face shield, or combined eye and respiratory protection.	flush with water, take to a doctor.

SPILLAGE	STORAGE	LABELING / NFPA
ventilate, under no circumstances spray liquid with water; (additional individual protection: breathing apparatus).	ventilate, refrigerate.	

NOTES
Density of liquid at boiling point 1.2kg/l. Critical temperature − 229° C. High atmospheric concentrations, e.g. in poorly ventilated spaces, can cause oxygen deficiency, with risk of unconsciousness. Use special insulated container.

NEOPENTYL GLYCOL

PHYSICAL PROPERTIES	IMPORTANT CHARACTERISTICS	
Boiling point, °C 204 Melting point, °C 120 Flash point, °C 107 Auto-ignition temperature, °C 388 Relative density (water = 1) 1.1 Relative vapor density (air = 1) 3.6 Relative density at 20 °C of saturated mixture vapor/air (air = 1) 1.0 Vapor pressure, mm Hg at 20 °C < 0.076 Solubility in water, g/100 ml 186 Explosive limits, vol% in air 1.4-18.8 Relative molecular mass 104.2	**COLORLESS TO WHITE HYGROSCOPIC CRYSTALS** Finely dispersed in air can cause dust explosion. Reacts with strong oxidants, with risk of fire and explosion.	
	TLV-TWA not available	
Gross formula: $C_5H_{12}O_2$	**Absorption route:** Can enter the body by inhalation or ingestion. Harmful atmospheric concentrations build up very slowly, if at all, on evaporation at approx. 20°C, but harmful concentrations of airborne particles can build up much more rapidly. **Immediate effects:** Irritates the eyes, skin and respiratory tract.	

HAZARDS/SYMPTOMS	PREVENTIVE MEASURES	FIRE EXTINGUISHING/FIRST AID
Fire: combustible.	keep away from open flame, no smoking.	water spray, powder.
Explosion: finely dispersed particles form explosive mixtures on contact with air.	keep dust from accumulating, closed system, explosion-proof electrical equipment and lighting.	
	KEEP DUST UNDER CONTROL	
Inhalation: cough, headache.	local exhaust or respiratory protection.	fresh air, rest, consult a doctor.
Skin: redness, pain.	gloves.	remove contaminated clothing, flush skin with water or shower.
Eyes: redness, pain, impaired vision.	goggles.	flush thoroughly with water (remove contact lenses if easy), take to a doctor.
Ingestion: cough, nausea.		rinse, mouth, give no liquids, refer to a doctor.

SPILLAGE	STORAGE	LABELING / NFPA
clean up spillage, flush away remainder with water; (additional individual protection: P2 respirator).	keep dry, separate from oxidants; ventilate at floor level.	

NOTES
Insufficient data on harmful effects to humans: exercise caution.

NICKEL CARBONYL

PHYSICAL PROPERTIES		IMPORTANT CHARACTERISTICS	
Boiling point, °C	43	**COLORLESS LIQUID WITH FAINT ODOR**	
Melting point, °C	− 19	Vapor is heavier than air and spreads at ground level, with risk of ignition at a distance. Do not use compressed air when filling, emptying or processing. Can ignite spontaneously on exposure to air and decomposes explosively when heated to 60°C. Decomposes on contact with acids, giving off → *carbon monoxide*. Reacts violently with oxidants, with risk of fire and explosion.	
Flash point, °C	< − 20		
Auto-ignition temperature, °C	60		
Relative density (water = 1)	1.3		
Relative vapor density (air = 1)	5.9		
Vapor pressure, mm Hg at 20°C	325	TLV-TWA 0.0001 ppm A1 0.007 mg/m³ A1	
Solubility in water, g/100 ml	0.02		
Explosive limits, vol% in air	2-34	**Absorption route:** Can enter the body by inhalation or ingestion or through the skin. Harmful atmospheric concentrations can build up very rapidly on evaporation at 20°C.	
Relative molecular mass	170.8	**Immediate effects:** Irritates the eyes, skin and respiratory tract. Inhalation of vapor/fumes can cause severe breathing difficulties (lung edema). In serious cases risk of seizures, unconsciousness and death.	
Gross formula:	C$_4$NiO$_4$		

HAZARDS/SYMPTOMS	PREVENTIVE MEASURES	FIRE EXTINGUISHING/FIRST AID
Fire: extremely flammable.	keep away from open flame and sparks, no smoking, avoid contact with oxidants.	powder, water spray, foam, carbon dioxide, (halons).
Explosion: forms explosive air-vapor mixtures, heating above 60°C can cause explosion.	sealed machinery, ventilation, explosion-proof electrical equipment and lighting.	in case of fire: keep tanks/drums cool by spraying with water, fight fire from sheltered location.
	STRICT HYGIENE	IN ALL CASES CALL A DOCTOR
Inhalation: sore throat, cough, severe breathing difficulties, headache, dizziness, vomiting.	ventilation, local exhaust or respiratory protection.	fresh air, rest, place in half-sitting position, take to hospital.
Skin: *is absorbed.*	gloves, protective clothing.	remove contaminated clothing, wash skin with soap and water, take to hospital.
Eyes: redness, pain.	face shield, or combined eye and respiratory protection.	flush with water, take to a doctor.
Ingestion: abdominal pain.		rinse mouth, take immediately to hospital.

SPILLAGE	STORAGE	LABELING / NFPA
evacuate area, call in an expert; (additional individual protection: CHEMICAL SUIT).	keep in cool, dark, fireproof place under inert gas, separate from oxidants and acids.	R: 11-26-40 S: 9-23-45 Flammable Toxic 4 3 3

NOTES
Odor limit is above TLV. Flush contaminated clothing with water (fire hazard). Lung edema symptoms usually develop several hours later and are aggravated by physical exertion: rest and hospitalization essential. As first aid, a doctor or authorized person should consider administering a corticosteroid spray. Use airtight packaging.

CAS-No: [7786-81-4]
nickel (II) sulfate
sulfuric acid, nickel salt

NiSO$_4$

NICKEL SULFATE

PHYSICAL PROPERTIES	IMPORTANT CHARACTERISTICS	
Decomposes below boiling point, °C 840 Relative density (water = 1) 3.7 Solubility in water good Relative molecular mass 154.8	**GREEN CRYSTALS** Decomposes when heated above 840°C, giving off corrosive vapors (→ *sulfur oxides*).	
	TLV-TWA	0.05 mg/m³ A1 ✷
	Absorption route: Can enter the body by inhalation or ingestion. Evaporation negligible at 20°C, but in finely dispersed form harmful atmospheric concentrations of airborne particles can build up rapidly. **Immediate effects:** Irritates the eyes and respiratory tract. **Effects of prolonged/repeated exposure:** Prolonged or repeated contact can cause eczema. Can cause asthma in case of hypersensitivity of the skin.	
Gross formula: NiO$_4$S		

HAZARDS/SYMPTOMS	PREVENTIVE MEASURES	FIRE EXTINGUISHING/FIRST AID
Fire: non-combustible.		in case of fire in immediate vicinity: use any extinguishing agent.
Inhalation: sore throat, metallic taste.	local exhaust or respiratory protection.	fresh air, rest, call a doctor if necessary.
Skin:	gloves.	remove contaminated clothing, flush skin with water or shower.
Eyes: redness, pain.	safety glasses.	flush with water, take to a doctor if necessary.
Ingestion: abdominal pain, diarrhea, nausea.		rinse mouth, call a doctor if necessary.

SPILLAGE	STORAGE	LABELING / NFPA
clean up spillage, carefully collect remainder; (additional individual protection: P2 respirator).		

NOTES
Besides anhydrous nickel sulfate, it can also contain 6 or 7 molecules of crystal water. Loss of crystal water when heated above 100°C. ✷ TLV equals that of nickel.

CAS-No: [54-11-5]
(S)-3-(1-methyl-2-pyrrolidinyl)pyridine
1-methyl-2-(3-pyridyl)pyrrolidine

$C_{10}H_{14}N_2$

NICOTINE

PHYSICAL PROPERTIES	IMPORTANT CHARACTERISTICS	
Boiling point, °C 247	**COLORLESS TO BROWN VISCOUS LIQUID**	
Melting point, °C − 80	Vapor mixes readily with air. Decomposes when heated or burned, giving off toxic (→ *carbon monoxide*) and corrosive (→ *nitrogen oxides*) vapors. Reacts with oxidants.	
Flash point, °C < 104		
Auto-ignition temperature, °C 240		
Relative density (water = 1) 1.0		
Relative vapor density (air = 1) 5.6	TLV-TWA (skin)	0.5 mg/m³
Relative density at 20 °C of saturated mixture		
vapor/air (air = 1) 1.0	**Absorption route:** Can enter the body by inhalation or ingestion or through the skin. Harmful atmospheric concentrations can build up fairly rapidly on evaporation at approx. 20°C - even more rapidly in aerosol form.	
Vapor pressure, mm Hg at 62 °C 0.99		
Solubility in water good	**Immediate effects:** Irritates the eyes, skin and respiratory tract. Can affect the nervous system. In serious cases risk of seizures, unconsciousness and death.	
Explosive limits, vol% in air 0.7-4.0		
Relative molecular mass 162.3	**Effects of prolonged/repeated exposure:** Prolonged or repeated contact can cause skin disorders.	
Log P octanol/water 1.2		
Gross formula: $C_{10}H_{14}N_2$		

HAZARDS/SYMPTOMS	PREVENTIVE MEASURES	FIRE EXTINGUISHING/FIRST AID
Fire: combustible.	keep away from open flame, no smoking.	powder, alcohol-resistant foam, water spray, carbon dioxide, (halons).
	STRICT HYGIENE	IN ALL CASES CALL IN A DOCTOR
Inhalation: perspiration, abdominal pain,	ventilation, local exhaust or respiratory protection.	fresh air, rest, artificial respiration if necessary, take to hospital.
Skin: *is absorbed*; see also 'Inhalation'.	gloves, protective clothing.	remove contaminated clothing, wash skin with soap and water, call a doctor.
Eyes: redness, pain.	face shield.	flush with water, take to a doctor if necessary.
Ingestion: see also 'Inhalation'.	do not eat, drink or smoke while working.	rinse mouth, immediately take to hospital.

SPILLAGE	STORAGE	LABELING / NFPA
collect leakage in sealable containers, soak up with sand or other inert absorbent and remove to safe place; (additional individual protection: breathing apparatus).	keep in fireproof place, separate from oxidants.	R: 26/27/28 S: 1-13-28-45 ☠ Toxic

NOTES

NITRATE FERTILIZERS

PHYSICAL PROPERTIES	IMPORTANT CHARACTERISTICS	
Auto-ignition temperature, °C ca. 150	**GRANULES** Granules are composed of nitrate compounds such as ammonium, calcium, potassium and sodium nitrate, with or without potassium salts and/or phosphates. Decomposes when heated or on contact with acids, strong reducing agents or oxidants, giving off toxic and corrosive vapors (→ *nitrous vapors*, → *chlorine*, → *hydrochloric acid*). In the case of fertilizers containing chlorides, e.g. potassium chloride, there is a risk of decomposition spreading through the entire mass (deflagration).	
	TLV-TWA not available	
	Absorption route: Can enter the body by inhalation or ingestion. **Immediate effects:** Corrosive to the eyes and skin. Vapors (from decomposition) are corrosive to the respiratory tract and can cause lung edema. In serious cases risk of unconsciousness and death.	

HAZARDS/SYMPTOMS	PREVENTIVE MEASURES	FIRE EXTINGUISHING/FIRST AID
Fire: combustible, mixtures containing chlorides can cause deflagration.	avoid contact with acids and strong reducing agents, keep away from open flame, no smoking.	large quantities of water, beware of possible suffocation.
	KEEP DUST UNDER CONTROL	
Inhalation: *corrosive*; sore throat, cough, severe breathing difficulties.	local exhaust or respiratory protection.	fresh air, rest, place in half-sitting position, take to hospital.
Skin: *corrosive*; redness, pain, serious burns.	gloves.	remove contaminated clothing, flush skin with water or shower, call a doctor.
Eyes: *corrosive*; redness, pain, impaired vision.	goggles.	flush with water, take to a doctor.
Ingestion: *corrosive*; sore throat, abdominal pain.		rinse mouth, immediately take to hospital.

SPILLAGE	STORAGE	LABELING / NFPA
clean up spillage, flush away remainder with water; (additional individual protection: P1 respirator).	keep in fireproof place, separate form combustible substances, reducing agents, oxidants, acids; avoid contact with electrical wiring, motors, portable lamps and steam lines.	

NOTES
Lung edema symptoms usually develop several hours later and are aggravated by physical exertion: rest and hospitalization essential. As first aid, a doctor or authorized person should consider administering a corticosteroid spray.

Transport Emergency Card TEC(R)-51G09

UN-number: 2067

NITRATING ACID
(mixture of sulfuric acid and nitric acid in various concentrations)

PHYSICAL PROPERTIES	IMPORTANT CHARACTERISTICS
Boiling point, °C 84-100 Melting point, °C −42 Relative density (water = 1) 1.5-1.8 Vapor pressure, mm Hg at 20 °C <42.6 Solubility in water ∞	**COLORLESS LIQUID WITH PUNGENT ODOR** Vapor mixes readily with air. Decomposes when heated, giving off toxic vapors. Strong oxidant which reacts violently with combustible substances and reducing agents, giving off → *nitrous vapors*. Strong acid which reacts violently with bases and is corrosive. Reacts violently with organic substances, solvents and many other substances.
	TLV-TWA ✴
	Absorption route: Can enter the body by inhalation or ingestion. Harmful atmospheric concentrations can build up very rapidly on evaporation at 20°C. **Immediate effects:** Corrosive to the eyes, skin and respiratory tract. Inhalation of vapor/fumes can cause severe breathing difficulties (lung edema). In serious cases risk of death.

HAZARDS/SYMPTOMS	PREVENTIVE MEASURES	FIRE EXTINGUISHING/FIRST AID
Fire: non-combustible, many chemical reactions can cause fire and explosion.	avoid contact with combustible substances.	in case of fire in immediate vicinity: use any extinguishing agent.
Explosion:		in case of fire: keep tanks/drums cool by spraying with water.
	STRICT HYGIENE	
Inhalation: *corrosive*; sore throat, cough, shortness of breath, severe breathing difficulties.	ventilation, local exhaust or respiratory protection.	fresh air, rest, place in half-sitting position, artificial respiration if necessary, take to hospital.
Skin: *corrosive*; redness, pain, serious burns.	gloves, protective clothing.	remove contaminated clothing, flush skin with water or shower, call a doctor or take directly to hospital.
Eyes: *corrosive*; redness, pain, impaired vision.	face shield, or combined eye and respiratory protection.	flush with water, take to a doctor.
Ingestion: *corrosive*; diarrhea, abdominal cramps, vomiting.		rinse mouth, take immediately to hospital.

SPILLAGE	STORAGE	LABELING / NFPA
collect leakage in sealable containers, soak up with sand or other inert absorbent and remove to safe place, flush away remainder with water; neutralize liquid with chalk or marl (CaCO₃), do not use sawdust or other combustible absorbents; (additional individual protection: CHEMICAL SUIT).	keep separate from combustible substances, reducing agents and bases.	R: 8-35 S: 23-26-30-36 Note B Oxidant Corrosive

NOTES
Flush contaminated clothing with water (fire hazard). ✴ TLV for sulfuric acid: TLV-TWA 1mg/m³ and STEL 3mg/m³; for nitric acid: TLV-TWA 2ppm (5mg/m³) and STEL 4ppm (10mg/m³). Lung edema symptoms usually develop several hours later and are aggravated by physical exertion: rest and hospitalization essential. As first aid, a doctor or authorized person should consider administering a corticosteroid spray. UNDER NO CIRCUMSTANCES add water to acid; when diluting ALWAYS add acid to water. Unbreakable packaging preferred; if breakable, keep in unbreakable container.

Transport Emergency Card TEC(R)-80G04/5 (R)

HI: 88; UN-number: 1796

NITRIC ACID
(up to 70%)

PHYSICAL PROPERTIES	IMPORTANT CHARACTERISTICS
Boiling point, °C 122 Melting point, °C −42 Relative density (water = 1) 1.4 Relative vapor density (air = 1) 2.2 Relative density at 20 °C of saturated mixture vapor/air (air = 1) 1.01 Vapor pressure, mm Hg at 20 °C 7.1 Solubility in water ∞ Relative molecular mass 63.0	**COLORLESS TO YELLOWISH-BROWN AQUEOUS SOLUTION OF NITRIC ACID IN VARIOUS CONCENTRATIONS UNDER 70%, WITH PUNGENT ODOR** Vapor mixes readily with air. Decomposes when heated or exposed to light, giving off → *nitrous vapors*. Strong oxidant which reacts violently with combustible substances and reducing agents. Strong acid which reacts violently with bases and is corrosive. Reacts violently with finely dispersed metals and many organic compounds, giving off → *nitrous vapors*.

	TLV-TWA	2 ppm	5.2 mg/m³
	STEL	4 ppm	10 mg/m³

Absorption route: Can enter the body by inhalation. Harmful atmospheric concentrations can build up fairly rapidly on evaporation at approx. 20°C - even more rapidly in aerosol form.
Immediate effects: Corrosive to the eyes, skin and respiratory tract. Inhalation of vapor/fumes can cause severe breathing difficulties (lung edema). In serious cases risk of death.

Gross formula: HNO$_3$

HAZARDS/SYMPTOMS	PREVENTIVE MEASURES	FIRE EXTINGUISHING/FIRST AID
Fire: non-combustible, many chemical reactions can cause fire and explosion.	avoid contact with combustible substances.	in case of fire in immediate vicinity: use any extinguishing agent.
Explosion: many chemical reactions can cause explosion.		in case of fire: keep tanks/drums cool by spraying with water.
Inhalation: *corrosive*; sore throat, cough, shortness of breath, severe breathing difficulties.	ventilation, local exhaust or respiratory protection.	fresh air, rest, place in half-sitting position, take to hospital.
Skin: *corrosive*; redness, pain, burns.	gloves, protective clothing.	remove contaminated clothing, flush skin with water or shower, call a doctor.
Eyes: *corrosive*; redness, pain, impaired vision.	face shield.	flush with water, take to a doctor.
Ingestion: *corrosive*; sore throat, abdominal pain.		rinse mouth, take immediately to hospital.

SPILLAGE	STORAGE	LABELING / NFPA
collect leakage in sealable containers, soak up with sand or other inert absorbent and remove to safe place, flush away remainder with water, do not use sawdust or other combustible absorbents; (additional individual protection: breathing apparatus).	keep in dark place, separate from combustible substances, reducing agents and strong bases; ventilate at floor level.	R: 35 S: 2-23-26-27 Note B Corrosive

NOTES
The measures on this card also apply to 50% nitric acid. Melting point of 50% nitric acid: −19°C; relative density: 1.3. Flush contaminated clothing with water (fire hazard). Lung edema symptoms usually develop several hours later and are aggravated by physical exertion: rest and hospitalization essential. As first aid, a doctor or authorized person should consider administering a corticosteroid spray. UNDER NO CIRCUMSTANCES add water to acid; when diluting ALWAYS add acid to water. Unbreakable packaging preferred; if breakable, keep in unbreakable container.

Transport Emergency Card TEC(R)-9A   **HI: 80; UN-number: 2031**

NITRIC ACID
(above 70%)

PHYSICAL PROPERTIES	IMPORTANT CHARACTERISTICS
Boiling point, °C 83 (100%) Melting point, °C −42 (100%) Relative density (water = 1) 1.5 Relative vapor density (air = 1) 2.2 Relative density at 20 °C of saturated mixture vapor/air (air = 1) 1.06 Vapor pressure, mm Hg at 20 °C 43 (100%) Solubility in water ∞ Relative molecular mass 63.0	**COLORLESS TO RED AQUEOUS SOLUTION OF NITRIC ACID IN VARIOUS CONCENTRATIONS ABOVE 70% WITH PUNGENT ODOR** Vapor mixes readily with air. Decomposes when heated or exposed to light, giving off toxic gas (→ *nitrous vapors*). Strong oxidant which reacts violently with combustible substances and reducing agents. Strong acid which reacts violently with bases and is corrosive. Reacts violently with finely dispersed metals, organic solvents and many other substances, with risk of fire and explosion.

TLV-TWA	2 ppm	5.2 mg/m³
STEL	4 ppm	10 mg/m³

Absorption route: Can enter the body by inhalation. Harmful atmospheric concentrations can build up very rapidly on evaporation at 20°C.
Immediate effects: Corrosive to the eyes, skin and respiratory tract. Inhalation of vapor/fumes can cause severe breathing difficulties (lung edema). In serious cases risk of death.

Gross formula: HNO₃

HAZARDS/SYMPTOMS	PREVENTIVE MEASURES	FIRE EXTINGUISHING/FIRST AID
Fire: non-combustible, but makes other substances more combustible; many chemical reactions can cause fire and explosion.	avoid contact with combustible substances, solvents and many other substances.	in case of fire in immediate vicinity: use any extinguishing agent.
Explosion: many chemical reactions can cause explosion.		in case of fire: keep tanks/drums cool by spraying with water.
	STRICT HYGIENE	
Inhalation: *corrosive*; sore throat, cough, shortness of breath, severe breathing difficulties.	ventilation, local exhaust or respiratory protection.	fresh air, rest, place in half-sitting position, take to hospital.
Skin: *corrosive*; redness, pain, burns.	gloves, protective clothing.	remove contaminated clothing, flush skin with water or shower, call a doctor.
Eyes: *corrosive*; redness, pain, impaired vision.	face shield, or combined eye and respiratory protection.	flush with water, take to a doctor.
Ingestion: *corrosive*; sore throat, abdominal pain.		rinse mouth, immediately take to hospital.

SPILLAGE	STORAGE	LABELING / NFPA
collect leakage in sealable containers, soak up with sand or other inert absorbent and remove to safe place, flush away remainder with water, do not use sawdust or other combustible absorbents; (additional individual protection: CHEMICAL SUIT).	keep in cool, dark place, separate from combustible substances, reducing agents and bases; ventilate at floor level.	R: 8-35 S: 23-26-36 Note B Oxidant Corrosive

NFPA diamond: 0 (top), 3 (left), 0 (right), oxy (bottom)

NOTES

In concentrations above 80% is referred to as fuming nitric acid. Physical properties of 100% nitric acid are given. Flush contaminated clothing with water (fire hazard). UN No. of red fuming nitric acid: 2032. Lung edema symptoms usually develop several hours later and are aggravated by physical exertion: rest and hospitalization essential. As first aid, a doctor or authorized person should consider administering a corticosteroid spray. UNDER NO CIRCUMSTANCES add water to acid; when diluting ALWAYS add acid to water. Unbreakable packaging preferred; if breakable, keep in unbreakable container. Packaging: dark glass, aluminum, stainless steel or suitable plastic.

Transport Emergency Card TEC(R)-9A   **HI: 885; UN-number: 2031**

1-amino-4-nitrobenzene
p-nitraniline
4-nitroaniline
4-nitrobenzeneamine

$NH_2C_6H_4NO_2$

p-NITROANILINE

PHYSICAL PROPERTIES	IMPORTANT CHARACTERISTICS	
Boiling point, °C 332 Melting point, °C 148 Flash point, °C 199 Relative density (water = 1) 1.4 Relative vapor density (air = 1) 4.8 Relative density at 20 °C of saturated mixture vapor/air (air = 1) 1.0 Vapor pressure, mm Hg at 20 °C 0.0015 Solubility in water none Relative molecular mass 138.1 Log P octanol/water 1.4	**YELLOW CRYSTALS** Vapor mixes readily with air. The melting of the crystals is accompanied by slight explosions. Decomposes when burned, giving off toxic vapors (→ *nitrous vapors*). Reacts with moisture, sulfuric acid and organic compounds.	
	TLV-TWA (skin) 1 ppm 6 mg/m³	
	Absorption route: Can enter the body by inhalation or ingestion or through the skin. Harmful atmospheric concentrations build up very slowly, if at all, on evaporation at approx. 20°C, but harmful concentrations of airborne particles can build up much more rapidly. **Immediate effects:** Irritates the eyes, skin and respiratory tract. In serious cases risk of death. **Effects of prolonged/repeated exposure:** Can affect the blood. Can affect the blood. Can cause liver damage.	
Gross formula: $C_6H_6N_2O_2$		

HAZARDS/SYMPTOMS	PREVENTIVE MEASURES	FIRE EXTINGUISHING/FIRST AID
Fire: combustible, many chemical reactions can cause fire and explosion.	keep away from open flames and sparks, no smoking; avoid contact with moisture or organic substances.	powder, large quantities of water.
Explosion: finely dispersed particles form explosive mixtures on contact with air.	keep dust from accumulating; sealed machinery, explosion-proof electrical equipment and lighting, grounding.	in case of fire: keep tanks/drums cool by spraying with water, fight fire from sheltered location.
	STRICT HYGIENE, KEEP DUST UNDER CONTROL	
Inhalation: sore throat, shortness of breath, dizziness, blue skin, feeling of weakness.	local exhaust or respiratory protection.	fresh air, rest, take to hospital.
Skin: *is absorbed*; redness; see also 'Inhalation'.	gloves, protective clothing.	remove contaminated clothing, wash skin with soap and water, call a doctor.
Eyes: redness, pain.	face shield, or combined eye and respiratory protection.	flush with water, take to a doctor if necessary.
Ingestion: abdominal pain, nausea; see also 'Inhalation'.	do not eat, drink or smoke while working.	rinse mouth, immediately take to hospital.

SPILLAGE	STORAGE	LABELING / NFPA
clean up spillage, carefully collect remainder and remove to safe place; (additional individual protection: P2 respirator).	keep in fireproof, dry place; keep separate form combustible and organic substances.	R: 23/24/25-33 S: 28-36/37-44 Note C ☠ Toxic ◇ 3/1/2

NOTES
Depending on the degree of exposure, regular medical checkups are advisable. Effects on blood due to formation of methemoglobin. Special first aid required in the event of poisoning: antidotes must be available (with instructions).

Transport Emergency Card TEC(R)-61G12B

HI: 60; UN-number: 1661

CAS-No: [98-95-3]
essence of Mirbane
oil of Mirbane
nitrobenzol

$C_6H_5NO_2$

NITROBENZENE

PHYSICAL PROPERTIES	IMPORTANT CHARACTERISTICS	
Boiling point, °C 210 Melting point, °C 6 Flash point, °C 88 Auto-ignition temperature, °C 480 Relative density (water = 1) 1.2 Relative vapor density (air = 1) 4.3 Relative density at 20 °C of saturated mixture vapor/air (air = 1) 1.0 Vapor pressure, mm Hg at 20 °C 0.15 Solubility in water, g/100 ml 0.2 Explosive limits, vol% in air 1.8-40 Electrical conductivity, pS/m 2x10⁴ Relative molecular mass 123.1 Log P octanol/water 1.9	**LIGHT YELLOW OILY LIQUID WITH CHARACTERISTIC ODOR** Vapor mixes readily with air. Toxic gases are a product of combustion (→ *nitrous vapors*). Can form explosive compounds with strong oxidants. Reacts violently with various organic and inorganic compounds. e.g. aluminum chloride with phenol, nitric acid, sodium chlorate and other strong oxidants, alkali metals, alcoholic potassium hydroxide, aniline with glycerine. Reacts with oxidants, reducing agents and combustible substances. Attacks many plastics.	
	TLV-TWA (skin) 1 ppm 5 mg/m³	
	Absorption route: Can enter the body by inhalation or ingestion or through the skin. Harmful atmospheric concentrations build up fairly slowly on evaporation at 20°C, but much more rapidly in aerosol form. **Immediate effects:** Irritates the eyes (causing light-sensitivity). In serious cases risk of death. **Effects of prolonged/repeated exposure:** Can affect the blood.	
Gross formula: $C_6H_5NO_2$		

HAZARDS/SYMPTOMS	PREVENTIVE MEASURES	FIRE EXTINGUISHING/FIRST AID
Fire: combustible.	keep away from open flame, no smoking; avoid contact with combustible substances.	powder, water spray, foam, carbon dioxide, (halons).
Explosion: above 88°C: forms explosive air-vapor mixtures.	above 88°C: sealed machinery, ventilation.	
	STRICT HYGIENE	
Inhalation: shortness of breath, dizziness, nausea, unconsciousness, blue skin, feeling of weakness.	ventilation, local exhaust or respiratory protection.	fresh air, rest, take to hospital.
Skin: *is absorbed*; see also 'Inhalation'.	gloves, protective clothing.	remove contaminated clothing, wash skin with soap and water, call a doctor.
Eyes: redness.	face shield.	flush with water, take to a doctor if necessary.
Ingestion: shortness of breath, nausea, dizziness, unconsciousness, feeling of weakness.	do not eat, drink or smoke while working.	rinse mouth, take immediately to hospital.

SPILLAGE	STORAGE	LABELING / NFPA
collect leakage in sealable containers, soak up with sand or other inert absorbent and remove to safe place.	keep separate from strong acids and bases, oxidants and combustible substances.	R: 26/27/28-33 S: 28-36/37-45 Toxic

NOTES

Depending on the degree of exposure, regular medical checkups are advisable. Effects on blood due to formation of methemoglobin. Special first aid required in the event of poisoning: antidotes must be available (with instructions). Especially dangerous in combination with nitric acid: when distilling from technical grade do not totally dehydrate because of risk of explosion. Symptoms usually develop several hours later: rest and hospitalization essential.

Transport Emergency Card TEC(R)-93

HI: 60; UN-number: 1662

NITROETHANE

PHYSICAL PROPERTIES	IMPORTANT CHARACTERISTICS	
Boiling point, °C 114 Melting point, °C −50 Flash point, °C 28 Auto-ignition temperature, °C 410 Relative density (water = 1) 1.1 Relative vapor density (air = 1) 2.6 Relative density at 20 °C of saturated mixture vapor/air (air = 1) 1.03 Vapor pressure, mm Hg at 20 °C 16 Solubility in water, g/100 ml at 20 °C 4.5 Explosive limits, vol% in air 3.4-? Electrical conductivity, pS/m 5 x 10⁷ Relative molecular mass 75.1 Log P octanol/water 0.2	**COLORLESS LIQUID WITH CHARACTERISTIC ODOR** Vapor mixes readily with air. Reacts with bases, acids or combination of amines and heavy metal oxides to form shock-sensitive compounds (salts, fulminates) or mixtures. Decomposes when heated to high temperatures (approx. 300°C) or burned, giving off toxic gases (→ *nitrous vapors*), with risk of explosion if temperature rises rapidly. Reacts with bases, oxidants and combustible substances, with risk of fire and explosion.	
	TLV-TWA 100 ppm 307 mg/m³	
	Absorption route: Can enter the body by inhalation or ingestion. Harmful atmospheric concentrations can build up fairly rapidly on evaporation at approx. 20°C - even more rapidly in aerosol form. **Immediate effects:** Irritates the eyes, skin and respiratory tract. Affects the nervous system.	
Gross formula: $C_2H_5NO_2$		

HAZARDS/SYMPTOMS	PREVENTIVE MEASURES	FIRE EXTINGUISHING/FIRST AID
Fire: flammable.	keep away from open flame and sparks, no smoking, avoid contact with combustible substances and oxidants.	water spray, alcohol-resistant foam, carbon dioxide, (halons).
Explosion: above 28°C: forms explosive air-vapor mixtures.	above 28°C: sealed machinery, ventilation, explosion-proof electrical equipment.	in case of fire: keep tanks/drums cool by spraying with water, fight fire from sheltered location.
Inhalation: cough, shortness of breath, headache.	ventilation, local exhaust or respiratory protection.	fresh air, rest, call a doctor.
Skin: redness.	gloves.	remove contaminated clothing, flush skin with water or shower.
Eyes: redness, pain.	acid goggles.	flush with water, take to doctor if necessary.
Ingestion: sore throat, abdominal pain.		rinse mouth, call a doctor or take to hospital.

SPILLAGE	STORAGE	LABELING / NFPA
collect leakage in sealable containers, soak up with sand or other inert absorbent and remove to safe place; (additional individual protection: breathing apparatus).	keep in fireproof place, separate from other substances.	R: 10-20/22 S: 9-25-41 ✖ Harmful

NOTES
Becomes shock-sensitive when contaminated with organic substances. Odor limit is above TLV.

Transport Emergency Card TEC(R)-30G36   **HI: 30; UN-number: 2842**

NITROGEN
(cylinder)

PHYSICAL PROPERTIES	IMPORTANT CHARACTERISTICS	
Boiling point, °C −196 Relative vapor density (air = 1) 0.97 Solubility in water none Relative molecular mass 28.0	**COLORLESS, ODORLESS COMPRESSED GAS** Gas mixes readily with air.	
	TLV-TWA not available	
	Immediate effects: Inhalation can cause severe breathing difficulties. In serious cases risk of unconsciousness.	
Gross formula: N₂		

HAZARDS/SYMPTOMS	PREVENTIVE MEASURES	FIRE EXTINGUISHING/FIRST AID
Fire: non-combustible.		in case of fire in immediate vicinity: use any extinguishing agent.
Explosion:		in case of fire: keep cylinders cool by spraying with water.
Inhalation: severe breathing difficulties, headache, dizziness, unconsciousness.	ventilation, local exhaust or respiratory protection.	fresh air, rest, artificial respiration if necessary, take to hospital.

SPILLAGE	STORAGE	LABELING / NFPA
ventilate; (additional individual protection: breathing apparatus).	if stored indoors, keep in cool, fireproof place; ventilate.	

NOTES
High atmospheric concentrations, e.g. in poorly ventilated spaces, can cause oxygen deficiency, with risk of unconsciousness.

Transport Emergency Card TEC(R)-20G01   **HI: 20; UN-number: 1066**

NITROGEN
(liquid, refrigerated)

PHYSICAL PROPERTIES	IMPORTANT CHARACTERISTICS	
Boiling point, °C − 196 Relative density (water = 1) see notes Relative vapor density (air = 1) 0.97 Relative molecular mass 28.0	**COLORLESS, EXTREMELY COLD LIQUID** Cold gas is heavier than air and can accumulate close to ground level, causing oxygen deficiency, with risk of unconsciousness.	
	TLV-TWA not available	
	Immediate effects: Cold liquid can cause frostbite.	
Gross formula: N_2		

HAZARDS/SYMPTOMS	PREVENTIVE MEASURES	FIRE EXTINGUISHING/FIRST AID
Fire: non-combustible.		in case of fire in immediate vicinity: use any extinguishing agent.
Inhalation: see Notes.	ventilation, local exhaust or respiratory protection; see Notes.	fresh air, rest, artificial respiration if necessary, call a doctor, take to hospital if necessary.
Skin: *in case of frostbite*: redness, pain, blisters, sores.	insulating gloves.	*in case of frostbite*: DO NOT remove clothing, flush skin with water or shower, take to a doctor.
Eyes: *in case of frostbite*: redness, pain, impaired vision.	face shield.	flush with water, take to a doctor.

SPILLAGE	STORAGE	LABELING / NFPA
ventilate, under no circumstances spray liquid with water; (additional individual protection: breathing apparatus).		(NFPA diamond: 3 / 0 / 0)

NOTES
High atmospheric concentrations, e.g. in poorly ventilated spaces, can cause oxygen deficiency, with risk of unconsciousness. Density of liquid at boiling point: 0.8kg/l.

Transport Emergency Card TEC(R)-112   **HI: 22; UN-number: 1977**

NITROGEN DIOXIDE
(cylinder)

PHYSICAL PROPERTIES	IMPORTANT CHARACTERISTICS
Boiling point, °C — 21 Melting point, °C — −11 Relative density (water = 1) — 1.45 Relative vapor density (air = 1) — 1.6 Vapor pressure, mm Hg at 20 °C — 767 Solubility in water — reaction Relative molecular mass — 46.0	**REDDISH-BROWN COMPRESSED LIQUEFIED GAS OR YELLOW LIQUID WITH PUNGENT ODOR** Nitrogen dioxide is a principal component of nitrous vapors. Decomposes when heated above 160° C, giving off → *nitric oxide* and → *oxygen*, with increased risk of fire. Strong oxidant which reacts violently with combustible substances and reducing agents. Reacts violently with anhydrous ammonia and chlorinated hydrocarbons. Reacts with water, forming → *nitric acid* and nitrous acid. Attacks many metals in the presence of moisture.

TLV-TWA	3 ppm	5.6 mg/m³
STEL	5 ppm	9.4 mg/m³

Absorption route: Can enter the body by inhalation. Harmful atmospheric concentrations can build up very rapidly if gas is released.
Immediate effects: Corrosive to the eyes, skin and respiratory tract. Inhalation of vapor/fumes can cause severe breathing difficulties (lung edema). In serious cases risk of death.

Gross formula: NO_2

HAZARDS/SYMPTOMS	PREVENTIVE MEASURES	FIRE EXTINGUISHING/FIRST AID
Fire: not combustible but enhances combustion of other substances.	avoid contact with combustible substances and reducing agents.	in case of fire in immediate vicinity: preferably no halons.
Explosion: mixtures with flammable vapors are explosive.		in case of fire: keep cylinders cool by spraying with water.
	STRICT HYGIENE	IN ALL CASES CALL IN A DOCTOR
Inhalation: *corrosive*; sore throat, cough, shortness of breath, severe breathing difficulties.	ventilation, local exhaust or respiratory protection.	fresh air, rest, place in half-sitting position, take to hospital.
Skin: *corrosive*; redness, pain, burns.	gloves, protective clothing.	remove contaminated clothing, flush skin with water or shower, call a doctor if necessary.
Eyes: *corrosive*; redness, pain, impaired vision.	face shield, or combined eye and respiratory protection.	flush with water, take to a doctor.

SPILLAGE	STORAGE	LABELING / NFPA
evacuate area, call in expert, ventilate, do not use sawdust or other combustible absorbents, combat gas cloud with water curtain, neutralize running water with chalk (CaCO₃) or soda; (additional individual protection: CHEMICAL SUIT).		

NOTES

Other components of nitrous vapors (NO_x): nitrous oxide (NO), nitrogen trioxide (N_2O_3) and nitrogen pentoxide (N_2O_5). Liquid in cylinder is nitrogen tetroxide (N_2O_4); gas contains 10% N_2O_4 at 20° C. Lung edema symptoms usually develop several hours later and are aggravated by physical exertion: rest and hospitalization essential. As first aid, a doctor or authorized person should consider administering a corticosteroid spray. Do not spray leaking cylinder with water (to avoid corrosion). Turn leaking cylinder so that leak is on top to prevent liquid nitrous oxide escaping. Cylinder can also contain nitrogen gas up to a pressure of 10atm. at 20°C.

laughing gas
nitrous (II) oxide

NITROGEN MONOXIDE
(cylinder)

PHYSICAL PROPERTIES	IMPORTANT CHARACTERISTICS	
Boiling point, °C −152 Melting point, °C −164 Relative density (water = 1) see notes Relative vapor density (air = 1) 1.03 Solubility in water, g/100 ml at 20 °C 0.006 Relative molecular mass 30.0	**COLORLESS, ODORLESS COMPRESSED GAS, TURNING BROWN ON EXPOSURE TO AIR** Nitrous vapors are a mixture of nitrous oxide and nitrogen dioxide. Gas mixes readily with air. Reacts violently with oxygen, phosphine, butadiene, ethylene oxide and vinyl methyl ether. Reacts with air to form → *nitrogen dioxide*. Attacks many metals in the presence of air or moisture.	
	TLV-TWA 25 ppm 31 mg/m³	
	Absorption route: Can enter the body by inhalation. Harmful atmospheric concentrations can build up very rapidly if gas is released. **Immediate effects:** Corrosive to the eyes, skin and respiratory tract. Inhalation of vapor/fumes can cause severe breathing difficulties (lung edema). In serious cases risk of death.	
Gross formula: NO		

HAZARDS/SYMPTOMS	PREVENTIVE MEASURES	FIRE EXTINGUISHING/FIRST AID
Fire: not combustible but enhances combustion of other substances.	keep away from open flame and sparks, no smoking.	in case of fire in immediate vicinity: use any extinguishing agent.
Explosion:		in case of fire: keep cylinders cool by spraying with water.
	STRICT HYGIENE	IN ALL CASES CALL IN A DOCTOR
Inhalation: *corrosive*; sore throat, cough, shortness of breath, severe breathing difficulties.	ventilation, local exhaust or respiratory protection.	fresh air, rest, place in half-sitting position, take to hospital.
Skin: *corrosive*; redness, pain, burns.	gloves.	remove contaminated clothing, flush skin with water or shower, call a doctor if necessary.
Eyes: *corrosive*; redness, pain, impaired vision.	acid goggles, or combined eye and respiratory protection.	flush with water, take to a doctor.

SPILLAGE	STORAGE	LABELING / NFPA
evacuate area, ventilate; (additional individual protection: breathing apparatus).	if stored indoors, keep in cool, fireproof place.	

NFPA diamond: top 0, left 3, right 0, bottom oxy

NOTES
Density of liquid at boiling point under 1atm: 1.3. Lung edema symptoms usually develop several hours later and are aggravated by physical exertion: rest and hospitalization essential. As first aid, a doctor or authorized person should consider administering a corticosteroid spray. Do not spray leaking cylinder with water (to avoid corrosion). Use cylinder with special fittings.

NITROGEN TRICHLORIDE

PHYSICAL PROPERTIES	IMPORTANT CHARACTERISTICS	
Boiling point (decomposes), °C ca. 70 Melting point, °C −40 Relative density (water = 1) 1.7 Relative vapor density (air = 1) 4.2 Relative density at 20 °C of saturated mixture vapor/air (air = 1) 1.6 Vapor pressure, mm Hg at 20 °C 152 Solubility in water none Relative molecular mass 120.4	**YELLOW VISCOUS LIQUID WITH PUNGENT ODOR** Vapor is heavier than air. Can decompose explosively when heated above 60°C or subjected to shock. Strong oxidant which reacts violently with combustible substances and reducing agents. Reacts violently with many substances, such as ammonia and hot water, with risk of fire and explosion.	
	TLV-TWA not available	
	Absorption route: Can enter the body by inhalation or ingestion. Harmful atmospheric concentrations can build up very rapidly on evaporation at 20°C. **Immediate effects:** Corrosive to the eyes, skin and respiratory tract. Inhalation of vapor/fumes can cause severe breathing difficulties (lung edema). In serious cases risk of death.	
Gross formula: Cl₃N		

HAZARDS/SYMPTOMS	PREVENTIVE MEASURES	FIRE EXTINGUISHING/FIRST AID
Fire: non-combustible, but makes other substances more combustible; many chemical reactions can cause fire and explosion.		
Explosion: heating, friction, light and contact with many substances can cause explosion.	do not subject to shocks or friction.	in case of fire: keep tanks/drums cool by spraying with water, but DO NOT spray substance with water, fight fire from protected location.
	STRICT HYGIENE	
Inhalation: *corrosive*; sore throat, cough, shortness of breath, severe breathing difficulties.	ventilation, local exhaust or respiratory protection.	fresh air, rest, place in half-sitting position, take to hospital.
Skin: *corrosive*; redness, pain.	gloves, protective clothing.	remove contaminated clothing, flush skin with water or shower, call a doctor.
Eyes: *corrosive*; redness, pain, impaired vision.	face shield, or combined eye and respiratory protection.	flush with water, take to a doctor.
Ingestion: *corrosive*; sore throat, pain with swallowing, severe breathing difficulties, blisters on lips and tongue, tongue pain.	do not eat, drink or smoke while working.	rinse mouth, immediately take to hospital.

SPILLAGE	STORAGE	LABELING / NFPA
evacuate area, call in an expert, render harmless with large quantities of bisulfite solution, flush away with water; (additional individual protection: CHEMICAL SUIT).	keep in cool, dark, fireproof place, separate from all other substances.	

NOTES
Flush contaminated clothing with water (fire hazard). Lung edema symptoms usually develop several hours later and are aggravated by physical exertion: rest and hospitalization essential. As first aid, a doctor or authorized person should consider administering a corticosteroid spray. Unbreakable packaging preferred; if breakable, keep in unbreakable container. Special precautions are required when handling this substance.

NITROMETHANE

PHYSICAL PROPERTIES	IMPORTANT CHARACTERISTICS
Boiling point, °C — 101 Melting point, °C — −29 Flash point, °C — 35 Auto-ignition temperature, °C — 415 Relative density (water = 1) — 1.1 Relative vapor density (air = 1) — 2.1 Relative density at 20 °C of saturated mixture vapor/air (air = 1) — 1.04 Vapor pressure, mm Hg at 20 °C — 28 Solubility in water, g/100 ml at 20 °C — 9.5 Explosive limits, vol% in air — 7.1-63 Electrical conductivity, pS/m — 5×10^5 Relative molecular mass — 61.0 Log P octanol/water — −0.1	**COLORLESS LIQUID WITH CHARACTERISTIC ODOR** Vapor mixes readily with air. Reacts with bases, acids or combination of amines and heavy metal oxides to form shock-sensitive compounds (salts, fulminates) or mixtures. Decomposes explosively when heated to high temperatures (approx. 300° C), giving off toxic gases (→ *nitrous vapors*); these are also a product of combustion. Reacts violently with very many compounds, with risk of fire and explosion. Attacks steel and copper slowly in the presence of moisture.
	TLV-TWA not available
	Absorption route: Can enter the body by inhalation or ingestion. Harmful atmospheric concentrations can build up fairly rapidly on evaporation at approx. 20° C - even more rapidly in aerosol form. **Immediate effects:** Irritates the eyes and respiratory tract. Affects the nervous system.
Gross formula: CH_3NO_2	

HAZARDS/SYMPTOMS	PREVENTIVE MEASURES	FIRE EXTINGUISHING/FIRST AID
Fire: flammable, many chemical reactions can cause fire and explosion.	keep away from open flame and sparks, no smoking, avoid contact with combustible substances and oxidants.	alcohol-resistant foam, carbon dioxide, (halons), NOT POWDER.
Explosion: above 35°C: forms explosive air-vapor mixtures; elevated temperature/pressure, friction or shock can cause explosion.	above 35°C: sealed machinery, ventilation, explosion-proof electrical equipment, do not subject to shock or friction.	in case of fire: keep tanks/drums cool by spraying with water, fight fire from sheltered location, wait until intact drums have cooled completely before approaching.
Inhalation: cough, shortness of breath, headache.	ventilation, local exhaust or respiratory protection.	fresh air, rest, call a doctor.
Skin: redness.	gloves.	remove contaminated clothing, flush skin with water or shower, refer to a doctor.
Eyes: redness, pain.	safety glasses.	flush with water, take to doctor if necessary.
Ingestion: abdominal pain, nausea.		rinse mouth, call a doctor or take to hospital.

SPILLAGE	STORAGE	LABELING / NFPA
collect leakage in sealable containers, soak up with sand or other inert absorbent and remove to safe place; (additional individual protection: breathing apparatus).	keep in fireproof place, separate from all other substances.	R: 5-10-22 S: 41 ✖ Harmful 3 / 1 / 4

NOTES
In confined spaces combustion can cause explosion. Gas bubbles greatly increase explosivity. Fight fire from safe distance. Depending on the degree of exposure, regular medical checkups are advisable.

Transport Emergency Card TEC(R)-30G08

HI: 30; UN-number: 1261

p-NITROPHENOL

PHYSICAL PROPERTIES	IMPORTANT CHARACTERISTICS	
Boiling point (decomposes), °C 279 Melting point, °C 113 Relative density (water = 1) 1.48 Relative vapor density (air = 1) 4.4 Vapor pressure, mm Hg at 49.3 °C 0.99 Solubility in water, g/100 ml at 25 °C 1.6 Relative molecular mass 139.1 Log P octanol/water 1.9	**COLORLESS TO LIGHT YELLOW CRYSTALS** Decomposes when heated or burned, giving off toxic vapors (→ *nitrous vapors*). Strong oxidant which reacts violently with combustible substances and reducing agents.	
	TLV-TWA not available	
	Absorption route: Can enter the body by inhalation or ingestion or through the skin. Harmful atmospheric concentrations build up very slowly, if at all, on evaporation at approx. 20°C, but harmful concentrations of airborne particles can build up much more rapidly. **Immediate effects:** Irritates the eyes, skin and respiratory tract.	
Gross formula: $C_6H_5NO_3$		

HAZARDS/SYMPTOMS	PREVENTIVE MEASURES	FIRE EXTINGUISHING/FIRST AID
Fire: combustible.	keep away from open flame, no smoking; avoid contact with combustible substances.	water spray, powder.
	STRICT HYGIENE, KEEP DUST UNDER CONTROL	
Inhalation: perspiration, thirst, fever, muscle weakness, severe fatigue.	local exhaust or respiratory protection.	fresh air, rest in a cool place, call a doctor or take to hospital.
Skin: *is absorbed*; redness, pain; see also 'Inhalation'.	gloves, protective clothing.	remove contaminated clothing, flush skin and wash with soap and water, call a doctor.
Eyes: redness, pain.	goggles, or combined eye and respiratory protection.	flush with water, take to a doctor if necessary.
Ingestion: abdominal pain; see also 'Inhalation'.		rinse mouth, immediately take to hospital.

SPILLAGE	STORAGE	LABELING / NFPA
clean up spillage, carefully collect remainder; (additional individual protection: P2 respirator).	keep separate from combustible substances and reducing agents.	R: 20/21/22-33 S: 28 ✖ Harmful

NFPA diamond: 1 (top), 3 (left), 2 (right)

NOTES
Depending on the degree of exposure, regular medical checkups are advisable.

1-NITROPROPANE

PHYSICAL PROPERTIES	IMPORTANT CHARACTERISTICS
Boiling point, °C 132 Melting point, °C 108 Flash point, °C 49 Auto-ignition temperature, °C 420 Relative density (water = 1) 1.0 Relative vapor density (air = 1) 3.1 Relative density at 20 °C of saturated mixture vapor/air (air = 1) 1.02 Vapor pressure, mm Hg at 20 °C 4.8 Solubility in water, g/100 ml 1.7 Explosive limits, vol% in air 2.2 Electrical conductivity, pS/m 10^7 Relative molecular mass 89.1 Log P octanol/water 0.7	**COLORLESS LIQUID** Vapor mixes readily with air. Reacts with salts of mercury and silver, acids, inorganic bases, amines and other substances to form shock-sensitive compounds. Can explode when heated, even in the absence of oxygen. Reacts with nitrous acid, forming propyl nitrolic acid; salts are explosive in dry state.

TLV-TWA	25 ppm	91 mg/m³

Absorption route: Can enter the body by inhalation or ingestion. Harmful atmospheric concentrations build up fairly slowly on evaporation at 20° C, but much more rapidly in aerosol form.
Immediate effects: Irritates the eyes, skin and respiratory tract. In serious cases risk of unconsciousness and death.
Effects of prolonged/repeated exposure: Can affect the blood. Can cause liver and kidney damage.

Gross formula: $C_3H_7NO_2$

HAZARDS/SYMPTOMS	PREVENTIVE MEASURES	FIRE EXTINGUISHING/FIRST AID
Fire: flammable, many chemical reactions can cause fire and explosion.	keep away from open flame and sparks, no smoking, avoid contact with other substances.	powder, alcohol-resistant foam, water spray, carbon dioxide, (halons).
Explosion: above 49°C: forms explosive air-vapor mixtures.	above 49°C: sealed machinery, ventilation, explosion-proof electrical equipment.	in case of fire: keep tanks/drums cool by spraying with water.
Inhalation: cough, headache, nausea, sleepiness, blue skin, feeling of weakness.	ventilation, local exhaust or respiratory protection.	fresh air, rest, call a doctor.
Skin: redness.	gloves.	remove contaminated clothing, flush skin with water or shower, refer to a doctor.
Eyes: redness, pain.	face shield.	flush with water, take to a doctor.
Ingestion: abdominal pain, nausea, diarrhea, unconsciousness, blue skin, feeling of weakness.		rinse mouth, immediately take to hospital.

SPILLAGE	STORAGE	LABELING / NFPA
collect leakage in sealable containers, soak up with sand or other inert absorbent and remove to safe place; (additional individual protection: breathing apparatus).	keep in fireproof place, separate from all other substances.	R: 10-20/21/22 S: 9 ✖ Harmful 3 / 1 2

NOTES
Odor limit is above TLV. Depending on the degree of exposure, regular medical checkups are advisable. Effects on blood due to formation of methemoglobin. Special first aid required in the event of poisoning: antidotes must be available (with instructions). Special first aid required in the event of poisoning: antidotes must be available (with instructions). Use airtight packaging.

Transport Emergency Card TEC(R)-30G36   **HI: 30; UN-number: 2608**

CAS-No: [79-46-9]
dimethylnitromethane
isonitropropane
nitroisopropane
2-NP

$CH_3CH(NO_2)CH_3$

2-NITROPROPANE

PHYSICAL PROPERTIES		IMPORTANT CHARACTERISTICS	
Boiling point, °C	120	**COLORLESS LIQUID WITH CHARACTERISTIC ODOR**	
Melting point, °C	−93	Vapor mixes readily with air. Reacts with bases or nitric acid to form shock-sensitive compounds.	
Flash point, °C	24	Decomposes when heated or burned, giving off toxic gases (→ *nitrous vapors*), with risk of	
Auto-ignition temperature, °C	425	explosion if rise in temperature is very rapid. Many other compounds - esp. bases, acids and	
Relative density (water = 1)	1.0	amines - promote its decomposition. Reacts violently with strong oxidants, bases, chlorosulfonic	
Relative vapor density (air = 1)	3.1	acid, oleum and combustible substances.	
Relative density at 20 °C of saturated mixture			
vapor/air (air = 1)	1.04	**TLV-TWA** 10 ppm 36 mg/m³	
Vapor pressure, mm Hg at 20 °C	13		
Solubility in water, g/100 ml	1.7	**Absorption route:** Can enter the body by inhalation or ingestion. Harmful atmospheric	
Explosive limits, vol% in air	2.6-11	concentrations can build up fairly rapidly on evaporation at approx. 20°C - even more rapidly in	
Electrical conductivity, pS/m	5 x 10⁷	aerosol form.	
Relative molecular mass	89.1	**Immediate effects:** Irritates the eyes, skin and respiratory tract. In serious cases risk of	
		unconsciousness and death.	
		Effects of prolonged/repeated exposure: Can affect the blood. Can cause liver and kidney	
		damage.	
Gross formula:	$C_3H_7NO_2$		

HAZARDS/SYMPTOMS	PREVENTIVE MEASURES	FIRE EXTINGUISHING/FIRST AID
Fire: flammable.	keep away from open flame and sparks, no smoking, avoid contact with other substances.	powder, alcohol-resistant foam, water spray, carbon dioxide, (halons).
Explosion: above 24°C: forms explosive air-vapor mixtures.	above 24°C: sealed machinery, ventilation, explosion-proof electrical equipment.	in case of fire: keep tanks/drums cool by spraying with water.
Inhalation: cough, headache, nausea, sleepiness, blue skin, feeling of weakness.	ventilation, local exhaust or respiratory protection.	fresh air, rest, call a doctor.
Skin: redness.	gloves.	remove contaminated clothing, flush skin with water or shower, refer to a doctor.
Eyes: redness, pain.	acid goggles.	flush with water, take to a doctor.
Ingestion: abdominal pain, diarrhea, nausea, unconsciousness, blue skin, feeling of weakness.		rinse mouth, take immediately to hospital.

SPILLAGE	STORAGE	LABELING / NFPA
collect leakage in sealable containers, soak up with sand or other inert absorbent and remove to safe place; (additional individual protection: breathing apparatus).	keep in fireproof place, separate from other substances.	R: 45-10-20/22 S: 53-9-44 Note E ☠ Toxic NFPA: 3 / 2 / 1

NOTES
Odor limit is above TLV. Depending on the degree of exposure, regular medical checkups are advisable. Effects on blood due to formation of methemoglobin. Special first aid required in the event of poisoning: antidotes must be available (with instructions).

Transport Emergency Card TEC(R)-30G13

HI: 30; UN-number: 2608

ortho-NITROTOLUENE

PHYSICAL PROPERTIES	IMPORTANT CHARACTERISTICS
Boiling point, °C — 222 Melting point, °C — −6 Flash point, °C — 106 Relative density (water = 1) — 1.2 Relative vapor density (air = 1) — 4.7 Relative density at 20 °C of saturated mixture vapor/air (air = 1) — 1.0 Vapor pressure, mm Hg at 20 °C — 0.11 Solubility in water — none Relative molecular mass — 137.1 Log P octanol/water — 2.3	**LIGHT YELLOW LIQUID WITH CHARACTERISTIC ODOR** Vapor mixes readily with air. Decomposes when heated or burned, giving off toxic gases (→ *nitrous vapors*). Reacts violently with oxidants, reducing agents, sodium hydroxide and strong acids, with risk of fire and explosion.
	TLV-TWA (skin)　　　　2 ppm　　　　11 mg/m³
	Absorption route: Can enter the body by inhalation or ingestion or through the skin. Harmful atmospheric concentrations build up fairly slowly on evaporation at 20° C, but much more rapidly in aerosol form. **Immediate effects:** Irritates the eyes, skin and respiratory tract. In serious cases risk of death. **Effects of prolonged/repeated exposure:** Prolonged or repeated contact can cause skin disorders. Can affect the blood. Can cause liver damage.
Gross formula:　　　　$C_7H_7NO_2$	

HAZARDS/SYMPTOMS	PREVENTIVE MEASURES	FIRE EXTINGUISHING/FIRST AID
Fire: combustible.	keep away from open flame, no smoking.	powder, water spray, foam, carbon dioxide, (halons).
Inhalation: shortness of breath, headache, dizziness, blue skin, feeling of weakness.	ventilation, local exhaust or respiratory protection.	fresh air, rest, take to hospital.
Skin: redness, pain; see also 'Inhalation'.	gloves, protective clothing.	remove contaminated clothing, wash skin with soap and water, call a doctor.
Eyes: redness, pain.	face shield.	flush with water, take to a doctor if necessary.
Ingestion: abdominal pain, nausea; see also 'Inhalation'.		rinse mouth, take immediately to hospital.

SPILLAGE	STORAGE	LABELING / NFPA
collect leakage in sealable containers, soak up with sand or other inert absorbent and remove to safe place.	keep separate from combustible substances, reducing agents, oxidants, strong acids and strong bases; ventilate at floor level.	R: 23/24/25-33 S: 28-37-44 Note C ☠️ Toxic　　　　　3 ◇ 1 / 1

NOTES
Depending on the degree of exposure, regular medical checkups are advisable. Effects on blood due to formation of methemoglobin. Special first aid required in the event of poisoning: antidotes must be available (with instructions). The measures on this card also apply to meta-nitrotoluene: boiling point 232° C; melting point 16° C; flash point 106° C.

Transport Emergency Card TEC(R)-61G06　　　　　　　　**HI: 60; UN-number: 1664**

CAS-No: [99-99-0]
4-methylnitrobenzene
p-methylnitrotoluene
4-nitrotoluene
4-nitrotoluol

$CH_3C_6H_4NO_2$

para-NITROTOLUENE

PHYSICAL PROPERTIES		IMPORTANT CHARACTERISTICS
Boiling point, °C	238	**LIGHT YELLOW CRYSTALS**
Melting point, °C	53	Vapor mixes readily with air. Decomposes when heated or burned, giving off toxic gases
Flash point, °C	106	(→ *nitrous vapors*). Reacts violently with oxidants, reducing agents and strong acids.
Relative density (water = 1)	1.2	
Relative vapor density (air = 1)	4.7	TLV-TWA (skin) 2 ppm 11 mg/m³
Relative density at 20 °C of saturated mixture		
vapor/air (air = 1)	1.0	**Absorption route:** Can enter the body by inhalation or ingestion or through the skin. Harmful
Vapor pressure, mm Hg at 20 °C	0.99	atmospheric concentrations build up fairly slowly on evaporation at 20°C, but much more rapidly
Solubility in water	none	in aerosol form.
Relative molecular mass	137.1	**Immediate effects:** Irritates the eyes, skin and respiratory tract. In serious cases risk of death.
Log P octanol/water	2.4	**Effects of prolonged/repeated exposure:** Prolonged or repeated contact can cause skin
		disorders. Can affect the blood. Can cause liver damage.
Gross formula:	$C_7H_7NO_2$	

HAZARDS/SYMPTOMS	PREVENTIVE MEASURES	FIRE EXTINGUISHING/FIRST AID
Fire: combustible.	keep away from open flame, no smoking.	powder, water spray, foam, carbon dioxide, (halons).
Inhalation: shortness of breath, headache, dizziness, blue skin, feeling of weakness.	local exhaust or respiratory protection.	fresh air, rest, take to hospital.
Skin: *is absorbed*; redness, pain; see also 'Inhalation'.	gloves, protective clothing.	remove contaminated clothing, wash skin with soap and water, call a doctor.
Eyes: redness, pain.	face shield.	flush with water, take to a doctor if necessary.
Ingestion: abdominal pain, nausea; see also 'Inhalation'.		rinse mouth, immediately take to hospital.

SPILLAGE	STORAGE	LABELING / NFPA
clean up spillage, carefully collect remainder; (additional individual protection: respirator with A/P2 filter).	keep separate from combustible substances, reducing agents, oxidants and sulfuric acid; ventilate at floor level.	R: 23/24/25-33 S: 28-37-44 Note C ☠ Toxic 3 1 1

NOTES
Depending on the degree of exposure, regular medical checkups are advisable. Effects on blood due to formation of methemoglobin. Special first aid required in the event of poisoning: antidotes must be available (with instructions). Special first aid required in the event of poisoning: antidotes must be available (with instructions). Residue from distillation can explode spontaneously.

Transport Emergency Card TEC(R)-879

HI: 60; UN-number: 1664

alcohol c-9
n-nonyl alcohol
octyl carbinol
pelargonic alcohol

$C_9H_{19}OH$

1-NONANOL

PHYSICAL PROPERTIES	IMPORTANT CHARACTERISTICS
Boiling point, °C 214 Melting point, °C −5 Flash point, °C 74 Auto-ignition temperature, °C ? Relative density (water = 1) 0.8 Relative vapor density (air = 1) 5.0 Relative density at 20 °C of saturated mixture vapor/air (air = 1) 1.0 Vapor pressure, mm Hg at 25 °C 0.76 Solubility in water none Explosive limits, vol% in air 0.8-6.1 Relative molecular mass 144.3 Gross formula: $C_9H_{20}O$	**COLORLESS LIQUID WITH CHARACTERISTIC ODOR** Vapor mixes readily with air. Reacts with strong oxidants. TLV-TWA not available **Absorption route:** Can enter the body by inhalation or ingestion or through the skin. Harmful atmospheric concentrations build up fairly slowly on evaporation at 20° C, but much more rapidly in aerosol form. **Immediate effects:** Irritates the eyes, skin and respiratory tract. Affects the nervous system. **Effects of prolonged/repeated exposure:** Can cause liver and kidney damage.

HAZARDS/SYMPTOMS	PREVENTIVE MEASURES	FIRE EXTINGUISHING/FIRST AID
Fire: combustible.	keep away from open flame, no smoking.	powder, AFFF, foam, carbon dioxide, (halons).
Explosion: above 74°C: forms explosive air-vapor mixtures.	above 74°C: sealed machinery, ventilation.	
Inhalation: sore throat, cough, dizziness, drowsiness.	ventilation, local exhaust or respiratory protection.	fresh air, rest, call a doctor.
Skin: redness.	gloves.	remove contaminated clothing, wash skin with soap and water.
Eyes: redness, pain.	safety glasses.	flush with water, take to a doctor if necessary.
Ingestion: abdominal pain, nausea, dizziness.		rinse mouth, call a doctor.

SPILLAGE	STORAGE	LABELING / NFPA
collect leakage in sealable containers, soak up with sand or other inert absorbent and remove to safe place.	keep separate from strong oxidants.	

NOTES
The isomers of 1-nonanol have lower boiling points.

$C_6H_4OHC_9H_{19}$

NONYLPHENOL

PHYSICAL PROPERTIES		IMPORTANT CHARACTERISTICS	
Boiling point, °C	295	**COLORLESS VISCOUS LIQUID WITH CHARACTERISTIC ODOR**	
Melting point, °C	2	Vapor mixes readily with air. Flow, agitation etc. can cause build-up of electrostatic charge due to liquid's low conductivity. Reacts violently with strong oxidants.	
Flash point, °C	141		
Relative density (water = 1)	0.95		
Relative vapor density (air = 1)	7.6	TLV-TWA not available	
Relative density at 20 °C of saturated mixture vapor/air (air = 1)	1.0		
Vapor pressure, mm Hg at 20 °C	0.76	**Absorption route:** Can enter the body by inhalation or ingestion or through the skin. Evaporation negligible at 20°C, but harmful atmospheric concentrations can build up rapidly in aerosol form.	
Solubility in water	none	**Immediate effects:** Corrosive to the eyes, skin and respiratory tract. Inhalation can cause lung edema. In serious cases risk of death.	
Electrical conductivity, pS/m	< 10⁴		
Relative molecular mass	220.3		
Gross formula:	$C_{15}H_{24}O$		

HAZARDS/SYMPTOMS	PREVENTIVE MEASURES	FIRE EXTINGUISHING/FIRST AID
Fire: combustible.	keep away from open flame, no smoking.	powder, AFFF, foam, carbon dioxide, (halons).
Inhalation: *corrosive*; sore throat, cough, severe breathing difficulties, unconsciousness.	ventilation.	fresh air, rest, place in half-sitting position, take to hospital.
Skin: *corrosive*; redness, pain, serious burns.	gloves.	remove contaminated clothing, flush skin with water or shower, call a doctor.
Eyes: *corrosive*; redness, pain, impaired vision.	face shield.	flush with water, take to a doctor.
Ingestion: *corrosive*; sore throat, abdominal pain, diarrhea, nausea.		rinse mouth, immediately take to hospital.

SPILLAGE	STORAGE	LABELING / NFPA
collect leakage in sealable containers, soak up with sand or other inert absorbent and remove to safe place.	keep separate from oxidants.	

NFPA: 1 / 2 / 0

NOTES
Lung edema symptoms usually develop several hours later and are aggravated by physical exertion: rest and hospitalization essential. As first aid, a doctor or authorized person should consider administering a corticosteroid spray.

Transport Emergency Card TEC(R)-61G08

OCTANE

PHYSICAL PROPERTIES		IMPORTANT CHARACTERISTICS	
Boiling point, °C	126	**COLORLESS LIQUID WITH CHARACTERISTIC ODOR**	
Melting point, °C	−57	Vapor mixes readily with air, forming explosive mixtures. Flow, agitation etc. can cause build-up of electrostatic charge due to liquid's low conductivity. Do not use compressed air when filling, emptying or processing.	
Flash point, °C	12		
Auto-ignition temperature, °C	210		
Relative density (water = 1)	0.7		
Relative vapor density (air = 1)	3.9	TLV-TWA 300 ppm 1400 mg/m³	
Relative density at 20 °C of saturated mixture vapor/air (air = 1)	1.04	STEL 375 ppm 1750 mg/m³	
Vapor pressure, mm Hg at 20 °C	11	**Absorption route:** Can enter the body by inhalation or ingestion or through the skin. Harmful atmospheric concentrations build up fairly slowly on evaporation at 20°C, but much more rapidly in aerosol form.	
Solubility in water	none		
Explosive limits, vol% in air	0.8-6.5	**Immediate effects:** Irritates the eyes, skin and respiratory tract. Liquid destroys the skin's natural oils. In substantial concentrations can impair consciousness. If liquid is swallowed, droplets can enter the lungs, with risk of pneumonia.	
Electrical conductivity, pS/m	<10⁴		
Relative molecular mass	114.2		
Gross formula:	C_8H_{18}		

HAZARDS/SYMPTOMS	PREVENTIVE MEASURES	FIRE EXTINGUISHING/FIRST AID
Fire: extremely flammable.	keep away from open flame and sparks, no smoking.	powder, AFFF, foam, carbon dioxide, (halons).
Explosion: forms explosive air-vapor mixtures.	sealed machinery, ventilation, explosion-proof electrical equipment and lighting, grounding.	in case of fire: keep tanks/drums cool by spraying with water.
Inhalation: severe breathing difficulties, drowsiness, sleepiness.	ventilation, local exhaust or respiratory protection.	fresh air, rest, artificial respiration if necessary, call a doctor.
Skin: redness.	gloves.	remove contaminated clothing, flush skin with water or shower.
Eyes: redness, pain.	safety glasses.	flush with water, take to a doctor if necessary.
Ingestion: abdominal pain, nausea.		rinse mouth, DO NOT induce vomiting, call a doctor or take to hospital.

SPILLAGE	STORAGE	LABELING / NFPA
collect leakage in sealable containers, soak up with sand or other inert absorbent and remove to safe place, do not flush into sewer.	keep in fireproof place.	R: 11 S: 9-16-29-33 Note C 🔥 Flammable

NFPA diamond: 3 / 0 / 0

NOTES
High atmospheric concentrations, e.g. in poorly ventilated spaces, can cause oxygen deficiency, with risk of unconsciousness. Commercial product is a mixture of isomers and often contains other hydrocarbons.

Transport Emergency Card TEC(R)-95

HI: 33; UN-number: 1262

CAS-No: [111-87-5]
caprylic alcohol
heptyl carbinol
1-hydroxyoctane
n-octanol
n-octyl alcohol

$C_8H_{17}OH$

1-OCTANOL

PHYSICAL PROPERTIES		IMPORTANT CHARACTERISTICS
Boiling point, °C	195	**COLORLESS VISCOUS LIQUID WITH CHARACTERISTIC ODOR**
Melting point, °C	−16	Vapor mixes readily with air. Reacts with strong oxidants.
Flash point, °C	81	
Auto-ignition temperature, °C	253	TLV-TWA not available
Relative density (water = 1)	0.8	
Relative vapor density (air = 1)	4.4	
Relative density at 20 °C of saturated mixture		**Absorption route:** Can enter the body by inhalation or ingestion or through the skin. Harmful atmospheric concentrations build up very slowly, if at all, on evaporation at approx. 20°C, but much more rapidly in aerosol form.
vapor/air (air = 1)	1.0	
Vapor pressure, mm Hg at 20 °C	0.023	
Solubility in water	none	**Immediate effects:** Irritates the eyes, skin and respiratory tract. Affects the nervous system. In serious cases risk of unconsciousness.
Explosive limits, vol% in air	0.2-30.3	
Electrical conductivity, pS/m	1.4x10⁷	
Relative molecular mass	130.2	
Log P octanol/water	3.2	
Gross formula:	$C_8H_{18}O$	

HAZARDS/SYMPTOMS	PREVENTIVE MEASURES	FIRE EXTINGUISHING/FIRST AID
Fire: combustible.	keep away from open flame, no smoking.	powder, AFFF, foam, carbon dioxide, (halons).
Explosion: above 81°C: forms explosive air-vapor mixtures.	above 81°C: sealed machinery, ventilation.	
Inhalation: cough, dizziness, nausea, sleepiness, high exposure can cause unconsciousness.	ventilation.	fresh air, rest, call a doctor.
Skin: *is absorbed*; redness, dizziness, sleepiness.	gloves.	remove contaminated clothing, wash skin with soap and water, refer to a doctor if necessary.
Eyes: redness, pain.	safety glasses.	flush with water, take to a doctor if necessary.
Ingestion: diarrhea, nausea, see also 'Inhalation'.		rinse mouth, call a doctor.

SPILLAGE	STORAGE	LABELING / NFPA
collect leakage in sealable containers, soak up with sand or other inert absorbent and remove to safe place.	keep separate from oxidants; ventilate at floor level.	2 1 0

NOTES
The synonym capryl alcohol is confusing; is also used for a mixture of isomers and 2-octanol. Alcohol consumption increases toxic effects.

Transport Emergency Card TEC(R)-30G30

HI: 30; UN-number: 1993

capryl alcohol
2-hydroxyoctane
methyl hexyl carbinol

$CH_3CHOH(CH_2)_5CH_3$

2-n-OCTANOL

PHYSICAL PROPERTIES	IMPORTANT CHARACTERISTICS
Boiling point, °C 180 Melting point, °C − 38 Flash point, °C 60 Auto-ignition temperature, °C ? Relative density (water = 1) 0.8 Relative vapor density (air = 1) 4.4 Relative density at 20 °C of saturated mixture vapor/air (air = 1) 1.0 Vapor pressure, mm Hg at 20 °C 0.19 Solubility in water, g/100 ml 0.1 Explosive limits, vol% in air 0.8-7.4 Relative molecular mass 130.2 Gross formula: $C_8H_{18}O$	**COLORLESS VISCOUS LIQUID WITH CHARACTERISTIC ODOR** Vapor mixes readily with air. Reacts with strong oxidants. TLV-TWA not available **Absorption route:** Can enter the body by inhalation or ingestion or through the skin. Harmful atmospheric concentrations build up very slowly, if at all, on evaporation at approx. 20°C, but much more rapidly in aerosol form. **Immediate effects:** Irritates the eyes, skin and respiratory tract. Affects the nervous system.

HAZARDS/SYMPTOMS	PREVENTIVE MEASURES	FIRE EXTINGUISHING/FIRST AID
Fire: combustible.	keep away from open flame, no smoking.	powder, AFFF, foam, carbon dioxide, (halons).
Explosion: above 60°C: forms explosive air-vapor mixtures.	above 60°C: sealed machinery, ventilation.	
Inhalation: sore throat, cough, dizziness, sleepiness.	ventilation.	fresh air, rest, call a doctor.
Skin: *is absorbed*; redness.	gloves.	remove contaminated clothing, flush skin with water or shower.
Eyes: redness, pain.	safety glasses.	flush with water, take to a doctor.
Ingestion: headache, nausea, dizziness, sleepiness.		rinse mouth, call a doctor.

SPILLAGE	STORAGE	LABELING / NFPA
collect leakage in sealable containers, soak up with sand or other inert absorbent and remove to safe place.	keep separate from oxidants; ventilate at floor level.	

NOTES
Capryl alcohol is a misnomer, previously used for a mixture of isomers of octanols; better to use the systematic names, e.g 2-octanol. Alcohol consumption increases toxic effects.

Transport Emergency Card TEC(R)-30G30

HI: 30; UN-number: 1993

para-OCTYL PHENOL

PHYSICAL PROPERTIES	IMPORTANT CHARACTERISTICS
Boiling point, °C 280 Melting point, °C 72 Flash point, °C see notes Relative density (water = 1) 0.9 Relative vapor density (air = 1) 7.1 Relative density at 20 °C of saturated mixture vapor/air (air = 1) 1.0 Solubility in water none Relative molecular mass 206.4	**WHITE OR LIGHT PINK POWDER** Decomposes when heated, giving off toxic vapors (→ *phenol*). Reacts with strong oxidants. In dry state can form electrostatic charge if stirred, transported pneumatically, poured etc.
	TLV-TWA not available
	Absorption route: Can enter the body by ingestion or through the skin. Evaporation negligible at 20°C, but harmful atmospheric concentrations can build up rapidly in aerosol form. **Immediate effects:** Corrosive to the eyes, skin and respiratory tract. Inhalation can cause lung edema. **Effects of prolonged/repeated exposure:** Frequent contact can cause decoloration (depigmentation) of the skin.
Gross formula: $C_{14}H_{22}O$	

HAZARDS/SYMPTOMS	PREVENTIVE MEASURES	FIRE EXTINGUISHING/FIRST AID
Fire: combustible.	keep away from open flame, no smoking.	water spray, powder.
Explosion: finely dispersed particles form explosive mixtures on contact with air.	keep dust from accumulating, sealed machinery, explosion-proof lighting, grounding.	
	KEEP DUST UNDER CONTROL	
Inhalation: *corrosive*; cough, shortness of breath, severe breathing difficulties.	local exhaust or respiratory protection.	fresh air, rest, place in half-sitting position, artificial respiration if necessary, take to hospital.
Skin: *corrosive, is absorbed*; redness, pain, serious burns.	gloves, protective clothing.	remove contaminated clothing, wash skin with soap and water, call a doctor.
Eyes: *corrosive*; redness, pain, impaired vision.	face shield, or combined eye and respiratory protection.	flush with water, take to a doctor.
Ingestion: *corrosive*; sore throat, diarrhea, abdominal cramps.		rinse mouth, take immediately to hospital.

SPILLAGE	STORAGE	LABELING / NFPA
clean up spillage, carefully collect remainder; (additional individual protection: P2 respirator).	ventilate at floor level.	

NOTES
Combustible, but flash point and explosive limits not given in the literature. Lung edema symptoms usually develop several hours later and are aggravated by physical exertion: rest and hospitalization essential. As first aid, a doctor or authorized person should consider administering a corticosteroid spray.

Transport Emergency Card TEC(R)-61G12C

HI: 60; UN-number: 2430

$$CH_3(CH_2)_7CH=CH(CH_2)_7CONH_2$$

OLEAMIDE

PHYSICAL PROPERTIES		IMPORTANT CHARACTERISTICS	
Melting point, °C	76	**IVORY-COLORED POWDER**	
Relative density (water = 1)	0.9	In dry state can form electrostatic charge if stirred, transported pneumatically, poured etc.	
Solubility in water	none	Decomposes when heated or burned, giving off → *nitrous vapors*. Reacts with oxidants.	
Relative molecular mass	281.5		
		TLV-TWA not available	
		Absorption route: Can enter the body by inhalation or ingestion. Evaporation negligible at 20°C, but harmful concentrations of airborne particles can build up rapidly.	
Gross formula:	$C_{18}H_{35}NO$		

HAZARDS/SYMPTOMS	PREVENTIVE MEASURES	FIRE EXTINGUISHING/FIRST AID
Fire: combustible.	keep away from open flame, no smoking.	water spray, powder.
Explosion: finely dispersed particles form explosive mixtures on contact with air.	keep dust from accumulating, sealed machinery, explosion-proof lighting, grounding.	
Inhalation:	local exhaust or respiratory protection.	fresh air, rest, call a doctor.
Skin:	gloves.	remove contaminated clothing, flush skin with water or shower, call a doctor if necessary.
Eyes:	goggles.	flush with water, take to a doctor if necessary.
Ingestion:	do not eat, drink or smoke while working.	rinse mouth, call a doctor.

SPILLAGE	STORAGE	LABELING / NFPA
clean up spillage, carefully collect remainder; (additional individual protection: P2 respirator).		

NOTES
Insufficient toxicological data on harmful effects to humans.

CAS-No: [112-80-1]
cis-9-octadecenoic acid
elaic acid
9,10-octadecenoic acid
oleinic acid
red oil

$$C_8H_{17}CH = CH(CH_2)_7COOH$$

OLEIC ACID

PHYSICAL PROPERTIES		IMPORTANT CHARACTERISTICS
Boiling point, °C	360	**COLORLESS TO LIGHT YELLOW LIQUID, TURNING DARK RED ON EXPOSURE TO AIR**
Melting point, °C	4	Decomposes when heated. Reacts with strong oxidants. Attacks aluminum.
Flash point, °C	189	
Auto-ignition temperature, °C	363	TLV-TWA not available
Relative density (water = 1)	0.9	
Vapor pressure, mm Hg at 176 °C	0.99	**Absorption route:** Can enter the body by ingestion. Evaporation negligible at 20°C, but unpleasant concentrations can build up in aerosol form.
Solubility in water	none	**Immediate effects:** Irritates the eyes and skin. Gastro-intestinal disorders likely if relatively large quantities are ingested.
Relative molecular mass	282.5	
Gross formula:	$C_{18}H_{34}O_2$	

HAZARDS/SYMPTOMS	PREVENTIVE MEASURES	FIRE EXTINGUISHING/FIRST AID
Fire: combustible.	keep away from open flame, no smoking.	powder, AFFF, foam, carbon dioxide, (halons).
Inhalation:	ventilation.	fresh air, rest, call a doctor if necessary.
Skin: redness, itching.	gloves.	wash skin with soap and water.
Eyes: redness, pain.	safety glasses.	flush with water, take to a doctor if necessary.

SPILLAGE	STORAGE	LABELING / NFPA
collect leakage in sealable containers, soak up with sand or other inert absorbent and remove to safe place.	keep separate from oxidants; ventilate at floor level.	

NOTES
The measures on this card also apply to palmitic acid and stearic acid.

OLEUM
(20% free SO_3)

PHYSICAL PROPERTIES	IMPORTANT CHARACTERISTICS
Boiling point, °C 138 Melting point, °C 2 Relative density (water = 1) 1.9 Relative vapor density (air = 1) 3.3 Relative density at 20 °C of saturated mixture vapor/air (air = 1) 1.01 Vapor pressure, mm Hg at 20 °C 2.1 Solubility in water reaction Relative molecular mass see notes	**COLORLESS TO BROWN HYGROSCOPIC VISCOUS SOLUTION OF SULFUR TRIOXIDE IN CONCENTRATED SULFURIC ACID WITH PUNGENT ODOR, FUMES ON EXPOSURE TO AIR** Strong oxidant which reacts violently with combustible substances and reducing agents. Strong acid which reacts violently with bases and is corrosive. Reacts violently with organic substances and water, evolving heat. Reacts with air to form corrosive vapors (which are heavier than air and spread at ground level). Attacks many metals, giving off flammable gas (→ *hydrogen*).

| | TLV-TWA | ✱ |

| | **Absorption route:** Can enter the body by inhalation or ingestion. Harmful atmospheric concentrations can build up very rapidly on evaporation at approx. 20°C.
Immediate effects: Corrosive to the eyes, skin and respiratory tract. Inhalation of vapor/fumes can cause severe breathing difficulties (lung edema). In serious cases risk of seizures and death. |

Gross formula: $O_3S.H_2O$

HAZARDS/SYMPTOMS	PREVENTIVE MEASURES	FIRE EXTINGUISHING/FIRST AID
Fire: non-combustible, many chemical reactions can cause fire and explosion.	avoid contact with combustible substances.	in case of fire in immediate vicinity: DO NOT USE WATER-BASED EXTINGUISHERS.
Explosion:		in case of fire: keep undamaged cylinders cool by spraying with water, but DO NOT spray substance with water.
	STRICT HYGIENE	IN ALL CASES CALL IN A DOCTOR
Inhalation: *corrosive*; sore throat, cough, shortness of breath, severe breathing difficulties.	ventilation, local exhaust or respiratory protection.	fresh air, rest, place in half-sitting position, take to hospital.
Skin: *corrosive*; redness, pain, serious burns.	gloves, protective clothing.	remove contaminated clothing, flush skin with water or shower, call a doctor.
Eyes: *corrosive*; redness, pain, impaired vision.	face shield, or combined eye and respiratory protection.	flush with water, take to a doctor.
Ingestion: *corrosive*; sore throat, abdominal pain, vomiting.		rinse mouth, immediately take to hospital.

SPILLAGE	STORAGE	LABELING / NFPA
evacuate area, call in an expert, collect leakage in sealable containers, flush away remainder with large quantities of water water, do not use sawdust or other combustible absorbents, combat gas cloud with water curtain; (additional individual protection: CHEMICAL SUIT).	keep dry, separate from combustible substances, reducing agents, hydrous solutions and bases; ventilate at floor level.	R: 14-35-37 S: 26-30 Note B Corrosive

NOTES

Relative molecular mass of H_2SO_4: 98.1; SO_3: 80.1. Lung edema symptoms usually develop several hours later and are aggravated by physical exertion: rest and hospitalization essential. As first aid, a doctor or authorized person should consider administering a corticosteroid spray. UNDER NO CIRCUMSTANCES add water to acid; when diluting ALWAYS add acid to water. Unbreakable packaging preferred; if breakable, keep in unbreakable container. ✱ TLV of sulfuric acid (TLV-TWA 1mg/m³ and STEL 3mg/m³) normally applied in practice.

Transport Emergency Card TEC(R)-10C

HI: X886; UN-number: 1831

OLEUM
(30% free SO_3)

PHYSICAL PROPERTIES	IMPORTANT CHARACTERISTICS
Boiling point, °C 116 Melting point, °C 21 Relative density (water = 1) 1.9 Relative vapor density (air = 1) 3.2 Relative density at 20 °C of saturated mixture vapor/air (air = 1) 1.01 Vapor pressure, mm Hg at 20 °C 3.6 Solubility in water reaction Relative molecular mass see notes	**COLORLESS TO BROWN HYGROSCOPIC VISCOUS SOLUTION OF SULFUR TRIOXIDE IN CONCENTRATED SULFURIC ACID, FUMING IN MOIST AIR, WITH PUNGENT ODOR** Strong oxidant which reacts violently with combustible substances and reducing agents. Strong acid which reacts violently with bases and is corrosive. Reacts with organic substances and water, evolving heat. Reacts with air to form corrosive vapors (which are heavier than air and spread at ground level). Attacks many metals, giving off flammable gas (→ *hydrogen*).

	TLV-TWA	✱

Absorption route: Can enter the body by inhalation or ingestion. Harmful atmospheric concentrations can build up very rapidly on evaporation at 20°C.
Immediate effects: Corrosive to the eyes, skin and respiratory tract. Inhalation of vapor/fumes can cause severe breathing difficulties (lung edema). In serious cases risk of seizures and death.

Gross formula: $O_3S.H_2O$

HAZARDS/SYMPTOMS	PREVENTIVE MEASURES	FIRE EXTINGUISHING/FIRST AID
Fire: many chemical reactions can cause fire and explosion.	avoid contact with combustible substances.	in case of fire in immediate vicinity: DO NOT USE WATER-BASED EXTINGUISHERS.
	STRICT HYGIENE	IN ALL CASES CALL IN A DOCTOR
Inhalation: *corrosive*; sore throat, cough, severe breathing difficulties.	ventilation, local exhaust or respiratory protection.	fresh air, rest, place in half-sitting position, take to hospital.
Skin: *corrosive*; redness, pain, serious burns.	gloves, protective clothing.	remove contaminated clothing, flush skin with water or shower, call a doctor.
Eyes: *corrosive*; redness, pain, impaired vision.	face shield, or combined eye and respiratory protection.	flush with water, take to a doctor.
Ingestion: *corrosive*; sore throat, abdominal pain, vomiting.		rinse mouth, immediately take to hospital.

SPILLAGE	STORAGE	LABELING / NFPA
evacuate area, call in an expert, collect leakage in sealable containers, flush away remainder with water, under no circumstances spray liquid with water, do not use sawdust or other combustible absorbents, combat gas cloud with water curtain; (additional individual protection: CHEMICAL SUIT).	keep dry, separate from combustible substances, reducing agents, hydrous solutions and bases; ventilate at floor level.	R: 14-35-37 S: 26-30 Note B Corrosive

NOTES

Relative molecular mass of H_2SO_4: 98.1; SO_3: 80.1. Lung edema symptoms usually develop several hours later and are aggravated by physical exertion: rest and hospitalization essential. As first aid, a doctor or authorized person should consider administering a corticosteroid spray. UNDER NO CIRCUMSTANCES add water to acid; when diluting ALWAYS add acid to water. ✱ For practical purposes TLV for sulfuric acid (TLV-TWA 1mg/m³, STEL 3mg/m³) can usually be applied.

Transport Emergency Card TEC(R)-10C

HI: X886; UN-number: 1831

OLEUM
(65% free SO_3)

PHYSICAL PROPERTIES	IMPORTANT CHARACTERISTICS
Boiling point, °C 60 Melting point, °C 5 Relative density (water = 1) 1.9 Relative vapor density (air = 1) 3.0 Relative density at 20 °C of saturated mixture vapor/air (air = 1) 1.3 Vapor pressure, mm Hg at 20 °C 131 Solubility in water reaction Relative molecular mass see notes	**COLORLESS TO BROWN HYGROSCOPIC VISCOUS SOLUTION OF SULFUR TRIOXIDE IN CONCENTRATED SULFURIC ACID, FUMING IN AIR, WITH PUNGENT ODOR** Vapor is heavier than air. Strong oxidant which reacts violently with combustible substances and reducing agents. Strong acid which reacts violently with bases and is corrosive. Reacts violently with organic substances and water, evolving heat. Reacts with air to form corrosive vapors (which are heavier than air and spread at ground level). Attacks many metals, giving off flammable gas (→ *hydrogen*).

TLV-TWA	✶

Absorption route: Can enter the body by inhalation or ingestion. Harmful atmospheric concentrations can build up very rapidly on evaporation at 20° C.
Immediate effects: Corrosive to the eyes, skin and respiratory tract. Inhalation of vapor/fumes can cause severe breathing difficulties (lung edema). In serious cases risk of seizures and death.

Gross formula: $O_3.S.H_2O$

HAZARDS/SYMPTOMS	PREVENTIVE MEASURES	FIRE EXTINGUISHING/FIRST AID
Fire: non-combustible, many chemical reactions can cause fire and explosion.	avoid contact with combustible substances.	in case of fire in immediate vicinity: DO NOT USE WATER-BASED EXTINGUISHERS.
Explosion:		in case of fire: keep undamaged cylinders cool by spraying with water, but DO NOT spray substance with water.
	STRICT HYGIENE	IN ALL CASES CALL IN A DOCTOR
Inhalation: *corrosive*; sore throat, cough, severe breathing difficulties.	ventilation, local exhaust or respiratory protection.	place in half-sitting position, administer amyl nitrite by inhalation and take to hospital.
Skin: *corrosive*; redness, pain, serious burns.	gloves, protective clothing.	remove contaminated clothing, flush skin with water or shower, call a doctor.
Eyes: *corrosive*; redness, pain, impaired vision.	face shield, or combined eye and respiratory protection.	flush with water, take to a doctor.
Ingestion: *corrosive*; sore throat, abdominal pain, vomiting.		rinse mouth, immediately take to hospital.

SPILLAGE	STORAGE	LABELING / NFPA
evacuate area, call in an expert, collect leakage in sealable containers, flush away remainder with water, under no circumstances spray liquid with water, do not use sawdust or other combustible absorbents, combat gas cloud with water curtain; (additional individual protection: CHEMICAL SUIT).	keep dry, separate from combustible substances, reducing agents, hydrous solutions and bases; ventilate at floor level.	R: 14-35-37 S: 26-30 Note B Corrosive

NOTES
Relative molecular mass of H_2SO_4: 98.1; SO_3: 80.1. Lung edema symptoms usually develop several hours later and are aggravated by physical exertion: rest and hospitalization essential. As first aid, a doctor or authorized person should consider administering a corticosteroid spray. UNDER NO CIRCUMSTANCES add water to acid; when diluting ALWAYS add acid to water. Unbreakable packaging preferred; if breakable, keep in unbreakable container. ✶ For practical purposes TLV for sulfuric acid (TLV-TWA 1mg/m³, STEL 3mg/m³) can usually be applied.

Transport Emergency Card TEC(R)-10C

HI: X886; UN-number: 1831

CAS-No: [143-28-2]
cis-9-octadecenol-1

$CH_3(CH_2)_7CH = CH(CH_2)_8OH$

OLEYL ALCOHOL

PHYSICAL PROPERTIES		IMPORTANT CHARACTERISTICS
Boiling point, °C	335	**COLORLESS OR LIGHT YELLOW VISCOUS LIQUID**
Melting point, °C	−7	Decomposes when heated, giving off acid, corrosive vapors. Reacts with oxidants.
Flash point, °C	170	
Relative density (water = 1)	0.85	TLV-TWA not available
Relative vapor density (air = 1)	9.3	
Relative density at 20 °C of saturated mixture vapor/air (air = 1)	1.0	**Absorption route:** Can enter the body by inhalation or ingestion. Harmful atmospheric concentrations build up very slowly, if at all, on evaporation at approx. 20°C, but much more rapidly in aerosol form.
Vapor pressure, mm Hg at 150 °C	<0.76	
Solubility in water	none	
Relative molecular mass	268.5	
Gross formula:	$C_{18}H_{36}O$	

HAZARDS/SYMPTOMS	PREVENTIVE MEASURES	FIRE EXTINGUISHING/FIRST AID
Fire: combustible.	keep away from open flame, no smoking.	powder, AFFF, foam, carbon dioxide, (halons).
Inhalation:	ventilation.	fresh air, rest, call a doctor if necessary.
Skin:	gloves.	remove contaminated clothing, wash skin with soap and water.
Eyes:	safety glasses.	flush with water, take to a doctor if necessary.
Ingestion:		rinse mouth, call a doctor if necessary.

SPILLAGE	STORAGE	LABELING / NFPA
collect leakage in sealable containers, soak up with sand or other inert absorbent and remove to safe place.	keep separate from oxidants; ventilate at floor level.	

NOTES
Insufficient toxicological data on harmful effects to humans.

OXALIC ACID

PHYSICAL PROPERTIES	IMPORTANT CHARACTERISTICS
Boiling point, °C see notes Melting point, °C 102 Relative density (water = 1) 1.7 Solubility in water, g/100 ml at 20 °C 14 Relative molecular mass 126.1 Log P octanol/water −0.7	**COLORLESS CRYSTALS** Decomposes when heated, giving off toxic gases (→ *carbon monoxide*). In aqueous solution is a medium strong acid and is corrosive. Reacts violently with silver and mercury. Reacts violently with strong oxidants, with risk of fire and explosion.

	TLV-TWA STEL	1 mg/m³ 2 mg/m³

	Absorption route: Can enter the body by inhalation or ingestion or through the skin. Evaporation negligible at 20°C, but harmful concentrations of airborne particles can build up rapidly. **Immediate effects:** Corrosive to the eyes, skin and respiratory tract. In serious cases risk of muscle cramps and death. **Effects of prolonged/repeated exposure:** Prolonged or repeated contact can cause skin disorders. Can cause liver and kidney damage.

Gross formula: C$_2$H$_2$O$_4$.2H$_2$O

HAZARDS/SYMPTOMS	PREVENTIVE MEASURES	FIRE EXTINGUISHING/FIRST AID
Fire: combustible.	keep away from open flame, no smoking.	water spray, powder.
	KEEP DUST UNDER CONTROL	
Inhalation: *corrosive*; sore throat, cough, shortness of breath.	local exhaust or respiratory protection.	fresh air, rest, place in half-sitting position, call a doctor.
Skin: *corrosive*; redness, pain, chemical burns, blisters.	gloves, protective clothing.	remove contaminated clothing, flush skin with water or shower, call a doctor.
Eyes: *corrosive*; redness, pain, impaired vision.	goggles.	flush with water, take to a doctor.
Ingestion: *corrosive*; sore throat, vomiting, unconsciousness, cramps.	do not eat, drink or smoke while working.	rinse mouth, take immediately to hospital.

SPILLAGE	STORAGE	LABELING / NFPA
clean up spillage, flush away remainder with water; (additional individual protection: P2 respirator).	keep separate from oxidants and strong bases.	R: 21/22 S: 2-24/25 ✖ Harmful 2 1 0

NOTES
Prolonged or repeated contact can cause gangrene. Apparent melting point due to loss of water of crystallization. Other apparent melting points: 187°C (anhydrous). Sublimation of anhydrous substance begins at 100°C. Log P octanol/water is estimated.

5-OXOHEXANENITRILE

PHYSICAL PROPERTIES		IMPORTANT CHARACTERISTICS	
Boiling point, °C	240	**CLEAR YELLOW LIQUID WITH CHARACTERISTIC ODOR**	
Flash point, °C	119	Vapor mixes readily with air. Toxic vapors are a product of combustion. Decomposes when	
Relative density (water = 1)	1.04	heated, giving off toxic gas ($\rightarrow$ *hydrogen cyanide*).	
Relative vapor density (air = 1)	3.9		
Relative density at 20 °C of saturated mixture		TLV-TWA	not available
vapor/air (air = 1)	1.0		
Vapor pressure, mm Hg at 20 °C	ca. 0.076	**Absorption route:** Can enter the body by inhalation or ingestion or through the skin. Insufficient	
Solubility in water	good	data on the rate at which harmful concentrations can build up.	
Relative molecular mass	111		
Gross formula:	C_6H_9NO		

HAZARDS/SYMPTOMS	PREVENTIVE MEASURES	FIRE EXTINGUISHING/FIRST AID
Fire: combustible.	keep away from open flame, no smoking.	powder, alcohol-resistant foam, water spray, carbon dioxide, (halons).
Inhalation: shortness of breath, headache, dizziness.	ventilation.	fresh air, rest, administer amyl nitrite by inhalation and take to hospital.
Skin: *is absorbed.*	gloves, protective clothing.	remove contaminated clothing, wash skin with soap and water, call a doctor.
Eyes:	safety glasses.	flush with water, take to a doctor if necessary.
Ingestion:		rinse mouth, call a doctor if necessary.

SPILLAGE	STORAGE	LABELING / NFPA
collect leakage in sealable containers, flush away remainder with water.	ventilate at floor level.	

NOTES
Do not take work clothing home. Insufficient toxicological data on harmful effects to humans.

OXYGEN
(cylinder)

PHYSICAL PROPERTIES	IMPORTANT CHARACTERISTICS	
Boiling point, °C　　　　　　　　− 183 Melting point, °C　　　　　　　　− 218 Relative vapor density (air = 1)　　1.1 Solubility in water, mg/l at 20 °C　ca. 10 Relative molecular mass　　　　32.0	**COLORLESS, ODORLESS COMPRESSED GAS** Reacts violently with combustible substances and reducing agents.	
	TLV-TWA　　　　　　　　not available	
Gross formula:　　　　　　　O_2		

HAZARDS/SYMPTOMS	PREVENTIVE MEASURES	FIRE EXTINGUISHING/FIRST AID
Fire: not combustible but enhances combustion of other substances and clothing.	avoid contact with combustible substances and reducing agents, VENTILATE, UNDER NO CIRCUMSTANCES BRING CLOTHING INTO CONTACT WITH OXYGEN, no smoking.	
Explosion:	KEEP MACHINERY AND LINES OIL/GREASE-FREE, under no circumstances aerate tanks with oxygen.	in case of fire: keep cylinders cool by spraying with water.
Inhalation:	ventilation.	
Skin:		first flush clothing with water, then REMOVE CLOTHING CONTAMINATED WITH OXYGEN (FIRE HAZARD).

SPILLAGE	STORAGE	LABELING / NFPA
ventilate.	if stored indoors, keep in cool, fireproof place, separate from combustible substances and reducing agents; ventilate.	 ◇ 0 / 0 0 / oxy

NOTES
Flush contaminated clothing with water (fire hazard). Many substances which are non-combustible in air (e.g. trichloroethylene, steel) will burn in pure oxygen. Even a slight rise in atmospheric oxygen level increases the flammability of all substances considerably. Close valve when not in use; check hoses and lines regularly; test for leaking connections with soap solution. Use cylinder with special fittings.

HI: 20; UN-number: 1072

OXYGEN
(liquid)

PHYSICAL PROPERTIES	IMPORTANT CHARACTERISTICS
Boiling point, °C — 183 Melting point, °C — 218 Relative density (water = 1) see notes Relative vapor density (air = 1) 1.1 Solubility in water none Relative molecular mass 32.0	**COLORLESS, EXTREMELY COLD LIQUID** Evaporation produces cold gas cloud which is heavier than surrounding air and spreads at ground level. Strong oxidant which reacts violently with combustible substances and reducing agents, with risk of fire and explosion.
	TLV-TWA not available
	Immediate effects: Can cause frostbite due to rapid evaporation.
Gross formula: O₂	

HAZARDS/SYMPTOMS	PREVENTIVE MEASURES	FIRE EXTINGUISHING/FIRST AID
Fire: non-combustible, but makes other substances more combustible.	keep away from open flame and sparks, no smoking, avoid contact with combustible substances and clothing.	in case of fire in immediate vicinity: use any extinguishing agent.
Explosion: contact with many substances can cause explosion.	KEEP MACHINERY AND LINES OIL/GREASE-FREE	
Inhalation:	ventilation.	
Skin: *in case of frostbite*: redness, pain, blisters.	insulated gloves, protective clothing.	*in case of frostbite*: flush clothing with water BEFORE REMOVING, then flush skin with water or shower, take to hospital.
Eyes: *in case of frostbite*: redness, pain, impaired vision.	face shield.	flush with water, take to doctor.

SPILLAGE	STORAGE	LABELING / NFPA
ventilate, under no circumstances spray liquid with water.	keep separate from combustible substances and reducing agents; ventilate at floor level.	R: 8-34 S: 21 🔥 Oxidant

NFPA diamond: 3 (left), 0 (top), 0 (right), oxy (bottom)

NOTES

Density of liquid at boiling point 1.1kg/l. Use special insulated container.

OZONE
(cylinder)

PHYSICAL PROPERTIES	IMPORTANT CHARACTERISTICS	
Boiling point, °C −112 Melting point, °C −193 Relative density (water = 1) 1.6 Relative vapor density (air = 1) 1.7 Vapor pressure, mm Hg at −12°C 41,800 Solubility in water, g/100 ml at 25°C 0.01 Relative molecular mass 48.0	**GAS DISSOLVED IN HALONS UNDER PRESSURE WITH PUNGENT ODOR** Gas is heavier than air. At room temperatur ozone decomposes slowly into oxygen; half-value time: 3 days at 20°C, 3 months at −50°C. Reacts with alkenes, aromates, ethers, rubber and many organic substances to form shock-sensitive compounds. Decomposes when heated, with risk of fire and explosion. Strong oxidant which reacts violently with combustible substances and reducing agents.	
	TLV-TWA	0.1 ppm 0.2 mg/m³ C
	Absorption route: Can enter the body by inhalation. Harmful atmospheric concentrations can build up very rapidly if gas is released. **Immediate effects:** Corrosive to the eyes and respiratory tract. Inhalation of vapor/fumes can cause severe breathing difficulties (lung edema). In serious cases risk of death. **Effects of prolonged/repeated exposure:** Prolonged exposure to vapor/fumes can cause lung disorders.	
Gross formula: O_3		

HAZARDS/SYMPTOMS	PREVENTIVE MEASURES	FIRE EXTINGUISHING/FIRST AID
Fire: non-combustible, but makes other substances more combustible; many chemical reactions can cause fire and explosion.	keep away from open flame and sparks, no smoking; avoid contact with combustible substances (clothing).	in case of fire in immediate vicinity: use any extinguishing agent.
Explosion: many chemical reactions can cause explosion.	special equipment, ventilation.	in case of fire: keep cylinders cool by spraying with water.
Inhalation: *corrosive*; sore throat, cough, shortness of breath, severe breathing difficulties.	ventilation, local exhaust or respiratory protection.	fresh air, rest, place in half-sitting position, take to hospital.
Skin:	gloves.	remove contaminated clothing, flush skin with water or shower.
Eyes: *corrosive*; redness, pain, impaired vision.	face shield, or combined eye and respiratory protection.	flush with water, take to a doctor.

SPILLAGE	STORAGE	LABELING / NFPA
evacuate area, call in an expert, ventilate, remove cylinder to safe place; (additional individual protection: breathing apparatus).	if stored indoors, keep in cool, fireproof place; ventilate at floor level.	⬦

NOTES

The ozon solution is frequently stored refrigerated in halons. Check with the supplier for the special measures which have to be applied for the application of ozone. The physical properties apply to the pure gas. When U.V. (ultraviolet) lamps are used the ozone concentration can sometimes exceed the TLV value. Lung edema symptoms usually develop several hours later and are aggravated by physical exertion: rest and hospitalization essential. Turn leaking cylinder so that leak is on top to prevent liquid ozone escaping. Use cylinder with special fittings.

$C_{(12-18)}H_{(26-38)}$

PARAFFIN
(oil)

PHYSICAL PROPERTIES	IMPORTANT CHARACTERISTICS	
Boiling point, °C <240	**COLORLESS ODORLESS VISCOUS LIQUID**	
Melting point, °C ca. 8	Vapor mixes readily with air.	
Flash point, °C ca. 115		
Auto-ignition temperature, °C ca. 200	TLV-TWA	2 mg/m³ ✶
Relative density (water = 1) ca. 0.8		
Relative vapor density (air = 1) 7.4		
Relative density at 20 °C of saturated mixture vapor/air (air = 1) 1	**Absorption route:** Can enter the body by inhalation or ingestion. Harmful atmospheric concentrations build up very slowly, if at all, on evaporation at approx. 20°C, but harmful concentrations of fumes can build up much more rapidly.	
Vapor pressure, mm Hg at 20 °C 0.23		
Solubility in water none		
Explosive limits, vol% in air 0.5-8		
Electrical conductivity, pS/m 1.6 x 10⁸		
Relative molecular mass 170-254		

HAZARDS/SYMPTOMS	PREVENTIVE MEASURES	FIRE EXTINGUISHING/FIRST AID
Fire: combustible.	keep away from open flame, no smoking.	powder, AFFF, foam, carbon dioxide, (halons).
Inhalation:	ventilation.	
Eyes:	safety glasses.	

SPILLAGE	STORAGE	LABELING / NFPA
collect leakage in sealable containers, soak up with sand or other inert absorbent and remove to safe place, allow remainder to set if possible and then clean up spillage.		

NOTES
Properties of solid form, paraffin wax, are similar. Can ignite on hot surface (200°C). ✶ TLV equals that of paraffin wax fume.

paraform
polyformaldehyde
polyoxymethylene

$HO(CH_2O)_nH$

PARAFORMALDEHYDE

PHYSICAL PROPERTIES	IMPORTANT CHARACTERISTICS
Decomposes below melting point, °C ? Flash point, °C 70 Auto-ignition temperature, °C 300 Relative density (water = 1) 1.4 Vapor pressure, mm Hg at 25 °C 1.4 Solubility in water poor Relative molecular mass variable	**WHITE POWDER WITH CHARACTERISTIC ODOR** Polymer of formaldehyde with variable composition (n = 8-100). Vapor mixes readily with air. In dry state can form electrostatic charge if stirred, transported pneumatically, poured etc. Decomposes when moderately heated, giving off toxic gas (→ *formaldehyde*). Reacts violently with strong oxidants, with risk of fire and explosion.

	TLV-TWA	✱

	Absorption route: Can enter the body by inhalation or ingestion or through the skin. Harmful atmospheric concentrations can build up fairly rapidly on evaporation at approx. 20°C; harmful concentrations of airborne particles can build up even more rapidly. **Immediate effects:** Irritates the skin. Corrosive to the eyes and respiratory tract. Inhalation of vapor/fumes can cause severe breathing difficulties (lung edema). In serious cases risk of death. **Effects of prolonged/repeated exposure:** Prolonged exposure can cause lung disorders (asthma). Prolonged or repeated contact can cause skin disorders. Can cause hypersensitivity.

Gross formula: $(CH_2O)_n.H_2O$

HAZARDS/SYMPTOMS	PREVENTIVE MEASURES	FIRE EXTINGUISHING/FIRST AID
Fire: combustible.	keep away from open flame, no smoking.	alcohol-resistant foam, powder, water spray, carbon dioxide.
Explosion: above 70°C: forms explosive air-vapor mixtures; finely dispersed particles form explosive mixtures on contact with air.	above 70°C: sealed machinery, ventilation, keep dust from accumulating, explosion-proof electrical equipment and lighting, grounding.	in case of fire: keep tanks/drums cool by spraying with water.
Inhalation: *corrosive*; sore throat, cough, shortness of breath, severe breathing difficulties.	local exhaust or respiratory protection.	fresh air, rest, place in half-sitting position, take to hospital.
Skin: redness, pain.	gloves.	remove contaminated clothing, wash skin with soap and water.
Eyes: *corrosive*; redness, pain, impaired vision.	goggles.	flush with water, take to doctor.
Ingestion: *corrosive*; sore throat, abdominal pain, vomiting.	do not eat, drink or smoke while working.	rinse mouth, take immediately to hospital.

SPILLAGE	STORAGE	LABELING / NFPA
clean up spillage, carefully collect remainder; (additional individual protection: breathing apparatus).	keep in cool, fireproof place, separate from oxidants.	1 / 2 / 0

NOTES
Decomposition temperature not given in the literature. ✱ TLV for formaldehyde (product of decomposition): TLV-TWA 1ppm (A2) and 1.2mg/m³ (A2); STEL 2ppm (A2) and 2.5mg/m³ (A2). Lung edema symptoms usually develop several hours later and are aggravated by physical exertion: rest and hospitalization essential. As first aid, a doctor or authorized person should consider administering a corticosteroid spray. Packaging: special material.

Transport Emergency Card TEC(R)-61G12

UN-number: 2213

PARALDEHYDE

PHYSICAL PROPERTIES	IMPORTANT CHARACTERISTICS
Boiling point, °C — 124 Melting point, °C — 12 Flash point, °C — 17 Auto-ignition temperature, °C — 235 Relative density (water = 1) — 0.99 Relative vapor density (air = 1) — 4.5 Relative density at 20 °C of saturated mixture vapor/air (air = 1) — 1.12 Vapor pressure, mm Hg at 20 °C — 25.8 Solubility in water, g/100 ml at 25 °C — 12 Explosive limits, vol% in air — 1.3-? Relative molecular mass — 132.2 Log P octanol/water — 0.7	**COLORLESS LIQUID WITH CHARACTERISTIC ODOR** Vapor is heavier than air and spreads at ground level, with risk of ignition at a distance. Do not use compressed air when filling, emptying or processing. Decomposes on contact with acids, forming → *acetaldehyde*. Reacts violently with oxidants.
	TLV-TWA not available
	Absorption route: Can enter the body by inhalation or ingestion. Harmful atmospheric concentrations can build up fairly rapidly on evaporation at approx. 20°C - even more rapidly in aerosol form. **Immediate effects:** Irritates the eyes, skin and respiratory tract. Affects the nervous system. In serious cases risk of unconsciousness.
Gross formula: C₆H₁₂O₃	

HAZARDS/SYMPTOMS	PREVENTIVE MEASURES	FIRE EXTINGUISHING/FIRST AID
Fire: extremely flammable.	keep away from open flame and sparks, no smoking.	powder, AFFF, foam, carbon dioxide, (halons).
Explosion: forms explosive air-vapor mixtures.	sealed machinery, ventilation, explosion-proof electrical equipment and lighting.	in case of fire: keep tanks/drums cool by spraying with water.
Inhalation: sore throat, headache, dizziness, nausea, sleepiness, unconsciousness.	ventilation, local exhaust or respiratory protection.	fresh air, rest, call a doctor.
Skin: redness.	gloves.	remove contaminated clothing, flush skin with water or shower.
Eyes: redness, pain.	face shield.	flush with water, take to a doctor if necessary.
Ingestion: abdominal pain, headache, nausea, sleepiness, unconsciousness.		rinse mouth, call a doctor or take to hospital.

SPILLAGE	STORAGE	LABELING / NFPA
collect leakage in sealable containers, soak up with sand or other inert absorbent and remove to safe place, flush away remainder with water; (additional individual protection: breathing apparatus).	keep in fireproof place, separate from oxidants and acids.	R: 11 S: 9-16-29-33 🔥 Flammable NFPA: 3 (top) 2 (left) 1 (right)

NOTES
Log P octanol/water is estimated. Alcohol consumption increases toxic effects.

CAS-No: [56-38-2]

O-O-diethyl-O-p-nitrophenylthiophospate
diethyl 4-nitrophenylphosphorothionate
diethyl p-nitrophenylthionophosphate
diethyl p-nitrophenylthiophospate
diethylparathion

$(C_2H_5O)_2\text{-}(P=S)\text{-}O\text{-}C_6H_4NO_2$

PARATHION

PHYSICAL PROPERTIES		IMPORTANT CHARACTERISTICS
		BROWN LIQUID WITH CHARACTERISTIC ODOR
Boiling point, °C	375	Decomposes when heated or burned, giving off toxic vapors (→ *nitrogen oxides*, → *oxides of*
Melting point, °C	6	*phosphorus*, → *oxides of sulfur*).
Flash point, °C	120	
Relative density (water = 1)	1.3	
Relative vapor density (air = 1)	10	TLV-TWA (skin) 0.1 mg/m³
Relative density at 20 °C of saturated mixture		
vapor/air (air = 1)	1.0	**Absorption route:** Can enter the body by inhalation or ingestion or through the skin or eyes.
Vapor pressure, mm Hg at 20 °C	3.6×10^{-5}	Harmful atmospheric concentrations build up very slowly, if at all, on evaporation at approx. 20°C,
Solubility in water	poor	but much more rapidly in aerosol form.
Relative molecular mass	291.3	**Immediate effects:** Affects the nervous system. Inhalation can cause severe breathing
Log P octanol/water	3.8	difficulties (lung edema). In serious cases risk of seizures, unconsciousness and death.
Gross formula: $C_{10}H_{14}NO_5PS$		

HAZARDS/SYMPTOMS	PREVENTIVE MEASURES	FIRE EXTINGUISHING/FIRST AID
Fire: combustible.	keep away from open flame, no smoking.	powder, water spray, foam, carbon dioxide, (halons).
	STRICT HYGIENE	
Inhalation: severe breathing difficulties, headache, dizziness, nausea, vomiting, cramps, unconsciousness.	ventilation, local exhaust or respiratory protection.	fresh air, rest, take to hospital.
Skin: *is absorbed.*	gloves, protective clothing.	remove contaminated clothing, wash skin with soap and water, take to hospital.
Eyes: *is absorbed;* redness, pain.	face shield, or combined eye and respiratory protection.	flush with water, take to doctor.
Ingestion: abdominal cramps, headache, severe breathing difficulties, nausea, dizziness, vomiting, unconsciousness.	do not eat, drink or smoke while working.	rinse mouth, take immediately to hospital.

SPILLAGE	STORAGE	LABELING / NFPA
collect leakage in sealable containers, soak up with sand or other inert absorbent and remove to safe place; (additional individual protection: breathing apparatus).	keep in cool, lockable storeroom; ventilate at floor level.	R: 26/27/28 S: 1-13-28-45 ☠ Toxic

NOTES
Blocks cholinesterase. Do not take work clothing home. Depending on the degree of exposure, regular medical checkups are advisable. Lung edema symptoms usually develop several hours later and are aggravated by physical exertion: rest and hospitalization essential. If symptoms of this form of lung edema present, DO NOT administer corticosteroid spray as first aid. Special first aid required in the event of poisoning. Parathion is an insecticide. Unbreakable packaging preferred; if breakable, keep in unbreakable container.

Transport Emergency Card TEC(R)-61G42/44

HI: 66; UN-number: 1668

CAS-No: [112-05-0]
n-nonoic acid
n-nonanoic acid
n-nonylic acid
1-octanecarboxylic acid
pelargic acid

$CH_3(CH_2)_7COOH$

PELARGONIC ACID

PHYSICAL PROPERTIES		IMPORTANT CHARACTERISTICS
Boiling point, °C	255	**COLORLESS LIQUID WITH PUNGENT ODOR**
Melting point, °C	12	Decomposes when heated, giving off acid vapors.
Flash point, °C	114	
Auto-ignition temperature, °C	405	TLV-TWA not available
Relative density (water = 1)	0.9	
Relative vapor density (air = 1)	5.4	
Relative density at 20 °C of saturated mixture		**Absorption route:** Can enter the body by inhalation or ingestion. Insufficient data on the rate at which harmful concentrations can build up.
vapor/air (air = 1)	1.00	**Immediate effects:** Irritates the skin and respiratory tract. Corrosive to the eyes. Inhalation can cause severe breathing difficulties.
Vapor pressure, mm Hg at 20 °C	0.023	
Solubility in water	none	
Explosive limits, vol% in air	1.2-?	**Effects of prolonged/repeated exposure:** Prolonged or repeated contact can cause skin disorders.
Relative molecular mass	158.2	
Gross formula:	$C_9H_{18}O_2$	

HAZARDS/SYMPTOMS	PREVENTIVE MEASURES	FIRE EXTINGUISHING/FIRST AID
Fire: combustible.	keep away from open flame, no smoking.	powder, AFFF, foam, carbon dioxide, (halons).
Inhalation: sore throat, cough, severe breathing difficulties, nausea.	ventilation.	fresh air, rest, place in half-sitting position, call a doctor.
Skin: redness, pain.	gloves.	remove contaminated clothing, wash skin with soap and water, call a doctor if necessary.
Eyes: *corrosive*; redness, pain, impaired vision.	acid goggles.	flush with water, take to a doctor.
Ingestion: abdominal pain, diarrhea, vomiting.		rinse mouth, call a doctor.

SPILLAGE	STORAGE	LABELING / NFPA
collect leakage in sealable containers, soak up with sand or other inert absorbent, clean up solid spillage and remove to safe place.		

NOTES
Gastro-intestional disorders unlikely unless large quantities are ingested.

PCP
penchlorol
penta
2,3,4,5,6-pentachlorophenol

PENTACHLOROPHENOL

PHYSICAL PROPERTIES	IMPORTANT CHARACTERISTICS
Boiling point (decomposes), °C 310 Melting point, °C 191 Relative density (water = 1) 2.0 Relative vapor density (air = 1) 9.1 Relative density at 20 °C of saturated mixture vapor/air (air = 1) 1.0 Vapor pressure, mm Hg at 20 °C 1.5x10^{-4} Solubility in water poor Relative molecular mass 266.5 Log P octanol/water 5.0	**WHITE CRYSTALS OR GRAYISH SOLID IN VARIOUS FORMS WITH CHARACTERISTIC ODOR** Decomposes on contact with water when heated above 190°C, giving off toxic and corrosive gases (→ *hydrochloric acid*).

TLV-TWA (skin)	0.5 mg/m³

Absorption route: Can enter the body by inhalation or ingestion or through the skin. Harmful atmospheric concentrations build up very slowly, if at all, on evaporation at approx. 20°C, but harmful concentrations of airborne particles can build up much more rapidly.
Immediate effects: Irritates the eyes and skin. Corrosive to the respiratory tract. Affects the nervous system. Inhalation can cause lung edema. In serious cases risk of death.
Effects of prolonged/repeated exposure: Can cause liver and kidney damage.

Gross formula: C_6HCl_5O

HAZARDS/SYMPTOMS	PREVENTIVE MEASURES	FIRE EXTINGUISHING/FIRST AID
Fire: non-combustible.		in case of fire in immediate vicinity: use any extinguishing agent.
	STRICT HYGIENE, KEEP DUST UNDER CONTROL	
Inhalation: headache, tiredness, sweating, feeling of weakness, muscle cramps, raised temperature.	sealed machinery, local exhaust or respiratory protection.	fresh air, rest, place in half sitting position, allow to cool off, take to hospital.
Skin: *is absorbed*; white spots, sometimes wounds.	gloves, protective clothing.	remove contaminated clothing, flush skin with water or shower, call a doctor.
Eyes: redness, pain.	face shield, or combined eye and respiratory protection.	flush with water, take to a doctor if necessary.
Ingestion: diarrhea, abdominal cramps, nausea, vomiting.	do not eat, drink or smoke while working.	rinse mouth, immediately take to hospital.

SPILLAGE	STORAGE	LABELING / NFPA
clean up spillage, carefully collect remainder and remove to safe place; (additional individual protection: P2 respirator).	ventilate at floor level.	R: 23/24/25 S: 28-36/39-44 ☠ Toxic

NFPA: 3 / 0 / 0

NOTES

Melting point technical grade is lower: approx. 170° C. Lung edema symptoms usually develop several hours later and are aggravated by physical exertion: rest and hospitalization essential. As first aid, a doctor or authorized person should consider administering a corticosteroid spray. The measures on this card also apply to sodium salt and pentachlorophenolate.

Transport Emergency Card TEC(R)-G1G52

HI: 60; UN-number: 2996

amyl hydride
n-pentane
normal pentane

C_5H_{12}

PENTANE

PHYSICAL PROPERTIES		IMPORTANT CHARACTERISTICS		
Boiling point, °C	36	**COLORLESS HIGHLY VOLATILE LIQUID WITH CHARACTERISTIC ODOR**		
Melting point, °C	−130	Vapor is heavier than air and spreads at ground level, with risk of ignition at a distance. Flow, agitation etc. can cause build-up of electrostatic charge due to liquid's low conductivity. Do not use compressed air when filling, emptying or processing.		
Flash point, °C	−40			
Auto-ignition temperature, °C	285			
Relative density (water = 1)	0.6			
Relative vapor density (air = 1)	2.5	TLV-TWA	600 ppm	1770 mg/m³
Relative density at 20 °C of saturated mixture vapor/air (air = 1)	1.8	STEL	750 ppm	2210 mg/m³
Vapor pressure, mm Hg at 20 °C	436			
Solubility in water, g/100 ml	0.03	**Absorption route:** Can enter the body by inhalation or ingestion. Harmful atmospheric concentrations can build up fairly rapidly on evaporation at approx. 20°C - even more rapidly in aerosol form.		
Explosive limits, vol% in air	1.4-8			
Minimum ignition energy, mJ	0.22	**Immediate effects:** Irritates the eyes, skin and respiratory tract. Liquid destroys the skin's natural oils. Affects the nervous system. If liquid is swallowed, droplets can enter the lungs, with risk of pneumonia.		
Electrical conductivity, pS/m	2x10⁴			
Relative molecular mass	72.2			
Gross formula:	C_5H_{12}			

HAZARDS/SYMPTOMS	PREVENTIVE MEASURES	FIRE EXTINGUISHING/FIRST AID
Fire: extremely flammable.	keep away from open flame and sparks, no smoking.	powder, AFFF, foam, carbon dioxide, (halons).
Explosion: forms explosive air-vapor mixtures.	sealed machinery, ventilation, explosion-proof electrical equipment and lighting, grounding, non-sparking hand tools.	in case of fire: keep tanks/drums cool by spraying with water.
Inhalation: sore throat, shortness of breath, dizziness, drowsiness.	ventilation, local exhaust or respiratory protection.	fresh air, rest, call a doctor.
Skin: redness.	gloves.	remove contaminated clothing, flush skin with water or shower.
Eyes: redness, pain.	safety glasses.	flush with water, take to a doctor if necessary.
Ingestion: abdominal pain, nausea.		rinse mouth, DO NOT induce vomiting, take to hospital.

SPILLAGE	STORAGE	LABELING / NFPA
evacuate area, call in an expert, collect leakage in sealable containers, soak up with sand or other inert absorbent and remove to safe place; (additional individual protection: breathing apparatus).	keep in cool, fireproof place.	R: 11 S: 9-16-29-33 Flammable 4 / 1 / 0

NOTES
Alcohol consumption increases toxic effects. High atmospheric concentrations, e.g. in poorly ventilated spaces, can cause oxygen deficiency, with risk of unconsciousness. Unbreakable packaging preferred; if breakable, keep in unbreakable container.

Transport Emergency Card TEC(R)-30G01

HI: 33; UN-number: 1265

$CH_3(CH_2)_2CHCH_2$

1-PENTENE

PHYSICAL PROPERTIES	IMPORTANT CHARACTERISTICS
Boiling point, °C 30 Melting point, °C − 165 Flash point, °C − 18 Auto-ignition temperature, °C 273 Relative density (water = 1) 0.6 Relative vapor density (air = 1) 2.4 Relative density at 20 °C of saturated mixture vapor/air (air = 1) 2.0 Vapor pressure, mm Hg at 18 °C 505 Solubility in water none Explosive limits, vol% in air 1.4-8.7 Electrical conductivity, pS/m < 10⁴ Relative molecular mass 70.1	**COLORLESS HIGHLY VOLATILE LIQUID WITH CHARACTERISTIC ODOR** Vapor is heavier than air and spreads at ground level, with risk of ignition at a distance. Can polymerize slowly. Flow, agitation etc. can cause build-up of electrostatic charge due to liquid's low conductivity. Do not use compressed air when filling, emptying or processing. Reacts violently with strong oxidants.

Electrical conductivity, pS/m $< 10^4$

	TLV-TWA not available
Gross formula: C_5H_{10}	**Absorption route:** Can enter the body by inhalation or ingestion. Harmful atmospheric concentrations can build up fairly rapidly on evaporation at approx. 20°C - even more rapidly in aerosol form. **Immediate effects:** Irritates the eyes, skin and respiratory tract. Affects the nervous system. In substantial concentrations can impair consciousness. If liquid is swallowed, droplets can enter the lungs, with risk of pneumonia. **Effects of prolonged/repeated exposure:** Can cause heart disorders.

HAZARDS/SYMPTOMS	PREVENTIVE MEASURES	FIRE EXTINGUISHING/FIRST AID
Fire: extremely flammable.	keep away from open flame and sparks, no smoking.	powder, AFFF, foam, carbon dioxide, (halons).
Explosion: forms explosive air-vapor mixtures.	sealed machinery, ventilation, explosion-proof electrical equipment and lighting, grounding, non-sparking hand tools.	in case of fire: keep tanks/drums cool by spraying with water.
Inhalation: headache, dizziness, drowsiness, unconsciousness.	ventilation, local exhaust or respiratory protection.	fresh air, rest, call a doctor.
Skin: redness.	gloves.	remove contaminated clothing, flush skin with water or shower.
Eyes: redness, pain.	safety glasses.	flush with water, take to a doctor if necessary.
Ingestion: abdominal pain, nausea.		rinse mouth, DO NOT induce vomiting, take immediately to hospital.

SPILLAGE	STORAGE	LABELING / NFPA
evacuate area, call in an expert, collect leakage in sealable containers, soak up with sand or other inert absorbent and remove to safe place, do not flush into sewer; (additional individual protection: breathing apparatus).	keep in cool, fireproof place.	◇ 4 / 1 / 0

NOTES
High atmospheric concentrations, e.g. in poorly ventilated spaces, can cause oxygen deficiency, with risk of unconsciousness. The measures on this card also apply to other pentenes.

Transport Emergency Card TEC(R)-125

HI: 33; UN-number: 1108

PERACETIC ACID
(40% solution in acetic acid)

PHYSICAL PROPERTIES	IMPORTANT CHARACTERISTICS
Boiling point, °C 105 Melting point, °C 0 Flash point, °C 41 Relative density (water = 1) 1.2 Relative vapor density (air = 1) 2.6 Relative density at 20 °C of saturated mixture vapor/air (air = 1) 1.04 Vapor pressure, mm Hg at 20 °C 20 Solubility in water ∞ Explosive limits, vol% in air ?-? Relative molecular mass 76.1	**COLORLESS LIQUID WITH PUNGENT ODOR** Vapor mixes readily with air, forming explosive mixtures. Can decompose explosively if subjected to shock. Can explode when heated to 110°C. Decomposes above 40°C, giving off → *oxygen*, with increased risk of fire. Strong oxidant which reacts violently with combustible substances and reducing agents. In aqueous solution is a strong acid which reacts violently with bases and is corrosive. Reacts violently with heavy metals and organic compounds.

		TLV-TWA	not available

	Absorption route: Can enter the body by inhalation or ingestion. Harmful atmospheric concentrations can build up fairly rapidly on evaporation at approx. 20°C - even more rapidly in aerosol form. **Immediate effects:** Corrosive to the eyes, skin and respiratory tract. Inhalation of vapor/fumes can cause severe breathing difficulties (lung edema). In serious cases risk of death.

Gross formula: C₂H₄O₃

HAZARDS/SYMPTOMS	PREVENTIVE MEASURES	FIRE EXTINGUISHING/FIRST AID
Fire: flammable, many chemical reactions can cause fire and explosion.	keep away from open flame and sparks, no smoking, avoid contact with hot surfaces (e.g. steam lines).	powder, alcohol-resistant foam, water spray, carbon dioxide, (halons).
Explosion: above 41°C: forms explosive air-vapor mixtures.	above 41°C: sealed machinery, ventilation, explosion-proof electrical equipment, do not subject to shock or friction.	in case of fire: keep tanks/drums cool by spraying with water, fight fire from sheltered location.
	STRICT HYGIENE, AVOID ALL CONTACT	
Inhalation: *corrosive*; sore throat, cough, shortness of breath, severe breathing difficulties.	ventilation, local exhaust or respiratory protection.	fresh air, rest, place in half-sitting position, take to hospital.
Skin: *corrosive*; redness, pain, burns.	gloves, protective clothing.	remove contaminated clothing, flush skin with water or shower, call a doctor if necessary.
Eyes: *corrosive*; redness, pain, impaired vision.	face shield.	flush with water, take to a doctor.
Ingestion: *corrosive*; abdominal pain, diarrhea, nausea.		rinse mouth, immediately take to hospital.

SPILLAGE	STORAGE	LABELING / NFPA
collect leakage in sealable containers, soak up with sand or other inert absorbent and remove to safe place, do not use sawdust or other combustible absorbents; (additional individual protection: breathing apparatus).	keep in fireproof, cool place; keep separate from combustible substances, reducing agents and strong bases.	R: 5-22-34 S: 3-27-36 Note B + D Oxidant Corrosive 3 2 4 oxy

NOTES
Explosive limits not given in the literature. Flush contaminated clothing with water (fire hazard). Lung edema symptoms usually develop several hours later and are aggravated by physical exertion: rest and hospitalization essential. As first aid, a doctor or authorized person should consider administering a °corticosteroid spray. Use airtight packaging.

HI: 539; UN-number: 2131

PERCHLORIC ACID
(solution in water 50-72%)

PHYSICAL PROPERTIES	IMPORTANT CHARACTERISTICS
Boiling point, °C 132-203 Melting point, °C − 18 Relative density (water = 1) 1.4-1.7 Solubility in water ∞ Relative molecular mass 100.5	**COLORLESS LIQUID WITH PUNGENT ODOR** Decomposes when moderately heated, giving off oxygen, with increased risk of fire. Strong oxidant which reacts violently with combustible substances and reducing agents. In aqueous solution is a strong acid which reacts violently with bases and is corrosive. Reacts violently with alcohols, glycols, ketones, ethers, hypophosphites, sulfoxides, metals and many other substances, with risk of fire and explosion. Can explode spontaneously as a result of dehydration, esp. by dehumidifying agents. Wood, paper, cotton, wool and other textiles can ignite spontaneously when contaminated by perchloric acid solutions.
	TLV-TWA not available
	Absorption route: Can enter the body by inhalation or ingestion. Evaporation negligible at 20° C, but harmful atmospheric concentrations can build up rapidly in aerosol form. **Immediate effects:** Corrosive to the eyes, skin and respiratory tract. Inhalation of vapor/fumes can cause severe breathing difficulties (lung edema). In serious cases risk of death.
Gross formula: ClHO$_4$	

HAZARDS/SYMPTOMS	PREVENTIVE MEASURES	FIRE EXTINGUISHING/FIRST AID
Fire: non-combustible, but makes other substances more combustible; many chemical reactions can cause fire and explosion.	avoid contact with organic substances.	in case of fire in immediate vicinity: do not use powder, halons or carbon dioxide; water spray preferred extinguishing agent.
Explosion: contact with many organic substances or heating above 75° C can cause explosion.		in case of fire: keep tanks/drums cool by spraying with water.
Inhalation: *corrosive*; sore throat, cough, severe breathing difficulties.	ventilation, local exhaust or respiratory protection.	fresh air, rest, place in half-sitting position, artificial respiration if necessary, take to hospital.
Skin: *corrosive*; redness, pain, serious burns.	gloves, protective clothing.	remove contaminated clothing, flush skin with water or shower, refer to a doctor.
Eyes: *corrosive*; redness, pain, impaired vision.	face shield, or combined eye and respiratory protection.	flush with water, take to doctor.
Ingestion: *corrosive*; sore throat, abdominal pain, diarrhea.		rinse mouth, take immediately to hospital.

SPILLAGE	STORAGE	LABELING / NFPA
collect leakage in sealable containers, soak up with sand or other inert absorbent and remove to safe place, flush away remainder with water, do not use sawdust or other combustible absorbents; (additional individual protection: breathing apparatus).	keep cool, separate from combustible substances and bases.	R: 5-8-35 S: 23-26-36 Note B Oxidant Corrosive 3 / 0 / 3 oxy

NOTES
Solutions containing more than 72% perchloric acid can decompose explosively if subjected to shock or heating. Becomes shock-sensitive when contaminated with organic substances. Flush contaminated clothing with water (fire hazard). Lung edema symptoms usually develop several hours later and are aggravated by physical exertion: rest and hospitalization essential. As first aid, a doctor or authorized person should consider administering a corticosteroid spray. UNDER NO CIRCUMSTANCES add water to acid; when diluting ALWAYS add acid to water. Packaging: special material. Unbreakable packaging preferred; if breakable, keep in unbreakable container.

Transport Emergency Card TEC(R)-716   HI: 558; UN-number: 1873

CAS-No: [127-18-4]
ethylene tetrachloride
perchloroethylene
perk
tetrachlorethylene

$Cl_2C = CCl_2$

PERCHLOROETHYLENE

PHYSICAL PROPERTIES	IMPORTANT CHARACTERISTICS
Boiling point, °C .... 121 Melting point, °C .... − 22 Relative density (water = 1) .... 1.6 Relative vapor density (air = 1) .... 5.8 Relative density at 20 °C of saturated mixture vapor/air (air = 1) .... 1.08 Vapor pressure, mm Hg at 20 °C ...... 13.7 Solubility in water, g/100 ml .... 0.015 Electrical conductivity, pS/m ..... 5.55x10^{10} Relative molecular mass .... 165.8 Log P octanol/water 2.6	**COLORLESS LIQUID WITH CHARACTERISTIC ODOR** Vapor mixes readily with air. Decomposes when heated above 150°C or exposed to UV light, giving off toxic gas (→ *phosgene*) and corrosive vapor (→ *hydrochloric acid*). Reacts violently with finely dispersed light metals and → *zinc*.
	TLV-TWA .... 50 ppm .... 379 mg/m³ STEL .... 200 ppm .... 1370 mg/m³
	Absorption route: Can enter the body by inhalation or ingestion or through the skin. Harmful atmospheric concentrations can build up fairly rapidly on evaporation at 20°C - even more rapidly in aerosol form. **Immediate effects:** Irritates the eyes, skin and respiratory tract. Liquid destroys the skin's natural oils. Affects the nervous system. In substantial concentrations can impair consciousness. Can cause liver damage.
Gross formula: .... C$_2$Cl$_4$	

HAZARDS/SYMPTOMS	PREVENTIVE MEASURES	FIRE EXTINGUISHING/FIRST AID
Fire: non-combustible.		in case of fire in immediate vicinity: use any extinguishing agent.
Inhalation: headache, dizziness, drowsiness.	ventilation, local exhaust or respiratory protection.	fresh air, rest, call a doctor.
Skin: *is absorbed*; redness.	gloves.	remove contaminated clothing, flush skin with water or shower.
Eyes: redness, pain.	safety glasses.	flush with water, take to a doctor if necessary.
Ingestion: abdominal pain, diarrhea, headache, nausea, dizziness.		rinse mouth, take immediately to hospital.

SPILLAGE	STORAGE	LABELING / NFPA
collect leakage in sealable containers, soak up with sand or other inert absorbent and remove to safe place; (additional individual protection: breathing apparatus).	keep in dark place, separate from light metals and zinc; ventilate at floor level.	R: 40 S: 23-36/37 ✖ Harmful

NOTES
Odor limit is above TLV. Alcohol consumption increases toxic effects. Under no circumstances use near flame or hot surface or when welding.

Transport Emergency Card TEC(R)-722

HI: 60; UN-number: 1897

PERMANGANATE/SULFURIC ACID SOLUTION

PHYSICAL PROPERTIES	IMPORTANT CHARACTERISTICS
Boiling point, °C > 100 Relative density (water = 1) > 1.2 Solubility in water ∞	**PURPLE TO BLACK AQUEOUS SOLUTION OF POTASSIUM PERMANGANATE AND SULFURIC ACID IN VARIOUS CONCENTRATIONS** Strong oxidant which reacts violently with combustible substances and reducing agents. Strong acid which reacts violently with bases and is corrosive. Reacts with hydrochloric acid and chlorides, giving off toxic → *chlorine* gas. Attacks many materials.

TLV-TWA	not available

Absorption route: Can enter the body by inhalation or ingestion. Evaporation negligible at 20°C, but harmful atmospheric concentrations can build up rapidly in aerosol form.
Immediate effects: Corrosive to the eyes, skin and respiratory tract. Inhalation of vapor/fumes can cause severe breathing difficulties (lung edema). In serious cases risk of death.

HAZARDS/SYMPTOMS	PREVENTIVE MEASURES	FIRE EXTINGUISHING/FIRST AID
Fire: non-combustible, many chemical reactions can cause fire and explosion.	avoid contact with combustible substances.	in case of fire in immediate vicinity: use any extinguishing agent.
Inhalation: *corrosive*; sore throat, cough, shortness of breath, severe breathing difficulties.	ventilation.	fresh air, rest.
Skin: *corrosive*; redness, pain, burns.	gloves, protective clothing.	remove contaminated clothing, flush skin with water or shower, call a doctor if necessary.
Eyes: *corrosive*; redness, pain, impaired vision.	face shield.	flush with water, take to a doctor.
Ingestion: *corrosive*; sore throat, abdominal pain, vomiting.		rinse mouth, immediately take to hospital.

SPILLAGE	STORAGE	LABELING / NFPA
collect leakage in sealable containers, flush away remainder with water, do not use sawdust or other combustible absorbents.	keep separate from combustible substances, reducing agents, strong bases, hydrochloric acid and chlorides.	

NOTES
Flush contaminated clothing with water (fire hazard). Lung edema symptoms usually develop several hours later and are aggravated by physical exertion: rest and hospitalization essential. As first aid, a doctor or authorized person should consider administering a corticosteroid spray. UNDER NO CIRCUMSTANCES add water to acid; when diluting ALWAYS add acid to water. Unbreakable packaging preferred; if breakable, keep in unbreakable container.

CAS-No: [8008-20-6 1)]
jet fuel
kerosine
light petroleum

PETROLEUM

PHYSICAL PROPERTIES		IMPORTANT CHARACTERISTICS
Boiling point, °C	150-300	**COLORLESS LIQUID WITH CHARACTERISTIC ODOR**
Melting point, °C	−20	Mixture of hydrocarbons; physical properties vary, depending on the composition. Vapor mixes
Flash point, °C	>39	readily with air. Flow, agitation etc. can cause build-up of electrostatic charge due to liquid's low
Auto-ignition temperature, °C	>220	conductivity.
Relative density (water = 1)	ca. 0.75	
Relative vapor density (air = 1)	4.5	
Relative density at 20 °C of saturated mixture		TLV-TWA not available
vapor/air (air = 1)	1.01	
Vapor pressure, mm Hg at 20 °C	<2.3	**Absorption route:** Can enter the body by inhalation or ingestion. Harmful atmospheric
Solubility in water	none	concentrations build up very slowly, if at all, on evaporation at approx. 20°C, but much more
Explosive limits, vol% in air	0.7-5	rapidly in aerosol form.
Electrical conductivity, pS/m	ca. 0.1	**Immediate effects:** Irritates the eyes. Liquid destroys the skin's natural oils. If liquid is swallowed,
Relative molecular mass	>130	droplets can enter the lungs, with risk of pneumonia.

HAZARDS/SYMPTOMS	PREVENTIVE MEASURES	FIRE EXTINGUISHING/FIRST AID
Fire: flammable.	keep away from open flame and sparks, no smoking.	powder, AFFF, foam, carbon dioxide, (halons).
Explosion: above 39°C: forms explosive air-vapor mixtures.	above 39°C: sealed machinery, ventilation, explosion-proof electrical equipment, grounding.	in case of fire: keep tanks/drums cool by spraying with water.
Inhalation: headache.	ventilation.	fresh air, rest.
Skin: redness.	gloves.	remove contaminated clothing, wash skin with soap and water.
Eyes: redness, pain.	safety glasses.	flush with water, take to a doctor if necessary.
Ingestion: abdominal pain, diarrhea, severe breathing difficulties.		rinse mouth, DO NOT induce vomiting, take immediately to hospital.

SPILLAGE	STORAGE	LABELING / NFPA
collect leakage in sealable containers, soak up with dry sand or other inert absorbent and remove to safe place.	keep in fireproof place.	R: 10

NFPA diamond: top 2, left 0, right 0

NOTES
Dyes are often added to comply with customs regulations (→ *furfural*). N.B. JP4 has a flash point below 22°C. CAS No. 'Kerosine, partial sulfonization' (68606-32-2); 'Kerosine (petroleum) hydrodesulfurized' (64742-81-0). The measures on this card also apply to lamp oil.

Transport Emergency Card TEC(R)-662

HI: 30; UN-number: 1223

PETROLEUM NAPHTHA

(aliphatic - initial boiling point <140°C; flash point <21°C)

PHYSICAL PROPERTIES	IMPORTANT CHARACTERISTICS
Boiling point, °C 24-140 Melting point, °C < −40 Flash point, °C <21 Auto-ignition temperature, °C 250 Relative vapor density (air = 1) 3 Relative density at 20 °C of saturated mixture vapor/air (air = 1) 1.1-2.1 Vapor pressure, mm Hg at 20 °C 30-608 Solubility in water none Explosive limits, vol% in air 0.6-8.0 Electrical conductivity, pS/m <10²	**COLORLESS LIQUID WITH CHARACTERISTIC ODOR** Mixture of aliphatic hydrocarbons with lower boiling point 24-140°C and upper boiling point generally below 150°C. The lower the boiling point, the higher the vapor pressure at 20°C and the lower the flash point. Vapor is heavier than air and spreads at ground level, with risk of ignition at a distance. Flow, agitation etc. can cause build-up of electrostatic charge due to liquid's low conductivity. Do not use compressed air when filling, emptying or processing.

TLV-TWA	not available

Absorption route: Can enter the body by inhalation or ingestion. Harmful atmospheric concentrations can build up fairly rapidly on evaporation at approx. 20°C - even more rapidly in aerosol form.
Immediate effects: Irritates the eyes, skin and respiratory tract. Liquid destroys the skin's natural oils. Affects the nervous system. If liquid is swallowed, droplets can enter the lungs, with risk of pneumonia.

HAZARDS/SYMPTOMS	PREVENTIVE MEASURES	FIRE EXTINGUISHING/FIRST AID
Fire: extremely flammable.	keep away from open flame and sparks, no smoking.	powder, AFFF, foam, carbon dioxide, (halons).
Explosion: forms explosive air-vapor mixtures.	sealed machinery, ventilation, explosion-proof electrical equipment and lighting, grounding.	in case of fire: keep tanks/drums cool by spraying with water.
Inhalation: headache, dizziness, nausea, drowsiness.	ventilation, local exhaust or respiratory protection.	fresh air, rest, call a doctor.
Skin: redness.	gloves.	remove contaminated clothing, wash skin with soap and water.
Eyes: redness, pain.	safety glasses, or combined eye and respiratory protection.	flush with water, take to a doctor if necessary.
Ingestion: abdominal pain, nausea, vomiting, severe breathing difficulties.		rinse mouth, DO NOT induce vomiting, rest, take immediately to hospital.

SPILLAGE	STORAGE	LABELING / NFPA
evacuate area, call in an expert, ventilate, collect leakage in sealable containers, soak up with sand or other inert absorbent and remove to safe place, do not flush into sewer; (additional individual protection: breathing apparatus).	keep in fireproof place.	R: 11 S: 9-16-29-33 🔥 Flammable

NOTES
Alcohol consumption increases toxic effects. The measures on this card also apply to petroleum naphthas 40/65, 62/82, 80/110, benzine, petroleum ether, petroleum spirits. Solvent naphtha is a trade name. Mixture can contain hexane.

Transport Emergency Card TEC(R)-754   HI: 33; UN-number: 1993

PETROLEUM NAPHTHA
(aliphatic - initial boiling point 120-180° C)

PHYSICAL PROPERTIES	IMPORTANT CHARACTERISTICS
Boiling point, °C 120-180 Melting point, °C < −40 Flash point, °C 21-55 Auto-ignition temperature, °C 210 Relative density (water = 1) 0.73-0.87 Relative vapor density (air = 1) 4.5 Vapor pressure, mm Hg at 20 °C 0.38-30 Solubility in water none Explosive limits, vol% in air 0.6-8.0 Electrical conductivity, pS/m <100	**COLORLESS LIQUID WITH CHARACTERISTIC ODOR** Mixture of aliphatic hydrocarbons with lower boiling point 120-180° C. The lower the boiling point, the higher the vapor pressure at 20° C and the lower the flash point. Flow, agitation etc. can cause build-up of electrostatic charge due to liquid's low conductivity. TLV-TWA not available **Absorption route:** Can enter the body by inhalation or ingestion. Harmful atmospheric concentrations build up fairly slowly on evaporation at 20° C, but much more rapidly in aerosol form. **Immediate effects:** Irritates the eyes, skin and respiratory tract. Liquid destroys the skin's natural oils. Affects the nervous system. If liquid is swallowed, droplets can enter the lungs, with risk of pneumonia.

HAZARDS/SYMPTOMS	PREVENTIVE MEASURES	FIRE EXTINGUISHING/FIRST AID
Fire: flammable.	keep away from open flame and sparks, no smoking.	powder, AFFF, foam, carbon dioxide, (halons).
Explosion: above 21°C: forms explosive air-vapor mixtures.	above 21°C: sealed machinery, ventilation, explosion-proof electrical equipment, grounding.	
Inhalation: sore throat, cough, dizziness, drowsiness.	ventilation, local exhaust or respiratory protection.	fresh air, rest, call a doctor.
Skin: redness.	gloves.	remove contaminated clothing, wash skin with soap and water.
Eyes: redness, pain.	safety glasses.	flush with water, take to a doctor if necessary.
Ingestion: sore throat, abdominal pain, diarrhea, severe breathing difficulties.		rinse mouth, DO NOT induce vomiting, call a doctor or take to hospital.

SPILLAGE	STORAGE	LABELING / NFPA
collect leakage in sealable containers, soak up with sand or other inert absorbent and remove to safe place; (additional individual protection: breathing apparatus).	keep in fireproof place.	R: 10

NFPA diamond: top 4, left 1, right 0

NOTES

Alcohol consumption increases toxic effects. White spirit, Shell Sol, solvent naphtha, turps (substitutes), petroleum thinner are trade names. Packaging: special material.

Transport Emergency Card TEC(R)-30G35

HI: 30; UN-number: 1993

CAS-No: [8052-41-3]
petroleum spirits
petroleum thinner
white spirit

PETROLEUM NAPHTHA

(aromatic - initial boiling point > 140°C < 190°C)

PHYSICAL PROPERTIES	IMPORTANT CHARACTERISTICS
Boiling point, °C 140-190 Melting point, °C < −40 Flash point, °C 25-72 Auto-ignition temperature, °C 400-500 Relative density (water = 1) ca. 0.9 Relative vapor density (air = 1) ca. 4.5 Relative density at 20 °C of saturated mixture vapor/air (air = 1) 1.0-1.1 Vapor pressure, mm Hg at 20 °C 2.3-30 Solubility in water none Explosive limits, vol% in air ca. 1-7.5 Electrical conductivity, pS/m 10^3	**COLORLESS LIQUID WITH CHARACTERISTIC ODOR** Mixture of aromatic hydrocarbons with lower boiling point 140-190°C. The lower the boiling point, the higher the vapor pressure at 20°C and the lower the flash point. Vapor mixes readily with air. Flow, agitation etc. can cause build-up of electrostatic charge due to liquid's low conductivity. TLV-TWA not available **Absorption route:** Can enter the body by inhalation or ingestion. Harmful atmospheric concentrations can build up fairly rapidly on evaporation at approx. 20°C - even more rapidly in aerosol form. **Immediate effects:** Irritates the eyes, skin and respiratory tract. Liquid destroys the skin's natural oils. Affects the nervous system. If liquid is swallowed, droplets can enter the lungs, with risk of pneumonia. **Effects of prolonged/repeated exposure:** Prolonged or repeated contact can cause skin disorders.

HAZARDS/SYMPTOMS	PREVENTIVE MEASURES	FIRE EXTINGUISHING/FIRST AID
Fire: flammable.	keep away from open flame and sparks, no smoking.	powder, AFFF, foam, carbon dioxide, (halons).
Explosion: above 25°C: forms explosive air-vapor mixtures.	above 25°C: sealed machinery, ventilation, explosion-proof electrical equipment, grounding.	in case of fire: keep tanks/drums cool by spraying with water.
Inhalation: headache, dizziness, sleepiness.	ventilation, local exhaust or respiratory protection.	fresh air, rest.
Skin: redness.	gloves.	remove contaminated clothing, flush skin with water or shower.
Eyes: redness, pain.	safety glasses.	flush with water, take to a doctor if necessary.
Ingestion: abdominal pain, nausea, severe breathing difficulties.		rinse mouth, DO NOT induce vomiting, call a doctor or take to hospital.

SPILLAGE	STORAGE	LABELING / NFPA
collect leakage in sealable containers, soak up with sand or other inert absorbent and remove to safe place; (additional individual protection: breathing apparatus).	keep in fireproof place.	R: 10

NOTES

Alcohol consumption increases toxic effects. Degree of toxicity depends on composition of mixture. Shell Sol and solvent naphtha are trade names. For data on commercial products see supplier's information. TLV depends on composition. Stoddard solvent (aromatic content < 20%) has TLV-TWA of 100ppm and 525mg/m³. Packaging: special material.

Transport Emergency Card TEC(R)-753

HI: 33; UN-number: 1993

CAS-No: [108-95-2]
carbolic acid
hydroxybenzene
oxybenzene

C_6H_5OH

PHENOL

PHYSICAL PROPERTIES		IMPORTANT CHARACTERISTICS
Boiling point, °C	182	**COLORLESS HYGROSCOPIC CRYSTALS WITH CHARACTERISTIC ODOR, TURNING PINK ON EXPOSURE TO AIR**
Melting point, °C	41	Vapor mixes readily with air. Reacts with oxidants. Attacks aluminum, lead and zinc.
Flash point, °C	78	
Auto-ignition temperature, °C	605	
Relative density (water = 1)	1.1	
Relative vapor density (air = 1)	3.2	TLV-TWA (skin) 5 ppm 19 mg/m³
Relative density at 20 °C of saturated mixture vapor/air (air = 1)	1.00	**Absorption route:** Can enter the body by inhalation or ingestion or - very rapidly - through the skin. Harmful atmospheric concentrations build up fairly slowly on evaporation at 20°C, but
Vapor pressure, mm Hg at 20°C	0.23	harmful concentrations of airborne particles can build up much more rapidly.
Solubility in water, g/100 ml	8	**Immediate effects:** Corrosive to the eyes, skin and respiratory tract. Affects the nervous system.
Explosive limits, vol% in air	1.3-9.5	Inhalation can cause severe breathing difficulties. In serious cases risk of unconsciousness and
Electrical conductivity, pS/m	2x10⁶	death.
Relative molecular mass	94.1	**Effects of prolonged/repeated exposure:** Can cause liver and kidney damage.
Log P octanol/water	1.5	
Gross formula:	C_6H_6O	

HAZARDS/SYMPTOMS	PREVENTIVE MEASURES	FIRE EXTINGUISHING/FIRST AID
Fire: combustible.	keep away from open flame, no smoking.	powder, water spray, foam, carbon dioxide, (halons).
Explosion: above 78°C: forms explosive air-vapor mixtures.	above 78°C: sealed machinery, ventilation.	
	STRICT HYGIENE, KEEP DUST UNDER CONTROL	ALL CASES URGENT - CALL A DOCTOR
Inhalation: *corrosive*; sore throat, cough, shortness of breath, severe breathing difficulties, dizziness, drowsiness, unconsciousness.	local exhaust or respiratory protection.	fresh air, rest, place in half-sitting position, take to hospital.
Skin: *corrosive, is absorbed*; redness, pain, serious burns; see also 'Inhalation'.	gloves, protective clothing.	remove contaminated clothing, flush skin thoroughly with water, take to hospital.
Eyes: *corrosive*; redness, pain, impaired vision.	face shield, or combined eye and respiratory protection.	flush with large quantities of water, take to a doctor.
Ingestion: *corrosive*; abdominal pain, diarrhea, vomiting; see also 'Inhalation'.	do not eat, drink or smoke while working.	rinse mouth, take immediately to hospital.

SPILLAGE	STORAGE	LABELING / NFPA
clean up spillage, carefully collect remainder and remove to safe place; (additional individual protection: respirator with A/P2 filter).	keep separate from oxidants; ventilate at floor level.	R: 24/25-34 S: 2-28-44 ☠ Toxic ◇ 2 / 3 0

NOTES

Often encountered in molten form (approx. 60°C), to permit transportation/processing as liquid: also becomes fluid on addition of 2% water. Wear appropriate protective clothing. Do not take work clothing home. Even in dilution can be absorbed through the skin. Do not touch contaminated portions of clothing or skin with bare hands. Depending on the degree of exposure, regular medical checkups are advisable. After thorough flushing with water, remaining phenol can be removed from skin with mixture of polyethylene glycol and alcohol (70:30), which must be available (with instructions).

Transport Emergency Card TEC(R)-10

HI: 68; UN-number: 1671

o-p-PHENOLSULFONIC ACID

PHYSICAL PROPERTIES		IMPORTANT CHARACTERISTICS	
Melting point, °C	50	**YELLOW HYGROSCOPIC CRYSTALS TURNING BROWN ON EXPOSURE TO AIR**	
Relative density (water = 1)	1.2	Decomposes when heated above 50°, giving off acid fumes. In aqueous solution is a strong acid which reacts violently with bases and is corrosive. Reacts with water and steam, evolving heat. Attacks many materials.	
Solubility in water	∞		
Relative molecular mass	174.2		
		TLV-TWA	not available
		Absorption route: Can enter the body by inhalation or ingestion. Insufficient data on the rate at which harmful concentrations can build up. **Immediate effects:** Corrosive to the eyes, skin and respiratory tract. Inhalation can cause lung edema. In serious cases risk of death.	
Gross formula:	$C_6H_6O_4S$		

HAZARDS/SYMPTOMS	PREVENTIVE MEASURES	FIRE EXTINGUISHING/FIRST AID
Fire: non-combustible.		in case of fire in immediate vicinity: use any extinguishing agent.
Inhalation: *corrosive*; sore throat, cough, severe breathing difficulties.	local exhaust or respiratory protection.	fresh air, rest, place in half-sitting position, take to hospital.
Skin: *corrosive*; redness, pain, serious burns.	gloves, protective clothing.	remove contaminated clothing, flush skin with water or shower, refer to a doctor.
Eyes: *corrosive*; redness, pain, impaired vision.	face shield.	flush with water, take to a doctor.
Ingestion: *corrosive*; sore throat, abdominal pain, diarrhea.	do not eat, drink or smoke while working.	rinse mouth, take immediately to hospital.

SPILLAGE	STORAGE	LABELING / NFPA
clean up spillage, flush away remainder with water, combat fumes with water curtain.	keep dry, separate from bases.	

NOTES

Lung edema symptoms usually develop several hours later and are aggravated by physical exertion: rest and hospitalization essential. As first aid, a doctor or authorized person should consider administering a corticosteroid spray. Applies to anhydrous form; with higher water of crystallization content, higher decomposition temperature.

Transport Emergency Card TEC(R)-80G01

HI: 80; UN-number: 1759

CAS-No: [92-84-2]
dibenzothiazine
dibenzo-1,4-thiazine
phenthiazine
thiodiphenylamine

$C_{12}H_9NS$

PHENOTHIAZINE

PHYSICAL PROPERTIES		IMPORTANT CHARACTERISTICS	
Boiling point (decomposes), °C	371	**YELLOW FLAKES, TURNING BROWN ON EXPOSURE TO AIR**	
Melting point, °C	185	Decomposes when heated to boiling point or on contact with acids, giving off toxic and corrosive vapors ($\rightarrow$ *sulfur dioxide* and $\rightarrow$ *nitrogen oxides*).	
Vapor pressure, mm Hg at 290 °C	40		
Solubility in water	poor		
Relative molecular mass	199.3	TLV-TWA (skin)	5 mg/m³ ✱
Log P octanol/water	4.2		
		Absorption route: Can enter the body by ingestion or through the skin. Evaporation negligible at 20°C, but harmful concentrations of airborne particles can build up rapidly.	
		Immediate effects: Irritates the skin - in some cases with intense itching - and respiratory tract; causes abdominal cramps and rapid rise in pulse rate, muscle spasms and muscle pains. Corrosive to the eyes.	
		Effects of prolonged/repeated exposure: Prolonged or repeated contact can cause skin disorders (photosensitivity). Can affect the blood. Can cause liver and kidney damage.	
Gross formula:	$C_{12}H_9NS$		

HAZARDS/SYMPTOMS	PREVENTIVE MEASURES	FIRE EXTINGUISHING/FIRST AID
Fire: combustible.	keep away from open flame, no smoking.	water spray, powder.
	KEEP DUST UNDER CONTROL	
Inhalation: sore throat, cough, shortness of breath.	local exhaust or respiratory protection.	fresh air, rest, call a doctor.
Skin: *is absorbed*; redness, pain, occasionally burns, itching.	gloves, protective clothing.	remove contaminated clothing, flush skin with water or shower, refer to a doctor.
Eyes: *corrosive*; redness, pain, impaired vision.	face shield, or combined eye and respiratory protection.	flush with water, take to a doctor.
Ingestion: abdominal cramps, vomiting, muscle spasms, muscle ache, cramps.		rinse mouth, call a doctor or take to hospital.

SPILLAGE	STORAGE	LABELING / NFPA
clean up spillage, carefully collect remainder; (additional individual protection: P2 respirator).	keep separate from strong acids.	

NOTES
Depending on the degree of exposure, regular medical checkups are advisable. No general toxicity on moderate excess exposure, but pinkish-red discoloration of hair and brown discoloration of fingernails. Temporary dyeing effect which lasts only as long as exposure. ✱ TLV based principally on skin symptoms.

CAS-No: [106-50-3]
p-aminoaniline
1,4-benzenediamine
1,4-diaminobenzene
PPD

$C_6H_4(NH_2)_2$

p-PHENYLENEDIAMINE

PHYSICAL PROPERTIES		IMPORTANT CHARACTERISTICS
Sublimation temperature, °C	267	**COLORLESS TO LIGHT RED CRYSTALS, DISCOLORING ON EXPOSURE TO AIR**
Melting point, °C	140	Vapor mixes readily with air. Toxic vapors (→ *nitrous vapors*) are a product of combustion.
Flash point, °C	156	Decomposes when heated, giving off → *nitrous vapors*. Strong reducing agent which reacts
Relative density (water = 1)	1.14	violently with oxidants.
Relative vapor density (air = 1)	3.7	
Relative density at 20 °C of saturated mixture		
vapor/air (air = 1)	1.0	
Vapor pressure, mm Hg at 20 °C	0.0076	**Absorption route:** Can enter the body by inhalation or ingestion or through the skin. Harmful
Solubility in water, g/100 ml at 24 °C	3.8	atmospheric concentrations build up very slowly, if at all, on evaporation at approx. 20° C, but
Explosive limits, vol% in air	1.5-?	harmful concentrations of airborne particles can build up much more rapidly.
Relative molecular mass	108.1	**Immediate effects:** Corrosive to the eyes, skin and respiratory tract. Inhalation of vapor/fumes
Log P octanol/water	−0.3	can cause severe breathing difficulties (lung edema).

TLV-TWA (skin) appears above "Absorption route": TLV-TWA (skin)   0.1 mg/m³

Effects of prolonged/repeated exposure: Prolonged or repeated contact can cause eczema. Hypersensitivity can asthma. Can cause liver and kidney damage. Can cause neurological disorders.

Gross formula: $C_6H_8N_2$

HAZARDS/SYMPTOMS	PREVENTIVE MEASURES	FIRE EXTINGUISHING/FIRST AID
Fire: combustible.	keep away from open flame, no smoking.	water spray, powder.
	STRICT HYGIENE, KEEP DUST UNDER CONTROL	
Inhalation: *corrosive*; sore throat, cough, shortness of breath, severe breathing difficulties.	local exhaust or respiratory protection.	fresh air, rest, place in half-sitting position, take to hospital.
Skin: *is absorbed, corrosive*; redness, pain, serious burns.	gloves, protective clothing.	remove contaminated clothing, flush skin with water or shower, refer to a doctor.
Eyes: *corrosive*; redness, pain, impaired vision.	face shield, or combined eye and respiratory protection.	flush with water, take to a doctor.
Ingestion: *corrosive*; diarrhea, abdominal cramps, vomiting.	do not eat, drink or smoke while working.	rinse mouth, immediately take to hospital.

SPILLAGE	STORAGE	LABELING / NFPA
clean up spillage, carefully collect remainder and remove to safe place; (additional individual protection: P3 respirator).	keep separate from oxidants.	R: 23/24/25-43 S: 28-44 Note C ☠ Toxic

NOTES

Depending on the degree of exposure, regular medical checkups are advisable. Lung edema symptoms usually develop several hours later and are aggravated by physical exertion: rest and hospitalization essential. As first aid, a doctor or authorized person should consider administering a corticosteroid spray. The measures on this card also apply to meta- and ortho-phenylenediamine.

Transport Emergency Card TEC(R)-748

HI: 60; UN-number: 1673

CAS-No: [122-98-5]
2-anilinoethanol
N-(2-hydroxyethyl)aniline
N-(2-hydroxyethyl)benzenamine
2-(phenylamino)ethanol

$C_6H_5NHCH_2CH_2OH$

N-PHENYLETHANOLAMINE

PHYSICAL PROPERTIES		IMPORTANT CHARACTERISTICS	
Boiling point, °C	285	**VISCOUS YELLOW TO REDDISH-BROWN LIQUID WITH CHARACTERISTIC ODOR**	
Melting point, °C	−30	Vapor mixes readily with air. Decomposes when heated or burned, giving off toxic and corrosive vapors. Reacts violently with strong oxidants and strong acids.	
Flash point, °C	148		
Auto-ignition temperature, °C	410		
Relative density (water = 1)	1.1		
Relative vapor density (air = 1)	4.7	TLV-TWA	not available
Relative density at 20 °C of saturated mixture vapor/air (air = 1)	10	**Absorption route:** Can enter the body by inhalation or ingestion. Harmful atmospheric concentrations build up very slowly, if at all, on evaporation at approx. 20°C, but much more rapidly in aerosol form.	
Vapor pressure, mm Hg at 20 °C	0.010		
Solubility in water, g/100 ml at 20 °C	4.6		
Relative molecular mass	137.2	**Immediate effects:** Irritates the eyes, skin and respiratory tract.	
Gross formula:	$C_8H_{11}NO$		

HAZARDS/SYMPTOMS	PREVENTIVE MEASURES	FIRE EXTINGUISHING/FIRST AID
Fire: combustible.	keep away from open flame, no smoking.	powder, AFFF, foam, carbon dioxide, (halons).
Inhalation: sore throat, cough, shortness of breath.	ventilation, local exhaust or respiratory protection.	fresh air, rest, call a doctor.
Skin: redness.	gloves.	remove contaminated clothing, flush skin with water or shower.
Eyes: redness, pain.	face shield.	flush with water, take to a doctor if necessary.
Ingestion: abdominal pain, nausea.	do not eat, drink or smoke while working.	rinse mouth, call a doctor.

SPILLAGE	STORAGE	LABELING / NFPA
collect leakage in sealable containers, soak up with sand or other inert absorbent and remove to safe place.	keep separate from oxidants and strong acids; ventilate at floor level.	

NFPA: 1 / 2 / 0

NOTES

ethoxybenzene
ethyl phenyl ether
phenetole
phenoxyethane

PHENYLETHYL ETHER

PHYSICAL PROPERTIES	IMPORTANT CHARACTERISTICS
Boiling point, °C 172 Melting point, °C −30 Flash point, °C 63 Relative density (water = 1) 0.97 Relative vapor density (air = 1) 4.2 Relative density at 20 °C of saturated mixture vapor/air (air = 1) 1.01 Vapor pressure, mm Hg at 20 °C 14 Solubility in water none Electrical conductivity, pS/m 1.7×10^6 Relative molecular mass 122.2	**COLORLESS VISCOUS LIQUID WITH CHARACTERISTIC ODOR** Vapor mixes readily with air. Presumed to be able to form peroxides. Reacts violently with strong oxidants. TLV-TWA not available **Absorption route:** Can enter the body by inhalation or ingestion or through the skin. Insufficient data on the rate at which harmful concentrations can build up. **Immediate effects:** Irritates the eyes, skin and respiratory tract. Liquid destroys the skin's natural oils.
Gross formula: $C_8H_{10}O$	

HAZARDS/SYMPTOMS	PREVENTIVE MEASURES	FIRE EXTINGUISHING/FIRST AID
Fire: combustible.	keep away from open flame, no smoking.	powder, AFFF, foam, carbon dioxide, (halons).
Explosion: above 63°C: forms explosive air-vapor mixtures.	above 63°C: sealed machinery, ventilation.	
Inhalation: sore throat, cough, sleepiness.	ventilation, local exhaust or respiratory protection.	fresh air, rest.
Skin: redness.	gloves.	remove contaminated clothing, flush skin with water or shower.
Eyes: redness, pain.	face shield.	flush with water, take to a doctor if necessary.
Ingestion: abdominal pain, nausea.		rinse mouth, rest, call a doctor.

SPILLAGE	STORAGE	LABELING / NFPA
collect leakage in non-sealable containers, soak up with sand or other inert absorbent and remove to safe place.		 2 0 0

NOTES
Before distilling check for peroxides; if found, render harmless.

Transport Emergency Card TEC(R)-30G15

CAS-No: [122-60-1]
1,2-epoxy-3-phenoxypropane
(2,3-epoxypropoxy)benzene
2,3-epoxypropylphenyl ether
PGE
(phenoxymethyl)oxirane

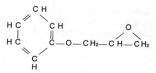

PHENYL GLYCIDYL ETHER

PHYSICAL PROPERTIES		IMPORTANT CHARACTERISTICS		
Boiling point, °C	245	**COLORLESS LIQUID**		
Melting point, °C	3.5	Vapor mixes readily with air. Presumed to be able to form peroxides. Reacts with many compounds.		
Flash point, °C	>80			
Relative density (water = 1)	1.1			
Relative vapor density (air = 1)	5.2	TLV-TWA	1 ppm	6.1 mg/m³
Relative density at 20 °C of saturated mixture vapor/air (air = 1)	1.0	**Absorption route:** Can enter the body by inhalation or ingestion or through the skin. Harmful atmospheric concentrations build up very slowly, if at all, on evaporation at approx. 20°C, but much more rapidly in aerosol form.		
Vapor pressure, mm Hg at 20 °C	9.9x10⁻³	**Immediate effects:** Irritates the eyes, skin and respiratory tract. Affects the nervous system.		
Solubility in water, g/100 ml at 20 °C	0.24	**Effects of prolonged/repeated exposure:** Prolonged or repeated contact can cause eczema.		
Relative molecular mass	150.2			
Gross formula:	$C_9H_{10}O_2$			

HAZARDS/SYMPTOMS	PREVENTIVE MEASURES	FIRE EXTINGUISHING/FIRST AID
Fire: combustible.	keep away from open flame, no smoking.	powder, AFFF, foam, carbon dioxide, (halons).
Explosion: above 80°C: forms explosive air-vapor mixtures.	above 80°C: sealed machinery, ventilation.	
Inhalation: drowsiness, restlessness.	ventilation.	fresh air, rest, call a doctor if necessary.
Skin: *is absorbed*; redness.	gloves.	remove contaminated clothing, flush skin with water or shower, call a doctor.
Eyes: redness, pain.	face shield.	flush with water, take to doctor.
Ingestion: nausea, vomiting, drowsiness, restlessness.		rinse mouth, take immediately to hospital.

SPILLAGE	STORAGE	LABELING / NFPA
collect leakage in sealable containers, soak up with sand or other inert absorbent and remove to safe place.	ventilate at floor level; add an inhibitor.	R: 21-43￼S: 24/25￼❌￼Harmful

NOTES
Before distilling check for peroxides; if found, render harmless. Alcohol consumption increases toxic effects.

α-**aminobenzeneacetic acid**
aminophenyl acetic acid

$C_6H_5CHNH_2COOH$

D(-)-α-PHENYLGLYCINE

PHYSICAL PROPERTIES		IMPORTANT CHARACTERISTICS	
Sublimation temperature, °C	256	**FINE WHITE CRYSTALS**	
Melting point (decomposes), °C	237	Decomposes when heated, giving off toxic vapors (→ *nitrogen oxides*).	
Flash point, °C	> 150		
Solubility in water, g/100 ml	0.3	TLV-TWA	not available
Minimum ignition energy, mJ	< 0.05		
Relative molecular mass	151.2		
		Absorption route: Can enter the body by inhalation or ingestion. Insufficient data on the rate at which harmful concentrations can build up.	
Gross formula:	$C_8H_9NO_2$		

HAZARDS/SYMPTOMS	PREVENTIVE MEASURES	FIRE EXTINGUISHING/FIRST AID
Fire: combustible.	keep away from open flame, no smoking.	water spray, powder.
Explosion: finely dispersed particles form explosive mixtures on contact with air.	non-sparking tools, keep dust from accumulating, sealed machinery, explosion-proof electrical equipment and lighting, grounding.	
Inhalation:	local exhaust or respiratory protection.	fresh air, rest, call a doctor if necessary.
Skin:	protective clothing.	remove contaminated clothing, flush skin with water or shower, call a doctor if necessary.
Eyes:	goggles, or combined eye and respiratory protection.	flush with water, take to a doctor if necessary.
Ingestion:		rinse mouth, call a doctor if necessary.

SPILLAGE	STORAGE	LABELING / NFPA
clean up spillage, carefully collect remainder; (additional individual protection: P2 respirator).		

NOTES
The measures on this card also apply to the other stereoisomers. Insufficient toxicological data on harmful effects to humans.

CAS-No: [39878-87-0]
phenylglycylchloride hydrochloride

$C_6H_5CH(NH_2.HCl)COCl$

α-PHENYLGLYCINE CHLORIDE-HYDROCHLORIDE

PHYSICAL PROPERTIES	IMPORTANT CHARACTERISTICS	
Melting point, °C 300 Solubility in water reaction Relative molecular mass 206.1	**WHITE TO LIGHT CRYSTALLINE POWDER WITH PUNGENT ODOR** Decomposes when heated, giving off toxic and corrosive vapors (→ *hydrochloric acid*). Reacts violently with water and alcohols evolving heat, forming corrosive vapors (→ *hydrochloric acid*). Reacts with air to form corrosive vapors (→ *hydrochloric acid*). Attacks many metals.	
	TLV-TWA not available	
	Absorption route: Can enter the body by inhalation or ingestion. Evaporation negligible at 20°C, but harmful concentrations of airborne particles can build up rapidly. **Immediate effects:** Irritates the eyes, skin and respiratory tract. Inhalation can cause lung edema. **Effects of prolonged/repeated exposure:** Prolonged or repeated contact can cause skin disorders.	
Gross formula: $C_8H_9Cl_2NO$		

HAZARDS/SYMPTOMS	PREVENTIVE MEASURES	FIRE EXTINGUISHING/FIRST AID
Fire: non-combustible.		in case of fire in immediate vicinity: use any extinguishing agent.
	KEEP DUST UNDER CONTROL	
Inhalation: *corrosive*; sore throat, cough, severe breathing difficulties.	local exhaust or respiratory protection.	fresh air, rest, place in half-sitting position, take to hospital.
Skin: *corrosive*; redness, pain, serious burns.	gloves, protective clothing.	remove contaminated clothing, flush skin with water or shower, refer to a doctor.
Eyes: *corrosive*; redness, pain, impaired vision.	face shield, or combined eye and respiratory protection.	flush with water, take to a doctor if necessary.
Ingestion: *corrosive*; sore throat, abdominal pain, diarrhea.	do not eat, drink or smoke while working.	rinse mouth, immediately take to hospital.

SPILLAGE	STORAGE	LABELING / NFPA
clean up spillage, carefully collect remainder; (additional individual protection: CHEMICAL SUIT).	keep in dry place, separate from alcohols.	

NOTES
Lung edema symptoms usually develop several hours later and are aggravated by physical exertion: rest and hospitalization essential. As first aid, a doctor or authorized person should consider administering a corticosteroid spray. The measures on this card also apply to the other stereoisomers. Use airtight packaging.

Transport Emergency Card TEC(R)-80G11

PHENYLHYDRAZINE

PHYSICAL PROPERTIES		IMPORTANT CHARACTERISTICS
Boiling point (decomposes), °C	243	**COLORLESS LIQUID OR CRYSTALS WITH CHARACTERISTIC ODOR, TURNING DARK ON EXPOSURE TO AIR**
Melting point, °C	20	Vapor mixes readily with air. Decomposes when heated or burned, giving off toxic vapors (→ *nitrous vapors*). Strong reducing agent which reacts violently with oxidants and lead tetroxide. Can ignite spontaneously in contact with cotton or sand. Reacts violently with methyl iodide and many other organic compounds, with risk of fire and explosion.
Flash point, °C	89	
Auto-ignition temperature, °C	174	
Relative density (water = 1)	1.1	
Relative vapor density (air = 1)	3.7	
Relative density at 20 °C of saturated mixture vapor/air (air = 1)	1.0	

		TLV-TWA (skin)	5 ppm	22 mg/m³ A2
Vapor pressure, mm Hg at 20 °C	0.076	STEL	10 ppm	44 mg/m³

PHYSICAL PROPERTIES (cont.)		IMPORTANT CHARACTERISTICS (cont.)
Solubility in water	moderate	**Absorption route:** Can enter the body by inhalation or ingestion or through the skin. Harmful atmospheric concentrations build up very slowly, if at all, on evaporation at approx. 20°C, but much more rapidly in aerosol form.
Relative molecular mass	108.1	**Immediate effects:** Inhalation of vapor/fumes can cause severe breathing difficulties (lung edema). In serious cases risk of death.
Log P octanol/water	1.3	**Effects of prolonged/repeated exposure:** Prolonged or repeated contact can cause eczema. Can cause liver and kidney damage. Has been found to cause a type of lung cancer in certain animal species under certain circumstances.
Gross formula:	$C_6H_8N_2$	

HAZARDS/SYMPTOMS	PREVENTIVE MEASURES	FIRE EXTINGUISHING/FIRST AID
Fire: combustible, many chemical reactions can cause fire and explosion.	keep away from open flame, no smoking, avoid contact with combustible substances and hot surfaces (steam lines).	powder, AFFF, foam, carbon dioxide, (halons).
Explosion: above 89°C: forms explosive air-vapor mixtures.	above 89°C: sealed machinery, ventilation.	
	STRICT HYGIENE	
Inhalation: *corrosive*; sore throat, cough, severe breathing difficulties, dizziness, blue skin, feeling of weakness.	ventilation.	fresh air, rest, place in half-sitting position, take to hospital.
Skin: *corrosive, is absorbed*; redness, pain, serious burns, dizziness, feeling of weakness.	gloves, protective clothing.	remove contaminated clothing, flush skin with water or shower, call a doctor if necessary.
Eyes: *corrosive*; redness, pain, impaired vision.	face shield, or combined eye and respiratory protection.	flush with water, take to a doctor.
Ingestion: *corrosive*; abdominal pain, dizziness, jaundice, vomiting, feeling of weakness.	do not eat, drink or smoke while working.	rinse mouth, take immediately to hospital.

SPILLAGE	STORAGE	LABELING / NFPA
collect leakage in sealable containers, clean up spillage, carefully collect remainder, do not soak up sand, sawdust etc.; (additional individual protection: breathing apparatus).	keep separate from oxidants, ventilate at floor level; add an inhibitor.	R: 23/24/25-36 S: 28-44 Toxic — NFPA diamond: 2 (top), 3 (left), 0 (right)

NOTES
Flush contaminated clothing with water (fire hazard). Depending on the degree of exposure, regular medical checkups are advisable. Lung edema symptoms usually develop several hours later and are aggravated by physical exertion: rest and hospitalization essential. As first aid, a doctor or authorized person should consider administering a corticosteroid spray. Contaminated clothing can ignite spontaneously when drying. Damages red blood cells, destroying hemoglobin and causing anemia (hemolytic anemia). Urine is dark, due to decomposition products of blood pigment.

Transport Emergency Card TEC(R)-61G06B

HI: 60; UN-number: 2572

CAS-No: [103-71-9]
carbanil
isocyanotobenzene
isocyanic acid, phenyl ester
phenylcarbimide
phenylcarbonimide

C_6H_5NCO

PHENYL ISOCYANATE

PHYSICAL PROPERTIES		IMPORTANT CHARACTERISTICS
Boiling point, °C	166	**COLORLESS LIQUID WITH PUNGENT ODOR**
Melting point, °C	−33	Vapor mixes readily with air. Decomposes when heated above boiling point or burned, giving off
Flash point, °C	51	toxic gases (→ *nitrous vapors*, → *carbon monoxide*). Reacts slowly with cold water and violently
Auto-ignition temperature, °C	300	with warm water, evolving heat and giving off → *carbon dioxide*. Reacts violently with alcohols,
Relative density (water = 1)	1.1	amines, strong acids and bases and strong oxidants. Attacks steel, copper and copper alloys.
Relative vapor density (air = 1)	4.1	
Vapor pressure, mm Hg at 20 °C	1.5	TLV-TWA not available
Solubility in water	reaction	
Explosive limits, vol% in air	?-?	**Absorption route:** Can enter the body by inhalation or ingestion or through the skin. Harmful
Relative molecular mass	119.1	atmospheric concentrations can build up very rapidly on evaporation at 20°C.
		Immediate effects: Corrosive to the eyes, skin and respiratory tract. Inhalation can cause severe
		breathing difficulties (asthma).
		Effects of prolonged/repeated exposure: Prolonged or repeated contact can cause skin
		disorders (eczema). Prolonged exposure to vapor/fumes can cause lung disorders.
Gross formula:	C_7H_5NO	

HAZARDS/SYMPTOMS	PREVENTIVE MEASURES	FIRE EXTINGUISHING/FIRST AID
Fire: flammable, many chemical reactions can cause fire and explosion.	keep away from open flame and sparks, no smoking, avoid contact with acids, bases, alcohols, amines and strong oxidants.	powder, carbon dioxide, (halons), DO NOT USE WATER-BASED EXTINGUISHERS.
Explosion: above 51°C: forms explosive air-vapor mixtures.	above 51°C: sealed machinery, ventilation, explosion-proof electrical equipment.	in case of fire: keep tanks/drums cool by spraying with water, but DO NOT spray substance with water.
	STRICT HYGIENE	IN ALL CASES CALL IN A DOCTOR
Inhalation: *corrosive*; sore throat, cough, shortness of breath, severe breathing difficulties.	ventilation, local exhaust or respiratory protection.	fresh air, rest, artificial respiration if necessary, take to hospital.
Skin: *corrosive, is absorbed*; redness, pain, burns.	gloves, protective clothing.	remove contaminated clothing, flush skin with water or shower, call a doctor if necessary.
Eyes: *corrosive*; redness, pain, impaired vision.	face shield.	flush with water, take to a doctor.
Ingestion: *corrosive*; diarrhea, abdominal cramps, nausea, vomiting.		rinse mouth, immediately take to hospital.

SPILLAGE	STORAGE	LABELING / NFPA
collect leakage in sealable containers, soak up with sand or other inert absorbent and remove to safe place, render harmless with dilute spirits of ammonia (50% water, 45% ethyl alcohol, 5% ammonia); (additional individual protection: breathing apparatus).	keep in dry, fireproof place, separate from oxidants, strong acids and strong bases.	

NOTES
Explosive limits not given in the literature. Reacts violently with water-based extinguishing agents. Depending on the degree of exposure, regular medical checkups are advisable. As first aid, a doctor or authorized person should consider administering a corticosteroid spray. Asthma symptoms usually develop several hours later and are aggravated by physical exertion: rest and hospitalization essential. Persons with history of asthma symptoms should under no circumstances come in contact again with substance. Packaging: special material.

Transport Emergency Card TEC(R)-626

HI: 63; UN-number: 2487

PHENYL (II) MERCURYBORATE

PHYSICAL PROPERTIES	IMPORTANT CHARACTERISTICS	
Boiling point (decomposes), °C .... ? Melting point, °C .... 112 Relative density (water = 1) .... > 1 Solubility in water .... very good Relative molecular mass 338.5	**WHITE ODORLESS POWDER** In dry state can form electrostatic charge if stirred, transported pneumatically, poured etc. Decomposes when heated or burned, giving off toxic vapors (incl. → *mercury*).	
	TLV-TWA (skin) STEL	0.01 mg/m³ ✱ 0.03 mg/m³
	Absorption route: Can enter the body by inhalation or ingestion or through the skin. Evaporation negligible at 20° C, but harmful atmospheric concentrations can build up rapidly in aerosol form. **Immediate effects:** Corrosive to the eyes, skin and respiratory tracts. In substantial concentrations can cause agitation, tingling in limbs, paralysis of the limbs. **Effects of prolonged/repeated exposure:** Prolonged or repeated contact can cause skin disorders (eczema).	
Gross formula: $C_6H_7BHgO_3$		

HAZARDS/SYMPTOMS	PREVENTIVE MEASURES	FIRE EXTINGUISHING/FIRST AID
Fire: combustible.	keep away from open flame, no smoking.	water spray, powder.
Explosion: finely dispersed particles form explosive mixtures on contact with air.	keep dust from accumulating; sealed machinery, explosion-proof electrical equipment and lighting, grounding.	
	KEEP DUST UNDER CONTROL	
Inhalation: *corrosive*; sore throat, cough, shortness of breath, nausea.	local exhaust or respiratory protection.	fresh air, rest, place in half-sitting position, call a doctor.
Skin: *corrosive, is absorbed*; redness, pain, serious burns.	gloves.	remove contaminated clothing, wash skin with soap and water, call a doctor if necessary.
Eyes: *corrosive*; redness, pain.	goggles, or combined eye and respiratory protection.	flush with water, take to a doctor.
Ingestion: *corrosive*; abdominal pain, diarrhea, vomiting.	do not eat, drink or smoke while working.	rinse mouth, take immediately to hospital.

SPILLAGE	STORAGE	LABELING / NFPA
dampen substance with water and clean up spillage, carefully collect remainder; (additional individual protection: P3 respirator).	keep in fireproof place.	R: 26/27/28-33 S: 2-13-28-36-45 Note A [☠ symbol] Toxic

NOTES
✱ TLV equals that of mercury. Do not take work clothing home. Unbreakable packaging preferred; if breakable, keep in unbreakable container.

N-PHENYL-1-NAPHTHYLAMINE

PHYSICAL PROPERTIES		IMPORTANT CHARACTERISTICS
Melting point, °C	62	**CRYSTALLINE WHITISH-YELLOW POWDER**
Flash point, °C	?	→ *Nitrous vapors* are a product of combustion. Decomposes when heated or on contact with oxidants, giving off toxic gases.
Relative density (water = 1)	1.2	
Vapor pressure, mm Hg at 335 °C	266	
Solubility in water	poor	TLV-TWA not available
Relative molecular mass	219.3	

Absorption route: Can enter the body by inhalation or ingestion or through the skin. Evaporation negligible at 20°C, but harmful atmospheric concentrations can build up rapidly in aerosol form.
Immediate effects: Irritates the eyes, skin and respiratory tract. In substantial concentrations can impair consciousness.
Effects of prolonged/repeated exposure: Can affect the blood. Can cause liver and kidney damage. Prolonged or repeated contact can cause skin disorders.

Gross formula: $C_{16}H_{13}N$

HAZARDS/SYMPTOMS	PREVENTIVE MEASURES	FIRE EXTINGUISHING/FIRST AID
Fire: combustible.	keep away from open flame, no smoking.	water spray, powder.
	STRICT HYGIENE, KEEP DUST UNDER CONTROL	
Inhalation: sore throat, shortness of breath, headache, dizziness, nausea, unconsciousness, blue skin, feeling of weakness.	local exhaust or respiratory protection.	fresh air, rest, call a doctor.
Skin: *is absorbed*; redness, pain; see also 'Inhalation'.	gloves, protective clothing.	remove contaminated clothing, wash skin with soap and water, call a doctor if necessary.
Eyes: redness, pain.	goggles.	flush with water, take to a doctor if necessary.
Ingestion: abdominal pain; see also 'Inhalation'.	do not eat, drink or smoke while working.	rinse mouth, take immediately to hospital.

SPILLAGE	STORAGE	LABELING / NFPA
clean up spillage, carefully collect remainder; (additional individual protection: P2 respirator).	keep separate from oxidants.	

NOTES
Combustible, but flash point not given in the literature. Alcohol consumption increases toxic effects. Depending on the degree of exposure, regular medical checkups are advisable. Effects on blood due to formation of methemoglobin. Special first aid required in the event of poisoning: antidotes must be available (with instructions).

CAS-No: [135-88-6]
2-anilinonaphthalene
N-(2-naphthyl)aniline
β-naphthylphenylamine
PBNA

$C_{10}H_7NHC_6H_5$

N-PHENYL-2-NAPHTHYLAMINE

PHYSICAL PROPERTIES	IMPORTANT CHARACTERISTICS	
Boiling point, °C　　　　　　　　395	**COLORLESS POWDER, TURNING GRAY, PINK IN LIGHT**	
Melting point, °C　　　　　　　　108	In dry state can cause electrostatic charge if stirred, transported pneumatically, poured etc.	
Flash point, °C　　　　　　see notes	Toxic → *nitrous vapors* are a product of combustion. Reacts violently with strong oxidants.	
Relative density (water = 1)　　　1.2		
Vapor pressure, mm Hg at 235 °C　15	TLV-TWA	A2
Solubility in water　　　　　　none		
Relative molecular mass　　　219.3	**Absorption route:** Can enter the body by inhalation or ingestion or through the skin. Evaporation negligible at 20° C, but harmful atmospheric concentrations can build up rapidly in aerosol form. **Immediate effects:** Irritates the eyes, skin and respiratory tract. Affects the nervous system. **Effects of prolonged/repeated exposure:** Prolonged or repeated contact can cause skin disorders. Can affect the blood. Can cause liver and kidney damage.	
Gross formula:　　　　$C_{16}H_{13}N$		

HAZARDS/SYMPTOMS	PREVENTIVE MEASURES	FIRE EXTINGUISHING/FIRST AID
Fire: combustible.	keep away from open flame, no smoking.	water spray, powder.
Explosion: finely dispersed particles can explode on contact with air.	keep dust from accumulating; sealed machinery, explosion-proof equipment and lighting, grounding.	in case of fire: keep tanks/drums cool by spraying with water.
	STRICT HYGIENE, KEEP DUST UNDER CONTROL	
Inhalation: sore throat, shortness of breath, headache, dizziness, nausea, unconsciousness, blue skin, feeling of weakness.	local exhaust or respiratory protection.	fresh air, rest, call a doctor.
Skin: *is absorbed*; redness, pain; see also 'Inhalation'.	gloves, protective clothing.	remove contaminated clothing, wash skin with soap and water, call a doctor if necessary.
Eyes: redness, pain.	goggles, or combined eye and respiratory protection.	flush with water, take to a doctor if necessary.
Ingestion: abdominal pain; see also 'Inhalation'.	do not eat, drink or smoke while working.	rinse mouth, take immediately to hospital.

SPILLAGE	STORAGE	LABELING / NFPA
clean up spillage, carefully collect remainder; (additional individual protection: P3 respirator).	keep separate from oxidants.	

NOTES
Combustible, but flash point not given in the literature. Alcohol consumption increases toxic effects. Depending on the degree of exposure, regular medical checkups are advisable. Effects on blood due to formation of methemoglobin. Special first aid required in the event of poisoning: antidotes must be available (with instructions).

CAS-No: [504-20-1]
2,6-dimethyl-2,5-heptadien-4-one
diisopropylidene acetone

$(CH_3)_2CCHCOCHC(CH_3)_2$

PHORONE

PHYSICAL PROPERTIES		IMPORTANT CHARACTERISTICS	
Boiling point, °C	198	**YELLOWISH-GREEN LIQUID OR CRYSTALS WITH CHARACTERISTIC ODOR**	
Melting point, °C	28	Vapor mixes readily with air. Reacts with acids, forming → *acetone*. Reacts with strong oxidants.	
Flash point, °C	85		
Relative density (water = 1)	0.9	**TLV-TWA** not available	
Relative vapor density (air = 1)	4.8		
Relative density at 20 °C of saturated mixture vapor/air (air = 1)	1.0	**Absorption route:** Can enter the body by inhalation or ingestion. Harmful atmospheric concentrations build up fairly slowly on evaporation at 20°C, but much more rapidly in aerosol form.	
Vapor pressure, mm Hg at 20 °C	0.38	**Immediate effects:** Irritates the eyes, skin and respiratory tract.	
Solubility in water	poor		
Explosive limits, vol% in air	0.8-3.8		
Relative molecular mass	138.2		
Gross formula:	$C_9H_{14}O$		

HAZARDS/SYMPTOMS	PREVENTIVE MEASURES	FIRE EXTINGUISHING/FIRST AID
Fire: combustible.	keep away from open flame, no smoking.	powder, AFFF, foam, carbon dioxide, (halons).
Explosion: above 85°C: forms explosive air-vapor mixtures.	above 85°C: sealed machinery, ventilation.	
Inhalation: sore throat, cough, shortness of breath.	ventilation, local exhaust or respiratory protection.	fresh air, rest, call a doctor.
Skin: redness, pain.	gloves.	remove contaminated clothing, flush skin with water or shower.
Eyes: redness, pain.	goggles.	flush with water, take to a doctor if necessary.
Ingestion: abdominal pain, nausea.		rinse mouth, call a doctor.

SPILLAGE	STORAGE	LABELING / NFPA
collect leakage in sealable containers, soak up with sand or other inert absorbent and remove to safe place; (additional individual protection: respirator with A/P2 filter).	keep separate from oxidants and strong acids; ventilate at floor level.	NFPA diamond: 2 (top), 2 (left), 0 (right)

NOTES
Insufficient toxicological data on harmful effects to humans.

CAS-No: [75-44-5]
carbonic dichloride
carbon oxychloride
carbonyl chloride
chloroformyl chloride

$COCl_2$

PHOSGENE
(cylinder)

PHYSICAL PROPERTIES		IMPORTANT CHARACTERISTICS	
Boiling point, °C	8.2	**COLORLESS COMPRESSED LIQUEFIED GAS WITH CHARACTERISTIC ODOR**	
Melting point, °C	− 118	Gas is heavier than air. Decomposes when heated above 300°C, giving off → *carbon monoxide*	
Relative density (water = 1)	1.4	and → *chlorine* gas. Reacts violently with anhydrous ammonia, strong oxidants and many other	
Relative vapor density (air = 1)	3.4	substances. Reacts slowly with water, forming → *hydrochloric acid* and → *carbon dioxide*.	
Vapor pressure, mm Hg at 20 °C	1178	Attacks many metals in moist environment.	
Solubility in water	reaction		
Relative molecular mass	98.9	TLV-TWA	0.1 ppm	0.40 mg/m³

Absorption route: Can enter the body by inhalation.
Immediate effects: Corrosive to the respiratory tract. Liquid can cause frostbite due to rapid evaporation. Inhalation can cause lung edema. In serious cases risk of death.

Gross formula: CCl_2O

HAZARDS/SYMPTOMS	PREVENTIVE MEASURES	FIRE EXTINGUISHING/FIRST AID
Fire: non-combustible.		
Explosion:		in case of fire: keep cylinders cool by spraying with water.
	STRICT HYGIENE	
Inhalation: *corrosive*; sore throat, cough, shortness of breath, severe breathing difficulties.	local exhaust or respiratory protection.	fresh air, rest, place in half-sitting position, take to hospital.
Skin: *in case of frostbite*: redness, pain, burns.	gloves, protective clothing.	*in case of frostbite*: DO NOT remove clothing, flush skin with water or shower, take to a doctor.
Eyes: *in case of frostbite*: redness, pain, impaired vision.	face shield, or combined eye and respiratory protection.	flush with water, take to a doctor.

SPILLAGE	STORAGE	LABELING / NFPA
evacuate area, call in an expert, ventilate, combat gas cloud with water curtain; (additional individual protection: CHEMICAL SUIT).	keep separate from all other substances.	0 / 4 / 1

NOTES

Odor limit is above TLV. Lung edema symptoms usually develop several hours later and are aggravated by physical exertion: rest and hospitalization essential. As first aid, a doctor or authorized person should consider administering a corticosteroid spray.

Transport Emergency Card TEC(R)-107

HI: 266; UN-number: 1076

PHOSPHINE
(cylinder)

PHYSICAL PROPERTIES	IMPORTANT CHARACTERISTICS
Boiling point, °C —88 Melting point, °C —133 Flash point, °C flammable gas Auto-ignition temperature, °C 100 Relative density (water = 1) 0.8 Relative vapor density (air = 1) 1.2 Vapor pressure, mm Hg at 20°C 31,844 Solubility in water, g/100 ml at 17°C 26 Explosive limits, vol% in air 1.8-? Relative molecular mass 34.0	**COLORLESS COMPRESSED LIQUEFIED GAS WITH CHARACTERISTIC ODOR** Gas is heavier than air and spreads at ground level, with risk of ignition at a distance. Vapors of → phosphoric acid are a product of combustion. Technical grades can ignite spontaneously on exposure to air when contaminated with diphosphine (P$_2$H$_4$), but not until 100°C when uncontaminated. Decomposes (with atmospheric oxygen excluded) when heated above 375°C, forming → phosphorus and → hydrogen, with risk of fire and explosion. Strong reducing agent which reacts violently with oxidants. Reacts violently with amines. Reacts spontaneously with oxidants such as chlorine. Reacts explosively with pure oxygen.

TLV-TWA	0.3 ppm	0.42 mg/m³	
STEL	1 ppm	1.4 mg/m³	

Absorption route: Can enter the body by inhalation. Harmful atmospheric concentrations can build up very rapidly if gas is released.
Immediate effects: Irritates the eyes, skin and respiratory tract. Can cause frostbite due to rapid evaporation. Inhalation can cause lung edema. Can affect the liver, heart and nervous system. Can cause death.

Gross formula: H$_3$P

HAZARDS/SYMPTOMS	PREVENTIVE MEASURES	FIRE EXTINGUISHING/FIRST AID
Fire: extremely flammable.	keep away from open flame and sparks, no smoking, avoid contact with hot surfaces (e.g. steam lines).	shut off supply; if impossible and no danger to surrounding area, allow fire to burn itself out; otherwise extinguish with powder, carbon dioxide, (halons).
Explosion: forms explosive air-gas mixtures.	sealed machinery, ventilation, explosion-proof electrical equipment and lighting, grounding.	in case of fire: keep cylinders cool by spraying with water, fight fire from sheltered location.
	STRICT HYGIENE	IN ALL CASES CALL IN A DOCTOR
Inhalation: headache, nausea, chest pain, drowsiness.	ventilation, local exhaust or respiratory protection.	fresh air, rest, place in half-sitting position, take to hospital.
Skin: in case of frostbite: redness, pain.	insulated gloves, protective clothing.	in case of frostbite: DO NOT remove clothing, flush skin with water or shower, refer to a doctor.
Eyes: in case of frostbite: redness, pain.	face shield, or combined eye and respiratory protection.	flush with water, take to a doctor.

SPILLAGE	STORAGE	LABELING / NFPA
evacuate area, call in an expert, render harmless with hypochlorite solution; (additional individual protection: CHEMICAL SUIT).	keep in fireproof place.	 3 4 2

NOTES
Clean lines after use with nitrogen. Lung edema symptoms usually develop several hours later and are aggravated by physical exertion: rest and hospitalization essential. As first aid, a doctor or authorized person should consider administering a corticosteroid spray. Turn leaking cylinder so that leak is on top to prevent liquid phosphine escaping. Relative density (water = 1) is at −90°C (liquid).

Transport Emergency Card TEC(R)-20G19

HI: 236; UN-number: 2199

PHOSPHORIC ACID
(25-85%)

PHYSICAL PROPERTIES		IMPORTANT CHARACTERISTICS	
Melting point, °C	11	**COLORLESS LIQUID**	
Relative density (water = 1)	1.2-1.7	Strong acid which reacts violently with bases and is corrosive. Attacks many metals, giving off flammable gas (→ *hydrogen*).	
Solubility in water	∞		
Electrical conductivity, pS/m	1.4×10^{21}		
Relative molecular mass	98		
		TLV-TWA	1 mg/m³
		STEL	3 mg/m³
		Absorption route: Can enter the body by inhalation or ingestion. Harmful atmospheric concentrations build up very slowly, if at all, on evaporation at approx. 20°C, but much more rapidly in aerosol form.	
		Immediate effects: Corrosive to the eyes, skin and respiratory tract. Inhalation of vapor/fumes can cause severe breathing difficulties (lung edema). In serious cases risk of death.	
Gross formula:	H_3O_4P		

HAZARDS/SYMPTOMS	PREVENTIVE MEASURES	FIRE EXTINGUISHING/FIRST AID
Fire: non-combustible.		in case of fire in immediate vicinity: use any extinguishing agent.
Inhalation: *corrosive*; sore throat, cough, shortness of breath, severe breathing difficulties.	local exhaust or respiratory protection.	fresh air, rest, place in half-sitting position, take to hospital.
Skin: *corrosive*; redness, pain, burns.	gloves, protective clothing.	remove contaminated clothing, flush skin with water or shower, take to a doctor.
Eyes: *corrosive*; redness, pain, impaired vision.	face shield.	flush with water, take to a doctor.
Ingestion: *corrosive*; sore throat, abdominal pain, nausea.		rinse mouth, immediately take to hospital.

SPILLAGE	STORAGE	LABELING / NFPA
collect leakage in sealable containers, flush away remainder with water.	keep separate from strong bases.	R: 34 S: 26 Note B Corrosive

NOTES

Crystallization temperature at 85% concentration: + 21°C; at 75% concentration: − 13°C. When kept above 75% the liquid is viscous. Lung edema symptoms usually develop several hours later and are aggravated by physical exertion: rest and hospitalization essential. As first aid, a doctor or authorized person should consider administering a corticosteroid spray.

Transport Emergency Card TEC(R)-82

HI: 80; UN-number: 1805

PHOSPHORIC ACID
(100%)

PHYSICAL PROPERTIES	IMPORTANT CHARACTERISTICS
Decomposes below boiling point, °C ca. 200 Melting point, °C 42 Relative density (water = 1) 1.9 Relative vapor density (air = 1) 3.4 Relative density at 20 °C of saturated mixture vapor/air (air = 1) 1.00 Vapor pressure, mm Hg at 20 °C 0.023 Solubility in water, g/100 ml 548 Relative molecular mass 98	**COLORLESS HYGROSCOPIC CRYSTALS** Vapor mixes readily with air. Decomposes when heated above approx. 200°C, forming pyrophosphorus and then toxic vapor (→ *phosphoric oxides*). In aqueous solution is a strong acid which reacts violently with bases and is corrosive. Attacks many metals, giving off flammable gas (→ *hydrogen*). Above 200°C also attacks earthenware and glass.

	TLV-TWA	1 mg/m³
	STEL	3 mg/m³

Absorption route: Can enter the body by inhalation or ingestion. Harmful atmospheric concentrations build up fairly slowly on evaporation at 20°C, but much more rapidly if vaporization occurs.
Immediate effects: Corrosive to the eyes, skin and respiratory tract. Inhalation of vapor/fumes can cause severe breathing difficulties (lung edema). In serious cases risk of death.

Gross formula: H_3O_4P

HAZARDS/SYMPTOMS	PREVENTIVE MEASURES	FIRE EXTINGUISHING/FIRST AID
Fire: non-combustible.		in case of fire in immediate vicinity: use any extinguishing agent.
	KEEP DUST UNDER CONTROL	
Inhalation: *corrosive*; sore throat, cough, severe breathing difficulties.	local exhaust or respiratory protection.	fresh air, rest, take to hospital.
Skin: *corrosive*; redness, pain, serious burns.	gloves, protective clothing.	remove contaminated clothing, flush skin with water or shower, take to a doctor.
Eyes: *corrosive*; redness, pain, impaired vision.	face shield.	flush with water, take to a doctor.
Ingestion: *corrosive*; sore throat, abdominal pain, nausea.		rinse mouth, call a doctor or take to hospital.

SPILLAGE	STORAGE	LABELING / NFPA
clean up spillage, flush away remainder with water; (additional individual protection: P2 respirator).	keep dry, separate from strong bases.	R: 34 S: 26 Note B Corrosive

NOTES

Lung edema symptoms usually develop several hours later and are aggravated by physical exertion: rest and hospitalization essential. As first aid, a doctor or authorized person should consider administering a corticosteroid spray. Does not attack stainless steel, copper, bronze or brass at room temperature. Solution is generally used. UNDER NO CIRCUMSTANCES add water to acid; when diluting ALWAYS add acid to water. Unbreakable packaging preferred; if breakable, keep in unbreakable container.

Transport Emergency Card TEC(R)-82

HI: 80; UN-number: 2834

PHOSPHORUS
(red)

PHYSICAL PROPERTIES	IMPORTANT CHARACTERISTICS
Sublimation temperature, °C 416 Melting point, °C 590 Auto-ignition temperature, °C 260 Relative density (water = 1) 2.3 Relative vapor density (air = 1) 4.3 Solubility in water none	**BROWNISH-RED SOLID IN VARIOUS FORMS** Cooling from vapor or liquid phase produces → *white phosphorus*. Corrosive gases are a product of combustion (e.g. → *phosphorus pentoxide*). Reacts with oxidants to form shock-sensitive mixtures.
	TLV-TWA not available
	Absorption route: Not harmful unless traces of → *white phosphorus* are present. Evaporation negligible at 20°C, but unpleasant concentrations of airborne particles can build up. **Immediate effects:** Irritates the eyes. Smoke from combustion is corrosive to the eyes, skin and respiratory organs, with risk of death from lung edema.
Gross formula: P_n	

HAZARDS/SYMPTOMS	PREVENTIVE MEASURES	FIRE EXTINGUISHING/FIRST AID
Fire: combustible, many chemical reactions can cause fire and explosion.	keep away from open flame and sparks, no smoking, avoid contact with oxidants.	large quantities of water.
Explosion: formation of shock-sensitive compounds with oxidants can cause explosion.	keep dust from accumulating, sealed machinery, explosion-proof electrical equipment and lighting, grounding.	
Inhalation:	ventilation.	
Skin:	gloves.	
Eyes: redness.	goggles.	flush with water, take to doctor if necessary.

SPILLAGE	STORAGE	LABELING / NFPA
clean up spillage; (additional individual protection: P2 respirator).	keep in cool, dry, fireproof place, separate from oxidants; under inert gas if packaging opened.	R: 11-16 S: 7-43 🔥 Flammable 1 / 1 / 1

NOTES
Black phosphorus is a graphite-like allotrope obtained under high pressure from white phosphorus. Above 550°C black phosphorus turns into red phosphorus. Melting point determined at pressure of 33mm Hg (triple point). Lung edema symptoms usually develop several hours later and are aggravated by physical exertion: rest and hospitalization essential. As first aid, a doctor or authorized person should consider administering a corticosteroid spray.

Transport Emergency Card TEC(R)-794

HI: 40; UN-number: 1338

PHOSPHORUS
(white)

PHYSICAL PROPERTIES	IMPORTANT CHARACTERISTICS
Boiling point, °C 280 Melting point, °C 44 Auto-ignition temperature, °C 30 Relative density (water = 1) 1.8 Relative vapor density (air = 1) 4.3 Relative density at 20 °C of saturated mixture vapor/air (air = 1) 1.0 Vapor pressure, mm Hg at 20 °C <0.076 Solubility in water none Electrical conductivity, pS/m 4 x 10^7 Relative molecular mass 123.9	**LIGHT YELLOW WAXY SOLID IN VARIOUS FORMS** Can polymerize to form → *red phosphorus* when heated above 250° C. Can ignite spontaneously on exposure to air, giving off corrosive clouds of fumes (incl. → *phosphorus pentoxide*). Kept under water to prevent spontaneous combustion. Strong reducing agent which reacts violently with oxidants. Reacts violently with halogens, sulfur and other substances, with risk of fire and explosion. Reacts with strong lye, giving off toxic gas (→ *phosphine*).
	TLV-TWA 0.1 mg/m³
Gross formula: P$_4$	**Absorption route:** Can enter the body by inhalation or ingestion or through the skin. Vapor is harmful to health; harmful concentrations do not build up if substance is kept under water. **Immediate effects:** Corrosive to the eyes, skin and respiratory tract (also applies to fumes from combustion). Inhalation can cause lung edema. In serious cases risk of death. Contact with the skin can cause deep, slow-healing burns due to ignition of substance. **Effects of prolonged/repeated exposure:** Repeated exposure to vapor can cause jaw disorders. Prolonged or repeated contact can cause eczema. Can cause liver, kidney and bone damage.

HAZARDS/SYMPTOMS	PREVENTIVE MEASURES	FIRE EXTINGUISHING/FIRST AID
Fire: combusts spontaneously on contact with air.	avoid contact with air; keep away from open flame and sparks, no smoking.	large quantities of water, water spray, sand.
	STRICT HYGIENE	
Inhalation: *corrosive*; sore throat, cough, severe breathing difficulties, nausea, vomiting, diarrhea, abdominal pain, unconsciousness.	ventilation, local exhaust or respiratory protection.	fresh air, rest, place in half-sitting position, take to hospital.
Skin: *corrosive*; redness, pain, serious burns.	gloves, protective clothing.	flush and only then remove clothing, flush skin with water or shower, take to hospital.
Eyes: *corrosive*; redness, pain, impaired vision.	face shield, or combined eye and respiratory protection.	flush with water, take to a doctor.
Ingestion: *corrosive*; sore throat, abdominal pain, diarrhea, nausea, vomiting, unconsciousness.	do not eat, drink or smoke while working.	rinse mouth, DO NOT induce vomiting, take immediately to hospital.

SPILLAGE	STORAGE	LABELING / NFPA
evacuate area, call in an expert, render small quantities harmless with copper sulfate solution, keep spillage damp (with water) and transfer to drum, do not flush into sewer; (additional individual protection: breathing apparatus).	keep in cool, dark, fireproof place under water, separate from oxidants, halogens and sulfur.	R: 17-26/28-35 S: 5-26-28-45 Flammable Toxic 3 4 2

NOTES
Black phosphorus is a graphite-like allotrope obtained under high pressure from white phosphorus. Above 550° C black phosphorus turns into red phosphorus. Flush contaminated clothing with water (fire hazard). Depending on the degree of exposure, regular medical checkups are advisable. Lung edema symptoms usually develop several hours later and are aggravated by physical exertion: rest and hospitalization essential. As first aid, a doctor or authorized person should consider administering a corticosteroid spray. Use airtight packaging. Unbreakable packaging preferred; if breakable, keep in unbreakable container.

Transport Emergency Card TEC(R)-714

HI: 46; UN-number: 1381

PHOSPHORUS OXYCHLORIDE

PHYSICAL PROPERTIES		IMPORTANT CHARACTERISTICS	
Boiling point, °C	107	**COLORLESS FUMING LIQUID WITH PUNGENT ODOR**	
Melting point, °C	1	Vapor is heavier than air. Both substance and products of decomposition are highly corrosive.	
Relative density (water = 1)	1.7	Decomposes when heated in the presence of oxygen, forming → *phosphorus oxides* and	
Relative vapor density (air = 1)	5.3	→ *chlorine*. Reacts violently with water and water vapor, evolving heat and forming → *phosphoric*	
Relative density at 20 °C of saturated mixture		*acid* and → *hydrochloric acid.*	
vapor/air (air = 1)	1.2		
Vapor pressure, mm Hg at 20 °C	27	TLV-TWA 0.1 ppm 0.63 mg/m³	
Solubility in water	reaction		
Relative molecular mass	153.4		
		Absorption route: Can enter the body by inhalation or ingestion. Harmful atmospheric concentrations can build up very rapidly on evaporation at 20°C.	
		Immediate effects: Corrosive to the eyes, skin and respiratory tract. Inhalation of vapor/fumes can cause severe breathing difficulties (lung edema). In serious cases risk of death.	
Gross formula:	Cl₃OP		

HAZARDS/SYMPTOMS	PREVENTIVE MEASURES	FIRE EXTINGUISHING/FIRST AID
Fire: non-combustible.		in case of fire in immediate vicinity: do not use water.
Inhalation: *corrosive*; sore throat, cough, severe breathing difficulties.	ventilation, local exhaust or respiratory protection.	fresh air, rest, place in half-sitting position, take to hospital.
Skin: *corrosive*; redness, pain, serious burns.	gloves, protective clothing.	remove contaminated clothing, flush skin with water or shower, call a doctor.
Eyes: *corrosive*; redness, pain, impaired vision.	face shield, or combined eye and respiratory protection.	flush with water, take to a doctor.
Ingestion: *corrosive*; sore throat, abdominal pain.		rinse mouth, immediately take to hospital.

SPILLAGE	STORAGE	LABELING / NFPA
evacuate area, call in an expert, collect leakage in sealable containers, soak up with sand or other inert absorbent and remove to safe place, under no circumstances spray liquid with water, combat gas cloud with water curtain; (additional individual protection: CHEMICAL SUIT).	keep dry, separate from bases; ventilate at floor level.	R: 34-37 S: 7/8-26 Corrosive

NOTES
Lung edema symptoms usually develop several hours later and are aggravated by physical exertion: rest and hospitalization essential. As first aid, a doctor or authorized person should consider administering a corticosteroid spray. Use airtight packaging.

Transport Emergency Card TEC(R)-539

HI: 80; UN-number: 1810

PHOSPHORUS PENTABROMIDE

PHYSICAL PROPERTIES	IMPORTANT CHARACTERISTICS
Melting point (decomposes), °C 100 Solubility in water reaction Relative molecular mass 430.5	**YELLOW CRYSTALS** Decomposes when heated, giving off toxic and corrosive gases. Highly corrosive, esp. in the presence of water and water vapor. Reacts violently with bases. Reacts violently with water and alcohols, forming corrosive vapors (→ *hydrogen bromide*). Reacts with air to form corrosive vapors (→ *hydrogen bromide*).
	TLV-TWA not available
	Absorption route: Can enter the body by inhalation or ingestion. **Immediate effects:** Corrosive to the eyes, skin and respiratory tract. Inhalation of vapor/fumes can cause severe breathing difficulties (lung edema). **Effects of prolonged/repeated exposure:** Prolonged or repeated contact with vapor and inhalation of particles can cause skin disorders (acne).
Gross formula: Br_5P	

HAZARDS/SYMPTOMS	PREVENTIVE MEASURES	FIRE EXTINGUISHING/FIRST AID
Fire: non-combustible, many chemical reactions can cause fire and explosion.	avoid contact with water, alcohols and bases.	DO NOT USE WATER-BASED EXTINGUISHERS.
Explosion:		in case of fire: keep tanks/drums cool by spraying with water, but DO NOT spray substance with water.
	STRICT HYGIENE, KEEP DUST UNDER CONTROL	IN ALL CASES CALL IN A DOCTOR
Inhalation: *corrosive*; sore throat, cough, shortness of breath, severe breathing difficulties.	local exhaust or respiratory protection.	fresh air, rest, place in half-sitting position, artificial respiration if necessary, take to hospital.
Skin: *corrosive*; redness, pain, serious burns.	gloves, protective clothing.	remove contaminated clothing, flush skin with water or shower, call a doctor or take to hospital.
Eyes: *corrosive*; redness, pain, impaired vision.	face shield, or combined eye and respiratory protection.	flush with water, take to a doctor.
Ingestion: *corrosive*; diarrhea, abdominal cramps, vomiting.	do not eat, drink or smoke while working.	rinse mouth, DO NOT induce vomiting, take immediately to hospital.

SPILLAGE	STORAGE	LABELING / NFPA
clean up spillage, flush away remainder with water; (additional individual protection: breathing apparatus).	keep dry, separate from alcohols and bases.	

NOTES
Depending on the degree of exposure, regular medical checkups are advisable. Lung edema symptoms usually develop several hours later and are aggravated by physical exertion: rest and hospitalization essential. As first aid, a doctor or authorized person should consider administering a corticosteroid spray. Use airtight packaging.

Transport Emergency Card TEC(R)-80G11

PHOSPHORUS PENTACHLORIDE

PHYSICAL PROPERTIES	IMPORTANT CHARACTERISTICS
Sublimation temperature, °C 162 Relative density (water = 1) 3.6 Relative vapor density (air = 1) 7.2 Relative density at 20°C of saturated mixture vapor/air (air = 1) 1.2 Vapor pressure, mm Hg at 30°C 0.084 Solubility in water reaction Relative molecular mass 208.2	**WHITE TO LIGHT YELLOW HYGROSCOPIC CRYSTALS WITH PUNGENT ODOR** Vapor is heavier than air. Decomposes when heated above 167°C, forming → *phosphorus trichloride* and → *chlorine*. The product of reaction with water is highly acidic, reacts violently with bases and is corrosive. Reacts violently with water, forming → *phosphoric acid* and → *hydrochloric acid*. Reacts with air to form corrosive vapors which spread at ground level. Attacks many metals, giving off flammable gas (→ *hydrogen*).

	TLV-TWA	0,1 ppm	0,85 mg/m³

Absorption route: Can enter the body by inhalation or ingestion. Harmful atmospheric concentrations can build up very rapidly on evaporation at 20°C.
Immediate effects: Corrosive to the eyes, skin and respiratory tract. Inhalation of vapor/fumes can cause severe breathing difficulties (lung edema). In serious cases risk of death.

Gross formula: Cl₅P

HAZARDS/SYMPTOMS	PREVENTIVE MEASURES	FIRE EXTINGUISHING/FIRST AID
Fire: non-combustible, many chemical reactions can cause fire and explosion.		in case of fire in immediate vicinity: DO NOT USE WATER-BASED EXTINGUISHERS.
	STRICT HYGIENE	IN ALL CASES CALL A DOCTOR
Inhalation: *corrosive*; sore throat, cough, shortness of breath, severe breathing difficulties.	ventilation, local exhaust or respiratory protection.	fresh air, rest, place in half-sitting position, take to hospital.
Skin: *corrosive*; redness, pain, serious burns.	gloves, protective clothing.	remove contaminated clothing, flush skin with water or shower, call a doctor.
Eyes: *corrosive*; redness, pain, impaired vision.	face shield, or combined eye and respiratory protection.	flush with water, take to a doctor.
Ingestion: *corrosive*; sore throat, abdominal pain, diarrhea, nausea, vomiting.		rinse mouth, take immediately to hospital.

SPILLAGE	STORAGE	LABELING / NFPA
evacuate area, call in an expert, clean up spillage, flush away remainder with water, combat gas cloud with water curtain; (additional individual protection: CHEMICAL SUIT).	keep dry, separate from strong bases.	R: 34-37 S: 7/8-26 Corrosive

NOTES

Triple point is above 760mm Hg, consequently sublimation at 162°C and melting/decomposition at 167°C, depending on the pressure. Depending on the degree of exposure, regular medical checkups are advisable. Lung edema symptoms usually develop several hours later and are aggravated by physical exertion: rest and hospitalization essential. As first aid, a doctor or authorized person should consider administering a corticosteroid spray.

Transport Emergency Card TEC(R)-80G11

HI: 80; UN-number: 1806

diphosphorus pentoxide
phosphorus oxide
phosphoric anhydride

PHOSPHORUS PENTOXIDE

PHYSICAL PROPERTIES	IMPORTANT CHARACTERISTICS
Sublimation temperature, °C 300 Melting point, °C 563 Relative density (water = 1) 2.4 Vapor pressure, mm Hg at 388 °C 0.99 Solubility in water reaction Relative molecular mass 142.0	**WHITE HIGHLY HYGROSCOPIC SOLID IN VARIOUS FORMS, DELIQUESCENT ON EXPOSURE TO AIR** In aqueous solution is a strong acid which reacts violently with bases and is corrosive. Reacts violently with water, forming → *phosphoric acid*. Reacts violently with substances which can give off moisture, e.g. wood, cotton, paper and straw, with risk of ignition. Attacks many metals.

PHYSICAL PROPERTIES (cont.)	TLV-TWA	1 mg/m³

	Absorption route: Can enter the body by inhalation or ingestion. Evaporation negligible at 20°C, but harmful concentrations of airborne particles can build up rapidly. **Immediate effects:** Corrosive to the eyes, skin and respiratory tract. Inhalation of vapor/fumes can cause severe breathing difficulties (lung edema).

Gross formula: O$_5$P$_2$

HAZARDS/SYMPTOMS	PREVENTIVE MEASURES	FIRE EXTINGUISHING/FIRST AID
Fire: non-combustible, many chemical reactions can cause fire and explosion.	avoid contact with combustible substances.	in case of fire in immediate vicinity: preferably no water-based extinguishers.
	STRICT HYGIENE	IN ALL CASES CALL IN A DOCTOR
Inhalation: *corrosive*; sore throat, cough, shortness of breath, severe breathing difficulties.	local exhaust or respiratory protection.	fresh air, rest, place in half-sitting position, take to hospital.
Skin: *corrosive*; redness, pain, burns.	gloves, protective clothing.	remove contaminated clothing, flush skin with water or shower, call a doctor if necessary.
Eyes: *corrosive*; redness, pain, impaired vision.	face shield, or combined eye and respiratory protection.	flush with water, take to a doctor.
Ingestion: *corrosive*; sore throat, abdominal pain, diarrhea, vomiting.	do not eat, drink or smoke while working.	rinse mouth, immediately take to hospital.

SPILLAGE	STORAGE	LABELING / NFPA
clean up spillage, flush away remainder with water; (additional individual protection: P2 respirator).	keep dry, separate from combustible substances, reducing agents and strong bases.	R: 35 S: 22-26 Corrosive

NOTES

Lung edema symptoms usually develop several hours later and are aggravated by physical exertion: rest and hospitalization essential. As first aid, a doctor or authorized person should consider administering a corticosteroid spray. Use airtight packaging.

Transport Emergency Card TEC(R)-80G03

HI: 80; UN-number: 1807

PHOSPHORUSTRIBROMIDE

PHYSICAL PROPERTIES	IMPORTANT CHARACTERISTICS
Boiling point, °C 173 Melting point, °C −40 Relative density (water = 1) 2.9 Relative vapor density (air = 1) 9.3 Relative density at 20 °C of saturated mixture vapor/air (air = 1) 1.02 Vapor pressure, mm Hg at 20 °C 2.2 Solubility in water reaction Relative molecular mass 270.7	**COLORLESS FUMING LIQUID WITH PUNGENT ODOR** Vapor mixes readily with air. Decomposes when heated, giving off toxic and corrosive vapors of → *bromine* and → *phosphorus oxide*. Reacts violently with reducing agents, ammonia and amines, with risk of fire and explosion. Reacts violently with water, giving off corrosive → *hydrogen bromide*. Reacts with moist air to form corrosive hydrogen bromide fumes which spread at floor level.

	TLV-TWA not available

	Absorption route: Can enter the body by inhalation or ingestion. Harmful atmospheric concentrations can build up fairly rapidly on evaporation at approx. 20°C - even more rapidly in aerosol form. **Immediate effects:** Corrosive to the eyes, skin and respiratory tract. Inhalation of vapor/fumes can cause severe breathing difficulties (lung edema). In serious cases risk of death.

Gross formula: Br₃P	

HAZARDS/SYMPTOMS	PREVENTIVE MEASURES	FIRE EXTINGUISHING/FIRST AID
Fire: non-combustible, many chemical reactions can cause fire and explosion.		in case of fire in immediate vicinity: preferably no water-based extinguishers.
	STRICT HYGIENE	IN ALL CASES CALL IN A DOCTOR
Inhalation: *corrosive*; sore throat, cough, shortness of breath, severe breathing difficulties.	ventilation, local exhaust or respiratory protection.	fresh air, rest, place in half-sitting position, take to hospital.
Skin: *corrosive*; redness, pain, serious burns.	gloves, protective clothing.	remove contaminated clothing, flush skin with water or shower, call a doctor.
Eyes: *corrosive*; redness, pain, impaired vision.	face shield.	flush with water, take to a doctor.
Ingestion: *corrosive*; sore throat, abdominal pain, diarrhea, vomiting.		rinse mouth, immediately take to hospital.

SPILLAGE	STORAGE	LABELING / NFPA
collect leakage in sealable containers, soak up with sand or other inert absorbent and remove to safe place; (additional individual protection: CHEMICAL SUIT).	keep dry; separate from combustible substances, reducing agents, strong bases, ammonia and amines, ventilate at floor level.	R: 14-34-37 S: 26 Corrosive

NOTES
Lung edema symptoms usually develop several hours later and are aggravated by physical exertion: rest and hospitalization essential. As first aid, a doctor or authorized person should consider administering a corticosteroid spray. Use airtight packaging.

Transport Emergency Card TEC(R)-80G10   **HI: 80; UN-number: 1808**

PHOSPHORUS TRICHLORIDE

PHYSICAL PROPERTIES	IMPORTANT CHARACTERISTICS
Boiling point, °C 76 Melting point, °C −112 Relative density (water = 1) 1.6 Relative vapor density (air = 1) 4.8 Relative density at 20 °C of saturated mixture vapor/air (air = 1) 1.5 Vapor pressure, mm Hg at 20 °C 97 Solubility in water reaction Relative molecular mass 137.3	**COLORLESS FUMING LIQUID WITH PUNGENT ODOR** Decomposes when heated, giving off toxic → *chlorine* gas. Reacts violently with alcohols, ammonia, bases, nitric acid and reducing agents, with risk of fire and explosion. Reacts with water, forming → *phosphine*, → *phosphorus acid* and → *hydrochloric acid* fumes. Reacts with air to form corrosive vapors (→ *hydrochloric acid*).

| | | TLV-TWA | 0.2 ppm | 1.1 mg/m³ |
| | | STEL | 0.5 ppm | 2.8 mg/m³ |

Absorption route: Can enter the body by inhalation or ingestion. Harmful atmospheric concentrations can build up very rapidly on evaporation at 20°C.
Immediate effects: Corrosive to the eyes, skin and respiratory tract. Inhalation of vapor/fumes can cause severe breathing difficulties (lung edema). In serious cases risk of death.

Gross formula: Cl₃P

HAZARDS/SYMPTOMS	PREVENTIVE MEASURES	FIRE EXTINGUISHING/FIRST AID
Fire: non-combustible, many chemical reactions can cause fire and explosion.	avoid contact with water, ammonia and many other substances.	in case of fire in immediate vicinity: do not use water-based extinguishers.
Explosion: contact with other substances can cause explosion.		in case of fire: keep tanks/drums cool by spraying with water, but DO NOT spray substance with water.
	STRICT HYGIENE	IN ALL CASES CALL IN A DOCTOR
Inhalation: *corrosive*; sore throat, cough, shortness of breath, severe breathing difficulties.	ventilation, local exhaust or respiratory protection.	fresh air, rest, place in half-sitting position, take to hospital.
Skin: *corrosive*; redness, pain, serious burns.	gloves, protective clothing.	remove contaminated clothing, flush skin with water or shower, call a doctor.
Eyes: *corrosive*; redness, pain, impaired vision.	face shield, or combined eye and respiratory protection.	flush with water, take to a doctor.
Ingestion: *corrosive*; sore throat, abdominal pain, vomiting.		rinse mouth, immediately take to hospital.

SPILLAGE	STORAGE	LABELING / NFPA
evacuate area, call in an expert, collect leakage in sealable containers, soak up with sand or other inert absorbent and remove to safe place, under no circumstances spray liquid with water, combat gas cloud with water curtain; (additional individual protection: CHEMICAL SUIT).	keep dry under inert gas, separate from all other substances; ventilate at floor level.	R: 34-37 S: 7/8-26 Corrosive 0 3　2 W

NOTES
Small leakages can be rendered harmless with large quantities of water; insufficient quantities usually cause ignition of resulting phosphine. Depending on the degree of exposure, regular medical checkups are advisable. Lung edema symptoms usually develop several hours later and are aggravated by physical exertion: rest and hospitalization essential. As first aid, a doctor or authorized person should consider administering a corticosteroid spray. Use airtight packaging. Insufficient data on harmful effects to humans: exercise great caution.

CAS-No: [88-99-3]
1,2-benzenedicarboxylic acid
o-benzene dicarboxylic acid
o-dicarboxybenzene
o-phthalic acid

$C_6H_4(COOH)_2$

PHTHALIC ACID

PHYSICAL PROPERTIES		IMPORTANT CHARACTERISTICS	
Melting point (decomposes), °C	191	**COLORLESS CRYSTALS**	
Flash point, °C	168	Decomposes when heated to approx. 190°C, giving off → *phthalic anhydride*. Reacts violently with strong oxidants. Reacts violently when heated with sodium nitrite, with risk of fire and explosion.	
Relative density (water = 1)	1.6		
Relative vapor density (air = 1)	5.7		
Solubility in water, g/100 ml	0.6		
Relative molecular mass	166.1	TLV-TWA	not available
Log P octanol/water	0.4		
		Absorption route: Can enter the body by inhalation or ingestion. **Immediate effects:** Irritates the eyes and respiratory tract.	
Gross formula:	$C_8H_6O_4$		

HAZARDS/SYMPTOMS	PREVENTIVE MEASURES	FIRE EXTINGUISHING/FIRST AID
Fire: combustible.	keep away from open flame, no smoking.	water spray, powder.
Explosion: finely dispersed particles form explosive mixtures on contact with air.	keep dust from accumulating, sealed machinery, explosion-proof electrical equipment and lighting, grounding.	
Inhalation: sore throat, cough.	local exhaust or respiratory protection.	fresh air, rest, place in half-sitting position, call a doctor.
Skin: redness.	gloves.	remove contaminated clothing, flush skin with water or shower, call a doctor if necessary.
Eyes: redness, pain.	goggles.	flush with water, take to a doctor if necessary.
Ingestion: abdominal cramps, vomiting.		rinse mouth, call a doctor or take to hospital.

SPILLAGE	STORAGE	LABELING / NFPA
clean up spillage, flush away remainder with water; (additional individual protection: P2 respirator).	keep separate from oxidants and strong acids.	1 / 0 / 1

NOTES
Explosive limits not given in the literature. The measures on this card also apply to metaphthalic acid (iso) and paraphthalic acid (tere). Melting point: isophthalic acid: 345°C; terephthalic acid: 348°C. Log P octanol/water is estimated.

CAS-No: [85-44-9]
1,3-isobenzofurandione
1,2-benzenedicarboxylic acid anhydride
phthalic acid anhydride

$C_6H_4(CO)_2O$

PHTHALIC ANHYDRIDE

PHYSICAL PROPERTIES	IMPORTANT CHARACTERISTICS
Boiling point, °C 295 Melting point, °C 131 Flash point, °C 152 Relative density (water = 1) 1.5 Relative vapor density (air = 1) 5.1 Relative density at 20 °C of saturated mixture vapor/air (air = 1) 1.0 Vapor pressure, mm Hg at 20 °C 2.3×10^{-3} Solubility in water, g/100 ml 0.6 Explosive limits, vol% in air 1.7-10.5 Relative molecular mass 148.1	**WHITE CRYSTALLINE FLAKES/NEEDLES OR POWDER** Vapor mixes readily with air. In dry state can form electrostatic charge if stirred, transported pneumatically, poured etc. Can ignite spontaneously on exposure to air (if fine powder is stirred up). Decomposes slowly on contact with water, forming → *phthalic acid*. Reacts with oxidants; reacts violently with nitric acid.

PHYSICAL PROPERTIES (cont.)	TLV-TWA	1 ppm	6.1 mg/m³

Gross formula: $C_8H_4O_3$	**Absorption route:** Can enter the body by inhalation or ingestion. Harmful atmospheric concentrations build up very slowly, if at all, on evaporation at approx. 20°C, but harmful concentrations of airborne particles can build up much more rapidly. **Immediate effects:** Corrosive to the eyes, skin and respiratory tract. Inhalation can cause lung edema. In serious cases risk of death. **Effects of prolonged/repeated exposure:** Prolonged or repeated contact can cause eczema and other skin disorders.

HAZARDS/SYMPTOMS	PREVENTIVE MEASURES	FIRE EXTINGUISHING/FIRST AID
Fire: combustible.	keep away from open flame, no smoking.	water spray, powder.
Explosion: finely dispersed particles can explode on contact with air.	keep dust from accumulating; use sealed machinery, explosion-proof equipment and lighting, grounding.	
Inhalation: *corrosive*; sore throat, cough, shortness of breath, severe breathing difficulties.	local exhaust or respiratory protection.	fresh air, rest, place in half-sitting position, take to hospital.
Skin: *corrosive*; redness, pain, burns.	gloves.	remove contaminated clothing, flush skin with water or shower, refer to a doctor.
Eyes: *corrosive*; redness, pain, impaired vision.	goggles, or combined eye and respiratory protection.	flush with water, take to a doctor.
Ingestion: *corrosive*; abdominal pain, diarrhea.	do not eat, drink or smoke while working.	rinse mouth, call a doctor or take to hospital.

SPILLAGE	STORAGE	LABELING / NFPA
clean up spillage, carefully collect remainder; (additional individual protection: P2 respirator).	keep dry, separate from nitric acid; ventilate at floor level.	R: 36/37/38 ✖ Irritant NFPA: 1 / 3 / 0

NOTES

Lung edema symptoms usually develop several hours later and are aggravated by physical exertion: rest and hospitalization essential. As first aid, a doctor or authorized person should consider administering a corticosteroid spray. Pneumatic transport requires explosion-proof equipment and grounding.

Transport Emergency Card TEC(R)-83

HI: 80; UN-number: 2214

CAS-No: [109-06-8]
2-methylpyridine

$NC_5H_4CH_3$

α-PICOLINE

PHYSICAL PROPERTIES	IMPORTANT CHARACTERISTICS
Boiling point, °C 129 Melting point, °C −67 Flash point, °C 27 Auto-ignition temperature, °C 535 Relative density (water = 1) 0.95 Relative vapor density (air = 1) 3.2 Relative density at 20 °C of saturated mixture vapor/air (air = 1) 1.03 Vapor pressure, mm Hg at 20 °C 9.12 Solubility in water very good Explosive limits, vol% in air 1.4-8.6 Electrical conductivity, pS/m 5.5x10⁷ Relative molecular mass 93.1 Log P octanol/water 1.1 Gross formula: C_6H_7N	**COLORLESS LIQUID WITH CHARACTERISTIC ODOR** Vapor mixes readily with air. Toxic gases are a product of combustion (incl. → *nitrous vapors*). Reacts with strong oxidants. Attacks copper and its alloys. TLV-TWA not available **Absorption route:** Can enter the body by inhalation or ingestion or through the skin. Harmful atmospheric concentrations can build up fairly rapidly on evaporation at approx. 20°C - even more rapidly in aerosol form. **Immediate effects:** Irritates the eyes, skin and respiratory tract. Liquid destroys the skin's natural oils. In substantial concentrations can impair consciousness.

HAZARDS/SYMPTOMS	PREVENTIVE MEASURES	FIRE EXTINGUISHING/FIRST AID
Fire: flammable.	keep away from open flame and sparks, no smoking.	powder, alcohol-resistant foam, large quantities of water, carbon dioxide, (halons).
Explosion: above 27°C: forms explosive air-vapor mixtures.	above 27°C: sealed machinery, ventilation, explosion-proof electrical equipment.	in case of fire: keep tanks/drums cool by spraying with water.
Inhalation: sore throat, cough, headache, sleepiness.	ventilation, local exhaust or respiratory protection.	fresh air, rest, call a doctor.
Skin: *is absorbed*; redness, blisters.	gloves, protective clothing.	remove contaminated clothing, wash skin with soap and water.
Eyes: redness, pain, impaired vision.	face shield, combined with respiratory protection.	flush with water, take to a doctor.
Ingestion: abdominal pain, diarrhea, nausea.		rinse mouth, call a doctor or take to hospital.

SPILLAGE	STORAGE	LABELING / NFPA
collect leakage in sealable containers, soak up with sand or other inert absorbent and remove to safe place; (additional individual protection: breathing apparatus).	keep in fireproof place.	R: 10-20/21/22-36/37 S: 26-36 ✖ Harmful

NFPA: 2 / 0 / 2

NOTES
Log P octanol/water is estimated.

Transport Emergency Card TEC(R)-832   HI: 30; UN-number: 2313

3-methylpyridine (β)
4-methylpyridine (γ)

CH₃
|
C—C
H
H C
N
C=C
H H

β, γ-PICOLINE

PHYSICAL PROPERTIES	IMPORTANT CHARACTERISTICS
Boiling point, °C 144 Melting point, °C − 18 Flash point, °C 40 Relative density (water = 1) 0.96 Relative vapor density (air = 1) 3.21 Relative density at 20 °C of saturated mixture vapor/air (air = 1) ca. 1.0 Vapor pressure, mm Hg at 20 °C ca. 4.6 Solubility in water ∞ Explosive limits, vol% in air 1.3-8.7 Relative molecular mass 93.1 Log P octanol/water 1.2	**COLORLESS LIQUID WITH CHARACTERISTIC ODOR** Vapor mixes readily with air. Decomposes when heated, giving off toxic vapors (→ *nitrous vapors*). Reacts with oxidants. TLV-TWA not available **Absorption route:** Can enter the body by inhalation or ingestion or through the skin. Harmful atmospheric concentrations can build up fairly rapidly on evaporation at approx. 20°C - even more rapidly in aerosol form. **Immediate effects:** Irritates the eyes, skin and respiratory tract. Liquid destroys the skin's natural oils. Affects the nervous system.
Gross formula: C₆H₇N	

HAZARDS/SYMPTOMS	PREVENTIVE MEASURES	FIRE EXTINGUISHING/FIRST AID
Fire: flammable.	keep away from open flame and sparks, no smoking.	powder, AFFF, foam, carbon dioxide, (halons).
Explosion: above 40°C: forms explosive air-vapor mixtures.	above 40°C: sealed machinery, ventilation, explosion-proof electrical equipment.	in case of fire: keep tanks/drums cool by spraying with water.
Inhalation: sore throat, cough, headache, sleepiness.	ventilation, local exhaust or respiratory protection.	fresh air, rest, call a doctor.
Skin: *is absorbed*; redness.	gloves.	remove contaminated clothing, wash skin with soap and water.
Eyes: redness, pain.	face shield.	flush with water, take to a doctor if necessary.
Ingestion: abdominal pain, diarrhea, nausea.		rinse mouth, call a doctor.

SPILLAGE	STORAGE	LABELING / NFPA
collect leakage in sealable containers, flush away remainder with water; (additional individual protection: breathing apparatus).	keep in fireproof place, separate from oxidants.	R: 10-20/22-24-36/37/38 S: 26-36-44 ☠ Toxic 2 / 2 / 0

NOTES
Melting point of γ-picoline: 3.6°C. Alcohol consumption increases toxic effects. CAS Nos: β-picoline: 108-99-0; γ-picoline: 108-89-4. Structure formula is for β-picoline.

CAS-No: [88-89-1]
carbazotic acid
nitroxanthic acid
picronitric acid
2,4,6-trinitrophenol

$(NO_2)_3(C_6H_2)OH$

PICRIC ACID

PHYSICAL PROPERTIES	IMPORTANT CHARACTERISTICS
Decomposes below boiling point, °C 300 Melting point, °C 122 Flash point, °C 150 Auto-ignition temperature, °C 300 Relative density (water = 1) 1.8 Relative vapor density (air = 1) 7.9 Relative density at 20°C of saturated mixture vapor/air (air = 1) 1.00 Vapor pressure, mm Hg at 20°C < <0.076 Solubility in water, g/100 ml 1.3 Relative molecular mass 229.1 Log P octanol/water 2.1	**YELLOW CRYSTALS** In dry state can form electrostatic charge if stirred, transported pneumatically, poured etc. Can decompose explosively if subjected to shock. Reacts with heavy metals, copper and zinc to form shock-sensitive compounds. Decomposes explosively when heated above 300°C, giving off toxic vapors (→ *carbon monoxide* and → *nitrous vapors*). Strong oxidant which reacts violently with combustible substances and reducing agents, with risk of fire and explosion. In aqueous solution is a strong acid which reacts violently with bases and is corrosive. Reacts with aluminum powder in the presence of water, with delayed spontaneous combustion. Attacks many metals, giving off flammable gas (→ *hydrogen*).

	TLV-TWA 0.1 mg/m³
	Absorption route: Can enter the body by inhalation or ingestion or through the skin. Evaporation negligible at 20°C, but harmful concentrations of airborne particles can build up rapidly. **Immediate effects:** Irritates the eyes, skin and respiratory tract. In substantial concentrations can cause muscle weakness, seizures and unconsciousness. **Effects of prolonged/repeated exposure:** Prolonged or repeated contact can cause eczema and yellow discoloration of skin. Can affect the blood. Can cause liver and kidney damage.
Gross formula: $C_6H_3N_3O_7$	

HAZARDS/SYMPTOMS	PREVENTIVE MEASURES	FIRE EXTINGUISHING/FIRST AID
Fire: extremely flammable, many chemical reactions can cause fire and explosion.	keep away from open flame and sparks, no smoking, avoid contact with combustible substances.	large quantities of water.
Explosion: finely dispersed particles form explosive mixtures on contact with air; subjecting to friction/shock or heating can cause explosion.	do not subject to shock or friction, keep dust from accumulating, sealed machinery, explosion-proof equipment and lighting, grounding. KEEP DUST UNDER CONTROL	in case of fire: keep tanks/drums cool by spraying with water, fight fire from sheltered location.
Inhalation: sore throat, cough.	local exhaust or respiratory protection.	fresh air, rest, call a doctor.
Skin: *is absorbed*; redness, pain, yellow skin, yellowing of white of the eye.	gloves.	remove contaminated clothing, flush skin with water or shower.
Eyes: redness, pain.	goggles, or combined eye and respiratory protection.	flush with water, take to a doctor if necessary.
Ingestion: abdominal pain, diarrhea, nausea, vomiting.	do not eat, drink or smoke while working.	rinse mouth, call a doctor or take immediately to hospital.

SPILLAGE	STORAGE	LABELING / NFPA
dampen and then clean up, carefully collect remainder and remove to safe place, do not use sawdust or other combustible absorbents; (additional individual protection: P3 respirator).	keep in fireproof place, separate from combustible substances, reducing agents and strong bases.	R: 2-4-23/24/25 S: 28-35-37-44 Explosive Toxic 3 4 4

NOTES
Flush contaminated clothing with water (fire hazard). Do not take work clothing home. Alcohol consumption increases toxic effects. Less dangerous in moist state (approx. 10% water); to prevent explosion mix with 20% water. Unbreakable packaging preferred; if breakable, keep in unbreakable container.

Transport Emergency Card TEC(R)-10G01

HI: 11; UN-number: 0154

CAS-No: [110-85-0]
1,4-diazacyclohexane
1,4-diethylenediamine
diethyleneimine
hexahydropyrazine
pyrazine hexahydride

PIPERAZINE

PHYSICAL PROPERTIES	IMPORTANT CHARACTERISTICS
Boiling point, °C 148 Melting point, °C 109 Flash point, °C 88 Auto-ignition temperature, °C 340 Relative density (water = 1) 1.1 Relative vapor density (air = 1) 3.0 Relative density at 20 °C of saturated mixture vapor/air (air = 1) 1.01 Vapor pressure, mm Hg at 20 °C 0.076 Solubility in water very good Relative molecular mass 86.1 Log P octanol/water − 1.2 Gross formula: $C_4H_{10}N_2$	**COLORLESS HYGROSCOPIC CRYSTALS WITH CHARACTERISTIC ODOR** Vapor mixes readily with air. Decomposes when heated or burned, giving off toxic vapors (→ *nitrous vapors*). In aqueous solution is a medium strong base which reacts violently with acids and corrodes aluminum, zinc etc. Reacts with carbon tetrachloride and nitrogen compounds. Reacts violently with strong oxidants, with risk of fire and explosion. Attacks copper, nickel and cobalt. TLV-TWA not available **Absorption route:** Can enter the body by inhalation or ingestion. Harmful atmospheric concentrations build up very slowly, if at all, on evaporation at approx. 20°C, but harmful concentrations of airborne particles can build up much more rapidly. **Immediate effects:** Corrosive to the eyes. In substantial concentrations can cause muscle weakness and tingling in arms and legs. Inhalation can cause severe breathing difficulties (asthma).

HAZARDS/SYMPTOMS	PREVENTIVE MEASURES	FIRE EXTINGUISHING/FIRST AID
Fire: combustible.	keep away from open flame, no smoking.	water spray, powder.
Explosion: above 88°C: forms explosive air-vapor mixtures.	above 88°C: sealed machinery, ventilation.	
Inhalation: sore throat, severe breathing difficulties; see also 'Ingestion'.	local exhaust or respiratory protection.	fresh air, rest, take to hospital.
Skin: *is absorbed*; redness, pain.	gloves, protective clothing.	remove contaminated clothing, flush skin with water or shower, refer to a doctor.
Eyes: *corrosive*; redness, pain, impaired vision.	face shield, or combined eye and respiratory protection.	flush with water, take to a doctor.
Ingestion: abdominal pain, diarrhea, headache, nausea, vomiting, disorientation, muscle weakness, seizures.		rinse mouth, call a doctor or take to hospital.

SPILLAGE	STORAGE	LABELING / NFPA
clean up spillage, flush away remainder with water, do not use sawdust or other combustible absorbents; (additional individual protection: P2 respirator).	keep dry, separate from oxidants and acids; ventilate at floor level.	R: 34 S: 26-36 Corrosive (NFPA: 2 / 2 / 0)

NOTES
Apparent melting point of hexahydrate: 44°C. Unbreakable packaging preferred; if breakable, keep in unbreakable container.

Transport Emergency Card TEC(R)-80G16

HI: 80; UN-number: 2579

CAS-No: [110-89-4]
azacyclohexane
cyclopentimine
hexahydropyridine
pentamethyleneimine

$CH_2(CH_2)_4NH$

PIPERIDINE

PHYSICAL PROPERTIES		IMPORTANT CHARACTERISTICS
Boiling point, °C	106	**COLORLESS LIQUID WITH CHARACTERISTIC ODOR**
Melting point, °C	-10	Vapor mixes readily with air, forming explosive mixtures. Do not use compressed air when filling,
Flash point, °C	16	emptying or processing. Decomposes when heated or burned, giving off toxic vapors (→ *nitrous*
Auto-ignition temperature, °C	?	*vapors*). Strong reducing agent which reacts violently with oxidants. In aqueous solution is a
Relative density (water = 1)	0.9	strong base which reacts violently with acids and corrodes aluminum, zinc etc.
Relative vapor density (air = 1)	2.9	
Relative density at 20 °C of saturated mixture		TLV-TWA not available
vapor/air (air = 1)	1.06	
Vapor pressure, mm Hg at 29 °C	40	**Absorption route:** Can enter the body by inhalation or ingestion. Harmful atmospheric
Solubility in water	∞	concentrations can build up fairly rapidly on evaporation at approx. 20°C - even more rapidly in
Explosive limits, vol% in air	?-?	aerosol form.
Relative molecular mass	85.2	**Immediate effects:** Corrosive to the eyes, skin and respiratory tract. Affects the nervous system.
Log P octanol/water	0.9	Inhalation of vapor/fumes can cause severe breathing difficulties (lung edema). In serious cases
		risk of death.
Gross formula:	$C_5H_{11}N$	

HAZARDS/SYMPTOMS	PREVENTIVE MEASURES	FIRE EXTINGUISHING/FIRST AID
Fire: extremely flammable.	keep away from open flame and sparks, no smoking.	powder, alcohol-resistant foam, large quantities of water, carbon dioxide, (halons).
Explosion: forms explosive air-vapor mixtures.	sealed machinery, ventilation, explosion-proof electrical equipment and lighting.	in case of fire: keep tanks/drums cool by spraying with water, fight fire from sheltered location.
Inhalation: *corrosive*; sore throat, cough, severe breathing difficulties, drowsiness.	ventilation, local exhaust or respiratory protection.	fresh air, rest, place in half-sitting position, artificial respiration if necessary, take to hospital.
Skin: *corrosive*; redness, pain, burns.	gloves, protective clothing.	remove contaminated clothing, flush skin with water or shower, refer to a doctor.
Eyes: *corrosive*; redness, pain, impaired vision.	face shield, or combined eye and respiratory protection.	flush with water, take to a doctor.
Ingestion: *corrosive*; abdominal pain, diarrhea, vomiting, sore throat.		rinse mouth, call a doctor or take to hospital.

SPILLAGE	STORAGE	LABELING / NFPA
collect leakage in sealable containers, flush away remainder with water, do not use sawdust or other combustible absorbents; (additional individual protection: CHEMICAL SUIT).	keep in fireproof place, separate from oxidants and acids.	R: 11-23/24-34 S: 16-26-27-44 Flammable Toxic NFPA: 2 / 3 / 3

NOTES

Technical grades have lower flash point. Auto-ignition temperature and explosive limits not given in the literature. Flush contaminated clothing with water (fire hazard). Lung edema symptoms usually develop several hours later and are aggravated by physical exertion: rest and hospitalization essential. As first aid, a doctor or authorized person should consider administering a corticosteroid spray. Unbreakable packaging preferred; if breakable, keep in unbreakable container.

Transport Emergency Card TEC(R)-30G31

HI: 338; UN-number: 2401

CAS-No: [75-98-9]
2,2-dimethylpropanoic acid
2,2-dimethylpropionic acid
neopentanoic acid
tert-pentanoic acid
trimethylacetic acid

$(CH_3)_3CCOOH$

PIVALIC ACID

PHYSICAL PROPERTIES	IMPORTANT CHARACTERISTICS
Boiling point, °C 164 Melting point, °C 34 Flash point, °C 64 Relative density (water = 1) 0.9 Relative vapor density (air = 1) 3.5 Relative density at 20 °C of saturated mixture vapor/air (air = 1) 1.0 Vapor pressure, mm Hg at 20 °C ca. 0.76 Vapor pressure, mm Hg at 70 °C 18 Solubility in water, g/100 ml 2.2 Relative molecular mass 102.1 Log P octanol/water 1.4	**HYGROSCOPIC COLORLESS OR WHITE CRYSTALS WITH PUNGENT ODOR** Decomposes when heated, giving off flammable, irritating vapors. In aqueous solution is a weak acid and slightly corrosive. Reacts with strong oxidants. TLV-TWA not available **Absorption route:** Can enter the body by inhalation or ingestion. Insufficient data on the rate at which harmful concentrations can build up. **Immediate effects:** Irritates the eyes, skin and respiratory tract.
Gross formula: $C_5H_{10}O_2$	

HAZARDS/SYMPTOMS	PREVENTIVE MEASURES	FIRE EXTINGUISHING/FIRST AID
Fire: combustible.	keep away from open flame, no smoking.	powder, AFFF, foam, carbon dioxide, (halons).
Explosion: finely dispersed particles form explosive mixtures on contact with air; above 55°C: forms explosive air-vapor mixtures.	keep dust from accumulating, sealed machinery, explosion-proof electrical equipment and lighting, grounding.	
	KEEP DUST UNDER CONTROL	
Inhalation: sore throat, cough.	local exhaust or respiratory protection.	fresh air, rest, call a doctor.
Skin: redness, pain.	gloves.	remove contaminated clothing, flush skin with water or shower, call a doctor.
Eyes: redness, pain, impaired vision.	face shield, or combined eye and respiratory protection.	flush with water, take to a doctor.
Ingestion: sore throat, abdominal pain.		rinse mouth, call a doctor or take to hospital.

SPILLAGE	STORAGE	LABELING / NFPA
clean up spillage, carefully collect remainder; (additional individual protection: respirator with A/P2 filter).	keep separate from strong bases; ventilate at floor level.	

NOTES
Log P octanol/water is estimated.

Transport Emergency Card TEC(R)-30G15

POTASSIUM

PHYSICAL PROPERTIES	IMPORTANT CHARACTERISTICS
Boiling point, °C 770 Melting point, °C 63 Auto-ignition temperature, °C 440 Relative density (water = 1) 0.86 Vapor pressure, mm Hg at 171 °C 0.76 Solubility in water reaction Relative atomic mass 39.1	**SOFT METALLIC LUMPS** Readily able to form peroxides. Can ignite spontaneously on exposure to air, esp. moist air, giving off corrosive potassium hydroxide fumes. Strong reducing agent which reacts violently with oxidants, chlorinated hydrocarbons and many other substances. Reacts violently with water, giving off flammable gas (→ *hydrogen*) and corrosive fumes (→ *potassium hydroxide*). Reacts with air to form a brown oxide layer.
	TLV-TWA not available
	Immediate effects: Corrosive to the eyes, skin and respiratory tract. Inhalation of vapor/fumes can cause severe breathing difficulties (lung edema).
Gross formula: K	

HAZARDS/SYMPTOMS	PREVENTIVE MEASURES	FIRE EXTINGUISHING/FIRST AID
Fire: extremely flammable, many chemical reactions can cause fire and explosion.	avoid contact with water and many other substances, keep away from open flame, no smoking.	dry sand, NO other extinguishing agents.
Explosion: hydrogen generation (with water) and many other reactions can cause explosion.		fight fire from sheltered location.
	PREVENT DISPERSION OF MIST (POTASSIUM HYDROXIDE)	
Inhalation: *corrosive*; sore throat, cough, severe breathing difficulties.	ventilation.	fresh air, rest, place in half-sitting position, artificial respiration if necessary, take to hospital.
Skin: *corrosive*; redness, pain, serious burns.	gloves.	remove contaminated clothing, flush skin with water or shower.
Eyes: *corrosive*; redness, pain, impaired vision.	face shield.	flush with water, take to a doctor.
Ingestion: *corrosive*; sore throat, abdominal pain, vomiting.		rinse mouth, call a doctor and take immediately to hospital.

SPILLAGE	STORAGE	LABELING / NFPA
evacuate area, call in an expert, cover spillage with dry sand and remove to safe place, render remainder harmless with spirits.	keep in dry fireproof place under oil or paraffin oil, separate from all other substances.	R: 14/15-34 S: 5-8-43 Flammable Corrosive

NOTES
Reacts violently with extinguishing agents such as water, bicarbonate and halons. Lung edema symptoms usually develop several hours later and are aggravated by physical exertion: rest and hospitalization essential. As first aid, a doctor or authorized person should consider administering a corticosteroid spray. Use airtight packaging. Unbreakable packaging preferred; if breakable, keep in unbreakable container.

Transport Emergency Card TEC(R)-711   HI: X423; UN-number: 2257

alum
aluminum potassium disulfate
aluminum potassium sulfate
potash alum

$KAl(SO_4)_2.12H_2O$

POTASSIUM ALUMINUM SULFATE

PHYSICAL PROPERTIES	IMPORTANT CHARACTERISTICS
Melting point, °C 92 Relative density (water = 1) 1.75 Solubility in water, g/100 ml at 20 °C 14 Relative molecular mass 474.4	**WHITE POWDER OR COLORLESS CRYSTALS** Decomposes when heated above 200°C, giving off corrosive vapors. In aqueous solution. Attacks many metals.
	TLV-TWA not available
	Absorption route: Can enter the body by ingestion. **Immediate effects:** Irritates the eyes and skin.
Gross formula: $AlKO_8S_2.12H_2O$	

HAZARDS/SYMPTOMS	PREVENTIVE MEASURES	FIRE EXTINGUISHING/FIRST AID
Fire: non-combustible.		in case of fire in immediate vicinity: use any extinguishing agent.
Inhalation:	local exhaust or respiratory protection.	
Skin: redness.	gloves.	remove contaminated clothing, flush skin with water or shower.
Eyes: redness, pain.	safety glasses.	flush with water, take to a doctor if necessary.
Ingestion: abdominal pain, diarrhea, vomiting.		rinse mouth, call a doctor.

SPILLAGE	STORAGE	LABELING / NFPA
clean up spillage, flush away remainder with water; (additional individual protection: P1 respirator).	keep cool.	

NOTES
Changes to anhydrous form at approx. 200°C.

POTASSIUM ANTIMONYLTARTRATE

PHYSICAL PROPERTIES	IMPORTANT CHARACTERISTICS	
Melting point (decomposes), °C 100 Relative density (water = 1) 2.6 Solubility in water, g/100 ml at 20 °C 8 Relative molecular mass 333.9	**COLORLESS CRYSTALS OR WHITE POWDER** Decomposes when heated above 100°C with loss of crystal water.	
	TLV-TWA	0.5 mg/m³ ✶
	Absorption route: Can enter the body by ingestion or through the skin. **Immediate effects:** Irritates the eyes, skin and respiratory tract. Affects the nervous system. **Effects of prolonged/repeated exposure:** Prolonged or repeated contact can cause skin disorders. Can cause asthmatic attacks. Can affect the blood. Can cause pneumonia and liver and kidney damage.	
Gross formula: $C_4H_4KO_7Sb \cdot \frac{1}{2}H_2O$		

HAZARDS/SYMPTOMS	PREVENTIVE MEASURES	FIRE EXTINGUISHING/FIRST AID
Fire: non-combustible.		in case of fire in immediate vicinity: use any extinguishing agent.
	STRICT HYGIENE, KEEP DUST UNDER CONTROL	
Inhalation: sore throat, cough, shortness of breath.	local exhaust or respiratory protection.	fresh air, rest, take to hospital.
Skin: *is absorbed*; redness, pain.	gloves.	remove contaminated clothing, flush skin with water or shower, call a doctor if necessary.
Eyes: redness, pain.	goggles.	flush with water, take to a doctor.
Ingestion: pain with swallowing, abdominal pain, diarrhea, vomiting.		rinse mouth, immediately take to hospital.

SPILLAGE	STORAGE	LABELING / NFPA
clean up spillage, carefully collect remainder; (additional individual protection: P2 respirator).		R: 20/22 S: 22 Note A ✖ Harmful

NOTES
✶ TLV equals that of antimony. Loss of crystalwater when heated up to 100°C. Depending on the degree of exposure, regular medical checkups are advisable. Effects on blood: leukopenia.

Transport Emergency Card TEC(R)-61G11

POTASSIUM BROMATE

PHYSICAL PROPERTIES	IMPORTANT CHARACTERISTICS
Boiling point (decomposes), °C — 370 Melting point, °C — 350 Relative density (water = 1) — 3.3 Solubility in water, g/100 ml — 7 Relative molecular mass — 167.0	**WHITE CRYSTALS OR CRYSTALLINE POWDER** Decomposes when heated above 370°C, giving off toxic and corrosive gas and oxygen, with increased risk of fire. Strong oxidant which reacts violently with combustible substances and reducing agents. Reacts violently with organic substances, e.g. grease (from fingers), metal powders, ammonium salts, sulfides, phosphorus, carbon and sulfur, with risk of fire and explosion.

TLV-TWA   not available

Absorption route: Can enter the body by inhalation or ingestion. Evaporation negligible at 20°C, but harmful concentrations of airborne particles can build up rapidly.
Immediate effects: If swallowed can affect the blood and cause blue coloration of skin - esp. lips - and mucous membranes. In serious cases risk of unconsciousness.
Effects of prolonged/repeated exposure: Can cause kidney and brain damage.

Gross formula: BrKO$_3$

HAZARDS/SYMPTOMS	PREVENTIVE MEASURES	FIRE EXTINGUISHING/FIRST AID
Fire: non-combustible, but makes other substances more combustible.		
Explosion: friction or shock can cause explosion when mixed with organic substances, various metals or non-metals.	keep dust from accumulating.	in case of fire: keep tanks/drums cool by spraying with water.
	KEEP DUST UNDER CONTROL	
Inhalation: cough, shortness of breath, headache, dizziness.	local exhaust or respiratory protection.	fresh air, rest, call a doctor and take to hospital.
Skin: redness.	gloves.	remove contaminated clothing, flush skin with water or shower.
Eyes: redness, pain.	goggles.	flush with water, take to doctor.
Ingestion: abdominal pain, dizziness, vomiting, unconsciousness, blue skin.		rinse mouth, take immediately to hospital.

SPILLAGE	STORAGE	LABELING / NFPA
clean up spillage, carefully collect remainder, do not use sawdust or other combustible absorbents, remove to safe place; (additional individual protection: P2 respirator).	keep in fireproof place, separate from combustible substances and reducing agents.	R: 9 S: 24/25-27 🔥 Oxidant

NOTES

Flush contaminated clothing with water (fire hazard). Becomes shock-sensitive when contaminated with organic substances. Depending on the degree of exposure, regular medical checkups are advisable. Special first aid required in the event of poisoning: antidotes must be available (with instructions). Decomposition produces corrosive gases which can cause lung edema. Lung edema symptoms usually develop several hours later and are aggravated by physical exertion: rest and hospitalization essential. Effects on blood due to formation of methemoglobin.

Transport Emergency Card TEC(R)-51G02

UN-number: 1484

POTASSIUM BROMIDE

PHYSICAL PROPERTIES		IMPORTANT CHARACTERISTICS	
Boiling point, °C	1435	**SLIGHTLY HYGROSCOPIC COLORLESS CRYSTALS OR WHITE GRANULES OR WHITE POWDER**	
Melting point, °C	730	Decomposes on contact with strong acids, giving off toxic vapors (→ *hydrogen bromide*). Reacts violently with bromotrifluoride, with risk of fire and explosion.	
Relative density (water = 1)	2.8		
Solubility in water, g/100 ml at 20 °C	66		
Relative molecular mass	119.0		
		TLV-TWA	not available
		Absorption route: Can enter the body by inhalation or ingestion. **Immediate effects:** In substantial concentrations can impair consciousness, can have tranquilizing effects and prevents epileptic attacks. Vapors given off on decomposition do have serious effects and can cause lung edema. **Effects of prolonged/repeated exposure:** Prolonged or repeated contact can cause skin disorders (bromine acne).	
Gross formula:	BrK		

HAZARDS/SYMPTOMS	PREVENTIVE MEASURES	FIRE EXTINGUISHING/FIRST AID
Fire: non-combustible.		in case of fire in immediate vicinity: use any extinguishing agent.
Inhalation: sore throat, cough, shortness of breath.	local exhaust or respiratory protection.	fresh air, rest, call a doctor.
Skin: redness.	gloves.	remove contaminated clothing, flush skin with water or shower.
Eyes: redness, pain.	safety glasses.	flush with water, take to a doctor if necessary.
Ingestion: pain with swallowing, abdominal pain, nausea, sleepiness.		rinse mouth, call a doctor.

SPILLAGE	STORAGE	LABELING / NFPA
clean up spillage, flush away remainder with water.	keep separate from strong acids.	

NOTES
Lung edema symptoms usually develop several hours later and are aggravated by physical exertion: rest and hospitalization essential. As first aid, a doctor or authorized person should consider administering a corticosteroid spray.

POTASSIUM CHLORATE

PHYSICAL PROPERTIES	IMPORTANT CHARACTERISTICS	
Decomposes below boiling point, °C **400** Melting point, °C **368** Relative density (water = 1) **2.3** Solubility in water, g/100 ml **7** Relative molecular mass **122.6**	**WHITE CRYSTALS OR WHITE POWDER** Decomposes explosively when heated above 400°C, giving off toxic, corrosive gases and → *oxygen*, with increased risk of fire. Strong oxidant which reacts violently with combustible substances and reducing agents. Reacts with strong acids, giving off toxic and explosive gas (→ *chlorine dioxide*). Forms explosive mixtures with many organic solids and metal powders.	
	TLV-TWA	not available
	Absorption route: Can enter the body by inhalation or ingestion. **Immediate effects:** Irritates the eyes, skin and respiratory tract. **Effects of prolonged/repeated exposure:** Can affect the blood. Can cause liver and kidney damage. Blueish-grey discoloration of skin - esp. lips - and mucous membranes. Red to brown urine. In serious cases risk of death.	
Gross formula: ClKO$_3$		

HAZARDS/SYMPTOMS	PREVENTIVE MEASURES	FIRE EXTINGUISHING/FIRST AID
Fire: not combustible but enhances combustion of other substances.	keep away from open flame and sparks, no smoking, avoid contact with combustible substances.	in case of fire in immediate vicinity: use any extinguishing agent.
Explosion: subjecting mixtures to friction or shock, even with only small quantities of organic material and limited ignition energy, can cause explosion.	keep dust from accumulating.	in case of fire: keep tanks/drums cool by spraying with water.
	KEEP DUST UNDER CONTROL	
Inhalation: sore throat, cough, blue skin.	local exhaust or respiratory protection.	fresh air, rest, call a doctor.
Skin: redness.	gloves.	remove contaminated clothing, flush skin with water or shower.
Eyes: redness, pain.	goggles.	flush with water, take to a doctor if necessary.
Ingestion: abdominal pain, diarrhea, nausea, feeling of weakness, blue skin, red-colored urine.		rinse mouth, immediately take to hospital.

SPILLAGE	STORAGE	LABELING / NFPA
clean up spillage, carefully collect remainder and remove to safe place, do not use sawdust or other combustible absorbents; (additional individual protection (against dust): P2 respirator).	keep in fireproof place, separate from combustible substances, reducing agents and strong acids.	R: 9-20/22 S: 2-13-16-27 🔥 Oxidant ✖ Harmful

NOTES
Flush contaminated clothing with water (fire hazard). Becomes shock-sensitive when contaminated with organic substances. Depending on the degree of exposure, regular medical checkups are advisable. Effects on blood due to formation of methemoglobin. Special first aid required in the event of poisoning: antidotes must be available (with instructions).

Transport Emergency Card TEC(R)-51G01

HI: 50; UN-number: 1485

POTASSIUM CHROMATE

PHYSICAL PROPERTIES	IMPORTANT CHARACTERISTICS	
Melting point, °C — 975 Relative density (water = 1) — 2.7 Solubility in water, g/100 ml at 20 °C — 64 Relative molecular mass — 194.2	**YELLOW CRYSTALS** Decomposes when heated, giving off → *oxygen*, with increased risk of fire. Strong oxidant which reacts violently with combustible substances and reducing agents, with risk of fire and explosion.	
	TLV-TWA	0.05 mg/m³ ✱
	Absorption route: Can enter the body by inhalation or ingestion. **Immediate effects:** Corrosive to the eyes, skin and respiratory tract. Inhalation of vapor can cause severe breathing difficulties (lung edema). **Effects of prolonged/repeated exposure:** Intermittent contact with limited quantities can cause skin disorders (eczema); prolonged or repeated contact can cause serious skin damage and characteristic chromium ulcers.	
Gross formula: CrK_2O_4		

HAZARDS/SYMPTOMS	PREVENTIVE MEASURES	FIRE EXTINGUISHING/FIRST AID
Fire: non-combustible, but makes other substances more combustible.	avoid contact with combustible substances.	in case of fire in immediate vicinity: use any extinguishing agent.
	STRICT HYGIENE, KEEP DUST UNDER CONTROL	
Inhalation: *corrosive*; sore throat, cough, severe breathing difficulties.	local exhaust or respiratory protection.	fresh air, rest, call a doctor.
Skin: *corrosive*; redness, pain, serious burns.	gloves, protective clothing.	remove contaminated clothing, flush skin with water or shower, refer to a doctor.
Eyes: *corrosive*; redness, pain, impaired vision.	face shield.	flush with water, take to a doctor.
Ingestion: *corrosive*; sore throat, abdominal pain, diarrhea, vomiting.	do not eat, drink or smoke while working.	rinse mouth, take immediately to hospital.

SPILLAGE	STORAGE	LABELING / NFPA
clean up spillage, carefully collect remainder; (additional individual protection: P3 respirator).	keep separate from combustible substances and reducing agents.	R: 36/37/38-43 S: 22-28 **✖** Irritant

NOTES
✱ TLV equals that of chromium (VI) compounds (water soluble). Flush contaminated clothing with water (fire hazard). Becomes shock-sensitive when contaminated with organic substances. In acid solution forms → *potassium dichromate*. Lung edema symptoms usually develop several hours later and are aggravated by physical exertion: rest and hospitalization essential. As first aid, a doctor or authorized person should consider administering a corticosteroid spray. Prolonged inhalation can cause perforation of the nasal septum, due to ulceration of the mucous membrane, corresponding to effect on the skin.

POTASSIUM CYANIDE

PHYSICAL PROPERTIES	IMPORTANT CHARACTERISTICS
Melting point, °C 635 Relative density (water = 1) 1.5 Solubility in water good Relative molecular mass 65.1	**WHITE HYGROSCOPIC CRYSTALS WITH CHARACTERISTIC ODOR** Reacts with acids or air to form highly toxic and flammable → *hydrogen cyanide*. In aqueous solution is a strong base which reacts violently with acids and corrodes aluminum, zinc etc. Reacts violently with strong oxidants, with risk of fire and explosion.
	TLV-TWA (skin) 5 mg/m³ ✳
	Absorption route: Can enter the body by inhalation or ingestion or through the skin. Evaporation negligible at 20°C, but harmful concentrations of airborne particles can build up rapidly. **Immediate effects:** Corrosive to the eyes, skin and respiratory tract. Impedes tissue respiration. Subject takes on a red hue. In serious cases risk of seizures and death.
Gross formula: CKN	

HAZARDS/SYMPTOMS	PREVENTIVE MEASURES	FIRE EXTINGUISHING/FIRST AID
Fire: not combustible but heating can cause formation of (combustible) → prussic acid.		in case of fire in immediate vicinity: preferably carbon dioxide.
	STRICT HYGIENE	
Inhalation: *corrosive*; sore throat, severe breathing difficulties, dizziness, cramps, unconsciousness, feeling of weakness.	local exhaust or respiratory protection.	fresh air, rest, artificial respiration if necessary, specific treatment, take to hospital.
Skin: *corrosive, is absorbed*; redness, pain, burns; see also 'Inhalation'.	gloves, protective clothing.	remove contaminated clothing, flush skin with water or shower, call a doctor.
Eyes: *corrosive*; redness, pain, impaired vision.	face shield.	flush with water, take to a doctor.
Ingestion: *corrosive*; pain with swallowing, stomachache; see also 'Inhalation'.	do not eat, drink or smoke while working.	rinse mouth, special treatment, take immediately to hospital.

SPILLAGE	STORAGE	LABELING / NFPA
clean up spillage, render remainder harmless with sodium hypochlorite, chloride of lime or ferrous sulfate and flush away with water; (additional individual protection: breathing apparatus).	keep dry, separate from oxidants and acids; ventilate.	R: 26/27/28-32 S: 1/2-7-28-29-45 Note A ☠ Toxic ◇ 3 0 0

NOTES
✳ TLV equals that of cyanides. Special first aid required in the event of poisoning: antidotes must be available (with instructions). Special treatment: administer oxygen (100%) if necessary, avoid mouth-to-mouth resuscitation if possible (risk to person assisting). Unbreakable packaging preferred; if breakable, keep in unbreakable container.

Transport Emergency Card TEC(R)-61G16   **HI: 66; UN-number: 1680**

POTASSIUMDICHROMATE

PHYSICAL PROPERTIES	IMPORTANT CHARACTERISTICS
Boiling point (decomposes), °C 500 Melting point, °C 398 Relative density (water = 1) 2.7 Solubility in water, g/100 ml at 20 °C 12.5 Relative molecular mass 294.2	**ORANGE-RED CRYSTALS** In aqueous solution attacks many materials, esp. in acidic environment. Strong oxidant which reacts violently with combustible substances and reducing agents, with risk of fire and explosion.

TLV-TWA   0,025 mg/m³ ✱
STEL   0,05 mg/m³

Absorption route: Can enter the body by inhalation or ingestion. Evaporation negligible at 20°C, but harmful concentrations of airborne particles can build up rapidly.
Immediate effects: Inhalation can cause lung edema. In serious cases risk of death.
Effects of prolonged/repeated exposure: Prolonged or repeated contact can cause skin disorders. Has been found to cause a type of lung cancer in certain animal species under certain circumstances.

Gross formula: $Cr_2K_2O_7$

HAZARDS/SYMPTOMS	PREVENTIVE MEASURES	FIRE EXTINGUISHING/FIRST AID
Fire: non-combustible, many chemical reactions can cause fire and explosion.	avoid contact with combustible substances.	in case of fire in immediate vicinity: use any extinguishing agent.
Inhalation: *corrosive*; sore throat, cough, shortness of breath, severe breathing difficulties.	local exhaust or respiratory protection.	fresh air, rest, place in half-sitting position, take to hospital.
Skin: *corrosive*; redness, pain, burns.	gloves.	remove contaminated clothing, flush skin with water or shower, call a doctor.
Eyes: *corrosive*; redness, pain, impaired vision.	goggles.	flush with water, take to a doctor.
Ingestion: sore throat, abdominal pain, vomiting.		rinse mouth, take immediately to hospital.

SPILLAGE	STORAGE	LABELING / NFPA
clean up spillage, carefully collect remainder; (additional individual protection: P3 respirator).	keep dry, separate from combustible substances and reducing agents.	R: 36/37/38-43 S: 22-28 ✖ Irritant

NOTES
✱ TLV equals that of chromium (VI) compounds (water soluble). Flush contaminated clothing with water (fire hazard). Depending on the degree of exposure, regular medical checkups are advisable. Lung edema symptoms usually develop several hours later and are aggravated by physical exertion: rest and hospitalization essential. As first aid, a doctor or authorized person should consider administering a corticosteroid spray. To acidify dichromate use sulfuric acid, NOT hydrochloric or nitric acid (chlorine/nitrous vapors given off). Packaging: special material.

Transport Emergency Card TEC(R)-51G02

CAS-No: [13746-66-2]
potassium hexacyanoferrate III
red prussiate of potash

$K_3Fe(CN)_6$

POTASSIUM FERRICYANIDE

PHYSICAL PROPERTIES	IMPORTANT CHARACTERISTICS	
Melting point (decomposes), °C ? Relative density (water = 1) 1.9 Solubility in water, g/100 ml at 20 °C 46 Relative molecular mass 329.3	**RED CRYSTALS OR POWDER** Decomposes slowly in aqueous solution, partly due to exposure on light. Decomposes when heated, forming toxic gas (→ *hydrocyanic acid*). When the substance decomposes, there is the odor of the typical bitter almond air of hydrocyanic acid. Reacts slowly withc acids, forming → *hydrocyanic acid*. Reacts violently with → *ammonia*, with risk of fire and explosion.	
	TLV-TWA not available	
	Absorption route: Can enter the body by ingestion. Evaporation negligible at 20° C, but harmful concentrations of airborne particles can build up rapidly.	
Gross formula: $C_6FeK_3N_6$		

HAZARDS/SYMPTOMS	PREVENTIVE MEASURES	FIRE EXTINGUISHING/FIRST AID
Fire: not combustible but enhances combustion of other substances.		in case of fire in immediate vicinity: use any extinguishing agent.
Explosion:		in case of fire: keep tanks/drums cool by spraying with water.
Inhalation:	local exhaust or respiratory protection.	fresh air, rest, refer to a doctor if necessary.
Skin:	gloves.	remove contaminated clothing, flush skin with water or shower.
Eyes:	goggles.	flush with water, take to a doctor if necessary.
Ingestion:		rinse mouth, call a doctor.

SPILLAGE	STORAGE	LABELING / NFPA
clean up spillage, carefully collect remainder; (additional individual protection: P2 respirator).	keep in dry, dark place, separate from strong acids.	

NOTES
Symptoms of hydrocyanic acid poisoning can occur due to decomposition in the stomach.

CAS-No: [14459-95-1]
yellow potassium prussiate
yellow prussiate of potash

$K_4Fe(CN)_6.3H_2O$

POTASSIUM FERROCYANIDE

PHYSICAL PROPERTIES	IMPORTANT CHARACTERISTICS	
Melting point, °C 70 Relative density (water = 1) 1.9 Solubility in water, g/100 ml at 12 °C 27.8 Relative molecular mass 422.4	**LIGHT YELLOW CRYSTALS** Decomposes when heated to red-hot, giving off toxic vapors (→ *hydrogen cyanide*). Reacts violently with cupric nitrate. Reacts with boiling acids, with risk of toxic gas being given off (→ *hydrogen cyanide*).	
	TLV-TWA not available	
	Absorption route: Can enter the body by ingestion. **Immediate effects:** Solid/fumes irritate(s) the eyes, skin and respiratory tract.	
Gross formula: $C_6FeK_4N_6.3H_2O$		

HAZARDS/SYMPTOMS	PREVENTIVE MEASURES	FIRE EXTINGUISHING/FIRST AID
Fire: non-combustible.		in case of fire in immediate vicinity: use any extinguishing agent.
Inhalation: sore throat, cough.	local exhaust or respiratory protection.	fresh air, rest.
Skin: redness.	gloves.	remove contaminated clothing, flush skin with water or shower.
Eyes: redness, pain.	safety glasses.	flush with water, take to a doctor if necessary.
Ingestion: diarrhea.		rinse mouth, call a doctor if necessary.

SPILLAGE	STORAGE	LABELING / NFPA
clean up spillage, flush away remainder with water; (additional individual protection: P1 respirator).	keep separate from copper nitrate.	

NOTES
Insufficient toxicological data on harmful effects to humans. Apparent melting point due to loss of water of crystallization. The measures on this card also apply to sodium ferrocyanide.

POTASSIUM FLUORIDE

PHYSICAL PROPERTIES		IMPORTANT CHARACTERISTICS	
Boiling point, °C	1505	**WHITE CRYSTALS**	
Melting point, °C	858	Decomposes when heated or on contact with acids, giving off toxic and corrosive gases	
Relative density (water = 1)	2.5	(→ *hydrogen fluoride*).	
Solubility in water, g/100 ml	92		
Relative molecular mass	58.1	TLV-TWA	2.5 mg/m³ ✱
		Absorption route: Can enter the body by inhalation or ingestion. **Immediate effects:** Corrosive to the eyes, skin and respiratory tract. Inhalation of vapor/fumes can cause severe breathing difficulties (lung edema). **Effects of prolonged/repeated exposure:** Can cause tooth and bone disorders.	
Gross formula:	FK		

HAZARDS/SYMPTOMS	PREVENTIVE MEASURES	FIRE EXTINGUISHING/FIRST AID
Fire: non-combustible.		in case of fire in immediate vicinity: use any extinguishing agent.
Inhalation: *corrosive*; sore throat, cough, severe breathing difficulties.	local exhaust or respiratory protection.	fresh air, rest, place in half-sitting position, take to hospital.
Skin: *corrosive*; redness, pain, serious burns.	gloves, protective clothing.	remove contaminated clothing, flush skin with water or shower, call a doctor if necessary.
Eyes: *corrosive*; redness, pain, impaired vision.	face shield, or combined eye and respiratory protection.	flush with water, take to a doctor.
Ingestion: *corrosive*; diarrhea, abdominal cramps, vomiting.	do not eat, drink or smoke while working.	rinse mouth, take immediately to hospital.

SPILLAGE	STORAGE	LABELING / NFPA
clean up spillage, carefully collect remainder; (additional individual protection: P3 respirator).	keep separate from acids.	R: 23/24/25 S: 1/2-26-44 ☠ Toxic

NOTES

✱ TLV equals that of fluorine. Lung edema symptoms usually develop several ours later and are aggravated by physical exertion: rest and hospitalization essential. As first aid, a doctor or authorized person should consider administering a corticosteroid spray. Packaging: special material.

Transport Emergency Card TEC(R)-61G11C   **HI: 60; UN-number: 1812**

CAS-No: [1310-58-3]
caustic potash
KOH
lye
potassa
potassium hydrate

POTASSIUM HYDROXIDE

PHYSICAL PROPERTIES		IMPORTANT CHARACTERISTICS	
Boiling point, °C	1320	**WHITE HYGROSCOPIC CRYSTALS OR PELLETS**	
Melting point, °C	360	In aqueous solution is a strong base which reacts violently with acids and corrodes tin, lead and	
Relative density (water = 1)	2.0	zinc. Dissolves in water, evolving large quantities of heat and possibly corrosive fumes. Attacks	
Solubility in water, g/100 ml at 20 °C	112	wool, leather and polyester fabric. Attacks many metals, giving off flammable gas ($\rightarrow$ *hydrogen*).	
Relative molecular mass	56.1		
		TLV-TWA	2 mg/m³ C
		Absorption route: Can enter the body by inhalation or ingestion. Evaporation negligible at 20°C, but harmful concentrations of airborne particles can build up rapidly. **Immediate effects:** Corrosive to the eyes, skin and respiratory tract. Inhalation of vapor/fumes can cause severe breathing difficulties (lung edema). Exposure to high concentrations can be fatal.	
Gross formula:	HKO		

HAZARDS/SYMPTOMS	PREVENTIVE MEASURES	FIRE EXTINGUISHING/FIRST AID
Fire: non-combustible.		in case of fire in immediate vicinity: preferably not water.
Inhalation: *corrosive*; sore throat, cough, shortness of breath, severe breathing difficulties.	local exhaust or respiratory protection.	fresh air, rest, place in half-sitting position, take to hospital.
Skin: *corrosive*; redness, pain, serious burns.	gloves, protective clothing.	remove contaminated clothing, flush skin with water or shower, refer to a doctor.
Eyes: *corrosive*; redness, pain, impaired vision.	face shield, or combined eye and respiratory protection.	flush with water, take to doctor if necessary.
Ingestion: *corrosive*; abdominal pain, diarrhea, vomiting.		rinse mouth, take immediately to hospital.

SPILLAGE	STORAGE	LABELING / NFPA
clean up spillage, flush away remainder with water; (additional individual protection: P2 respirator).	keep dry, separate from acids.	R: 35 S: 2-26-37/39 Corrosive 3 / 0 / 1

NOTES
When dissolving use exhaust and add solid to water in small quantities to avoid heating and formation of fumes. TLV is maximum and must not be exceeded. Lung edema symptoms usually develop several hours later and are aggravated by physical exertion: rest and hospitalization essential. As first aid, a doctor or authorized person should consider administering a corticosteroid spray.

Transport Emergency Card TEC(R)-123

HI: 80; UN-number: 1813

POTASSIUM IODATE

PHYSICAL PROPERTIES	IMPORTANT CHARACTERISTICS	
Melting point (decomposes), °C 560 Relative density (water = 1) 3.9 Solubility in water, g/100 ml at 0 °C 5 Relative molecular mass 214.0	**WHITE CRYSTALS OR WHITE POWDER** Decomposes slowly when melted, giving off → *oxygen*, with increased risk of fire. Strong oxidant which reacts violently with combustible substances and reducing agents. Reacts violently with aluminum, carbon, sulfur, sulfides and many organic substances, with risk of fire and explosion.	
	TLV-TWA not available	
	Absorption route: Can enter the body by inhalation or ingestion. Evaporation negligible at 20°C, but harmful concentrations of airborne particles can build up rapidly.	
Gross formula: IKO$_3$		

HAZARDS/SYMPTOMS	PREVENTIVE MEASURES	FIRE EXTINGUISHING/FIRST AID
Fire: not combustible but enhances combustion of other substances.	avoid contact with combustible substances.	in case of fire in immediate vicinity: use any extinguishing agent.
Inhalation:	local exhaust or respiratory protection.	fresh air, rest.
Skin:	gloves.	remove contaminated clothing, flush skin with water or shower, call a doctor.
Eyes:	goggles.	flush with water, take to a doctor.
Ingestion:		rinse mouth, call a doctor.

SPILLAGE	STORAGE	LABELING / NFPA
clean up spillage, flush away remainder with water; (additional individual protection: P2 respirator).	keep separate from combustible substances and reducing agents.	

NOTES
Flush contaminated clothing with water (fire hazard). Insufficient toxicological data on harmful effects to humans.

POTASSIUM IODIDE

PHYSICAL PROPERTIES	IMPORTANT CHARACTERISTICS
Boiling point, °C 1330 Melting point, °C 686 Relative density (water = 1) 3.1 Solubility in water, g/100 ml 144 Relative molecular mass 166.0	**WHITE CRYSTALS OR WHITE POWDER, TURNING YELLOW ON EXPOSURE TO AIR** Decomposes on exposure to air and light, forming iodine. Corrosive in aqueous solution. Reacts violently with alkali metals, fluoroperchloric acid, bromo and chlorotrifluorides, with risk of fire and explosion.
	TLV-TWA not available
	Absorption route: Can enter the body by ingestion. **Immediate effects:** Irritates the eyes.
Gross formula: IK	

HAZARDS/SYMPTOMS	PREVENTIVE MEASURES	FIRE EXTINGUISHING/FIRST AID
Fire: non-combustible.		in case of fire in immediate vicinity: use any extinguishing agent.
Inhalation:	local exhaust or respiratory protection.	
Skin:	gloves.	
Eyes: redness, pain.	safety glasses.	flush with water, take to a doctor if necessary.

SPILLAGE	STORAGE	LABELING / NFPA
clean up spillage, flush away remainder with water; (additional individual protection: P2 respirator).	keep in dry, dark place.	

NOTES
In the case of people who are allergic to iodine, absorption of substance can cause allergic reactions (incl. allergic eczema, distress). Use airtight packaging.

POTASSIUM NITRATE

PHYSICAL PROPERTIES	IMPORTANT CHARACTERISTICS
Boiling point (decomposes), °C 400 Melting point, °C 337 Relative density (water = 1) 2.1 Solubility in water, g/100 ml 35 Relative molecular mass 101.1	**COLORLESS CRYSTALS, WHITE POWDER** Decomposes when heated above 400°C, giving off → *potassium nitrite* and oxygen, with increased risk of fire. When heated above 900°C forms → *nitrous vapors* and potassium peroxide, which reacts violently with water. Strong oxidant which reacts violently with combustible substances and reducing agents, with risk of fire and explosion. Reacts violently with a large number of substances, incl. calcium sulfide, sodium acetate and trichloroethylene, with risk of fire and explosion. Reacts with strong acids, giving off toxic vapors (→ *nitrous vapors*).
	TLV-TWA not available
Gross formula: KNO$_3$	**Absorption route:** Can enter the body by inhalation or ingestion. Evaporation negligible at 20°C, but unpleasant concentrations of airborne particles can build up. **Immediate effects:** Irritates the eyes, skin and respiratory tract. In substantial concentrations can cause shaking, seizures etc. **Effects of prolonged/repeated exposure:** Can affect the blood.

HAZARDS/SYMPTOMS	PREVENTIVE MEASURES	FIRE EXTINGUISHING/FIRST AID
Fire: not combustible but enhances combustion of other substances.	combustible substances and reducing agents.	in case of fire in immediate vicinity: use any extinguishing agent.
Explosion: mixing with organic materials can cause explosion.		in case of fire: keep tanks/drums cool by spraying with water.
	KEEP DUST UNDER CONTROL	
Inhalation: cough, shortness of breath.	local exhaust or respiratory protection.	fresh air, rest, call a doctor.
Skin: redness.	gloves.	remove contaminated clothing, flush skin with water or shower.
Eyes: redness, pain.	goggles.	flush with water, take to a doctor.
Ingestion: abdominal cramps, feeling of weakness, blue skin.		rinse mouth, call a doctor or take to hospital.

SPILLAGE	STORAGE	LABELING / NFPA
clean up spillage, flush away remainder with water; (additional individual protection: P2 respirator).	keep dry, separate from combustible substances, reducing agents and strong acids.	

NOTES

Flush contaminated clothing with water (fire hazard). Effects on blood due to formation of methemoglobin. Special first aid required in the event of poisoning: antidotes must be available (with instructions).

POTASSIUM NITRITE

PHYSICAL PROPERTIES	IMPORTANT CHARACTERISTICS
Melting point, °C 441 Relative density (water = 1) 1.9 Solubility in water, g/100 ml at 20 °C 290 Relative molecular mass 85.1	**HYGROSCOPIC WHITE OR LIGHT YELLOW CRYSTALS OR STICKS** Decomposes when heated above 350 °C and on contact with acids, giving off toxic vapors (→ *nitrous vapors*). Decomposition can be explosive at temperatures above 540°C. Reacts violently with ammonium salts, amines, many organic substances and reducing agents, with risk of fire and explosion.
	TLV-TWA not available
	Absorption route: Can enter the body by inhalation or ingestion. **Immediate effects:** In serious cases risk of death. **Effects of prolonged/repeated exposure:** Can affect the blood.
Gross formula: KNO$_2$	

HAZARDS/SYMPTOMS	PREVENTIVE MEASURES	FIRE EXTINGUISHING/FIRST AID
Fire: not combustible but enhances combustion of other substances.	keep away from open flame, no smoking; avoid contact with combustible substances.	in case of fire in immediate vicinity: use any extinguishing agent.
Explosion: many chemical reactions can cause explosion.		in case of fire: keep tanks/drums cool by spraying with water.
	KEEP DUST UNDER CONTROL	
Inhalation: shortness of breath, headache, dizziness, see also: 'Ingestion'.	local exhaust or respiratory protection.	fresh air, rest, call a doctor.
Skin:	gloves.	remove contaminated clothing, flush skin with water or shower.
Eyes: redness, pain.	face shield.	flush with water, take to a doctor if necessary.
Ingestion: abdominal pain, nausea, dizziness, vomiting, blue skin, feeling of weakness.	do not eat, drink or smoke while working.	rinse mouth, immediately take to hospital.

SPILLAGE	STORAGE	LABELING / NFPA
clean up spillage, carefully collect remainder.	keep dry, separate from combustible substances, reducing agents and acids.	R: 8-25 S: 44 Oxidant Toxic

NOTES
Depending on the degree of exposure, regular medical checkups are advisable. Effects on blood due to formation of methemoglobin. Special first aid required in the event of poisoning: antidotes must be available (with instructions).

Transport Emergency Card TEC(R)-51G02   **UN-number: 1488**

POTASSIUM PERIODATE

PHYSICAL PROPERTIES	IMPORTANT CHARACTERISTICS	
Melting point, °C 582 Relative density (water = 1) 3.6 Solubility in water, g/100 ml 0.4 Relative molecular mass 230	**COLORLESS CRYSTALS** Decomposes in flame or on hot surface, giving off toxic vapors. Decomposes when heated, giving off oxygen, with increased risk of fire. Strong oxidant which reacts violently with combustible substances and reducing agents.	
	TLV-TWA not available	
	Absorption route: Can enter the body by inhalation or ingestion. **Immediate effects:** Irritates the eyes, skin and respiratory tract.	
Gross formula: IKO$_4$		

HAZARDS/SYMPTOMS	PREVENTIVE MEASURES	FIRE EXTINGUISHING/FIRST AID
Fire: non-combustible, but makes other substances more combustible; many chemical reactions can cause fire and explosion.	avoid contact with combustible substances.	in case of fire in immediate vicinity: use any extinguishing agent.
Inhalation: sore throat, cough, shortness of breath.	local exhaust or respiratory protection.	fresh air, rest.
Skin: irritation.	gloves.	remove contaminated clothing, flush skin with water or shower, call a doctor.
Eyes: redness.	face shield.	flush with water, take to a doctor.
Ingestion: sore throat, abdominal pain.		rinse mouth, call a doctor.

SPILLAGE	STORAGE	LABELING / NFPA
clean up spillage, flush away remainder with water.	keep separate from combustible substances and reducing agents.	

NOTES
Insufficient toxicological data on harmful effects to humans.

Transport Emergency Card TEC(R)-51G02

POTASSIUM PERMANGANATE

PHYSICAL PROPERTIES	IMPORTANT CHARACTERISTICS	
Decomposes below melting point, °C 240 Relative density (water = 1) 2.7 Solubility in water, g/100 ml at 20 °C 6.4 Relative molecular mass 158.0	**DARK PURPLE ODORLESS CRYSTALS** Decomposes when heated above 240°C, giving off oxygen, with increased risk of fire. Strong oxidant which reacts violently with combustible substances and reducing agents. Reacts explosively on contact with acetic anhydride, anhydrous ammonia, hydrogen peroxide, sulfur, phosphorus etc. and many organic substances. Reacts with hydrochloric acid, giving off toxic gas (→ *chlorine*). Reacts with concentrated sulfuric acid to form manganese heptaoxide, which can decompose explosively if moderately heated or subjected to shock.	
	TLV-TWA	5 mg/m³ ✳
	Absorption route: Can enter the body by inhalation or ingestion. Evaporation negligible at 20°C, but harmful concentrations of airborne particles (powder) can build up rapidly. **Immediate effects:** Corrosive to the eyes, skin and respiratory tract. Inhalation of vapor/fumes can cause severe breathing difficulties (lung edema). In serious cases risk of death.	
Gross formula: KMnO$_4$		

HAZARDS/SYMPTOMS	PREVENTIVE MEASURES	FIRE EXTINGUISHING/FIRST AID
Fire: non-combustible, but makes other substances more combustible.	combustible substances.	in case of fire in immediate vicinity: use any extinguishing agent.
Explosion: reactions with many substances can cause explosion.		
	KEEP DUST UNDER CONTROL	
Inhalation: *corrosive*; sore throat, cough, shortness of breath, severe breathing difficulties.	local exhaust or respiratory protection.	fresh air, rest, place in half-sitting position, take to hospital.
Skin: *corrosive*; redness, pain, serious burns.	gloves.	remove contaminated clothing, flush skin with water or shower, refer to a doctor.
Eyes: *corrosive*; redness, pain, impaired vision.	face shield.	flush with water, take to doctor.
Ingestion: *corrosive*; sore throat, abdominal pain, vomiting.		rinse mouth, take immediately to hospital.

SPILLAGE	STORAGE	LABELING / NFPA
clean up spillage, flush away remainder with water; (additional individual protection: P2 respirator).	keep separate from combustible substances, reducing agents, hydrochloric acid and hydrogen peroxide.	R: 8-22 S: 2 🔥 Oxidant ✖ Harmful

NOTES
✳ TLV equals that of manganese (dust and compounds). Flush contaminated clothing with water (fire hazard). To acidify permanganate solutions use sulfuric acid, not hydrochloric or nitric acid (→ *chlorine/nitrous* vapors given off). Lung edema symptoms usually develop several hours later and are aggravated by physical exertion: rest and hospitalization essential. As first aid, a doctor or authorized person should consider administering a corticosteroid spray.

POTASSIUMPERSULFATE

PHYSICAL PROPERTIES	IMPORTANT CHARACTERISTICS	
Melting point (decomposes), °C 100 Relative density (water = 1) 2.5 Solubility in water, g/100 ml at 20 °C 5 Relative molecular mass 270.3	**WHITE CRYSTALS** Decomposes when heated above 100°C or in solution when heated above 50°C, giving off → *hydrochloric acid* and corrosive → *sulfuric acid* vapors. Strong oxidant which reacts violently with combustible substances and reducing agents. Reacts violently with strong alkalines such as caustic sodium and caustic potash.	
	TLV-TWA not available	
	Absorption route: Can enter the body by inhalation or ingestion. Evaporation negligible at 20°C, but harmful concentrations of airborne particles (powder) can build up rapidly. **Immediate effects:** Corrosive to the eyes, skin and respiratory tract. Inhalation can cause lung edema. **Effects of prolonged/repeated exposure:** Prolonged or repeated contact can cause eczema.	
Gross formula: $K_2O_8S_2$		

HAZARDS/SYMPTOMS	PREVENTIVE MEASURES	FIRE EXTINGUISHING/FIRST AID
Fire: not combustible but enhances combustion of other substances.	keep away from open flame, no smoking; avoid contact with combustible substances.	in case of fire in immediate vicinity: use any extinguishing agent.
Explosion: reaction with reducing agents can cause explosion.		
	KEEP DUST UNDER CONTROL	
Inhalation: *corrosive*; sore throat, cough, shortness of breath, severe breathing difficulties.	local exhaust or respiratory protection.	fresh air, rest, take to hospital.
Skin: *corrosive*; redness, pain, serious burns.	gloves.	remove contaminated clothing, flush skin with water or shower, take to hospital.
Eyes: *corrosive*; redness, pain, impaired vision.	face shield.	flush with water, take to a doctor.
Ingestion: *corrosive*; abdominal pain, diarrhea, nausea, vomiting.	do not eat, drink or smoke while working.	rinse mouth, immediately take to hospital.

SPILLAGE	STORAGE	LABELING / NFPA
clean up spillage, flush away remainder with water, do not use sawdust or other combustible absorbents; (additional individual protection: P2 respirator).	keep separate from combustible substances, reducing agents and strong bases.	

NOTES
Flush contaminated clothing with water (fire hazard). Lung edema symptoms usually develop several hours later and are aggravated by physical exertion: rest and hospitalization essential. As first aid, a doctor or authorized person should consider administering a corticosteroid spray.

Transport Emergency Card TEC(R)-51G02

UN-number: 1492

POTASSIUM PYROSULFITE

PHYSICAL PROPERTIES	IMPORTANT CHARACTERISTICS	
Melting point (decomposes), °C 190 Relative density (water = 1) 2.3 Solubility in water moderate Relative molecular mass 222.3	**COLORLESS CRYSTALS OR WHITE POWDER WITH PUNGENT ODOR** Decomposes when heated or on contact with acids, giving off toxic vapor (→ *sulfur dioxide*). Reacts with air to form potassium sulfate.	
	TLV-TWA not available	
	Absorption route: Can enter the body by inhalation or ingestion. Evaporation negligible at 20°C, but harmful concentrations of airborne particles can build up rapidly. **Immediate effects:** Irritates the eyes, skin and respiratory tract.	
Gross formula: $K_2O_5S_2$		

HAZARDS/SYMPTOMS	PREVENTIVE MEASURES	FIRE EXTINGUISHING/FIRST AID
Fire: non-combustible.		in case of fire in immediate vicinity: use any extinguishing agent.
Inhalation: sore throat, cough, shortness of breath.	local exhaust or respiratory protection.	fresh air, rest, call a doctor.
Skin: redness, pain.	gloves.	remove contaminated clothing, flush skin with water or shower, refer to a doctor.
Eyes: redness, pain.	safety glasses.	flush with water, take to a doctor if necessary.
Ingestion: sore throat, abdominal pain.		rinse mouth, call a doctor.

SPILLAGE	STORAGE	LABELING / NFPA
clean up spillage, flush away remainder with water; (additional individual protection: P2 respirator).	keep dry, separate from acids.	

NOTES
Although the substance is normally non-combustible, strong friction (e.g. grinding) can cause ignition. The measures on this card also apply to sodium bisulfite. Use airtight packaging.

potassium sulfocyanate
potassium sulfocyanide
potassium thiocyanate

POTASSIUM RHODANIDE

PHYSICAL PROPERTIES	IMPORTANT CHARACTERISTICS	
Boiling point (decomposes), °C 500 Melting point, °C 173 Relative density (water = 1) 1.9 Solubility in water, g/100 ml at 20 °C 217 Relative molecular mass 97.2	**COLORLESS HYGROSCOPIC CRYSTALS** Decomposes when heated, giving off toxic vapors of cyanides.	
	TLV-TWA not available	
	Absorption route: Can enter the body by inhalation or ingestion. Evaporation negligible at 20° C, but harmful concentrations of airborne particles (powder) can build up rapidly. **Immediate effects:** In substantial concentrations can impair consciousness. **Effects of prolonged/repeated exposure:** Prolonged or repeated contact can cause skin disorders. Can lower blood pressure.	
Gross formula: CKNS		

HAZARDS/SYMPTOMS	PREVENTIVE MEASURES	FIRE EXTINGUISHING/FIRST AID
Fire: non-combustible.		in case of fire in immediate vicinity: use any extinguishing agent.
Inhalation: dizziness, feeling of weakness.		fresh air, rest, call a doctor if necessary.
Skin: redness.	gloves.	remove contaminated clothing, flush skin with water or shower.
Eyes: redness, pain.	safety glasses.	flush with water, take to a doctor.
Ingestion: headache, dizziness, nausea, vomiting, feeling of weakness.		rinse mouth, call a doctor or take to hospital.

SPILLAGE	STORAGE	LABELING / NFPA
clean up spillage, carefully collect remainder.	keep dry.	

NOTES

POTASSIUM SULFIDE

PHYSICAL PROPERTIES	IMPORTANT CHARACTERISTICS
Melting point, °C 912 Relative density (water = 1) 1.8 Solubility in water good Relative molecular mass 110.3	**WHITE TO REDDISH-BROWN HYGROSCOPIC CRYSTALS WITH CHARACTERISTIC ODOR** Can ignite spontaneously on exposure to air. Toxic → *sulfur dioxide* is a product of combustion. Can explode if heated rapidly, or subjected to friction or shock. Reacts with acids, giving off toxic and flammable → *hydrogen sulfide gas*. Reacts violently with oxidants, giving off toxic → *sulfur dioxide*.

TLV-TWA	not available

Absorption route: Can enter the body by inhalation or ingestion. Evaporation negligible at 20°C, but harmful concentrations of airborne particles can build up rapidly.
Immediate effects: Corrosive to the eyes, skin and respiratory tract.

Gross formula: K$_2$S

HAZARDS/SYMPTOMS	PREVENTIVE MEASURES	FIRE EXTINGUISHING/FIRST AID
Fire: extremely flammable.	keep away from open flame, no smoking.	large quantities of water, water spray.
Explosion: finely dispersed particles can explode on contact with air.	do not subject to shock or friction, keep dust from accumulating, sealed machinery, explosion-proof equipment and lighting, grounding.	in case of fire: keep tanks/drums cool by spraying with water.
Inhalation: *corrosive*; sore throat, cough, shortness of breath.	local exhaust or respiratory protection.	fresh air, rest, place in half-sitting position, call a doctor.
Skin: *corrosive*; redness, pain, burns.	gloves, protective clothing.	remove contaminated clothing, flush skin with water or shower, refer to a doctor.
Eyes: *corrosive*; redness, pain, impaired vision.	face shield, or combined eye and respiratory protection.	flush with water, take to a doctor.
Ingestion: *corrosive*; sore throat, abdominal pain, diarrhea, vomiting.	do not eat, drink or smoke while working.	rinse mouth, take immediately to hospital.

SPILLAGE	STORAGE	LABELING / NFPA
clean up spillage, render remainder harmless with ferrous (III) chloride solution (caution: forms → *hydrogen sulfide*) and then render harmless with soda, flush away with water; (additional individual protection: breathing apparatus).	keep in cool, dry, fireproof place, separate from oxidants and acids.	R: 31-34 S: 26 Corrosive 1 / 3 0

NOTES
Potassium sulphide pentahydrate (K$_2$S.5H$_2$O) is non-combustible and non-explosive. Use airtight packaging.

Transport Emergency Card TEC(R)-80G09

HI: 43; UN-number: 1382

CAS-No: [463-49-0]
allene
dimethylenemethane

$CH_2 = C = CH_2$

PROPADIENE
(cylinder)

PHYSICAL PROPERTIES	IMPORTANT CHARACTERISTICS
Boiling point, °C −34 Melting point, °C −136 Flash point, °C flammable gas Relative density (water = 1) 0.6 Relative vapor density (air = 1) 1.4 Vapor pressure, mm Hg at 20 °C 5320 Solubility in water none Explosive limits, vol% in air 3.1-12 Relative molecular mass 40.1	**COLORLESS COMPRESSED LIQUEFIED GAS WITH CHARACTERISTIC ODOR** Gas is heavier than air and spreads at ground level, with risk of ignition at a distance, and can accumulate close to ground level, causing oxygen deficiency, with risk of unconsciousness. Flow, agitation etc. can cause build-up of electrostatic charge due to liquid's low conductivity. Reacts violently with oxidants and many other substances.
	TLV-TWA not available
	Absorption route: Can enter the body by inhalation or through the skin. Can saturate the air if released, with risk of suffocation. **Immediate effects:** Irritates the eyes, skin and respiratory tract. Liquid can cause frostbite due to rapid evaporation. Affects the nervous system.
Gross formula: C_3H_4	

HAZARDS/SYMPTOMS	PREVENTIVE MEASURES	FIRE EXTINGUISHING/FIRST AID
Fire: extremely flammable.	keep away from open flame and sparks, no smoking, avoid contact with oxidants.	shut off supply; if impossible and no danger to surrounding area, allow fire to burn itself out; otherwise extinguish with powder, carbon dioxide, (halons).
Explosion: forms explosive air-gas mixtures.	sealed machinery, ventilation, explosion-proof electrical equipment and lighting, grounding when pumping etc. in liquid form, non-sparking hand tools.	in case of fire: keep cylinders cool by spraying with water, fight fire from sheltered location.
Inhalation: sore throat, cough, shortness of breath, headache, unconsciousness.	ventilation, local exhaust or respiratory protection.	fresh air, rest, call a doctor.
Skin: *in case of frostbite*: redness, pain, burns.	insulating gloves.	*in case of frostbite*: DO NOT remove clothing, flush skin with water or shower, refer to a doctor if necessary.
Eyes: *in case of frostbite*: redness, pain, impaired vision.	face shield.	flush with water, take to a doctor if necessary.

SPILLAGE	STORAGE	LABELING / NFPA
evacuate area, call in an expert, ventilate, under no circumstances spray liquid with water; (additional individual protection: breathing apparatus).	keep in cool, fireproof place.	

NOTES
High atmospheric concentrations, e.g. in poorly ventilated spaces, can cause oxygen deficiency, with risk of unconsciousness. Turn leaking cylinder so that leak is on top to prevent liquid propadiene escaping. Material used for lines must not exceed 63% copper content (to avoid formation of methyl acetylene).

Transport Emergency Card TEC(R)-20G11

UN-number: 2200

CAS-No: [74-98-6]
dimethylmethane
liquified petroleum gas
propyl hydride

C_3H_8

PROPANE
(cylinder)

PHYSICAL PROPERTIES	IMPORTANT CHARACTERISTICS
Boiling point, °C −42 Melting point, °C −187 Flash point, °C flammable gas Auto-ignition temperature, °C 470 Relative density (water = 1) 0.5 Relative vapor density (air = 1) 1.6 Vapor pressure, mm Hg at 20 °C 6840 Solubility in water none Explosive limits, vol% in air 2.1-9.5 Minimum ignition energy, mJ 0.25 Electrical conductivity, pS/m 50 Relative molecular mass 44 Gross formula: C_3H_8	**COLORLESS, ODORLESS COMPRESSED LIQUEFIED GAS** Gas is heavier than air and spreads at ground level, with risk of ignition at a distance, and can accumulate close to ground level, causing oxygen deficiency, with risk of unconsciousness. Flow, agitation etc. can cause build-up of electrostatic charge due to liquid's low conductivity. TLV-TWA not available **Absorption route:** Can saturate the air if released, with risk of suffocation. **Immediate effects:** Liquid can cause frostbite due to rapid evaporation.

HAZARDS/SYMPTOMS	PREVENTIVE MEASURES	FIRE EXTINGUISHING/FIRST AID
Fire: extremely flammable.	keep away from open flame and sparks, no smoking.	shut off supply; if impossible and no danger to surrounding area, allow fire to burn itself out; otherwise extinguish with powder, carbon dioxide, (halons).
Explosion: forms explosive air-gas mixtures.	sealed machinery, ventilation, explosion-proof electrical equipment and lighting, grounding when pumping etc. in liquid form, non-sparking hand tools.	in case of fire: keep cylinders cool by spraying with water, fight fire from sheltered location.
Inhalation: shortness of breath, headache, drowsiness, unconsciousness.	ventilation, local exhaust or respiratory protection.	fresh air, rest, artificial respiration if necessary, call a doctor.
Skin: *in case of frostbite*: redness, pain, blisters.	insulating gloves.	*in case of frostbite*: DO NOT remove clothing, flush skin with water or shower, refer to a doctor.
Eyes: redness, pain, impaired vision.	face shield.	flush with water, take to a doctor if necessary.

SPILLAGE	STORAGE	LABELING / NFPA
evacuate area, call in an expert, ventilate, under no circumstances spray liquid with water; (additional individual protection: breathing apparatus).	keep in cool, fireproof place.	4 1 1

NOTES
High atmospheric concentrations, e.g. in poorly ventilated spaces, can cause oxygen deficiency, with risk of unconsciousness. Technical propane contains 20% propene, which causes an increase in pressure. Turn leaking cylinder so that leak is on top to prevent liquid propane escaping.

Transport Emergency Card TEC(R)-27A

HI: 23; UN-number: 1978

CAS-No: [71-23-8]
ethyl carbinol
1-hydroxy propane
1-propanol
propyl alcohol
n-propyl alcohol

$CH_3CH_2CH_2OH$

n-PROPANOL

PHYSICAL PROPERTIES		IMPORTANT CHARACTERISTICS		
Boiling point, °C	97	**COLORLESS LIQUID WITH CHARACTERISTIC ODOR**		
Melting point, °C	−126	Vapor mixes readily with air, forming explosive mixtures. Do not use compressed air when filling, emptying or processing. Reacts violently with strong oxidants, with risk of fire and explosion. Reacts with alkaline-earth and alkali metals, giving off → *hydrogen*.		
Flash point, °C	15			
Auto-ignition temperature, °C	405			
Relative density (water = 1)	0.8			
Relative vapor density (air = 1)	2.1	TLV-TWA (skin)	200 ppm	492 mg/m³
Relative density at 20 °C of saturated mixture vapor/air (air = 1)	1.02	STEL	250 ppm	614 mg/m³
Vapor pressure, mm Hg at 20 °C	14	**Absorption route:** Can enter the body by inhalation or ingestion or through the skin. Harmful atmospheric concentrations build up fairly slowly on evaporation, but much more rapidly in aerosol form.		
Solubility in water	∞			
Explosive limits, vol% in air	2.1-13.5			
Electrical conductivity, pS/m	9.2 x 10⁵	**Immediate effects:** Irritates the eyes, skin and respiratory tract. Liquid destroys the skin's natural oils. In substantial concentrations can impair consciousness. In serious cases risk of unconsciousness.		
Relative molecular mass	60.1			
Log P octanol/water	0.3			
Gross formula:	C_3H_8O			

HAZARDS/SYMPTOMS	PREVENTIVE MEASURES	FIRE EXTINGUISHING/FIRST AID
Fire: extremely flammable.	keep away from open flame and sparks, no smoking.	powder, alcohol-resistant foam, large quantities of water, carbon dioxide, (halons).
Explosion: forms explosive air-vapor mixtures.	sealed machinery, ventilation, explosion-proof electrical equipment and lighting.	in case of fire: keep tanks/drums cool by spraying with water.
Inhalation: cough, shortness of breath, drowsiness, unconsciousness.	ventilation, local exhaust or respiratory protection.	fresh air, rest, call a doctor if necessary.
Skin: *is absorbed*; redness.	gloves.	remove contaminated clothing, flush skin with water or shower, call a doctor if necessary.
Eyes: redness, pain.	safety glasses.	flush with water, take to doctor.
Ingestion: abdominal pain, drowsiness.		rinse mouth, call a doctor if necessary.

SPILLAGE	STORAGE	LABELING / NFPA
collect leakage in sealable containers, soak up with sand or other inert absorbent and remove to safe place.	keep in fireproof place, separate from oxidants.	R: 11 S: 7-16 Note C 🔥 Flammable

NOTES
Alcohol consumption increases toxic effects.

Transport Emergency Card TEC(R)-543

HI: 33; UN-number: 1274

PROPARGYL BROMIDE

PHYSICAL PROPERTIES		IMPORTANT CHARACTERISTICS
Boiling point, °C	89	**COLORLESS TO LIGHT YELLOW LIQUID WITH PUNGENT ODOR**
Melting point, °C	−61	Vapor is heavier than air and spreads at ground level, with risk of ignition at a distance. Flow, agitation etc. can cause build-up of electrostatic charge due to liquid's low conductivity. Do not use compressed air when filling, emptying or processing. Toxic bromide vapors are a product of combustion. Can deflagrate (burn explosively) when heated. Can decompose explosively if subjected to shock. Forms shock-sensitive compounds with chloropicrin. Heating above 220° C and local ignition can cause the substance to explode. Can cause explosion on contact with copper, copper alloys, silver or mercury. Reacts violently with oxidants and many other substances, with risk of fire and explosion.
Flash point, °C	10	
Relative density (water = 1)	1.6	
Relative vapor density (air = 1)	4.1	
Relative density at 20 ° C of saturated mixture		
vapor/air (air = 1)	1.2	
Vapor pressure, mm Hg at 20 °C	54	
Solubility in water	none	
Explosive limits, vol% in air	3-?	
Relative molecular mass	119.0	
		TLV-TWA not available
		Absorption route: Can enter the body by inhalation or ingestion. Harmful atmospheric concentrations can build up very rapidly on evaporation at 20° C. **Immediate effects:** Corrosive to the eyes, skin and respiratory tract. Inhalation of vapor/fumes can cause severe breathing difficulties (lung edema). In serious cases risk of death.
Gross formula:	C_3H_3Br	

HAZARDS/SYMPTOMS	PREVENTIVE MEASURES	FIRE EXTINGUISHING/FIRST AID
Fire: extremely flammable, many chemical reactions can cause fire and explosion.	keep away from open flame and sparks, no smoking; avoid contact with hot surfaces because of detonation at 220° C (steam lines), avoid contact with copper.	powder, water spray, foam, carbon dioxide, (halons).
Explosion: forms explosive air-vapor mixtures, subjecting to shock/friction can cause explosion.	sealed machinery, ventilation, explosion-proof electrical equipment and lighting, grounding, non-sparking tools, do not subject to shocks or friction.	in case of fire: keep tanks/drums cool by spraying with water, fight fire from sheltered location.
Inhalation: *corrosive*; cough, severe breathing difficulties.	ventilation, local exhaust or respiratory protection.	fresh air, rest, place in half-sitting position, take to hospital.
Skin: *corrosive*; redness, pain, burns.	gloves, protective clothing.	remove contaminated clothing, flush skin with water or shower, refer to a doctor.
Eyes: *corrosive*; redness, pain, impaired vision.	face shield, or combined eye and respiratory protection.	flush with water, take to a doctor.
Ingestion: *corrosive*; abdominal cramps, pain with swallowing.		rinse mouth, immediately take to hospital.

SPILLAGE	STORAGE	LABELING / NFPA
evacuate area, call in an expert, collect leakage in non-sealable containers, soak up with sand or other inert absorbent and remove to safe place; do not use sawdust or any other combustible absorbent; (additional individual protection: CHEMICAL SUIT).	since it is an explosive substance keep in a cool place and separate from oxidants.	3 / 4 / 4

NOTES
Propargylbromide often occurs as a mixture with toluene, in which form it is virtually non-explosive. In confined spaces combustion can cause explosion. Lung edema symptoms usually develop several hours later and are aggravated by physical exertion: rest and hospitalization essential. As first aid, a doctor or authorized person should consider administering a corticosteroid spray. Unbreakable packaging preferred; if breakable, keep in unbreakable container.

UN-number: 2345

PROPARGYL BROMIDE
(in toluene)

PHYSICAL PROPERTIES	IMPORTANT CHARACTERISTICS
Boiling point, °C ca.90 Flash point, °C 4 Solubility in water none Explosive limits, vol% in air 1.2-7 Relative molecular mass 119.0	**COLORLESS 20-30% SOLUTION OF PROPARGYL BROMIDE IN TOLUENE WITH PUNGENT ODOR** Vapor is heavier than air and spreads at ground level, with risk of ignition at a distance. Flow, agitation etc. can cause build-up of electrostatic charge due to the liquid's low conductivity. Do not use compressed air when filling, emptying or processing. Toxic bromide vapors are a product of combustion. Can deflagrate (burn explosively when) heated. Contact with copper or its alloys, mercury or silver can cause explosion. Reacts violently with strong oxidants and many other substances, with risk of fire and explosion.

	TLV-TWA	not available

	Absorption route: Can enter the body by inhalation or ingestion or through the skin. Harmful atmospheric concentrations can build up very rapidly on evaporation at 20°C. **Immediate effects:** Irritates the eyes, skin and respiratory tract. Liquid destroys the skin's natural oils. Inhalation of vapor/fumes can cause severe breathing difficulties (lung edema). In serious cases risk of death.
Gross formula: C_3H_3Br	

HAZARDS/SYMPTOMS	PREVENTIVE MEASURES	FIRE EXTINGUISHING/FIRST AID
Fire: extremely flammable.	keep away from open flame, sparks and hot surfaces (e.g. steam lines), no smoking.	powder, AFFF, foam, carbon dioxide, (halons).
Explosion: forms explosive air-vapor mixtures.	sealed machinery, ventilation, explosion-proof equipment and lighting, grounding.	in case of fire: keep tanks/drums cool by spraying with water, fight fire from sheltered location.
Inhalation: *corrosive*; sore throat, cough, shortness of breath, severe breathing difficulties, nausea.	ventilation, local exhaust or respiratory protection.	fresh air, rest, place in half-sitting position, take to hospital.
Skin: *corrosive, is absorbed*; redness, pain, burns.	gloves, protective clothing.	remove contaminated clothing, flush skin with water or shower, take to a doctor.
Eyes: *corrosive*; redness, pain, impaired vision.	face shield, or combined eye and respiratory protection.	flush with water, take to a doctor.
Ingestion: *corrosive*; abdominal pain, diarrhea, vomiting.		rinse mouth, take immediately to hospital.

SPILLAGE	STORAGE	LABELING / NFPA
collect leakage in sealable containers, soak up with dry sand or other inert absorbent and remove to safe place, do not use sawdust or other combustible absorbents; (additional individual protection: breathing apparatus).	keep in fireproof place, separate from oxidants.	3 / 4 / 4

NOTES
Explosive limits given are those of toluene. Lung edema symptoms usually develop several hours later and are aggravated by physical exertion: rest and hospitalization essential. As first aid, a doctor or authorized person should consider administering a corticosteroid spray. TLV-TWA for toluene is 100ppm and 377mg/m³, STEL is 150ppm and 565mg/m³; limit for solution in toluene is probably lower.

Transport Emergency Card TEC(R)-30G32

PROPARGYL CHLORIDE

PHYSICAL PROPERTIES		IMPORTANT CHARACTERISTICS	
Boiling point, °C	57	**COLORLESS LIQUID WITH PUNGENT ODOR**	
Melting point, °C	−78	Vapor is heavier than air and spreads at ground level, with risk of ignition at a distance. Do not use compressed air when filling, emptying or processing. Can deflagrate (burn explosively) when heated. Can decompose explosively if subjected to shock. Decomposes when heated or burned, giving off corrosive vapors (→ *hydrochloric acid*). Reacts with oxidants.	
Flash point, °C	> 15		
Auto-ignition temperature, °C	295		
Relative density (water = 1)	1.03		
Relative vapor density (air = 1)	2.6		
Relative density at 20 °C of saturated mixture		TLV-TWA not available	
vapor/air (air = 1)	1.4		
Vapor pressure, mm Hg at 20 °C	ca. 194	**Absorption route:** Can enter the body by inhalation or ingestion. Harmful atmospheric concentrations can build up rapidly on evaporation at 20°C.	
Solubility in water	none		
Explosive limits, vol% in air	?-?	**Immediate effects:** Corrosive to the eyes, skin and respiratory tract. Inhalation of vapor/fumes can cause severe breathing difficulties (lung edema). In serious cases risk of death.	
Relative molecular mass	74.5		
Gross formula:	C_3H_3Cl		

HAZARDS/SYMPTOMS	PREVENTIVE MEASURES	FIRE EXTINGUISHING/FIRST AID
Fire: extremely flammable, many chemical reactions can cause fire and explosion.	keep away from open flame and sparks, no smoking.	powder, AFFF, foam, carbon dioxide, (halons).
Explosion: forms explosive air-vapor mixtures; heating, shock and jolts can cause explosion.	sealed machinery, ventilation, explosion-proof electrical equipment and lighting, non-sparking tools, do not subject to shocks or friction.	in case of fire: keep tanks/drums cool by spraying with water, fight fire from sheltered location.
Inhalation: *corrosive*; sore throat, cough, shortness of breath, severe breathing difficulties.	ventilation, local exhaust or respiratory protection.	fresh air, rest, place in half-sitting position, take to hospital.
Skin: *corrosive*; redness, pain, burns.	gloves, protective clothing.	remove contaminated clothing, flush skin with water or shower, refer to a doctor.
Eyes: redness, pain, impaired vision.	face shield, or combined eye and respiratory protection.	flush with water, take to a doctor.
Ingestion: *corrosive*, abdominal pain, diarrhea, nausea.		rinse mouth, immediately take to hospital.

SPILLAGE	STORAGE	LABELING / NFPA
evacuate area, call in an expert, collect leakage in non-sealable containers, soak up with sand or other inert absorbent and remove to safe place, do not use sawdust or other combustible absorbents; (additional individual protection: CHEMICAL SUIT).	keep in cool, fireproof place, separate from oxidants.	

NOTES
Explosive limits not given in the literature. Lung edema symptoms usually develop several hours later and are aggravated by physical exertion: rest and hospitalization essential. As first aid, a doctor or authorized person should consider administering a corticosteroid spray.

Transport Emergency Card TEC(R)-61G03

PROPARGYL CHLORIDE
(in toluene)

PHYSICAL PROPERTIES	IMPORTANT CHARACTERISTICS
Boiling point, °C ca. 100 Flash point, °C 4 Auto-ignition temperature, °C <510 Relative density (water = 1) ca. 0.94 Relative vapor density (air = 1) 3.2 Relative density at 20°C of saturated mixture vapor/air (air = 1) ca. 1.1 Vapor pressure, mm Hg at 20°C ca. 23 Solubility in water none Explosive limits, vol% in air ?-?	**COLORLESS 20-30% SOLUTION OF PROPARGYL CHLORIDE IN TOLUENE WITH PUNGENT ODOR** Vapor is heavier than air and spreads at ground level, with risk of ignition at a distance. Flow, agitation etc. can cause build-up of electrostatic charge due to liquid's low conductivity. Do not use compressed air when filling, emptying or processing. Corrosive vapors are a product of combustion. Reacts violently with strong oxidants.
	TLV-TWA not available
	Absorption route: Can enter the body by inhalation or ingestion or through the skin. Harmful atmospheric concentrations can build up very rapidly on evaporation at 20°C. **Immediate effects:** Corrosive to the eyes, skin and respiratory tract. Liquid destroys the skin's natural oils. Affects the nervous system. Inhalation of vapor/fumes can cause severe breathing difficulties (lung edema).
Gross formula: C_3H_3Cl	

HAZARDS/SYMPTOMS	PREVENTIVE MEASURES	FIRE EXTINGUISHING/FIRST AID
Fire: extremely flammable, many chemical reactions can cause fire and explosion.	keep away from open flame and sparks, no smoking.	powder, AFFF, foam, carbon dioxide, (halons).
Explosion: forms explosive air-vapor mixtures.	sealed machinery, ventilation, explosion-proof electrical equipment and lighting, grounding.	in case of fire: keep tanks/drums cool by spraying with water.
Inhalation: *corrosive*; sore throat, cough, shortness of breath, severe breathing difficulties, headache, drowsiness.	ventilation, local exhaust or respiratory protection.	fresh air, rest, place in half-sitting position, take to hospital.
Skin: *corrosive*; redness, pain, burns.	gloves, protective clothing.	remove contaminated clothing, flush skin with water or shower, refer to a doctor.
Eyes: *corrosive*; redness, pain, impaired vision.	face shield, or combined eye and respiratory protection.	flush with water, take to a doctor.
Ingestion: *corrosive*; abdominal pain, headache, nausea, drowsiness.		rinse mouth, DO NOT induce vomiting, take immediately to hospital.

SPILLAGE	STORAGE	LABELING / NFPA
collect leakage in sealable containers, soak up with sand or other inert absorbent, do not use sawdust or other combustible absorbents; (additional individual protection: breathing apparatus).	keep in fireproof place, separate from oxidants.	

NOTES

In regard to the physical properties are primarily derived from → *toluene*. See also → *propargyl chloride*. Explosive limits not given in the literature. TLV for toluene: TLV-TWA 100ppm and 377mg/m³, STEL 150ppm and 565mg/m³; the permissible concentration is lower for the solution than toluene. Lung edema symptoms usually develop several hours later and are aggravated by physical exertion: rest and hospitalization essential. As first aid, a doctor or authorized person should consider administering a corticosteroid spray.

Transport Emergency Card TEC(R)-30G02

PROPENE TRIMER

PHYSICAL PROPERTIES	IMPORTANT CHARACTERISTICS	
Boiling point, °C 132-143 Melting point, °C −76 Flash point, °C 25 Auto-ignition temperature, °C 260 Relative density (water = 1) 0.7 Relative vapor density (air = 1) 4.35 Relative density at 20 °C of saturated mixture vapor/air (air = 1) 1.0 Vapor pressure, mm Hg at 20 °C ca. 6.1 Solubility in water none Explosive limits, vol% in air 0.5-4.0 Relative molecular mass 126.2	**COLORLESS LIQUID WITH CHARACTERISTIC ODOR, MIXTURE OF BRANCHED C₉-OLEFINS** Vapor mixes readily with air, forming explosive mixtures. Able to form peroxides and polymerize. Flow, agitation etc. can cause build-up of electrostatic charge due to liquid's low conductivity. Reacts violently with strong oxidants.	
	TLV-TWA not available	
	Absorption route: Can enter the body by inhalation or ingestion. Harmful atmospheric concentrations build up fairly slowly on evaporation, but much more rapidly in aerosol form. **Immediate effects:** Irritates the eyes, skin and respiratory tract. Liquid destroys the skin's natural oils. Affects the nervous system. In serious cases risk of unconsciousness. If liquid is swallowed, droplets can enter the lungs, with risk of pneumonia.	
Gross formula: C_9H_{18}		

HAZARDS/SYMPTOMS	PREVENTIVE MEASURES	FIRE EXTINGUISHING/FIRST AID
Fire: flammable.	keep away from open flame and sparks, no smoking.	powder, AFFF, foam, carbon dioxide, (halons).
Explosion: above 25°C: forms explosive air-vapor mixtures.	above 25°C: sealed machinery, ventilation, explosion-proof electrical equipment, grounding.	in case of fire: keep tanks/drums cool by spraying with water.
Inhalation: cough, dizziness, drowsiness.	ventilation, local exhaust or respiratory protection.	fresh air, rest, call a doctor, administer oxygen if necessary.
Skin: redness.	gloves.	remove contaminated clothing, flush skin with water or shower, refer to doctor if necessary.
Eyes: redness, pain.	safety glasses.	flush with water, take to a doctor if necessary.
Ingestion: nausea.		rinse mouth, DO NOT induce vomiting, take immediately to hospital.

SPILLAGE	STORAGE	LABELING / NFPA
collect leakage in sealable containers, soak up with sand or other inert absorbent and remove to safe place.	keep in fireproof place; add an inhibitor.	

NOTES
Before distilling check for peroxides; if found, render harmless. Alcohol consumption increases toxic effects.

Transport Emergency Card TEC(R)-151

HI: 30; UN-number: 2057

CAS-No: [57-57-8]
hydracrylic acid beta-lactone
2-oxetanone
propanolide
1,3-propiolactone

β-PROPIOLACTONE

<table>
<tr><th colspan="2">PHYSICAL PROPERTIES</th><th colspan="2">IMPORTANT CHARACTERISTICS</th></tr>
</table>

PHYSICAL PROPERTIES		IMPORTANT CHARACTERISTICS	

PHYSICAL PROPERTIES

Decomposes below boiling point, °C	155
Melting point, °C	−33
Flash point, °C	74
Relative density (water = 1)	1.1
Relative vapor density (air = 1)	2.5
Relative density at 20 °C of saturated mixture vapor/air (air = 1)	1.00
Vapor pressure, mm Hg at 20°C	2.3
Solubility in water, g/100 ml	67
Explosive limits, vol% in air	2.9-?
Relative molecular mass	72.1

Gross formula: $C_3H_4O_2$

IMPORTANT CHARACTERISTICS

COLORLESS LIQUID WITH CHARACTERISTIC ODOR
Vapor mixes readily with air. Can polymerize when moderately heated; stable at temperatures below 5°C. Decomposes when heated, giving off corrosive gases. Reacts with many substances, evolving heat. Reacts slowly with water, forming 3-hydroxypropionic acid.

TLV-TWA	0.5 ppm	1.5 mg/m³ A2

Absorption route: Can enter the body by inhalation or ingestion or through the skin. Harmful atmospheric concentrations can build up fairly rapidly on evaporation at approx. 20°C - even more rapidly in aerosol form.
Immediate effects: Corrosive to the eyes, skin and respiratory tract. Inhalation can cause lung edema. In serious cases risk of death.
Effects of prolonged/repeated exposure: Prolonged or repeated contact can cause skin disorders. Has been found to cause a type of cancer in certain animal species under certain circumstances.

HAZARDS/SYMPTOMS	PREVENTIVE MEASURES	FIRE EXTINGUISHING/FIRST AID
Fire: combustible.	keep away from open flame, no smoking.	powder, alcohol-resistant foam, water spray, carbon dioxide, (halons).
Explosion: above 74°C: forms explosive air-vapor mixtures.	above 74°C: sealed machinery, ventilation.	in case of fire: keep tanks/drums cool by spraying with water.
	STRICT HYGIENE	
Inhalation: *corrosive*; sore throat, cough, shortness of breath, severe breathing difficulties.	ventilation, local exhaust or respiratory protection.	fresh air, rest, place in half-sitting position, take to hospital.
Skin: *corrosive*; redness, pain, burns.	gloves, protective clothing.	remove contaminated clothing, flush skin with water or shower, refer to a doctor, take to hospital if necessary.
Eyes: *corrosive*; redness, pain, impaired vision.	face shield.	flush with water, take to a doctor.
Ingestion: *corrosive*; abdominal cramps, nausea.		rinse mouth, immediately take to hospital.

SPILLAGE	STORAGE	LABELING / NFPA
collect leakage in sealable containers, soak up with sand or other inert absorbent and remove to safe place; (additional individual protection: breathing apparatus).	keep dry under refrigeration, separate from oxidants; ventilate at floor level.	R: 45-26-36/38 S: 53-45 Note E ☠ Highly toxic

NFPA diamond: 2 / 0 / 0

NOTES
Depending on the degree of exposure, regular medical checkups are advisable. Lung edema symptoms usually develop several hours later and are aggravated by physical exertion: rest and hospitalization essential. As first aid, a doctor or authorized person should consider administering a corticosteroid spray. Unbreakable packaging preferred; if breakable, keep in unbreakable container.

Transport Emergency Card TEC(R)-61G07

CAS-No: [123-38-6]
propanal
propionic aldehyde
propyl aldehyde

C_2H_5CHO

PROPIONALDEHYDE

PHYSICAL PROPERTIES	IMPORTANT CHARACTERISTICS
Boiling point, °C 49 Melting point, °C −81 Flash point, °C < −20 Auto-ignition temperature, °C 207 Relative density (water = 1) 0.8 Relative vapor density (air = 1) 2.0 Relative density at 20 °C of saturated mixture vapor/air (air = 1) 1.3 Vapor pressure, mm Hg at 20 °C 261 Solubility in water, g/100 ml at 20 °C 20 Explosive limits, vol% in air 2.3-21 Minimum ignition energy, mJ 0.4 Electrical conductivity, pS/m 10^{10} Relative molecular mass 58.1 Log P octanol/water 0.4 Gross formula: C_3H_6O	**COLORLESS LIQUID WITH PUNGENT ODOR** Vapor is heavier than air and spreads at ground level, with risk of ignition at a distance. Do not use compressed air when filling, emptying or processing. Decomposes when heated or exposed to UV radiation - esp. in the presence of iodine, hydrogen sulfide, bases or acids - giving off toxic → *carbon monoxide* and flammable → *ethane*. Can be self-igniting when finely dispersed, e.g. on textiles or other absorbent. Reacts violently with acids and bases, and extremely violently with oxidants, with risk of fire and explosion. TLV-TWA not available **Absorption route:** Can enter the body by inhalation or ingestion or through the skin. Harmful atmospheric concentrations can build up fairly rapidly on evaporation at approx. 20°C - even more rapidly in aerosol form. **Immediate effects:** Corrosive to the eyes, skin and respiratory tract. Can cause liver damage. Inhalation of vapor/fumes can cause severe breathing difficulties (lung edema). In serious cases risk of seizures and death.

HAZARDS/SYMPTOMS	PREVENTIVE MEASURES	FIRE EXTINGUISHING/FIRST AID
Fire: extremely flammable.	keep away from open flame and sparks, no smoking, avoid contact with oxidants.	powder, alcohol-resistant foam, large quantities of water, carbon dioxide, (halons).
Explosion: forms explosive air-vapor mixtures.	sealed machinery, ventilation, explosion-proof electrical equipment and lighting, non-sparking hand tools.	in case of fire: keep tanks/drums cool by spraying with water.
Inhalation: *corrosive*; sore throat, cough, shortness of breath, severe breathing difficulties.	ventilation, local exhaust or respiratory protection.	fresh air, rest, place in half-sitting position, take to hospital.
Skin: *corrosive*; redness, pain, serious burns.	gloves, protective clothing.	remove contaminated clothing, flush skin with water or shower, call a doctor.
Eyes: *corrosive*; redness, pain, impaired vision.	face shield, or combined eye and respiratory protection.	flush with water, take to a doctor.
Ingestion: *corrosive*; sore throat, abdominal pain, diarrhea, vomiting.		rinse mouth, take immediately to hospital.

SPILLAGE	STORAGE	LABELING / NFPA
evacuate area, collect leakage in sealable containers, do not use sawdust or other combustible absorbents, flush away with large quantities of water; (additional individual protection: breathing apparatus).	keep in dark, fireproof place, separate from oxidants, acids and bases.	R: 11-36/37/38 S: 9-16-29 ⬢ Flammable ✖ Irritant NFPA: 3 / 2 / 2

NOTES
Log P octanol/water is estimated. Flush contaminated clothing with water (fire hazard). Lung edema symptoms usually develop several hours later and are aggravated by physical exertion: rest and hospitalization essential. As first aid, a doctor or authorized person should consider administering a corticosteroid spray.

Transport Emergency Card TEC(R)-692

HI: 33; UN-number: 1275

CAS-No: [79-09-4]
ethanecarboxylic acid
ethylformic acid
metacetonic acid
methylacetic acid
propanoic acid

CH_3CH_2COOH

PROPIONIC ACID

PHYSICAL PROPERTIES		IMPORTANT CHARACTERISTICS
Boiling point, °C	141	**COLORLESS LIQUID WITH PUNGENT ODOR**
Melting point, °C	− 22	Vapor mixes readily with air. Flow, agitation etc. can cause build-up of electrostatic charge due to
Flash point, °C	52	liquid's low conductivity. Reacts with amines and bases, evolving heat. Medium strong acid which
Auto-ignition temperature, °C	465	reacts with bases and is corrosive. Reacts violently with strong oxidants. Attacks many metals,
Relative density (water = 1)	0.99	giving off flammable gas (→ *hydrogen*).
Relative vapor density (air = 1)	2.6	

	TLV-TWA	10 ppm	30 mg/m³

Relative density at 20 °C of saturated mixture
vapor/air (air = 1)   1.01
Vapor pressure, mm Hg at 20 °C   3.0
Solubility in water   ∞
Explosive limits, vol% in air   2.1-12
Electrical conductivity, pS/m   < 3x10⁴
Relative molecular mass   74.1
Log P octanol/water   0.3

Absorption route: Can enter the body by inhalation or ingestion. Harmful atmospheric concentrations can build up fairly rapidly on evaporation at approx. 20° C - even more rapidly in aerosol form.
Immediate effects: Corrosive to the eyes, skin and respiratory tract.

Gross formula: $C_3H_6O_2$

HAZARDS/SYMPTOMS	PREVENTIVE MEASURES	FIRE EXTINGUISHING/FIRST AID
Fire: flammable.	keep away from open flame and sparks, no smoking.	powder, alcohol-resistant foam, water spray, carbon dioxide, (halons).
Explosion: above 52°C: forms explosive air-vapor mixtures.	above 52°C: sealed machinery, ventilation, explosion-proof electrical equipment, grounding.	in case of fire: keep tanks/drums cool by spraying with water.
Inhalation: *corrosive*; sore throat, cough, shortness of breath.	ventilation, local exhaust or respiratory protection.	fresh air, rest, place in half-sitting position, call a doctor.
Skin: *corrosive*; redness, pain.	gloves, protective clothing.	remove contaminated clothing, flush skin with water or shower, call a doctor.
Eyes: *corrosive*; redness, pain, impaired vision.	face shield.	flush with water, take to a doctor.
Ingestion: *corrosive*; sore throat, abdominal pain, vomiting.		rinse mouth, call a doctor or take to hospital.

SPILLAGE	STORAGE	LABELING / NFPA
collect leakage in sealable containers, flush away remainder with water; (additional individual protection: breathing apparatus).	keep in fireproof place, separate from oxidants and strong bases.	R: 34 S: 2-23-26 Note B Corrosive NFPA: 2 / 2 / 0

NOTES

HI: 80; UN-number: 1848

755

PROPIONIC ANHYDRIDE

PHYSICAL PROPERTIES	IMPORTANT CHARACTERISTICS
Boiling point, °C 167 Melting point, °C −45 Flash point, °C 74 Relative density (water = 1) 1.02 Relative vapor density (air = 1) 4.5 Relative density at 20 °C of saturated mixture vapor/air (air = 1) 1.0 Vapor pressure, mm Hg at 20 °C 0.99 Solubility in water reaction Relative molecular mass 130.1	**COLORLESS LIQUID WITH PUNGENT ODOR** Vapor mixes readily with air. Decomposes on contact with water or moisture, giving off corrosive → *propionic acid*. Reacts violently with acids, bases and oxidants, with risk of fire and explosion. TLV-TWA not available **Absorption route:** Can enter the body by inhalation or ingestion. Harmful atmospheric concentrations can build up fairly rapidly on evaporation at approx. 20°C - even more rapidly in aerosol form. **Immediate effects:** Corrosive to the eyes, skin and respiratory tract. Inhalation of vapor/fumes can cause severe breathing difficulties (lung edema). In serious cases risk of death.
Gross formula: C$_6$H$_{10}$O$_3$	

HAZARDS/SYMPTOMS	PREVENTIVE MEASURES	FIRE EXTINGUISHING/FIRST AID
Fire: combustible.	keep away from open flame, no smoking.	powder, carbon dioxide, (halons).
Explosion: above 74°C: forms explosive air-vapor mixtures.	above 74°C: sealed machinery, ventilation.	
Inhalation: *corrosive*; sore throat, cough, shortness of breath, severe breathing difficulties.	ventilation, local exhaust or respiratory protection.	fresh air, rest, place in half-sitting position, take to hospital.
Skin: *corrosive*; redness, pain, serious burns.	gloves, protective clothing.	remove contaminated clothing, flush skin with water or shower, refer to a doctor.
Eyes: *corrosive*; redness, pain, impaired vision.	face shield.	flush with water, take to a doctor.
Ingestion: *corrosive*; sore throat, abdominal pain.		rinse mouth, immediately take to hospital.

SPILLAGE	STORAGE	LABELING / NFPA
collect leakage in sealable containers, flush away remainder with water; (additional individual protection: breathing apparatus).	keep dry.	R: 34 S: 26 Corrosive 2 / 2 / 1

NOTES
Fight large-scale fires with large quantities of water. Lung edema symptoms usually develop several hours later and are aggravated by physical exertion: rest and hospitalization essential. As first aid, a doctor or authorized person should consider administering a corticosteroid spray.

Transport Emergency Card TEC(R)-80G20C   **HI: 80; UN-number: 2496**

cyanoethane
ethyl cyanide
propanenitrile
propionic nitrile
propylnitrile

$C_2H_5C \equiv N$

PROPIONITRILE

PHYSICAL PROPERTIES		IMPORTANT CHARACTERISTICS
Boiling point, °C	97	**COLORLESS LIQUID WITH CHARACTERISTIC ODOR**
Melting point, °C	−93	Vapor mixes readily with air, forming explosive mixtures. Decomposes when heated or burned,
Flash point, °C	2	giving off toxic vapors (→ *nitrous vapors* and → *hydrogen cyanide*). Reacts violently with
Relative density (water = 1)	0.8	oxidants, with risk of fire and explosion. Reacts with warm water, steam and acids, giving off
Relative vapor density (air = 1)	1.9	flammable and toxic vapors (→ *hydrogen cyanide*).
Relative density at 20 °C of saturated mixture		
vapor/air (air = 1)	1.04	
Vapor pressure, mm Hg at 20 °C	37	TLV-TWA not available
Solubility in water, g/100 ml at 40 °C	11.9	
Explosive limits, vol% in air	3.1-?	**Absorption route:** Can enter the body by inhalation of vapor, through the skin or by ingestion.
Electrical conductivity, pS/m	8.5x10⁶	Harmful atmospheric concentrations can build up very rapidly on evaporation at 20°C.
Relative molecular mass	55.1	**Immediate effects:** Irritates the eyes, skin and respiratory tract. Impedes tissue respiration. Can
Log P octanol/water	0.1	affect the nervous system. At high exposure levels can cause unconsciousness and death.
		Effects can be delayed. Keep under medical observation.
		Effects of prolonged/repeated exposure: Liquid destroys the skin's natural oils. Can cause
		birth defects.
Gross formula:	C_3H_5N	

HAZARDS/SYMPTOMS	PREVENTIVE MEASURES	FIRE EXTINGUISHING/FIRST AID
Fire: extremely flammable.	keep away from open flame and sparks, no smoking.	powder, AFFF, foam, carbon dioxide, (halons).
Explosion: forms explosive air-vapor mixtures.	sealed machinery, ventilation, explosion-proof electrical equipment and lighting; do not use compressed air when filling, emptying or processing.	in case of fire: keep tanks/drums cool by spraying with water.
	STRICT HYGIENE, AVOID EXPOSURE OF PREGNANT WOMEN	IN ALL CASES CALL IN A DOCTOR
Inhalation: blue lips or fingernails, disorientation, dizziness, headache, unconsciousness, severe breathing difficulties.	ventilation, local exhaust or respiratory protection.	fresh air, rest, artificial respiration if necessary, special treatment, take immediately to hospital.
Skin: *is absorbed*; redness; see also 'Inhalation'.	gloves, protective clothing.	remove contaminated clothing, flush skin with water or shower, consult a doctor.
Eyes: redness, pain.	face shield, or combined eye and respiratory protection.	flush thoroughly with water (remove contact lenses if easy), take to a doctor.
Ingestion: abdominal cramps; see also 'Inhalation'.		rinse mouth, special treatment, take immediately to hospital.

SPILLAGE	STORAGE	LABELING / NFPA
evacuate area, call in an expert, ventilate, collect leakage in sealable containers, soak up with sand or other inert absorbent and remove to safe place; (additional individual protection: breathing apparatus).	keep in fireproof place, separate from oxidants and strong acids.	3 / 4 / 1

NOTES
Special first aid required in the event of poisoning: antidotes must be available (with instructions). Unbreakable packaging preferred; if breakable, keep in unbreakable container.

Transport Emergency Card TEC(R)-30G45

HI: 336; UN-number: 2404

C_2H_5COCl

propanoyl chloride
propionic chloride

PROPIONYL CHLORIDE

PHYSICAL PROPERTIES	IMPORTANT CHARACTERISTICS
Boiling point, °C ... 80 Melting point, °C ... −94 Flash point, °C ... 12 Relative density (water = 1) ... 1.1 Relative vapor density (air = 1) ... 3.2 Relative density at 20 °C of saturated mixture vapor/air (air = 1) ... 1.2 Vapor pressure, mm Hg at 20 °C ... 62 Solubility in water ... reaction Relative molecular mass ... 92.5	**COLORLESS LIQUID WITH PUNGENT ODOR** Vapor is heavier than air and spreads at ground level, with risk of ignition at a distance. Flow, agitation etc. can cause build-up of electrostatic charge due to liquid's low conductivity. Do not use compressed air when filling, emptying or processing. Decomposes when heated or burned, giving off toxic gases (→ *phosgene* and → *hydrochloric acid*). Reacts violently with water, alcohols and many other compounds, giving off corrosive gases (→ *hydrochloric acid*). Reacts with air to form corrosive vapors (acid fumes) which are heavier than air and spread at ground level.
	TLV-TWA not available
	Absorption route: Can enter the body by inhalation or ingestion. Harmful atmospheric concentrations can build up very rapidly on evaporation at 20°C. **Immediate effects:** Corrosive to the eyes, skin and respiratory tract. Inhalation of vapor/fumes can cause severe breathing difficulties (lung edema). In serious cases risk of death.
Gross formula: C_3H_5ClO	

HAZARDS/SYMPTOMS	PREVENTIVE MEASURES	FIRE EXTINGUISHING/FIRST AID
Fire: extremely flammable, many chemical reactions can cause fire and explosion.	keep away from open flame and sparks, no smoking, avoid contact with other substances.	powder, carbon dioxide, (halons), DO NOT USE WATER-BASE EXTINGUISHERS.
Explosion: forms explosive air-vapor mixtures.	sealed machinery, ventilation, explosion-proof electrical equipment and lighting, grounding.	in case of fire: keep tanks/drums cool by spraying with water, but DO NOT spray substance with water.
	STRICT HYGIENE	IN ALL CASES CALL IN A DOCTOR
Inhalation: *corrosive*; sore throat, cough, shortness of breath, severe breathing difficulties.	ventilation, local exhaust or respiratory protection.	fresh air, rest, place in half-sitting position, artificial respiration, take to hospital.
Skin: *corrosive*; redness, pain, burns.	gloves, protective clothing.	remove contaminated clothing, flush skin with water or shower, refer to a doctor.
Eyes: *corrosive*; redness, pain, impaired vision.	face shield, or combined eye and respiratory protection.	flush with water, take to a doctor.
Ingestion: *corrosive*; sore throat, abdominal pain, diarrhea.		rinse mouth, call a doctor or take to hospital.

SPILLAGE	STORAGE	LABELING / NFPA
evacuate area, call in an expert, ventilate, collect leakage in sealable containers, soak up with sand or other inert absorbent and remove to safe place; (additional individual protection: CHEMICAL SUIT).	keep in dry, fireproof place, separate from alcohols.	R: 11-14-34 S: 9-16-26 Flammable Corrosive 3 / 3 / 1

NOTES
Reacts violently with extinguishing agents containing water. Lung edema symptoms usually develop several hours later and are aggravated by physical exertion: rest and hospitalization essential. As first aid, a doctor or authorized person should consider administering a corticosteroid spray. Use airtight packaging.

Transport Emergency Card TEC(R)-30G31

HI: 338; UN-number: 1815

CAS-No: [109-60-4]
acetic acid, n-propyl ester
1-acetoxypropane
n-propyl acetate
1-propyl acetate

$CH_3COOC_3H_7$

PROPYL ACETATE

PHYSICAL PROPERTIES	IMPORTANT CHARACTERISTICS
Boiling point, °C 102 Melting point, °C −92 Flash point, °C 14 Auto-ignition temperature, °C 450 Relative density (water = 1) 0.9 Relative vapor density (air = 1) 3.5 Relative density at 20 °C of saturated mixture vapor/air (air = 1) 1.08 Vapor pressure, mm Hg at 20 °C 25 Solubility in water, g/100 ml at 20 °C 2 Explosive limits, vol% in air 1.7-8 Electrical conductivity, pS/m 2.2×10^7 Relative molecular mass 102.1 Log P octanol/water 1.5 Gross formula: $C_5H_{10}O_2$	**COLORLESS LIQUID WITH CHARACTERISTIC ODOR** Vapor mixes readily with air, forming explosive mixtures. Do not use compressed air when filling, emptying or processing. TLV-TWA 200 ppm 885 mg/m³ STEL 250 ppm 1040 mg/m³ **Absorption route:** Can enter the body by inhalation or ingestion. Harmful atmospheric concentrations can build up fairly rapidly on evaporation at 20° C - even more rapidly in aerosol form. **Immediate effects:** Irritates the eyes, skin and respiratory tract. Liquid destroys the skin's natural oils. Affects the nervous system.

HAZARDS/SYMPTOMS	PREVENTIVE MEASURES	FIRE EXTINGUISHING/FIRST AID
Fire: extremely flammable.	keep away from open flame and sparks, no smoking.	powder, AFFF, foam, carbon dioxide, (halons).
Explosion: forms explosive air-vapor mixtures.	sealed machinery, ventilation, explosion-proof electrical equipment and lighting.	in case of fire: keep tanks/drums cool by spraying with water.
Inhalation: cough, shortness of breath, dizziness.	ventilation, local exhaust or respiratory protection.	fresh air, rest, call a doctor.
Skin: redness.	gloves.	remove contaminated clothing, flush skin with water or shower.
Eyes: redness, pain.	safety glasses.	flush with water, take to a doctor if necessary.
Ingestion: sore throat, abdominal pain, diarrhea.		rinse mouth, call a doctor or take to hospital.

SPILLAGE	STORAGE	LABELING / NFPA
collect leakage in sealable containers, soak up with sand or other inert absorbent and remove to safe place; (additional individual protection: breathing apparatus).	keep in fireproof place.	R: 11 S: 16-23-29-33 Note C Flammable 3 1 0

NOTES
Log P octanol/water is estimated. Alcohol consumption increases toxic effects.

CAS-No: [107-10-8]
1-aminopropane
monopropylamine
propanamine
n-propylamine
TEA

$CH_3CH_2CH_2NH_2$

PROPYLAMINE

PHYSICAL PROPERTIES	IMPORTANT CHARACTERISTICS
Boiling point, °C 48 Melting point, °C − 83 Flash point, °C < − 20 Auto-ignition temperature, °C 318 Relative density (water = 1) 0.7 Relative vapor density (air = 1) 2.0 Relative density at 20 °C of saturated mixture vapor/air (air = 1) 1.3 Vapor pressure, mm Hg at 20 °C 250 Solubility in water ∞ Explosive limits, vol% in air 2.0-10.4 Relative molecular mass 59.1 Log P octanol/water 0.2 Gross formula: C_3H_9N	**COLORLESS HYGROSCOPIC LIQUID WITH CHARACTERISTIC ODOR** Vapor (gas) is heavier than air and spreads at ground level, with risk of ignition at a distance. Do not use compressed air when filling, emptying or processing. Decomposes when heated or burned, giving off toxic gases (→ *nitrous vapors*). Medium strong base which reacts with acids and corrodes aluminum, zinc etc. Reacts violently with concentrated acids, nitroparaffins, halogenated hydrocarbons, alcohols and many other substances. Reacts violently with strong oxidants and mercury, with risk of fire and explosion. Attacks copper and its alloys. TLV-TWA not available **Absorption route:** Can enter the body by inhalation or ingestion or through the skin. Harmful atmospheric concentrations can build up very rapidly on evaporation at 20°C. **Immediate effects:** Corrosive to the eyes, skin and respiratory tract. Inhalation can cause lung edema. In serious cases risk of death. **Effects of prolonged/repeated exposure:** Prolonged or repeated contact can cause skin disorders.

HAZARDS/SYMPTOMS	PREVENTIVE MEASURES	FIRE EXTINGUISHING/FIRST AID
Fire: extremely flammable.	keep away from open flame and sparks, no smoking.	powder, alcohol-resistant foam, large quantities of water, carbon dioxide, (halons).
Explosion: forms explosive air-vapor mixtures.	sealed machinery, ventilation, explosion-proof electrical equipment and lighting.	in case of fire: keep tanks/drums cool by spraying with water.
Inhalation: *corrosive*; sore throat, cough, shortness of breath, severe breathing difficulties.	ventilation, local exhaust or respiratory protection.	fresh air, rest, take to hospital.
Skin: *corrosive*; redness, pain, serious burns.	gloves, protective clothing.	remove contaminated clothing, flush skin with water or shower, call a doctor.
Eyes: *corrosive*; redness, pain, impaired vision.	face shield, or combined eye and respiratory protection.	flush with water, take to a doctor.
Ingestion: *corrosive*; sore throat, abdominal cramps, vomiting.		rinse mouth, immediately take to hospital.

SPILLAGE	STORAGE	LABELING / NFPA
evacuate area, call in an expert, collect leakage in sealable containers, soak up with sand or other inert absorbent and remove to safe place; (additional individual protection: CHEMICAL SUIT).	keep in fireproof place, separate from other substances.	◇ 3 / 3 0

NOTES
Log P octanol/water is estimated. Lung edema symptoms usually develop several hours later and are aggravated by physical exertion: rest and hospitalization essential. As first aid, a doctor or authorized person should consider administering a corticosteroid spray. Unbreakable packaging preferred; if breakable, keep in unbreakable container.

Transport Emergency Card TEC(R)-30G31   **HI: 338; UN-number: 1277**

CAS-No: [115-07-1]
methylethene
methylethylene
propene

$CH_2 = CHCH_3$

PROPYLENE
(cylinder)

PHYSICAL PROPERTIES		IMPORTANT CHARACTERISTICS	
Boiling point, °C	−48	**COLORLESS COMPRESSED LIQUEFIED GAS**	
Melting point, °C	−185	Gas is heavier than air and spreads at ground level, with risk of ignition at a distance, and can accumulate close to ground level, causing oxygen deficiency, with risk of unconsciousness. Able to form peroxides and thus to polymerize. Flow, agitation etc. can cause build-up of electrostatic charge due to liquid's low conductivity. Reacts violently with strong oxidants and many other compounds, with risk of fire and explosion.	
Flash point, °C	−108		
Auto-ignition temperature, °C	460		
Relative density (water = 1)	0.5		
Relative vapor density (air = 1)	1.5		
Vapor pressure, mm Hg at 20 °C	7600		
Solubility in water	none	**TLV-TWA** not available	
Explosive limits, vol% in air	2.0-11.1		
Minimum ignition energy, mJ	0.28	**Absorption route:** Can enter the body by inhalation. Harmful atmospheric concentrations can build up very rapidly if gas is released. Can saturate the air if released, with risk of suffocation. **Immediate effects:** Liquid can cause frostbite due to rapid evaporation. Affects the nervous system.	
Relative molecular mass	42.1		
Gross formula:	C_3H_6		

HAZARDS/SYMPTOMS	PREVENTIVE MEASURES	FIRE EXTINGUISHING/FIRST AID
Fire: extremely flammable.	keep away from open flame and sparks, no smoking.	shut off supply; if impossible and no danger to surrounding area, allow fire to burn itself out; otherwise extinguish with powder, carbon dioxide, (halons).
Explosion: forms explosive air-gas mixtures.	sealed machinery, ventilation, explosion-proof electrical equipment and lighting, grounding, non-sparking hand tools.	in case of fire: keep cylinders cool by spraying with water.
Inhalation: headache, dizziness, drowsiness, unconsciousness.	ventilation, local exhaust or respiratory protection.	fresh air, rest, call a doctor if necessary.
Skin: *in case of frostbite:* redness, pain, blisters.	insulating gloves.	*in case of frostbite:* DO NOT remove clothing, flush skin with water or shower, take to a doctor.
Eyes: *in case of frostbite:* redness, pain, impaired vision.	face shield.	flush with water, take to a doctor.

SPILLAGE	STORAGE	LABELING / NFPA
evacuate area, call in an expert, ventilate, under no circumstances spray liquid with water; (additional individual protection: breathing apparatus).	keep in cool, fireproof place.	4 / 1 / 1

NOTES

Before distilling check for peroxides; if found, render harmless. On rendering explosive gas/vapor-mixtures inert, see chapter 'Tables and formulas'. High atmospheric concentrations, e.g. in poorly ventilated spaces, can cause oxygen deficiency, with risk of unconsciousness. Turn leaking cylinder so that leak is on top to prevent liquid propene escaping.

Transport Emergency Card TEC(R)-137

HI: 23; UN-number: 1077

CAS-No: [57-55-6]
2-hydroxypropanol
methylethyleneglycol
methylethylglycol
methyl glycol
1,2-propanediol

CH₃CHOHCH₂OH

PROPYLENE GLYCOL

PHYSICAL PROPERTIES		IMPORTANT CHARACTERISTICS	
Boiling point, °C	188	**COLORLESS HYGROSCOPIC VISCOUS LIQUID**	
Melting point, °C	ca. −60	Vapor mixes readily with air. Reacts violently with strong oxidants, with risk of fire.	
Flash point, °C	99		
Auto-ignition temperature, °C	ca. 400	TLV-TWA	not available
Relative density (water = 1)	1.04		
Relative vapor density (air = 1)	2.6	**Absorption route:** Can enter the body by inhalation or ingestion. Harmful atmospheric	
Relative density at 20 °C of saturated mixture		concentrations build up very slowly, if at all, on evaporation at approx. 20°C, but much more	
vapor/air (air = 1)	1.0	rapidly in aerosol form.	
Vapor pressure, mm Hg at 20 °C	0.099	**Immediate effects:** Irritates the eyes and respiratory tract.	
Solubility in water	∞		
Explosive limits, vol% in air	2.5-12.6		
Relative molecular mass	76.1		
Log P octanol/water	− 1.4		
Gross formula:	$C_3H_8O_2$		

HAZARDS/SYMPTOMS	PREVENTIVE MEASURES	FIRE EXTINGUISHING/FIRST AID
Fire: combustible.	keep away from open flame, no smoking.	powder, alcohol-resistant foam, water spray, carbon dioxide, (halons).
Explosion: above 99°C: forms explosive air-vapor mixtures.	above 99°C: sealed machinery, ventilation.	
Inhalation: sore throat, cough, headache.	ventilation.	fresh air, rest.
Skin: redness, pain.	gloves.	remove contaminated clothing, flush skin with water or shower.
Eyes: redness, pain.	acid goggles.	flush with water, take to a doctor if necessary.
Ingestion: abdominal pain, nausea.		rinse mouth, call a doctor.

SPILLAGE	STORAGE	LABELING / NFPA
collect leakage in sealable containers, flush away remainder with water.	keep dry, separate from oxidants; ventilate at floor level.	

NOTES
Log P octanol/water is estimated.

CAS-No: [107-98-2]
1-methoxy-2-hydroxypropane
2-methoxy-1-methylethanol
1-methoxy-2-propanol
PGME
propylene glycol methyl ether

$CH_3OCH_2CHOHCH_3$

PROPYLENE GLYCOL METHYL ETHER

PHYSICAL PROPERTIES		IMPORTANT CHARACTERISTICS		
Boiling point, °C	120	**COLORLESS LIQUID WITH CHARACTERISTIC ODOR**		
Melting point, °C	−97	Vapor mixes readily with air. Presumed to be able to form peroxides. Reacts with strong oxidants.		
Flash point, °C	36			
Auto-ignition temperature, °C	286	TLV-TWA	100 ppm	369 mg/m³
Relative density (water = 1)	0.9	STEL	150 ppm	553 mg/m³
Relative vapor density (air = 1)	3.1			
Relative density at 20 °C of saturated mixture vapor/air (air = 1)	1.02	**Absorption route:** Can enter the body by inhalation or ingestion or through the skin. Harmful atmospheric concentrations build up fairly slowly on evaporation at 20°C, but much more rapidly in aerosol form.		
Vapor pressure, mm Hg at 20 °C	ca. 9.1			
Solubility in water	∞	**Immediate effects:** Irritates the eyes, skin and respiratory tract. Affects the nervous system.		
Explosive limits, vol% in air	1.9-13.1			
Relative molecular mass	90.1			
Gross formula:	$C_4H_{10}O_2$			

HAZARDS/SYMPTOMS	PREVENTIVE MEASURES	FIRE EXTINGUISHING/FIRST AID
Fire: flammable.	keep away from open flame and sparks, no smoking.	powder, alcohol-resistant foam, water spray, carbon dioxide, (halons).
Explosion: above 36°C: forms explosive air-vapor mixtures.	above 36°C: sealed machinery, ventilation, explosion-proof electrical equipment.	in case of fire: keep tanks/drums cool by spraying with water.
Inhalation: headache, sleepiness.	ventilation, local exhaust or respiratory protection.	fresh air, rest, call a doctor.
Skin: redness.	gloves.	remove contaminated clothing, flush skin with water or shower.
Eyes: redness.	safety glasses.	flush with water, take to a doctor if necessary.
Ingestion: abdominal pain; see also 'Inhalation'.		rinse mouth, call a doctor.

SPILLAGE	STORAGE	LABELING / NFPA
collect leakage in sealable containers, flush away remainder with water.	keep in fireproof place.	R: 10 S: 24

NFPA: 3 / 0 / 0

NOTES
Before distilling check for peroxides; if found, render harmless. Dowanol is a trade name.

Transport Emergency Card TEC(R)-30G35

HI: 30; UN-number: 1993

763

CAS-No: [75-56-9]
1,2-epoxypropane
epoxypropane
methyl ethylene oxide
methyloxirane
1,2-propylene oxide

PROPYLENE OXIDE

PHYSICAL PROPERTIES	IMPORTANT CHARACTERISTICS	
Boiling point, °C 34 Melting point, °C −112 Flash point, °C −37 Auto-ignition temperature, °C 550 Relative density (water = 1) 0.9 Relative vapor density (air = 1) 2.0 Relative density at 20 °C of saturated mixture vapor/air (air = 1) 1.6 Vapor pressure, mm Hg at 20 °C 433 Solubility in water, g/100 ml 40 Explosive limits, vol% in air 1.9-37 Minimum ignition energy, mJ 0.13 Relative molecular mass 58.1	**COLORLESS HIGHLY VOLATILE LIQUID WITH CHARACTERISTIC ODOR** Vapor is heavier than air and spreads at ground level, with risk of ignition at a distance. Presumed to be able to form peroxides and thus to polymerize. Bases, acids or metal halides can cause violent polymerization. Do not use compressed air when filling, emptying or processing. Reacts violently with oxidants, acids, bases, anhydrides, chlorine and ammonia, with risk of fire and explosion.	
	TLV-TWA	29 ppm 48 mg/m³
Gross formula: C₃H₆O	**Absorption route:** Can enter the body by inhalation or ingestion. Harmful atmospheric concentrations can build up very rapidly on evaporation at 20°C. **Immediate effects:** Corrosive to the eyes, skin and respiratory tract. Even in aqueous solution (up to 1%) can cause severe chemical burns on contact with skin. Affects the nervous system. Inhalation can cause lung edema. **Effects of prolonged/repeated exposure:** Prolonged or repeated contact can cause skin disorders.	

HAZARDS/SYMPTOMS	PREVENTIVE MEASURES	FIRE EXTINGUISHING/FIRST AID
Fire: extremely flammable.	keep away from open flame and sparks, no smoking.	powder, alcohol-resistant foam, large quantities of water, carbon dioxide, (halons).
Explosion: forms explosive air-vapor mixtures.	sealed machinery, ventilation, explosion-proof electrical equipment and lighting, non-sparking tools.	in case of fire: keep tanks/drums cool by spraying with water.
Inhalation: *corrosive*; cough, severe breathing difficulties, dizziness, vomiting.	ventilation, local exhaust or respiratory protection.	fresh air, rest, place in half-sitting position, take to hospital.
Skin: *corrosive*; redness, pain, burns.	gloves, protective clothing.	remove contaminated clothing, flush skin with water or shower, refer to a doctor.
Eyes: *corrosive*; redness, pain, impaired vision.	face shield, or combined eye and respiratory protection.	flush with water, take to doctor.
Ingestion: *corrosive*; abdominal pain, abdominal cramps.		rinse mouth, take immediately to hospital.

SPILLAGE	STORAGE	LABELING / NFPA
evacuate area, call in an expert, collect leakage in sealable containers, soak up with sand or other inert absorbent and remove to safe place; (additional individual protection: breathing apparatus).	keep in cool, fireproof place, separate from all other substances.	R: 45-12-20/21/22-36/37/38 S: 53-3/7/9-16-33-44 Note E 🔥 Highly flammable ☠ Toxic

NOTES
Before distilling check for peroxides; if found, render harmless. Odor limit is above TLV. Lung edema symptoms usually develop several hours later and are aggravated by physical exertion: rest and hospitalization essential. As first aid, a doctor or authorized person should consider administering a corticosteroid spray. Unbreakable packaging preferred; if breakable, keep in unbreakable container. Flash point of 1% aqueous solution: 23°C.

Transport Emergency Card TEC(R)-158

HI: 33; UN-number: 1280

CAS-No: [110-86-1]
azabenzene
azine

PYRIDINE

PHYSICAL PROPERTIES		IMPORTANT CHARACTERISTICS
Boiling point, °C	115	**COLORLESS TO LIGHT YELLOW LIQUID WITH NAUSEOUS ODOR**
Melting point, °C	−42	Vapor mixes readily with air, forming explosive mixtures. Do not use compressed air when filling,
Flash point, °C	17	emptying or processing. Reacts violently with strong oxidants.
Auto-ignition temperature, °C	482	
Relative density (water = 1)	0.98	

TLV-TWA	5 ppm	16 mg/m³

Relative density at 20 °C of saturated mixture
vapor/air (air = 1)   1.03

Absorption route: Can enter the body by inhalation or ingestion or through the skin. Harmful atmospheric concentrations can build up fairly rapidly on evaporation at approx. 20°C - even more rapidly in aerosol form.

Vapor pressure, mm Hg at 20 °C   25
Solubility in water   ∞
Explosive limits, vol% in air   1.8-12.4
Electrical conductivity, pS/m   4 x 10⁶
Relative molecular mass   79.1
Log P octanol/water   0.7

Immediate effects: Irritates the eyes, skin and respiratory tract. In substantial concentrations can impair consciousness.
Effects of prolonged/repeated exposure: Can cause liver and kidney damage. Can make the skin hypersensitive to sunlight.

Gross formula: C_5H_5N

HAZARDS/SYMPTOMS	PREVENTIVE MEASURES	FIRE EXTINGUISHING/FIRST AID
Fire: extremely flammable.	keep away from open flame and sparks, no smoking.	powder, alcohol-resistant foam, large quantities of water, carbon dioxide, (halons).
Explosion: forms explosive air-vapor mixtures.	sealed machinery, ventilation, explosion-proof electrical equipment and lighting.	in case of fire: keep tanks/drums cool by spraying with water.
Inhalation: cough, shortness of breath, headache, dizziness, nausea, vomiting, sleeplessness.	ventilation, local exhaust or respiratory protection.	fresh air, rest, call a doctor.
Skin: *is absorbed*; redness.	gloves.	remove contaminated clothing, flush skin with water or shower, call a doctor.
Eyes: redness, pain, impaired vision.	face shield, or combined eye and respiratory protection.	flush with water, take to doctor if necessary.
Ingestion: abdominal pain, diarrhea, vomiting.		rinse mouth, call a doctor or take to hospital.

SPILLAGE	STORAGE	LABELING / NFPA
collect leakage in sealable containers, soak up with sand or other inert absorbent and remove to safe place; (additional individual protection: breathing apparatus).	keep in fireproof place, separate from oxidants.	R: 11-20/21/22 S: 26-28

Flammable   Harmful

NFPA: 3 / 2 / 0

NOTES

Depending on the degree of exposure, regular medical checkups are advisable.

CAS-No: [120-80-9]
1,2-benzenediol
catechol
1,2-dihydroxybenzene

$C_6H_4(OH)_2$

PYROCATECHOL

PHYSICAL PROPERTIES		IMPORTANT CHARACTERISTICS	
Boiling point, °C	246	**COLORLESS CRYSTALS WITH CHARACTERISTIC ODOR, TURNING BROWN ON EXPOSURE TO AIR OR LIGHT**	
Melting point, °C	105		
Flash point, °C	127	Reacts with oxidants. In dry state can form electrostatic charge if stirred, transported pneumatically, poured etc.	
Relative density (water = 1)	1.3		
Vapor pressure, mm Hg at 20 °C	0.15		
Solubility in water, g/100 ml at 20 °C	43	TLV-TWA 5 ppm 23 mg/m³	
Relative molecular mass	110.1		
Log P octanol/water	0.9	**Absorption route:** Can enter the body by inhalation or ingestion or through the skin. Harmful atmospheric concentrations build up fairly slowly on evaporation at 20°C, but harmful concentrations of airborne particles can build up much more rapidly. **Immediate effects:** Corrosive to the eyes, skin and respiratory tract. In substantial concentrations can impair consciousness and raise blood pressure. Inhalation can cause lung edema. In serious cases risk of seizures, unconsciousness and death. **Effects of prolonged/repeated exposure:** Can affect the blood. Can cause liver, kidney and brain damage.	
Gross formula: $C_6H_6O_2$			

HAZARDS/SYMPTOMS	PREVENTIVE MEASURES	FIRE EXTINGUISHING/FIRST AID
Fire: combustible.	keep away from open flame, no smoking.	water spray, powder.
Explosion: finely dispersed particles form explosive mixtures on contact with air.	keep dust from accumulating, sealed machinery, explosion-proof lighting, grounding.	
	STRICT HYGIENE	IN ALL CASES CALL IN A DOCTOR
Inhalation: *corrosive*; sore throat, cough, severe breathing difficulties, cramps, unconsciousness.	local exhaust or respiratory protection.	fresh air, rest, place in half-sitting position, take to hospital.
Skin: *corrosive, is absorbed*; redness, pain, serious burns; see also 'Inhalation'.	gloves, protective clothing.	remove contaminated clothing, flush skin with water or shower, take to hospital.
Eyes: *corrosive*; redness, pain, impaired vision.	face shield, or combined eye and respiratory protection.	flush with water, take to a doctor.
Ingestion: *corrosive*; abdominal pain, diarrhea, vomiting; see also 'Inhalation'.	do not eat, drink or smoke while working.	rinse mouth, take immediately to hospital.

SPILLAGE	STORAGE	LABELING / NFPA
clean up spillage, carefully collect remainder; (additional individual protection: respirator with A/P2 filter).	keep separate from strong oxidants; ventilate at floor level.	R: 21/22-36/38 S: 22-26-37 ❌ Harmful

NOTES
Remarkably volatile, despite being a solid. Depending on the degree of exposure, regular medical checkups are advisable. Lung edema symptoms usually develop several hours later and are aggravated by physical exertion: rest and hospitalization essential. As first aid, a doctor or authorized person should consider administering a corticosteroid spray. Use airtight packaging.

Transport Emergency Card TEC(R)-61G12

CAS-No: [87-66-1]
1,2,3-benzenetriol
pyrogallic acid
1,2,3-trihydroxybenzene

$C_6H_3(OH)_3$

PYROGALLOL

PHYSICAL PROPERTIES	IMPORTANT CHARACTERISTICS
Boiling point, °C 309 Melting point, °C 131 Flash point, °C see notes Relative density (water = 1) 1.5 Vapor pressure, mm Hg at 168 °C 9.9 Solubility in water, g/100 ml at 20 °C 60 Relative molecular mass 126.1 Log P octanol/water 0.3	**WHITE CRYSTALS, TURNING GRAY ON EXPOSURE TO AIR OR LIGHT** Reacts with oxidants. In dry state can form electrostatic charge if stirred, transported pneumatically, poured etc.
	TLV-TWA not available
	Absorption route: Can enter the body by inhalation or ingestion. Harmful atmospheric concentrations build up very slowly, if at all, on evaporation at 20°C, but harmful **Immediate effects:** Irritates the eyes, skin and respiratory tract. Affects the nervous system. **Effects of prolonged/repeated exposure:** Prolonged or repeated contact can cause skin disorders. Can affect the blood. Can cause liver and kidney damage.
Gross formula: $C_6H_6O_3$	

HAZARDS/SYMPTOMS	PREVENTIVE MEASURES	FIRE EXTINGUISHING/FIRST AID
Fire: combustible.	keep away from open flame, no smoking.	water spray, powder.
Explosion: finely dispersed particles form explosive mixtures on contact with air.	keep dust from accumulating, sealed machinery, explosion-proof lighting, grounding.	
	KEEP DUST UNDER CONTROL	
Inhalation: cough, shortness of breath, nausea, feeling of weakness.	local exhaust or respiratory protection.	fresh air, rest, call a doctor.
Skin: redness, pain.	gloves.	remove contaminated clothing, flush skin with water or shower, call a doctor.
Eyes: redness, pain.	goggles.	flush with water, take to a doctor.
Ingestion: diarrhea, vomiting, unconsciousness; see also 'Inhalation'.	do not eat, drink or smoke while working.	rinse mouth, immediately take to hospital.

SPILLAGE	STORAGE	LABELING / NFPA
clean up spillage, carefully collect remainder; (additional individual protection: P2 respirator).	keep separate from strong oxidants; ventilate at floor level.	R: 20/21/22 ❌ Harmful

NOTES
Remarkably volatile, despite being a solid. Combustible, but flash point and explosive limits not given in the literature. Log P octanol/water is estimated. Depending on the degree of exposure, regular medical checkups are advisable.

PYROPHOSPHORIC ACID

PHYSICAL PROPERTIES	IMPORTANT CHARACTERISTICS	
Melting point, °C ca. 60 Solubility in water, g/100 ml at 23 °C 709 Relative molecular mass 178.0	**COLORLESS HYGROSCOPIC CRYSTALS OR VITREOUS SOLID** Decomposes when heated, giving off corrosive → *phosphoric oxide* vapors. In aqueous solution is a strong acid which reacts violently with bases and is corrosive. In water is changed to → *phosphoric acid*, evolving heat.	
	TLV-TWA not available	
	Absorption route: Can enter the body by inhalation or ingestion. Evaporation negligible at 20°C, but harmful concentrations of airborne particles can build up rapidly. **Immediate effects:** Corrosive to the eyes, skin and respiratory tract. Inhalation of finely dispersed substance or solution in aerosol form can cause severe breathing difficulties (lung edema).	
Gross formula: $H_4O_7P_2$		

HAZARDS/SYMPTOMS	PREVENTIVE MEASURES	FIRE EXTINGUISHING/FIRST AID
Fire: non-combustible.		in case of fire in immediate vicinity: use any extinguishing agent.
Inhalation: *corrosive*; sore throat, cough, shortness of breath.	local exhaust or respiratory protection.	fresh air, rest, place in half-sitting position, call a doctor.
Skin: *corrosive*; redness, pain, burns.	gloves, protective clothing.	remove contaminated clothing, flush skin with water or shower, refer to a doctor.
Eyes: *corrosive*; redness, pain, impaired vision.	face shield.	flush with water, take to a doctor.
Ingestion: *corrosive*; sore throat, abdominal pain, diarrhea.		rinse mouth, call a doctor or take to hospital.

SPILLAGE	STORAGE	LABELING / NFPA
clean up spillage, flush away remainder with water; (additional individual protection: P2 respirator).	keep dry, separate from strong bases.	

NOTES
TLV for → *phosphoric acid* can usually be applied. Lung edema symptoms usually develop several hours later and are aggravated by physical exertion: rest and hospitalization essential. As first aid, a doctor or authorized person should consider administering a corticosteroid spray.

CAS-No: [106-51-4]
p-benzoquinone
1,4-benzoquinone
1,4-cyclohexadienedione
2,5-cyclohexane-1,4 dione
p-quinone

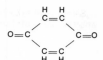

QUINONE

PHYSICAL PROPERTIES	IMPORTANT CHARACTERISTICS
Melting point (decomposes), °C 116 Flash point, °C ca. 40 Auto-ignition temperature, °C 435 Relative density (water = 1) 1.3 Relative vapor density (air = 1) 3.7 Relative density at 20 °C of saturated mixture vapor/air (air = 1) 1.0 Vapor pressure, mm Hg at 20 °C 0.019 Solubility in water moderate Relative molecular mass 108.1 Log P octanol/water 0.2	**YELLOW CRYSTALS WITH PUNGENT ODOR** Can sublimate even at room temperature. Finely dispersed in air can cause dust explosion. In dry state can form electrostatic charge if stirred, transported pneumatically, poured etc. Above 60°C when moist is self-heating and decomposes, giving off toxic gases (→ *carbon monoxide*). Weak oxidant which can react violently with some combustible substances and reducing agents. Reacts violently with strong bases.
	TLV-TWA 0.1 ppm 0.44 mg/m³
	Absorption route: Can enter the body by inhalation or ingestion or through the skin. Harmful atmospheric concentrations can build up fairly rapidly on evaporation at approx. 20°C; harmful concentrations of airborne particles can build up even more rapidly. **Immediate effects:** Corrosive to the eyes, skin and respiratory tract. Inhalation of particles/vapor can cause lung edema. Can affect the sclera, causing corneal edema, ulceration and scarring. Exposure to high concentrations can be fatal. **Effects of prolonged/repeated exposure:** Can affect the sclera, causing pigmentation of the sclera, corneal deformation and impaired vision. Prolonged or repeated contact can cause skin disorders.
Gross formula: $C_6H_4O_2$	

HAZARDS/SYMPTOMS	PREVENTIVE MEASURES	FIRE EXTINGUISHING/FIRST AID
Fire: combustible.	keep away from open flame, no smoking.	water spray, powder, carbon dioxide.
Explosion: above 40°C: forms explosive air-vapor mixtures; finely dispersed particles form explosive mixtures on contact with air.	keep dust from accumulating, closed system, explosion-proof electrical equipment and lighting, prevent buildup of electrostatic charges by grounding, etc.	
	STRICT HYGIENE, KEEP DUST UNDER CONTROL	
Inhalation: *corrosive*; sore throat, cough, shortness of breath, severe breathing difficulties.	local exhaust or respiratory protection.	fresh air, rest, place in half-sitting position, artificial respiration if necessary, take to hospital.
Skin: *corrosive, is absorbed*; redness, pain.	gloves.	remove contaminated clothing, flush skin with water or shower, consult a doctor.
Eyes: *corrosive*; redness, pain, impaired vision, prolonged exposure can cause loss of sight.	safety glasses, or combined eye and respiratory protection.	flush thoroughly with water (remove contact lenses if easy), take to a doctor.
Ingestion: *corrosive*; abdominal pain, diarrhea, nausea, vomiting.	do not eat, drink or smoke while working.	rinse mouth, DO NOT induce vomiting, take immediately to hospital.

SPILLAGE	STORAGE	LABELING / NFPA
clean up spillage, carefully collect remainder; (additional individual protection: breathing apparatus).	keep separate from combustible substances, reducing agents and strong bases; ventilate at floor level.	R: 23/25-36/37/38 S: 26-28-44 ☠ Toxic

NOTES
Physical properties (e.g. flash point) depend largely on humidity. One source gives flash point as 38-93°C. Lung edema symptoms usually develop several hours later and are aggravated by physical exertion: rest and hospitalization essential. As first aid, a doctor or authorized person should consider administering a corticosteroid spray.

Transport Emergency Card TEC(R)-61G12B

HI: 60; UN-number: 2587

RANEY NICKEL

PHYSICAL PROPERTIES	IMPORTANT CHARACTERISTICS
Auto-ignition temperature, °C ? Relative density (water = 1) 1.7-1.8 Solubility in water none Relative atomic mass 58.7 (as Ni)	**GRAYISH-BLACK SPONGY POWDER** Composed mainly of finely dispersed nickel, with a small percentage of aluminum. In active state contains hydrogen bonded to surface of metal. Can ignite spontaneously on exposure to air. In storage slowly releases → *hydrogen*, with risk of fire and explosion. Strong reducing agent which reacts violently with oxidants. Reacts violently with acids, giving off flammable gas (→ *hydrogen*).

	TLV-TWA — 0.05 mg/m³ A1 ✻

Absorption route: Can enter the body by inhalation or ingestion. Harmful concentrations of airborne particles can build up rapidly.
Immediate effects: Irritates the eyes and respiratory tract.
Effects of prolonged/repeated exposure: Prolonged or repeated contact can cause eczema. Has been found to cause a type of lung and nasal cancer under certain circumstances.

Gross formula: Ni

HAZARDS/SYMPTOMS	PREVENTIVE MEASURES	FIRE EXTINGUISHING/FIRST AID
Fire: extremely flammable, ignites on contact with air.	AVOID CONTACT WITH AIR	dry sand, special extinguishing powder, large quantities of water.
Explosion: finely dispersed particles form explosive mixtures on contact with air.	keep dust from accumulating, sealed machinery, explosion-proof electrical equipment and lighting, grounding.	
Inhalation: cough.	local exhaust or respiratory protection.	fresh air, rest.
Skin: redness, pain.	gloves.	remove contaminated clothing, flush skin with water or shower.
Eyes: redness, pain.	goggles.	flush with water, take to a doctor.

SPILLAGE	STORAGE	LABELING / NFPA
keep wet (under water if possible), call in an expert; (additional individual protection: P2 respirator).	keep in fireproof place under water or inert gas, separate from strong acids.	⬦

NOTES
✻ TLV equals that of nickel (metal). Flush contaminated clothing with water (fire hazard). Prepared by treating a nickel-aluminum alloy with a strong base. Special packaging.

lead oxide red
lead tetroxide

RED LEAD

PHYSICAL PROPERTIES	IMPORTANT CHARACTERISTICS	
Decomposes below melting point, °C 500 Relative density (water = 1) 9.1 Solubility in water none Relative molecular mass 685.6	**CLEAR RED POWDER** Decomposes when heated above 500°C, giving off → *oxygen*. Toxic → *lead* vapors can form when heated with organic substances (e.g. paint). Reacts violently with aluminum and zinc powder various reducing agents.	
	TLV-TWA 0.15 mg/m³ ✱	
	Absorption route: Can enter the body by inhalation or ingestion. Evaporation at 20°C negligible, but harmful atmospheric concentrations can build up rapidly in aerosol form. **Immediate effects:** Irritates the eyes and respiratory tract. **Effects of prolonged/repeated exposure:** Can affect the blood. Can cause kidney damage.	
Gross formula: O$_4$Pb$_3$		

HAZARDS/SYMPTOMS	PREVENTIVE MEASURES	FIRE EXTINGUISHING/FIRST AID
Fire: not combustible but enhances combustion of other substances.		in case of fire in immediate vicinity: use any extinguishing agent.
	STRICT HYGIENE, KEEP DUST UNDER CONTROL	
Inhalation: sore throat, cough, headache, dizziness, constipation, abdominal cramps.	local exhaust or respiratory protection.	fresh air, rest, call a doctor.
Skin: redness.	gloves.	remove contaminated clothing, flush skin with water or shower.
Eyes: redness, pain.	goggles.	flush with water, take to a doctor.
Ingestion: headache, nausea, constipation, abdominal cramps.	do not eat, drink or smoke while working.	rinse mouth, immediately take to hospital.

SPILLAGE	STORAGE	LABELING / NFPA
clean up spillage, carefully collect remainder; (additional individual protection: P2 respirator).	keep separate from stron reducing agents.	R: 20/22-23 S: 13-20/21 Note A ✖ Harmful

NOTES
✱ TLV equals that of lead. The primer that contains red lead as effective steel conservation agent is often referred to as red lead. Do not take work clothing home. When processing and removing the primer (even after years) all the above mentioned measures apply. Special first aid required in the event of poisoning: antidotes must be available (with instructions). This also applies to heating, welding and cutting of materials with red lead. Lead level in blood is highly recommended as a yardstick for exposure.

Transport Emergency Card TEC(R)-221

HI: 60; UN-number: 2291

CAS-No: [108-46-3]
1,3-benzenediol
1,3-dihydroxybenzene
m-dihydroxybenzene
m-hydroquinone
3-hydroxyphenol

$C_6H_4(OH)_2$

RESORCINOL

PHYSICAL PROPERTIES		IMPORTANT CHARACTERISTICS	
Boiling point, °C	280	**WHITE CRYSTALS, TURNING PINK ON EXPOSURE TO AIR**	
Melting point, °C	110	Reacts violently with strong oxidants. Can form electrostatic charge if stirred, transported pneumatically, poured etc.; finely dispersed in air can cause dust explosion.	
Flash point, °C	127		
Auto-ignition temperature, °C	600		
Relative density (water = 1)	1.3	TLV-TWA	10 ppm	45 mg/m³
Relative vapor density (air = 1)	3.8	STEL	20 ppm	90 mg/m³
Relative density at 20 °C of saturated mixture vapor/air (air = 1)	1.0	**Absorption route:** Can enter the body by inhalation or ingestion or through the skin. Harmful atmospheric concentrations build up very slowly, if at all, on evaporation at approx. 20°C, but harmful concentrations of airborne particles can build up much more rapidly.	
Vapor pressure, mm Hg at 20 °C	0.0076		
Solubility in water, g/100 ml at 20 °C	140		
Explosive limits, vol% in air	1.4–?	**Immediate effects:** Irritates the eyes, skin and respiratory tract. Affects the nervous system. In serious cases risk of seizures and death.	
Relative molecular mass	110.1		
Log P octanol/water	0.8	**Effects of prolonged/repeated exposure:** Prolonged or repeated contact can cause skin disorders. Can affect the blood. Can cause liver, kidney and heart damage.	
Gross formula:	$C_6H_6O_2$		

HAZARDS/SYMPTOMS	PREVENTIVE MEASURES	FIRE EXTINGUISHING/FIRST AID
Fire: combustible.	keep away from open flame, no smoking.	water spray, powder.
Explosion: finely dispersed particles form explosive mixtures on contact with air.	keep dust from accumulating, explosion-proof electrical equipment and lighting, grounding.	
	STRICT HYGIENE	
Inhalation: sore throat, cough, shortness of breath, dizziness, cramps, unconsciousness, blue feeling of weakness.	local exhaust or respiratory protection.	fresh air, rest, take to hospital.
Skin: *is absorbed*; redness; see also 'Inhalation'.	gloves, protective clothing.	remove contaminated clothing, wash skin with soap and water, call a doctor.
Eyes: redness, pain, impaired vision.	face shield, or combined eye and respiratory protection.	flush with water, take to doctor.
Ingestion: abdominal pain; see also 'Inhalation'.		rinse mouth, induce vomiting, take immediately to hospital.

SPILLAGE	STORAGE	LABELING / NFPA
clean up spillage, carefully collect remainder; (additional individual protection: P2 respirator).	keep separate from strong oxidants.	R: 22-36/38 S: 26 ✖ Harmful

NOTES

Depending on the degree of exposure, regular medical checkups are advisable. Effects on blood due to formation of methemoglobin. Special first aid required in the event of poisoning: antidotes must be available (with instructions).

Transport Emergency Card TEC(R)-61G12C

HI: 60; UN-number: 2876

CAS-No: [69-72-7]
2-hydroxybenzoic acid

HOC_6H_4COOH

SALICYLIC ACID

PHYSICAL PROPERTIES		IMPORTANT CHARACTERISTICS	
Melting point, °C	157	**WHITE POWDER OR CRYSTALS, DISCOLORING ON EXPOSURE TO SUNLIGHT**	
Flash point, °C	157	In dry state can form electrostatic charge if stirred, transported pneumatically, poured etc.	
Auto-ignition temperature, °C	570	Decomposes when heated, giving off toxic → *phenol* vapors and → *carbon dioxide*. Reacts with	
Relative density (water = 1)	1.4	strong oxidants, iron salts, lead acetate and iodine.	
Vapor pressure, mm Hg at 114 °C	0.99		
Solubility in water, g/100 ml at 20 °C	0.2	TLV-TWA not available	
Relative molecular mass	138.1		
Log P octanol/water	2.2		
		Absorption route: Can enter the body by inhalation or ingestion or through the skin. Evaporation negligible at 20°C, but harmful concentrations of airborne particles can build up rapidly. **Immediate effects:** Irritates the eyes, skin and respiratory tract. Affects the nervous system. In serious cases risk of seizures, unconsciousness and death. **Effects of prolonged/repeated exposure:** Prolonged or repeated contact can cause eczema.	
Gross formula: $C_7H_6O_3$			

HAZARDS/SYMPTOMS	PREVENTIVE MEASURES	FIRE EXTINGUISHING/FIRST AID
Fire: combustible.	keep away from open flame, no smoking.	water spray, powder.
Explosion: finely dispersed particles form explosive mixtures on contact with air.	keep dust from accumulating, sealed machinery, explosion-proof electrical equipment and lighting, grounding.	
Inhalation: sore throat, cough, headache, dizziness, ringing in the ears, rapid pulse.	local exhaust or respiratory protection.	fresh air, rest, call a doctor.
Skin: *is absorbed*; redness, pain.	gloves.	remove contaminated clothing, flush skin with water or shower, call a doctor.
Eyes: redness, pain.	goggles, or combined eye and respiratory protection.	flush with water, take to a doctor if necessary.
Ingestion: abdominal pain, nausea, vomiting.	do not eat, drink or smoke while working.	rinse mouth, take immediately to hospital.

SPILLAGE	STORAGE	LABELING / NFPA
clean up spillage, carefully collect remainder; (additional individual protection: P2 respirator).	keep separate from oxidants.	

NOTES

SELENIUM DIOXIDE

PHYSICAL PROPERTIES		IMPORTANT CHARACTERISTICS	
Sublimation temperature, °C	315	**WHITE HYGROSCOPIC POWDER OR NEEDLES**	
Relative density (water = 1)	4	Reacts with many substances, giving off toxic vapors (selenium). Attacks many metals.	
Vapor pressure, mm Hg at 70 °C	13		
Solubility in water, g/100 ml at 14 °C	38.4	TLV-TWA	0.2 mg/m³ ✹
Relative molecular mass	111		
		Absorption route: Can enter the body by inhalation or ingestion or through the skin. Harmful atmospheric concentrations can build up fairly rapidly on evaporation at approx. 20° C; harmful concentrations of airborne particles can build up even more rapidly.	
		Immediate effects: Corrosive to the eyes, skin and respiratory tract. Inhalation can cause lung edema. In serious cases risk of seizures, unconsciousness and death.	
		Effects of prolonged/repeated exposure: Can cause liver and kidney damage.	
Gross formula:	O$_2$Se		

HAZARDS/SYMPTOMS	PREVENTIVE MEASURES	FIRE EXTINGUISHING/FIRST AID
Fire: non-combustible.		
	STRICT HYGIENE	IN ALL CASES CALL IN A DOCTOR
Inhalation: *corrosive*; sore throat, cough, severe breathing difficulties, nausea, exhalation smells of garlic.	local exhaust or respiratory protection.	fresh air, rest, place in half-sitting position, take to hospital.
Skin: *corrosive*; redness, pain, burns.	gloves, protective clothing.	remove contaminated clothing, flush skin with water or shower, call a doctor.
Eyes: *corrosive*; redness, pain, impaired vision.	face shield, or combined eye and respiratory protection.	flush with water, take to a doctor.
Ingestion: *corrosive*; sore throat, abdominal pain, diarrhea, nausea, vomiting; see also 'Inhalation'.	do not eat, drink or smoke while working.	rinse mouth, immediately take to hospital.

SPILLAGE	STORAGE	LABELING / NFPA
call in an expert, clean up spillage, carefully collect remainder and remove to safe place; (additional individual protection: breathing apparatus).	keep dry; ventilate at floor level.	R: 23/25 - 33 S: 20/21 - 28 - 44 Note A ☠ Toxic

NOTES
✹ TLV equals that of selenium. Lung edema symptoms usually develop several hours later and are aggravated by physical exertion: rest and hospitalization essential. As first aid, a doctor or authorized person should consider administering a corticosteroid spray. Unbreakable airtight packaging preferred; if breakable, keep in unbreakable container.

HI: 60; UN-number: 2811

SELENIUM HYDRIDE
(cylinder)

PHYSICAL PROPERTIES	IMPORTANT CHARACTERISTICS
Boiling point, °C −42 Melting point, °C −64 Flash point, °C flammable gas Relative density (water = 1) 2.0 Relative vapor density (air = 1) 2.8 Vapor pressure, mm Hg at 21 °C 7296 Solubility in water, g/100 ml at 20 °C 0.7 Explosive limits, vol% in air ?-? Relative molecular mass 81.0	**COLORLESS COMPRESSED LIQUEFIED GAS WITH CHARACTERISTIC ODOR** Gas is heavier than air and spreads at ground level, with risk of ignition at a distance. Corrosive, toxic vapors are a product of combustion. Decomposes when heated above 160°C, forming selenium and → *hydrogen*. Strong reducing agent which reacts violently with oxidants, with risk of fire and explosion.

TLV-TWA	0.05 ppm	0.16 mg/m³ ✱

Absorption route: Can enter the body by inhalation. Harmful atmospheric concentrations can build up very rapidly if gas is released.
Immediate effects: Corrosive to the eyes, skin and respiratory tract. Can cause frostbite due to rapid evaporation. Inhalation of vapor/fumes can cause severe breathing difficulties (lung edema).

Gross formula: H$_2$Se

HAZARDS/SYMPTOMS	PREVENTIVE MEASURES	FIRE EXTINGUISHING/FIRST AID
Fire: extremely flammable.	keep away from open flame and sparks, no smoking.	shut off supply; if impossible and no danger to surrounding area, allow fire to burn itself out; otherwise extinguish with powder, carbon dioxide, (halons).
Explosion: forms explosive air-gas mixtures.	sealed machinery, ventilation, explosion-proof electrical equipment and lighting. STRICT HYGIENE	in case of fire: keep cylinders cool by spraying with water.
Inhalation: *corrosive*; sore throat, cough, severe breathing difficulties, nausea, feeling of weakness.	ventilation, local exhaust or respiratory protection.	fresh air, rest, place in half-sitting position, take to hospital.
Skin: *in case of frostbite*: redness, pain, sores.	insulating gloves.	*in case of frostbite*: DO NOT remove clothing, flush skin with water or shower, take to hospital.
Eyes: *corrosive, in case of frostbite*: redness, pain, impaired vision.	face shield, or combined eye and respiratory protection.	flush with water, take to a doctor.

SPILLAGE	STORAGE	LABELING / NFPA
evacuate area, call in an expert, ventilate, under no circumstances spray liquid with water; (additional individual protection: breathing apparatus).	keep in cool, fireproof place.	

NOTES
Explosive limits not given in the literature. ✱ TLV equals that of selenium. Lung edema symptoms usually develop several hours later and are aggravated by physical exertion: rest and hospitalization essential. As first aid, a doctor or authorized person should consider administering a corticosteroid spray. Turn leaking cylinder so that leak is on top to prevent liquid hydrogen selenide escaping.

Transport Emergency Card TEC(R)-20G12

HI: 236; UN-number: 2202

SELENIUM OXYCHLORIDE

PHYSICAL PROPERTIES	IMPORTANT CHARACTERISTICS	
Boiling point, °C — 176 Melting point, °C — 8.5 Relative density (water = 1) — 2.4 Relative vapor density (air = 1) — 5.1 Relative density at 20 °C of saturated mixture vapor/air (air = 1) — 1.0 Vapor pressure, mm Hg at 35 °C — 0.99 Solubility in water — reaction Relative molecular mass — 165.9	**COLORLESS OR YELLOW LIQUID** Reacts with water to form → *hydrochloric acid*. Reacts with air to form corrosive vapors (→ *hydrochloric acid*). Reacts violently with phosphorus, antimony and alkali metals.	
	TLV-TWA	0.2 mg/m³ ✱
	Absorption route: Can enter the body by inhalation or ingestion or through the skin. Harmful atmospheric concentrations can build up rapidly on evaporation at approx. 20°C - even more rapidly in aerosol form. **Immediate effects:** Corrosive to the eyes, skin and respiratory tract. Inhalation of vapor/fumes can cause severe breathing difficulties (lung edema). In serious cases risk of seizures, unconsciousness and death. **Effects of prolonged/repeated exposure:** Can cause liver and kidney damage.	
Gross formula: ClOSe		

HAZARDS/SYMPTOMS	PREVENTIVE MEASURES	FIRE EXTINGUISHING/FIRST AID
Fire: non-combustible.	avoid contact with antimony, phosphorus and light metals.	in case of fire in immediate vicinity: preferably no water or water-based extinguishers.
	STRICT HYGIENE	IN ALL CASES CALL IN A DOCTOR
Inhalation: *corrosive*; sore throat, cough, severe breathing difficulties, nausea, exhalation smells of garlic.	ventilation, local exhaust or respiratory protection.	fresh air, rest, place in half-sitting position, take to hospital.
Skin: *corrosive*; redness, pain, serious burns.	gloves, protective clothing.	remove contaminated clothing, flush skin with water or shower, call a doctor.
Eyes: *corrosive*; redness, pain, impaired vision.	face shield, or combined eye and respiratory protection.	flush with water, take to a doctor.
Ingestion: *corrosive*; sore throat, abdominal pain, diarrhea, nausea, vomiting; see also 'Inhalation'.	do not eat, drink or smoke while working.	rinse mouth, immediately take to hospital.

SPILLAGE	STORAGE	LABELING / NFPA
call in an expert, collect leakage in sealable containers, soak up with sand or other inert absorbent and remove to safe place; (additional individual protection: breathing apparatus).	keep dry, separate from antimony, phosphorus and light metals; ventilate at floor level.	R: 23/25-33 S: 20/21-28-44 Note A ☠ Toxic

NOTES
✱ TLV equals that of selenium. Lung edema symptoms usually develop several hours later and are aggravated by physical exertion: rest and hospitalization essential. As first aid, a doctor or authorized person should consider administering a corticosteroid spray. Use airtight packaging.

UN-number: 2879

SELENIUM TRIOXIDE

PHYSICAL PROPERTIES	IMPORTANT CHARACTERISTICS	
Melting point, °C 118 Relative density (water = 1) 3.6 Solubility in water reaction Relative molecular mass 126.9	**YELLOWISH-WHITE HYGROSCOPIC POWDER** Strong oxidant which reacts violently with combustible substances and reducing agents. Reacts violently with water, giving off corrosive selenic acid. Corrosive to metals in the presence of moisture.	
	TLV-TWA	0.2 mg/m³ ✱
Gross formula: O$_3$Se	**Absorption route:** Can enter the body by inhalation or ingestion or through the skin. Evaporation negligible at 20°C, but harmful concentrations of airborne particles can build up rapidly. **Immediate effects:** Corrosive to the eyes, skin and respiratory tract. Inhalation can cause lung edema. In serious cases risk of seizures, unconsciousness and death. **Effects of prolonged/repeated exposure:** Can cause liver and kidney damage.	

HAZARDS/SYMPTOMS	PREVENTIVE MEASURES	FIRE EXTINGUISHING/FIRST AID
Fire: not combustible but enhances combustion of other substances.	avoid contact with combustible substances.	in case of fire in immediate vicinity: preferably no water-based extinguishers.
Explosion: many chemical reactions can cause explosion.		
	STRICT HYGIENE	
Inhalation: *corrosive*; sore throat, cough, severe breathing difficulties, nausea, exhalation smells of garlic.	local exhaust or respiratory protection.	fresh air, rest, place in half-sitting position, take to hospital.
Skin: *corrosive*; redness, pain, burns.	gloves, protective clothing.	remove contaminated clothing, flush skin with water or shower, call a doctor.
Eyes: *corrosive*; redness, pain, impaired vision.	face shield, or combined eye and respiratory protection.	flush with water, take to a doctor.
Ingestion: *corrosive*; sore throat, abdominal pain, diarrhea, nausea, vomiting; see also 'Inhalation'.	do not eat, drink or smoke while working.	rinse mouth, take immediately to hospital.

SPILLAGE	STORAGE	LABELING / NFPA
clean up spillage, carefully collect remainder; (additional individual protection: P2 respirator).	keep dry, separate from combustible substances.	R: 23/25-33 S: 20/21-28-44 Note A ☠ Toxic

NOTES
✱ TLV equals that of selenium. Lung edema sumptoms usually develop several hours later and are aggravated by physical exertion: rest and hospitalization essential. As first aid, a doctor or authorized person should consider administering a corticosteroid spray.

Transport Emergency Card TEC(R)-61G11

SELENOUS ACID

PHYSICAL PROPERTIES	IMPORTANT CHARACTERISTICS	
Decomposes below melting point, °C 70 Relative density (water = 1) 3.0 Vapor pressure, mm Hg at 15 °C 2.1 Solubility in water, g/100 ml at 20 °C 167 Relative molecular mass 129.0	**TRANSPARENT HYGROSCOPIC CRYSTALS** Decomposes when heated or exposed to air, forming → *selenium oxides* and water.	
	TLV-TWA	0.2 mg/m³ ✳
	Absorption route: Can enter the body by inhalation or ingestion. Harmful atmospheric concentrations can build up rapidly on evaporation at 20°C. **Immediate effects:** Irritates the eyes, skin and respiratory tract. Inhalation of vapor/fumes can cause severe breathing difficulties (lung edema). In serious cases risk of seizures, unconsciousness and death. **Effects of prolonged/repeated exposure:** Can cause liver and kidney damage.	
Gross formula: H_2O_3Se		

HAZARDS/SYMPTOMS	PREVENTIVE MEASURES	FIRE EXTINGUISHING/FIRST AID
Fire: non-combustible.		
Inhalation: *corrosive*; sore throat, cough, severe breathing difficulties, nausea, exhalation smells of garlic.	local exhaust or respiratory protection.	fresh air, rest, place in half-sitting position, take to hospital.
Skin: *corrosive*; redness, pain, serious burns.	gloves, protective clothing.	remove contaminated clothing, flush skin with water or shower, refer to a doctor.
Eyes: *corrosive*; redness, pain, impaired vision.	face shield, or combined eye and respiratory protection.	flush with water, take to a doctor.
Ingestion: *corrosive*; sore throat, abdominal pain, nausea, vomiting; see also 'Inhalation'.	do not eat, drink or smoke while working.	rinse mouth, take immediately to hospital.

SPILLAGE	STORAGE	LABELING / NFPA
call in an expert, clean up spillage, carefully collect remainder; (additional individual protection: breathing apparatus).	keep dry; ventilate at floor level.	R: 23/25-33 S: 20/21-28-44 Note A ☠ Toxic

NOTES
✳ TLV equals that of selenium. Lung edema symptoms usually develop several hours later and are aggravated by physical exertion: rest and hospitalization essential. As first aid, a doctor or authorized person should consider administering a corticosteroid spray. Use airtight packaging.

Transport Emergency Card TEC(R)-80G38

CAS-No: [7631-86-9]
quartz
quartz powder

SiO_2

SILICON DIOXIDE

PHYSICAL PROPERTIES	IMPORTANT CHARACTERISTICS
Boiling point, °C 2230 Melting point, °C 1610 Relative density (water = 1) 2.6 Solubility in water none Relative molecular mass 60.1 Gross formula: O_2Si	**WHITE POWDER** Silicon dioxide is a component of among others, sand, granite, scouring powder and clay. TLV-TWA 0.1 mg/m³ ✱ **Absorption route:** Can enter the body by inhalation. Evaporation negligible at 20°C, but harmful atmospheric concentrations can build up rapidly in aerosol form. **Immediate effects:** Irritates the eyes. **Effects of prolonged/repeated exposure:** Prolonged contact and inhalation of dust particles can cause (irreversible) lung disorders (silicosis).

HAZARDS/SYMPTOMS	PREVENTIVE MEASURES	FIRE EXTINGUISHING/FIRST AID
Fire: non-combustible.		in case of fire in immediate vicinity: use any extinguishing agent.
Inhalation: cough, irritation of throat.	local exhaust or respiratory protection.	fresh air, rest.
Eyes: redness, pain.	goggles, or combined eye and respiratory protection.	flush with water, take to a doctor if necessary.

SPILLAGE	STORAGE	LABELING / NFPA
clean up spillage, flush away remainder with water; (additional individual protection: P2 respirator).		

NOTES
Dutch silicosis legislation has laid down strigent rules for safe handling of silicon dioxide. ✱ MAC value as respirable dust.

SILICON OXIDE

PHYSICAL PROPERTIES	IMPORTANT CHARACTERISTICS	
Sublimation temperature, °C 1880 Melting point, °C > 1702 Relative density (water = 1) 2.1 Solubility in water none Relative molecular mass 44.1	**BLACK POWDER** The substance is made up of hard and durable particles. Finely dispersed in air can cause dust explosion. In dry state can form electrostatic charge if stirred, transported pneumatically, poured etc. Can ignite spontaneously on exposure to air. Strong reducing agent which reacts violently with oxidants and can consume the oxygen in the air.	
	TLV-TWA not available	
	Absorption route: The substance is not absorbed by the body. Evaporation negligible at 20°C, but harmful concentrations of airborne particles can build up rapidly. **Immediate effects:** Irritates the eyes and respiratory tract. **Effects of prolonged/repeated exposure:** Prolonged or repeated contact with dust can cause lung disorders.	
Gross formula: OSi		

HAZARDS/SYMPTOMS	PREVENTIVE MEASURES	FIRE EXTINGUISHING/FIRST AID
Fire: extremely flammable.	keep away from open flame and sparks, no smoking, keep exposure of freshly prepared powder to a minimum.	special powder extinguisher, dry sand, NO other extinguishing agents.
Explosion: finely dispersed particles form explosive mixtures on contact with air.	keep dust from accumulating, closed system, explosion-proof electrical equipment and lighting, prevent buildup of electrostatic charges by grounding, etc.	
	KEEP DUST UNDER CONTROL	
Inhalation: cough.	local exhaust or respiratory protection.	fresh air, rest.
Skin:	gloves.	
Eyes: irritation, redness.	acid goggles.	flush thoroughly with water (remove contact lenses if easy), take to a doctor.
Ingestion:	do not eat, drink or smoke while working.	

SPILLAGE	STORAGE	LABELING / NFPA
clean up spillage, carefully collect remainder; (additional individual protection: P2 respirator).	keep in fireproof place, separate from oxidants.	◇

NOTES
Unstable: disproportions slowly to Si and SiO_2, causing color to change to dark brown. Insufficient toxicological data on harmful effects to humans.

SILICON TETRACHLORIDE

PHYSICAL PROPERTIES		IMPORTANT CHARACTERISTICS
Boiling point, °C	58	**COLORLESS FUMING LIQUID WITH PUNGENT ODOR**
Melting point, °C	−70	Vapor is heavier than air. Reacts violently with water, water vapor and alcohol, forming
Relative density (water = 1)	1.5	→ *hydrochloric acid* and → *hydrochloric acid* fumes and evolving heat. Reacts with air to form
Relative vapor density (air = 1)	5.9	corrosive vapors (→ *hydrochloric acid*). Attacks many metals in the presence of water, giving off
Relative density at 20 °C of saturated mixture		flammable gas (→ *hydrogen*).
vapor/air (air = 1)	2.2	
Vapor pressure, mm Hg at 20 °C	196	TLV-TWA not available
Solubility in water	reaction	
Relative molecular mass	170	**Absorption route:** Can enter the body by inhalation or ingestion or through the skin. Harmful
		concentrations build up very rapidly on evaporation at approx. 20°C.
		Immediate effects: Corrosive to the eyes, skin and respiratory tract. Inhalation of vapor/fumes
		can cause severe breathing difficulties (lung edema). In serious cases risk of death.
Gross formula:	Cl$_4$Si	

HAZARDS/SYMPTOMS	PREVENTIVE MEASURES	FIRE EXTINGUISHING/FIRST AID
Fire: non-combustible.		in case of fire in immediate vicinity: preferably no water.
Explosion:		in case of fire: keep tanks/drums cool by spraying with water, but DO NOT spray substance with water.
	STRICT HYGIENE	
Inhalation: *corrosive*; sore throat, cough, shortness of breath, severe breathing difficulties.	ventilation, local exhaust or respiratory protection.	fresh air, rest, place in half-sitting position, take to hospital.
Skin: *corrosive*; redness, pain, burns.	gloves, protective clothing.	remove contaminated clothing, flush skin with water or shower, take to a doctor.
Eyes: *corrosive*; redness, pain, impaired vision.	face shield, or combined eye and respiratory protection.	flush with water, take to a doctor.
Ingestion: *corrosive*; diarrhea, nausea.		rinse mouth, take immediately to hospital.

SPILLAGE	STORAGE	LABELING / NFPA
render harmless with moist sodium bicarbonate and flush away remainder with water; (additional individual protection: CHEMICAL SUIT).	keep cool and dry, separate from aqueous solutions and alcohols; ventilate.	R: 14-36/37/38 S: 7/8-26 ✖ Irritant NFPA: 3 / 0 / 2 / W

NOTES
Lung edema symptoms usually develop several hours later and are aggravated by physical exertion: rest and hospitalization essential. As first aid, a doctor or authorized person should consider administering a corticosteroid spray. Use airtight packaging.

Transport Emergency Card TEC(R)-609

HI: 80; UN-number: 1818

SILVER NITRATE

PHYSICAL PROPERTIES	IMPORTANT CHARACTERISTICS	
Decomposes below boiling point, °C 444 Melting point, °C 212 Relative density (water = 1) 4.3 Solubility in water, g/100 ml at 20 °C 219 Relative molecular mass 169.9	**COLORLESS TO GRAY CRYSTALS** Decomposes on contact with organic contaminants when exposed to light. Decomposes when heated, giving off → *nitrous vapors*. Reacts with anhydrous ammonia to form shock-sensitive compounds. Oxidant, corrosive.	
	TLV-TWA	0.01 mg/m³ ✶
	Absorption route: Can enter the body by inhalation or ingestion. Evaporation negligible at 20°C, but harmful concentrations of airborne particles can build up rapidly. **Immediate effects:** Irritates the eyes, skin and respiratory tract. In serious cases risk of seizures. **Effects of prolonged/repeated exposure:** Can affect the blood if swallowed.	
Gross formula: AgNO$_3$		

HAZARDS/SYMPTOMS	PREVENTIVE MEASURES	FIRE EXTINGUISHING/FIRST AID
Fire: non-combustible, but makes other substances more combustible.	avoid contact with combustible substances and reducing agents.	in case of fire in immediate vicinity: use any extinguishing agent.
Explosion: old ammoniacal silver solutions can cause explosion.		in case of fire: keep tanks/drums cool by spraying with water.
	KEEP DUST UNDER CONTROL	
Inhalation: sore throat, cough, shortness of breath.	local exhaust or respiratory protection.	fresh air, rest, call a doctor.
Skin: redness, pain.	gloves.	remove contaminated clothing, flush skin with water or shower.
Eyes: redness, pain.	face shield.	flush with water, take to doctor.
Ingestion: abdominal pain, feeling of weakness, muscle cramps, blue skin.	do not eat, drink or smoke while working.	rinse mouth, call a doctor or take to hospital.

SPILLAGE	STORAGE	LABELING / NFPA
clean up spillage, carefully collect remainder; (additional individual protection: P3 respirator).	keep in dark place, separate from combustible substances and reducing agents.	R: 34 S: 2-26 Corrosive

NOTES
✶ TLV equals that of silver (soluble compounds). Effects on blood due to formation of methemoglobin. Special first aid required in the event of poisoning: antidotes must be available (with instructions).

Transport Emergency Card TEC(R)-51G02

UN-number: 1493

SODIUM

PHYSICAL PROPERTIES	IMPORTANT CHARACTERISTICS
Boiling point, °C 883 Melting point, °C 98 Auto-ignition temperature, °C see notes Relative density (water = 1) 0.97 Vapor pressure, mm Hg at 400 °C 1.2 Solubility in water reaction Relative atomic mass 23.0	**LIGHT SILVER-WHITE SOFT METAL, TURNING GRAY ON EXPOSURE TO AIR** Readily able to form peroxides on contact with moist air. Can ignite spontaneously on exposure to air, esp. moist air. Strong reducing agent which reacts violently with oxidants, with risk of fire and explosion. Reacts violently with water and water vapor, giving off flammable gas (→ *hydrogen*) and corrosive fumes (→ *sodium hydroxide*), with risk of fire and explosion. Reacts violently with many compounds, esp. halogens, halogenated hydrocarbons, acids, carbon dioxide, haloids and oxides of heavy metals, with risk of fire and explosion.
	TLV-TWA not available
	Absorption route: Can enter the body by inhalation or ingestion. Evaporation negligible at 20°C, but harmful atmospheric concentrations of fumes can build up rapidly on contact with water. **Immediate effects:** Corrosive to the eyes, skin and respiratory tract. Inhalation can cause lung edema. Can affect the cornea, causing impairment of vision. Exposure to high concentrations can be fatal.
Gross formula: Na	

HAZARDS/SYMPTOMS	PREVENTIVE MEASURES	FIRE EXTINGUISHING/FIRST AID
Fire: extremely flammable, many chemical reactions can cause fire and explosion.	keep away from open flame and sparks, no smoking, avoid contact with water and other substances.	special powder extinguisher, dry sand, NO other extinguishing agents.
Explosion: contact with water or moisture (hydrogen formation) or violent reactions can cause explosion.	prevent contact with water, moisture and many other substances.	
	KEEP DUST UNDER CONTROL, STRICT HYGIENE	
Inhalation: *corrosive*; cough, shortness of breath, sore throat, severe breathing difficulties.	ventilation.	fresh air, rest, place in half-sitting position, artificial respiration if necessary, take immediately to hospital.
Skin: *corrosive*; burns, pain.	gloves.	remove contaminated clothing, flush skin with water or shower, consult a doctor.
Eyes: *corrosive*; redness, pain, impaired vision, serious burns.	face shield.	flush thoroughly with water (remove contact lenses if easy), take to a doctor.
Ingestion: *corrosive*; abdominal pain, nausea, sore throat.		rinse mouth, DO NOT induce vomiting, take immediately to hospital.

SPILLAGE	STORAGE	LABELING / NFPA
evacuate area, call in an expert, collect dry spillage, cover remainder with oil, paraffin oil or dry sand and render harmless with spirits.	keep in dry fireproof place under oil or paraffin oil, separate from all other substances.	R: 14-15-34 S: 5-8-43 Flammable Corrosive 3 / 3 / 2 / W

NOTES

Autoignition temperature in dry air: 115°C. Reacts violently with extinguishing agents such as water, foam, carbon dioxide and halons. Lung edema symptoms usually develop several hours later and are aggravated by physical exertion: rest and hospitalization essential. As first aid, a doctor or authorized person should consider administering a suitable spray. Unbreakable packaging preferred; if breakable, keep in unbreakable container.

Transport Emergency Card TEC(R)-197/710   **HI: X423; UN-number: 1428**

SODIUM ACETATE

PHYSICAL PROPERTIES	IMPORTANT CHARACTERISTICS	
Decomposes below boiling point, °C ? Melting point, °C 324 Auto-ignition temperature, °C 607 Relative density (water = 1) 1.53 Solubility in water very good Relative molecular mass 82	**WHITE HYGROSCOPIC CRYSTALS** Decomposes when heated and on contact with strong acids, giving off corrosive vapors (→ *acetic acid*).	
	TLV-TWA not available	
	Absorption route: Can enter the body by inhalation or ingestion. Evaporation negligible at 20°C, but harmful concentrations of airborne particles can build up rapidly. **Immediate effects:** Irritates the eyes, skin and respiratory tract.	
Gross formula: $C_2H_3NaO_2$		

HAZARDS/SYMPTOMS	PREVENTIVE MEASURES	FIRE EXTINGUISHING/FIRST AID
Fire: combustible.	keep away from open flame, no smoking.	water spray, powder.
Explosion: finely dispersed particles form explosive mixtures on contact with air.	keep dust from accumulating; sealed machinery, explosion-proof electrical equipment and lighting, grounding.	
Inhalation: cough, shortness of breath.	local exhaust or respiratory protection.	fresh air, rest, place in half-sitting position, call a doctor.
Skin: redness, pain.	gloves.	remove contaminated clothing, flush skin with water or shower, refer to a doctor.
Eyes: redness, pain.	goggles.	flush with water, take to a doctor if necessary.
Ingestion: abdominal pain, vomiting.		rinse mouth, call a doctor.

SPILLAGE	STORAGE	LABELING / NFPA
clean up spillage, flush away remainder with water.	keep dry, separate from acids.	

NOTES
Decomposition temperature not given in the literature. The measures on this card also apply to potassium acetate.

SODIUM ALUMINATE

PHYSICAL PROPERTIES	IMPORTANT CHARACTERISTICS
Melting point, °C 1800 Relative density (water = 1) > 1.5 Solubility in water good Relative molecular mass 82.0	**WHITE HYGROSCOPIC POWDER** In aqueous solution is a strong base which reacts violently with acids and corrodes aluminum, zinc etc.
	TLV-TWA not available
	Absorption route: Can enter the body by ingestion. Harmful atmospheric concentrations build up very slowly, if at all, on evaporation at approx. 20°C, but much more rapidly in aerosol form. **Immediate effects:** Corrosive to the eyes and skin.
Gross formula: AlNaO$_2$	

HAZARDS/SYMPTOMS	PREVENTIVE MEASURES	FIRE EXTINGUISHING/FIRST AID
Fire: non-combustible.		in case of fire in immediate vicinity: use any extinguishing agent.
Inhalation:	local exhaust or respiratory protection.	
Skin: redness, pain.	gloves.	remove contaminated clothing, flush skin with water or shower, call a doctor.
Eyes: redness, pain, impaired vision.	face shield.	flush with water, take to a doctor if necessary.
Ingestion: sore throat, abdominal pain, diarrhea.		rinse mouth, call a doctor or take to hospital.

SPILLAGE	STORAGE	LABELING / NFPA
clean up spillage, flush away remainder with water; (additional individual protection: P2 respirator).	keep dry.	

NOTES

SODIUM AMIDE

PHYSICAL PROPERTIES	IMPORTANT CHARACTERISTICS
Boiling point, °C 400 Melting point, °C 210 Relative density (water = 1) 1.4 Solubility in water reaction Relative molecular mass 39.0	**WHITE CRYSTALS OR CRYSTALLINE POWDER WITH CHARACTERISTIC ODOR** Readily able to form peroxides, making it shock-sensitive. In dry state can form electrostatic charge if stirred, transported pneumatically, poured etc. Very strong base and reacts readily with very weak acids. Reacts violently with water, forming toxic and corrosive vapors (→ *ammonia*) and corrosive → *sodium hydroxide*. Reacts with oxidants and reacts violently with halogenated hydrocarbons. Reacts with air to form corrosive and toxic vapor (→ *ammonia*) and very explosive yellowish-brown oxidation products.
	TLV-TWA not available
	Absorption route: Can enter the body by inhalation or ingestion. Evaporation negligible at 20°C, but harmful concentrations of airborne particles can build up rapidly. **Immediate effects:** Corrosive to the eyes, skin and respiratory tract. Inhalation can cause lung edema. In serious cases risk of death.
Gross formula: H$_2$NNa	

HAZARDS/SYMPTOMS	PREVENTIVE MEASURES	FIRE EXTINGUISHING/FIRST AID
Fire: combustible.	keep away from open flame and sparks, no smoking.	special powder extinguisher, dry sand, NO other extinguishing agents, DO NOT USE WATER-BASED EXTINGUISHERS.
Explosion: contact with water or exposure to air can cause explosion.	do not subject to shock or friction, avoid contact with moisture.	
	STRICT HYGIENE, KEEP DUST UNDER CONTROL	IN ALL CASES CALL IN A DOCTOR
Inhalation: *corrosive*; sore throat, shortness of breath, severe breathing difficulties.	local exhaust or respiratory protection.	fresh air, rest, place in half-sitting position, take to hospital.
Skin: *corrosive*; redness, pain, serious burns.	gloves, protective clothing.	remove contaminated clothing, flush skin with water or shower, call a doctor if necessary.
Eyes: *corrosive*; redness, pain, impaired vision.	face shield, or combined eye and respiratory protection.	flush with water, take to a doctor.
Ingestion: *corrosive*; abdominal cramps, diarrhea, vomiting.		rinse mouth, immediately take to hospital.

SPILLAGE	STORAGE	LABELING / NFPA
evacuate area, call in an expert, clean up spillage, carefully collect remainder and remove to safe place; (additional individual protection: breathing apparatus).	keep in dry, fireproof place under inert gas or under dry pentane, toluene or xylene; separate from oxidants or acids.	

NOTES
Reacts violently with extinguishing agents such as water, foam and halons. Lung edema symptoms usually develop several hours later and are aggravated by physical exertion: rest and hospitalization essential. As first aid, a doctor or authorized person should consider administering a corticosteroid spray. The measures on this card also apply to potassium amide. Do not stock the substance but prepare shortly before use. When explosive compounds develop on exposure to air (yellow-brown discoloration), destroy the substance by covering with → *toluene*, then slowly adding (anhydrous) → *ethanol* while stirring, and finely neutralizing the substance with diluted → *hydrochloric acid*. Use airtight packaging.

Transport Emergency Card TEC(R)-43G06

SODIUM AZIDE

PHYSICAL PROPERTIES	IMPORTANT CHARACTERISTICS
Melting point (decomposes), °C 275 Relative density (water = 1) 1.8 Solubility in water, g/100 ml at 17 °C 72 Relative molecular mass 65.0	**COLORLESS CRYSTALS OR WHITE POWDER** Reacts with copper, lead, silver and carbon disulfide to form particularly shock-sensitive compounds. Decomposes, possibly explosively, when heated above melting point. Reacts with acids, forming toxic and explosive hydrogen azide. In aqueous solution is a weak base.

	TLV-TWA	0.11 ppm	0.29 mg/m³ C

Absorption route: Can enter the body by inhalation or ingestion. Evaporation negligible at 20°C, but harmful concentrations of airborne particles can build up rapidly.
Immediate effects: Irritates the eyes, respiratory tract and mucous membranes. Affects the nervous system. In serious cases risk of seizures, unconsciousness and death.
Effects of prolonged/repeated exposure: Affects the blood vessels.

Gross formula: N$_3$Na

HAZARDS/SYMPTOMS	PREVENTIVE MEASURES	FIRE EXTINGUISHING/FIRST AID
Fire: combustible.	keep away from open flame and sparks, no smoking.	dry sand, special extinguishing powder; in case of fire in immediate vicinity: use no halons.
Explosion: shock, friction or heating can cause explosion if substance contaminated with azides of heavy metals.	do not subject to shock or friction, keep dust from accumulating, sealed machinery, explosion-proof equipment and lighting, grounding.	
	STRICT HYGIENE, KEEP DUST UNDER CONTROL	
Inhalation: sore throat, cough, shortness of breath, dizziness, feeling of weakness.	ventilation, local exhaust or respiratory protection.	fresh air, rest, place in half-sitting position, take to hospital.
Skin: redness, pain.	gloves.	remove contaminated clothing, flush skin with water or shower, refer to a doctor.
Eyes: redness, pain, impaired vision.	goggles, or combined eye and respiratory protection.	flush with water, take to a doctor.
Ingestion: abdominal pain, dizziness, cramps.	do not eat, drink or smoke while working.	rinse mouth, immediately take to hospital.

SPILLAGE	STORAGE	LABELING / NFPA
clean up spillage, flush away remainder with water; (additional individual protection: P3 respirator).	keep in cool, fireproof place, separate from acids.	R: 28-32 S: 28 ☠ Toxic

NOTES

TLV is maximum and must not be exceeded. Depending on the degree of exposure, regular medical checkups are advisable. Packaging: special material.

NaHCO$_3$

bicarbonate of soda
double carbonate soda
sodium hydrogen carbonate

SODIUM BICARBONATE

PHYSICAL PROPERTIES	IMPORTANT CHARACTERISTICS	
Decomposes below melting point, °C 50 Relative density (water = 1) 2.2 Solubility in water, g/100 ml at 20 °C 9 Relative molecular mass 84.0	**WHITE CRYSTALLINE POWDER** Decomposes when heated above 50°C, giving off → *carbon dioxide* gas and → *sodium carbonate*. Reacts violently with strong acids, giving off → *carbon dioxide* gas.	
	TLV-TWA not available	
	Absorption route: Can enter the body by inhalation or ingestion. Evaporation negligible at 20° C, but unpleasant concentrations of airborne particles can build up. **Immediate effects:** Irritates the eyes and respiratory tract.	
Gross formula: CHNaO$_3$		

HAZARDS/SYMPTOMS	PREVENTIVE MEASURES	FIRE EXTINGUISHING/FIRST AID
Fire: non-combustible.		in case of fire in immediate vicinity: use any extinguishing agent.
Inhalation: sore throat, cough.	local exhaust or respiratory protection.	fresh air, rest.
Skin:	gloves.	
Eyes: redness, pain.	goggles.	flush with water, take to a doctor if necessary.
Ingestion: abdominal pain.		rinse mouth, call a doctor if necessary.

SPILLAGE	STORAGE	LABELING / NFPA
clean up spillage, flush away remainder with water; (additional individual protection: P1 respirator).	keep separate from acids.	

NOTES

CAS-No: [7631-90-5]
sodium acid sulfite
sodium hydrogen sulfite

NaHSO$_3$

SODIUM BISULFITE

PHYSICAL PROPERTIES	IMPORTANT CHARACTERISTICS	
Decomposes below melting point, °C ? Relative density (water = 1) 1.5 Solubility in water, g/100 ml 30 Relative molecular mass 104.1	**WHITE CRYSTALLINE POWDER WITH PUNGENT ODOR** Decomposes when heated or on contact with acids, giving off corrosive and toxic gases (→ *oxides of sulfur*). Reacts violently with acids, giving off toxic gas (→ *oxides of sulfur*). Reacts with oxidants. Attacks many metals.	
	TLV-TWA	5 mg/m³
	Absorption route: Can enter the body by inhalation or ingestion. Harmful concentrations can build up fairly rapidly due to emission of → *sulfur dioxide*. **Immediate effects:** Irritates the eyes, skin and respiratory tract.	
Gross formula: HNaO$_3$S		

HAZARDS/SYMPTOMS	PREVENTIVE MEASURES	FIRE EXTINGUISHING/FIRST AID
Fire: non-combustible.		in case of fire in immediate vicinity: use any extinguishing agent.
Inhalation: sore throat, cough, shortness of breath.	local exhaust or respiratory protection.	fresh air, rest, call a doctor.
Skin: redness.	gloves.	remove contaminated clothing, flush skin with water or shower.
Eyes: redness, pain.	goggles.	flush with water, take to doctor if necessary.
Ingestion: abdominal pain, nausea.		rinse mouth, call a doctor if necessary.

SPILLAGE	STORAGE	LABELING / NFPA
clean up spillage, flush away remainder with water; (additional individual protection: P2 respirator).	keep cool, separate from acids; ventilate at floor level.	

NOTES
Decomposition temperature not given in the literature. Sometimes referred to as sodium hydrosulfite, but this name should not be used as it is also used (incorrectly) for sodium dithionite (Na$_2$S$_2$O$_4$).

UN-number: 2693

SODIUM BOROHYDRIDE

PHYSICAL PROPERTIES	IMPORTANT CHARACTERISTICS	
Decomposes below boiling point, °C 300 Melting point, °C 36 Auto-ignition temperature, °C 250 Relative density (water = 1) 1.07 Solubility in water reaction Relative molecular mass 37.9	**WHITE HYGROSCOPIC POWDER** Decomposes when heated, giving off toxic vapors. Decomposes on contact with water or moist air, giving off → *hydrogen*, with risk of fire and explosion. Aqueous solutions with sodium hydroxide are stable. Reacts violently with strong oxidants, aldehydes and ketones. Reacts violently with acids, giving off → *diborane* and → *hydrogen*, with risk of fire and explosion. Vapors from a saturated solution in dimethylformamide can ignite spontaneously.	
	TLV-TWA not available	
	Absorption route: Can enter the body by inhalation or ingestion. Harmful atmospheric concentrations build up very slowly, if at all, on evaporation at approx. 20°C, but much more rapidly in aerosol form. **Immediate effects:** Irritates the eyes, skin and respiratory tract. Affects the nervous system. In serious cases risk of death. **Effects of prolonged/repeated exposure:** Can cause brain damage.	
Gross formula: BH$_4$Na		

HAZARDS/SYMPTOMS	PREVENTIVE MEASURES	FIRE EXTINGUISHING/FIRST AID
Fire: combustible.	keep away from open flame, no smoking.	powder, carbon dioxide, (halons).
	KEEP DUST UNDER CONTROL	
Inhalation: sore throat, cough, shortness of breath, severe breathing difficulties.	local exhaust or respiratory protection.	fresh air, rest, place in half-sitting position, call a doctor or take to hospital.
Skin: redness.	gloves.	remove contaminated clothing, flush skin with water or shower, refer to a doctor.
Eyes: redness, pain.	goggles.	flush with water, take to a doctor if necessary.
Ingestion: sore throat, abdominal pain, headache, dizziness.	do not eat, drink or smoke while working.	rinse mouth, immediately take to hospital.

SPILLAGE	STORAGE	LABELING / NFPA
clean up spillage, carefully collect remainder; (additional individual protection: P2 respirator).	keep dry, separate from oxidants and acids.	◇

NOTES
Use airtight packaging.

Transport Emergency Card TEC(R)-43G04

UN-number: 1426

SODIUM BROMATE

PHYSICAL PROPERTIES	IMPORTANT CHARACTERISTICS
Melting point (decomposes), °C 380 Relative density (water = 1) 3.4 Solubility in water, g/100 ml 36 Relative molecular mass 150.9	**WHITE CRYSTALS OR POWDER** Decomposes when heated above 380°C, giving off → *oxygen*, with increased risk of fire. Strong oxidant which reacts violently with combustible substances and reducing agents, e.g. textiles, oil, fat, grease (from fingers), sugar, sawdust, ammonium salts, carbon, phosphorus, metal powders and sulfides, with risk of fire and explosion.

Continued as a combined block:

	TLV-TWA not available
	Absorption route: Can enter the body by inhalation or ingestion. Evaporation negligible at 20°C, but unpleasant concentrations of airborne particles can build up. **Immediate effects:** Irritates the eyes, skin and respiratory tract. Affects the nervous system. In serious cases risk of unconsciousness. Can affect the blood if swallowed. **Effects of prolonged/repeated exposure:** Can affect the blood. Can cause kidney and brain damage.
Gross formula: BrNaO$_3$	

HAZARDS/SYMPTOMS	PREVENTIVE MEASURES	FIRE EXTINGUISHING/FIRST AID
Fire: not combustible but enhances combustion of other substances.		in case of fire in immediate vicinity: preferably large quantities of water.
Explosion:	keep dust from accumulating.	
	KEEP DUST UNDER CONTROL	
Inhalation: cough, shortness of breath, headache, dizziness, blue skin.	local exhaust or respiratory protection.	fresh air, rest, call a doctor and take to hospital.
Skin: redness.	gloves.	remove contaminated clothing, flush skin with water or shower.
Eyes: redness, pain.	goggles.	flush with water, take to a doctor.
Ingestion: abdominal pain, dizziness, vomiting.		rinse mouth, immediately take to hospital.

SPILLAGE	STORAGE	LABELING / NFPA
clean up spillage, flush away remainder with water; (additional individual protection: P2 respirator).	keep in fireproof place, separate from combustible substances and reducing agents.	

NOTES
Flush contaminated clothing with water (fire hazard). Increased fire/explosion risk when contaminated with combustible substances. Depending on the degree of exposure, regular medical checkups are advisable. Effects on blood due to formation of methemoglobin. Special first aid required in the event of poisoning: antidotes must be available (with instructions).

Transport Emergency Card TEC(R)-51G02   **UN-number: 1494**

CAS-No: [497-19-8]
carbonic acid, disodium salt
disodium carbonate
soda
soda ash
soda monohydrate

Na_2CO_3

SODIUM CARBONATE

PHYSICAL PROPERTIES	IMPORTANT CHARACTERISTICS
Boiling point (decomposes), °C 1600 Melting point, °C 851 Relative density (water = 1) 2.5 Solubility in water, g/100 ml at 20 °C 22 Relative molecular mass 106.0	**WHITE HYGROSCOPIC CRYSTALS OR WHITE HYGROSCOPIC POWDER** In aqueous solution is a strong base which reacts violently with acids and corrodes aluminum, zinc etc. Reacts violently with acids, giving off → *carbon dioxide*.
	TLV-TWA not available
Gross formula: CNa_2O_3	**Absorption route:** Can enter the body by inhalation or ingestion. Evaporation negligible at 20°C, but unpleasant concentrations of airborne particles can build up. **Immediate effects:** Irritates the skin and respiratory tract. Corrosive to the eyes.

HAZARDS/SYMPTOMS	PREVENTIVE MEASURES	FIRE EXTINGUISHING/FIRST AID
Fire: non-combustible.		in case of fire in immediate vicinity: use any extinguishing agent.
Inhalation: cough, shortness of breath.	local exhaust or respiratory protection.	fresh air, rest, call a doctor.
Skin: redness.	gloves.	remove contaminated clothing, flush skin with water or shower.
Eyes: *corrosive*; redness, pain, impaired vision.	goggles.	flush with water, take to a doctor.
Ingestion: sore throat, abdominal pain.		rinse mouth, call a doctor.

SPILLAGE	STORAGE	LABELING / NFPA
clean up spillage, flush away remainder with water; (additional individual protection: P2 respirator).	keep dry, separate from acids.	R: 36 S: 22-26 ✖ Irritant

NOTES
The measures on this card also apply to potassium carbonate. Solubility of potassium carbonate in water at 20°C; 112g/100ml.

SODIUM CHLORATE

PHYSICAL PROPERTIES	IMPORTANT CHARACTERISTICS
Decomposes below boiling point, °C 300 Melting point, °C 248 Relative density (water = 1) 2.5 Solubility in water, g/100 ml at 20 °C 100 Relative molecular mass 106.4	**WHITE CRYSTALS** Reacts with organic contaminants to form shock-sensitive mixtures. Decomposes when heated above 300°C, giving off oxygen, with increased risk of fire. Strong oxidant which reacts violently with combustible substances and reducing agents. Reacts with strong acids, giving off → *carbon dioxide*.

TLV-TWA	not available

Absorption route: Can enter the body by inhalation or ingestion. Evaporation negligible at 20°C, but harmful concentrations of airborne particles can build up rapidly.
Immediate effects: Irritates the eyes, skin and respiratory tract. In serious cases risk of death.
Effects of prolonged/repeated exposure: Can affect the blood. Can cause liver and kidney damage.

Gross formula: ClNaO₃

HAZARDS/SYMPTOMS	PREVENTIVE MEASURES	FIRE EXTINGUISHING/FIRST AID
Fire: non-combustible, but makes other substances more combustible; many chemical reactions can cause fire and explosion.	avoid contact with combustible substances.	
Explosion: very close contact with textiles, oil, fat, sugar, salts of ammonia, carbon, sulfur, metal powders, sulfides etc. can cause explosion.		
	KEEP DUST UNDER CONTROL	
Inhalation: sore throat, cough, headache, dizziness, blue skin, feeling of weakness.	local exhaust or respiratory protection.	fresh air, rest, call a doctor.
Skin: redness.	gloves.	remove contaminated clothing, flush skin with water or shower.
Eyes: redness, pain.	goggles.	flush with water, take to a doctor.
Ingestion: diarrhea, abdominal cramps, vomiting, unconsciousness, blue skin.		rinse mouth, immediately take to hospital.

SPILLAGE	STORAGE	LABELING / NFPA
clean up spillage, flush away remainder with water, do not use sawdust or other combustible absorbents; (additional individual protection: P2 respirator).	keep dry, separate from combustible substances and reducing agents.	R: 9-20/22 S: 2-13-16-27 Oxidant Harmful

NOTES
Flush contaminated clothing with water (fire hazard). Becomes shock-sensitive when contaminated with organic substances. Depending on the degree of exposure, regular medical checkups are advisable. Effects on blood due to formation of methemoglobin. Special first aid required in the event of poisoning: antidotes must be available (with instructions).

Transport Emergency Card TEC(R)-47   **HI: 50; UN-number: 1495**

SODIUM CHLORIDE

PHYSICAL PROPERTIES		IMPORTANT CHARACTERISTICS	
Boiling point, °C	1413	**WHITE HYGROSCOPIC CRYSTALS OR POWDER**	
Melting point, °C	801	When strongly heated gives off irritating vapors. Attacks many metals and building materials.	
Relative density (water = 1)	2.2		
Solubility in water, g/100 ml	37	TLV-TWA	not available
Relative molecular mass	58.5		
Log P octanol/water	−3.0		
		Absorption route: Can enter the body by ingestion.	
		Immediate effects: Irritates the eyes.	
		Effects of prolonged/repeated exposure: Prolonged or repeated contact can cause skin disorders (ulcers).	
Gross formula:	ClNa		

HAZARDS/SYMPTOMS	PREVENTIVE MEASURES	FIRE EXTINGUISHING/FIRST AID
Fire: non-combustible.		in case of fire in immediate vicinity: use any extinguishing agent.
Skin:	gloves.	
Eyes: redness, pain.	safety glasses.	flush with water, take to a doctor if necessary.

SPILLAGE	STORAGE	LABELING / NFPA
clean up spillage, flush away remainder with water.		

NOTES
The solution in water can also be called a brine. Log P octanol/water is estimated. The measures on this card also apply to potassium chloride.

SODIUM CHLORITE

PHYSICAL PROPERTIES	IMPORTANT CHARACTERISTICS	
Decomposes below boiling point, ° C 180 Solubility in water, g/100 ml 74 Relative molecular mass 90.4	**WHITE CRYSTALS OR CRYSTALLINE POWDER, SLIGHTLY HYGROSCOPIC** Decomposes when heated above 150° C, giving off toxic gases and oxygen, with increased risk of fire. Strong oxidant which reacts violently with combustible substances and reducing agents. Reacts with acid, giving off toxic gas (→ *chlorine dioxide*).	
	TLV-TWA not available	
	Absorption route: Can enter the body by inhalation or ingestion. Evaporation negligible at 20° C, but harmful concentrations of airborne particles can build up rapidly. **Immediate effects:** Corrosive to the eyes, skin and respiratory tract. Inhalation can cause lung edema. In serious cases risk of death.	
Gross formula: ClNaO$_2$		

HAZARDS/SYMPTOMS	PREVENTIVE MEASURES	FIRE EXTINGUISHING/FIRST AID
Fire: non-combustible, but makes other substances more combustible.		in case of fire in immediate vicinity: preferably large quantities of water.
Explosion:	keep dust from accumulating.	
Inhalation: *corrosive*; sore throat, cough, shortness of breath, severe breathing difficulties.	local exhaust or respiratory protection.	fresh air, rest, place in half-sitting position, take to hospital.
Skin: *corrosive*; redness, pain, serious burns.	gloves, protective clothing.	flush clothing with water before removing, then flush skin with water or shower, call a doctor.
Eyes: *corrosive*; redness, pain, impaired vision.	face shield, or combined eye and respiratory protection.	flush with water, take to doctor if necessary.
Ingestion: *corrosive*; abdominal pain, vomiting.		rinse mouth, call a doctor or take to hospital.

SPILLAGE	STORAGE	LABELING / NFPA
clean up spillage, flush away remainder with water, do not use sawdust or other combustible absorbents; (additional individual protection: P2 respirator).	keep in fireproof place, separate from combustible substances, reducing agents and acids.	 1 1 1 oxy

NOTES
Flush contaminated clothing with water (fire hazard). Risk of fire and explosion greatly increased when contaminated with combustible substances. Lung edema symptoms usually develop several hours later and are aggravated by physical exertion: rest and hospitalization essential. As first aid, a doctor or authorized person should consider administering a corticosteroid spray.

CAS-No: [143-33-9]   NaCN
cyanide of sodium
hydrocyanic, sodium salt
NaCN

SODIUM CYANIDE

PHYSICAL PROPERTIES	IMPORTANT CHARACTERISTICS	
Boiling point, °C — 1496 Melting point, °C — 563 Relative density (water = 1) — 1.6 Solubility in water, g/100 ml at 20 °C — 37 Relative molecular mass — 49	**WHITE HYGROSCOPIC SOLID** In aqueous solution is a strong base which reacts violently with acids and corrodes aluminum, zinc etc. Reacts violently with acids (even CO_2), giving off toxic → *hydrogen cyanide*. Reacts violently with hot, strong oxidants.	
	TLV-TWA (skin)	5 mg/m³ ✶
	Absorption route: Can enter the body by inhalation or ingestion or through the skin. Evaporation negligible at 20°C, but harmful concentrations of airborne particles can build up rapidly. **Immediate effects:** Corrosive to the eyes, skin and respiratory tract. Impedes tissue respiration. In serious cases risk of seizures and death.	
Gross formula: — CNNa		

HAZARDS/SYMPTOMS	PREVENTIVE MEASURES	FIRE EXTINGUISHING/FIRST AID
Fire: non-combustible.		in case of fire in immediate vicinity: use any extinguishing agent.
	STRICT HYGIENE	IN ALL CASES CALL A DOCTOR
Inhalation: *corrosive*; sore throat, severe breathing difficulties, feeling of weakness.	local exhaust or respiratory protection.	fresh air, rest, place in half-sitting position, special treatment, take to hospital.
Skin: *corrosive, is absorbed*; redness, pain, burns; see also 'Inhalation'.	gloves, protective clothing.	remove contaminated clothing, flush skin with water or shower, call a doctor.
Eyes: *corrosive*; redness, pain, impaired vision.	face shield.	flush with water, take to a doctor if necessary.
Ingestion: *corrosive*; sore throat, abdominal pain; see also 'Inhalation'.	do not eat, drink or smoke while working.	rinse mouth, special treatment, take immediately to hospital.

SPILLAGE	STORAGE	LABELING / NFPA
call in an expert, clean up spillage, carefully collect remainder, render harmless with large quantities of sodium hypochlorite solution; (additional individual protection: breathing apparatus).	keep dry, separate from acids; ventilate.	R: 26/27/28-32 S: 1/2-7-28-29-45 Note A ☠ Toxic

NOTES
✶ TLV equals that of cyanides. Special first aid required in the event of poisoning: antidotes must be available (with instructions). Special treatment: administer oxygen (100%) if necessary, avoid mouth-to-mouth resuscitation if possible (risk to person assisting). Unbreakable packaging preferred; if breakable, keep in unbreakable container.

Transport Emergency Card TEC(R)-61G15   **UN-number: 1689**

CAS-No: [143-33-9]
cyanide of sodium
hydrocyanic acid, sodium salt
NaCN

NaCN

SODIUM CYANIDE

(30% solution)

PHYSICAL PROPERTIES	IMPORTANT CHARACTERISTICS	
Boiling point, °C　　　　　　　112 Solubility in water　　　　　　∞ Relative molecular mass　　as NaCN 49	**COLORLESS LIQUID WITH CHARACTERISTIC ODOR** In aqueous solution is a strong base which reacts violently with acids and corrodes aluminum, zinc etc. Reacts violently with acids, giving off toxic → *hydrogen cyanide*. Reacts violently with hot, strong oxidants. Reacts with air to form very small quantities of → *hydrogen cyanide*.	
	TLV-TWA (skin)	5 mg/m³ ✳
	Absorption route: Can enter the body by inhalation or ingestion or through the skin. Evaporation negligible at 20°C, but harmful atmospheric concentrations can build up rapidly in aerosol form. **Immediate effects:** Corrosive to the eyes, skin and respiratory tract. Impedes tissue respiration. In serious cases risk of seizures and death.	
Gross formula:　　　　　　CNNa		

HAZARDS/SYMPTOMS	PREVENTIVE MEASURES	FIRE EXTINGUISHING/FIRST AID
Fire: non-combustible.		in case of fire in immediate vicinity: use any extinguishing agent.
	STRICT HYGIENE	
Inhalation: *corrosive*; sore throat, severe breathing difficulties, feeling of weakness, unconsciousness.	ventilation.	fresh air, rest, place in half-sitting position, special treatment, take to hospital.
Skin: *corrosive, is absorbed*; redness, pain, burns; see also 'Inhalation'.	gloves, protective clothing.	remove contaminated clothing, flush skin with water or shower, take to hospital.
Eyes: *corrosive*; redness, pain, impaired vision.	face shield, or combined eye and respiratory protection.	flush with water, take to a doctor.
Ingestion: *corrosive*; abdominal pain, sore throat; see also 'Inhalation'.	do not eat, drink or smoke while working.	rinse mouth, special treatment, take immediately to hospital.

SPILLAGE	STORAGE	LABELING / NFPA
call in an expert, collect leakage in sealable containers, soak up with sand or other inert absorbent and remove to safe place, render harmless with large quantities of sodium hypochlorite solution; (additional individual protection: breathing apparatus).	keep separate from oxidants and acids; ventilate.	R: 26/27/28-32 S: 1/2-7-28-29-45 Note A ☣ Toxic

NFPA diamond: 3 (left), 0 (top), 0 (right)

NOTES
✳ TLV equals that of cyanides. Special first aid required in the event of poisoning: antidotes must be available (with instructions). Special treatment: administer oxygen (100%) if necessary, avoid mouth-to-mouth resuscitation if possible (risk to person assisting). Unbreakable packaging preferred; if breakable, keep in unbreakable container.

HI: 66; UN-number: 1935

SODIUM DICHROMATE

PHYSICAL PROPERTIES	IMPORTANT CHARACTERISTICS	
Decomposes below boiling point, °C 400 Melting point, °C 357 Relative density (water = 1) 2.3 Solubility in water, g/100 ml at 20 °C 270 Relative molecular mass 262	**ORANGE-RED HYGROSCOPIC CRYSTALS** In aqueous solution attacks many materials, esp. in the presence of acids. Strong oxidant which reacts violently with combustible substances and reducing agents, with risk of fire and explosion.	
	TLV-TWA (skin)	0.05 mg/m³ ✱
	Absorption route: Can enter the body by inhalation or ingestion. **Immediate effects:** Corrosive to the eyes, skin and respiratory tract. Inhalation can cause lung edema. In serious cases risk of death. **Effects of prolonged/repeated exposure:** Prolonged or repeated contact can cause skin disorders. Has been found to cause a type of lung cancer in certain animal species under certain circumstances.	
Gross formula: Cr$_2$Na$_2$O$_7$.2H$_2$O		

HAZARDS/SYMPTOMS	PREVENTIVE MEASURES	FIRE EXTINGUISHING/FIRST AID
Fire: non-combustible, many chemical reactions can cause fire and explosion.	avoid contact with combustible substances.	in case of fire in immediate vicinity: use any extinguishing agent.
	STRICT HYGIENE	
Inhalation: *corrosive*; sore throat, cough, shortness of breath, severe breathing difficulties.	local exhaust or respiratory protection.	fresh air, rest, place in half-sitting position, take to hospital.
Skin: *corrosive*; redness, pain, burns.	gloves.	remove contaminated clothing, flush skin with water or shower, call a doctor.
Eyes: *corrosive*; redness, pain, impaired vision.	goggles.	flush with water, take to a doctor.
Ingestion: *corrosive*; sore throat, abdominal pain, vomiting.	do not eat, drink or smoke while working.	rinse mouth, immediately take to hospital.

SPILLAGE	STORAGE	LABELING / NFPA
clean up spillage, carefully collect remainder; (additional individual protection: P3 respirator).	keep dry, separate from combustible substances and reducing agents.	R: 36/37/38-43 S: 22-28 ✖ Irritant

NOTES
✱ TLV equals that of chromium (VI) compounds. Flush contaminated clothing with water (fire hazard). Depending on the degree of exposure, regular medical checkups are advisable. Lung edema symptoms usually develop several hours later and are aggravated by physical exertion: rest and hospitalization essential. As first aid, a doctor or authorized person should consider administering a corticosteroid spray. To acidify dichromate use sulfuric acid, NOT hydrochloric or nitric acid (chlorine/nitrous vapors given off). Packaging: special material.

Transport Emergency Card TEC(R)-51G02

UN-number: 1497

SODIUMDIETHYLDITHIOCARBAMATE

PHYSICAL PROPERTIES	IMPORTANT CHARACTERISTICS	
Decomposes below boiling point, °C ? Melting point, °C 95 Relative density (water = 1) 1.1 Solubility in water good Relative molecular mass 171.3	**WHITE HYGROSCOPIC CRYSTALS** Decomposes on contact with acids, giving off toxic and flammable → *carbon disulfide*. Decomposes when heated or burned, giving off toxic and corrosive vapors (→ *nitrogen oxides* and → *sulfur oxides*).	
	TLV-TWA not available	
	Absorption route: Can enter the body by inhalation or ingestion. Evaporation negligible at 20°C, but unpleasant concentrations of airborne particles can build up. **Immediate effects:** Irritates the eyes, skin and respiratory tract.	
Gross formula: $C_5H_{10}NNaS_2$		

HAZARDS/SYMPTOMS	PREVENTIVE MEASURES	FIRE EXTINGUISHING/FIRST AID
Fire: combustible.	keep away from open flame, no smoking.	water spray, powder.
Inhalation: sore throat, cough, shortness of breath.	local exhaust or respiratory protection.	fresh air, rest, call a doctor.
Skin: redness, pain.	gloves.	remove contaminated clothing, flush skin with water or shower.
Eyes: redness, pain.	goggles.	flush with water, take to a doctor if necessary.
Ingestion: abdominal pain, nausea.	do not eat, drink or smoke while working.	rinse mouth, call a doctor or take to hospital.

SPILLAGE	STORAGE	LABELING / NFPA
clean up spillage, carefully collect remainder; (additional individual protection: P2 respirator).	keep dry; keep separate from acids, ventilate at floor level.	

NOTES
Decomposition temperature not given in the literature. Insufficient toxicological data on harmful effects to humans. Used as inhibitor in → *diethylether* to prevent formation of peroxides. Usually the substance is used with 3 molecules of crystal water, which deliquesce on exposure to air.

Transport Emergency Card TEC(R)-61G47

SODIUMDITHIONITE

PHYSICAL PROPERTIES	IMPORTANT CHARACTERISTICS
Melting point (decomposes), °C 52 Flash point, °C see notes Relative density (water = 1) 1.6 Solubility in water, g/100 ml at 20 °C 22 Relative molecular mass 210.1 Gross formula: $Na_2O_4S_2$	**GRANULAR WHITE OR LIGHT YELLOW POWDER WITH SLIGHTLY PUNGENT ODOR** Can ignite spontaneously on exposure to moist air. Can produce thermal explosion when heated. Decomposes when heated above 100°C and on contact with small amount of water, giving off toxic and corrosive vapors (→ *sulfur dioxides*), with risk of fire and explosion. Strong reducing agent which reacts violently with oxidants. Under certain conditions reacts with lye, forming flammable gas (→ *hydrogen*). Reacts with acids, forming → *sulfur dioxide*. TLV-TWA not available **Absorption route:** Can enter the body by inhalation or ingestion or through the skin. Evaporation negligible at 20°C, but harmful concentrations of airborne particles can build up rapidly. **Immediate effects:** Irritates the eyes, skin and respiratory tract.

HAZARDS/SYMPTOMS	PREVENTIVE MEASURES	FIRE EXTINGUISHING/FIRST AID
Fire: combustible, even on contact with moisture.	keep away from open flame, no smoking; avoid contact with air, moist air or oxidants.	large quantities of water, carbon dioxide, dry sand, special powder extinguisher, foam is ineffective.
Inhalation: sore throat, cough, shortness of breath.	local exhaust or respiratory protection.	fresh air, rest, call a doctor.
Skin: redness.	gloves, protective clothing.	remove contaminated clothing, flush skin with water or shower.
Eyes: redness, pain.	goggles, or combined eye and respiratory protection.	flush with water, take to a doctor if necessary.
Ingestion: abdominal pain, nausea.	do not eat, drink or smoke while working.	rinse mouth, call a doctor.

SPILLAGE	STORAGE	LABELING / NFPA
if necessary render harmless spillage with solution of soda and sodium hypochlorite, clean up spillage, flush away remainder with water, do not use sawdust or other combustible absorbents; (additional individual protection: breathing apparatus).	keep in fireproof, dry place; keep separate from oxidants and combustible substances.	R: 7-22-31 S: 7/8-26-28-43 ✖ Harmful NFPA: 3 / 1 / 2

NOTES
Apparent melting point due to loss of water of crystallization. Flush contaminated clothing with water (fire hazard). Do not take work clothing home. When dissolving in water always add sodiumdithionite to the water, do not add water to the substance. In sealed containers the moist product can heat up spontaneously, with risk of explosion. Sodium thionite is sometimes mistakenly named sodium hydrosulfite or sodium hyposulfite. Conite is a trade name. Use airtight packaging.

Transport Emergency Card TEC(R)-747   **HI: 43; UN-number: 1384**

$C_{12}H_{25}C_6H_4SO_3Na$

SODIUMDODECYLBENZENESULFONATE

PHYSICAL PROPERTIES	IMPORTANT CHARACTERISTICS	
Boiling point, °C ? Solubility in water good Relative molecular mass 348.5	**WHITE TO YELLOW POWDER** Decomposes when heated, giving off toxic gas (→ *sulfur dioxide*).	
	TLV-TWA not available	
	Absorption route: Can enter the body by inhalation or ingestion. Evaporation negligible at 20°C, but harmful concentrations of airborne particles can build up rapidly. **Immediate effects:** Irritates the eyes, skin and respiratory tract.	
Gross formula: $C_{18}H_{29}NaO_3S$		

HAZARDS/SYMPTOMS	PREVENTIVE MEASURES	FIRE EXTINGUISHING/FIRST AID
Fire: non-combustible.		in case of fire in immediate vicinity: use any extinguishing agent.
Inhalation: cough, shortness of breath.	local exhaust or respiratory protection.	fresh air, rest, call a doctor if necessary.
Skin: redness.	gloves.	remove contaminated clothing, flush skin with water or shower.
Eyes: redness, pain.	goggles.	flush with water, take to a doctor if necessary.
Ingestion: abdominal pain, nausea.		rinse mouth, call a doctor.

SPILLAGE	STORAGE	LABELING / NFPA
clean up spillage, flush away remainder with water; (additional individual protection: P2 respirator).		

NOTES

SODIUM FLUORIDE

PHYSICAL PROPERTIES	IMPORTANT CHARACTERISTICS	
Boiling point, °C 1700 Melting point, °C 990 Relative density (water = 1) 2.8 Solubility in water, g/100 ml at 18 °C 4.2 Relative molecular mass 42.0	**WHITE CRYSTALS OR POWDER** Decomposes in flame or on hot surface, giving off toxic gas ($\rightarrow$ *hydrogen fluoride*). In aqueous solution corrosive, e.g. to glass. Reacts with acids, forming corrosive and toxic $\rightarrow$ *hydrogen fluoride*.	
	TLV-TWA	2.5 mg/m³ ✱
	Absorption route: Can enter the body by inhalation or ingestion. Evaporation negligible at 20°C, but harmful concentrations of airborne particles can build up rapidly. **Immediate effects:** Irritates the skin. Corrosive to the eyes and respiratory tract. Inhalation can cause lung edema. In serious cases risk of death. **Effects of prolonged/repeated exposure:** Can cause kidney damage.	
Gross formula: FNa		

HAZARDS/SYMPTOMS	PREVENTIVE MEASURES	FIRE EXTINGUISHING/FIRST AID
Fire: non-combustible.		in case of fire in immediate vicinity: use any extinguishing agent.
	KEEP DUST UNDER CONTROL	
Inhalation: *corrosive*; sore throat, cough, shortness of breath, severe breathing difficulties.	local exhaust or respiratory protection.	fresh air, rest, place in half-sitting position, take to hospital.
Skin: redness, pain.	gloves.	remove contaminated clothing, flush skin with water or shower, refer to a doctor if necessary.
Eyes: *corrosive*; redness, pain, impaired vision.	face shield.	flush with water, take to a doctor.
Ingestion: *corrosive*; abdominal pain, diarrhea, vomiting.	do not eat, drink or smoke while working.	rinse mouth, immediately take to hospital.

SPILLAGE	STORAGE	LABELING / NFPA
clean up spillage, carefully collect remainder; (additional individual protection: P2 respirator).	keep separate from acids.	R: 23/24/25 S: 1/2-26-44 ☠ Toxic

NFPA diamond: 0 / 2 / 0

NOTES
✱ TLV equals that of fluoride. Depending on the degree of exposure, regular medical checkups are advisable. Lung edema symptoms usually develop several hours later and are aggravated by physical exertion: rest and hospitalization essential. As first aid, a doctor or authorized person should consider administering a corticosteroid spray.

Transport Emergency Card TEC(R)-899

HI: 60; UN-number: 1690

SODIUM FORMATE

PHYSICAL PROPERTIES	IMPORTANT CHARACTERISTICS	
Decomposes below boiling point, °C > 300 Melting point, °C 253 Relative density (water = 1) 1.9 Solubility in water, g/100 ml 77 Relative molecular mass 68.0	**DELIQUESCENT HYGROSCOPIC CRYSTALS OR WHITE CRYSTALLINE POWDER WITH CHARACTERISTIC ODOR** Decomposes when heated between 300-450°C, giving off combustible and toxic gases (→ *hydrogen* and → *carbon monoxide*). Reacts with strong acids, forming corrosive → *formic acid.*	
	TLV-TWA not available	
	Absorption route: Can enter the body by inhalation or ingestion or through the skin. Evaporation negligible at 20°C, but unpleasant concentrations of airborne particles can build up. **Immediate effects:** Irritates the respiratory tract. Corrosive to the eyes and skin.	
Gross formula: CHNaO₂		

HAZARDS/SYMPTOMS	PREVENTIVE MEASURES	FIRE EXTINGUISHING/FIRST AID
Fire: non-combustible.		in case of fire in immediate vicinity: use any extinguishing agent.
Inhalation: sore throat, cough, shortness of breath.	local exhaust or respiratory protection.	fresh air, rest, call a doctor.
Skin: *corrosive*; redness, pain, burns.	gloves.	remove contaminated clothing, flush skin with water or shower, refer to a doctor.
Eyes: *corrosive*; redness, pain, impaired vision.	goggles.	flush with water, take to a doctor.
Ingestion: sore throat, abdominal cramps, vomiting.	do not eat, drink or smoke while working.	rinse mouth, call a doctor or take to hospital.

SPILLAGE	STORAGE	LABELING / NFPA
(additional individual protection: P2 respirator).	keep dry, separate from strong acids.	

NOTES

SODIUM HEXAMETAPHOSPHATE

PHYSICAL PROPERTIES	IMPORTANT CHARACTERISTICS	
Melting point, °C 630 Relative density (water = 1) 2 Solubility in water good Relative molecular mass 611.8	**WHITE HYGROSCOPIC POWDER** Decomposes when heated, giving off toxic phosphorus oxides and sodium oxide.	
	TLV-TWA not available	
	Absorption route: Can enter the body by inhalation or ingestion. Evaporation negligible at 20°C, but unpleasant atmospheric concentrations can build up rapidly in aerosol form. **Immediate effects:** Irritates the eyes, skin and respiratory tract.	
Gross formula: NaO_3P		

HAZARDS/SYMPTOMS	PREVENTIVE MEASURES	FIRE EXTINGUISHING/FIRST AID
Fire: non-combustible.		in case of fire in immediate vicinity: use any extinguishing agent.
Inhalation: sore throat, cough.	local exhaust or respiratory protection.	fresh air, rest, call a doctor.
Skin: redness.	gloves.	remove contaminated clothing, flush skin with water or shower.
Eyes: redness.	goggles.	flush with water, take to a doctor if necessary.
Ingestion: diarrhea, nausea, vomiting.		rinse mouth, call a doctor.

SPILLAGE	STORAGE	LABELING / NFPA
clean up spillage, flush away remainder with water; (additional individual protection: P1 respirator).		

NOTES
Sodium hexametaphosphate is a decalcifier. Available as Calgon in shops. Sodium hexamethaphosphate is not a 'hexamer', but a mixture of polymeric phosphates.

caustic flake
caustic soda
lye
sodium hydrate

NaOH

SODIUM HYDROXIDE

PHYSICAL PROPERTIES		IMPORTANT CHARACTERISTICS	
Boiling point, °C	1390	**WHITE HYGROSCOPIC PELLETS OR FLAKES, WHITE SOLID IN VARIOUS FORMS**	
Melting point, °C	318	Strong base which reacts violently with acids and corrodes aluminum, zinc etc. Also reacts violently with halogenated hydrocarbons and nitrogen compounds, with risk of fire and explosion. Reacts violently with water (e.g. when dissolved), evolving large quantities of heat, with risk of formation of corrosive fumes.	
Relative density (water = 1)	2.1		
Vapor pressure, mm Hg at 700 °C	0.76		
Solubility in water, g/100 ml at 25 °C	111		
Relative molecular mass	40		
		TLV-TWA	2 mg/m³ C
		Absorption route: Can enter the body by inhalation or ingestion. Evaporation negligible at 20° C, but harmful concentrations of airborne particles can build up rapidly. **Immediate effects:** Corrosive to the eyes, skin and respiratory tract; effects often delayed. Inhalation can cause lung edema.	
Gross formula:	HNaO		

HAZARDS/SYMPTOMS	PREVENTIVE MEASURES	FIRE EXTINGUISHING/FIRST AID
Fire: non-combustible.		in case of fire in immediate vicinity: use any extinguishing agent.
Inhalation: *corrosive*; sore throat, cough, severe breathing difficulties.	local exhaust or respiratory protection.	fresh air, rest, place in half-sitting position, take to hospital.
Skin: *corrosive*; redness, pain, serious burns.	gloves.	remove contaminated clothing, flush skin with water or shower, refer to a doctor.
Eyes: *corrosive*; redness, pain, impaired vision.	face shield.	flush with water, take to a doctor.
Ingestion: *corrosive*; sore throat, diarrhea, abdominal cramps, vomiting.		rinse mouth, take immediately to hospital.

SPILLAGE	STORAGE	LABELING / NFPA
clean up spillage, flush away remainder with water; (additional individual protection: P2 respirator).	keep separate from acids.	R: 35 S: 2-26-37/39 Corrosive 0 / 3 / 1

NOTES
Aqueous solution (up to 70%) is referred to as caustic soda solution. TLV is maximum and must not be exceeded. Lung edema symptoms usually develop several hours later and are aggravated by physical exertion: rest and hospitalization essential. As first aid, a doctor or authorized person should consider administering a corticosteroid spray. When dissolving use exhaust and add solid to water in small quantities to avoid heating and formation of fumes.

Transport Emergency Card TEC(R)-121

HI: 80; UN-number: 1823

CAS-No: [7681-52-9]
bleach
liquid bleach
hypochlorous acid, sodium salt
NaOCl
sodium oxychloride

NaClO

SODIUM HYPOCHLORITE
(solution with 150g/l active chlorine)

PHYSICAL PROPERTIES	IMPORTANT CHARACTERISTICS
Relative density (water = 1) 1.2 Solubility in water ∞ Relative molecular mass 74.4 NaClO	**CLEAR YELLOWISH-GREEN SOLUTION OF SODIUM HYPOCHLORITE WITH 12.5% ACTIVE CHLORINE (150G/L)** Decomposes when heated, giving off toxic gas (→ *chlorine*). Decomposes when exposed to sunlight, giving off oxygen, with increased risk of fire. Strong oxidant which reacts violently with combustible substances and reducing agents, with risk of fire and explosion. In aqueous solution is a strong base which reacts violently with acids and corrodes aluminum, zinc etc. Attacks all metals.
	TLV-TWA not available
	Absorption route: Can enter the body by inhalation or ingestion. Insufficient data on the rate at which harmful concentrations can build up. **Immediate effects:** Corrosive to the skin, mucous membranes, eyes and respiratory tract. Inhalation of vapor/fumes can cause severe breathing difficulties (lung edema).
Gross formula: ClNaO	

HAZARDS/SYMPTOMS	PREVENTIVE MEASURES	FIRE EXTINGUISHING/FIRST AID
Fire: non-combustible, but makes other substances more combustible; many chemical reactions can cause fire and explosion.	avoid contact with combustible substances.	in case of fire in immediate vicinity: use any extinguishing agent.
Explosion:		in case of fire: keep tanks/drums cool by spraying with water.
Inhalation: *corrosive*; sore throat, cough, severe breathing difficulties.	ventilation, local exhaust or respiratory protection.	fresh air, rest, place in half-sitting position, take to hospital.
Skin: redness, pain, blisters, sores.	gloves, protective clothing.	remove contaminated clothing, flush skin with water or shower, refer to a doctor.
Eyes: *corrosive*; redness, pain, impaired vision.	face shield.	flush with water, take to doctor.
Ingestion: *corrosive*; sore throat, abdominal cramps, nausea.		rinse mouth, take immediately to hospital.

SPILLAGE	STORAGE	LABELING / NFPA
collect leakage in sealable containers, flush away remainder with water.	keep in cool, dark place, separate from combustible substances, reducing agents and acids.	R: 31-34 S: 2-28 Corrosive

NOTES
Flush contaminated clothing with water (fire hazard). Crystallizes out below − 20° C. Lung edema symptoms usually develop several hours later and are aggravated by physical exertion: rest and hospitalization essential. As first aid, a doctor or authorized person should consider administering a corticosteroid spray. Packaging: special material.

Transport Emergency Card TEC(R)-45

HI: 85; UN-number: 1791

SODIUM HYPOCHLORITE (SOLUTION)

PHYSICAL PROPERTIES	IMPORTANT CHARACTERISTICS
Melting point, °C −6 Relative density (water = 1) 1.1 Solubility in water ∞ Relative molecular mass 74.4	**CLEAR LIGHT YELLOW LIQUID WITH CHARACTERISTIC ODOR** Aqueous solution of sodium hypochlorite with active chlorine content of 100g/l (bleach) or 50g/l (Javelle water). Decomposes when heated, giving off toxic gas (→ *chlorine*). Decomposes when exposed to sunlight, giving off oxygen, with increased risk of fire. Strong oxidant which reacts violently with combustible substances and reducing agents, with risk of fire and explosion. In aqueous solution is a strong base which reacts violently with acids and corrodes aluminum, zinc etc. Attacks many metals.

TLV-TWA	not available

Absorption route: Can enter the body by inhalation or ingestion. Harmful atmospheric concentrations build up very slowly, if at all, on evaporation at approx. 20°C, but much more rapidly in aerosol form.
Immediate effects: Irritates the skin and mucous membranes. Corrosive to the eyes and respiratory tract. Inhalation of vapor/fumes can cause severe breathing difficulties (lung edema).

Gross formula: ClNaO

HAZARDS/SYMPTOMS	PREVENTIVE MEASURES	FIRE EXTINGUISHING/FIRST AID
Fire: non-combustible.		in case of fire in immediate vicinity: preferably no carbon dioxide.
Explosion:		in case of fire: keep tanks/drums cool by spraying with water.
Inhalation: *corrosive*; sore throat, cough, severe breathing difficulties.	ventilation, local exhaust or respiratory protection.	fresh air, rest, place in half-sitting position, take to hospital.
Skin: redness, pain.	gloves, protective clothing.	remove contaminated clothing, flush skin with water or shower, call a doctor if necessary.
Eyes: *corrosive*; redness, pain, impaired vision.	face shield.	flush with water, take to a doctor.
Ingestion: *corrosive*; sore throat, abdominal pain, vomiting.		rinse mouth, take immediately to hospital.

SPILLAGE	STORAGE	LABELING / NFPA
collect leakage in sealable containers, flush away remainder with water.	keep in cool, dark place, separate from acids.	R: 31-36/38 S: 2-25 Note B ✖ Irritant

NOTES
Flush contaminated clothing with water (fire hazard). Lung edema symptoms usually develop several hours later and are aggravated by physical exertion: rest and hospitalization essential. As first aid, a doctor or authorized person should consider administering a corticosteroid spray. Packaging: special material.

Transport Emergency Card TEC(R)-45

HI: 85; UN-number: 1791

SODIUM IODIDE

PHYSICAL PROPERTIES		IMPORTANT CHARACTERISTICS	
Boiling point, °C	1304	**COLORLESS TO WHITE HYGROSCOPIC CRYSTALS, TURNING BROWN ON EXPOSURE TO AIR** Decomposes on contact with concentrated strong acids, giving off toxic vapors. Reacts violently with perchloric acid and bromotrifluoride. Also reacts with other oxidants, giving off → *iodium* vapors.	
Melting point, °C	661		
Relative density (water = 1)	3.7		
Solubility in water, g/100 ml	185		
Relative molecular mass	149.9		
		TLV-TWA	not available
		Absorption route: Can enter the body by ingestion. Evaporation negligible at 20°C, but unpleasant concentrations of airborne particles can build up. **Immediate effects:** Irritates the eyes and skin.	
Gross formula:	INa		

HAZARDS/SYMPTOMS	PREVENTIVE MEASURES	FIRE EXTINGUISHING/FIRST AID
Fire: non-combustible.		in case of fire in immediate vicinity: use any extinguishing agent.
Inhalation:	ventilation.	
Skin: redness, pain.	gloves.	
Eyes: redness, pain.	goggles.	flush with water, take to a doctor if necessary.

SPILLAGE	STORAGE	LABELING / NFPA
clean up spillage, flush away remainder with water.	keep in dry, dark place, keep separate from oxidants.	1 / 2 / 0

NOTES

SODIUM METHYLATE

PHYSICAL PROPERTIES	IMPORTANT CHARACTERISTICS
Melting point (decomposes), °C 126 Auto-ignition temperature, °C 70 Solubility in water reaction Relative molecular mass 54.0	**WHITE POWDER** Can ignite spontaneously on exposure to air. Decomposes when heated above 126°C, giving off flammable and toxic vapors. Strong reducing agent which reacts violently with oxidants, with risk of fire and explosion. Reacts explosively with a mixture of *chloroform* and *methanol*. Reacts violently with water, giving off → *methanol* and → *caustic soda*, and reacts with acids. Reacts with light metals, giving off → *hydrogen* gas, with risk of fire and explosion.
	TLV-TWA not available
	Absorption route: Can enter the body by inhalation or ingestion. Can cause burns on contact with moist skin. **Immediate effects:** Corrosive to the eyes, skin and respiratory tract. Inhalation can cause lung edema.
Gross formula: CH₃NaO	

HAZARDS/SYMPTOMS	PREVENTIVE MEASURES	FIRE EXTINGUISHING/FIRST AID
Fire: extremely flammable, many chemical reactions can cause fire and explosion, esp. on contact with moisture.	keep away from open flame and sparks, no smoking; avoid contact with water (or moisture) and with hot surfaces (steam lines).	powder, dry sand, DO NOT USE WATER-BASED EXTINGUISHERS.
Explosion: vapor formation can cause explosion.		in case of fire: keep tanks/drums cool by spraying with water, but DO NOT spray substance with water, fight fire from protected location.
Inhalation: *corrosive*; sore throat, cough, severe breathing difficulties.	local exhaust or respiratory protection.	fresh air, rest, place in half-sitting position, take to hospital.
Skin: *corrosive*; redness, pain, serious burns.	gloves, protective clothing.	first wipe skin clean, remove contaminated clothing, flush skin with water or shower, call a doctor if necessary.
Eyes: *corrosive*; redness, pain, impaired vision.	face shield, or combined eye and respiratory protection.	flush with water, take to a doctor.
Ingestion: *corrosive*; sore throat, abdominal cramps, diarrhea, vomiting.		rinse mouth, take immediately to hospital.

SPILLAGE	STORAGE	LABELING / NFPA
clean up spillage, carefully collect remainder, under no circumstances spray substance with water, do not use sawdust or other combustible absorbents; (additional individual protection: breathing apparatus).	keep in dry, fireproof place, separate from oxidants, strong acids, water and hydrous solutions/mixtures.	R: 11-14-34 S: 8-16-26-43 Note A Flammable Corrosive

NOTES
Solution in methanol has flash point of approx. 23°C. Flush contaminated clothing with water (fire hazard). Reacts violently with extinguishing agents such as water. Lung edema symptoms usually develop several hours later and are aggravated by physical exertion: rest and hospitalization essential. As first aid, a doctor or authorized person should consider administering a corticosteroid spray. Reacts with water in the body to form caustic soda and methyl alcohol: medical data for caustic soda are given. Use airtight packaging. Unbreakable packaging preferred; if breakable, keep in unbreakable container.

UN-number: 1431

SODIUM NITRATE

PHYSICAL PROPERTIES	IMPORTANT CHARACTERISTICS	
Decomposes below boiling point, °C 380 Melting point, °C 307 Relative density (water = 1) 2.3 Solubility in water, g/100 ml at 20 °C 90 Relative molecular mass 85.0	**WHITE HYGROSCOPIC CRYSTALS** Decomposes when heated above 380° C, giving off toxic gases (→ *nitrous vapors*) and → *oxygen*, with increased risk of fire. Strong oxidant which reacts violently with combustible substances and reducing agents, with risk of fire and explosion. Reacts with strong acids, giving off → *nitrous vapors*.	
	TLV-TWA	not available
	Absorption route: Can enter the body by inhalation or ingestion. Evaporation negligible at 20° C, but harmful atmospheric concentrations can build up rapidly if heated or in the form of airborne particles. **Immediate effects:** Irritates the eyes, skin and respiratory tract. **Effects of prolonged/repeated exposure:** Can affect the blood. Affects the blood vessels. In serious cases risk of death.	
Gross formula: NNaO$_3$		

HAZARDS/SYMPTOMS	PREVENTIVE MEASURES	FIRE EXTINGUISHING/FIRST AID
Fire: not combustible but enhances combustion of other substances.		in case of fire in immediate vicinity: use any extinguishing agent.
Explosion: mixing with combustible solids can cause explosion.		in case of fire: keep tanks/drums cool by spraying with water.
	KEEP DUST UNDER CONTROL	
Inhalation: cough, shortness of breath.	local exhaust or respiratory protection.	fresh air, rest, call a doctor.
Skin: redness.	gloves.	remove contaminated clothing, flush skin with water or shower.
Eyes: redness, pain.	goggles.	flush with water, take to a doctor.
Ingestion: abdominal cramps, nausea, vomiting, feeling of weakness, diarrhea, blue skin, muscle cramps.		rinse mouth, call a doctor or take to hospital.

SPILLAGE	STORAGE	LABELING / NFPA
clean up spillage, flush away remainder with water; (additional individual protection: P2 respirator).	keep dry, separate from combustible substances, reducing agents and strong acids.	

NOTES
Flush contaminated clothing with water (fire hazard). Effects on blood due to formation of methemoglobin. Special first aid required in the event of poisoning: antidotes must be available (with instructions).

Transport Emergency Card TEC(R)-51G01

HI: 50; UN-number: 1498

SODIUM NITRITE

PHYSICAL PROPERTIES	IMPORTANT CHARACTERISTICS
Decomposes below boiling point, °C　　320 Melting point, °C　　271 Relative density (water = 1)　　2.2 Solubility in water, g/100 ml at 15 °C　　82 Relative molecular mass　　69.0	**LIGHT YELLOW OR WHITE HYGROSCOPIC CRYSTALS, GRANULES OR POWDER** Decomposes explosively when heated above 538°C. Also decomposes when heated to below 538°C, giving off → *oxygen*. Either a strong oxidant or a strong reducing agent, depending on the conditions. Reacts violently with combustible substances and reducing agents, with risk of fire and explosion. Reacts with acids, giving off toxic gas (→ *nitrous vapors*). Reacts violently when heated with salts of ammonia, thiosulfate, urea or cyanides, with risk of fire and explosion.
	TLV-TWA　　　　　　　　　　not available
	Absorption route: Can enter the body by inhalation or ingestion or through the skin. Evaporation negligible at 20°C, but harmful concentrations of airborne particles can build up rapidly. **Immediate effects:** In serious cases risk of death. **Effects of prolonged/repeated exposure:** Affects the blood vessels and can affect the blood.
Gross formula:　　　　　　NNaO$_2$	

HAZARDS/SYMPTOMS	PREVENTIVE MEASURES	FIRE EXTINGUISHING/FIRST AID
Fire: non-combustible, many chemical reactions can cause fire and explosion.	avoid contact with combustible substances.	in case of fire in immediate vicinity: use any extinguishing agent.
Explosion: heating with other substances can cause explosion.		
	KEEP DUST UNDER CONTROL	
Inhalation: shortness of breath, headache, dizziness, blue skin, feeling of weakness.	local exhaust or respiratory protection.	fresh air, rest, call a doctor or take to hospital.
Skin: *is absorbed*; see also 'Inhalation'.	gloves.	remove contaminated clothing, flush skin with water or shower, call a doctor.
Eyes: redness, pain.	goggles.	flush with water, take to a doctor if necessary.
Ingestion: abdominal pain; see also 'Inhalation'.	do not eat, drink or smoke while working.	rinse mouth, call a doctor or take to hospital.

SPILLAGE	STORAGE	LABELING / NFPA
clean up spillage, flush away remainder with water, do not use sawdust or other combustible absorbents; (additional individual protection: P3 respirator).	keep dry, separate from combustible substances, reducing agents and acids.	R: 8-25 S: 44 🔥　　　☠ Oxidant　　Toxic

NOTES
Flush contaminated clothing with water (fire hazard). Effects on blood due to formation of methemoglobin. Special first aid required in the event of poisoning: antidotes must be available (with instructions). Use airtight packaging.

SODIUMPERBORATE

PHYSICAL PROPERTIES	IMPORTANT CHARACTERISTICS
Boiling point, °C — see notes Melting point, °C — 63 Solubility in water, g/100 ml at 15 °C — 2.5 Relative molecular mass — 153.9	**COLORLESS CRYSTALS** Decomposes when heated above 60°C or on contact with moist air, giving off oxygen. Strong oxidant which reacts violently with combustible substances and reducing agents. The solution in water is alkaline.
	TLV-TWA not available
	Absorption route: Can enter the body by inhalation or ingestion. Evaporation negligible at 20°C, but harmful concentrations of airborne particles can build up rapidly. **Immediate effects:** Affects the nervous system. Affects metabolism. **Effects of prolonged/repeated exposure:** Can affect the blood.
Gross formula: $BH_2NaO_4.3H_2O$	

HAZARDS/SYMPTOMS	PREVENTIVE MEASURES	FIRE EXTINGUISHING/FIRST AID
Fire: not combustible but enhances combustion of other substances.	avoid contact with combustible substances.	in case of fire in immediate vicinity: use any extinguishing agent.
Explosion:		in case of fire: keep tanks/drums cool by spraying with water.
	KEEP DUST UNDER CONTROL	
Inhalation: sore throat, cough.	local exhaust or respiratory protection.	fresh air, rest.
Skin: redness.	gloves.	remove contaminated clothing, flush skin with water or shower.
Eyes: redness, pain.	goggles.	flush with water, take to a doctor if necessary.
Ingestion: diarrhea, stomachache, vomiting, unconsciousness, rash.		rinse mouth, immediately take to hospital.

SPILLAGE	STORAGE	LABELING / NFPA
clean up spillage, flush away remainder with water; (additional individual protection: P2 respirator).	keep cool and dry, separate from combustible substances and reducing agents.	◇◇

NOTES
Boils at 130-150°C after solution in crystal water. Depending on the degree of exposure, regular medical checkups are advisable.

SODIUM PEROXIDE

PHYSICAL PROPERTIES	IMPORTANT CHARACTERISTICS
Boiling point (decomposes), °C 657 Melting point (decomposes), °C 460 Relative density (water = 1) 2.8 Solubility in water reaction Relative molecular mass 78	**LIGHT YELLOW CRYSTALS** When heated gives off oxygen, with increased risk of fire. Strong oxidant which reacts violently with combustible substances, reducing agents and light metals, with risk of fire and explosion. Reacts violently with water, evolving heat and forming a strong base (→ *sodium hydroxide*) and → *oxygen*.
	TLV-TWA not available
	Absorption route: Can enter the body by inhalation or ingestion. **Immediate effects:** Corrosive to the eyes, skin and respiratory tract. Inhalation can cause lung edema.
Gross formula: Na$_2$O$_2$	

HAZARDS/SYMPTOMS	PREVENTIVE MEASURES	FIRE EXTINGUISHING/FIRST AID
Fire: non-combustible, many chemical reactions can cause fire and explosion.	avoid contact with combustible substances and reducing agents.	dry sand, dry foam, powder, DO NOT USE WATER-BASED EXTINGUISHERS, DO NOT use carbon dioxide.
Explosion:		in case of fire: keep tanks/drums cool by spraying with water, but DO NOT spray substance with water.
	STRICT HYGIENE, KEEP DUST UNDER CONTROL	
Inhalation: *corrosive*; sore throat, shortness of breath, severe breathing difficulties.	local exhaust or respiratory protection.	fresh air, rest, place in half-sitting position, artificial respiration if necessary, take to hospital.
Skin: *corrosive*; redness, pain, serious burns.	gloves, protective clothing.	remove contaminated clothing, flush skin with water or shower, call a doctor.
Eyes: *corrosive*; redness, pain, impaired vision.	face shield, or combined eye and respiratory protection.	flush with water, take to a doctor.
Ingestion: *corrosive*; sore throat, vomiting.	do not eat, drink or smoke while working.	rinse mouth, immediately take to hospital.

SPILLAGE	STORAGE	LABELING / NFPA
clean up spillage and remove to safe place; (additional individual protection: P2 respirator).	keep cool and dry, separate from combustible substances, reducing agents and light metals.	R: 8-35 S: 8-27-39 Oxidant Corrosive 3 — 1 / 0 / oxy

NOTES

Flush contaminated clothing with water (fire hazard). Lung edema symptoms usually develop several hours later and are aggravated by physical exertion: rest and hospitalization essential. As first aid, a doctor or authorized person should consider administering a corticosteroid spray.

Transport Emergency Card TEC(R)-51G02   **HI: 50; UN-number: 1504**

SODIUM PERSULPHATE

PHYSICAL PROPERTIES	IMPORTANT CHARACTERISTICS
Melting point (decomposes), °C 100 Relative density (water = 1) 2.0 Solubility in water very good Relative molecular mass 238.1	**COLORLESS CRYSTALS OR WHITE CRYSTALLINE POWDER** Decomposes when heated or on exposure to air, giving off corrosive vapors (→ *sulfur dioxide* and → *sodium peroxide*). Reacts violently with combustible substances and reducing agents, with risk of fire and explosion.
	TLV-TWA not available
	Absorption route: Can enter the body by inhalation or ingestion. **Immediate effects:** Corrosive to the eyes, skin and respiratory tract. Inhalation can cause lung edema. **Effects of prolonged/repeated exposure:** Prolonged or repeated contact can cause skin disorders.
Gross formula: $Na_2O_8S_2$	

HAZARDS/SYMPTOMS	PREVENTIVE MEASURES	FIRE EXTINGUISHING/FIRST AID
Fire: non-combustible, many chemical reactions can cause fire and explosion.	avoid contact with combustible substances.	in case of fire in immediate vicinity: use any extinguishing agent.
Explosion: many chemical reactions can cause explosion.		
	KEEP DUST UNDER CONTROL	
Inhalation: *corrosive*; sore throat, cough, shortness of breath, severe breathing difficulties.	local exhaust or respiratory protection.	fresh air, rest, take to hospital.
Skin: *corrosive*; redness, pain, serious burns.	gloves.	remove contaminated clothing, flush skin with water or shower, call a doctor.
Eyes: *corrosive*; redness, pain, impaired vision.	goggles, or combined eye and respiratory protection.	flush with water, take to a doctor.
Ingestion: *corrosive*; abdominal pain, diarrhea, nausea, vomiting.		rinse mouth, immediately take to hospital.

SPILLAGE	STORAGE	LABELING / NFPA
clean up spillage, carefully collect remainder; (additional individual protection: P2 respirator).	keep separate from combustible substances and reducing agents.	

NOTES
Flush contaminated clothing with water (fire hazard). Lung edema symptoms usually develop several hours later and are aggravated by physical exertion: rest and hospitalization essential. As first aid, a doctor or authorized person should consider administering a corticosteroid spray.

Transport Emergency Card TEC(R)-51G02

UN-number: 1505

SODIUM PYROPHOSPHATE

PHYSICAL PROPERTIES	IMPORTANT CHARACTERISTICS	
Melting point, °C 880 Relative density (water = 1) 2.5 Solubility in water, g/100 ml at 25 °C 6.7 Relative molecular mass 256.9	**WHITE CRYSTALS OR WHITE POWDER** Decomposes slowly on contact with water, forming → *disodium phoshate*.	
	TLV-TWA	5 mg/m³
	Absorption route: Can enter the body by inhalation or ingestion. Evaporation negligible at 20°C, but in finely dispersed form harmfull concentrations of airborne particles can build up. **Immediate effects:** Irritates the eyes, skin and respiratory tract.	
Gross formula: $Na_4O_7P_2$		

HAZARDS/SYMPTOMS	PREVENTIVE MEASURES	FIRE EXTINGUISHING/FIRST AID
Fire: non-combustible.		in case of fire in immediate vicinity: use any extinguishing agent.
Inhalation: sore throat, cough, shortness of breath.	local exhaust or respiratory protection.	fresh air, rest, call a doctor.
Skin: redness, pain.	gloves.	remove contaminated clothing, flush skin with water or shower, refer to a doctor if necessary.
Eyes: redness, pain, impaired vision.	goggles.	flush with water, take to a doctor.
Ingestion: abdominal pain, nausea.		rinse mouth, call a doctor or take to hospital.

SPILLAGE	STORAGE	LABELING / NFPA
clean up spillage, flush away remainder with water; (additional individual protection: P2 respirator).		

NOTES
The substance also exists with 10 molecules of crystal water; apparent melting point at 94°C due to loss of water of crystallization.

SODIUM RHODANIDE

PHYSICAL PROPERTIES	IMPORTANT CHARACTERISTICS	
Melting point (decomposes), °C 287 Relative density (water = 1) 1.7 Solubility in water, g/100 ml at 21 °C 139 Relative molecular mass 81.1	**COLORLESS HYGROSCOPIC CRYSTALS** Decomposes when heated, giving off toxic vapors (→ *nitrous vapors*, → *oxides of sulfur* and → *cyanides*). Decomposes when exposed to light. Reacts violently with strong oxidants.	
	TLV-TWA not available	
	Absorption route: Can enter the body by ingestion. **Immediate effects:** Inhalation of vapor/fumes can cause lung edema. **Effects of prolonged/repeated exposure:** Affects the central nervous system and blood pressure.	
Gross formula: , CNNaS		

HAZARDS/SYMPTOMS	PREVENTIVE MEASURES	FIRE EXTINGUISHING/FIRST AID
Fire: non-combustible.		in case of fire in immediate vicinity: use any extinguishing agent.
Inhalation:	local exhaust or respiratory protection.	
Skin: redness.	gloves.	remove contaminated clothing, flush skin with water or shower.
Eyes: redness, pain.	safety glasses.	flush with water, take to a doctor.
Ingestion: headache, nausea, vomiting, feeling of weakness, dizziness.		rinse mouth, call a doctor or take to hospital.

SPILLAGE	STORAGE	LABELING / NFPA
clean up spillage, carefully collect remainder; (additional individual protection: P2 respirator).	keep in dry, dark place, separate from oxidants.	

NOTES

SODIUM SELENITE

PHYSICAL PROPERTIES	IMPORTANT CHARACTERISTICS	
Boiling point, °C　　　　　　? Melting point, °C　　　　　　? Relative molecular mass　　263	**WHITE CRYSTALS** Decomposes when heated, forming sodium oxide and → *selenium oxides*.	
	TLV-TWA (skin)	0.2 mg/m³ ✷
	Absorption route: Can enter the body by inhalation or ingestion. **Immediate effects:** Corrosive to the eyes, skin and respiratory tract. Inhalation can cause lung edema. In serious cases risk of death.	
Gross formula:　　　　$Na_2O_3Se.5H_2O$		

HAZARDS/SYMPTOMS	PREVENTIVE MEASURES	FIRE EXTINGUISHING/FIRST AID
Fire: non-combustible.		in case of fire in immediate vicinity: use any extinguishing agent.
	KEEP DUST UNDER CONTROL	
Inhalation: *corrosive*; sore throat, cough, severe breathing difficulties.	local exhaust or respiratory protection.	fresh air, rest, place in half-sitting position, take to hospital.
Skin: *corrosive*; redness, pain, serious burns.	gloves, protective clothing.	remove contaminated clothing, flush skin with water or shower, refer to a doctor.
Eyes: *corrosive*; redness, pain, impaired vision.	close-fitting safety glasses.	flush with water, take to a doctor.
Ingestion: *corrosive*; sore throat, abdominal pain, vomiting, metallic taste.	do not eat, drink or smoke while working.	rinse mouth, immediately take to hospital.

SPILLAGE	STORAGE	LABELING / NFPA
call in expert, clean up spillage, carefully collect remainder; (additional individual protection: P3 respirator).		R: 23/25-33 S: 20/21-28-44 Note A ☠ Toxic

NOTES

✷ TLV equals that of selenium. Lung edema symptoms usually develop several hours later and are aggravated by physical exertion: rest and hospitalization essential. As first aid, a doctor or authorized person should consider administering a corticosteroid spray.

SODIUM SILICATE

PHYSICAL PROPERTIES	IMPORTANT CHARACTERISTICS
Melting point, °C 1088 Relative density (water = 1) 2.4 Solubility in water good Relative molecular mass 122.1	**WHITISH-GREEN POWDER OR VITREOUS LUMPS** In aqueous solution is a strong base which reacts violently with acids and corrodes aluminum, zinc etc. Reacts violently with fluorine.
	TLV-TWA not available
	Absorption route: Can enter the body by inhalation or ingestion. Evaporation negligible at 20°C, but harmful concentrations of airborne particles can build up rapidly. **Immediate effects:** Irritates the eyes, skin and respiratory tract.
Gross formula: Na$_2$O$_3$Si	

HAZARDS/SYMPTOMS	PREVENTIVE MEASURES	FIRE EXTINGUISHING/FIRST AID
Fire: non-combustible.		in case of fire in immediate vicinity: use any extinguishing agent.
Inhalation: sore throat, cough, shortness of breath.	local exhaust or respiratory protection.	fresh air, rest, call a doctor.
Skin: redness, pain.	gloves.	remove contaminated clothing, flush skin with water or shower, refer to a doctor.
Eyes: redness, pain, impaired vision.	face shield.	flush with water, take to a doctor.
Ingestion: sore throat, abdominal pain, nausea, vomiting.		rinse mouth, immediately take to hospital.

SPILLAGE	STORAGE	LABELING / NFPA
clean up spillage, flush away remainder with water; (additional individual protection: P2 respirator).		

NOTES
Lumps of sodium metasilicate dissolve slowly in water.

Transport Emergency Card TEC(R)-80G09

SODIUM SILICATE
(solution in water)

PHYSICAL PROPERTIES	IMPORTANT CHARACTERISTICS	
Solubility in water ∞ Relative molecular mass 122.1	**COLORLESS OR LIGHT GREEN VISCOUS LIQUID** In aqueous solution is a strong base which reacts violently with acids and corrodes aluminum, zinc etc.	
	TLV-TWA not available	
	Absorption route: Can enter the body by inhalation or ingestion. Harmful atmospheric concentrations build up very slowly, if at all, on evaporation at approx. 20°C, but much more rapidly in aerosol form. **Immediate effects:** Irritates the eyes, skin and respiratory tract.	
Gross formula: Na_2O_3Si		

HAZARDS/SYMPTOMS	PREVENTIVE MEASURES	FIRE EXTINGUISHING/FIRST AID
Fire: non-combustible.		in case of fire in immediate vicinity: use any extinguishing agent.
Inhalation: sore throat, cough, shortness of breath.		fresh air, rest, call a doctor.
Skin: redness, pain.	gloves, protective clothing.	remove contaminated clothing, flush skin with water or shower.
Eyes: redness, pain.	face shield.	flush with water, take to a doctor if necessary.
Ingestion: sore throat, abdominal pain, nausea, vomiting.		rinse mouth, call a doctor.

SPILLAGE	STORAGE	LABELING / NFPA
collect leakage in sealable containers, flush away remainder with water.	keep separate from strong acids.	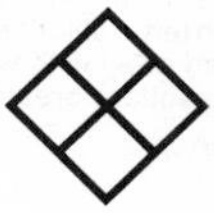

NOTES

CAS-No: [1313-82-2]
disodium monosulfide
disodium sulfide
sodium monosulfide

$Na_2S.9H_2O$

SODIUM SULFIDE

PHYSICAL PROPERTIES	IMPORTANT CHARACTERISTICS
Melting point, °C 50 Relative density (water = 1) 1.4 Solubility in water, g/100 ml at 25 °C 18 Relative molecular mass 240.2	**WHITE TO YELLOW HYGROSCOPIC CRYSTALS WITH CHARACTERISTIC ODOR** Anhydrous sodium sulfide can ignite spontaneously on exposure to air. Can explode if heated rapidly. Decomposes when burned, giving off toxic → *sulfur dioxide*, or on contact with acids, giving off toxic gas (→ *hydrogen sulfide*). In aqueous solution is a strong base which reacts violently with acids and corrodes aluminum, zinc etc. Reacts violently with strong oxidants, giving off toxic gas (→ *sulfur dioxide*). Attacks steel.
	TLV-TWA not available
	Absorption route: Can enter the body by inhalation or ingestion. Insufficient data on the rate at which harmful concentrations can build up. **Immediate effects:** Corrosive to the eyes, skin and respiratory tract.
Gross formula: $Na_2S.9H_2O$	

HAZARDS/SYMPTOMS	PREVENTIVE MEASURES	FIRE EXTINGUISHING/FIRST AID
Fire: anhydrous substance is combustible.		large quantities of water.
Explosion: rapid heating or friction can cause explosion.	do not subject to shocks or friction.	
	KEEP DUST UNDER CONTROL	IN ALL CASES CALL IN A DOCTOR
Inhalation: *corrosive*; sore throat, cough, shortness of breath.	local exhaust or respiratory protection.	fresh air, rest, place in half-sitting position, call a doctor.
Skin: *corrosive*; redness, pain, burns.	gloves, protective clothing.	remove contaminated clothing, flush skin with water or shower, refer to a doctor.
Eyes: *corrosive*; redness, pain, impaired vision.	face shield, or combined eye and respiratory protection.	flush with water, take to a doctor.
Ingestion: *corrosive*; sore throat, abdominal pain, diarrhea, nausea, vomiting.	do not eat, drink or smoke while working.	rinse mouth, take immediately to hospital.

SPILLAGE	STORAGE	LABELING / NFPA
clean up spillage, render remainder harmless with ferric chloride solution and then add soda, flush away with water; (additional individual protection: breathing apparatus).	keep dry, separate from oxidants and acids.	R: 31-34 S: 26 Corrosive 3 — 1 / 1

NOTES
Apparent melting point due to loss of water of crystallization. Other apparent melting points: anhydrous: 920°C; $Na_2S.5H_2O$: 120°C. Use airtight packaging.

Transport Emergency Card TEC(R)-80G13

HI: 43; UN-number: 1385

SODIUM SULFITE (ANHYDROUS)

PHYSICAL PROPERTIES	IMPORTANT CHARACTERISTICS	
Decomposes below melting point, °C ca. 600 Relative density (water = 1) 2.6 Solubility in water, g/100 ml at 20 °C 23 Relative molecular mass 126.1	**WHITE CRYSTALS OR POWDER** Decomposes when heated above 600°C, giving off toxic gas (→ *sulfur dioxide*). Reacts with strong acids, giving off → *sulfur dioxide*. *Reacts with oxidants.*	
	TLV-TWA not available	
	Absorption route: Can enter the body by inhalation or ingestion. Evaporation negligible at 20°C, but harmful concentrations of airborne particles can build up rapidly. **Immediate effects:** Corrosive to the eyes, skin and respiratory tract. Inhalation can cause lung edema. In serious cases risk of death.	
Gross formula: Na_2O_3S		

HAZARDS/SYMPTOMS	PREVENTIVE MEASURES	FIRE EXTINGUISHING/FIRST AID
Fire: non-combustible.		in case of fire in immediate vicinity: use any extinguishing agent.
Inhalation: *corrosive*; sore throat, cough, shortness of breath.	local exhaust or respiratory protection.	fresh air, rest, place in half-sitting position, take to hospital.
Skin: *corrosive*; redness, pain, burns.	gloves.	remove contaminated clothing, flush skin with water or shower, refer to a doctor if necessary.
Eyes: *corrosive*; redness, pain, impaired vision.	goggles.	flush with water, take to a doctor.
Ingestion: *corrosive*; abdominal pain, diarrhea, vomiting.		rinse mouth, take to hospital if necessary.

SPILLAGE	STORAGE	LABELING / NFPA
clean up spillage, flush away remainder with water; (additional individual protection: P2 respirator).	keep dry, separate from acids.	

NOTES
Other apparent melting points: with 7H₂O: > 150°C. Lung edema symptoms usually develop several hours later and are aggravated by physical exertion: rest and hospitalization essential. As first aid, a doctor or authorized person should consider administering a corticosteroid spray.

SODIUM THIOSULFATE

PHYSICAL PROPERTIES	IMPORTANT CHARACTERISTICS	
Melting point, °C 48 Relative density (water = 1) 1.7 Solubility in water, g/100 ml at 0 °C 79 Relative molecular mass 248.2	**COLORLESS CRYSTALS OR WHITE POWDER** Reacts with chlorine in solution, forming sodium hydrosulfate. Decomposes when heated, giving off sodium oxide, → *sulfur dioxide* and → *sulfur trioxide*. Reacts with acids, giving off → *sulfur dioxide*. Reacts violently with oxidants, e.g. nitrates and nitrites, with risk of explosion. Reacts with halogens.	
	TLV-TWA not available	
	Absorption route: Can enter the body by ingestion. **Immediate effects:** Irritates the eyes, nose and throat.	
Gross formula: $Na_2O_3S_2.5H_2O$		

HAZARDS/SYMPTOMS	PREVENTIVE MEASURES	FIRE EXTINGUISHING/FIRST AID
Fire: non-combustible.		in case of fire in immediate vicinity: use any extinguishing agent.
Eyes: redness, irritation.	safety glasses.	flush with water, take to a doctor if necessary.
Ingestion: in large quantities: diarrhea.		

SPILLAGE	STORAGE	LABELING / NFPA
clean up spillage, flush away remainder with water.		

NOTES
Used in photographic fixing baths. Also referred to *incorrectly* as *sodium hyposulfite*. Apparent melting point due to loss of water of crystallization. Other apparent melting points: 100°C (on loss of all water of crystallization).

SODIUM TRICHLOROACETATE

PHYSICAL PROPERTIES	IMPORTANT CHARACTERISTICS	
Decomposes below melting point, °C 140 Relative density (water = 1) 0.9 Solubility in water, g/100 ml at 20 °C 153 Relative molecular mass 185.4	**WHITE TO LIGHT GRAY GRANULATE OR POWDER** Decomposes when heated above 140°C, giving off toxic and corrosive vapors. In aqueous solution decomposes when heated above 40°C, giving off → *chloroform*. Attacks many metals.	
	TLV-TWA not available	
	Absorption route: Can enter the body by inhalation or ingestion. Insufficient data on the rate at which harmful concentrations can build up. **Immediate effects:** Irritates the eyes, skin and respiratory tract.	
Gross formula: $C_2Cl_3NaO_2$		

HAZARDS/SYMPTOMS	PREVENTIVE MEASURES	FIRE EXTINGUISHING/FIRST AID
Fire: non-combustible.		in case of fire in immediate vicinity: use any extinguishing agent.
Inhalation: sore throat, cough, shortness of breath.	local exhaust or respiratory protection.	fresh air, rest, call a doctor.
Skin: itching, redness, pain.	gloves.	remove contaminated clothing, flush skin with water or shower.
Eyes: redness, pain.	safety glasses.	flush with water, take to a doctor if necessary.
Ingestion: abdominal pain, nausea.		rinse mouth, call a doctor or take to hospital.

SPILLAGE	STORAGE	LABELING / NFPA
clean up spillage, flush away remainder with water; (additional individual protection: P2 respirator).		R: 22 S: 24/25 ✖ Harmful

NOTES
Konesta is a trade name.

$Na_5P_3O_{10}$

SODIUM TRIPOLYPHOSPHATE

PHYSICAL PROPERTIES	IMPORTANT CHARACTERISTICS	
Melting point (decomposes), °C 622 Relative density (water = 1) < 1.5 Solubility in water, g/100 ml at 25 °C 20 Relative molecular mass 367.9	**WHITE HYGROSCOPIC POWDER** Decomposes when heated, forming toxic phosphorus oxides and sodium oxide.	
	TLV-TWA not available	
	Absorption route: Can enter the body by inhalation or ingestion. Evaporation negligible at 20° C, but unpleasant atmospheric concentrations can build up rapidly in aerosol form. **Immediate effects:** Irritates the eyes, skin and respiratory tract.	
Gross formula: $Na_5O_{10}P_3$		

HAZARDS/SYMPTOMS	PREVENTIVE MEASURES	FIRE EXTINGUISHING/FIRST AID
Fire: non-combustible.		in case of fire in immediate vicinity: use any extinguishing agent.
Inhalation: sore throat, cough, shortness of breath.	local exhaust or respiratory protection.	fresh air, rest, call a doctor.
Skin: redness, pain.	gloves.	remove contaminated clothing, flush skin with water or shower.
Eyes: redness, pain.	goggles.	flush with water, take to a doctor.
Ingestion: sore throat, abdominal pain, diarrhea, vomiting.		rinse mouth, immediately take to hospital.

SPILLAGE	STORAGE	LABELING / NFPA
clean up spillage, flush away remainder with water; (additional individual protection: P1 respirator).		

NOTES

Transport Emergency Card TEC(R)-80G09

CAS-No: [100-42-5]
cinnamene
cinnamol
ethenylbenzene
styrol
vinylbenzene

$C_6H_5CH=CH_2$

STYRENE

PHYSICAL PROPERTIES		IMPORTANT CHARACTERISTICS	
Boiling point, °C	145	**COLORLESS LIQUID WITH CHARACTERISTIC ODOR**	
Melting point, °C	−31	Able to form peroxides when exposed to light or air or on contact with acids. Able to polymerize violently when heated or on contact with rust, acids or peroxides, with risk of fire and explosion. Flow, agitation etc. can cause build-up of electrostatic charge due to liquid's low conductivity. Reacts violently with strong oxidants, with risk of fire and explosion. Attacks copper slowly.	
Flash point, °C	31		
Auto-ignition temperature, °C	490		
Relative density (water = 1)	0.9		
Relative vapor density (air = 1)	3.6		
Relative density at 20 °C of saturated mixture vapor/air (air = 1)	1.02	TLV-TWA (skin) 50 ppm 213 mg/m³	
Vapor pressure, mm Hg at 20 °C	4.6	STEL 100 ppm 426 mg/m³	
Solubility in water, g/100 ml at 20 °C	0.03		
Explosive limits, vol% in air	1.1-8	**Absorption route:** Can enter the body by inhalation or ingestion. Harmful atmospheric concentrations build up fairly slowly on evaporation at 20°C, but much more rapidly in aerosol form.	
Relative molecular mass	104.2		
		Immediate effects: Irritates the eyes, skin and respiratory tract.	
		Effects of prolonged/repeated exposure: Frequent high-level exposure can cause diseases of the nervous system, with muscle disorders, feeling of weakness in the legs and tingling in fingers and toes.	
Gross formula:	C_8H_8		

HAZARDS/SYMPTOMS	PREVENTIVE MEASURES	FIRE EXTINGUISHING/FIRST AID
Fire: flammable.	keep away from open flame and sparks, no smoking, avoid contact with strong oxidants.	powder, AFFF, foam, carbon dioxide, (halons).
Explosion: above 31°C: forms explosive air-vapor mixtures.	above 31°C: sealed machinery, ventilation, explosion-proof electrical equipment, grounding.	in case of fire: keep tanks/drums cool by spraying with water, fight fire from sheltered location.
Inhalation: cough, headache, dizziness, nausea, vomiting, sleepiness.	ventilation, local exhaust or respiratory protection.	fresh air, rest, call a doctor.
Skin: redness.	gloves.	remove contaminated clothing, wash skin with soap and water, call a doctor.
Eyes: redness, pain.	face shield.	flush with water, take to a doctor if necessary.
Ingestion: sore throat, abdominal pain, stomachache, headache, dizziness, vomiting, drowsiness.		rinse mouth, DO NOT induce vomiting, call a doctor or take to hospital.

SPILLAGE	STORAGE	LABELING / NFPA
collect leakage in sealable containers, soak up with sand or other inert absorbent and remove to safe place.	keep in cool, dark, fireproof place, separate from oxidants; add an inhibitor.	R: 10-20-36/38 S: 23 Note D ✖ Harmful

NFPA: 3 / 2 / 2

NOTES
Before distilling check for peroxides; if found, render harmless. Polymerization inhibitor can be rendered ineffective by contact with lye etc. Alcohol consumption increases toxic effects.

Transport Emergency Card TEC(R)-101

HI: 39; UN-number: 2055

CAS-No: [96-09-3]
ethenyl benzene oxide
phenylethylene oxide

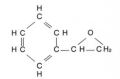

STYRENE OXIDE

PHYSICAL PROPERTIES	IMPORTANT CHARACTERISTICS
Boiling point, °C 194 Melting point, °C − 37 Flash point, °C 74 Auto-ignition temperature, °C 435 Relative density (water = 1) 1.05 Relative vapor density (air = 1) 4.1 Relative density at 20 °C of saturated mixture vapor/air (air = 1) 1.0 Vapor pressure, mm Hg at 20 °C 0.30 Solubility in water, g/100 ml 0.1 Relative molecular mass 120.1	**COLORLESS LIQUID** Vapor mixes readily with air, forming explosive mixtures. Acids, bases and some salts can cause violent polymerization. Reacts with strong oxidants and many other compounds.
	TLV-TWA not available
	Absorption route: Can enter the body by inhalation or ingestion or through the skin. Insufficient data on the rate at which harmful concentrations can build up. **Immediate effects:** Irritates the skin and respiratory tract. Corrosive to the eyes. **Effects of prolonged/repeated exposure:** Prolonged or repeated contact can cause skin disorders. Frequent high-level exposure can cause liver damage and nerve tissue disease, combined with muscle disorders, feeling of weakness in the legs and tingling in fingers and toes.
Gross formula: C_8H_8O	

HAZARDS/SYMPTOMS	PREVENTIVE MEASURES	FIRE EXTINGUISHING/FIRST AID
Fire: combustible.	keep away from open flame, no smoking.	powder, AFFF, foam, carbon dioxide, (halons).
Explosion: above 74°C: forms explosive air-vapor mixtures.	above 74°C: sealed machinery, ventilation, grounding.	
Inhalation: sore throat, cough, shortness of breath.	ventilation, local exhaust or respiratory protection.	fresh air, rest, call a doctor if necessary.
Skin: redness.	gloves, protective clothing.	remove contaminated clothing, flush skin with water or shower.
Eyes: *corrosive*; redness, pain, impaired vision.	face shield.	flush with water, take to a doctor.
Ingestion: sore throat, abdominal pain, nausea, stomachache, drowsiness.		rinse mouth, call a doctor.

SPILLAGE	STORAGE	LABELING / NFPA
collect leakage in sealable containers, soak up with sand or other inert absorbent and remove to safe place.	keep separate from acids and bases.	R: 45-21-36 S: 53-44 Note E ☠ Toxic

NFPA diamond: 2 / 0 / 2

NOTES
Aqueous solutions as weak as 1% cause skin hypersensitivity. Persons who have developed hypersensitivity due to direct contact with dilute liquid are often hypersensitive to vapor as well.

Transport Emergency Card TEC(R)-30G15

1,2-ethanecarboxylic acid
butanedioic acid

$HOOC(CH_2)_2COOH$

SUCCINIC ACID

PHYSICAL PROPERTIES	IMPORTANT CHARACTERISTICS
Boiling point (decomposes), °C 235 Melting point, °C 188 Relative density (water = 1) 1.6 Relative vapor density (air = 1) 4.0 Relative density at 20 °C of saturated mixture vapor/air (air = 1) 1.0 Vapor pressure, mm Hg at 20 °C <0.076 Solubility in water, g/100 ml at 20 °C 8 Relative molecular mass 118.1 Log P octanol/water −0.6 Gross formula: $C_4H_6O_4$	**WHITE CRYSTALS** Decomposes when heated to boiling point, forming succinic anhydride. Reacts with strong oxidants and bases. TLV-TWA not available **Absorption route:** Can enter the body by inhalation or ingestion. Evaporation negligible at 20°C, but unpleasant concentrations of airborne particles can build up. **Immediate effects:** Irritates the eyes, skin and respiratory tract.

HAZARDS/SYMPTOMS	PREVENTIVE MEASURES	FIRE EXTINGUISHING/FIRST AID
Fire: combustible.	keep away from open flame, no smoking.	water spray, powder.
	KEEP DUST UNDER CONTROL	
Inhalation: sore throat, cough.	local exhaust or respiratory protection.	fresh air, rest, place in half-sitting position, call a doctor if necessary.
Skin: redness, pain.	gloves.	remove contaminated clothing, flush skin with water or shower.
Eyes: redness, pain.	safety glasses.	flush with water, take to a doctor.
Ingestion: sore throat, abdominal pain, diarrhea, vomiting.		rinse mouth, consult a doctor.

SPILLAGE	STORAGE	LABELING / NFPA
clean up spillage, flush away remainder with water; (additional individual protection: P1 respirator).	keep separate from strong bases.	

NOTES

SULFAMIC ACID

PHYSICAL PROPERTIES	IMPORTANT CHARACTERISTICS	
Decomposes below melting point, °C 206 Relative density (water = 1) 2.12 Solubility in water, g/100 ml at 20 °C 17.5 Relative molecular mass 97.1	**COLORLESS CRYSTALS OR WHITE POWDER** Decomposes when heated, giving off → *nitrogen* and corrosive → *sulfur oxide* gases. In aqueous solution is a medium strong acid which reacts violently with bases and is corrosive. Reacts violently with chlorine and nitric acid. The aqueous solution hydrolizes to ammoniumbisulfate when subjected to rise of temperature.	
	TLV-TWA not available	
	Absorption route: Can enter the body by inhalation or ingestion. **Immediate effects:** Corrosive to the eyes, skin and respiratory tract. Inhalation can cause lung edema. In serious cases risk of death.	
Gross formula: H_3NO_3S		

HAZARDS/SYMPTOMS	PREVENTIVE MEASURES	FIRE EXTINGUISHING/FIRST AID
Fire: non-combustible.		in case of fire in immediate vicinity: use any extinguishing agent, see 'Notes'.
Explosion:		in case of fire: keep tanks/drums cool by spraying with water.
Inhalation: *corrosive*; sore throat, cough, shortness of breath, lung edema.	local exhaust or respiratory protection.	fresh air, rest, place in half-sitting position, call a doctor.
Skin: *corrosive*; redness, pain.	gloves, protective clothing.	remove contaminated clothing, flush skin with water or shower.
Eyes: *corrosive*; redness, pain, impaired vision.	face shield, or combined eye and respiratory protection.	flush with water, take to a doctor.
Ingestion: *corrosive*; diarrhea, abdominal cramps, mouth and tongue pain, sore throat, nausea, stomachache.		rinse mouth, call a doctor or take to hospital.

SPILLAGE	STORAGE	LABELING / NFPA
clean up spillage, flush away remainder with water; (additional individual protection: P2 respirator).	keep dry, separate from strong bases.	R: 36/38 S: 2-26-28 ✖ Irritant

NOTES
Lung edema symptoms usually develop several hours later and are aggravated by physical exertion: rest and hospitalization essential. As first aid, a doctor or authorized person should consider administering a corticosteroid spray. If extinguished with water it must be expected that the flowing water will become acidic.

CAS-No: [121-57-3]
p-aminobenzenesulfonic acid
p-anilinesulfonic acid

$NH_2C_6H_4SO_3H.H_2O$

p-SULFANILIC ACID
(monohydrate)

PHYSICAL PROPERTIES	IMPORTANT CHARACTERISTICS
Decomposes below melting point, °C 288 Relative density (water = 1) 1.5 Solubility in water, g/100 ml at 20 °C 1 Relative molecular mass 191.2	**COLORLESS CRYSTALS, DISINTEGRATING IN LIGHT** Decomposes when heated above 288°C or on contact with strong acids, giving off → *nitrous vapors* and → *sulfur trioxide*. Slightly acidic in aqueous solution.
	TLV-TWA not available
	Absorption route: Can enter the body by inhalation or ingestion. Evaporation negligible at 20°C, but harmful concentrations of airborne particles can build up rapidly. **Immediate effects:** Corrosive to the eyes, skin and respiratory tract. Inhalation of vapors resulting from combustion or reaction with acids can cause lung edema. In serious cases risk of death. **Effects of prolonged/repeated exposure:** Repeated exposure to high concentrations can cause liver and kidney disease and can affect the blood.
Gross formula: $C_6H_7NO_3S.H_2O$	

HAZARDS/SYMPTOMS	PREVENTIVE MEASURES	FIRE EXTINGUISHING/FIRST AID
Fire: non-combustible.		in case of fire in immediate vicinity: use any extinguishing agent.
Inhalation: *corrosive*; sore throat, cough, shortness of breath, lung edema.	local exhaust or respiratory protection.	fresh air, rest, place in half-sitting position, take to hospital.
Skin: *corrosive*; redness, pain.	gloves.	remove contaminated clothing, flush skin with water or shower, call a doctor if necessary.
Eyes: *corrosive*; redness, pain, impaired vision.	goggles.	flush with water, take to a doctor.
Ingestion: *corrosive*; sore throat, abdominal pain, diarrhea, nausea, vomiting, mouth and tongue pain, stomachache.		rinse mouth, take immediately to hospital.

SPILLAGE	STORAGE	LABELING / NFPA
clean up spillage, carefully collect remainder; (additional individual protection: P2 respirator).		R: 20/21/22 S: 25-28 ✖ Harmful

NOTES
Changes to anhydrous form at 100°C. Lung edema symptoms usually develop several hours later and are aggravated by physical exertion: rest and hospitalization essential. As first aid, a doctor or authorized person should consider administering a corticosteroid spray.

SULFUR
(molten)

PHYSICAL PROPERTIES		IMPORTANT CHARACTERISTICS
Boiling point, °C	445	**YELLOW TO REDDISH-BROWN LIQUID WITH TEMPERATURE OF APPROX. 140 °C AND CHARACTERISTIC ODOR**
Melting point, °C	114	
Flash point, °C	<205	Harmful vapors are produced in considerable quantities on exposure to air at 140°C. Toxic
Auto-ignition temperature, °C	ca. 235	→ *hydrogen sulfide* can be given off due to contaminants. Flow, agitation etc. can cause build-up
Relative density (water = 1)	2.07	of electrostatic charge due to liquid's low conductivity. Toxic → *sulfur dioxide* is a product of
Relative vapor density (air = 1)	8.9	combustion. Reacts with iron, forming pyrophoric compounds. Reacts violently with strong
Vapor pressure, mm Hg at 184 °C	0.99	oxidants. Liquid reacts with (1) air, giving off toxic → *sulfur dioxide*, and (2) hydrogen, giving off
Solubility in water	none	toxic → *hydrogen sulfide*. Attacks copper, silver and mercury.
Explosive limits, g/m³ in air	35-1400	
Relative molecular mass	256.5	

TLV-TWA not available

Absorption route: Can enter the body by inhalation or ingestion. Harmful atmospheric concentrations can build up rapidly at a temperature of 140°C.
Immediate effects: Irritates the respiratory tract. In substantial concentrations can impair consciousness. In serious cases risk of paralysis of the respiratory organs.

Gross formula: S₈

HAZARDS/SYMPTOMS	PREVENTIVE MEASURES	FIRE EXTINGUISHING/FIRST AID
Fire: combustible.	keep away from open flame, no smoking.	steam.
Explosion: finely dispered particles can explode on contact with air.	sealed machinery, explosion-proof electrical equipment and lighting, grounding.	under no circumstances direct tight stream of water onto hot liquid.
Inhalation: sore throat, cough, shortness of breath, headache.	ventilation, local exhaust or respiratory protection.	fresh air, rest, call a doctor.
Skin: *in case of burns*: redness, pain, serious burns.	heat-insulated gloves, protective clothing.	*in case of burns*: DO NOT remove clothing, flush skin with water or shower, take to a doctor if necessary.
Eyes: *in case of burns*: redness, pain, impaired vision.	face shield.	flush with water, take to a doctor.
Ingestion: *in case of burns*: mouth and throat pain, abdominal pain, diarrhea; also nausea, drowsiness, unconsciousness.		rinse mouth, take immediately to hospital.

SPILLAGE	STORAGE	LABELING / NFPA
allow leakage to set, clean up solid and remove to safe place, under no circumstances spray substance with water; (additional individual protection: breathing apparatus).		1 / 2 / 0

NOTES

Formation of hydrogen sulfide depends largely on contaminants.

Transport Emergency Card TEC(R)-115   HI: 44; UN-number: 2448

SULFUR
(solid)

PHYSICAL PROPERTIES		IMPORTANT CHARACTERISTICS
Boiling point, °C	445	**YELLOW SOLID IN VARIOUS FORMS**
Melting point, °C	114	In dry state can form electrostatic charge if stirred, transported pneumatically, poured etc. Toxic
Flash point, °C	168-207	→ *sulfur dioxide* is a product of combustion. Reacts violently with strong oxidants, with risk of fire
Auto-ignition temperature, °C	ca. 235	and explosion. Attacks copper, silver and mercury.
Relative density (water = 1)	2.07	
Vapor pressure, mm Hg at 20 °C	< 0.00076	TLV-TWA not available
Solubility in water	none	
Explosive limits, g/m³ in air	35-1400	**Absorption route:** Can enter the body by inhalation or ingestion. Evaporation negligible at 20°C,
Relative molecular mass	256.5	but harmful concentrations of airborne particles can build up rapidly.
		Immediate effects: Irritates the eyes, skin and respiratory tract. In substantial concentrations
		can impair consciousness.
Gross formula:	S$_8$	

HAZARDS/SYMPTOMS	PREVENTIVE MEASURES	FIRE EXTINGUISHING/FIRST AID
Fire: combustible.	keep away from open flame, no smoking.	water spray, powder.
Explosion: finely dispersed particles can explode on contact with air.	keep dust from accumulating; sealed machinery, explosion-proof electrical equipment and lighting, grounding.	
Inhalation: sore throat, cough, headache.	local exhaust or respiratory protection.	fresh air, rest, call a doctor.
Skin: redness.	gloves.	remove contaminated clothing, flush skin with water or shower.
Eyes: redness, pain.	goggles.	flush with water, take to a doctor if necessary.
Ingestion: sore throat, nausea, headache, drowsiness, unconsciousness, abdominal pain, diarrhea.	do not eat, drink or smoke while working.	rinse mouth, take immediately to hospital.

SPILLAGE	STORAGE	LABELING / NFPA
clean up spillage, flush away remainder with water; (additional individual protection: P2 respirator).	keep in fireproof place, separate from oxidants.	◇ 1 / 1 X 0

NOTES
Auto-ignition temperature: dust cloud 190°C, dust layer 220°C.

Transport Emergency Card TEC(R)-115A

HI: 40; UN-number: 1350

SULFUR BROMIDE

PHYSICAL PROPERTIES		IMPORTANT CHARACTERISTICS
Boiling point, °C	?	**YELLOW TO RED FUMING LIQUID WITH PUNGENT ODOR**
Melting point, °C	− 40	Decomposes when heated or burned, giving off toxic and corrosive gases. Reacts with water and
Flash point, °C	see notes	oxidants, giving off toxic and corrosive gases (→ *hydrogen bromide* and → *sulfur dioxide*).
Relative density (water = 1)	2.6	Reacts with air to form corrosive fumes. Attacks many metals, esp. in the presence of moisture.
Relative vapor density (air = 1)	7.7	
Solubility in water	reaction	TLV-TWA — not available
Relative molecular mass	224.0	
		Absorption route: Can enter the body by inhalation or ingestion. Harmful atmospheric concentrations can build up rapidly on evaporation at approx. 20°C - even more rapidly in aerosol form.
		Immediate effects: Corrosive to the eyes, skin and respiratory tract. Inhalation of vapor/fumes can cause severe breathing difficulties (lung edema). In serious cases risk of death.
Gross formula:	Br_2S_2	

HAZARDS/SYMPTOMS	PREVENTIVE MEASURES	FIRE EXTINGUISHING/FIRST AID
Fire: combustible.	keep away from open flame, no smoking.	powder, carbon dioxide, DO NOT USE WATER-BASED EXTINGUISHERS.
	STRICT HYGIENE	IN ALL CASES CALL IN A DOCTOR
Inhalation: *corrosive*; sore throat, cough, shortness of breath, severe breathing difficulties.	ventilation, local exhaust or respiratory protection.	fresh air, rest, place in half-sitting position, take to hospital.
Skin: *corrosive*; redness, pain, serious burns.	gloves, protective clothing.	remove contaminated clothing, flush skin with water or shower, take to a doctor or take immediately to hospital.
Eyes: *corrosive*; redness, pain, impaired vision.	face shield, or combined eye and respiratory protection.	flush with water, take to a doctor.
Ingestion: *corrosive*; sore throat, abdominal pain, diarrhea, vomiting.		rinse mouth, take immediately to hospital.

SPILLAGE	STORAGE	LABELING / NFPA
collect leakage in sealable containers, soak up with dry sand or other inert absorbent and remove to safe place; (additional individual protection: CHEMICAL SUIT).	keep dry, separate from strong oxidants; ventilate at floor level.	

NOTES
Combustible, but flash point and explosive limits not given in the literature. Lung edema symptoms usually develop several hours later and are aggravated by physical exertion: rest and hospitalization essential. As first aid, a doctor or authorized person should consider administering a corticosteroid spray. Use airtight packaging.

Transport Emergency Card TEC(R)-80G10

SULFUR CHLORIDE

PHYSICAL PROPERTIES	IMPORTANT CHARACTERISTICS
Boiling point, °C 138 Melting point, °C −77 Flash point, °C 118 Auto-ignition temperature, °C 233 Relative density (water = 1) 1.68 Relative vapor density (air = 1) 4.66 Relative density at 20 °C of saturated mixture vapor/air (air = 1) 1.05 Vapor pressure, mm Hg at 20 °C 9.9 Solubility in water reaction Relative molecular mass 135.0	**YELLOW TO RED VISCOUS LIQUID WITH PUNGENT ODOR** Decomposes when heated, giving off toxic and corrosive gases. Decomposes on contact with water, giving off → *hydrogen chloride* and → *sulfur dioxide*. Reacts violently with oxidants. Reacts with air to form corrosive vapors (→ *hydrochloric acid*). Highly corrosive, esp. in the presence of water.

TLV-TWA	1 ppm	5.5 mg/m³ C

Absorption route: Can enter the body by inhalation or ingestion. Harmful atmospheric concentrations can build up very rapidly on evaporation at 20°C.
Immediate effects: Corrosive to the eyes, skin and respiratory tract. Inhalation of vapor/fumes can cause severe breathing difficulties (lung edema). In serious cases risk of death.

Gross formula: Cl_2S_2

HAZARDS/SYMPTOMS	PREVENTIVE MEASURES	FIRE EXTINGUISHING/FIRST AID
Fire: combustible.	keep away from open flame, no smoking.	powder, carbon dioxide, DO NOT USE WATER-BASED EXTINGUISHERS.
	STRICT HYGIENE	IN ALL CASES CALL IN A DOCTOR
Inhalation: *corrosive*; sore throat, cough, shortness of breath, severe breathing difficulties.	ventilation, local exhaust or respiratory protection.	fresh air, rest, place in half-sitting position, take to hospital.
Skin: *corrosive*; redness, pain, serious burns.	gloves, protective clothing.	remove contaminated clothing, flush skin with water or shower, take to doctor or take immediately to hospital.
Eyes: *corrosive*; redness, pain, impaired vision.	face shield, or combined eye and respiratory protection.	flush with water, take to a doctor.
Ingestion: *corrosive*; sore throat, abdominal pain, diarrhea, vomiting.		rinse mouth, take immediately to hospital.

SPILLAGE	STORAGE	LABELING / NFPA
collect leakage in sealable containers, flush away remainder with water, combat gas cloud with water curtain; (additional individual protection: CHEMICALSUIT).	keep dry, separate from oxidants; ventilate at floor level.	R: 14-34-37 S: 26 Corrosive 2 1 1

NOTES
Lung edema symptoms usually develop several hours later and are aggravated by physical exertion: rest and hospitalization essential. As first aid, a doctor or authorized person should consider administering a corticosteroid spray. Use airtight packaging.

SULFUR DICHLORIDE

PHYSICAL PROPERTIES	IMPORTANT CHARACTERISTICS
Boiling point (decomposes), °C 59 Melting point, °C −78 Relative density (water = 1) 1.6 Relative vapor density (air = 1) 3.6 Relative density at 20 °C of saturated mixture vapor/air (air = 1) 1.5 Vapor pressure, mm Hg at 20 °C ca. 152 Solubility in water reaction Relative molecular mass 103 Gross formula: Cl$_2$S	**BROWN FUMING LIQUID WITH PUNGENT ODOR** Vapor is heavier than air. Decomposes when heated above boiling point, giving off toxic and corrosive gases. Reacts violently with water, alcohols, amines and bases. Reacts with air to form corrosive fumes (→ *hydrochloric acid*). TLV-TWA not available **Absorption route:** Can enter the body by inhalation or ingestion. Harmful atmospheric concentrations can build up very rapidly on evaporation at 20°C. **Immediate effects:** Corrosive to the eyes, skin and respiratory tract. Inhalation of vapor/fumes can cause severe breathing difficulties (lung edema). In serious cases risk of death.

HAZARDS/SYMPTOMS	PREVENTIVE MEASURES	FIRE EXTINGUISHING/FIRST AID
Fire: non-combustible.		in case of fire in immediate vicinity: preferably not water.
	STRICT HYGIENE	IN ALL CASES CALL IN A DOCTOR
Inhalation: *corrosive*; sore throat, cough, shortness of breath, severe breathing difficulties.	ventilation, local exhaust or respiratory protection.	fresh air, rest, place in half-sitting position, take to hospital.
Skin: *corrosive*; redness, pain, serious burns.	gloves, protective clothing.	remove contaminated clothing, flush skin with water or shower, take to a doctor or take immediately to hospital.
Eyes: *corrosive*; redness, pain, impaired vision.	face shield, or combined eye and respiratory protection.	flush with water, take to a doctor.
Ingestion: *corrosive*; sore throat, abdominal pain, diarrhea, chest pain, vomiting.		rinse mouth, immediately take to hospital.

SPILLAGE	STORAGE	LABELING / NFPA
evacuate area, call in an expert, collect leakage in sealable containers, soak up with dry sand or other inert absorbent and remove to safe place; (additional individual protection: CHEMICAL SUIT).	keep dry, separate from strong bases, alcohols and amines; ventilate at floor level.	R: 14-34-37 S: 26 ⬛ Corrosive

NOTES
Lung edema symptoms usually develop several hours later and are aggravated by physical exertion: rest and hospitalization essential. As first aid, a doctor or authorized person should consider administering a corticosteroid spray. Use airtight packaging. Unbreakable packaging preferred; if breakable, keep in unbreakable container.

Transport Emergency Card TEC(R)-80G10

HI: X88; UN-number: 1828

SULFUR DIOXIDE
(cylinder)

PHYSICAL PROPERTIES		IMPORTANT CHARACTERISTICS	
Boiling point, °C	− 10	**COLORLESS COMPRESSED LIQUEFIED GAS WITH PUNGENT ODOR**	
Melting point, °C	− 76	Gas is heavier than air. In aqueous solution is a medium strong acid. Reacts violently with anhydrous ammonia, bases, amines and chlorine. Dissolves in many organic solvents. Does not attack most materials in dry state.	
Relative density (water = 1)	1.4		
Relative vapor density (air = 1)	2.3		
Vapor pressure, mm Hg at 20 °C	2508		
Solubility in water, g/100 ml at 20 °C	10.5	TLV-TWA 2 ppm 5.2 mg/m³	
Relative molecular mass	64.0	STEL 5 ppm 13 mg/m³	
		Absorption route: Can enter the body by inhalation. Harmful atmospheric concentrations can build up very rapidly if gas is released.	
		Immediate effects: Corrosive to the eyes, skin and respiratory tract. Liquid can cause frostbite due to rapid evaporation. Inhalation can cause lung edema. In serious cases risk of death.	
Gross formula:	O$_2$S		

HAZARDS/SYMPTOMS	PREVENTIVE MEASURES	FIRE EXTINGUISHING/FIRST AID
Fire: non-combustible.		in case of fire in immediate vicinity: use any extinguishing agent.
Explosion:		in case of fire: keep cylinders cool by spraying with water.
Inhalation: *corrosive*; sore throat, cough, shortness of breath, severe breathing difficulties.	ventilation, local exhaust or respiratory protection.	fresh air, rest, place in half-sitting position, take to hospital.
Skin: *corrosive, in case of frostbite*: redness, pain, burns.	insulating gloves.	*in case of frostbite*: DO NOT remove clothing, flush skin with water or shower, take to a doctor.
Eyes: *corrosive*; redness, pain, impaired vision.	face shield, or combined eye and respiratory protection.	flush with water, take to a doctor.

SPILLAGE	STORAGE	LABELING / NFPA
evacuate area, call in an expert, ventilate, under no circumstances spray liquid with water, combat gas cloud with water curtain; (additional individual protection: CHEMICAL SUIT).	if stored indoors, keep in cool, fireproof place.	0 / 3 0

NOTES
Lung edema symptoms usually develop several hours later and are aggravated by physical exertion: rest and hospitalization essential. As first aid, a doctor or authorized person should consider administering a corticosteroid spray. Do not spray leaking cylinder with water (to avoid corrosion). Turn leaking cylinder so that leak is on top to prevent liquid sulfur dioxide escaping; remove to safe place if possible. Use cylinder with special fittings.

Transport Emergency Card TEC(R)-15

HI: 26; UN-number: 1079

SULFUR HEXAFLUORIDE
(cylinder)

PHYSICAL PROPERTIES		IMPORTANT CHARACTERISTICS
Sublimation temperature, °C	−64	**COLORLESS COMPRESSED LIQUEFIED GAS**
Melting point, °C	−51	Gas is heavier than air and can accumulate close to ground level, causing oxygen deficiency, with
Relative density (water = 1)	1.5	risk of unconsciousness. Decomposes when heated above 500°C, giving off toxic gases
Relative vapor density (air = 1)	5.1	(fluorine and sulfur compounds). In the presence of moisture above 200°C hydrolysis can take
Vapor pressure, mm Hg at 20 °C	15,960	place.
Solubility in water	none	
Relative molecular mass	146.1	

	TLV-TWA	1000 ppm	5970 mg/m³

Absorption route: Can enter the body by inhalation. Can saturate the air if released, with risk of suffocation.
Immediate effects: Liquid can cause frostbite due to rapid evaporation.

Gross formula: F$_6$S

HAZARDS/SYMPTOMS	PREVENTIVE MEASURES	FIRE EXTINGUISHING/FIRST AID
Fire: non-combustible.		in case of fire in immediate vicinity: use any extinguishing agent.
Explosion:		in case of fire: keep cylinders cool by spraying with water.
Inhalation: headache, drowsiness.	ventilation, local exhaust or respiratory protection.	fresh air, rest.
Skin: *in case of frostbite*: redness, pain, sores.	insulating gloves.	*in case frostbite*: DO NOT remove clothing, flush skin with water or shower, take to a doctor.
Eyes: *in case of frostbite*: redness, pain, impaired vision.	face shield.	flush with water, take to a doctor.

SPILLAGE	STORAGE	LABELING / NFPA
ventilate, under no circumstances spray liquid with water; (additional individual protection: breathing apparatus).	if stored indoors, keep in cool, fireproof place.	

NOTES
High atmospheric concentrations, e.g. in poorly ventilated spaces, can cause oxygen deficiency, with risk of unconsciousness. Turn leaking cylinder so that leak is on top to prevent liquid sulfur hexafluoride escaping. Impure sulfur hexafluoride may contain sulfur compounds with lower fluorine content. These are toxic and react with water vapor forming corrosive and toxic → *hydrogen fluoride*. Note especially the possible presence of sulfur pentafluoride (TLV-TWA: C 0.01ppm and C 0.10mg/m³).

Transport Emergency Card TEC(R)-860

HI: 20; UN-number: 1080

SULFURIC ACID
(ca. 98%)

PHYSICAL PROPERTIES	IMPORTANT CHARACTERISTICS	
Boiling point, °C 290 Melting point, °C 11 Relative density (water = 1) 1.83 Relative vapor density (air = 1) 3.4 Relative density at 20 °C of saturated mixture vapor/air (air = 1) 1 Vapor pressure, mm Hg at 20 °C < 0.0076 Solubility in water ∞ Relative molecular mass 98.1	**COLORLESS OILY HYGROSCOPIC LIQUID WITH PUNGENT ODOR** Vapor mixes readily with air. Decomposes when heated, giving off toxic vapors. Strong oxidant which reacts violently with combustible substances and reducing agents. Strong acid which reacts violently with bases and is corrosive. Reacts violently with organic substances, solvents and many other substances, with risk of fire and explosion. Reacts with metals, giving off → *hydrogen* gas, with risk of fire and explosion. Concentrated acid does not attack steel.	
	TLV-TWA 1 mg/m³ STEL 3 mg/m³	
	Absorption route: Can enter the body by inhalation or ingestion. Harmful atmospheric concentrations build up very slowly, if at all, on evaporation at approx. 20°C, but much more rapidly in aerosol form. **Immediate effects:** Corrosive to the eyes, skin and respiratory tract. Inhalation of vapor/fumes can cause severe breathing difficulties (lung edema).	
Gross formula: H_2O_4S		

HAZARDS/SYMPTOMS	PREVENTIVE MEASURES	FIRE EXTINGUISHING/FIRST AID
Fire: non-combustible, many chemical reactions can cause fire and explosion.	avoid contact with combustible substances, solvents and many other substances.	in case of fire in immediate vicinity: DO NOT USE WATER-BASED EXTINGUISHERS.
Explosion: many chemical reactions can cause explosion.		
	STRICT HYGIENE	IN ALL CASES CALL IN A DOCTOR
Inhalation: *corrosive*; sore throat, cough, shortness of breath, severe breathing difficulties.	ventilation, local exhaust or respiratory protection.	fresh air, rest, place in half-sitting position, take to hospital.
Skin: *corrosive*; redness, pain, burns.	gloves, protective clothing.	remove contaminated clothing, flush skin with water or shower, refer to a doctor.
Eyes: *corrosive*; redness, pain, impaired vision.	face shield, or combined eye and respiratory protection.	flush with water, take to a doctor.
Ingestion: *corrosive*; sore throat, abdominal pain.		rinse mouth, immediately take to hospital.

SPILLAGE	STORAGE	LABELING / NFPA
collect leakage in sealable containers, soak up with dry sand or other inert absorbent and remove to safe place, flush away remainder with water, under no circumstances spray liquid with water, do not use sawdust or other combustible absorbents; (additional individual protection: CHEMICAL SUIT).	keep separate from combustible substances, reducing agents and bases; ventilate at floor level.	R: 35 S: 2-26-30 Note B Corrosive

NOTES
Occurs in various dilutions. 96%: boiling point and melting point 330°C and −15°C respectively; 64%: 142°C and −32°C. Lung edema symptoms usually develop several hours later and are aggravated by physical exertion: rest and hospitalization essential. As first aid, a doctor or authorized person should consider administering a corticosteroid spray. UNDER NO CIRCUMSTANCES add water to acid; when diluting ALWAYS add acid to water. Unbreakable packaging preferred; if breakable, keep in unbreakable container.

Transport Emergency Card TEC(R)-10B

HI: 80; UN-number: 1830

SULFURIC ACID
(solution in water up to 98%)

PHYSICAL PROPERTIES	IMPORTANT CHARACTERISTICS
Melting point, °C 93%: −32 Relative density (water = 1) 1-1.8 Relative vapor density (air = 1) 3.4 Solubility in water ∞ Relative molecular mass 98.1	**COLORLESS LIQUID** Above 60% solution is a strong oxidant which reacts with many organic compounds and attacks clothing. Strong acid which reacts violently with bases and is corrosive. Reacts violently with many substances, evolving heat. Attacks base metals (except lead), giving off flammable gas (→ *hydrogen*).

TLV-TWA STEL	1 mg/m³ 3 mg/m³

Absorption route: Can enter the body by inhalation or ingestion. Evaporation negligible at 20°C, but harmful atmospheric concentrations can build up rapidly in aerosol form.
Immediate effects: Corrosive to the eyes, skin and respiratory tract. Inhalation can cause lung edema. In serious cases risk of death.

Gross formula: H₂O₄S

HAZARDS/SYMPTOMS	PREVENTIVE MEASURES	FIRE EXTINGUISHING/FIRST AID
Fire: non-combustible, many chemical reactions can cause fire and explosion.	avoid contact with metals and organic substances.	in case of fire in immediate vicinity: use any extinguishing agent.
Explosion: many chemical reactions can cause explosion.		
		IN ALL CASES CALL A DOCTOR
Inhalation: *corrosive*; sore throat, cough, shortness of breath, severe breathing difficulties.	ventilation, local exhaust or respiratory protection.	fresh air, rest, place in half-sitting position, artificial respiration if necessary, take to hospital.
Skin: *corrosive*; redness, pain, serious burns.	gloves, protective clothing.	remove contaminated clothing, flush skin with water or shower, call a doctor if necessary.
Eyes: *corrosive*; redness, pain, impaired vision.	face shield.	flush with water, take to a doctor.
Ingestion: *corrosive*; sore throat, abdominal pain, vomiting, diarrhea.		rinse mouth, take immediately to hospital.

SPILLAGE	STORAGE	LABELING / NFPA
collect leakage in sealable containers, neutralize spillage and flush away with water (risk of heating and acid spattering).	keep separate from combustible substances, reducing agents and strong bases.	R: 35 S: 2-26-30 Note B Corrosive

NFPA diamond: Health 3, Fire 0, Reactivity 2, W

NOTES
Lung edema symptoms usually develop several hours later and are aggravated by physical exertion: rest and hospitalization essential. As first aid, a doctor or authorized person should consider administering a corticosteroid spray. UNDER NO CIRCUMSTANCES add water to acid; when diluting ALWAYS add acid to water. Melting point 65%: −44°C; 78% −8°C. Packaging: special material.

Transport Emergency Card TEC(R)-10a (⩾75%)

HI: 80; UN-number: 1830

γ-SULFURTRIOXIDE

PHYSICAL PROPERTIES		IMPORTANT CHARACTERISTICS
Boiling point, °C	45	**COLORLESS HYGROSCOPIC LIQUID OR ICE LIKE HYGROSCOPIC CRYSTALS WITH PUNGENT ODOR**
Melting point, °C	17	
Relative density (water = 1)	1.9	Vapor is heavier than air and reacts with moist air to form white, corrosive fumes (→ *sulfuric acid*).
Relative vapor density (air = 1)	2.8	Without inhibitor the substance transforms slowly to α-isomer (see notes). Strong oxidant which
Relative density at 20 °C of saturated mixture		reacts violently with combustible substances and reducing agents, with risk of fire and explosion.
vapor/air (air = 1)	ca. 2	Reacts extremely violently with water, forming → *sulfuric acid*). Removes water form many
Vapor pressure, mm Hg at 25 °C	439	organic substances, with strong heat development with risk of fire. Attacks many metals in the
Solubility in water	reaction	presence of moisture, giving off flammable gas (→ *hydrogen*).
Relative molecular mass	80.1	

TLV-TWA not available

Absorption route: Can enter the body by inhalation or ingestion.
Immediate effects: Corrosive to the eyes, skin and respiratory tract. Inhalation of vapor/fumes can cause severe breathing difficulties (lung edema). In serious cases risk of death.

Gross formula: O$_3$S

HAZARDS/SYMPTOMS	PREVENTIVE MEASURES	FIRE EXTINGUISHING/FIRST AID
Fire: not combustible but enhances combustion of other substances.	avoid contact with combustible substances.	
Explosion: many chemical reactions can cause explosion.		in case of fire: keep tanks/drums cool by spraying with water, but DO NOT spray substance with water.
Inhalation: *corrosive*; sore throat, cough, severe breathing difficulties.	ventilation, local exhaust or respiratory protection.	fresh air, rest, place in half-sitting position, take to hospital.
Skin: *corrosive*; redness, pain, serious burns.	gloves, protective clothing.	remove contaminated clothing, flush skin with water or shower, call a doctor.
Eyes: *corrosive*; redness, pain, impaired vision.	face shield, or combined eye and respiratory protection.	flush with water, take to a doctor.
Ingestion: *corrosive*; abdominal pain, vomiting, pain with swallowing.		rinse mouth, immediately take to hospital.

SPILLAGE	STORAGE	LABELING / NFPA
evacuate area, call in an expert, clean up spillage or collect leakage in sealable containers, soak up with dry sand or other inert absorbent and remove to safe place, under no circumstances spray liquid with water, do not use sawdust or other combustible absorbents; (additional individual protection: CHEMICAL SUIT).	keep in dry place at temperatur between 17 and 25°C, separate form combustible substances, reducing agents and bases.	

NOTES

When moisture is attracted the substance readily converts to sulfuric acid. The γ-isomer converts into the β- and α-isomer, which melt 33° C and 62° C. Melting of β-isomer, but especially of α-isomer is associated with an explosive increase in the vapor pressure. Thus, do not allow to melt in glass. The stabilized commercial grades Sulfan A and B contain primalrily β- and α-isomers; Sulfan C is not stabilized. Lung edema symptoms usually develop several hours later and are aggravated by physical exertion: rest and hospitalization essential. As first aid, a doctor or authorized person should consider administering a corticosteroid spray. UNDER NO CIRCUMSTANCES add water to sulfur trioxide. Use airtight packaging. Unbreakable packaging preferred; if breakable, keep in unbreakable container.

SULFURYL CHLORIDE

PHYSICAL PROPERTIES	IMPORTANT CHARACTERISTICS	
Boiling point, °C .... 69 Melting point, °C .... − 54 Relative density (water = 1) .... 1.7 Relative vapor density (air = 1) .... 4.7 Relative density at 20 °C of saturated mixture vapor/air (air = 1) .... 1.5 Vapor pressure, mm Hg at 18 °C ... 101 Solubility in water .... reaction Relative molecular mass 135.0	**COLORLESS TO YELLOW LIQUID WITH PUNGENT ODOR** Vapor is heavier than air. Decomposes when heated above 160° C, giving off corrosive and toxic gases (→ *sulfur dioxide* and → *chlorine*). Reacts with water, forming → *hydrochloric acid* vapor and → *sulfuric acid*. Reacts violently with alkali metals, bases, phosphorus and many organic compounds. Reacts with air to form corrosive fumes. Attacks many metals in the presence of moisture, giving off flammable gas (→ *hydrogen*).	
	TLV-TWA not available	
	Absorption route: Can enter the body by inhalation or ingestion. Harmful atmospheric concentrations can build up very rapidly on evaporation at 20°C. **Immediate effects:** Corrosive to the eyes, skin and respiratory tract. Inhalation can cause lung edema. In serious cases risk of death.	
Gross formula: Cl$_2$O$_2$S		

HAZARDS/SYMPTOMS	PREVENTIVE MEASURES	FIRE EXTINGUISHING/FIRST AID
Fire: non-combustible.		in case of fire in immediate vicinity: preferably no water-based extinguishers.
Explosion:		in case of fire: keep tanks/drums cool by spraying with water, but DO NOT spray substance with water.
	STRICT HYGIENE	
Inhalation: *corrosive*; sore throat, cough, shortness of breath, severe breathing difficulties.	ventilation, local exhaust or respiratory protection.	fresh air, rest, place in half-sitting position, artificial respiration if necessary, take to hospital.
Skin: *corrosive*; redness, pain, burns.	gloves, protective clothing.	remove contaminated clothing, flush skin with water or shower, refer to doctor if necessary.
Eyes: *corrosive*; redness, pain, impaired vision.	face shield, or combined eye and respiratory protection.	flush with water, take to an eye doctor.
Ingestion: *corrosive*; sore throat, abdominal pain, blisters on lips and mouth, tongue pain, chest pain, stomachache.		rinse mouth, call a doctor and take to hospital.

SPILLAGE	STORAGE	LABELING / NFPA
evacuate area, call in an expert, collect leakage in sealable containers, carefully neutralize spillage with sodium bicarbonate and flush away with water; (additional individual protection: CHEMICAL SUIT).	keep dry, separate from strong bases; ventilate at floor level.	R: 14-34-37 S: 26 Corrosive

NOTES
Lung edema symptoms usually develop several hours later and are aggravated by physical exertion: rest and hospitalization essential. As first aid, a doctor or authorized person should consider administering a corticosteroid spray. Forms a stable hydrate with small quantities of ice-cold water. Use airtight packaging. Unbreakable packaging preferred; if breakable, keep in unbreakable container.

Transport Emergency Card TEC(R)-681   **HI: X88; UN-number: 1834**

NFPA: 3 / 0 / 1

french chalk
soapstone
steatite
talcum

$3MgO.4SiO_2.H_2O$

TALC

PHYSICAL PROPERTIES	IMPORTANT CHARACTERISTICS	
Relative density (water = 1) 2.8 Solubility in water none Relative molecular mass 379.3	**WHITE TO GRAYISH-WHITE POWDER**	
	TLV-TWA 2 mg/m³ ✱	
	Absorption route: Fine airborne particles can be inhaled (e.g. from a talcum powder dispenser). Evaporation negligible at 20°C, but harmful concentrations of airborne particles can build up rapidly.	
Gross formula: $Mg_3O_{11}Si_4.H_2O$		

HAZARDS/SYMPTOMS	PREVENTIVE MEASURES	FIRE EXTINGUISHING/FIRST AID
Fire: non-combustible.		in case of fire in immediate vicinity: use any extinguishing agent.
Inhalation: prickly cough, tightness of the chest.	local exhaust or respiratory protection.	
Eyes: redness.	goggles.	
Ingestion: cough, tickling in throat.		

SPILLAGE	STORAGE	LABELING / NFPA
clean up spillage; (additional individual protection: P2 respirator).		

NOTES
✱ TLV as respirable dust and it applies to talc without asbestos fibers. Industrial talcs, esp. containing fibers, can sometimes include asbestos fibers: these could cause lung disorders if inhaled. In this case the rules for → *asbestos* apply. No special precautions required for pure talc.

L-TARTARIC ACID

PHYSICAL PROPERTIES		IMPORTANT CHARACTERISTICS	
Melting point, °C	170	**WHITE CRYSTALS**	
Flash point, °C	210	Decomposes when heated above 210°C, giving off flammable vapors. In aqueous solution is a medium strong acid. Reacts with bases and is corrosive. Reacts with strong oxidants.	
Auto-ignition temperature, °C	425		
Relative density (water = 1)	1.8		
Vapor pressure, mm Hg at 20 °C	<0.076	TLV-TWA	not available
Solubility in water, g/100 ml at 20 °C	139		
Relative molecular mass	150.1	**Absorption route:** Can enter the body by inhalation or ingestion. Evaporation negligible at 20°C, but harmful concentrations of airborne particles can build up rapidly.	
Log P octanol/water	−1.7	**Immediate effects:** Irritates the eyes, skin and respiratory tract.	
Gross formula:	C$_4$H$_6$O$_6$		

HAZARDS/SYMPTOMS	PREVENTIVE MEASURES	FIRE EXTINGUISHING/FIRST AID
Fire: combustible.	keep away from open flame, no smoking.	water spray, powder.
Inhalation: sore throat, cough, shortness of breath.	local exhaust or respiratory protection.	fresh air, rest.
Skin: redness, pain.	gloves.	remove contaminated clothing, flush skin with water or shower.
Eyes: redness, pain.	goggles.	flush with water, take to a doctor if necessary.
Ingestion: sore throat, abdominal pain.		rinse mouth, call a doctor.

SPILLAGE	STORAGE	LABELING / NFPA
clean up spillage, flush away remainder with water; (additional individual protection: P2 respirator).		

NOTES
Log P octanol/water is estimated. Salts are referred to as tartrates. The measures on this card also apply to D-tartaric acid, LD-tartaric acid and meso-tartaric acid.

CAS-No: [558-13-4]
carbon tetrabromide
methane tetrabromide

CBr_4

TETRABROMOMETHANE

PHYSICAL PROPERTIES		IMPORTANT CHARACTERISTICS	
Boiling point, °C	190	**COLORLESS CRYSTALS**	
Melting point, °C	90	Decomposes in flame or on hot surface, giving off toxic vapors.	
Relative density (water = 1)	3.4		
Relative vapor density (air = 1)	11.6	TLV-TWA — 0.1 ppm — 1.4 mg/m³	
Relative density at 20 °C of saturated mixture		STEL — 0.3 ppm — 4.1 mg/m³	
vapor/air (air = 1)	1.01		
Vapor pressure, mm Hg at 20 °C	ca. 0.53	**Absorption route:** Can enter the body by inhalation or ingestion. Harmful atmospheric concentrations can build up fairly rapidly on evaporation at approx. 20°C; harmful concentrations of airborne particles can build up even more rapidly.	
Solubility in water	none		
Relative molecular mass	331.7		
		Immediate effects: Corrosive to the eyes, skin and respiratory tract. Exposure to very low concentrations cause watering of the eyes. Watering of the eyes warns sufficiently to avoid high *acute* exposure, but insufficiently to avoid high *chronic* exposure. High concentrations can cause unconsciousness. Inhalation of vapor/fumes can cause severe breathing difficulties (lung edema).	
		Effects of prolonged/repeated exposure: Can cause liver and kidney damage.	
Gross formula:	Br_4C		

HAZARDS/SYMPTOMS	PREVENTIVE MEASURES	FIRE EXTINGUISHING/FIRST AID
Fire: non-combustible.		in case of fire in immediate vicinity: use any extinguishing agent.
	STRICT HYGIENE, KEEP DUST UNDER CONTROL	
Inhalation: profuse watering of the eyes, *corrosive*; sore throat, cough, severe breathing difficulties, drowsiness, sleepiness, unconsciousness.	ventilation, local exhaust or respiratory protection.	fresh air, rest, place in half-sitting position, take to hospital.
Skin: *corrosive*; redness, pain, burns.	gloves.	remove contaminated clothing, flush skin with water or shower, call a doctor.
Eyes: *corrosive*; redness, pain, impaired vision.	face shield.	flush with water, take to a doctor.
Ingestion: profuse watering of the eyes, *corrosive*; sore throat, abdominal pain, diarrhea, drowsiness, sleepiness, unconsciousness.	do not eat, drink or smoke while working.	rinse mouth, immediately take to hospital.

SPILLAGE	STORAGE	LABELING / NFPA
clean up spillage, carefully collect remainder, remove to safe place in sealed containers; (additional individual protection: breathing apparatus).	ventilate at floor level.	

NOTES
Lung edema symptoms usually develop several hours later and are aggravated by physical exertion: rest and hospitalization essential. As first aid, a doctor or authorized person should consider administering a corticosteroid spray. Under no circumstances use near flame or hot surface or when welding.

Transport Emergency Card TEC(R)-61G11C

HI: 60; UN-number: 2516

CAS-No: [79-34-5]
acetylene tetrachloride
TCE
tetrachloroethane
sym-tetrachloroethane

$CHCl_2CHCl_2$

1,1,2,2-TETRACHLOROETHANE

PHYSICAL PROPERTIES	IMPORTANT CHARACTERISTICS	
Boiling point, °C 146 Melting point, °C −43 Relative density (water = 1) 1.6 Relative vapor density (air = 1) 5.8 Relative density at 20 °C of saturated mixture vapor/air (air = 1) 1.03 Vapor pressure, mm Hg at 20 °C 5.0 Solubility in water, g/100 ml at 25 °C 0.3 Electrical conductivity, pS/m 4.5 x 10⁵ Relative molecular mass 167.9	**COLORLESS LIQUID WITH CHARACTERISTIC ODOR** Vapor mixes readily with air. Decomposes in flame or on hot surface, giving off → *hydrochloric acid* gas. Decomposes slowly on exposure to light, air or moisture, forming → *hydrochloric acid*. Reacts violently with light metals and sodium amide, with risk of fire and explosion. On contact with bases forms explosive dichloroacetylene.	

TLV-TWA (skin)	1 ppm	6.9 mg/m³

Absorption route: Can enter the body by inhalation or ingestion or through the skin. Harmful atmospheric concentrations can build up fairly rapidly on evaporation at approx. 20°C - even more rapidly in aerosol form.
Immediate effects: Irritates the eyes and respiratory tract. Liquid destroys the skin's natural oils. Affects the nervous system. In serious cases risk of unconsciousness and death.
Effects of prolonged/repeated exposure: Can cause liver and kidney damage.

Gross formula: $C_2H_2Cl_4$

HAZARDS/SYMPTOMS	PREVENTIVE MEASURES	FIRE EXTINGUISHING/FIRST AID
Fire: non-combustible.		in case of fire in immediate vicinity: use any extinguishing agent.
Inhalation: cough, shortness of breath, dizziness, nausea, drowsiness, unconsciousness.	ventilation, local exhaust or respiratory protection.	fresh air, rest, place in half-sitting position, take to hospital.
Skin: *is absorbed*; redness.	gloves, protective clothing.	remove contaminated clothing, wash skin with soap and water, call a doctor.
Eyes: redness, pain.	face shield, or combined eye and respiratory protection.	flush with water, take to a doctor if necessary.
Ingestion: sore throat, abdominal pain, vomiting; see also 'Inhalation'.		rinse mouth, immediately take to hospital.

SPILLAGE	STORAGE	LABELING / NFPA
collect leakage in sealable containers, soak up with sand or other inert absorbent and remove to safe place; (additional individual protection: breathing apparatus).	keep in dry, dark place; ventilate at floor level.	R: 26/27 S: 2-38-45 ☠ Toxic

NOTES
Alcohol consumption increases toxic effects. Depending on the degree of exposure, regular medical checkups are advisable. Under no circumstances use near flame or hot surface or when welding.

Transport Emergency Card TEC(R)-719

HI: 60; UN-number: 1702

TETRACHLOROPHENOL
(various isomers)

PHYSICAL PROPERTIES	IMPORTANT CHARACTERISTICS
Boiling point (decomposes), °C　　ca. 290 Melting point, °C　　50-70 Relative density (water = 1)　　1.7 Relative vapor density (air = 1)　　8.0 Relative density at 20 °C of saturated mixture vapor/air (air = 1)　　1.0 Vapor pressure, mm Hg at 20 °C　　<0.076 Solubility in water　　poor Relative molecular mass　　231.9	**GRAY CRYSTALS OR BROWN SOLID IN VARIOUS FORMS WITH CHARACTERISTIC ODOR** Vapor mixes readily with air. Decomposes when heated above 280°C, giving off toxic and corrosive vapors (→ *hydrochloric acid*).
	TLV-TWA　　　　　　　　not available
	Absorption route: Can enter the body by inhalation or ingestion. Harmful atmospheric concentrations build up very slowly, if at all, on evaporation at approx. 20°C, but harmful concentrations of airborne particles can build up much more rapidly. **Immediate effects:** Irritates the eyes, skin and respiratory tract. Affects the nervous system and metabolism. In serious cases risk of seizures and death. **Effects of prolonged/repeated exposure:** Can cause liver and kidney damage.
Gross formula:　　　　$C_6H_2Cl_4O$	

HAZARDS/SYMPTOMS	PREVENTIVE MEASURES	FIRE EXTINGUISHING/FIRST AID
Fire: non-combustible.		in case of fire in immediate vicinity: use any extinguishing agent.
	KEEP DUST UNDER CONTROL	
Inhalation: sore throat, cough, fever, muscle weakness, shaking, muscle cramps.	local exhaust or respiratory protection.	fresh air, rest, take to hospital.
Skin: redness, pain.	gloves, protective clothing.	remove contaminated clothing, flush skin with water or shower, refer to a doctor.
Eyes: redness, pain.	face shield.	flush with water, take to a doctor.
Ingestion: abdominal pain, diarrhea, vomiting; see also 'Inhalation'.	do not eat, drink or smoke while working.	rinse mouth, immediately take to hospital.

SPILLAGE	STORAGE	LABELING / NFPA
claean up spillage, carefully collect remainder, then remove to safe place; (additional individual protection: P2 respirator).		R: 25-36/38 S: 26-28-37-44 Toxic

NOTES
This card applies to all isomers of tetrachlorophenol. ✳ Labeling applies to 2, 3, 4, 6 tetrachlorophenol.

Transport Emergency Card TEC(R)-61G11

HI: 60; UN-number: 2020

$H_2NCH_2(CH_2NHCH_2)_3CH_2NH_2$

TETRAETHYLENEPENTAMINE

PHYSICAL PROPERTIES		IMPORTANT CHARACTERISTICS	
Boiling point, °C	320	**YELLOW VISCOUS LIQUID WITH CHARACTERISTIC ODOR**	
Melting point, °C	− 30	Vapor mixes readily with air. Toxic and corrosive vapors are a product of combustion (→ *nitrous*	
Flash point, °C	170	*vapors*). Medium strong base. Reacts with chlorinated hydrocarbons. Reacts with oxidants.	
Auto-ignition temperature, °C	321	Promotes spontaneous decomposition of nitrogen compounds. Attacks copper, nickel and	
Relative density (water = 1)	1.0	cobalt.	
Relative vapor density (air = 1)	6.5		
Relative density at 20 °C of saturated mixture		TLV-TWA	not available
vapor/air (air = 1)	1.0		
Vapor pressure, mm Hg at 20 °C	< 0.076	**Absorption route:** Can enter the body by inhalation or ingestion. Harmful atmospheric	
Solubility in water	∞	concentrations build up fairly slowly on evaporation at 20°C, but much more rapidly in aerosol	
Relative molecular mass	189.3	form.	
		Immediate effects: Corrosive to the eyes, skin and respiratory tract. Inhalation of vapor/fumes	
		can cause severe breathing difficulties (lung edema). In serious cases risk of death.	
Gross formula:	$C_8H_{23}N_5$		

HAZARDS/SYMPTOMS	PREVENTIVE MEASURES	FIRE EXTINGUISHING/FIRST AID
Fire: combustible.	keep away from open flame, no smoking.	water spray, alcohol-resistant foam, powder, carbon dioxide, DO NOT USE HALONS.
Inhalation: *corrosive*; sore throat, cough, shortness of breath, severe breathing difficulties.	ventilation, local exhaust or respiratory protection.	fresh air, rest, place in half-sitting position, take to hospital.
Skin: *corrosive*; redness, pain, serious burns.	gloves, protective clothing.	remove contaminated clothing, flush skin with water or shower, call a doctor if necessary.
Eyes: *corrosive*; redness, pain, impaired vision.	face shield.	flush with water, take to a doctor.
Ingestion: *corrosive*; abdominal pain, diarrhea, vomiting.		rinse mouth, take immediately to hospital.

SPILLAGE	STORAGE	LABELING / NFPA
collect leakage in sealable containers, flush away remainder with water.	keep separate from strong acids and nitrogen compounds; ventilate at floor level.	R: 21/22-34-43 S: 26-36/37/39 Corrosive

NOTES

Commercial product contains approx. 40-50% linear isomer; non-lineair isomer and cyclic pentaethylene pentamine also present. Physical properties relate to commercial products. Lung edema symptoms usually develop several hours later and are aggravated by physical exertion: rest and hospitalization essential. As first aid, a doctor or authorized person should consider administering a corticosteroid spray. Reacts violently with extinguishing agents such as halons. Unbreakable packaging preferred; if breakable, keep in unbreakable container.

Transport Emergency Card TEC(R)-80G16

HI: 80; UN-number: 2320

CAS-No: [78-00-2]
lead tetraethyl
TEL
tetraethyle plumb
tetraethylplumbane

$(C_2H_5)_4Pb$

TETRAETHYL LEAD

PHYSICAL PROPERTIES	IMPORTANT CHARACTERISTICS
Decomposes below boiling point, °C 125-150 Melting point, °C − 136 Flash point, °C 80 Auto-ignition temperature, °C see notes Relative density (water = 1) 1.65 Relative vapor density (air = 1) 11.2 Relative density at 20 °C of saturated mixture vapor/air (air = 1) 1.0 Vapor pressure, mm Hg at 20 °C 0.23 Solubility in water poor Explosive limits, vol% in air 1.8-? Relative molecular mass 323.4 Gross formula: $C_8H_{20}Pb$	**COLORLESS VISCOUS LIQUID WITH CHARACTERISTIC ODOR** Vapor mixes readily with air. Can decompose explosively at 320° C if heated rapidly. Decomposes when heated above 125° C and exposed to light, giving off toxic and flammable vapors. Reacts violently with oxidants, with risk of fire and explosion. Reacts with fats. Attacks rubber. TLV-TWA (skin) 0.1 mg/m³ ✱ **Absorption route:** Can enter the body by inhalation or ingestion or through the skin. Harmful atmospheric concentrations can build up very rapidly on evaporation at 20° C. **Immediate effects:** Irritates the eyes, skin and respiratory tract. Affects the nervous system. In serious cases risk of death.

HAZARDS/SYMPTOMS	PREVENTIVE MEASURES	FIRE EXTINGUISHING/FIRST AID
Fire: combustible.	keep away from open flame, no smoking.	powder, AFFF, foam, carbon dioxide, (halons); in case of fire in immediate vicinity: use any extinguishing agent.
Explosion: above 80°C: forms explosive air-vapor mixtures; rapid heating can cause explosion.	above 80°C: sealed machinery, ventilation.	
Inhalation: headache, dizziness, cramps, unconsciousness.	ventilation, local exhaust or respiratory protection.	fresh air, rest, take to hospital.
Skin: *is absorbed*; redness.	gloves, protective clothing.	remove contaminated clothing, wash skin with soap and water, take to hospital.
Eyes: pain, impaired vision.	face shield, or combined eye and respiratory protection.	flush with water, take to a doctor if necessary.
Ingestion: headache, dizziness, cramps, unconsciousness.	do not eat, drink or smoke while working.	rinse mouth, immediately take to hospital.

SPILLAGE	STORAGE	LABELING / NFPA
evacuate area, call in an expert, collect leakage in sealable containers, soak up with sand or other inert absorbent and remove to safe place, render remainder harmless with potassium permanganate solution (70g/l), do not flush into sewer; (additional individual protection: CHEMICAL SUIT).	keep in fireproof place, separate from oxidants.	R: 26/27/28-33 S: 13-26-36/37-45 Note A ☠ Toxic 2 / 3 3

NOTES

✱ TLV equals that of lead. Technical grades are often referred to as antiknock compounds: these contain varying quantities of dichloroethane and dibromoethane. Depending on the degree of exposure, regular medical checkups are advisable. Unbreakable packaging preferred; if breakable, keep in unbreakable container.

Transport Emergency Card TEC(R)-157

HI: 66; UN-number: 1649

CAS-No: [109-99-9]
butylene oxide
diethylene oxide
1,4-epoxybutane
furanidine
tetramethylene oxide

TETRAHYDROFURAN

PHYSICAL PROPERTIES		IMPORTANT CHARACTERISTICS	
Boiling point, °C	66	**COLORLESS LIQUID WITH CHARACTERISTIC ODOR**	
Melting point, °C	− 107	Vapor is heavier than air and spreads at ground level, with risk of ignition at a distance. Able to form peroxides. Do not use compressed air when filling, emptying or processing. Reacts with oxidants. Attacks some plastics.	
Flash point, °C	− 17		
Auto-ignition temperature, °C	230		
Relative density (water = 1)	0.9		
Relative vapor density (air = 1)	2.5	TLV-TWA	200 ppm	590 mg/m³
Relative density at 20 °C of saturated mixture		STEL	250 ppm	737 mg/m³
vapor/air (air = 1)	1.3		
Vapor pressure, mm Hg at 20 °C	152	**Absorption route:** Can enter the body by inhalation or ingestion. Harmful atmospheric concentrations can build up fairly rapidly on evaporation at approx. 20°C - even more rapidly in aerosol form.	
Solubility in water	∞		
Explosive limits, vol% in air	1.5-12.4		
Minimum ignition energy, mJ	0.54	**Immediate effects:** Irritates the eyes, skin and respiratory tract. Affects the nervous system. Can impair consciousness.	
Relative molecular mass	72.1		
Gross formula:	C_4H_8O		

HAZARDS/SYMPTOMS	PREVENTIVE MEASURES	FIRE EXTINGUISHING/FIRST AID
Fire: extremely flammable.	keep away from open flame and sparks, no smoking.	powder, alcohol-resistant foam, water spray, carbon dioxide, (halons).
Explosion: forms explosive air-vapor mixtures; formation of (explosive) peroxides can cause explosion.	sealed machinery, ventilation, explosion-proof electrical equipment and lighting, non-sparking tools.	in case of fire: keep tanks/drums cool by spraying with water.
Inhalation: sore throat, cough, shortness of breath, headache, drowsiness.	ventilation, local exhaust or respiratory protection.	fresh air, rest, call a doctor.
Skin: redness, pain.	gloves.	remove contaminated clothing, flush skin with water or shower.
Eyes: redness, pain.	face shield.	flush with water, take to a doctor if necessary.
Ingestion: sore throat, abdominal pain, drowsiness.		rinse mouth, call a doctor or take to hospital.

SPILLAGE	STORAGE	LABELING / NFPA
collect leakage in sealable containers, soak up with sand or other inert absorbent and remove to safe place; (additional individual protection: breathing apparatus).	keep in fireproof place, separate from oxidants; add an inhibitor.	R: 11-19-36/37 S: 16-29-33 Flammable Irritant

NOTES
Impurities (→ *furan*) increase toxicity. Before distilling check for peroxides; if found, render harmless. Alcohol consumption increases toxic effects. Commercial grade is usually stabilized with 4-methyl-2,6-di-t-butylphenol or hydroquinone ferrosulfate. Use airtight packaging.

Transport Emergency Card TEC(R)-601

HI: 33; UN-number: 2056

CAS-No: [97-99-4]
tetrahydro-2-furanmethanol
tetrahydrofurfuryl carbinol
THFA

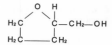

TETRAHYDROFURFURYL ALCOHOL

PHYSICAL PROPERTIES		IMPORTANT CHARACTERISTICS
Boiling point, °C	178	**COLORLESS OILY HYGROSCOPIC LIQUID WITH CHARACTERISTIC ODOR**
Melting point, °C	< −80	Vapor mixes readily with air. Presumed to be able to form peroxides. Reacts violently with strong
Flash point, °C	74	oxidants, with risk of fire and explosion.
Auto-ignition temperature, °C	280	
Relative density (water = 1)	1.1	TLV-TWA not available
Relative vapor density (air = 1)	3.5	
Relative density at 20 °C of saturated mixture		**Absorption route:** Can enter the body by inhalation or ingestion or through the skin. Insufficient
vapor/air (air = 1)	1.0	data available on the rate at which harmful concentrations can build up.
Vapor pressure, mm Hg at 20 °C	0.23	**Immediate effects:** Irritates the eyes, skin and respiratory tract. In serious cases risk of
Solubility in water	∞	unconsciousness.
Explosive limits, vol% in air	1.5-9.7	**Effects of prolonged/repeated exposure:** Can cause liver and kidney damage.
Relative molecular mass	102.1	
Gross formula:	$C_5H_{10}O_2$	

HAZARDS/SYMPTOMS	PREVENTIVE MEASURES	FIRE EXTINGUISHING/FIRST AID
Fire: combustible.	keep away from open flame, no smoking.	powder, alcohol-resistant foam, water spray, carbon dioxide, (halons).
Explosion: above 74°C: forms explosive air-vapor mixtures.	above 74°C: sealed machinery, ventilation.	
Inhalation: sore throat, cough, headache, dizziness, drowsiness, unconsciousness.	ventilation, local exhaust or respiratory protection.	fresh air, rest, place in half-sitting position, call a doctor.
Skin: *is absorbed*; redness, pain.	gloves, protective clothing.	remove contaminated clothing, flush skin with water or shower, call a doctor.
Eyes: redness, pain.	safety glasses.	flush with water, take to a doctor if necessary.
Ingestion: abdominal pain; see also 'Inhalation'.		rinse mouth, call a doctor or take to hospital.

SPILLAGE	STORAGE	LABELING / NFPA
collect leakage in sealable containers, flush away remainder with water.	keep separate from oxidants.	R: 36 S: 39 ✖ Irritant

NOTES
Before distilling check for peroxides; if found, render harmless. Alcohol consumption increases toxic effects.

1,2,3,4-TETRAHYDRONAPHTHALENE

PHYSICAL PROPERTIES	IMPORTANT CHARACTERISTICS
Boiling point, °C 207 Melting point, °C − 36 Flash point, °C 77 Auto-ignition temperature, °C 425 Relative density (water = 1) 0.98 Relative vapor density (air = 1) 4.6 Relative density at 20 °C of saturated mixture vapor/air (air = 1) 1.0 Vapor pressure, mm Hg at 20 °C 0.30 Solubility in water none Explosive limits, vol% in air 0.8-5.0 Relative molecular mass 132	**COLORLESS LIQUID WITH CHARACTERISTIC ODOR** Vapor mixes readily with air. Can form peroxides on prolonged exposure to air. Flow, agitation etc. can cause build-up of electrostatic charge due to liquid's low conductivity. TLV-TWA not available **Absorption route:** Can enter the body by inhalation or ingestion. Harmful atmospheric concentrations build up very slowly, if at all, on evaporation at approx. 20° C, but much more rapidly in aerosol form. **Immediate effects:** Irritates the eyes, skin and respiratory tract. If liquid is swallowed, droplets can enter the lungs, with risk of pneumonia. **Effects of prolonged/repeated exposure:** Prolonged or repeated contact can cause skin disorders. Can cause liver and kidney damage.
Gross formula: $C_{10}H_{12}$	

HAZARDS/SYMPTOMS	PREVENTIVE MEASURES	FIRE EXTINGUISHING/FIRST AID
Fire: combustible.	keep away from open flame, no smoking.	powder, AFFF, foam, carbon dioxide, (halons).
Explosion: above 77°C: forms explosive air-vapor mixtures.	above 77°C: sealed machinery, ventilation, explosion-proof electrical equipment, grounding.	in case of fire: keep tanks/drums cool by spraying with water.
Inhalation: headache, dizziness, sleepiness, unconsciousness.	ventilation, local exhaust or respiratory protection.	fresh air, rest, call a doctor.
Skin: redness, pain.	gloves.	remove contaminated clothing, flush skin with water or shower.
Eyes: redness, pain.	face shield.	flush with water, take to a doctor if necessary.
Ingestion: abdominal pain, vomiting; see also 'Inhalation'.		rinse mouth, DO NOT induce vomiting, take to hospital.

SPILLAGE	STORAGE	LABELING / NFPA
collect leakage in sealable containers, soak up with sand or other inert absorbent and remove to safe place.		NFPA diamond: 2 (top), 1 (left), 0 (right)

NOTES
Before distilling check for peroxides; if found, render harmless. Alcohol consumption increases toxic effects. Depending on the degree of exposure, regular medical checkups are advisable.

Transport Emergency Card TEC(R)-30G15

CAS-No: [54668-9]
isopropyl titanate
titanium isopropylate
TPT

$Ti(C_3H_7O)_4$

TETRA-ISOPROPYLTITANATE

PHYSICAL PROPERTIES	IMPORTANT CHARACTERISTICS
Boiling point, °C — 220 Melting point, °C — 15-20 Flash point, °C — 22 Relative density (water = 1) — 0.96 Relative vapor density (air = 1) — 9.8 Relative density at 20 °C of saturated mixture vapor/air (air = 1) — 1.00 Vapor pressure, mm Hg at 20 °C — ca. 0.076 Solubility in water — reaction Explosive limits, vol% in air — ?-? Relative molecular mass — 284.2 Gross formula: $C_{12}H_{28}O_4Ti$	**COLORLESS TO YELLOW LIQUID** Vapor mixes readily with air, giving off fumes. Decomposes on contact with water vapor, water and steam, giving off flammable and toxic vapors. Reacts violently with strong oxidants. TLV-TWA not available **Absorption route:** Can enter the body by inhalation or ingestion. Insufficient data on the rate at which harmful concentrations can build up. **Immediate effects:** Irritates the eyes, skin and respiratory tract.

HAZARDS/SYMPTOMS	PREVENTIVE MEASURES	FIRE EXTINGUISHING/FIRST AID
Fire: flammable.	keep away from open flame and sparks, no smoking.	powder, carbon dioxide, (halons), DO NOT USE WATER-BASED EXTINGUISHERS.
Explosion: above 22°C: forms explosive air-vapor mixtures.	above 22°C: sealed machinery, ventilation, explosion-proof electrical equipment.	in case of fire: keep tanks/drums cool by spraying with water, but DO NOT spray substance with water.
Inhalation: sore throat, cough.	ventilation, local exhaust or respiratory protection.	fresh air, rest, call a doctor.
Skin: redness.	gloves.	remove contaminated clothing, flush skin with water or shower.
Eyes: redness, pain.	safety glasses.	flush with water, take to a doctor if necessary.
Ingestion: abdominal pain.		rinse mouth, call a doctor.

SPILLAGE	STORAGE	LABELING / NFPA
collect leakage in sealable containers, soak up with sand or other inert absorbent and remove to safe place.	keep in dry, fireproof place, separate from oxidants.	

NOTES
Explosive limits not given in the literature. Reacts violently with extinguishing agents such as water and foam. Insufficient toxicological data on harmful effects to humans. Use airtight packaging.

Transport Emergency Card TEC(R)-30G08

851

TETRAMETHYLLEAD

PHYSICAL PROPERTIES	IMPORTANT CHARACTERISTICS
Boiling point (decomposes), °C 110 Melting point, °C −28 Flash point, °C <21 Relative density (water = 1) 2.0 Relative vapor density (air = 1) 9.2 Relative density at 20 °C of saturated mixture vapor/air (air = 1) 1.27 Vapor pressure, mm Hg at 20 °C 30 Solubility in water poor Explosive limits, vol% in air 1.8-? Relative molecular mass 267.3	**COLORLESS LIQUID WITH CHARACTERISTIC ODOR** Vapor is heavier than air and spreads at ground level, with risk of ignition at a distance. Decomposes when heated above 110°C, giving off toxic and flammable gases. Decomposes explosively when heated above 320°C. Reacts with strong oxidants, with risk of fire and explosion. Attacks lipids and rubber.
	TLV-TWA (skin) 0.15 mg/m³ ✱
	Absorption route: Can enter the body by inhalation or ingestion or through the skin. Harmful atmospheric concentrations can build up very rapidly on evaporation at 20°C. **Immediate effects:** Irritates the eyes, skin and respiratory tract. In substantial concentrations can cause agitation, sleeplessness and hallucination. In serious cases risk of death.
Gross formula: C$_4$H$_{12}$Pb	

HAZARDS/SYMPTOMS	PREVENTIVE MEASURES	FIRE EXTINGUISHING/FIRST AID
Fire: extremely flammable.	keep away from open flame and sparks, no smoking.	powder, water spray, foam, carbon dioxide, (halons), in case of fire, in immediate vicinity: use any extinguishing agent.
Explosion: forms explosive air-vapor mixtures.	sealed machinery, ventilation, explosion-proof electrical equipment and lighting.	in case of fire: keep tanks/drums cool by spraying with water.
	STRICT HYGIENE	
Inhalation: headache, dizziness, cramps, unconsciousness.	ventilation, local exhaust or respiratory protection.	fresh air, rest, take to hospital.
Skin: *is absorbed*; redness.	gloves, protective clothing.	remove contaminated clothing, wash skin with soap and water, take to hospital.
Eyes: redness, pain.	face shield, or combined eye and respiratory protection.	flush with water, take to a doctor.
Ingestion: diarrhea, abdominal cramps, headache, dizziness, unconsciousness.	do not eat, drink or smoke while working.	rinse mouth, immediately take to hospital.

SPILLAGE	STORAGE	LABELING / NFPA
evacuate area, call in an expert, collect leakage in sealable containers, soak up with sand or other inert absorbent and remove to safe place; (additional individual protection: CHEMICAL SUIT).	keep in dark, fireproof place, separate from oxidants.	R: 26/27/28-33 S: 13-26-36/37-45 Note A ☠ Toxic 3 / 3 / 3

NOTES
Depending on the degree of exposure, regular medical checkups are advisable. Technical grades are usually called anti-knock compounds and contain various quantities of dichloroethane and dibromoethane. Unbreakable packaging preferred; if breakable, keep in unbreakable container.

Transport Emergency Card TEC(R)-157

HI: 663; UN-number: 1649

TETRAPHENYLTIN

PHYSICAL PROPERTIES		IMPORTANT CHARACTERISTICS	
Boiling point, °C	> 420	**WHITE POWDER**	
Melting point, °C	226	Decomposes when heated, giving off toxic vapors. Reacts with strong oxidants.	
Flash point, °C	232		
Relative density (water = 1)	1.5	TLV-TWA (skin)	0.1 mg/m³ ✷
Solubility in water	none	STEL	0.2 mg/m³
Relative molecular mass	427.1		

Absorption route: Can enter the body by inhalation or ingestion or through the skin. Evaporation negligible at 20°C, but harmful concentrations of airborne particles can build up rapidly.
Immediate effects: Irritates the eyes, skin and respiratory tract. Affects the nervous system. In serious cases risk of death.
Effects of prolonged/repeated exposure: Prolonged or repeated contact can cause skin disorders. Inhalation of particles can cause lung disorders. Can cause liver and kidney damage.

Gross formula: $C_{24}H_{20}Sn$

HAZARDS/SYMPTOMS	PREVENTIVE MEASURES	FIRE EXTINGUISHING/FIRST AID
Fire: combustible.	keep away from open flame, no smoking.	water spray, powder.
	STRICT HYGIENE, KEEP DUST UNDER CONTROL	
Inhalation: headache, dizziness, abdominal cramps.	local exhaust or respiratory protection.	fresh air, rest, call a doctor.
Skin: *is absorbed*; redness, pain, inflammation of the follicles.	gloves.	remove contaminated clothing, flush skin with water or shower, refer to a doctor.
Eyes: redness, pain.	goggles, or combined eye and respiratory protection.	flush with water, take to a doctor.
Ingestion: abdominal cramps, headache, dizziness.	do not eat, drink or smoke while working.	rinse mouth, immediately take to hospital.

SPILLAGE	STORAGE	LABELING / NFPA
clean up spillage, carefully collect remainder; (additional individual protection: P3 respirator).	keep separate from strong oxidants.	1 / 3 / 0

NOTES
✷ TLV equals that of tin (organic compounds). Depending on the degree of exposure, regular medical checkups are advisable.

THIOACETIC ACID

PHYSICAL PROPERTIES	IMPORTANT CHARACTERISTICS
Boiling point, °C 87-93 Melting point, °C < − 17 Flash point, °C 18-21 Relative density (water = 1) 1.07 Relative vapor density (air = 1) 2.6 Relative density at 20 °C of saturated mixture vapor/air (air = 1) 1.06 Vapor pressure, mm Hg at 20 °C ca. 35.7 Solubility in water good Explosive limits, vol% in air ?-? Relative molecular mass 76.1 Gross formula: C$_2$H$_4$OS	**COLORLESS TO LIGHT YELLOW FUMING LIQUID WITH PUNGENT ODOR** Vapor mixes readily with air, forming explosive mixtures. Do not use compressed air when filling, emptying or processing. Toxic gases are a product of combustion (→ *sulfur dioxide*). Medium strong acid which reacts violently with bases and is corrosive. Reacts violently with strong oxidants, with risk of fire and explosion. Reacts slowly with water, forming → *acetic acid* and → *hydrogen sulfide*. Reacts with air to form strong, irritating vapors. TLV-TWA not available **Absorption route:** Can enter the body by inhalation or ingestion. Harmful atmospheric concentrations can build up very rapidly on evaporation at approx. 20°C. **Immediate effects:** Irritates the eyes, skin and respiratory tract. **Effects of prolonged/repeated exposure:** Prolonged or repeated contact can cause skin disorders.

HAZARDS/SYMPTOMS	PREVENTIVE MEASURES	FIRE EXTINGUISHING/FIRST AID
Fire: extremely flammable.	keep away from open flame and sparks, no smoking.	powder, alcohol-resistant foam, large quantities of water, carbon dioxide, (halons).
Explosion: forms explosive air-vapor mixtures.	sealed machinery, ventilation, explosion-proof electrical equipment and lighting.	in case of fire: keep tanks/drums cool by spraying with water.
Inhalation: sore throat, cough.	ventilation, local exhaust or respiratory protection.	fresh air, rest, call a doctor.
Skin: redness, pain.	gloves.	remove contaminated clothing, flush skin with water or shower, refer to a doctor if necessary.
Eyes: redness, pain.	face shield, or combined eye and respiratory protection.	flush with water, take to a doctor if necessary.
Ingestion: abdominal pain, diarrhea.		rinse mouth, call a doctor.

SPILLAGE	STORAGE	LABELING / NFPA
collect leakage in sealable containers, soak up with sand or other inert absorbent and remove to safe place; (additional individual protection: breathing apparatus).	keep in fireproof place, separate from oxidants and strong bases.	

NOTES
Explosive limits not given in the literature. Unbreakable packaging preferred; if breakable, keep in unbreakable container.

Transport Emergency Card TEC(R)-30G34

HI: 33; UN-number: 2436

CAS-No: [111-48-8]
β-bis-hydroxyethyl sulfide
dihydroxyethyl sulfide
thiodiethylene glycol

$(CH_2CH_2OH)_2S$

THIODIGLYCOL

PHYSICAL PROPERTIES	IMPORTANT CHARACTERISTICS
Boiling point, °C 282 Melting point, °C − 10 Flash point, °C 160 Auto-ignition temperature, °C 245 Relative density (water = 1) 1.2 Relative vapor density (air = 1) 4.2 Relative density at 20 °C of saturated mixture vapor/air (air = 1) 1.0 Vapor pressure, mm Hg at 20 °C < 0.076 Solubility in water ∞ Relative molecular mass 122.2	**COLORLESS VISCOUS LIQUID WITH CHARACTERISTIC ODOR** Vapor mixes readily with air. Decomposes when heated or burned, giving off toxic gases (→ *sulfur dioxide*). Reacts violently with hydrochloric acid and many other substances, giving off toxic vapors.
	TLV-TWA not available
	Absorption route: Can enter the body by inhalation or ingestion. Harmful atmospheric concentrations build up fairly slowly on evaporation at 20°C, but much more rapidly in aerosol form. **Immediate effects:** Irritates the eyes, skin and respiratory tract. Affects the nervous system. Can cause kidney damage.
Gross formula: $C_4H_{10}O_2S$	

HAZARDS/SYMPTOMS	PREVENTIVE MEASURES	FIRE EXTINGUISHING/FIRST AID
Fire: combustible.	keep away from open flame, no smoking.	powder, alcohol-resistant foam, water spray, carbon dioxide, (halons).
Inhalation: cough, shortness of breath, dizziness, sleepiness.	ventilation, local exhaust or respiratory protection.	fresh air, rest, call a doctor.
Skin: redness, pain.	gloves.	remove contaminated clothing, flush skin with water or shower, call a doctor.
Eyes: redness, pain.	face shield.	flush with water, take to a doctor.
Ingestion: abdominal pain, nausea, dizziness, sleepiness, unconsciousness.		rinse mouth, call a doctor or take to hospital.

SPILLAGE	STORAGE	LABELING / NFPA
collect leakage in sealable containers, soak up with sand or other inert absorbent and remove to safe place.	keep separate from acids; ventilate at floor level.	R: 36 ✖ Irritant (NFPA diamond: 2 — 1 — 0)

NOTES

THIONYL CHLORIDE

PHYSICAL PROPERTIES	IMPORTANT CHARACTERISTICS	
Boiling point, °C 79 Melting point, °C − 105 Relative density (water = 1) 1.6 Relative vapor density (air = 1) 4.1 Relative density at 20 °C of saturated mixture vapor/air (air = 1) 1.4 Vapor pressure, mm Hg at 20 °C 94 Solubility in water reaction Relative molecular mass 119	**COLORLESS TO YELLOW LIQUID WITH PUNGENT ODOR** Vapor is heavier than air. Decomposes when heated above boiling point, giving off → *chlorine*, → *sulfur chloride* and → *sulfur dioxide*. Decomposes when exposed to light, giving off → *sulfur dioxide*. On contact with water forms → *sulfuric acid* and → *hydrochloric acid*. Reacts violently with water, alcohols and many organic compounds, with risk of fire and explosion. Reacts with air to form corrosive fumes (→ *hydrochloric acid* and → *sulfur dioxide*). Attacks many materials.	
	TLV-TWA	1 ppm	4.9 mg/m³ C
	Absorption route: Can enter the body by inhalation or ingestion. Harmful atmospheric concentrations can build up very rapidly on evaporation at 20°C. **Immediate effects:** Corrosive to the eyes, skin and respiratory tract. Inhalation can cause lung edema. In serious cases risk of death.	
Gross formula: Cl$_2$OS		

HAZARDS/SYMPTOMS	PREVENTIVE MEASURES	FIRE EXTINGUISHING/FIRST AID
Fire: non-combustible.		in case of fire in immediate vicinity: preferably not water.
Explosion: many chemical reactions can cause explosion.	avoid contact with alcohols and other organic compounds.	in case of fire: keep tanks/drums cool by spraying with water, but DO NOT spray substance with water.
	STRICT HYGIENE	
Inhalation: *corrosive*; sore throat, shortness of breath, severe breathing difficulties.	ventilation, local exhaust or respiratory protection.	fresh air, rest, place in half-sitting position, artificial respiration if necessary, take to hospital.
Skin: *corrosive*; redness, pain, burns.	gloves, protective clothing.	remove contaminated clothing, flush skin with water or shower, refer to a doctor if necessary.
Eyes: *corrosive*; redness, pain, impaired vision.	face shield, or combined eye and respiratory protection.	flush with water, take to a doctor if necessary.
Ingestion: *corrosive*; sore throat, abdominal pain.		rinse mouth, take to hospital.

SPILLAGE	STORAGE	LABELING / NFPA
collect leakage in sealable containers, soak up with dry sand or other inert absorbent and remove to safe place, under no circumstances spray liquid with water; (additional individual protection: CHEMICAL SUIT).	keep in dry, dark place, separate from bases, alcohols and many other organic compounds.	R: 14-34-37 S: 26 Corrosive NFPA: 0 / 3 / 2 / W

NOTES
TLV is maximum and must not be exceeded. Lung edema symptoms usually develop several hours later and are aggravated by physical exertion: rest and hospitalization essential. As first aid, a doctor or authorized person should consider administering a corticosteroid spray. Use airtight packaging. Unbreakable packaging preferred; if breakable, keep in unbreakable container.

Transport Emergency Card TEC(R)-680

HI: X88; UN-number: 1836

$$\begin{array}{c} S \\ HC \diagup \diagdown CH \\ \| \quad \| \\ HC\underline{\quad\quad}CH \end{array}$$

THIOPHENE

PHYSICAL PROPERTIES	IMPORTANT CHARACTERISTICS
Boiling point, °C 84 Melting point, °C −38 Flash point, °C −9 Auto-ignition temperature, °C 395 Relative density (water = 1) 1.1 Relative vapor density (air = 1) 2.9 Relative density at 20 °C of saturated mixture vapor/air (air = 1) 1.16 Vapor pressure, mm Hg at 20 °C 61 Solubility in water none Explosive limits, vol% in air 1.5-12.5 Minimum ignition energy, mJ 0.4-0.6 Electrical conductivity, pS/m < 10⁴ Relative molecular mass 84.1 Gross formula: C_4H_4S	**COLORLESS LIQUID WITH CHARACTERISTIC ODOR** Vapor is heavier than air and spreads at ground level, with risk of ignition at a distance. Flow, agitation etc. can cause build-up of electrostatic charge due to liquid's low conductivity. Do not use compressed air when filling, emptying or processing. Toxic gases are a product of combustion (→ *sulfur dioxide*). Reacts violently with strong oxidants. Reacts violently with concentrated nitric acid and solid calcium hypochlorite. TLV-TWA not available **Absorption route:** Can enter the body by inhalation or ingestion. Harmful atmospheric concentrations can build up fairly rapidly on evaporation at approx. 20°C - even more rapidly in aerosol form. **Immediate effects:** Irritates the eyes and skin.

HAZARDS/SYMPTOMS	PREVENTIVE MEASURES	FIRE EXTINGUISHING/FIRST AID
Fire: extremely flammable, many chemical reactions can cause fire and explosion.	keep away from open flame and sparks, no smoking, avoid contact with oxidants.	powder, AFFF, foam, carbon dioxide, (halons).
Explosion: forms explosive air-vapor mixtures.	sealed machinery, ventilation, explosion-proof electrical equipment and lighting, grounding, non-sparking hand tools.	in case of fire: keep tanks/drums cool by spraying with water.
Inhalation: sore throat, cough.	ventilation, local exhaust or respiratory protection.	fresh air, rest.
Skin: redness.	gloves.	remove contaminated clothing, wash skin with soap and water.
Eyes: redness.	safety glasses.	flush with water, take to a doctor if necessary.
Ingestion:		rinse mouth, call a doctor or take immediately to hospital.

SPILLAGE	STORAGE	LABELING / NFPA
collect leakage in sealable containers, soak up with sand or other inert absorbent and remove to safe place; (additional individual protection: breathing apparatus).	keep in fireproof place, separate from oxidants.	3 2 0

NOTES
Insufficient toxicological data on harmful effects to humans.

THIOPHENOL

PHYSICAL PROPERTIES	IMPORTANT CHARACTERISTICS	
Boiling point, °C 168 Melting point, °C − 15 Flash point, °C 22 Relative density (water = 1) 1.07 Relative vapor density (air = 1) 3.8 Relative density at 20 °C of saturated mixture vapor/air (air = 1) 1.0 Vapor pressure, mm Hg at 20 °C 1.1 Solubility in water none Explosive limits, vol% in air ?-? Relative molecular mass 110.2	**COLORLESS LIQUID WITH CHARACTERISTIC ODOR** Vapor mixes readily with air. Decomposes when heated, giving off toxic vapors (→ *sulfur dioxide*). Reacts with acids, giving off toxic vapors (sulfur compounds).	
	TLV-TWA	0.5 ppm	2.3 mg/m³
	Absorption route: Can enter the body by inhalation or ingestion. Harmful atmospheric concentrations can build up fairly rapidly on evaporation at approx. 20°C - even more rapidly in aerosol form. **Immediate effects:** Corrosive to the eyes, skin and respiratory tract. Liquid destroys the skin's natural oils. Affects the nervous system. Inhalation of vapor/fumes can cause severe breathing difficulties (lung edema). In serious cases risk of death.	
Gross formula: C_6H_6S		

HAZARDS/SYMPTOMS	PREVENTIVE MEASURES	FIRE EXTINGUISHING/FIRST AID
Fire: flammable.	keep away from open flame and sparks, no smoking.	powder, AFFF, foam, carbon dioxide, (halons).
Explosion: above 22°C: forms explosive air-vapor mixtures.	above 22°C: sealed machinery, ventilation, explosion-proof electrical equipment.	in case of fire: keep tanks/drums cool by spraying with water.
	STRICT HYGIENE	
Inhalation: *corrosive*; sore throat, cough, shortness of breath, severe breathing difficulties, nausea, vomiting.	ventilation, local exhaust or respiratory protection.	fresh air, rest, place in half-sitting position, take to hospital.
Skin: *corrosive*; redness, pain, burns.	gloves, protective clothing.	remove contaminated clothing, flush skin with water or shower, call a doctor.
Eyes: *corrosive*; redness, pain, impaired vision.	face shield.	flush with water, take to a doctor.
Ingestion: *corrosive*; sore throat, abdominal pain, nausea, vomiting.		rinse mouth, take immediately to hospital.

SPILLAGE	STORAGE	LABELING / NFPA
collect leakage in sealable containers, soak up with sand or other inert absorbent and remove to safe place; (additional individual protection: breathing apparatus).	keep in fireproof place, separate from acids.	

NOTES
Technical grades are usually contaminated with → *carbon disulfide*, which lowers flash point. Explosive limits not given in the literature. Lung edema symptoms usually develop several hours later and are aggravated by physical exertion: rest and hospitalization essential. As first aid, a doctor or authorized person should consider administering a corticosteroid spray.

Transport Emergency Card TEC(R)-61G04

HI: 663; UN-number: 2337

TIN
(powder)

PHYSICAL PROPERTIES	IMPORTANT CHARACTERISTICS	
Boiling point, °C .... 2270 Melting point, °C .... 232 Relative density (water = 1) ... 7.3 Solubility in water .... none Relative molecular mass ... 118.7	**WHITE POWDER** At low temperatures changes slowly to α-crystalline form (gray tin). Reacts violently with strong oxidants and carbon tetrachloride. Reacts slowly with strong lyes, and violently with strong acids, giving off flammable gas (→ *hydrogen*). Oxidizes slowly to tin dioxide on exposure to moist air.	
	TLV-TWA	2 mg/m³
	Absorption route: Can enter the body by inhalation or ingestion. Harmful concentrations of airborne particles can build up rapidly. **Immediate effects:** Irritates the eyes and respiratory tract.	
Gross formula: Sn		

HAZARDS/SYMPTOMS	PREVENTIVE MEASURES	FIRE EXTINGUISHING/FIRST AID
Fire: combustible.	keep away from open flame, no smoking.	dry sand, special extinguishing powder, no other extinguishing agents.
Inhalation: cough.	local exhaust or respiratory protection.	fresh air, rest.
Eyes: redness, pain.	goggles.	flush with water, take to a doctor if necessary.
Ingestion: abdominal pain.		rinse mouth, call a doctor if necessary.

SPILLAGE	STORAGE	LABELING / NFPA
first dampen and then clean up spillage, carefully collect remainder; (additional individual protection: P2 respirator).	keep separate from oxidants and strong acids.	

NOTES
Burning powder reacts violently with extinguishing agents such as carbon dioxide, bicarbonate, water and halons.

TIN DICHLORIDE
(dihydrate)

PHYSICAL PROPERTIES	IMPORTANT CHARACTERISTICS	
Boiling point (decomposes), °C 652 Melting point, °C 38 Relative density (water = 1) 2.7 Solubility in water ∞ Relative molecular mass 225.7	**COLORLESS CRYSTALS** Decomposes when heated, giving off corrosive vapors (→ *hydrochloric acid*). Strong reducing agent which reacts violently with oxidants. Reacts with oxygen in air, forming insoluble tin oxydichloride.	
	TLV-TWA	2 mg/m³ ✱
	Absorption route: Can enter the body by inhalation or ingestion. Evaporation negligible at 20°C, but harmful concentrations of airborne particles can build up rapidly. **Immediate effects:** Irritates the eyes, skin and respiratory tract. Inhalation of finely dispersed substance can cause severe breathing difficulties. Contact with vapor or inhalation of particles can cause lung disorders. **Effects of prolonged/repeated exposure:** Can cause liver and kidney damage.	
Gross formula: $Cl_2Sn.2H_2O$		

HAZARDS/SYMPTOMS	PREVENTIVE MEASURES	FIRE EXTINGUISHING/FIRST AID
Fire: non-combustible.		in case of fire in immediate vicinity: use any extinguishing agent.
	KEEP DUST UNDER CONTROL	
Inhalation: sore throat, cough, shortness of breath.	local exhaust or respiratory protection.	fresh air, rest, call a doctor.
Skin: redness.	gloves.	remove contaminated clothing, flush skin with water or shower, refer to a doctor.
Eyes: redness, pain.	goggles.	flush with water, take to a doctor.
Ingestion: abdominal pain, diarrhea, nausea, vomiting.		rinse mouth, call a doctor.

SPILLAGE	STORAGE	LABELING / NFPA
clean up spillage, carefully collect remainder; (additional individual protection: P2 respirator).	keep separate from oxidants.	

NOTES

✱ TLV equals that of tin. Not to be confused with tin tetrachloride $SnCl_4$. Apparent melting point due to loss of water of crystallization. Anhydrous tin (II) chloride melts at 246°C. Depending on the degree of exposure, regular medical checkups are advisable.

TINTETRACHLORIDE

PHYSICAL PROPERTIES		IMPORTANT CHARACTERISTICS
Boiling point, °C	114	**COLORLESS HYGROSCOPIC LIQUID WITH PUNGENT ODOR**
Melting point, °C	− 33	Vapor is heavier than air. Reacts violently with water, forming corrosive → *hydrochloric acid* and → *tin oxide*. Reacts with moist air to form corrosive → *hydrochloric acid* vapors.
Relative density (water = 1)	2.2	
Relative vapor density (air = 1)	8.9	
Relative density at 20 °C of saturated mixture vapor/air (air = 1)	1.19	

TLV-TWA	2 mg/m³ ✳	

Absorption route: Can enter the body by inhalation or ingestion. Harmful atmospheric concentrations can build up very rapidly on evaporation at 20°C.
Immediate effects: Corrosive to the eyes, skin and respiratory tract. Affects the nervous system. Inhalation of vapor/fumes can cause severe breathing difficulties (lung edema). In serious cases risk of death.
Effects of prolonged/repeated exposure: Can cause liver, kidney and brain damage.

Physical properties (continued):

Vapor pressure, mm Hg at 20 °C	18
Solubility in water	reaction
Relative molecular mass	260.5

Gross formula: Cl$_4$Sn

HAZARDS/SYMPTOMS	PREVENTIVE MEASURES	FIRE EXTINGUISHING/FIRST AID
Fire: non-combustible.		in case of fire in immediate vicinity: preferably not water.
Inhalation: *corrosive*; sore throat, cough, shortness of breath, severe breathing difficulties, headache, sleepiness.	ventilation, local exhaust or respiratory protection.	fresh air, rest, place in half-sitting position, take to hospital.
Skin: *corrosive*; redness, pain, burns.	gloves, protective clothing.	remove contaminated clothing, flush skin with water or shower, refer to a doctor.
Eyes: *corrosive*; redness, pain, impaired vision.	face shield, or combined eye and respiratory protection.	flush with water, take to a doctor.
Ingestion: *corrosive*; sore throat, abdominal pain, diarrhea, headache, sleepiness.		rinse mouth, immediately take to hospital.

SPILLAGE	STORAGE	LABELING / NFPA
collect leakage in sealable containers, soak up with dry sand or other inert absorbent and remove to safe place, under no circumstances spray liquid with water; (additional individual protection: CHEMICAL SUIT).	keep dry; ventilate at floor level.	R: 34-37 S: 7/8-26 Corrosive NFPA: 3 / 0 / 1

NOTES
✳ TLV equals that of tin. Not to be confused with tin(II)chloride SnCl$_2$2H$_2$O. Depending on the degree of exposure, regular medical checkups are advisable. Lung edema symptoms usually develop several hours later and are aggravated by physical exertion: rest and hospitalization essential. As first aid, a doctor or authorized person should consider administering a corticosteroid spray. Use airtight packaging. Unbreakable packaging preferred; if breakable, keep in unbreakable container.

anatase
rutile
titanium white

TITANIUM DIOXIDE

<table>
<tr><td colspan="2">PHYSICAL PROPERTIES</td><td colspan="3">IMPORTANT CHARACTERISTICS</td></tr>
<tr>
<td>
Boiling point, °C
Melting point, °C
Relative density (water = 1)
Solubility in water
Relative molecular mass
</td>
<td>
2500-3000
1840
4.3
none
79.9
</td>
<td colspan="3">
WHITE POWDER
The principal types are anatase and rutile.
</td>
</tr>
<tr>
<td colspan="2"></td>
<td colspan="2">TLV-TWA</td>
<td>10 mg/m³</td>
</tr>
<tr>
<td colspan="2"></td>
<td colspan="3">
Absorption route: Can enter the body by inhalation or ingestion. Evaporation negligible at 20°C, but unpleasant concentrations of airborne particles can build up.
Immediate effects: Irritates the respiratory tract.
</td>
</tr>
<tr>
<td>Gross formula:</td>
<td>O$_2$Ti</td>
<td colspan="3"></td>
</tr>
<tr>
<td colspan="2">HAZARDS/SYMPTOMS</td>
<td colspan="2">PREVENTIVE MEASURES</td>
<td>FIRE EXTINGUISHING/FIRST AID</td>
</tr>
<tr>
<td colspan="2">**Fire:** non-combustible.</td>
<td colspan="2"></td>
<td>in case of fire in immediate vicinity: use any extinguishing agent.</td>
</tr>
<tr>
<td colspan="2">**Inhalation:** cough.</td>
<td colspan="2">local exhaust or respiratory protection.</td>
<td>fresh air, rest.</td>
</tr>
<tr>
<td colspan="2">**Eyes:** redness.</td>
<td colspan="2">goggles.</td>
<td>flush with water, take to a doctor if necessary.</td>
</tr>
<tr>
<td colspan="2">SPILLAGE</td>
<td colspan="2">STORAGE</td>
<td>LABELING / NFPA</td>
</tr>
<tr>
<td colspan="2">clean up spillage, flush away with water; (additional individual protection: P1 respirator).</td>
<td colspan="2"></td>
<td></td>
</tr>
<tr>
<td colspan="5">NOTES</td>
</tr>
<tr>
<td colspan="5"></td>
</tr>
</table>

TITANIUM TETRACHLORIDE

PHYSICAL PROPERTIES		IMPORTANT CHARACTERISTICS	
Boiling point, °C	136	**COLORLESS HYGROSCOPIC LIQUID WITH PUNGENT ODOR**	
Melting point, °C	−24	Vapor mixes readily with air. Reacts violently with water and alcohols, giving off corrosive → *hydrochloric acid*. Reacts with air to form corrosive → *hydrochloric acid* fumes.	
Relative density (water = 1)	1.7		
Relative vapor density (air = 1)	6.5		
Relative density at 20 °C of saturated mixture vapor/air (air = 1)	1.07	TLV-TWA	not available
Vapor pressure, mm Hg at 20 °C	9652		
Solubility in water	reaction	**Absorption route:** Can enter the body by inhalation or ingestion. Harmful atmospheric concentrations can build up fairly rapidly on evaporation at approx. 20°C - even more rapidly in aerosol form.	
Relative molecular mass	189.7		
		Immediate effects: Corrosive to the eyes, skin and respiratory tract. Inhalation of vapor/fumes can cause severe breathing difficulties (lung edema). In serious cases risk of seizures.	
Gross formula:	Cl$_4$Ti		

HAZARDS/SYMPTOMS	PREVENTIVE MEASURES	FIRE EXTINGUISHING/FIRST AID
Fire: non-combustible.		in case of fire in immediate vicinity: do not use water.
	STRICT HYGIENE	
Inhalation: *corrosive*; sore throat, cough, shortness of breath, severe breathing difficulties.	ventilation, local exhaust or respiratory protection.	fresh air, rest, place in half-sitting position, artificial respiration if necessary, take to hospital.
Skin: *corrosive*; redness, pain, burns.	gloves, protective clothing.	remove contaminated clothing, flush skin with water or shower, refer to a doctor.
Eyes: *corrosive*; redness, pain, impaired vision.	face shield, or combined eye and respiratory protection.	flush with water, take to a doctor.
Ingestion: *corrosive*; diarrhea, abdominal cramps, vomiting.		rinse mouth, take immediately to hospital.

SPILLAGE	STORAGE	LABELING / NFPA
evacuate area, collect leakage in sealable containers, soak up with sand or other inert absorbent and remove to safe place, under no circumstances spray liquid with water; (additional individual protection: CHEMICAL SUIT).	keep dry; ventilate at floor level.	R: 14-34-36/37 S: 7/8-26 Corrosive

NOTES

For practical purposes TLV for hydrochloric acid should be applied. Lung edema symptoms usually develop several hours later and are aggravated by physical exertion: rest and hospitalization essential. As first aid, a doctor or authorized person should consider administering a corticosteroid spray. Use airtight packaging. Unbreakable packaging preferred; if breakable, keep in unbreakable container.

Transport Emergency Card TEC(R)-733

HI: 80; UN-number: 1838

TITANIUM TRICHLORIDE

PHYSICAL PROPERTIES	IMPORTANT CHARACTERISTICS	
Melting point (decomposes), °C 440 Relative density (water = 1) 2.6 Relative molecular mass 154.3	**PURPLE POWDER** Decomposes when heated above 440°C, giving off → *hydrochloric acid* fumes. Strong reducing agent which reacts violently with oxidants. Reacts violently with water, giving off → *hydrochloric acid* fumes.	
	TLV-TWA not available	
	Absorption route: Can enter the body by inhalation or ingestion. **Immediate effects:** Corrosive to the eyes, skin and respiratory tract. Inhalation of vapor/fumes can cause severe breathing difficulties (lung edema). In serious cases risk of death.	
Gross formula: Cl₃Ti		

HAZARDS/SYMPTOMS	PREVENTIVE MEASURES	FIRE EXTINGUISHING/FIRST AID
Fire: combustible, many chemical reactions can cause fire and explosion.	avoid contact with oxidants and water vapor.	dry sand, special powder extinguisher, DO NOT USE WATER.
Inhalation: *corrosive*; sore throat, cough, shortness of breath, severe breathing difficulties.	local exhaust or respiratory protection.	fresh air, rest, place in half-sitting position, artificial respiration if necessary, take to hospital.
Skin: *corrosive*; redness, pain, serious burns.	gloves, protective clothing.	remove contaminated clothing, flush skin with water or shower, refer to a doctor.
Eyes: *corrosive*; redness, pain, impaired vision.	goggles, or combined eye and respiratory protection.	flush with water, take to a doctor.
Ingestion: *corrosive*; diarrhea, abdominal cramps, vomiting.	do not eat, drink or smoke while working.	rinse mouth, immediately take to hospital.

SPILLAGE	STORAGE	LABELING / NFPA
clean up spillage, carefully collect remainder; (additional individual protection: breathing apparatus).	keep dry and under inert gas, separate from oxidants.	

NOTES
Lung edema symptoms usually develop several hours later and are aggravated by physical exertion: rest and hospitalization essential. As first aid, a doctor or authorized person should consider administering a corticosteroid spray. Use airtight packaging.

Transport Emergency Card TEC(R)-80 (not pyrophoric)

HI: 80; UN-number: 2441

CAS-No: [108-88-3]
methylbenzene
methylbenzol
phenylmethane
toluol

$C_6H_5CH_3$

TOLUENE

PHYSICAL PROPERTIES		IMPORTANT CHARACTERISTICS	
Boiling point, °C	111	**COLORLESS LIQUID WITH CHARACTERISTIC ODOR**	
Melting point, °C	−95	Vapor mixes readily with air, forming explosive mixtures. Flow, agitation etc. can cause build-up of electrostatic charge due to liquid's low conductivity. Do not use compressed air when filling, emptying or processing. Reacts violently with strong oxidants, with risk of fire and explosion.	
Flash point, °C	4		
Auto-ignition temperature, °C	535		
Relative density (water = 1)	0.9		
Relative vapor density (air = 1)	3.2	TLV-TWA (skin) 100 ppm 377 mg/m³	
Relative density at 20 °C of saturated mixture vapor/air (air = 1)	1.06	STEL 150 ppm 565 mg/m³	
Vapor pressure, mm Hg at 20 °C	22	**Absorption route:** Can enter the body by inhalation or ingestion. Harmful atmospheric concentrations can build up fairly rapidly on evaporation at approx. 20°C - even more rapidly in aerosol form.	
Solubility in water	none		
Explosive limits, vol% in air	1.2 - 7		
Electrical conductivity, pS/m	1	**Immediate effects:** Irritates the eyes, skin and respiratory tract. Liquid destroys the skin's natural oils. Affects the nervous system. In serious cases risk of unconsciousness. If liquid is swallowed, droplets can enter the lungs, with risk of pneumonia.	
Relative molecular mass	92.1		
Log P octanol/water	2.6		
Gross formula:	C_7H_8		

HAZARDS/SYMPTOMS	PREVENTIVE MEASURES	FIRE EXTINGUISHING/FIRST AID
Fire: extremely flammable.	keep away from open flame and sparks, no smoking.	powder, AFFF, foam, carbon dioxide, (halons).
Explosion: forms explosive air-vapor mixtures.	sealed machinery, ventilation, explosion-proof electrical equipment and lighting, grounding.	in case of fire: keep tanks/drums cool by spraying with water.
Inhalation: headache, dizziness, nausea, drowsiness.	ventilation, local exhaust or respiratory protection.	fresh air, rest, call a doctor.
Skin: redness.	gloves.	remove contaminated clothing, flush skin with water or shower, call a doctor if necessary.
Eyes: redness.	safety glasses.	flush with water, take to a doctor if necessary.
Ingestion: abdominal cramps, headache, dizziness, drowsiness.	do not eat, drink or smoke while working.	rinse mouth, DO NOT induce vomiting, take immediately to hospital.

SPILLAGE	STORAGE	LABELING / NFPA
collect leakage in sealable containers, soak up with sand or other inert absorbent and remove to safe place; (additional individual protection: breathing apparatus).	keep in fireproof place, separate from oxidants.	R: 11-20 S: 16-25-29-33 Flammable Harmful

NOTES
Alcohol consumption increases toxic effects.

Transport Emergency Card TEC(R)-31

HI: 33; UN-number: 1294

CAS-No: [584-84-9]
2,4-diisocyanate-1-methylbenzene
2,4-diisocyanatotoluene
2,4-TDI
toluylene-2,4-diisocyanate

$CH_3C_6H_3(NCO)_2$

TOLUENE-2,4-DIISOCYANATE

PHYSICAL PROPERTIES		IMPORTANT CHARACTERISTICS
Boiling point, °C	251	**CLEAR TO LIGHT YELLOW LIQUID OR CRYSTALS WITH PUNGENT ODOR**
Melting point, °C	20	Vapor mixes readily with air. Decomposes when heated, giving off toxic gases. Reacts violently
Flash point, °C	127	with strong oxidants, amines, alcohols, acids, warm water etc., with risk of fire and explosion.
Auto-ignition temperature, °C	620	

TLV-TWA: 0.005 ppm / 0.036 mg/m³ C
STEL: 0.02 ppm / 0.14 mg/m³

Relative density (water = 1)	1.2
Relative vapor density (air = 1)	6
Relative density at 20 °C of saturated mixture vapor/air (air = 1)	1.0
Vapor pressure, mm Hg at 20 °C	0.030
Solubility in water	reaction
Explosive limits, vol% in air	0.9-9.5
Relative molecular mass	174

Absorption route: Can enter the body by inhalation or ingestion or through the skin. Harmful atmospheric concentrations can build up fairly rapidly on evaporation at approx. 20°C - even more rapidly in aerosol form.
Immediate effects: Corrosive to the eyes, skin and respiratory tract. Inhalation can cause severe breathing difficulties (asthma).
Effects of prolonged/repeated exposure: Prolonged or repeated contact can cause allergic disorders of the skin and respiratory tract.

Gross formula: $C_9H_6N_2O_2$

HAZARDS/SYMPTOMS	PREVENTIVE MEASURES	FIRE EXTINGUISHING/FIRST AID
Fire: combustible, many chemical reactions can cause fire and explosion.	keep away from open flame, no smoking, avoid contact with amines, alcohols, acids and warm water, etc.	powder, alcohol-resistant foam, water spray, carbon dioxide, (halons).
	STRICT HYGIENE	
Inhalation: *corrosive*; sore throat, cough, shortness of breath, severe breathing difficulties.	ventilation, local exhaust or respiratory protection.	fresh air, rest, place in half-sitting position, call a doctor or take to hospital.
Skin: *corrosive, is absorbed*; redness, pain.	gloves, protective clothing.	remove contaminated clothing, flush skin with water or shower, take to hospital.
Eyes: *corrosive*; redness, pain, impaired vision.	face shield, or combined eye and respiratory protection.	flush with water, take to doctor.
Ingestion: *corrosive*; sore throat, abdominal pain, diarrhea.	do not eat, drink or smoke while working.	rinse mouth, call a doctor.

SPILLAGE	STORAGE	LABELING / NFPA
collect leakage in sealable containers, soak up with sand or other inert absorbent and remove to safe place, render remainder harmless with dilute spirits of ammonia; (additional individual protection: breathing apparatus).	keep in dry, fireproof place, separate from oxidants, amines, alcohols and acids.	R: 26-36/37/38-42 S: 26-28-38-45 Toxic 1 / 3 2

NOTES

At low temperatures TDI reactions develop slowly. Odor limit is above TLV. TLV is maximum and must not be exceeded. Dilute spirits of ammonia: 50% water, 45% ethyl alcohol, 5% concentrated ammonia. Depending on the degree of exposure, regular medical checkups are advisable. Asthma symptoms usually develop several hours later and are aggravated by physical exertion: rest and hospitalization essential. Persons with history of asthma symptoms should under no circumstances come in contact again with substance. Use airtight packaging made of special material.

Transport Emergency Card TEC(R)-173

HI: 60; UN-number: 2078

CAS-No: [104-15-4]
4-methylbenzenesulfonic acid
p-methylbenzenesulfonic acid
4-toluenesulfonic acid
tosic acid

$CH_3C_6H_4SO_3H$

p-TOLUENESULFONIC ACID

PHYSICAL PROPERTIES		IMPORTANT CHARACTERISTICS	
Melting point, °C	106	**COLORLESS HYGROSCOPIC CRYSTALS WITH PUNGENT ODOR**	
Flash point, °C	180	Decomposes when heated or burned, giving off toxic and corrosive vapors ($\rightarrow$ *sulfur oxide*). In aqueous solution is a strong acid which reacts violently with bases and is corrosive. Attacks many metals, giving off flammable gas ($\rightarrow$ *hydrogen*).	
Relative density (water = 1)	1.2		
Vapor pressure, mm Hg at 140 °C	21		
Solubility in water, g/100 ml	67		
Relative molecular mass	172.2		
		TLV-TWA	not available
		Absorption route: Can enter the body by inhalation or ingestion. Insufficient data on the rate at which harmful concentrations can build up. **Immediate effects:** Irritates the eyes, skin and respiratory tract.	
Gross formula:	$C_7H_8O_3S$		

HAZARDS/SYMPTOMS	PREVENTIVE MEASURES	FIRE EXTINGUISHING/FIRST AID
Fire: combustible.	keep away from open flame, no smoking.	powder, alcohol-resistant foam, water spray, carbon dioxide, (halons).
Inhalation: cough, shortness of breath.	ventilation, local exhaust or respiratory protection.	fresh air, rest, place in half-sitting position, call a doctor.
Skin: redness, pain.	gloves, protective clothing.	remove contaminated clothing, flush skin with water or shower, refer to a doctor.
Eyes: redness, pain, impaired vision.	face shield, or combined eye and respiratory protection.	flush with water, take to a doctor.
Ingestion: abdominal cramps, vomiting.		rinse mouth, call a doctor or take to hospital.

SPILLAGE	STORAGE	LABELING / NFPA
clean up spillage, flush away remainder with water; (additional individual protection: P2 respirator).	keep dry, separate from strong bases; ventilate at floor level.	1 / 3 / 1

NOTES
p-Tolenesulfonic acid can contain variable quantities of sulfuric acid. Depending on the concentration of sulfuric acid the substance must be labelled if (1) it contains more than 5% sulfuric acid as corrosive and (2) it contains 5% or less sulfuric acid as harmful/irritating.

Transport Emergency Card TEC(R)-80G09C

HI: 80; UN-number: 2585

p-TOLUENESULFONYL CHLORIDE

PHYSICAL PROPERTIES	IMPORTANT CHARACTERISTICS	
Decomposes below boiling point, °C ... ? Melting point, °C ... 70 Flash point, °C ... 110 Auto-ignition temperature, °C ... 492 Relative density (water = 1) ... 1.3 Vapor pressure, mm Hg at 88 °C ... 0.99 Solubility in water ... reaction Relative molecular mass ... 190.7 Gross formula: $C_7H_7ClO_2S$	**COLORLESS CRYSTALS** Decomposes when heated, giving off toxic vapors. Reacts with water and water vapor, forming → *hydrochloric acid* and → *p-toluenesulfonic acid* and giving off corrosive fumes. Reacts violently with bases. TLV-TWA not available **Absorption route:** Can enter the body by inhalation or ingestion. Evaporation negligible at 20°C, but at temperatures above melting point harmful atmospheric concentrations can build up due to vaporization. **Immediate effects:** Corrosive to the eyes, skin and respiratory tract. Inhalation of vapor/fumes can cause severe breathing difficulties (lung edema). In serious cases risk of death. **Effects of prolonged/repeated exposure:** Prolonged contact can cause lung disorders.	

HAZARDS/SYMPTOMS	PREVENTIVE MEASURES	FIRE EXTINGUISHING/FIRST AID
Fire: combustible.	keep away from open flame, no smoking.	powder, carbon dioxide, (halons).
	STRICT HYGIENE	
Inhalation: *corrosive*; sore throat, cough, shortness of breath, severe breathing difficulties.	local exhaust or respiratory protection.	fresh air, rest, take to hospital.
Skin: *corrosive*; redness, pain, serious burns.	gloves, protective clothing.	remove contaminated clothing, flush skin with water or shower, refer to a doctor.
Eyes: *corrosive*; redness, pain, impaired vision.	face shield.	flush with water, take to a doctor.
Ingestion: *corrosive*; sore throat, abdominal pain, vomiting.	do not eat, drink or smoke while working.	rinse mouth, take immediately to hospital.

SPILLAGE	STORAGE	LABELING / NFPA
clean up spillage, soak up with sand or other inert absorbent and remove to safe place; (additional individual protection: breathing apparatus).	keep dry, separate from strong bases.	

NOTES
Decomposition temperature not given in the literature. Lung edema symptoms usually develop several hours later and are aggravated by physical exertion: rest and hospitalization essential. As first aid, a doctor or authorized person should consider administering a corticosteroid spray.

Transport Emergency Card TEC(R)-80G11

CAS-No: [95-53-4]
1-amino-2-methylbenzene
o-aminotoluene
2-methylaniline
2-methylbenzenamine
o-toluidine

$CH_3C_6H_4NH_2$

1,2-TOLUIDINE

PHYSICAL PROPERTIES		IMPORTANT CHARACTERISTICS	
Boiling point, °C	200	**COLORLESS LIQUID WITH CHARACTERISTIC ODOR, TURNING YELLOWISH-BROWN ON EXPOSURE TO AIR**	
Melting point, °C	− 16		
Flash point, °C	85	Vapor mixes readily with air. Toxic gases are a product of combustion (→ *nitrous vapors*). Reacts with strong oxidants. Reacts violently with strong acids.	
Auto-ignition temperature, °C	480		
Relative density (water = 1)	1.0		
Relative vapor density (air = 1)	3.7		
Relative density at 20 °C of saturated mixture vapor/air (air = 1)	1.0	TLV-TWA (skin) 2 ppm 8.8 mg/m³ A2	
Vapor pressure, mm Hg at 20 °C	0.15		
Solubility in water, g/100 ml at 25 °C	1.5	**Absorption route:** Can enter the body by inhalation or ingestion or through the skin. Harmful atmospheric concentrations build up fairly slowly on evaporation at 20°C, but much more rapidly in aerosol form.	
Explosive limits, vol% in air	1.5-?		
Electrical conductivity, pS/m	3.8 x 10⁷	**Immediate effects:** Affects the nervous system.	
Relative molecular mass	107.2	**Effects of prolonged/repeated exposure:** Can affect the blood. Can cause kidney and bladder damage.	
Log P octanol/water	1.3		
Gross formula:	C_7H_9N		

HAZARDS/SYMPTOMS	PREVENTIVE MEASURES	FIRE EXTINGUISHING/FIRST AID
Fire: combustible.	keep away from open flame, no smoking.	powder, AFFF, foam, carbon dioxide, (halons).
Explosion: above 85°C: forms explosive air-vapor mixtures.	above 85°C: sealed machinery, ventilation.	
	STRICT HYGIENE	
Inhalation: headache, dizziness, blue skin, feeling of weakness.	ventilation, local exhaust or respiratory protection.	fresh air, rest, call a doctor.
Skin: *is absorbed*; see also 'Inhalation'.	gloves, protective clothing.	remove contaminated clothing, wash skin with soap and water, call a doctor.
Eyes: redness, pain.	face shield.	flush with water, take to a doctor if necessary.
Ingestion: headache, dizziness, blue skin, feeling of weakness.		rinse mouth, immediately take to hospital.

SPILLAGE	STORAGE	LABELING / NFPA
collect leakage in sealable containers, soak up with sand or other inert absorbent and remove to safe place.	keep separate from strong acids; ventilate at floor level.	R: 23/24/25-33 S: 28-36/37-44 Note C ☠ Toxic 2 / 3 / 0

NOTES
Odor limit is above TLV. Alcohol consumption increases toxic effects. Depending on the degree of exposure, regular medical checkups are advisable. Effects on blood due to formation of methemoglobin. Special first aid required in the event of poisoning: antidotes must be available (with instructions). The measures on this card also apply to meta-toluidine.

Transport Emergency Card TEC(R)-669

HI: 60; UN-number: 1708

CAS-No: [102-70-5]
tri-2-propenylamine
N,N-di-2-propenyl-2-propen-1-amine
TAA

$(CH_2\!=\!CHCH_2)_3N$

TRIALLYLAMINE

PHYSICAL PROPERTIES		IMPORTANT CHARACTERISTICS
Boiling point, °C	150	**COLORLESS LIQUID WITH CHARACTERISTIC ODOR**
Melting point, °C	−70	Vapor mixes readily with air. Decomposes when heated, giving off flammable and toxic vapors. Strong reducing agent which reacts violently with oxidants. Strong base which reacts violently with acids and corrodes aluminum, zinc etc.
Flash point, °C	39	
Auto-ignition temperature, °C	?	
Relative density (water = 1)	0.8	
Relative vapor density (air = 1)	4.7	
Relative density at 20 °C of saturated mixture vapor/air (air = 1)	1.01	**TLV-TWA** not available
Vapor pressure, mm Hg at 20 °C	ca. 3.0	**Absorption route:** Can enter the body by inhalation or ingestion. Harmful atmospheric concentrations can build up fairly rapidly on evaporation at approx. 20°C - even more rapidly in aerosol form.
Solubility in water, g/100 ml	0.2	
Explosive limits, vol% in air	?-?	**Immediate effects:** Corrosive to the eyes, skin and respiratory tract. Inhalation can cause lung edema. In serious cases risk of death.
Relative molecular mass	137.2	
Gross formula:	$C_9H_{15}N$	

HAZARDS/SYMPTOMS	PREVENTIVE MEASURES	FIRE EXTINGUISHING/FIRST AID
Fire: flammable.	keep away from open flame and sparks, no smoking.	powder, AFFF, foam, carbon dioxide, (halons).
Explosion: above 39°C: forms explosive air-vapor mixtures.	above 39°C: sealed machinery, ventilation, explosion-proof electrical equipment.	in case of fire: keep tanks/drums cool by spraying with water.
Inhalation: *corrosive*; sore throat, cough, shortness of breath, severe breathing difficulties.	ventilation, local exhaust or respiratory protection.	fresh air, rest, place in half-sitting position, artificial respiration if necessary, take to hospital.
Skin: *corrosive*; redness, burns.	gloves, protective clothing.	remove contaminated clothing, flush skin with water or shower, call a doctor.
Eyes: *corrosive*; redness, pain, impaired vision.	face shield, or combined eye and respiratory protection.	flush with water, take to doctor.
Ingestion: *corrosive*; sore throat, abdominal pain, vomiting.		rinse mouth, take immediately to hospital.

SPILLAGE	STORAGE	LABELING / NFPA
collect leakage in sealable containers, soak up with sand or other inert absorbent and remove to safe place; (additional individual protection: breathing apparatus).	keep in fireproof place, separate from oxidants and acids.	◇

NOTES
Explosive limits not given in the literature. Lung edema symptoms usually develop several hours later and are aggravated by physical exertion: rest and hospitalization essential. As first aid, a doctor or authorized person should consider administering a corticosteroid spray.

Transport Emergency Card TEC(R)-30G36

HI: 30; UN-number: 2610

CAS-No: [102-82-9]
N,N-dibutyl-1-butaneamine
TNBA
tri-n-butylamine
tris-n-butylamine

$(C_4H_9)_3N$

TRIBUTYLAMINE

PHYSICAL PROPERTIES		IMPORTANT CHARACTERISTICS	
Boiling point, °C	214	**COLORLESS TO LIGHT YELLOW HYGROSCOPIC LIQUID WITH AMINE ODOR**	
Melting point, °C	−70	Vapor mixes readily with air. Decomposes when heated or burned, giving off toxic vapors ($\rightarrow$ *nitrous vapors*). Strong base which reacts violently with acids and corrodes aluminum, zinc etc. Reacts violently with mercury, with risk of fire and explosion. Reacts with oxidants.	
Flash point, °C	63		
Relative density (water = 1)	0.78		
Relative density at 20 °C of saturated mixture			
vapor/air (air = 1)	1.0	TLV-TWA	not available
Vapor pressure, mm Hg at 100 °C	ca. 0.15		
Solubility in water, g/100 ml at 18 °C	0.004	**Absorption route:** Can enter the body by inhalation or ingestion. Harmful atmospheric concentrations build up fairly slowly on evaporation at approx. 20° C, but much more rapidly in aerosol form.	
Relative molecular mass	185.3		
Log P octanol/water	1.5		
		Immediate effects: Corrosive to the eyes, skin and respiratory tract. Inhalation can cause lung edema.	
Gross formula:	$C_{12}H_{27}N$		

HAZARDS/SYMPTOMS	PREVENTIVE MEASURES	FIRE EXTINGUISHING/FIRST AID
Fire: combustible.	keep away from open flame, no smoking.	powder, AFFF, foam, carbon dioxide, (halons).
Explosion: above 63°C: forms explosive air-vapor mixtures.	above 63°C: sealed machinery, ventilation.	
Inhalation: *corrosive*; sore throat, cough, shortness of breath, severe breathing difficulties.	ventilation, local exhaust or respiratory protection.	fresh air, rest, place in half-sitting position, take to hospital.
Skin: *corrosive*; redness, pain, burns.	gloves, protective clothing.	remove contaminated clothing, flush skin with water or shower, refer to a doctor.
Eyes: *corrosive*; redness, pain, impaired vision.	face shield.	flush with water, take to a doctor.
Ingestion: *corrosive*; sore throat, abdominal pain, diarrhea, nausea.		rinse mouth, take to hospital.

SPILLAGE	STORAGE	LABELING / NFPA
collect leakage in sealable containers, flush away remainder with water.	keep separate from strong acids and mercury; ventilate at floor level.	2 / 3 0

NOTES

Log P octanol/water is estimated. Lung edema symptoms usually develop several hours later and are aggravated by physical exertion: rest and hospitalization essential. As first aid, a doctor or authorized person should consider administering a corticosteroid spray. Unbreakable packaging preferred; if breakable, keep in unbreakable container.

Transport Emergency Card TEC(R)-80G16

HI: 80; UN-number: 2542

TRIBUTYL PHOSPHATE

PHYSICAL PROPERTIES	IMPORTANT CHARACTERISTICS	
Boiling point (decomposes), °C — 290 Melting point, °C — < −80 Flash point, °C — 146 Relative density (water = 1) — 0.98 Relative vapor density (air = 1) — 9.2 Relative density at 20 °C of saturated mixture vapor/air (air = 1) — 1.00 Vapor pressure, mm Hg at 20°C — < <0.076 Solubility in water, g/100 ml — 0.6 Relative molecular mass — 266.3	**COLORLESS, VISCOUS, ALMOST ODORLESS LIQUID** Vapor mixes readily with air. Toxic and corrosive → *phosphorus oxide* vapors are a product of combustion. Decomposes when heated above 290°C or on contact with warm water, forming → *phosphoric acid* and → *butanol*. Reacts violently with strong oxidants.	
	TLV-TWA	0.2 ppm	2.2 mg/m³
	Absorption route: Can enter the body by inhalation or ingestion. Harmful atmospheric concentrations build up very slowly, if at all, on evaporation at approx. 20°C, but much more rapidly in aerosol form. **Immediate effects:** Corrosive to the eyes, skin and respiratory tract.	
Gross formula: $C_{12}H_{27}O_4P$		

HAZARDS/SYMPTOMS	PREVENTIVE MEASURES	FIRE EXTINGUISHING/FIRST AID
Fire: combustible.	keep away from open flame, no smoking.	powder, AFFF, foam, carbon dioxide, (halons).
Explosion: aerosol forms explosive mixtures with air.		
Inhalation: *corrosive*; sore throat, cough, shortness of breath.	ventilation.	fresh air, rest, place in half-sitting position, call a doctor.
Skin: *corrosive*; redness, pain, burns.	gloves.	remove contaminated clothing, flush skin with water or shower, call a doctor if necessary.
Eyes: *corrosive*; redness, pain, impaired vision.	face shield.	flush with water, take to a doctor.
Ingestion: *corrosive*; abdominal pain, diarrhea, vomiting.	do not eat, drink or smoke while working.	rinse mouth, take to hospital.

SPILLAGE	STORAGE	LABELING / NFPA
collect leakage in sealable containers, soak up with sand or other inert absorbent and remove to safe place.	keep separate from oxidants; ventilate at floor level.	R: 22 S: 25 ✖ Harmful

NOTES

CAS-No: [76-03-9]
TCA
trichlorethanoic acid
trichloroethanoic acid

Cl₃CCOOH

TRICHLOROACETIC ACID

PHYSICAL PROPERTIES	IMPORTANT CHARACTERISTICS
Boiling point, ° C 198 Melting point, ° C 58 Relative density (water = 1) 1.6 Relative vapor density (air = 1) 5.6 Relative density at 20 ° C of saturated mixture vapor/air (air = 1) 1.0 Vapor pressure, mm Hg at 20 ° C 0.076 Solubility in water, g/100 ml at 20 ° C 120 Relative molecular mass 163.4 Log P octanol/water 1.1	**WHITE HYGROSCOPIC CRYSTALS WITH PUNGENT ODOR** Vapor mixes readily with air. Decomposes when heated, giving off toxic (→ *phosgene*, → *carbon monoxide*) and corrosive (→ *hydrochloric acid*) vapors. In aqueous solution decomposes when moderately heated, giving off → *chloroform* and → *carbon dioxide*. In aqueous solution is a strong acid which reacts violently with bases and is corrosive.

	TLV-TWA	1 ppm	6.7 mg/m³

Absorption route: Can enter the body by inhalation or ingestion. Harmful atmospheric concentrations build up fairly slowly on evaporation, but harmful concentrations of airborne particles can build up much more rapidly.
Immediate effects: Corrosive to the eyes, skin and respiratory tract. Inhalation can cause lung edema. In serious cases risk of death.

Gross formula: C₂HCl₃O₂

HAZARDS/SYMPTOMS	PREVENTIVE MEASURES	FIRE EXTINGUISHING/FIRST AID
Fire: non-combustible.		in case of fire in immediate vicinity: use any extinguishing agent.
	KEEP DUST UNDER CONTROL	
Inhalation: *corrosive*; sore throat, cough, shortness of breath, severe breathing difficulties.	local exhaust or respiratory protection.	fresh air, rest, place in half-sitting position, take to hospital.
Skin: *corrosive*; redness, pain, burns.	gloves, protective clothing.	remove contaminated clothing, flush skin with water or shower, refer to a doctor.
Eyes: *corrosive*; redness, pain, impaired vision.	face shield.	flush with water, take to a doctor.
Ingestion: *corrosive*; sore throat, abdominal pain, vomiting.		rinse mouth, call a doctor or take to hospital.

SPILLAGE	STORAGE	LABELING / NFPA
render harmless with soda, flush away remainder with water; (additional individual protection: respirator with A/P2 filter).	keep dry, separate from strong bases; ventilate at floor level.	R: 35 S: 24/25-26 Corrosive

NOTES
Log P octanol/water is estimated. Lung edema symptoms usually develop several hours later and are aggravated by physical exertion: rest and hospitalization essential. As first aid, a doctor or authorized person should consider administering a corticosteroid spray. Dilute aqueous solutions decompose slowly. Unbreakable packaging preferred; if breakable, keep in unbreakable container.

Transport Emergency Card TEC(R)-80G21B

HI: 80; UN-number: 1839

1,2,4-TRICHLOROBENZENE

PHYSICAL PROPERTIES		IMPORTANT CHARACTERISTICS	
Boiling point, °C	213	**COLORLESS LIQUID OR WHITE CRYSTALS WITH CHARACTERISTIC ODOR**	
Melting point, °C	17	Decomposes when heated, giving off toxic and corrosive gases (→ *phosgene* and	
Flash point, °C	105	→ *hydrochloric acid*). Reacts violently with strong oxidants.	
Relative density (water = 1)	1.5		
Relative vapor density (air = 1)	6.3	TLV-TWA	5 ppm	37 mg/m³ C
Relative density at 20 °C of saturated mixture			
vapor/air (air = 1)	1.0	**Absorption route:** Can enter the body by inhalation or ingestion or through the skin. Harmful	
Vapor pressure, mm Hg at 20 °C	0.30	atmospheric concentrations can build up fairly slowly on evaporation at approx. 20°C, but more	
Solubility in water	none	rapidly in aerosol form.	
Relative molecular mass	181.5	**Immediate effects:** Corrosive to the eyes, skin and respiratory tract. In substantial	
		concentrations can impair consciousness. Inhalation of vapor/fumes can cause severe breathing	
		difficulties (lung edema). In serious cases risk of unconsciousness and death. If liquid is	
		swallowed, droplets can enter the lungs, with risk of pneumonia.	
		Effects of prolonged/repeated exposure: Can cause liver and kidney damage.	
Gross formula:	$C_6H_3Cl_3$		

HAZARDS/SYMPTOMS	PREVENTIVE MEASURES	FIRE EXTINGUISHING/FIRST AID
Fire: combustible.	keep away from open flame, no smoking.	water spray, foam, powder, carbon dioxide, (halons).
	KEEP DUST UNDER CONTROL	
Inhalation: *corrosive*; sore throat, cough, severe breathing difficulties, drowsiness.	local exhaust or respiratory protection.	fresh air, rest, place in half-sitting position, take to hospital.
Skin: *corrosive*; redness, pain, burns.	gloves.	remove contaminated clothing, flush skin with water or shower, call a doctor.
Eyes: *corrosive*; redness, pain, impaired vision.	face shield, or combined eye and respiratory protection.	flush with water, take to a doctor.
Ingestion: *corrosive*; sore throat, abdominal pain, vomiting.	do not eat, drink or smoke while working.	rinse mouth, call a doctor or take to hospital.

SPILLAGE	STORAGE	LABELING / NFPA
Allow spillage to set as far as possible and clean up, soak up liquid with dry sand or other inert absorbent and remove to safe place, carefully collect remainder; (additional individual protection: breathing apparatus or respirator with A/P2 filter).		(NFPA diamond: top 1, left 2, right 0)

NOTES
TLV is maximum and must not be exceeded. Depending on the degree of exposure, regular medical checkups are advisable. Lung edema symptoms usually develop several hours later and are aggravated by physical exertion: rest and hospitalization essential. As first aid, a doctor or authorized person should consider administering a corticosteroid spray. The measures on this card also apply to 1,2,3-trichlorobenzene and 1,3,5-trichlorobenzene. Physical properties of 1,2,3-trichlorobenzene: flash point: 113°C; melting point: 52°C; vapor pressure at 40°C: 0.99mm Hg. Physical properties of 1,3,5-trichlorobenzene: flash point: 107°C; melting point: 63°C; vapor pressure at 78°C: 9.9mm Hg.

Transport Emergency Card TEC(R)-61G06C

HI: 60; UN-number: 2321

CAS-No: [79-00-5]
ethane trichloride
1,1,2-trichlorethane
β-trichloroethane
vinyl trichloride

$CHCl_2\,CH_2Cl$

1,1,2-TRICHLOROETHANE

PHYSICAL PROPERTIES		IMPORTANT CHARACTERISTICS	
Boiling point, °C	114	**COLORLESS LIQUID WITH CHARACTERISTIC ODOR**	
Melting point, °C	− 36	Vapor mixes readily with air. Decomposes in flame or on hot surface, giving off toxic and corrosive vapors (→ *hydrochloric acid*). Reacts violently with metal powders and sodium amide, with risk of fire and explosion. Attacks plastics and aluminum.	
Flash point, °C	none		
Relative density (water = 1)	1.4		
Relative vapor density (air = 1)	4.6		
Relative density at 20 °C of saturated mixture vapor/air (air = 1)	1.08	TLV-TWA (skin) 10 ppm 55 mg/m³	
Vapor pressure, mm Hg at 20 °C	19		
Solubility in water, g/100 ml at 20 °C	0.4	**Absorption route:** Can enter the body by inhalation or ingestion or through the skin. Harmful atmospheric concentrations can build up fairly rapidly on evaporation at approx. 20°C - even more rapidly in aerosol form.	
Relative molecular mass	133.4		
		Immediate effects: Irritates the eyes, skin and respiratory tract. Liquid destroys the skin's natural oils. In substantial concentrations can impair consciousness.	
		Effects of prolonged/repeated exposure: Can cause liver and kidney damage.	
Gross formula:	$C_2H_3Cl_3$		

HAZARDS/SYMPTOMS	PREVENTIVE MEASURES	FIRE EXTINGUISHING/FIRST AID
Fire: non-combustible; see Notes.		in case of fire in immediate vicinity: use any extinguishing agent.
Inhalation: cough, headache, dizziness, drowsiness.	ventilation, local exhaust or respiratory protection.	fresh air, rest, call a doctor.
Skin: *is absorbed*; redness, pain.	gloves, protective clothing.	remove contaminated clothing, flush skin and wash thoroughly with soap and water, call a doctor.
Eyes: redness, pain.	face shield.	
Ingestion: sore throat, abdominal pain, nausea; see also 'Inhalation'.		rinse mouth, immediately take to hospital.

SPILLAGE	STORAGE	LABELING / NFPA
collect leakage in sealable containers, soak up with sand or other inert absorbent and remove to safe place; (additional individual protection: breathing apparatus).	keep separate from metal powders; ventilate at floor level.	R: 20/21/22 S: 9 ✖ Harmful

NFPA diagram: 3 (left), 1 (top), 0 (right)

NOTES
Under certain conditions can form flammable vapor-air mixtures (12-?% by vol.). Addition of small quantities of combustible substances or raising of oxygen content increases flammability considerably. Alcohol consumption increases toxic effects. Depending on the degree of exposure, regular medical checkups are advisable. Under no circumstances use near flame or hot surface or when welding. Packaging: not aluminum.

Transport Emergency Card TEC(R)-61G06B

CAS-No: [79-01-6]
ethinyl trichloride
trichloroethene
tce

$ClCH = CCl_2$

TRICHLOROETHYLENE

<table>
<tr><td colspan="2">PHYSICAL PROPERTIES</td><td colspan="3">IMPORTANT CHARACTERISTICS</td></tr>
<tr>
<td>
Boiling point, °C
Melting point, °C
Flash point, °C
Auto-ignition temperature, °C
Relative density (water = 1)
Relative vapor density (air = 1)
Relative density at 20 °C of saturated mixture
vapor/air (air = 1)
Vapor pressure, mm Hg at 20 °C
Solubility in water, g/100 ml at 20 °C
Electrical conductivity, pS/m
Relative molecular mass

Gross formula:
</td>
<td>
87
− 86
none
410
1.5
4.5

1.3
58.5
0.1
800
131.4

C_2HCl_3
</td>
<td colspan="3">
COLORLESS LIQUID WITH CHARACTERISTIC ODOR
Vapor is heavier than air. Flow, agitation etc. can cause build-up of electrostatic charge due to liquid's low conductivity. Decomposes in flame or on hot surface, giving off toxic gas (→ *phosgene*) and corrosive vapors (→ *hydrochloric acid*); also on exposure to UV radiation or welding. Reacts violently with alkali metals and metal powders. Reacts violently with concentrated lye, giving off dichloroacetylene, with risk of fire and explosion.
<hr>
TLV-TWA   50 ppm   269 mg/m³
STEL   200 ppm   1070 mg/m³
<hr>
Absorption route: Can enter the body by inhalation or ingestion or through the skin. Harmful atmospheric concentrations can build up fairly rapidly on evaporation at approx. 20°C - even more rapidly in aerosol form.
Immediate effects: Irritates the eyes, skin and respiratory tract. Liquid destroys the skin's natural oils. In substantial concentrations can cause impairment of consciousness and heart rate disorders. In serious cases risk of unconsciousness and death.
Effects of prolonged/repeated exposure: Can cause liver and kidney damage.
</td>
</tr>
<tr>
<td colspan="2">HAZARDS/SYMPTOMS</td>
<td>PREVENTIVE MEASURES</td>
<td colspan="2">FIRE EXTINGUISHING/FIRST AID</td>
</tr>
<tr>
<td colspan="2">**Fire:** non-combustible.</td>
<td></td>
<td colspan="2">in case of fire in immediate vicinity: use any extinguishing agent.</td>
</tr>
<tr>
<td colspan="2">**Explosion:**</td>
<td></td>
<td colspan="2">in case of fire: keep tanks/drums cool by spraying with water.</td>
</tr>
<tr>
<td colspan="2">**Inhalation:** cough, severe breathing difficulties, dizziness, drowsiness, unconsciousness.</td>
<td>ventilation, local exhaust or respiratory protection.</td>
<td colspan="2">fresh air, rest, artificial respiration if necessary, take to hospital.</td>
</tr>
<tr>
<td colspan="2">**Skin:** redness.</td>
<td>gloves.</td>
<td colspan="2">remove contaminated clothing, flush skin with water or shower, call a doctor.</td>
</tr>
<tr>
<td colspan="2">**Eyes:** redness, pain.</td>
<td>close-fitting safety glasses.</td>
<td colspan="2">flush with water, take to a doctor if necessary.</td>
</tr>
<tr>
<td colspan="2">**Ingestion:** cough, abdominal pain, diarrhea, dizziness, unconsciousness.</td>
<td></td>
<td colspan="2">rinse mouth, call a doctor or take to hospital.</td>
</tr>
<tr>
<td colspan="2">SPILLAGE</td>
<td>STORAGE</td>
<td colspan="2">LABELING / NFPA</td>
</tr>
<tr>
<td colspan="2">collect leakage in sealable containers, soak up with sand or other inert absorbent and remove to safe place; (additional individual protection: breathing apparatus).</td>
<td>ventilate at floor level.</td>
<td colspan="2">R: 40
S: 23-36/37

✖
Harmful

    2
2   0</td>
</tr>
<tr>
<td colspan="5">NOTES</td>
</tr>
<tr>
<td colspan="5">Can form flammable air-vapor mixtures under certain circumstances (7.9-90% by vol). Explosion can be initiated only by strong energy source (e.g. welding torch). Addition of small quantities of combustible substances or raising of oxygen content increases flammability considerably. Odor limit is above TLV. Alcohol consumption increases toxic effects. Depending on the degree of exposure, regular medical checkups are advisable. Technical grades can contain small quantities of *epichlorohydrin*. Under no circumstances use near flame or hot surface or when welding.</td>
</tr>
</table>

Transport Emergency Card TEC(R)-723

HI: 60; UN-number: 1710

CAS-No: [87-90-1]
trichlorocyanuric acid
1,3,5-trichloroisocyanuric acid
1,3,5-trichloro-1,3,5-triazinetrione

TRICHLOROISOCYANURIC ACID

PHYSICAL PROPERTIES	IMPORTANT CHARACTERISTICS
Decomposes below melting point, °C 225 Solubility in water, g/100 ml at 25 °C 1.2 Relative molecular mass 232.5	**WHITE HYGROSCOPIC CRYSTALS WITH CHARACTERISTIC ODOR** Decomposes when heated above 200°C, giving off toxic gases (*chlorine* and → *carbon monoxide*). HCIO on contact with water. Reacts with anhydrous ammonia, amines, urea and other nitrogenous substances, forming explosive → *nitrogen trichloride*.

TLV-TWA	not available

Absorption route: Can enter the body by inhalation or ingestion. Evaporation negligible at 20°C, but harmful concentrations of airborne particles can build up rapidly.
Immediate effects: Irritates the eyes and respiratory tract. Can cause damage to gastro-intestinal tract if ingested. In substantial concentrations can cause liver and kidney damage. In serious cases risk of death.

Gross formula: $C_3Cl_3N_3O_3$

HAZARDS/SYMPTOMS	PREVENTIVE MEASURES	FIRE EXTINGUISHING/FIRST AID
Fire: non-combustible, but makes other substances more combustible: many chemical reactions can cause fire and explosion.	avoid contact with combustible substances and substances containing nitrogen.	in case of fire in immediate vicinity: use any extinguishing agent.
Explosion: reaction with nitrogen compounds can cause explosion.		
	KEEP DUST UNDER CONTROL	
Inhalation: sore throat, cough, nausea.	local exhaust or respiratory protection.	fresh air, rest.
Skin: redness.	gloves.	remove contaminated clothing, flush skin with water or shower.
Eyes: redness, pain, impaired vision.	goggles.	flush with water, take to a doctor.
Ingestion: abdominal pain, abdominal cramps, nausea, feeling of weakness.	do not eat, drink or smoke while working.	rinse mouth, call a doctor or take to hospital.

SPILLAGE	STORAGE	LABELING / NFPA
clean up spillage, flush away remainder with water; (additional individual protection: P2 respirator).	keep dry, separate from combustible substances, reducing agents and nitrogen compounds.	R: 8-22-31-36/37 S: 8-26-41 🔥 Oxidant ✖ Harmful NFPA: 3 / 0 / 2 oxy

NOTES
Flush contaminated clothing with water (fire hazard).

CAS-No: [75-69-4]
fluorocarbon 11
halocarbon 11
propellant 11
R 11

CFCl$_3$

TRICHLOROMONOFLUOROMETHANE

PHYSICAL PROPERTIES	IMPORTANT CHARACTERISTICS	
Boiling point, °C 24 Melting point, °C − 111 Relative density (water = 1) 1.5 Relative vapor density (air = 1) 4.8 Relative density at 20 °C of saturated mixture vapor/air (air = 1) 4.4 Vapor pressure, mm Hg at 20 °C 684 Solubility in water, g/100 ml at 20 °C 0.1 Relative molecular mass 137.4	**COLORLESS, ALMOST ODORLESS HIGHLY VOLATILE LIQUID** Vapor is heavier than air and can accumulate close to ground level, causing oxygen deficiency, with risk of unconsciousness. Decomposes in flame or on hot surface, giving off toxic gas. Reacts violently with alkali metals, metal powders and alkali amides. Attacks magnesium and zinc.	
	TLV-TWA	1000 ppm	5600 mg/m³ C
	Absorption route: Can enter the body by inhalation. Harmful atmospheric concentrations can build up fairly rapidly on evaporation at approx. 20°C - even more rapidly in aerosol form. **Immediate effects:** Affects the heart. Liquid can cause frostbite due to rapid evaporation.	
Gross formula: CCl$_3$F		

HAZARDS/SYMPTOMS	PREVENTIVE MEASURES	FIRE EXTINGUISHING/FIRST AID
Fire: non-combustible.		in case of fire in immediate vicinity: use any extinguishing agent.
Explosion:		in case of fire: keep cylinders or tanks/drums cool by spraying with water.
Inhalation:	ventilation, local exhaust or respiratory protection.	fresh air, rest, call a doctor if necessary.
Skin: *in case of frostbite:* redness, pain, sores.	gloves.	*in case of frostbite:* DO NOT remove clothing, flush skin with water or shower, refer to a doctor.
Eyes: *in case of frostbite:* redness, pain, impaired vision.	safety glasses.	*in case of frostbite:* flush with water, take to a doctor.

SPILLAGE	STORAGE	LABELING / NFPA
ventilation, collect leakage in sealable containers, soak up with sand or other inert absorbent and remove to safe place; (additional individual protection: breathing apparatus).	mostly in cylinders; if stored indoors, keep in cool, fireproof place.	

NOTES
High atmospheric concentrations, e.g. in poorly ventilated spaces, can cause oxygen deficiency, with risk of unconsciousness. Arcton 11, Freon 11 and Frigen 11 are trade names. TLV is maximum and must not be exceeded. Under no circumstances use near flame or hot surface or when welding. Turn leaking cylinder so that leak is on top to prevent liquid trichloromonofluoromethane escaping.

phenachlor
1,3,5-trichloro-2-hydroxybenzene

$C_6H_2Cl_3OH$

2,4,6-TRICHLOROPHENOL

PHYSICAL PROPERTIES		IMPORTANT CHARACTERISTICS
Boiling point, °C	245	**LIGHT YELLOW CRYSTALLINE POWDER OR FLAKES WITH CHARACTERISTIC ODOR**
Melting point, °C	68	Vapor mixes readily with air. Decomposes when heated, giving off toxic vapors (incl. → *hydrochloric acid*). Reacts with strong oxidants.
Flash point, °C	?	
Auto-ignition temperature, °C	?	
Relative density (water = 1)	15	TLV-TWA not available
Relative vapor density (air = 1)	6.9	
Relative density at 20 °C of saturated mixture vapor/air (air = 1)	1.0	**Absorption route:** Can enter the body by inhalation or ingestion. Harmful atmospheric concentrations build up fairly slowly on evaporation at 20°C, but much more rapidly in aerosol form.
Vapor pressure, mm Hg at 20°C	0.027	**Immediate effects:** Irritates the eyes, skin and respiratory tract. Affects the nervous system. Affects metabolism.
Solubility in water	none	**Effects of prolonged/repeated exposure:** Prolonged or repeated contact can cause skin disorders. Can cause liver and kidney damage.
Relative molecular mass	197.5	
Log P octanol/water	3.4	
Gross formula: $C_6H_3Cl_3O$		

HAZARDS/SYMPTOMS	PREVENTIVE MEASURES	FIRE EXTINGUISHING/FIRST AID
Fire: non-combustible.		in case of fire in immediate vicinity: use any extinguishing agent.
	KEEP DUST UNDER CONTROL	
Inhalation: sore throat, cough, tiredness, sweating, thirst, muscle cramps, rise in temperature, feeling of weakness.	local exhaust or respiratory protection.	fresh air, rest, help cool down, take to hospital.
Skin: redness, pain.	gloves, protective clothing.	remove contaminated clothing, wash skin with soap and water, refer to a doctor.
Eyes: redness, pain.	goggles, or combined eye and respiratory protection.	flush with water, take to a doctor if necessary.
Ingestion: abdominal pain, diarrhea, vomiting; see also 'Inhalation'.		rinse mouth, immediately take to hospital.

SPILLAGE	STORAGE	LABELING / NFPA
clean up spillage, carefully collect remainder; (additional individual protection: P2 respirator).	keep separate from oxidants.	R: 22-36/38 S: 26-28 ☒ Harmful

NOTES
Moderately soluble in hot water. The measures on this card also apply to 2,3,5-trichlorophenol. Under certain circumstances trichlorophenols - esp. alkaline salts - can change into highly toxic tetrachlorodibenzodioxins.

Transport Emergency Card TEC(R)-61G01

HI: 60; UN-number: 2020

2,4,5-TRICHLOROPHENOXY-ACETIC ACID

PHYSICAL PROPERTIES		IMPORTANT CHARACTERISTICS
Melting point, °C	153	**WHITE CRYSTALS**
Relative density (water = 1)	1.8	Decomposes when heated, giving off toxic gases.
Solubility in water	none	
Relative molecular mass	255.5	

TLV-TWA 10 mg/m³

Absorption route: Can enter the body by inhalation or ingestion. Evaporation negligible at 20°C, but harmful concentrations of airborne particles can build up rapidly.
Immediate effects: Irritates the eyes and respiratory tract. Affects the nervous system. Affects carbohydrate metabolism.
Effects of prolonged/repeated exposure: Can cause liver and kidney damage.

Gross formula: $C_8H_5Cl_3O_3$

HAZARDS/SYMPTOMS	PREVENTIVE MEASURES	FIRE EXTINGUISHING/FIRST AID
Fire: non-combustible.		in case of fire in immediate vicinity: use any extinguishing agent.
	KEEP DUST UNDER CONTROL, AVOID EXPOSURE OF PREGNANT WOMEN AS FAR AS POSSIBLE	
Inhalation: sore throat, cough, drowsiness, muscle weakness, cramps.	local exhaust or respiratory protection.	fresh air, rest, take to hospital.
Skin: redness.	gloves.	remove contaminated clothing, wash skin with soap and water.
Eyes: redness, pain.	safety glasses, or combined eye and respiratory protection.	flush with water, take to a doctor if necessary.
Ingestion: abdominal pain, muscle weakness, cramps, drowsiness.		rinse mouth, immediately take to hospital.

SPILLAGE	STORAGE	LABELING / NFPA
clean up spillage, carefully collect remainder; (additional individual protection: P2 respirator).		R: 20/21/22-40 S: 2-13 ✖ Harmful

NOTES
Depending on the degree of exposure, regular medical checkups are advisable. 2,4,5-T is a pesticide.

Transport Emergency Card TEC(R)-61G55

TRICHLOROSILANE

PHYSICAL PROPERTIES	IMPORTANT CHARACTERISTICS
Boiling point, °C 32 Melting point, °C − 127 Flash point, °C − 25 Auto-ignition temperature, °C 185 Relative density (water = 1) 1.3 Relative vapor density (air = 1) 4.7 Relative density at 20 °C of saturated mixture vapor/air (air = 1) 3.3 Vapor pressure, mm Hg at 20 °C 505 Solubility in water reaction Explosive limits, vol% in air 1.2-90.5 Relative molecular mass 135.5 Gross formula: Cl₃HSi	**COLORLESS, FUMING, HIGHLY VOLATILE LIQUID WITH PUNGENT ODOR** Vapor is heavier than air and spreads at ground level, with risk of ignition at a distance. Do not use compressed air when filling, emptying or processing. Toxic and corrosive vapors are a product of combustion (incl. → *hydrochloric acid*). Can deflagrate (burn explosively) when heated. Reacts violently with acids, oxidants, and many other substances. Reacts violently with water, steam and moist air, giving off flammable (→ *hydrogen*) and corrosive (→ *hydrochloric acid*) gases. Attacks light metals. TLV-TWA not available **Absorption route:** Can enter the body by inhalation or ingestion. Harmful atmospheric concentrations can build up very rapidly on evaporation at 20°C. **Immediate effects:** Corrosive to the eyes, skin and respiratory tract. Inhalation of vapor/fumes can cause severe breathing difficulties (lung edema). In serious cases risk of death. **Effects of prolonged/repeated exposure:** Prolonged or repeated contact can cause skin disorders.

HAZARDS/SYMPTOMS	PREVENTIVE MEASURES	FIRE EXTINGUISHING/FIRST AID
Fire: extremely flammable, many chemical reactions can cause fire and explosion.	keep away from open flame and sparks, no smoking, avoid contact with hot surfaces (e.g. steam lines).	powder, dry sand, DO NOT USE WATER-BASED EXTINGUISHERS.
Explosion: forms explosive air-vapor mixtures, many chemical reactions can cause explosion.	sealed machinery, ventilation, explosion-proof electrical equipment and lighting, grounding.	in case of fire: keep tanks/drums cool by spraying with water, but DO NOT spray substance with water.
	STRICT HYGIENE	
Inhalation: *corrosive*; sore throat, cough, shortness of breath, severe breathing difficulties.	ventilation, local exhaust or respiratory protection.	fresh air, rest, place in half-sitting position, take to hospital.
Skin: *corrosive*; redness, pain, serious burns.	gloves, protective clothing.	remove contaminated clothing, flush skin with water or shower, refer to a doctor.
Eyes: *corrosive*; redness, pain, impaired vision.	face shield, or combined eye and respiratory protection.	flush with water, take to a doctor.
Ingestion: *corrosive*; sore throat, abdominal pain, diarrhea.		rinse mouth, take immediately to hospital.

SPILLAGE	STORAGE	LABELING / NFPA
evacuate area, call in an expert, collect leakage in sealable containers, sprinkle area of spillage thoroughly with soda and flush away with water after waiting 10 minutes; (additional individual protection: CHEMICAL SUIT).	keep in cool, dry, fireproof place, separate from all other substances.	R: 15-17 S: 24/25-43 🔥 Flammable NFPA: 4 / 3 / 2 / W

NOTES
Reacts violently with extinguishing agents such as water and foam. Lung edema symptoms usually develop several hours later and are aggravated by physical exertion: rest and hospitalization essential. As first aid, a doctor or authorized person should consider administering a corticosteroid spray. Use airtight packaging. Unbreakable packaging preferred; if breakable, keep in unbreakable container.

Transport Emergency Card TEC(R)-611

HI: X338; UN-number: 1295

TRIDECANOL
(mixtures of isomers)

PHYSICAL PROPERTIES	IMPORTANT CHARACTERISTICS
Boiling point, °C 250-270 Melting point, °C 33 Flash point, °C 82-115 Auto-ignition temperature, °C 260 Relative density (water = 1) 0.8 Relative vapor density (air = 1) 6.9 Relative density at 20 °C of saturated mixture vapor/air (air = 1) 1.0 Vapor pressure, mm Hg at 20 °C ca. 76 Solubility in water none Relative molecular mass 200.4	**WHITE SOLID IN VARIOUS FORMS** Vapor mixes readily with air. Reacts with strong oxidants. TLV-TWA not available **Absorption route:** Can enter the body by inhalation or ingestion or through the skin. **Immediate effects:** Irritates the eyes, skin and respiratory tract.
Gross formula: $C_{13}H_{28}O$	

HAZARDS/SYMPTOMS	PREVENTIVE MEASURES	FIRE EXTINGUISHING/FIRST AID
Fire: combustible.	keep away from open flame, no smoking.	powder, water spray, foam, carbon dioxide, (halons).
Explosion: above 82°C: forms explosive air-vapor mixtures.	above 82°C: sealed machinery, ventilation.	
Inhalation: sore throat, cough.	local exhaust or respiratory protection.	fresh air, rest, call a doctor if necessary.
Skin: redness.	gloves.	remove contaminated clothing, flush skin with water or shower.
Eyes: redness, pain.	goggles.	flush with water, take to a doctor if necessary.
Ingestion: abdominal pain, drowsiness, unconsciousness.		rinse mouth, call a doctor.

SPILLAGE	STORAGE	LABELING / NFPA
allow spillage to set, clean up spillage, carefully collect remainder; (additional individual protection: P1 respirator).		

NOTES
Depending on the isomer ratio the substance can exist as a liquid. Insufficient toxicological data on harmful effects to humans.

CAS-No: [102-71-6]
2,2',2''-nitrilo-triethanol
daltogen
TEA
tri(2-hydroxyethyl)amine

$(C_2H_4OH)_3N$

TRIETHANOLAMINE

PHYSICAL PROPERTIES		IMPORTANT CHARACTERISTICS
Boiling point, °C	360	**COLORLESS VISCOUS LIQUID OR CRYSTALS, TURNING BROWN ON EXPOSURE TO AIR**
Melting point, °C	18	Toxic vapors are a product of combustion (→ *nitrous vapors*). Decomposes when heated, giving
Flash point, °C	179	off toxic vapors (→ *nitrous vapors*). Reacts violently with strong oxidants. In aqueous solution is a
Auto-ignition temperature, °C	324	medium strong base.
Relative density (water = 1)	1.1	
Relative vapor density (air = 1)	5.1	
Relative density at 20 °C of saturated mixture		TLV-TWA not available
vapor/air (air = 1)	1.0	
Vapor pressure, mm Hg at 20 °C	0.020	**Absorption route:** Can enter the body by inhalation or ingestion or through the skin.
Solubility in water	∞	**Immediate effects:** Irritates the eyes, skin and respiratory tract. Inhalation of vapor/fumes can
Explosive limits, vol% in air	1.2-?	cause severe breathing difficulties (lung edema).
Relative molecular mass	149.2	**Effects of prolonged/repeated exposure:** Prolonged or repeated contact can cause eczema.
Log P octanol/water	− 1.5	
Gross formula:	$C_6H_{15}NO_3$	

HAZARDS/SYMPTOMS	PREVENTIVE MEASURES	FIRE EXTINGUISHING/FIRST AID
Fire: combustible.	keep away from open flame, no smoking.	powder, alcohol-resistant foam, water spray, carbon dioxide, (halons).
Inhalation: sore throat, cough, diarrhea.	ventilation.	fresh air, rest, take to hospital.
Skin: *is absorbed*; redness, pain.	gloves.	remove contaminated clothing, flush skin with water or shower, refer to a doctor.
Eyes: redness, pain.	face shield.	flush with water, take to a doctor.
Ingestion: sore throat, abdominal pain, diarrhea, stomachache, nausea, vomiting.		rinse mouth, call a doctor or take to hospital.

SPILLAGE	STORAGE	LABELING / NFPA
collect leakage in sealable containers, flush away remainder with water.	keep separate from oxidants; ventilate at floor level.	1 / 2 / 1

NOTES
Log P octanol/water is estimated. Lung edema symptoms usually develop several hours later and are aggravated by physical exertion: rest and hospitalization essential. As first aid, a doctor or authorized person should consider administering a corticosteroid spray.

TRIETHYLALUMINUM

PHYSICAL PROPERTIES	IMPORTANT CHARACTERISTICS	
Boiling point, °C 187 Melting point, °C −46 Flash point, °C −52 Relative density (water = 1) 0.83 Relative vapor density (air = 1) 3.9 Vapor pressure, mm Hg at 20 °C <0.76 Solubility in water reaction Relative molecular mass 114.1	**COLORLESS FUMING LIQUID** Ignites spontaneously on exposure to air. Aluminum oxide vapors are a product of combustion. Strong reducing agent which reacts violently with oxidants. Reacts violently with water, alcohols, phenols, amines, carbon dioxide, oxides of sulfur, nitrogen oxides, halogenated hydrocarbons and many other substances, with risk of fire and explosion. Reacts with air to form irritating vapors.	
	TLV-TWA	2 mg/m³
Gross formula: $C_6H_{15}Al$	**Absorption route:** Can enter the body by inhalation or ingestion. Harmful atmospheric concentrations can build up fairly rapidly on evaporation at approx. 20°C - even more rapidly in aerosol form. **Immediate effects:** Corrosive to the eyes, skin and respiratory tract. Inhalation of vapor/fumes can cause severe breathing difficulties (lung edema). In serious cases risk of death.	

HAZARDS/SYMPTOMS	PREVENTIVE MEASURES	FIRE EXTINGUISHING/FIRST AID
Fire: extremely flammable, many chemical reactions can cause fire and explosion.	keep away from open flame and sparks, no smoking, avoid contact with air, water and many other substances.	special extinguishing powder, dry sand, NO other extinguishing agents, can reignite after being extinguished.
Explosion: contact with water and many other substances can cause explosion.		in case of fire: keep undamaged cylinders cool by spraying with water, but DO NOT spray substance with water.
	STRICT HYGIENE	
Inhalation: *corrosive*; sore throat, cough, severe breathing difficulties.	completely sealed machinery, ventilation, local exhaust or respiratory protection.	fresh air, rest, place in half-sitting position, take to hospital.
Skin: *corrosive*; serious burns, slow healing.	gloves, protective clothing.	remove contaminated clothing, flush skin with water or shower, refer to a doctor.
Eyes: *corrosive*; redness, pain, impaired vision.	face shield, or combined eye and respiratory protection.	flush with water, take to a doctor.
Ingestion: *corrosive*; sore throat, abdominal pain, diarrhea.		rinse mouth, take immediately to hospital.

SPILLAGE	STORAGE	LABELING / NFPA
evacuate area, call in an expert, dilute leakage immediately with diesel/machine oil, soak up/cover with dry sand/table salt and remove to safe place; (additional individual protection: CHEMICAL SUIT).	keep in dry, fireproof place under dry nitrogen, separate from all other substances.	R: 14-17-34 S: 16-43 Note A ◊ Flammable ▨ Corrosive

NFPA: 4 (top), 3 (left), 3 (right), W (bottom)

NOTES
Usually sold as 15-30% solution in hydrocarbons, which does not ignite spontaneously on exposure to air. For 15% solution see relevant entry. Reacts violently with extinguishing agents such as water, foam, halons and ordinary powder. Lung edema symptoms usually develop several hours later and are aggravated by physical exertion: rest and hospitalization essential. As first aid, a doctor or authorized person should consider administering a corticosteroid spray. Use cylinder with special fittings.

Transport Emergency Card TEC(R)-42G02

HI: X333; UN-number: 1102

TRIETHYLALUMINUM

(15% solution in hexane)

PHYSICAL PROPERTIES		IMPORTANT CHARACTERISTICS
Boiling point, °C	69	**COLORLESS FUMING LIQUID**
Melting point, °C	−95	Vapor is heavier than air and spreads at ground level, with risk of ignition at a distance. Flow,
Flash point, °C	−26	agitation etc. can cause build-up of electrostatic charge due to liquid's low conductivity. Do not
Auto-ignition temperature, °C	240	use compressed air when filling, emptying or processing. Once solvent has evaporated (e.g. on
Relative density (water = 1)	0.8	clothing) can ignite spontaneously on exposure to air. Aluminum oxide vapors are a product of
Relative vapor density (air = 1)	3.0	combustion. Strong reducing agent which reacts violently with
Relative density at 20 °C of saturated mixture		water, alcohols, phenols, amines, halogenated hydrocarbons, carbon dioxide, nitrogen oxide,
vapor/air (air = 1)	1.3	sulfur oxide and many other substances, with risk of fire and explosion.
Vapor pressure, mm Hg at 20 °C	122	

Solubility in water	reaction	TLV-TWA
Explosive limits, vol% in air	1.1-7.5	
Minimum ignition energy, mJ	0.24	
Electrical conductivity, pS/m	100	

TLV-TWA   2 mg/m³ ✶

Absorption route: Can enter the body by inhalation or ingestion. Harmful atmospheric concentrations can build up fairly rapidly on evaporation at approx. 20°C - even more rapidly in aerosol form.
Immediate effects: Corrosive to the eyes, skin and respiratory tract. Inhalation of vapor/fumes can cause severe breathing difficulties (lung edema). In serious cases risk of death. Can cause metal fume fever.

Gross formula: $C_6H_{15}Al$

HAZARDS/SYMPTOMS	PREVENTIVE MEASURES	FIRE EXTINGUISHING/FIRST AID
Fire: extremely flammable, many chemical reactions can cause fire and explosion.	keep away from open flame and sparks, no smoking, avoid contact with water, alcohols, phenols and many other substances.	special extinguishing powder, dry sand, NO other extinguishing agents, can reignite after being extinguished.
Explosion: forms explosive air-vapor mixtures; contact with water, alcohols, phenols and many other substances can cause explosion.	sealed machinery, ventilation, explosion-proof electrical equipment and lighting, grounding, non-sparking hand tools.	in case of fire: keep tanks/drums cool by spraying with water, but DO NOT spray substance with water, fight fire from protected location.
	STRICT HYGIENE	
Inhalation: *corrosive*; sore throat, cough, severe breathing difficulties, metal fume fever.	sealed machinery, ventilation, local exhaust or respiratory protection.	fresh air, rest, place in half-sitting position, take to hospital.
Skin: *corrosive*; serious burns, slow healing.	gloves, protective clothing.	remove contaminated clothing, flush skin with water or shower, refer to a doctor.
Eyes: *corrosive*; redness, pain, impaired vision.	face shield, or combined eye and respiratory protection.	flush with water, take to a doctor.
Ingestion: *corrosive*; sore throat, abdominal pain, diarrhea, mouth pain.		rinse mouth, immediately take to hospital.

SPILLAGE	STORAGE	LABELING / NFPA
evacuate area, call in an expert, collect leakage in sealable containers, soak up remainder with dry sand/table salt and remove to safe place, under no circumstances spray liquid with water; (additional individual protection: CHEMICAL SUIT).	keep in dry, fireproof place under inert gas, separate from all other substances.	R: 14-17-34 S: 16-43 Note A Flammable Corrosive

NOTES

Physical properties (boiling point, flash point, autoignition point, explosive limits, vapor pressure, relative density, conductivity and ignition energy) are based on hexane. For pure substance: → *triethylaluminium*. Reacts violently with extinguishing agents such as water, foam, halons, carbon dioxide and ordinary powder. Remove contaminated clothing immediately and flush with water (risk of spontaneous combustion). Lung edema symptoms usually develop several hours later and are aggravated by physical exertion: rest and hospitalization essential. As first aid, a doctor or authorized person should consider administering a corticosteroid spray. ✶ For the vapor usually the TLV of → *hexane* can be used.

TRIETHYLAMINE

PHYSICAL PROPERTIES		IMPORTANT CHARACTERISTICS
Boiling point, °C	89	**COLORLESS LIQUID WITH CHARACTERISTIC ODOR**
Melting point, °C	−115	Vapor is heavier than air and spreads at ground level, with risk of ignition at a distance. Do not use
Flash point, °C	−17	compressed air when filling, emptying or processing. Toxic vapors are a product of combustion
Auto-ignition temperature, °C	230	(→ *nitrous vapors* and → *carbon monoxide*). Decomposes when heated, giving off flammable and
Relative density (water = 1)	0.7	toxic gases. Strong base which reacts violently with acids and corrodes aluminum, zinc etc.
Relative vapor density (air = 1)	3.5	Reacts violently with strong oxidants, nitroparaffins and nitrogen tetroxide, with risk of fire and
Relative density at 20 °C of saturated mixture		explosion.
vapor/air (air = 1)	1.2	
Vapor pressure, mm Hg at 20 °C	53	TLV-TWA (skin) 10 ppm 41 mg/m³
Solubility in water, g/100 ml at 20 °C	17	STEL 15 ppm 62 mg/m³
Explosive limits, vol% in air	1.2-8.0	
Minimum ignition energy, mJ	0.75	**Absorption route:** Can enter the body by inhalation or ingestion or through the skin. Harmful
Relative molecular mass	101.2	atmospheric concentrations can build up very rapidly on evaporation at 20° C.
Log P octanol/water	1.4	**Immediate effects:** Corrosive to the eyes, skin and respiratory tract. Local effect is more
		significant than toxic effects. Inhalation can cause lung edema. In serious cases risk of death.
		Can affect the central nervous system.
Gross formula:	$C_6H_{15}N$	

HAZARDS/SYMPTOMS	PREVENTIVE MEASURES	FIRE EXTINGUISHING/FIRST AID
Fire: extremely flammable.	keep away from open flame and sparks, no smoking, avoid contact with strong oxidants.	powder, AFFF, foam, carbon dioxide, (halons).
Explosion: forms explosive air-vapor mixtures.	sealed machinery, ventilation, explosion-proof electrical equipment and lighting.	in case of fire: keep tanks/drums cool by spraying with water.
	STRICT HYGIENE	
Inhalation: *corrosive*; sore throat, cough, shortness of breath, severe breathing difficulties, headache.	ventilation, local exhaust or respiratory protection.	fresh air, rest, place in half-sitting position, take to hospital.
Skin: *is absorbed, corrosive*; redness, pain, serious burns.	gloves, protective clothing.	remove contaminated clothing, flush skin with water or shower, call a doctor.
Eyes: *corrosive*; redness, pain, impaired vision.	face shield, or combined eye and respiratory protection.	flush with water, take to a doctor.
Ingestion: *corrosive*; sore throat, abdominal pain, stomachache.		rinse mouth, take to hospital.

SPILLAGE	STORAGE	LABELING / NFPA
evacuate area, call in an expert, collect leakage in sealable containers, soak up with sand or other inert absorbent and remove to safe place; (additional individual protection: breathing apparatus).	keep in fireproof place, separate from oxidants and strong acids.	R: 11-36/37 S: 16-26-29 Flammable Irritant 3 2 0

NOTES

Solubility decreases with increasing temperature. Lung edema symptoms usually develop several hours later and are aggravated by physical exertion: rest and hospitalization essential. As first aid, a doctor or authorized person should consider administering a corticosteroid spray. Unbreakable packaging preferred; if breakable, keep in unbreakable container.

Transport Emergency Card TEC(R)-658

HI: 338; UN-number: 1296

TRIETHYLBORINE

PHYSICAL PROPERTIES	IMPORTANT CHARACTERISTICS
Boiling point, °C — 95 Melting point, °C — −93 Flash point, °C — < −93 Auto-ignition temperature, °C — <0 Relative density (water = 1) — 0.7 Relative vapor density (air = 1) — 3.4 Relative density at 20 °C of saturated mixture vapor/air (air = 1) — 1.12 Vapor pressure, mm Hg at 20 °C — 38 Solubility in water — none Explosive limits, vol% in air — ?-? Relative molecular mass — 98.0	**COLORLESS LIQUID** Vapor is heavier than air and spreads at ground level, with risk of spontaneous combustion. Do not use compressed air when filling, emptying or processing. Can ignite spontaneously on exposure to air. Decomposes when heated or burned, giving off toxic gases. Strong reducing agent which reacts violently with oxidants and many other substances. Reacts slowly with alcohols and carbon acids.
	TLV-TWA not available
Gross formula: $C_6H_{15}B$	**Absorption route:** Can enter the body by inhalation or ingestion. Insufficient data on the rate at which harmful concentrations can build up. Harmful concentrations likely to build up very rapidly on evaporation. **Immediate effects:** Corrosive to the eyes, skin and respiratory tract. Inhalation of vapor/fumes can cause severe breathing difficulties (lung edema).

HAZARDS/SYMPTOMS	PREVENTIVE MEASURES	FIRE EXTINGUISHING/FIRST AID
Fire: extremely flammable, combusts spontaneously on contact with air.	keep away from open flame and sparks, no smoking; avoid contact with air, oxidants and hot surfaces (e.g. steam lines).	powder, carbon dioxide or foam; DO NOT USE HALONS.
Explosion: forms explosive air-vapor mixtures, many chemical reactions can cause explosion.	completely sealed machinery.	in case of fire: keep tanks/drums cool by spraying with water, fight fire from sheltered location.
Inhalation: *corrosive*; sore throat, cough, severe breathing difficulties.	ventilation, local exhaust or respiratory protection.	fresh air, rest, place in half-sitting position, take to hospital.
Skin: *corrosive*; redness, pain, serious burns.	gloves, protective clothing.	remove contaminated clothing, flush skin with water or shower, take to hospital.
Eyes: *corrosive*; redness, pain, impaired vision.	face shield, or combined eye and respiratory protection.	flush with water, take to an eye doctor.
Ingestion: *corrosive*; sore throat, abdominal pain, diarrhea; soreness of lips, mouth, esophagus and stomach.		rinse mouth, take immediately to hospital.

SPILLAGE	STORAGE	LABELING / NFPA
evacuate area, call in an expert, soak up in ten-fold quantity of diesel oil and remove to safe place, do not use sawdust or other combustible absorbents, do not flush into sewer; (additional individual protection: breathing apparatus).	keep in fireproof place under inert gas, separate from all other substances.	(NFPA diamond: blue 1, red 3, yellow 3, white W̶)

NOTES
Explosive limits not given in the literature. Flush contaminated clothing with water (fire hazard); treat with special extinguishing powder and do not reuse. Reacts violently with extinguishing agents such as halons. Lung edema symptoms usually develop several hours later and are aggravated by physical exertion: rest and hospitalization essential. As first aid, a doctor or authorized person should consider administering a corticosteroid spray. Use original packaging; keep airtight.

CAS-No: [112-27-6]
1,2-bis(2-hydroxyethoxy)ethane
3,6-dioxaoctane-1,8-diol
2,2'-ethylenedioxydiethanol
TEG

$(CH_2OHCH_2OCH_2)_2$

TRIETHYLENE GLYCOL

PHYSICAL PROPERTIES		IMPORTANT CHARACTERISTICS	
Boiling point, °C	287	**COLORLESS, VISCOUS HYGROSCOPIC LIQUID**	
Melting point, °C	−5	Vapor mixes readily with air. Reacts with strong oxidants.	
Flash point, °C	170		
Auto-ignition temperature, °C	323	TLV-TWA	not available
Relative density (water = 1)	1.1		
Relative vapor density (air = 1)	5.2		
Relative density at 20 °C of saturated mixture		**Absorption route:** Can enter the body by inhalation or ingestion. Harmful atmospheric concentrations build up very slowly, if at all, on evaporation at approx. 20°C, but much more rapidly in aerosol form.	
vapor/air (air = 1)	1.00		
Vapor pressure, mm Hg at 20 °C	0.099		
Solubility in water	∞		
Explosive limits, vol% in air	0.9-9.2		
Electrical conductivity, pS/m	8.4×10^6		
Relative molecular mass	150.2		
Log P octanol/water	−1.7		
Gross formula:	$C_6H_{14}O_4$		

HAZARDS/SYMPTOMS	PREVENTIVE MEASURES	FIRE EXTINGUISHING/FIRST AID
Fire: combustible.	keep away from open flame, no smoking.	powder, alcohol-resistant foam, water spray, carbon dioxide, (halons).
Inhalation:	ventilation.	
Skin:	gloves.	flush skin with water or shower.
Eyes:	safety glasses.	flush with water, take to a doctor if necessary.
Ingestion: headache, nausea.		call a doctor if necessary.

SPILLAGE	STORAGE	LABELING / NFPA
collect leakage in sealable containers, flush away remainder with water.		 1 / 1 / 0

NOTES
Log P octanol/water is estimated.

CAS-No: [112-50-5]
ethyltriglycol
ethoxytriethylene glycol
triethylene glycol ethyl ether
triglycol monoethyl ether
3,6,9-trioxaundecan-1-ol

$C_2H_5(OCH_2CH_2)_3OH$

TRIETHYLENE GLYCOL MONOETHYL ETHER

PHYSICAL PROPERTIES		IMPORTANT CHARACTERISTICS	
Boiling point, °C	255	**COLORLESS, HYGROSCOPIC VISCOUS LIQUID WITH CHARACTERISTIC ODOR**	
Melting point, °C	−19	Vapor mixes readily with air.	
Flash point, °C	110		
Relative density (water = 1)	1.02	TLV-TWA not available	
Relative vapor density (air = 1)	6.1		
Relative density at 20 °C of saturated mixture vapor/air (air = 1)	1.01	**Absorption route:** Can enter the body by inhalation or ingestion or through the skin. Harmful atmospheric concentrations build up very slowly, if at all, on evaporation at approx. 20°C, but much more rapidly in aerosol form.	
Vapor pressure, mm Hg at 20 °C	0.051		
Solubility in water	∞		
Relative molecular mass	178.2	**Immediate effects:** Irritates the eyes. Liquid destroys the skin's natural oils.	
Gross formula: $C_8H_{18}O_4$			

HAZARDS/SYMPTOMS	PREVENTIVE MEASURES	FIRE EXTINGUISHING/FIRST AID
Fire: combustible.	keep away from open flame, no smoking.	powder, alcohol-resistant foam, water spray, carbon dioxide, (halons).
Inhalation: sore throat, cough.	ventilation.	fresh air, rest.
Skin: *is absorbed*; redness, rawness.	gloves.	remove contaminated clothing, flush skin with water or shower.
Eyes: redness, pain, impaired vision.	safety glasses.	flush with water, take to an eye doctor.
Ingestion: headache, nausea.		call a doctor.

SPILLAGE	STORAGE	LABELING / NFPA
collect leakage in sealable containers, flush away remainder with water.	keep dry.	

NOTES
Trioxytol is a tradename.

CAS-No: [112-24-3]
N,N'-bis(2-aminoethyl)ethylenediamine
3,6-diazaoctane-1,8-diamine
TETA
TRIEN

$H_2NCH_2(CH_2NHCH_2)_2CH_2NH_2$

TRIETHYLENETETRAMINE

PHYSICAL PROPERTIES	IMPORTANT CHARACTERISTICS
Boiling point, °C 277 Melting point, °C −35 Flash point, °C 143 Auto-ignition temperature, °C 335 Relative density (water = 1) 1.0 Relative vapor density (air = 1) 5.0 Relative density at 20 °C of saturated mixture vapor/air (air = 1) 1.0 Vapor pressure, mm Hg at 20 °C < < 0.076 Solubility in water ∞ Explosive limits, vol% in air 0.7-7.2 Relative molecular mass 146.2 Log P octanol/water − 1.7 Gross formula: $C_6H_{18}N_4$	**YELLOW VISCOUS LIQUID WITH CHARACTERISTIC ODOR** Vapor mixes readily with air. Toxic and corrosive vapors are a product of combustion (→ *nitrous vapors*). Medium strong base. Reacts with chlorinated hydrocarbons. Reacts violently with strong oxidants. Promotes spontaneous decomposition of nitrogen compounds. Attacks copper and its alloys, nickel and cobalt. TLV-TWA not available **Absorption route:** Can enter the body by inhalation or ingestion or through the skin. Harmful atmospheric concentrations build up very slowly, if at all, on evaporation at approx. 20° C, but much more rapidly in aerosol form. **Immediate effects:** Corrosive to the eyes, skin and respiratory tract. Can cause liver damage. Inhalation of vapor/fumes can cause severe breathing difficulties (lung edema). **Effects of prolonged/repeated exposure:** Prolonged or repeated contact can cause skin disorders. Repeated contact can cause asthma. Local effect is more significant than toxic effect.

HAZARDS/SYMPTOMS	PREVENTIVE MEASURES	FIRE EXTINGUISHING/FIRST AID
Fire: combustible.	keep away from open flame, no smoking.	water spray, alcohol-resistant foam, powder, carbon dioxide, DO NOT USE HALONS.
	STRICT HYGIENE	
Inhalation: *corrosive*; sore throat, cough, shortness of breath, severe breathing difficulties, headache.	ventilation, local exhaust or respiratory protection.	fresh air, rest, place in half-sitting position, take to hospital.
Skin: *corrosive, is absorbed*; redness, pain, burns.	gloves, protective clothing.	remove contaminated clothing, flush skin with water or shower, refer to a doctor.
Eyes: *corrosive*; redness, pain, impaired vision.	face shield.	flush with water, take to an eye doctor.
Ingestion: *corrosive*; sore throat, abdominal pain, vomiting, mouth pain, stomachache.		rinse mouth, take to hospital.

SPILLAGE	STORAGE	LABELING / NFPA
collect leakage in sealable containers, flush away remainder with water.	keep separate from oxidants, strong acids and nitrogen compounds, ventilate at floor level.	R: 21-34-43 S: 26-36/37/39 [corrosive symbol] Corrosive

NOTES
The commercial product contains approx. 75% linear isomer, with approx. 5% branched isomer and approx. 20% cyclic tetraethylenetetramine. Physical properties given relate to the commercial product. Log P octanol/water is estimated. Lung edema symptoms usually develop several hours later and are aggravated by physical exertion: rest and hospitalization essential. As first aid, a doctor or authorized person should consider administering a corticosteroid spray. Due to its viscosity liquid becomes virtually solid at approx. 10° C. Unbreakable packaging preferred; if breakable, keep in unbreakable container.

Transport Emergency Card TEC(R)-659

HI: 80; UN-number: 2259

TRIFLUOROACETIC ACID

PHYSICAL PROPERTIES	IMPORTANT CHARACTERISTICS
Boiling point, °C 72 Melting point, °C −15 Relative density (water = 1) 1.5 Relative vapor density (air = 1) 3.9 Relative density at 20 °C of saturated mixture vapor/air (air = 1) 1.3 Vapor pressure, mm Hg at 20 °C 80 Solubility in water good Relative molecular mass 114	**COLORLESS HYGROSCOPIC LIQUID WITH PUNGENT ODOR** Vapor is heavier than air. Decomposes when heated and on contact with strong acids, giving off toxic vapors. Strong acid which reacts violently with bases and is corrosive. Attacks many metals.
	TLV-TWA not available
	Absorption route: Can enter the body by inhalation or ingestion or through the skin. Harmful atmospheric concentrations can build up very rapidly on evaporation at 20°C. **Immediate effects:** Corrosive to the eyes, skin and respiratory tract. Inhalation of vapor/fumes can cause severe breathing difficulties (lung edema). In serious cases risk of death.
Gross formula: C$_2$HF$_3$O$_2$	

HAZARDS/SYMPTOMS	PREVENTIVE MEASURES	FIRE EXTINGUISHING/FIRST AID
Fire: non-combustible.		in case of fire in immediate vicinity: use any extinguishing agent.
Explosion:		in case of fire: keep tanks/drums cool by spraying with water.
Inhalation: *corrosive*; sore throat, cough, severe breathing difficulties, chest pain.	ventilation, local exhaust or respiratory protection.	fresh air, rest, place in half-sitting position, artificial respiration if necessary, take to hospital.
Skin: *is absorbed, corrosive*; redness, pain, serious burns.	gloves, protective clothing.	remove contaminated clothing, flush skin with water or shower, take to hospital.
Eyes: *corrosive*; redness, pain, impaired vision.	face shield, or combined eye and respiratory protection.	flush with water, take to an eye doctor.
Ingestion: *corrosive*; sore throat, abdominal pain, diarrhea, pain with swallowing, stomachache, chest pain, blisters on lips and tongue.		take immediately to hospital.

SPILLAGE	STORAGE	LABELING / NFPA
collect leakage in sealable containers, flush away remainder with water, soak up with chalk or soda and remove to safe place; (additional individual protection: CHEMICAL SUIT).	keep separate from strong bases.	R: 20-35 S: 9-26-27-28 Note B ⬛ Corrosive

NOTES
Lung edema symptoms usually develop several hours later and are aggravated by physical exertion: rest and hospitalization essential. As first aid, a doctor or authorized person should consider administering a corticosteroid spray. Unbreakable packaging preferred; if breakable, keep in unbreakable container.

Transport Emergency Card TEC(R)-80G17   HI: 88; UN-number: 2699

TRIFLUOROMETHANE

(cylinder)

PHYSICAL PROPERTIES	IMPORTANT CHARACTERISTICS
Boiling point, °C −82 Melting point, °C −163 Relative density (water = 1) 1.44 Relative vapor density (air = 1) 2.4 Vapor pressure, mm Hg at 20°C 33,896 Solubility in water none Relative molecular mass 70.0	**COLORLESS COMPRESSED LIQUEFIED GAS** Vapor is heavier than air and can accumulate close to ground level, causing oxygen deficiency, with risk of unconsciousness. Decomposes when heated or on contact with metals, giving off corrosive and toxic vapors. Reacts violently with alkaline-earth and alkali metals. Attacks magnesium and its alloys.
	TLV-TWA not available
	Absorption route: Can enter the body by inhalation. **Immediate effects:** Liquid can cause frostbite due to rapid evaporation. In substantial concentrations can impair consciousness.
Gross formula: CHF$_3$	

HAZARDS/SYMPTOMS	PREVENTIVE MEASURES	FIRE EXTINGUISHING/FIRST AID
Fire: non-combustible.		in case of fire in immediate vicinity: use any extinguishing agent.
Explosion:		in case of fire: keep cylinders cool by spraying with water.
Inhalation: headache, dizziness, sleepiness, unconsciousness.	ventilation, local exhaust or respiratory protection.	fresh air, rest, call a doctor if necessary.
Skin: *in case of frostbite*: redness, pain, sores.	insulating gloves.	*in case of frostbite*: DO NOT remove clothing, flush skin with water or shower, call a doctor if necessary.
Eyes: *in case of frostbite*: redness, pain, impaired vision.	face shield, or combined eye and respiratory protection.	flush with water, take to a doctor.

SPILLAGE	STORAGE	LABELING / NFPA
evacuate area, call in an expert, ventilate; (additional individual protection: breathing apparatus).	if stored indoors, keep in cool, fireproof place.	

NOTES

High atmospheric concentrations, e.g. in poorly ventilated spaces, can cause oxygen deficiency, with risk of unconsciousness. Freon 23 and Frigen 23 are trade names. Trifluoromethane is used among others as propellant in spray cans and as an extinguishing agent. After use as extinguishing agent provide good ventilation. Under no circumstances use near flame or hot surface or when welding. Turn leaking cylinder so that leak is on top to prevent liquid trifluoromethane escaping.

Transport Emergency Card TEC(R)-644

HI: 20; UN-number: 1984

CAS-No: [100-99-2]
TIBAL

$(C_4H_9)_3Al$

TRIISOBUTYLALUMINUM

PHYSICAL PROPERTIES	IMPORTANT CHARACTERISTICS	
Boiling point (decomposes), °C 212 Melting point, °C 6 Relative density (water = 1) 0.78 Relative vapor density (air = 1) 6.3 Vapor pressure, mm Hg at 20 °C <0.76 Solubility in water reaction Relative molecular mass 198.3	**COLORLESS FUMING LIQUID** Ignites spontaneously on exposure to air. Aluminum oxide vapors are a product of combustion. Strong reducing agent which reacts violently with oxidants. Reacts violently with water, alcohols, phenols, amines, carbon dioxide, oxides of sulfur, nitrogen oxides, halogens, halogenated hydrocarbons and many other substances, with risk of fire and explosion. Reacts with air to form irritating vapors.	
	TLV-TWA	2 mg/m³
Gross formula: $C_{12}H_{27}Al$	**Absorption route:** Can enter the body by inhalation or ingestion. Harmful atmospheric concentrations can build up fairly rapidly on evaporation at approx. 20°C - even more rapidly in aerosol form. **Immediate effects:** Corrosive to the eyes, skin and respiratory tract. Inhalation of vapor/fumes can cause severe breathing difficulties (lung edema). In serious cases risk of death.	

HAZARDS/SYMPTOMS	PREVENTIVE MEASURES	FIRE EXTINGUISHING/FIRST AID
Fire: extremely flammable, many chemical reactions can cause fire and explosion.	keep away from open flame and sparks, no smoking, avoid contact with water, air and many other substances.	special extinguishing powder, dry sand, NO other extinguishing agents, can reignite after being extinguished.
Explosion: contact with water and many other substances can cause explosion.		in case of fire: keep undamaged cylinders cool by spraying with water, but DO NOT spray substance with water.
	STRICT HYGIENE	
Inhalation: *corrosive*; sore throat, cough, severe breathing difficulties.	completely sealed machinery, ventilation, local exhaust or respiratory protection.	fresh air, rest, place in half-sitting position, take to hospital.
Skin: *corrosive*; serious burns, slow healing.	gloves, protective clothing.	remove contaminated clothing, flush skin with water or shower, refer to a doctor.
Eyes: *corrosive*; redness, pain, impaired vision.	face shield, or combined eye and respiratory protection.	flush with water, take to a doctor.
Ingestion: *corrosive*; sore throat, abdominal pain, diarrhea.		rinse mouth, take immediately to hospital.

SPILLAGE	STORAGE	LABELING / NFPA
evacuate area, call in an expert, dilute leakage with diesel/machine oil, soak up/cover with dry sand/table salt and remove to safe place, under no circumstances spray liquid with water; (additional individual protection: CHEMICAL SUIT).	keep in dry, fireproof place under dry nitrogen, separate from all other substances.	R: 14-17-34 S: 16-43 Note A Flammable Corrosive 4/3/3 W

NOTES
Usually sold as 15-30% solution in hydrocarbons, which does not ignite spontaneously on exposure to air. For 15% solution see relevant entry. Reacts violently with extinguishing agents such as carbon dioxide, halons, water, foam, and ordinary powder. Lung edema symptoms usually develop several hours later and are aggravated by physical exertion: rest and hospitalization essential. As first aid, a doctor or authorized person should consider administering a corticosteroid spray. Use cylinder with special fittings.

Transport Emergency Card TEC(R)-42G02

HI: X333; UN-number: 1930

TRIISOBUTYLALUMINUM

(15% solution in hexane)

PHYSICAL PROPERTIES		IMPORTANT CHARACTERISTICS
Boiling point, °C	69	**COLORLESS FUMING LIQUID**
Melting point, °C	−95	Vapor is heavier than air and spreads at ground level, with risk of ignition at a distance. Flow, agitation etc. can cause build-up of electrostatic charge due to liquid's low conductivity. Do not use compressed air when filling, emptying or processing. Once solvent has evaporated (e.g. on clothing) can ignite spontaneously. Aluminum oxide vapors are a product of combustion. Strong reducing agent which reacts violently with oxidants. Reacts violently with water, alcohols, amines, phenols, halogenated hydrocarbons, carbon dioxide, nitrogen oxides, oxides of sulfur and many other compounds.
Flash point, °C	−26	
Auto-ignition temperature, °C	240	
Relative density (water = 1)	0.74	
Relative vapor density (air = 1)	3.0	
Relative density at 20 °C of saturated mixture vapor/air (air = 1)	1.3	
Vapor pressure, mm Hg at 20 °C	121.6	
Solubility in water	reaction	TLV-TWA 2 mg/m³ ✳
Explosive limits, vol% in air	1.1-7.5	
Minimum ignition energy, mJ	0.24	**Absorption route:** Can enter the body by inhalation or ingestion. Harmful atmospheric concentrations can build up fairly rapidly on evaporation at approx. 20°C - even more rapidly in aerosol form.
Electrical conductivity, pS/m	100	**Immediate effects:** Corrosive to the eyes, skin and respiratory tract. Inhalation of vapor/fumes can cause severe breathing difficulties (lung edema). In serious cases risk of death.
Gross formula:	$C_{12}H_{27}Al$	

HAZARDS/SYMPTOMS	PREVENTIVE MEASURES	FIRE EXTINGUISHING/FIRST AID
Fire: extremely flammable, many chemical reactions can cause fire and explosion.	keep away from open flame and sparks, no smoking, avoid contact with water, phenols, alcohols and many other substances.	special extinguishing powder, dry sand, NO other extinguishing agents, can reignite after being extinguished.
Explosion: forms explosive air-vapor mixtures; contact with water, alcohols, phenols and many other substances can cause explosion.	sealed machinery, ventilation, explosion-proof electrical equipment and lighting, grounding, non-sparking tools.	in case of fire: keep undamaged cylinders cool by spraying with water, but DO NOT spray substance with water.
	STRICT HYGIENE	
Inhalation: corrosive; sore throat, cough, severe breathing difficulties.	sealed machinery, ventilation, local exhaust or respiratory protection.	fresh air, rest, place in half-sitting position, take to hospital.
Skin: corrosive; redness, serious burns, slow healing.	gloves, protective clothing.	remove contaminated clothing, flush skin with water or shower, refer to a doctor.
Eyes: corrosive; redness, pain, impaired vision.	face shield, or combined eye and respiratory protection.	flush with water, take to a doctor.
Ingestion: corrosive; sore throat, abdominal pain, diarrhea.		rinse mouth, immediately take to hospital.

SPILLAGE	STORAGE	LABELING / NFPA
evacuate area, call in an expert, collect leakage in sealable containers, soak up remainder with dry sand/table salt and remove to safe place, under no circumstances spray liquid with water; (additional individual protection: CHEMICAL SUIT).	keep in dry, fireproof place under inert gas, separate from all other substances.	R: 14-17-34 S: 16-43 Note A 🔥 Flammable ☣ Corrosive NFPA: 4 / 3 / 3 / W

NOTES

Physical properties (boiling point, flash point, autoignition point, explosive limits, vapor pressure, relative density, conductivity and ignition energy) are based on hexane. For pure triisobutyl aluminum see relevant card. ✳ For TLV of *hexane* see relevant card. Reacts violently with extinguishing agents such as water, foam, halons, carbon dioxide and ordinary powder. Remove contaminated clothing immediately (risk of spontaneous combustion). Lung edema symptoms usually develop several hours later and are aggravated by physical exertion: rest and hospitalization essential. As first aid, a doctor or authorized person should consider administering a corticosteroid spray. Use cylinder with special fittings.

Transport Emergency Card TEC(R)-42G04

HI: X333; UN-number: 3051

CAS-No: [552-30-7]
1,2,3-benzenetricarboxylic acid-1,2-anhydride
TMA

TRIMELLITIC ANHYDRIDE

PHYSICAL PROPERTIES		IMPORTANT CHARACTERISTICS	
Boiling point, °C	ca. 240	**WHITE CRYSTALS OR POWDER**	
Melting point, °C	ca. 165	Vapor mixes readily with air. In dry state can form electrostatic charge if stirred, transported pneumatically, poured etc. Reacts violently with oxidants.	
Flash point, °C	see notes		
Relative density (water = 1)	1.6		
Relative vapor density (air = 1)	5.7	TLV-TWA 0.005 ppm 0.039 mg/m³	
Vapor pressure, mm Hg at 20 °C	< 0.0076		
Solubility in water	poor	**Absorption route:** Can enter the body by inhalation or ingestion. Evaporation negligible at 20° C, but harmful concentrations of airborne particles can build up rapidly.	
Relative molecular mass	192.2		
		Immediate effects: Corrosive to the eyes, skin and respiratory tract. Inhalation can cause lung edema and asthma. Exposure to high concentrations can be fatal. Keep under medical observation.	
		Effects of prolonged/repeated exposure: Prolonged or repeated exposure can cause asthmatic reaction.	
Gross formula:	$C_9H_4O_5$		

HAZARDS/SYMPTOMS	PREVENTIVE MEASURES	FIRE EXTINGUISHING/FIRST AID
Fire: combustible.	keep away from open flame, no smoking.	water spray, powder.
Explosion: finely dispersed particles form explosive mixtures on contact with air.	keep dust from accumulating, sealed machinery, explosion-proof electrical equipment and lighting, grounding.	
	KEEP DUST UNDER CONTROL, STRICT HYGIENE	IN ALL CASES CALL IN A DOCTOR
Inhalation: *corrosive*; burning sensation, cough, difficulty breathing, sore throat, headache, nausea.	local exhaust or respiratory protection.	fresh air, rest, place in half-sitting position, artificial respiration if necessary, take immediately to hospital.
Skin: *corrosive*; redness, burning sensation, pain.	gloves.	remove contaminated clothing, flush skin with water or shower, consult a doctor.
Eyes: *corrosive*; redness, pain, impaired vision.	goggles, or combined eye and respiratory protection.	flush thoroughly with water (remove contact lenses if easy), take to a doctor.
Ingestion: *corrosive*; abdominal pain, burning sensation, nausea, sore throat.		rinse mouth, DO NOT induce vomiting, give no liquids, take immediately to hospital.

SPILLAGE	STORAGE	LABELING / NFPA
clean up spillage, carefully collect remainder; (additional individual protection: P2 respirator).	keep dry, separate from oxidants; ventilate at floor level.	R: 36/37/38-42 S: 22-28 ✖ Irritant

NOTES

Reacts slowly with water to form trimellitic acid. Combustible, but flash point and explosive limits not given in the literature. Lung edema symptoms usually develop several hours later and are aggravated by physical exertion: rest and hospitalization essential. As first aid, a doctor or authorized person should consider administering a suitable spray. Asthma symptoms usually develop after some delay: rest and hospitalization essential. Persons with history of asthma symptoms should under no circumstances come in contact again with substance.

CAS-No: [75-24-1]
TMA

$(CH_3)_3Al$

TRIMETHYL ALUMINUM

PHYSICAL PROPERTIES		IMPORTANT CHARACTERISTICS
Boiling point, °C	126	**COLORLESS FUMING LIQUID**
Melting point, °C	15.4	Ignites spontaneously on exposure to air. Aluminum oxide vapors are a product of combustion. Strong reducing agent which reacts violently with oxidants. Reacts violently with water, alcohols, phenols, amines, carbon dioxide, oxides of sulfur, nitrogen oxides, halogens, halogenated hydrocarbons and many other substances, with risk of fire and explosion. Reacts with air to form irritating vapors.
Relative density (water = 1)	0.75	
Relative vapor density (air = 1)	ca. 2.5	
Vapor pressure, mm Hg at 20 °C	6.4	
Solubility in water	reaction	
Relative molecular mass	72.1	
		TLV-TWA 2 mg/m³
		Absorption route: Can enter the body by inhalation or ingestion. Harmful atmospheric concentrations can build up fairly rapidly on evaporation at approx. 20°C - even more rapidly in aerosol form.
		Immediate effects: Corrosive to the eyes, skin and respiratory tract. Inhalation of vapor/fumes can cause severe breathing difficulties (lung edema). In serious cases risk of death.
Gross formula:	C_3H_9Al	

HAZARDS/SYMPTOMS	PREVENTIVE MEASURES	FIRE EXTINGUISHING/FIRST AID
Fire: extremely flammable, many chemical reactions can cause fire and explosion.	keep away from open flame and sparks, no smoking, avoid contact with air, water and many other substances.	special extinguishing powder, dry sand, NO other extinguishing agents, can reignite after being extinguished.
Explosion: contact with water and many other substances can cause explosion.		in case of fire: keep undamaged cylinders cool by spraying with water, but DO NOT spray substance with water.
	STRICT HYGIENE	
Inhalation: *corrosive*; sore throat, cough, severe breathing difficulties.	completely sealed machinery, ventilation, local exhaust or respiratory protection.	fresh air, rest, place in half-sitting position, take to hospital.
Skin: *corrosive*; serious burns, slow healing.	gloves, protective clothing.	remove contaminated clothing, flush skin with water or shower, refer to a doctor.
Eyes: *corrosive*; redness, pain, impaired vision.	face shield, or combined eye and respiratory protection.	flush with water, take to a doctor.
Ingestion: *corrosive*; sore throat, abdominal pain, diarrhea.		rinse mouth, take immediately to hospital.

SPILLAGE	STORAGE	LABELING / NFPA
evacuate area, call in an expert, dilute leakage immediately with diesel/machine oil, soak up/cover with dry sand/table salt and remove to safe place, under no circumstances spray liquid with water; (additional individual protection: CHEMICAL SUIT).	keep in dry, fireproof place under dry nitrogen, separate from all other substances.	R: 14-17-34 S: 16-43 Note A Flammable Corrosive

NOTES
Usually sold as 15-30% solution in hydrocarbons, which does not ignite spontaneously on exposure to air. For 15% solution see relevant entry. Reacts violently with extinguishing agents such as water, foam, halons, carbon dioxide and ordinary powder. Lung edema symptoms usually develop several hours later and are aggravated by physical exertion: rest and hospitalization essential. As first aid, a doctor or authorized person should consider administering a corticosteroid spray. Use cylinder with special fittings.

Transport Emergency Card TEC(R)-42G02

HI: X333; UN-number: 1103

TRIMETHYLALUMINUM
(15% solution in hexane)

PHYSICAL PROPERTIES	IMPORTANT CHARACTERISTICS
Boiling point, °C 69 Melting point, °C − 95 Flash point, °C − 26 Auto-ignition temperature, °C 240 Relative density (water = 1) 0.71 Relative vapor density (air = 1) 3.0 Relative density at 20 °C of saturated mixture vapor/air (air = 1) 1.3 Vapor pressure, mm Hg at 20 °C 121.6 Solubility in water reaction Explosive limits, vol% in air 1.1-7.5 Minimum ignition energy, mJ 0.24 Electrical conductivity, pS/m 100 Gross formula: C$_3$H$_9$Al	**COLORLESS FUMING LIQUID** Vapor is heavier than air and spreads at ground level, with risk of ignition at a distance. Flow, agitation etc. can cause build-up of electrostatic charge due to liquid's low conductivity. Do not use compressed air when filling, emptying or processing. Once solvent has evaporated (e.g. on clothing) ignites spontaneously on exposure to air. Aluminum oxide vapors are a product of combustion. Strong reducing agent which reacts violently with oxidants. Reacts violently with water, alcohols, phenols, amines, halogenated hydrocarbons and many other substances, with risk of fire and explosion.

TLV-TWA	2 mg/m³ ✱	

Absorption route: Can enter the body by inhalation or ingestion. Harmful atmospheric concentrations can build up fairly rapidly on evaporation at approx. 20°C - even more rapidly in aerosol form.
Immediate effects: Corrosive to the eyes, skin and respiratory tract. Inhalation of vapor/fumes can cause severe breathing difficulties (lung edema). In serious cases risk of death. Can cause metal fume fever.

HAZARDS/SYMPTOMS	PREVENTIVE MEASURES	FIRE EXTINGUISHING/FIRST AID
Fire: extremely flammable, many chemical reactions can cause fire and explosion.	keep away from open flame and sparks, no smoking, avoid contact with water, alcohols, phenols and many other substances.	special extinguishing powder, dry sand, NO other extinguishing agents, can reignite after being extinguished.
Explosion: forms explosive air-vapor mixtures, contact with water and many other substances can cause explosion.	sealed machinery, ventilation, explosion-proof electrical equipment and lighting, grounding, non-sparking tools.	in case of fire: keep undamaged cylinders cool by spraying with water, but DO NOT spray substance with water.
	STRICT HYGIENE	
Inhalation: *corrosive*; sore throat, cough, severe breathing difficulties, metal fume fever.	sealed machinery, ventilation, local exhaust or respiratory protection.	fresh air, rest, place in half-sitting position, take to hospital.
Skin: *corrosive*; serious burns, slow healing.	gloves, protective clothing.	remove contaminated clothing, flush skin with water or shower, refer to a doctor.
Eyes: *corrosive*; redness, pain, impaired vision.	face shield, or combined eye and respiratory protection.	flush with water, take to a doctor.
Ingestion: *corrosive*; abdominal pain, diarrhea, mouth pain.		rinse mouth, immediately take to hospital.

SPILLAGE	STORAGE	LABELING / NFPA
evacuate area, call in an expert, collect leakage in sealable containers, soak up remainder with dry sand/table salt and remove to safe place, under no circumstances spray liquid with water; (additional individual protection: CHEMICAL SUIT).	keep in dry, fireproof place under inert gas, separate from all other substances.	R: 14-17-34 S: 16-43 Note A 🔥 Flammable ☒ Corrosive NFPA diamond: 3 / 3 / W

NOTES
Physical properties (boiling point, flash point, autoignition point, explosive limits, vapor pressure, relative density, conductivity and ignition energy) are based on hexane. For pure trimethyl aluminum see relevant card. ✱ For TLV of *hexane* see relevant card. Reacts violently with extinguishing agents such as water, foam, halons, carbon dioxide and ordinary powder. Remove contaminated clothing immediately and flush with water (risk of spontaneous combustion). Lung edema symptoms usually develop several hours later and are aggravated by physical exertion: rest and hospitalization essential. As first aid, a doctor or authorized person should consider administering a corticosteroid spray. Use cylinder with special fittings.

Transport Emergency Card TEC(R)-42G02

HI: X333; UN-number: 1103

TRIMETHYLAMINE
(cylinder)

PHYSICAL PROPERTIES		IMPORTANT CHARACTERISTICS	
Boiling point, °C	3	**COLORLESS COMPRESSED LIQUEFIED GAS WITH PUNGENT ODOR**	
Melting point, °C	− 117	Vapor is heavier than air and spreads at ground level, with risk of ignition at a distance. → *nitrous*	
Flash point, °C	− 18	*vapors* are a product of combustion. Reacts with mercury to form shock-sensitive compounds.	
Auto-ignition temperature, °C	190	Strong base which reacts violently with acids and corrodes aluminum, zinc etc. Reacts violently	
Relative density (water = 1)	0.6	with oxidants. Attacks copper and its alloys.	
Relative vapor density (air = 1)	2.0		
Vapor pressure, mm Hg at 20 °C	1444	TLV-TWA	10 ppm	24 mg/m³
Solubility in water, g/100 ml at 20 °C	160	STEL	15 ppm	36 mg/m³
Explosive limits, vol% in air	2-11.6		
Electrical conductivity, pS/m	2.2x10⁴	**Absorption route:** Can enter the body by inhalation or ingestion. Harmful atmospheric	
Relative molecular mass	59.1	concentrations can build up very rapidly if gas is released.	
Log P octanol/water	0.3	**Immediate effects:** Corrosive to the eyes, skin and respiratory tract. Can cause frostbite due to	
		rapid evaporation. Inhalation of vapor/fumes can cause severe breathing difficulties (lung	
		edema).	
Gross formula:	C_3H_9N		

Note: TLV-TWA 10 ppm 24 mg/m³; STEL 15 ppm 36 mg/m³ shown in the characteristics column.

HAZARDS/SYMPTOMS	PREVENTIVE MEASURES	FIRE EXTINGUISHING/FIRST AID
Fire: extremely flammable.	keep away from open flame and sparks, no smoking, avoid contact with hot surfaces (e.g. steam lines).	shut off supply; if impossible and no danger to surrounding area, allow fire to burn itself out; otherwise extinguish with powder, carbon dioxide, (halons).
Explosion: forms explosive air-gas mixtures.	sealed machinery, ventilation, explosion-proof electrical equipment and lighting, grounding when pumping etc. in liquid form.	in case of fire: keep cylinders cool by spraying with water.
Inhalation: *corrosive*; sore throat, cough, shortness of breath, severe breathing difficulties.	ventilation, local exhaust or respiratory protection.	fresh air, rest, place in half-sitting position, call a doctor or take to hospital.
Skin: *corrosive*; redness, pain.	insulated gloves, protective clothing.	remove contaminated clothing, flush skin with water or shower, refer to a doctor if necessary.
Eyes: *corrosive*; redness, pain, impaired vision.	acid goggles, or combined eye and respiratory protection.	flush with water, take to a doctor.
Ingestion: *corrosive*; abdominal pain, diarrhea, vomiting.		rinse mouth, immediately take to hospital.

SPILLAGE	STORAGE	LABELING / NFPA
evacuate area, call in an expert, ventilate, under no circumstances spray liquid with water, combat gas cloud with water curtain; (additional individual protection: CHEMICAL SUIT).	keep in cool, fireproof place, separate from oxidants, acids and mercury.	 NFPA diamond: 4 (top), 3 (left), 0 (right)

NOTES
Flush contaminated clothing with water (fire hazard). Lung edema symptoms usually develop several hours later and are aggravated by physical exertion: rest and hospitalization essential. As first aid, a doctor or authorized person should consider administering a corticosteroid spray. Do not spray leaking cylinder with water (to avoid corrosion). Turn leaking cylinder so that leak is on top to prevent liquid trimethylamine escaping. Use cylinder with special fittings.

Transport Emergency Card TEC(R)-206

HI: 236; UN-number: 1083

CAS-No: [75-50-3]
N,N-dimethylmethanamine
TMA
trimethylamine

$(CH_3)_3N$

TRIMETHYLAMINE
(40% aqueous solution)

PHYSICAL PROPERTIES	IMPORTANT CHARACTERISTICS
Boiling point, °C 30 Melting point, °C +2 Flash point, °C < −16 Auto-ignition temperature, °C 190 Relative density (water = 1) 0.7 Relative vapor density (air = 1) 2.0 Relative density at 20 °C of saturated mixture vapor/air (air = 1) 1.6 Vapor pressure, mm Hg at 20 °C 505 Solubility in water ∞ Explosive limits, vol% in air 2-11.6 Relative molecular mass 59.1 Log P octanol/water 0.3 Gross formula: C_3H_9N	**COLORLESS AQUEOUS SOLUTION OF TRIMETHYLAMINE WITH PUNGENT ODOR** Vapor is heavier than air and spreads at ground level, with risk of ignition at a distance. Do not use compressed air when filling, emptying or processing. → *Nitrous vapors* are a product of combustion. Reacts with mercury to form shock-sensitive compounds. Strong base which reacts violently with acids and corrodes aluminum, zinc etc. Reacts violently with oxidants and many other compounds. Attacks copper and its alloys. TLV-TWA 10 ppm 24 mg/m³ STEL 15 ppm 36 mg/m³ **Absorption route:** Can enter the body by inhalation or ingestion. Harmful atmospheric concentrations can build up very rapidly on evaporation at 20°C. **Immediate effects:** Corrosive to the eyes, skin and respiratory tract. Inhalation of vapor/fumes can cause severe breathing difficulties (lung edema).

HAZARDS/SYMPTOMS	PREVENTIVE MEASURES	FIRE EXTINGUISHING/FIRST AID
Fire: extremely flammable.	keep away from open flame and sparks, no smoking, avoid contact with hot surfaces (e.g. steam lines).	powder, alcohol-resistant foam, water spray, carbon dioxide, (halons).
Explosion: forms explosive air-vapor mixtures.	sealed machinery, ventilation, explosion-proof electrical equipment and lighting.	in case of fire: keep tanks/drums cool by spraying with water.
Inhalation: *corrosive*; sore throat, cough, shortness of breath, severe breathing difficulties.	ventilation, local exhaust or respiratory protection.	fresh air, rest, place in half-sitting position, call a doctor.
Skin: *corrosive*; redness, pain, serious burns.	gloves, protective clothing.	remove contaminated clothing, flush skin with water or shower, call a doctor.
Eyes: *corrosive*; redness, pain, impaired vision.	face shield, or combined eye and respiratory protection.	flush with water, take to a doctor.
Ingestion: *corrosive*; abdominal pain, diarrhea, vomiting.		rinse mouth, call a doctor or take immediately to hospital.

SPILLAGE	STORAGE	LABELING / NFPA
evacuate area, call in an expert, collect leakage in sealable containers and remove to safe place, render spillage harmless with sodium bisulfate solution and flush away with water; (additional individual protection: CHEMICAL SUIT).	keep in cool, fireproof place, separate from oxidants, acids and mercury.	R: 13-36/37 S: 16-26-29 Note C 🔥 Flammable ✖ Irritant NFPA: 3 / 4 / 0

NOTES
Flush contaminated clothing with water (fire hazard). Lung edema symptoms usually develop several hours later and are aggravated by physical exertion: rest and hospitalization essential. As first aid, a doctor or authorized person should consider administering a corticosteroid spray. Unbreakable packaging preferred; if breakable, keep in unbreakable container.

Transport Emergency Card TEC(R)-206

HI: 236; UN-number: 1083

$(CH_3)_3SiCl$

TRIMETHYLCHLOROSILANE

PHYSICAL PROPERTIES	IMPORTANT CHARACTERISTICS
Boiling point, °C 57 Melting point, °C −50 Flash point, °C −28 Auto-ignition temperature, °C 417 Relative density (water = 1) 0.85 Relative vapor density (air = 1) 3.75 Relative density at 20°C of saturated mixture vapor/air (air = 1) 1.7 Vapor pressure, mm Hg at 20°C 200 Solubility in water reaction Explosive limits, vol% in air 2-6 Minimum ignition energy, mJ 0.9 Relative molecular mass 108.7	**COLORLESS LIQUID WITH PUNGENT ODOR** Vapor is heavier than air and spreads at ground level, with risk of ignition at a distance. Corrosive → *hydrochloric acid* fumes are a product of combustion. Reacts violently with water, ketones, alcohols, amines and many other substances, giving off toxic and corrosive gases, with risk of fire and explosion. Reacts with air to form corrosive hydrochloric acid fumes.
	TLV-TWA not available
Gross formula: C_3H_9ClSi	**Absorption route:** Can enter the body by inhalation or ingestion. Harmful atmospheric concentrations can build up very rapidly on evaporation at 20°C. **Immediate effects:** Corrosive to the eyes, skin and respiratory tract. Inhalation of vapor/fumes can cause severe breathing difficulties (lung edema). In serious cases risk of death.

HAZARDS/SYMPTOMS	PREVENTIVE MEASURES	FIRE EXTINGUISHING/FIRST AID
Fire: extremely flammable.	keep away from open flame and sparks, no smoking.	AFFF, powder, carbon dioxide, (halons), DO NOT use water.
Explosion: forms explosive air-vapor mixtures, many chemical reactions can cause explosion.	sealed machinery, ventilation, explosion-proof electrical equipment and lighting.	in case of fire: keep tanks/drums cool by spraying with water, but DO NOT spray substance with water.
Inhalation: *corrosive*; sore throat, cough, shortness of breath, severe breathing difficulties.	ventilation, local exhaust or respiratory protection.	fresh air, rest, place in half-sitting position, take to hospital.
Skin: *corrosive*; redness, pain, serious burns.	gloves, protective clothing.	remove contaminated clothing, flush skin with water or shower, refer to a doctor.
Eyes: *corrosive*; redness, pain, impaired vision.	face shield, or combined eye and respiratory protection.	flush with water, take to a doctor.
Ingestion: *corrosive*; sore throat, abdominal pain, diarrhea.		rinse mouth, take immediately to hospital.

SPILLAGE	STORAGE	LABELING / NFPA
evacuate area, call in an expert, collect leakage in sealable containers, soak up with calcium carbonate or soda and flush away with large quantities of water after waiting 10 minutes; (additional individual protection: CHEMICAL SUIT).	keep in dry, fireproof place, separate from all other substances.	 3 / 3 2 / W

NOTES
Lung edema symptoms usually develop several hours later and are aggravated by physical exertion: rest and hospitalization essential. As first aid, a doctor or authorized person should consider administering a corticosteroid spray. Reacts violently with extinguishing agents such as water and foam. Use airtight packaging. Unbreakable packaging preferred; if breakable, keep in unbreakable container.

Transport Emergency Card TEC(R)-30G40

HI: X338; UN-number: 1298

CAS-No: [504-63-2]
2-deoxyglycerol
1,3-dihydroxypropane
1,3-propanediol
β-propylene glycol
1,3-propylene glycol

HOCH₂CH₂CH₂OH

TRIMETHYLENE GLYCOL

PHYSICAL PROPERTIES		IMPORTANT CHARACTERISTICS	
Boiling point, °C	211	**COLORLESS, ODORLESS, VISCOUS HYGROSCOPIC LIQUID**	
Flash point, °C	>74	Vapor mixes readily with air. Reacts with strong oxidants.	
Auto-ignition temperature, °C	400		
Relative density (water = 1)	1.05	TLV-TWA	not available
Relative vapor density (air = 1)	2.6		
Relative density at 20 °C of saturated mixture		**Absorption route:** Can enter the body by inhalation or ingestion. Harmful atmospheric concentrations build up very slowly, if at all, on evaporation at approx. 20°C, but much more rapidly in aerosol form.	
vapor/air (air = 1)	1.0		
Vapor pressure, mm Hg at 20 °C	ca. 0.076		
Solubility in water	∞	**Immediate effects:** Irritates the eyes and respiratory tract. Liquid destroys the skin's natural oils.	
Relative molecular mass	76.1		
Log P octanol/water	−1.6		
Gross formula:	C₃H₈O₂		

HAZARDS/SYMPTOMS	PREVENTIVE MEASURES	FIRE EXTINGUISHING/FIRST AID
Fire: combustible.	keep away from open flame, no smoking.	powder, alcohol-resistant foam, water spray, carbon dioxide, (halons).
Explosion: above 74°C: forms explosive air-vapor mixtures.	above 74°C: sealed machinery, ventilation.	
Inhalation: sore throat, cough.	ventilation.	fresh air, rest.
Skin: redness, itching.	gloves.	remove contaminated clothing, flush skin with water or shower.
Eyes: redness, pain.	acid goggles.	flush with water, take to a doctor if necessary.
Ingestion: abdominal pain, nausea.		rinse mouth, call a doctor or take to hospital.

SPILLAGE	STORAGE	LABELING / NFPA
collect leakage in sealable containers, flush away remainder with water.	keep dry, separate from oxidants; ventilate at floor level.	

NOTES
Log P octanol/water is estimated.

CAS-No: [78-30-8]
o-cresyl phosphate
o-tolyl phosphate
TOCP

$(CH_3C_6H_4O)_3P=O$

TRIORTHOCRESYL PHOSPHATE

PHYSICAL PROPERTIES	IMPORTANT CHARACTERISTICS	
Decomposes below boiling point, °C 410 Melting point, °C −28 Flash point, °C 225 Auto-ignition temperature, °C 385 Relative density (water = 1) 1.2 Relative vapor density (air = 1) 12.7 Relative density at 20 °C of saturated mixture vapor/air (air = 1) 1.00 Vapor pressure, mm Hg at 20 °C 1.5x10⁻⁶ Solubility in water, g/100 ml at 25 °C 0.01 Relative molecular mass 368.4	**COLORLESS VISCOUS LIQUID** Decomposes when heated above 400°C or burned, giving off toxic vapors (→ *phosphorus oxides*). Moisture causes hydrolysis and formation of → *phosphoric acid*. Reacts violently with strong oxidants.	
	TLV-TWA (skin)	0.1 mg/m³
	Absorption route: Can enter the body by inhalation or ingestion or through the skin. Evaporation negligible at 20°C, but harmful atmospheric concentrations can build up rapidly in aerosol form. **Immediate effects:** Can cause nerve disorders. In serious cases risk of permanent injury or death.	
Gross formula: $C_{21}H_{21}O_4P$		

HAZARDS/SYMPTOMS	PREVENTIVE MEASURES	FIRE EXTINGUISHING/FIRST AID
Fire: combustible.	keep away from open flame, no smoking.	powder, water spray, foam, carbon dioxide, (halons).
Inhalation: headache, nausea, vomiting, pain in arms and legs, paralysis (arms and legs).	ventilation.	fresh air, rest, call a doctor.
Skin: *is absorbed*; see also 'Inhalation'.	gloves, protective clothing.	remove contaminated clothing, wash skin with soap and water, refer to a doctor.
Eyes: redness, pain.	face shield.	flush with water, take to a doctor if necessary.
Ingestion: abdominal pain, nausea; see also 'Inhalation'.	do not eat, drink or smoke while working.	rinse mouth, immediately take to hospital.

SPILLAGE	STORAGE	LABELING / NFPA
collect leakage in sealable containers, soak up with sand or other inert absorbent and remove to safe place.	keep dry, separate from oxidants; ventilate at floor level.	R: 23/24/25-39 S: 20/21-28-44 Note C ☠ Toxic

NOTES
Odor limit is above TLV. TLV is for ortho-isomer. Depending on the degree of exposure, regular medical checkups are advisable. Symptoms usually develop after some time and are accelerated by physical exertion: observation in hospital often necessary. Commercial product is a mixture of ortho, meta and para-isomers; ortho-isomer is most toxic.

Transport Emergency Card TEC(R)-61G06B (> 3% ortho-isomeer)   **HI: 60; UN-number: 2574**

CAS-No: [110-88-3]
metaformaldehyde
trioxymethylene

1,3,5-TRIOXANE

PHYSICAL PROPERTIES	IMPORTANT CHARACTERISTICS	
Boiling point, °C 115 Melting point, °C 64 Flash point, °C 45 Auto-ignition temperature, °C 410 Relative density (water = 1) 1.2 Relative vapor density (air = 1) 3.1 Relative density at 20 °C of saturated mixture vapor/air (air = 1) 1.03 Vapor pressure, mm Hg at 25 °C 13 Solubility in water, g/100 ml at 18 °C 17 Explosive limits, vol% in air 3.6-29 Relative molecular mass 90.1	**WHITE CRYSTALS OR LUMPS WITH CHARACTERISTIC ODOR** Vapor mixes readily with air. Presumed to be able to form peroxides, which can ignite due to shock or friction. In dry state can form electrostatic charge if stirred, transported pneumatically, poured etc. → *Formaldehyde* can form in case of incomplete combustion. Can produce thermal explosion when heated. Decomposes on contact with acids, forming → *formaldehyde*. Reacts violently with strong oxidants, with risk of fire and explosion.	
	TLV-TWA not available	
Gross formula: $C_3H_6O_3$	**Absorption route:** Can enter the body by inhalation or ingestion or through the skin. Insufficient data on the rate at which harmful concentrations can build up. **Immediate effects:** Corrosive to the eyes, skin and respiratory tract. In substantial concentrations can impair consciousness.	

HAZARDS/SYMPTOMS	PREVENTIVE MEASURES	FIRE EXTINGUISHING/FIRST AID
Fire: flammable.	keep away from open flame and sparks, no smoking.	water spray, powder.
Explosion: above 45°C: forms explosive air-vapor mixtures; finely dispersed particles form explosive mixtures on contact with air.	keep dust from accumulating; sealed machinery, explosion-proof electrical equipment and lighting, grounding.	in case of fire: keep tanks/drums cool by spraying with water.
Inhalation: *corrosive*; sore throat, cough, shortness of breath.	local exhaust or respiratory protection.	fresh air, rest, call a doctor.
Skin: *corrosive, is absorbed*; redness, pain, burns.	gloves.	remove contaminated clothing.
Eyes: *corrosive*; redness, pain, impaired vision.	face shield.	flush with water, take to a doctor.
Ingestion: *corrosive*, sore throat, abdominal pain, vomiting, sleepiness.		rinse mouth, immediately take to hospital.

SPILLAGE	STORAGE	LABELING / NFPA
clean up spillage, flush away remainder with water; (additional individual protection: respirator with A/P2 filter).	keep in fireproof place, separate from oxidants and acids.	R: 22 S: 24/25 ❎ Harmful NFPA: 2 / 2 / 0

NOTES

Transport Emergency Card TEC(R)-80G09

TRIPHENYL PHOSPHATE

PHYSICAL PROPERTIES	IMPORTANT CHARACTERISTICS	
Boiling point, °C 370 Melting point, °C 49 Flash point, °C 220 Vapor pressure, mm Hg at 20 °C < <0.076 Solubility in water none Relative molecular mass 326.3	**WHITE CRYSTALS OR FLAKES** Decomposes when heated or burned, giving off toxic gases (→ *phosphorus pentoxide*).	
	TLV-TWA	3 mg/m³
	Absorption route: Can enter the body by inhalation or ingestion. Harmful atmospheric concentrations build up very slowly, if at all, on evaporation at approx. 20°C, but much more rapidly in aerosol form. **Immediate effects:** Irritates the eyes and respiratory tract. **Effects of prolonged/repeated exposure:** Prolonged exposure can cause paralysis.	
Gross formula: $C_{18}H_{15}O_4P$		

HAZARDS/SYMPTOMS	PREVENTIVE MEASURES	FIRE EXTINGUISHING/FIRST AID
Fire: non-combustible.		in case of fire in immediate vicinity: preferably powder or water spray.
	KEEP DUST UNDER CONTROL	
Inhalation: hot vapor: sore throat, cough, shortness of breath.	local exhaust or respiratory protection.	fresh air, rest, call a doctor.
Skin:	gloves.	remove contaminated clothing, flush skin with water or shower.
Eyes: redness, pain.	goggles, or combined eye and respiratory protection.	flush with water, take to a doctor if necessary.
Ingestion: sore throat, abdominal pain.		call a doctor.

SPILLAGE	STORAGE	LABELING / NFPA
clean up spillage, carefully collect remainder; (additional individual protection: P2 respirator).		1 / 2 / 0

NOTES
Although combustible, combustion ceases when ignition source is removed. Effect on nervous system: slight blocking of cholinesterase.

TRIPHENYLPHOSPHITE

PHYSICAL PROPERTIES	IMPORTANT CHARACTERISTICS
Boiling point, °C 360 Melting point, °C 25 Flash point, °C 218 Relative density (water = 1) 1.2 Vapor pressure, mm Hg at 20 °C <0.076 Solubility in water none Relative molecular mass 310.3	**LIGHT YELLOW VISCOUS LIQUID, CRYSTALS OR PASTE WITH CHARACTERISTIC ODOR** Decomposes when heated or burned, giving off toxic gases (→ *phosphorus pentoxide*). TLV-TWA not available **Absorption route:** Can enter the body by ingestion or through the skin. Evaporation negligible at 20°C, but harmful concentrations of airborne particles can build up rapidly. **Immediate effects:** Irritates the eyes, skin and respiratory tract. In serious cases risk of seizures, unconsciousness and death. **Effects of prolonged/repeated exposure:** Prolonged or repeated contact can cause skin disorders. Can cause paralysis and brain damage.
Gross formula: $C_{18}H_{15}O_3P$	

HAZARDS/SYMPTOMS	PREVENTIVE MEASURES	FIRE EXTINGUISHING/FIRST AID
Fire: combustible.	keep away from open flame, no smoking.	powder, water spray, foam, carbon dioxide, (halons).
Inhalation: hot vapor: sore throat, cough, shortness of breath.	local exhaust or respiratory protection.	fresh air, rest, call a doctor.
Skin: *is absorbed*; redness, pain.	gloves.	remove contaminated clothing, flush skin with water or shower, refer to a doctor.
Eyes: redness, pain, impaired vision.	goggles, or combined eye and respiratory protection.	flush with water, take to an eye doctor.
Ingestion: drowsiness, feeling of weakness, trembling, fever, paralysis.		rinse mouth, immediately take to hospital.

SPILLAGE	STORAGE	LABELING / NFPA
first allow to set, then clean up spillage, carefully collect remainder; (additional individual protection: P2 respirator)		R: 36/38 S: 28 ✖ Irritant

NOTES

Blocks cholinesterase. Symptoms of paralysis often appear only after a few days.

CAS-No: [102-69-2]
N,N-dipropyl-1-propanamine
TNPA
tri-n-propylamine

$(C_3H_7)_3N$

TRIPROPYLAMINE

PHYSICAL PROPERTIES	IMPORTANT CHARACTERISTICS
Boiling point, °C — 156 Melting point, °C — −94 Flash point, °C — 34 Auto-ignition temperature, °C — 180 Relative density (water = 1) — 0.75 Relative vapor density (air = 1) — 4.9 Relative density at 20 °C of saturated mixture vapor/air (air = 1) — 1.01 Vapor pressure, mm Hg at 20° C — 2.5 Solubility in water, g/100 ml at 20 °C — 0.26 Explosive limits, vol% in air — 0.7-5.6 Relative molecular mass — 143.3 Log P octanol/water — 2.8	**COLORLESS LIQUID WITH CHARACTERISTIC ODOR** Vapor mixes readily with air. Decomposes when heated, giving off flammable and toxic gases (→ nitrous vapors). Medium strong base. Reacts violently with strong oxidants and many organic compounds, with risk of fire and explosion. Attacks copper and its alloys.
	TLV-TWA not available
	Absorption route: Can enter the body by inhalation or ingestion. Harmful atmospheric concentrations can build up fairly rapidly on evaporation at approx. 20° C - even more rapidly in aerosol form. **Immediate effects:** Corrosive to the eyes, skin and respiratory tract. Inhalation of vapor/fumes can cause severe breathing difficulties (lung edema). In serious cases risk of death.
Gross formula: $C_9H_{21}N$	

HAZARDS/SYMPTOMS	PREVENTIVE MEASURES	FIRE EXTINGUISHING/FIRST AID
Fire: flammable.	keep away from open flame and sparks, no smoking, avoid contact with hot surfaces (e.g. steam lines).	powder, AFFF, foam, carbon dioxide, (halons).
Explosion: above 34°C: forms explosive air-vapor mixtures.	above 34°C: sealed machinery, ventilation, explosion-proof electrical equipment.	in case of fire: keep tanks/drums cool by spraying with water.
Inhalation: *corrosive*; sore throat, cough, shortness of breath, severe breathing difficulties.	ventilation, local exhaust or respiratory protection.	fresh air, rest, place in half-sitting position.
Skin: *corrosive*; redness, pain, burns.	gloves, protective clothing.	remove contaminated clothing, flush skin with water or shower, call a doctor.
Eyes: *corrosive*; redness, pain, impaired vision.	face shield.	flush with water, take to a doctor.
Ingestion: *corrosive*; sore throat, abdominal pain.		rinse mouth, immediately take to hospital.

SPILLAGE	STORAGE	LABELING / NFPA
collect leakage in sealable containers, soak up with sand or other inert absorbent and remove to safe place; (additional individual protection: breathing apparatus).	keep in fireproof place, separate from oxidants and strong acids.	 2 2 0

NOTES
Lung edema symptoms usually develop several hours later and are aggravated by physical exertion: rest and hospitalization essential. As first aid, a doctor or authorized person should consider administering a corticosteroid spray. The measures on this card also apply to triisopropylamine.

Transport Emergency Card TEC(R)-163

HI: 83; UN-number: 2260

TRIPROPYLENE GLYCOL

PHYSICAL PROPERTIES	IMPORTANT CHARACTERISTICS
Boiling point, °C 268 Flash point, °C 141 Relative density (water = 1) 1.02 Relative vapor density (air = 1) 6.6 Relative density at 20 °C of saturated mixture vapor/air (air = 1) 1.0 Vapor pressure, mm Hg at 20 °C <0.076 Solubility in water ∞ Relative molecular mass 192.3 Gross formula: C$_9$H$_{20}$O$_4$	**COLORLESS LIQUID** Vapor mixes readily with air. Reacts with strong oxidants. TLV-TWA not available **Absorption route:** Can enter the body by inhalation or ingestion. Harmful atmospheric concentrations build up very slowly, if at all, on evaporation at approx. 20°C, but much more rapidly in aerosol form. **Immediate effects:** Affects the nervous system.

HAZARDS/SYMPTOMS	PREVENTIVE MEASURES	FIRE EXTINGUISHING/FIRST AID
Fire: combustible.	keep away from open flame, no smoking.	powder, alcohol-resistant foam, water spray, carbon dioxide, (halons).
Inhalation: agitation.	ventilation.	fresh air, rest, call a doctor.
Skin: redness.	gloves.	remove contaminated clothing, flush skin with water or shower.
Eyes: redness, pain.	safety glasses.	flush with water, take to a doctor if necessary.
Ingestion: abdominal pain; see also 'Inhalation'.		rinse mouth, rest, call a doctor.

SPILLAGE	STORAGE	LABELING / NFPA
collect leakage in sealable containers, flush away remainder with water.	keep separate from oxidants; ventilate at floor level.	

NOTES

TRISODIUM PHOSPHATE

PHYSICAL PROPERTIES	IMPORTANT CHARACTERISTICS	
Melting point, °C 73 Relative density (water = 1) 1.6 Solubility in water, g/100 ml at 20 °C 121 Relative molecular mass 380.2	**COLORLESS CRYSTALS** In aqueous solution is a medium strong base which reacts violently with acids and corrodes aluminum, zinc etc. Reacts with air to form disodium phosphate and → *sodium carbonate*.	
	TLV-TWA not available	
	Absorption route: Can enter the body by inhalation or ingestion. Evaporation negligible at 20°C, but harmful concentrations of airborne particles can build up rapidly. **Immediate effects:** Irritates the eyes, skin and respiratory tract.	
Gross formula: $Na_3O_4P.12H_2O$		

HAZARDS/SYMPTOMS	PREVENTIVE MEASURES	FIRE EXTINGUISHING/FIRST AID
Fire: non-combustible.		in case of fire in immediate vicinity: use any extinguishing agent.
Inhalation: sore throat, cough, shortness of breath.	local exhaust or respiratory protection.	fresh air, rest, call a doctor.
Skin: redness.	gloves.	remove contaminated clothing, flush skin with water or shower.
Eyes: redness, pain.	goggles.	flush with water, take to a doctor if necessary.
Ingestion: abdominal pain, diarrhea.		rinse mouth, call a doctor.

SPILLAGE	STORAGE	LABELING / NFPA
clean up spillage, flush away remainder with water; (additional individual protection: P2 respirator).	keep separate from strong acids.	◇

NOTES
Apparent melting point due to loss of water of crystallization.

$C_{10}H_{16}$

TURPENTINE

PHYSICAL PROPERTIES		IMPORTANT CHARACTERISTICS
Boiling point, °C	165 (mean)	**COLORLESS LIQUID WITH CHARACTERISTIC ODOR**
Melting point, °C	− 55	Vapor mixes readily with air. Mixture of hydrocarbons (mainly diterpenes) with boiling range
Flash point, °C	35	150-180°C. Flow, agitation etc. can cause build-up of electrostatic charge due to liquid's low
Auto-ignition temperature, °C	255	conductivity. Attacks rubber. Reacts violently with strong oxidants.
Relative density (water = 1)	0.9	
Relative vapor density (air = 1)	4.7	

TLV-TWA	100 ppm	556 mg/m³

Relative density at 20 °C of saturated mixture vapor/air (air = 1) 1.01
Vapor pressure, mm Hg at 20 °C 1.9
Solubility in water poor
Explosive limits, vol% in air 0.8-6
Electrical conductivity, pS/m < 1.7x10⁶
Relative molecular mass 136

Absorption route: Can enter the body by inhalation or ingestion or through the skin. Harmful atmospheric concentrations build up fairly slowly on evaporation, but much more rapidly in aerosol form.
Immediate effects: Irritates the eyes, skin and respiratory tract. Inhalation of vapor/fumes can cause severe breathing difficulties (lung edema).
Effects of prolonged/repeated exposure: Frequent prolonged contact can cause hypersensitivity. Prolonged or repeated contact can cause skin disorders. Loss of blood through urine and/or bone marrow damage can result in anemia. Can cause kidney and bladder damage.

Gross formula: $C_{10}H_{16}$

HAZARDS/SYMPTOMS	PREVENTIVE MEASURES	FIRE EXTINGUISHING/FIRST AID
Fire: flammable.	keep away from open flame and sparks, no smoking.	powder, AFFF, foam, carbon dioxide, (halons).
Explosion: above 35°C: forms explosive air-vapor mixtures.	above 35°C: sealed machinery, ventilation, explosion-proof electrical equipment, grounding.	in case of fire: keep tanks/drums cool by spraying with water.
Inhalation: sore throat, cough, severe breathing difficulties, headache, dizziness, nausea, chest pain, drowsiness.	ventilation, local exhaust or respiratory protection.	fresh air, rest, take to hospital.
Skin: *is absorbed*; redness, pain.	gloves.	remove contaminated clothing, wash skin with soap and water.
Eyes: redness, pain, impaired vision.	safety glasses, or combined eye and respiratory protection.	flush with water, take to a doctor.
Ingestion: sore throat, cough, abdominal pain, stomachache, blood in the urine, muscle weakness, cramps, impaired vision, drowsiness, unconsciousness.		rinse mouth, DO NOT induce vomiting, take immediately to hospital.

SPILLAGE	STORAGE	LABELING / NFPA
collect leakage in sealable containers, soak up with sand or other inert absorbent and remove to safe place.	keep in fireproof place, separate from oxidants.	R: 10-20/21/22 S: 2 ✖ Harmful

NFPA diamond: 3 / 1 / 0

NOTES

Alcohol consumption increases toxic effects. Lung edema symptoms usually develop several hours later and are aggravated by physical exertion: rest and hospitalization essential. As first aid, a doctor or authorized person should consider administering a corticosteroid spray. Odor threshold approx. 100ppm. Throat irritation begins at approx. 125ppm; eye and nose irritation at approx. 175ppm.

Transport Emergency Card TEC(R)-21

HI: 30; UN-number: 1299

CAS-No: [57-13-6]
carbamide
carbamimidic acid
carbonyl diamide
carbonyl diamine
isourea

$(NH_2)_2CO$

UREA

PHYSICAL PROPERTIES	IMPORTANT CHARACTERISTICS	
Decomposes below boiling point, °C ? Melting point, °C 133 Relative density (water = 1) 1.3 Solubility in water, g/100 ml 100 Relative molecular mass 60.6	**WHITE CRYSTALS OR WHITE POWDER** Decomposes in flame or on hot surface, giving off toxic vapors (→ *nitrous vapors*). Decomposes when heated, giving off suffocating gas (→ *anhydrous ammonia*). Hydrolyzes slowly on contact with air. Reacts with hypochlorite, phosphorus pentachloride and many other chlorinating agents, giving off explosive → *nitrogen trichloride*. Reacts with oxidants. Reacts with nitrites, with risk of fire and explosion.	
	TLV-TWA not available	
	Absorption route: Can enter the body by inhalation or ingestion or through the skin. **Immediate effects:** Irritates the eyes and respiratory tract; on prolonged or repeated contact also the skin.	
Gross formula: CH_4N_2O		

HAZARDS/SYMPTOMS	PREVENTIVE MEASURES	FIRE EXTINGUISHING/FIRST AID
Fire: non-combustible, many chemical reactions can cause fire and explosion.		in case of fire in immediate vicinity: use any extinguishing agent.
Inhalation: sore throat, cough, shortness of breath.	local exhaust or respiratory protection.	fresh air, rest, call a doctor if necessary.
Skin: redness.	gloves.	remove contaminated clothing, flush skin with water or shower.
Eyes: redness, pain.	goggles.	flush with water, take to a doctor if necessary.
Ingestion: sore throat, abdominal pain.		rinse mouth, call a doctor if necessary.

SPILLAGE	STORAGE	LABELING / NFPA
clean up spillage, flush away remainder with water; (additional individual protection: P1 respirator).	keep dry.	

NOTES
Can be made to burn.

VANADIUM OXYTRICHLORIDE

PHYSICAL PROPERTIES	IMPORTANT CHARACTERISTICS	
Boiling point, °C 127 Melting point, °C −77 Relative density (water = 1) 1.8 Relative vapor density (air = 1) 6.0 Relative density at 20 °C of saturated mixture vapor/air (air = 1) 1.03 Vapor pressure, mm Hg at 20 °C ca. 6.1 Solubility in water reaction Relative molecular mass 173.3	**COLORLESS TO YELLOW LIQUID WITH PUNGENT ODOR** Vapor mixes readily with air. Reacts violently with water, forming → *hydrochloric acid*. Reacts with bases and many organic compounds. Reacts with air to form corrosive mixtures (→ *hydrochloric acid*). Attacks many metals in the presence of moisture.	
	TLV-TWA	0.05 mg/m³ ✳
	Absorption route: Can enter the body by inhalation or ingestion. Harmful atmospheric concentrations can build up very rapidly on evaporation at 20°C. **Immediate effects:** Irritates the eyes, skin and respiratory tract. Can cause nerve damage. In substantial concentrations can cause cardiovascular disorders. Inhalation can cause severe breathing difficulties (asthma).	
Gross formula: Cl$_3$OV		

HAZARDS/SYMPTOMS	PREVENTIVE MEASURES	FIRE EXTINGUISHING/FIRST AID
Fire: non-combustible.		in case of fire in immediate vicinity: do not use water.
	STRICT HYGIENE	
Inhalation: sore throat, cough, shortness of breath.	ventilation, local exhaust or respiratory protection.	fresh air, rest, place in half-sitting position, call a doctor.
Skin: redness.	gloves, protective clothing.	remove contaminated clothing, wash skin with soap and water.
Eyes: redness, pain.	face shield, or combined eye and respiratory protection.	flush with water, take to a doctor.
Ingestion: abdominal pain, diarrhea, nausea, dizziness.		rinse mouth, take to hospital.

SPILLAGE	STORAGE	LABELING / NFPA
collect leakage in sealable containers, soak up with sand or other inert absorbent and remove to safe place, under no circumstances spray liquid with water; (additional individual protection: CHEMICAL SUIT).	keep dry, separate from strong bases; ventilate.	◇

NOTES

✳ TLV equals that of vanadium (V$_2$O$_5$, respirable dust and fume). Asthma symptoms usually develop several hours later and are aggravated by physical exertion: rest and hospitalization essential. Persons with history of asthma symptoms should under no circumstances come in contact again with substance. Use airtight packaging. Unbreakable packaging preferred; if breakable, keep in unbreakable container.

Transport Emergency Card TEC(R)-80G10

HI: 80; UN-number: 2443

VANADIUMPENTOXIDE

PHYSICAL PROPERTIES	IMPORTANT CHARACTERISTICS
Boiling point (decomposes), °C 1750 Melting point, °C 690 Relative density (water = 1) 3.4 Solubility in water, g/100 ml at 20 °C 0.8 Relative molecular mass 181.9	**YELLOW TO RED CRYSTALS** Vanadium pentoxide acts as a catalyst in many oxidation reactions. In aqueous solution reacts aciditic. The solubility in acids and bases is better than in water.

| | TLV-TWA | 0.05 mg/m³ ✳ |

	Absorption route: Can enter the body by inhalation or ingestion. Evaporation negligible at 20°C, but harmful concentrations of airborne particles can build up rapidly. **Immediate effects:** Irritates the eyes, skin and respiratory tract. Inhalation of dust or fumes can cause severe breathing difficulties (lung edema). **Effects of prolonged/repeated exposure:** Prolonged and repeated contact with dust can cause lung disorders.
Gross formula: O_5V_2	

HAZARDS/SYMPTOMS	PREVENTIVE MEASURES	FIRE EXTINGUISHING/FIRST AID
Fire: non-combustible, but enhances combustion of other substances catalytically.	avoid contact with combustible substances.	in case of fire in immediate vicinity: use any extinguishing agent.
	KEEP DUST UNDER CONTROL	
Inhalation: sore throat, cough, stuffed nose, shortness of breath, headache.	ventilation, local exhaust or respiratory protection.	fresh air, rest, place in half sitting position, call a doctor, take to hospital if necessary.
Skin: redness, itching.	gloves.	remove contaminated clothing, wash skin thoroughly with soap and water.
Eyes: redness, pain, profuse watering of the eyes.	face shield, or combined eye and respiratory protection.	flush with water, take to a doctor.
Ingestion: metallic taste, abdominal cramps, vomiting.		rinse mouth, call a doctor or take immediately to hospital.

SPILLAGE	STORAGE	LABELING / NFPA
clean up spillage, carefully collect remainder; (additional individual protection: P2 respirator).	separate from combustible substances.	R: 20 S: 22 ✖ Harmful

NOTES
✳ TLV equals that of vanadium (respirable dust and fume). If vanadium pentoxide develops use P3 respirator (class 2c). There are two grades available in trade: the spray dried and the normal (fused) form. Depending on the degree of exposure, regular medical checkups are advisable. Lung edema symptoms ussually develop several hours later and are aggravated by physical exertion: rest and hospitalization essential. As first aid, a doctor or authorized person should consider administering a corticosteroid spray.

VANADIUMSULFIDE

PHYSICAL PROPERTIES	IMPORTANT CHARACTERISTICS
Decomposes below melting point, °C ? Relative density (water = 1) 4.2 Solubility in water reaction Relative molecular mass 83	**BLACK CRYSTALS OR POWDER** Toxic gases (→ *sulfur dioxide*) and toxic fumes (→ *vanadium pentoxide*) are a product of combustion or heating. Reacts violently with acids and water (vapor), giving off toxic gas (→ *hydrogen sulfide*). Reacts violently with strong oxidants.
	TLV-TWA not available
	Absorption route: Can enter the body by inhalation or ingestion. Evaporation negligible at 20°C, but harmful concentrations of airborne particles can build up rapidly. **Immediate effects:** Irritates the eyes and respiratory tract. Inhalation of vapor/fumes can cause severe breathing difficulties (lung edema).
Gross formula: VS	

HAZARDS/SYMPTOMS	PREVENTIVE MEASURES	FIRE EXTINGUISHING/FIRST AID
Fire: combustible.	keep away from open flame, no smoking.	AFFF, powder, carbon dioxide, DO NOT USE OTHER WATER-BASED EXTINGUISHERS.
Explosion: finely dispersed particles form explosive mixtures on contact with air.	keep dust from accumulating; sealed machinery, explosion-proof electrical equipment and lighting, grounding.	in case of fire: keep tanks/drums cool by spraying with water, but DO NOT spray substance with water.
	KEEP DUST UNDER CONTROL	
Inhalation: sore throat, cough, shortness of breath, headache.	local exhaust or respiratory protection.	fresh air, rest, place in half-sitting position, call a doctor or take to hospital.
Skin: redness.	gloves.	remove contaminated clothing, wash skin with soap and water.
Eyes: redness, pain.	acid goggles.	flush with water, take to a doctor.
Ingestion: sore throat, cough, diarrhea, shortness of breath, nausea.	do not eat, drink or smoke while working.	rinse mouth, call a doctor or take to hospital.

SPILLAGE	STORAGE	LABELING / NFPA
clean up spillage, carefully collect remainder; (additional individual protection: breathing apparatus).	keep dry, separate from oxidants and acids.	

NOTES
Depending on the degree of exposure, regular medical checkups are advisable. Lung edema symptoms usually develop several hours later and are aggravated by physical exertion: rest and hospitalization essential. As first aid, a doctor or authorized person should consider administering a corticosteroid spray.

Transport Emergency Card TEC(R)-61G12

UN-number: 2575

VANADIUM TETRACHLORIDE

PHYSICAL PROPERTIES	IMPORTANT CHARACTERISTICS
Boiling point, °C 148 Melting point, °C −28 Relative density (water = 1) 1.8 Relative vapor density (air = 1) 6 Relative density at 20 °C of saturated mixture vapor/air (air = 1) 1.04 Vapor pressure, mm Hg at 20 °C 5.9 Solubility in water reaction Relative molecular mass 192.8	**REDDISH-BROWN LIQUID WITH PUNGENT ODOR** Vapor mixes readily with air. Decomposes when exposed to light, giving off toxic → *chlorine* gas. Reacts violently with water, forming corrosive vanadium trichloride, vanadium oxydichloride and → *hydrochloric acid* and evolving heat. Reacts with air to form corrosive → *vanadium pentoxide* and → *hydrochloric acid* fumes. Attacks many metals (e.g. iron, steel, aluminum).
	TLV-TWA 0.05 mg/m³ ✱
	Absorption route: Can enter the body by inhalation or ingestion. Harmful atmospheric concentrations can build up fairly rapidly on evaporation at approx. 20°C - even more rapidly in aerosol form. **Immediate effects:** Corrosive to the eyes, skin and respiratory tract. Inhalation of vapor/fumes can cause severe breathing difficulties (lung edema). In serious cases risk of death.
Gross formula: Cl_4V	

HAZARDS/SYMPTOMS	PREVENTIVE MEASURES	FIRE EXTINGUISHING/FIRST AID
Fire: non-combustible.		in case of fire in immediate vicinity: do not use water.
	STRICT HYGIENE	
Inhalation: *corrosive*; sore throat, cough, severe breathing difficulties.	ventilation, local exhaust or respiratory protection.	fresh air, rest, place in half-sitting position, take to hospital.
Skin: *corrosive*; redness, pain, burns.	gloves, protective clothing.	remove contaminated clothing, flush skin with water or shower, refer to a doctor.
Eyes: *corrosive*; redness, pain, impaired vision.	face shield, or combined eye and respiratory protection.	flush with water, take to a doctor.
Ingestion: *corrosive*; sore throat, abdominal pain, diarrhea.		rinse mouth, take immediately to hospital.

SPILLAGE	STORAGE	LABELING / NFPA
evacuate area, collect leakage in sealable containers, soak up with sand or other inert absorbent and remove to safe place, under no circumstances spray liquid with water; (additional individual protection: CHEMICAL SUIT).	keep in dry, dark place; ventilate at floor level.	◇

NOTES
✱ TLV equals that of vanadium (V_2O_5, respirable dust and fume). For practical purposes TLV for hydrochloric acid should be applied. Lung edema symptoms usually develop several hours later and are aggravated by physical exertion: rest and hospitalization essential. As first aid, a doctor or authorized person should consider administering a corticosteroid spray. Use airtight packaging. Unbreakable packaging preferred; if breakable, keep in unbreakable container.

Transport Emergency Card TEC(R)-80G10   **HI: 88; UN-number: 2444**

VASELINE

PHYSICAL PROPERTIES	IMPORTANT CHARACTERISTICS	
Melting point, °C ca. 40 Flash point, °C 185 Relative density (water = 1) 0.9 Solubility in water none	**WHITE OR YELLOW PASTE** Mixture of higher hydrocarbons.	
	TLV-TWA not available	

HAZARDS/SYMPTOMS	PREVENTIVE MEASURES	FIRE EXTINGUISHING/FIRST AID
Fire: combustible. **Ingestion:** in large quantities: abdominal pain, diarrhea.	keep away from open flame, no smoking.	powder, AFFF, foam, carbon dioxide, (halons). DO NOT induce vomiting, call a doctor.

SPILLAGE	STORAGE	LABELING / NFPA
clean up spillage, soak up with sand or other inert absorbent.		

NOTES
Vaseline is a basic material for many ointments.

CAS-No: [108-05-4]
acetic acid ethenyl ester
acetic acid vinyl ester
ethenyl acetate
ethenyl ethanoate
vinyl acetate monomer

$CH_3COOCH=CH_2$

VINYL ACETATE

PHYSICAL PROPERTIES		IMPORTANT CHARACTERISTICS	
Boiling point, °C	72	**COLORLESS LIQUID WITH CHARACTERISTIC ODOR**	
Melting point, °C	−100	Vapor is heavier than air and spreads at ground level, with risk of ignition at a distance. Able to	
Flash point, °C	−8	form peroxides. Readily able to polymerize, evolving heat. Do not use compressed air when	
Auto-ignition temperature, °C	385	filling, emptying or processing. Reacts violently with strong oxidants.	
Relative density (water = 1)	0.9		
Relative vapor density (air = 1)	3.0	TLV-TWA 10 ppm 35 mg/m³	
Relative density at 20 °C of saturated mixture		STEL 20 ppm 70 mg/m³	
vapor/air (air = 1)	1.2		
Vapor pressure, mm Hg at 20 °C	85.9	**Absorption route:** Can enter the body by inhalation or ingestion. Harmful atmospheric	
Solubility in water, g/100 ml at 20 °C	2.5	concentrations can build up very rapidly on evaporation at 20°C.	
Explosive limits, vol% in air	2.6-13.4	**Immediate effects:** Irritates the eyes, skin and respiratory tract. In substantial concentrations	
Minimum ignition energy, mJ	0.7	can impair consciousness.	
Relative molecular mass	86.1		
Gross formula:	$C_4H_6O_2$		

HAZARDS/SYMPTOMS	PREVENTIVE MEASURES	FIRE EXTINGUISHING/FIRST AID
Fire: extremely flammable.	keep away from open flame and sparks, no smoking.	powder, AFFF, foam, carbon dioxide, (halons).
Explosion: forms explosive air-vapor mixtures.	sealed machinery, ventilation, explosion-proof electrical equipment and lighting, grounding, non-sparking hand tools.	in case of fire: keep tanks/drums cool by spraying with water.
Inhalation: sore throat, hoarseness, shortness of breath.	ventilation, local exhaust or respiratory protection.	fresh air, rest, place in half-sitting position, call a doctor.
Skin: redness, pain.	gloves.	remove contaminated clothing, flush skin with water or shower.
Eyes: redness, pain.	face shield, or combined eye and respiratory protection.	flush with water, take to a doctor if necessary.
Ingestion: sore throat, abdominal pain, vomiting.		rinse mouth, call a doctor or take to hospital.

SPILLAGE	STORAGE	LABELING / NFPA
collect leakage in sealable containers, soak up with sand or other inert absorbent and remove to safe place; (additional individual protection: breathing apparatus).	keep in cool, fireproof place, separate from oxidants; add an inhibitor.	R: 11 S: 16-23-29-33 Note D 🔥 Flammable 3 2 2

NOTES
With diphenylamine as inhibitor can be stored indefinitely; with hydroquinone max. 60 days. Added inhibitor can itself cause strong irritation. Alcohol consumption increases toxic effects.

Transport Emergency Card TEC(R)-3

HI: 339; UN-number: 1301

VINYL BROMIDE
(cylinder)

PHYSICAL PROPERTIES	IMPORTANT CHARACTERISTICS
Boiling point, °C 16 Melting point, °C − 138 Flash point, °C flammable gas Relative density (water = 1) 1.5 Relative vapor density (air = 1) 3.7 Vapor pressure, mm Hg at 20 °C 912 Solubility in water poor Explosive limits, vol% in air 5.6-13.5 Relative molecular mass 107.0	**COLORLESS COMPRESSED LIQUEFIED GAS WITH CHARACTERISTIC ODOR** Gas is heavier than air and spreads at ground level, with risk of ignition at a distance. Light, peroxides and hydrogen sulfide can cause violent polymerization. Flow, agitation etc. can cause build-up of electrostatic charge due to liquid's low conductivity. Do not use compressed air when filling, emptying or processing. Toxic gases are a product of combustion (incl. → *bromine* and → *hydrogen*). Strong reducing agent which reacts violently with oxidants. Reacts violently with chlorine, fluorine and acetylene.

TLV-TWA	5 ppm	22 mg/m³ A2

Absorption route: Can enter the body by inhalation or ingestion or through the skin. Harmful atmospheric concentrations can build up very rapidly if gas is released.
Immediate effects: Irritates the eyes, skin and respiratory tract. Liquid destroys the skin's natural oils. Can cause frostbite due to rapid evaporation. Affects the nervous system. In serious cases risk of unconsciousness.

Gross formula: C_2H_3Br

HAZARDS/SYMPTOMS	PREVENTIVE MEASURES	FIRE EXTINGUISHING/FIRST AID
Fire: extremely flammable.	keep away from open flame and sparks, no smoking.	shut off supply; if impossible and no danger to surrounding area, allow fire to burn itself out; otherwise extinguish with powder, carbon dioxide, (halons).
Explosion: forms explosive air-gas mixtures.	sealed machinery, ventilation, explosion-proof electrical equipment and lighting, grounding when pumping etc. in liquid form.	in case of fire: keep cylinders cool by spraying with water.
	STRICT HYGIENE	IN ALL CASES CALL IN A DOCTOR
Inhalation: sore throat, cough, dizziness, unconsciousness.	ventilation, local exhaust or respiratory protection.	fresh air, rest, call a doctor.
Skin: *is absorbed*; redness; see also 'Inhalation'.	insulated gloves, protective clothing.	remove contaminated clothing, flush skin with water or shower, call a doctor.
Eyes: redness, pain.	face shield, or combined eye and respiratory protection.	flush with water, take to a doctor.
Ingestion: abdominal pain, nausea; see also 'Inhalation'.		rinse mouth, immediately take to hospital.

SPILLAGE	STORAGE	LABELING / NFPA
evacuate area, call in an expert, ventilate, under no circumstances spray liquid with water; (additional individual protection: breathing apparatus).	keep in cool, fireproof place.	0 / 2 / 1

NOTES
Alcohol consumption increases toxic effects. Do not spray leaking cylinder with water (to avoid corrosion). Turn leaking cylinder so that leak is on top to prevent liquid vinyl bromide escaping.

CAS-No: [75-01-4]
chloroethylene
ethylene monochloride
monochloroethene
VC
vinyl chloride monomer

$H_2C = CHCl$

VINYL CHLORIDE
(cylinder)

PHYSICAL PROPERTIES	IMPORTANT CHARACTERISTICS
Boiling point, °C − 14 Melting point, °C − 154 Flash point, °C flammable gas Auto-ignition temperature, °C 415 Relative density (water = 1) 0.90 Relative vapor density (air = 1) 2.2 Vapor pressure, mm Hg at 20 °C 2584 Solubility in water, g/100 ml at 25 °C 0.3 Explosive limits, vol% in air 3.6-31 Minimum ignition energy, mJ <0.3 Relative molecular mass 62.5	**COLORLESS COMPRESSED LIQUEFIED GAS WITH CHARACTERISTIC ODOR** Gas is heavier than air and spreads at ground level, with risk of ignition at a distance. Able to form peroxides and thus to polymerize. Also able to polymerize when moderately heated and exposed to air and light. Toxic and corrosive gases are a product of combustion (→ *phosgene* and → *hydrochloric acid*). Reacts violently with strong oxidants.

TLV-TWA	5 ppm	13 mg/m³ A1	

Absorption route: Can enter the body by inhalation or through the skin. Harmful atmospheric concentrations can build up very rapidly if gas is released.
Immediate effects: Irritates the eyes. Can cause frostbite due to rapid evaporation. Affects the nervous system. Prolonged exposure to vapor can cause disorders in various organs due to damage to connective tissue and/or growth of tumors.
Effects of prolonged/repeated exposure: Prolonged or repeated contact can cause skin and bone disorders. Can affect the blood. Can cause liver, kidney and heart damage. Has been found to cause a type of cancer in humans under certain circumstances.

Gross formula: C_2H_3Cl

HAZARDS/SYMPTOMS	PREVENTIVE MEASURES	FIRE EXTINGUISHING/FIRST AID
Fire: extremely flammable.	keep away from open flame and sparks, no smoking.	shut off supply; if impossible and no danger to surrounding area, allow fire to burn itself out; otherwise extinguish with powder, carbon dioxide, (halons).
Explosion: forms explosive air-gas mixtures.	sealed machinery, ventilation, explosion-proof electrical equipment and lighting, non-sparking tools.	in case of fire: keep cylinder cool by spraying with water, fight fire from sheltered location.
	AVOID ALL CONTACT	
Inhalation: headache, dizziness, nausea, sleepiness.	ventilation, local exhaust or respiratory protection.	fresh air, rest, call a doctor.
Skin: *in case of frostbite:* redness, pain, burns.	insulated gloves.	*in case of frostbite:* DO NOT remove clothing, flush skin with water or shower.
Eyes: redness, pain.	face shield, or combined eye and respiratory protection.	flush with water, take to a doctor.

SPILLAGE	STORAGE	LABELING / NFPA
evacuate area, call in an expert, ventilate, under no circumstances spray liquid with water; (additional individual protection: breathing apparatus).	keep in cool, fireproof place; add an inhibitor (for long-term storage).	 4 2 2

NOTES
Before distilling check for peroxides; if found, render harmless. Depending on the degree of exposure, regular medical checkups are advisable. Turn leaking cylinder so that leak is on top to prevent liquid vinyl chloride escaping.

Transport Emergency Card TEC(R)-150

HI: 239; UN-number: 1086

CAS-No: [75-02-5]
fluoroethene
fluoroethylene
monofluoroethylene

$CH_2 = CHF$

VINYL FLUORIDE
(cylinder)

PHYSICAL PROPERTIES	IMPORTANT CHARACTERISTICS
Boiling point, °C −72 Melting point, °C −160 Flash point, °C flammable gas Auto-ignition temperature, °C 460 Relative density (water = 1) 0.7 Relative vapor density (air = 1) 1.6 Vapor pressure, mm Hg at 21 °C 19.4 Solubility in water poor Explosive limits, vol% in air 2.6-29 Relative molecular mass 46.0	**COLORLESS COMPRESSED LIQUEFIED GAS** Gas is heavier than air and spreads at ground level, with risk of ignition at a distance. Under certain conditions can polymerize, evolving large quantities of heat. Flow, agitation etc. can cause build-up of electrostatic charge due to liquid's low conductivity. Decomposes when heated or burned, giving off toxic gases → *hydrogen fluoride*. Reacts violently with oxidants.
	TLV-TWA not available
	Absorption route: Can enter the body by inhalation. Harmful atmospheric concentrations can build up very rapidly if gas is released. **Immediate effects:** Irritates the eyes and respiratory tract. Liquid can cause frostbite due to rapid evaporation. Affects the nervous system. In serious cases risk of unconsciousness.
Gross formula: C_2H_3F	

HAZARDS/SYMPTOMS	PREVENTIVE MEASURES	FIRE EXTINGUISHING/FIRST AID
Fire: extremely flammable.	keep away from open flame and sparks, no smoking.	shut off supply; if impossible and no danger to surrounding area, allow fire to burn itself out; otherwise extinguish with powder, carbon dioxide, (halons).
Explosion: forms explosive air-gas mixtures.	sealed machinery, ventilation, explosion-proof electrical equipment and lighting, grounding.	in case of fire: keep cylinders cool by spraying with water.
Inhalation: cough, shortness of breath, dizziness.	ventilation, local exhaust or respiratory protection.	fresh air, rest, call a doctor.
Skin: *in case of frostbite*: redness, pain, sores.	insulating gloves.	*in case of frostbite*: DO NOT remove clothing, flush skin with water or shower, call a doctor if necessary.
Eyes: *in case of frostbite*: redness, pain, impaired vision.	safety glasses, or combined eye and respiratory protection.	flush with water, take to a doctor.

SPILLAGE	STORAGE	LABELING / NFPA
evacuate area, call in an expert, ventilate, under no circumstances spray liquid with water; (additional individual protection: breathing apparatus).	keep in cool, fireproof place, separate from oxidants.	4 / 1 2

NOTES
High atmospheric concentrations, e.g. in poorly ventilated spaces, can cause oxygen deficiency, with risk of unconsciousness. Do not spray leaking cylinder with water (to avoid corrosion). Turn leaking cylinder so that leak is on top to prevent liquid vinyl fluoride escaping.

Transport Emergency Card TEC(R)-682

HI: 239; UN-number: 1860

919

CAS-No: [107-25-5]
methoxyethene
methoxyethylene
methyl vinyl ether
MVE

$CH_2 = CHOCH_3$

VINYL METHYL ETHER
(cylinder)

PHYSICAL PROPERTIES	IMPORTANT CHARACTERISTICS
Boiling point, °C 6 Melting point, °C − 122 Flash point, °C flammable gas Relative density (water = 1) 0.8 Relative vapor density (air = 1) 20 Vapor pressure, mm Hg at 20 °C 1117 Solubility in water poor Explosive limits, vol% in air 2.6-39 Relative molecular mass 58.1 Gross formula: C_3H_6O	**COLORLESS COMPRESSED LIQUEFIED GAS WITH CHARACTERISTIC ODOR** Gas is heavier than air and spreads at ground level, with risk of ignition at a distance. Able to form peroxides and polymerize. Flow, agitation etc. can cause build-up of electrostatic charge due to liquid's low conductivity. Decomposes on contact with dilute acids or water, giving off → *acetaldehyde* and → *methanol*. Reacts violently with strong oxidants and acids, with risk of fire and explosion. TLV-TWA not available **Absorption route:** Can enter the body by inhalation. Harmful atmospheric concentrations can build up very rapidly if gas is released. **Immediate effects:** Irritates the eyes, skin and respiratory tract. Liquid can cause frostbite due to rapid evaporation. Affects the nervous system. In serious cases risk of seizures, unconsciousness and death. **Effects of prolonged/repeated exposure:** Prolonged or repeated contact can cause skin disorders. Can cause liver, kidney and brain damage.

HAZARDS/SYMPTOMS	PREVENTIVE MEASURES	FIRE EXTINGUISHING/FIRST AID
Fire: extremely flammable.	keep away from open flame and sparks, no smoking.	shut off supply; if impossible and no danger to surrounding area, allow fire to burn itself out; otherwise extinguish with powder, carbon dioxide, (halons).
Explosion: forms explosive air-gas mixtures.	sealed machinery, ventilation, explosion-proof electrical equipment and lighting, grounding when pumping etc. in liquid form.	in case of fire: keep cylinders cool by spraying with water.
Inhalation: sore throat, drowsiness, unconsciousness.	ventilation, local exhaust or respiratory protection.	fresh air, rest, call a doctor.
Skin: *in case of frostbite*: redness, pain, sores.	insulated gloves.	*in case of frostbite*: DO NOT remove clothing, flush skin with water or shower, call a doctor if necessary.
Eyes: *in case of frostbite*: redness, pain.	face shield, or combined eye and respiratory protection.	flush with water, take to a doctor.

SPILLAGE	STORAGE	LABELING / NFPA
ventilate, under no circumstances spray liquid with water; (additional individual protection: breathing apparatus).	if stored indoors, keep in cool, fireproof place.	R: 13 S: 9-16-33 Note D 🔥 Flammable NFPA: 4 / 2 / 2

NOTES
Do not spray leaking cylinder with water (to avoid corrosion). Turn leaking cylinder so that leak is on top to prevent liquid vinyl methyl ether escaping.

Transport Emergency Card TEC(R)-684

HI: 236; UN-number: 1087

5-VINYL-2-NORBORNENE

PHYSICAL PROPERTIES	IMPORTANT CHARACTERISTICS
Boiling point, °C — 141 Melting point, °C — −80 Flash point, °C — 28 Relative density (water = 1) — 0.9 Relative vapor density (air = 1) — 4.2 Relative density at 20 °C of saturated mixture vapor/air (air = 1) — 1.02 Vapor pressure, mm Hg at 20 °C — 4.0 Solubility in water — none Explosive limits, vol% in air — ?-? Relative molecular mass — 120.0	**COLORLESS LIQUID WITH CHARACTERISTIC ODOR** Vapor mixes readily with air. Able to polymerize. Flow, agitation etc. can cause build-up of electrostatic charge due to liquid's low conductivity. Reacts violently with strong oxidants. TLV-TWA not available **Absorption route:** Can enter the body by inhalation or ingestion. Harmful atmospheric concentrations can build up fairly rapidly on evaporation at approx. 20°C - even more rapidly in aerosol form. **Immediate effects:** Irritates the eyes, skin and respiratory tract.
Gross formula: C_9H_{12}	

HAZARDS/SYMPTOMS	PREVENTIVE MEASURES	FIRE EXTINGUISHING/FIRST AID
Fire: flammable.	keep away from open flame and sparks, no smoking.	powder, AFFF, foam, carbon dioxide, (halons).
Explosion: above 28°C: forms explosive air-vapor mixtures.	above 28°C: sealed machinery, ventilation, explosion-proof electrical equipment, grounding.	in case of fire: keep tanks/drums cool by spraying with water.
Inhalation: sore throat, cough, shortness of breath.	ventilation, local exhaust or respiratory protection.	fresh air, rest, call a doctor if necessary.
Skin: redness.	gloves.	remove contaminated clothing, flush skin with water or shower.
Eyes: redness, pain.	safety glasses.	flush with water, take to a doctor if necessary.
Ingestion: abdominal pain, nausea, vomiting.		rinse mouth, call a doctor if necessary.

SPILLAGE	STORAGE	LABELING / NFPA
collect leakage in sealable containers, soak up with sand or other inert absorbent and remove to safe place; (additional individual protection: breathing apparatus).	keep in cool, fireproof place, separate from oxidants; add an inhibitor.	

NOTES
Insufficient toxicological data on harmful effects to humans. Explosive limits not given in the literature. For safety purposes TLV for 5-ethylidene-2-norborene is usually a good guide (TLV-TWA: C 5ppm and C 25mg/m³).

Transport Emergency Card TEC(R)-30G08

CAS-No: [100-69-6]

2-VINYLPYRIDINE

PHYSICAL PROPERTIES	IMPORTANT CHARACTERISTICS
Boiling point, °C 159 Flash point, °C see notes Relative density (water = 1) 0.98 Relative vapor density (air = 1) 3.6 Relative density at 20 °C of saturated mixture vapor/air (air = 1) 1.01 Vapor pressure, mm Hg at 20 °C ca. 1.5 Solubility in water, g/100 ml 2.5 Relative molecular mass 105.1 Gross formula: C_7H_7N	**COLORLESS LIQUID WITH CHARACTERISTIC ODOR** Vapor mixes readily with air. Able to form peroxides and thus to polymerize. Decomposes when heated or burned, giving off toxic vapors. Reacts violently with strong oxidants. TLV-TWA not available **Absorption route:** Can enter the body by inhalation or ingestion or through the skin. Harmful atmospheric concentrations can build up fairly rapidly on evaporation at approx. 20°C - even more rapidly in aerosol form. **Immediate effects:** Corrosive to the eyes, skin and respiratory tract. In substantial concentrations can impair consciousness. Inhalation of vapor/fumes can cause severe breathing difficulties (lung edema). In serious cases risk of death.

HAZARDS/SYMPTOMS	PREVENTIVE MEASURES	FIRE EXTINGUISHING/FIRST AID
Fire: combustible.	keep away from open flame, no smoking.	powder, AFFF, foam, carbon dioxide, (halons).
Inhalation: *corrosive*; sore throat, cough, shortness of breath, severe breathing difficulties.	ventilation, local exhaust or respiratory protection.	fresh air, rest, place in half-sitting position, take to hospital.
Skin: *corrosive, is absorbed*; redness, pain, burns.	gloves, protective clothing.	remove contaminated clothing, flush skin with water or shower, call a doctor.
Eyes: *corrosive*; redness, pain, impaired vision.	face shield.	flush with water, take to a doctor.
Ingestion: *corrosive*; sore throat, abdominal pain, diarrhea, vomiting.		rinse mouth, immediately take to hospital.

SPILLAGE	STORAGE	LABELING / NFPA
collect leakage in sealable containers, soak up with sand or other inert absorbent and remove to safe place; (additional individual protection: breathing apparatus).	keep in fireproof place, separate from strong oxidants; add an inhibitor.	

NOTES
Combustible, but flash point and explosive limits not given in the literature. Before distilling check for peroxides; if found, render harmless. Lung edema symptoms usually develop several hours later and are aggravated by physical exertion: rest and hospitalization essential. As first aid, a doctor or authorized person should consider administering a corticosteroid spray. Commercial products usually contain inhibitor.

Transport Emergency Card TEC(R)-80G08

HI: 639; UN-number: 3073

$CH_2 = CHC_6H_4CH_3$

m- AND p-VINYLTOLUENE

PHYSICAL PROPERTIES		IMPORTANT CHARACTERISTICS	
Boiling point, °C	170	**COLORLESS LIQUID WITH CHARACTERISTIC ODOR**	
Melting point, °C	−77	Vapor mixes readily with air. Able to form peroxides and thus to polymerize. The presence of metallic salts enhances polymerization. Strong reducing agent which reacts violently with oxidants.	
Flash point, °C	53		
Auto-ignition temperature, °C	490		
Relative density (water = 1)	0.9		
Relative vapor density (air = 1)	4.1	TLV-TWA	50 ppm	242 mg/m³
Relative density at 20 °C of saturated mixture		STEL	100 ppm	483 mg/m³
vapor/air (air = 1)	1.0		
Vapor pressure, mm Hg at 20 °C	ca. 1.5	**Absorption route:** Can enter the body by inhalation or ingestion or through the skin. Harmful atmospheric concentrations build up fairly slowly on evaporation at 20°C, but much more rapidly in aerosol form.	
Solubility in water	poor		
Explosive limits, vol% in air	0.8-11	**Immediate effects:** Irritates the eyes, skin and respiratory tract. Liquid destroys the skin's natural oils. In substantial concentrations can impair consciousness.	
Relative molecular mass	118.2		
Gross formula:	C_9H_{10}		

HAZARDS/SYMPTOMS	PREVENTIVE MEASURES	FIRE EXTINGUISHING/FIRST AID
Fire: flammable.	keep away from open flame and sparks, no smoking.	powder, AFFF, foam, carbon dioxide, (halons).
Explosion: above 53°C: forms explosive air-vapor mixtures.	above 53°C: sealed machinery, ventilation, explosion-proof electrical equipment.	in case of fire: keep tanks/drums cool by spraying with water.
Inhalation: sore throat, headache, dizziness, drowsiness.	ventilation, local exhaust or respiratory protection.	fresh air, rest, call a doctor.
Skin: *is absorbed*; redness, pain.	gloves, protective clothing.	remove contaminated clothing, flush skin with water or shower, call a doctor.
Eyes: redness, pain.	safety glasses.	flush with water, take to a doctor if necessary.
Ingestion: abdominal pain, nausea, vomiting.		rinse mouth, DO NOT induce vomiting, take immediately to hospital.

SPILLAGE	STORAGE	LABELING / NFPA
collect leakage in sealable containers, soak up with sand or other inert absorbent and remove to safe place.	keep in fireproof, cool and dark place, separate from oxidants, add an inhibitor (tert. butylcatechol, 50ppm).	(NFPA diamond: top 2, left 2, right 1)

NOTES
Not to be confused with α-*methylstyrene*. Alcohol consumption increases toxic effects.

XENON
(cylinder)

PHYSICAL PROPERTIES	IMPORTANT CHARACTERISTICS
Boiling point, °C −108 Melting point, °C −112 Relative density (water = 1) 1.5 Relative vapor density (air = 1) 5.4 Solubility in water poor Relative atomic mass 131.3 Log P octanol/water 1.4	**COLORLESS, ODORLESS COMPRESSED GAS OR COMPRESSED LIQUEFIED GAS** Gas is heavier than air and can accumulate close to ground level, causing oxygen deficiency, with risk of unconsciousness.
	TLV-TWA not available
	Absorption route: Can saturate the air if released, with risk of suffocation. **Immediate effects:** Liquid can cause frostbite due to rapid evaporation. Inhalation can cause severe breathing difficulties.
Gross formula: Xe	

HAZARDS/SYMPTOMS	PREVENTIVE MEASURES	FIRE EXTINGUISHING/FIRST AID
Fire: non-combustible.		in case of fire in immediate vicinity: use any extinguishing agent.
Explosion:		in case of fire: keep cylinders cool by spraying with water.
Inhalation: severe breathing difficulties, headache, dizziness, unconsciousness.	ventilation, local exhaust or respiratory protection.	fresh air, rest, artificial respiration if necessary, take to hospital.

SPILLAGE	STORAGE	LABELING / NFPA
ventilate; (additional individual protection: breathing apparatus).	if stored indoors, keep in cool, fireproof place.	

NOTES
Critical temperature 16.6°C. Log P octanol/water is estimated. High atmospheric concentrations, e.g. in poorly ventilated spaces, can cause oxygen deficiency, with risk of unconsciousness.

Transport Emergency Card TEC(R)-20G16   **HI: 20; UN-number: 2036**

XENON
(liquid)

PHYSICAL PROPERTIES	IMPORTANT CHARACTERISTICS	
Boiling point, °C − 108.1 Melting point, °C − 112 Relative density (water = 1) see notes Vapor pressure, mm Hg at − 108 °C 760 Solubility in water poor Relative atomic mass 131.3 Log P octanol/water 1.4	**COLORLESS, ODORLESS, EXTREMELY COLD LIQUID** *Cold* vapor is heavier than air.	
	TLV-TWA not available	
	Absorption route: If liquid is released, air can become saturated due to evaporation, with risk of suffocation. **Immediate effects:** Cold liquid can cause frostbite.	
Gross formula: Xe		

HAZARDS/SYMPTOMS	PREVENTIVE MEASURES	FIRE EXTINGUISHING/FIRST AID
Fire: non-combustible.		in case of fire in immediate vicinity: use any extinguishing agent, but DO NOT spray liquid with water.
Inhalation: see 'Notes'.	ventilation, local exhaust or respiratory protection.	fresh air, rest, artificial respiration if necessary, call a doctor, take to hospital if necessary.
Skin: *in case of frostbite*: redness, pain, blisters, sores.	insulated gloves.	*in case of frostbite*: DO NOT remove clothing, flush skin with water or shower.
Eyes: *in case of frostbite*: redness, pain, impaired vision.	face shield, or combined eye and respiratory protection.	flush with water, take to a doctor.

SPILLAGE	STORAGE	LABELING / NFPA
ventilate, under no circumstances spray liquid with water; (additional individual protection: breathing apparatus).	keep cool; ventilate.	

NOTES
Density of liquid at boiling point 3.05kg/l. Critical temperature 16.6° C. Log P octanol/water is estimated. High atmospheric concentrations, e.g. in poorly ventilated spaces, can cause oxygen deficiency, with risk of unconsciousness. Use special insulated drums.

Transport Emergency Card TEC(R)-20G22   **HI: 22; UN-number: 2591**

1,3-dimethylbenzene
1,3-xylene
m-xylol

m-XYLENE

PHYSICAL PROPERTIES		IMPORTANT CHARACTERISTICS	
Boiling point, °C	139	**COLORLESS LIQUID WITH CHARACTERISTIC ODOR**	
Melting point, °C	−48	Vapor mixes readily with air. Flow, agitation etc. can cause build-up of electrostatic charge due to liquid's low conductivity. Reacts violently with strong oxidants.	
Flash point, °C	25		
Auto-ignition temperature, °C	525		
Relative density (water = 1)	0.86	TLV-TWA	100 ppm	434 mg/m³
Relative vapor density (air = 1)	3.6	STEL	150 ppm	651 mg/m³

Relative density at 20°C of saturated mixture vapor/air (air = 1)	1.02
Vapor pressure, mm Hg at 20°C	6.1
Solubility in water	none
Explosive limits, vol% in air	1.1-7.0
Electrical conductivity, pS/m	<10⁻¹
Relative molecular mass	106.2
Log P octanol/water	3.2

Absorption route: Can enter the body by inhalation or ingestion or through the skin. Harmful atmospheric concentrations build up fairly slowly on evaporation at 20°C, but much more rapidly in aerosol form.
Immediate effects: Irritates the eyes, skin and respiratory tract. Liquid destroys the skin's natural oils. Affects the nervous system. In serious cases risk of unconsciousness. If liquid is swallowed, droplets can enter the lungs, with risk of pneumonia. The risk of poisoning from absorption through the skin is practically zero.
Effects of prolonged/repeated exposure: Prolonged or repeated contact can cause skin disorders.

Gross formula: C_8H_{10}

HAZARDS/SYMPTOMS	PREVENTIVE MEASURES	FIRE EXTINGUISHING/FIRST AID
Fire: flammable.	keep away from open flame and sparks, no smoking.	powder, AFFF, foam, carbon dioxide, (halons).
Explosion: above 25°C: forms explosive air-vapor mixtures.	above 25°C: sealed machinery, ventilation, explosion-proof electrical equipment, grounding.	in case of fire: keep tanks/drums cool by spraying with water.
Inhalation: cough, headache, dizziness, nausea, unconsciousness.	ventilation, local exhaust or respiratory protection.	fresh air, rest, call a doctor.
Skin: redness, pain.	gloves.	remove contaminated clothing, flush skin with water or shower.
Eyes: redness, pain, impaired vision.	face shield.	flush with water, take to a doctor if necessary.
Ingestion: cough, headache, nausea, dizziness, vomiting, unconsciousness.		rinse mouth, DO NOT induce vomiting, take immediately to hospital.

SPILLAGE	STORAGE	LABELING / NFPA
collect leakage in sealable containers, soak up with sand or other inert absorbent and remove to safe place.	keep in fireproof place, separate from oxidants.	R: 10-20/21-38 S: 25 ✖ Harmful

NFPA: 3 / 2 / 0

NOTES

Commercial product xylene is a mixture of isomers, principally *meta-xylene*. Alcohol consumption increases toxic effects. The measures on this card also apply to ortho-xylene. Physical properties of ortho-xylene; boiling point 144°C; melting point −25°C; flash point: 30°C; autoignition temperature: 465°C.

Transport Emergency Card TEC(R)-33

HI: 30; UN-number: 1307

p-XYLENE

PHYSICAL PROPERTIES		IMPORTANT CHARACTERISTICS		
Boiling point, °C	138	**COLORLESS LIQUID WITH CHARACTERISTIC ODOR**		
Melting point, °C	13	Vapor mixes readily with air. Flow, agitation etc. can cause build-up of electrostatic charge due to		
Flash point, °C	25	liquid's low conductivity.		
Auto-ignition temperature, °C	525			
Relative density (water = 1)	0.86	TLV-TWA	100 ppm	434 mg/m³
Relative vapor density (air = 1)	3.6	STEL	150 ppm	651 mg/m³
Relative density at 20 °C of saturated mixture				
vapor/air (air = 1)	1.02	**Absorption route:** Can enter the body by inhalation or ingestion or through the skin. Harmful		
Vapor pressure, mm Hg at 20 °C	6.2	atmospheric concentrations build up fairly slowly on evaporation at 20° C, but much more rapidly		
Solubility in water	none	in aerosol form.		
Explosive limits, vol% in air	1.1-7.0	**Immediate effects:** Irritates the eyes, skin and respiratory tract. Liquid destroys the skin's natural		
Electrical conductivity, pS/m	< 10⁻¹	oils. Affects the nervous system. In serious cases risk of unconsciousness. If liquid is swallowed,		
Relative molecular mass	106.2	droplets can enter the lungs, with risk of pneumonia.		
Log P octanol/water	3.2	**Effects of prolonged/repeated exposure:** Prolonged or repeated contact can cause skin		
		disorders.		
Gross formula:	C_8H_{10}			

HAZARDS/SYMPTOMS	PREVENTIVE MEASURES	FIRE EXTINGUISHING/FIRST AID
Fire: flammable.	keep away from open flame and sparks, no smoking.	powder, AFFF, foam, carbon dioxide, (halons).
Explosion: above 25°C: forms explosive air-vapor mixtures.	above 25°C: sealed machinery, ventilation, explosion-proof electrical equipment, grounding.	in case of fire: keep tanks/drums cool by spraying with water.
Inhalation: cough, headache, dizziness, nausea, unconsciousness.	ventilation, local exhaust or respiratory protection.	fresh air, rest, call a doctor.
Skin: redness, pain.	gloves.	remove contaminated clothing, flush skin with water or shower.
Eyes: redness, pain, impaired vision.	face shield.	flush with water, take to a doctor if necessary.
Ingestion: cough, headache, dizziness, nausea, vomiting, unconsciousness.		rinse mouth, DO NOT induce vomiting, call a doctor or take to hospital.

SPILLAGE	STORAGE	LABELING / NFPA
collect leakage in sealable containers, soak up with sand or other inert absorbent and remove to safe place.	keep in fireproof place, separate from oxidants.	R: 10-20/21-38 S: 25 ✖ Harmful

NFPA: 3 / 2 / 0

NOTES
Alcohol consumption increases toxic effects. Commercial product xylene is a mixture of isomers, principally *meta-xylene*.

dimethylaniline
dimethylbenzeneamine

$(CH_3)_2C_6H_3NH_2$

XYLIDINE
(mixture of o-, m-, and p-isomers)

PHYSICAL PROPERTIES	IMPORTANT CHARACTERISTICS
Boiling point, °C .... 215-223 Melting point, °C .... − 15/ + 51 Flash point, °C .... 97 Auto-ignition temperature, °C ... see notes Relative density (water = 1) ... 0.97-0.99 Relative vapor density (air = 1) ... 4.18 Relative density at 20 °C of saturated mixture vapor/air (air = 1) .... 1.0 Vapor pressure, mm Hg at 20 °C ... 152 Solubility in water .... poor Relative molecular mass ...... 121.2 Log P octanol/water 1.8	**COLORLESS TO BROWN LIQUID OR SOLID WITH CHARACTERISTIC ODOR** Vapor mixes readily with air. Decomposes when heated, giving off toxic gases. Reacts violently with strong oxidants and strong acids. Forms explosive chloramines in contact with hypochlorites.

	TLV-TWA (skin) 0.5 ppm 2.5 mg/m³ A2

Absorption route: Can enter the body by inhalation or ingestion or through the skin. Harmful atmospheric concentrations build up fairly slowly on evaporation at 20° C, but much more rapidly in aerosol form.
Immediate effects: Irritates the eyes, skin and respiratory tract. Affects the nervous system.
Effects of prolonged/repeated exposure: At high exposure levels can cause lung, liver and kidney disorders. Can affect the blood.

Gross formula: $C_8H_{11}N$

HAZARDS/SYMPTOMS	PREVENTIVE MEASURES	FIRE EXTINGUISHING/FIRST AID
Fire: combustible.	keep away from open flame, no smoking.	powder, AFFF, foam, carbon dioxide, (halons).
Explosion: above 97° C: forms explosive air-vapor mixtures.	above 97°C: sealed machinery, ventilation.	
	STRICT HYGIENE	**IN ALL CASES CALL IN A DOCTOR**
Inhalation: shortness of breath, headache, dizziness, drowsiness, blue skin, feeling of weakness.	ventilation, local exhaust or respiratory protection.	fresh air, rest, artificial respiration if necessary, take to hospital.
Skin: *is absorbed*; redness; see also 'Inhalation'.	gloves, protective clothing.	remove contaminated clothing, wash skin with soap and water, take to hospital.
Eyes: redness, pain.	face shield.	flush with water, take to a doctor if necessary.
Ingestion: abdominal pain, nausea; see also 'Inhalation'.	do not eat, drink or smoke while working.	rinse mouth, take immediately to hospital.

SPILLAGE	STORAGE	LABELING / NFPA
collect leakage in sealable containers, soak up with sand or other inert absorbent and remove to safe place.	keep separate from oxidants and strong acids.	R: 23/24/25-33 S: 28-36/37-44 Note C ☠ Toxic

NFPA diamond: 3 (left), 1 (top), 0 (right)

NOTES

Boiling point, melting point and relative density depend on isomers. Flash point of 2-3-isomer and vapor pressure at 20° C of 2-6-isomer are given. Log P octanol/water is estimated. Alcohol consumption increases toxic effects. Depending on the degree of exposure, regular medical checkups are advisable. Effects on blood due to formation of methemoglobin. Special first aid required in the event of poisoning: antidotes must be available (with instructions).

Transport Emergency Card TEC(R)-679

HI: 60; UN-number: 1711

ZINC
(powder, pyrophoric)

PHYSICAL PROPERTIES	IMPORTANT CHARACTERISTICS
Boiling point, °C 907 Melting point, °C 420 Auto-ignition temperature, °C 460 Relative density (water = 1) 7.1 Vapor pressure, mm Hg at 478 °C 0.99 Solubility in water reaction Explosive limits, g/m³ in air 500-? Minimum ignition energy, mJ 960 Relative atomic mass 65.3	**LIGHT GRAY POWDER** In dry state can form electrostatic charge if stirred, transported pneumatically, poured etc. Can ignite spontaneously on exposure to air. Strong reducing agent which reacts violently with oxidants. Reacts with water and reacts violently with acids and bases, giving off flammable gas (→ *hydrogen*). Reacts violently with sulfur, halogenated hydrocarbons and many other substances, with risk of fire and explosion.

	TLV-TWA not available

Absorption route: Can enter the body by inhalation or ingestion. Evaporation negligible at 20° C, but harmfull concentrations of airborne particles can build up rapidly.
Immediate effects: Inhalation of dust and vapor can cause metal fume fever.

Gross formula: Zn

HAZARDS/SYMPTOMS	PREVENTIVE MEASURES	FIRE EXTINGUISHING/FIRST AID
Fire: extremely flammable.	keep away from open flame and sparks, no smoking.	dry sand, special powder extinguishers, NO OTHER extinguishing agents.
Explosion: finely dispersed particles form explosive mixtures on contact with air.	keep dust from accumulating; sealed machinery, explosion-proof electrical equipment and lighting, grounding.	
Inhalation: cough, fever.	local exhaust or respiratory protection.	fresh air, rest, call a doctor.
Skin:	gloves.	
Eyes:	goggles.	
Ingestion: possibly nausea.		

SPILLAGE	STORAGE	LABELING / NFPA
clean up spillage, carefully collect remainder; (additional individual protection: P2 respirator).	keep in fireproof, dry place, separate from oxidants, acids and bases.	R: 15-17 S: 7/8-43 🔥 Highly flammable

NOTES
Flush contaminated clothing with water (fire hazard). Reacts violently with extinguishing agents such as water, powder, foam and carbon dioxide. Use airtight packaging. Stabilized zinc powder is hardly combustible.

Transport Emergency Card TEC(R)-42G03

HI: 43; UN-number: 1383

ZINC CHLORIDE

PHYSICAL PROPERTIES		IMPORTANT CHARACTERISTICS	
Boiling point, °C	732	**COLORLESS HYGROSCOPIC CRYSTALS OR POWDER**	
Melting point, °C	290	Decomposes when heated, giving off toxic vapors. In aqueous solution is a medium strong acid	
Relative density (water = 1)	2.9	which reacts violently with bases and is corrosive.	
Vapor pressure, mm Hg at 428 °C	0.99		
Solubility in water, g/100 ml at 25 °C	432	TLV-TWA	1 mg/m³ ✱
Relative molecular mass	136.3	STEL	2 mg/m³
		Absorption route: Can enter the body by inhalation or ingestion. Evaporation negligible at 20°C, but harmful concentrations of airborne particles can build up rapidly.	
		Immediate effects: Corrosive to the eyes, skin and respiratory tract. Inhalation can cause lung edema. In serious cases risk of death.	
Gross formula:	Cl$_2$Zn		

HAZARDS/SYMPTOMS	PREVENTIVE MEASURES	FIRE EXTINGUISHING/FIRST AID
Fire: non-combustible.		in case of fire in immediate vicinity: use any extinguishing agent.
Inhalation: *corrosive*; sore throat, cough, shortness of breath, severe breathing difficulties.	local exhaust or respiratory protection.	fresh air, rest, place in half-sitting position, take to hospital.
Skin: *corrosive*; redness, pain, serious burns.	gloves, protective clothing.	remove contaminated clothing, flush skin with water or shower, take to hospital.
Eyes: *corrosive*; redness, pain, impaired vision.	face shield.	flush with water, take to a doctor.
Ingestion: *corrosive*; sore throat, abdominal pain, diarrhea, vomiting.		rinse mouth, take immediately to hospital.

SPILLAGE	STORAGE	LABELING / NFPA
clean up spillage, carefully collect remainder; (additional individual protection: P2 respirator).	keep dry, separate from strong bases.	R: 34 S: 7/8-28 Corrosive

NOTES
✱ TLV equals that of zinc chloride fume. When heated (e.g. in soldering) can give off fumes. Lung edema symptoms usually develop several hours later and are aggravated by physical exertion: rest and hospitalization essential. As first aid, a doctor or authorized person should consider administering a corticosteroid spray. Use airtight packaging.

Transport Emergency Card TEC(R)-80G11   **HI: 80; UN-number: 2331**

ZINC NITRATE
(hexahydrate)

PHYSICAL PROPERTIES	IMPORTANT CHARACTERISTICS	
Decomposes below boiling point, °C 140 Melting point, °C 36 Relative density (water = 1) 2.0 Solubility in water, g/100 ml 200 Relative molecular mass 297.5	**COLORLESS CRYSTALS OR LUMPS** Decomposes when heated above 140°C, giving off toxic gases (→ *nitrous vapors* and → *zinc oxide* fumes. Strong oxidant which reacts violently with combustible substances and reducing agents. Reacts violently with carbon, copper powder, metal sulfides, phophorus and sulfur.	
	TLV-TWA not available	
	Absorption route: Can enter the body by inhalation or ingestion. Evaporation negligible at 20°C, but in finely dispersed form harmfull concentrations of airborne particles can build up. **Immediate effects:** Corrosive to the eyes, skin and respiratory tract. Inhalation can cause lung edema. In serious cases risk of unconsciousness. **Effects of prolonged/repeated exposure:** Can affect the blood.	
Gross formula: $N_2O_6Zn.6H_2O$		

HAZARDS/SYMPTOMS	PREVENTIVE MEASURES	FIRE EXTINGUISHING/FIRST AID
Fire: not combustible but enhances combustion of other substances.	avoid contact with combustible substances.	in case of fire in immediate vicinity: use any extinguishing agent.
Inhalation: *corrosive*; sore throat, cough, severe breathing difficulties.	local exhaust or respiratory protection.	fresh air, rest, place in half-sitting position, artificial respiration if necessary, take to hospital.
Skin: *corrosive*; redness, pain, burns.	gloves, protective clothing.	remove contaminated clothing, flush skin with water or shower, refer to a doctor.
Eyes: *corrosive*; redness, pain, impaired vision.	face shield.	flush with water, take to a doctor.
Ingestion: *corrosive*; sore throat, abdominal pain, nausea, vomiting, blue skin, feeling of weakness.		rinse mouth, take to hospital.

SPILLAGE	STORAGE	LABELING / NFPA
clean up spillage, carefully collect remainder; (additional individual protection: P2 respirator).	keep separate from combustible substances and reducing agents.	◇

NOTES
Apparent melting point due to loss of water of crystallization; melting point of anhydrous zinc nitrate at approx 110°C. Flush contaminated clothing with water (fire hazard). Lung edema symptoms usually develop several hours later and are aggravated by physical exertion: rest and hospitalization essential. As first aid, a doctor or authorized person should consider administering a corticosteroid spray. Special first aid required in the event of poisoning: antidotes must be available (with instructions).

Transport Emergency Card TEC(R)-51G02   **UN-number: 1514**

calamine
zincite
zinc monoxide
zincoid
zinc white

ZnO

ZINC OXIDE

PHYSICAL PROPERTIES	IMPORTANT CHARACTERISTICS	
Melting point, °C > 1800 Relative density (water = 1) 5.5 Solubility in water none Relative molecular mass 81.4	**WHITE POWDER** Decomposes on contact with acids, in which it dissolves.	
	TLV-TWA 5 mg/m³ ✱ STEL 10 mg/m³	
	Absorption route: Can enter the body by inhalation or ingestion. Harmful concentrations of airborne particles can build up rapidly. **Immediate effects:** Irritates the eyes and respiratory tract.	
Gross formula: OZn		

HAZARDS/SYMPTOMS	PREVENTIVE MEASURES	FIRE EXTINGUISHING/FIRST AID
Fire: non-combustible.		in case of fire in immediate vicinity: use any extinguishing agent.
Inhalation: sore throat, cough, headache.	local exhaust or respiratory protection.	fresh air, rest, call a doctor if necessary.
Skin:	gloves.	
Eyes: redness, pain.	goggles.	flush with water, take to a doctor if necessary.
Ingestion: diarrhea or constipation.		rinse mouth, call a doctor if necessary.

SPILLAGE	STORAGE	LABELING / NFPA
clean up spillage; (additional individual protection: P2 respirator).		

NOTES
✱ TLV equals that of zinc (fume). Zinc oxide fumes given off when burning or welding materials containing zinc can cause metal fume fever if inhaled.

CAS-No: [557-05-1]
stearic acid, zinc salt
zinc distearate
zinc octadecanoate

$Zn(C_{18}H_{35}O_2)_2$

ZINC STEARATE

PHYSICAL PROPERTIES	IMPORTANT CHARACTERISTICS	
Melting point, °C 120-130 Flash point, °C 277 Auto-ignition temperature, °C 420 Relative density (water = 1) 1.1 Solubility in water none Explosive limits, g/m³ in air 30-? Relative molecular mass 632.3	**WHITE POWDER** In dry state can form electrostatic charge if stirred, transported pneumatically, poured etc. Harmful vapors are a product of combustion. Decomposes when heated or burned, giving off acid vapors and → *zinc oxide*. Exposure can cause metal fume fever.	
	TLV-TWA	10 mg/m³
	Absorption route: Can enter the body by inhalation or ingestion. Evaporation negligible at 20°C, but unpleasant concentrations of airborne particles can build up. **Immediate effects:** Irritates the eyes and respiratory tract. **Effects of prolonged/repeated exposure:** Prolonged exposure to particles can cause lung disorders.	
Gross formula: $C_{36}H_{70}O_4Zn$		

HAZARDS/SYMPTOMS	PREVENTIVE MEASURES	FIRE EXTINGUISHING/FIRST AID
Fire: combustible.	keep away from open flame, no smoking.	water spray, powder.
Explosion: finely dispersed particles form explosive mixtures on contact with air.	keep dust from accumulating, sealed machinery, explosion-proof electrical equipment and lighting, grounding.	
Inhalation: sore throat, cough, shortness of breath.	local exhaust or respiratory protection.	fresh air, rest, call a doctor if necessary.
Skin: redness.	gloves.	remove contaminated clothing, flush skin with water or shower.
Eyes: redness, pain.	goggles.	flush with water, take to a doctor if necessary.
Ingestion: diarrhea, abdominal cramps.		rinse mouth, call a doctor if necessary.

SPILLAGE	STORAGE	LABELING / NFPA
clean up spillage, carefully collect remainder; (additional individual protection: P1 respirator).		

NOTES

ZINC SULFATE
(heptahydrate)

PHYSICAL PROPERTIES	IMPORTANT CHARACTERISTICS	
Decomposes below melting point, °C 500 Relative density (water = 1) 1.97 Solubility in water, g/100 ml 166 Relative molecular mass 287.5	**COLORLESS CRYSTALS** Decomposes when heated, giving off toxic gas (→ *sulfur trioxide*) and → *zinc oxide* fume. Reacts with strong bases, evolving heat.	
	TLV-TWA not available	
	Absorption route: Can enter the body by inhalation or ingestion. Evaporation negligible at 20°C, but harmful concentrations of airborne particles can build up rapidly. **Immediate effects:** Corrosive to the eyes, skin and respiratory tract. Inhalation can cause lung edema. In serious cases risk of death.	
Gross formula: O$_4$SZn.7H$_2$O		

HAZARDS/SYMPTOMS	PREVENTIVE MEASURES	FIRE EXTINGUISHING/FIRST AID
Fire: non-combustible.		in case of fire in immediate vicinity: use any extinguishing agent.
Inhalation: *corrosive*; sore throat, cough, shortness of breath, severe breathing difficulties.	local exhaust or respiratory protection.	fresh air, rest, place in half-sitting position, take to hospital.
Skin: *corrosive*; redness, pain.	gloves.	remove contaminated clothing, flush skin with water or shower, refer to a doctor.
Eyes: *corrosive*; redness, pain, impaired vision.	face shield.	flush with water, take to a doctor if necessary.
Ingestion: *corrosive*; diarrhea, abdominal cramps, vomiting.		rinse mouth, take to hospital.

SPILLAGE	STORAGE	LABELING / NFPA
clean up spillage, flush away remainder with water; (additional individual protection: P2 respirator).		

NOTES
Deliquesces at 100°C. Transition to anhydrous state takes place at 238°C. Lung edema symptoms usually develop several hours later and are aggravated by physical exertion: rest and hospitalization essential. As first aid, a doctor or authorized person should consider administering a corticosteroid spray.

ZIRCONIUM CARBONATE

PHYSICAL PROPERTIES	IMPORTANT CHARACTERISTICS	
Relative density (water = 1) > 2 Solubility in water none Relative molecular mass 431.7	**WHITE AMORPHOUS POWDER** Dissolves in acids, releasing carbon dioxide gas.	
	TLV-TWA 5 mg/m^3 ✱ STEL 10 mg/m^3	
	Absorption route: Can enter the body by inhalation or ingestion. Evaporation negligible at 20° C, but harmful concentrations of airborne particles can build up rapidly. **Immediate effects:** Irritates the eyes and respiratory tract.	
Gross formula: CO$_4$Zr		

HAZARDS/SYMPTOMS	PREVENTIVE MEASURES	FIRE EXTINGUISHING/FIRST AID
Fire: non-combustible.		in case of fire in immediate vicinity: use any extinguishing agent.
Inhalation: cough, shortness of breath.	local exhaust or respiratory protection.	fresh air, rest.
Eyes: redness, pain.	goggles.	flush with water, take to a doctor if necessary.
Ingestion: see 'Notes'.		

SPILLAGE	STORAGE	LABELING / NFPA
clean up spillage, carefuly collect remainder; (additional individual protection: P2 respirator).		◇

NOTES
✱ TLV equals that of zirconium. Theoretically, zirconium carbonate dissolves in gastric acid, but it is converted into an insoluble colloidal zirconium compound in the gastrointestinal tract, leading to poor resorption.

EXPLANATORY NOTES

The information contained in the sheets has been standardized so as to facilitate comparison of the relative hazards of different substances. The editorial policy followed is clarified in the Explanatory Notes below, where the terms, measurements and standard phrases used on the sheets are expanded upon. The layout of the Explanatory Notes is based on that of the sheets.

N.B.
(i)   When the name of a substance oppears in italics, it refers to the relevant sheet(s) (see indexes).

(ii)   A < B: A is less than B (or B is greater than A).

C ≥ D: C is greater than or equal to D (or D is less than or equal to C).

< < : much less than.

Contents:

NAME

PHYSICAL PROPERTIES	IMPORTANT CHARACTERISTICS	
§ 2	Physical/Chemical Data: § 3.1	
	TLV-TWA § 3.2	
	Absorption route: § 3.3 **Acute inhalation risk:** § 3.3 **Immediate effects:** § 3.4 **Effects of prolonged/repeated exposure:** § 3.5	
Gross formula:		

HAZARDS/SYMPTOMS	PREVENTIVE MEASURES	FIRE EXTINGUISHING/FIRST AID
Fire: § 4.1.1	§ 5.1.1	§ 6.1.1
Explosion: § 4.1.2	§ 5.1.2	§ 6.1.2
	Special Warning: § 5.2	§ 6.2
Inhalation: § 4.2	§ 5.2.1	§ 6.2.1
Skin: § 4.2	§ 5.2.2	§ 6.2.2
Eyes: § 4.2	§ 5.2.3	§ 6.2.3
Ingestion: § 4.2	§ 5.2.4	§ 6.2.4

SPILLAGE	STORAGE	LABELING / NFPA
§ 7	§ 8	§ 9

NOTES
§ 10

Transport Emergency Card TEC(R)-....　　　　　　　　　　　　　　　　　　**HI-no.: ...; UN-no.: ...**

Layout of Chemical Safety Sheet; numbers refer to Explanatory Notes

1. HEADER

Name (main entry)	The IUPAC (International Union of Pure and Applied Chemistry) guidelines have been taken into account in the choice of names; trivial names have been used where these are more frequently encountered in the chemical industry.
Synonyms	Alternative names (incl. IUPAC) are listed at the top left of each sheet.
Cylinder	This designation (under the main heading) indicates that the sheet treats the form of the substance stored in cylinders.
Liquefied	This designation (under the main heading) indicates a liquefied gas at atmospheric pressure, e.g. in a Dewar vessel.
Liquid, refrigerated	This designation (under the main heading) indicates a liquefied gas under refrigeration.
Powder	This designation (under the main heading) indicates that the sheet concerns the substance in this form only.
CAS	Chemical Abstracts Service Registry Number. Since the majority of substances go by several different names (incl. trade names), CAS numbers are being used increasingly (in addition to the names of the substances) to avoid confusion. Source: *EPA Toxic Substances Control Act; Chemical Substances Inventory.*
Formula	If necessary the structural formula of the substance is provided, although in most cases a more detailed version of the molecular formula is given.

2. PHYSICAL PROPERTIES

?	Value unknown.
Boiling point, ...°C	The boiling point of the anhydrous substance in whole degrees Celsius at atmospheric pressure (760mm Hg), unless otherwise stated.
Sublimation point, ...°C	The substance passes directly from the solid to the gaseous phase on moderate heating (at 760mm Hg); sublimation occurs at the temperature given.
Decomposes below boiling point, ...°C	The substance decomposes below the boiling point (at 760mm Hg/1 atm); decomposition occurs at the temperature given. If the temperature is not known precisely, this is indicated as 'see Notes'.
Boiling point (decomposes), ...°C	The substance decomposes while boiling (at 760mm Hg); given in whole °C.

| *Melting point, ...°C* | In whole °C at atmospheric pressure (760mm Hg). In the case of substances containing water of crystallization, the apparent melting point is given in the *Notes*. Melting/solidifying points which lie far below 0°C are not given, since (i) they may be imprecise, and (ii) it matters little, for considerations of safety, whether a substance melts at -60 or -100°C. |

| *Decomposes below melting point, ...°C* | The substance decomposes before reaching the melting point (at 760mm Hg/1atm); decomposition occurs at the temperature given. If the temperature is not known precisely, this is indicated as 'see Notes'. |

| *Melting point (decomposes), ...°C* | The substance decomposes while melting (at 760mm Hg/1atm). |

| *Flash point, ...°C* | In whole °C. One definition is: The lowest temperature (at 760mm Hg/1atm) at which a liquid gives off flammable vapor in sufficient quantity to ignite when mixed with air at or near the surface of the liquid on application of a flame or spark . Although this description is clear, its practical usefulness is severely limited; it is therefore customary to give the method of measurement along with the flash point. Moreover, the values given by various authors may differ due the presence of impurities. When the precise flash point is needed, we recommend establishing the flash point of a technical-grade sample of the product by testing. |

| *Flash point 'none'* | When the explosive limits can be determined, but not the flash point. |

| *Flash point 'flammable gas'* | Flammable gases are identified as such; no flash point is given. |

| *Autoignition point, ...°C* | In whole °C. Also called the ignition temperature. A distinction is made between the flash point and the autoignition point. To determine the flash point, an air-vapor mixture is ignited by, e.g., a gas flame. The same mixture can, however, ignite 'spontaneously' at higher temperatures. According to DIN 51794, the autoignition point is the lowest temperature of a glass surface at which droplets of a combustible liquid falling onto this surface will undergo spontaneous combustion. Due to the catalyzing effects of metals, contaminants etc., ignition may in practice occur at a lower temperature than the one given. The autoignition point should be taken into consideration when selecting electrical equipment for areas where explosive air-vapor mixtures can form. |

Classification of electrical equipment by temperature

temperature classification of electrical equipment	maximum permissible surface temperature of electrical equipment (in °C)	ignition temp. of gas/vapor (in °C)
T1	> 300°C and ≤ 450°C	> 450°C
T2	> 200°C and ≤ 300°C	> 300°C
T3	> 135°C and ≤ 200°C	> 200°C
T4	> 100°C and ≤ 135°C	> 135°C
T5	> 85°C and ≤ 135°C	> 100°C
T6	≤ 85°C	> 85°C

Relative density (water = 1)	Rounded off to the nearest tenth (or hundredth when between 0.95 and 1.15). For considerations of safety, it is important to know whether the substance floats on water or sinks. In the case of compressed liquefied gases, the density of the liquid is given; this is used to determine the capacity of gas cylinders etc. The density given for liquids, solids and liquefied compressed gases is assumed to be at normal ambient temperature; entries deviating from this policy are mentioned in the *Notes*. In the case of compressed gases there is no liquid state and hence no value is given. In the case of refrigerated compressed liquefied gases, the density of the liquid at boiling point (at atmospheric pressure) is given in the *Notes*.
Relative vapor density at or above boiling point (air = 1)	The density of the pure vapor given off when the liquid boils. In practice, the relative vapor density is usable for gases and, if necessary, for liquids with a boiling point below 30°C. The value is rounded off to the nearest tenth (or hundredth when between 0.9 and 1.1).
Relative density at 20°C of saturated air-vapor mixture (air = 1)	Relative density of the air-vapor mixture given off above a liquid at normal ambient temperature and atmospheric pressure, i.e. 20°C and 760mm Hg. For considerations of safety, it is important to know whether the air-vapor mixture is heavier or lighter than air under the same conditions. When the relative density of the saturated vapor is between 0.9 and 1.1, mixing with air occurs readily and the value is rounded off to the nearest hundredth; outside these limits, to the nearest tenth.
Vapor pressure, ...mm Hg ...°C	In the case of gases in cylinders the pressure given is reached at the temperature stated.
	The vapor pressure of liquids and solids is given, together with the temperature at which the stated pressure is reached. Vapor pressure values are usually taken from the literature; lacking these, the estimated values are given (see *Tables and Formulas*). These latter readings are preceded by 'ca.'(=approx.), since they sometimes deviate from the actual value. The vapor pressure of substances with a boiling point of 350°C or higher is generally not given, since the values are usually extremely low (i.e. vapor formation is negligible).
Solubility in water	Solubility in water is normally given in g/100ml at 20°C. In the case of gases, a (partial) pressure above the solution of 760mm Hg is assumed. When the degree of solubility in water is not known precisely, it is given as follows:

none	: <0.1	g/100ml
poor	: 0.1-1	g/100ml
moderate	: 1-10	g/100ml
good	: 10-100	g/100ml
very good	: >100	g/100ml
∞	: combines with water in any concentration to form one phase.	

If a substance reacts readily with water, this is indicated by the term 'reaction.'

940

| Explosive limits, % by vol. in air | The explosive limits describe the range within which a mixture of air with a gas, vapor, fumes or powder undergoes spontaneous combustion or explodes. The explosive limits of gases and of the vapors of liquids and solids are given in % by volume. Vapor pressure, flash point and lower explosive limit are interrelated. No explosive limits are given for powdered substances and fumes, since these depend on particle size. In general, the explosive limits of powders are between 0.02 and a few kg/m^3. |

| Mininum ignition energy, mJ | The minimum amount of energy needed to ignite a combustible air-vapor mixture of the substance. When this value is less than 0.6milliJoules, further information is given under *Preventive Measures*. |

| Specific conductivity, pS/m | A measurement of electrical conductivity in liquids, given in picoSiemens per meter at temperatures between 15°C and 25°C. It can vary widely if contaminants are present, however, and the values given cannot be regarded as anything but rough estimates. For the other specifications in use (in addition to pS/m or $pS.m^{-1}$) see *Tables and Formulas*. A static charge can build up in liquids when the specific electrical conductivity is less than $10^4 pS.m^{-1}$. In these cases, more information is given under the headings *Preventive Measures* and *Important Characteristics*. |

| Relative molecular mass | The sum of the relative atomic masses of the composite atoms of a molecule of the substance. |

| Relative atomic mass | The mass of 1 atom of the substance, divided by 1/12 of the mass of an atom of carbon. |

| Log P octanol/water | The octanol/water coefficient indicates the ratio of the concentration in octanol to the concentration in water when the substance is dissolved in a mixture of these two liquids. It is customary to state the logarithm of the coefficient as log P octanol/water. If an estimated value is given, mention is made of this fact in the Notes. The octanol/water coefficient is used to determine the substance's potential for harming the environment. The higher the value, the higher the chance of the substance accumulating in living tissue (especially in fats), in particular at values greater than 3.0. |

3. IMPORTANT CHARACTERISTICS

3.1 Physical/Chemical Data
3.2 ACGIH Threshold Limit Values
3.3 Absorption Route/Acute Inhalation Risk
3.4 Immediate Effects
3.5 Effects of Prolonged/Repeated Exposure

3.1 *Physical/Chemical Data*

| GAS/LIQUID/SOLID | When the boiling point is below 15°C the substance is referred to as a 'gas,' and when it is between 15°C and 30°C as a 'gas or liquid.' When the melting point is below 15°C the substance is referred to as a 'liquid,' and when it is between 15°C and 30°C as a 'liquid or solid.' |

| CHARACTERISTIC ODOR | Many substances have a characteristic odor. Although descriptions of the respective odors exist, these have been omitted from the sheets because of their subjective nature. See also *Notes*. |

941

Vapor (gas) is heavier than air and spreads at ground level, with risk of ignition at a distance	Applies to substances which form an air-vapor mixture with a relative density greater than 1.1 and a flash point below 21°C. When such a substance is released, it spreads at ground level and forms an explosive mixture which can be ignited even at a considerable distance from the source.
Vapor (gas) mixes readily with air, forming explosive mixtures	Applies to substances which form an air-vapor mixture with a relative density between 0.9 and 1.1 and a flash point below 21°C. Although less dangerous than heavier vapors, the possibility of explosion should not be discounted.
Gas is lighter than air	Applies to combustible gases with a relative density less than or equal to 0.9. When such a gas is released, it collects at the highest point of a building (where exhaust fans must be located).
... form peroxides	Applies to ethers etc. Peroxides can form in the storage area. Residues from evaporation or heating can explode spontaneously. In some cases the peroxides thus formed can cause polymerization. Peroxides are normally rendered harmless with iron (II) sulfate or by passing the liquid over a bed of activated aluminum oxide (see Data Sheet 1-655-rev. 82 of the National Safety Council).
... polymerize/ polymerization	Polymerization is a chemical reaction whereby the molecules of a substance combine to form larger molecules. Heat is usually released in the process and can cause fire and/or explosion.
Flow, agitation etc. can cause build-up of electrostatic charge due to liquid's low conductivity	Pumping, stirring, filtering etc. of a liquid with a specific conductivity of less than $10^4 pS.m^{-1}$ can generate electrostatic charges. This is even more likely to occur if the liquid is contaminated with particles of a liquid, gas or solid (in mixture or suspension). The charge generated is picked up by conductors (pumps, drums, lines, pipes etc.) and, if strong enough, a spark can jump the gap between these conductors and earth, igniting any combustible air-vapor mixture in the vicinity. Grounding these conductors and connnecting them one to another is one preventive measure. Electrostatic charges are also thought to be the ignition source responsible for some explosions involving dust/fumes. In these cases, sparks are caused either by inadequate connections between - or grounding of - the conductors, or by a discharge from the dust/fume cloud. The conductivity of dust clouds/fumes, on the other hand, plays but a small role in their ability to build up a charge.
In dry state can form electrostatic charge if stirred, transported pneumatically, poured, etc.	There is a danger of dust explosions: special safety precautions are required (see *Preventive Measures*).
Finely dispersed in air can cause dust explosion	Applies to flammable powdered or finely granulated substances (diameter <0.5mm), which can deflagrate (burn explosively) on ignition, even if damp. In an enclosed space, combustion can lead to a very violent dust explosion.

942

Do not use compressed air when filling, emptying or processing	In order to prevent the formation/spreading of flammable air-vapor mixtures/fumes, in the case of extremely flammable liquids. These liquids should be transported by gravity, pumping or, if necessary, by propelling with inert gas.
Can deflagrate (burn explosively) when heated	Applies to substances which can be made to react by general or local heating, even without the addition of air. In a confined space, combustion can lead to a sudden, violent explosion.
Can deplete atmospheric oxygen	In a confined space the substance can cause a dangerous depletion of atmospheric oxygen.
Acid/base	Strong acids and bases, especially in concentrated form, rapidly attack the skin and eyes, even causing blindness. Many of them also react violently with numerous substances and attack many metals. Medium strong acids and bases pose a lesser - but by no means negligible - danger.

3.2 ACGIH Threshold Limit Values

Threshold Limit Values (TLVs) refer to airborne concentrations of substances and represent conditions under which it is believed that nearly all workers may be repeatedly exposed day after day without adverse health effects. Because of wide variation in individual susceptibility, however, a small percentage of workers may experience discomfort from some substances at concentrations at or below the threshold limit; a smaller percentage may be affected more seriously by aggravation of a pre-existing condition or by development of an occupational illness. Smoking of tobacco is harmful for several reasons. Smoking may act to enhance the biological effects of chemicals encountered in the workplace and may reduce the body's defense mechanisms against toxic substances.

TLVs are based on the best available information from industrial experience, from experimental human and animal studies, and, when possible, from a combination of the three. The basis on which the values are established may differ from substance to substance; protection against impairment of health may be a guiding factor for some, whereas reasonable freedom from irritation, narcosis, nuisance, or other forms of stress may form the basis for others.
The amount and nature of the information available for establishing a TLV varies from substance to substance; consequently, the precision of the estimated TLV is also subject to variation and the latest TLV *Documentation* should be consulted in order to assess the extent of the data available for a given substance.
These limits are intended for use in the practice of industrial hygiene as guidelines or recommendations in the control of potential health hazards and for no other use, e.g., in the evaluation or control of community air pollution nuisances; in estimating the toxic potential of continuous, uninterrupted exposures or other extended work periods; as proof or disproof of an existing disease or physical condition; or adoption by countries whose working conditions differ from those in the United States of America and where substances and processes differ. These limits *are not* fine lines between safe and dangerous concentration nor are they a relative index of toxicity. They should not be used by anyone untrained in the discipline industrial hygiene.

943

The TLVs, as issued by the American Conference of Governmental Industrial Hygienists, are recommendations and should be used as guidelines for good practices. In spite of the fact that serious injury is not believed likely as a result of exposure to the threshold limit concentrations, the best practice is to maintain concentrations of all atmospheric contaminants as low as is practical. The American Conference of Governmental Industrial Hygienists disclaims liability with respect to the use of TLVs.

Three categories of Threshold Limit Values (TLVs) are specified herein, as follows:

TLV-TWA

Threshold Limit Value - Time-Weighted Average (TLV-TWA) is the time-weighted average concentration for a normal 8-hour workday and a 40-hour workweek, to which nearly all workers may be repeatedly exposed, day after day, without adverse affects.

TLV-STEL

Threshold Limit Value - Short-Term Exposure Limit (STEL) is the concentration to which workers can be exposed continuously for a short period of time without suffering from (1) irritation, (2) chronic or irreversible tissue damage, or (3) narcosis of sufficient degree to increase the likelihood of accidental injury, impair self-rescue or materially reduce work efficiency, and provided that the daily TLV-TWA is not exceeded. It is not a separate independent exposure limit; rather, it supplements the time-weighted average (TWA) limit where there are recognized acute effects from a substance whose toxic effects are primarily of chronic nature. STELs are recommended only where toxic effects have been reported from high short-term exposures in either humans or animals. A STEL is defined as a 15-minute TWA exposure which should not be exceeded at any time during a workday even if the 8-hour TWA is within the TLV-TWA. Exposures above the TLV-TWA up to the STEL should not be longer than 15 minutes and should not occur more than four times per day. There should be at least 60 minutes between successive exposures in this range. An averaging period other than 15 minutes may be recommended when this is warranted by observed biological effects.

C

Threshold Limit Value - Ceiling (TLV-C) is the concentration that should not be exceeded during any part of the working exposure.

In conventional industrial hygiene practice if instantaneous monitoring is not feasible, then the TLV-C can be assessed by sampling over a 15-minute period except for those substances that may cause immediate irritation when exposures are short.

For some substances, e.g., irritant gases, only one category, the TLV-Ceiling, may be relevant. For other substances, one or two categories may be relevant, depending on their physiologic action. It is important to observe that if any one of these types of TLVs is exceedeed, a potential hazard from that substance is presumed to exist.

The Chemical Substances TLV Committee holds to the opinion that TLVs based on physical irritation should be considered no less binding than those based on physical impairment. There is increasing evidence that physical irritation may initiate, promote, or accelerate physical impairment through interaction with other chemical or biological agents.

| Conversion of TLVs in ppm to mg/m³ | TLVs for gases and vapors are usually established in terms of parts per million of substance in air by volume (ppm). For convenience to the user, these TLVs are also listed in terms of milligrams of substance per cubic meter of air (mg/m³). The conversion is based on 760 torr barometric pressure and 25°C (77°F), giving a conversion equation of: |

$$\text{TLV in mg/m}^3 = \frac{(\text{TLV in ppm})\ (\text{gram molecular weight of the substance})}{24.45}$$

Resulting values are rounded to two significant figures below 100 and three significant figures above 100. This is not done to give any converted value a greater precision than that of the original TLV, but to avoid increasing or decreasing the TLV significantly merely by the conversion of units.

| *Skin* | The notation 'skin' refers to the potential contribution to the overall exposure by the cutaneous route including mucous membranes and eye, either by airborne or, more particularly, by direct contact with the substance. Vehicles can alter skin absorbtion. |

| *A1* | Indicates that the substance is a *Confirmed Human Carcinogen*: a substance, or substance associated with industrial processes, recognized to have carcinogenic potential. |

| *A2* | Indicates that the substance is a *Suspected Human Carcinogen*: a chemical substance, or a substance associated with industrial processes, which is suspect of inducing cancer, based on either limited epidemiological or demonstration of carcinogesis in one or more animal species by appropriate methods. |

3.3   Absorption Route/Acute Inhalation Risk

| *Can enter the body by ...* | Identifying the absorption route is critical since, although liquids and solids are generally absorbed by ingestion, a number of substances can also be absorbed through the mucous membranes or through unbroken skin. Gases, vapors, smoke, fumes and fine particles usually enter the body by inhalation, the absorption route most commonly encountered in the workplace. The *acute inhalation risk* of gases and vapors given on the sheet relates to the STEL (Short Term Exposure Limit). For details see the TLV list of the American Conference of Governmental Industrial Hygienists. |

| *(a) Harmful atmospheric concentrations can build up very rapidly if gas is released* *(b) Can saturate the air if released, with risk of suffocation* | (a) applies to gases with a STEL ≤ 5000ppm, (b) to gases which have a STEL > 5000ppm or are listed as causing oxygen displacement. |

Harmful atmospheric	The following phrases are used to complete this sentence:
concentrations build up	– very slowly, if at all
... on evaporation at	– fairly slowly
approx. 20°C	– fairly rapidly
	– very rapidly.

Harmful atmospheric concentrations build up ... on evaporation at approx. 20°C

These describe the acute inhalation risk from substances that can give off harmful vapors, indicating the speed with which the STEL (see above) is reached under standardized conditions of evaporation. The more rapid the accumulation, the greater the risk of inhaling a harmful concentration of vapor released into the air. The corresponding values in the RIR index are shown below (for the method of calculation, see the chapter *Tables and Formulas*).

RIR index

< 12	*very slowly, if at all*
12-120	*fairly slowly*
120-4000	*fairly rapidly*
> 4000	*very rapidly*

N.B.

(i) Even with substances which evaporate slowly – whether liquids (in aerosol form) or solids (as airborne particles) – a harmful atmospheric concentration can be reached very rapidly.

(ii) Where neither the STEL nor the TLV for a substance is known, a conservative estimate is given, based on other toxicological data.

Evaporation negligible at 20°C, but (a) harmful ... concentrations ... can build up rapidly (b) unpleasant concentrations ... can build up ...

For substances whose evaporation is negligible, *one* of the phrases applies;

(a) refers to substances classified as harmful or toxic (class 2b or 2c; see 'Supplementary information', Respiratory protection).

(b) to 'inert' dusts (class 2a).

3.4 Immediate Effects

3.4.1 Acute Local Effects

... the eyes, skin and respiratory tract

On the sheets the terms 'irritates' or 'corrosive to' are used, depending on the severity of the effects.

946

Inhalation ... can cause severe breathing difficulties (lung edema)	Inhalation of highly corrosive substances, e.g. many acids and bases – inhaled as gas, vapor, fumes or dust – can cause lung disorders (e.g. lung edema = fluid in the lungs) a considerable time after they are inhaled; this can lead to severe breathing difficulties. A person who has inhaled an appreciable amount of such a substance should have complete rest (preferably in a half-sitting position) and be taken to hospital, even if no symptoms are present as yet. Administering a corticosteroid spray should be considered; this, however, should be done only by a doctor or other authorized person. The sheets for all substances which can cause lung edema contain this warning in the *Notes.* Severe breathing difficulties can also be caused by disorders other than lung edema, e.g. oxygen deficiency and asthma.
Can cause frostbite due to rapid evaporation	Substances can produce harmful effects by physical as well as chemical means. This is especially true of liquefied gases, which can evaporate so rapidly as to cause a drastic lowering of the temperature of the eyes or skin, or even frostbite.

3.4.2   Acute General Effects

Deals with those effects which can occur during or shortly after exposure and at a place other than the place of exposure.

Substance affects ...	Self-explanatory.
Can cause ... damage Can affect ...	Damage to, or changes in, an organ can occur even in the absence of immediate symptoms (which may develop later).
If liquid is swallowed, droplets can enter the lungs, with risk of pneumonia	Some substances which display little toxicity if ingested can be aspirated (entering the lungs as a result of choking, vomiting etc.), subsequently causing pneumonia.

3.5   Effects of Prolonged/Repeated Exposure

Prolonged or repeated contact can cause skin disorders	As already indicated (under *Can cause ... damage*), non-acute effects can be delayed. Repeated exposure to certain substances can also be harmful, leading to skin/lung disorders, many of them (e.g. eczema) occasioned by an allergic reaction. Persons found to be allergic should avoid all further contact with the substance.
Liquid destroys the skin's natural oils	Many liquids which have no acute effects on the skin and do not cause allergic reactions, can destroy the skin's oils on prolonged or repeated exposure, leaving the skin raw, dry and red. After coming in contact with such a substance (sometimes used as a skin cleaner), the skin should be washed with soap and water and then oiled with lanoline or vaseline.
Prolonged/repeated exposure (to vapor/particles) can cause lung disorders	Lung disorders include: chronic bronchitis, asthma, emphysema. 'Repeated' refers to contact on more than one occasion, even if no acute symptoms have presented. 'Prolonged and repeated' is used for exposure which must exceed a given limit in order to cause disorders.

Has been found to cause a type of ... cancer in humans under certain circumstances	This substance has been found to be carcinogenic in humans, although – as the qualification 'under certain circumstances' indicates – exposure need not necessarily cause cancer. The risk depends greatly on the concentration and duration of the exposure; the absorption route is also critical. The company doctor/safety officer needs to be aware of any situations which pose a cancer risk to employees and should take all necessary precautions.
Has been found to cause a type of ... cancer in certain animal species under certain circumstances	This substance has been shown to be carcinogenic in one or more animal species, depending on the circumstances. Although it is not known whether these substances are also carcinogenic in humans, they must be considered highly suspect, and the same caution should be exercized as when handling those chemicals known to cause cancer in humans. The company doctor/safety officer needs to be aware of any situations which pose a cancer risk to employees and should take all necessary safety precautions.
Can cause hereditary genetic damage	Applies to substances known to be mutagens (capable of inducing gene mutation) in humans, or which should be considered as such because of their properties.
Suspected of being able to cause hereditary genetic damage	Applies to substances whose properties give reason for concern as to possible mutagenic effects in humans, although these are as yet unproven.

4. HAZARDS/SYMPTOMS

4.1   Fire/Explosion Hazard
4.2   Immediate Symptoms

4.1   Fire/Explosion Hazard

Fire

Extremely flammable	Applies to substances with a flash point below 21°C (including flammable gases) and, in particular, to solids which can ignite 'spontaneously' (on exposure to air).
Flammable	Applies to substances with a flash point above 21°C but below 55°C, and to solids which ignite readily, even at low oxygen levels.
Combustible	Applies to all other combustible liquids and solids, including those with a flash point above 100°C.
Non-combustible but enhances combustibility of other substances	Applies to substances which readily give off oxygen, such as perchlorates, peroxides and other oxidants.
Non-combustible	Applies to substances which cease to burn in air of normal composition at normal pressure when the ignition source is removed. Such substances can sometimes be *made to burn*, however; drawing a line between these and combustible substances is often difficult. Where necessary, further information is given in the *Notes.*

948

But contact with moisture can cause fire and explosion	Preceded by 'non-combustible', this applies to solids and liquids which form flammable gases on contact with water or moist air but do not themselves react with oxygen.

Explosion

Forms explosive air-gas mixtures	Applies to flammable gases.
Forms explosive air-vapor mixtures	Applies to liquids with a flash point below 21°C. Explosive air-vapor mixtures are very likely to form if the liquid is released at normal ambient temperature (indoors or outdoors).
Above ... °C: forms explosive air-vapor mixtures	Applies to substances with a flash point between 21°C and 100°C.
Finely dispersed particles form explosive mixtures on contact with air	Applies to combustible powdered substances. N.B. Fumes from combustible liquids are usually also explosive.

4.2   Immediate Symptoms

The medical data are classified by absorption route: inhalation, skin, eyes, ingestion. Under these categories, some of the most important acute symptoms caused by excessive contact are listed. Although it is assumed that no one will knowingly eat or drink chemicals, symptoms stemming from this manner of absorption are listed where the substance is highly toxic. Solids or liquids which adhere to the skin (hands) or clothes are especially hazardous. Anyone experiencing one of the listed symptoms during or following exposure should report to the company doctor/first aid station.

5.   PREVENTIVE MEASURES

5.1   Preventing Fire/Explosion
5.2   Preventing Exposure/Contact

5.1   Preventing Fire/Explosion

Fire

Keep away from open flame and sparks, no smoking	Applies to flammable gases and liquids with a flash point below 55°C, and to solids which (i) ignite readily even at low oxygen levels, or (ii) can form flammable gases on contact with moisture.
Keep away from open flame, no smoking	Applies to all other combustible substances. 'Open flame' includes hot surfaces with a temperature above the autoignition point of the substance.
Avoid contact with ...	Applies to substances which react violently with the other substance(s) listed (with risk of fire/explosion). Sheets for substances with a low autignition point also bear a warning regarding contact with hot surfaces.

Sealed machinery, ventilation, explosion-proof electrical equipment and lighting	Applies to liquids with a flash point below 21°C and to flammable gases. The recommendations state the standard safety precautions which must be taken to prevent the formation (and ignition) of explosive air-gas/vapor mixtures. N.B. (i) Specifying machinery so as to ensure that the composition of the air-gas/vapor mixtures remains within safe limits falls outside the parameters of this book, although some hints are given in the chapter *Tables and Formulas, Rendering explosive mixtures inert.* (ii) 'Explosion-proof electrical equipment and lighting' means electrical equipment and lighting suitable for use in an area where there is a danger of explosion.
Above ... °C: sealed machinery, ventilation, explosion-proof electrical equipment	Applies to all liquids with a flash point between 21°C and 55°C. See also preceding entry.
Above ... °C: sealed machinery, ventilation	Applies to all liquids with a flash point between 55°C and 100°C.
Grounding, grounding when pumping etc. in liquid form	Applies to combustible liquids and liquefied gases in which, due to flow, friction etc., electrostatic charges can build up (conductivity less than $10^4 pS.m^-1$). Measures to limit the build-up of electrostatic charges or prevent discharges (including grounding of all machinery at all times) are essential.
Non-sparking hand tools	Applies to substances with a minimum ignition energy of less than 0.6mJ.
Do not subject to shock or friction	Shock or friction can cause the substance to decompose explosively.
Keep dust from accumulating, sealed, machinery, explosion-proof electrical equipment and lighting, grounding	Applies to finely dispersed particles which form explosive mixtures on contact with air. In some cases, installation of special equipment to prevent dust explosions will be required. The risk of dust explosions can be reduced by taking measures to prevent the build-up of electrostatic charges (consult an expert).

5.2   Preventing Exposure/Contact

Measures for preventing excessive exposure are listed, in order of increasing protection offered, for each of the possible absorption routes; in other words, the first precaution listed applies under all circumstances, unless the safety officer recommends otherwise. In practice, the safety officer will advise as to which of the precautions are most suitable. These should be incorporated in any industrial process at the planning stage, priority being given to minimizing contact between personnel and dangerous substances. Individual protection should be seen as an additional precaution, not a first line of defense. The assessment of hazard needs to take both the toxicity of the substance and the risk of absorption into consideration.

Risk of absorption is determined by:
- physical properties, e.g. vapor pressure, speed of evaporation, boiling point, solubility;
- particle size, in the case of solids;
- the layout and size of the workplace and the possibility of dispersion by air currents. Rapid dispersion can pose a danger to other personnel. High local concentrations can be reached if the substance does not mix well with air;
- the conditions under which the substance is used.

Keep dust under control/prevent formation of fumes

Harmful fumes, powder or dust can form during processing.

Strict hygiene

When handling chemicals hygienic procedures should be strictly observed. 'Strict hygiene' indicates that the compilers consider this to be a particularly dangerous substance, demanding special care.

Avoid all contact

Substance is highly toxic. Note that the acute symptoms listed may not appear immediately.

Avoid exposure of pregnant women as far as possible

Only indicated when the substance is a teratogen (capable of interfering with the development of a fetus, causing birth defects, when absorbed by a pregnant woman). Pregnant women and women of childbearing age should not be allowed to come in contact with quantities sufficient to have teratogenic effects; they should be advised of the risks involved before being allowed to work with such substances. The company doctor/safety officer should also be consulted as to the safest way of handling these substances.

5.2.1 Inhalation

Ventilation

Applies to substances which generally present no risk of serious harmful effects due to inhalation of the gas or vapor, and where the TLV cannot be exceeded under normal working conditions.

Local exhaust

Ventilation may offer insufficient protection, or may even be undesirable.

Respiratory protection

Necessary when ventilation and exhaust systems are inadequate to prevent the build-up of harmful concentrations.

Respiratory protection means:
- breathing apparatus;
- respirators;
- respirators with absorbing filters/canisters of the appropriate type.

N.B. When processing suspensions or working with a liquid/solution in aerosol form (e.g. when spraying paint), a simple respirator (or face mask) is totally inadequate.

5.2.2 Skin

Gloves

Absorption through – and corrosion of – the skin can usually be prevented by wearing the appropriate type of rubber/plastic gloves. For some chemicals, however, there is as yet no material that offers adequate protection; in these cases suitably sealed equipment must be used. Work gloves of leather or heavy cloth should under no circumstances be used when working with liquids or fine powders, since they provide only limited protection against abrasions and cuts. For information on glove selection, see the chapter *Recommendations on Gloves and Other Protective Clothing.*

Heat-insulated gloves/insulated gloves

To prevent frostbite or burns when handling cold substances (liquefied compressed gases or liquids with a boiling point below 20° C) or hot substances (e.g. molten sulfur). If the substance can penetrate or damage the skin, a glove material incorporating rubber or plastic must be found which is sufficiently resistant to the substance.

Protective clothing

Ideally, the workplace should be so arranged that normal work clothes offer ample protection, although this will not always be possible. 'Protective clothing' is indicated only if incidental contact with the substance through normal work clothes can have serious consequences.

This is the case, for instance:
- with liquids that burn the skin, e.g. strong oxidants and concentrated strong acids and bases;
- when the absorption of a substance (through the skin) carries serious risks;
- with substances that can cause allergies.

Advice as to what protective clothing, if any, should be worn under given working conditions can best be obtained from the safety officer. For advice on choice of materials, see the chapter *Recommendations on Gloves and Other Protective Clothing.*
N.B. When wearing boots, pant legs should be worn over the tops of the boots.

5.2.3 Eyes

Contact lenses

In industrial situations, contact lenses usually offer no eye protection, and can even pose an added risk when the eyes are exposed to chemical vapors. All recommendations as to eye protection (goggles, face shield etc.) still apply, therefore, even if the worker wears contact lenses. The company doctor/safety officer can advise in each individual case whether contact lenses are appropriate.

Safety glasses

To be worn when the only threat posed by incidental contact with liquids and solids is mechanical injury to eyes/skin.

Acid goggles

To be worn when handling liquids which are hazardous to the eyes, but not to the facial skin in the event of incidental contact.

| Goggles | To be worn when handling solids and other substances which are hazardous to the eyes (other than causing mechanical injury) or powders which cause extreme eye irritation, but when incidental contact with facial skin poses little danger. |

| Face shield | To be worn when handling liquids and solids which are not prone to give off dust/particles but pose a danger to both eyes and facial skin.
N.B. A face shield offers insufficient protection if droplets can spatter up behind it, e.g. if an object should fall into a container of liquid. |

| Combined eye and respiratory protection | An alternative to other forms of eye protection in work situations where:
— there is a very substantial risk of inhaling a harmful concentration of vapor;
— inhalation of fine particles (powder) must be avoided;
— all skin contact must be avoided.
The combination can consist of:
— breathing apparatus;
— full-face mask with external air supply or filter canister.
For more details on these forms of individual protection, see *Supplementary Information, Respiratory protection.* |

5.2.4   Ingestion

| Do not eat, drink or smoke while working | EATING, DRINKING OR SMOKING IN THE WORKPLACE SHOULD BE STRICTLY PROHIBITED.
Under normal circumstances, a recommendation not to eat, drink or smoke while working will be unnecessary, since this can be considered normal working procedure; as there is a great risk of food contamination while working with harmful powders or viscous liquids, however, the sheet may bear a special warning to this effect. Washing the hands before eating, drinking or smoking is strongly recommended. |

6.   FIRE EXTINGUISHING/FIRST AID

6.1   Fire/Explosion
6.2   First Aid

6.1   Fire/Explosion

Extinguishing Agents

Extinguishing agents have been selected and listed in order of preference on the basis of their usability by laymen under varying conditions and their effectiveness. In some cases extinguishing agents are proscribed because of possible dangerous chemical reactions. Instructions for fire department personnel are given in the *Notes.* In exceptional cases, a special extinguishing agent may be indicated; any further information necessary is given in the *Notes.*

| Water | Works primarily by cooling the burning substance, in particular due to the cooling action of evaporation, the steam formed in the process serving to expel oxygen from the area. |

Water spray	The more even distribution of water occasioned by this method increases the cooling and blanketing effect.
AFFF	Acronym for Aqueous Film Forming Foam, a liquid which, when mixed with water, produces a blanketing film. For details of function see *Foam*.
Foam	Refers to mechanically-produced foam, which works by interfering with the transfer of heat between the flame and the burning substance, thereby impeding evaporation.
Alcohol-resistant foam	Alcohols, ketones and esters can break down many types of foam, thus cancelling out the extinguishing action; fires involving these materials need to be fought with special foams which do not break down as readily. AFFF-ATC also falls under this category.
Powder	Refers particularly to powders based on bicarbonate or ammonium phosphate. The chain reaction which keeps the fire burning is curbed (a process termed 'negative catalysis').
Halons	Halons (i.e. BCF and BTM) are compounds of fluorine, chlorine and/or bromine with hydrocarbons such as methane and ethane; their extinguishing action is also due to 'negative catalysis'.
	N.B. The use of CFCs (chlorofluorocarbons - this includes halons 1211/BCF and 1301/BTM) is being curtailed by the Montreal Protocol on Substances that deplete the Ozone Layer. Given the possible environmental consequences, it is not certain what position governments will take on halon extinguishers; the editors advise limiting their use, employing alternative extinguishing agents where feasible and restricting as far as possible the testing of halon extinguishing systems.
Carbon dioxide	The extinguishing action of carbon dioxide is based on oxygen displacement; it is only suitable for fighting small fires (in their early stages) when there is little or no wind.
In case of fire in immediate vicinity: use any extinguishing agent/do not use ...	The selection of an extinguishing agent is based, first and foremost, on the burning substances. In some cases non-combustible substances in the immediate vicinity can rule out the use of certain extinguishing agents which react with the substance, substantially increasing the risk of fire, explosion or poisoning. Any such limitations are mentioned under this heading.

Prevention of fire/explosion

Shut off supply; if impossible and no danger to surrounding area, allow fire to burn itself out; otherwise extinguish with ...	Applies to combustible gases. The best extinguishing method is to shut off the supply. Should this not be possible, letting the fire burn itself out 'under surveillance' is far preferable to putting it out: if this has to be done without shutting off the supply, the resulting air-gas mixture could explode if it reignites.
In case of fire: keep ... cool by spraying with water	Cylinders, tanks and drums exposed to radiant heat as the result of a fire in the vicinity must be kept cool to avoid rupture/collapse due to increased pressure or localized overheating.

In case of fire: keep ... cool by spraying with water, but do not spray substance with water	Indicates that the substance reacts dangerously with water.
Fight fire from sheltered location	Indicates danger of explosion.

6.2 First Aid

Potential first-aiders are referred to the instructional literature and to the instructions on *First aid in the event of poisoning* in the chapter *Supplementary Information,* which complement the information given on the sheets.

In all cases call a doctor	Whenever a substance can cause serious effects, regardless of the absorption route, a doctor should be called or consulted.
Call a doctor	Call in a doctor if one is available on the premises; otherwise take the person to the doctor/emergency room (show the relevant sheet(s)).
Take to/refer to a doctor	The patient should be treated without delay. If there is no doctor available on the premises, take the person to a doctor (take along the relevant sheet(s)).
Take to hospital	1. Engage the help of the company infirmary or doctor on the premises, explaining that the sheet(s) recommend hospitalization. Leave further decisions to the health professional(s). 2. In all other cases, arrange transport to the hospital (take along the relevant sheet(s)).

6.2.1 Inhalation

Fresh air, rest	Applies in all cases where inhalation of a substance has produced disorders or symptoms.
Half-sitting position	Applies in those cases where inhalation of an irritating or corrosive substance has produced shortness of breath, or where there is danger of lung edema. Often the most pleasant position for the patient.
Artificial respiration if necessary	Applies in serious cases, where the person has virtually or entirely stopped breathing, or severe breathing difficulties carry the risk of suffocation. 'Artificial respiration' means 'mouth-to-mouth resuscitation' (place a handkerchief over the patient's mouth). Although administration of oxygen would normally be indicated in many of these cases, this has been intentionally omitted, since (i) when incorrectly administered, it often does more harm than good, and (ii) oxygen should be administered only by qualified medical and paramedical personnel. Oxygen equipment should be available in the workplace, however, for use by qualified persons.

Remove contaminated clothing, flush with water	In most cases it is sensible to remove contaminated clothing and shoes *before* washing the skin with water, thus ensuring that contact with the skin ceases as quickly as possible. It is preferable, however, to remove clothing while flushing with water or showering.
Flush with water, remove contaminated clothing	If the skin and clothing have been contaminated with oxidants and there is a danger of the clothing catching fire, the clothing should be flushed with water (e.g. under the shower) before it is removed.
Flush with water, do not remove clothing	In the case of burns or frostbite, clothing should not be removed because of the increased risk of infection from broken blisters. Clothing and skin should, of course, be flushed with water.
Wash with soap and water	Actively clean the skin with water. Do not wash if the skin is broken or if this is likely to occur during washing.
Flush with water	Allow stream of water to clean the skin (preferably by showering).

6.2.3 Eyes

Flush with water	For a period of fifteen minutes. Remove contact lenses if this can be done with ease.
Take to an eye doctor	Applies to substances which can damage the cornea or cause other eye disorders. In all cases first flush the eye(s) with water for a period of fifteen minutes, and then take person to an eye doctor.

6.2.4. Ingestion

Rinse mouth	Especially in cases where mouth and throat may be affected.
Drink water	To dilute the substance in the stomach. Warning: UNDER NO CIRCUMSTANCES GIVE LIQUIDS TO AN UNCONSCIOUS PERSON.
Do not induce vomiting	Because of the danger of inhaling stomach contents when vomiting, it is safer to take the victim to a doctor, who will then make a decision on the best means of emptying the stomach.

7. SPILLAGE

The more dangerous the substance, the more important it is to have contingency plans to deal with any release, including the following damage-control measures:
- means of collecting leakages;
- sand or other suitable absorbent to confine/soak up materials;
- means of sealing leaking containers;
- ventilation.

The following aids should therefore be kept on hand:
- for individual protection: safety glasses, face shields, special clothing, aprons, boots, gloves, respirators etc.;
- containers (in which to place other leaking drums etc.);
- neutralizing agents;
- sand or other suitable absorbents.

The Chemical Manufacturers' Association gives detailed cleanup procedures for many substances in the publication *MCA Laboratory Waste Manual,* Washington, 1975.

Evacuate area, call in an expert	Applies to a number of very hazardous substances which, if released, would pose a serious danger to anyone in the vicinity. An expert should be called in as soon as possible.
Ventilation	Applies when (i) dissipation by evaporation is possible and poses no risks, or (ii) when other dispersion methods pose even greater risks.
Flush away remainder with water	Applies to solids and liquids which (i) cannot cause serious environmental pollution in small quantities, and (ii) pose no danger of explosion in the sewers.
Clean up spillage, soak up with sand or other inert absorbent and remove to safe place	Applies to liquids which must not be flushed into the sewer, (i) because they are extremely flammable (flash point below 21°C), (ii) because they pose a danger to health, or (iii) because of the threat of serious environmental pollution. 'Safe place' means one where (i) there is no risk of explosion, (ii) no one can be exposed to the substance, and (iii) there is no risk of environmental pollution.
Carefully collect remainder	Applies when dispersion of the substance must be prevented under all circumstances.
Under no circumstances spray liquid with water	Applies to substances which react violently with water and to extremely cold liquids; adding water greatly increases the speed of evaporation.
Do not flush into sewer	An additional warning for extremely flammable liquids which float on water and thus pose a serious danger of causing an explosion in the sewers.

N.B. Specific, self-explanatory instructions on cleanup may also be found under *Spillage.* The special equipment used for cleaning up laboratory spillages can be obtained from the chemical suppliers.

Additional individual protection	Applies in situations where the cleanup of substantial quantities requires more protection than that listed under *Preventive Measures.*
Breathing apparatus	Recommended when working with toxic gases, liquids or solids where harmful concentrations of vapor can build up rapidly.

N.B. Breathing apparatus should not be worn without prior approval from a doctor; regular practice with the unit is necessary.

Chemical suit	Recommended when gas, vapor or fumes can cause damage by entering the respiratory organs and attacking unbroken skin.

| Respirator | Recommended when the airborne particles of a substance can be harmful to health; the choice of respirator depends on the type of substance and the working conditions. For details see *Supplementary Information, Respiratory protection.* The respirator listed on the sheet is to be worn when (i) the ambient air contains a minimum of 19% oxygen and is free from toxic gases, and (ii) considerable freedom of movement is required (i.e. not in emergency situations). |

Respirator with ... filter Recommended when working with harmful powders which, even in limited quantities, can give off harmful vapors. The respirator + filter consists of a gas-absorbing canister and a dust mask.

8. STORAGE

It goes without saying that the various federal/state/local laws/ordinances concerning the proper storage of chemicals should be investigated and complied with. Even if a substance is not covered under the law, it is recommended that access be limited to those who are qualified to handle the substance professionally. The more dangerous the substance, the higher the risk from injudicious handling of the material.

Keep in fireproof place Applies to flammable substances and combustible substances in cylinders, and to solids that fall into the highest combustibility category or give off combustible gases in the presence of moisture.

If stored indoors, keep in fireproof place When cylinders are stored in a building the storage place must be fireproof, even if the gas is not combustible.

Keep separate from ... Applies in those cases where the substance reacts violently with, or gives off toxic gases on contact with, the substance(s) listed. Information on reactivity is given under *Important Characteristics.*

Keep cool Applies to substances which (i) are stored in cylinders, (ii) are highly volatile, (iii) can decompose or polymerize when moderately heated, or (iv) can form peroxides. In most cases, a room which never reaches temperatures above 25°C will be suitable.

Keep dry Applies to hygroscopic substances and substances which can react with water and/or moist air.

Keep in dark place Applies to substances which can react, polymerize or form peroxides when exposed to light.

Ventilate at floor level Applies to substances where the relative density of the saturated air-vapor mixture is greater than 0.9.

N.B. This phrase is omitted when 'keep in fireproof place' is indicated, since this latter requirement includes ventilation at floor level.

Under inert gas Applies to substances which react or polymerize on contact with air. Nitrogen or a noble gas will normally be suitable. If only carbon dioxide is available consult an expert before using.

Add an inhibitor, only if stabilized	Applies to substances which (i) can polymerize, (ii) can react violently while in storage, or (iii) can form hazardous compounds (e.g. with atmospheric oxygen to form peroxides). The selection of an inhibitor must always be left to an expert, who must also be consulted if there is uncertainty as to whether an inhibitor has been added.

9. LABELING/NFPA

Labeling is a measure taken to alert workers, consumers and other parties to the undesirable effects of dangerous substances. The labeling procedure followed on the sheets was established by the EEC and is compulsory. The substances so labeled are contained in the EEC list of Dangerous Substances. Suppliers are responsible for the choice of appropriate labeling for substances not included in the EEC list which, on the basis of their specific properties, must be provided with a warning label. The warning label consists of: danger symbol(s); indications of special hazards (Risk Phrases); safety recommendations (Safety Phrases); in some cases Note(s) (A-F). For editorial reasons, only the numbers of the Risk and Safety Phrases are listed on the sheets, along with the code for any applicable Note (A-F).

Within the European Economic Community, dangerous substances and preparations must, by law, be provided with a warning label bearing all the above indications in the language of the country in which the product is located (EEC Directive 67/548).

The EEC labeling system applies to the delivery of dangerous substances and not to their transportation; cylinders are also not covered by these regulations.

Key

E	Explosive
O	Oxidizing
T^+	Very toxic
T	Toxic
X_n	Harmful
C	Corrosive
X_i	Irritant
F^+	Extremely flammable
F	Highly flammable

Meaning of Notes A-F

Where there is more than one possible classification for substances listed in Annex I of EEC Directive 67/548 (i.e. where substances are designated generically, e.g. as '... compounds' or '... salt'), select the category which represents the highest degree of hazard.

Note A

The name of the substance stated on the label must be included in Annex I. Where a substance is listed in Annex I generically (e.g. as '... compounds' or '... salt'), the manufacturer, or other person marketing it, must state the chemical name on the label, e.g. $BeCl_2$ must be labeled as 'beryllium chloride'.

Note B

Certain substances (acids, bases etc.) are sold in aqueous solutions of varying concentrations; since the various concentrations pose different hazards, each must be provided with an appropriate warning label. Where a substance is listed in Annex I generically (e.g. as 'nitric acid ...%'), the manufacturer, or other person marketing the substance, must state the percentage concentration on the label, e.g. nitric acid 45%. Percentages always refer to weight ratio, unless otherwise stated. Other data (e.g. specific gravity, Baumé values) or descriptions (e.g. ice-cold, fuming concentrate) are permitted.

Note C

Some organic substances and derivatives on the market are composed of either one specific isomer or a mixture of isomers. Where these are listed in Annex I generically (e.g. as 'hexane'), the manufacturer, or other person marketing the substance, must clearly list on the label whether the product contains:
a.   one specific isomer, or
b.   a mixture of isomers.

Examples:
a.   orthoxylene
b.   xylene (mixture of isomers)

Note D

Some substances which polymerize readily or decompose spontaneously are generally encountered in stabilized form, the form under which they are listed in Annex I. Such substances are occasionally marketed in non-stabilized form however, in which case the manufacturer, or other person marketing them, must state the name of the substance on the label, followed by the designation '(not stabilized)'.
Example: methacrylic acid (not stabilized).

Note E

In the case of E-substances so designated, the R-phrases R20-28 - and any combinations thereof - must be preceded by the word 'also.'

Examples:
R23: Toxic by inhalation.
R27/28: Very toxic in contact with skin and if swallowed.

Note F

The substance may contain a stabilizer. If the stabilizer modifies hazardous properties of the substance as listed on the Annex I label, a label must be compiled in accordance with the regulations for the resulting hazardous preparation.

List of EEC Risk (R) and Safety (S) phrases for Classification and Labeling

Risk (R) phrases

R1	Explosive when dry
R2	Risk of explosion by shock, friction, fire or other sources of ignition
R3	Extreme risk of explosion by shock, friction, fire or other sources of ignition
R4	Forms very sensitive explosive metallic compounds
R5	Heating may cause an explosion
R6	Explosive with or without contact with air
R7	May cause fire
R8	Contact with combustible material may cause fire
R9	Explosive when mixed with combustible material
R10	Flammable
R11	Highly flammable
R12	Extremely flammable
R13	Extremely flammable liquefied gas
R14	Reacts violently with water
R15	Contact with water liberates highly flammable gases
R16	Explosive when mixed with oxidizing substances
R17	Spontaneously flammable in air
R18	In use, may form flammable/explosive vapour-air mixture
R19	May form explosive peroxides
R20	Harmful by inhalation
R21	Harmful in contact with skin
R22	Harmful if swallowed
R23	Toxic by inhalation
R24	Toxic in contact with skin
R25	Toxic if swallowed
R26	Very toxic by inhalation
R27	Very toxic in contact with skin
R28	Very toxic if swallowed
R29	Contact with water liberates toxic gas
R30	Can become highly flammable in use
R31	Contact with acids liberates toxic gases
R32	Contact with acids liberates very toxic gas
R33	Danger of cumulative effects
R34	Causes burns
R35	Causes severe burns
R36	Irritating to eyes
R37	Irritating to respiratory system
R38	Irritating to skin
R39	Danger of very serious irreversible effects
R40	Possible risk of irreversible effects
R41	Risk of serious damage to eyes
R42	May cause sensitization by inhalation
R43	May cause sensitization by skin contact
R44	Risk of explosion if heated under confinement
R45	May cause cancer
R46	May cause heritable genetic damage
R47	May cause birth defects
R48	Danger of serious damage to health by prolonged exposure

Combination risk (R) phrases

R14/15	Reacts violently with water, liberating highly flammable gases
R15/29	Contact with water liberates toxic, highly flammable gas
R20/21	Harmful by inhalation and in contact with skin
R20/21/22	Harmful by inhalation, in contact with skin and if swallowed
R20/22	Harmful by inhalation and if swallowed
R21/22	Harmful in contact with skin and if swallowed
R23/24	Toxic by inhalation and in contact with skin
R23/24/25	Toxic by inhalation, in contact with skin and if swallowed
R23/25	Toxic by inhalation and if swallowed
R24/25	Toxic in contact with skin and if swallowed
R26/27	Very toxic by inhalation and in contact with skin
R26/27/28	Very toxic by inhalation, in contact with skin and if swallowed
R26/28	Very toxic by inhalation and if swallowed
R27/28	Very toxic in contact with skin and if swallowed
R36/37	Irritating to eyes and respiratory system
R36/37/38	Irritating to eyes, respiratory system and skin
R36/38	Irritating to eyes and skin
R37/38	Irritating to respiratory system and skin
R42/43	May cause sensitization by inhalation and skin contact

Safety (S) phrases

S1	Keep locked up
S2	Keep out of reached of children
S3	Keep in a cool place
S4	Keep away from living quarters
S5	Keep contents under ... (appropriate liquid to be specified by the manufacturer)
S6	Keep under ... (inert gas to be specified by the manufacturer)
S7	Keep container tightly closed
S8	Keep container dry
S9	Keep container in a well-ventilated place
S12	Do not keep the container sealed
S13	Keep away from food, drink and animal feedingstuffs
S14	Keep away from ...; (incompatible materials by the manufacturer)
S15	Keep away from heat
S16	Keep away from sources of ignition - No smoking
S17	Keep away from combustible material
S18	Handle and open container with care
S20	When using do not eat or drink
S21	When using do not smoke
S22	Do not breathe dust
S23	Do not breathe gas/fumes/vapour/spray (appropriate wording to be specified by the manufacturer)
S24	Avoid contact with skin
S25	Avoid contact with eyes
S26	In case of contact with eyes, rinse immediately with plenty of water and seek medical advice
S27	Take off immediately all contaminated clothing
S28	After contact with skin, wash immediately with plenty of ... (to be specified by the manufacturer)
S29	Do not empty into drains
S30	Never add water to this product

S33	Take precautionary measures against static discharges
S34	Avoid shock and friction
S35	This material and its container must be disposed of in a safe way
S36	Wear suitable protective clothing
S37	Wear suitable gloves
S38	In case of insufficient ventilation wear suitable respiratory equipment
S39	Wear eye/face protection
S40	To clean the floor and all objects contaminated by this material use... (to be specified by the manufacturer)
S41	In case of fire and/or explosion do not breathe fumes
S42	During fumigation/spraying wear suitable respiratory equipment (appropriate wording to be specified by the manufacturer)
S43	In case of fire use ... (indicate in the space the precise type of fire-fighting equipment. If water increases the rids add: Never use water)
S43a	In case of fire use sand, earth, dry chemical or foam
S43b	In case of fire use sand, earth, dry chemical or alcohol type foam
S44	If you feel unwell seek medical advice (show the label where possible)
S45	In case of accident or if you feel unwell seek medical advice immediately (show the label where possible)
S46	If swallowed, seek medical advice immediately and show this container or label
S47	Keep at a temperature not exceeding... C (to be specified by the manufacturer)
S48	Keep wetted with ... (appropriate material to be specified by the manufacturer)
S49	Keep only in the original container
S50	Do not mix with... (to be specified by the manufacturer)
S51	Use only in well-ventilated areas
S52	Not recommended for interior use on large surface areas
S53	Avoid exposure – obtain special instructions before use

Combination safety (S) phrases

S1/2	Keep locked up and out of reach of children
S3/7/9	Keep container tightly closed, in a cool, well ventilated place
S3/9	Keep in a cool, well ventilated place
S3/9/14	Keep in a cool, well ventilated place away from ... (incompatible materials to be indicated by the manufacturer)
S3/9/14/49	Keep only in the original container in a cool, well ventilated place away from ... (incompatible materials to be indicated by the manufacturer)
S3/9/49	Keep only in the original container in a cool, well ventilated place
S3/14	Keep in a cool place away from ... (incompatible materials to be indicated by the manufacturer
S7/8	Keep container tightly closed and dry
S7/9	Keep container tightly closed and in a well ventilated place
S20/21	When using do not eat, drink or smoke
S24/25	Avoid contact with skin and eyes
S36/37	Wear suitable protective clothing and gloves
S36/37/39	Wear suitable protective clothing, gloves and eye/face protection
S36/39	Wear suitable protective clothing and eye/face protection
S37/39	Wear suitable gloves and eye/face protection
S47/49	Keep only in the original container at a temperature not exceeding... °C (to be specified by the manufacturer)

NFPA

The NFPA system is intended to provide basic information to fire fighting, emergency, and other personnel, enabling them to more easily decide whether to evacuate the area or to commence emergency control procedures. It is also intended to provide them with information to assist in selecting fire fighting tactics and emergency procedures. The system identifies the hazards of a material in terms of three principal categories: 'health', 'flammability', and 'reactivity'. It indicates the degree of severity by a numerical rating that ranges from four (4), indicating severe hazard, to zero (0), indicating no hazard.

H: Health hazards
F: Flammability hazards
R: Reactivity (instability) hazards

H-rating

4 Materials that on very short exposure could cause death or major residual injury.

3 Materials that on short exposure could cause serious temporary or residual injury.

2 Materials that on intense or continued but not chronic exposure could cause temporary incapacitation or possible residual injury.

1 Materials that on exposure would cause irritation but only minor residual injury.

0 Materials that on exposure under fire conditions would offer no hazard beyond that of ordinary combustible material.

F-rating

4 Materials that will rapidly or completely vaporize at amosphereic pressure and normal ambient temperature, or that are readily dispersed in air and that will burn readily.

3 Liquids and solids that can be ignited under almost ambient temperature conditions.

2 Materials that must be moderately heated or exposed to relatively high ambient temperatures before ignition can occur.

1 Materials that must be preheated before ignition can occur.

0 Materials that will not burn.

R-rating

4 Materials that in themselves are readily capable of detonation or of explosive decomposition or reaction at normal temperatures and pressures.

3 Materials that in themselves are capable of detonation or explosive decomposition or reaction but require a strong initiating source or which must be heated under confinement before initiation or which react explosively with water.

2 Materials that readily undergo violent chemical change at elevated temperatures and pressures or which react violently with water or which may form explosive mixtures with water.

1 Materials that in themselves are normally stable, but which can become unstable at elevated temperatures and pressures.

0 Materials that in themselves are normally stable, even under fire exposure conditions, and which are not reactive with water.

OXY	Sometimes in the lower part of the NFPA diamond the word OXY appears. This indicates that the substance is a strong oxidant.

W

This symbol should be interpreted in conjunction with the reactivity ratings:

R = 3 Apart form the hazards mentioned under 3, the substance may react explosively with water. Safe guarding against the effects of an explosion is necessary in case water-based extinguishing agents are used.

R = 2 Apart from the hazards mentioned under 2, the substance may react violently with water or form explosive compounds.

R = 1 Apart from the hazards mentioned under 1, the substance may react with water but not violently.

W does not occur in conjunction with R = 4 or R = 0.

N.B.: The NFPA sometimes comes to an estimation of hazards that differs from that of other institutions, e.g. the editors of this book. Also when estimating health hazards in case of fire sometimes the effects of the combustion products of the substance are taken into account.

10. NOTES

Addition of small quantities of combustible substances or raising oxygen content enhances combustibility considerably

Many 'non-combustible' substances, e.g. trichloroethylene, chloroform, are combustible and can explode on contact with air when (i) the atmospheric oxygen level is above normal (such as when cleaning oxygen lines), or (ii) the liquid is contaminated with a combustible liquid.

Flush contaminated clothing with water (fire hazard)

Applies to strong oxidants and reducing agents. When clothing is contaminated with these substances there is increased risk of fire.

Odor gives inadequate warning of exceeded TLV

The identifiability of a substance by smell is expressed as the odor limit, the atmospheric concentration of the substance at which 50% of test subjects can *just* smell it. Unfortunately, research findings differ, and this contributes an element of uncertainty. Whether someone can smell a substance under certain circumstances depends on many factors, such as individual sensitivity, illnesses (e.g. colds), smoking habits and the presence of other odors (e.g. in mixtures). It is thus rarely, if ever, advisable to judge air purity in the workplace on the basis of the sense of smell. Consequently, the sheets always state 'odor gives inadequate warning of exceeded TLV' unless the safety margin between the TLV and the odor limit is such that there is a 99% certainty that a level in excess of the TLV will be detected under normal circumstances - which excludes anyone with a cold! Even when there is an ample safety margin, it is undesirable to rely on the sense of smell. See also chapter *Tables and Formulas*.

Effects on blood due to formation of methemoglobin	Certain substances, if absorbed into the bloodstream, can affect oxygen transport by transforming hemoglobin, the oxygen carrier, into methemoglobin. If the hemoglobin level falls too low, respiration is impaired.

Alcohol consumption increases toxic effects

The consequences of excessive exposure to certain substances can be exacerbated by consumption of alcohol before, during or after work.

Depending on the degree of exposure, regular medical checkups are advisable

The company doctor and the safety officer should consult with each other as to the conditions under which a subject exposed to the substance is to report for regular medical checkups.

Symptoms usually develop after some time: rest and hospitalization essential

Since with certain substances there is a definite time-lag between the initial exposure and the appearance of the first symptoms, the subject should be placed under observation in hospital so that medical assistance will be available immediately should it be needed.

Insufficient data on harmful effects to humans

In the opinion of the medical committee, too few data are available regarding the effects of exposure to this substance in the work environment for any responsible comments to be made on symptoms, prevention and first aid.

High atmospheric concentrations, e.g. in poorly ventilated spaces, can cause oxygen deficiency, with risk of unconsciousness

Applies to gases (including those known to cause asphyxiation) which have few, if any, immediate harmful effects at room temperature, and to some liquids with a high vapor pressure. These substances, when released, can cause dangerously low oxygen levels in the workplace. The atmospheric oxygen level decreases with a rise in pressure: at a vapor pressure of 30mm Hg it drops from the normal 20.7% by vol. to 20.0%; at 60mm Hg it drops to 19.0%, which is already dangerously low. Although the vapor pressure saturation point will not be reached in most cases, because there is normally at least some ventilation, saturated vapor pressure values of 30mm Hg or higher must be seen as an increasingly serious cause for concern. A degree of preparedness is essential, particularly when working with cryogenic liquids such as liquid nitrogen, since accidents with these substances can very rapidly cause oxygen deficiency in confined spaces. When this danger exists, always work with breathing apparatus or an oxygen mask, since oxygen deficiency can without warning lead to unconsciousness or even death.

N.B. DO NOT USE A RESPIRATOR/DUST MASK!

Under no circumstances use near flame or hot surface or when welding

Indicates that high temperatures or ultraviolet rays can produce toxic substances. This applies particularly to non-combustible chlorine compounds (which in most cases produce phosgene).

Use cylinder with special fittings

Indicates special handling requirements.

Packaging	Original packaging is generally the best. Changing the packaging can even pose serious risks, as in the case of organic peroxides. Reclosable packaging is normally required, because: – dispersion (by evaporation etc.) is always undesirable; – open packaging can lead to contamination, with the risk of undesirable chemical reactions.
Use airtight packaging	Indicates a substance which can produce hazardous effects on contact with air.
Packaging: special material	This indication is present only when special packaging requirements (materials for drums etc.) cannot be assumed to be generally known; the tendency of acids, bases and water-based liquids to attack metals and the unsuitability of many plastics for organic liquids are general knowledge and thus not mentioned. The chemical suppliers can usually give advice on the correct materials.
Unbreakable packaging preferred; if breakable, keep in unbreakable container	Indicates a very hazardous substance. The breakable packaging is protected by the unbreakable; the 'container' must also be able to collect any leakages.
Transport emergency card, TEC(R)	Listing of emergency measures to be taken following an accident during transportation (compiled by the European Council of Chemical Industry Federations).
HI	The Hazard Identification code must be used (together with the UN number – see below) when transporting dangerous substances. HI is given above the UN number on the rectangular orange signboard affixed to the back of tanker trucks etc.

Key   *first number (immediate danger):*
2 = gas
3 = flammable liquid
4 = flammable solid
5 = oxidant or organic peroxide
6 = toxic substance
7 = radioactive substance
8 = corrosive substance

second and third numbers (attendant danger)
0 = no meaning
1 = danger of explosion
2 = danger of gas release
3 = danger of ignition
5 = danger of oxidation
6 = danger of poisoning
8 = danger of corrosion
9 = danger of violent reaction due to polymerization or decomposition

An X before HI means avoid contact with water.

Two consecutive identical numbers indicate increased danger, for example:

22 = refrigerated gas
33 = extremely flammable liquid (flash point below 21°C)
66 = extremely toxic substance
266 = extremely toxic gas
88 = highly corrosive substance

UN number

Number assigned by the UN to substances; used primarily for transportation purposes. Source: *United National Transport of Dangerous Goods;* recommendations prepared by the Committee of Experts on the Transport of Dangerous Goods, second revised edition, New York 1980, ISBN 0-94394-04-2.

TABELS AND FORMULAS

Conversion factors   (rounded off)

1 N.m^{-2}	= 1 pascal (Pa)
1 bar	= 100,000 Pa
	= 100 kPa (kilopascal)
1 mbar	= 0.750 mm Hg
	= 0.1 kPa
	= 1hPa (hectopascal)
1mm Hg	= 1.333 mbar
1 phys. atmosphere (atm)	= 760mm Hg
	= 1.013 bar
1 tech. atmosphere (at)	= 1 kg force/cm^2
	= 1 kp/cm^3 = 10 m column of water
	= 0.981 bar
1 kg force	= 1 metric kilopound
	= 9.81 N
1 psi	= 69 mbar
1 joule (J)	= 1 N.m
	= 1 watt-second = 0.24 calorie (cal)
1 cal	= 4.2 J
1 Mho/cm (Ω^{-1} cm^{-1})	= 10^{14} pS/m
1 Siemens/cm (S.cm^{-1})	= 10^{14} pS/m
1 Siemens/m (S.m^{-1})	= 10^{12} pS/m

Conversion of Fahrenheit to Centigrade to Kelvin

$(°F - 32) \times 5/9 = °C$ $273.2 + °C = K$

Relative atomic mass values of some elements

Aluminium	Al	27.0	Magnesium	Mg	24.3
Antimony	Sb	121.8	Manganese	Mn	54.9
Arsenic	As	74.9	Mercury	Hg	200.6
Barium	Ba	137.3	Molybdenum	Mo	95.9
Boron	B	10.8	Nickel	Ni	58.7
Bromine	Br	79.9	Nitrogen	N	14.0
Cadmium	Cd	112.4	Oxygen	O	16.0
Calcium	Ca	40.1	Phosphorus	P	31.0
Carbon	C	12.0	Potassium	K	39.1
Chlorine	Cl	35.5	Selenium	Se	79.0
Chromium	Cr	52.0	Silicon	Si	28.1
Cobalt	Co	58.9	Silver	Ag	107.9
Copper	Cu	63.5	Sodium	Na	23.0
Fluorine	F	19.0	Sulfur	S	32.1
Hydrogen	H	1.0	Tin	Sn	118.7
Iodine	I	126.9	Titanium	Ti	47.9
Iron	Fe	55.8	Vanadium	V	50.9
Lead	Pb	207.2	Zinc	Zn	65.4

Composition of clean, dray air

oxygen	20.93% by vol.
nitrogen	78.10% by vol.
carbon dioxide	0.03% by vol.
noble gases etc.	0.94% by vol.

Air usually contains approx. 1% water vapor by vol. (which corresponds to 40% relative humidity at 20°C); oxygen content is then 20.7% by vol.

Relative density

From the point of view of safety, it is important to know whether a gas or vapor is lighter or heavier than air.

In the case of *gases*, the following applies:
Taking the specific density of air at a given temperature to be 1, the relative density of a gas at the same temperature is:

$$\text{rel. density of gas} = \frac{M}{29.0}$$

where M is the relative molecular mass of the gas and 29.0 is the mean molecular mass of air.

This measurement is most often referred to as the relative vapor density; in the case of liquids, however, it applies only at temperatures above boiling point.

Liquids at room temperature (approx. 20°) give off an air-vapor mixture whose density relative to pure air depends not only on the relative molecular mass of the liquid but also on the saturated vapor pressure. This vapor density is given on the sheets as: 'Relative density at 20°C of saturated air-vapor mixture (air = 1)'. It is calculated using the following formula:

$$d_m = \frac{1013 - p_{20}}{1013} + \frac{1000\ Mp_{20}}{22.4 \cdot 1013 \cdot 1293}$$

(1293 = specific mass of air in g/m^3 at 0°C and 1013mbar)

which reduces to:
$$d_m = 1 + 34.1 \cdot 10^{-6} \cdot p_{20} \cdot (M-29.0)$$

where:

d_m	=	relative density of the saturated air-vapor mixture at 20°C
p_{20}	=	saturated vapor pressure of the liquid in mbar at 20°C
M	=	relative molecular mass of the liquid

Atmospheric concentrations of gases and vapors

These are expressed in % by vol., ppm or mg/m^3.

1% by vol.	=	1 unit by volume of gas/vapor per 1000 units of contaminated air
	=	10,000 ppm

1ppm	= 1 unit by volume of gas/vapor per million units of contaminated air
	= $1cm^3$ (ml) of gas/vapor per m^3 contaminated air
	= 10^{-4}% by vol. of contaminated air
$1mg/m^3$	= 1mg of gas or vapor per m^3 contaminated air

The ratio between ppm and mg/m^3 at 20°C and 1013mbar is as follows:

$$1 \text{ ppm} = \frac{M}{24} \text{ mg/m}^3 \qquad\qquad 1 \text{ mg/m}^3 = \frac{24}{M} \text{ ppm}$$

where M is the relative molecular mass of the substance.

At other temperatures, the ratio is as follows:

$$1 \text{ ppm} = \frac{273\,M}{22.4\,T} \text{ mg/m}^3$$

where T is the temperature in °K; 22.4 is the molar volume in l of a gas at 0°C and 1013mbar.

Vapor pressure and vapor concentration

Where the saturated vapor pressure of a substance at a given temperature is known, the vapor concentration can be approximated as follows:

$$C = \frac{M}{22.4} \cdot \frac{p}{1013} \cdot \frac{273}{T} \cdot 10^6 \text{ where}$$

C = concentration in mg/m^3 of air
M = relative molecular mass
p = vapor pressure in mbar at temperature T
T = temperature in °K

At 20°C the formula becomes:

$$C_{20} = 41\,M \cdot p_{20}$$

The ratio between vapor pressure and % by vol. in air is:

$$\% \text{ by vol.} = \frac{100p}{1013}$$

At and atmospheric pressure of 1013mbar, a partial vapor pressur of 1mbar corresponds to a concentration of 987ppm.

Approximating the saturated vapor pressure of liquids

Where no value can be found in the literature, the following formula can be used:

$$\log p_{20} = 2.8808 - \frac{(a_n t_k + b_n)(t_k - 20)}{296.1 - 0.15\,t_k}$$

where:

p_{20} = saturated vapor pressure at 20°C in mm Hg(!)
t_k = boiling point in °C at atmospheric pressure

a_n and b_n = physical constants depending on the value of n for the substance as given in the table below.

n	a_n	b_n
1	0.0021	4.31
2	0.0021	4.54
3	0.0021	4.77
4	0.0022	5.00
5	0.0023	5.22
6	0.0023	5.44
7	0.0023	5.67
8	0.0023	5.90

The values of n of a limited number of substances are given in the literature. Other substances can be classified by chemical 'family', as explained below.

Halogen compounds: take halogen atoms to be 'H' and determine the value of n. If a substance is difficult to classify, select n = 4.

n = 2
- hydrocarbons and substances in which there are proportionately few elements present other than C and H
- ethers
- silicons
- sulfides

n = 3
- aldehydes (alkanes)
- epoxy compounds
- esters
- ketones
- nitrogen compounds

n = 4
- phenols (including multivalent phenols)
- lower esters (with relatively high proportion of 0)

n = 5
- carboxylic acids
- acid anhydrides

n = 7
- alcohols
- glycols

Estimation of flash point and explosive limits

The best way to determine the flash point is by experiment. If this is not practicable and the literature gives no figures, it can be approximated, provided the boiling point is known, using the method described in the *Journal of Chemical Education* 50 (1973); A 85-9; *Journal Fire and Flammability* 8 (1977) 38-40; *Chem. Eng.*, December 1983, p. 88.
For methods of approximating flash points of mixtures, see *Fire and Materials* (1976) (1) 134-9.

Rendering explosive mixtures inert

The explosive limits of an explosive air-gas/vapor mixture approach each other as the proportion of inert gas in the mixture increases. At a certain percentage of inert gas they coincide; at this point the mixture is no longer explosive. The narrowing of the explosive range varies from one inert gas to another, and the damping effect of carbon dioxide is generally greater than that of nitrogen.

Where the percentage of combustible gas/vapor is not known, the following table may prove useful in determining the risk of explosion:

Gas/vapor	Inert gas	max. % O_2	min. % w/o air	min. % with air
Hydrogen	Carbon dioxide	5.9	57	91
Carbon monoxide		5.9	40	68
Ethylene		11.7		
Ethane		13.3	32	88
1,3-Butadiene		13.0		
Benzene		13.9		
Cyclopropane		13.9		
Propylene		14.1	~29	~89
Propane		14.2	29	89
Pentane		14.4		
Hexane		14.5	29	94
Butane		14.5	28	90
Methane		14.6	23	77
Gasoline		~14.5	~29	~93
Hydrogen	Nitrogen	5.0	71	95
Carbon monoxide		5.6	58	80
Ethylene		10.0		
Ethane		11.0	44	93
1,3-Butadiene		10.4		
Benzene		11.2		
Cyclopropane		11.7		
Propylene		11.5		
Propane		11.8	42	94
Pentane		11.5		
Hexane		12.1	41	96
Butane		12.1	40	95
Methane		12.1	36	86
Gasoline		~11.8	~42	~96
Gasoline	Exhaust gases	~12.9	~36	~94
	$CHCl_2F$ (R 21)	~17.2	~16	~86
	CCl_2F_2 (R 12)	~17.8	~12	~78
	CCl_3F (R 11)	~18.1	~11	~77

~ = approximate value

max. % O_2: % by vol. of oxygen in the overall mixture (combustible gas/vapor + air + inert gas) below which the mixture is not explosive.

min. % w/o air: % by vol. of inert gas in the overall mixture (combustible gas/vapor + air + inert gas) above which the mixture is not explosive.

min. % with air: % by vol. of inert gas in the overall mixture (combustible gas/vapor + air + inert gas) above which the mixture is not explosive and no explosive mixture can form upon dilution with air.

The required safety margin depends, inter alia, on the accuracy of the methods used to measure oxygen content and volume of inert gas, and on the degree of mixing of the gases. The values of 'min. % with air' are, of course, high. Where the composition of the overall mixture is known, the risk of explosion should be determined by means of a graph in order to avoid needless dilution.

References:

1.   Zabetakis, M.G. 'Flammability characteristics of combustible gases and vapors,' *U.S. Bureau of Mines Bulletin* 627 (1964) Washington DC.
2.   Nabert, K. Schön, G. *Sicherheitstechnische Kennzahlen brennbare Gase und Dämpfe,* Braunschweig: Deutscher Eichverlag, 1970 (2. erweiterte Auflage).

Classification of acids and bases according to strength

Acids and bases are defined as 'strong' when they totally *dissociate* in water. This chemical classification is too narrow as regards possible damage to the skin or eyes and is in need of further gradation. An acid is classified as strong on the sheets when the estimated pH of the saturated aqueous solution (max. 10 mole) is less than or equal to 0.2 at 20°C ($pH_s \leq 0.2$). An acid is described as 'medium strong' when the pH is between 0.2 and 2.0. Since human skin and eyes are more sensitive to bases than to acids, the following has been adopted: strong bases have a pH_s greater than or equal to 13.0, medium strong bases a pH_s which is between 11.0 and 13.0.

Calculation of the pH_s
acids: $pH_s = (pK_a + pC_s)/2$
bases: $pH_s = 14 - (pK_b + pC_s)/2$

where:
pK = strength exponent of acid (pK_a) or base (pK_b) at 20°C
pC_s = concentration exponent of saturated aqueous solution/10 mole solution at 20°C in mol./l, as calculated using the formula:

$$pC_s = -\log(1000 . D.S.) / (M(100 D + S))$$

where:
S = solubility in g per 100ml water at 20°C
D = relative density of substance (water = 1)
M = relative molecular mass

Values where $pC_s < -1$ are taken to be -1 because of the upper limit of 10 mole.

Values for pK can be found, inter alia, in Weast, *Handbook of Chemistry and Physics.*
N.B. The literature sometimes gives pK_a values for basis instead of pK_b values; the latter can then be calculated using the formula $pK_a + pK_b = 14.0$.

Index of relative inhalation risk (RIR index)

The TLV is one of various indices of permissible exposure levels to chemicals; another is the 'Short Term Exposure Limit' (STEL), which represents a 15-minute concentration which should not be exceeded in an 8-hour workday. Its aim is particularly to prevent the acute effects of short-term exposure. STEL values can be found, along with a more precise definition, in *Threshold Limit Values,* published by the ACGIH (Cincinnati, USA).

The *RIR index* indicates how quickly the STEL is reached on evaporation under normal conditions at 20°C.

Calculation of the RIR-index:
a.   if $p_{20} < 200$ mbar: $RIR = C_s/STEL$

b.   if $p_{20} \geq 200$ mbar: $RIR = (10^6/STEL) \ln (10^6/(10^6 - C_s))$

where:
p_{20} = saturated vapor pressure of the substance in mbar at 20°C
C_s = saturated vapor concentration in ppm at 20°C. Calculation: $C_s = 10^6/1013)p_{20}$
$\ln$ = natural logarithm
STEL = The Short Term Exposure Limit in ppm (as listed in the the most recent ACGIH report).

If no STEL value for a given substance is available, use:
−   the ceiling value (C) of the TLV (ACGIH), or
−   the TLV time-weighted average (TWA), multiplied by 3.
    If TLV > 500ppm, multiply by 2.

In some cases, if neither the STEL nor the TLV is available, a STEL value can be estimated using other toxological data to give an indication of the order of magnitude of the RIR index; extreme caution is, however, advised. Even when the STEL value is known, the calculated RIR index should be considered in connection with other data on the substance.

Meaning of RIR under the following standard conditions:
−   surface area of liquid: $1m^2$
−   air speed over surface of the liquid: 0.1m/s
−   volume within which evaporation occurs at 20°C without ventilation: $100m^3$

Where the vapor can be assumed to mix homogeneously in air, the time (in minutes) used to calculate the STEL value is as follows:

$$t_{STEL} = 3714/RIR$$

Taken in conjunction with the classification on the sheets this gives the following table:

RIR-index	t_{STEL} in minutes
< 12	> 310
12 − 120	310 − 31
120 − 4000	31 − 1
⩾ 4000	< 1

The RIR-index for gases is infinitely large and t_{STEL}, consequently, 0 minutes.

Odor Safety Factor

The sense of smell can sometimes be of use in ascertaining that the TLV of a hazardous substance has been exceeded before it is too late. J.E. Amoore and E. Hautala (*Journal of Applied Toxicology* 3, No. 6 (1983) 272ff) have developed an Odor Safety Factor classification with five categories; the have also compiled a list of mean odor limits for 214 volatile substances based on various sources. These limits indicate the atmospheric concentration at which 50% of the test subjects, having been given prior warning, were just able to identify the odor.

The Odor Safety Factor is defined as:

$$OSF = \frac{TLV-TWA \ (ppm)}{odor \ limit \ (ppm)}$$

Thus the OSF is 1 when a concentration equal to the TLV-value is detected by 50% of the subjects.

Amoore and Hautala also investigated the effects of distraction (e.g. by work-related tasks) on odor identification and discovered that, under these conditions, the OSF reached 26 before 50% of subjects detected that the TLV had been attained. In the case of subjects given prior warning, 99% could detect attainment of the TLV at OSF = 26.

OSF values are divided into the following categories.

Categorie	OSF	Description
A	⩾ 550	More than 90% of exposed persons able to detect attainment of TLV by smell, even when distracted by work environment
B	26-550	50-90% of distracted subjects able to detect
C	1-26	Less than 50% of the distracted subjects able to detect
D	0.18-1	10-50% of subjects given prior warning able to detect
E	< 0.18	Less than 10% of subjects given prior warning able to detect

The OSF can be calculated simply from the TLV and odor limit of a substance; the table can then be used to ascertain the appropriate OSF category. Oxygen-depleting gases have no TLV-value, but an Odor Safety Factor can nonetheless be calculated by using a value of 14,000ppm instead of the TLV. When an extraneous gas reaches an atmospheric concentration of 14,000ppm, the oxygen level is already a dangerously low 18%.

Only when working with substances in category A or B is there a reasonable to good chance of detecting the odor before it is too late; consequently sheets for substances where the OSF is less than 26 bear the warning that the odor gives insufficient warning of concentrations in excess of the TLV. Most volatile substances discussed in the Amoore and Hautala article fall into category C, D or E. In other words, the odor by itself usually gives insufficient warning, particularly in situations where several chemicals are present, masking one another's odor. There are other reasons why the sense of smell – however sensitive – must be regarded as inadequate to the task of monitoring concentrations of dangerous gases and vapors: some substances are odor-free; habituation or deadening of the olfactory nerves can occur; variations from the average olfactory sensibility etc. Given that exceeding the TLV of some chemicals can quickly lead to acutely dangerous situations, we must conclude that the sense of smell is of limited value when it comes to reliably detecting harmful vapors and gases.

RECOMMENDATIONS FOR GLOVES AND OTHER PROTECTIVE CLOTHING

When working with chemicals, it is usually necessary to prevent contact with the skin. Although in some cases the skin can be sufficiently resistant to the substance to prevent absorption, in most cases contact leads to skin disorders. A substance can also penetrate the skin without attacking it, being absorbed by the blood and doing harm elsewhere in the body.

When handling relatively harmless dry solids, leather work gloves are usually the best choice, except for persons whose skin is sensitive to the chromium residue left over from the tanning process. In these cases, leather gloves tanned with vegetable products or lined plastic gloves should be used. In all other cases, especially when handling liquids, normal working gloves are inadequate. It is advisable to choose the glove material carefully, keeping in mind (a) the nature of the work and (b) the chemicals against which protection is required.

(a)   Nature of work
–   If gloves have to be worn continuously, skin irritation and even skin disorders can occur. As a preventive measure, it may be necessary to wear lined work gloves, or a pair of cotton gloves inside the work gloves.
–   The more prolonged and/or repeated the contact with chemicals, the more durable and impermeable the gloves need to be; gauntlets are recommended for prolonged/repeated contact.
(b)   The chemicals can:
–   attack the glove material, dissolving or weakening it;
–   be soaked up by the glove material, causing it to swell and come into contact with the skin;
–   permeate the glove material without being soaked up, still making contact with the skin.

Much more is known about the durability of glove materials than their impermeability, and this should be realized when consulting tables on glove material durability. When working with extremely dangerous liquids, the potential for absorption of the substance by the glove material makes the use of disposable gloves advisable. The impermeability of a glove material depends on its thickness and the nature of the liquid; the following glove materials are listed in increasing order of impermeability (to most organic liquids): polyvinyl chloride, natural rubber, neoprene, nitrile rubber, butyl rubber, polyvinyl alcohol. Polyvinyl alcohol is permeable to lower alcohols, however, whereas butyl rubber is not. Polyvinyl chloride, butyl rubber, natural rubber and neoprene are the least permeable to aqueous liquids. Polyvinyl alcohol is unusable in an aqueous milieu, since it dissolves.

For guidance on the selection of protective clothing, see *Guidelines for the Selection of Chemical Protective Clothing,* published by the American Conference of Governmental Industrial Hygienists (ACGIH). This contains, inter alia, two tables on the various applications of 13 types of material used in gloves and protective clothing, particularly as regards durability and permeability. Using these tables, a *List of Recommendations* has been drawn up for those substances for which there is a sheet. The list is intended to provide initial suggestions; the final choice should be made on the basis of on-site trials. An expert should always be consulted.

Use of safety gloves
–   One or more persons should be placed in charge of supplying gloves and given special instructions.
–   Cotton gloves should be provided to wear inside the safety gloves.
–   Gloves should be stored in a cool place, out of the sun, to prevent premature deterioration.
–   Before using always check gloves (even new ones) for defects. They can be checked for holes by inflating them with air and putting them under water (this should not be done with PVA gloves).

977

- Before putting on the gloves, wash hands (and cut fingernails, if necessary).
- Do not remove gloves until the exterior has been thoroughly cleaned (do not use water on PVA gloves).
- Re-use of gloves is permitted only if they are verifiably clean and free of absorbed chemicals (by washing inside and outside with soap and warm water and drying out). Do not re-use if gloves have stiffened or show other signs of deterioration.
- The above applies mutatis mutandis to other protective clothing.

References:

- ACGIH Inc. *Guidelines for the Selection of Chemical Protective Clothing.* Cincinnati, 1987.
- Estlander, T. *Protective gloves.* Proceedings of fourth Finnish-Soviet joint symposium on industrial hygiene. Helsinki: Institute of Occupational Health, 1984.

Notes on List of Recommendations

Names of chemicals
Usually the same as on the sheets; use the Index to find synonyms at the end of the table.

Types of material
Butyl/Neoprene
Butyl Rubber (Butyl)
Chlorinated Polyethylene (CPE)
Natural Rubber
Nitrile Rubber (Nitrile)
Nitrile Rubber + Polyvinyl Chloride (Nitrile + PVC)
Polyethylene (PE)
Polyvinyl Alcohol (PV Alcohol)
Polyvinyl Chloride (PVC)
Viton
Viton/Neoprene

Some other materials:
- Polyurethane
- PV Acetate
- Styrene-Butadiene Rubber (SBR)
- Saranex
- Teflon

N.B. Teflon-based and similar materials fall into the category of chlorinated compounds.

Recommendations
The recommendations on the various materials are based on research data on durability and permeability (penetration time), as well as on the classification system used by the various suppliers (3 = excellent; 2 = good; 1 = fair; 0 = poor). In the list of recommendations this is indicated as follows:

Single and double, upper and lower case "r's" and "n's" are used to convey the recommendations. Briefly, RR, R, rr, and r indicate various degrees of good resistance and NN, N, nn, and n indicate various degrees of poor chemical resistance. Double characters indicate that there are test data to support the recommendations, and single characters indicate that only qualitative information was available. Upper case characters indicate concensus and a relatively large amount of information, whereas lower case indicates a relatively small amount of information or inconsistencies.

For further details please refer to *Guidelines for the Selection of Chemical Protective Clothing,* published by the American Conference of Governmental Industrial Hygienists; Cincinnati, 1987.

LIST OF RECOMMENDATIONS
for gloves and other protective clothing when working with chemicals

Name	BUTYL	CPE	VITON/NEOPRENE	NATURAL RUBBER	NEOPRENE	NITRILE+PVC	NITRILE	PE	PV ALCOHOL	PVC	VITON	BUTYL/NEOPRENE	Other materials
Acetaldehyde	RR	NN		NN	NN	NN	NN	NN	nn	NN	NN		TEFLON(rr) SILVERSHIELD(rr)
Acetic Acid	R	rr		nn	RR	nn	RR	nn	n	NN	rr		NEOP+NAT RUB (RR) NEOP/NAT RUB (RR)
Acetic Acid, < 30%					r			rr		n			
Acetic Acid, 30-70%		r		rr	rr		rr			rr			POLYURETHANE(r)
Acetic Acid, > 70%				nn	n		rr			rr			SBR(r)
Acetic Anhydride	rr	rr		NN	nn					NN	n		TEFLON(rr)
Acetone	RR	NN		NN	NN	nn	NN	NN	NN	NN	NN		TEFLON(rr) PV ACETATE(rr) SILVERSHIELD(rr)
Acetonitrile	RR	rr	nn	NN	NN		NN	NN	rr	NN	rr	rr	TEFLON(rr) SARANEX(rr) VIT/CLORBUTL(rr) SILVERSHIELD(rr)
Acetophenone			n									r	TEFLON(rr)
Acetyl Chloride	n	r								n			TEFLON(rr)
Acrolein	rr	NN					NN		nn		NN		VIT/CLORBUTL(rr)
Acrylic Acid		r								n			TEFLON(rr)
Acrylonitrile	n	nn		N	r	n	n	NN		N	n		POLYURETHANE(r)
Allyl Alcohol	RR	rr	rr		RR			rr	nn	NN	rr	rr	TEFLON(rr) SARANEX(rr)
Allylamine	nn			NN					nn	NN			
Allyl Chloride	r	rr								n			TEFLON(rr) PV ACETATE(rr)
Allyl Glycidyl Ether	rr			rr	nn		rr		rr	rr			NEOP/NAT RUB(rr)
Ammonia			n		r					r	n	r	
Ammonium Fluoride													NEOP+NAT RUB(r)
Ammonium Fluoride, 30-70%				rr	rr		rr			rr			
Ammonium Hydroxide	R	r		rr	rr	NN	rr	NN	n	NN	r		SBR(R)
Ammonium Hydroxide, <30%				RR	RR		RR			NN	rr		
Ammonium Hydroxide, 30-70%				rr	rr	r	rr	NN		R			

Name	BUTYL	CPE	VITON/NEOPRENE	NATURAL RUBBER	NEOPRENE	NITRILE+PVC	NITRILE	PE	PV ALCOHOL	PVC	VITON	BUTYL/NEOPRENE	Other materials
Ammonium Sulfate	r			R	R	r	r			R			POLYURETHANE(R)
Amyl Acetate	R	n	n	nn	nn	nn	nn	NN	rr	nn	n	n	NEOP+NAT RUB(r)
Amyl Alcohol	rr		r	NN	RR	NN	nn	nn	rr	NN	rr	r	NEOPRENE+SBR(R) POLYURETHANE(R) SBR(R) SBR/NEOPRENE(R)
Aniline	RR	r	rr	NN	NN	NN	nn	NN	RR	NN	NN	rr	TEFLON(rr) NEOP+NAT RUB(rr) SARANEX(rr) SILVERSHIELD(rr)
Arsenic Trichloride			n									n	
Asphalt			r									n	
Benzaldehyde	rr	n	n	nn	nn	n	nn	NN	RR	N	n	r	SBR(r)
Benzene	NN	nn	rr	NN	NN	NN	NN	NN	NN	NN	nn	rr	PV ACETATE(rr) SILVERSHIELD(rr)
Benzenesulfonic Acid				rr			rr						
Benzethonium Chloride	rr			rr	rr					rr			
Benzonitrile	rr			NN					rr	nn			
Benzoyl Chloride	rr	n	r		nn				rr	NN	nn	n	
Benzyl Alcohol	rr		r	r	R	r	r			R	rr	r	SBR(R) SBR/NEOPRENE(r)
Benzyl Benzoate			r									r	
Benzyl Chloride	n	nn	r	N	N	n	r			N		n	TEFLON(rr)
Bis(2-Ethylhexyl) Phthalate	rr		r	rr	rr		RR		nn	NN	rr	r	
Boric Acid	rr				rr		rr			r	rr		
Bromine								NN					
Bromine Trifluoride			n									n	
Bromoacetonitrile	rr			NN					rr		rr		
Bromobenzene	nn		r				nn		rr		rr	n	
Bromochloromethane			r									r	
2-Bromoethanol	rr			NN						NN	rr		
1-Bromo-2-propanol	rr			NN					rr		rr		
3-Bromo-1-propanol	rr				rr				rr		rr		
Butadiene	rr			NN	nn					NN	rr		
Butane	n			N	R	r	n			N	r		NEOPRENE+SBR(r) POLYURETHANE(R)
tert-Butanol	rr		r	NN	rr					NN		r	

Name	BUTYL	CPE	VITON/NEOPRENE	NATURAL RUBBER	NEOPRENE	NITRILE+PVC	NITRILE	PE	PV ALCOHOL	PVC	VITON	BUTYL/NEOPRENE	Other materials
Butene				n	r		r			n			
Butyl Acetate	rr	r		NN	NN	nn	NN	NN	rr	NN	nn		TEFLON(rr) SILVERSHIELD(rr)
Butyl Acrylate			n									n	TEFLON(rr)
Butyl Alcohol	R	r		nn	RR	nn	RR	RR	nn	nn	r		TEFLON(rr) POLYURETHANE(r)
Butylamine	nn	NN	n	NN	nn					NN		n	TEFLON(rr)
iso-Butylamine	nn	rr	n		NN				nn	NN		n	VIT/CLORBUTL(rr)
sec-Butylamine	nn				nn		nn			NN			
tert-Butylamine	nn				nn		nn			NN			
n-Butyl Benzoate			r									r	
n-Butyl Carbitol			r									r	
Butyl Cellosolve			r				NN		rr	rr		n	NAT+NEOP+NBR(r)
Butyl Cellosolve Acetate													NAT+NEOP+NBR(r)
n-Butyl Chloride							nn		rr	nn	rr		
Butylene					r		r			r			SBR(r)
tert-Butyl Mercaptan			n									n	
Butyl Oleate			r									r	
n-Butyl Phthalate	RR	n	r	nn	NN	rr	NN	rr	RR	N	rr	r	SBR(r) SILVERSHIELD(rr) SBR/NEOPRENE(r)
Butyl Stearate			r									r	
o-tert-Butyl Toluene	RR				rr		rr		rr		RR		SILVERSHIELD(rr)
Butyraldehyde	nn		n	R	nn	r	r		nn	R	nn	r	TEFLON(rr) SBR(r) SBR/NEOPRENE(r)
Butyric Acid		r								r			
Calcium Bisulfate, 30-70%		r										r	
Calcium Chloride	r			R	R	r	r			R			
Calcium Hydroxide					r					r			
Calcium Hypochlorite	r			R	R	r	r			R			
Carbitol			r									r	
Carbon Disulfide	NN	NN		N	N	n	NN	NN	RR	N	RR		
Carbontetrabromide			r									n	

Name	BUTYL	CPE	VITON/NEOPRENE	NATURAL RUBBER	NEOPRENE	NITRILE+PVC	NITRILE	PE	PV ALCOHOL	PVC	VITON	BUTYL/NEOPRENE	Other materials
Carbon Tetrachloride	N	nn	r	NN	NN	NN	N	NN	RR	NN	rr	n	TEFLON(rr) PV ACETATE(rr) SILVERSHIELD(rr)
Chlorine	n	r	rr	n	rr		n	NN		NN	rr	rr	SARANEX(rr)
Chloroacetic Acid								NN					SARANEX(rr)
Chloroacetone	r		n	n	R	n				N		r	
Chloroacetonitrile	rr			NN				rr			rr		
Chlorobenzene	nn	n	r	N	NN	n	nn	NN	nn	NN	RR	n	TEFLON(rr)
2-Chloro-1,3-butadiene			r		NN		NN		rr	NN		n	
Chlorodibromomethane	nn								nn	NN	rr		
Chlorododecane			r									n	
Chloroform	N	NN	r	NN	NN	n	NN	NN	RR	NN	rr	n	TEFLON(rr) POLYURETHANE(r) VIT/CLORBUTL(rr)
3-Chloro-2-methylpropene	nn								NN	NN	rr		
Chloronaphthalenes (all isomers)			r	N	n		n		rr	n	rr	r	SILVERSHIELD(RR)
2-Chloro-2-nitropropane	rr			NN					rr	nn			
1-Chloro-2-propanol	rr			NN						NN	rr		
3-Chloro-1-propanol	rr								nn	NN	rr		
Chlorosulfonic Acid					n			rr			n	n	SARANEX(rr)
o-Chlorotoluene							nn				rr		
p-Chlorotoluene							nn				rr		
Chromic Acid	n	r		NN	N	RR	N	rr		RR	r		
Chromic Acid, <30%													NEOP+NAT RUB(rr)
Chromic Acid, 30-70%					n	n	r	n	n	nn			
Cimethyl Sulfoxide		rr		RR	RR	rr	nn	rr		NN			VIT/CLORBUTL(rr) NEOP/NAT RUB(rr)
Citric Acid	r			R	R	R	R			R	r		POLYURETHANE(R) SBR(R)
Citric Acid, < 30%					rr	rr	rr	rr	rr	nn	rr		
Copper Chloride					n	r	r			r			
Copper Sulfate					n	R	r			R			
Creosote	n		r	N	rr	r	R			r	rr	r	SBR(r)
Creosote, Wood					rr						rr		
m-Cresol					nn	rr	rr	rr		NN			TEFLON(rr) NEOP/NAT RUB(rr)

Name	BUTYL	CPE	VITON/NEOPRENE	NATURAL RUBBER	NEOPRENE	NITRILE + PVC	NITRILE	PE	PV ALCOHOL	PVC	VITON	BUTYL/NEOPRENE	Other materials
Cresols	n	r	r					nn		n		n	SARANEX(rr)
Crotonaldehyde	rr	nn			nn				NN		NN		TEFLON(rr)
Cumene		rr	r									n	
Cupric Chloride	r			n	R	r				R			
Cupric Sulfate	r			n	R	r	r			R			
Cyclohexane	N	r	r	NN	NN	n	RR	NN	nn	NN	RR	n	TEFLON(rr)
Cyclohexanol	n	r	r	NN	NN	nn	RR	rr	RR	rr	RR	r	SILVERSHIELD(rr)
Cyclohexanone	rr	n	n				nn		rr	n	nn	n	SILVERSHIELD(rr)
Cyclohexylamine	nn			NN	nn		nn						
Cymene			r									n	
Decahydronaphthalene			r									n	
Decanal (all isomers)	rr		rr		rr		rr			NN	rr	rr	SARANEX(rr)
Decane			r									n	
Diacetin			r									r	
Diacetone Alcohol	r		n	r	R	r				R		r	SBR(r) SBR/NEOPRENE(r)
Diallyamine	nn								rr	NN	rr		
1,3-Diaminopropane	nn			NN	rr					NN			
Di-n-amylamine					nn		rr			NN	rr		
Dibenzyl Ether	r		n	N	R	r	r			R		r	
Dibenzyl Sebacate			r									r	
1,2-Dibromo-3-chloropropane		n								n			
Dibutylamine			n				nn		rr	NN	rr	n	
Dibutylether			n									n	
Dibutyl Sebacate			r									n	
Dichloroacetyl Chloride	nn								rr	NN	rr		
Dichlorobenzene	n	nn								n			
1,2-Dichlorobenzene			r				nn				rr	n	
1,3-Dichlorobenzene							nn				rr		
Dichlorobromomethane	nn									NN	rr		VITON/BUTYL(rr)
1,4-Dichloro-2-butene	nn	NN			NN		NN	NN	rr	NN	RR-		SARANEX(rr)
Dichloroethane			r	n	n		n			n		n	TEFLON(rr)
Dichloroethylene (all isomers)		n								n			
1,2-Dichloroethylene							NN		nn	NN	nn		

Name	BUTYL	CPE	VITON/NEOPRENE	NATURAL RUBBER	NEOPRENE	NITRILE+PVC	NITRILE	PE	PV ALCOHOL	PVC	VITON	BUTYL/NEOPRENE	Other materials
cis-Dichloroethylene	nn								NN	NN	rr		
trans-1,2-Dichloroethylene	NN								rr	NN	rr		
2,2'-Dichloroethyl Ether	nn	rr											TEFLON(rr)
Dichloropropane-Dichloropropene													TEFLON(rr)
Dichloropropane (all isomers)		nn											TEFLON(RR)
1,3-Dichloropropene	nn								rr	NN	rr		
2,3-Dichloro-1-propene	nn								rr	NN	rr		
Dicyclohexylamine			n									n	
Diethanolamine	rr			n	rr	n	nn			r	rr		TEFLON(rr) NEOP+NAT RUB(r)
Diethylaminoethanol	rr						nn		rr		rr		
Diethylamine	nn		n	NN	NN	n	NN	NN	n	NN	nn	n	SBR(r) SILVERSHIELD(rr)
Diethyl Benzene			r									n	
Diethylene Glycol			r		r					r		r	
Diethylenetriamine	rr				nn					nn	rr		
Di-(2-Ethylhexyl)-Sebacate			r									r	
Diethyl Oxalate			r									r	
Diethyl Sebacate			r									r	
Diisobutylamine					nn		rr		rr		rr		
Diisobutylene			r									n	
Diisobutyl Ketone	n	n		nn	nn	n	rr	NN	rr	N			
Diisobutyl Ketone, >70%	RR						RR		rr		RR		SILVERSHIELD(rr)
Diisoctyl Phthalate			r									r	
Diisooctyl Sebacate			r									r	
Diisopropylamine					nn		rr			NN	rr		TEFLON(rr)
Diisopropyl Benzene (all isomers)			r									n	
Diisopropyl Ketone			n									n	
N,N-Dimethylacetamide		nn		nn	nn		nn						SARANEX(rr)
Dimethylaminopropylamine	nn			NN	nn					NN			
Dimethylamine	rr			NN	nn				nn	NN			
Dimethylanaline			n									n	
α, α-Dimethylbenzyl Hydroperoxide													TEFLON(rr)
Dimethylbutylamine	nn						nn		nn	NN			

ame	BUTYL	CPE	VITON/NEOPRENE	NATURAL RUBBER	NEOPRENE	NITRILE+PVC	NITRILE	PE	PV ALCOHOL	PVC	VITON	BUTYL/NEOPRENE	Other materials
Dimethylethanolamine	rr			NN	nn		nn						
Dimethylformamide	RR			nn	NN	n	NN	nn	NN	N	NN		TEFLON(rr) VIT/CLORBUTL(rr) SILVERSHIELD(rr)
1,1-Dimethylhydrazine	RR	nn	n	NN	NN		NN		NN	NN	nn	n	CHLOROBUTYL(rr)
1,2-Dimethylhydrazine			n									n	
Dimethyl Phthalate			r									r	
Dimethylvinylchloride							NN		rr	NN	rr		
Di-n-octyl Phthalate				n	r	nn		NN		nn			NEOP+NAT RUB(r)
1,4-Dioxane	RR	r		NN	NN	n	NN	NN	n	NN	NN		TEFLON(rr)
1,3-Dioxolane			n									n	
Dipentene			r									n	
Dipropylamine	nn												TEFLON(rr) POLYCARBONAT(rr)
Di-sec-Octyl Phthalate			r								r		
Disulfur Dichloride			r									n	
Divinyl Benzene	RR						rr		rr		RR		SILVERSHIELD(rr)
Epichlorohydrin	nn	n		NN	nn		nn	NN	NN	nn	nn		TEFLON(RR)
1,2-Epoxybutane	nn				NN				rr		NN		
Ethane	n	r						rr		r			
Ethanol, 30-70%								rr					
Ethanol, > 70%								rr					
Ethanolamine	rr	r	n	RR	RR	RR	RR	rr	rr	rr	n	n	
2-Ethoxyethyl Acetate		n		n	rr		n			n		r	NAT+NEOP+NBR(r)
Ethyl Acetate	rr	nn	n	NN	NN	nn	NN	NN	n	nn	n	n	SILVERSHIELD(rr)
Ethyl Acetoacetate		n										r	
Ethyl Alcohol	R			NN	NN	nn	RR	NN	rr	NN	r		TEFLON(rr) NEOP/NAT RUB(rr)
Ethylamine, 30-70%	rr						rr						TEFLON(rr)
Ethyl Acrylate	nn	NN	nn		NN			NN	rr	NN	NN	rr	TEFLON(rr) SARANEX(rr)
Ethyl Benzene		r			n					nn	n	n	TEFLON(rr)
Ethyl Benzoate		r										r	
Ethyl Bromide					NN				rr	NN	nn		
Ethyl-n-butylamine							nn		rr	NN	nn		

Name	BUTYL	CPE	VITON/NEOPRENE	NATURAL RUBBER	NEOPRENE	NITRILE+PVC	NITRILE	PE	PV ALCOHOL	PVC	VITON	BUTYL/NEOPRENE	Other materials
Ethyl Cellosolve	rr		r	NN	NN		NN		NN	NN		n	NAT+NEOP+NBR(R)
Ethyl Chlorocarbonate			r									n	
Ethyl Cyanide	nn			NN					rr	NN			
Ethylene Acetate				n	n		n			n			
Ethylene Chlorohydrin	rr		r		rr				rr		rr	n	
Ethylene Dibromide	nn	nn	rr	NN	NN		NN	NN	RR	NN	nn	NN	TEFLON(RR)
Ethylene Dichloride	nn	nn	r	NN	NN	n	NN	NN	nn	NN	rr	n	SILVERSHIELD(rr)
Ethylenediamine	rr	nn	n	NN	rr			nn		NN		n	TEFLON(rr) SARANEX(rr)
Ethylene Glycol	R	r	r	RR	rr	RR	RR	RR	rr	nn	r	r	TEFLON(rr) NEOP+NAT RUB(rr)
Ethylene Oxide	r	r			n		NN			n	n		
Ethyl Ether	NN	r	n	NN	NN	nn	NN	NN	RR	nn	NN	n	TEFLON(rr) SBR(R) SILVERSHIELD(rr)
Ethyl Formate	r		n	r	R	n	R			N		n	SBR(R) SBR/NEOPRENE(r)
2-Ethylhexanoic Acid					rr		rr			rr			
2-Ethyl-1-Hexanol	rr				rr				rr		rr		
Ethylenimine	nn					nn	NN						
Ethyl Mercaptan			n									n	
Ethyl Methacrylate	nn	nn					nn		rr	NN		r	
Ethyl Silicate			r									r	
Ethylidene Dichloride	nn								rr	NN	nn		
Ferric Chloride	r			r	R	r				R			
Fluorine			n	r	r					r		n	
Fluorobenzene			r									n	
Fluoroboric Acid			r									r	
Fluorosilicic Acid			r									r	
Formaldehyde, < 37%	RR	rr	r	NN	NN	nn	NN	RR	n	NN	RR	r	TEFLON(rr) SBR(R) SILVERSHIELD(rr)
Formamide, 30%													NEOP+NAT RUB(r)
Formic Acid	R	r		R	R	R	r	NN		R	n		POLYURETHANE(R) SBR(R)
Formic Acid, 30-70%										r			

Name	BUTYL	CPE	VITON/NEOPRENE	NATURAL RUBBER	NEOPRENE	NITRILE+PVC	NITRILE	PE	PV ALCOHOL	PVC	VITON	BUTYL/NEOPRENE	Other materials
Formic Acid, > 70%				rr	rr	nn	rr	nn	n	nn			NEOP + NAT RUB(rr)
Freon 11	n		r	n	R		n			n	r	n	
Freon 12	r			n	R		r			n	r		
Freon 21				n	n		n			n			
Freon 22					r	r	n			n			
Freon 112		r										n	
Freon 11482		r										n	
Freon TF	n		r	NN	RR	nn	RR	NN	nn	NN		r	
Freon TMC	rr	rr		NN	NN		NN			rr	rr		CHLOROBUTYL(rr)
Furan	nn		n						rr	NN	nn	n	
Furfural	RR	r	n	NN	NN	nn	NN	NN	rr	N	rr	r	TEFLON(rr) SILVERSHIELD(rr)
2-Furylmethanol			n									r	
Gasoline	nn	r	rr	N	rr	NN	rr	NN	rr	NN	r	nn	POLYURETHANE(R)
Glutaraldehyde	rr				rr					rr	rr		
Glycerin Triacetate			r									r	
Glycerol	r		r	r	R	r	R			r		r	
Glycols			r									r	
Halothane	nn									rr	NN	nn	
Heptane					NN	nn	rr	RR	rr	NN	rr		
Hexachlorocyclopentadiene	nn						nn			rr		rr	
Hexamethylphosphoamide	rr				nn		nn	nn			nn		
Hexanal			r									r	
Hexane	NN	rr		NN	NN	NN	NN	NN	RR	NN	RR		TEFLON(rr) POLYURETHANE(R) VIT/CLORBUTL(rr) SILVERSHIELD(rr)
1-Hexene			r									n	
Hexyl Alcohol			r									r	
Hydrazine	rr	n		r	rr		rr			rr	n		CHLOROBUTYL(rr)
Hydrazine, 30-70%	rr			rr	rr	rr	RR	rr	n	rr			SILVERSHIELD(rr)
Hydrobromic Acid	r			r	R	r				R			SBR(r) SBR/NEOPRENE(r)
Hydrobromic Acid, 30-70%			r									r	
Hydrochloric Acid	nn	rr	rr	rr	RR	RR	rr			NN	rr	rr	

Name	BUTYL	CPE	VITON/NEOPRENE	NATURAL RUBBER	NEOPRENE	NITRILE+PVC	NITRILE	PE	PV ALCOHOL	PVC	VITON	BUTYL/NEOPRENE	Other materials
Hydrochloric Acid, <30%	r			RR	rr	rr	rr	rr	n	rr	r		
Hydrochloric Acid, 30-70%	r		r	RR	RR	n	RR	nn	n	nn	rr	r	NEOP/NAT RUB(RR)
Hydrochloric Acid, >70%				rr									NEOP+NAT RUB(rr) NEOP/NAT RUB(rr)
Hydrocyanic Acid	rr				r			rr		nn	r		
Hydrofluoric Acid	nn	r	rr	N	N	n	n	rr		NN	rr	rr	POLYCARBONAT(rr) SARANEX(rr) NEOPRENE+PVC(rr)
Hydrofluoric Acid, <30%				r	R		r			r			NAT+NEOP+NBR(r)
Hydrofluoric Acid, 30-70%			r	RR	rr	NN	nn	rr	n	nn	r	r	NEOP+NAT RUB(RR) SBR(r) NEOP/NAT RUB(rr)
Hydrofluoric Acid, >70%				rr									NEOP+NAT RUB(rr) NEOP/NAT RUB(rr)
Hydrogen Peroxide	nn	rr	r	r	R	r	n			nn	r	r	
Hydrogen Peroxide, <30%													CHLOROBUTYL(rr)
Hydrogen Peroxide, 30-70%				RR	NN		rr	rr	n	rr			NEOP+NAT RUB(rr) NEOP/NAT RUB(rr)
Hydrogen Peroxide, >70%			r								n		
Hydrogen Phosphide				nn	NN			nn		nn			
Hydrogen Sulfide	r	r		n	R	r				R			SBR(r)
Hydroquinone	n			R	R	rr	r	rr	n	rr	r		SBR(r)
Hydroquinone, <30%				rr	rr		rr			rr			
Hylene					r					r			POLYURETHANE(r)
Iminobispropylamine	rr			NN	nn						rr		
Iodine Pentafluoride			n								n		
b-Ionone	rr							rr					
Isoamyl Acetate	NN			NN	NN		nn	NN		NN			
Isoamylnitrile					nn		rr		rr		rr		
Isobutyl Acrylate	nn						nn		rr	NN			
Isobutyl Alcohol	rr		r	nn	NN	NN	RR	NN	n	NN	rr	r	NEOP+NAT RUB(r) SBR(r)
Isobutylene								rr					
Isobutyl Nitrite	nn						nn			NN	nn		
Isobutyraldehyde	rr				nn				NN	NN			
Isooctane	n	r	r	NN	NN	nn	NN	nn	nn	NN	r	n	

Name	BUTYL	CPE	VITON/NEOPRENE	NATURAL RUBBER	NEOPRENE	NITRILE+PVC	NITRILE	PE	PV ALCOHOL	PVC	VITON	BUTYL/NEOPRENE	Other materials
Isophorone			n									r	
Isoprene					nn		nn		rr		rr		
Isopropyl Acetate			n									n	
Isopropylamine	nn				nn					NN	NN		TEFLON(rr) PV ACETATE(rr)
Isopropyl Alcohol	rr	rr	r	NN	RR	nn	RR	NN		nn	r	r	CHLOROBUTYL(rr) TEFLON(RR)
Isopropyl Chloride			r								r		
Isopropyl Ether		rr	n	NN	nn		rr		rr	nn	rr	n	
Isopropylmethacrylate	nn						nn		rr	NN			
JP-4, Jet Fuel					n						r		
Kerosene	n		r	nn	N	rr	rr	nn	rr	nn	r	n	POLYURETHANE(r)
Lactic Acid	R	r	r	R	R	R	R			R	r	r	POLYURETHANE(R)
Lactic Acid, < 30%					r					r			
Lactic Acid, > 70%				rr	rr	rr	rr	rr	rr	rr			
Lauric Acid				n	r					r			
Lauric Acid, 30-70%				rr	rr		rr			nn			
d-Limonene	rr		r		rr		rr		rr			n	
Linoleic Acid			r	n	r					r		r	
Maleic Acid	n			n	R	R	n		n	R	r		SBR(r) SBR/NEOPRENE(r)
Malic Acid			r	r						r			
Maleic Acid, > 70%				rr	rr	rr	rr	rr		rr			
Mercury			r		r					r		r	
Mesityl Oxide		nn	n									r	VIT/CLORBUTL(rr)
Methacrylonitrile	rr			NN					nn	NN			
Methane								rr					
Methanesulfonic Acid					rr					rr			
Methanol	rr	rr	rr	NN	NN	nn	NN	nn	NN	NN	nn	rr	TEFLON(rr) SBR(R) VIT/CLORBUTL(rr) PV ACETATE(rr)
Methanol, < 30%								rr					
Methanol, > 70%								rr					
4-Methoxy-4-methyl-2-pentanone	rr				rr				rr		nn		
Methyl Acetate	rr		n	NN	r	n	n	NN	nn	NN		n	

Name	BUTYL	CPE	VITON/NEOPRENE	NATURAL RUBBER	NEOPRENE	NITRILE+PVC	NITRILE	PE	PV ALCOHOL	PVC	VITON	BUTYL/NEOPRENE	Other materials
Methyl Acrylate	rr		n	NN	nn				rr			n	TEFLON(rr)
Methylacrylic Acid			n									r	
Methylamine	r			nn	rr		rr		n	rr			SBR(r)
Methylamine, 30-70%	rr					nn	rr	NN		NN	rr		SILVERSHIELD(rr)
3-Methylaminopropylamine	rr			NN	nn					NN			
Methyl Aniline			n									n	
Methyl Bromide	n			N	r		r	nn		N		r	SARANEX(rr) SBR(r)
Methyl n-Butyl Ketone			r									n	
Methyl Cellosolve	rr				n	R	r	nn		R			NAT+NEOP+NBR(r)
Methyl Cellosolve Acetate													NAT+NEOP+NBR(r)
Methyl Cellulose					n	r		r		r			POLYURETHANE(r)
Methyl Chloride	n			N	n	n	n	n		N			
Methyl Chloroacetate													SARANEX(rr)
Methyl Chloroform	NN	n	r	NN	NN	NN	NN	NN	RR	NN	rr	n	TEFLON(rr) SILVERSHIELD(rr)
Methyl Chloroformate	nn												
Methyl Cyclopentane			r									n	
n-Methylethanolamine	rr			NN	rr								CELLUL ACRYL(rr)
Methyl Ethyl Ketone	RR	nn	NN	NN	NN	NN	NN	NN	nn	NN	NN	NN	CHLOROBUTYL(rr) POLYURETHANE(r) SBR(r)
Methyl Ethyl Ketone Peroxide	rr			nn	rr						rr		
Methyl Formate			n									n	
Methylhydrazine	NN	nn								nn	rr		CHLOROBUTYL(rr) CR 39(rr)
Methyl Iodide	nn			NN	NN		NN	NN	rr		rr		
Methyl Isobutyl Ketone	NN	n	n	NN	NN	nn	NN	NN	rr	nn	nn	n	TEFLON(rr) SBR(r)
Methyl Isocyanate	nn			NN	NN				rr		NN		
Methyl Methacrylate	nn	n	n	NN	n	n	n	NN	RR	NN		n	TEFLON(rr)
Methyl Oleate			r								r		
N-Methyl-2-pyrrolidone				NN	NN		NN			NN			
Methyl Salicylate			n									r	
Methyl-vinyl-ketone		nn											
Methylene Bromide					n	n	n	n	NN	rr	n		

Name	BUTYL	CPE	VITON/NEOPRENE	NATURAL RUBBER	NEOPRENE	NITRILE+PVC	NITRILE	PE	PV ALCOHOL	PVC	VITON	BUTYL/NEOPRENE	Other materials
Methylene Chloride	NN	nn	r	NN	NN	nn	NN	NN	nn	NN	nn	n	SILVERSHIELD(rr)
Mineral Oil			r									n	
Mineral Spirits				N	rr	NN	rr	NN	rr	NN			SBR(r)
Monoisopropanolamine	rr				rr					rr	rr		
Morpholine	RR			nn	N	r	NN		RR	n	RR		SILVERSHIELD(rr)
Naphtha, V.M.&P	N	rr	r	N	nn	NN	rr	NN	rr	NN	r	n	POLYURETHANE(R)
Natural Gas, Liquified			r									n	
Nickel Carbonyl	r	r								r			
Nitric Acid	n	rr	rr	n	rr	nn	n	rr		NN	rr	rr	NEOP+NAT RUB(rr) NAT+NEOP+NBR(r)
Nitric Acid, <30%	r	r		rr	rr	rr	rr	nn		rr	r		NEOP+NAT RUB(rr) NEOP/NAT RUB(rr)
Nitric Acid, 30-70%				rr	n			nn	n	rr	r		NEOP+NAT RUB(rr) NEOP/NAT RUB(rr)
Nitric Acid, >70%	n	n		nn	n	NN	nn	nn	n	NN	rr		SARANEX(rr) NEOP/NAT RUB(rr)
Nitric Acid, Fuming Red	nn	nn		rr	rr		rr		NN	NN	rr		CHLOROBUTYL(rr) NEOP/NAT RUB(rr)
Nitric Oxide	r	r							r				
Nitrobenzene	N	RR		NN	NN	n	NN		RR	N	RR		TEFLON(rr) VIT/CLORBUTL(rr) SILVERSHIELD(rr)
Nitroethane	rr		n	NN	nn				rr			r	NEOP+NAT RUB(r)
Nitrogen Dioxide			n									n	
Nitrogen Tetroxide	nn	rr	n					rr		nn	nn	n	CHLOROBUTYL(rr) CR 39(rr)
Nitroglycerine	r	r											
Nitromethane	rr			NN	rr		nn	rr	NN	n	n		
Nitropropane	RR			n	nn	NN	NN	NN	NN	nn	nn		SILVERSHIELD(rr)
1-Nitropropane	rr						nn		rr		nn		TEFLON(rr)
2-Nitropropane	rr			NN	nn				rr				
n-Nitrosodimethylamine		nn											
Nitrotoluene					r					r			SBR(r) SBR/NEOPRENE(r)
o-Nitrotoluene	nn												
p-Nitrotoluene	rr												POLYCARBONAT(rr)
Nonylphenol					rr		rr						

ame	BUTYL	CPE	VITON/NEOPRENE	NATURAL RUBBER	NEOPRENE	NITRILE+PVC	NITRILE	PE	PV ALCOHOL	PVC	VITON	BUTYL/NEOPRENE	Other materials
Octadecane			r									n	
n-Octane			r	nn	rr	rr	rr			nn		n	
1-Octanol			r									r	
n-Octanol	r		r	nn	RR	rr	RR		rr	RR		r	SBR(R) SBR/NEOPRENE(r)
Oleic Acid	n	r	r	nn	rr	rr	rr	rr	rr	rr	r	r	NEOP+NAT RUB(r) POLYURETHANE(r) SBR/NEOPRENE(r)
Oxalic Acid	rr	r		RR	RR	RR	RR	rr	n	RR	rr		SBR(r)
Palmitic Acid				NN	rr		nn			rr	r		
Pentachlorophenol					NN		rr		NN	rr			
Pentane	n			NN	NN	NN	NN	NN	nn	NN	RR		SILVERSHIELD(rr)
Pentene					r		r			n			SBR(r)
Perchloric Acid	r		r	N	rr	rr	rr	rr		rr	r	r	
Perchloric Acid, 30-70%				n	rr	r	rr		n	rr			
Phenol	R	nn		NN	nn	n	NN	rr	nn	NN	n		TEFLON(rr) NEOP+NAT RUB(rr) NEOP/NAT RUB(rr)
Phenol, <30%								rr					
Phenol, >70%	RR				NN	nn	NN	rr		NN	RR		
Phenolphthalein				rr	rr		rr			rr			
Phenyl Ethyl Ether			n									n	
Phenylenediamine				n	r					n			
Phenyl Glycidyl Ether	rr				rr	nn		nn	rr	nn			
Phosphoric Acid	r				rr	rr	rr	rr	rr	n	rr		SARANEX(rr)
Phosphoric Acid, <30%	r	r			r					r	r		
Phosphoric Acid, 30-70%					r						r		
Phosphoric Acid, >70%	r	r		RR	rr	rr	rr	rr		rr	r		NEOP+NAT RUB(rr) NEOP/NAT RUB(rr)
Phosphorus Oxychloride		nn			NN	nn				NN	nn		
Phosphorus Trichloride			r									r	
Picric Acid	r			R	R	r	R			n	r	r	
Pinene (all isomers)			r									n	
1-Piperazineethaneamine	rr												
Piperidine					n							n	

Name	BUTYL	CPE	VITON/NEOPRENE	NATURAL RUBBER	NEOPRENE	NITRILE+PVC	NITRILE	PE	PV ALCOHOL	PVC	VITON	BUTYL/NEOPRENE	Other materials
Polychlorinated Biphenyls (PCB's)	nn	n		NN	rr			NN	rr	n	rr		TEFLON(rr) SARANEX(RR)
Potassium Bromide				r	r					r			
Potassium Chloride										r			
Potassium Cyanide										r			
Potassium Dichromate	r			r	R	r				R			SBR(r) SBR/NEOPRENE(r)
Potassium Hydroxide	r				R	R	r	R		R	n		NEOPRENE+SBR(r) SBR(r) SBR/NEOPRENE(r)
Potassium Hydroxide, <30%					r						r		
Potassium Hydroxide, 30-70%	r	r	r	RR	RR	RR	RR	rr	n	RR		r	NEOP+NAT RUB(rr) SBR(r)
Propane	N			N	R	R	n	rr		r	r		POLYURETHANE(R)
β-Propiolactone	rr			nn	nn		nn	NN		nn	nn		
Propionaldehyde	rr				nn				NN		NN		
Propionic Acid								NN					TEFLON(rr)
Propionic Anhydride								NN					
Propyl Acetate	rr		n	NN	n	n	NN	NN	rr	N	n	n	SILVERSHIELD(rr)
Propyl Alcohol	r		r	nn	RR	NN	RR	NN	n	nn	r	r	TEFLON(rr) PV ACETATE(rr)
n-Propylamine	nn	NN											TEFLON(rr) PV ACETATE(rr)
Propyl Chloride													PV ACETATE(rr)
Propylene Dichloride	nn								rr	NN	rr		
Propylenediamine	rr				rr					nn	rr		
Propylene Glycol					rr		rr	RR		rr			NEOP/NAT RUB(rr)
Propylene Oxide	rr		n	NN	n	n	n	NN	NN	n	NN	n	TEFLON(rr)
1,3-Propylene Oxide	nn			NN					NN		NN		
Propylmethacrylate	nn						nn		rr	NN			
Pyridine	r			NN	NN		NN	rr			n		
Pyrrole			n									n	
Silver Nitrate				r	r					n			
Sodium Chloride										r			
Sodium Cyanide, < 30%								rr					
Sodium Cyanide, 30-70%								NN					

Name	BUTYL	CPE	VITON/NEOPRENE	NATURAL RUBBER	NEOPRENE	NITRILE+PVC	NITRILE	PE	PV ALCOHOL	PVC	VITON	BUTYL/NEOPRENE	Other materials
Sodium Hydroxide	n	rr		R	R	rr	R	rr		rr			VIT/CLORBUTL(rr) SILVERSHIELD(rr)
Sodium Hydroxide, <30%								rr					NEOP+NAT RUB(rr)
Sodium Hydroxide, 30-70%	rr	r	rr	RR	RR	RR	RR	RR	n	RR	RR	rr	SARANEX(RR)
Sodium Hypochlorite, 30-70%					rr	rr	rr	rr		rr			
Sodium Silicate			r									r	
Sodium Sulfide					r					r			
Sodium Thiosulfate					r					r			
Stannous Chloride	r				r	R	r			R			POLYURETHANE(r) SBR(r) SBR/NEOPRENE(r)
Stearic Acid	r			R	R	r	R			R			POLYURETHANE(R) SBR(R)
Styrene	n	rr	r	NN	NN	NN	NN	NN	rr	NN	r	n	TEFLON(rr) VIT/CLORBUTL(rr)
Sulfuric Acid	n	RR	rr	N	rr	nn	n	rr		NN	rr	rr	SARANEX(RR)
Sulfuric Acid, <30%	r	r		r	R	rr	R	RR		rr	r		SARANEX(rr)
Sulfuric Acid, 30-70%	r	rr		rr	rr		rr	rr		rr	r		SARANEX(rr)
Sulfuric Acid, >70%	r	r		n	n	nn	N	RR	n	NN	rr		TEFLON(rr) SARANEX(rr) NEOP/NAT RUB(rr)
Sulfur Monochloride			r									n	
Tannic Acid	R	r		R	R	rr	R	rr		rr	r		SBR(R)
Tannic Acid, 30-70%					rr	rr	rr			rr			
Tannic Acid, > 70%					rr	rr	nn						
Terpineol			r									n	
Tetrabutyl Orthotitanate			r									r	
1,1,1,2-Tetrachloroethane	nn								rr	NN	rr		
1,1,2,2-Tetrachloroethane	nn	n		NN	NN		nn	NN	rr	NN	rr		TEFLON(rr)
Tetrachloroethylene	NN	rr	r	NN	NN	NN	nn	NN	nn	NN	RR	n	SILVERSHIELD(rr)
Tetrachlorodifluoroethane			r									n	
Tetraethylenepentamine	rr			nn	nn						rr		
Tetrafluoroetylene	rr				rr				rr		rr		
Tetrahydrofuran	NN	NN	NN	NN	NN	n	NN	NN	n	NN	NN	NN	TEFLON(rr)
1,2,3,4-Tetrahydronaphthalene			r									n	
N,N,N',N'-Tetramethylenediamine	nn						nn			NN	nn		

Name	BUTYL	CPE	VITON/NEOPRENE	NATURAL RUBBER	NEOPRENE	NITRILE+PVC	NITRILE	PE	PV ALCOHOL	PVC	VITON	BUTYL/NEOPRENE	Other materials
Thionyl Chloride			r									n	
Thiophenol	nn								rr				
Titanium Tetrachloride			r									n	
Toluene Diisocyanate	RR	rr		NN	n		rr	rr	RR	nn	RR		TEFLON(RR) SILVERSHIELD(RR)
Toluene	NN	r	rr	NN	NN	nn	NN	NN	NN	NN	nn		TEFLON(rr) VIT/CLORBUTL(rr) SILVERSHIELD(rr)
p-Toluenesulfonic Acid		rr			rr					rr			
o-Toluidine													TEFLON(rr)
Triallylamine					nn		rr			NN	rr		
Tributyl Phosphate			r									r	
Trichloroacetaldehyde	rr								rr	NN	rr		
Trichloroacetonitrile	rr				rr				rr		rr		
1,2,4-Trichlorobenzene	NN			NN	rr			NN	rr		NN		TEFLON(rr)
1,1,2-Trichloroethane	nn			NN	NN		NN	NN	nn	NN	rr		TEFLON(rr)
Trichloroethylene	NN	nn		NN	NN	NN	NN	NN	NN	NN	nn		CHLOROBUTYL(r) TEFLON(RR) SILVERSHIELD(rr)
1,2,3-Trichloropropane	nn						nn		rr		rr		
Tricresyl Phosphate	rr	r	r	nn	nn	rr	n	rr	RR	rr	rr	r	SBR/NEOPRENE(r)
Triethanolamine	r	r	r	N	R	rr	R	rr		rr	n	r	
Triethanolamine, > 70%			rr		rr	r	rr		rr	rr			
Triethyl Aluminum			n									n	
Triethylamine		rr	n		nn		RR			NN	rr	n	
Triethylenetetramine	rr				rr		nn				rr		
Trifluoroethanol			rr		rr		NN	RR		NN			NEOP+NAT RUB(rr) NEOP/NAT RUB(rr)
Triisooctyl Phosphate			r									r	
2,2,4-Trimethylpentane			r									n	
2,2,4-Trimethyl-1-pentene			r									n	
2,4,6-Trinitrotoluene	n			N	N	r				R			
Trioctyl Phosphine			r									r	
Tri-n-propylamine					nn		rr		rr		rr		
Turpentine	N	r	r	N	N	N	nn	nn	rr	N	r	n	TEFLON(rr) POLYURETHANE(r)

Name	BUTYL	CPE	VITON/NEOPRENE	NATURAL RUBBER	NEOPRENE	NITRILE + PVC	NITRILE	PE	PV ALCOHOL	PVC	VITON	BUTYL/NEOPRENE	Other materials
Urea				r	r	r	r			r			
Valeronitrile	rr			NN	nn				rr				
Vinyl Acetate	n	r								n			TEFLON(rr)
Vinyl Chloride	n	rr					NN			n	rr		SILVERSHIELD(rr)
4-Vinyl-1-cyclohexane	nn						nn		nn		rr		
Vinylidene Fluoride	rr			NN	NN					NN	rr		
Xylene	n	n	r	NN	NN	NN	NN	NN	RR	NN	rr	n	TEFLON(rr)
m-Xylene	NN				nn		nn		rr		RR		
o-Xylene	nn	nn					nn		rr		rr		
p-Xylene	nn						nn		rr	NN	rr		

SUPPLEMENTARY INFORMATION

Respiratory protection

When working with chemicals which may be harmful if inhaled, measures should be taken to prevent gases, vapors and finely dispersed particles being released into the ambient air, so as to obviate the use of additional individual protection under normal working conditions. Situations can arise, however, where individual protection is essential; an expert should make the selection. Here we shall confine ourselves to explaining some of the terms which appear on the sheets.

Types of respiratory protection
There are two means of providing an air supply:
- by purifying the ambient air (using a respirator);
- by obtaining air from an external source.

A variety of mouthpieces, masks and helmets can be used. A chemical suit protects the entire body.

Protection against gases and vapors
In most cases a face mask or helmet connected to an external source is preferable. This system, however, has drawbacks: either freedom of movement is limited by the supply line, or the operator has to be trained in the use of air tanks which are cumbersome and have only a limited capacity. Under certain conditions a respirator can be used; the filters used in respirators, however, are of limited capacity.

Respirators may be used only if:
- the oxygen level in the ambient air does not fall below a minimum of 19%;
- the atmospheric concentration of contaminants is no more than 0.1% (= 1,000ppm) when using low-capacity filters (filters with renewable filter pads, grade 1 or 2), or 1% (= 10,000ppm) when using high-capacity filters (canisters, grade 3);
- the contamination does not take the form of special category substances (see below);
- the correct filter is used.

Gas filters must be used once only, and for a limited duration; an expert should always be consulted on questions of filter life. As already stated, the filter must be appropriate for the relevant contaminant. In the EEC filters are coded by letter and color:

A – brown : for vapors/gases of organic solvents etc. with a boiling point above 65°C
B – gray : for acid gases
E – yellow : for sulfur dioxide
K – green : for anhydrous ammonia

In addition, the following filter canisters are available:

X   for organic substances with a boiling point below 65°C
Hg   for mercury
CO   for carbon monoxide
NO   for nitrogen oxides
ABEK   universal filter

Nose-and-mouth masks with low-capacity filters should be used only for light work requiring great freedom of movement, where the concentration of contaminants does not exceed 1000ppm.

Protection against dust
Questions to be asked:
- Is protection required only against airborne particles or also against gas/vapor?
- Is the dust harmful only when inhaled or also on contact with the skin or eyes?

Where respiratory protection only is required - and there is a minimum of 19% atmospheric oxygen - a good-quality dust mask may be used (for exceptions see below under *Special category substances*).

Dust can be divided into three categories:

2a	2b	2c
Inert airborne dust with TLV 10mg/m^3	Harmful substance with TLV 0.1-10mg/m^3 (except for asbestos)	Toxic substance with TLV <0.1mg/m^3 Asbestos Human carcinogens Spores, bacteria, viruses, proteolytic enzymes

N.B. If a solid also gives off vapors, a dust filter alone may not be sufficient.

Dust filters should be selected in accordance with the table below.

Type of dust filter	Substance category		
	2a	2b	2c
P1	yes	no	no
P2	yes	yes	no
P3	yes	yes	yes

P3 filters must not be combined with half-face masks or other types of mask which do not fit tightly. If the substance is also harmful to the eyes and skin, breathing apparatus should be used, or, in special cases, a chemical suit.

Protection against dust and gas/vapor
If protection against gas/vapor as well as dust is required, it should first be decided whether a filter will provide sufficient protection against the gas/vapor. If so, it must be combined with the appropriate dust filter for the dust category. The applications of such combined filters are limited; in most such cases a supply of air from an external source will be necessary.

When using either filters or an external air supply, the mask – or other headpiece – must fit tightly.

Special category substances
The substances requiring special respiratory protection are as follows:
- anomalous substances;
- teratogens (substances which cause fetal malformation);
- mutagens (substances which can cause mutations in offspring);
- proven human carcinogens.

Anomalous substances (e.g. methyl alcohol, methyl chloride and methyl bromide) are so poorly absorbed by filters - or saturate filters so quickly - that a respirator can never afford sufficient protection; an external air supply should always be used when handling these substances.

First aid in the event of poisoning

The sheets list all the first-aid measures which can safely be carried out by untrained persons. What follows is not intended as a substitute for the first-aid manuals: the measures described are no more than initial steps, which will always need to be followed by treatment at the hands of trained first-aiders, paramedics, doctors etc. Prompt first aid, however, can save lives!

Anyone wishing to help a suspected victim of poisoning should first consider the following points.

1. Am I endangering myself?
Consider yourself first! Do you also run the risk of inhaling toxic or corrosive gases, or of getting toxic or corrosive substances on your skin or in your eyes? Take steps to ensure that you don't also become a victim. Depending on the situation, you may need to use respiratory protection, a face shield, gloves etc.

2. Are other people in danger?
Consider others besides yourself and the victim. If there is any danger, deal with that first.

Only after resolving these two questions should you give your attention to aiding the person.

3. Get the victim out of danger as quickly as possible.
Fresh air is essential for anyone poisoned by inhalation (of gas, vapor, fumes, dust).

4. Next consider the general condition of the victim.
A.   Is he/she breathing?
B.   Is he/she conscious?
C.   How is his/her circulation?

A. Respiration
If the victim is not breathing (or scarcely breathing), mouth-to-mouth resuscitation must be given as soon as possible and maintained until he or she begins breathing independently.
–   Before starting, place a handkerchief or similar piece of cloth over the victim's mouth.
–   Tilt the head far enough back to provide an unobstructed air passage from mouth to lungs.

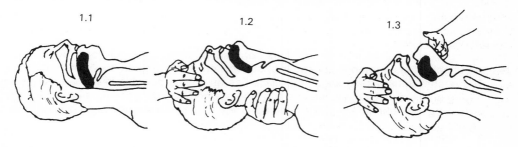

Figure 1: (1.1) Air passage is obstructed by retracted tongue (in black), due to relaxation of lower jaw muscles; (1.2) 'head back' and (1.3) 'chin up' movement to remove obstruction

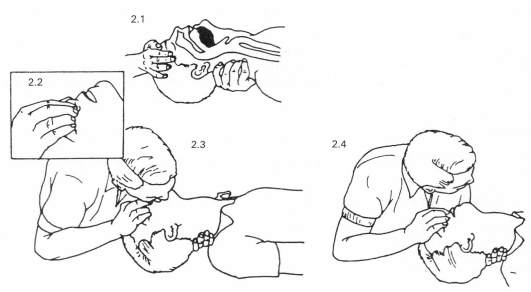

Figure 2: Mouth-to-mouth resuscitation
2.1　'Head back - chin up'
2.2　Pinch nostrils
2.3　Exhale forcefully into victim's mouth
2.4　Watch and listen for victim's breathing
Repeat the series 2.2, 2.3, 2.4 about 12 times a minute until the victim begins to breathe independently.

B. Unconsciousness
The victim is unconscious if he or she does not respond when spoken to or nudged.
- An unconscious person is in danger of suffocation, if left lying on his or her back.
- The unconscious person should be placed as quickly as possible in the recovery position. If lying on the right side, the left arm and left knee and hip joint should be bent, with the right arm along the right flank and the right leg extended. Ensure that the head is tilted back (see Figure 3).
- Tight clothing (e.g. collar, belt) should be loosened.
- Loose objects in the mouth (e.g. dentures, chewing gum) should be removed. To do this, open the victim's mouth and squeeze the cheeks between the teeth with the thumb and forefinger of one hand while removing the object with the other hand.

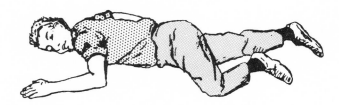

Figure 3: Recovery position

C. Circulation
The victim's circulation must be maintained as far as possible.
- An untrained helper can only see to it that the victim is kept warm.
- Especially important is to place a jacket, blanket or other insulation under the victim.

1000

−   Heart massage techniques are beyond the scope of this book; the reader is referred to the first-aid manuals and handbooks of clinical toxicology.

5. Call for help (doctor, nurse, paramedic)

Respiration, unconsciousness and circulation are matters of life and death; if you are alone, first attend to these before calling for expert assistance. If mouth-to-mouth resuscitation is required, this should take precedence over calling in a professional. If there are other people around, send them to get help while you start with the first aid.

When reporting the incident, be specific:
−   Location of incident.
−   Precise nature of incident.
−   Number of victims.
−   Location of victims.
−   Likely cause (toxic substance).
−   General condition of victims.
−   If victim(s) not breathing, request oxygen equipment.

In larger establishments: call the company doctor/infirmary. In establishments which do not employ on-site health professionals, it will depend on any arrangements made with local doctors whether a doctor is called in or the victim is taken to the hospital. The victim should be taken immediately to the hospital, however, if this is specified on the card (send along the card(s)).

6. First aid measures listed on the cards

(a) Do not give the victim water to drink unless he or she can hold the glass.
UNDER NO CIRCUMSTANCES GIVE LIQUIDS TO AN UNCONSCIOUS PERSON: THIS CAN CAUSE SUFFOCATION!
(b) When flushing the victim's eyes with water, under no circumstances direct water straight from the faucet into the eyes. Use a special eye fountain or cup if possible.
N.B. Inducing vomiting is not recommended because of the risk of swallowing or inhaling the vomit.

Summary

When giving first aid, follow the order below.
1.   Consider yourself.
2.   Consider other people.
3.   Get the victim out of danger.
4.   Help the victim:
     (a) if not breathing: give artificial respiration;
     (b) if unconscious: place in recovery position;
     (c) keep him or her warm.
5.   Report the incident: location, nature, substance, general condition of victim.
6.   Under no circumstances induce vomiting if victim is unconscious.
7.   Under no circumstances give liquids to an unconscious victim.

INDEX

The index lists the names of the chemicals - main entries (upper case), synonyms (lower case), trade names (in italics) etc. - alphabetically. The following prefixes and interpolations are treated as part of the name: bis, cyclo, iso and the numerical designations mono, di (bi), tri, penta, hexa, octa etc. Cyclopentane will thus be found under 'c' and pentachlorophenol under 'p'; in fluorodinitrobenzene, 'di' is part of the name.

The following prefixes and interpolations are not treated as part of the name:
ortho- (o-), meta- (m-), para- (p-), alpha- (α), beta- (β), gamma- (γ) etc.; primary- (prim-), secondary- (sec-), tertiary- (tert-), cis-, trans-, dextro- (d-), levo- (l-), normal (n-), N- (= bond to nitrogen atom).

Ortho-cresol (o-cresol) should thus be sought under 'c', tert-amyl alcohol under 'a' (but iso-amyl alcohol under 'i').

As stated in the Introduction, the cards themselves are organized alphabetically by main entry, using the above rules.

ALPHABETICAL INDEX

Name	CAS-number	EEC-number	RTECS number	HI number	UN-number	Page
A						
acetal	105-57-7	605-015-00-1	AB2800000	33	1088	301
ACETALDEHYDE	75-07-0	605-003-00-6	AB1925000	33	1089	001
β-acetaldehyde oxime	107-29-9		AB2975000		2332	002
β-ACETALDOXIME	107-29-9		AB2975000		2332	002
bis(acetato)cobalt	71-48-7		AG3480000			227
ACETIC ACID (100%)	64-19-7	607-002-01-3	AF1225000	83	2789	003
ACETIC ACID (85% in water)	64-19-7	607-002-00-6	AF1340000	83	2789	004
acetic acid, amyl ester	628-63-7	607-130-00-2	AJ1925000	30	1104	056
acetic acid anhydride	108-24-7	607-008-00-9	AK1925000	83	1715	005
acetic acid, benzyl ester	140-11-4		AF5075000			104
acetic acid, 2-butoxyethyl ester	112-07-2	607-038-00-2	KJ8925000			409
acetic acid, n-butyl ester	123-86-4	607-025-00-1	AF7350000	30	1123	139
acetic acid dimethylamide	127-19-5	616-011-00-4	AB7700000			337
acetic acid ethenyl ester	108-05-4	607-023-00-0	AK0875000	339	1301	916
acetic acid 2-ethoxyethyl ester	111-15-9	607-037-00-7	KK8225000	30	1172	383
acetic acid ethyl ester	141-78-6	607-022-00-5	AH5425000	33	1173	384
acetic acid, iso-butyl ester	110-19-0	607-026-00-7	AI4025000	33	1213	500
acetic acid, isopentyl ester	123-92-2		NS9800000	30	1104	494
acetic acid methyl ester	79-20-9	607-021-00-x	AI9100000	33	1231	571
acetic acid, 1-methylethyl ester	108-21-4	607-024-00-6	AI4930000	33	1220	511
acetic acid, 2-methylpropyl ester	110-19-0	607-026-00-7	AI4025000	33	1213	500
acetic acid, pentyl ester	628-63-7	607-130-00-2	AJ1925000	30	1104	056
acetic acid, n-propyl ester	109-60-4	607-024-00-6	AJ3675000	33	1276	759
acetic acid vinyl ester	108-05-4	607-023-00-0	AK0875000	339	1301	916
acetic aldehyde	75-07-0	605-003-00-6	AB1925000	33	1089	001
ACETIC ANHYDRIDE	108-24-7	607-008-00-9	AK1925000	83	1715	005
acetic bromide	506-96-7		AO5955000	80	1716	012
acetic sec butylester	105-46-4	607-026-00-7	AF7380000	30	1123	140
acetic ether	141-78-6	607-022-00-5	AH5425000	33	1173	384
ACETONE	67-64-1	606-001-00-8	AL3150000	33	1090	006
ACETONE CYANOHYDRIN	75-86-5	608-004-00-x	OD9275000	66	1541	007
ACETONITRILE	75-05-8	608-001-00-3	AL7700000	336	1648	008
ACETONYLACETONE	110-13-4		MO3150000			009
acetonyl chloride	78-95-5		UC700000	60	1695	203
ACETOPHENONE	98-86-2		AM5250000			010
acetoxyethane	141-78-6	607-022-00-5	AH5425000	33	1173	384
acetoxymethylbenzene	140-11-4		AF5075000			104
2-acetoxypentane	626-38-0	607-130-00-2	AJ2100000	30	1104	057
1-acetoxypropane	109-60-4	607-024-00-6	AJ3675000	33	1276	759
2-acetoxypropane	108-21-4	607-024-00-6	AI4930000	33	1220	511
α-acetoxytoluene	140-11-4		AF5075000			104
ACETYL ACETONE	123-54-6	606-029-00-0	SA1925000	30	2310	011
acetylaldehyde	75-07-0	605-003-00-6	AB1925000	33	1089	001
acetyl anhydride	108-24-7	607-008-00-9	AK1925000	83	1715	005
acetylbenzene	98-86-2		AM5250000			010
ACETYLBROMIDE	506-96-7		AO5955000	80	1716	012
ACETYL CAPROLACTAM	1888-91-1		CM3599500			013
ACETYL CHLORIDE	75-36-5	607-011-00-5	AO6390000	X338	1717	014
acetyldimethylamine	127-19-5	616-011-00-4	AB7700000			337
ACETYLENE	74-86-2	601-015-00-0	AO9600000	239	1001	015
ACETYLENE (cylinder)	74-86-2	601-015-00-0	AO9600000	239	1001	016
acetylene tetrachloride	79-34-5	602-015-00-3	KI8575000	60	1702	844
acetyl ether	108-24-7	607-008-00-9	AK1925000	83	1715	005
acetyl ethylene	78-94-4		EM9800000	339	1251	619
acetyl hexanolactam	1888-91-1		CM3599500			013
acetyl ketene	674-82-8	606-017-00-5	RQ8225000	39	2521	335

Name	CAS-number	EEC-number	RTECS number	HI number	UN-number	Page
acetyl oxide	108-24-7	607-008-00-9	AK1925000	83	1715	005
acraldehyde	107-02-8	605-008-00-3	AS1050000	336	1092	017
acroleic acid	79-10-7	607-061-00-8	AS4375000	89	2218	019
ACROLEIN	107-02-8	605-008-00-3	AS1050000	336	1092	017
acroleine	107-02-8	605-008-00-3	AS1050000	336	1092	017
acrylaldehyde	107-02-8	605-008-00-3	AS1050000	336	1092	017
ACRYLAMIDE	79-06-1	616-003-00-0	AS3325000	60	2074	018
ACRYLIC ACID	79-10-7	607-061-00-8	AS4375000	89	2218	019
acrylic acid n-butyl ester	141-32-2	607-062-00-3	UD3150000	39	2348	141
acrylic acid, ethyl ester	140-88-5	607-032-00-x	AT0700000	339	1917	386
acrylic acid, methyl ester	96-33-3	607-034-00-0	AT2800000	339	1919	575
acrylic aldehyde	107-02-8	605-008-00-3	AS1050000	336	1092	017
acrylic amide	79-06-1	616-003-00-0	AS3325000	60	2074	018
ACRYLONITRILE	107-13-1	608-003-00-4	AT5250000	336	1093	020
ADIPIC ACID	124-04-9	607-144-00-9	AU8400000			021
adipic acid dinitrile	111-69-3		AV2625000	60	2205	022
ADIPONITRILE	111-69-3		AV2625000	60	2205	022
adipylnitrile	111-69-3		AV2625000	60	2205	022
AGE	106-92-3	603-038-00-1	RR0875000	30	2219	030
AIR (liquid, refrigerated)				225	1003	023
alcohol (80-90% ethanol)	64-17-5	603-002-00-5	KQ6300000	33	1170	583
alcohol c-9	143-08-8		RB1575000			654
Aldol	107-89-1		ES3150000	60	2839	486
alkaline zirconium carbonate	15667-84-2					935
allene (cylinder)	463-49-0		BA0400000		2200	745
ALLYL ALCOHOL	107-18-6	603-015-00-6	BA5075000	663	1098	024
ALLYLAMINE	107-11-9	612-046-00-4	BA5425000	336	2334	025
ALLYL BROMIDE	106-95-6		UC7090000	336	1099	026
ALLYL CHLORIDE	107-05-1	602-029-00-x	UC7350000	336	1100	027
allylene (cylinder)	74-99-7		UK4250000	239	1060	573
allyl 2,3-epoxypropyl ether	106-92-3	603-038-00-1	RR0875000	30	2219	030
ALLYL ETHER	557-40-4		KN7525000	336	2360	028
ALLYL ETHYL ETHER	557-31-3		KM9120000	336	2335	029
ALLYL GLYCIDYL ETHER	106-92-3	603-038-00-1	RR0875000	30	2219	030
1-allyloxy-2,3-epoxypropane	106-92-3	603-038-00-1	RR0875000	30	2219	030
ALLYL PHENYL ETHER	1746-13-0		DA8575000			031
alum	7784-24-9		WS5690000			723
ALUMINUM (powder)	7429-90-5	013-002-00-1	BD0330000	43	1396	032
aluminum alum	10043-01-3		BD1700000			037
ALUMINUM CHLORIDE (anhydrous)	7446-70-0	013-003-00-7	BD0525000	80	1726	033
aluminum lithium hydride	16853-85-3	001-002-00-4	BD0100000		1410	532
ALUMINUM NITRATE (nonahydrate)	13473-90-0		BD1040000		1438	034
ALUMINUM OXIDE	1344-28-1		BD1200000			035
ALUMINUM PHOSPHIDE	20859-73-8	015-004-00-8	BD1400000	66	1397	036
aluminum potassium disulfate	7784-24-9		WS5690000			723
aluminum potassium sulfate	7784-24-9		WS5690000			723
aluminum powder (powder)	7429-90-5	013-002-00-1	BD0330000	43	1396	032
ALUMINUM SULFATE	10043-01-3		BD1700000			037
aluminum trichloride (anhydrous)	7446-70-0	013-003-00-7	BD0525000	80	1726	033
aluminum trinitrate (nonahydrate)	13473-90-0		BD1040000		1438	034
aluminum trisulfate	10043-01-3		BD1700000			037
alunogenite	10043-01-3		BD1700000			037
aminic acid	64-18-6	607-001-00-1	LQ4900000	80	1779	433
AMINOACETIC ACID	56-40-6		MB7600000			038
p-aminoaniline	106-50-3	612-028-00-6	SS8050000	60	1673	690
amino benzene	62-53-3	612-008-00-7	BW6650000	60	1547	065
α-aminobenzeneacetic acid	875-74-1					694
p-aminobenzenesulfonic acid (monohydrate)	121-57-3	612-014-00-x	WP3895500			829
1-aminobutane	109-73-9	612-005-00-0	EO2975000	338	1125	144
aminocaproic lactam	105-60-2		CM3675000			189
1-amino-2-chlorobenzene	95-51-2	612-010-00-8	BX0525000	60	2019	206
1-amino-4-chlorobenzene	106-47-8	612-010-00-8	BX0700000	60	2018	207
aminocyclohexane	108-91-8	612-050-00-6	GX0700000	83	2357	255
2-amino-2,2-dimethylethanol	124-68-5	6034-070-00-6	UA5950000			499
1-amino-2,4-dinitrobenzene	97-02-9	612-040-00-1	BX9100000	60	1596	353
aminoethane (cylinder)	75-04-7	612-002-00-4	KH2100000	236	1036	391
1-aminoethane (cylinder)	75-04-7	612-002-00-4	KH2100000	236	1036	391

Name	CAS-number	EEC-number	RTECS number	HI number	UN-number	Page
aminoethyl alcohol	141-43-5	603-030-00-8	KJ5775000	80	2491	379
2-aminoethyl alcohol	141-43-5	603-030-00-8	KJ5775000	80	2491	379
2-((2-aminoethyl)amino)ethanol	111-41-1		KJ6300000			487
aminoethylethandiamine	111-40-0	612-058-00-x	IE1225000	80	2079	314
N-aminoethylethanol amine	111-41-1		KJ6300000			487
N,N'-bis(2-aminoethyl)ethylenediamine	112-24-3	612-059-00-5	YE6650000	80	2259	890
N-AMINOETHYLPIPERAZINE	140-31-8	612-065-00-8	TK8050000	80	2815	039
aminohexahydrobenzene	108-91-8	612-050-00-6	GX0700000	83	2357	255
1-amino-2-hydroxypropane	78-96-6	603-082-00-1	UA5775000			510
2-aminoisobutane	75-64-9		EO3330000			145
2-aminoisobutanol	124-68-5	6034-070-00-6	UA5950000			499
aminomethane (in water, 40%)	74-89-5	612-001-00-9	PF6300000	338	1235	581
1-amino-2-methylbenzene	95-53-4	612-024-00-4	XU2975000	60	1708	869
2-amino-2-methylpropane	75-64-9		EO3330000			145
2-amino-2-methyl-1-propanol	124-68-5	6034-070-00-6	UA5950000			499
1-amino-4-nitrobenzene	100-01-6	612-012-00-9	BY7000000	60	1661	640
m-aminophenol, see o-AMINOPHENOL						040
o-AMINOPHENOL	95-55-6	612-033-00-3	SJ4950000	60	2512	040
p-aminophenol, see o-AMINOPHENOL						040
aminophenyl acetic acid	875-74-1					694
1-aminopropane	107-10-8		UH9100000	338	1277	760
2-aminopropane	75-31-0	612-007-00-1	NT8400000	338	1221	513
1-amino-propanol-2	78-96-6	603-082-00-1	UA5775000			510
3-amino-1-propene	107-11-9	612-046-00-4	BA5425000	336	2334	025
3-aminopropylene	107-11-9	612-046-00-4	BA5425000	336	2334	025
aminosulfonic acid	5329-14-6	016-026-00-0	WO5950000			828
o-aminotoluene	95-53-4	612-024-00-4	XU2975000	60	1708	869
ammonia (cylinder)	7664-41-7	007-001-00-5	BO0875000	268	1005	042
AMMONIA (25% solution of ammonia in water)	1336-21-6	007-001-02-x	BQ9625000	80	2672	041
AMMONIA (ANHYDROUS) (cylinder)	7664-41-7	007-001-00-5	BO0875000	268	1005	042
ammoniac (cylinder)	7664-41-7	007-001-00-5	BO0875000	268	1005	042
AMMONIUM BICARBONATE	1066-33-7		BO8600000			043
AMMONIUMBICHROMATE	7789-09-5	024-003-00-1	HX7650000		1439	044
AMMONIUMBISULFIDE	12124-99-1		BS4900000		2683	045
AMMONIUM CHLORIDE	12125-02-9	017-014-00-8	BP4550000			046
ammoniumdichromate	7789-09-5	024-003-00-1	HX7650000		1439	044
AMMONIUMFLUORIDE	12125-01-8	009-006-00-8	BQ6300000	60	2505	047
ammoniumfluorosilicate	16919-19-0		GQ9450000	60	2854	053
ammonium hydrogen carbonate	1066-33-7		BO8600000			043
ammonium hydrogen sulfide	12124-99-1		BS4900000		2683	045
AMMONIUM NITRATE	6484-52-2		BR9050000	50	1942	048
AMMONIUM OXALATE (monohydrate)	6009-70-7	607-007-00-3		60	2449	049
AMMONIUM PERSULFATE	7727-54-0				1444	050
sec-AMMONIUM PHOSPHATE	7783-28-0					051
AMMONIUM RHODANIDE	1762-95-4		XK7875000			052
AMMONIUMSILICOFLUORIDE	16919-19-0		GQ9450000	60	2854	053
AMMONIUM SULFATE	7783-20-2		BS4500000			054
AMMONIUM SULFATE (40% solution)	7783-20-2		BS4500000			055
ammonium sulfide	12124-99-1		BS4900000		2683	045
amosite (brown)	12001-29-5		CI6478500		2212	078
amosite (brown)	12001-29-5		CI6478500		2590	078
amygdalic acid	90-64-2		OO6300000			546
n-AMYLACETATE	628-63-7	607-130-00-2	AJ1925000	30	1104	056
sec-AMYL ACETATE	626-38-0	607-130-00-2	AJ2100000	30	1104	057
amyl acetic ester	628-63-7	607-130-00-2	AJ1925000	30	1104	056
prim-n-AMYL ALCOHOL	71-41-0	603-006-00-7	SB9800000	30	1105	058
sec-amyl alcohol	6032-29-7	603-006-00-7	SA4900000	30	1105	059
sec-n-AMYL ALCOHOL	6032-29-7	603-006-00-7	SA4900000	30	1105	059
t-amyl alcohol	75-85-4	603-007-00-2		33	1105	060
tert-AMYL ALCOHOL	75-85-4	603-007-00-2		33	1105	060
amyl carbinol	111-27-3	603-059-00-6	MQ4025000	30	1987	466
AMYLENE-DIMER	16736-42-8				1142	061
amylene hydrate	75-85-4	603-007-00-2		33	1105	060
AMYL ETHER	693-65-2		SC2900000			062
amyl ethyl ketone	541-85-5	606-020-00-1	MJ7350000	30	2271	392
amyl hydride	109-66-0	601-006-00-1	RZ9450000	33	1265	677
amylmethyl carbinol	543-49-7		MJ2975000	30	1987	455

Name	CAS-number	EEC-number	RTECS number	HI number	UN-number	Page
AMYL NITRITE	110-46-3		NT0187500		1113	063
p-tert-AMYLPHENOL	80-46-6		SM6825000	60	2430	064
AN	107-13-1	608-003-00-4	AT5250000	336	1093	020
anatase	13463-67-7		XR2275000			862
ANILINE	62-53-3	612-008-00-7	BW6650000	60	1547	065
aniline chloride	142-04-1	612-009-00-2	CX0875000		1548	066
ANILINE HYDROCHLORIDE	142-04-1	612-009-00-2	CX0875000		1548	066
aniline salt	142-04-1	612-009-00-2	CX0875000		1548	066
p-anilinesulfonic acid (monohydrate)	121-57-3	612-014-00-x	WP3895500			829
2-anilinoethanol	122-98-5		KJ7175000			691
anilinomethane	100-61-8	612-015-00-5	BY4550000	60	2294	582
2-anilinonaphthalene	135-88-6		QM4555000			700
ANISOLE	100-66-3		BZ8050000	30	2222	067
anthion	7727-21-2		SE0400000		1492	741
ANTHRACENE	120-12-7		CA9350000			068
anthracin	120-12-7		CA9350000			068
antimony (III) oxide	1309-64-4		CC5650000	60	1549	072
ANTIMONY PENTACHLORIDE	7647-18-9	051-002-00-3	CC5075000	80	1730	069
antimony pentafluoride, see ANTIMONY PENTACHLORIDE						069
antimony perchloride	7647-18-9	051-002-00-3	CC5075000	80	1730	069
antimony potassium tartrate	11071-15-1	051-003-00-9	CC7350000			724
ANTIMONY TRIBROMIDE	7789-61-9	051-003-00-9	CC4400000	60		070
ANTIMONY TRICHLORIDE	10025-91-9	051-001-00-8	CC4900000	80	1733	071
ANTIMONY TRIOXIDE	1309-64-4		CC5650000	60	1549	072
aragonite	471-34-1		FF9335000			176
Arcton 11	75-69-4		PB6125000			878
Arcton 12 (cylinder)	75-71-8		PA8200000	20	1028	286
Arcton 21 (cylinder)	75-43-4		PA8400000	20	1029	292
argon compressed (cylinder)	7440-37-1		CF2300000	20	1006	073
ARGON (cylinder)	7440-37-1		CF2300000	20	1006	073
argon gas (cylinder)	7440-37-1		CF2300000	20	1006	073
ARGON (liquid, cooled)	7440-37-1		CF2300000	22	1951	074
arsenic chloride	7784-34-1	033-002-00-5	CG1750000	66	1560	075
ARSENIC TRICHLORIDE	7784-34-1	033-002-00-5	CG1750000	66	1560	075
arsenic trihydride (cylinder)	7784-42-1	033-002-00-5	CG6475000	236	2188	077
ARSENIC TRIOXIDE	1327-53-3	033-003-00-0	CG3325000	60	1561	076
arsenious chloride	7784-34-1	033-002-00-5	CG1750000	66	1560	075
arsenous anhydride	1327-53-3	033-003-00-0	CG3325000	60	1561	076
ARSINE (cylinder)	7784-42-1	033-002-00-5	CG6475000	236	2188	077
artificial malachite (basic)	12069-69-1		GL6910000			235
ASBESTOS	12001-29-5		CI6478500		2212	078
ASBESTOS	12001-29-5		CI6478500		2590	078
ATE	97-93-8	013-004-00-2	BD2050000	X333	1102	884
ATE (15% solution in hexane)	97-93-8	013-004-00-2	BD2050000	X333	1102	885
ATM (15% solution in hexane)	75-24-1	013-004-00-2	BD2204000	X333	1103	897
azabenzene	110-86-1	613-002-00-7	UR8400000	336	1282	765
azacyclohexane	110-89-4	613-027-00-3	TM3500000	338	2401	720
3-azapentane-1,5-diamine	111-40-0	612-058-00-x	IE1225000	80	2079	314
azine	110-86-1	613-002-00-7	UR8400000	336	1282	765
AZOBENZENE	103-33-3	611-001-00-6	CN1400000			079
azobenzide	103-33-3	611-001-00-6	CN1400000			079
azobenzol	103-33-3	611-001-00-6	CN1400000			079
azobisbenzene	103-33-3	611-001-00-6	CN1400000			079

B

Name	CAS-number	EEC-number	RTECS number	HI number	UN-number	Page
BARIUM	7440-39-3		CQ8370000		1400	080
BARIUM ACETATE	543-80-6	056-002-00-7	AF4550000		1564	081
BARIUM AZIDE (anhydrous)	18810-58-7	056-002-00-7	CQ8500000		0224	082
BARIUM CARBONATE	513-77-9	056-002-00-7	CQ8600000	60	1564	083
BARIUM CHLORATE (monohydrate)	13477-00-4	017-003-00-8	FN9770000		1445	084
BARIUM CHLORIDE	10361-37-2	056-002-00-7	CQ8750000	60	1564	085
BARIUM DIPHENYLAMINE SULFONATE	1300-92-1	056-002-00-7			1564	086
barium hydrate (octahydrate)	17194-00-2	056-002-00-7		60	1564	087
BARIUM HYDROXIDE (octahydrate)	17194-00-2	056-002-00-7		60	1564	087

Name	CAS-number	EEC-number	RTECS number	HI number	UN-number	Page
BARIUM NITRATE	10022-31-8	056-002-00-7	CQ9625000	50	1446	088
BARIUM PERCHLORATE (trihydrate)	13465-95-7	017-007-00-x	SC7550000		1447	089
BARIUM PEROXIDE	1304-29-6	056-001-00-1	CR0175000	50	1449	090
BARIUM SULFIDE	21109-95-5	016-002-00-x		60	1564	091
BARIUM SULFITE	7787-39-5	056-002-00-7			1564	092
BCME	542-88-1	603-046-00-5	KN1575000		2249	287
1,2,3-benzenetricarboxylic acid-1,2-anhy	552-30-7	607-097-00-4				895
BENZAL CHLORIDE	98-87-3	602-058-00-1	CZ5075000	68	1886	093
BENZALDEHYDE	100-52-7	605-012-00-5	CU4375000	30	1990	094
BENZALDEHYDE CYANOHYDRIN	532-28-5		OO8400000			095
benzenamine	62-53-3	612-008-00-7	BW6650000	60	1547	065
BENZENE	71-43-2	601-020-00-8	CY1400000	33	1114	096
benzenecarbonal	100-52-7	605-012-00-5	CU4375000	30	1990	094
benzenecarbonyl chloride	98-88-4	607-012-00-0	DM6600000	80	1736	103
benzene carboxaldehyde	100-52-7	605-012-00-5	CU4375000	30	1990	094
benzenecarboxylic acid	65-85-0		DG0875000			099
benzene chloride	108-90-7	602-033-00-1	CZ0175000	30	1134	621
1,4-benzenediamine	106-50-3	612-028-00-6	SS8050000	60	1673	690
o-benzene dicarboxylic acid	88-99-3		TH9625000			714
1,2-benzenedicarboxylic acid	88-99-3		TH9625000			714
1,2-benzenedicarboxylic acid anhydride	85-44-9	607-009-00-4	TI3150000	80	2214	715
1,2-benzenedicarboxylic acid dimethyl ester	131-11-3		TI1575000			349
1,4-benzene dicarboxylic methyl ester	120-61-6		WZ1225000			352
1,2-benzenediol	120-80-9	604-016-00-4	UX1050000			766
1,3-benzenediol	108-46-3	604-010-00-1	VG9625000	60	2876	772
1,4-benzenediol	123-31-9	604-005-00-4	MX3500000	60	2662	485
benzeneformic acid	65-85-0		DG0875000			099
γ-benzene hexachloride	58-89-9	602-043-00-6	GV4900000	60	2761	459
benzene hexahydride	110-82-7	601-017-00-1	GU6300000	33	1145	250
benzenemethylal	100-52-7	605-012-00-5	CU4375000	30	1990	094
benzenesulfonic chloride	98-09-9		DB8750000	80	2225	097
benzenesulfonyl chloride	98-09-9		DB8750000	80	2225	097
BENZENESULFONYL CHLORIDE	98-09-9		DB8750000	80	2225	097
benzenethiol	108-98-5		DC0525000	663	2337	858
1,2,3-benzenetriol	87-66-1	604-009-00-6	UX2800000			767
benzenyl chloride	98-07-7		XT9275000	80	2226	101
benzenyl trichloride	98-07-7		XT9275000	80	2226	101
BENZIDINE	92-87-5	612-042-00-2	DC9625000	60	1885	098
benzine, see PETROLEUM NAPHTHA						684
benzoglycolic acid	90-64-2		OO6300000			546
benzohydroquinone	123-31-9	604-005-00-4	MX3500000	60	2662	485
BENZOIC ACID	65-85-0		DG0875000			099
benzoic acid benzylester	120-51-4	607-085-00-9	DG4200000			106
benzoic acid ethyl ester	93-89-0		DH0200000			395
benzoic acid, methyl ester	93-58-3		DH3850000		2938	584
benzoic acid peroxide (50%)	94-36-0	617-008-00-0	DM8575000	11	2087	272
benzoic aldehyde	100-52-7	605-012-00-5	CU4375000	30	1990	094
benzol	71-43-2	601-020-00-8	CY1400000	33	1114	096
BENZONITRILE	100-47-0	608-012-00-3	DI2450000	60	2224	100
1,2-benzopyrone	91-64-5		GN4200000			244
p-benzoquinone	106-51-4	606-013-00-3	DK2625000	60	2587	769
1,4-benzoquinone	106-51-4	606-013-00-3	DK2625000	60	2587	769
2-benzothiazolethiol	149-30-4		DL6475000			551
BENZOTRICHLORIDE	98-07-7		XT9275000	80	2226	101
BENZOTRIFLUORIDE	98-08-8	602-056-00-7	XT9450000	33	2338	102
BENZOYL CHLORIDE	98-88-4	607-012-00-0	DM6600000	80	1736	103
benzoyl peroxide (50%)	94-36-0	617-008-00-0	DM8575000	11	2087	272
benzoyl superoxide (50%)	94-36-0	617-008-00-0	DM8575000	11	2087	272
BENZYL ACETATE	140-11-4		AF5075000			104
BENZYL ALCOHOL	100-51-6	603-057-00-5	DN3150000			105
BENZYL BENZOATE	120-51-4	607-085-00-9	DG4200000			106
BENZYL BROMIDE	100-39-0	602-057-00-2	XS7965000	60	1737	107
BENZYL BUTYL ETHER	588-67-0					108
BENZYL CHLORIDE	100-44-7	602-037-00-3	XS8925000	68	1738	109
BENZYLCYANIDE	140-29-4		AM1400000	60	2470	110
benzyl ethanoate	140-11-4		AF5075000			104
BENZYL ETHER	103-50-4		DQ6125000			111

Name	CAS-number	EEC-number	RTECS number	HI number	UN-number	Page
BENZYL ETHYL ETHER	539-30-0					112
BENZYL FORMATE	104-57-4		LQ5400000			113
BERYLLIUM	7440-41-7	004-001-00-7	DS1750000	60	1567	114
BERYLLIUM OXIDE	1304-56-9	004-002-00-2	DS4025000		1566	115
n-BGE	2426-08-6	603-039-00-7	TX4200000	30	1993	155
bicarbonate of soda	144-55-8		VZ0950000			788
bicarburetted hydrogen (cylinder)	74-85-1	601-010-00-3	KU5340000	23	1962	380
bichloracetic acid	79-43-6	607-066-00-5	AG6125000	80	1764	282
bichromate sulfuric acid solution (Sol. of potassium/sodium dichromate in sulfuric acid of various concentrations)				88	2240	297
bicyclo(4,4,0)decane	91-17-8		QJ3150000	30	1147	261
BICYCLOHEPTADIENE	121-46-0		RB6535000	33	2251	116
bicyclo-2,2,1-hepta-2,5-diene	121-46-0		RB6535000	33	2251	116
bicyclopentadiene	77-73-6		PC1050000	30	2048	299
biethylene (cylinder)	106-99-0	601-013-00-x	EI9275000	239	1010	130
(1,1'-bifenyl)-4,4'-diamine	92-87-5	612-042-00-2	DC9625000	60	1885	098
bimethyl (cylinder)	74-84-0	601-002-00-x	KH3800000	23	1035	377
biphenyl	92-52-4		DU8050000			360
1,1'-biphenyl	92-52-4		DU8050000			360
biphenyl oxide	101-84-8		KN8970000			362
biphenylsulfonic acid						363
bis(2-chloroethyl)ether	111-44-4	603-029-00-2	KN0875000	63	1916	291
bis-chloromethyl ether	542-88-1	603-046-00-5	KN1575000		2249	287
biscyclopentadiene	77-73-6		PC1050000	30	2048	299
bis(2-ethoxyethyl)ether	112-36-7		KN3160000			311
bis(2-ethylhexyl) phthalate	117-81-7		TI0350000			358
Bisfenol A	80-05-7		SL6300000			489
1,2-bis(2-hydroxyethoxy)ethane	112-27-6		YE4550000			888
β-bis-hydroxyethyl sulfide	111-48-8	603-081-00-6	KM2975000			855
2,2-bis(4-hydroxyfenyl)propane	80-05-7		SL6300000			489
bis(2-hydroxypropyl) ether	110-98-5		UB8785000			366
bismuthbromide	778-58-8					117
BISMUTHTRIBROMIDE	778-58-8					117
bivinyl (cylinder)	106-99-0	601-013-00-x	EI9275000	239	1010	130
bleach (solution with 150g/l active chlorine)	7681-52-9	017-011-00-1	NH3486300	85	1791	806
blue copperas	7758-98-7		GL8800000	60	1645	240
blue powder (powder, pyrophoric)	7440-66-6	030-001-00-1	ZG8600000	43	1383	929
blue vitriol	7758-98-7		GL8800000	60	1645	240
BORAX	1303-96-4		VZ2275000			118
BORIC ACID	10043-35-3		ED4550000			119
boric hydride (cylinder)	19287-45-7		HQ9275000	236		273
BORIUMTRIFLUORIDE (cylinder)	7637-07-2	005-001-00-x	ED2275000	286	1008	120
2-bornanone	76-22-2		EX1225000			187
boron chloride (cylinder)	10294-34-5	005-002-00-5	ED1925000	286	1741	121
boron ethane (cylinder)	19287-45-7		HQ9275000	236		273
BORON TRICHLORIDE (cylinder)	10294-34-5	005-002-00-5	ED1925000	286	1741	121
BORON TRIFLUORIDE ETHERATE	109-63-7		KX7375000	83	2604	122
BrCN	506-68-3		GT2100000		1889	124
bromide	7758-02-3		TS7650000			726
BROMINE	7726-95-6	035-001-00-5	EF9100000	886	1744	123
BROMINE CYANIDE	506-68-3		GT2100000		1889	124
BROMO ACETONE	598-31-2		UC0525000	60	1569	125
o-, m-, p-BROMOANILINE						126
BROMOBENZENE	108-86-1	602-060-00-9	CY9000000	30	2514	127
1-BROMO-3-CHLOROPROPANE	109-70-6		TX4113000	60	2688	128
2-bromo-2-chloro-1,1,1-trifluoroethane	151-67-7		KH6550000			451
bromocyan	506-68-3		GT2100000		1889	124
bromocyanide	506-68-3		GT2100000		1889	124
bromocyanogen	506-68-3		GT2100000		1889	124
bromoethane	74-96-4		KH6475000	60	1891	396
bromoethene (cylinder)	593-60-2	602-024-00-2	KU8400000	236	1085	917
bromoethylene (cylinder)	593-60-2	602-024-00-2	KU8400000	236	1085	917
bromomethane (cylinder)	74-83-9	602-002-00-3	PA4900000	26	1062	586
bromopropanone	598-31-2		UC0525000	60	1569	125
1-bromo-2-propene	106-95-6		UC7090000	336	1099	026
3-bromopropene	106-95-6		UC7090000	336	1099	026

Name	CAS-number	EEC-number	RTECS number	HI number	UN-number	Page
3-bromopropylene	106-95-6		UC7090000	336	1099	026
3-bromo-1-propyne	106-96-7		UK4375000		2345	748
3-bromopropyne (in toluene)	106-96-7		UK4375000			749
3-bromo-1-propyne (in toluene)	106-96-7		UK4375000			749
α-bromotoluene	100-39-0	602-057-00-2	XS7965000	60	1737	107
BRUCINE (quarternary hydrate)	357-57-3	614-006-00-1	EH8925000		1570	129
brucine hydrate (quarternary hydrate)	357-57-3	614-006-00-1	EH8925000		1570	129
1,3-BUTADIENE (cylinder)	106-99-0	601-013-00-x	EI9275000	239	1010	130
n-butanal	123-72-8	605-006-00-2	ES2275000	33	1129	164
BUTANE (cylinder)	106-97-8	601-004-00-0	EJ4200000	23	1011	131
n-butane (cylinder)	106-97-8	601-004-00-0	EJ4200000	23	1011	131
1,3-butanediamine	590-88-5		EJ6700000			266
1,4-butanedicarboxylic acid	124-04-9	607-144-00-9	AU8400000			021
butanedioic acid	110-15-6		WM4900000			827
cis-butanedioic acid	141-82-2		OO0175000			543
butanedioic acid diethyl ester	123-25-1		WM7400000			319
1,2-BUTANEDIOL	0584-03-2					132
1,3-BUTANEDIOL	107-88-0		EK0440000			133
1,4-BUTANEDIOL	110-63-4		EK0525000			134
2,3-BUTANEDIOL	513-85-9		EK0532000			135
2,3-butanedione	431-03-8		EK2625000	33	2346	263
butanenitrile	109-74-0	608-005-00-5	ET8750000	336	2411	166
butanoic acid butyl ester	109-21-7	607-031-00-4	ES8120000		1993	148
n-butanoic acid	107-92-6	607-135-00-x	ES5425000	80	2820	165
n-butanol	71-36-3	603-004-00-6	EO1400000	30	1120	142
sec-BUTANOL	78-92-2	603-004-00-6	EO1750000	30	1120	136
t-butanol	75-65-0	603-005-00-1	EO1925000	30	1120	143
1-butanol	71-36-3	603-004-00-6	EO1400000	30	1120	142
2-butanol	78-92-2	603-004-00-6	EO1750000	30	1120	136
2-butanone oxime	96-29-7	616-014-00-0	EL9275000			598
2-butanoneperoxide (solution with approx. 10 % active oxygen content)	1338-23-4		EL9450000	539	2250	597
2-butanoneperoxide (solution with approx. 10 % active oxygen content)	1338-23-4		EL9450000	539	2563	597
2-butanone	78-93-3	606-002-00-3	EL6475000	33	1193	596
Butanox (solution with approx. 10 % active oxygen content)	1338-23-4		EL9450000	539	2250	597
Butanox (solution with approx. 10 % active oxygen content)	1338-23-4		EL9450000	539	2563	597
butanoyl chloride	141-75-3	607-136-00-5	EU5523000	338	2353	167
1-BUTENE (cylinder)	106-98-9	601-012-00-4	EM2893000	23	1012	137
2-butene, see 1-BUTENE						137
cis-butenedioic acid	110-16-7	607-095-00-3	OM9625000			545
cis-butenedioic anhydride	108-31-6	607-096-00-9	ON3675000	80	2215	544
1,2-butene oxide	106-88-7		EK3675000	339	3022	152
trans-2-butenoic acid	3724-65-0		GQ2800000	89	2823	243
butenone	78-94-4		EM9800000	339	1251	619
2-butenone	78-94-4		EM9800000	339	1251	619
3-buten-2-one	78-94-4		EM9800000	339	1251	619
1-butoxybutane	142-96-1	603-054-00-9	EK5425000	30	1149	153
2-butoxyethanol acetate	112-07-2	607-038-00-2	KJ8925000			409
2-BUTOXYETHANOL	111-76-2	603-014-00-0	KJ8575000	60	2369	138
2-(2-butoxyethoxy)ethanol	112-34-5		KJ9100000			312
2-butoxyethyl-acetate	112-07-2	607-038-00-2	KJ8925000			409
butyl acetate	123-86-4	607-025-00-1	AF7350000	30	1123	139
n-BUTYL ACETATE	123-86-4	607-025-00-1	AF7350000	30	1123	139
sec-BUTYLACETATE	105-46-4	607-026-00-7	AF7380000	30	1123	140
1-butyl acetate	123-86-4	607-025-00-1	AF7350000	30	1123	139
butylacetic acid	142-62-1		MO5250000	80	2829	188
n-BUTYL ACRYLATE	141-32-2	607-062-00-3	UD3150000	39	2348	141
n-BUTYL ALCOHOL	71-36-3	603-004-00-6	EO1400000	30	1120	142
tert-BUTYL ALCOHOL	75-65-0	603-005-00-1	EO1925000	30	1120	143
n-butylaldehyde	123-72-8	605-006-00-2	ES2275000	33	1129	164
BUTYLAMINE	109-73-9	612-005-00-0	EO2975000	338	1125	144
n-butylamine	109-73-9	612-005-00-0	EO2975000	338	1125	144
tert-BUTYLAMINE	75-64-9		EO3330000			145
butyl benzoate	136-60-7		DG4925000			146

Name	CAS-number	EEC-number	RTECS number	HI number	UN-number	Page
n-BUTYL BENZOATE	136-60-7		DG4925000			146
p-t-butylbenzoic acid	98-73-7		DG4708000			147
p-tert-BUTYLBENZOIC ACID	98-73-7		DG4708000			147
4-tert-butylbenzoic acid	98-73-7		DG4708000			147
butyl benzyl ether	588-67-0					108
N-butyl-1-butanamine	111-92-2	612-049-00-0	HR7780000	83	2248	275
2-butylbutanoic acid	149-57-5		MO7700000			415
n-BUTYLBUTYRATE	109-21-7	607-031-00-4	ES8120000		1993	148
n-butyl carbinol	71-41-0	603-006-00-7	SB9800000	30	1105	058
butyl carbitol	112-34-5		KJ9100000			312
Butyl Carbitol	112-34-5		KJ9100000			312
p-tert-BUTYLCATECHOL	98-29-3		UX1400000		2921	149
butyl 'Cellosolve' acetate	112-07-2	607-038-00-2	KJ8925000			409
Butyl Cellosolve	111-76-2	603-014-00-0	KJ8575000	60	2369	138
BUTYL CHLORIDE	109-69-3		EJ6300000	33	1127	150
n-butyl chloride	109-69-3		EJ6300000	33	1127	150
tert-BUTYLCHLORIDE	507-20-0		TX5040000	33	1127	151
Butyl Diethoxol	112-34-5		KJ9100000			312
Butyl Dioxitol	112-34-5		KJ9100000			312
α-butylene (cylinder)	106-98-9	601-012-00-4	EM2893000	23	1012	137
β-butylene, see 1-BUTENE						137
1,2-butylene glycol	0584-03-2					132
1,3-butylene glycol	107-88-0		EK0440000			133
1,4-butylenegylcol	110-63-4		EK0525000			134
2,3-butylene glycol	513-85-9		EK0532000			135
butylene oxide	109-99-9	603-025-00-0	LU5950000	33	2056	848
1,2-BUTYLENE OXIDE	106-88-7		EK3675000	339	3022	152
2,3-butylene oxide, see 1,2-BUTYLENE OXIDE						152
butyl ethanoate	123-86-4	607-025-00-1	AF7350000	30	1123	139
BUTYL ETHER	142-96-1	603-054-00-9	EK5425000	30	1149	153
t-butyl ether, see BUTYL ETHER						153
Butyl Ethoxol	111-76-2	603-014-00-0	KJ8575000	60	2369	138
butylethylacetaldehyde	123-05-7		MN7525000	30	1191	414
butylethylacetic acid	149-57-5		MO7700000			415
n-butyl ethyl ether	628-81-9		KN4725000	33	1179	397
n-BUTYLFORMATE	592-84-7	607-017-00-8	LQ5500000	33	1128	154
n-BUTYL GLYCIDYL ETHER	2426-08-6	603-039-00-7	TX4200000	30	1993	155
BUTYL GLYCOLATE	7397-62-8					156
butyl hydride (cylinder)	106-97-8	601-004-00-0	EJ4200000	23	1011	131
butylhydroxytoluene	128-37-0		GO7875000			277
n-BUTYLLITHIUM	109-72-8			X333	2445	157
n-BUTYLLITHIUM (15% solution in hexane)	109-72-8			X323	2445	158
BUTYL METHACRYLATE	97-88-1	607-033-00-5	OZ3675000	39	2227	159
butyl 2-methacrylate	97-88-1	607-033-00-5	OZ3675000	39	2227	159
tert-butyl methyl ether	1634-04-4		KN5250000	33	2398	587
butyl 2-methyl-2-propanoate	97-88-1	607-033-00-5	OZ3675000	39	2227	159
bis(tert-butyl)peroxide	110-05-4	617-001-00-2	ER2450000	539	2102	278
tert-butyl peroxide	110-05-4	617-001-00-2	ER2450000	539	2102	278
butyl 2-propenoate	141-32-2	607-062-00-3	UD3150000	39	2348	141
n-BUTYL PROPIONATE	590-01-2	607-029-00-3	UE8245000	30	1914	160
4-tert-butylpyrocatechol	98-29-3		UX1400000		2921	149
BUTYL TITANATE	5593-70-4		XR1585000			161
p-tert-BUTYLTOLUENE	98-51-1		XS8400000	30	2667	162
4-tert-butyltoluene	98-51-1		XS8400000	30	2667	162
2-BUTYNE	503-17-3		GQ7210000	339	1144	163
BUTYRALDEHYDE	123-72-8	605-006-00-2	ES2275000	33	1129	164
BUTYRIC ACID	107-92-6	607-135-00-x	ES5425000	80	2820	165
n-butyric acid	107-92-6	607-135-00-x	ES5425000	80	2820	165
butyric alcohol	71-36-3	603-004-00-6	EO1400000	30	1120	142
N-BUTYRONITRILE	109-74-0	608-005-00-5	ET8750000	336	2411	166
butyroyl chloride	141-75-3	607-136-00-5	EU5523000	338	2353	167
BUTYRYL CHLORIDE	141-75-3	607-136-00-5	EU5523000	338	2353	167

C

Name	CAS-number	EEC-number	RTECS number	HI number	UN-number	Page
cadaverine	462-94-2		SA0200000			267
CADMIUM	7440-43-9		EU9800000			168
CADMIUM ACETATE	543-90-8	048-001-00-5	EU9810000	60	2570	169
CADMIUM HYDROXIDE	21041-95-2	048-001-00-5			2570	170
CADMIUM OXIDE	1306-19-0	048-002-00-0	EV1925000		2570	171
CADMIUM SULFIDE	1306-23-6		EV3150000		2570	172
cajeputene	138-86-3	601-029-00-7	OS8100000	30	2052	530
calamine	1314-13-2		ZH4810000			932
calcite	471-34-1		FF9335000			176
CALCIUM	7440-70-2	020-001-00-x	EV8040000	X423	1401	173
calcium acetylide	75-20-7	006-004-00-9	EV9400000	423	1402	175
CALCIUM BROMATE (monohydrate)	10102-75-7				1450	174
CALCIUM CARBIDE	75-20-7	006-004-00-9	EV9400000	423	1402	175
CALCIUM CARBONATE	471-34-1		FF9335000			176
CALCIUM CHLORATE (monohydrate)	10137-74-3		FN9800000		1452	177
CALCIUM CHLORIDE	10043-52-4	017-013-00-2	EV9800000		1453	178
calcium dicarbide	75-20-7	006-004-00-9	EV9400000	423	1402	175
CALCIUM HYDRIDE	7789-78-8	001-004-00-5	EW2440000	X423	1404	179
CALCIUM HYPOCHLORITE	7778-54-3	017-012-00-7	NH3485000		1748	180
CALCIUM NITRATE	10124-37-5		EW2985000		1454	181
CALCIUM NITRITE	13780-06-8				2627	182
CALCIUM OXIDE	1305-78-8		EW3100000		1910	183
calcium oxychloride	7778-54-3	017-012-00-7	NH3485000		1748	180
CALCIUM PHOSPHIDE	1305-99-3	015-003-00-2	EW3870000	43	1360	184
CALCIUM SULFIDE	20548-54-3	016-004-00-0				185
Calgon	10124-56-8		OY3675000			804
CALOMEL	10112-91-1	080-003-00-1	OV8740000		2025	186
calx	1305-78-8		EW3100000		1910	183
CAMPHOR	76-22-2		EX1225000			187
CAPROIC ACID	142-62-1		MO5250000	80	2829	188
CAPROLACTAM	105-60-2		CM3675000			189
ε-caprolactam	105-60-2		CM3675000			189
capronic acid	142-62-1		MO5250000	80	2829	188
caproyl alcohol	111-27-3	603-059-00-6	MQ4025000	30	1987	466
capryl alcohol	123-96-6		RH0795000	30	1993	658
caprylic alcohol	111-87-5		RH6550000	30	1993	657
carbamaldehyde	75-12-7		LQ0525000			432
carbamide	57-13-6		YR6250000			910
carbamimidic acid	57-13-6		YR6250000			910
carbanil	103-71-9		DA3675000	63	2487	697
carbaryl	63-25-2	006-011-00-7	FC5950000			610
carbazotic acid	88-89-1	609-009-00-x	TJ7875000	11	0154	718
carbide	75-20-7	006-004-00-9	EV9400000	423	1402	175
carbinol	67-56-1	603-001-00-x	PC1400000	336	1230	568
carbitol	111-90-0		KK8750000			313
Carbitol solvent	111-90-0		KK8750000			313
carbolic acid	108-95-2	604-001-00-2	SJ3325000	68	1671	687
carbon bisulfide	75-15-0	006-003-00-3	FF6650000	336	1131	192
CARBON DIOXIDE (cylinder)	124-38-9		FF6400000	20	1013	190
carbon dioxide gas (cylinder)	124-38-9		FF6400000	20	1013	190
CARBON DIOXIDE (solid)	124-38-9		FF6400000	22	1845	191
CARBON DISULFIDE	75-15-0	006-003-00-3	FF6650000	336	1131	192
carbon hexachloride	67-72-1		KI4025000			460
carbonic acid, cyclic ethylene ester	96-49-1		FF9550000			402
carbonic acid diethyl ester	105-58-8		FF9800000	30	2366	309
carbonic acid, disodium salt	497-19-8	011-005-00-2	VZ4050000			792
carbonic acid, lithium salt	554-13-2		OJ5800000			533
carbonic anhydride (solid)	124-38-9		FF6400000	22	1845	191
carbonic dichloride (cylinder)	75-44-5	006-002-00-8	SY5600000	266	1076	702
carbonic oxide (cylinder)	630-08-0	006-001-00-2	FG3500000	236	1016	193
CARBON MONOXIDE (cylinder)	630-08-0	006-001-00-2	FG3500000	236	1016	193
carbon oil	71-43-2	601-020-00-8	CY1400000	33	1114	096
carbon oxide (cylinder)	630-08-0	006-001-00-2	FG3500000	236	1016	193
carbon oxychloride (cylinder)	75-44-5	006-002-00-8	SY5600000	266	1076	702
carbon sulfide	75-15-0	006-003-00-3	FF6650000	336	1131	192
carbon tet	56-23-5	602-008-00-5	FG4900000	60	1846	194
carbon tetrabromide	558-13-4		FG4725000	60	2516	843

Name	CAS-number	EEC-number	RTECS number	HI number	UN-number	Page
CARBON TETRACHLORIDE	56-23-5	602-008-00-5	FG4900000	60	1846	194
carbonyl chloride (cylinder)	75-44-5	006-002-00-8	SY5600000	266	1076	702
carbonyl diamide	57-13-6		YR6250000			910
carbonyl diamine	57-13-6		YR6250000			910
CARBORUNDUM	409-21-2		VW0450000			195
carboxybenzene	65-85-0		DG0875000			099
catechol	120-80-9	604-016-00-4	UX1050000			766
caustic flake	1310-73-2	011-002-00-6	WB4900000	80	1823	805
caustic potash	1310-58-3	019-002-00-8	TT2100000	80	1813	734
caustic soda	1310-73-2	011-002-00-6	WB4900000	80	1823	805
CAUSTIC SODA SOLUTION (33%)	1310-73-2	011-002-01-3	WB4900000	80	1824	196
Cellosolve	110-80-5	603-012-00-x	KK8050000	30	1171	382
Cellosolve acetate	111-15-9	607-037-00-7	KK8225000	30	1172	383
Cereclor	63449-39-8		RV0490000			200
CHA	108-91-8	612-050-00-6	GX0700000	83	2357	255
chalk	471-34-1		FF9335000			176
CHINOLINE	92-22-5		VA9275000	60	2656	197
chloracetyl chloride	79-04-9	607-080-00-1	AO6475000	X80	1752	205
CHLORAL	75-87-6		FM7870000	60	2075	198
CHLORAL HYDRATE	302-17-0	605-014-00-6	FM8750000	60	2811	199
CHLORINATED PARAFFIN	63449-39-8		RV0490000			200
CHLORINE (cylinder)	7782-50-5	017-001-00-7	FO2100000	266	1017	201
chlorine cyanide (cylinder)	506-77-4		GT2275000	236	1589	247
CHLORINE DIOXIDE	10049-04-4		FO3000000			202
chlorine oxide	10049-04-4		FO3000000			202
chlorine (IV) oxide	10049-04-4		FO3000000			202
chlorine peroxide	10049-04-4		FO3000000			202
chloroacetic acid	79-11-8	607-003-00-1	AF8575000	80	1751	620
α-chloroacetic acid	79-11-8	607-003-00-1	AF8575000	80	1751	620
chloroacetic acid chloride	79-04-9	607-080-00-1	AO6475000	X80	1752	205
chloroacetic acid, ethyl ester	105-39-5	607-070-00-7	AF9110000	63	1181	399
CHLOROACETONE	78-95-5		UC700000	60	1695	203
CHLOROACETONITRILE	107-14-2	608-008-00-1	AL8225000	60	2668	204
CHLOROACETYL CHLORIDE	79-04-9	607-080-00-1	AO6475000	X80	1752	205
3-chloroallyl chloride	542-75-6	602-030-00-5	UC8310000	30	2047	296
p-chloroaminobenzene	106-47-8	612-010-00-8	BX0700000	60	2018	207
4-chloro-1-aminobenzene	106-47-8	612-010-00-8	BX0700000	60	2018	207
m-chloroaniline, see o-CHLOROANILINE						206
o-CHLOROANILINE	95-51-2	612-010-00-8	BX0525000	60	2019	206
p-CHLOROANILINE	106-47-8	612-010-00-8	BX0700000	60	2018	207
2-chloroaniline	95-51-2	612-010-00-8	BX0525000	60	2019	206
3-chloroaniline, see o-CHLOROANILINE						206
α-chlorobenzaldehyde	98-88-4	607-012-00-0	DM6600000	80	1736	103
2-chlorobenzenamine	95-51-2	612-010-00-8	BX0525000	60	2019	206
chlorobenzene	108-90-7	602-033-00-1	CZ0175000	30	1134	621
4-chlorobenzeneamine	106-47-8	612-010-00-8	BX0700000	60	2018	207
chlorobutadiene	126-99-8	602-036-00-8	EI9625000	336	1991	219
2-chloro-1,3 butadiene	126-99-8	602-036-00-8	EI9625000	336	1991	219
1-chlorobutane	109-69-3		EJ6300000	33	1127	150
chloro(chloromethoxy)methane	542-88-1	603-046-00-5	KN1575000		2249	287
4-chloro-m-cresol	59-50-7	604-014-00-3	GO7100000	60	2669	208
p-chlorocresol	59-50-7	604-014-00-3	GO7100000	60	2669	208
p-CHLORO-m-CRESOL	59-50-7	604-014-00-3	GO7100000	60	2669	208
4-chloro-o-cresoxyacetic acid	94-74-6	607-051-00-3	AG1575000			592
chlorocyan (cylinder)	506-77-4		GT2275000	236	1589	247
chlorocyanide (cylinder)	506-77-4		GT2275000	236	1589	247
chlorocyanogen (cylinder)	506-77-4		GT2275000	236	1589	247
2-chloro-1,3-diene	126-99-8	602-036-00-8	EI9625000	336	1991	219
chlorodimethyl ether	107-30-2	603-075-00-3	KN6650000	336	1239	211
1-chloro-2,4-dinitrobenzene	97-00-7	610-003-00-4	CZ0525000	60	1577	355
1-chloro-2,3-epoxypropane	106-89-8	603-026-00-6	TX4900000	63	2023	376
chloroethane (cylinder)	75-00-3	602-009-00-0	KH7525000	236	1037	398
chloroethanoic acid	79-11-8	607-003-00-1	AF8575000	80	1751	620
2-chloroethanol	107-07-3	603-028-00-7	KK0875000	60	1135	209
β-chloroethyl alcohol	107-07-3	603-028-00-7	KK0875000	60	1135	209
2-CHLOROETHYL ALCOHOL	107-07-3	603-028-00-7	KK0875000	60	1135	209
chloroethylene (cylinder)	75-01-4	602-023-00-7	KU9625000	239	1086	918

Name	CAS-number	EEC-number	RTECS number	HI number	UN-number	Page
CHLOROFORM	67-66-3	602-006-00-4	FS9100000	60	1888	210
chloroformyl chloride (cylinder)	75-44-5	006-002-00-8	SY5600000	266	1076	702
4-chloro-1-hydroxybenzene	106-48-9	604-008-00-0	SK2800000	60	2020	217
1-chloro-2-hydroxybenzene	95-57-8	604-008-00-0	SK2625000	68	2021	623
2-chloro-1-hydroxybenzene	95-57-8	604-008-00-0	SK2625000	68	2021	623
4-chloro-1-hydroxy-3-methylbenzene	59-50-7	604-014-00-3	GO7100000	60	2669	208
2-chloro-5-hydroxytoluene	59-50-7	604-014-00-3	GO7100000	60	2669	208
1-chloro-2-ketopropane	78-95-5		UC700000	60	1695	203
chloromethane (cylinder)	74-87-3	602-001-00-7	PA6300000	236	1063	589
(chloromethyl)benzene	100-44-7	602-037-00-3	XS8925000	68	1738	109
chloromethylcyanide	107-14-2	608-008-00-1	AL8225000	60	2668	204
chloromethyl ether	542-88-1	603-046-00-5	KN1575000		2249	287
CHLOROMETHYL METHYL ETHER	107-30-2	603-075-00-3	KN6650000	336	1239	211
chloromethyl methyl ketone	78-95-5		UC700000	60	1695	203
chloromethyloxirane	106-89-8	603-026-00-6	TX4900000	63	2023	376
4-chloro-3-methylphenol	59-50-7	604-014-00-3	GO7100000	60	2669	208
4-chloro-2-methylphenoxyacetic acid	94-74-6	607-051-00-3	AG1575000			592
1-chloro-2-methylpropane	513-36-0			33	1127	503
2-chloro-2-methylpropane	507-20-0		TX5040000	33	1127	151
α-chloronaphtalene	90-13-1		QJ2100000			212
1-CHLORONAPHTALENE	90-13-1		QJ2100000			212
2-chloronaphtalene, see 1-CHLORONAPHTALENE						212
CHLORONITROANILINE	121-87-9	610-006-00-0	BX1400000	60	2237	213
p-chloro-o-nitroaniline, see CHLORONITROANILINE						213
2-chloro-4-nitroaniline	121-87-9	610-006-00-0	BX1400000	60	2237	213
m-chloronitrobenzene, see o-CHLORONITROBENZENE						214
o-CHLORONITROBENZENE	88-73-3		CZ0875000	60	1578	214
p-CHLORONITROBENZENE	100-00-5	610-005-00-5	CZ1050000	60	1578	215
1-chloro-2-nitrobenzene	88-73-3		CZ0875000	60	1578	214
1,1-CHLORONITROETHANE	598-92-5		KH7875000			216
1-chloro-2-oxopropane	78-95-5		UC700000	60	1695	203
o-chlorophenol	95-57-8	604-008-00-0	SK2625000	68	2021	623
p-CHLOROPHENOL	106-48-9	604-008-00-0	SK2800000	60	2020	217
2-chlorophenol	95-57-8	604-008-00-0	SK2625000	68	2021	623
4-chlorophenol	106-48-9	604-008-00-0	SK2800000	60	2020	217
2-chlorophenylamine	95-51-2	612-010-00-8	BX0525000	60	2019	206
4-chlorophenylamine	106-47-8	612-010-00-8	BX0700000	60	2018	207
CHLOROPICRIN	76-06-2	610-001-00-3	PB6300000	66	1580	218
chloroprene	126-99-8	602-036-00-8	EI9625000	336	1991	219
β-chloroprene	126-99-8	602-036-00-8	EI9625000	336	1991	219
2-CHLOROPRENE	126-99-8	602-036-00-8	EI9625000	336	1991	219
2-chloropropane	75-29-6	602-018-00-x	TX4410000	33	2356	514
1-chloro-2-propanone	78-95-5		UC700000	60	1695	203
3-chloro-1-propene	107-05-1	602-029-00-x	UC7350000	336	1100	027
3-chloropropenyl chloride	542-75-6	602-030-00-5	UC8310000	30	2047	296
α-CHLOROPROPIONIC ACID	598-78-7	607-139-00-1	UE8575000	80	2511	220
2-chloropropionic acid	598-78-7	607-139-00-1	UE8575000	80	2511	220
chloropropylene oxide	106-89-8	603-026-00-6	TX4900000	63	2023	376
3-chloro-1,2-propylene oxide	106-89-8	603-026-00-6	TX4900000	63	2023	376
3-chloropropylene	107-05-1	602-029-00-x	UC7350000	336	1100	027
3-chloro-1-propyne	624-65-7					750
3-chloro-1-propyne (in toluene)	624-65-7					751
CHLOROSULFONIC ACID	7790-94-5	016-017-00-1	FX5730000	88	1754	221
Chlorothane	71-55-6	602-013-00-2	KJ2975000	60	2831	590
Chlorothene	71-55-6	602-013-00-2	KJ2975000	60	2831	590
4-chloro-o-toloxyacetic acid	94-74-6	607-051-00-3	AG1575000			592
α-chlorotoluene	100-44-7	602-037-00-3	XS8925000	68	1738	109
chlorotrifluoromethane (cylinder)	75-72-9		PA6410000	20	1022	624
1,4-chloronitrobenzene	100-00-5	610-005-00-5	CZ1050000	60	1578	215
chromic acid	1333-82-0	024-001-00-0	GB6650000	50	1463	222
chromic acid (Sol. of potassium/sodium dichromate in sulfuric acid of various concentrations)				88	2240	297
chromic anhydride	1333-82-0	024-001-00-0	GB6650000	50	1463	222
chromium (VI) oxide	1333-82-0	024-001-00-0	GB6650000	50	1463	222
CHROMIUM TRIOXIDE	1333-82-0	024-001-00-0	GB6650000	50	1463	222
chrysotile (white)	12001-29-5		CI6478500		2212	078

Name	CAS-number	EEC-number	RTECS number	HI number	UN-number	Page
chrysotile (white)	12001-29-5		CI6478500		2590	078
cinnamene	100-42-5	601-026-00-0	WL3675000	39	2055	825
CINNAMIC ALDEHYDE	104-55-2		GD6475000			223
cinnamol	100-42-5	601-026-00-0	WL3675000	39	2055	825
cinnamyl aldehyde	104-55-2		GD6475000			223
CITRIC ACID	77-92-9		GE7350000			224
CITRONELLA (mixture of essential oils)	8000-29-1		GE8750000			225
citronella oil (mixture of essential oils)	8000-29-1		GE8750000			225
ClCN (cylinder)	506-77-4		GT2275000	236	1589	247
ClO2	10049-04-4		FO3000000			202
CMME	107-30-2	603-075-00-3	KN6650000	336	1239	211
CO (cylinder)	630-08-0	006-001-00-2	FG3500000	236	1016	193
coal naphtha	71-43-2	601-020-00-8	CY1400000	33	1114	096
COBALT (powder)	7440-48-4					226
cobalt acetate	71-48-7		AG3480000			227
COBALT (II) ACETATE	71-48-7		AG3480000			227
COBALT (II) CHLORIDE	7646-79-9		GF9800000			228
cobalt diacetate	71-48-7		AG3480000			227
cobalt muriate	7646-79-9		GF9800000			228
cobalt nitrate, hexahydrate	10026-22-9		QU7355500			229
COBALT (II) NITRATE	10026-22-9		QU7355500			229
cobaltous acetate	71-48-7		AG3480000			227
cobaltous chloride	7646-79-9		GF9800000			228
cobalt sulfate	10124-43-3		GG3100000			230
COBALT (II) SULFATE	10124-43-3		GG3100000			230
COLCHICINE	64-86-8	614-005-00-6	GH0700000			231
COLLODION (nitrocellulose solution in ether)	9004-70-0		QW0975200	33	2059	232
Conite	7775-14-6	016-028-00-1	JP2100000	43	1384	800
COPPER (powder)	7440-50-8		GL5325000			233
COPPER (II) ACETATE	142-71-2		AG3480000			234
COPPER (II) CARBONATE (basic)	12069-69-1		GL6910000			235
COPPER (I) CHLORIDE	7758-89-6	029-001-00-4	GL6990000			236
COPPER (II) CHLORIDE	7447-39-4		GL7237000	60	2802	237
COPPER (II) HYDROXIDE	20427-59-2		GL7600000			238
COPPER (II) NITRATE	3251-23-8		QU7400000			239
COPPER (II) SULFATE	7758-98-7		GL8800000	60	1645	240
m-cresol, see o-CRESOL						241
o-CRESOL	95-48-7	604-004-00-9	GO6300000	60	2076	241
p-cresol, see o-CRESOL						241
o-cresylic acid	95-48-7	604-004-00-9	GO6300000	60	2076	241
o-cresyl phosphate	78-30-8	015-015-00-8	TD0175000	60	2574	902
crocidolite (blue)	12001-29-5		CI6478500		2212	078
crocidolite (blue)	12001-29-5		CI6478500		2590	078
CROTONALDEHYDE	123-73-9	605-009-00-9	GP9625000	33	1143	242
CROTONIC ACID	3724-65-0		GQ2800000	89	2823	243
crotonylene	503-17-3		GQ7210000	339	1144	163
CUMARINE	91-64-5		GN4200000			244
CUMENE	98-82-8	601-024-00-x	GR8575000	30	1918	245
CUMENE HYDROPEROXIDE (70-80%)	80-15-9	617-002-00-8	MX2450000	539	2116	246
α-cumenyl hydroperoxide (70-80%)	80-15-9	617-002-00-8	MX2450000	539	2116	246
cumol	98-82-8	601-024-00-x	GR8575000	30	1918	245
cupric acetate	142-71-2		AG3480000			234
cupric carbonate (basic)	12069-69-1		GL6910000			235
cupric chloride	7447-39-4		GL7237000	60	2802	237
cupric hydroxide	20427-59-2		GL7600000			238
cupric nitrate	3251-23-8		QU7400000			239
cuprochloride	7758-89-6	029-001-00-4	GL6990000			236
cyanide of sodium (30% solution)	143-33-9	006-007-00-5	VZ7525000	66	1935	797
cyanide of sodium	143-33-9	006-007-00-5	VZ7525000		1689	796
cyanoethane	107-12-0		UF9625000	336	2404	757
cyanoethylene	107-13-1	608-003-00-4	AT5250000	336	1093	020
cyanogen bromide	506-68-3		GT2100000		1889	124
CYANOGEN CHLORIDE (cylinder)	506-77-4		GT2275000	236	1589	247
cyanoguanidine	461-58-5		ME9950000			298
cyanomethane	75-05-8	608-001-00-3	AL7700000	336	1648	008
2-cyano-2-propanol	75-86-5	608-004-00-x	OD9275000	66	1541	007
2-cyano-1-propene	126-98-7	608-010-00-2	UD1400000			566

Name	CAS-number	EEC-number	RTECS number	HI number	UN-number	Page
v-cyanotoluene	140-29-4		AM1400000	60	2470	110
cyanuric acid	108-80-5		XZ1800000			505
cyanuric acid chloride	108-77-0	613-009-00-5	XZ1400000	80	2670	248
CYANURIC CHLORIDE	108-77-0	613-009-00-5	XZ1400000	80	2670	248
cyanuryl chloride	108-77-0	613-009-00-5	XZ1400000	80	2670	248
CYCLOHEPTATRIENE	544-25-2		GU3675000		2603	249
1,4-cyclohexadienedione	106-51-4	606-013-00-3	DK2625000	60	2587	769
cyclohexanamine	108-91-8	612-050-00-6	GX0700000	83	2357	255
CYCLOHEXANE	110-82-7	601-017-00-1	GU6300000	33	1145	250
2,5-cyclohexane-1,4 dione	106-51-4	606-013-00-3	DK2625000	60	2587	769
CYCLOHEXANOL	108-93-0	603-009-00-3	GV7875000			251
CYCLOHEXANONE	108-94-1	606-010-00-7	GW1050000	30	1915	252
CYCLOHEXANON OXIME	100-64-1		GW1925000			253
cyclohexatriene	71-43-2	601-020-00-8	CY1400000	33	1114	096
CYCLOHEXENE	110-83-8		GW2500000	33	2256	254
CYCLOHEXYLAMINE	108-91-8	612-050-00-6	GX0700000	83	2357	255
cyclohexylmethane	108-87-2	601-018-00-7	GV6125000	33	2296	593
1,3-CYCLOPENTADIENE	542-92-7		GY1000000			256
1,3-cyclopentadiene, dimer	77-73-6		PC1050000	30	2048	299
CYCLOPENTANE	287-92-3	601-030-00-2	GY2390000	33	1146	257
CYCLOPENTENE	142-29-0		GY5950000	33	2246	258
cyclopentimine	110-89-4	613-027-00-3	TM3500000	338	2401	720
CYCLOPROPANE (cylinder)	75-19-4	601-016-00-6	GZ0690000	23	1027	259
m-cymene, see p-CYMENE						260
o-cymene, see p-CYMENE						260
p-CYMENE	99-87-6		GZ5950000	30	2046	260

D

Name	CAS-number	EEC-number	RTECS number	HI number	UN-number	Page
2,4-D	94-75-7	607-039-00-8	AG6825000	60	1609	294
DAA	124-02-7		UC6650000	338	2359	264
DAA	123-42-2	603-016-00-1	SA9100000	33	1148	262
daltogen	102-71-6		KL9275000			883
DBA	111-92-2	612-049-00-0	HR7780000	83	2248	275
DBCP	96-12-8	602-021-00-6	TX8750000	60	2872	274
DBP	84-74-2		TI0875000			279
DCA	79-43-6	607-066-00-5	AG6125000	80	1764	282
1,1-DCE	75-35-4	602-025-00-8	KV9275000	339	1303	290
DCPD	77-73-6		PC1050000	30	2048	299
DEA	111-42-2	603-071-00-1	KL2975000	80	1719	300
DEAC	96-10-6		BD0558000	X333	2221	302
DEAC (15% solution in hexane)	96-10-6		BD0558000	X333	2220	303
DEAH	871-27-2			X323	2220	304
DEAH (15% solution in hexane)	87-27-2			X333	2220	305
deanol	108-01-0	603-047-00-0	KK6125000	30	2051	339
decadiene	16736-42-8				1142	061
DECAHYDRONAPHTHALENE (cis/trans)	91-17-8		QJ3150000	30	1147	261
decalin	91-17-8		QJ3150000	30	1147	261
2-deoxyglycerol	504-63-2		TY2010000			901
DEP	84-66-2		TI1050000			318
DETA	111-40-0	612-058-00-x	IE1225000	80	2079	314
detergent alkylate	123-01-3		CZ9540000			372
DGE	2238-07-5		KN2350000			324
diacetic ester	141-97-9		AK5250000			385
DIACETONE ALCOHOL	123-42-2	603-016-00-1	SA9100000	33	1148	262
DIACETYL	431-03-8		EK2625000	33	2346	263
diacetyl methane	123-54-6	606-029-00-0	SA1925000	30	2310	011
DIALLYLAMINE	124-02-7		UC6650000	338	2359	264
diallyl ether	557-40-4		KN7525000	336	2360	028
DIALLYL PHTALATE	131-17-9	607-086-00-4	CZ4200000			265
dialuminum sulfate	10043-01-3		BD1700000			037
diamide (100%)	302-01-2		MU7175000	886	2029	468
diamine (100%)	302-01-2		MU7175000	886	2029	468
1,4-diaminobenzene	106-50-3	612-028-00-6	SS8050000	60	1673	690
1,3-DIAMINOBUTANE	590-88-5		EJ6700000			266

Name	CAS-number	EEC-number	RTECS number	HI number	UN-number	Page
dichloroethanoic acid	79-43-6	607-066-00-5	AG6125000	80	1764	282
1,1-DICHLOROETHENE	75-35-4	602-025-00-8	KV9275000	339	1303	290
1,1-dichloroethylene	75-35-4	602-025-00-8	KV9275000	339	1303	290
sym-DICHLOROETHYL ETHER	111-44-4	603-029-00-2	KN0875000	63	1916	291
dichloroethyl ether	111-44-4	603-029-00-2	KN0875000	63	1916	291
dichlorofluoromethane (cylinder)	75-43-4		PA8400000	20	1029	292
dichloromethane	75-09-02	602-004-00-3	PA8050000	60	1593	594
(dichloromethyl)benzene	98-87-3	602-058-00-8	CZ5075000	68	1886	093
DICHLOROMONOFLUOROMETHANE (cylinder)	75-43-4		PA8400000	20	1029	292
2,4-DICHLOROPHENOL	120-83-2	604-011-00-7	SK8575000	60	2021	293
2,4-DICHLOROPHENOXYACETIC ACID	94-75-7	607-039-00-8	AG6825000	60	1609	294
1,2-DICHLOROPROPANE	78-87-5	602-020-00-0	TX9625000	33	1279	295
1,3-dichloropropane, see 1,2-DICHLOROPROPANE						295
2,2-dichloropropane, see 1,2-DICHLOROPROPANE						295
dichloropropene	542-75-6	602-030-00-5	UC8310000	30	2047	296
1,3-dichloro-1-propene	542-75-6	602-030-00-5	UC8310000	30	2047	296
1,3-DICHLOROPROPENE	542-75-6	602-030-00-5	UC8310000	30	2047	296
1,3-dichloropropylene	542-75-6	602-030-00-5	UC8310000	30	2047	296
dichloropropyl ether, see sym-DICHLOROETHYL ETHER						291
α,α-dichlorotoluene	98-87-3	602-058-00-8	CZ5075000	68	1886	093
DICHROMATE/SULFURIC ACID SOLUTION (Sol. of potassium/sodium dichromate in sulfuric acid of various concentrations)				88	2240	297
DICYANDIAMIDE	461-58-5		ME9950000			298
1,4-dicyanobutane	111-69-3		AV2625000	60	2205	022
DICYCLOPENTADIENE	77-73-6		PC1050000	30	2048	299
di-(2,3-epoxypropyl)ether	2238-07-5		KN2350000			324
diesel oil	64741-44-2		LX3296000	30	1202	440
diesel oil, see GAS OIL						440
DIETHANOLAMINE	111-42-2	603-071-00-1	KL2975000	80	1719	300
1,1-diethoxyethane	105-57-7	605-015-00-1	AB2800000	33	1088	301
1,2-diethoxy ethane	629-14-1		KI14100000	30	1153	315
DIETHYLACETAL	105-57-7	605-015-00-1	AB2800000	33	1088	301
DIETHYLALUMINUM CHLORIDE	96-10-6		BD0558000	X333	2221	302
DIETHYLALUMINUMCHLORIDE (15% solution in hexane)	96-10-6		BD0558000	X333	2220	303
DIETHYLALUMINUM HYDRIDE	871-27-2			X323	2220	304
DIETHYLALUMINUM HYDRIDE (15% solution in hexane)	87-27-2			X333	2220	305
DIETHYLAMINE	109-89-7	612-003-00-x	HZ8750000	338	1154	306
DIETHYLAMINOETHANOL	100-37-8	603-048-00-6	KK5075000	30	2686	307
N,N-DIETHYLANILINE	91-66-7	612-054-00-8	BX3400000	60	2432	308
diethyl carbitol	112-36-7		KN3160000			311
DIETHYL CARBONATE	105-58-8		FF9800000	30	2366	309
1,4-diethylenediamine	110-85-0	612-057-00-4	TK7800000	80	2579	719
1,4-diethylene dioxide	123-91-1	603-024-00-5	JG8225000	33	1165	359
diethylene ether	123-91-1	603-024-00-5	JG8225000	33	1165	359
DIETHYLENE GLYCOL	111-46-6		ID5950000			310
diethylene glycol butyl ether	112-34-5		KJ9100000			312
diethylene glycol n-butyl ether	112-34-5		KJ9100000			312
DIETHYLENE GLYCOL DIETHYL ETHER	112-36-7		KN3160000			311
diethylene glycol ethyl ether	111-90-0		KK8750000			313
DIETHYLENE GLYCOL MONOBUTYL ETHER	112-34-5		KJ9100000			312
DIETHYLENE GLYCOL MONOETHYL ETHER	111-90-0		KK8750000			313
diethylene glycol monomethyl ether, see DIETHYLENE GLYCOL MONOETHYL ETHER						313
diethylene imidoxide	110-91-8	613-028-00-9	QD6475000	30	2054	626
diethyleneimine	110-85-0	612-057-00-4	TK7800000	80	2579	719
diethylene oxide	109-99-9	603-025-00-0	LU5950000	33	2056	848
diethylene oximide	110-91-8	613-028-00-9	QD6475000	30	2054	626
DIETHYLENETRIAMINE	111-40-0	612-058-00-x	IE1225000	80	2079	314
diethylenimide oxide	110-91-8	613-028-00-9	QD6475000	30	2054	626
N,N-diethylethanamine	121-44-8	612-004-00-5	YE0175000	338	1296	886
N,N-diethylethanolamine	100-37-8	603-048-00-6	KK5075000	30	2686	307
diethyl ether	60-29-7	603-022-00-4	KI5775000	33	1155	381
DIETHYL GLYCOL	629-14-1		KI14100000	30	1153	315
di(2-ethylhexyl) phthalate	117-81-7		TI0350000			358
DIETHYL KETONE	96-22-0	606-006-00-5	SA8050000	33	1156	316

Name	CAS-number	EEC-number	RTECS number	HI number	UN-number	Page
diethyl malonate	105-53-3		OO0700000			422
diethyl 4-nitrophenylphosphorothionate	56-38-2	015-034-00-1	TF4550000	66	1668	674
diethyl p-nitrophenylthionophosphate	56-38-2	015-034-00-1	TF4550000	66	1668	674
diethyl p-nitrophenylthiophospate	56-38-2	015-034-00-1	TF4550000	66	1668	674
diethylolamine	111-42-2	603-071-00-1	KL2975000	80	1719	300
O-O-diethyl-O-p-nitrophenylthiophospate	56-38-2	015-034-00-1	TF4550000	66	1668	674
DIETHYL OXALATE	95-92-1	607-147-00-5	RQ8225000	60	2525	317
diethyl oxide	60-29-7	603-022-00-4	KI5775000	33	1155	381
diethylparathion	56-38-2	015-034-00-1	TF4550000	66	1668	674
diethylphenylamine	91-66-7	612-054-00-8	BX3400000	60	2432	308
p-DIETHYL PHTHALATE	84-66-2		TI1050000			318
DIETHYL SUCCINATE	123-25-1		WM7400000			319
DIETHYLSULFATE	64-67-5	016-027-00-6	WS7875000	60	1594	320
DIETHYL SULFIDE	352-93-2		LC7200000	336	2375	321
diethyl sulphate	64-67-5	016-027-00-6	WS7875000	60	1594	320
difenylol propane	80-05-7		SL6300000			489
difluorochloromethane (cylinder)	75-45-6		PA6390000	20	1018	622
difluoroethane (cylinder)	75-37-6		KI1410000	23	1030	322
1,1-DIFLUOROETHANE (cylinder)	75-37-6		KI1410000	23	1030	322
1,1-difluoroethene (cylinder)	75-38-7		KW0560000	239	1959	323
1,1-DIFLUOROETHYLENE (cylinder)	75-38-7		KW0560000	239	1959	323
difluoro-1,1 ethylene (cylinder)	75-38-7		KW0560000	239	1959	323
1,3-diformal propane	111-30-8		MA2450000			442
diformyl	107-22-2	605-016-00-7	MD2625000			449
DIGLYCIDYL ETHER	2238-07-5		KN2350000			324
diglycol	111-46-6		ID5950000			310
dihydro-2,5-dioxofuran	108-31-6	607-096-00-9	ON3675000	80	2215	544
dihydrooxirene (cylinder)	75-21-8	603-023-00-x	KX2450000	236	1040	412
2,3-DIHYDROPYRAN	110-87-2		UP7700000	33	2376	325
m-dihydroxybenzene	108-46-3	604-010-00-1	VG9625000	60	2876	772
p-dihydroxybenzene	123-31-9	604-005-00-4	MX3500000	60	2662	485
1,2-dihydroxybenzene	120-80-9	604-016-00-4	UX1050000			766
1,3-dihydroxybenzene	108-46-3	604-010-00-1	VG9625000	60	2876	772
1,2-dihydroxybutane	0584-03-2					132
2,3-dihydoxybutane	513-85-9		EK0532000			135
2,2'-dihydroxydipropyl ether	110-98-5		UB8785000			366
1,2-dihydroxyethane	107-21-1	603-027-00-1	KW2975000			406
di(2-hydroxyethyl)amine	111-42-2	603-071-00-1	KL2975000	80	1719	300
dihydroxyethyl sulfide	111-48-8	603-081-00-6	KM2975000			855
2,5-dihydroxyhexane	2935-44-6		MO2275000			465
2,4-dihydroxy-2-methylpentane	107-41-5	603-053-00-3	SA0810000			467
1,3-dihydroxypropane	504-63-2		TY2010000			901
dihydroxysuccinic acid	87-69-4		WW8750000			842
DIISOBUTYL ALUMINUM CHLORIDE	1779-25-5		BD0560000	X333	2221	326
DIISOBUTYL ALUMINUM CHLORIDE (15% solution in hexane)	1779-25-5		BD0560000	X333	2220	327
DIISOBUTYL ALUMINUM HYDRIDE	1191-15-7		BD0710000	X333	3051	328
DIISOBUTYL ALUMINUM HYDRIDE (15% solution in hexane)	1191-15-7		BD0710000	X333	2220	329
DIISOBUTYLCARBINOL	108-82-7		MJ3325000	30	1993	330
DIISOBUTYLENE	25167-70-8	601-031-00-8	SB2717000	33	2050	331
DIISOBUTYL KETONE	108-83-8	606-005-00-x	MJ5775000	30	1157	332
diisobutyl phenol	1806-26-4			60	2430	659
2,4-diisocyanate-1-methylbenzene	584-84-9	615-006-00-4	CZ6300000	60	2078	866
2,4-diisocyanatotoluene	584-84-9	615-006-00-4	CZ6300000	60	2078	866
DIISOOCTYL PHTHALATE	27554-26-3		TI1300000			333
di-sec-octyl phthalate	117-81-7		TI0350000			358
DIISOPROPYL CARBINOL	600-36-2					334
diisopropyl ether	108-20-3	603-045-00-x	TZ4300000	33	1159	515
diisopropylidene acetone	504-20-1		MI5500000			701
diisopropyl oxide	108-20-3	603-045-00-x	TZ4300000	33	1159	515
DIKETENE	674-82-8	606-017-00-5	RQ8225000	39	2521	335
dilithium carbonate	554-13-2		OJ5800000			533
dimazine	57-14-7	007-012-00-5	MV2450000	338	1163	348
1,1-dimethoxy ethane	534-15-6	605-007-00-8	AB2825000	33	2377	336
1,2-dimethoxyethane	110-71-4	603-031-00-3	KI1451000			408
dimethoxymethane	109-87-5		PA8750000	33	1234	576

Name	CAS-number	EEC-number	RTECS number	HI number	UN-number	Page
2,3-dimethoxystrichnidin-10-one (quarternary hydrate)	357-57-3	614-006-00-1	EH8925000		1570	129
2,3-dimethoxystrychnine (quarternary hydrate)	357-57-3	614-006-00-1	EH8925000		1570	129
dimethyl (cylinder)	74-84-0	601-002-00-x	KH3800000	23	1035	377
DIMETHYLACETAL	534-15-6	605-007-00-8	AB2825000	33	2377	336
dimethylacetamide	127-19-5	616-011-00-4	AB7700000			337
N,N-DIMETHYL ACETAMIDE	127-19-5	616-011-00-4	AB7700000			337
dimethylacetonylcarbinol	123-42-2	603-016-00-1	SA9100000	33	1148	262
dimethylacetylene	503-17-3		GQ7210000	339	1144	163
DIMETHYLAMINE (40% aqueous solution)	124-40-3	612-001-00-9	IP8750000	338	1160	338
dimethylamino benzene	121-69-7	612-016-00-0	BX4725000	60	2253	342
dimethylaminoethanol	108-01-0	603-047-00-0	KK6125000	30	2051	339
2-DIMETHYLAMINOETHANOL	108-01-0	603-047-00-0	KK6125000	30	2051	339
3-DIMETHYLAMINOPROPYLAMINE	109-55-7	612-061-00-6	TX7525000			340
DIMETHYLAMMONIUMCHLORIDE	506-59-2		IQ0220000			341
dimethylaniline (mixture of o-, m-, and p-isomers)	1300-73-8	612-027-00-0	ZE8575000	60	1711	928
dimethylaniline	121-69-7	612-016-00-0	BX4725000	60	2253	342
N,N-DIMETHYLANILINE	121-69-7	612-016-00-0	BX4725000	60	2253	342
dimethylbenzeneamine (mixture of o-, m-, and p-isomers)	1300-73-8	612-027-00-0	ZE8575000	60	1711	928
N,N-dimethylbenzeneamine	121-69-7	612-016-00-0	BX4725000	60	2253	342
dimethyl benzeneorthodicarboxylate	131-11-3		TI1575000			349
1,3-dimethylbenzene	108-38-3	601-039-00-1	ZE2275000	30	1307	926
1,4-dimethylbenzene	106-42-3	601-040-00-7	ZE2625000	30	1307	927
α,α-dimethylbenzyl hydroperoxide (70-80%)	80-15-9	617-002-00-8	MX2450000	539	2116	246
1,3-dimethylbutanol	108-11-2	603-008-00-8	SA7350000	30	2053	604
dimethylcarbinol	67-63-0	603-003-00-0	NT8050000	33	1219	512
DIMETHYL CARBONATE	616-38-6	607-013-00-6	FG0450000	33	1161	343
dimethylchloroether	107-30-2	603-075-00-3	KN6650000	336	1239	211
1,4-DIMETHYLCYCLOHEXANE- trans	589-90-2	601-019-00-2	GV0200000	33	2263	344
DIMETHYLDICHLOROSILANE	75-78-5	014-003-00-x	VV3150000	X338	1162	345
1,1'-dimethyldiethylene glycol	110-98-5		UB8785000			366
dimethylenediamine	107-15-3	612-006-00-6	KH8575000	83	1604	404
dimethylenemethane (cylinder)	463-49-0		BA0400000		2200	745
1,1-dimethylethane (cylinder)	75-28-5	601-004-00-0	TZ4300000	23	1969	497
1,1-dimethylethanol	75-65-0	603-005-00-1	EO1925000	30	1120	143
dimethylethanolamine	108-01-0	603-047-00-0	KK6125000	30	2051	339
DIMETHYL ETHER (cylinder)	115-10-6	603-019-00-8	PM4780000	23	1033	346
4-(1,1-dimethylethyl)benzoic acid	98-73-7		DG4708000			147
dimethylethylcarbinol	75-85-4	603-007-00-2		33	1105	060
di-(1-methylethyl)ether	108-20-3	603-045-00-x	TZ4300000	33	1159	515
bis(1,1-dimethylethyl)peroxide	110-05-4	617-001-00-2	ER2450000	539	2102	278
dimethyl formaldehyde	67-64-1	606-001-00-8	AL3150000	33	1090	006
DIMETHYLFORMAMIDE	68-12-2	616-001-00-x	LQ2100000	30	2265	347
N,N-dimethylforrmamide	68-12-2	616-001-00-x	LQ2100000	30	2265	347
2,6-dimethyl-2,5-heptadien-4-one	504-20-1		MI5500000			701
2,6-dimethyl-4-heptanol	108-82-7		MJ3325000	30	1993	330
2,6-dimethyl-4-heptanone	108-83-8	606-005-00-x	MJ5775000	30	1157	332
dimethylhydrazine	57-14-7	007-012-00-5	MV2450000	338	1163	348
asym-dimethylhydrazine	57-14-7	007-012-00-5	MV2450000	338	1163	348
N,N-dimethylhydrazine	57-14-7	007-012-00-5	MV2450000	338	1163	348
1,1-DIMETHYLHYDRAZINE	57-14-7	007-012-00-5	MV2450000	338	1163	348
N,N-dimethyl-2-hydroxyethylamine	108-01-0	603-047-00-0	KK6125000	30	2051	339
dimethyl ketone	67-64-1	606-001-00-8	AL3150000	33	1090	006
N,N-dimethylmethanamine (cylinder)	75-50-3	612-001-00-9	PA0350000	236	1083	898
N,N-dimethylmethanamine (40% aqueous solution)	75-50-3	612-001-00-9	PA0350000	236	1083	899
dimethylmethane (cylinder)	74-98-6	601-003-00-5	TX2275000	23	1978	746
dimethylnitromethane	79-46-9	609-002-00-1	TZ5250000	30	2608	651
2,4-dimethylpentanol-3	600-36-2					334
dimethylphenylamine	121-69-7	612-016-00-0	BX4725000	60	2253	342
DIMETHYL PHTHALATE	131-11-3		TI1575000			349
para-dimethylphthalate	120-61-6		WZ1225000			352
2,2-dimethyl-1,3-propanediol	126-30-7		TY5775000			632
2,2 dimethylpropane, see BUTANE						131
2,2-dimethylpropane, see ISOBUTANE						497
2,2-dimethylpropanoic acid	75-98-9		TO7700000			721

Name	CAS-number	EEC-number	RTECS number	HI number	UN-number	Page
2,2-dimethylpropionic acid	75-98-9		TO7700000			721
(dimethyl-1,1- propyl)-4 phenol	80-46-6		SM6825000	60	2430	064
4-(1,1-dimethylpropyl)phenol	80-46-6		SM6825000	60	2430	064
2,6-dimethylpyridine	108-48-5		OK9700000			536
DIMETHYL SULFATE	77-78-1	016-023-00-4	WS8225000	66	1595	350
DIMETHYL SULFOXIDE	67-68-5		PV6210000			351
dimethyl sulphate	77-78-1	016-023-00-4	WS8225000	66	1595	350
DIMETHYL TEREPHTHALATE	120-61-6		WZ1225000			352
dinitro-2,4 aniline	97-02-9	612-040-00-1	BX9100000	60	1596	353
2,4-DINITROANILINE	97-02-9	612-040-00-1	BX9100000	60	1596	353
2,4-dinitrobenzeneamine	97-02-9	612-040-00-1	BX9100000	60	1596	353
3,5-dinitrobenzoic chloride	99-33-2		DM6637000			354
3,5-DINITROBENZOYLCHLORIDE	99-33-2		DM6637000			354
DINITROCHLOROBENZENE	97-00-7	610-003-00-4	CZ0525000	60	1577	355
1,3-dinitro-4-chlorobenzene	97-00-7	610-003-00-4	CZ0525000	60	1577	355
dinitrofluorobenzene	70-34-8		CZ7800000			430
2,4-DINITROPHENOL	25550-58-7	609-016-00-8	SL2625000		0076	356
2,4-dinitrophenylamine	97-02-9	612-040-00-1	BX9100000	60	1596	353
DINONYLPHENOL	1323-65-5					357
DIOCTYL PHTHALATE	117-81-7		TI0350000			358
di-sec-octyl phthalate	117-81-7		TI0350000			358
diolamine	111-42-2	603-071-00-1	KL2975000	80	1719	300
DIOP	27554-26-3		TI1300000			333
dioxan	123-91-1	603-024-00-5	JG8225000	33	1165	359
1,4-DIOXANE	123-91-1	603-024-00-5	JG8225000	33	1165	359
3,6-dioxaoctane-1,8-diol	112-27-6		YE4550000			888
Dioxitol	111-90-0		KK8750000			313
dioxolone-2	96-49-1		FF9550000			402
dioxyethylene ether	123-91-1	603-024-00-5	JG8225000	33	1165	359
4,8-dioxy-undecane diol,-1,11	24800-44-0		YK6825000			907
dipentene	138-86-3	601-029-00-7	OS8100000	30	2052	530
DIPHENYL	92-52-4		DU8050000			360
1,2-diphenyldiazene	103-33-3	611-001-00-6	CN1400000			079
diphenyldiimide	103-33-3	611-001-00-6	CN1400000			079
4,4'-diphenylenediamine	92-87-5	612-042-00-2	DC9625000	60	1885	098
diphenyl ether	101-84-8		KN8970000			362
DIPHENYLMETHANE-4,4'-DIISOCYANATE	101-68-8	615-005-01-6	NQ9350000	60	2489	361
4-4'-diphenylmethane diisocyanate	101-68-8	615-005-01-6	NQ9350000	60	2489	361
DIPHENYL OXIDE	101-84-8		KN8970000			362
DIPHENYLSULFONIC ACID						363
diphosphoric acid	2466-09-3					768
diphosphorus pentoxide	1314-56-3	015-010-00-0	TH3945000	80	1807	711
di-2-propenylamine	124-02-7		UC6650000	338	2359	264
DI-n-PROPYLALUMINUM HYDRIDE	2036-15-9			X333	2221	364
DI-n-PROPYLALUMINUM HYDRIDE (15 % solution in hexane)	2036-15-9			X333	2220	365
DIPROPYLENE GLYCOL	110-98-5		UB8785000			366
DIPROPYLENE TRIAMINE	56-18-8	612-063-00-7	JL9450000	80	2269	367
dipropylmethane	142-82-5	601-008-00-2	MI7700000	33	1206	454
N,N-dipropyl-1-propanamine	102-69-2		TX1575000	83	2260	906
disodium acid phosphate	7558-79-4		WC4500000			368
disodium carbonate	497-19-8	011-005-00-2	VZ4050000			792
disodium hydrogen phosphate	7558-79-4		WC4500000			368
disodium monosulfide	1313-82-2	016-009-00-8	WE1905000	43	1385	820
disodium orthophosphate	7558-79-4		WC4500000			368
DISODIUMPHOSPHATE	7558-79-4		WC4500000			368
disodium sulfide	1313-82-2	016-009-00-8	WE1905000	43	1385	820
disulfur dichloride	10025-67-9	016-012-00-4	WS4300000	88	1828	833
dithiocarbonic anhydride	75-15-0	006-003-00-3	FF6650000	336	1131	192
divanadiumpentoxide	1314-62-1	023-001-00-8	YW2450000	60	2862	912
divinyl (cylinder)	106-99-0	601-013-00-x	EI9275000	239	1010	130
1,4-DIVINYLBENZENE (inhibited)	1321-74-0		CZ9370000			369
1,4-divinylbenzene (isomers), see 1,4-DIVINYLBENZENE						369
DMA (40% aqueous solution)	124-40-3	612-001-00-9	IP8750000	338	1160	338
DMAC	127-19-5	616-011-00-4	AB7700000			337
DME (cylinder)	115-10-6	603-019-00-8	PM4780000	23	1033	346
DMF	68-12-2	616-001-00-x	LQ2100000	30	2265	347

Name	CAS-number	EEC-number	RTECS number	HI number	UN-number	Page
DMFA	68-12-2	616-001-00-x	LQ2100000	30	2265	347
DMP	131-11-3		TI1575000			349
DMS	77-78-1	016-023-00-4	WS8225000	66	1595	350
DMSO	67-68-5		PV6210000			351
DMT	120-61-6		WZ1225000			352
DNBA	111-92-2	612-049-00-0	HR7780000	83	2248	275
DNPAH	2036-15-9			X333	2221	364
DNPAH (15 % solution in hexane)	2036-15-9			X333	2220	365
1-DODECANE	112-41-4					370
1-dodecanethiol	112-55-0		JR3155000	30	1228	373
dodecanoic acid	143-07-7		OE9800000			522
DODECANOYL PEROXIDE	105-74-8		OF2625000		2124	371
DODECYLBENZENE	123-01-3		CZ9540000			372
α-dodecylene	112-41-4					370
dodecyl mercaptan	112-55-0		JR3155000	30	1228	373
n-DODECYL MERCAPTAN	112-55-0		JR3155000	30	1228	373
tert-DODECYL MERCAPTAN	2466-19-1			30	1228	374
dolomite	471-34-1		FF9335000			176
DOP	117-81-7		TI0350000			358
double carbonate soda	144-55-8		VZ0950000			788
Dowanol	107-98-2	603-064-00-3	UB7700000	30	1993	763
Dowtherm	101-84-8		KN8970000			362
2,4-DP, see 2,4-DICHLOROPHENOXYACETIC ACID						294
dracylic acid	65-85-0		DG0875000			099
dry ice (solid)	124-38-9		FF6400000	22	1845	191
DTBP	110-05-4	617-001-00-2	ER2450000	539	2102	278
DVB (inhibited)	1321-74-0		CZ9370000			369

E

Name	CAS-number	EEC-number	RTECS number	HI number	UN-number	Page
EA (cylinder)	75-04-7	612-002-00-4	KH2100000	236	1036	391
EADC (15% solution in hexane)	563-43-9		BD0705000	X333	2221	388
EADC	563-43-9		BD0705000	X333	2220	387
EAK	541-85-5	606-020-00-1	MJ7350000	30	2271	392
EASC	12075-68-2		BD1950000	X333	2221	389
EASC (15% solution in hexane)	12075-68-2		BD1950000	X333	2220	390
EB	100-41-4	601-023-00-4	DA0700000	33	1175	394
EDB	106-93-4	602-010-00-6	KH9275000	60	1605	405
EDTA	60-00-4		AH4025000			375
EGME	109-86-4	603-011-00-4	KL5775000	30	1188	600
EGMEA	110-49-6	607-036-00-1	KL5950000	30	1189	601
elaic acid	112-80-1		RG2275000			661
enanthic acid	111-14-8		MJ1575000			456
ENB	16219-75-3		RB9450000			419
EPICHLOROHYDRIN	106-89-8	603-026-00-6	TX4900000	63	2023	376
epihydrin alcohol	556-52-5	603-063-00-8	UB4375000			447
epoxy butane	106-88-8		EK3675000	339	3022	152
1,4-epoxybutane	109-99-9	603-025-00-0	LU5950000	33	2056	848
1,2-epoxy-3-chloropropane	106-89-8	603-026-00-6	TX4900000	63	2023	376
1,2-epoxyethane (cylinder)	75-21-8	603-023-00-x	KX2450000	236	1040	412
1,2-epoxy-3-phenoxypropane	122-60-1	603-067-00-x	TZ3675000			693
epoxypropane	75-56-9	603-055-00-4	TZ2975000	33	1280	764
1,2-epoxypropane	75-56-9	603-055-00-4	TZ2975000	33	1280	764
2,3-epoxy-1-propanol	556-52-5	603-063-00-8	UB4375000			447
(2,3-epoxypropoxy)benzene	122-60-1	603-067-00-x	TZ3675000			693
2,3-epoxypropyl butyl ether	2426-08-6	603-039-00-7	TX4200000	30	1993	155
2,3-epoxypropylphenyl ether	122-60-1	603-067-00-x	TZ3675000			693
ethanal	75-07-0	605-003-00-6	AB1925000	33	1089	001
ethanamine (cylinder)	75-04-7	612-002-00-4	KH2100000	236	1036	391
ETHANE (cylinder)	74-84-0	601-002-00-x	KH3800000	23	1035	377
ethanecarboxylic acid	79-09-4	607-089-00-0	UE5950000	80	1848	755
1,2-ethanecarboxylic acid	110-15-6		WM4900000			827
1,2-ethanediamine	107-15-3	612-006-00-6	KH8575000	83	1604	404
ethane dichloride	107-06-2	602-012-00-7	KI0525000	33	1184	289
1,2-ethanediol	107-21-1	603-027-00-1	KW2975000			406

1022

Name	CAS-number	EEC-number	RTECS number	HI number	UN-number	Page
1,1'-(1,2-ethanediylbis(oxy))bisbutane	112-48-1		KH9450000			407
ethane isocyanate	109-90-0		NQ8825000	336	2481	421
ethane nitrile	75-05-8	608-001-00-3	AL7700000	336	1648	008
oxybis-1,1'-ethane	60-29-7	603-022-00-4	KI5775000	33	1155	381
ethanethiol	75-08-1	016-022-00-9	KI9625000	336	2363	423
ethanethiolic acid	507-09-5		AJ5600000	33	2436	854
ethane trichloride	79-00-5	602-014-00-8	FG4900000			875
ethanoic acid (100%)	64-19-7	607-002-01-3	AF1225000	83	2789	003
ethanoic acid (85% in water)	64-19-7	607-002-00-6	AF1340000	83	2789	004
ETHANOL	64-17-5	603-002-00-5	KQ6300000	33	1170	378
ETHANOLAMINE	141-43-5	603-030-00-8	KJ5775000	80	2491	379
1-ethanol-2-thiol	60-24-2		KL5600000	60	2966	552
ethanoyl chloride	75-36-5	607-011-00-5	AO6390000	X338	1717	014
ethene (liquid, refrigerated)	74-85-1	601-010-00-3	KU5340000	223	1038	401
ETHENE (cylinder)	74-85-1	601-010-00-3	KU5340000	23	1962	380
ethenyl acetate	108-05-4	607-023-00-0	AK0875000	339	1301	916
ethenylbenzene	100-42-5	601-026-00-0	WL3675000	39	2055	825
ethenyl benzene oxide	96-09-3	603-084-00-2	CZ9625000			826
ethenyl ethanoate	108-05-4	607-023-00-0	AK0875000	339	1301	916
ETHER	60-29-7	603-022-00-4	KI5775000	33	1155	381
ethinyl trichloride	79-01-6	602-027-00-9	KX4550000	60	1710	876
?ethoxol acetate (cylinder)	74-84-0	601-002-00-x	KH3800000	23	1035	377
ethoxybenzene	103-73-1		SI7700000			692
2-ethoxybutane	628-81-9		KN4725000	33	1179	397
ethoxy carbonyl ethylene	140-88-5	607-032-00-x	AT0700000	339	1917	386
ethoxydiglycol	111-90-0		KK8750000			313
2-ETHOXYETHANOL	110-80-5	603-012-00-x	KK8050000	30	1171	382
2-(2-ethoxyethoxy) ethanol	111-90-0		KK8750000			313
2-ETHOXYETHYL ACETATE	111-15-9	607-037-00-7	KK8225000	30	1172	383
2-ethoxyethyl ether	112-36-7		KN3160000			311
(ethoxymethyl)benzene	539-30-0					112
ethoxytriethylene glycol	112-50-5		KK8950000			889
ETHYL ACETATE	141-78-6	607-022-00-5	AH5425000	33	1173	384
ethylacetic acid	107-92-6	607-135-00-x	ES5425000	80	2820	165
ethyl acetic ester	141-78-6	607-022-00-5	AH5425000	33	1173	384
ETHYL ACETOACETATE	141-97-5		AK5250000			385
ethylacetone	107-87-9		SA7875000	33	1249	613
ethyl acetylacetate	141-97-9		AK5250000			385
ETHYL ACRYLATE	140-88-5	607-032-00-x	AT0700000	339	1917	386
ethyl alcohol	64-17-5	603-002-00-5	KQ6300000	33	1170	378
ethyl alcohol (80-90% ethanol)	64-17-5	603-002-00-5	KQ6300000	33	1170	583
ethyl aldehyde	75-07-0	605-003-00-6	AB1925000	33	1089	001
ETHYLALUMINUM DICHLORIDE	563-43-9		BD0705000	X333	2220	387
ETHYLALUMINUM DICHLORIDE (15% solution in hexane)	563-43-9		BD0705000	X333	2221	388
ETHYLALUMINUM SESQUICHLORIDE	12075-68-2		BD1950000	X333	2221	389
ETHYL ALUMINUM SESQUICHLORIDE (15% solution in hexane)	12075-68-2		BD1950000	X333	2220	390
ETHYLAMINE (cylinder)	75-04-7	612-002-00-4	KH2100000	236	1036	391
ETHYL AMYL KETONE	541-85-5	606-020-00-1	MJ7350000	30	2271	392
ethylaniline	103-69-5	612-053-00-2	BX9780000	60	2272	393
N-ETHYLANILINE	103-69-5	612-053-00-2	BX9780000	60	2272	393
ETHYLBENZENE	100-41-4	601-023-00-4	DA0700000	33	1175	394
N-ethylbenzeneamine	103-69-5	612-053-00-2	BX9780000	60	2272	393
ETHYL BENZOATE	93-89-0		DH0200000			395
ethylbenzol	100-41-4	601-023-00-4	DA0700000	33	1175	394
ETHYL BROMIDE	74-96-4		KH6475000	60	1891	396
ETHYL-n-BUTYL ETHER	628-81-9		KN4725000	33	1179	397
2-ethylcaproic acid	149-57-5		MO7700000			415
ethyl carbinol	71-23-8	603-003-00-0	UH8225000	33	1274	747
ethyl carbonate	105-58-8		FF9800000	30	2366	309
ETHYL CHLORIDE (cylinder)	75-00-3	602-009-00-0	KH7525000	236	1037	398
ETHYL CHLOROACETATE	105-39-5	607-070-00-7	AF9110000	63	1181	399
ethyl α-chloroacetate	105-39-5	607-070-00-7	AF9110000	63	1181	399
ethyl 2-chloroacetate	105-39-5	607-070-00-7	AF9110000	63	1181	399
ethylchlorocarbonate	541-41-3	607-020-00-4	LQ6125000	336	1182	400
ethyl chloroethanoate	105-39-5	607-070-00-7	AF9110000	63	1181	399

Name	CAS-number	EEC-number	RTECS number	HI number	UN-number	Page
ETHYL CHLOROFORMATE	541-41-3	607-020-00-4	LQ6125000	336	1182	400
ethyl cyanide	107-12-0		UF9625000	336	2404	757
ethyl diglyme	112-36-7		KN3160000			311
ethylene (cylinder)	74-85-1	601-010-00-3	KU5340000	23	1962	380
ETHYLENE (liquid, refrigerated)	74-85-1	601-010-00-3	KU5340000	223	1038	401
ethylene alcohol	107-21-1	603-027-00-1	KW2975000			406
ethylene bromide	106-93-4	602-010-00-6	KH9275000	60	1605	405
ETHYLENE CARBONATE	96-49-1		FF9550000			402
cyclic ethylene carbonate	96-49-1		FF9550000			402
ethylenecarboxamide	79-06-1	616-003-00-0	AS3325000	60	2074	018
ethylenecarboxylic acid	79-10-7	607-061-00-8	AS4375000	89	2218	019
ethylene chlorhydrine	107-07-3	603-028-00-7	KK0875000	60	1135	209
ethylene chloride	107-06-2	602-012-00-7	KI0525000	33	1184	289
ethylene chlorohydrin	107-07-3	603-028-00-7	KK0875000	60	1135	209
ETHYLENE CYANOHYDRIN	109-78-4		MU5250000			403
ethylene diamine	107-15-3	612-006-00-6	KH8575000	83	1604	404
ETHYLENEDIAMINE	107-15-3	612-006-00-6	KH8575000	83	1604	404
1,2-ethylenediamine	107-15-3	612-006-00-6	KH8575000	83	1604	404
ethylenediaminetetraacetic acid	60-00-4		AH4025000			375
ETHYLENE DIBROMIDE	106-93-4	602-010-00-6	KH9275000	60	1605	405
1,2-ethylene dibromide	106-93-4	602-010-00-6	KH9275000	60	1605	405
trans-ethylenedicarboxylic acid	110-17-8	607-146-00-x	LS9625000			436
ethylene dichloride	107-06-2	602-012-00-7	KI0525000	33	1184	289
1,2-ethylene dichloride	107-06-2	602-012-00-7	KI0525000	33	1184	289
ethylene diglycol	111-46-6		ID5950000			310
2,2'-ethylenedioxydiethanol	112-27-6		YE4550000			888
ethylene fluoride (cylinder)	75-37-6		KI1410000	23	1030	322
ETHYLENE GLYCOL	107-21-1	603-027-00-1	KW2975000			406
ethylene glycol n-butyl ether	111-76-2	603-014-00-0	KJ8575000	60	2369	138
ethylene glycol carbonate	96-49-1		FF9550000			402
ETHYLENE GLYCOL DIBUTYL ETHER	112-48-1		KH9450000			407
ethylene glycol diethyl ether	629-14-1		KI14100000	30	1153	315
ETHYLENE GLYCOL DIMETHYL ETHER	110-71-4	603-031-00-3	KI1451000			408
ethylene glycol ethyl ether acetate	111-15-9	607-037-00-7	KK8225000	30	1172	383
Ethylene glycol monoethyl ether acetate	111-15-9	607-037-00-7	KK8225000	30	1172	383
ethylene glycol monobutyl ether	111-76-2	603-014-00-0	KJ8575000	60	2369	138
ETHYLENE GLYCOL MONOBUTYL ETHER ACETATE	112-07-2	607-038-00-2	KJ8925000			409
ethylene glycol monoethyl ether	110-80-5	603-012-00-x	KK8050000	30	1171	382
ethylene glycol monoisopropyl ether	109-59-1	603-013-00-5	KL5075000			516
ethylene glycol monomethyl ether	109-86-4	603-011-00-4	KL5775000	30	1188	600
ethylene glycol monomethyl ether acetate	110-49-6	607-036-00-1	KL5950000	30	1189	601
ETHYLENE GLYCOL MONOPHENYL ETHER	122-99-6		KM0350000			410
ethylene hexachloride	67-72-1		KI4025000			460
ETHYLENEIMINE	151-56-4	613-001-00-1	KX5075000	336	1185	411
ethylene monochloride (cylinder)	75-01-4	602-023-00-7	KU9625000	239	1086	918
ETHYLENE OXIDE (cylinder)	75-21-8	603-023-00-x	KX2450000	236	1040	412
ethylene tetrachloride	127-18-4	602-028-00-4	KX3850000	60	1897	681
ethyl ethanoate	141-78-6	607-022-00-5	AH5425000	33	1173	384
ethyl ether	60-29-7	603-022-00-4	KI5775000	33	1155	381
ethylethylene oxide	106-88-7		EK3675000	339	3022	152
ETHYL FORMATE	109-94-4	607-015-00-7	LQ8400000	33	1190	413
ethylformic acid	79-09-4	607-089-00-0	UE5950000	80	1848	755
2-ethylhexaldehyde	123-05-7		MN7525000	30	1191	414
2-ETHYLHEXANAL	123-05-7		MN7525000	30	1191	414
2-ETHYLHEXOIC ACID	149-57-5		MO7700000			415
2-ETHYLHEXYL ACRYLATE	103-11-7	607-107-00-7	AT0855000			416
2-ETHYLHEXYLAMINE-1	104-75-6		MQ5250000	83	2276	417
ETHYLHEXYL GLYCIDYL ETHER						418
ethyl hydride (cylinder)	74-84-0	601-002-00-x	KH3800000	23	1035	377
ethyl hydrosulfide	75-08-1	016-022-00-9	KI9625000	336	2363	423
ethylic acid (100%)	64-19-7	607-002-01-3	AF1225000	83	2789	003
ethylic acid (85% in water)	64-19-7	607-002-00-6	AF1340000	83	2789	004
ethylidene chloride	75-34-3	602-011-00-1	KI0175000	336	2362	288
ethylidene dichloride	75-34-3	602-011-00-1	KI0175000	336	2362	288
1,1-ethylidene dichloride	75-34-3	602-011-00-1	KI0175000	336	2362	288
ethylidene difluoride (cylinder)	75-37-6		KI1410000	23	1030	322

Name	CAS-number	EEC-number	RTECS number	HI number	UN-number	Page
ethylidene fluoride (cylinder)	75-37-6		KI1410000	23	1030	322
5-ETHYLIDENE-2-NORBORNENE	16219-75-3		RB9450000			419
ETHYL IODIDE	75-03-6		KI4750000			420
ETHYL ISOCYANATE	109-90-0		NQ8825000	336	2481	421
ETHYL MALONATE	105-53-3		OO0700000			422
ETHYLMERCAPTAN	75-08-1	016-022-00-9	KI9625000	336	2363	423
ethylmethanoate	109-94-4	607-015-00-7	LQ8400000	33	1190	413
ethyl methyl carbinol	78-92-2	603-004-00-6	EO1750000	30	1120	136
ethylmethylketone	78-93-3	606-002-00-3	EL6475000	33	1193	596
ethyl methyl ketone peroxide (solution with approx. 10 % active oxygen content)	1338-23-4		EL9450000	539	2250	597
ethyl methyl ketone peroxide (solution with approx. 10 % active oxygen content)	1338-23-4		EL9450000	539	2563	597
ethyl methyl ketoxime	96-29-7	616-014-00-0	EL9275000			598
ethyl monochloroacetate	105-39-5	607-070-00-7	AF9110000	63	1181	399
ethyl nitril	75-05-8	608-001-00-3	AL7700000	336	1648	008
ethylolamine	141-43-5	603-030-00-8	KJ5775000	80	2491	379
ethyl orthosilicate	78-10-4	014-005-0	VV9450000	30	1292	425
ethyl oxide	60-29-7	603-022-00-4	KI5775000	33	1155	381
ethyloxirane	106-88-7		EK3675000	339	3022	152
ethyl 3-oxobutanoate	141-97-9		AK5250000			385
ethyl phthalate	84-66-2		TI1050000			318
ethylpropanoate	105-37-3	607-028-00-8	UF3675000	33	1195	424
ethyl propenoate	140-88-5	607-032-00-x	AT0700000	339	1917	386
ethyl 2-propenoate	140-88-5	607-032-00-x	AT0700000	339	1917	386
ETHYL PROPIONATE	105-37-3	607-028-00-8	UF3675000	33	1195	424
ETHYL SILICATE	78-10-4	014-005-0	VV9450000	30	1292	425
ethyl succinate	123-25-1		WM7400000			319
ethyl sulfate	64-67-5	016-027-00-6	WS7875000	60	1594	320
ethyl sulfhydrate	75-08-1	016-022-00-9	KI9625000	336	2363	423
ethyl sulfide	352-93-2		LC7200000	336	2375	321
ethylthioethane	352-93-2		LC7200000	336	2375	321
ethyltriglycol	112-50-5		KK8950000			889
ethyne	74-86-2	601-015-00-0	AO9600000	239	1001	015
ethyne (cylinder)	74-86-2	601-015-00-0	AO9600000	239	1001	016
ethanoic anhydrate	108-24-7	607-008-00-9	AK1925000	83	1715	005
ethyl phenyl ether	103-73-1		SI7700000			692

F

Name	CAS-number	EEC-number	RTECS number	HI number	UN-number	Page
FERROUS (III) CHLORIDE	7705-08-0		LJ9100000	80	1773	426
FERROUS (II) SULFATE	7782-63-0		NO8500000			427
FLUOBORIC ACID (25-78% solution)	16872-11-0		ED2685000	80	1775	428
FLUORINE (cylinder)	7782-41-4	009-001-00-0	LM6475000		1045	429
fluoroboric acid (25-78% solution)	16872-11-0		ED2685000	80	1775	428
fluorocarbon 11	75-69-4		PB6125000			878
fluorocarbon 12 (cylinder)	75-71-8		PA8200000	20	1028	286
1-FLUORO-2,4-DINITROBENZENE	70-34-8		CZ7800000			430
fluoroethene (cylinder)	75-02-5		YZ7351000	239	1860	919
fluoroethylene (cylinder)	75-02-5		YZ7351000	239	1860	919
fluoroform (cylinder)	75-46-7		PB6900000	20	1984	892
fluorohydric acid (cylinder)	7664-39-3	009-002-00-6	MW7875000	886	1052	479
formal	109-87-5		PA8750000	33	1234	576
FORMALDEHYDE (37% solution in water with 10% methanol)	50-00-0	605-001-00-5	LP8925000	80	2209	431
formaldehyde dimethylacetal	109-87-5		PA8750000	33	1234	576
formalin (37% solution in water with 10% methanol)	50-00-0	605-001-00-5	LP8925000	80	2209	431
FORMAMIDE	75-12-7		LQ0525000			432
FORMIC ACID	64-18-6	607-001-00-1	LQ4900000	80	1779	433
formic acid butylester	592-84-7	607-017-00-8	LQ5500000	33	1128	154
formic acid ethyl ester	109-94-4	607-015-00-7	LQ8400000	33	1190	413
formic acid, methylester	107-31-3	607-014-00-1	LQ8925000	33	1243	599
formic acid, sodium salt	141-53-7		LR0350000			803
formic benzyl ester	104-57-4		LQ5400000			113

Name	CAS-number	EEC-number	RTECS number	HI number	UN-number	Page
formmonomethylamide	123-39-7		LQ3000000			625
formonitrile (cylinder)	74-90-8	006-006-00-x	MW6825000	663	1051	478
formonitrile (20% solution in water)	74-90-8	006-006-00-x	MW6825000	663	1613	473
N-formyldimethylamine	68-12-2	616-001-00-x	LQ2100000	30	2265	347
formylic acid	64-18-6	607-001-00-1	LQ4900000	80	1779	433
formyl trichloride	67-66-3	602-006-00-4	FS9100000	60	1888	210
french chalk	14807-96-6		RB1575000			841
Freon 11	75-69-4		PB6125000			878
freon 12 (cylinder)	75-71-8		PA8200000	20	1028	286
Freon 12 (cylinder)	75-71-8		PA8200000	20	1028	286
Freon 21 (cylinder)	75-43-4		PA8400000	20	1029	292
Freon 22 (cylinder)	75-45-6		PA6390000	20	1018	622
Freon 23 (cylinder)	75-46-7		PB6900000	20	1984	892
Frigen 11	75-69-4		PB6125000			878
Frigen 12 (cylinder)	75-71-8		PA8200000	20	1028	286
Frigen 21 (cylinder)	75-43-4		PA8400000	20	1029	292
Frigen 22 (cylinder)	75-45-6		PA6390000	20	1018	622
frigen 23 (cylinder)	75-46-7		PB6900000	20	1984	892
FUEL OIL	68476-31-3			30	1202	434
FUEL OIL (heating oil)	8006-61-9		LX3300000	30	1202	435
FUMARIC ACID	110-17-8	607-146-00-x	LS9625000			436
fuming sulfuric acid (20% free SO₃)	8014-95-7	016-019-00-2	WS5605000	X886	1831	662
fural	98-01-1	605-010-00-4	LT7000000	30	1199	437
2-furaldehyde	98-01-1	605-010-00-4	LT7000000	30	1199	437
2-furancarbinol	98-00-0	603-018-00-2	LH9100000	60	2874	438
2-furancarboxaldehyde	98-01-1	605-010-00-4	LT7000000	30	1199	437
2,5-furandione	108-31-6	607-096-00-9	ON3675000	80	2215	544
furanidine	109-99-9	603-025-00-0	LU5950000	33	2056	848
2-furanmethanol	98-00-0	603-018-00-2	LH9100000	60	2874	438
FURFURAL	98-01-1	605-010-00-4	LT7000000	30	1199	437
furfuralalcohol	98-00-0	603-018-00-2	LH9100000	60	2874	438
furfuraldehyde	98-01-1	605-010-00-4	LT7000000	30	1199	437
FURFURYL ALCOHOL	98-00-0	603-018-00-2	LH9100000	60	2874	438
2-furyl carbinol	98-00-0	603-018-00-2	LH9100000	60	2874	438
F-21 (cylinder)	75-43-4		PA8400000	20	1029	292

G

Name	CAS-number	EEC-number	RTECS number	HI number	UN-number	Page
GALLIC ACID	149-91-7		LW7525000			439
GAS OIL	64741-44-2		LX3296000	30	1202	440
GASOLINE	8006-61-9	650-001-01-8	LX3300000	33	1203	441
gaultheria oil	119-36-8		VO4725000			615
GDME	110-71-4	603-031-00-3	KI1451000			408
Gilotherm	101-84-8		KN8970000			362
glutaral	111-30-8		MA2450000			442
GLUTARALDEHYDE	111-30-8		MA2450000			442
GLUTARIC ACID	110-94-1		MA3740000			443
glutaric dialdehyde	111-30-8		MA2450000			442
glycerin	56-81-5		MA8050000			445
GLYCERIN TRIACETATE	102-76-1		AK3675000			444
GLYCEROL	56-81-5		MA8050000			445
GLYCEROLTRIBUTYRATE	60-01-5		ET7350000			446
GLYCIDOL	556-52-5	603-063-00-8	UB4375000			447
glycidyl alcohol	556-52-5	603-063-00-8	UB4375000			447
glycidyl ethylhexyl ether						418
glycine	56-40-6		MB7600000			038
glycinol	141-43-5	603-030-00-8	KJ5775000	80	2491	379
glycocol	56-40-6		MB7600000			038
glycol	107-21-1	603-027-00-1	KW2975000			406
glycol carbonate	96-49-1		FF9550000			402
glycol ether	111-46-6		ID5950000			310
glycolic butyl ester	7397-62-8					156
GLYCOLLIC ACID	79-14-1		MC5250000			448
GLYOXAL	107-22-2	605-016-00-7	MD2625000			449
Graham's salt	10124-56-8		OY3675000			804

Name	CAS-number	EEC-number	RTECS number	HI number	UN-number	Page
HYDRAZINESOLUTION (15%)	7803-57-8	007-008-01-0	NC2975000	86	2030	469
hydrazinobenzene	100-63-0	612-023-00-9	MV8925000	60	2572	696
hydrazomethane	60-34-4		MV5600000	338	1244	602
hydrobromic acid (47% solution in water)	10035-10-6	035-002-01-8	MW3850000	80	1788	476
hydrobromic acid (cylinder)	10035-10-6	035-002-00-0	MW3850000	286	1048	475
HYDROCARBON-SOLVENTS (aliphatic - initial boiling point > 180°C)				30	1993	470
HYDROCARBON-SOLVENTS (aromatic - initial boiling point > 180°C)				30	1202	471
HYDROCHLORIC ACID (ca. 36%)	7647-01-0	017-002-01-x	MW4025000	80	1789	472
anhydrous hydrochloric acid (cylinder)	7647-01-0	017-002-00-2	MW4025000	286	1050	477
hydrochloric ether (cylinder)	75-00-3	602-009-00-0	KH7525000	236	1037	398
hydrochloride (cylinder)	7647-01-0	017-002-00-2	MW4025000	286	1050	477
hydrocyanic acid (cylinder)	74-90-8	006-006-00-x	MW6825000	663	1051	478
hydrocyanic acid, sodium salt (30% solution)	143-33-9	006-007-00-5	VZ7525000	66	1935	797
HYDROCYANIC ACID (20% solution in water)	74-90-8	006-006-00-x	MW6825000	663	1613	473
hydrocyanic, sodium salt	143-33-9	006-007-00-5	VZ7525000		1689	796
anhydrous hydrofluoric acid (cylinder)	7664-39-3	009-002-00-6	MW7875000	886	1052	479
HYDROGEN (cylinder)	1333-74-0	001-001-00-9	MW8900000	23	1049	474
hydrogen arsenide (cylinder)	7784-42-1	033-002-00-5	CG6475000	236	2188	077
HYDROGEN BROMIDE (cylinder)	10035-10-6	035-002-00-0	MW3850000	286	1048	475
HYDROGEN BROMIDE (47% solution in water)	10035-10-6	035-002-01-8	MW3850000	80	1788	476
hydrogen carboxylic acid	64-18-6	607-001-00-1	LQ4900000	80	1779	433
HYDROGEN CHLORIDE (cylinder)	7647-01-0	017-002-00-2	MW4025000	286	1050	477
aqueous hydrogen chloride (ca. 36%)	7647-01-0	017-002-01-x	MW4025000	80	1789	472
hydrogen, compressed (cylinder)	1333-74-0	001-001-00-9	MW8900000	23	1049	474
hydrogen cyanide (20% solution in water)	74-90-8	006-006-00-x	MW6825000	663	1613	473
HYDROGEN CYANIDE (cylinder)	74-90-8	006-006-00-x	MW6825000	663	1051	478
HYDROGEN FLUORIDE (cylinder)	7664-39-3	009-002-00-6	MW7875000	886	1052	479
HYDROGEN FLUORIDE (hydrogen fluoride solution 30-80%)	7664-39-3	009-003-00-1	MW7875000	886	1790	480
anhydrous hydrogen fluoride (cylinder)	7664-39-3	009-002-00-6	MW7875000	886	1052	479
hydrogen gas (cylinder)	1333-74-0	001-001-00-9	MW8900000	23	1049	474
molecular hydrogen (cylinder)	1333-74-0	001-001-00-9	MW8900000	23	1049	474
HYDROGEN PEROXIDE (10%)	7722-84-1		MX089000	85	2984	481
HYDROGEN PEROXIDE (ca. 35%)	7722-84-1	008-003-01-6	MX0899000	85	2014	482
HYDROGEN PEROXIDE (50%)	7722-84-1	008-003-01-6	MX0899500	85	2014	483
hydrogen phosphide (cylinder)	7803-51-2		SY7525000	236	2199	703
hydrogen selenide (cylinder)	7783-07-5	034-002-00-8	MX1050000	236	2202	775
HYDROGEN SULFIDE (cylinder)	7783-06-4	016-001-00-4	MX1225000	236	1053	484
hydrogen tetrafluoroborate (25-78% solution)	16872-11-0		ED2685000	80	1775	428
hydroquinol	123-31-9	604-005-00-4	MX3500000	60	2662	485
m-hydroquinone	108-46-3	604-010-00-1	VG9625000	60	2876	772
HYDROQUINONE	123-31-9	604-005-00-4	MX3500000	60	2662	485
hydroxyacetic acid	79-14-1		MC5250000			448
o-hydroxyaniline	95-55-6	612-033-00-3	SJ4950000	60	2512	040
o-hydroxyanisole	90-05-1		SL7525000			450
α-hydroxy-α-toluic acid	90-64-2		OO6300000			546
hydroxybenzene	108-95-2	604-001-00-2	SJ3325000	68	1671	687
2-hydroxybenzoic acid	69-72-7		VO0525000			773
3-HYDROXYBUTANAL	107-89-1		ES3150000	60	2839	486
1-hydroxybutane	71-36-3	603-004-00-6	EO1400000	30	1120	142
2-hydroxybutane	78-92-2	603-004-00-6	EO1750000	30	1120	136
hydroxy-tert-butylamine	124-68-5	6034-070-00-6	UA5950000			499
β-hydroxybutyraldehyde	107-89-1		ES3150000	60	2839	486
2-hydroxychlorobenzene	95-57-8	604-008-00-0	SK2625000	68	2021	623
4-hydroxychlorobenzene	106-48-9	604-008-00-0	SK2800000	60	2020	217
1-hydroxy-2,4-dichlorobenzene	120-83-2	604-011-00-7	SK8575000	60	2021	293
3-hydroxy-1,2-epoxypropane	556-52-5	603-063-00-8	UB4375000			447
hydroxyethanoic acid	79-14-1		MC5250000			448
1-(2-hydroxyethylamino)-2-aminoethane	111-41-1		KJ6300000			487
2-(2-hydroxyethylamino)ethylamine	111-41-1		KJ6300000			487
N-(2-hydroxyethyl)aniline	122-98-5		KJ7175000			691
(1-hydroxyethyl)benzene	98-85-1		DO9275000			585
N-(2-hydroxyethyl)benzenamine	122-98-5		KJ7175000			691
β-hydroxyethyldimethylamine	108-01-0	603-047-00-0	KK6125000	30	2051	339
HYDROXYETHYLETHYLENEDIAMINE	111-41-1		KJ6300000			487

Name	CAS-number	EEC-number	RTECS number	HI number	UN-number	Page
ISOBUTYLACETATE	110-19-0	607-026-00-7	AI4025000	33	1213	500
isobutylalcohol	78-83-1	603-004-00-6	NP9625000	30	1212	498
isobutyl aluminum chloride (15% solution in hexane)	1888-87-5			X333	2220	502
isobutyl aluminum chloride	1888-87-5			X333	2221	501
ISOBUTYL ALUMINUM DICHLORIDE	1888-87-5			X333	2221	501
ISOBUTYL ALUMINUM DICHLORIDE (15% solution in hexane)	1888-87-5			X333	2220	502
ISOBUTYL CHLORIDE	513-36-0			33	1127	503
ISOBUTYLENE (cylinder)	115-11-7	601-012-00-4	UD0890000	23	1055	504
isobutyl ketone	108-83-8	606-005-00-x	MJ5775000	30	1157	332
iso-butyl methyl ketone	108-10-1	606-004-00-4	SA9275000	33	1245	605
isobutyraldehyde, see BUTYRALDEHYDE						164
isobutyric acid, see BUTYRIC ACID						165
isocyanotobenzene	103-71-9		DA3675000	63	2487	697
isocyanatomethane	624-83-9	615-001-00-7	NQ9450000	336	2480	606
isocyanic acid methyl ester	624-83-9	615-001-00-7	NQ9450000	336	2480	606
isocyanic acid, phenyl ester	103-71-9		DA3675000	63	2487	697
ISOCYANURIC ACID	108-80-5		XZ1800000			505
isodecanol	25339-17-7		NR0960000			506
ISODECYLALCOHOL	25339-17-7		NR0960000			506
isonitropropane	79-46-9	609-002-00-1	TZ5250000	30	2608	651
isononylcarbinol	108-82-7		MJ3325000	30	1993	330
ISOOCTANE	540-84-1	601-009-00-8	SA3320000	33	1262	507
ISOPENTANE	78-78-4	601-006-00-1	EK4430000	33	1265	508
isopentyl acetate	123-92-2		NS9800000	30	1104	494
isopentyl alcohol	123-51-3	603-006-00-7	EL5425000	30	1201	495
isopentyl nitrite	110-46-3		NT0187500		1113	063
ISOPRENE	78-79-5	601-014-00-5	NT4037000	339	1218	509
isopropanol	67-63-0	603-003-00-0	NT8050000	33	1219	512
ISOPROPANOLAMINE	78-96-6	603-082-00-1	UA5775000			510
isopropene cyanide	126-98-7	608-010-00-2	UD1400000			566
isopropenylbenzene	98-83-9	601-027-00-6	WL5075300	30	2303	617
4-isopropenyl-1-methyl-1-cyclohexene	138-86-3	601-029-00-7	OS8100000	30	2052	530
isopropenylnitril	126-98-7	608-010-00-2	UD1400000			566
2-isopropoxyethanol	109-59-1	603-013-00-5	KL5075000			516
2-isopropoxypropane	108-20-3	603-045-00-x	TZ4300000	33	1159	515
ISOPROPYL ACETATE	108-21-4	607-024-00-6	AI4930000	33	1220	511
isopropylacetone	108-10-1	606-004-00-4	SA9275000	33	1245	605
ISOPROPYL ALCOHOL	67-63-0	603-003-00-0	NT8050000	33	1219	512
ISOPROPYLAMINE	75-31-0	612-007-00-1	NT8400000	338	1221	513
isopropyl benzene	98-82-8	601-024-00-x	GR8575000	30	1918	245
isopropylcarbinol	78-83-1	603-004-00-6	NP9625000	30	1212	498
Isopropyl Cellosolve	109-59-1	603-013-00-5	KL5075000			516
ISOPROPYL CHLORIDE	75-29-6	602-018-00-x	TX4410000	33	2356	514
ISOPROPYL ETHER	108-20-3	603-045-00-x	TZ4300000	33	1159	515
ISOPROPYL GLYCOL	109-59-1	603-013-00-5	KL5075000			516
isopropylideneacetone	141-79-7	606-009-00-1	SB4200000	30	1229	562
Isopropyloxitol	109-59-1	603-013-00-5	KL5075000			516
isopropyl titanate	54668-9		NT8060000			851
isourea	57-13-6		YR6250000			910
isovalerone	108-83-8	606-005-00-x	MJ5775000	30	1157	332

J

jet fuel	8008-20-6 1)	650-001-02-5	OA5500000	30	1223	683

K

kerosine	8008-20-6 1)	650-001-02-5	OA5500000	30	1223	683
ketone propane	67-64-1	606-001-00-8	AL3150000	33	1090	006
KOH	1310-58-3	019-002-00-8	TT2100000	80	1813	734
Konesta	650-51-1	607-005-00-2	AJ9100000			823

1030

Name	CAS-number	EEC-number	RTECS number	HI number	UN-number	Page
MAGNESIUMHYDROXIDE	1309-42-8					539
MAGNESIUM NITRATE	13446-18-9		OM3756000		1474	540
magnesium (II) nitrate	13446-18-9		OM3756000		1474	540
MAGNESIUM OXIDE	1309-48-4		OM3850000			541
MAGNESIUM SULFATE	7487-88-9		OM4500000			542
maleic acid	110-16-7	607-095-00-3	OM9625000			545
MALEIC ACID	141-82-2		OO0175000			543
MALEIC ANHYDRIDE	108-31-6	607-096-00-9	ON3675000	80	2215	544
maleinic acid	141-82-2		OO0175000			543
MALEINIC ACID	110-16-7	607-095-00-3	OM9625000			545
malenic acid	110-16-7	607-095-00-3	OM9625000			545
malonic ester	105-53-3		OO0700000			422
MANDELIC ACID	90-64-2		OO6300000			546
mandelonitrile	532-28-5		OO8400000			095
MANGANESE ACETATE	638-38-0		AI5770000			547
manganese binoxide	1313-13-9	025-001-00-3	OP0350000			548
manganese black	1313-13-9	025-001-00-3	OP0350000			548
MANGANESE DIOXIDE	1313-13-9	025-001-00-3	OP0350000			548
manganese (II) acetate	638-38-0		AI5770000			547
MANGANOUS SULFATE (monohydrate)	10034-96-5		OP0893500			549
marble	471-34-1		FF9335000			176
MASC (15% solution in hexane)	12542-85-7		BD1970000	X333	2221	580
MASC	12542-85-7		BD1970000	X333	2220	579
MBK	591-78-6	606-030-00-6	MP1400000			588
MBT	149-30-4		DL6475000			551
MCA	79-11-8	607-003-00-1	AF8575000	80	1751	620
MCP	94-74-6	607-051-00-3	AG1575000			592
MCPA	94-74-6	607-051-00-3	AG1575000			592
MCPA	94-74-6	607-051-00-3	AG1575000			592
MEK	78-93-3	606-002-00-3	EL6475000	33	1193	596
MEK peroxide (solution with approx. 10 % active oxygen content)	1338-23-4		EL9450000	539	2250	597
MEK peroxide (solution with approx. 10 % active oxygen content)	1338-23-4		EL9450000	539	2563	597
MEK-oxime	96-29-7	616-014-00-0	EL9275000			598
MEKP (solution with approx. 10 % active oxygen content)	1338-23-4		EL9450000	539	2250	597
MEKP (solution with approx. 10 % active oxygen content)	1338-23-4		EL9450000	539	2563	597
MELAMINE	108-78-1		OS0700000			550
dl-p-mentha-1,8-diene	138-86-3	601-029-00-7	OS8100000	30	2052	530
2-MERCAPTOBENZOTHIAZOLE	149-30-4		DL6475000			551
1-mercaptododecane	112-55-0		JR3155000	30	1228	373
mercaptoethane	75-08-1	016-022-00-9	KI9625000	336	2363	423
2-MERCAPTOETHANOL	60-24-2		KL5600000	60	2966	552
mercaptomethane (cylinder)	74-93-1	016-021-00-3	PB4375000	236	1064	607
mercuric acetate	1600-27-7	080-002-00-6	AI8575000	60	1629	555
mercuric bromide	7789-47-1	080-002-00-6	OV7415000		1634	556
mercuric chloride	7487-94-7	080-002-00-6	OV9100000	60	1624	557
mercuric nitrate	10045-94-0	080-002-00-6	OW8225000		1625	558
MERCURIC (II) OXIDE (yellow and red)	21908-53-2	080-002-00-6	OW8750000	60	1645	553
mercuric rhodanide	592-85-8	080-002-00-6	XL1550000		1646	560
mercuricsulfate	7783-35-9	080-002-00-6	OX0500000		1645	559
mercuric thiocyanate	592-85-8	080-002-00-6	XL1550000		1646	560
mercurochloride	10112-91-1	080-003-00-1	OV8740000		2025	186
MERCURY	7439-97-6	080-001-00-0	OV4550000		2809	554
MERCURY (II) ACETATE	1600-27-7	080-002-00-6	AI8575000	60	1629	555
MERCURY (II) BROMIDE	7789-47-1	080-002-00-6	OV7415000		1634	556
mercury (I) chloride	10112-91-1	080-003-00-1	OV8740000		2025	186
MERCURY (II) CHLORIDE	7487-94-7	080-002-00-6	OV9100000	60	1624	557
mercury iodide, see MERCURY (II) BROMIDE						556
MERCURY (II) NITRATE	10045-94-0	080-002-00-6	OW8225000		1625	558
mercury oxide (yellow and red)	21908-53-2	080-002-00-6	OW8750000	60	1645	553
MERCURY (II) SULFATE	7783-35-9	080-002-00-6	OX0500000		1645	559
MERCURY (II) THIOCYANATE	592-85-8	080-002-00-6	XL1550000		1646	560
merrillite (powder, pyrophoric)	7440-66-6	030-001-00-1	ZG8600000	43	1383	929
MESITYLENE	108-67-8	601-025-00-5	OX6825000	30	2325	561

Name	CAS-number	EEC-number	RTECS number	HI number	UN-number	Page
MESITYL OXIDE	141-79-7	606-009-00-1	SB4200000	30	1229	562
MESYL CHLORIDE	124-63-0					563
metacetaldehyde	9002-91-9	605-005-00-7	XF9900000		1332	564
metacetone	96-22-0	606-006-00-5	SA8050000	33	1156	316
metacetonic acid	79-09-4	607-089-00-0	UE5950000	80	1848	755
metaformaldehyde	110-88-3	605-002-00-0	YK0350000			903
METALDEHYDE	9002-91-9	605-005-00-7	XF9900000		1332	564
METAPHOSPHORIC ACID	37267-86-0					565
α-methacrylic acid	79-41-4	607-088-00-5	OZ2975000	89	2531	570
β-methacrylic acid	3724-65-0		GQ2800000	89	2823	243
methacrylic acid, methyl ester	80-62-6	607-035-00-6	OZ5075000	339	1247	608
METHACRYLONITRILE	126-98-7	608-010-00-2	UD1400000			566
methanal (37% solution in water with 10% methanol)	50-00-0	605-001-00-5	LP8925000	80	2209	431
methanamide	75-12-7		LQ0525000			432
methanamine (in water, 40%)	74-89-5	612-001-00-9	PF6300000	338	1235	581
METHANE (cylinder)	74-82-8	601-001-00-4	PA1490000	23	1971	567
methanecarbonitril	75-05-8	608-001-00-3	AL7700000	336	1648	008
methanecarboxylic acid (85% in water)	64-19-7	607-002-00-6	AF1340000	83	2789	004
methanecarboxylic acid (100%)	64-19-7	607-002-01-3	AF1225000	83	2789	003
methane gas (cylinder)	74-82-8	601-001-00-4	PA1490000	23	1971	567
liquefied methane (liquid, refrigerated)	74-82-8		PA1490000	223	1972	629
methanesulfonyl chloride	124-63-0					563
methane tetrabromide	558-13-4		FG4725000	60	2516	843
methanethiol (cylinder)	74-93-1	016-021-00-3	PB4375000	236	1064	607
methane trichloride	67-66-3	602-006-00-4	FS9100000	60	1888	210
methanoic acid	64-18-6	607-001-00-1	LQ4900000	80	1779	433
METHANOL	67-56-1	603-001-00-x	PC1400000	336	1230	568
methenamine	100-97-0		MN4725000		1328	463
methenyl chloride	67-66-3	602-006-00-4	FS9100000	60	1888	210
methoxybenzene	100-66-3		BZ8050000	30	2222	067
1-methoxybutane-1,3-dione	105-45-3		AK5775000			572
METHOXYETHANE (cylinder)	540-67-0	603-020-00-3	KO0260000		1039	569
2-methoxyethanol	109-86-4	603-011-00-4	KL5775000	30	1188	600
methoxyethene (cylinder)	107-25-5	603-021-00-9	KO2300000	236	1087	920
2-methoxyethyl acetate	110-49-6	607-036-00-1	KL5950000	30	1189	601
methoxyethylene (cylinder)	107-25-5	603-021-00-9	KO2300000	236	1087	920
1-methoxy-2-hydroxypropane	107-98-2	603-064-00-3	UB7700000	30	1993	763
methoxymethyl chloride	107-30-2	603-075-00-3	KN6650000	336	1239	211
2-methoxy-1-methylethanol	107-98-2	603-064-00-3	UB7700000	30	1993	763
2-methoxy-2-methyl propane	1634-04-4		KN5250000	33	2398	587
2-methoxyphenol	90-05-1		SL7525000			450
1-methoxypropane	557-17-5		KO2280000	33	2612	612
1-methoxy-2-propanol	107-98-2	603-064-00-3	UB7700000	30	1993	763
METHYACRYLIC ACID	79-41-4	607-088-00-5	OZ2975000	89	2531	570
1-methylethyl acetate	108-21-4	607-024-00-6	AI4930000	33	1220	511
METHYL ACETATE	79-20-9	607-021-00-x	AI9100000	33	1231	571
methylacetic acid	79-09-4	607-089-00-0	UE5950000	80	1848	755
METHYL ACETOACETATE	105-45-3		AK5775000			572
methyl acetone	78-93-3	606-002-00-3	EL6475000	33	1193	596
methyl acetylacetate	105-45-3		AK5775000			572
METHYL ACETYLENE (cylinder)	74-99-7		UK4250000	239	1060	573
METHYL ACETYLENE-PROPADIENE MIXTURE (cylinder)				239	1060	574
METHYL ACRYLATE	96-33-3	607-034-00-0	AT2800000	339	1919	575
2-methylacrylonitrile	126-98-7	608-010-00-2	UD1400000			566
METHYLAL	109-87-5		PA8750000	33	1234	576
methyl alcohol	67-56-1	603-001-00-x	PC1400000	336	1230	568
methyl aldehyde (37% solution in water with 10% methanol)	50-00-0	605-001-00-5	LP8925000	80	2209	431
METHYLALUMINUM DICHLORIDE	917-65-7			X333	2221	577
METHYLALUMINUM DICHLORIDE (15% solution in hexane)	917-65-7			X333	2220	578
METHYLALUMINUM SESQUICHLORIDE	12542-85-7		BD1970000	X333	2220	579
METHYLALUMINUM SESQUICHLORIDE (15% solution in hexane)	12542-85-7		BD1970000	X333	2221	580
METHYLAMINE (in water, 40%)	74-89-5	612-001-00-9	PF6300000	338	1235	581

Name	CAS-number	EEC-number	RTECS number	HI number	UN-number	Page
N-methylaminobenzene	100-61-8	612-015-00-5	BY4550000	60	2294	582
N-methylaminoethanol	109-83-1	603-080-00-0	KL6650000			595
methylamyl alcohol	108-11-2	603-008-00-8	SA7350000	30	2053	604
methyl-n-amyl carbinol	543-49-7		MJ2975000	30	1987	455
N-METHYLANILINE	100-61-8	612-015-00-5	BY4550000	60	2294	582
2-methylaniline	95-53-4	612-024-00-4	XU2975000	60	1708	869
METHYLATED SPIRIT (80-90% ethanol)	64-17-5	603-002-00-5	KQ6300000	33	1170	583
2-methylbenzenamine	95-53-4	612-024-00-4	XU2975000	60	1708	869
methylbenzene	108-88-3	601-021-00-3	XS5250000	33	1294	865
N-methylbenzeneamine	100-61-8	612-015-00-5	BY4550000	60	2294	582
4-methylbenzenesulfonic acid	104-15-4	016-030-00-2	XT6300000	80	2585	867
p-methylbenzenesulfonic acid	104-15-4	016-030-00-2	XT6300000	80	2585	867
METHYL BENZOATE	93-58-3		DH3850000		2938	584
methylbenzol	108-88-3	601-021-00-3	XS5250000	33	1294	865
α-METHYLBENZYL ALCOHOL	98-85-1		DO9275000			585
METHYL BROMIDE (cylinder)	74-83-9	602-002-00-3	PA4900000	26	1062	586
2-methyl-1,3-butadiene	78-79-5	601-014-00-5	NT4037000	339	1218	509
3-methyl-1,3-butadiene	78-79-5	601-014-00-5	NT4037000	339	1218	509
2-methylbutane	78-78-4	601-006-00-1	EK4430000	33	1265	508
2-methyl-2-butanol	75-85-4	603-007-00-2		33	1105	060
3-methyl-1-butanol	123-51-3	603-006-00-7	EL5425000	30	1201	495
1-methylbutyl acetate	626-38-0	607-130-00-2	AJ2100000	30	1104	057
3-methyl-1-butyl acetate	123-92-2		NS9800000	30	1104	494
2-methyl butylacrylate	97-88-1	607-033-00-5	OZ3675000	39	2227	159
1-methyl-4-tert-butylbenzene	98-51-1		XS8400000	30	2667	162
METHYL-tert-BUTYLETHER	1634-04-4		KN5250000	33	2398	587
3-methyl butyl formate	110-45-2	607-018-00-3	NT0185000	30	1109	496
METHYL BUTYL KETONE	591-78-6	606-030-00-6	MP1400000			588
methyl-n-butyl ketone	591-78-6	606-030-00-6	MP1400000			588
methyl carbonate	616-38-6	607-013-00-6	FG0450000	33	1161	343
methylcarbylamine	624-83-9	615-001-00-7	NQ9450000	336	2480	606
methylcatechol	90-05-1		SL7525000			450
Methyl Cellosolve	109-86-4	603-011-00-4	KL5775000	30	1188	600
Methyl Cellosolve acetate	110-49-6	607-036-00-1	KL5950000	30	1189	601
2-(2-methyl-4-chlorophenoxy)propionic acid, see 2-METHYL-4-CHLOROPHENOXYACETIC ACID						592
METHYL CHLORIDE (cylinder)	74-87-3	602-001-00-7	PA6300000	236	1063	589
methyl chloroacetate, see ETHYL CHLOROACETATE						399
methyl chlorocarbonate	79-22-1	607-019-00-9	FG3675000	336	1238	591
METHYL CHLOROFORM	71-55-6	602-013-00-2	KJ2975000	60	2831	590
METHYL CHLOROFORMATE	79-22-1	607-019-00-9	FG3675000	336	1238	591
2-METHYL-4-CHLOROPHENOXYACETIC ACID	94-74-6	607-051-00-2	AG1575000			592
methyl cyanide	75-05-8	608-001-00-3	AL7700000	336	1648	008
METHYLCYCLOHEXANE	108-87-2	601-018-00-7	GV6125000	33	2296	593
methyl 1,1-dimethylethyl ether	1634-04-4		KN5250000	33	2398	587
methylene acetone	78-94-4		EM9800000	339	1251	619
methylene bisphenyl isocyanate	101-68-8	615-005-01-6	NQ9350000	60	2489	361
METHYLENE CHLORIDE	75-09-02	602-004-00-3	PA8050000	60	1593	594
methylene dimethyl ether	109-87-5		PA8750000	33	1234	576
4,4'-methylenediphenyl diisocyanate	101-68-8	615-005-01-6	NQ9350000	60	2489	361
methylene oxide (37% solution in water with 10% methanol)	50-00-0	605-001-00-5	LP8925000	80	2209	431
methylene bis-phenyl isocyanate	101-68-8	615-005-01-6	NQ9350000	60	2489	361
methyl ethanoate	79-20-9	607-021-00-x	AI9100000	33	1231	571
methylethanolamine	109-83-1	603-080-00-0	KL6650000			595
N-METHYLETHANOLAMINE	109-83-1	603-080-00-0	KL6650000			595
methylethene (cylinder)	115-07-1	601-011-00-9	UC6740000	23	1077	761
(1-methylethenyl)benzene	98-83-9	601-027-00-6	WL5075300	30	2303	617
methyl ether (cylinder)	115-10-6	603-019-00-8	PM4780000	23	1033	346
Methylethoxlacetate	110-49-6	607-036-00-1	KL5950000	30	1189	601
1-methylethylamine	75-31-0	612-007-00-1	NT8400000	338	1221	513
1-methylethyl benzene	98-82-8	601-024-00-x	GR8575000	30	1918	245
methylethylene (cylinder)	115-07-1	601-011-00-9	UC6740000	23	1077	761
methylethyleneglycol	57-55-6		TY2000000			762
methyl ethylene oxide	75-56-9	603-055-00-4	TZ2975000	33	1280	764
methyl ethyl ether (cylinder)	540-67-0	603-020-00-3	KO0260000		1039	569
methylethylglycol	57-55-6		TY2000000			762
4,4'-(1-methyl ethylidene)bisphenol	80-05-7		SL6300000			489

1034

Name	CAS-number	EEC-number	RTECS number	HI number	UN-number	Page
methylethylketone	78-93-3	606-002-00-3	EL6475000	33	1193	596
METHYL ETHYL KETONE	78-93-3	606-002-00-3	EL6475000	33	1193	596
METHYL ETHYL KETONE PEROXIDE (solution with approx. 10 % active oxygen content)	1338-23-4		EL9450000	539	2250	597
METHYL ETHYL KETONE PEROXIDE (solution with approx. 10 % active oxygen content)	1338-23-4		EL9450000	539	2563	597
METHYL ETHYL KETONOXIME	96-29-7	616-014-00-0	EL9275000			598
methylethylmethane (cylinder)	106-97-8	601-004-00-0	EJ4200000	23	1011	131
1-(1,1-dimethylethyl)-4-methylbenzene	98-51-1		XS8400000	30	2667	162
METHYL FORMATE	107-31-3	607-014-00-1	LQ8925000	33	1243	599
methyl glycol	57-55-6		TY2000000			762
METHYL GLYCOL	109-86-4	603-011-00-4	KL5775000	30	1188	600
METHYL GLYCOL ACETATE	110-49-6	607-036-00-1	KL5950000	30	1189	601
methyl glycol monoacetate	110-49-6	607-036-00-1	KL5950000	30	1189	601
5-methyl-3-heptanone	541-85-5	606-020-00-1	MJ7350000	30	2271	392
methyl hexyl carbinol	123-96-6		RH0795000	30	1993	658
methyl hydrate	67-56-1	603-001-00-x	PC1400000	336	1230	568
METHYL HYDRAZINE	60-34-4		MV5600000	338	1244	602
N-methylhydrazine	60-34-4		MV5600000	338	1244	602
1-methylhydrazine	60-34-4		MV5600000	338	1244	602
methyl hydride (cylinder)	74-82-8	601-001-00-4	PA1490000	23	1971	567
2-methyl-2-p-hydroxyphenylbutane	80-46-6		SM6825000	60	2430	064
METHYL IODIDE	74-88-4	602-005-00-9	PA9450000	60	2644	603
methyl isobutenyl ketone	141-79-7	606-009-00-1	SB4200000	30	1229	562
METHYLISOBUTYL CARBINOL	108-11-2	603-008-00-8	SA7350000	30	2053	604
METHYL ISOBUTYL KETONE	108-10-1	606-004-00-4	SA9275000	33	1245	605
METHYL ISOCYANATE	624-83-9	615-001-00-7	NQ9450000	336	2480	606
methyl ketone	67-64-1	606-001-00-8	AL3150000	33	1090	006
2-methyllactonitril	75-86-5	608-004-00-x	OD9275000	66	1541	007
METHYL MERCAPTANE (cylinder)	74-93-1	016-021-00-3	PB4375000	236	1064	607
METHYL METHACRYLATE	80-62-6	607-035-00-6	OZ5075000	339	1247	608
N-methylmethanamine	109-89-7	612-003-00-x	HZ8750000	338	1154	306
N-methylmethanamine (40% aqueous solution)	124-40-3	612-001-00-9	IP8750000	338	1160	338
methylmethane (cylinder)	74-84-0	601-002-00-x	KH3800000	23	1035	377
methyl methanoate	107-31-3	607-014-00-1	LQ8925000	33	1243	599
methyl-α-methylacrylate	80-62-6	607-035-00-6	OZ5075000	339	1247	608
methyl-2-methyl-2-propenoate	80-62-6	607-035-00-6	OZ5075000	339	1247	608
methyl 2-methyl propanoate	80-62-6	607-035-00-6	OZ5075000	339	1247	608
N-METHYL MORPHOLINE	109-02-4		QE5775000	83	2535	609
4-methyl morpholine	109-02-4		QE5775000	83	2535	609
N-METHYL-1-NAPHTHYLCARBAMATE	63-25-2	006-011-00-7	FC5950000			610
o-methylnitrobenzene	88-72-2	609-006-00-3	XT3150000	60	1664	652
2-methylnitrobenzene	88-72-2	609-006-00-3	XT3150000	60	1664	652
4-methylnitrobenzene	99-99-0	609-006-00-3	CZ9540000	60	1664	653
p-methylnitrotoluene	99-99-0	609-006-00-3	CZ9540000	60	1664	653
methylol	67-56-1	603-001-00-x	PC1400000	336	1230	568
methyl orthosilicate	681-84-5		VV9800000	336	2606	616
methyloxirane	75-56-9	603-055-00-4	TZ2975000	33	1280	764
Methyl Oxitol	109-86-4	603-011-00-4	KL5775000	30	1188	600
methyl 3-oxobutanoate	105-45-3		AK5775000			572
methyl 3-oxobutyrate	105-45-3		AK5775000			572
Methyl Oxitol acetate	110-49-6	607-036-00-1	KL5950000	30	1189	601
2-methyl-2,4-pentanediol	107-41-5	603-053-00-3	SA0810000			467
2-methylpentane-2,4-diol	107-41-5	603-053-00-3	SA0810000			467
4-methyl-2-pentanol	108-11-2	603-008-00-8	SA7350000	30	2053	604
4-methyl-2-pentanone	108-10-1	606-004-00-4	SA9275000	33	1245	605
4-methyl-3-penten-2-one	141-79-7	606-009-00-1	SB4200000	30	1229	562
4-methyl-2-pentyl alcohol	108-11-2	603-008-00-8	SA7350000	30	2053	604
2-methylphenol	95-48-7	604-004-00-9	GO6300000	60	2076	241
methylphenylamine	100-61-8	612-015-00-5	BY4550000	60	2294	582
N-methylphenylamine	100-61-8	612-015-00-5	BY4550000	60	2294	582
methylphenylcarbinol	98-85-1		DO9275000			585
methylphenyl ether	100-66-3		BZ8050000	30	2222	067
1-methyl-1-phenylethene	98-83-9	601-027-00-6	WL5075300	30	2303	617
methylphenylketone	98-86-2		AM5250000			010
methyl phthalate	131-11-3		TI1575000			349
2-methyl-2-propanamine	75-64-9		EO3330000			145

Name	CAS-number	EEC-number	RTECS number	HI number	UN-number	Page
2-methylpropane (cylinder)	75-28-5	601-004-00-0	TZ4300000	23	1969	497
2-methyl-2-propanol	75-65-0	603-005-00-1	EO1925000	30	1120	143
2-methyl-1-propanol	78-83-1	603-004-00-6	NP9625000	30	1212	498
2-methylpropene (cylinder)	115-11-7	601-012-00-4	UD0890000	23	1055	504
2-methyl-2-propenenitrile	126-98-7	608-010-00-2	UD1400000			566
methyl propenoate	96-33-3	607-034-00-0	AT2800000	339	1919	575
methyl-2-propenoate	96-33-3	607-034-00-0	AT2800000	339	1919	575
2-methylpropenoic acid	79-41-4	607-088-00-5	OZ2975000	89	2531	570
METHYLPROPIONATE	554-12-1	607-027-00-2	UF5970000	33	1248	611
2-methylpropyl acetate	110-19-0	607-026-00-7	AI4025000	33	1213	500
methyl propyl carbinol	6032-29-7	603-006-00-7	SA4900000	30	1105	059
β-methylpropyl ethanoate	110-19-0	607-026-00-7	AI4025000	33	1213	500
METHYL PROPYL ETHER	557-17-5		KO2280000	33	2612	612
METHYLPROPYLKETONE	107-87-9		SA7875000	33	1249	613
3-methylpyridine (β)	1)	613-037-00-8		30	2313	717
4-methylpyridine (γ)	1)	613-037-00-8		30	2313	717
2-methylpyridine	109-06-8	613-036-00-2	TJ4900000	30	2313	716
1-methyl-2-(3-pyridyl)pyrrolidine	54-11-5	614-001-00-4	QS5250000	60	1654	635
N-methylpyrrolidone	872-50-4	606-021-00-7	UY5790000			614
1-methylpyrollidone	872-50-4	606-021-00-7	UY5790000			614
1-methyl-2-pyrollidone	872-50-4	606-021-00-7	UY5790000			614
1-METHYL-2-PYROLLIDINONE	872-50-4	606-021-00-7	UY5790000			614
(S)-3-(1-methyl-2-pyrrolidinyl)pyridine	54-11-5	614-001-00-4	QS5250000	60	1654	635
METHYL SALICYLATE	119-36-8		VO4725000			615
METHYL SILICATE	681-84-5		VV9800000	336	2606	616
α-METHYLSTYRENE	98-83-9	601-027-00-6	WL5075300	30	2303	617
m- and p-methylstyrene				39	2618	923
3- and 4-methylstyrene				39	2618	923
methyl sulfhydrate (cylinder)	74-93-1	016-021-00-3	PB4375000	236	1064	607
methyltrichloromethane	71-55-6	602-013-00-2	KJ2975000	60	2831	590
METHYLTRICHLOROSILANE	75-79-6	014-004-00-5	VV4550000	X338	1250	618
methyl vinyl ether (cylinder)	107-25-5	603-021-00-9	KO2300000	236	1087	920
METHYL VINYL KETONE	78-94-4		EM9800000	339	1251	619
MIBC	108-11-2	603-008-00-8	SA7350000	30	2053	604
MIBK	108-10-1	606-004-00-4	SA9275000	33	1245	605
MIC	624-83-9	615-001-00-7	NQ9450000	336	2480	606
milk acid	598-82-3		OD2800000			519
MIPA	75-31-0	612-007-00-1	NT8400000	338	1221	513
essence of Mirbane	98-95-3	609-003-00-7	DA6475000	60	1662	641
oil of Mirbane	98-95-3	609-003-00-7	DA6475000	60	1662	641
mixed acid (mixture of sulfuric acid and nitric acid in various concentrations)		007-005-00-7		88	1796	637
MMH	60-34-4		MV5600000	338	1244	602
monobromobenzene	108-86-1	602-060-00-9	CY9000000	30	2514	127
monobromoethane	74-96-4		KH6475000	60	1891	396
monobromomethane (cylinder)	74-83-9	602-002-00-3	PA4900000	26	1062	586
MONOCHLOROACETIC ACID	79-11-8	607-003-00-1	AF8575000	80	1751	620
monochloroacetyl chloride	79-04-9	607-080-00-1	AO6475000	X80	1752	205
MONOCHLOROBENZENE	108-90-7	602-033-00-1	CZ0175000	30	1134	621
MONOCHLORODIFLUOROMETHANE (cylinder)	75-45-6		PA6390000	20	1018	622
monochlorodimethyl ether	107-30-2	603-075-00-3	KN6650000	336	1239	211
monochloroethane (cylinder)	75-00-3	602-009-00-0	KH7525000	236	1037	398
monochloroethanoic acid	79-11-8	607-003-00-1	AF8575000	80	1751	620
2-monochloroethanol	107-07-3	603-028-00-7	KK0875000	60	1135	209
monochloroethene (cylinder)	75-01-4	602-023-00-7	KU9625000	239	1086	918
monochloromethane (cylinder)	74-87-3	602-001-00-7	PA6300000	236	1063	589
o-MONOCHLOROPHENOL	95-57-8	604-008-00-0	SK2625000	68	2021	623
MONOCHLOROTRIFLUOROMETHANE (cylinder)	75-72-9		PA6410000	20	1022	624
monoethylamine (cylinder)	75-04-7	612-002-00-4	KH2100000	236	1036	391
monoethylene glycol	107-21-1	603-027-00-1	KW2975000			406
monofluoroethylene (cylinder)	75-02-5		YZ7351000	239	1860	919
monoisopropanolamine	78-96-6	603-082-00-1	UA5775000			510
monomethylamine (in water, 40%)	74-89-5	612-001-00-9	PF6300000	338	1235	581
monomethylethanolamine	109-83-1	603-080-00-0	KL6650000			595
MONOMETHYLFORMAMIDE	123-39-7		LQ3000000			625
monopropylamine	107-10-8		UH9100000	338	1277	760

Name	CAS-number	EEC-number	RTECS number	HI number	UN-number	Page
MORPHOLINE	110-91-8	613-028-00-9	QD6475000	30	2054	626
motor fuel	8006-61-9	650-001-01-8	LX3300000	33	1203	441
MTBE	1634-04-4		KN5250000	33	2398	587
muriatic acid (ca. 36%)	7647-01-0	017-002-01-x	MW4025000	80	1789	472
muriatic ether (cylinder)	75-00-3	602-009-00-0	KH7525000	236	1037	398
MVE (cylinder)	107-25-5	603-021-00-9	KO2300000	236	1087	920

N

Name	CAS-number	EEC-number	RTECS number	HI number	UN-number	Page
NaCN	143-33-9	006-007-00-5	VZ7525000		1689	796
NaCN (30% solution)	143-33-9	006-007-00-5	VZ7525000	66	1935	797
NaOCl (solution with 150g/l active chlorine)	7681-52-9	017-011-00-1	NH3486300	85	1791	806
NAPHTHALENE	91-20-3		QJ0525000	40	1334	627
1-naphthalenol methyl carbamate	63-25-2	006-011-00-7	FC5950000			610
naphthalin	91-20-3		QJ0525000	40	1334	627
naphthene	91-20-3		QJ0525000	40	1334	627
N-(2-naphthyl)aniline	135-88-6		QM4555000			700
1-naphthyl-N-methylcarbamate	63-25-2	006-011-00-7	FC5950000			610
β-naphthylphenylamine	135-88-6		QM4555000			700
narcylene (cylinder)	74-86-2	601-015-00-0	AO9600000	239	1001	016
natural gas, see METHANE						567
NATURAL GAS (cylinder)	74-82-8		PA1490000	223	1972	628
NATURAL GAS (liquid, refrigerated)	74-82-8		PA1490000	223	1972	629
NEON (cylinder)	7440-01-9		QP4450000	22	1913	630
NEON (liquid, refrigerated)	7440-01-9		QP4450000	20	1065	631
neon, compressed (cylinder)	7440-01-9		QP4450000	22	1913	630
neon gas (cylinder)	7440-01-9		QP4450000	22	1913	630
neopentane, see BUTANE						131
neopentane, see ISOBUTANE						497
neopentanoic acid	75-98-9		TO7700000			721
NEOPENTYL GLYCOL	126-30-7		TY5775000			632
NICKEL CARBONYL	13463-39-3	028-001-00-1	QR6300000	663	1259	633
NICKEL SULFATE	7786-81-4		QR9350000			634
nickel (II) sulfate	7786-81-4		QR9350000			634
nickel tetracarbonyl	13463-39-3	028-001-00-1	QR6300000	663	1259	633
NICOTINE	54-11-5	614-001-00-4	QS5250000	60	1654	635
niobe oil	93-58-3		DH3850000		2938	584
p-nitraniline	100-01-6	612-012-00-9	BY7000000	60	1661	640
NITRATE FERTILIZERS					2067	636
NITRATING ACID (mixture of sulfuric acid and nitric acid in various concentrations)		007-005-00-7		88	1796	637
NITRIC ACID (up to 70%)	7697-37-2	007-004-01-9	QU5775000	80	2031	638
NITRIC ACID (above 70%)	7697-37-2	007-004-00-1	QU5900000	885	2031	639
nitric acid (50%), see NITRIC ACID						638
2,2',2''-nitrilo-triethanol	102-71-6		KL9275000			883
p-NITROANILINE	100-01-6	612-012-00-9	BY7000000	60	1661	640
4-nitroaniline	100-01-6	612-012-00-9	BY7000000	60	1661	640
NITROBENZENE	98-95-3	609-003-00-7	DA6475000	60	1662	641
4-nitrobenzeneamine	100-01-6	612-012-00-9	BY7000000	60	1661	640
nitrobenzol	98-95-3	609-003-00-7	DA6475000	60	1662	641
nitrocarbol	75-52-5	609-036-00-7	PA9800000	30	1261	648
nitrocellulose (sol) (nitrocellulose solution in ether)	9004-70-0		QW0975200	33	2059	232
m-nitrochlorobenzene, see o-CHLORONITROBENZENE						214
o-nitrochlorobenzene	88-73-3		CZ0875000	60	1578	214
p-nitrochlorobenzene	100-00-5	610-005-00-5	CZ1050000	60	1578	215
1,1-nitrochloroethane	598-92-5		KH7875000			216
nitrochloroform	76-06-2	610-001-00-3	PB6300000	66	1580	218
NITROETHANE	79-24-3	609-035-00-1	KI5600000	30	2842	642
NITROGEN (cylinder)	7727-37-9		QW9700000	20	1066	643
NITROGEN (liquid, refrigerated)	7727-37-9		QW9700000	22	1977	644
nitrogen chloride	10025-85-1		QW9740000			647
NITROGEN DIOXIDE (cylinder)	10102-44-0	007-002-00-0	QW9800000	265	1067	645
nitrogen monoxide (cylinder)	10024-97-2		QX1350000	225	1070	521
NITROGEN MONOXIDE (cylinder)	10102-43-9		QX0525000			646

Name	CAS-number	EEC-number	RTECS number	HI number	UN-number	Page
nitrogen peroxide (cylinder)	10102-44-0	007-002-00-0	QW9800000	265	1067	645
nitrogen tetroxide (cylinder)	10102-44-0	007-002-00-0	QW9800000	265	1067	645
NITROGEN TRICHLORIDE	10025-85-1		QW9740000			647
nitroisopropane	79-46-9	609-002-00-1	TZ5250000	30	2608	651
NITROMETHANE	75-52-5	609-036-00-7	PA9800000	30	1261	648
p-NITROPHENOL	100-02-7	609-015-00-2	SM2275000	60	1663	649
4-nitrophenol	100-02-7	609-015-00-2	SM2275000	60	1663	649
1-NITROPROPANE	108-03-2	609-001-00-6	TZ5075000	30	2608	650
2-NITROPROPANE	79-46-9	609-002-00-1	TZ5250000	30	2608	651
m-nitrotoluene, see ortho-NITROTOLUENE						652
ortho-NITROTOLUENE	88-72-2	609-006-00-3	XT3150000	60	1664	652
para-NITROTOLUENE	99-99-0	609-006-00-3	CZ9540000	60	1664	653
2-nitrotoluene	88-72-2	609-006-00-3	XT3150000	60	1664	652
4-nitrotoluene	99-99-0	609-006-00-3	CZ9540000	60	1664	653
4-nitrotoluol	99-99-0	609-006-00-3	CZ9540000	60	1664	653
nitrotrichloromethane	76-06-2	610-001-00-3	PB6300000	66	1580	218
nitrous oxide (cylinder)	10024-97-2		QX1350000	225	1070	521
nitrous (II) oxide (cylinder)	10102-43-9		QX0525000			646
nitrous vapor (cylinder)	10102-44-0	007-002-00-0	QW9800000	265	1067	645
nitroxanthic acid	88-89-1	609-009-00-x	TJ7875000	11	0154	718
NMP	872-50-4	606-021-00-7	UY5790000			614
n-nonanoic acid	112-05-0		RA6650000			675
1-NONANOL	143-08-8		RB1575000			654
n-nonoic acid	112-05-0		RA6650000			675
n-nonyl alcohol	143-08-8		RB1575000			654
n-nonylic acid	112-05-0		RA6650000			675
NONYLPHENOL	25154-52-3		SM5600000			655
2,5-norbordiene	121-46-0		RB6535000	33	2251	116
Norway saltpeter	6484-52-2		BR9050000	50	1942	048
2-NP	79-46-9	609-002-00-1	TZ5250000	30	2608	651

O

Name	CAS-number	EEC-number	RTECS number	HI number	UN-number	Page
cis-9-octadecenoic acid	112-80-1		RG2275000			661
9,10-octadecenoic acid	112-80-1		RG2275000			661
cis-9-octadecenol-1	143-28-2		RG4120000			665
OCTANE	111-65-9	601-009-00-8	RG8400000	33	1262	656
1-octanecarboxylic acid	112-05-0		RA6650000			675
n-octane	111-65-9	601-009-00-8	RG8400000	33	1262	656
n-octanol	111-87-5		RH6550000	30	1993	657
1-OCTANOL	111-87-5		RH6550000	30	1993	657
2-n-OCTANOL	123-96-6		RH0795000	30	1993	658
n-octyl alcohol	111-87-5		RH6550000	30	1993	657
octyl carbinol	143-08-8		RB1575000			654
para-OCTYL PHENOL	1806-26-4			60	2430	659
olamine	141-43-5	603-030-00-8	KJ5775000	80	2491	379
OLEAMIDE	301-02-0					660
olefiant gas (cylinder)	74-85-1	601-010-00-3	KU5340000	23	1962	380
OLEIC ACID	112-80-1		RG2275000			661
oleinic acid	112-80-1		RG2275000			661
OLEUM (20% free SO$_3$)	8014-95-7	016-019-00-2	WS5605000	X886	1831	662
OLEUM (30% free SO$_3$)	8014-95-7	016-019-00-2	WS5605000	X886	1831	663
OLEUM (65% free SO$_3$)	8014-95-7	016-019-00-2	WS5605000	X886	1831	664
OLEYL ALCOHOL	143-28-2		RG4120000			665
orthoboric acid	10043-35-3		ED4550000			119
1-oxa-4-azacyclohexane	110-91-8	613-028-00-9	QD6475000	30	2054	626
oxacyclopropane (cylinder)	75-21-8	603-023-00-x	KX2450000	236	1040	412
OXALIC ACID	144-62-7	607-006-00-8	RO2450000			666
oxalic acid diethyl ester	95-92-1	607-147-00-5	RQ8225000	60	2525	317
oxalic aldehyde	107-22-2	605-016-00-7	MD2625000			449
oxammonium	7803-49-8		MV8050000			490
oxane (cylinder)	75-21-8	603-023-00-x	KX2450000	236	1040	412
2-oxetanone	57-57-8	606-031-00-1	RQ7350000			753
2,2'-oxibisethanol	111-46-6		ID5950000			310
oxirane (cylinder)	75-21-8	603-023-00-x	KX2450000	236	1040	412

Name	CAS-number	EEC-number	RTECS number	HI number	UN-number	Page
2-oxobornane	76-22-2		EX1225000			187
3-oxobutanoic acid ethyl ester	141-97-9		AK5250000			385
2-oxohexamethyleneimine	105-60-2		CM3675000			189
5-OXOHEXANENITRILE	10412-98-3					667
oxomethane (37% solution in water with 10% methanol)	50-00-0	605-001-00-5	LP8925000	80	2209	431
oxybenzene	108-95-2	604-001-00-2	SJ3325000	68	1671	687
1,1'-oxybisbenzene	101-84-8		KN8970000			362
1,1'-Oxybisbenzene	101-84-8		KN8970000			362
2,2'-oxybispropane	108-20-3	603-045-00-x	TZ4300000	33	1159	515
2,2'-oxydiethanol	111-46-6		ID5950000			310
OXYGEN (cylinder)	7782-44-7		RS2060000	20	1072	668
OXYGEN (liquid)	7782-44-7		RS2060000	22	1073	669
1,1'-oxydi-2-propanol	110-98-5		UB8785000			366
OZONE (cylinder)	10028-15-6		RS8225000			670

P

Name	CAS-number	EEC-number	RTECS number	HI number	UN-number	Page
palmitic acid, see OLEIC ACID						661
PARAFFIN (oil)	8002-74-2		WV6300000			671
paraform	30525-89-4		RV0540000		2213	672
PARAFORMALDEHYDE	30525-89-4		RV0540000		2213	672
PARALDEHYDE	123-63-7	605-004-00-1	YK0525000	30	1264	673
paranaphthalene	120-12-7		CA9350000			068
PARATHION	56-38-2	015-034-00-1	TF4550000	66	1668	674
PBNA	135-88-6		QM4555000			700
PCP	87-86-5	604-002-00-8	SM6300000	60	2996	676
PDCB	106-46-7	602-035-00-2	CZ4550000	60	1592	285
pebble lime	1305-78-8		EW3100000		1910	183
pelargic acid	112-05-0		RA6650000			675
PELARGONIC ACID	112-05-0		RA6650000			675
pelargonic alcohol	143-08-8		RB1575000			654
penchlorol	87-86-5	604-002-00-8	SM6300000	60	2996	676
penta	87-86-5	604-002-00-8	SM6300000	60	2996	676
PENTACHLOROPHENOL	87-86-5	604-002-00-8	SM6300000	60	2996	676
pentachlorophenolate, see PENTACHLOROPHENOL						676
2,3,4,5,6-pentachlorophenol	87-86-5	604-002-00-8	SM6300000	60	2996	676
pentamethylene	287-92-3	601-030-00-2	GY2390000	33	1146	257
pentamethylenediamine	462-94-2		SA0200000			267
pentamethyleneimine	110-89-4	613-027-00-3	TM3500000	338	2401	720
PENTANE	109-66-0	601-006-00-1	RZ9450000	33	1265	677
pentanedial	111-30-8		MA2450000			442
pentane-1,5-diamine	462-94-2		SA0200000			267
pentanedioic acid	110-94-1		MA3740000			443
2,4-pentanedione	123-54-6	606-029-00-0	SA1925000	30	2310	011
n-pentane	109-66-0	601-006-00-1	RZ9450000	33	1265	677
normal pentane	109-66-0	601-006-00-1	RZ9450000	33	1265	677
tert-pentanoic acid	75-98-9		TO7700000			721
1-pentanol acetate	628-63-7	607-130-00-2	AJ1925000	30	1104	056
2-pentanol acetate	626-38-0	607-130-00-2	AJ2100000	30	1104	057
tert-pentanol	75-85-4	603-007-00-2		33	1105	060
1-pentanol	71-41-0	603-006-00-7	SB9800000	30	1105	058
2-pentanol	6032-29-7	603-006-00-7	SA4900000	30	1105	059
3-pentanone	96-22-0	606-006-00-5	SA8050000	33	1156	316
pentaphene	80-46-6		SM6825000	60	2430	064
1-PENTENE	109-67-1			33	1108	678
pentiformic acid	142-62-1		MO5250000	80	2829	188
n-pentyl acetate	628-63-7	607-130-00-2	AJ1925000	30	1104	056
2-pentyl acetate	626-38-0	607-130-00-2	AJ2100000	30	1104	057
pentyl alcohol	71-41-0	603-006-00-7	SB9800000	30	1105	058
sec-pentyl alcohol	6032-29-7	603-006-00-7	SA4900000	30	1105	059
t-pentyl alcohol	75-85-4	603-007-00-2		33	1105	060
pentyl carbinol	111-27-3	603-059-00-6	MQ4025000	30	1987	466
pentyl ether	693-65-2		SC2900000			062
pentylformic acid	142-62-1		MO5250000	80	2829	188

Name	CAS-number	EEC-number	RTECS number	HI number	UN-number	Page
PERACETIC ACID (40% solution in acetic acid)	79-21-0	607-094-00-8	SD8927000	539	2131	679
PERCHLORIC ACID (solution in water 50-72%)	7601-90-3	017-006-00-4	SC7501100	558	1873	680
perchlorobenzene	118-74-1		DA2975000	60	2729	457
perchlorobutadiene	87-68-3		EJ0700000	60	2279	458
perchloroethane	67-72-1		KI4025000			460
perchloroethylene	127-18-4	602-028-00-4	KX3850000	60	1897	681
PERCHLOROETHYLENE	127-18-4	602-028-00-4	KX3850000	60	1897	681
perchloromethane	56-23-5	602-008-00-5	FG4900000	60	1846	194
perk	127-18-4	602-028-00-4	KX3850000	60	1897	681
PERMANGANATE/SULFURIC ACID SOLUTION						682
peroxyacetic acid (40% solution in acetic acid)	79-21-0	607-094-00-8	SD8927000	539	2131	679
PETROLEUM	8008-20-6 1)	650-001-02-5	OA5500000	30	1223	683
petroleum ether, see PETROLEUM NAPHTHA						684
petroleum naphta 40/65, see PETROLEUM NAPHTHA						684
petroleum naphta 62/82, see PETROLEUM NAPHTHA						684
petroleum naphta 80/110, see PETROLEUM NAPHTHA						684
PETROLEUM NAPHTHA (aliphatic - initial boiling point <140°C; flash point <21°C)	8002-05-9	650-001-01-8	SE7175000	33	1993	684
PETROLEUM NAPHTHA (aliphatic - initial boiling point 120-180°C)		650-001-02-5		30	1993	685
PETROLEUM NAPHTHA (aromatic - initial boiling point >140°C <190°C)	8052-41-3	650-001-02-5	WJ8925000	33	1993	686
petroleum spirit, see PETROLEUM NAPHTHA						684
petroleum spirits (aromatic - initial boiling point >140°C <190°C)	8052-41-3	650-001-02-5	WJ8925000	33	1993	686
petroleum spirits (aliphatic - initial boiling point 120-180°C)		650-001-02-5		30	1993	685
petroleum thinner (aliphatic - initial boiling point 120-180°C)		650-001-02-5		30	1993	685
petroleum thinner (aromatic - initial boiling point >140°C <190°C)	8052-41-3	650-001-02-5	WJ8925000	33	1993	686
PGE	122-60-1	603-067-00-x	TZ3675000			693
PGME	107-98-2	603-064-00-3	UB7700000	30	1993	763
phenachlor	88-06-2	604-012-00-2	SN1575000	60	2020	879
1-phenethyl alcohol	98-85-1		DO9275000			585
phenetole	103-73-1		SI7700000			692
PHENOL	108-95-2	604-001-00-2	SJ3325000	68	1671	687
o-p-PHENOLSULFONIC ACID	98-67-9		DB6970000	80	1759	688
PHENOTHIAZINE	92-84-2		SN5075000			689
phenoxyethane	103-73-1		SI7700000			692
2-phenoxyethanol	122-99-6		KM0350000			410
(phenoxymethyl)oxirane	122-60-1	603-067-00-x	TZ3675000			693
phenthiazine	92-84-2		SN5075000			689
phenylacetonitrile	140-29-4		AM1400000	60	2470	110
phenyl allyl ether	1746-13-0		DA8575000			031
phenylamine	62-53-3	612-008-00-7	BW6650000	60	1547	065
2-(phenylamino)ethanol	122-98-5		KJ7175000			691
phenylbenzene	92-52-4		DU8050000			360
phenyl bromide	108-86-1	602-060-00-9	CY9000000	30	2514	127
phenylcarbimide	103-71-9		DA3675000	63	2487	697
phenylcarbinol	100-51-6	603-057-00-5	DN3150000			105
phenylcarbonimide	103-71-9		DA3675000	63	2487	697
phenylcarboxylic acid	65-85-0		DG0875000			099
phenyl cellosolve	122-99-6		KM0350000			410
phenyl chloride	108-90-7	602-033-00-1	CZ0175000	30	1134	621
phenylchloroform	98-07-7		XT9275000	80	2226	101
phenyl cyanide	100-47-0	608-012-00-3	DI2450000	60	2224	100
p-PHENYLENEDIAMINE	106-50-3	612-028-00-6	SS8050000	60	1673	690
m-phenylenediamine, see p-PHENYLENEDIAMINE						690
o-phenylenediamine, see p-PHENYLENEDIAMINE						690
phenylethane	100-41-4	601-023-00-4	DA0700000	33	1175	394
N-PHENYLETHANOLAMINE	122-98-5		KJ7175000			691
1-phenylethanone	98-86-2		AM5250000			010
phenyl ether	101-84-8		KN8970000			362
phenylethylene oxide	96-09-3	603-084-00-2	CZ9625000			826
PHENYLETHYL ETHER	103-73-1		SI7700000			692
PHENYL GLYCIDYL ETHER	122-60-1	603-067-00-x	TZ3675000			693

Name	CAS-number	EEC-number	RTECS number	HI number	UN-number	Page
POTASSIUM ALUMINUM SULFATE	7784-24-9		WS5690000			723
potassium amide, see SODIUM AMIDE						786
POTASSIUM ANTIMONYLTARTRATE	11071-15-1	051-003-00-9	CC7350000			724
POTASSIUM BROMATE	7758-01-2	035-003-00-6	EF8725000		1484	725
POTASSIUM BROMIDE	7758-02-3		TS7650000			726
potassium carbonate, see SODIUM CARBONATE						792
POTASSIUM CHLORATE	3811-04-9	017-004-00-3	FO0350000	50	1485	727
potassium chloride, see SODIUM CHLORIDE						794
POTASSIUM CHROMATE	7789-00-6	024-006-00-8	GB2940000			728
POTASSIUM CYANIDE	151-50-8	006-007-00-5	TS8750000	66	1680	729
POTASSIUMDICHROMATE	7778-50-9	024-002-00-6	HX7680000			730
POTASSIUM FERRICYANIDE	13746-66-2		LJ8225000			731
POTASSIUM FERROCYANIDE	14459-95-1		LJ8219000			732
POTASSIUM FLUORIDE	7789-23-3	009-005-00-2	TT0700000	60	1812	733
potassium hexacyanoferrate III	13746-66-2		LJ8225000			731
potassium hydrate	1310-58-3	019-002-00-8	TT2100000	80	1813	734
POTASSIUM HYDROXIDE	1310-58-3	019-002-00-8	TT2100000	80	1813	734
POTASSIUM IODATE	7758-05-6		NN1350000			735
POTASSIUM IODIDE	7681-11-0		TT2975000			736
potassium metabisulfite	16731-55-8		TT4920000			742
POTASSIUM NITRATE	7757-79-1		TT3700000		1486	737
POTASSIUM NITRITE	7758-09-0	007-011-00-x	TT3750000		1488	738
POTASSIUM PERIODATE	7790-21-8					739
POTASSIUM PERMANGANATE	7722-64-7	025-002-00-9	SD6475000	50	1490	740
POTASSIUMPERSULFATE	7727-21-2		SE0400000		1492	741
POTASSIUM PYROSULFITE	16731-55-8		TT4920000			742
POTASSIUM RHODANIDE	333-20-0		XL1925000			743
POTASSIUM SULFIDE	1312-73-8	016-006-00-1	TT6000000	43	1382	744
potassium sulfocyanate	333-20-0		XL1925000			743
potassium sulfocyanide	333-20-0		XL1925000			743
potassium thiocyanate	333-20-0		XL1925000			743
PPD	106-50-3	612-028-00-6	SS8050000	60	1673	690
PROPADIENE (cylinder)	463-49-0		BA0400000		2200	745
propanal	123-38-6	605-018-00-8	UE0350000	33	1275	754
2-propanal	107-02-8	605-008-00-3	AS1050000	336	1092	017
propanamine	107-10-8		UH9100000	338	1277	760
PROPANE (cylinder)	74-98-6	601-003-00-5	TX2275000	23	1978	746
2-propane carbonic acid	3724-65-0		GQ2800000	89	2823	243
1-propanecarboxylic acid	107-92-6	607-135-00-x	ES5425000	80	2820	165
1,2-propanediamine	78-90-0		TX6650000	83	2258	269
1,3-propanediamine	109-76-2		TX6825000			270
1,3-propane dicarboxylic acid	110-94-1		MA3740000			443
1,2-propanediol	57-55-6		TY2000000			762
1,3-propanediol	504-63-2		TY2010000			901
propanenitrile	107-12-0		UF9625000	336	2404	757
1,2,3-propanetriol	56-81-5		MA8050000			445
1,2,3-propane triol triacetate	102-76-1		AK3675000			444
propanoic acid	79-09-4	607-089-00-0	UE5950000	80	1848	755
propanoic acid butyl ester	590-01-2	607-029-00-3	UE8245000	30	1914	160
propanoic acid ethyl ester	105-37-3	607-028-00-8	UF3675000	33	1195	424
propanoic acid methyl ester	554-12-1	607-027-00-2	UF5970000	33	1248	611
n-PROPANOL	71-23-8	603-003-00-0	UH8225000	33	1274	747
1-propanol	71-23-8	603-003-00-0	UH8225000	33	1274	747
2-propanol	67-63-0	603-003-00-0	NT8050000	33	1219	512
propanolide	57-57-8	606-031-00-1	RQ7350000			753
2-propanone	67-64-1	606-001-00-8	AL3150000	33	1090	006
propanoyl chloride	79-03-8	607-093-00-2	UG6657000	338	1815	758
PROPARGYL BROMIDE	106-96-7		UK4375000		2345	748
PROPARGYL BROMIDE (in toluene)	106-96-7		UK4375000			749
PROPARGYL CHLORIDE	624-65-7					750
PROPARGYL CHLORIDE (in toluene)	624-65-7					751
propellant 11	75-69-4		PB6125000			878
propellant 12 (cylinder)	75-71-8		PA8200000	20	1028	286
propenamide	79-06-1	616-003-00-0	AS3325000	60	2074	018
2-propenamide	107-11-9	612-046-00-4	BA5425000	336	2334	025
propene (cylinder)	115-07-1	601-011-00-9	UC6740000	23	1077	761
propeneacid	79-10-7	607-061-00-8	AS4375000	89	2218	019

Name	CAS-number	EEC-number	RTECS number	HI number	UN-number	Page
2-propenenitrile	107-13-1	608-003-00-4	AT5250000	336	1093	020
PROPENE TRIMER	13987-04-7			30	2057	752
propenoic acid	79-10-7	607-061-00-8	AS4375000	89	2218	019
propenoic acid amide	79-06-1	616-003-00-0	AS3325000	60	2074	018
2-propenoic acid butyl ester	141-32-2	607-062-00-3	UD3150000	39	2348	141
2-propenoic acid, ethyl ester	140-88-5	607-032-00-x	AT0700000	339	1917	386
2-propenoic acid, methyl ester	96-33-3	607-034-00-0	AT2800000	339	1919	575
1-propen-3-ol	107-18-6	603-015-00-6	BA5075000	663	1098	024
2-propen-1-ol	107-18-6	603-015-00-6	BA5075000	663	1098	024
2-propenyl alcohol	107-18-6	603-015-00-6	BA5075000	663	1098	024
2-propenylamine	107-11-9	612-046-00-4	BA5425000	336	2334	025
(2-propenyloxy)benzene	1746-13-0		DA8575000			031
((2-propenyloxy)methyl)oxirane	106-92-3	603-038-00-1	RR0875000	30	2219	030
N,N-di-2-propenyl-2-propen-1-amine	102-70-5		XX5950000	30	2610	870
N-2-propenyl-2-propen-1-amine	124-02-7		UC6650000	338	2359	264
β-PROPIOLACTONE	57-57-8	606-031-00-1	RQ7350000			753
1,3-propiolactone	57-57-8	606-031-00-1	RQ7350000			753
PROPIONALDEHYDE	123-38-6	605-018-00-8	UE0350000	33	1275	754
propione	96-22-0	606-006-00-5	SA8050000	33	1156	316
PROPIONIC ACID	79-09-4	607-089-00-0	UE5950000	80	1848	755
propionic aldehyde	123-38-6	605-018-00-8	UE0350000	33	1275	754
PROPIONIC ANHYDRIDE	123-62-6	607-010-00-x	UF9100000	80	2496	756
propionic chloride	79-03-8	607-093-00-2	UG6657000	338	1815	758
propionic ether	105-37-3	607-028-00-8	UF3675000	33	1195	424
propionic nitrile	107-12-0		UF9625000	336	2404	757
PROPIONITRILE	107-12-0		UF9625000	336	2404	757
PROPIONYL CHLORIDE	79-03-8	607-093-00-2	UG6657000	338	1815	758
PROPYL ACETATE	109-60-4	607-024-00-6	AJ3675000	33	1276	759
n-propyl acetate	109-60-4	607-024-00-6	AJ3675000	33	1276	759
1-propyl acetate	109-60-4	607-024-00-6	AJ3675000	33	1276	759
2-propyl acetate	108-21-4	607-024-00-6	AI4930000	33	1220	511
propyl alcohol	71-23-8	603-003-00-0	UH8225000	33	1274	747
n-propyl alcohol	71-23-8	603-003-00-0	UH8225000	33	1274	747
sec-propyl alcohol	67-63-0	603-003-00-0	NT8050000	33	1219	512
propyl aldehyde	123-38-6	605-018-00-8	UE0350000	33	1275	754
PROPYLAMINE	107-10-8		UH9100000	338	1277	760
n-propylamine	107-10-8		UH9100000	338	1277	760
2-propylamine	75-31-0	612-007-00-1	NT8400000	338	1221	513
propyl carbinol	71-36-3	603-004-00-6	EO1400000	30	1120	142
propyl cyanide	109-74-0	608-005-00-5	ET8750000	336	2411	166
α,β-propylene dichloride	78-87-5	602-020-00-0	TX9625000	33	1279	295
PROPYLENE (cylinder)	115-07-1	601-011-00-9	UC6740000	23	1077	761
propylene chloride	78-87-5	602-020-00-0	TX9625000	33	1279	295
propylenediamine	78-90-0		TX6650000	83	2258	269
propylene dichloride	78-87-5	602-020-00-0	TX9625000	33	1279	295
PROPYLENE GLYCOL	57-55-6		TY2000000			762
β-propylene glycol	504-63-2		TY2010000			901
PROPYLENE GLYCOL METHYL ETHER	107-98-2	603-064-00-3	UB7700000	30	1993	763
propylene glycol methyl ether	107-98-2	603-064-00-3	UB7700000	30	1993	763
1,3-propylene glycol	504-63-2		TY2010000			901
PROPYLENE OXIDE	75-56-9	603-055-00-4	TZ2975000	33	1280	764
propylene trimer	13987-04-7			30	2057	752
1,2-propylene oxide	75-56-9	603-055-00-4	TZ2975000	33	1280	764
propylformic acid	107-92-6	607-135-00-x	ES5425000	80	2820	165
propyl hydride (cylinder)	74-98-6	601-003-00-5	TX2275000	23	1978	746
propylnitrile	107-12-0		UF9625000	336	2404	757
propyne (cylinder)	74-99-7		UK4250000	239	1060	573
mixture of propyne and allene (cylinder)				239	1060	574
1-propyne (cylinder)	74-99-7		UK4250000	239	1060	573
Prussic acid (20% solution in water)	74-90-8	006-006-00-x	MW6825000	663	1613	473
Prussic acid (cylinder)	74-90-8	006-006-00-x	MW6825000	663	1051	478
pyrazine hexahydride	110-85-0	612-057-00-4	TK7800000	80	2579	719
PYRIDINE	110-86-1	613-002-00-7	UR8400000	336	1282	765
PYROCATECHOL	120-80-9	604-016-00-4	UX1050000			766
pyrocatechol monomethyl ether	90-05-1		SL7525000			450
pyrogallic acid	87-66-1	604-009-00-6	UX2800000			767
PYROGALLOL	87-66-1	604-009-00-6	UX2800000			767

Name	CAS-number	EEC-number	RTECS number	HI number	UN-number	Page
pyrophoric aluminum powder (powder)	7429-90-5	013-002-00-1	BD0330000	43	1396	032
PYROPHOSPHORIC ACID	2466-09-3					768

Q

Name	CAS-number	EEC-number	RTECS number	HI number	UN-number	Page
quartz	7631-86-9		VV7310000			779
quartz powder	7631-86-9		VV7310000			779
quicklime	1305-78-8		EW3100000		1910	183
quicksilver	7439-97-6	080-001-00-0	OV4550000		2809	554
quinoline	92-22-5		VA9275000	60	2656	197
QUINONE	106-51-4	606-013-00-3	DK2625000	60	2587	769
p-quinone	106-51-4	606-013-00-3	DK2625000	60	2587	769

R

Name	CAS-number	EEC-number	RTECS number	HI number	UN-number	Page
RANEY NICKEL	7440-02-0		QR5950000		1378	770
raney nickel catalyst	7440-02-0		QR5950000		1378	770
RED LEAD	1314-41-6	082-001-00-6	OG5425000	60	2291	771
red oil	112-80-1		RG2275000			661
red prussiate of potash	13746-66-2		LJ8225000			731
refrigerant 12 (cylinder)	75-71-8		PA8200000	20	1028	286
regular gasoline	8006-61-9	650-001-01-8	LX3300000	33	1203	441
RESORCINOL	108-46-3	604-010-00-1	VG9625000	60	2876	772
rock salt	7647-14-5		VZ4725000			794
rutile	13463-67-7		XR2275000			862
R 11	75-69-4		PB6125000			878
R 12 (cylinder)	75-71-8		PA8200000	20	1028	286
R13 (cylinder)	75-72-9		PA6410000	20	1022	624
R 22 (cylinder)	75-45-6		PA6390000	20	1018	622

S

Name	CAS-number	EEC-number	RTECS number	HI number	UN-number	Page
sal ammoniac	12125-02-9	017-014-00-8	BP4550000			046
SALICYLIC ACID	69-72-7		VO0525000			773
saltpeter	7757-79-1		TT3700000		1486	737
SELENIUM DIOXIDE	7446-08-4	034-002-00-8	VS8575000	60	2811	774
SELENIUM HYDRIDE (cylinder)	7783-07-5	034-002-00-8	MX1050000	236	2202	775
selenium (IV) oxide	7446-08-4	034-002-00-8	VS8575000	60	2811	774
selenium (VI) oxide	13768-86-0	034-002-00-8				777
SELENIUM OXYCHLORIDE	7791-23-3	034-002-00-8	VS7000000		2879	776
SELENIUM TRIOXIDE	13768-86-0	034-002-00-8				777
SELENOUS ACID	7783-00-8	034-002-00-8	VS7175000			778
Shell Sol (aliphatic - initial boiling point 120-180° C)		650-001-02-5		30	1993	685
Shell Sol (aromatic - initial boiling point > 140° C < 190° C)	8052-41-3	650-001-02-5	WJ8925000	33	1993	686
Shell Sol (aliphatic - initial boiling point > 180° C)				30	1993	470
Shell Sol (aromatic - initial boiling point > 180° C)				30	1202	471
silicic acid tetraethyl ester	78-10-4	014-005-0	VV9450000	30	1292	425
silicic acid tetramethyl ester	681-84-5		VV9800000	336	2606	616
silicochloroform	10025-78-2	014-001-00-9	VV5950000	X338	1295	881
silicon carbide	409-21-2		VW0450000			195
SILICON DIOXIDE	7631-86-9		VV7310000			779
SILICON OXIDE	10097-28-6					780
silicon-II-oxide	10097-28-6					780
SILICON TETRACHLORIDE	10026-04-7	014-002-00-4	VW0525000	80	1818	781
silundum	409-21-2		VW0450000			195
SILVER NITRATE	7761-88-8	047-001-00-2	VW4725000		1493	782
soapstone	14807-96-6		RB1575000			841
soda	497-19-8	011-005-00-2	VZ4050000			792
soda ash	497-19-8	011-005-00-2	VZ4050000			792

Name	CAS-number	EEC-number	RTECS number	HI number	UN-number	Page
sodamide	7782-92-5		VY2775000			786
soda monohydrate	497-19-8	011-005-00-2	VZ4050000			792
soda niter	7631-99-4		WC5600000	50	1498	810
SODIUM	7440-23-5	011-001-00-0	VY0686000	X423	1428	783
SODIUM ACETATE	127-09-3		AJ4375000			784
sodium acid sulfite	7631-90-5		VZ2000000		2693	789
SODIUM ALUMINATE	1302-42-7		BD1600000	80	2812	785
sodium meta-aluminate	1302-42-7		BD1600000	80	2812	785
SODIUM AMIDE	7782-92-5		VY2775000			786
SODIUM AZIDE	26628-22-8	011-004-00-7	VY8050000	60	1687	787
SODIUM BICARBONATE	144-55-8		VZ0950000			788
sodium bichromate	10588-01-9	024-004-00-7	HX7700000		1497	798
sodium bisulfite, see POTASSIUM PYROSULFITE						742
SODIUM BISULFITE	7631-90-5		VZ2000000		2693	789
SODIUM BOROHYDRIDE	16940-66-2		ED3325000		1426	790
SODIUM BROMATE	7789-38-0		EF8750000		1494	791
SODIUM CARBONATE	497-19-8	011-005-00-2	VZ4050000			792
SODIUM CHLORATE	7775-09-9	017-005-00-9	FO0525000	50	1495	793
SODIUM CHLORIDE	7647-14-5		VZ4725000			794
SODIUM CHLORITE	7758-19-2		VZ4800000		1496	795
SODIUM CYANIDE	143-33-9	006-007-00-5	VZ7525000		1689	796
SODIUM CYANIDE (30% solution)	143-33-9	006-007-00-5	VZ7525000	66	1935	797
SODIUM DICHROMATE	10588-01-9	024-004-00-7	HX7700000		1497	798
SODIUMDIETHYLDITHIOCARBAMATE	148-18-5		EZ6475000			799
sodiumdiethylthiocarbamate	148-18-5		EZ6475000			799
SODIUMDITHIONITE	7775-14-6	016-028-00-1	JP2100000	43	1384	800
SODIUMDODECYLBENZENESULFONATE	25155-30-0		DB6825000			801
sodium ferrocyanide, see POTASSIUM FERROCYANIDE						732
SODIUM FLUORIDE	7681-49-4	009-004-00-7	WB0350000	60	1690	802
SODIUM FORMATE	141-53-7		LR0350000			803
SODIUM HEXAMETAPHOSPHATE	10124-56-8		OY3675000			804
sodium hydrate	1310-73-2	011-002-00-6	WB4900000	80	1823	805
sodium hydrogen carbonate	144-55-8		VZ0950000			788
sodium hydrogen sulfite	7631-90-5		VZ2000000		2693	789
SODIUM HYDROXIDE	1310-73-2	011-002-00-6	WB4900000	80	1823	805
sodium hydroxide solution	1310-73-2	011-002-01-3	WB4900000	80	1824	196
SODIUM HYPOCHLORITE (solution with 150g/l active chlorine)	7681-52-9	017-011-00-1	NH3486300	85	1791	806
SODIUM HYPOCHLORITE (SOLUTION)	7681-52-9	017-011-01-9	NH3486300	85	1791	807
SODIUM IODIDE	7681-82-5		WB6475000			808
sodiummetaborate-peroxi-hydrate	1113-47-9					812
anhydrous sodium metasilicate	6834-92-0		VV9275000			818
sodium methoxide	124-41-4	603-040-00-2	PC3570000		1431	809
SODIUM METHYLATE	124-41-4	603-040-00-2	PC3570000		1431	809
sodium monosulfide	1313-82-2	016-009-00-8	WE1905000	43	1385	820
SODIUM NITRATE	7631-99-4		WC5600000	50	1498	810
SODIUM NITRITE	7632-00-0	007-010-00-4	RA1225000		1500	811
sodium oxychloride (solution with 150g/l active chlorine)	7681-52-9	017-011-00-1	NH3486300	85	1791	806
SODIUMPERBORATE	1113-47-9					812
SODIUM PEROXIDE	1313-60-6	011-003-00-1		50	1504	813
sodium peroxydisulphate	7775-27-2		SE0525000		1505	814
SODIUM PERSULPHATE	7775-27-2		SE0525000		1505	814
sodium polymetaphosphate	10124-56-8		OY3675000			804
SODIUM PYROPHOSPHATE	7722-88-5		UX7350000			815
sodium rhodanide, see POTASSIUM RHODANIDE						743
SODIUM RHODANIDE	540-72-7		XL2275000			816
sodium salt, see PENTACHLOROPHENOL						676
SODIUM SELENITE	10102-18-8	034-002-00-8	VS7350000		2630	817
SODIUM SILICATE	6834-92-0		VV9275000			818
SODIUM SILICATE (solution in water)	1344-09-8					819
sodium silicofluoride, see AMMONIUMSILICOFLUORIDE						053
SODIUM SULFIDE	1313-82-2	016-009-00-8	WE1905000	43	1385	820
SODIUM SULFITE (ANHYDROUS)	7757-83-7		WE2150000			821
sodium TCA	650-51-1	607-005-00-2	AJ9100000			823
sodium tetraborate	1303-96-4		VZ2275000			118
sodium tetrahydroborate	16940-66-2		ED3325000		1426	790

Name	CAS-number	EEC-number	RTECS number	HI number	UN-number	Page
sodium thiocyanate	540-72-7		XL2275000			816
SODIUM THIOSULFATE	10102-17-7		WE6660000			822
SODIUM TRICHLOROACETATE	650-51-1	607-005-00-2	AJ9100000			823
SODIUM TRIPOLYPHOSPHATE	7758-29-4		YK4570000			824
solvent naphtha (aliphatic - initial boiling point 120-180°C)		650-001-02-5		30	1993	685
Solvent naphtha (aliphatic - initial boiling point <140°C; flash point <21°C)	8002-05-9	650-001-01-8	SE7175000	33	1993	684
Solvent naphtha (aromatic - initial boiling point >140°C <190°C)	8052-41-3	650-001-02-5	WJ8925000	33	1993	686
stannous chloride	7646-78-8	050-001-00-5	XP8750000	80	1827	861
stannous chloride (dihydrate)	7772-99-8		XP8700000			860
stearic acid, see OLEIC ACID						661
stearic acid, zinc salt	557-05-1		ZH5200000			933
steatite	14807-96-6		RB1575000			841
styrallyl alcohol	98-85-1		DO9275000			585
STYRENE	100-42-5	601-026-00-0	WL3675000	39	2055	825
STYRENE OXIDE	96-09-3	603-084-00-2	CZ9625000			826
styrol	100-42-5	601-026-00-0	WL3675000	39	2055	825
sublimate	7487-94-7	080-002-00-6	OV9100000	60	1624	557
SUCCINIC ACID	110-15-6		WM4900000			827
SULFAMIC ACID	5329-14-6	016-026-00-0	WO5950000			828
sulfan-B	7446-11-9		WS4830000	X88	1829	839
p-SULFANILIC ACID (monohydrate)	121-57-3	612-014-00-x	WP3895500			829
suficyl bis(methane)	67-68-5		PV6210000			351
sulfocarbolic acid	98-67-9		DB6970000	80	1759	688
SULFUR (molten)	7704-34-9		WS4250000	44	2448	830
SULFUR (solid)	7704-34-9		WS4250000	40	1350	831
SULFUR BROMIDE	13172-31-1					832
SULFUR CHLORIDE	10025-67-9	016-012-00-4	WS4300000	88	1828	833
SULFUR DICHLORIDE	10545-99-0	016-013-00-x	WS4500000	X88	1828	834
SULFUR DIOXIDE (cylinder)	7446-09-5	016-011-00-9	WS4550000	26	1079	835
sulfur fluoride (cylinder)	2551-62-4		WS4900000	20	1080	836
SULFUR HEXAFLUORIDE (cylinder)	2551-62-4		WS4900000	20	1080	836
SULFURIC ACID (ca. 98%)	7664-93-9	016-020-00-8	WS5600000	80	1830	837
sulfuric acid diethyl ester	64-67-5	016-027-00-6	WS7875000	60	1594	320
sulfuric acid dimethyl ester	77-78-1	016-023-00-4	WS8225000	66	1595	350
sulfuric acid, fuming (65% free SO$_3$)	8014-95-7	016-019-00-2	WS5605000	X886	1831	664
sulfuric acid, fuming (30% free SO$_3$)	8014-95-7	016-019-00-2	WS5605000	X886	1831	663
sulfuric acid, nickel salt	7786-81-4		QR9350000			634
SULFURIC ACID (solution in water up to 98%)	7664-93-9	016-020-00-8	WS5600000	80	1830	838
sulfuric chlorohydrin	7790-94-5	016-017-00-1	FX5730000	88	1754	221
sulfur monobromide	13172-31-1					832
sulfur monochloride	10025-67-9	016-012-00-4	WS4300000	88	1828	833
sulfurous anhydride (cylinder)	7446-09-5	016-011-00-9	WS4550000	26	1079	835
sulfurous oxide (cylinder)	7446-09-5	016-011-00-9	WS4550000	26	1079	835
sulfurous oxychloride	7719-09-7	016-015-00-0	XM5150000	X88	1836	856
γ-SULFURTRIOXIDE	7446-11-9		WS4830000	X88	1829	839
SULFURYL CHLORIDE	7791-25-5	016-016-00-6	WT4870000	X88	1834	840
sweet-birch oil	119-36-8		VO4725000			615

T

Name	CAS-number	EEC-number	RTECS number	HI number	UN-number	Page
TAA	102-70-5		XX5950000	30	2610	870
table salt	7647-14-5		VZ4725000			794
TALC	14807-96-6		RB1575000			841
talcum	14807-96-6		RB1575000			841
tar camphor	91-20-3		QJ0525000	40	1334	627
D-tartaric acid, see L-TARTARIC ACID						842
L-TARTARIC ACID	87-69-4		WW8750000			842
LD-tartaric acid, see L-TARTARIC ACID						842
meso-tartaric acid, see L-TARTARIC ACID						842
TBBA	98-73-7		DG4708000			147
TBP	126-73-8	015-014-00-2	TC7700000			872
TBT	5593-70-4		XR1585000			161

1046

Name	CAS-number	EEC-number	RTECS number	HI number	UN-number	Page
TCA	76-03-9	607-004-00-7	AJ7875000	80	1839	873
tce	79-01-6	602-027-00-9	KX4550000	60	1710	876
TCE	79-34-5	602-015-00-3	KI8575000	60	1702	844
TCP (various isomers)	58-90-2	604-013-00-8	SM9275000	60	2020	845
2,4-TDI	584-84-9	615-006-00-4	CZ6300000	60	2078	866
TEA	107-10-8		UH9100000	338	1277	760
TEA	102-71-6		KL9275000			883
TEA	97-93-8	013-004-00-2	BD2050000	X333	1102	884
TEG	112-27-6		YE4550000			888
TEL	78-00-2	082-002-00-3	TP4550000	66	1649	847
TETA	112-24-3	612-059-00-5	YE6650000	80	2259	890
TETRABROMOMETHANE	558-13-4		FG4725000	60	2516	843
tetrabutyl titanate	5593-70-4		XR1585000			161
tetracarbonyl nickel	13463-39-3	028-001-00-1	QR6300000	663	1259	633
tetrachlorethylene	127-18-4	602-028-00-4	KX3850000	60	1897	681
tetrachloroethane	79-34-5	602-015-00-3	KI8575000	60	1702	844
sym-tetrachloroethane	79-34-5	602-015-00-3	KI8575000	60	1702	844
1,1,2,2-TETRACHLOROETHANE	79-34-5	602-015-00-3	KI8575000	60	1702	844
tetrachloromethane	56-23-5	602-008-00-5	FG4900000	60	1846	194
TETRACHLOROPHENOL (various isomers)	58-90-2	604-013-00-8	SM9275000	60	2020	845
tetrachlorosilane	10026-04-7	014-002-00-4	VW0525000	80	1818	781
tetraethoxysilane	78-10-4	014-005-0	VV9450000	30	1292	425
TETRAETHYLENEPENTAMINE	112-57-2	612-060-00-0	KH8585000	80	2320	846
tetraethyle plumb	78-00-2	082-002-00-3	TP4550000	66	1649	847
TETRAETHYL LEAD	78-00-2	082-002-00-3	TP4550000	66	1649	847
tetraethyl orthosilicate	78-10-4	014-005-0	VV9450000	30	1292	425
tetraethylplumbane	78-00-2	082-002-00-3	TP4550000	66	1649	847
tetraethyl silicate	78-10-4	014-005-0	VV9450000	30	1292	425
1,2,3,4-tetrahydrobenzene	110-83-8		GW2500000	33	2256	254
tetrahydro-p-dioxin	123-91-1	603-024-00-5	JG8225000	33	1165	359
TETRAHYDROFURAN	109-99-9	603-025-00-0	LU5950000	33	2056	848
tetrahydro-2-furanmethanol	97-99-4	603-061-00-7	LU2450000			849
TETRAHYDROFURFURYL ALCOHOL	97-99-4	603-061-00-7	LU2450000			849
tetrahydrofurfuryl carbinol	97-99-4	603-061-00-7	LU2450000			849
1,2,3,4-TETRAHYDRONAPHTHALENE	119-64-2		QK3850000			850
tetrahydro-1,4-oxazine	110-91-8	613-028-00-9	QD6475000	30	2054	626
(S)-N-(5,6,7,9-tetrahydro-1,2,3,10-	64-86-8	614-005-00-6	GH0700000			231
TETRA-ISOPROPYLTITANATE	54668-9		NT8060000			851
tetralin	119-64-2		QK3850000			850
tetramethoxysilane	681-84-5		VV9800000	336	2606	616
tetramethylene cyanide	111-69-3		AV2625000	60	2205	022
tetramethylenediamine, see 1,3-DIAMINOBUTANE						266
tetramethylene oxide	109-99-9	603-025-00-0	LU5950000	33	2056	848
TETRAMETHYLLEAD	75-74-1	082-002-00-3	TP4725000	663	1649	852
tetramethyl orthosilicate	681-84-5		VV9800000	336	2606	616
TETRAPHENYLTIN	595-90-4					853
tetrasodiumpyrophosphate	7722-88-5		UX7350000			815
Thermex	101-84-8		KN8970000			362
THFA	97-99-4	603-061-00-7	LU2450000			849
thiacetic acid	507-09-5		AJ5600000	33	2436	854
3-thiapentane	352-93-2		LC7200000	336	2375	321
THIOACETIC ACID	507-09-5		AJ5600000	33	2436	854
1,1'-thiobisethane	352-93-2		LC7200000	336	2375	321
thiodiethylene glycol	111-48-8	603-081-00-6	KM2975000			855
THIODIGLYCOL	111-48-8	603-081-00-6	KM2975000			855
thiodiphenylamine	92-84-2		SN5075000			689
2-thioethanol	60-24-2		KL5600000	60	2966	552
thioethyl alcohol	75-08-1	016-022-00-9	KI9625000	336	2363	423
thioethylene glycol	60-24-2		KL5600000	60	2966	552
thioethylether	352-93-2		LC7200000	336	2375	321
thiofuran	110-02-1		XM7350000	33	2414	857
thioglycol	60-24-2		KL5600000	60	2966	552
thiomethanol (cylinder)	74-93-1	016-021-00-3	PB4375000	236	1064	607
THIONYL CHLORIDE	7719-09-7	016-015-00-0	XM5150000	X88	1836	856
THIOPHENE	110-02-1		XM7350000	33	2414	857
THIOPHENOL	108-98-5		DC0525000	663	2337	858
TIBAL	100-99-2	013-004-00-2	BD2203500	X333	1930	893

Name	CAS-number	EEC-number	RTECS number	HI number	UN-number	Page
TIBAL (15% solution in hexane)	100-99-2	013-004-00-2	BD2203500	X333	3051	894
TIN (powder)	7440-31-5		XP7320000			859
TIN DICHLORIDE (dihydrate)	7772-99-8		XP8700000			860
TINTETRACHLORIDE	7646-78-8	050-001-00-5	XP8750000	80	1827	861
tintetraphenyl	595-90-4					853
titanium butylate	5593-70-4		XR1585000			161
titanium (IV) chloride	7550-45-0	022-001-00-5	XR1925000	80	1838	863
titanium (III) chloride	7705-07-9		XR1924000	80	2441	864
TITANIUM DIOXIDE	13463-67-7		XR2275000			862
titanium isopropylate	54668-9		NT8060000			851
TITANIUM TETRACHLORIDE	7550-45-0	022-001-00-5	XR1925000	80	1838	863
TITANIUM TRICHLORIDE	7705-07-9		XR1924000	80	2441	864
titanium white	13463-67-7		XR2275000			862
titanous chloride	7705-07-9		XR1924000	80	2441	864
TMA	552-30-7	607-097-00-4				895
TMA (40% aqueous solution)	75-50-3	612-001-00-9	PA0350000	236	1083	899
TMA	75-24-1	013-004-00-2	BD2204000	X333	1103	896
TMA (cylinder)	75-50-3	612-001-00-9	PA0350000	236	1083	898
TML	75-74-1	082-002-00-3	TP4725000	663	1649	852
TNBA	102-82-9		YA0350000	80	2542	871
TNPA	102-69-2		TX1575000	83	2260	906
TOCP	78-30-8	015-015-00-8	TD0175000	60	2574	902
TOLUENE	108-88-3	601-021-00-3	XS5250000	33	1294	865
TOLUENE-2,4-DIISOCYANATE	584-84-9	615-006-00-4	CZ6300000	60	2078	866
toluene hexahydride	108-87-2	601-018-00-7	GV6125000	33	2296	593
p-toluenesulfochloride	98-59-9					868
p-TOLUENESULFONIC ACID	104-15-4	016-030-00-2	XT6300000	80	2585	867
4-toluenesulfonic acid	104-15-4	016-030-00-2	XT6300000	80	2585	867
p-TOLUENESULFONYL CHLORIDE	98-59-9					868
toluene trichloride	98-07-7		XT9275000	80	2226	101
toluene trifluoride	98-08-8	602-056-00-7	XT9450000	33	2338	102
meta-toluidine, see 1,2-TOLUIDINE						869
o-toluidine	95-53-4	612-024-00-4	XU2975000	60	1708	869
1,2-TOLUIDINE	95-53-4	612-024-00-4	XU2975000	60	1708	869
α-tolunitrile	140-29-4		AM1400000	60	2470	110
toluol	108-88-3	601-021-00-3	XS5250000	33	1294	865
toluylene-2,4-diisocyanate	584-84-9	615-006-00-4	CZ6300000	60	2078	866
o-tolyl phosphate	78-30-8	015-015-00-8	TD0175000	60	2574	902
Topanol OC	128-37-0		GO7875000			277
tosic acid	104-15-4	016-030-00-2	XT6300000	80	2585	867
tosyl chloride	98-59-9					868
toxilic acid	110-16-7	607-095-00-3	OM9625000			545
TPP	115-86-6		TC8400000			904
TPT	54668-9		NT8060000			851
triacetine	102-76-1		AK3675000			444
TRIALLYLAMINE	102-70-5		XX5950000	30	2610	870
1,3,5-triazine-2,4,6-triamine	108-78-1		OS0700000			550
TRIBUTYLAMINE	102-82-9		YA0350000	80	2542	871
TRIBUTYL PHOSPHATE	126-73-8	015-014-00-2	TC7700000			872
tributyrine	60-01-5		ET7350000			446
1,2,3-trichloorbenzene, see 1,2,4-TRICHLOROBENZENE						874
1,3,5-trichloorbenzene, see 1,2,4-TRICHLOROBENZENE						874
1,1,2-trichlorethane	79-00-5	602-014-00-8	FG4900000			875
trichlorethanoic acid	76-03-9	607-004-00-7	AJ7875000	80	1839	873
trichloroacetaldehyde	75-87-6		FM7870000	60	2075	198
trichloroacetaldehyde monohydrate	302-17-0	605-014-00-6	FM8750000	60	2811	199
TRICHLOROACETIC ACID	76-03-9	607-004-00-7	AJ7875000	80	1839	873
unsym-trichlorobenzene	120-82-1		DC2100000	60	2321	874
1,2,4-TRICHLOROBENZENE	120-82-1		DC2100000	60	2321	874
trichlorocyanidine	108-77-0	613-009-00-5	XZ1400000	80	2670	248
trichlorocyanuric acid	87-90-1	613-031-00-5	XZ1925000		2468	877
trichloroethanal	75-87-6		FM7870000	60	2075	198
trichloroethane	71-55-6	602-013-00-2	KJ2975000	60	2831	590
β-trichloroethane	79-00-5	602-014-00-8	FG4900000			875
1,1,1-trichloroethane	71-55-6	602-013-00-2	KJ2975000	60	2831	590
1,1,2-TRICHLOROETHANE	79-00-5	602-014-00-8	FG4900000			875
trichloroethanoic acid	76-03-9	607-004-00-7	AJ7875000	80	1839	873

1048

Name	CAS-number	EEC-number	RTECS number	HI number	UN-number	Page
trichloroethene	79-01-6	602-027-00-9	KX4550000	60	1710	876
TRICHLOROETHYLENE	79-01-6	602-027-00-9	KX4550000	60	1710	876
trichloroethylidene glycol	302-17-0	605-014-00-6	FM8750000	60	2811	199
trichloroform	67-66-3	602-006-00-4	FS9100000	60	1888	210
1,3,5-trichloro-2-hydroxybenzene	88-06-2	604-012-00-2	SN1575000	60	2020	879
TRICHLOROISOCYANURIC ACID	87-90-1	613-031-00-5	XZ1925000		2468	877
1,3,5-trichloroisocyanuric acid	87-90-1	613-031-00-5	XZ1925000		2468	877
trichloromethane	67-66-3	602-006-00-4	FS9100000	60	1888	210
TRICHLOROMONOFLUOROMETHANE	75-69-4		PB6125000			878
trichloronitromethane	76-06-2	610-001-00-3	PB6300000	66	1580	218
2,3,5-trichlorophenol, see 2,4,6-TRICHLOROPHENOL						879
2,4,6-TRICHLOROPHENOL	88-06-2	604-012-00-2	SN1575000	60	2020	879
2,4,5-TRICHLOROPHENOXY-ACETIC ACID	93-76-5	607-041-00-9	AJ8400000			880
TRICHLOROSILANE	10025-78-2	014-001-00-9	VV5950000	X338	1295	881
1,3,5-trichloro-1,3,5-triazinetrione	87-90-1	613-031-00-5	XZ1925000		2468	877
2,4,6-trichloro-1,3,5-triazine	108-77-0	613-009-00-5	XZ1400000	80	2670	248
TRIDECANOL (mixtures of isomers)	112-70-9		YD4200000			882
tridecylalcohol (mixtures of isomers)	112-70-9		YD4200000			882
TRIEN	112-24-3	612-059-00-5	YE6650000	80	2259	890
TRIETHANOLAMINE	102-71-6		KL9275000			883
TRIETHYLALUMINUM	97-93-8	013-004-00-2	BD2050000	X333	1102	884
TRIETHYLALUMINUM (15% solution in hexane)	97-93-8	013-004-00-2	BD2050000	X333	1102	885
TRIETHYLAMINE	121-44-8	612-004-00-5	YE0175000	338	1296	886
triethylborane	97-94-9		ED2100000			887
TRIETHYLBORINE	97-94-9		ED2100000			887
TRIETHYLENE GLYCOL	112-27-6		YE4550000			888
triethylene glycol ethyl ether	112-50-5		KK8950000			889
TRIETHYLENE GLYCOL MONOETHYL ETHER	112-50-5		KK8950000			889
TRIETHYLENETETRAMINE	112-24-3	612-059-00-5	YE6650000	80	2259	890
TRIFLUOROACETIC ACID	76-05-1	607-091-00-1	AJ9625000	88	2699	891
trifluoroborane (cylinder)	7637-07-2	005-001-00-x	ED2275000	286	1008	120
TRIFLUOROMETHANE (cylinder)	75-46-7		PB6900000	20	1984	892
trifluoromethylbenzene	98-08-8	602-056-00-7	XT9450000	33	2338	102
triglycol monoethyl ether	112-50-5		KK8950000			889
1,2,3-trihydroxybenzene	87-66-1	604-009-00-6	UX2800000			767
3,4,5-trihydroxybenzoic acid	149-91-7		LW7525000			439
trihydroxycyanidine	108-80-5		XZ1800000			505
tri(2-hydroxyethyl)amine	102-71-6		KL9275000			883
2,4,6-trihydroxy-1,3,5-triazine	108-80-5		XZ1800000			505
TRIISOBUTYLALUMINUM	100-99-2	013-004-00-2	BD2203500	X333	1930	893
TRIISOBUTYLALUMINUM (15% solution in hexane)	100-99-2	013-004-00-2	BD2203500	X333	3051	894
triisopropylamine, see TRIPROPYLAMINE						906
TRIMELLITIC ANHYDRIDE	552-30-7	607-097-00-4				895
trimethylacetic acid	75-98-9		TO7700000			721
TRIMETHYL ALUMINUM	75-24-1	013-004-00-2	BD2204000	X333	1103	896
TRIMETHYLALUMINUM (15% solution in hexane)	75-24-1	013-004-00-2	BD2204000	X333	1103	897
trimethylamine (40% aqueous solution)	75-50-3	612-001-00-9	PA0350000	236	1083	899
TRIMETHYLAMINE (cylinder)	75-50-3	612-001-00-9	PA0350000	236	1083	898
TRIMETHYLAMINE (40% aqueous solution)	75-50-3	612-001-00-9	PA0350000	236	1083	899
trimethylaminomethane	75-64-9		EO3330000			145
1,3,5-trimethylbenzene	108-67-8	601-025-00-5	OX6825000	30	2325	561
trimethylbenzol	108-67-8	601-025-00-5	OX6825000	30	2325	561
1,7,7-trimethylbicycol(2,2,1)heptanone-2	76-22-2		EX1225000			187
trimethyl carbinol	75-65-0	603-005-00-1	EO1925000	30	1120	143
trimethylcarbinylamine	75-64-9		EO3330000			145
TRIMETHYLCHLOROSILANE	75-77-4		VV2710000	X338	1298	900
trimethylene (cylinder)	75-19-4	601-016-00-6	GZ0690000	23	1027	259
trimethylene chlorobromide	109-70-6		TX4113000	60	2688	128
1,3-trimethylene diamine	109-76-2		TX6825000			270
TRIMETHYLENE GLYCOL	504-63-2		TY2010000			901
trimethylmethane (cylinder)	75-28-5	601-004-00-0	TZ4300000	23	1969	497
2,2,4-trimethylpentane	540-84-1	601-009-00-8	SA3320000	33	1262	507
2,4,4-trimethylpentene-1	25167-70-8	601-031-00-8	SB2717000	33	2050	331
2,4,4-trimethylpentene-2, see DIISOBUTYLENE						331
2,4,4-trimethylpentene-2, see COPPER (II) ACETATE						234
2,4,6,-trimethyl-1,3,5-trioxane	123-63-7	605-004-00-1	YK0525000	30	1264	673

Name	CAS-number	EEC-number	RTECS number	HI number	UN-number	Page
-trimethyltrimethylene glyco	107-41-5	603-053-00-3	SA0810000			467
tri-n-butylamine	102-82-9		YA0350000	80	2542	871
2,4,6-trinitrophenol	88-89-1	609-009-00-x	TJ7875000	11	0154	718
TRIORTHOCRESYL PHOSPHATE	78-30-8	015-015-00-8	TD0175000	60	2574	902
1,3,5-TRIOXANE	110-88-3	605-002-00-0	YK0350000			903
3,6,9-trioxaundecan-1-ol	112-50-5		KK8950000			889
trioxymethylene	110-88-3	605-002-00-0	YK0350000			903
Trioxytol	112-50-5		KK8950000			889
TRIPHENYL PHOSPHATE	115-86-6		TC8400000			904
TRIPHENYLPHOSPHITE	101-02-0	015-105-00-7	TH1575000			905
tri-2-propenylamine	102-70-5		XX5950000	30	2610	870
TRIPROPYLAMINE	102-69-2		TX1575000	83	2260	906
tri-n-propylamine	102-69-2		TX1575000	83	2260	906
TRIPROPYLENE GLYCOL	24800-44-0		YK6825000			907
tris-n-butylamine	102-82-9		YA0350000	80	2542	871
trisodium orthophosphate	7601-54-9		TC9490000			908
TRISODIUM PHOSPHATE	7601-54-9		TC9490000			908
TSP	7601-54-9		TC9490000			908
TURPENTINE	8006-64-2	650-002-00-6	YO8400000	30	1299	909
turps (aliphatic - initial boiling point 120-180°C)		650-001-02-5		30	1993	685
turps substitute (aliphatic - initial boiling point 120-180°C)		650-001-02-5		30	1993	685
2,4,5-T	93-76-5	607-041-00-9	AJ8400000			880

U

Name	CAS-number	EEC-number	RTECS number	HI number	UN-number	Page
UDMH	57-14-7	007-012-00-5	MV2450000	338	1163	348
UREA	57-13-6		YR6250000			910
uromaline	90-64-2		OO6300000			546

V

Name	CAS-number	EEC-number	RTECS number	HI number	UN-number	Page
vanadium (IV) chloride	7632-51-1		YW2625000	88	2444	914
vanadiummonosulfide	11130-24-8				2575	913
VANADIUM OXYTRICHLORIDE	7727-18-6		YW2975000	80	2443	911
VANADIUMPENTOXIDE	1314-62-1	023-001-00-8	YW2450000	60	2862	912
VANADIUMSULFIDE	11130-24-8				2575	913
VANADIUM TETRACHLORIDE	7632-51-1		YW2625000	88	2444	914
vanadyl trichloride	7727-18-6		YW2975000	80	2443	911
VASELINE	8009-03-8					915
VC (cylinder)	75-01-4	602-023-00-7	KU9625000	239	1086	918
vinegar (100%)	64-19-7	607-002-01-3	AF1225000	83	2789	003
vinegar (85% in water)	64-19-7	607-002-00-6	AF1340000	83	2789	004
VINYL ACETATE	108-05-4	607-023-00-0	AK0875000	339	1301	916
vinyl acetate monomer	108-05-4	607-023-00-0	AK0875000	339	1301	916
vinyl amide	79-06-1	616-003-00-0	AS3325000	60	2074	018
vinylbenzene	100-42-5	601-026-00-0	WL3675000	39	2055	825
VINYL BROMIDE (cylinder)	593-60-2	602-024-00-2	KU8400000	236	1085	917
vinyl carbinol	107-18-6	603-015-00-6	BA5075000	663	1098	024
VINYL CHLORIDE (cylinder)	75-01-4	602-023-00-7	KU9625000	239	1086	918
vinyl chloride monomer (cylinder)	75-01-4	602-023-00-7	KU9625000	239	1086	918
vinyl cyanide	107-13-1	608-003-00-4	AT5250000	336	1093	020
vinylethylene (cylinder)	106-99-0	601-013-00-x	EI9275000	239	1010	130
VINYL FLUORIDE (cylinder)	75-02-5		YZ7351000	239	1860	919
vinylformic acid	79-10-7	607-061-00-8	AS4375000	89	2218	019
vinylidene chloride	75-35-4	602-025-00-8	KV9275000	339	1303	290
vinylidene dichloride	75-35-4	602-025-00-8	KV9275000	339	1303	290
vinylidene difluoride (cylinder)	75-38-7		KW0560000	239	1959	323
vinylidene fluoride (cylinder)	75-38-7		KW0560000	239	1959	323
vinylidine chloride	75-35-4	602-025-00-8	KV9275000	339	1303	290
VINYL METHYL ETHER (cylinder)	107-25-5	603-021-00-9	KO2300000	236	1087	920
5-VINYL-2-NORBORNENE	3048-64-4		RC0350000			921
2-VINYLPYRIDINE	100-69-6		UU1040000	639	3073	922

Name	CAS-number	EEC-number	RTECS number	HI number	UN-number	Page
vinylstyrene (inhibited)	1321-74-0		CZ9370000			369
m- AND p-VINYLTOLUENE				39	2618	923
vinyl trichloride	79-00-5	602-014-00-8	FG4900000			875

W

Name	CAS-number	EEC-number	RTECS number	HI number	UN-number	Page
water glass (solution in water)	1344-09-8					819
white spirit (aliphatic - initial boiling point 120-180°C)		650-001-02-5		30	1993	685
white spirit (aromatic - initial boiling point > 140°C < 190°C)	8052-41-3	650-001-02-5	WJ8925000	33	1993	686
white tar	91-20-3		QJ0525000	40	1334	627
wintergreen oil	119-36-8		VO4725000			615
wood alcohol	67-56-1	603-001-00-x	PC1400000	336	1230	568
wool fat	8006-54-0					520

X

Name	CAS-number	EEC-number	RTECS number	HI number	UN-number	Page
XENON (cylinder)	7440-63-3		ZE1280000	20	2036	924
XENON (liquid)	7440-63-3		ZE1280000	22	2591	925
m-XYLENE	108-38-3	601-039-00-1	ZE2275000	30	1307	926
o-xylene, see m-XYLENE						926
p-XYLENE	106-42-3	601-040-00-7	ZE2625000	30	1307	927
1,3-xylene	108-38-3	601-039-00-1	ZE2275000	30	1307	926
XYLIDINE (mixture of o-, m-, and p-isomers)	1300-73-8	612-027-00-0	ZE8575000	60	1711	928
m-xylol	108-38-3	601-039-00-1	ZE2275000	30	1307	926

Y

Name	CAS-number	EEC-number	RTECS number	HI number	UN-number	Page
yellow potassium prussiate	14459-95-1		LJ8219000			732
yellow prussiate of potash	14459-95-1		LJ8219000			732

Z

Name	CAS-number	EEC-number	RTECS number	HI number	UN-number	Page
ZINC (powder, pyrophoric)	7440-66-6	030-001-00-1	ZG8600000	43	1383	929
ZINC CHLORIDE	7646-85-7	030-003-00-2	ZH1400000	80	2331	930
zinc distearate	557-05-1		ZH5200000			933
zincite	1314-13-2		ZH4810000			932
zinc monoxide	1314-13-2		ZH4810000			932
ZINC NITRATE (hexahydrate)	7779-88-6		ZH4772000		1514	931
zinc octadecanoate	557-05-1		ZH5200000			933
zincoid	1314-13-2		ZH4810000			932
ZINC OXIDE	1314-13-2		ZH4810000			932
ZINC STEARATE	557-05-1		ZH5200000			933
ZINC SULFATE (heptahydrate)	7733-02-0		ZH5260000			934
zinc white	1314-13-2		ZH4810000			932
ZIRCONIUM CARBONATE	15667-84-2					935

UN-NUMBER INDEX